JAY WITHGOTT • SCOTT BRENNAN

ENVIRONMENT

THE SCIENCE BEHIND THE STORIES

Fourth Edition

Benjamin Cummings

Boston   Columbus   Indianapolis   New York   San Francisco   Upper Saddle River
Amsterdam   Cape Town   Dubai   London   Madrid   Milan   Munich   Paris   Montréal   Toronto
Delhi   Mexico City   São Paulo   Sydney   Hong Kong   Seoul   Singapore   Taipei   Tokyo

VP/Editor-in-Chief: Beth Wilbur
Executive Director of Development: Deborah Gale
Executive Editor: Chalon Bridges
Project Editor: Nora Lally-Graves
Developmental Editor: Nora Lally-Graves
Editorial Assistant: Rachel Brickner
Marketing Manager: Lauren Garritson
Editorial Media Manager: Tim Flem
Mastering Media Producer: Wendy Romaniecki
Managing Editor, Environmental Science: Gina Cheselka

Project Manager, Science: Beth Sweeten
Art Production Manager: Connie Long
Operations Specialist: Maura Zeldivar
Photo Researchers: Maureen Raymond, Dian Lofton
Composition: PreMediaGlobal, Inc.
Production Editor: Kelly Keeler, PreMediaGlobal, Inc.
Illustrations: Imagineering
Interior Design: Maureen Eide
Cover Design: Yvo Riezebos Design
Cover Photograph: Photolibrary/Dave Jacobs

Photo credits follow the glossary.

Printed in the United States
10 9 8 7 6 5 4 3

Library of Congress Cataloging-in-Publication Data
Withgott, Jay.
Environment: the science behind the stories / Jay Withgott, Scott Brennan. — 4th ed.
    p. cm.
  Includes bibliographical references and index.
  ISBN 978-0-321-71534-0 (pbk.)
1. Environmental sciences.   I. Brennan, Scott.   II. Title.
GE105.B74 2011
333.7—dc22

2010029039

ISBN-10: 0-321-71534-9/ISBN-13: 978-0-321-71534-0 (Student edition)
ISBN-10: 0-321-72155-1/ISBN-13: 978-0-321-72155-6 (Books á la Carte)

Benjamin Cummings
is an imprint of

www.pearsonhighered.com

ABOUT THE AUTHORS

JAY H. WITHGOTT is a science and environmental writer with a background in scientific research and teaching. He holds degrees from Yale University, the University of Arkansas, and the University of Arizona. As a researcher, he has published scientific papers on topics in ecology, evolution, animal behavior, and conservation biology in journals including *Proceedings of the National Academy of Sciences, Proceedings of the Royal Society of London B, Evolution,* and *Animal Behavior.* He has taught university-level laboratory courses in ecology, ornithology, vertebrate diversity, anatomy, and general biology.

As a science writer, Jay has authored articles for a variety of journals and magazines including *Science, New Scientist, BioScience, Smithsonian, Current Biology, Conservation in Practice,* and *Natural History.* He combines his scientific expertise with his past experience as a reporter and editor for daily newspapers to make science accessible and engaging for general audiences.

Jay lives with his wife, biologist Susan Masta, in Portland, Oregon.

SCOTT BRENNAN has taught environmental science, ecology, resource policy, and journalism at Western Washington University and at Walla Walla Community College. He has also worked as a journalist, photographer, and consultant.

ABOUT OUR SUSTAINABILITY INITIATIVES

This book is carefully crafted to minimize environmental impact. The materials used to manufacture this book originated from sources committed to responsible forestry practices. The paper is Forest Stewardship Council™ (FSC®) certified. The printing, binding, cover, and paper come from facilities that minimize waste, energy consumption, and the use of harmful chemicals.

Pearson closes the loop by recycling every out-of-date text returned to our warehouse. We pulp the books, and the pulp is used to produce items such as paper coffee cups and shopping bags. In addition, Pearson aims to become the first climate neutral educational publishing company.

The future holds great promise for reducing our impact on Earth's environment, and Pearson is proud to be leading the way. We strive to publish the best books with the most up-to-date and accurate content, and to do so in ways that minimize our impact on Earth.

MIX
Paper from
responsible sources
FSC
www.fsc.org
FSC® C005928

Benjamin Cummings
is an imprint of

iii

CONTENTS

iv

LETTER TO THE STUDENT

Dear Student,

You are coming of age at a unique and momentous time in history. Within your lifetime, our global society must chart a promising course for a sustainable future—or it will risk peril. The stakes could not be higher, and the path we take will depend largely on how we choose to interact with our environment.

Today we live long lives enriched with astonishing technologies, in societies more free, just, and equal than ever before. We enjoy wealth on a scale our ancestors could hardly have dreamed of. Yet we have purchased these wonderful things at a price. By tapping into Earth's resources and ecological services, we are depleting our planet's bank account and running up its credit card. Never before has Earth held so many people making so many demands upon it. We are altering our planet's land, air, water, nutrient cycles, biodiversity, and climate at dizzying speeds. More than ever before, the future of our society rests with how we treat the world around us.

Environmental science helps show us how Earth's systems function and how we influence these systems. It gives us a big-picture understanding of the world and our place within it. Studying environmental science helps us comprehend the problems we create, and it can reveal ways to fix those problems. Environmental science is not just some subject you learn in college; it's something that relates to everything around you for your entire life!

We have written this book because today's students will shape tomorrow's world. At this unique moment in history, students of your generation are key to achieving a sustainable future for our civilization. Your decisions and actions will make all the difference. Conversely, you will want to read this book because your world is being shaped by the phenomena you will learn about in your environmental science course. Environmental science is your key to basic literacy in the issues of the 21st century, and the more and sooner you can learn, the better off you will be. As you gain an understanding of how people depend on nature, how our society relies on what our environment provides, and how science can help guide us toward a sustainable future, it will help you and it will help our society.

In this book we try to focus on providing hope and solutions. The many environmental challenges that face us can seem overwhelming, but we want you to feel encouraged and motivated. Remember that each dilemma is also an opportunity; for every problem that human carelessness has managed to create, human ingenuity can devise an answer. Now is the time for innovation and creativity and the fresh perspectives that a new generation can offer. Your own ideas and energy *will* make a difference. You are the solution!

JAY WITHGOTT AND SCOTT BRENNAN

LETTER TO THE INSTRUCTOR

Dear Instructor,

You have taken on one of our most important jobs: educating today's students—the citizens and leaders of tomorrow—on the fundamentals of the world around them and on the most vital issues of our time. We wrote this book to assist you in this endeavor, because we feel that the crucial role of environmental science in today's world makes it imperative to engage, educate, and inspire a broad audience of students.

In *Environment: The Science behind the Stories*, we try to implement the very best in modern teaching approaches and to clarify how science can inform human efforts to create a sustainable society. We also aim to maintain a balanced approach and to encourage critical thinking as we flesh out the social debate over environmental issues. And as we assess the challenges facing our society and our planet, we focus on providing hope and solutions.

In crafting the fourth edition of this text, we incorporated the most current information from this fast-moving field and streamlined our presentation to make learning straightforward and appealing. This edition contains more information in fewer pages in a more readable style. We examined every line with care to make sure all content is accurate, clear, and up-to-date. Moreover, a number of major changes are new to this edition.

NEW TO THIS EDITION

This fourth edition contains a number of major changes that together enhance the effectiveness of our presentation while strengthening our ongoing commitment to teaching science in an engaging and accessible way.

➤ **EnvisionIt** A full-page photo essay appearing once in each of our 17 Issues chapters, the *EnvisionIt* feature is designed to draw in visual learners and to help all students envision vital concepts. Some *EnvisionIt* pages help students visualize the scale of a phenomenon. Others bring vibrant life to the human side of an environmental issue or show students how issues affect people in cultures throughout the world. Each page concludes with a list of things students can personally do to help address the issue being presented. Because today's students are so visually driven, we expect this new feature will help to better engage them in the book and in your course.

➤ **Enhanced geology coverage** We have enhanced our coverage of the geosciences throughout the book. Expanded coverage of the fundamentals of geology is moved forward and now drives our heavily revised Chapter 2, along with newly added coverage of geologic hazards. In addition, we have added an entirely new chapter on minerals and mining (Chapter 23). This chapter highlights the significance of mineral resources in our lives, the techniques and impacts of mining, and the importance of using mineral resources as sustainably as possible.

➤ **New *Case Studies* and *Science behind the Stories*** We have introduced eight new *Central Case Studies* and 17 new *Science behind the Story* features, replacing existing ones. Thus, fully one-third of these hallmark features of our book are new to this edition, giving you a more current and exciting selection of scientific studies to highlight and stories to tell.

➤ **FSC-certified paper** One of these brand-new *Central Case Studies* (Chapter 12) describes how our book is using FSC-certified sustainable paper from the Upper Peninsula of Michigan. Students learn about the process used to make the very textbook they are holding in their hands while they learn how certification programs and corporate responsibility can help drive sustainability efforts.

➤ **Enhanced photos, art, and graphs** We have greatly strengthened our visual presentation of material throughout the text, seeking new and effective ways of teaching visually with improved photos, artwork, and graphs. This edition includes 300 new photos and 130 new graphs and illustrations. In addition, 135 existing figures have been revised to reflect current data or for better clarity or pedagogy.

➤ **Chapter sequence** We have modified the sequence of chapters in our Foundations section to better match the preferences of most instructors, placing ethics, economics, and policy after the natural sciences but retaining these as vital foundations of environmental science.

➤ **An exciting new online platform** With this edition we are thrilled to gain a new online learning and assessment platform called *MasteringEnvironmentalScience*. Powerful yet easy to use, you can employ *Mastering* to assess student learning outside the classroom. After consulting with instructors, we apportioned our features and activities between the printed book and the *Mastering* platform in a way that makes best use of the advantages of each medium. As a result, you will find certain popular features from previous editions of the print book now on *Mastering*. These include (1) *Interpreting Graphs and Data* exercises, which, along with the interactive *GraphIt!* program, guide students in exploring how to present and interpret data and how to create graphs; (2) interactive *Causes and Consequences* exercises, which let students probe the causes behind major issues, their consequences, and possible solutions; and (3) *Viewpoints*, paired essays authored by invited experts who present divergent points of view on topical questions, providing a taste of informed arguments directly from individuals actively involved in work— and

debate—on environmental issues. Quiz questions lead students to critically think through points made in the essays.

➤ **Video Field Trips** Our favorite brand-new item on the *MasteringEnvironmentalScience* site is a series of five *Video Field Trips*. These brief videos are a wonderful resource, especially for courses unable to take students into the field. With these videos you can kick off your class period with a short visit to a landfill, wastewater treatment plant, or organic farm!

➤ **Currency and coverage of topical issues** By incorporating the most recent data possible, we've aimed to live up to our book's hard-won reputation for currency. We've also enhanced coverage of issues currently gaining prominence. To reflect the central role of energy issues today, we have bolstered our already-strong three chapters on energy. Likewise, we have taken care to update and strengthen our comprehensive chapter on climate change, as well as the many ways in which climate change appears throughout the book as it touches on other issues. This edition also benefits from extra attention to sustainable agriculture, green-collar jobs in renewable energy, carbon capture and storage, carbon taxation and cap-and-trade approaches, and a variety of topics of special interest to students, ranging from the impacts of bottled water to the use of forensics in conservation biology. Throughout, we have maintained our applications of the ecological footprint concept and our use of sustainability as an organizing theme. Our campus sustainability coverage (Chapter 24) shows students how their peers across North America are applying principles and lessons from environmental science to forge sustainable solutions on their own campuses.

EXISTING FEATURES

We have also retained the major features that made the first three editions of our book unique and that are proving so successful in classrooms across North America:

➤ **Central Case Studies integrated throughout the text** Your feedback to us has confirmed that telling compelling stories about real people and real places is the best way to capture students' interest. Narratives also help teach abstract concepts by giving students a tangible framework with which to incorporate new ideas. Many textbooks serve up case studies in isolated boxes, but we have chosen to integrate each chapter's *Central Case Study* into the main text, weaving information and elaboration throughout the chapter. In this way, the concrete realities of the case study help to illustrate the topics we cover. We are gratified that students and instructors using our book have consistently applauded this approach, and we hope it continues to bring about a new level of effectiveness in environmental science education.

➤ **The Science behind the Story** Because we strive to engage students in the scientific process of testing and discovery, we discuss the scientific method and the social context of science in our opening chapter, and we describe hundreds of real-life studies throughout the text.

We also feature in each chapter two *The Science behind the Story* boxes, which elaborate on particular studies, guiding readers through details of the research. In this way we show not merely *what* scientists discovered, but *how* they discovered it.

➤ **Weighing the Issues** The multifaceted issues in environmental science often lack black-and-white answers, so students need critical-thinking skills to help navigate the gray areas at the juncture of science, policy, and ethics. Our *Weighing the Issues* questions aim to help you develop these skills in your students. These questions serve as stopping points for students to reflect upon what they have read, wrestle with complex dilemmas, and engage in spirited classroom discussion.

➤ **Diverse end-of-chapter features** The four features that conclude each chapter target particular student needs. *Reviewing Objectives* summarizes each chapter's main points and relates them to the learning objectives presented at the chapter's opening, enabling students to confirm that they have understood the most crucial ideas and to review concepts by turning to specified page numbers. *Testing Your Comprehension* provides concise study questions on main topics in each chapter, while *Seeking Solutions* encourages broader creative thinking aimed at finding solutions. "Think It Through" questions within this section place students in a scenario and empower them to make decisions to resolve problems. Lastly, *Calculating Ecological Footprints* enables students to quantify the environmental impacts of their own choices and then measure how individual impacts scale up to the societal level.

➤ **An emphasis on solutions** For many students in environmental science courses, the deluge of environmental problems can cause them to feel that there is no hope or that they cannot personally make a difference. We have aimed to counter this impression by drawing out innovative solutions being implemented or considered around the world. While being careful not to paint too rosy a picture of the challenges that lie ahead, we try to instill hope and encourage action. To recognize the efforts of faculty and students toward encouraging sustainable practices on campus and in the community, Pearson Education will continue to award a Sustainable Solutions Award to the campuses which best exemplify the principles of sustainability. Go online at **www.masteringenvironmentalscience.com** for entry details and for profiles of previous Sustainable Solutions Award winners.

Environment: The Science behind the Stories has grown from our experiences in teaching, research, and writing. Jay Withgott has synthesized and presented science to a wide readership. His experience in distilling and making accessible the fruits of scientific inquiry shapes our book's content and presentation. We have been guided in our efforts by input from the hundreds of instructors across North America who have served as reviewers and advisors. The participation of so many learned and thoughtful experts has improved this volume in countless ways.

We sincerely hope that our efforts are worthy of the immense importance of our subject matter. We invite you to let us know how well we have achieved our goals and where you feel we have fallen short. Please write the authors in care of Chalon Bridges (chalon.bridges@pearson.com) at Pearson Education. We value your feedback and want to know how we can serve you better.

JAY WITHGOTT AND SCOTT BRENNAN

INSTRUCTOR SUPPLEMENTS

Instructor Resource Center on DVD 0-321-72138-1 with TestGen

This powerful media package is organized chapter-by-chapter and includes all teaching resources in one convenient location. You'll find Video Field Trips, PowerPoint presentations, Active Lecture questions to facilitate class discussions (for use with or without clickers), and an image library that includes all art and tables from the text.

Included on the IRDVD, the test bank includes hundreds of multiple-choice questions plus unique graphing, and scenario-based questions to test students' critical-thinking abilities.

Instructor Guide 0-321-72137-3

This comprehensive resource provides chapter outlines, key terms, and teaching tips for lecture and classroom activities.

Blackboard Premium for Environment 0-321-72141-1

Blackboard Open Access 0-321-72143-8

MasteringEnvironmentalScience™ for Environment: The Science behind the Stories 0-321-72147-0

New to Environmental Science but used by over a million science students, the Mastering platform is the most effective and widely used online tutorial, homework, and assessment system for the sciences.

Where current issues meet current science

Integrated Central Case Studies begin and are woven throughout each chapter, highlighting real people and real places in order to bring current environmental issues to life.

◀ **NEW! 30% of the Case Studies** in the book are completely new, including new case studies on sustainable paper (Chapter 12) and the mining of tantalum for use in cell phones and other digital devices (Chapter 23).

CENTRAL CASE STUDY

Certified Sustainable Paper in Your Textbook

"As a company and as individuals, we must be able to look ourselves in the mirror and know that we are truly doing what is right and good for the environment."

—Rick Willett, president and CEO, NewPage Corporation

"FSC is the high bar in forest certification, because its standards protect forests of significant conservation value and consistently deliver meaningful improvements in forest management on the ground."

—Kerry Cesareo, deputy director, WWF-US Forests Program

As you turn the pages of this textbook, you are handling paper made from trees that were grown, managed, harvested, and processed using certified sustainable practices.

If you were to trace the paper in this book back to its origin, you would find yourself standing in a diverse mixed forest of aspen, birch, beech, maple, spruce, and pine in the Upper Peninsula of Michigan. This sparsely populated region near the Canadian border, flanked by Wisconsin, Lake Michigan, and Lake Superior, remains heavily forested, despite supplying timber to our society for nearly 200 years.

The trees cut to make this book's paper were selected for harvest based on a sustainable management plan designed to avoid depleting the forest of its mature trees or degrading the ecological functions the forest performs. The logs were then transported to a nearby pulp and paper mill at Escanaba, a community of 13,000 people on the shore of Lake Michigan. The mill is Escanaba's largest employer, providing jobs to about 1,100 residents.

Worker examines paper being produced at the mill in Escanaba, Michigan

At the Escanaba mill, the wood is chipped and then fed into a digester, where the chips are cooked with chemicals to break down the wood's molecular bonds. Lignin in the wood (the polymer that gives wood its sturdiness) is separated out and discarded [...] paper is ma [...] wood's cellu [...] Millworkers [...] and screen [...] and mix the [...] water and a [...] smooth, wa [...] onto a movi [...] water drains [...] rollers press [...] into thin she [...] newly forme [...] dried on he [...] made ready [...] variety of co [...] wound into [...] to be cut int [...]

In this w [...] mill produces about 770,000 tons [...] year. It does its best to recycle ch [...] used in the process, and it combu [...] to help power the mill.

At every stage in this process [...] third-party inspectors from the Fo [...] Council (FSC) examine the practic [...]

▶ The **Central Case Studies** draw students in at the start of each chapter and continue to unfold throughout the chapter text, making the science more understandable and interesting to learn.

FIGURE 12.19 ▲ Progress worldwide toward sustainable forestry is mixed, as shown by six leading indicators. On the one hand, forest uses are shifting toward conservation, and timber production is increasingly coming from plantations. On the other hand, global forest area is still shrinking. Data from U.N. Food and Agriculture Organization, 2010. *Global forest resources assessment 2010.* Data reflect averages as reported by FAO for most-recent years, ending in 2010 but extending back variably to 2005, 2000, or 1990.

Chart axis labels: Loss / Gain; Total forest; Primary forest; Forest for timber production; Forest for soil/water protection; Forest for biodiversity conservation; Productive plantations. X-axis: Average annual net change in forest area (million hectares per year), -6 -4 -2 0 2 4 6.

FIGURE 12.20 ▲ An inspector marks wood with the FSC logo after confirming that it was harvested in accord with FSC criteria for sustainable harvesting. A consumer movement centered on independent certification of sustainable wood products is allowing consumer choice to promote sustainable forestry practices.

to pulping to production—have met strict standards. FSC certification rests on 10 general principles (**TABLE 12.2**) and 56 more-detailed criteria.

The number of FSC-certified forests, companies, and products is growing quickly. Five percent of the world's forests managed for timber production are now FSC certified. This totaled over 128 million ha (320 million acres) in 80 nations as of spring 2010. FSC-labeled products accounted for over $20 billion in sales in 2008. About 1,000 operations are certified, and over 16,000 companies are chain-of-custody certified. Some of the

growth of certification is tied to the increasing construction of sustainable, or green, buildings (pp. 000, 000).

This expanding participation means that FSC-certified products are becoming easier for consumers to find. Consumer demand for sustainable wood has been great enough that many major retail businesses such as Home Depot now sell sustainable wood. The purchasing decisions of such retailers are influencing the practices of timber companies. You can look for the logos of certifying organizations (**FIGURE 12.20**) on forest products where they are sold. By asking retailers whether they carry certified wood, you make businesses aware of your preferences and help drive demand for these products. Through our requests and our purchases, we as consumers can influence practices in forests from Michigan to Malaysia to Manaus. Sustainable forestry is more costly for the timber industry, but if certification standards are kept adequately strong, then consumer choice in the marketplace can be a powerful driver of sustainable forestry practices.

PARKS AND PROTECTED AREAS

As our world fills with more people consuming more goods, the conservation and sustainable management of resources from forests and other ecosystems becomes ever more important. So does our need to preserve land and functional ecosystems by setting aside tracts of undisturbed land intended to remain forever undeveloped.

Preservation has been part of the American psyche ever since John Muir rallied support for saving scenic lands in the Sierras (p. 143). For ethical reasons as well as pragmatic ecological and economic ones, Americans and people worldwide have chosen to set aside areas of land in perpetuity to be protected from development. Today 12% of the world's land area is designated for preservation in various types of parks, reserves, and protected areas.

TABLE 12.2 Ten Principles of Forest Stewardship Council (FSC) Certification

To receive FSC certification, forest product companies must:

- ▶ Comply with all laws and treaties.
- ▶ Show uncontested, clearly defined, long-term land rights.
- ▶ Recognize and respect indigenous peoples' rights.
- ▶ Maintain or enhance long-term social and economic well-being of forest workers and local communities, and respect workers' rights.
- ▶ Use and share benefits derived from the forest equitably.
- ▶ Reduce environmental impacts of logging and maintain the forest's ecological functions.
- ▶ Continuously update an appropriate management plan.
- ▶ Monitor and assess forest condition, management activities, and social and environmental impacts.
- ▶ Maintain forests of high conservation value.
- ▶ Promote restoration and conservation of natural forests.

EnvisionIt photo essays engage students by helping them visualize environmental issues and consider potential solutions.

ENVISIONIT

Each year Earth loses 18 million acres of forest – an area the size of South Carolina.

Satellite photo of Rondonia, Brazil, 1975

As roads are built into the Amazon rainforest, small settlers and large corporations cut the forest and convert the land for cattle ranching and soy production. Development follows wherever roads are built.

Satellite photo of same area, 2001

Forest loss drives biodiversity loss, climate change, erosion, and other problems.

Cattle on burned and cleared land

A soy farm leaves a tiny forest fragment

YOU CAN MAKE A DIFFERENCE

➤ Buy certified sustainable wood products.

➤ Eat less meat (raising livestock requires more land than raising crops).

➤ Contribute to forest protection efforts in your region and around the world.

Often people can farm the land for only a few years before the soil gives out and they have to cut more forest.

Settlers in the Amazon

320

▲ **NEW! EnvisionIt Photo Essays** appear in every chapter from 8-24 and cover current topics like deforestation and the Deepwater Horizon oil spill.

Where current issues **meet current science**

Science behind the Story boxes highlight how scientists pursue questions and test predictions.

The SCIENCE behind the Story

Using Forensics to Uncover Illegal Whaling

As any television buff knows, forensic science is a crucial tool in solving mysteries and fighting crime. In recent years conservation biologists have been using forensics to unearth secrets and catch bad guys in the multi-billion-dollar illegal global wildlife trade. One such detective story comes from the Pacific Ocean and Japan.

The meat from whales has long been a delicacy in Japan and other nations. Whaling ships decimated populations of most species of whales in the 20th century through overhunting, and the International Whaling Commission (IWC) outlawed commercial whaling worldwide beginning in 1986. Yet whale meat continues to be sold at market to wealthy consumers today (see photo). This meat comes legally from several sources:

1. From scientific hunts. Japan and several other nations negotiated with the IWC to continue to hunt limited numbers of whales for research purposes, and this meat may be sold afterwards.
2. From whales killed accidentally when caught in fishing nets meant for other animals (bycatch; p. 000).
3. Possibly from stockpiles frozen before the IWC's moratorium.

However, conservation biologists long suspected that much of the whale meat on the market was actually caught illegally for the purpose of selling for food and that fleets from Japan and other nations were killing more whales than international law allowed. Once DNA sequencing technology

Dr. C. Scott Baker of Oregon State University

was developed, scientists could use this tool to find out.

The detectives in this story are conservation geneticists C. Scott Baker, Stephen Palumbi, Frank Cipriano, and their colleagues. For close to two decades they have been traveling to Asia on what has amounted to top-secret grocery shopping trips.

Minke whales **(a)** from the western Pacific Ocean may end up as meat in Japanese or Korean markets **(b)**.

It began in 1993, when Baker and Palumbi bought samples of whale meat—all labeled simply as *kujira*, the generic Japanese term for whale meat—from a number of markets in Japan and sequenced DNA from these samples. Law forbids the export of whale meat, so the researchers had to run their analyses in their hotel rooms with portable genetic kits. Once they were back home in the United States, they compared their data with sequences from known whale species.

By analyzing which samples matched which, they concluded that they had sampled meat from nine minke whales, four fin whales, one humpback whale, and two dolphins. Moreover, because subspecies of whales from different oceans differ genetically, the researchers were able to analyze the genetic variation in their samples and learn that one fin whale came from the Atlantic whereas the other three were from the Pacific, and that eight of the nine minke whales came from the Southern Hemisphere.

Because several of these species and/or subspecies were off-limits to hunting, the data suggested that some of the meat had been hunted, processed, or traded illegally. Baker and Palumbi concluded in a 1994 paper in *Science* that "the existence of legal whaling serves as a cover for the sale of illegal whale products." They urged that the international community monitor catches more closely.

Two year[...] and Cipriano[...]

304

◄ **NEW!** **30% of the Science behind the Stories** are completely new, including the use of science to uncover illegal whaling (Chapter 11) and chemicals in bottled water (Chapter 15).

The Science behind the Story boxes ► carefully walk you through the scientific process—not only what scientists know, but how they discovered it.

The SCIENCE behind the Story

Is It Better in a Bottle?

Quick—Which is safer and healthier for you to drink, tap water or bottled water?

If you said bottled water, you're not alone. Most people think bottled water is safer and healthier, which is why sales have doubled each decade for the past 20 years. But is bottled water really as pure and clean as its marketers want us to think?

It's hard to know the answer, because companies are not required to tell us anything about the quality of the water in their bottles, or even where the water comes from. Municipalities that provide tap water to their residents need to submit regular reports to the Environmental [Protec]tion Agency describing their

Martin Wagner and Jörg Oehlmann, Johann Wolfgang Goethe University, Germany

Group each published reports detailing how consumers cannot get basic information about the bottled water they purchase. The Environmental Working Group (EWG), based in Washington, D.C., surveyed the labels and websites of 188 brands of bottled water. Only 2 of these brands disclosed information comparable to th[...] required of munici-pal tap [water...] the FDA told

fertilizer, and various industrial compounds such as solvents and plasticizers (**first figure**). Each brand contained 8 contaminants on average, and two brands had levels of chemicals that exceeded legal limits in California and industry safety guidelines.

Two brands showed the chemical composition of standard municipal water treatment—including chlorine, fluoride, and other by-products of disinfection—indicating that these were identical to tap water. Indeed, an estimated 25–44% of bottled water *is* tap water. Corporations can essentially just turn on the faucet, filter the water, bottle it, and sell it to us at marked-up prices.

EW[...]

Help students learn how to **read graphs** and **understand data**.

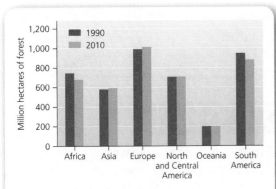

FIGURE 12.6 ▲ Africa and South America are experiencing deforestation as they attempt to develop, extract resources, and provide new agricultural land for their growing populations. In Europe, forested area is slowly increasing as some formerly farmed areas are abandoned and allowed to grow back into forest. Data for North and Central America reflect a balance of forest regrowth in North America and forest loss in Central America. In Asia, natural forests are being lost, but China's extensive planting of tree plantations (here counted as forests) to combat desertification has increased forest cover for Asia since 1990. Data from U.N. Food and Agriculture Organization, 2010. *Global forest resources assessment 2010.*

▲ We use the most current data available and cite sources, so you can trace the information presented.

▼ Unique **Graphing Appendix** shows students how to read and understand the most common types of graphs.

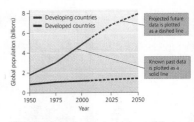

KEY CONCEPT: Projections

Besides showing observed data, we can use graphs to show data that is predicted for the future, based on models, simulations, or extrapolations from past data. Often, projected future data on a line graph are shown with dashed lines, as in **FIGURE A.4**, to indicate that they are less certain than data that has already been observed.

FIGURE A.4 ▲ Past and projected population growth for developing and developed countries. (Figure 8.22, page 217).

▼ **Calculating Ecological Footprint** activities at the end of every chapter let students evaluate the impact of actions.

CALCULATING ECOLOGICAL FOOTPRINTS

In the United States, a common dream is to own a suburban home with a weed-free green lawn. Nationwide, Americans tend about 40.5 million acres of lawn grass. But conventional lawn care involves applying fertilizers, pesticides, and irrigation water, and using gasoline or electricity for mowing and other care—all of which raise environmental and health concerns. Using the figures for a typical lawn in the table, calculate the total amount of fertilizer, water, and gasoline used in lawn care across the nation each year.

	Acreage of lawn	Fertilizer used (lbs)	Water used (gal)	Gasoline used (gal)
For the typical 1/4-acre lawn	0.25	37	16,000	4.9
For all lawns in your hometown				
For all lawns in the United States	40,500,000			

Data from Chameides, B., 2008. http://www.nicholas.duke.edu/thegreengrok/lawns.

Make Learning Part of the Grade®

www.masteringenvironmentalscience.com

MasteringEnvironmentalScience helps students arrive better prepared for lectures through a variety of auto-graded activities.

VIDEO field trips

◀ **Video Field Trips** give students a fascinating behind-the-scenes tour of coal-fired power plants, wastewater treatment facilities, landfills, farms, and more. Each Video Field Trip includes assessment questions to assign as homework.

BioFlix®

◀ **BioFlix® activities** are highly interactive and use 3D animations.

Topics include:
- Cellular Respiration
- Photosynthesis
- Mechanisms of Evolution
- Carbon Cycle
- Population Ecology

Current Events quizzes prompt students ▶ to read recent *New York Times* articles.

The New York Times

Wildlife Toll Mounts as BP Oil Inundates Gulf Coast Marshes

By APRIL REESE of Greenwire
Published: June 7, 2010

VENICE, La. -- After several hours motoring through the bays and passes that web across the Mississippi River Delta, Bob Ford, a wildlife biologist with the Fish and Wildlife Service, spots a few brown pelicans and frigate birds perched on the remnants of a hurricane-ravaged barge platform in Redfish Bay.

✉ SIGN IN TO E-MAIL

🖨 PRINT

▼ The **Pearson eText** gives students access to the text whenever and wherever they can access the Internet. The eText pages look exactly like the printed text, and include powerful interactive and customization functions.

Highlight Function
lets students highlight what they want to remember.

Interactive Glossary
provides pop-up definitions and terms.

Instructors can share
their notes and highlights with students and can also hide chapters.

Annotation Function
allows students to take notes.

Google-based
search function.

Hyperlinks to quizzes, tests, activities, and animations.

Zoom in and out for better viewing.

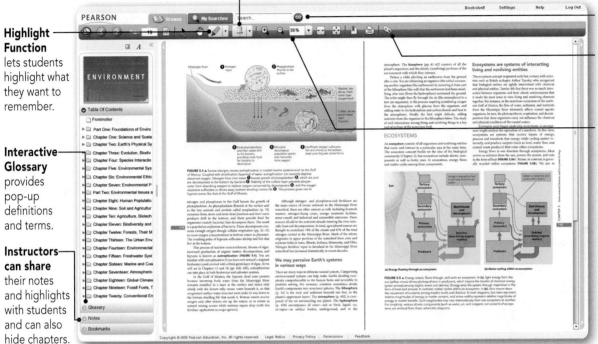

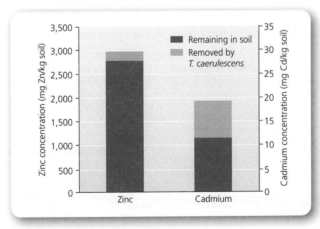

◀ **Interpreting Graphs and Data**
activities use real research examples to help students understand how to graph data and interpret graphs.

Additional Study Tools on MasteringEnvironmentalScience™:

- **Viewpoints** activities offer a taste of the debates surrounding environmental issues through a pair of informed arguments on each topic—directly from individuals actively working on the issues. Each is followed by assignable quiz questions that check reading comprehension and encourage students to think critically.

- **Causes & Consequences** exercises help students determine the causes, consequences, and potential solutions to environmental issues.

- **Reading Quizzes** help your students keep on track with your reading assignments and test their true understanding of the content.

- **Self Study Area** offers a 24/7 study resource.

Make Learning Part of the Grade®

www.masteringenvironmentalscience.com

The Mastering platform is the most effective and widely used online tutorial, homework, and assessment system for the sciences.

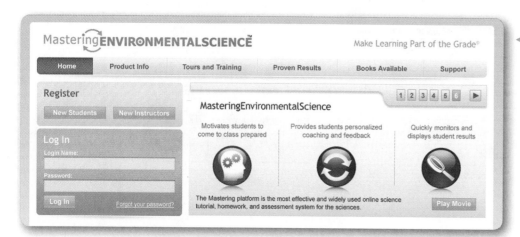

- Over **one million** active student users
- 99.8% server reliability

▼ **The MasteringEnvironmentalScience gradebook** provides you with quick results and easy-to-interpret insights into student performance.

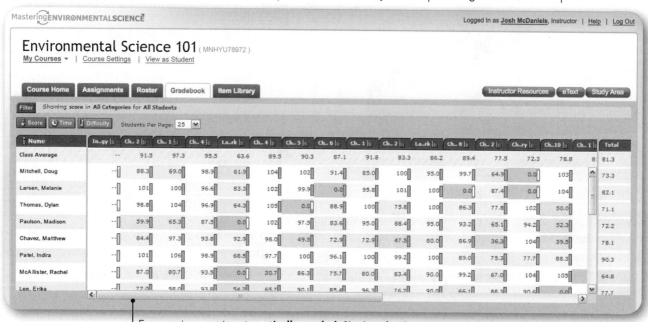

Every assignment is **automatically graded**. Shades of red highlight vulnerable students and challenging assignments.

data on student performance

▼ **Student performance snapshots** provide quick results and unique insight.

Customize Content
Questions and answers can be easily edited.

Pre-Lecture Quizzes
Assign reading quizzes before lecture to help students arrive prepared.

At-A-Glance Statistics
Reveal data for your class as well as national results.

Color Bar Data
Green indicates correct answers, Red shows the percentage of students who requested an answer. Orange indicates the average number of wrong answers per student.

Wrong Answers Summary
Gives unique insight into your students' misunderstandings and allows for just-in-time teaching adjustments.

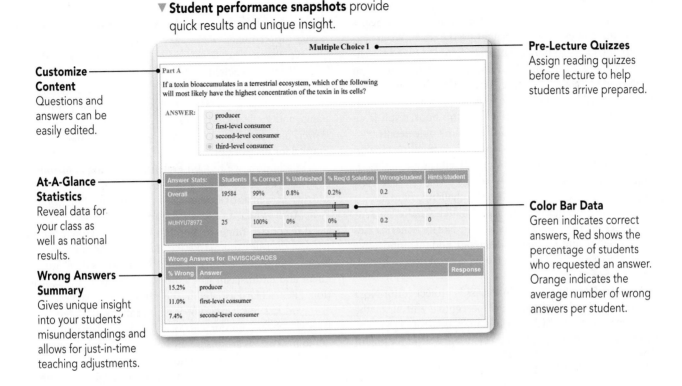

Multiple Choice 1

Part A

If a toxin bioaccumulates in a terrestrial ecosystem, which of the following will most likely have the highest concentration of the toxin in its cells?

ANSWER:
- producer
- first-level consumer
- second-level consumer
- third-level consumer

Answer Stats:	Students	% Correct	% Unfinished	% Req'd Solution	Wrong/student	Hints/student
Overall	19584	99%	0.8%	0.2%	0.2	0
MUHYU78972	25	100%	0%	0%	0.2	0

Wrong Answers for ENVISCIGRADES

% Wrong	Answer	Response
15.2%	producer	
11.0%	first-level consumer	
7.4%	second-level consumer	

Continuously Improving Content

MasteringEnvironmentalScience™ offers a dynamic pool of assignable content that improves with student usage. Detailed analysis of student performance statistics—including time spent, answers submitted, and hints used—ensures the highest quality content.

1. We conduct a **thorough analysis** of each problem by reviewing the color bar data that has been generated through classroom use with real students. Data includes difficulty level, number of attempts, number of hints, and confusing or distracting factors.

2. We **make corrections** to improve problems that have ambiguous answer choices, ineffectual distractors, or tricky language.

3. We **continue to strengthen** each problem by gathering color bar data from real students…

…and then the process repeats. This ongoing process collects input from every student who works with MasteringEnvironmentalScience.

ACKNOWLEDGMENTS

A textbook is the product of *many* more minds and hearts than one might guess from the names on the cover. We are exceedingly fortunate to be supported and guided by a tremendous publishing team and by a small army of experts in environmental science who have generously shared their time and expertise. Although we alone, as authors, bear responsibility for any inaccuracies, the strengths of this book result from the collective labor and dedication of innumerable people.

We would first like to thank our executive editor, Chalon Bridges. Chalon's commitment to quality education and publishing is simply inspirational. By her own example she constantly motivates our team to relish the challenge of making a successful and well-received book still better. Chalon's extensive interaction with instructors across North America has helped us define and refine our pedagogy and our innovative features. Every aspect of this book owes a great deal to her astute guidance and vision.

Also crucial to this edition's success is project editor Nora Lally-Graves. Nora's insightful and perceptive editing of the text was matched only by her skillful management of publishing logistics—and we greatly appreciate her sound judgment in both arenas. Her deadline whip-cracking (always with an encouraging word) kept the project on track despite the best authorial attempts to throw it off course.

We wish to thank our Editor-in-Chief Beth Wilbur and Editorial Director Frank Ruggirello for their support of the project through its several editions.

Beth Sweeten managed production of the book under the guidance of Gina Cheselka. Connie Long handled the art program, and Maureen Eide designed the new text interior. Sally Peyrefitte once again provided meticulous copy-editing of our text. We send a big thanks to production editor Kelly Keeler and the rest of the staff at PreMediaGlobal for their great work in putting this fourth edition together.

This edition's enhanced photo program benefits from extensive photo research by Maureen Raymond and Dian Lofton. For input and management on photos we thank Travis Amos and Elaine Soares.

We thank executive director of development Deborah Gale, as well as editorial assistant Rachel Brickner. Project editor Leata Holloway helped get the edition going in its early stages.

We thank Kayla Rihani, Julie Stoughton, Steven Frankel, Tim Flem, Lee Ann Doctor, Wendy Romaniecki, Pete Ratkevich, Karen Sheh, Tania Mlawer, and David Chavez for their work on the Mastering website and our media supplements.

We thank Eric Flagg for his tremendous Video Field Trips. We are so excited about them!

Our thanks go to Danielle DuCharme for updating our Instructor's Guide, to Thomas Pliske for his help with Test Bank, and to Heidi Marcum for doing the revision of the PowerPoint lectures. We would also like to thank Steven Frankel for his work on the clickers.

In addition, we remain grateful for lasting contributions to the book's earlier editions by Susan Teahan, and Mary Ann Murray, as well as by Etienne Benson, Russell Chun, Jonathan Frye, April Lynch, and Kristy Manning.

Of course, none of this has any impact on education without the sales and marketing staff to get the book into your hands. Marketing manager Lauren Garritson is dedicating her talent and enthusiasm to the book's promotion and distribution. Lauren Harp assisted in the edition's early stages.

And last but surely not least, the many sales representatives who help communicate our vision and deliver our product to instructors are absolutely vital, and we deeply appreciate their tireless work and commitment.

In the lists that follow, we acknowledge the many instructors and outside experts who have helped us maximize the quality and accuracy of our presentation through their chapter reviews, feature reviews, class tests, or other services. We wish to extend a special thank-you to Dr. Ninian Stein of Wheaton College for forwarding extensive student reviews of our third edition, as well as Dr. Steven Rudnick of UMass-Boston, Dr. Lorne Wolfe of Georgia Southern University, Dr. Todd Tracy of Northwestern College, and Ms. Christine Brady of Cal Poly Pomona for previous student reviews. If the thoughtfulness and thoroughness of our many reviewers are any indication, we feel confident that the teaching of environmental science is in excellent hands!

Lastly, Jay gives loving thanks to his wife, Susan Masta, who has endured this book's writing and revision with tremendous patience and sacrifice and has provided support and sustenance throughout.

We dedicate this book to today's students, who will shape tomorrow's world.

JAY WITHGOTT AND SCOTT BRENNAN

REVIEWERS

Pre-Revision Reviewers for the Fourth Edition

Isoken T. Aighewi, *University of Maryland–Eastern Shore*
Jon Barbour, *University of Colorado–Denver*
Lisa K. Bonneau, *Metropolitan Community College–Blue River*
Bruno Borsari, *Winona State University*
Hauke Busch, *Augusta State University*
Kelly Cartwright, *College of Lake County*
Mandy L. Comes, *Rockingham Community College*
Roger del Moral, *University of Washington*
Kathy McCann Evans, *Reading Area Community College*
Marcia L. Gillette, *Indiana University–Kokomo*
Nisse Goldberg, *Jacksonville University*
Jason Hlebakos, *Mount San Jacinto College*
James M. Hutcheon, *Asian University for Women*
Juana Ibanez, *University of New Orleans*
Gail F. Johnston, *Lindenwood University*
David M. Kargbo, *Temple University*
Susan Karr, *Carson–Newman College*
Jerry H. Kavouras, *Lewis University*
Kevin King, *Clinton Community College*
Diana Kropf-Gomez, *Richland College (DCCCD)*
Kurt Leuschner, *College of the Desert*
Kenneth Mantai, *State University of New York at Fredonia*
Michael D. Marlen, *Southwestern Illinois College*
Robert M. L. McKay, *Bowling Green State University*
Alberto Mestas-Nunez, *Texas A&M University–Corpus Christi*
Michael Nicodemus, *Abilene Christian University*
Tommy Parker, *University of Louisville*
Barry Perlmutter, *College of Southern Nevada*
Patricia Bolin Ratliff, *Eastern Oklahoma State College*
Virginia Rivers, *Truckee Meadows Community College*
John Rueter, *Portland State University*
Brian R. Shmaefsky, *Lone Star College–Kingwood*
Roy Sofield, *Chattanooga State Technical Community College*
Annelle Soponis, *Reading Area Community College*
Douglas Spieles, *Denison University*
Keith S. Summerville, *Drake University*
Bill Stewart, *Middle Tennessee State University.*
Jamey Thompson, *Hudson Valley Community College*
Thomas Tyning, *Berkshire Community College*
Peter Weishampel, *Northland College*
Jeffrey Wilcox, *University of North Carolina at Asheville*
Shaun Wilson, *East Carolina University*
Brian G. Wolff, *Minnesota State Colleges and Universities*
Jessica Wooten, *Franklin University*

Chapter Reviewers for the Fourth Edition

Matthew Abbott, *Des Moines Area Community College*
Deniz Z. Altin, *Georgia Perimeter College*
Mark W. Anderson, *The University of Maine*
Lisa K. Bonneau, *Metropolitan Community College–Blue River*
John S. Campbell, *Northwest College*
Myra Carmen Hall, *Georgia Perimeter College*
Kelly Cartwright, *College of Lake County*
Tait Chirenje, *The Richard Stockton College of New Jersey*
Donna Cohen, *MassBay Community College*
Michael L. Denniston, *Georgia Perimeter College*
Brad Fiero, *Pima Community College*

Marcia Gillette, *Indiana University–Kokomo*
Stanley J. Kabala, *Duquesne University*
Myung-Hoon Kim, *Georgia Perimeter College*
Ned J. Knight, *Linfield College*
Penelope M. Koines, *University of Maryland*
Erica Kosal, *North Carolina Wesleyan College*
James D. Kubicki, *The Pennsylvania State University*
Carlton Lee Rockett, *Bowling Green State University*
Dan McNally, *Bryant University*
Kiran Misra, *Edinboro University of Pennsylvania*
Bruce Olszewski, *San Jose State University*
Thomas E. Pliske, *Florida International University*
Virginia Rivers, *Truckee Meadows Community College*
Ajgaonkar Sandiford, *College of DuPage*
Brian R. Shmaefsky, *Lone Star College*
Linda Sigismondi, *University of Rio Grande*
Ninian R. Stein, *Wheaton College*
Dion C. Stewart, *Georgia Perimeter College*
Julie Stoughton, *University of Nevada–Reno*
Keith Summerville, *Drake University*
Keith S. Summerville, *Drake University*
Jamey Thompson, *Hudson Valley Community College*
Todd Tracy, *Northwestern College*

Consulting Reviewers for the Fourth Edition

Shamim Ahsan, *Metropolitan State College of Denver*; Laura Beaton, *York College, City University of New York*; Dale Burnside, *Lenoir-Rhyne University*; Amy Ellwein, *University of New Mexico*; Bob Dennison, *Lead teacher at HISD and Robert E. Lee High School*; Michael L. Denniston, *Georgia Perimeter College*; Doreen Dewell, *Whatcom Community College* Brad Chandler, *Palo Alto College*; Neil Ingraham, *California State University, Fresno*; Jason Janke, *Metropolitan State College of Denver*; Linda Jensen-Carey, *Southwestern Michigan College*; Myung-Hoon Kim, *Georgia Perimeter College*; Ned J. Knight, *Linfield College*; Andrew Lapinski, *Reading Area Community College*; Kimberly Marsella, *Skidmore College*; Benjamin Neimark, *Temple University*; Shana Petermann, *Minnesota State Community and Technical College–Moorhead*; Craig D. Phelps, *Rutgers University*; Thomas E. Pliske, *Florida International University*; Alison Purcell, *Humboldt State University*; Irene Rossell, *University of North Carolina*; Elizabeth Shrader, *Community College of Baltimore County*; Patricia L. Smith, *Valencia Community College*; Debra Socci, *Seminole Community College*; Richard Stevens, *Louisiana State University*; Todd Tracy, *Northwestern College*; John F. Weishampel, *University of Central Florida*; Kelly Wessell, *Tompkins Cortland Community College*; Karen Zagula, *Waketech Community College*

Focus-Group Advisors for the Fourth Edition

Marsha Fanning, *Lenoir Rhyne College*; Scott Gleeson, *University of Kentucky*; Don Hyder, *San Juan College*; Kurt Leuschner, *College of the Desert*; Robert McKay, *Bowling Green State University*; Neal Phillip, *Bronx Community College*; Gerald Pollack, *Georgia Perimeter College*; Norm Strobel, *Bluegrass Community Technical College*; Rudi Thompson, *University of North Texas*; Luanne Clark, *Lansing Community College*; Matthew Laposata, *Kennesaw State University*; Heidi Marcum, *Baylor University*; Frank Phillips, *McNeese State University*; Kayla Rihani, *Northeast Illinois University*; Gary Silverman, *Bowling Green State University*; Debra Socci, *Seminole Community College*; Ninian Stein, *University of Massachusetts, Boston*; Kathy Evans, *Reading Area*

Community College; Heidi Marcum, *Baylor University*; Brian Maurer, *Michigan State University*; John Pleasants, *Iowa State University*; Mike Priano, *Westchester Community College*; Daniel Ratcliff, *Rose State College*; James Riley, *University of Arizona*; Kimberly Schulte, *Georgia Perimeter College*; Annelle Soponis, *Reading Area Community College*; Todd Tarrant, *Michigan State University*; John Weishampel, *University of Central Florida*; Narinder Bansal, *Ohlone College*

Suppliers of Student Reviews

Christine Brady, *California State Polytechnic University, Pomona*
Steven Rudnick, *University of Massachusetts–Boston*
Ninian R. Stein, *Wheaton College*
Todd Tracy, *Northwestern College*
Lorne Wolfe, *Georgia Southern University*

Reviewers for Previous Editions

David Aborne, *University of Tennessee-Chattanooga*; Jeffrey Albert, *Watson Institute of International Studies*; Shamim Ahsan, *Metropolitan State College of Denver*; John V. Aliff, *Georgia Perimeter College*; Mary E. Allen, *Hartwick College*; Dula Amarasiriwardena, *Hampshire College*; Gary I. Anderson, *Santa Rosa Junior College*; Corey Andries, *Albuquerque Technical Vocational Institute*; David M. Armstrong, *University of Colorado-Boulder*; David L. Arnold, *Ball State University*; Joseph Arruda, *Pittsburg State University*; Thomas W. H. Backman, *Linfield College*; Timothy J. Bailey, *Pittsburg State University*; Stokes Baker, *University of Detroit*; Kenneth Banks, *University of North Texas*; Reuben Barret, *Prairie State College*; Morgan Barrows, *Saddleback College*; Henry Bart, *LaSalle University*; James Bartalome, *University of California-Berkeley*; Marilynn Bartels, *Black Hawk College*; David Bass, *University of Central Oklahoma*; Christy Bazan, *Illinois State University*; Christopher Beals, *Volunteer State Community College*; Hans T. Beck, *Northern Illinois University*; Richard Beckwitt, *Framingham State College*; Barbara Bekken, *Virginia Polytechnic Institute and State University*; Elizabeth Bell, *Santa Clara University*; Timothy Bell, *Chicago State University*; David Belt, *Johnson County Community College*; Gary Beluzo, *Holyoke Community College*; Terrence Bensel, *Allegheny College*; Bob Bennett, *University of Arkansas*; William B. N. Berry, *University of California, Berkeley*; Kristina Beuning, *University of Wisconsin–Eau Claire*; Peter Biesmeyer, *North Country Community College*; Donna Bivans, *Pitt Community College*; Grady Price Blount, *Texas A&M University-Corpus Christi*; Marsha Bollinger, *Winthrop University*; Richard D. Bowden, *Allegheny College*; Frederick J. Brenner, *Grove City College*; Nancy Broshot, *Linfield College*; Bonnie L. Brown, *Virginia Commonwealth University*; David Brown, *California State University-Chico*; Evert Brown, *Casper College*; Hugh Brown, *Ball State University*; J. Christopher Brown, *University of Kansas*; Dan Buresh, *Sitting Bull College*; Lee Burras, *Iowa State University*; Christina Buttington, *University of Wisconsin–Milwaukee*; Charles E. Button, *University of Cincinnati, Clermont College*; John S. Campbell, *Northwest College*; Mike Carney, *Jenks High School*; Kelly S. Cartwright, *College of Lake County*; Jon Cawley, *Roanoke College*; Michelle Cawthorn, *Georgia Southern University*; Linda Chalker-Scott, *University of Washington*; Brad S. Chandler, *Palo Alto College*; Paul Chandler, *Ball State University*; David A. Charlet, *Community College of Southern Nevada*; Sudip Chattopadhyay, *San Francisco State University*; Tait Chirenje, *Richard Stockton College*; Richard Clements, *Chattanooga State Technical Community College*; Kenneth E. Clifton, *Lewis and Clark College*; Reggie Cobb, *Nash Community College*; John E. Cochran, *Columbia Basin College*; Luke W. Cole, *Center on Race, Poverty, and the Environment*; Thomas L. Crisman, *University of Florida*; Jessica Crowe, *South Georgia College*; Ann Cutter, *Randolph Community College*; Gregory A. Dahlem, *Northern Kentucky University*; Randi Darling, *Westfield State College*; Mary E. Davis, *University of Massachusetts, Boston*; Thomas A. Davis, *Loras College*; Lola M. Deets, *Pennsylvania State University-Erie*; Ed DeGrauw, *Portland Community College*; Roger del Moral, *University of Washington*; Craig Diamond, *Florida State University*; Darren Divine, *Community College of Southern Nevada*; Stephanie Dockstader, *Monroe Community College*; Toby Dogwiler, *Winona State University*; Jeffrey Dorale, *University of Iowa*; Tracey Dosch, *Waubonsee Community College*; Michael L. Draney, *University of Wisconsin–Green Bay*; Iver W.

Duedall, *Florida Institute of Technology*; Dee Eggers, *University of North Carolina-Asheville*; Jane Ellis, *Presbyterian College*; JodyLee Estrada Duek, *Pima Community College*; Jeffrey R. Dunk, *Humboldt State University*; Jean W. Dupon, *Menlo College*; Robert M. East, Jr., *Washington & Jefferson College*; Margaret L. Edwards-Wilson, *Ferris State University*; Anne H. Ehrlich, *Stanford University*; Thomas R. Embich, *Harrisburg Area Community College*; Kenneth Engelbrecht, *Metropolitan State College of Denver*; Bill Epperly, *Robert Morris College*; Corey Etchberger, *Johnson County Community College*; W. F. J. Evans, *Trent University*; Paul Fader, *Freed Hardeman University*; Joseph Fail, *Johnson C. Smith University*; Bonnie Fancher, *Switzerland County High School*; Jiasong Fang, *Iowa State University*; Leslie Fay, *Rock Valley College*; Debra A. Feikert, *Antelope Valley College*; M. Siobhan Fennessy, *Kenyon College*; Francette Fey, *Macomb Community College*; Steven Fields, *Winthrop University*; Brad Fiero, *Pima Community College*; Dane Fisher, *Pfeiffer University*; David G. Fisher, *Maharishi University of Management*; Linda M. Fitzhugh, *Gulf Coast Community College*; Doug Flournoy, *Indian Hills Community College-Ottumwa*; Johanna Foster, *Johnson County Community College*; Chris Fox, *Catonsville Community College*; Nancy Frank, *University of Wisconsin–Milwaukee*; Steven Frankel, *Northeastern Illinois University*; Arthur Fredeen, *University of Northern British Columbia*; Chad Freed, *Widener University*; Robert Frye, *University of Arizona*; Laura Furlong, *Northwestern College*; Navida Gangully, *Oak Ridge High School*; Sandi B. Gardner, *Triton College*; Kristen S. Genet, *Anoka Ramsey Community College*; Stephen Getchell, *Mohawk Valley Community College*; Marcia Gillette, *Indiana University-Kokomo*; Sue Glenn, *Gloucester County College*; Thad Godish, *Ball State University*; Michele Goldsmith, *Emerson College*; Jeffrey J. Gordon, *Bowling Green State University*; John G. Graveel, *Purdue University*; Jack Greene, *Millikan High School*; Cheryl Greengrove, *University of Washington*; Amy R. Gregory, *University of Cincinnati, Clermont College*; Carol Griffin, *Grand Valley State University*; Carl W. Grobe, *Westfield State College*; Sherri Gross, *Ithaca College*; David E. Grunklee, *Hawkeye Community College*; Judy Guinan, *Radford University*; Gian Gupta, *University of Maryland, Eastern Shore*; Mark Gustafson, *Texas Lutheran University*; Daniel Guthrie, *Claremont College*; Sue Habeck, *Tacoma Community College*; David Hacker, *New Mexico Highlands University*; Greg Haenel, *Elon University*; Mark Hammer, *Wayne State University*; Grace Hanners, *Huntingtown High School*; Michael Hanson, *Bellevue Community College*; Alton Harestad, *Simon Fraser University*; Barbara Harvey, *Kirkwood Community College*; David Hassenzahl, *University of Nevada Las Vegas*; Jill Haukos, *South Plains College*; Keith Hench, *Kirkwood Community College*; George Hinman, *Washington State University*; Joseph Hobbs, *University of Missouri-Columbia*; Jason Hoeksema, *Cabrillo College*; Curtis Hollabaugh, *University of West Georgia*; Robert D. Hollister, *Grand Valley State University*; David Hong, *Diamond Bar High School*; Catherine Hooey, *Pittsburgh State University*; Kathleen Hornberger, *Widener University*; Debra Howell, *Chabot College*; April Huff, *North Seattle Community College*; Pamela Davey Huggins, *Fairmont State University*; Barbara Hunnicutt, *Seminole Community College*; Jonathan E. Hutchins, *Buena Vista University*; Daniel Hyke, *Alhambra High School*; Juana Ibanez, *University of New Orleans*; Walter Illman, *University of Iowa*; Daniel Ippolito, *Anderson University*; Bonnie Jacobs, *Southern Methodist University*; Nan Jenks-Jay, *Middlebury College*; Stephen R. Johnson, *William Penn University*; Gina Johnston, *California State University, Chico*; Paul Jurena, *University of Texas–San Antonio*; Richard R. Jurin, *University of Northern Colorado*; Thomas M. Justice, *McLennan Community College*; Stanley S. Kabala, *Duquesne University*; Brian Kaestner, *Saint Mary's Hall*; Steve Kahl, *Plymouth State University*; Carol Kearns, *University of Colorado-Boulder*; Richard R. Keenan, *Providence Senior High School*; Dawn G. Keller, *Hawkeye Community College*; John C. Kinworthy, *Concordia University*; Cindy Klevickis, *James Madison University*; Ned J. Knight, *Linfield College*; David Knowles, *East Carolina University*; Penelope M. Koines, *University of Maryland*; Alexander Kolovos, *University of North Carolina-Chapel Hill*; Erica Kosal, *North Carolina Wesleyan College*; Steven Kosztya, *Baldwin Wallace College*; Robert J. Koester, *Ball State University*; Tom Kozel, *Anderson College*; Robert G. Kremer, *Metropolitan State College of Denver*; Jim Krest, *University of South Florida–South Florida*; Sushma Krishnamurthy, *Texas A&M International University*; Jerome Kruegar, *South Dakota State University*; James Kubicki, *Penn State University*;

Frank T. Kuserk, *Moravian College*; Diane M. LaCole, *Georgia Perimeter College*; Troy A. Ladine, *East Texas Baptist University*; William R. Lammela, *Nazareth College*; Vic Landrum, *Washburn University*; Tom Langen, *Clarkson University*; Andrew Lapinski, *Reading Area Community College*; Michael T. Lares, *University of Mary*; Kim D. B. Largen, *George Mason University*; John Latto, *University of California-Berkeley*; Lissa Leege, *Georgia Southern University*; James Lehner, *Taft School*; Stephen D. Lewis, *California State University, Fresno*; Chun Liang, *Miami University*; John Logue, *University of South Carolina-Sumter*; John F. Looney, Jr., *University of Massachusetts-Boston*; Joseph Luczkovich, *East Carolina University*; Linda Lusby, *Acadia University*; Richard A. Lutz, *Rutgers University*; Jennifer Lyman, *Rocky Mountain College*; Les M. Lynn, *Bergen Community College*; Timothy F. Lyon, *Ball State University*; Sue Ellen Lyons, *Holy Cross School*; Ian R. MacDonald, *Texas A&M University*; James G. March, *Washington and Jefferson College*; Blase Maffia, *University of Miami*; Robert L. Mahler, *University of Idaho*; Keith Malmos, *Valencia Community College*; Kenneth Mantai, *State University of New York-Fredonia*; Anthony J.M. Marcattilio, *St. Cloud State University*; Heidi Marcum, *Baylor University*; Nancy Markee, *University of Nevada-Reno*; Patrick S. Market, *University of Missouri-Columbia*; Steven R. Martin, *Humboldt State University*; John Mathwig, *College of Lake County*; Allan Matthias, *University of Arizona*; Robert Mauck, *Kenyon College*; Bill Mautz, *University of New Hampshire*; Debbie McClinton, *Brevard Community College*; Paul McDaniel, *University of Idaho*; Jake McDonald, *University of New Mexico*; Gregory McIsaac, *Cornell University*; Dan McNally, *Bryant University*; Richard McNeil, *Cornell University*; Julie Meents, *Columbia College*; Mike L. Meyer, *New Mexico Highlands University*; Steven J. Meyer, *University of Wisconsin-Green Bay*; Patrick Michaels, *Cato Institute*; Christopher Migliaccio, *Miami Dade Community College*; Matthew R. Milnes, *University of California, Irvine*; Kiran Misra, *Edinboro University of Pennsylvania*; Mark Mitch, *New England College*; Lori Moore, *Northwest Iowa Community College*; Paul Montagna, *University of Texas-Austin*; Brian W. Moores, *Randolph-Macon College*; James T. Morris, *University of South Carolina*; Sherri Morris, *Bradley University*; Mary Murphy, *Penn State Abington*; William M. Murphy, *California State University-Chico*; Carla S. Murray, *Carl Sandburg College*; Rao Mylavarapu, *University of Florida*; Jane Nadel-Klein, *Trinity College*; Muthena Naseri, *Moorpark College*; Michael J. Neilson, *University of Alabama-Birmingham*; Richard A. Niesenbaum, *Muhlenberg College*; Moti Nissani, *Wayne State University*; Richard B. Norgaard, *University of California-Berkeley*; John Novak, *Colgate University*; Mark P. Oemke, *Alma College*; Niamh O'Leary, *Wells College*; Bruce Olszewski, *San Jose State University*; Brian O'Neill, *Brown University*; Nancy Ostiguy, *Penn State University*; David R. Ownby, *Stephen F. Austin State University*; Eric Pallant, *Allegheny College*; Philip Parker, *University of Wisconsin-Platteville*; Brian Peck, *Simpson College*; Brian D. Peer, *Simpson College*; Clayton Penniman, *Central Connecticut State*; Christopher Pennuto, *Buffalo State College*; Donald J. Perkey, *University of Alabama-Huntsville*; Shana Petermann, *Minnesota State Community and Technical College-Moorhead*; Raymond Pierotti, *University of Kansas*; Craig D. Phelps, *Rutgers University*; Frank X. Phillips, *McNeese State University*; Elizabeth Pixley, *Monroe Community College*; Thomas E. Pliske, *Florida International University*; Daryl Prigmore, *University of Colorado*; Avram G. Primack, *Miami University of Ohio*; Sarah Quast, *Middlesex Community College*; Loren A. Raymond, *Appalachian State University*; Barbara Reynolds, *University of North Carolina-Asheville*; Thomas J. Rice, *California Polytechnic State University*; Samuel K. Riffell, *Mississippi State University*; Gary Ritchison, *Eastern Kentucky University*; Roger Robbins, *East Carolina University*; Tom Robertson, *Portland Community College, Rock Creek Campus*; Mark Robson, *University of Medicine and Dentistry of New Jersey*; Carlton Lee Rockett, *Bowling Green State University*; Angel M. Rodriguez, *Broward Community College*; Deanne Roquet, *Lake Superior College*; Armin Rosencranz, *Stanford University*; Robert E. Roth, *The Ohio State University*; George E. Rough, *South Puget Sound Community College*; Steven Rudnick, *University of Massachusetts-Boston*; John Rueter, *Portland State University*; Christopher T. Ruhland, *Minnesota State University*; Shamili A. Sandiford; *College of DuPage*; Robert Sanford, *University of Southern Maine*; Ronald Sass, *Rice University*; Carl Schafer, *University of Connecticut*; Jeffery A. Schneider, *State University of New York-Oswego*; Edward G. Schultz, III, *Valencia Community College*; Mark Schwartz, *University of California-Davis*; Jennifer Scrafford, *Loyola College*; Richard Seigel, *Towson University*; Julie Seiter, *University of Nevada-Las Vegas*; Wendy E. Sera, *NDAA's National Ocean Service*; Maureen Sevigny, *Oregon Institute of Technology*; Rebecca Sheesley, *University of Wisconsin-Madison*; Pamela Shlachtman, *Miami Palmetto Senior High School*; Brian Shmaefsky, *Kingwood College*; William Shockner, *Community College of Baltimore County*; Christian V. Shorey, *University of Iowa*; Robert Sidorsky, *Northfield Mt. Hermon High School*; Linda Sigismondi, *University of Rio Grande*; Jeffrey Simmons, *West Virginia Wesleyan College*; Cynthia Simon, *University of New England*; Jan Simpkin, *College of Southern Idaho*; Michael Singer, *Wesleyan University*; Diane Sklensky, *Le Moyne College*; Ben Smith, *Palos Verdes Peninsula High School*; Mark Smith, *Chaffey College*; Patricia L. Smith, *Valencia Community College*; Sherilyn Smith, *Le Moyne College*; Debra Socci, *Seminole Community College*; Roy Sofield, *Chattanooga State Technical Community College*; Douglas J. Spieles, *Denison University*; Ravi Srinivas, *University of St. Thomas*; Bruce Stallsmith, *University of Alabama-Huntsville*; Jon G. Stanley, *Metropolitan State College of Denver*; Jeff Steinmetz, *Queens University of Charlotte*; Richard J. Strange, *University of Tennessee*; Robert Strikwerda, *Indiana University-Kokomo*; Richard Stringer, *Harrisburg Area Community College*; Andrew Suarez, *University of Illinois*; Keith S. Summerville, *Drake University*; Ronald Sundell, *Northern Michigan University*; Bruce Sundrud, *Harrisburg Area Community College*; Jim Swan, *Albuquerque Technical Vocational Institute*; Mark L. Taper, *Montana State University*; Max R. Terman, *Tabor College*; Julienne Thomas, *Robert Morris College*; Patricia Terry, *University of Wisconsin-Green Bay*; Jamey Thompson, *Hudson Valley Community College*; Todd Tracy, *Northwestern College*; Amy Treonis, *Creighton University*; Adrian Treves, *Wildlife Conservation Society*; Frederick R. Troeh, *Iowa State University*; Virginia Turner, *Robert Morris College*; Michael Tveten, *Pima Community College*; Charles Umbanhowar, *St. Olaf College*; G. Peter van Walsum, *Baylor University*; Callie A. Vanderbilt, *San Juan College*; Elichia A. Venso, *Salisbury University*; Rob Viens, *Bellevue Community College*; Michael Vorwerk, *Westfield State College*; Caryl Waggett, *Allegheny College*; Maud M. Walsh, *Louisiana State University*; Daniel W. Ward, *Waubonsee Community College*; Darrell Watson, *The University of Mary Hardin Baylor*; Phillip L. Watson, *Ferris State University*; Lisa Weasel, *Portland State University*; Kathryn Weatherhead, *Hilton Head High School*; John F. Weishampel, *University of Central Florida*; Barry Welch, *San Antonio College*; James W.C. White, *University of Colorado*; Susan Whitehead, *Becker College*; Richard D. Wilk, *Union College*; Donald L. Williams, *Park University*; Justin Williams, *Sam Houston University*; Ray E. Williams, *Rio Hondo College*; Roberta Williams, *University of Nevada-Las Vegas*; Dwina Willis, *Freed-Hardeman University*; Tom Wilson, *University of Arizona*; James Winebrake, *Rochester Institute of Technology*; Danielle Wirth, *Des Moines Area Community College*; Lorne Wolfe, *Georgia Southern University*; Marjorie Wonham, *University of Alberta*; Wes Wood, *Auburn University*; Jeffrey S. Wooters, *Pensacola Junior College*; Joan G. Wright, *Truckee Meadows Community College*; Michael Wright, *Truckee Meadows Community College*; S. Rebecca Yeomans, *South Georgia College*; Lynne Zeman, *Kirkwood Community College*; Zhihong Zhang, *Chatham College*.

Class Testers for Previous Editions

David Aborne, *University of Tennessee-Chattanooga*; Reuben Barret, *Prairie State College*; Morgan Barrows, *Saddleback College*; Henry Bart, *LaSalle University*; James Bartalome, *University of California-Berkeley*; Christy Bazan, *Illinois State University*; Richard Beckwitt, *Framingham State College*; Elizabeth Bell, *Santa Clara University*; Peter Biesmeyer, *North Country Community College*; Donna Bivans, *Pitt Community College*; Evert Brown, *Casper College*; Christina Buttington, *University of Wisconsin-Milwaukee*; Tait Chirenje, *Richard Stockton College*; Reggie Cobb, *Nash Community College*; Ann Cutter, *Randolph Community College*; Lola Deets, *Pennsylvania State University-Erie*; Ed DeGrauw, *Portland Community College*; Stephanie Dockstader, *Monroe Community College*; Dee Eggers, *University of North Carolina-Asheville*; Jane Ellis, *Presbyterian College*; Paul Fader, *Freed Hardeman University*; Joseph Fail, *Johnson C. Smith University*; Brad Fiero, *Pima Community College, West*

Campus; Dane Fisher, *Pfeiffer University;* Chad Freed, *Widener University;* Sue Glenn, *Gloucester County College;* Sue Habeck, *Tacoma Community College;* Mark Hammer, *Wayne State University;* Michael Hanson, *Bellevue Community College;* David Hassenzahl, *Oakland Community College;* Kathleen Hornberger, *Widener University;* Paul Jurena, *University of Texas–San Antonio;* Dawn Keller, *Hawkeye Community College;* David Knowles, *East Carolina University;* Erica Kosal, *Wesleyan College;* John Logue, *University of Southern Carolina Sumter;* Keith Malmos, *Valencia Community College;* Nancy Markee, *University of Nevada–Reno;* Bill Mautz, *University of New Hampshire;* Julie Meents, *Columbia College;* Stephen Getchell, *Mohawk Valley Community College;* Lori Moore, *Northwest Iowa Community College;* Elizabeth Pixley, *Monroe Community College;* John Novak, *Colgate University;* Brian Peck, *Simpson College;* Sarah Quast, *Middlesex Community College;* Roger Robbins, *East Carolina University;* Mark Schwartz, *University of California–Davis;* Julie Seiter, *University of Nevada–Las Vegas;* Brian Shmaefsky, *Kingwood College;* Diane Sklensky, *Le Moyne College;* Mark Smith, *Fullerton College;* Patricia Smith, *Valencia Community College East;* Sherilyn Smith, *Le Moyne College;* Jim Swan, *Albuquerque Technical Vocational Institute;* Amy Treonis, *Creighton University;* Darrell Watson, *The University of Mary Hardin Baylor;* Barry Welch, *San Antonio College;* Susan Whitehead, *Becker College;* Roberta Williams, *University of Nevada–Las Vegas;* Justin Williams, *Sam Houston University;* Tom Wilson, *University of Arizona.*

PART ONE

FOUNDATIONS OF ENVIRONMENTAL SCIENCE

Climber on "The Diving Board" at Half Dome in Yosemite National Park.

Our Island, Earth

1 SCIENCE AND SUSTAINABILITY: AN INTRODUCTION TO ENVIRONMENTAL SCIENCE

UPON COMPLETING THIS CHAPTER, YOU WILL BE ABLE TO:

- Define the term *environment* and describe the field of environmental science
- Explain the importance of natural resources and ecosystem services to our lives
- Discuss the effects of population growth and resource consumption

- Characterize the interdisciplinary nature of environmental science
- Understand the scientific method and the process of science
- Diagnose and illustrate some of the pressures on the global environment
- Articulate the concepts of sustainability and sustainable development

OUR ISLAND, EARTH

Viewed from space, our home planet resembles a small blue marble suspended in a vast inky-black void. Earth may seem enormous to us as we go about our lives on its surface, but the astronaut's view suggests that Earth and its systems are finite and limited. From this perspective, it becomes clear that as our population, technological powers, and consumption of resources increase, so does our capacity to alter our planet and damage the very systems that keep us alive.

Our environment surrounds us

A photograph of Earth offers a revealing perspective, but it cannot convey the complexity of our environment. Our **environment** consists of all the living and nonliving things around us. It includes the continents, oceans, clouds, and ice caps you can see in the photo of Earth from space, as well as the animals, plants, forests, and farms that comprise the landscapes surrounding us. In a more inclusive sense, it encompasses our built environment as well—the structures, urban centers, and living spaces that people have created. In its broadest sense, our environment also includes the complex webs of social relationships and institutions that shape our daily lives.

People commonly use the term *environment* in the first, most narrow sense—to mean a nonhuman or "natural" world apart from human society. This usage is unfortunate, because it masks the vital fact that people exist within the environment and are part of nature. As one of many species on Earth, we share with others the same dependence on a healthy, functioning planet. The limitations of language make it all too easy to speak of "people and nature," or "humans and the environment," as though they are separate and do not interact. However, the fundamental insight of environmental science is that we are part of the "natural" world and that our interactions with its other parts matter a great deal.

Environmental science explores our interactions with the world

Understanding our relationship with the world around us is vital because we depend utterly on our environment for air, water, food, shelter, and everything else essential for living. Moreover, we modify our environment. Many of our actions have enriched our lives, bringing us better health; longer life spans; and greater material wealth, mobility, and leisure time—but they have also often degraded the natural systems that sustain us. Impacts such as air and water pollution, soil erosion, and species extinction compromise our well-being, pose risks to human life, and jeopardize our ability to build a society that will survive and thrive in the long term.

Environmental science is the study of how the natural world works, how our environment affects us, and how we affect our environment. We need to understand our interactions with our environment in order to devise solutions to our most pressing challenges. It can be daunting to reflect on the sheer magnitude of environmental dilemmas that confront us today, but these problems also bring countless opportunities for creative solutions.

Environmental scientists study the issues most centrally important to our world and its future. Right now, global conditions are changing more quickly than ever. Right now, through science, we are gaining knowledge more rapidly than ever. And right now, the window of opportunity for acting to solve problems is still open. With such bountiful challenges and opportunities, this particular moment in history is indeed an exciting time to be alive—and to be studying environmental science.

We rely on natural resources

An island by definition is finite and bounded, and its inhabitants must cope with limitations in the materials they need. On our island, Earth, human beings, like all living things, ultimately face environmental constraints. Specifically, there are limits to many of our **natural resources**, the various substances and energy sources that we take from our environment and that we need to survive. Natural resources that are replenished over short periods are known as **renewable natural resources**. Some renewable resources, such as sunlight, wind, and wave energy, are perpetually renewed and essentially inexhaustible. Others, such as timber, water, and soil, renew themselves over months, years, or decades. In contrast, resources such as mineral ores and crude oil are in finite supply and are formed much more slowly than we use them. These are known as **nonrenewable natural resources**. Once we deplete them, they are no longer available.

We can view the renewability of natural resources as a continuum (**FIGURE 1.1**). Renewable resources such as timber, water, and soil can be depleted if we use them faster than they are replenished. For example, pumping groundwater faster than it is restored can deplete underground aquifers and turn lush landscapes into deserts. Populations of animals and plants we take from the wild may vanish if we overharvest them.

We rely on ecosystem services

If we think of natural resources as "goods" produced by nature, then it is also true that Earth's natural systems provide "services" on which we depend. Our planet's ecological systems purify air and water, cycle nutrients, regulate climate, pollinate plants, and receive and recycle our waste. Such essential services are commonly called **ecosystem services**. Ecosystem services arise from the normal functioning of natural systems, and although these processes are not meant for our benefit, we could not survive without them. Later in this book we will examine the countless and profound ways that ecosystem services support our lives and civilization (pp. 121–122, 160–161).

Just as we can deplete natural resources if we take too many of them, we can degrade ecosystem services by depleting resources, destroying habitat, or generating pollution. In recent years, our depletion of nature's goods and our disruption of nature's services have both intensified, driven by rising affluence and a human population that grows larger every day.

Population growth amplifies our impact

For nearly all of human history, less than a million people populated Earth at any one time. Today our population has grown beyond 6.9 *billion* people—several thousand times more! **FIGURE 1.2** shows just how recently and suddenly this monumental change has come about.

Two phenomena triggered remarkable increases in population size. The first was our transition from a hunter-gatherer

Renewable natural resources

- Sunlight
- Wind energy
- Wave energy
- Geothermal energy

- Fresh water
- Forest products
- Agricultural crops
- Soils

Nonrenewable natural resources

- Crude oil
- Natural gas
- Coal
- Copper, aluminum, and other metals

FIGURE 1.1 ▲ Natural resources lie along a continuum from perpetually renewable to nonrenewable. Perpetually renewable, or inexhaustible, resources, such as sunlight and wind energy, will always be there for us. Renewable resources such as timber, soils, and fresh water may be replenished on intermediate time scales, if we are careful not to deplete them. Nonrenewable resources, such as oil and coal, exist in limited amounts that could one day be gone.

lifestyle to an agricultural way of life. This change began around 10,000 years ago and is known as the **agricultural revolution**. As people began to grow crops, domesticate animals, and live sedentary lives on farms and in villages, they found it easier to meet their nutritional needs. As a result, they began to live longer and to produce more children.

The second notable phenomenon, known as the **industrial revolution**, began in the mid-1700s. It entailed a shift from rural life, animal-powered agriculture, and handcrafted goods to an urban society (**FIGURE 1.3**) provisioned by the mass production of factory-made goods and powered by **fossil fuels** (nonrenewable energy sources including oil, coal, and natural gas; pp. 533–545). In the wake of industrialization, our technological advances have brought improvements in sanitation and medicine, and we have enhanced agricultural production with fossil-fuel-powered equipment and synthetic pesticides and fertilizers (pp. 254–255).

The factors driving population growth have brought us better lives in many ways. But as our world fills up with people, population growth has begun to threaten our well-being. We must ask how well the planet can accommodate 7 billion of us—or the 9 billion forecast for 2050. Already our sheer numbers, unparalleled in history, are putting unprecedented stress on natural systems and the availability of resources.

Resource consumption exerts social and environmental pressures

Population growth is unquestionably at the root of many environmental problems. However, the growth in our exploitation and consumption of resources is also to blame. The industrial revolution enhanced the material affluence of many of the world's people by considerably increasing our consumption of natural resources and manufactured goods.

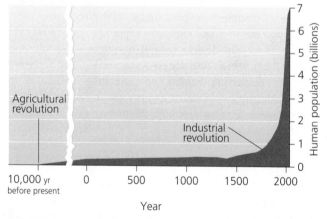

FIGURE 1.2 ▲ For almost all of human history, our population was low and relatively stable. It increased after the agricultural revolution and then skyrocketed as a result of the industrial revolution. Data compiled from U.S. Census Bureau, U.N. Population Division, and other sources.

FIGURE 1.3 ▲ As our population swells and as nations industrialize, people are streaming to cities, creating congested urban areas such as this one in Java, Indonesia.

The "tragedy of the commons" When publicly accessible resources are open to unregulated exploitation, they inevitably become overused and, as a result, are damaged or depleted. So argued the late Garrett Hardin of the University of California at Santa Barbara in his 1968 essay in the journal *Science*, titled "The Tragedy of the Commons."

Basing his argument on a scenario described in a 19th-century pamphlet, Hardin explained that in a public pasture (or "common") open to unregulated grazing, each person who grazes animals will be motivated by self-interest to increase the number of his or her animals in the pasture. Because no single person owns the pasture, no one has incentive to expend effort taking care of it, and everyone takes what he or she can until the resource is depleted. This is known as the **tragedy of the commons**. Ultimately, overgrazing will cause the pasture's food production to collapse.

Some have argued that private ownership best addresses this problem. Others point to cases in which people sharing a common resource have voluntarily organized and cooperated to enforce its responsible use. Still others maintain that the dilemma justifies government regulation of the private use of resources held in common by the public, from grazing land and forests to clean air and water.

WEIGHING THE ISSUES

The Tragedy of the Commons Imagine you make your living by fishing. You are free to boat anywhere and set out as many lines and traps as you like, and your catches have been good. However, the fishing grounds are getting crowded, and you find yourself competing with more and more people for fewer and fewer fish. Catches decline year by year, leaving you and the other fishers with catches too meager to support your families. Some call for dividing the waters and selling access to individuals plot by plot. Others implore the government to regulate how many fish can be caught. Still others want to team up, set quotas themselves, and prevent newcomers from entering the market. What do you think is the best way to combat this tragedy of the commons and restore the fishery, and why?

Our ecological footprint As global affluence has increased, human society has consumed more and more of the planet's limited resources. We can quantify resource consumption using the concept of the "ecological footprint," developed in the 1990s by environmental scientists Mathis Wackernagel and William Rees. An **ecological footprint** expresses environmental impact in terms of the cumulative area of biologically productive land and water required to provide the resources a person or population consumes and to dispose of or recycle the waste the person or population produces (**FIGURE 1.4**). It measures the total area of Earth's biologically productive surface that a given person or population "uses" once all direct and indirect impacts are totaled up.

For humanity as a whole, Wackernagel and his colleagues calculate that our ecological footprint now surpasses Earth's productive capacity by about 30%. That is, we are running a global deficit, depleting renewable resources by using them 30% faster than they are being replenished. This is essentially

like drawing the principal out of a bank account rather than living off the interest. This global deficit has been termed **overshoot**, because we have overshot, or surpassed, Earth's capacity to sustainably support us (**FIGURE 1.5**). Moreover,

FIGURE 1.4 ▲ An "ecological footprint" represents the total area of biologically productive land and water needed to produce the resources and dispose of the waste for a given person or population. The footprint of an average citizen of an affluent nation is much larger than the physical area in which the person lives day to day. Adapted from an illustration by Philip Testemale in Wackernagel, M., and W. Rees, 1996. *Our ecological footprint: Reducing human impact on the Earth.* Gabriola Island, British Columbia: New Society Publishers.

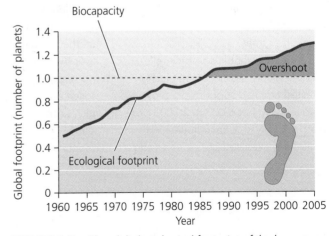

FIGURE 1.5 ▲ The global ecological footprint of the human population is over 2.5 times larger than it was a half-century ago and now exceeds what Earth can bear in the long run, scientists have calculated. Data indicate that we have already overshot Earth's biocapacity—its capacity to support us—by 30%; that is, we are using renewable natural resources 30% faster than they are being replenished. Data from WWF International. 2008. *Living planet report 2008.* Published in October 2008 by WWF-World Wide Fund for Nature. © 2008 WWF.

The SCIENCE behind the Story

The Lesson of Easter Island

Easter Island is one of the most remote spots on the globe, located in the Pacific Ocean 3,750 km (2,325 mi) from South America and 2,250 km (1,395 mi) from the nearest inhabited island. When the first European explorers reached the island (today called Rapa Nui) in 1722, they found a barren landscape populated by fewer than 2,000 people, who lived in caves and eked out a marginal existence from a few meager crops. However, explorers also noted that the desolate island featured hundreds of gigantic statues of carved stone, evidence that a sophisticated civilization had once inhabited the island.

Historians and anthropologists long wondered how people without wheels or ropes, on an island without trees, could have moved statues 10 m (33 ft) high weighing 90 metric tons (99 tons) as far as 10 km (6.2 mi) from the quarries where they were chiseled to the coastal sites where they were erected. The explanation, scientists discovered, was that the island did not always lack trees.

Indeed, scientific research tells us that the island had once been lushly forested and had supported a prosperous society of 6,000 to 30,000 people. Tragically, this once-flourishing civilization overused its resources and

The immense moai (statues) of Easter Island

cut down all its trees, destroying itself in a downward spiral of starvation and conflict. Today Easter Island stands as a parable and a warning for what can happen when a population consumes too much of the limited resources that support it.

To solve the mystery of Easter Island's past, scientists have used various methods. Some, such as British scientist John Flenley, have excavated sediments from the bottom of the island's lakes, drilling cores deep into the mud and examining ancient grains of pollen preserved there. Pollen grains vary from one plant species to another, so scientists can reconstruct, layer by layer, the history of vegetation in a region through time. By analyzing pollen grains under scanning electron microscopes, Flenley and other researchers found that when Polynesian people arrived (likely between A.D. 300

and 900), the island was covered with a species of palm tree related to the Chilean wine palm, a tall and thick-trunked tree.

Adding to this evidence, archaeologists located ancient palm nut casings in caves and crevices, and a geologist found carbon-lined channels in the soil that matched root channels typical of the Chilean wine palm. Scientists deciphering the island people's script on stone tablets discerned characters etched in the form of palm trees.

By studying pollen and the remains of wood from charcoal, scientists, among them French archaeologist Catherine Orliac, found that at least 21 other species of plants, many of them trees, had also been common but are now completely gone. The island had clearly supported a diverse forest. However, starting around A.D. 750, tree populations declined and ferns and grasses became more common, according to pollen analysis from one lake site. By A.D. 950, the trees were largely gone, and around A.D. 1400 overall pollen levels plummeted, indicating a dearth of vegetation.

The same sequence of events occurred two centuries later at the other two lake sites, which were higher and more remote from village areas.

people from wealthy nations such as the United States have much larger ecological footprints than do people from poorer nations. If all the world's people consumed resources at the rate of U.S. citizens, we would need the equivalent of four and one-half planet Earths.

Environmental science can help us avoid past mistakes

It remains to be seen what consequences resource consumption and population growth will have for today's

The haunting statues of Easter Island (Rapa Nui) were erected by a sophisticated civilization that collapsed after depleting its resource base and devastating its island environment.

David Steadman showed that at least 31 species of birds nested on Easter Island and provided food for the islanders. Today, only one native bird species is left.

Remains from charcoal fires aged using radioisotopes of carbon (p. 26) show that besides crops and birds, early islanders feasted on the bounty of the sea, including porpoises, fish, sharks, turtles, octopus, and shellfish. But analysis of islanders' diets in the later years indicated that the people consumed little seafood. With the trees gone, the islanders could no longer build the great double canoes their proud Polynesian ancestors had used for centuries to fish and travel among islands.

As resources declined, archaeologists found, the islanders began keeping their main domesticated food animal, chickens, in stone fortresses with entrances designed to prevent theft. The once prosperous and peaceful civilization fell into clan warfare, as revealed by unearthed weapons, skeletons, and skulls with head wounds.

Is the story of Easter Island as unique and isolated as the island itself, or does it hold lessons for our world today? Like the Easter Islanders, we are all stranded together on an island with limited resources. Earth may be vastly larger and richer in resources than Easter Island, but Earth's human population is also much greater.

The Easter Islanders must have seen that they were depleting their resources, but it seems that they could not stop. Whether we can learn from the history of Easter Island and act more wisely to conserve the resources on our island, Earth, is entirely up to us. ∎

Researchers first hypothesized that the forest loss was due to climate change, but evidence instead supported the hypothesis that the people had gradually denuded their own island.

The trees provided fuelwood, building material for houses and canoes, fruit to eat, fiber for clothing—and, presumably, logs with which to move the stone statues. By hiring groups of men to recreate the feat, anthropologists have experimentally tested hypotheses about how the islanders moved their monoliths down from the quarries. The methods that have worked involve using numerous tree trunks as rollers or sleds, along with great quantities of rope. The only likely source of rope on the island would have been the fibrous inner bark of the hauhau tree, a species that today is near extinction.

With the trees gone, soil would have eroded away—a phenomenon confirmed by data from the lake bottoms, where large quantities of sediment accumulated. Faster runoff of rainwater would have meant less fresh water available for drinking. Runoff and erosion would have degraded the islanders' agricultural land, lowering yields of bananas, sugarcane, and sweet potatoes. Reduced agricultural production would have led to starvation and population decline.

Archaeological evidence supports a scenario of environmental degradation and civilization decline. Analysis of 6,500 bones by archaeologist

global society, but we have historical evidence that civilizations can crumble when pressures from population and consumption overwhelm resource availability. Easter Island is a classic case (see **THE SCIENCE BEHIND THE STORY**, above).

Many great civilizations have fallen after degrading their environments, and each has left devastated landscapes in its wake. Historians have concluded that environmental degradation contributed in part to the fall of the Greek and Roman empires; the Angkor civilization of Southeast Asia;

and the Maya, Anasazi, and other civilizations of the New World. In Iraq and other regions of the Middle East, areas that are barren desert today were lush enough to support the origin of agriculture when great ancient civilizations thrived there. In his 2005 book *Collapse*, scientist and author Jared Diamond synthesized existing research and formulated general reasons why civilizations succeed and persist, or fail and collapse. Success and persistence, he argued, depend largely on how societies interact with their environments and on how they respond to problems.

In today's globalized society, the stakes are higher than ever because our environmental impacts are global. If we cannot forge sustainable solutions to our problems, then the resulting societal collapse will be global. Fortunately, environmental science holds keys to building a better world. By studying environmental science, you will learn to evaluate the changes happening around us and to think critically and creatively about actions to take in response.

THE NATURE OF ENVIRONMENTAL SCIENCE

Environmental scientists aim to comprehend how Earth's natural systems function, how these systems affect people, and how we are influencing those systems. Many environmental scientists are motivated by a desire to develop solutions to environmental problems. These solutions (such as new technologies, policy decisions, or resource management strategies) are *applications* of environmental science. The study of such applications and their consequences is, in turn, also part of environmental science.

Environmental science is an interdisciplinary pursuit

Studying our interactions with our environment is a complex endeavor that requires expertise from many disciplines, including ecology, earth science, chemistry, biology, geography, economics, political science, demography, ethics, and others. Environmental science is thus an **interdisciplinary** field—one that borrows techniques from multiple disciplines and brings their research results together into a broad synthesis (**FIGURE 1.6**).

Traditional established disciplines are valuable because their scholars delve deeply into topics, uncovering new knowledge and developing expertise in particular areas. In contrast, interdisciplinary fields are valuable because their practitioners consolidate and synthesize the specialized knowledge from many different disciplines and make sense of it in a broad context to better serve the multifaceted interests of society.

Environmental science is especially broad because it encompasses not only the **natural sciences** (disciplines that examine the natural world), but also the **social sciences** (disciplines that address human interactions and institutions). The natural sciences provide us the means to obtain and interpret information about the world around us. However, to fully comprehend our interactions with our

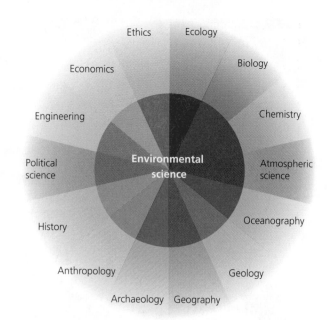

FIGURE 1.6 ▲ Environmental science is an interdisciplinary pursuit, involving input from many different established fields of study across the natural sciences and social sciences.

environment, we also need to weigh values and understand human behavior—and this requires the social sciences. Most environmental science programs focus predominantly on the natural sciences, whereas programs that incorporate the social sciences extensively often use the term **environmental studies**. Whichever approach one takes, these fields reflect many diverse perspectives and sources of knowledge.

Just as an interdisciplinary approach to studying issues can help us better understand them, an integrated approach to addressing environmental problems can produce effective solutions for society. As one example, we used to add lead to gasoline to make cars run more smoothly, even though researchers knew that lead emissions from tailpipes caused health problems, including brain damage and premature death. In 1970 air pollution was severe, and motor vehicles accounted for 78% of U.S. lead emissions. Over the following years, environmental scientists, engineers, medical researchers, and policymakers all merged their knowledge and skills into a process that eventually resulted in a ban on leaded gasoline. By 1996 all gasoline sold in the United States was unleaded, and the nation's largest source of atmospheric lead emissions had been completely eliminated.

People vary in their perception of environmental problems

Environmental science arose in the latter half of the 20th century as people sought to better understand environmental problems, their origins, and their solutions. However, the perception of what constitutes a problem may vary from one person to another, or from one situation to another. A person's age, gender, class, race, nationality, employment, income, and educational background can all affect whether he or she considers a given environmental condition or change to be a "problem."

For instance, people today are more likely to view the spraying of the pesticide DDT as a problem than they did in the 1950s, because today more is known about the health risks of pesticides (**FIGURE 1.7**). However, a person living today in a malaria-infested village in Africa or India may welcome the use of DDT if it kills mosquitoes that transmit malaria, because he or she may view malaria as a more immediate health threat. Thus an African and an American who have each knowledgeably assessed the pros and cons may, because of differences in their circumstances, differ in their judgment of DDT's severity as an environmental problem.

People may also vary in their awareness of problems. For example, in many cultures women are responsible for collecting water and fuelwood. As a result, they are often the first to perceive environmental degradation that affects these resources, whereas men simply might not "see" the problem. As another example, in most societies information about environmental health risks tends to reach wealthy people more readily than poor people. Thus, who you are, where you live, and what you do influences how you perceive your environment, how change affects you, and how you react to change. In Chapter 6 we will examine the diversity of human values and philosophies and

FIGURE 1.8 ▲ Environmental scientists play roles very different from those of environmental activists, such as the protestors shown here. Although many environmental scientists search for solutions to environmental problems, they generally aim to keep their research rigorously objective and free from advocacy.

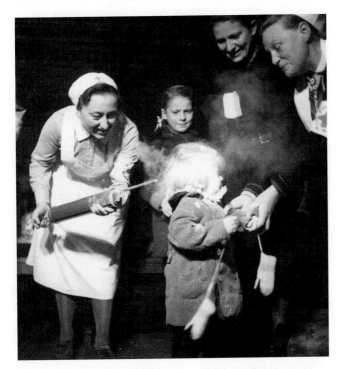

FIGURE 1.7 ▲ How a person or a society defines an environmental problem can vary with time and circumstance. In 1945, health hazards of the pesticide DDT were not yet known, so children were doused with the chemical to treat head lice. Today, knowing of its toxicity to people and wildlife, many developed nations have banned DDT. However, in developing countries where malaria is a threat, DDT is welcomed as a means of eradicating mosquitoes that transmit the disease.

consider their consequences for how we define environmental problems.

Environmental science is not the same as environmentalism

Although many environmental scientists are interested in solving problems, it would be incorrect to confuse environmental science with environmentalism, or environmental activism. They are *not* the same. **Environmentalism** is a social movement dedicated to protecting the natural world—and, by extension, people—from undesirable changes brought about by human actions (**FIGURE 1.8**). Environmental science, in contrast, is the pursuit of knowledge about the workings of the environment and our interactions with it.

Although environmental scientists study many of the same issues environmentalists care about, as scientists they aim to maintain an objective approach in how they conduct their work. Like any other human endeavor, science can never be entirely free of social or political influence—and like all people, scientists are motivated by personal values and interests. Yet although personal values and social concerns may shape how scientists choose their goals, scientists try to avoid undue influence from outside pressures and preconceptions when carrying out their work. Remaining free from bias, and open to whatever conclusions the data reveal, is a hallmark of the effective scientist.

THE NATURE OF SCIENCE

Modern scientists describe **science** as a systematic process for learning about the world and testing our understanding of it. The term *science* is also commonly used to refer to the accumulated body of knowledge that arises from this dynamic process of questioning, observation, testing, and discovery.

Knowledge gained from science can be applied to address societal needs. Among the applications of science is its use in developing technology and in informing policy and management decisions (**FIGURE 1.9**). Many scientists are motivated by the potential for developing such useful applications, whereas others are motivated simply by a desire to understand how the world works.

Why does science matter? The late astronomer and author Carl Sagan wrote the following in his 1995 treatise, *The Demon Haunted World: Science as a Candle in the Dark:*

> We've arranged a global civilization in which the most crucial elements—transportation, communications, and all other industries; agriculture, medicine, education, entertainment, protecting the environment; and even the key democratic institution of voting—profoundly depend on science and technology. . . . Science is an attempt, largely successful, to understand the world, to get a grip on things, to get hold of ourselves, to steer a safe course.

Sagan and many other thinkers have argued that science is essential if we hope to sort fact from fiction and develop solutions to the challenges we face. Moreover, making science accessible and understandable to as many people as possible is also vital if we are to make informed decisions as a society.

Scientists test ideas by critically examining evidence

Science is all about asking and answering questions. Scientists examine how the world works by making observations, taking measurements, and designing tests to determine whether ideas are supported by evidence. If a particular explanation is testable and resists repeated attempts to disprove it, scientists will come to accept it as a true explanation. Scientific inquiry thus consists of an incremental approach to the truth.

The effective scientist thinks critically and does not simply accept the conventional wisdom that he or she hears from others. The scientist becomes excited by novel ideas but is a skeptic and judges ideas by the strength of evidence that supports them. In these ways, scientists are good role models for the rest of us, because every one of us can benefit from learning to think critically in our everyday lives.

Science advances in different ways

A great deal of scientific work is **observational science** or **descriptive science**, types of research in which scientists gather basic information about organisms, materials, systems, or processes that are not well known or that cannot be manipulated in experiments. In these approaches, researchers explore new frontiers of knowledge by observing and measuring phenomena to gain a better understanding of the world. Such research is common in traditional fields such as astronomy, paleontology, and taxonomy, as well as in newer and expanding fields such as molecular biology and genomics.

Once enough general information is known about a subject, scientists can begin posing more specific questions that ask how and why things are the way they are. At this point they may pursue **hypothesis-driven science**, research that proceeds in a more targeted and structured manner, using experiments to test hypotheses within a framework traditionally known as the scientific method.

The scientific method is a traditional approach to research

The **scientific method** is a technique for testing ideas with observations. There is nothing mysterious or intimidating about the scientific method; it is merely a formalized version

(a) Prescribed burning

(b) Chevy Volt, an electric hybrid car

FIGURE 1.9 ▲ Scientific knowledge can be applied in policy and management decisions and in engineering and technology. Prescribed burning **(a)**, shown here in the Ouachita National Forest, Arkansas, is a management practice to restore healthy forests that is informed by scientific research into forest ecology. Energy-efficient automobiles such as the Chevy Volt **(b)**, an electric car due for release in late 2010, are technological advances made possible by materials and energy research.

of the way any of us might naturally use logic to resolve a question.

Because science is an active, creative process, an innovative researcher may find good reason to stray from the traditional scientific method when a particular situation demands it. Moreover, scientists in different fields approach their work differently because they deal with dissimilar types of information. A natural scientist, such as a chemist, will conduct research quite differently from a social scientist, such as a sociologist. However, scientists of all persuasions broadly agree on fundamental elements of the process of scientific inquiry.

The scientific method relies on the following assumptions:

► The universe functions in accordance with fixed natural laws that do not change from time to time or from place to place.

► All events arise from some cause or causes and, in turn, cause other events.

► We can use our senses and reasoning abilities to detect and describe natural processes that underlie the cause-and-effect relationships we observe in nature.

As practiced by individual researchers or research teams, the scientific method (**FIGURE 1.10**) typically consists of the steps outlined below.

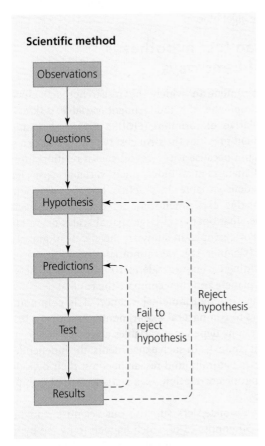

Scientific method

Observations → Questions → Hypothesis → Predictions → Test → Results

Fail to reject hypothesis

Reject hypothesis

FIGURE 1.10 ▲ The scientific method is the traditional experimental approach that scientists use to learn how the world works. This diagram is a simplified generalization that, although useful for instructive purposes, cannot convey the true dynamic and creative nature of science. Moreover, researchers from different disciplines may pursue their work in ways that vary legitimately from this model.

Make observations Advances in science typically begin with the observation of some phenomenon that the scientist wishes to explain. Observations set the scientific method in motion and also function throughout the process.

Ask questions Curiosity is a fundamental human characteristic—just observe the explorations of a young child in a new environment. Babies want to touch, taste, watch, and listen to anything that catches their attention, and as soon as they can speak, they begin asking questions. Scientists, in this respect, are kids at heart. Why are certain plants or animals less common today than they once were? Why are storms becoming more severe and flooding more frequent? What is causing excessive growth of algae in local ponds? When pesticides poison fish or frogs, does that indicate that people may also be affected? All of these are questions environmental scientists ask.

Develop a hypothesis Scientists attempt to answer their questions by devising explanations that they can test. A **hypothesis** is a statement that attempts to explain a phenomenon or answer a scientific question. For example, a scientist investigating the question of why algae are growing excessively in local ponds might observe chemical fertilizers being applied on farm fields nearby. The scientist might then state a hypothesis as follows: "Agricultural fertilizers running into ponds cause the amount of algae in the ponds to increase." Sometimes a scientist begins with a *null hypothesis*, which postulates no relationship between variables (such as between fertilizer and algal growth), and tests this against *alternative hypotheses* that propose particular relationships.

Make predictions The scientist next uses his or her hypothesis to generate **predictions**, specific statements that can be directly and unequivocally tested. In our algae example, a researcher might predict: "If agricultural fertilizers are added to a pond, the quantity of algae in the pond will increase."

Test the predictions Scientists test predictions one at a time by gathering evidence that could potentially refute the prediction and thus disprove the hypothesis (**FIGURE 1.11**). The strongest form of evidence comes from experimentation. An **experiment** is an activity designed to test the validity of a prediction or a hypothesis. It involves manipulating **variables**, or conditions that can change.

For example, a scientist could test the prediction linking algal growth to fertilizer by selecting two identical ponds and adding fertilizer to one while leaving the other in its natural state. In this example, fertilizer input is an **independent variable**, a variable the scientist manipulates, whereas the quantity of algae that results is the **dependent variable**, one that depends on the fertilizer input. If the two ponds are identical except for a single independent variable (fertilizer input), then any differences that arise between the ponds can be attributed to that variable. Such an experiment is known as a **controlled experiment** because the scientist controls for the effects of all variables except the one whose effect he or she is testing. In our example, the pond left unfertilized serves as a **control**, an unmanipulated point of comparison for the manipulated **treatment** pond.

FIGURE 1.11 ▲ Here Dr. Jennifer Smith of the Scripps Institution of Oceanography in San Diego uses a quadrat with a digital camera to photograph sites along a transect of a coral reef at a remote atoll in the South Pacific. Data from analysis of the photos will help her test hypotheses about how human impacts affect the condition and community structure of coral reefs.

Whenever possible, it is best to replicate one's experiment; that is, to stage multiple tests of the same comparison of control and treatment. Our scientist could perform a replicated experiment on, say, 10 pairs of ponds, adding fertilizer to one of each pair.

Analyze and interpret results Scientists record **data**, or information, from their studies. They particularly value quantitative data (information expressed using numbers), because numbers provide precision and are easy to compare. The scientist running the fertilization experiment, for instance, might quantify the area of water surface covered by algae in each pond or might measure the dry weight of algae in a certain volume of water taken from each.

However, even with the precision that numbers provide, a scientist's results may not be clear-cut. Data from treatments and controls may vary only slightly, or replicates may yield different results. The researcher must therefore analyze the data using statistical tests. With these mathematical methods, scientists can determine objectively and precisely the strength and reliability of patterns they find.

Some research, especially in the social sciences, involves data that is qualitative, or not expressible in terms of numbers. Research involving historical texts, personal interviews, surveys, case studies, or descriptive observation of behavior can include qualitative data on which quantitative statistical analysis may not be possible.

If experiments disprove a hypothesis, the scientist will reject it and may formulate a new hypothesis to replace it. If experiments fail to disprove a hypothesis, this result lends support to the hypothesis but does not *prove* it is correct. The scientist may choose to generate new predictions to test the hypothesis in different ways and further assess its likelihood of being true. Thus, the scientific method loops back on itself, often giving rise to repeated rounds of hypothesis revision, prediction, and testing (see Figure 1.10). If repeated tests fail to reject a particular hypothesis, evidence in favor of it accumulates, and the researcher may eventually conclude that the hypothesis is well supported.

Ideally, the scientist would want to test all possible explanations for the question of interest. For instance, our researcher might propose two alternative hypotheses for why algae increase in fertilized ponds: Is it because the chemical nutrients spur algal growth directly, or is it because fertilizer contamination decreases the numbers of fish or invertebrate animals that eat algae? It is possible, of course, that both hypotheses could be correct and that each may explain some portion of the initial observation that local ponds were experiencing algal blooms.

We can test hypotheses in different ways

An experiment in which the researcher actively chooses and manipulates the independent variable is known as a **manipulative experiment** (**FIGURE 1.12A**). A manipulative experiment provides the strongest type of evidence a scientist can obtain, because it can reveal causal relationships, showing that changes in an independent variable cause changes in a dependent variable. In practice, however, we cannot run manipulative experiments for all questions, especially for processes that operate at large spatial scales or on long time scales. For example, in studying the effects of global climate change (Chapter 18), we cannot run a manipulative experiment adding carbon dioxide to 10 treatment planets and 10 control planets and then compare the results!

Thus, in environmental science, it is common for researchers to run **natural experiments** (**FIGURE 1.12B**), which compare how dependent variables are expressed in naturally different contexts. In such experiments, the independent variable varies naturally, and researchers test their hypotheses by searching for **correlation**, or a statistical relationship among variables.

For instance, let's suppose our scientist studying algae surveys 50 ponds, 25 of which happen to be fed by fertilizer runoff from nearby farm fields and 25 of which are not. Let's say he or she finds seven times more algal growth in the fertilized ponds than in the unfertilized ponds. The scientist would conclude that algal growth is correlated with fertilizer input; that is, that one tends to increase along with the other.

This type of evidence is weaker than the causal demonstration that manipulative experiments can provide, but sometimes

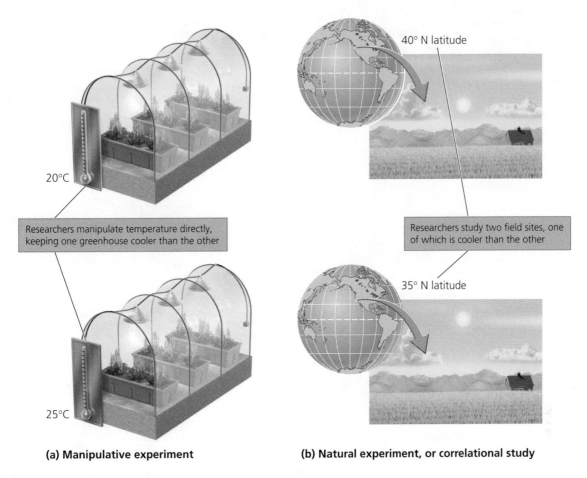

20°C

Researchers manipulate temperature directly, keeping one greenhouse cooler than the other

25°C

40° N latitude

Researchers study two field sites, one of which is cooler than the other

35° N latitude

(a) Manipulative experiment

(b) Natural experiment, or correlational study

FIGURE 1.12 ▲ A researcher wishing to test how temperature affects the growth of a crop might run a manipulative experiment **(a)**, in which the crop is grown in two identical greenhouses, one kept at 20 °C (68 °F) and the other kept at 25 °C (77 °F). Alternatively, the researcher might run a "natural experiment" **(b)**, in which he or she compares the growth of the crop in two fields at different latitudes, a cool northerly location and a warm southerly one. Because it would be difficult to hold all variables besides temperature constant in the natural experiment, the researcher might want to study a number of northern and southern fields and correlate data on temperature and crop growth.

a natural experiment is the only feasible approach for a subject of immense scale, such as an ecosystem or a planet. Because of their large scale and complexity, many questions in environmental science must be addressed with correlative data. As such, environmental scientists cannot always provide clear-cut, black-and-white answers to questions from policymakers and the public. Nonetheless, good correlative studies can make for strong science, and they preserve the real-world complexity that manipulative experiments often sacrifice.

The scientific process does not stop with the scientific method

Scientific work takes place within the context of a community of peers. To have impact, a researcher's work must be published and made accessible to this community. Thus, the scientific method is embedded within a larger process involving the scientific community as a whole (**FIGURE 1.13**).

Peer review   When a researcher's work is done and the results analyzed, he or she writes up the findings and submits them to a journal for publication. The journal's editor asks

several other scientists who specialize in the subject area to examine the manuscript, provide comments and criticism (generally anonymously), and judge whether the work merits publication in the journal. This procedure, known as **peer review**, is an essential part of the scientific process.

Peer review is a valuable guard against faulty research contaminating the literature on which all scientists rely. However, because scientists are human and may have their own personal biases and agendas, politics can sometimes creep into the review process. Fortunately, just as individual scientists strive to remain objective in conducting their research, the scientific community does its best to ensure fair review of all work. Winston Churchill once called democracy the worst form of government, except for all the others that had been tried. The same might be said about peer review; it is an imperfect system, yet no one has come up with a better one.

Conference presentations   Scientists frequently present their work at professional conferences, where they interact with colleagues and receive informal comments on their research. Such feedback can help improve a scientist's work before it is submitted for publication.

FIGURE 1.13 ▶ The scientific method (inner yellow box) followed by individual researchers or research teams exists within the context of the overall process of science at the level of the scientific community (outer green box). This process includes peer review and publication of research, acquisition of funding, and the elaboration of theory through the cumulative work of many researchers.

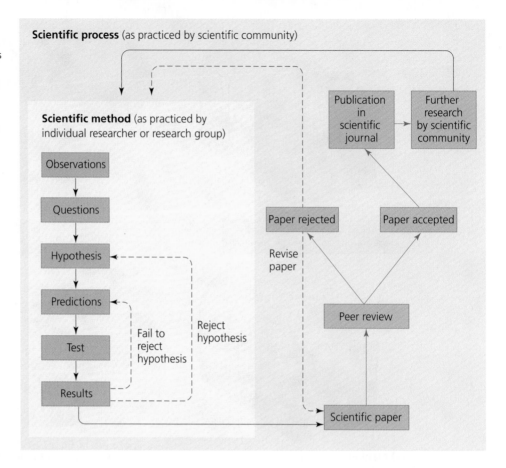

Grants and funding To fund their research, most scientists are forced to spend enormous amounts of time writing grant applications to request money from private foundations or from government agencies such as the National Science Foundation. Grant applications undergo peer review just as scientific papers do, and competition for funding is generally intense.

Scientists' reliance on funding sources can occasionally lead to conflicts of interest. A researcher who obtains data showing his or her funding source in an unfavorable light may be reluctant to publish the results for fear of losing funding—or worse yet, may be tempted to doctor the results. This situation can arise, for instance, when an industry funds research to test its products for safety or environmental impact. Few scientists succumb to these temptations, but some funding sources have been known to pressure their researchers for certain results. This is why as a student or informed citizen, when critically assessing a scientific study, you should always try to find out where the researchers obtained funding.

WEIGHING THE ISSUES

Follow the Money Let us say you are a research scientist, and you want to study the impacts of chemicals released by pulp-and-paper mills on nearby lakes. Obtaining research funding has been difficult. Then a representative from a large pulp-and-paper company contacts you. The company also is interested in how its chemical effluents affect water bodies, and it would like to fund your research. What are the benefits and drawbacks of this offer? Would you accept the offer?

Repeatability Even when a hypothesis appears to explain an observed phenomenon, scientists are inherently wary of accepting it. The careful scientist may test a hypothesis repeatedly in various ways before submitting the findings for publication. Following publication, other scientists may attempt to reproduce the results in their own experiments and analyses.

Theories If a hypothesis survives repeated testing by numerous research teams and continues to predict experimental outcomes and observations accurately, it may potentially be incorporated into a theory. A **theory** is a widely-accepted, well-tested explanation of one or more cause-and-effect relationships that has been extensively validated by a great amount of research. Whereas a hypothesis is a simple explanatory statement that may be disproven by a single experiment, a theory consolidates many related hypotheses that have been supported by a large body of experimental and observational data.

Note that scientific use of the word *theory* differs from popular usage of the word. In everyday language when we say something is "just a theory," we are suggesting it is a speculative idea without much substance. Scientists, however, mean just the opposite when they use the term. To them, a theory is a conceptual framework that effectively explains a phenomenon and has undergone extensive and rigorous testing, such that confidence in it is extremely strong.

For example, Darwin's theory of evolution by natural selection (pp. 52–55) has been supported and elaborated by many thousands of studies over 150 years of intensive research. Research has shown repeatedly and in great detail how plants and animals change over generations, or evolve, to express characteristics that best promote survival and reproduction. Because of its strong support and explanatory power, evolutionary theory is the central unifying principle of modern biology. Other prominent scientific theories include atomic theory, cell theory, the big bang theory, plate tectonics, and general relativity.

Applications Knowledge gained from scientific research may be applied to help fulfill society's needs and address society's problems. As discussed earlier (see Figure 1.9), scientific research informs and facilitates new technologies, engineering approaches, policy decisions, and management strategies. However, even when research is able to provide clear information, deciding on the optimal social response to a problem can still be difficult. Moreover, many predicaments addressed by environmental science are so-called *wicked problems*: problems complex enough to have no simple solution and whose very nature changes over time. As a result, we need scientists to continue studying such problems as they evolve.

Science goes through "paradigm shifts"

As the scientific community accumulates data in any given area of research, interpretations may change. Thomas Kuhn's influential 1962 book *The Structure of Scientific Revolutions* argued that science goes through periodic revolutions: dramatic upheavals in thought, in which one scientific **paradigm**, or dominant view, is abandoned for another. For example, before the 16th century, scientists believed that Earth was at the center of the universe. Their data on the movements of planets fit that concept somewhat well—yet the idea eventually was disproved by Nicolaus Copernicus, who showed that placing the sun at the center of the solar system explained the planetary data much better.

Another paradigm shift occurred in the 1960s, when geologists accepted plate tectonics (pp. 35–37). By this time, evidence for the movement of continents and the action of tectonic plates had accumulated and become overwhelmingly convincing. Such paradigm shifts demonstrate the strength and vitality of science, showing it to be a process that refines and improves itself through time.

Understanding how science works is vital to assessing how scientific ideas and interpretations change through time as information accrues. This process is especially relevant in environmental science—a young field that is changing rapidly as we learn vast amounts of new information, as human impacts on the planet multiply, and as we gather lessons from the consequences of our actions.

SUSTAINABILITY AND THE FUTURE OF OUR WORLD

Throughout this book you will encounter environmental scientists asking questions, testing hypotheses, conducting experiments, gathering and analyzing data, and drawing conclusions about environmental processes and the causes and consequences of environmental change. Environmental scientists who aim to understand the condition of our environment and the consequences of our impacts are addressing the most centrally important issues of our time.

Achieving sustainable solutions is vital

The primary challenge in our increasingly populated world is how to live within our planet's means, such that Earth and its resources can sustain us—and all life on Earth—for the future. This is the challenge of **sustainability**, a guiding principle of modern environmental science. Sustainability means leaving our children and grandchildren a world as rich and full as the world we live in now. It means conserving Earth's resources so that our descendants may enjoy them as we have. It means developing solutions that work in the long term. Sustainability requires maintaining fully functioning ecological systems, because we cannot sustain human civilization without sustaining the natural systems that nourish it.

We can think of our planet's resources as a bank account. If we deplete resources, we draw down the bank account. However, we can choose instead to use the interest and leave the principal intact so that we can continue using the interest far into the future. Presently we are drawing down Earth's **natural capital**—its accumulated wealth of resources. Recall (p. 5) that researchers estimate that we are withdrawing our planet's natural capital 30% faster than it is being replenished. To live off nature's interest—its replenishable resources—is sustainable. To draw down resources faster than they are replaced is to eat into nature's capital—the bank account for our planet and our civilization—and we cannot get away with this for long.

Sustainability is a concept you will encounter throughout this book, infused through issue after issue. Our final chapter (Chapter 24) rounds out our discussion by presenting a broad summary of approaches to sustainability—on college and university campuses and in the world at large. Along the way, we will explore a wide array of issues in our far-reaching search for sustainable solutions.

Population and consumption drive environmental impact

We modify our environment in diverse ways, but the steep and sudden rise in human population (Chapter 8) has amplified nearly all of our impacts. We add about 80 million people to the planet each year—that's over 200,000 per day. Today, the rate of population growth is slowing, but our absolute numbers continue to increase.

Our consumption of resources has risen even faster than our population. The modern rise in affluence has been a positive development for humanity, and our conversion of the planet's natural capital has made life more pleasant for us so far. However, like rising population, rising per capita consumption magnifies the demands we make on our environment.

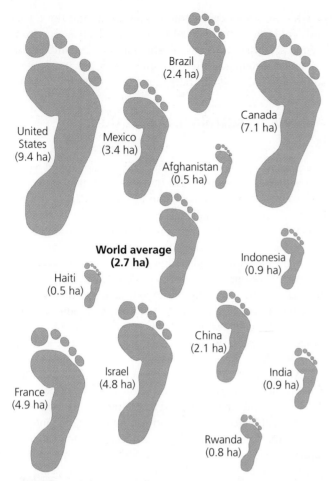

FIGURE 1.14 ▲ The citizens of some nations have much larger ecological footprints than the citizens of others. Shown here are ecological footprints for average citizens of several developed and developing nations, along with the world's average per capita footprint of 2.7 hectares. One hectare (ha) = 2.47 acres. Data are for 2005, from Global Footprint Network, 2008.

Moreover, the world's citizens have not benefited equally from our overall rise in affluence. Today the 20 wealthiest nations boast over 55 times the per capita income of the 20 poorest nations—nearly three times the gap that existed just four decades ago. The ecological footprint of the average citizen of a developed nation such as the United States is considerably larger than that of the average resident of a developing country (**FIGURE 1.14**). Within the United States, the richest 10% of people claim fully half the income, and the richest 1% claim nearly a quarter of all income.

WEIGHING THE ISSUES

Ecological Footprints What do you think accounts for the variation in sizes of per capita ecological footprints among societies? Do you think that nations with larger footprints have a moral obligation to reduce their environmental impact, so as to leave more resources available for nations with smaller footprints?

We face challenges with agriculture, pollution, and biodiversity

Our dramatic growth in population and consumption is due in part to our success in expanding and intensifying the production of food (Chapters 9 and 10). Since the origin of agriculture and the industrial revolution, progressively more powerful technologies have enabled us to grow more and more food per unit of land. These advances in agriculture must be counted as one of humanity's great achievements, but they have come at some cost. We have converted nearly half the planet's land surface for agriculture, and our extensive use of chemical fertilizers and pesticides poisons organisms and alters natural systems. Erosion, climate change, and poorly managed irrigation destroy 5–7 million ha (hectares; 12.5–17.5 million acres) of productive cropland each year. Together, agriculture, urban sprawl, and various other land uses have substantially affected most of the landscape of the United States and all other nations (**FIGURE 1.15**).

Meanwhile, pollution from our farms, industries, households, and vehicles dirties our land, water, and air. Outdoor air pollution, indoor air pollution, and water pollution contribute to the deaths of millions of people each year (Chapters 15–17). Environmental toxicologists are chronicling the impacts on people and wildlife of the many synthetic chemicals and other pollutants we emit into the environment (Chapter 14).

Perhaps our most pressing pollution challenge is to address the looming specter of global climate change (Chapter 18). Scientists have firmly concluded that human activity is altering the composition of the atmosphere and that these changes are affecting Earth's climate. Since the start of the industrial revolution, our combustion of fossil fuels and our deforestation of the landscape have boosted atmospheric carbon dioxide concentrations by 39%, to a level not present in at least 800,000 years, and probably 20 million years. Carbon dioxide and several other gases absorb and emit radiation and warm Earth's surface, which in turn causes glacial melting; sea level rise; changes in rainfall patterns; increased storms; and impacts on crops, forests, wildlife, and health.

The combined impact of human actions—climate change, overharvesting, pollution, the introduction of non-native species, and particularly habitat alteration—has driven many species toward extinction (Chapter 11). Today Earth's biological diversity, or **biodiversity**, the cumulative number and diversity of living things (pp. 55, 282–287), is declining dramatically (**FIGURE 1.16**). Biologists say we are setting in motion a mass extinction event (pp. 59–61, 288–290) comparable to only five others documented in all of Earth's history. Biologist Edward O. Wilson has warned that the loss of biodiversity is our most threatening dilemma because extinction is irreversible; once a species has become extinct, it is lost forever.

The most comprehensive scientific assessment of the condition of the world's ecological systems and their ability to continue supporting our civilization was completed in 2005, when over 2,000 of the world's leading environmental scientists from nearly 100 nations completed the **Millennium Ecosystem Assessment**. The main findings of this exhaustive project are summarized in **TABLE 1.1**. The Millennium Ecosystem Assessment makes it clear that our degradation

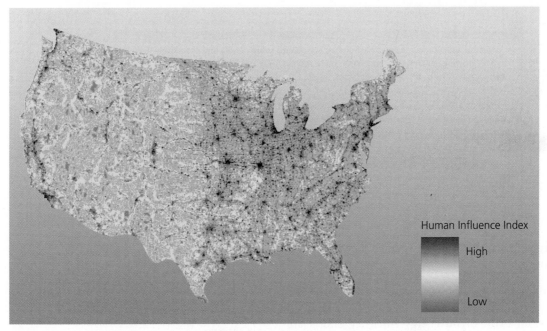

FIGURE 1.15 ▲ Human activity has heavily influenced much of the United States, especially in cities, on farms, and across the eastern portion of the nation. This map summarizes influence on terrestrial ecosystems by human settlement, roads and transportation networks, nighttime light pollution, and agriculture and other land use. It demonstrates that we live in a highly modified environment and suggests we would be wise to carefully nurture natural systems and manage remaining resources. Used by permission of the Center for International Earth Science Information Network (CIESIN), The Earth Institute, Columbia University. © 2008.

of the world's environmental systems is having negative impacts on all of us, but that with care and diligence we can still turn many of these trends around.

Our energy choices will influence our future enormously

Our reliance on fossil fuels to power our civilization has intensified virtually every impact we have on our environment, from habitat alteration to air pollution to climate change. Fossil fuels have also brought us the material affluence we enjoy. By exploiting the richly concentrated energy in coal, oil, and natural gas, we have been able to power the machinery of the industrial revolution, produce the chemicals that boost agricultural yields, run the vehicles and transportation networks of our mobile society, and manufacture and distribute our countless consumer products. The lifestyles we lead today are a direct result of the availability of fossil fuels (Chapter 19).

However, in extracting fossil fuels, we are splurging on a one-time bonanza. Scientists calculate that we have depleted roughly half the world's oil supplies and that we are in for a rude awakening soon, once supply begins to decline while demand continues to rise (pp. 539–544). We are also approaching peak production of natural gas, and coal is also nonrenewable and in finite supply. How we handle the imminent crises of fossil fuel depletion will largely determine the nature of our lives in the 21st century.

Sustainable solutions abound

Humanity's challenge is to develop solutions that enhance our quality of life while protecting and restoring the environment

FIGURE 1.16 ▲ Human activities are pushing many organisms, including the panda, toward extinction. Efforts to save endangered species and reduce the loss of biodiversity include many approaches, but all require that adequate areas of appropriate habitat be preserved in the wild.

that supports us. Fortunately, many workable solutions are at hand. For instance:

- ▶ A multitude of renewable energy sources (Chapters 20 and 21) are being developed to take the place of fossil fuels (**FIGURE 1.17**), and energy-efficiency efforts are gaining ground.

- ▶ In response to agricultural impacts, scientists and others have developed and promoted soil conservation, high-efficiency irrigation, and organic agriculture.

- ▶ Legislation and technological advances have reduced the pollution emitted by industry and automobiles in wealthier countries.

- ▶ Amid deep concern over the state of global biodiversity, advances in conservation biology (pp. 300–309) are helping to protect habitat, slow extinction, and safeguard endangered species.

- ▶ Recycling is helping to mitigate our waste disposal problems (Chapter 22).

- ▶ Nations, states, municipalities, businesses, and individuals are beginning to take steps to reduce emissions of the greenhouse gases that drive climate change.

18 **FIGURE 1.17 ▼** We can develop clean and renewable energy sources, for our sustainable use now and in the future. Just as a flowering plant gathers energy from the sun, we can use rooftop panels like these to harness solar energy.

These are but a few of the many efforts we will examine while exploring sustainable solutions in the course of this book.

Are things getting better or worse?

Despite the myriad challenges we face, some people maintain that the general conditions of human life and the environment are in fact getting better, not worse. A well-known proponent of this view, Danish statistician Bjorn Lomborg, wrote the following in his 2001 book *The Skeptical Environmentalist*:

> We are not running out of energy or natural resources. There will be more and more food per head of the world's population. Fewer and fewer people are starving. In 1900 we lived for an average of 30 years; today we live for 67. . . . The air and water around us are becoming less and less polluted. Mankind's lot has actually improved in terms of practically every measurable indicator.

Furthermore, some people maintain that we will find ways to make Earth's natural resources meet all of our needs indefinitely and that human ingenuity will see us through any difficulty. Such views are sometimes characterized as **Cornucopian**. In Greek mythology, *cornucopia*—literally "horn of plenty"—is the name for a magical goat's horn that overflowed with grain, fruit, and flowers. In contrast, people who predict doom and disaster have been called **Cassandras**, after the mythical princess of Troy with the gift of prophecy, whose dire predictions were not believed.

At least three questions are worth asking each time you are confronted with seemingly conflicting statements from Cornucopians and Cassandras:

1. Do the impacts being debated pertain only to people or also to other organisms and to natural systems?

2. Are the proponents of each view thinking in the short term or in the long term?

3. Are they considering all costs and benefits relevant for the question at hand, or only some?

As you proceed through this book and encounter countless contentious issues, consider how a person's perception of them may be influenced by these three factors.

Sustainable development involves environmental protection, economic well-being, and social equity

Environmental protection often is portrayed as threatening people's economic and social needs, but environmental scientists have long recognized that our civilization cannot exist without an intact and functional natural environment. In recent years, people of all persuasions have increasingly realized how environmental quality supports human quality of life.

Moreover, we now recognize that it is society's poorer people who suffer the most from environmental degradation. This realization has led advocates of environmental protection, economic development, and social justice to begin working together toward common goals. This cooperative approach has given rise to the modern drive for sustainable development.

Economists employ the term *development* to describe the use of natural resources for economic advancement (as opposed to simple subsistence, or survival). More broadly, development involves making purposeful changes intended to improve the quality of human life. Construction of homes, schools, hospitals, power plants, factories, and transportation networks are all examples of development. **Sustainable development** is the use of resources in a manner that satisfies our current needs but does not compromise the future availability of resources. The United Nations (p. 189) defines sustainable development as development that "meets the needs of the present without sacrificing the ability of future generations to meet their own needs." This definition is taken from the United Nations–sponsored Brundtland Commission (named after its chair, Norwegian prime minister Gro Harlem Brundtland), which published an influential 1987 report titled *Our Common Future*.

Prior to the Brundtland Report, most people aware of human impact on the environment might have thought "sustainable development" to be an oxymoron—a phrase that contradicts itself. Environmental advocates had long pointed out that development often so degrades the natural environment that it threatens the very improvements for human life that were intended. Conversely, many people remain under the impression that protecting the environment is incompatible with serving people's economic needs.

People also vary in how they interpret the phrase "sustainable development." Some businesspeople use it to mean ever-increasing economic gain, regardless of social or ecological well-being—whereas most environmental advocates use it to mean a novel kind of development that values and prioritizes environmental protection. Proponents of a school of thought called "weak sustainability" feel that we can allow natural capital to decline over time as long as

FIGURE 1.18 ▲ Former South African President Thabo Mbeki hugs a boy who performed in the welcoming ceremony of the United Nations–sponsored World Summit on Sustainable Development, held in Johannesburg, South Africa, in 2002, where 10,000 delegates from 200 nations set sustainable development goals. Sustainability requires that each generation leave enough resources for future generations to live as well or better.

human-made capital increases to compensate for it. In contrast, proponents of "strong sustainability" insist that human-made capital cannot substitute for natural capital and that therefore we must not allow natural capital to degrade.

In part because of the ambiguity resulting from differing interpretations of the term, just about everyone claims to support sustainable development these days. Sustainable development efforts are being undertaken by governments, businesses, industries, organizations, and individuals everywhere—from students on campus (Chapter 24) to international representatives at the United Nations (**FIGURE 1.18**). Many of these efforts truly are innovative and call for business and development interests to satisfy a **triple bottom line**, in which the goal is not simply the maximization of profit or economic advancement, but also environmental protection and the promotion of social equity. Solutions that meet environmental, economic, and social goals simultaneously, addressing the triple bottom line, can help pave the way for a truly sustainable society that promotes economic and social well-being while limiting environmental impact.

These aims require us to make an ethical commitment to our fellow citizens and to future generations. They also require that we apply knowledge from the sciences to help us devise ways to limit our impact and maintain the functioning environmental systems on which we depend. The question "How can we develop in a sustainable way?" may well be the single most important question in the world today. Environmental science holds the key to addressing this question.

➤ CONCLUSION

Finding effective ways of living peacefully, healthfully, and sustainably on our diverse and complex planet will require a thorough scientific understanding of both natural and social systems. Environmental science helps us understand our intricate relationship with our environment and informs our attempts to solve and prevent environmental problems. Identifying a problem is the first step in devising a solution to it. Many of the trends detailed in this book may cause us worry, but others give us reason to hope. Solving environmental problems can move us toward health, longevity, peace, and prosperity. Science in general, and environmental science in particular, can aid us in our efforts to develop balanced and workable solutions to the many challenges we face today and to create a better world for ourselves and our children.

REVIEWING OBJECTIVES

You should now be able to:

DEFINE THE TERM *ENVIRONMENT* AND THE FIELD OF ENVIRONMENTAL SCIENCE

- Our environment consists of everything around us, including living and nonliving things. (p. 3)
- Humans are part of the environment and are not separate from nature. (p. 3)
- Environmental science is the study of how the natural world works, how our environment affects us, and how we affect our environment. (p. 3)

EXPLAIN THE IMPORTANCE OF NATURAL RESOURCES AND ECOSYSTEM SERVICES TO OUR LIVES

- Some resources are inexhaustible or perpetually renewable, others are nonrenewable, and still others are renewable if we do not overexploit them. (pp. 3–4)
- Ecosystem services are benefits we receive from the processes and normal functioning of natural systems. (p. 3)
- Resources and ecosystem services are essential to human life and civilization, yet we are depleting and degrading them. (p. 3)

DISCUSS THE EFFECTS OF POPULATION GROWTH AND RESOURCE CONSUMPTION

- Rapid growth of the human population magnifies our environmental impacts. (pp. 3–4)
- The tragedy of the commons can lead to resource depletion. (p. 5)
- The ecological footprint concept quantifies a person's or nation's resource consumption in terms of area of biologically productive land and water. (pp. 5–6)

CHARACTERIZE THE INTERDISCIPLINARY NATURE OF ENVIRONMENTAL SCIENCE

- Environmental science employs approaches and insights from numerous disciplines in the natural sciences and social sciences. (p. 8)

UNDERSTAND THE SCIENTIFIC METHOD AND THE PROCESS OF SCIENCE

- Science is a process of using observations to test ideas. (p. 10)
- The scientific method consists of a series of steps that include making observations, formulating questions, stating a hypothesis, generating predictions, testing predictions, and analyzing results obtained from the tests. (pp. 10–12)
- There are different ways to test questions scientifically (for example, with manipulative experiments to determine causation or with natural experiments and correlation). (pp. 12–13)
- Scientific research occurs within a larger process that includes peer review of work, journal publication, and interaction with colleagues. (pp. 13–15)
- Science goes through paradigm shifts. This openness to change is what gives science its strength. (p. 15)

DIAGNOSE AND ILLUSTRATE SOME OF THE PRESSURES ON THE GLOBAL ENVIRONMENT

- Rising human population and intensifying per capita consumption magnify human impacts on the environment. (pp. 15–16)
- Human activities such as industrial agriculture and fossil fuel use are having diverse impacts, including resource depletion, air and water pollution, climate change, habitat destruction, and biodiversity loss. (pp. 16–17)
- We are striving to develop sustainable solutions to our environmental challenges. (pp. 18–19)

ARTICULATE THE CONCEPTS OF SUSTAINABILITY AND SUSTAINABLE DEVELOPMENT

- Sustainability means living within the planet's means, such that Earth's resources can sustain us—and other species—for the future. (p. 15)
- Sustainable development means pursuing environmental, economic, and social goals in a coordinated way, and it is the most important pursuit in our society today. (p. 19)

TESTING YOUR COMPREHENSION

1. What do renewable resources and nonrenewable resources have in common? How are they different? Identify two renewable and two nonrenewable resources.

2. How and why did the agricultural revolution affect human population size? How and why did the industrial revolution affect human population size? Explain what environmental impacts have resulted.

3. What is *the tragedy of the commons*? Explain how the concept might apply to an unregulated industry that is a source of water pollution.

4. What is *environmental science*? Name several disciplines involved in environmental science.

5. What are the two meanings of *science*? Name three applications of science.

6. Describe the scientific method. What is its typical sequence of steps?

7. Explain the difference between correlation and causation, and state how these concepts relate to manipulative and natural experiments.

8. What needs to occur before a researcher's results are published? Why is this process important?

9. Give examples of three major environmental problems in the world today, along with their causes. How are these problems interrelated? Can you name a potential solution for each?

10. How can *sustainable development* be defined? What is meant by the *triple bottom line*? Why is it important to pursue sustainable development?

SEEKING SOLUTIONS

1. Many resources are renewable if we use them in moderation but can become nonrenewable if we overexploit them. Order the following resources on a continuum of renewability (see Figure 1.1), from most renewable to least renewable: soils, timber, fresh water, food crops, and biodiversity. What factors influenced your choices? For each resource, what might constitute overexploitation, and what might constitute sustainable use?

2. Why do you think the Easter Islanders did not or could not stop themselves from stripping their island of all its trees? What similarities do you perceive between the history of Easter Island and the modern history of our society? What differences do you see between their predicament and ours?

3. What environmental problem do *you* feel most acutely yourself? Do you think there are people in the world who do not view your issue as an environmental problem? Who might they be, and why might they take a different view?

4. If the human population were to stabilize tomorrow and never surpass 7 billion people, would that solve our environmental problems? Which types of problems might be alleviated, and which might continue to worsen?

5. Consider the historic expansion of agriculture and our ability to feed increasing numbers of people, as described in this chapter. Now ask yourself, "Are things getting better or worse?" Ask this question from four points of view: (1) the human perspective, (2) the perspective of other organisms, (3) a short-term perspective, and (4) a long-term perspective. Do your answers to this question change? If so, how?

6. **THINK IT THROUGH** You have become head of a major funding agency that disburses funding to researchers pursuing work in environmental science. You must give your staff several priorities to determine what types of scientific research to fund. What environmental problems would you most like to see addressed with research? Describe the research you think would need to be completed so that workable solutions to these problems can be developed. Would more than science be needed to develop sustainable solutions?

CALCULATING ECOLOGICAL FOOTPRINTS

Mathis Wackernagel and his many colleagues at the Global Footprint Network (www.footprintnetwork.org) have continued to refine the method of calculating ecological footprints—the amount of biologically productive land and water required to produce the energy and natural resources we consume and to absorb the wastes we generate. According to their most recent data, there are nearly 2.1 hectares (5.2 acres) available for every person in the world,

yet we use on average more than 2.7 ha (6.7 acres) per person, creating a global ecological deficit, or overshoot (p. 5), of about 30%.

Compare the ecological footprints of each nation listed in the table. Calculate their proportional relationships to the world population's average ecological footprint and to the area available globally to meet our ecological demands.

Nation	Ecological footprint (hectares per person)	Proportion relative to world average footprint	Proportion relative to world area available
Bangladesh	0.6	0.2 (0.6 ÷ 2.7)	0.3 (0.6 ÷ 2.1)
Tanzania	1.1		
Colombia	1.8		
Thailand	2.1		
Mexico	3.4		
Sweden	5.1		
United States	9.4		
World average	2.7	1.0 (2.7 ÷ 2.7)	1.29 (2.7 ÷ 2.1)
Your personal footprint (see Question 4)			

Data from *Living planet report 2008*. WWF International, Zoological Society of London, and Global Footprint Network.

1. Why do you think the ecological footprint for people in Bangladesh is so small?

2. Why is it so large for people in the United States?

3. Based on the data in the table, how do you think average per capita income affects ecological footprints?

4. Go to an online footprint calculator such as the one at http://www.myfootprint.org or http://www.footprintnetwork.org/en/index.php/GFN/page/personal_footprint, and take the test to determine your own personal ecological footprint. Enter the value you obtain in the table, and calculate the other values as you did for each nation. How does your footprint compare to those of the average person in the United States? How does it compare to that of people from other nations? Name three actions you could take to reduce your footprint. (*Note*: Save this number—you will calculate your footprint again in Chapter 24 at the end of your course!)

Mastering**ENVIRONMENTALSCIENCE**™

Go to **www.masteringenvironmentalscience.com** for practice quizzes, Pearson eText, videos, current events, and more.

Geothermal power plant at The Geysers in northern California

2 EARTH'S PHYSICAL SYSTEMS: MATTER, ENERGY, AND GEOLOGY

UPON COMPLETING THIS CHAPTER, YOU WILL BE ABLE TO:

- Explain the fundamentals of matter and chemistry and apply them to real-world situations
- Differentiate among forms of energy and explain the basics of energy flow
- Distinguish photosynthesis, cellular respiration, and chemosynthesis, and summarize their importance to living things

- Explain how plate tectonics and the rock cycle shape the landscape around us and the earth beneath our feet
- List major types of geologic hazards and describe ways to mitigate their impacts

Clean Green Energy Beneath Our Feet: The Geysers in California

NORTH AMERICA

The Geysers

California

Atlantic Ocean

Pacific Ocean

"You drill a well [at The Geysers] and just pure steam comes pouring out of the ground and you don't have to do anything except run it through a turbine."

—Dr. David Blackwell of Southern Methodist University's Geothermal Lab

"Enhanced Geothermal Systems could be the 'killer app' of the energy world."

—Dan Reicher, Director of Climate and Energy Initiatives for Google.org, which is funding geothermal energy research

In 1847, explorer William Bell Elliott was hunting grizzly bears in the hills of northern California when he stumbled into a bizarre valley full of hissing jets of steam, sputtering sulfurous vents, and bubbling hot springs. He likened the scene to "the ruins of a recently burned city" and "the gates of hell."

A century and a half later, the spouting vapors in this region that he named "The Geysers" are providing renewable energy for modern residents of the region. For in these hills north of San Francisco, just beyond the wineries of the Napa and Sonoma valleys, rocks deep underground get so hot that water turns to steam. By tapping into these reservoirs of trapped steam, engineers at a complex of geothermal power plants—also called The Geysers—are producing enough electricity to power nearly a million homes.

Geothermal energy (from the Greek *geo* for "Earth" and *therme* for "heat") is thermal energy from below Earth's surface. At the site of The Geysers, magma (molten rock) rises relatively close to the surface, keeping the underground rock unusually hot. A *geyser* is a location where hot water and steam spout out of the ground, forced up naturally by pressure. There are, strictly speaking, no geysers at the site of The Geysers— only *fumaroles*, spouts of cteam or other gases. Yet since 1921, people have tapped this area for its geothermal

An engineer inspects turbine blades at a geothermal power plant at The Geysers

energy by extracting steam created by the underground heat and using it to turn turbines to generate electricity. Today the 22 power plants at The Geysers generate more geothermal power than anywhere else in the world.

However, there was a problem. The power plants were drawing up steam more quickly than it was being replenished underground, and although the plants returned some water to the ground, they were gradually depleting the subterranean reservoirs. Power production peaked in 1987 and then began declining as the groundwater supply dwindled.

In response, the managers began to import treated wastewater from surrounding communities and inject it into the ground. The idea was to replenish the aquifer to sustain the steam supply, while also doing a service for the communities. These efforts—the first of their kind in the world—seem to be working. The injected water has replenished the groundwater supply, and power generation is proceeding apace at stable levels.

Another problem has emerged, though: earthquakes. Extracting steam from underground and pumping water into the ground changes subterranean pressures in ways we cannot accurately predict—and changing pressures can cause earthquakes. Indeed, since the 1970s the area has been beset with flurries of small earthquakes (called microearthquakes). No major quake has struck the region yet, but many people fear one could.

In recent years, the company that owns most of the plants has embarked on a multi-million-dollar effort to drill new wells and enhance production. California's government has mandated that the state's utilities obtain fully one-third of their energy from renewable sources by the year 2020. The Geysers already provide one-fourth of California's renewable energy, so they will be a vital part of this effort.

In 2009, engineers from one company were drilling deeply into the ground at The Geysers, pursuing a new strategy called Enhanced Geothermal Systems (EGS). EGS seeks to harness more power from more places by drilling more deeply, pumping in fluid that cracks the rocks and becomes readily heated, and then drawing up the heated fluid. However, in Europe some EGS projects have been shut down after they appeared to generate earthquakes. At The Geysers, the project proceeded amid public concern over earthquakes but was suspended when the drill got stuck less than halfway to its intended depth.

As we try to move away from fossil fuels and begin powering our society with clean renewable energy, we are looking to the sun, to the wind, to the waves, and to the geothermal energy right beneath our feet. But if the price of abundant geothermal power is more earthquakes, that's a price we might not want to pay. We can hope and expect that more scientific study, creative technology, and public input can help us surmount the challenges that lie between our fossil fuel economy of today and the renewable energy future of tomorrow.

MATTER, CHEMISTRY, AND THE ENVIRONMENT

Geothermal heating arises from the interplay of rock, water, and heat. It involves geology, chemistry, and energy, and it is a process that takes place on a large scale. Environmental scientists regularly study these types of processes to understand how our planet works. Because all large-scale processes are made up of small-scale components, however, environmental science—the broadest of scientific fields—must also study small phenomena. At the smallest scale, an understanding of matter itself helps us to fully appreciate all the processes of our world.

All material in the universe that has mass and occupies space—solid, liquid, and gas alike—is termed **matter**. In our quick tour of matter in the pages that follow, we examine types of matter and some of the important ways they interact—phenomena that together we term **chemistry**. Once you examine any environmental issue, you will likely discover chemistry playing a central role. Chemistry is crucial for understanding how gases such as carbon dioxide and methane contribute to global climate change, how pollutants such as sulfur dioxide and nitric oxide cause acid rain, and how pesticides and other compounds we release into the environment affect the health of wildlife and people. Chemistry is central, too, in understanding water pollution and wastewater treatment, hazardous waste and its cleanup and disposal, atmospheric ozone depletion, and most energy issues. Moreover, countless applications of chemistry can help us address environmental problems.

Matter is conserved

To appreciate the chemistry involved in environmental science, we must begin with a grasp of the fundamentals. Matter may be transformed from one type of substance into others, but it cannot be created or destroyed. This principle is referred to as the **law of conservation of matter**. In environmental science, this principle helps us understand that the amount of matter stays constant as it is recycled in nutrient cycles and ecosystems (pp. 115, 122). It also makes clear that we cannot simply wish away the matter (such as waste and pollution) that we want to eliminate. Any piece of garbage or drop of spilled oil or billow of smokestack pollution or canister of nuclear waste that we dispose of will not simply disappear. Instead, we will need to take responsible steps to mitigate its impacts.

Atoms and elements are chemical building blocks

The geothermally heated water that rises up as steam at The Geysers is made up of **hydrogen** and **oxygen**, each of which is an element. An **element** is a fundamental type of matter, a chemical substance with a given set of properties, which cannot be broken down into substances with other properties. Chemists currently recognize 92 elements occurring in nature, as well as more than 20 others that they have created in the lab.

TABLE 2.1 Earth's Most Abundant Chemical Elements, by Mass

Earth's crust	Oceans	Air	Organisms
Oxygen (O), 49.5%	Oxygen (O), 88.3%	Nitrogen (N), 78.1%	Oxygen (O), 65.0%
Silicon (Si), 25.7%	Hydrogen (H), 11.0%	Oxygen (O), 21.0%	Carbon (C), 18.5%
Aluminum (Al), 7.4%	Chlorine (Cl), 1.9%	Argon (Ar), 0.9%	Hydrogen (H), 9.5%
Iron (Fe), 4.7%	Sodium (Na), 1.1%	Other, <0.1%	Nitrogen (N), 3.3%
Calcium (Ca), 3.6%	Magnesium (Mg), 0.1%		Calcium (Ca), 1.5%
Sodium (Na), 2.8%	Sulfur (S), 0.1%		Phosphorus (P), 1.0%
Potassium (K), 2.6%	Calcium (Ca), <0.1%		Potassium (K), 0.4%
Magnesium (Mg), 2.1%	Potassium (K), <0.1%		Sulfur (S), 0.3%
Other, 1.6%	Bromine (Br), <0.1%		Other, 0.5%

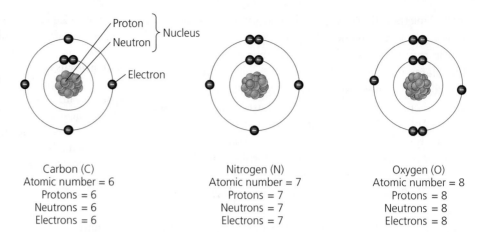

FIGURE 2.1 ▶ In an atom, protons and neutrons stay in the nucleus, and electrons move around the nucleus. Each chemical element has its own particular number of protons. Carbon possesses 6 protons, nitrogen 7, and oxygen 8. These schematic diagrams are meant to clearly show and compare numbers of electrons for these three elements. In reality, however, electrons do *not* orbit the nucleus in rings as shown; they move through space in more complex ways.

Carbon (C)
Atomic number = 6
Protons = 6
Neutrons = 6
Electrons = 6

Nitrogen (N)
Atomic number = 7
Protons = 7
Neutrons = 7
Electrons = 7

Oxygen (O)
Atomic number = 8
Protons = 8
Neutrons = 8
Electrons = 8

Besides hydrogen and oxygen, elements especially abundant on our planet include **silicon** (in Earth's crust), **nitrogen** (in the air), and **carbon** (in living organisms) (**TABLE 2.1**). Each element is assigned an abbreviation, or chemical symbol (for instance, H for hydrogen, and O for oxygen). The *periodic table of the elements* (see **APPENDIX C**) organizes the elements according to their chemical properties and behavior.

Atoms are the smallest units that maintain the chemical properties of the element. Atoms of each element hold a defined number of **protons** (positively charged particles) in the atom's nucleus (its dense center), and this number is called the element's *atomic number*. (Elemental carbon, for instance, has six protons in its nucleus; thus, its atomic number is 6.) Most atoms also contain **neutrons** (particles lacking electric charge) in their nuclei, and an element's *mass number* denotes the combined number of protons and neutrons in the atom. An atom's nucleus is surrounded by negatively charged particles known as **electrons**, which balance the positive charge of the protons (**FIGURE 2.1**).

Isotopes Although all atoms of a given element contain the same number of protons, they do not necessarily contain the same number of neutrons. Atoms of the same element with differing numbers of neutrons are referred to as **isotopes** (**FIGURE 2.2A**). Thus, isotopes of a given element have the same atomic number yet different mass numbers. Isotopes are denoted by their elemental symbol preceded by the mass number. For example, ^{14}C (carbon-14) is an isotope of carbon with eight neutrons (and six protons) in the nucleus rather than the six neutrons (and six protons) of ^{12}C (carbon-12), the most abundant carbon isotope.

Some isotopes are **radioactive** and "decay," changing their chemical identity as they shed subatomic particles and emit high-energy radiation. Such *radioisotopes* decay into lighter and lighter radioisotopes, until they become stable isotopes (isotopes that are not radioactive). The ongoing radioactive decay of elements such as uranium, potassium, and thorium deep within Earth emits heat that travels slowly upward toward Earth's surface, heating rock and water along the way and eventually being released to the atmosphere. It is this process that leads to geothermal heating at The Geysers and elsewhere.

Each radioisotope decays at a rate determined by that isotope's **half-life**, the amount of time it takes for one-half the atoms to give off radiation and decay. Different radioisotopes have very different half-lives, ranging from fractions of a second to billions of years. The radioisotope uranium-235 (^{235}U) is our society's source of energy for commercial nuclear power (pp. 564–576). Uranium-235 decays into a series of daughter isotopes, eventually forming lead-207 (^{207}Pb), and has a half-life of about 700 million years.

Ions Atoms may also gain or lose electrons, thereby becoming **ions**, electrically charged atoms or combinations of atoms (**FIGURE 2.2B**). Ions are denoted by their elemental symbol followed by their ionic charge. For instance, a common ion used by mussels and clams to form shells is Ca^{2+}, a calcium atom that has lost two electrons and so has a charge of positive 2.

Atoms bond to form molecules and compounds

Atoms can bond together and form **molecules**, combinations of two or more atoms. Common molecules containing only a single element include those of hydrogen (H_2) and oxygen (O_2), each of which exists as a gas at room temperature. A molecule composed of atoms of two or more different elements is

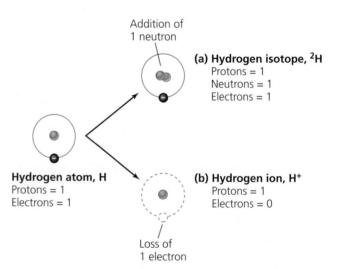

Addition of 1 neutron

(a) Hydrogen isotope, ^2H
Protons = 1
Neutrons = 1
Electrons = 1

Hydrogen atom, H
Protons = 1
Electrons = 1

(b) Hydrogen ion, H^+
Protons = 1
Electrons = 0

Loss of 1 electron

FIGURE 2.2 ▲ The mass number of hydrogen is 1, because a typical atom of this element contains one proton and no neutrons. Deuterium (hydrogen-2 or ^2H), a different isotope of hydrogen **(a)**, contains a neutron as well as a proton and thus has greater mass than a typical hydrogen atom; its mass number is 2. Shown in **(b)** is the hydrogen ion, H^+. By losing its electron, it gains a positive charge.

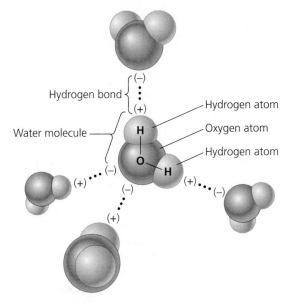

FIGURE 2.3 ▲ Water is a unique compound that has several properties crucial for life. Hydrogen bonds give water cohesion by enabling water molecules to adhere loosely to one another.

called a **compound**. **Water** is a compound; it is composed of two hydrogen atoms bonded to one oxygen atom, and it is denoted by the chemical formula H_2O. Another compound is **carbon dioxide**, consisting of one carbon atom bonded to two oxygen atoms; its chemical formula is CO_2.

Atoms bond together because of an attraction for one another's electrons. Because the strength of this attraction varies among elements, atoms may be held together in different ways. Two atoms of hydrogen share electrons equally as they bind together to form hydrogen gas, H_2. With water, in contrast, oxygen attracts electrons more strongly than does hydrogen. In compounds in which the strength of attraction is sufficiently unequal, an electron may be transferred from one atom to another, creating oppositely charged ions. Such associations are called *ionic compounds*, or *salts*. Table salt (NaCl) contains ionic bonds between positively charged sodium ions (Na^+), each of which donated an electron, and

negatively charged chloride ions (Cl^-), each of which received an electron.

Elements, molecules, and compounds can also come together in mixtures without chemically bonding or reacting. Homogeneous mixtures of substances are called *solutions*, a term most often applied to liquids, but also applicable to some gases and solids. Air in the atmosphere is a solution formed of constituents such as nitrogen, oxygen, water vapor, carbon dioxide, **methane** (CH_4), and **ozone** (O_3). Ocean water, plant sap, petroleum, and metal alloys (p. 646) such as brass are all solutions.

Water's chemistry facilitates life

Water has unique properties that give it an amazing capacity to support life. A water molecule has a partial negative charge at its oxygen end and partial positive charges at its hydrogen ends. This arrangement allows water molecules to adhere to one another in a type of weakly attractive interaction called a hydrogen bond (**FIGURE 2.3**). The loose connections of hydrogen bonds give water several properties that help to support life and stabilize Earth's climate:

▶ Water's cohesion (think of how a water droplet holds together) facilitates the transport of nutrients and waste in organisms.

▶ Heating weakens hydrogen bonds before it speeds molecular motion, so water can absorb a great deal of heat with only small changes in its temperature. This capacity to resist change helps stabilize water bodies, organisms, and climate systems.

▶ Water molecules in ice are farther apart than in liquid water (**FIGURE 2.4**), so ice is less dense. This enables ice to float on water, insulating lakes and ponds and preventing them from freezing solid in winter.

▶ Water molecules bond well with ions and other partially charged molecules, so water can hold in solution, or dissolve, many other molecules, including chemicals vital for life.

FIGURE 2.4 ◀ Ice floats in liquid water because ice is less dense. In ice, molecules are connected by stable hydrogen bonds, forming a spacious crystal lattice. In liquid water, hydrogen bonds frequently break and reform, and the molecules are closer together and less organized.

Hydrogen ions determine acidity

In any aqueous solution, a small number of water molecules split apart, each forming a hydrogen ion (H^+) and a hydroxide ion (OH^-). The product of hydrogen and hydroxide ion concentrations is always the same; as one increases, the other decreases. Pure water contains equal numbers of these ions. Solutions in which the H^+ concentration is greater than the OH^- concentration are **acidic**, whereas solutions in which the OH^- concentration exceeds the H^+ concentration are **basic**, or alkaline.

The **pH** scale (**FIGURE 2.5**) quantifies the acidity or alkalinity of solutions. It runs from 0 to 14; pure water has a hydrogen ion concentration of 10^{-7} and a pH of 7. Solutions with pH less than 7 are acidic and those with pH greater than 7 are basic. Because the pH scale is logarithmic, each step on the scale represents a tenfold difference in hydrogen ion concentration. Thus, a substance with pH of 6 contains 10 times as many hydrogen ions as a substance with pH of 7. Figure 2.5 shows pH for a number of common substances. Water can vary tremendously in pH, depending on what chemicals it holds in solution. For instance, geothermally heated water may hold a variety of chemicals dissolved from surrounding rock. Industrial air pollution makes precipitation more acidic (pp. 482–486), and the pH of rain throughout most of the eastern United States now averages well below 5.

Matter is composed of organic and inorganic compounds

Beyond their need for water, living things also depend on organic compounds, which they create and of which they are created. **Organic compounds** consist of carbon atoms (and generally hydrogen atoms) joined by bonds, and they may also include other elements, such as nitrogen, oxygen, sulfur, and phosphorus. Inorganic compounds, in contrast, lack carbon–carbon bonds. When carbon atoms bond together in long chains, the resulting molecules may be called **polymers**.

Carbon's unusual ability to bond together in chains, rings, and other structures to build elaborate molecules has resulted in millions of different organic compounds. One class of such compounds that is important in environmental science is the **hydrocarbons**, which consist solely of atoms of carbon and hydrogen (although other elements may enter these compounds as impurities). Fossil fuels and the many petroleum products we make from them (pp. 533, 541) consist largely of hydrocarbons.

The simplest hydrocarbon is methane (CH_4), the key component of natural gas; it has one carbon atom bonded to four hydrogen atoms (**FIGURE 2.6A**). Adding another carbon atom and two more hydrogen atoms gives us ethane (C_2H_6), the next-simplest hydrocarbon (**FIGURE 2.6B**). The smallest (and therefore lightest-weight) hydrocarbons (those consisting of four or fewer carbon atoms) exist in a gaseous state at moderate temperatures and pressures. Larger (therefore heavier) hydrocarbons are liquids, whereas those containing over 20 carbon atoms are normally solids. Some hydrocarbons from petroleum are known to pose health hazards to wildlife and people. For example, some polycyclic aromatic hydrocarbons, or PAHs (**FIGURE 2.6C**), can evaporate from spilled oil and gasoline and can mix with water, putting the eggs and young of fish and other aquatic creatures at risk. PAHs also occur in particulate form in various combustion products, including cigarette smoke, wood smoke, and charred meat.

Macromolecules are building blocks of life

Some polymers—proteins, nucleic acids, and carbohydrates—play key roles as building blocks of life. Along with lipids (chemically diverse compounds that do not dissolve in water, such as fats, oils, and waxes), these types of molecules are referred to as **macromolecules** because of their large sizes.

Proteins consist of long chains of organic molecules called *amino acids*. The many types of proteins serve various functions. Some help produce tissues and provide structural support; for example, animals use proteins to generate skin, hair, muscles, and tendons. Some proteins help store energy, whereas others transport substances. Some act in the immune system to defend the organism against foreign attackers. Still others are

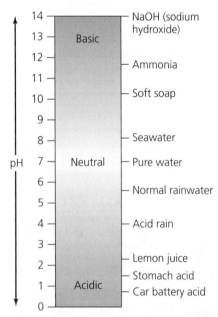

FIGURE 2.5 ▲ The pH scale measures how acidic or basic (alkaline) a solution is. The pH of pure water is 7, the midpoint of the scale. Acidic solutions have higher hydrogen ion concentrations and lower pH, whereas basic solutions have lower hydrogen ion concentrations and higher pH.

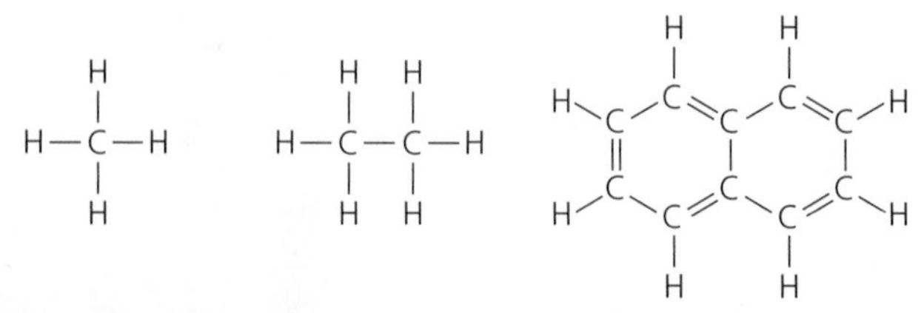

(a) Methane, CH_4   **(b) Ethane, C_2H_6**   **(c) Naphthalene, $C_{10}H_8$ (a polycyclic aromatic hydrocarbon)**

FIGURE 2.6 ▲ The simplest hydrocarbon is methane **(a)**. Many hydrocarbons consist of linear chains of carbon atoms with hydrogen atoms attached; the shortest of these is ethane **(b)**. Volatile hydrocarbons with multiple rings, such as naphthalene **(c)**, are called polycyclic aromatic hydrocarbons (PAHs).

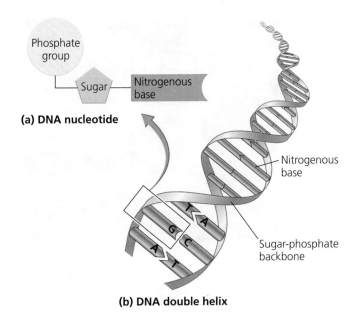

Phosphate
group

Sugar

Nitrogenous
base

(a) DNA nucleotide

Nitrogenous
base

Sugar-phosphate
backbone

(b) DNA double helix

FIGURE 2.7 ▲ Nucleic acids encode genetic information in the sequence of nucleotides **(a)**, small molecules that pair together like rungs of a ladder. DNA includes four types of nucleotides, each with a different nitrogenous base: adenine (A), guanine (G), cytosine (C), and thymine (T). Adenine (A) pairs with thymine (T), and cytosine (C) pairs with guanine (G). In RNA, thymine is replaced by uracil (U). DNA twists into the shape of a double helix **(b)**.

hormones, molecules that act as chemical messengers within an organism. Proteins can also serve as *enzymes*, molecules that catalyze, or promote, certain chemical reactions.

Nucleic acids direct the production of proteins. The two nucleic acids—**deoxyribonucleic acid (DNA)** and **ribonucleic acid (RNA)**—carry the hereditary information for organisms and are responsible for passing traits from parents to offspring. Nucleic acids are composed of series of nucleotides, each of which contains a sugar molecule, a phosphate group, and a nitrogenous base. DNA contains four types of nucleotides and can be pictured as a ladder twisted into a spiral, giving the molecule a shape called a double helix (**FIGURE 2.7**). Regions of DNA coding for particular proteins that perform particular functions are called **genes**.

Carbohydrates include simple sugars that are three to seven carbon atoms long. Glucose ($C_6H_{12}O_6$) fuels living cells and serves as a building block for complex carbohydrates, such as starch. Plants use starch to store energy, and animals eat plants to acquire starch. Plants and animals also use complex carbohydrates to build structure. Insects and crustaceans form hard shells from the carbohydrate chitin. Cellulose, the most abundant organic compound on Earth, is a complex carbohydrate found in the cell walls of leaves, bark, stems, and roots.

We create synthetic polymers

The polymers in nature that are so vital to our survival have inspired chemists. These scientists have taken the polymer concept and run with it, creating innumerable types of synthetic (human-made) polymers, which we call **plastics**. Polyethylene, polypropylene, polyurethane, and polystyrene are just a few of the many synthetic polymers in our manufactured products today. (We often know types of plastic by brand names, such as Nylon, Teflon, and Kevlar.) Plastics, many of them derived from hydrocarbons in petroleum, are all around us in our everyday lives, from furniture to food containers to fiber optics to fleece jackets (see Figure 19.12, p. 541).

We value synthetic polymers because they are long-lasting and resist chemical breakdown. Ironically, this is also the reason that they are such a serious source of waste and pollution. In future chapters we will see how synthetic pollutants that resist breakdown can cause problems for wildlife and human health (pp. 378–392), water quality (pp. 420–424), marine animals (pp. 445–446), and waste management (pp. 619–621). Fortunately, chemists, policymakers, and citizens are finding ways to design less-polluting substances and to recycle materials more effectively.

ENERGY: AN INTRODUCTION

The fact that plastics and so many of our consumer products are derived from the hydrocarbons of petroleum is just one reason why our society would benefit from conserving its dwindling petroleum supplies. As we shall see (Chapters 19–21), shifting to renewable sources of energy such as geothermal energy will cut down on pollution, mitigate climate change, and help assure us long-lasting energy supplies into the future.

But what, exactly, is energy? **Energy** is the capacity to change the position, physical composition, or temperature of matter—in other words, a force that can accomplish work. Energy powers everything from the chemical reactions within our bodies to the geologic forces that shape our planet. Energy is involved in every physical, chemical, and biological process.

As we shall see shortly, geologic events involve some of the most dramatic releases of energy in nature; think of volcanoes erupting, earthquakes shaking the land, and landslides sending material down mountain slopes. Energy is omnipresent in living things, as well. A sparrow in flight expends energy to propel its body through the air (change of position). When the sparrow lays an egg, its body uses energy to create the calcium-based eggshell and color it with pigment (change in composition). The sparrow sitting on its nest transfers energy from its body to heat the developing chicks inside its eggs (change of temperature).

Energy comes in different forms

Energy manifests itself in different ways and can be converted from one form to another. Two major forms of energy that scientists commonly distinguish are **potential energy**, energy of position; and **kinetic energy**, energy of motion. Consider underground water that is heated by thermal energy from deep within Earth. Its pressure rises, but horizontal layers of rock may prevent the heated and pressurized water from moving upward toward the surface. This causes the water to accumulate potential energy. If the water finds a way to the surface, the potential energy is converted to kinetic energy in the form of motion as the water rushes upward, perhaps spouting above the surface as a geyser.

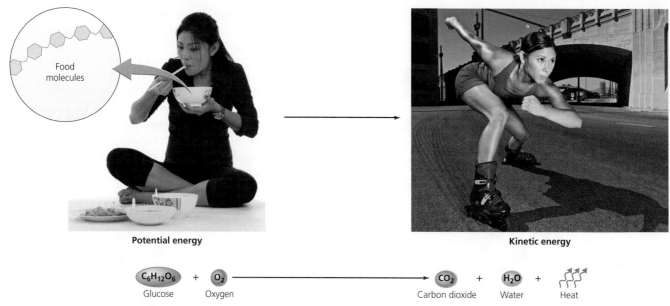

Potential energy

Kinetic energy

$$C_6H_{12}O_6 + O_2 \longrightarrow CO_2 + H_2O + Heat$$
Glucose   Oxygen   Carbon dioxide   Water   Heat

FIGURE 2.8 ▲ Energy is released when potential energy is converted to kinetic energy. Potential energy stored in sugars (such as glucose) in the food we eat **(a)**, combined with oxygen, becomes kinetic energy when we exercise **(b)**, releasing carbon dioxide, water, and heat as by-products.

Energy conversions take place at the atomic level every time a chemical bond is broken or formed. Chemical energy is essentially potential energy stored in the bonds between atoms. Bonds differ in their amounts of chemical energy, depending on the atoms they hold together. Converting a molecule with high-energy bonds (such as the carbon–carbon bonds of fossil fuels) into molecules with lower-energy bonds (such as the bonds in water or carbon dioxide) releases energy by changing potential energy into kinetic energy and produces motion, action, or heat. Just as automobile engines split the hydrocarbons of gasoline to release chemical energy and generate movement, our bodies split glucose molecules in our food for the same purpose (**FIGURE 2.8**).

Besides occurring as chemical energy, potential energy can occur as nuclear energy (the energy that holds atomic nuclei together) or mechanical energy (such as is stored in a compressed spring). Kinetic energy can also express itself in different forms, including thermal energy, light energy, sound energy, or electrical energy—all of which involve the movement of atoms, subatomic particles, molecules, or objects.

Energy is always conserved but can change in quality

Although energy can change from one form to another, it cannot be created or destroyed. Just as matter is conserved (p. 25), the total energy in the universe remains constant and thus is said to be conserved. Scientists refer to this principle as the **first law of thermodynamics**. When heated underground water surges toward the surface, the kinetic energy of its movement will equal the potential energy it had held while underground. Likewise, we obtain energy from the food we eat and then expend it in exercise, apply it toward maintaining our body and all its functions, or store it in fat. We do not somehow create additional energy or end up with less than the food gives us. Any particular system in nature can

temporarily increase or decrease in energy, but the total amount in the universe remains constant.

Although the overall amount of energy is conserved in any conversion of energy, the **second law of thermodynamics** states that the nature of energy will change from a more-ordered state to a less-ordered state, as long as no force counteracts this tendency. That is, systems tend to move toward increasing disorder, or *entropy*. For instance, after death every organism undergoes decomposition and loses its structure. A log of firewood—the highly organized and structurally complex product of many years of tree growth—transforms in the campfire to a residue of carbon ash, smoke, and gases such as carbon dioxide and water vapor, as well as the light and the heat of the flame (**FIGURE 2.9**). With the help of oxygen, the complex biological polymers that make up the wood are converted into a disorganized assortment of rudimentary molecules and heat and light energy.

If the second law of thermodynamics specifies that systems tend to move toward disorder, then how does any system ever hold together? The order of an object or system can be increased by the input of energy from outside the system. This is precisely what living organisms do. Organisms maintain their structure and function by consuming energy. They represent a constant struggle to maintain order and combat the natural tendency toward disorder.

The nature of an energy source helps determine how easily people can harness it. Sources such as fossil fuels and the electricity we produce in power plants contain concentrated energy that we can readily release. It is relatively easy for us to gain large amounts of energy efficiently from such sources. In contrast, sunlight and the heat stored in ocean water are more diffuse. Each day the world's oceans absorb heat energy from the sun equivalent to that of 250 billion barrels of oil—more than 3,000 times as much as our global society uses in a year. However, because this energy is spread across such vast areas, it is difficult to harness efficiently.

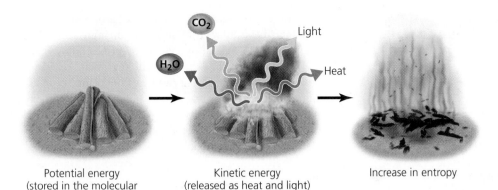

Potential energy
(stored in the molecular
bonds of wood)

Kinetic energy
(released as heat and light)

Increase in entropy

In each attempt we make to harness energy, some portion escapes. We can express our degree of success in capturing energy in terms of the *energy conversion efficiency*, the ratio of useful output of energy to the amount that we need to input. When we burn gasoline in an automobile engine, only about 16% of the energy released is used to power the automobile, and the rest of the energy is converted to heat and escapes without being used (see Figure 18.26, p. 519). Incandescent light bulbs are even less efficient; only 5% of their energy is converted to light, while the rest escapes as heat. Viewed in this context, the 7–15% efficiency of most existing geothermal power plants does not look bad—and engineers are attempting to increase this efficiency.

Light energy from the sun powers most living systems

The energy that powers Earth's biological systems comes primarily from the sun. The sun releases radiation from large portions of the electromagnetic spectrum, although our atmosphere filters much of this out and we see only some of this radiation as visible light (**FIGURE 2.10**). Most of the sun's energy is reflected, or else absorbed and re-emitted, by the atmosphere, land, or water (see Figure 18.1, p. 496). Solar energy drives winds, ocean currents, weather, and climate patterns. A small amount (less than 1% of the total) powers plant growth, and a still smaller amount flows from plants into the organisms that eat them and the organisms that decompose dead organic matter. A minuscule percentage of this energy is eventually deposited below ground in the chemical bonds in fossil fuels.

Some organisms use the sun's radiation directly to produce their own food. Such organisms, called **autotrophs** or **primary producers**, include green plants, algae, and cyanobacteria (a type of bacteria named for their cyan, or blue-green, color). Autotrophs turn light energy from the sun into chemical energy in a process called **photosynthesis** (**FIGURE 2.11**). In photosynthesis, sunlight powers a series of chemical reactions that convert carbon dioxide and water into sugars, transforming diffuse energy from the sun into concentrated energy the organism can use. Photosynthesis is an example of moving toward a state of lower entropy, and as such it requires a substantial input of outside energy.

Photosynthesis produces food for plants and animals

Photosynthesis occurs within cell organelles called *chloroplasts*, where the light-absorbing pigment *chlorophyll* (which is what makes plants green) uses solar energy (light energy from the sun) to initiate a series of chemical reactions called *light reactions*. During these reactions, water molecules split and react to form hydrogen ions (H^+) and molecular oxygen (O_2), thus creating the oxygen that we breathe. The light reactions also produce small, high-energy molecules that are used to fuel reactions in the *Calvin cycle*. In these reactions, carbon atoms from carbon dioxide are linked together to manufacture sugars.

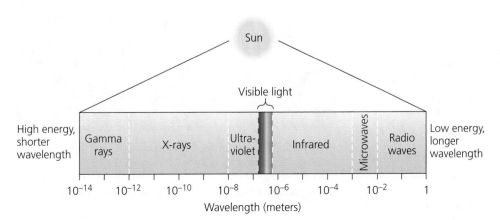

Sun

Visible light

High energy, shorter wavelength

Gamma rays

X-rays

Ultra-violet

Infrared

Microwaves

Radio waves

Low energy, longer wavelength

10^{-14} 10^{-12} 10^{-10} 10^{-8} 10^{-6} 10^{-4} 10^{-2} 1

Wavelength (meters)

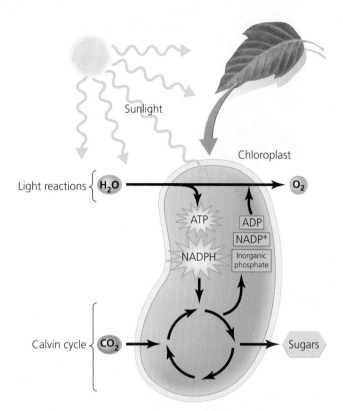

FIGURE 2.11 ▲ In photosynthesis, autotrophs such as plants, algae, and cyanobacteria use sunlight to convert carbon dioxide and water into sugars and oxygen. This schematic diagram summarizes the complex sets of chemical reactions that take place within chloroplasts. In the light reactions, water is converted to oxygen in the presence of sunlight, creating high-energy molecules (ATP and NADPH). These molecules help drive reactions in the Calvin cycle, in which carbon dioxide is used to produce sugars. Molecules of ADP, NADP⁺, and inorganic phosphate created in the Calvin cycle in turn help power the light reactions, creating an endless loop.

Photosynthesis is a complex process, but the overall reaction can be summarized with the following equation:

$$6CO_2 + 12H_2O + \text{the sun's energy} \longrightarrow C_6H_{12}O_6 + 6O_2 + 6H_2O$$
$$\text{(sugar)}$$

The numbers preceding each molecular formula indicate how many of each molecule are involved in the reaction. Note that the sums of the numbers on each side of the equation for each element are equal; that is, there are 6 C, 24 H, and 24 O on each side. This illustrates how chemical equations are balanced, with each atom recycled and matter conserved. No atoms are lost; they are simply rearranged among molecules. Note also that water appears on both sides of the equation. The reason is that for every 12 water molecules that are input and split in the process, 6 water molecules are newly created. We can streamline the photosynthesis equation by showing only the net loss of 6 water molecules:

$$6CO_2 + 6H_2O + \text{the sun's energy} \longrightarrow C_6H_{12}O_6 \text{ (sugar)} + 6O_2$$

Thus in photosynthesis, water, carbon dioxide, and light energy from the sun are transformed to produce sugar (glucose) and oxygen. To accomplish this, green plants draw up water from the ground through their roots, absorb carbon dioxide from the air through their leaves, and harness sunlight. With these ingredients, they create sugars for their growth and maintenance, and they release oxygen as a by-product.

Animals, in turn, depend on the sugars and oxygen from photosynthesis. Animals survive by eating plants, or by eating animals that have eaten plants, and by taking in oxygen (**FIGURE 2.12**). In fact, it is thought that animals appeared on Earth's surface only after the planet's atmosphere had been supplied with oxygen by cyanobacteria, the earliest autotrophs.

Cellular respiration releases chemical energy

Organisms make use of the chemical energy created by photosynthesis in a process called **cellular respiration**, which is vital to life. To release the chemical energy of glucose, cells employ oxygen to convert glucose back into its original starting materials, water and carbon dioxide. Note that this is the process illustrated in both Figures 2.8 and 2.12. The energy released during this process is used to form chemical bonds or to perform other tasks within cells. The net equation for cellular respiration is the exact opposite of that for photosynthesis:

$$C_6H_{12}O_6\text{(sugar)} + 6O_2 \longrightarrow 6CO_2 + 6H_2O + \text{energy}$$

FIGURE 2.12 ▲ When an animal such as a deer eats the leaves of a plant, it consumes the sugars the plant has produced through photosynthesis. The animal gains energy from those sugars, along with oxygen, through the process of cellular respiration.

32

However, the energy released per glucose molecule in respiration is only two-thirds of the energy input per glucose molecule in photosynthesis—a prime example of the second law of thermodynamics. Cellular respiration is a continuous process occurring in all living things, and is essential to life. Thus, it occurs in the autotrophs that create glucose and also in **heterotrophs**, organisms that gain their energy by feeding on other organisms. Heterotrophs include most animals, as well as the fungi and microbes that decompose organic matter.

Geothermal energy also powers Earth's systems

Although the sun is life's primary energy source, it is not the only one for our planet. A minor additional energy source is the gravitational pull of the moon, which in conjunction with the sun's pull causes ocean tides. As we have seen, a more significant additional energy source is geothermal heating emanating from inside Earth, powered primarily by radioactivity (p. 26). Radiation from radioisotopes deep inside our planet heats the inner Earth, and this heat gradually makes its way to the surface. There it heats magma that erupts from volcanoes, drives plate tectonics (p. 35), and warms water, which in some locations shoots out of the ground in the form of geysers (**FIGURE 2.13**).

Although we harness geothermal energy in some locations for our own use today, geothermal energy was powering biological communities long before people appeared on Earth. On the ocean floor, jets of geothermally heated water—essentially underwater geysers—gush into the icy-cold depths. In one of the more amazing scientific discoveries of recent decades, scientists realized that these **hydrothermal vents** can host entire communities of specialized organisms that thrive in the extreme high-temperature, high-pressure conditions. Gigantic clams, immense tubeworms, and odd mussels,

FIGURE 2.13 ▼ This geyser in the Black Rock Desert of Nevada propels scalding water into the air, powered by geothermal energy from deep below ground. The bright colors of the rocks are from colonies of bacteria that thrive in the hot, mineral-laden water.

(a) Hydrothermal vent

(b) Giant tubeworms

FIGURE 2.14 ▲ Hydrothermal vents on the ocean floor **(a)** send spouts of hot, mineral-rich water into the cold blackness of the deep sea. Specialized biological communities thrive in these unusual conditions. Odd creatures such as giant tubeworms **(b)** survive thanks to bacteria that produce food from hydrogen sulfide through the process of chemosynthesis.

shrimps, crabs, and fish all flourish in the seemingly hostile environment near scalding water that shoots out of tall chimneys of encrusted minerals (**FIGURE 2.14**).

These locations are so deep underwater that they completely lack sunlight, so the energy flow of these communities cannot be fueled through photosynthesis. Instead, bacteria in deep-sea vents use the chemical-bond energy of hydrogen sulfide (H_2S) to transform inorganic carbon into organic carbon compounds in a process called **chemosynthesis**:

$$6CO_2 + 6H_2O + 3H_2S \rightarrow C_6H_{12}O_6 \text{ (sugar)} + 3H_2SO_4$$

Chemosynthesis occurs in various ways, but note how this particular reaction for chemosynthesis closely resembles the photosynthesis reaction. These two processes use different energy sources, but each uses water and carbon dioxide to produce sugar and a by-product, and each produces potential energy that is later released during respiration. Energy from chemosynthesis passes through the deep-sea-vent animal community as heterotrophs such as clams, mussels, and shrimp gain nutrition from chemoautotrophic bacteria. Hydrothermal vent communities excited scientists because they were novel and unexpected, and they showed just how much we still have to learn about the workings of our planet.

GEOLOGY: THE PHYSICAL BASIS FOR ENVIRONMENTAL SCIENCE

If we want to understand how our planet functions, a good way to start is to examine the rocks, soil, and sediments beneath our feet. The physical processes that take place at and below Earth's surface shape the landscape around us and lay the foundation for most environmental systems and for life. Understanding the physical nature of our planet also benefits our society, for without the study of Earth's rocks and the processes that shape them, we would have no energy from geothermal sources or from fossil fuels! The harnessing of geothermal energy provides just one example of how we draw on resources and processes from beneath the surface of our planet and put them to use in our everyday lives.

Our planet is dynamic, and this dynamism is what motivates **geology**, the study of Earth's physical features, processes, and history. A human lifetime is just a blink of an eye in the long course of geologic time, and the Earth we experience is merely a snapshot in our changing planet's long history. We can begin to grasp this long-term dynamism as we consider two processes of fundamental importance—plate tectonics and the rock cycle. After examining the geologic processes that shape our planet, we will close our chapter by exploring the catastrophic hazards we sometimes face from these processes.

Earth consists of layers

Most geologic processes take place near Earth's surface, but our planet consists of multiple layers (**FIGURE 2.15**). At Earth's center is a dense **core** consisting mostly of iron, solid in the inner core and molten in the outer core. Surrounding the core is a thick layer of less dense, elastic rock called the **mantle**. A portion of the upper mantle called the *asthenosphere* contains especially soft rock, melted in some areas. The harder rock above the asthenosphere is what we know as the **lithosphere**. The lithosphere includes both the uppermost mantle and the entirety of Earth's third major layer, the **crust**, the thin, brittle, low-density layer of rock that covers Earth's surface. The intense heat in the inner Earth rises from core to mantle to crust, and it eventually dissipates at the surface. In regions where the asthenosphere's molten rock approaches to within a few miles of the surface, we can harness geothermal energy by drilling boreholes into the crust (**FIGURE 2.16**).

In fact, we don't even need to drill deep or find geysers to take advantage of geothermal energy. Anywhere in the world, the soil and rock just below the surface is fairly constant in temperature, cooler than the air above the surface in summer and warmer than the air in winter. Because of this, you can use a geothermal heat pump (also called a ground-source heat pump) to heat and cool a home (Figure 21.18, p. 607). Over

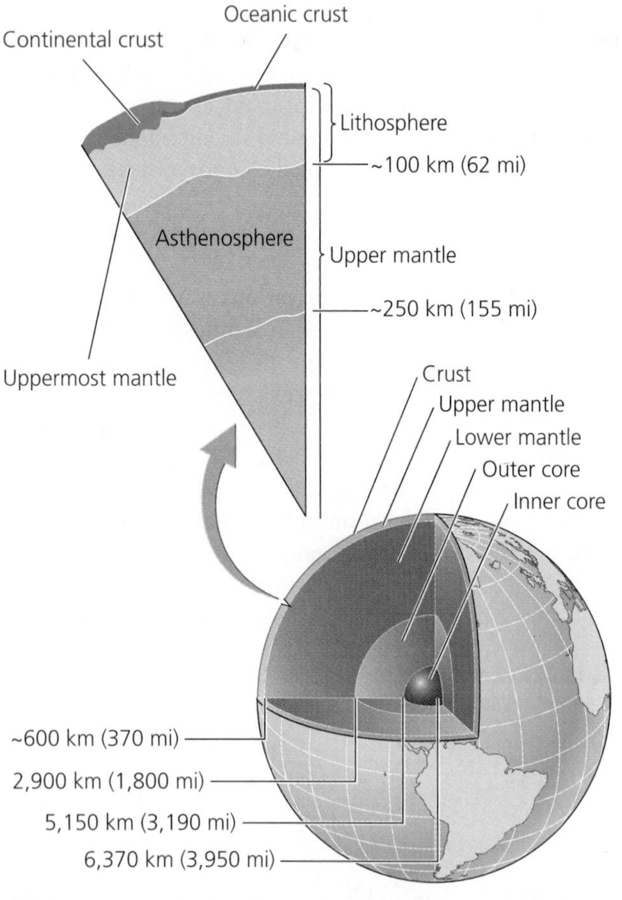

FIGURE 2.15 ▲ Earth's three primary layers—core, mantle, and crust—are themselves layered. The inner core of solid iron is surrounded by an outer core of molten iron, and the rocky mantle includes the molten *asthenosphere* near its upper edge. At Earth's surface, dense and thin oceanic crust abuts lighter, thicker continental crust. The *lithosphere* consists of the crust and the uppermost mantle above the asthenosphere.

Continental crust
Oceanic crust
Lithosphere
~100 km (62 mi)
Asthenosphere
Upper mantle
~250 km (155 mi)
Uppermost mantle
Crust
Upper mantle
Lower mantle
Outer core
Inner core
~600 km (370 mi)
2,900 km (1,800 mi)
5,150 km (3,190 mi)
6,370 km (3,950 mi)

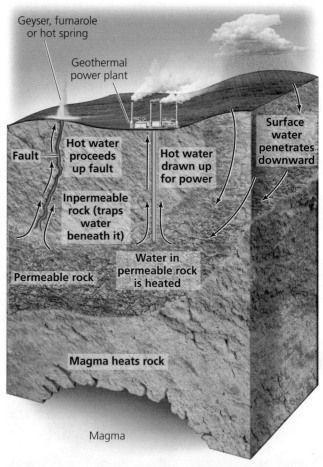

Geyser, fumarole or hot spring
Geothermal power plant
Fault
Hot water proceeds up fault
Hot water drawn up for power
Surface water penetrates downward
Inpermeable rock (traps water beneath it)
Permeable rock
Water in permeable rock is heated
Magma heats rock
Magma

FIGURE 2.16 ▲ A body of magma near Earth's surface heats water within permeable rock layers. Trapped by impermeable rock layers above it, this heated, pressurized water can escape in liquid form or as steam through faults in the crust, reaching the surface as a geyser, fumarole, or hot spring. Precipitation and surface water replenish the permeable layer over time as water percolates downward. We can harness geothermal energy for electrical power by drilling into permeable layers to release the trapped water or steam.

600,000 U.S. homes now use these highly efficient systems. We will learn more about these and the other methods of geothermal heating in our discussion of renewable energy sources in Chapter 21.

The heat from the inner layers of Earth also drives convection currents that flow in loops in the mantle, pushing the mantle's soft rock cyclically upward (as it warms) and downward (as it cools), like a gigantic conveyor belt system. As the mantle material moves, it drags large plates of lithosphere along its surface. This movement of lithospheric plates is known as **plate tectonics**, a process of extraordinary importance to our planet.

Plate tectonics shapes Earth's geography

Our planet's surface consists of about 15 major tectonic plates, which fit together like pieces of a jigsaw puzzle (**FIGURE 2.17**). Imagine peeling an orange and then placing the pieces of peel back onto the fruit; the ragged pieces of peel are like the lithospheric plates riding atop Earth's surface. However, the plates are thinner relative to the planet's size, more like the skin of an apple. These plates move at rates of roughly 2–15 cm (1–6 in.) per year. This slow movement has influenced Earth's climate and life's evolution throughout our planet's history as the continents combined, separated, and recombined in various configurations. By studying ancient rock formations throughout the world, geologists have determined that at least twice, all landmasses were joined together in a "supercontinent." Scientists have dubbed the one that occurred about 225 million years ago *Pangaea* (see Figure 2.17).

There are three types of plate boundaries

The processes that occur at the boundaries between plates have major consequences. We can categorize these boundaries into three types: divergent, transform, and convergent plate boundaries.

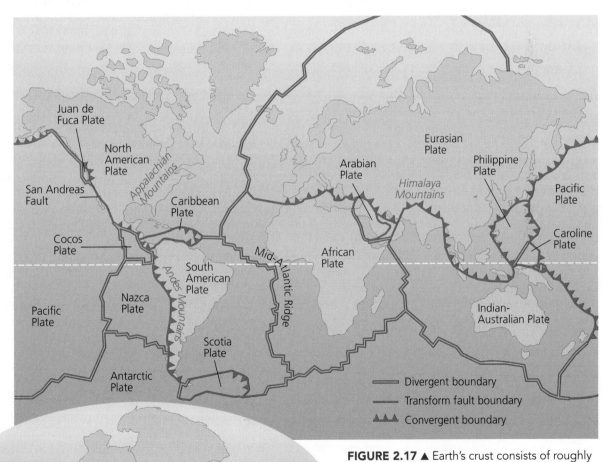

FIGURE 2.17 ▲ Earth's crust consists of roughly 15 major plates (above) that move very slowly by the process of plate tectonics. Today's continents were joined together in the landmass Pangaea (left) about 225 million years ago.

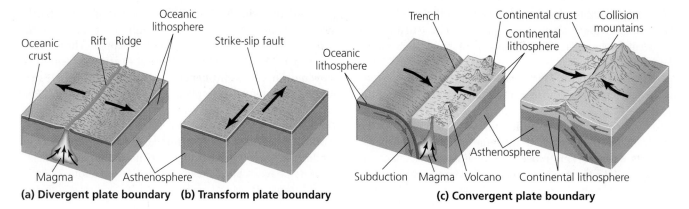

Oceanic crust | Rift | Ridge | Oceanic lithosphere | Strike-slip fault | Oceanic lithosphere | Trench | Continental crust | Continental lithosphere | Collision mountains

Magma | Asthenosphere

Subduction | Magma | Volcano | Asthenosphere | Continental lithosphere

(a) Divergent plate boundary **(b) Transform plate boundary**   **(c) Convergent plate boundary**

FIGURE 2.18 ▲ Different types of boundaries between tectonic plates generate different geologic processes. At a divergent plate boundary, such as a mid-ocean ridge on the seafloor **(a)**, the two plates move gradually away from the boundary in the manner of conveyor belts, and magma from beneath the crust may extrude as lava. At a transform plate boundary **(b)**, two plates slide alongside one another, creating friction that leads to earthquakes. Where plates collide at a convergent plate boundary **(c)**, one plate is subducted beneath another, leading to volcanism. If continental crust from two plates collides, the buckling of rock can form mountain ranges.

At **divergent plate boundaries**, tectonic plates push apart from one another as **magma** (rock heated to a molten, liquid state) rises upward to the surface, creating new lithosphere as it cools (**FIGURE 2.18A**). An example is the Mid-Atlantic Ridge, part of a 74,000-km (46,000-mi) system of divergent plate boundaries slicing across the floors of the world's oceans.

Where two plates meet, they may slip and grind alongside one another, forming a **transform plate boundary** (**FIGURE 2.18B**). This movement creates friction that generates earthquakes (pp. 40–41) along strike-slip faults. *Faults* are fractures in Earth's crust, and at strike-slip faults each landmass moves horizontally in opposite directions. The Pacific Plate and the North American Plate slide past one another along California's San Andreas Fault, which runs north-south to the west of The Geysers. Southern California is slowly inching its way northward along the San Andreas Fault, and the site of Los Angeles will eventually reach that of San Francisco.

Convergent plate boundaries, where two plates converge or come together, can give rise to different outcomes (**FIGURE 2.18C**). As plates of newly formed lithosphere push outward from divergent plate boundaries, this oceanic lithosphere gradually cools, becoming denser. Eventually (after millions of years) it becomes denser than the asthenosphere beneath it, and dives downward into the asthenosphere in a process called **subduction**. As the lithospheric plate descends, it slides beneath a neighboring plate that is less dense, forming a convergent plate boundary. The subducted plate is heated and pressurized as it sinks, and water vapor escapes, helping to melt rock (by lowering its melting temperature). The molten rock rises, and this magma may erupt through the surface at volcanoes (pp. 42–43).

When one plate of oceanic lithosphere is subducted beneath another plate of oceanic lithosphere, the resulting volcanism may form arcs of islands, such as Japan and the Aleutian Islands of Alaska. Subduction zones may also create deep trenches, such as the Mariana Trench, our planet's

deepest abyss. When oceanic lithosphere slides beneath continental lithosphere, this leads to the formation of volcanic mountain ranges that parallel coastlines (a process shown in the left-hand drawing of Figure 2.18c). The Cascades in the Pacific Northwest, where Mount Saint Helens erupted violently in 1980 (see Figure 17.11a, p. 469) and renewed its activity in 2004, are fueled by magma from subduction. Another example is South America's Andes Mountains, where the Nazca Plate slides beneath the South American Plate.

When two plates of continental lithosphere meet, the continental crust on both sides resists subduction and instead crushes together, bending, buckling, and deforming layers of rock from both plates in a **continental collision** (shown in the right-hand drawing of Figure 2.18c). Portions of the accumulating masses of buckled crust are forced upward as they are pressed together, and mountain ranges result. The Himalayas, the world's highest mountains, result from the Indian-Australian Plate's collision with the Eurasian Plate beginning 40–50 million years ago, and these mountains are still rising today as these plates converge. The Appalachian Mountains of the eastern United States, once the world's highest mountains themselves, resulted from a more-ancient collision with the edge of what is today Africa.

Tectonics produces Earth's landforms

In these ways, the processes of plate tectonics create the landforms around us. Tectonic movements build mountains; shape the geography of oceans, islands, and continents; and give rise to earthquakes and volcanoes. They also help determine the locations of geothermal energy resources. The Geysers in California are located above a region of subduction, which is why magma rises to within 4 mi (6.4 km) of the surface.

Another location with exceptional geothermal energy resources is the nation of Iceland, an island built when magma from underwater volcanoes along the Mid-Atlantic Ridge was extruded above the ocean surface and cooled. Iceland possesses numerous volcanoes and geysers, including the Eyjafjallajokull volcano whose eruption disrupted air travel in Europe for days in 2010. In Iceland, geothermally heated water is piped throughout towns and cities, and nine out of ten people heat their homes with geothermal energy.

The topography created by tectonic processes in turn shapes climate by altering patterns of rainfall, wind, ocean currents, heating, and cooling—all of which affect rates of weathering and erosion and the ability of plants and animals to inhabit different regions. Thus, the locations of biomes (pp. 96–102) are influenced by plate tectonics. Moreover, tectonics has affected the history of life's evolution; the convergence of landmasses into supercontinents such as Pangaea is thought to have contributed to widespread extinctions by reducing the area of species-rich coastal regions and by creating an arid continental interior with extreme temperature swings.

Only in the last several decades have scientists learned about plate tectonics—this environmental system of such fundamental importance was completely unknown to humanity just half a century ago. Amazingly, our civilization was sending people to the moon by the time we were coming to understand the movement of land under our very feet. We are still learning: Only now are scientists creating a comprehensive map of what lies underneath the United States (see **THE SCIENCE BEHIND THE STORY**, pp. 38–39).

The rock cycle alters rock

Just as plate tectonics shows geology's dynamism at a large scale, the rock cycle shows it at a smaller one. We tend to think of rock as pretty solid stuff. Yet over geologic time, rocks and the minerals that comprise them are heated, melted,

cooled, broken down, and reassembled in a very slow process called the **rock cycle** (**FIGURE 2.19**).

A **rock** is any solid aggregation of minerals. A **mineral**, in turn, is any naturally occurring solid element or inorganic compound with a crystal structure, a specific chemical composition, and distinct physical properties. The type of rock in a given region affects soil characteristics and thereby influences the region's plant community. Understanding the rock cycle enables us to better appreciate the formation and conservation of soils, mineral resources, fossil fuels, groundwater sources, geothermal energy sources, and other natural resources—all of which we discuss in later chapters.

Igneous rock All rocks can melt. At high enough temperatures, rock will enter the molten, liquid state called magma. If magma is released through the lithosphere (as in a volcanic eruption), it may flow or spatter across Earth's surface as **lava**. Rock that forms when magma or lava cools is called **igneous rock** (from the Latin *ignis*, meaning "fire") (**FIGURE 2.19A**).

Igneous rock comes in two main classes because magma can solidify in different ways. When magma cools slowly and solidifies while it is below Earth's surface, it forms *intrusive* igneous rock. This process created the famous rock formations at Yosemite National Park (**FIGURE 2.20A**). Granite is the best-known type of intrusive rock. A slow cooling process allows minerals of different types to aggregate into large crystals, giving granite its multicolored, coarse-grained appearance. In contrast, when molten rock is ejected from a volcano, it cools quickly, so minerals have little time to grow into coarse crystals. This class of igneous

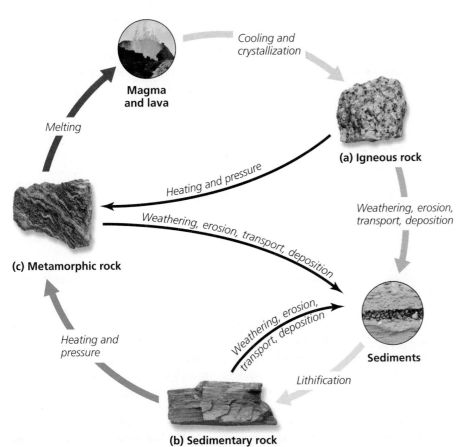

FIGURE 2.19 ◀ In the rock cycle, igneous rock **(a)** is formed when rock melts and the resulting magma or lava then cools. Sedimentary rock **(b)** is formed when rock is weathered and eroded and the resulting sediments are compressed to form new rock. Metamorphic rock **(c)** is formed when rock is subjected to intense heat and pressure underground. Through these processes (shown by arrows of different colors), each type of rock can be converted into either of the other two types.

The SCIENCE behind the Story

Mapping the Unknown Beneath Us

If a team of scientists knocked on your door wanting to install instruments in your backyard to measure signals from earthquakes from around the world, would you let them? They could be at your door any day now, if you happen to be at one of many predetermined grid points in a bold program called USArray.

What goes on beneath Earth's surface affects us all—through earthquakes, volcanoes, and geothermal heating—yet the underground workings of our planet remain largely a mystery. Through the years, scientists and engineers have developed ways to infer what the subsurface realm looks like and to determine just what is brewing down there.

Today, the most ambitious such program ever is underway—an attempt to comprehensively map the lithosphere beneath the entire United States. A large consortium of researchers has embarked on USArray, a 15-year program to map the subsurface geology of the United States. The program aims to image the structure of the crust and upper mantle and to

Technician installing a seismometer for USArray

analyze how earthquakes and other geologic processes relate to the plate tectonic movements going on beneath our feet.

Central to this effort is the *seismometer*. This instrument measures and records the seismic waves generated by earthquakes (or explosions or other sources of vibrations). It produces visual displays called seismograms to help researchers visualize these signals.

Worldwide, hundreds of earthquakes occur each day. Most are too small for us to notice, but seismometers can sense even the weakest ones from great distances. When an earthquake occurs somewhere in the world, seismic waves travel outward in all

directions. Seismometers in various locations record these waves—and when multiple seismic stations record the same earthquake, comparing seismograms from the stations can reveal the location and magnitude of the earthquake through the process of triangulation.

Researchers use this same approach to infer the underground structure of rock layers. A grid of multiple seismometers triangulating on the waves emitted by an earthquake can give three-dimensional information about the way those waves travel underground. Seismic waves travel more quickly through cooler and denser materials. Thus, if it takes longer for a seismic wave to reach a station than expected from the distance alone, that information indicates that underground features in the direction of the quake are warmer or less dense than average. Once the speeds of waves from enough earthquakes are recorded from all directions, the station can map the characteristics of rock in all directions around the station.

rock is called *extrusive* igneous rock, and its most common representative is basalt, the principal rock type of the Hawaiian Islands (**FIGURE 2.20B**).

Sedimentary rock All exposed rock weathers away with time. The relentless forces of wind, water, freezing, and thawing eat away at rocks, stripping off one tiny grain (or large chunk) after another. Through weathering (pp. 227–228) and erosion (p. 231), particles of rock blown by wind or washed away by water come to rest downhill, downstream, or downwind from their sources, eventually forming **sediments**.

Alternatively, some sediments form chemically from the precipitation of substances out of solution.

Sediment layers accumulate over time, causing the weight and pressure of overlying layers to increase. **Sedimentary rock** (**FIGURE 2.19B**) is formed as sediments are physically pressed together (*compaction*) and as dissolved minerals seep through sediments and act as a kind of glue, binding sediment particles together (*cementation*). The formation of rock through these processes of compaction and cementation is termed *lithification*. Examples of sedimentary rock include sandstone, made of cemented

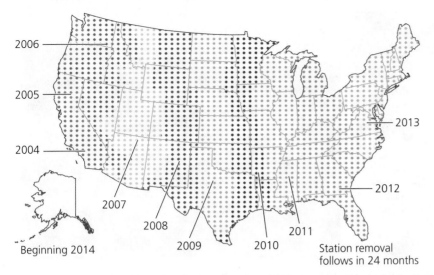

2006
2005
2004
2007
2008
2009
2010
2011
2012
2013
Beginning 2014
Station removal follows in 24 months

USArray stations are moving eastward across the United States. Symbols show stations, and bands of color indicate the year in which the area is covered.

To map the lithosphere under the United States, seismologists and geologists with the USArray project are moving a "transportable array" of 400 seismometers across the lower 48 U.S. states from west to east over the course of 10 years (see figure). Seismometers are spaced 70 km (43 mi) apart, and each stays in place for 2 years. Technicians then move the westernmost stations eastward, leapfrogging the rest, and the entire array shifts in a band across the country. After finishing along the Atlantic coast in the year 2013, having completed 1,600 data collection points, the grid will move to Alaska. Data from the seismometers are transmitted continuously to a facility at the University of California–San Diego,

and then to a national management center.

To zero in on regions of special interest, researchers also use a "flexible array," transporting a pool of 2,100 seismometers to targeted areas they want to study more carefully. Spacing the seismometers more closely in a tighter grid allows for data to be collected at higher resolution. With this higher-quality data, researchers can examine the depth of fault systems, the extent of magma chambers underneath volcanoes, geophysical patterns beneath mountain ranges and lake beds, and the deep structure of rock at the boundaries of tectonic plates.

The transportable and the flexible arrays are tied together by

a "reference network" of permanent seismometer stations operated by the U.S. Geological Survey (USGS). These are widely spaced (about 300 km [186 mi] apart) in a grid throughout the country for the duration of the program and beyond.

To accomplish all this, researchers need the cooperation of private land-owners willing to situate the instruments on their land. College students during the summer are helping to recruit landowners, who can sign up or recommend locations online at the project's website.

USArray is one of three projects in a larger program called Earthscope, funded by the National Science Foundation. In the second project, researchers are mapping regions that are being stretched and strained along the tectonic boundary between the Pacific and North American plates. In the third project, researchers have drilled a 3-km (1.9-mi) hole straight into the San Andreas Fault, midway between San Francisco and Los Angeles, in order to directly observe what is going on inside an active fault. Together, all these data will help scientists map the underground geology of the United States and learn how subsurface processes interact. With this knowledge, we will better understand earthquakes, and perhaps one day we will be able to predict them. ∎

sand particles; shale, comprised of still smaller mud particles; and limestone, formed as dissolved calcite precipitates from water or as calcite from marine organisms settles to the bottom.

These processes also create the fossils of organisms (p. 58) and the fossil fuels we use for energy (p. 533). Because sedimentary layers pile up in chronological order (**FIGURE 2.20C**), geologists and paleontologists can assign relative dates to fossils they find in sedimentary rock, and thereby they make inferences about Earth's history (see **THE SCIENCE BEHIND THE STORY**, pp. 44–45).

Metamorphic rock Geologic forces may bend, uplift, compress, or stretch rock. When any type of rock is subjected to great heat or pressure, it may alter its form, becoming **metamorphic rock** (from the Greek for "changed form") (**FIGURE 2.19C**). The forces that metamorphose rock generally occur deep underground, at temperatures lower than the rock's melting point but high enough to change its appearance and physical properties. Metamorphic rock (**FIGURE 2.20D**) includes rock such as slate, formed when shale is subjected to heat and pressure, and marble, formed when limestone is heated and pressurized.

(a) Igneous rock: Granite at Yosemite National Park

(b) Igneous rock: Basalt from magma in Hawaii

(c) Sedimentary rock: Sandstone in Arizona

(d) Metamorphic rock: Gneiss in Utah

FIGURE 2.20 ▲ The towering rock formations of Yosemite National Park **(a)** are made of granite, a type of intrusive igneous rock, whereas the lava flows on Hawaii **(b)** form basalt, a type of extrusive igneous rock. This layered formation in Paria Canyon, Arizona **(c)** is an example of sandstone, a type of sedimentary rock. This gneiss (pronounced "nice") at Antelope Island, Utah **(d)**, is a type of metamorphic rock.

GEOLOGIC AND NATURAL HAZARDS

By driving plate tectonics, Earth's geothermal heating gives rise to creative forces that shape our planet—yet some of the consequences of tectonic movement can also pose hazards to us. Earthquakes and volcanoes are examples of geologic hazards. We can see how such hazards relate to tectonic processes by examining a map of the circum-Pacific belt, or "ring of fire" (**FIGURE 2.21**). Compare the route of the circum-Pacific belt to some of the tectonic plate boundaries shown in Figure 2.17, and note how closely they match. Nine out of 10 earthquakes and over half the world's volcanoes occur along this 40,000-km (25,000-mi) arc of subduction zones and fault systems. Like many locations along the circum-Pacific belt, The Geysers has experienced earthquakes and volcanism frequently in its past.

Earthquakes result from movement at plate boundaries and faults

Along tectonic plate boundaries, and in other places where faults occur, Earth may relieve built-up pressure in fits and starts. Each release of energy causes what we know as an **earthquake**. Most earthquakes are barely perceptible, but occasionally they are powerful enough to do tremendous damage to human life and property (**FIGURE 2.22; TABLE 2.2**). Damage is generally greatest where soils are loose or saturated with water—areas of cities built atop landfill are particularly susceptible. For instance, during the 1989 Loma Prieta earthquake that shook San Francisco and Oakland while their baseball teams were playing in the World Series, one of the areas hardest hit was San Francisco's Marina District, a neighborhood built atop soil and debris dumped into the bay, including rubble from the city's 1906 earthquake.

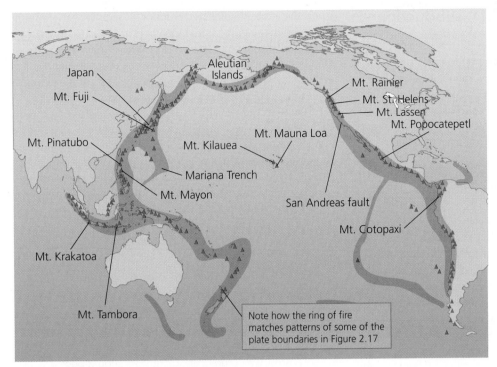

FIGURE 2.21 ◄ Most of our planet's volcanoes and earthquakes occur along the circum-Pacific belt, or "ring of fire," the system of subduction zones and other plate boundaries that encircles the Pacific Ocean. In this map, red symbols indicate major volcanoes, and gray-shaded areas indicate areas of greatest earthquake risk. Compare the distribution of these hazards with the tectonic plate boundaries shown in Figure 2.17.

Japan
Mt. Fuji
Mt. Pinatubo
Mt. Kilauea
Mt. Mauna Loa
Mariana Trench
Mt. Mayon
Mt. Krakatoa
Mt. Tambora
Aleutian Islands
Mt. Rainier
Mt. St. Helens
Mt. Lassen
Mt. Popocatepetl
San Andreas fault
Mt. Cotopaxi

Note how the ring of fire matches patterns of some of the plate boundaries in Figure 2.17

TABLE 2.2	Examples of Large or Recent Earthquakes		
Location	**Year**	**Fatalities**	**Magnitude[1]**
Shaanxi Province, China	1556	830,000	~8
Lisbon, Portugal	1755	70,000[2]	8.7
San Francisco, California	1906	3,000	7.8
Kwanto, Japan	1923	143,000	7.9
Anchorage, Alaska	1964	128[2]	9.2
Tangshan, China	1976	255,000+	7.5
Michoacan, Mexico	1985	9,500	8.0
Loma Prieta, California	1989	63	6.9
Northridge, California	1994	60	6.7
Kobe, Japan	1995	5,502	6.9
Northern Sumatra	2004	228,000[2]	9.1
Kashmir, Pakistan	2005	86,000	7.6
Sichuan Province, China	2008	50,000+	7.9
Port-au-Prince, Haiti	2010	236,000	7.0
Maule, Chile	2010	500	8.8

[1] Measured by moment magnitude; each full unit is roughly 32 times as powerful as the preceding full unit.
[2] Includes deaths from the resulting tsunami.

Systems (pp. 25, 607), in which engineers drill deeply into rock, intentionally fracture it, pump water into the system of fractures, and then extract the water or steam once it is heated. This approach shows great promise for expanding our use of geothermal energy, but some efforts have been stalled or cancelled once earthquakes resulted.

To minimize damage from earthquakes, engineers have developed ways to protect buildings from shaking. They do this by strengthening structural components while also designing points at which a structure can move and sway harmlessly with ground motion. Just as a flexible tree trunk bends in a storm while a brittle one breaks, buildings with built-in flexibility are more likely to withstand an earthquake's violent shaking. Such designs are an important part of new building codes in California, Japan, and other quake-prone regions, and many older structures are being retrofitted to meet these codes.

FIGURE 2.22 ▼ The 2010 earthquake in Haiti devastated the capital city of Port-au-Prince and killed an estimated 230,000 people.

Some regions with active geothermal heating such as The Geysers experience frequent earthquakes of very low intensity (microearthquakes). Extracting steam for power and pumping water to replenish the supply at The Geysers have caused an increase in these quakes by altering the pressures belowground. Earthquakes can be one unfortunate side effect of the new approach called Enhanced Geothermal

Volcanoes arise from rifts, subduction zones, or hotspots

Where molten rock, hot gas, or ash erupt through Earth's surface, a **volcano** is formed, often creating a mountain over time as cooled lava accumulates. As we have seen, lava can extrude along mid-ocean ridges, or over subduction zones as one tectonic plate dives beneath another. Lava may also be emitted at *hotspots*, localized areas where plugs of molten rock from the mantle erupt through the crust. As a tectonic plate moves across a hotspot, repeated eruptions from this source may create a linear series of volcanoes. The Hawaiian Islands provide an example of this process (**FIGURE 2.23A**).

At some volcanoes, lava flows slowly downhill, such as at Mount Kilauea in Hawaii (**FIGURE 2.23B**), which has been erupting continuously since 1983! At other times, a volcano may let loose large amounts of ash and cinder in a sudden explosion, such as during the 1980 eruption of Mount Saint Helens (see Figure 17.11a, p. 469). Sometimes a volcano can unleash a *pyroclastic flow*—a fast-moving cloud of toxic gas, ash, and rock fragments that races down the slopes at speeds up to 725 km/hr (450 mph), enveloping everything in its path. Such a flow buried the inhabitants of the ancient Roman cities of Pompeii and Herculaneum in A.D. 79, when Mount Vesuvius erupted.

Besides affecting people, volcanic eruptions exert environmental impacts (**TABLE 2.3**). Ash blocks sunlight, while sulfur emissions lead to a sulfuric acid haze that blocks radiation and cools the atmosphere. Large eruptions—such as that of Mount Pinatubo in the Philippines in 1991 (p. 468)—can depress temperatures throughout the world. When Indonesia's Mount Tambora erupted in 1815, it cooled average global temperatures by 0.4–0.7 °C (0.7–1.3 °F), enough to cause crop failures worldwide and make 1816 "the year without a summer."

One of the world's largest volcanoes—so large it is called a "supervolcano"—lies in the United States. The entire basin of Yellowstone National Park is an ancient supervolcano that has at times erupted so massively as to cover large parts of the continent deeply in ash. Although another eruption is not expected imminently, the region is still geothermally active; the park hosts fully two-thirds of the world's geysers.

WEIGHING THE ISSUES

Your Risk from Natural Hazards What types of natural hazards (earthquakes? flooding? landslides? fires? hurricanes? tornadoes?) occur in the area where you live? Name three things you personally can do to minimize your risk from these hazards. If a natural disaster strikes, should people be allowed to rebuild in the same areas if those areas are prone to experience the hazard again? Discuss one example you are familiar with from news accounts or personal experience.

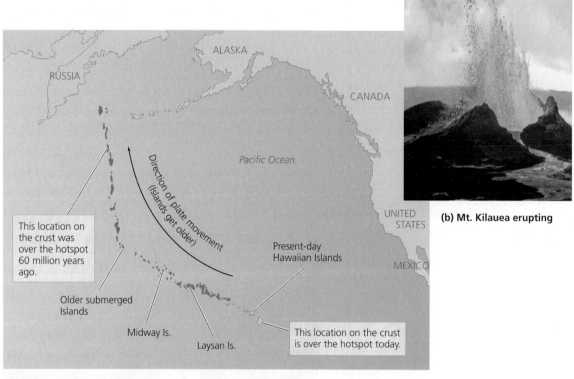

(b) Mt. Kilauea erupting

(a) Current and former Hawaiian islands, formed as crust moves over a volcanic hot spot

FIGURE 2.23 ▲ The Hawaiian Islands **(a)** have been formed by repeated eruptions from a hotspot of magma in the mantle as the Pacific Plate passes over the hotspot. The Big Island of Hawaii is most recently formed, and it is still volcanically active. The other islands are older and have already begun eroding away. To their northwest stretches a long series of former islands, now submerged. In the future, a new island will one day rise above the sea to the southeast of today's Big Island. The active volcano Kilauea **(b)**, on the Big Island's southeast coast, is the youngest of Hawaii's volcanoes, currently located above the edge of the hotspot.

TABLE 2.3 Examples of Notable Volcanic Eruptions

Location	Year	Impacts	Magnitude[2]
Yellowstone Caldera, Wyoming, United States	640,000 B.P.[1]	Most recent "mega-eruption" at site of Yellowstone National Park	8
Mount Mazama, Oregon, United States	6,870 B.P.	Created Crater Lake	7
Mount Vesuvius, Italy	A.D. 79	Buried Pompeii and Herculaneum	5
Mount Tambora, Indonesia	1815	Created "year without a summer"; killed at least 70,000 people	7
Krakatau, Indonesia	1883	Killed over 36,000 people; heard 5,000 km (3,000 mi) away; affected weather for 5 years	6
Mount Saint Helens, Washington, United States	1980	Blew top off mountain; sent ash 19 km (12 mi) into sky and into 11 U.S. states; 57 people killed	5
Kilauea, Hawaii, United States	1983–present	Continuous lava flow	1
Mount Pinatubo, Philippines	1991	Sulfuric aerosols lowered world temperature 0.5 °C (0.9 °F)	6
Eyjafjallajokull, Iceland	2010	Ash cloud disrupted air travel throughout Europe	1

[1]B.P. = years before the present.
[2]Measured by the Volcanic Explosivity Index, which ranges from 0 (least powerful) to 8 (most powerful).

Landslides are a form of mass wasting

On a smaller scale than volcanoes or earthquakes, another type of geologic hazard, the **landslide**, occurs when large amounts of rock or soil collapse and flow downhill. Landslides are a severe and often sudden manifestation of the more general phenomenon of **mass wasting**, the downslope movement of soil and rock due to gravity. Mass wasting occurs naturally, and heavy rains may saturate soils and trigger mudslides of soil, rock, and water. However, mass wasting can also be brought about by human land use practices that expose or loosen soil, making slopes more prone to collapse.

Most often, mass wasting erodes unstable hillsides, damaging property one structure at a time (**FIGURE 2.24**). Occasionally,

mass wasting events can be colossal and deadly; mudslides that followed the torrential rainfall of Hurricane Mitch in Nicaragua and Honduras in 1998 killed over 11,000 people. Mudslides caused when volcanic eruptions melt snow and send huge volumes of destabilized mud racing downhill are called *lahars*, and these are particularly dangerous. A lahar buried the entire town of Armero, Colombia, in 1985 following an eruption, killing 21,000 people.

Tsunamis can follow earthquakes, volcanoes, or landslides

Earthquakes, volcanic eruptions, and large coastal landslides can all displace huge volumes of ocean water instantaneously and trigger a **tsunami**, an immense swell, or wave, of

FIGURE 2.24 ◀ Landslides are a frequent occurrence in sloping areas along the California coast, particularly after heavy winter rains saturate soils. Homes built on unstable slopes, such as these in Laguna Beach, Orange County, can be damaged or destroyed when slopes give way.

The SCIENCE behind the Story

Have We Brought on a New Geologic Epoch?

Geologist Jan Zalasiewicz, University of Leicester

Have the impacts of human beings on Earth been so profound as to warrant creating a new time period in the geologic record?

Scientists are conservative when it comes to revising the fundamental building blocks of their disciplines—so when a group of geologists suggested naming the current portion of our planet's history the "Anthropocene" (from the Greek word *anthropos*, meaning "human"), scientists sat up and took notice.

The idea began to catch on after being proposed by Nobel Prize–winning chemist Paul Crutzen (p. 480) in 2000. In 2008, a team of 21 scientists from the Stratigraphy Commission of the Geological Society of London formally advocated the proposal in a paper in *GSA Today*, a journal of the Geological Society of America (GSA).

To understand this provocative idea, let's first turn to the geologic timescale (**APPENDIX D**) in the back of this book. The scale on the left side of that page shows the full span of Earth's history—all 4.5 *billion* years of it. The scale on the right magnifies the most recent 543 million years. Geologists have subdivided this stretch of time into three eras and 10 periods. The most recent period, the Quaternary period, occupies a thin slice of time at the top of the scale, because this

period began "only" 1.8 million years ago. Take a moment to reflect on the mind-boggling depth of geologic time.

Geologists divide this immensely long timescale using evidence from *stratigraphy*, the study of *strata*, or layers, of sedimentary rock. (As we have seen, sedimentary rock is laid down in chronological sequence, so by studying it researchers can infer how conditions changed through time.) Where scientists find evidence for major and sudden changes in the physical, chemical, or biological conditions present on Earth between one set of layers and the next, they create a boundary between geologic time periods. For instance, fossil evidence for mass extinctions (pp. 59–61, 288–290) determines several boundaries, such as that between the Permian and Triassic periods.

We live in the Holocene epoch, the most recent slice of the Quaternary period. The Holocene

epoch began about 11,500 years ago with a warming trend that melted glaciers and brought Earth out of its most recent ice age. Since then, Earth's climate has been remarkably constant, and this constancy provided our species with the long-term stability we needed to develop agriculture and civilization.

However, Crutzen and the 21 scientists led by geologist Jan Zalasiewicz of the University of Leicester, U.K., note that since the industrial revolution (p. 4), human activity has had major impacts on Earth's basic processes. The question for geologists is: Have those effects been strong enough to warrant naming a new geologic epoch after ourselves?

In their GSA paper, the British geologists reviewed a broad sweep of published scientific evidence and advanced several reasons to rename the past 200 or so years the *Anthropocene*.

First, humans have caused a sharp increase in erosion worldwide (**see top panel of figure**). By clearing forest and raising crops, we have sent immense amounts of soil downwind and downstream, from continents into oceans. This rapid deposition of sediment in the oceans will be noticeable in the stratigraphic record far into the future as today's sediments become

water that can travel thousands of miles across oceans. The world's attention was drawn to this hazard on December 26, 2004, when a massive tsunami, triggered by an earthquake off Sumatra, devastated the coastlines of countries all around the Indian Ocean, including Indonesia, Thailand, Sri Lanka, India, and several African nations. Roughly 228,000 people were killed, 1–2 million were displaced, and whole communities were destroyed (**FIGURE 2.25**). Coral reefs, coastal forests, and wetlands were damaged, and salt water

contaminated soil and aquifers, making it difficult to restore the affected areas.

Those of us who live in the United States and Canada should not consider tsunamis to be something that occurs only in faraway places. A large tsunami struck North America's Pacific coast in 1700 following a huge earthquake in the Pacific Northwest, and one following the Alaskan earthquake of 1964 drowned over 100 people from Alaska to California. Residents of Boston, New York, Washington D.C., and other cities along

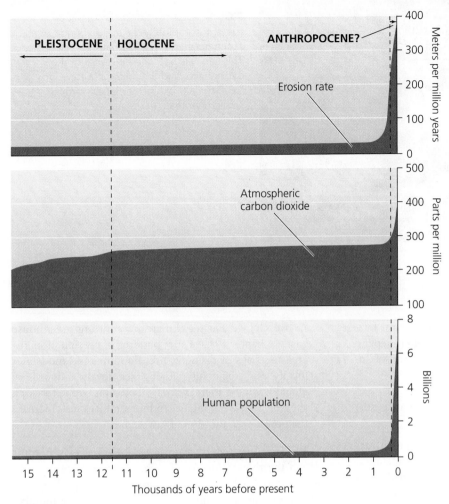

Global soil erosion rates (top) and atmospheric carbon dioxide concentrations (middle) have increased sharply in just the last few hundred years, along with human population (bottom). These patterns have persuaded some geologists that we should recognize a new epoch in Earth history and call it the Anthropocene. Adapted from Zalasiewicz, J. et al. 2008. Are we now living in the Anthropocene? *GSA Today* 18 (2): 4-8, Figure 1.

compacted into tomorrow's sedimentary rock layers.

Second, our species has rapidly altered the composition of the atmosphere by emitting greenhouse gases **(see middle panel of figure)** as a result of deforestation, agriculture, and, especially, our combustion of coal, oil, and natural gas. As we shall see in Chapter 18 (pp. 500–501), we have already brought carbon dioxide and methane to their highest levels in at least 800,000 years.

As greenhouse gas concentrations rise, so does temperature. Earth's temperature has risen 0.7 °C (1.3 °F) in the past century and is predicted to rise by 1.8–4.0 °C (3.2–7.2 °F) in the current century (pp. 504–505). Rising temperatures are melting polar ice, and the influx of meltwater into the oceans, combined with the fact that warmed water expands in volume, means that sea level rises. Global sea level rose 17 cm (6.7 in.) in the last century and will likely rise by much more in this century (pp. 507–511). Moreover, increasing atmospheric carbon dioxide acidifies ocean water (pp. 438, 440–441), which kills off coral reefs (and will even likely dissolve some of the geologic strata being formed).

All these changes—along with the hunting, pollution, and habitat disturbance we are inflicting on Earth's biotic communities—are causing extinctions of animals and plants (pp. 290–296). If we step back to view our impacts in deep geologic time, it becomes clear that we are setting in motion a rapid new mass extinction event.

Our impacts—intensified by our recent explosion in population **(see bottom panel of figure)**—are happening in the blink of an eye, geologically speaking. The entire period of our society's existence so far could end up represented in as little as a millimeter of rock in the future, so the changes we are bringing about now would appear sudden to a geologist of the far future.

Many scientists resist the idea of renaming our current epoch because they question whether the proposal holds true to the tradition of defining time periods based on changes actually seen in the geologic record. But supporters of the idea maintain that today's changes will be readily visible in future stratigraphy and that we should recognize this unprecedented time of rapid change in Earth's history by denoting it the Anthropocene. ■

the Atlantic coast could be at risk if an unstable portion of a Canary Island volcano that researchers are monitoring were to one day slump into the sea.

Since the 2004 tsunami, nations and international agencies have stepped up efforts to develop systems to give coastal residents advance warning of approaching tsunamis. In addition, we can lessen the impacts of tsunamis if we preserve coral reefs and mangrove forests (pp. 439–444), which help protect coastlines by absorbing wave energy.

We can worsen or mitigate the impacts of natural hazards

Aside from geologic hazards, people face other types of natural hazards. Heavy rains can lead to flooding that ravages lowlying areas near rivers and streams (p. 411). Coastal erosion can eat away at beaches (p. 509). Wildfire can threaten life and property in fire-prone areas (p. 328). Tornadoes and hurricanes (p. 468) can cause extensive damage and loss of life.

FIGURE 2.25 ◄ The December 2004 tsunami completely destroyed Banda Aceh, at the northern tip of Sumatra in Indonesia. This site was near the epicenter of the earthquake, and the influx of water—up to 30 m (100 ft) high in places—arrived within minutes.

Although we refer to such phenomena as "natural hazards," the magnitude of their impacts upon us often depend on choices we make. We tend to worsen the impacts of so-called natural hazards in various ways:

▶ As our population grows, more people live in areas susceptible to natural disasters.

▶ Many of us choose to live in areas that we deem attractive but that are also prone to hazards. For instance, coastlines are vulnerable to tsunamis and erosion by storms, and mountainous areas are prone to volcanoes and mass wasting.

▶ We use and engineer landscapes around us in ways that can increase the frequency or severity of natural hazards. Damming and diking rivers to control floods can sometimes lead to catastrophic flooding (p. 411), and suppressing natural wildfire puts forests at risk of larger, truly damaging fires (p. 509). Mining practices (pp. 649–653), clearing forests for agriculture, and clear-cutting on slopes (p. 326) can each induce mass wasting, increase runoff, compact soil, and change drainage patterns.

▶ As we change Earth's climate by emitting greenhouse gases (Chapter 19), we alter patterns of precipitation, increasing risks of drought, fire, flooding, and mudslides locally and regionally. Rising sea levels induced by global warming increase coastal erosion. Some research suggests that warming ocean temperatures may increase the power and duration of hurricanes.

We can often reduce or mitigate the impacts of hazards through the thoughtful use of technology, engineering, and policy, informed by a solid understanding of geology and ecology. Examples of this, as we've noted, include building earthquake-resistant structures, designing early warning systems for tsunamis and volcanoes, and conserving reefs and shoreline vegetation to protect against tsunamis and coastal erosion. In addition, better forestry, agriculture, and mining practices can help prevent mass wasting. Zoning regulations, building codes, and insurance incentives that discourage development in areas prone to landslides, floods, fires, and storm surges can help keep us out of harm's way and can decrease taxpayer expense in cleaning up after natural disasters. Finally, mitigating global climate change may help reduce the frequency of natural hazards in many regions.

➤ CONCLUSION

Geothermal heating provides one window into the broader phenomena of the chemical basis of matter, the nature of energy, and the geologic processes that shape our planet. These physical phenomena provide the basis on which much of environmental science is built.

An understanding of matter is essential for all science. Moreover, chemistry can provide tools for finding solutions to environmental problems, whether one wants to analyze agricultural practices, manage water resources, reform energy policy, conduct toxicological studies, or find ways to mitigate global climate change.

Likewise, an understanding of energy is both of fundamental scientific importance and of considerable practical

relevance. Since the industrial revolution we have powered our societies largely with fossil fuels, but as these fuels dwindle and as climate change worsens, we will need to shift to renewable energy sources such as geothermal energy.

Physical processes of geology such as plate tectonics and the rock cycle are centrally important because they shape Earth's terrain and form the foundation for living systems that overlie the landscape. Geologic processes also generate phenomena that can threaten our lives and property, including earthquakes, volcanoes, landslides, and tsunamis. Matter, energy, and geology are in some way tied to nearly every significant process in environmental science.

REVIEWING OBJECTIVES

You should now be able to:

EXPLAIN THE FUNDAMENTALS OF MATTER AND CHEMISTRY AND APPLY THEM TO REAL-WORLD SITUATIONS

- Understanding chemistry provides a powerful tool for understanding environmental science and developing solutions to environmental problems. (p. 25)
- Atoms can form molecules and compounds, and changes at the atomic level can result in alternate forms of elements, such as ions and isotopes. (pp. 25–27)
- Water's chemistry facilitates life. (p. 27)
- The pH scale quantifies acidity and alkalinity. (p. 28)
- Living things depend on organic compounds, which are carbon-based. (p. 28)
- Macromolecules, including proteins, nucleic acids, carbohydrates, and lipids, are key building blocks of life. (pp. 28–29)
- Chemists have designed synthetic polymers (such as plastics) based on natural ones. (p. 29)

DIFFERENTIATE AMONG FORMS OF ENERGY AND EXPLAIN THE BASICS OF ENERGY FLOW

- Energy can convert from one form to another; for instance, from potential to kinetic energy, and vice versa. Chemical energy is potential energy in the bonds between atoms. (pp. 29–30)
- The total amount of energy in the universe is conserved; it cannot be created or lost. (p. 30)
- Systems tend to increase in entropy, or disorder, unless energy is added to build or maintain order and complexity. (pp. 30–31)
- Earth's systems are powered mainly by radiation from the sun, geothermal heating from the planet's core, and gravitational interactions among Earth, the sun, and the moon. (pp. 31, 33)

DISTINGUISH PHOTOSYNTHESIS, CELLULAR RESPIRATION, AND CHEMOSYNTHESIS, AND SUMMARIZE THEIR IMPORTANCE TO LIVING THINGS

- In photosynthesis, autotrophs use carbon dioxide, water, and solar energy to produce oxygen and the sugars they need. (pp. 31–32)
- In cellular respiration, organisms extract energy from sugars by converting them in the presence of oxygen into carbon dioxide and water. (pp. 32–33)
- In chemosynthesis, specialized autotrophs use carbon dioxide, water, and chemical energy from minerals to produce sugars. (p. 33)

EXPLAIN HOW PLATE TECTONICS AND THE ROCK CYCLE SHAPE THE LANDSCAPE AROUND US AND THE EARTH BENEATH OUR FEET

- Earth's geology is dynamic, and a human lifetime is a blink of an eye in the long course of geologic time. (pp. 33–34, 44–45)
- Earth consists of distinct layers that differ in composition, temperature, density, and other characteristics. (p. 34)
- Plate tectonics is a fundamental system that shapes Earth's physical geography, as well as producing earthquakes and volcanoes. (pp. 35–37)
- Tectonic plates meet at three types of boundaries: divergent, transform, and convergent. (pp. 35–36)
- Matter is cycled within the lithosphere, and rocks transform from one type to another. (pp. 37–40)

LIST MAJOR TYPES OF GEOLOGIC HAZARDS AND DESCRIBE WAYS TO MITIGATE THEIR IMPACTS

- The circum-Pacific belt, or "ring of fire," spawns most of the world's volcanoes and earthquakes. (pp. 40–41)
- Earthquakes result from movement at faults and plate boundaries. We cannot prevent them, but we can build structures and cities in safer ways. (pp. 40–41)
- Volcanoes arise from heating by magma at rifts, subduction zones, or hotspots. (pp. 42–43)
- Landslides and other forms of mass wasting can occur on small or large scales; damage can be minimized by understanding their risks. (p. 43)
- Tsunamis can flood coastlines and cause immense damage. Early warning systems will be key in minimizing future losses. (pp. 43–46)
- We often worsen impacts from natural hazards, but we can reduce them through better land use practices. (p. 46).

TESTING YOUR COMPREHENSION

1. What are the basic building blocks of matter? Provide several examples using chemicals common in Earth's physical or biological systems.

2. How does an ion differ from an isotope? Now differentiate among an atom, a molecule, and a compound.

3. Describe two major forms of energy, and give examples of each.

4. State the first law of thermodynamics. Now compare it to the second law of thermodynamics.

5. Describe the three major sources of energy that power Earth's environmental systems.

6. What substances are produced by photosynthesis? By cellular respiration? By chemosynthesis?

7. Name the primary layers that make up our planet. Which portions does the lithosphere include?

8. Describe what occurs at a divergent plate boundary. What happens at a transform plate boundary? Compare and contrast the types of processes that can occur at a convergent plate boundary.

9. Name the three main types of rocks, and describe how each type may be converted to the others via the rock cycle.

10. What causes earthquakes? What is a tsunami, and what causes them? How does a Hawaiian volcano such as Kilauea differ from a volcano in the Cascades of North America such as Mount Saint Helens?

SEEKING SOLUTIONS

1. Think of an example of an environmental problem not mentioned in this chapter that a good knowledge of chemistry might help us solve. How could chemistry help us address the problem?

2. Think about the ways we harness and use energy sources in our society—both renewable sources such as geothermal energy and nonrenewable sources such as coal, oil, and natural gas. What implications does the first law of thermodynamics have for our energy usage? How is the second law of thermodynamics relevant to our use of energy?

3. Referring to the chemical reactions for photosynthesis and respiration, provide an argument for why increasing amounts of carbon dioxide in the atmosphere due to global climate change (Chapter 18) might potentially increase amounts of oxygen in the atmosphere. Now give an argument for why it might potentially decrease amounts of atmospheric oxygen. What would you need to know to determine which of these two outcomes might occur?

4. Describe how plate tectonics accounts for the formation of (a) mountains, (b) volcanoes, and (c) earthquakes. Why do you think it took so long for scientists to discover an environmental system of such fundamental importance as plate tectonics?

5. For each of the following "natural hazards," describe one thing that can be done to minimize or mitigate their impacts on our lives and property:

▸ Earthquakes

▸ Landslides

▸ Flooding

6. **THINK IT THROUGH** You live in a community just outside a national park famous for its geysers, and you sit on the board of your regional electric utility. The utility is considering drilling for heated underground water and steam and constructing a geothermal power plant in your town. Many of your fellow citizens are enthusiastic about the idea of clean renewable energy. However, some fear that the project could deplete the steam reservoirs that power the geysers and bring visitors to the national park. Others fear that earthquakes may be caused. Others worry that salts in the underground water could corrode pipes and render the plant unworkable. As a member of the utility board, what specific information would you insist on obtaining from geologists and other scientists before you cast your vote on the project? What assurances from scientific research would you feel you need before voting in favor of the project?

CALCULATING ECOLOGICAL FOOTPRINTS

In ecological systems, a rough rule of thumb is that when energy is transferred from plants to plant-eaters or from prey to predator, the efficiency is only about 10% (p. 85). Much of this inefficiency is a consequence of the second law of thermodynamics. Another way to think of this is that eating 1 Calorie of meat from an animal is the ecological equivalent of eating 10 Calories of plant material.

Humans are considered omnivores because we can eat both plants and animals. The choices we make about what to eat have significant ecological consequences. With this in mind, calculate the ecological energy requirements for four different diets, each of which provides a total of 2,000 dietary Calories per day.

Diet	Source of Calories	Number of Calories consumed	Ecologically equivalent Calories	Total ecologically equivalent Calories
100% plant	Plant			
0% animal	Animal			
90% plant	Plant	1,800	1,800	3,800
10% animal	Animal	200	2,000	
50% plant	Plant			
50% animal	Animal			
0% plant	Plant			
100% animal	Animal			

1. How many ecologically equivalent Calories would it take to support you for a year, for each of the four diets listed?

2. How does the ecological impact from a diet consisting strictly of animal products (e.g., milk, other dairy products, eggs, and meat) compare with that of a strictly vegetarian diet? How many additional ecologically equivalent Calories do you consume each day by including as little as 10% of your Calories from animal sources?

3. What percentages of the Calories in your own diet do you think come from plant versus animal sources? Estimate the ecological impact of your diet, relative to a strictly vegetarian one.

4. Describe some challenges of providing food for the growing human population, especially as people in many poorer nations develop a taste for an American-style diet rich in animal protein and fat.

Mastering ENVIRONMENTAL SCIENCE™

Go to **www.masteringenvironmentalscience.com** for practice quizzes, Pearson eText, videos, current events, and more.

Monteverde cloud forest, Costa Rica

3 EVOLUTION, BIODIVERSITY, AND POPULATION ECOLOGY

UPON COMPLETING THIS CHAPTER, YOU WILL BE ABLE TO:

- Explain the process of natural selection and cite evidence for this process
- Describe the ways in which evolution influences biodiversity
- Discuss reasons for species extinction and mass extinction events
- List the levels of ecological organization

- Outline the characteristics of populations that help predict population growth
- Assess logistic growth, carrying capacity, limiting factors, and other fundamental concepts in population ecology
- Identify efforts and challenges involved in the conservation of biodiversity

CENTRAL **CASE STUDY**

Striking Gold in a Costa Rican Cloud Forest

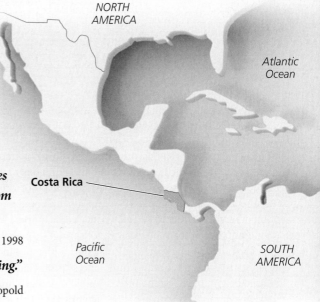

NORTH AMERICA

Atlantic Ocean

Costa Rica

Pacific Ocean

SOUTH AMERICA

"What a terrible feeling to realize that within my own lifetime, a species of such unusual beauty, one that I had discovered, should disappear from our planet."

—Dr. Jay M. Savage, describing the Golden Toad in 1998

"To keep every cog and wheel is the first precaution of intelligent tinkering."

—Aldo Leopold

During a 1963 visit to Central America, biologist Jay M. Savage heard rumors of a previously undocumented toad living in Costa Rica's mountainous Monteverde region. The elusive amphibian, according to local residents, was best known for its color: a brilliant golden yellow-orange. Savage was told the toad was hard to find because it appeared only during the early part of the region's rainy season.

Golden toads at Monteverde

Monteverde means "green mountain" in Spanish, and the name could not be more appropriate. The village of Monteverde sits beneath the verdant slopes of the Cordillera de Tilarán, mountains that receive over 400 cm (157 in.) of annual rainfall. Some of the lush forests above Monteverde, which begin at an altitude of around 1,500 m (4,920 ft, just under a mile high), are known as *cloud forests* because much of their moisture arrives with low-moving clouds that blow inland from the Caribbean Sea.

Monteverde's cloud forest was not fully explored at the time of Savage's first visit, and researchers who had been there described the area as pristine, with a rich bounty of ferns, liverworts, mosses, clinging vines, orchids, and other organisms that thrive in cool, misty environments. Savage knew that such conditions create ideal habitat for many toads and other amphibians.

In May of 1964, Savage organized an expedition into the muddy mountains above Monteverde to try to document the existence of the previously unknown toad species in its natural habitat. Late in the afternoon of May 14, he and his colleagues found what they were looking for. Approaching the mountain's crest, they spotted bright orange patches on the forest's black floor. In one area only 5 m (16 ft) in diameter, they counted 200 male golden toads searching for females at a breeding congregation. Savage gave the creature the scientific name *Bufo periglenes* (literally, "the brilliant toad").

The discovery received international attention, making a celebrity of the tiny toad and making a travel destination of its mountain home, which was soon protected within the Monteverde Cloud Forest Preserve.

At the time, no one knew that Monteverde's ecosystem was about to be transformed. No one foresaw that the oceans and atmosphere would begin warming because of global climate change (Chapter 18) and cause Monteverde's moisture-bearing clouds to rise, drying the forest. No one anticipated the spread of a lethal fungal disease called chytridiomycosis, caused by the pathogen *Batrachochytrium dendrobatidis*, that would infect frogs and toads even in the most pristine environments. And no one could guess that the glamorous, newly discovered golden toad would vanish from the Earth in less than 25 years.

EVOLUTION AS THE WELLSPRING OF EARTH'S BIODIVERSITY

Although the golden toad was new to science (and countless types of organisms still await discovery), we understand quite well how our world became populated with the remarkable diversity of life we see today. Science shows us that our planet has progressed from a stark world inhabited solely by microbes to a lush cornucopia of 1.8 million (and likely millions more) species (**FIGURE 3.1**). A **species** is a particular type of organism or, more precisely, a population or group of populations whose members share certain characteristics and can freely breed with one another and produce fertile offspring. A **population** is a group of individuals of a particular species that live in a particular area. Over eons of time, our planet's species and populations have been molded by the process of biological evolution, giving us the vibrant abundance of life that enriches Earth today.

Evolution in the broad sense means change over time, and biological evolution consists of genetic change in populations of organisms across generations. These changes in genes (p. 29) often lead to modifications in the appearance, functioning, or behavior of organisms from generation to generation through time. Biological evolution results from random genetic changes, and it may proceed randomly or may be directed by natural selection. **Natural selection** is the process by which inherited characteristics that enhance survival and reproduction are passed on more frequently to future generations than those that do not, thus altering the genetic makeup of populations through time.

Evolution by natural selection is one of the best-supported and most illuminating concepts in all of science. From a scientific standpoint, evolutionary theory is indispensable, because it is the foundation of modern biology. Understanding evolution is also vital for a full appreciation of environmental science. Perceiving how species adapt to their environments and change over time is crucial for comprehending ecology and learning the history of life. Evolutionary processes influence many aspects of environmental science, including pesticide resistance, agriculture, medicine, and environmental health.

Natural selection shapes organisms and diversity

In 1858, **Charles Darwin** and **Alfred Russel Wallace** each independently proposed the concept of natural selection as a mechanism for evolution and as a way to explain the great variety of living things. Both Darwin and Wallace were exceptionally keen naturalists from England who had studied plants and animals in such exotic locales as the Galápagos Islands (Darwin) and the Malay Archipelago (Wallace). In the century and a half since then, countless thousands of scientists have refined our understanding of natural selection and evolution.

Natural selection is a simple concept that offers an astonishingly powerful explanation for patterns evident in nature. The idea of natural selection follows logically from a few straightforward premises that are readily apparent to anyone who observes the life around us (**TABLE 3.1**). One is that organisms face a constant struggle to survive and reproduce. Another is that organisms tend to produce more offspring than can survive. A third is that individuals of a species vary in their characteristics. Although not known in Darwin's and Wallace's time, we now know that variation is due to differences in genes, the environments within which genes are expressed, and the interactions between genes and environment. As a result of this variation, some individuals within a species will happen to be better suited to their environment than others and thus will be able to reproduce more effectively.

Many characteristics are passed from parent to offspring through the genes, and a parent that produces many offspring

(a) Resplendent quetzal

(b) Heliconia flower

(c) Harlequin frog

(d) Scutellerid bug

FIGURE 3.1 ▲ Much of our planet's biological diversity resides in tropical rainforests. Monteverde's cloud-forest community includes organisms such as this **(a)** resplendent quetzal (*Pharomachrus mocinno*), **(b)** heliconia (*Heliconia wagneriana*), **(c)** harlequin frog (*Atelopus varius*), and **(d)** scutellerid bug (*Pachycoris torridus*).

TABLE 3.1	The Logic of Natural Selection

▸ Organisms struggle to survive and reproduce.

▸ Organisms tend to produce more offspring than can survive.

▸ Individuals vary in their characteristics.

▸ Many characteristics are inherited by offspring from parents.

Therefore,

▸ Some individuals will be better suited to their environment than other individuals.

▸ Some individuals will produce more offspring or offspring of higher quality than others, thus transmitting more genes to future generations.

▸ Future generations will contain more genes, and thus more characteristics, of the better-reproducing individuals. As a result, characteristics evolve across generations through time.

will pass on more genes to the next generation than a parent that produces few or no offspring. In the next generation, therefore, the genes of better-adapted individuals will be more prevalent than those of individuals that are less well adapted. From one generation to another through time, characteristics, or *traits*, that lead to better and better reproductive success in a given environment will evolve in the population. This process is termed **adaptation**, and a trait that promotes success is called an **adaptation** or an **adaptive trait**.

Natural selection acts on genetic variation

For an organism to pass a trait along to future generations, genes in the organism's DNA (p. 29) must code for the trait. In an organism's lifetime, its DNA will be copied millions of times by millions of cells. In all this copying and recopying, sometimes a mistake is made. Accidental changes in DNA, called **mutations**, give rise to genetic variation among individuals. If a mutation occurs in a sperm or egg cell, it may be passed on to the next generation. Most mutations have little effect, but some can be deadly, whereas others can be beneficial. Those that are not lethal provide the genetic variation on which natural selection acts.

Genetic variation is also generated as organisms mix their genetic material through sexual reproduction. When organisms reproduce sexually, a portion of each parent's genes contributes to the genes of the offspring. This process produces novel combinations of genes, generating variation among individuals.

Natural selection can act on genetic variation and alter organismal characteristics through time in three main ways (**FIGURE 3.2**). Selection that drives a feature in one direction rather than another—for example, toward larger or smaller, faster or slower—is called *directional selection*. In contrast, *stabilizing selection* favors intermediate traits, in essence preserving the status quo. Under *disruptive selection*, traits diverge from their starting condition in two or more directions.

The generation of variation and natural selection's action upon it require a great deal of time. A species cannot simply adapt at once every time environmental conditions change. For instance, the warming of our global climate today (Chapter 18) is occurring too rapidly for most species to adapt, and we will likely lose many species to extinction.

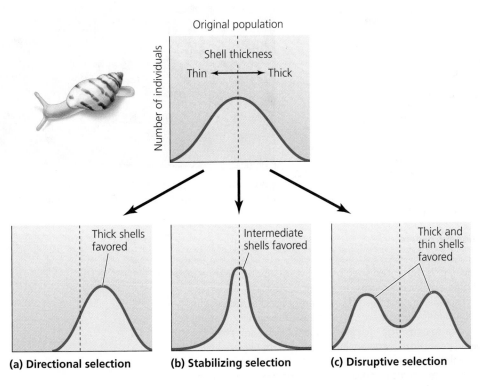

(a) Directional selection   **(b) Stabilizing selection**   **(c) Disruptive selection**

FIGURE 3.2 ◀ Selection can act in three ways. Consider snails living in a tropical cloud forest, and assume that we begin with a population of snails with shells of different thicknesses (top graph). Because shells protect snails against predators, snails with thick shells may be favored over those with thin shells, through *directional selection* **(a)** Alternatively, suppose that a shell that is too thin breaks easily, whereas a shell that is too thick wastes resources that are better used for feeding or reproduction. In such a case, *stabilizing selection* **(b)** could favor snails with shells that are neither too thick nor too thin. Under *disruptive selection* **(c)** extreme traits are favored. For example, if thin-shelled snails are so resource-efficient that they out-reproduce intermediate-shelled snails, and thick-shelled snails are so well protected from predators that they also out-reproduce intermediate-shelled snails, then each "extreme" strategy works more effectively than a compromise between the two.

Selective pressures from the environment influence adaptation

Environmental conditions determine what pressures natural selection will exert, and these selective pressures affect which members of a population will survive and reproduce. Over many generations, this results in the evolution of traits that enable success within the environment in question. Closely related species that live in very different environments and thus experience very different selective pressures tend to diverge in their traits as the differing pressures drive the evolution of different adaptations (**FIGURE 3.3A**). Conversely, sometimes very unrelated species may have similar traits as a result of adapting to selective pressures from similar environments; this is called *convergent evolution* (**FIGURE 3.3B**).

However, environments change over time, and organisms may move to new places and encounter new conditions. In either case, a trait that promotes success at one time or location may not do so at another. The population of golden toads that had adapted to the moist conditions of Monteverde's cloud forest did not persist after Monteverde's climate started to become drier 25 years ago. Varying environmental conditions in time and space make adaptation a moving target.

In all these ways, variable genes and variable environments interact as species engage in a perpetual process of adapting to the changing conditions around them. During

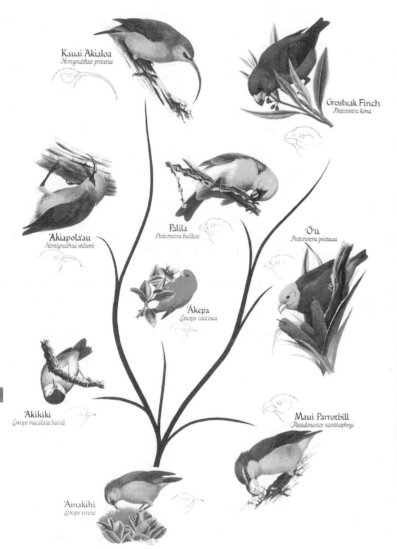

(a) Divergent evolution of Hawaiian honeycreepers.

(b) Convergent evolution of a cactus in Arizona (top) and a euphorb (spurge) in the Canary Islands (bottom).

FIGURE 3.3 ▲ Natural selection can cause closely related species to diverge in appearance (a) as they adapt to selective pressures from different environments. In the group of birds known as Hawaiian honeycreepers, closely related species have adapted to different food resources and habitats, as indicated by the diversity in their plumage colors and the shapes of their bills. Natural selection can also cause distantly related species to converge in appearance (b) if the selective pressures of their environments are similar. A classic case of such convergent evolution involves many cacti of the Americas and euphorbs of Africa. Plants in each family independently adapted to arid environments through the evolution of tough succulent stems to hold water, thorns to keep thirsty animals away, and photosynthetic stems without leaves to reduce surface area and water loss.

this process, natural selection does not simply weed out unfit individuals. It also helps to elaborate and diversify traits that in the long term may help lead to the formation of new species and whole new types of organisms.

Evidence of natural selection is all around us

The results of natural selection are all around us, visible in every adaptation of every organism. In addition, countless lab experiments (mostly with fast-reproducing organisms such as bacteria and fruit flies) have demonstrated rapid evolution of traits.

The evidence for selection that may be most familiar to us is that which Darwin himself cited prominently in his work 150 years ago: our breeding of domesticated animals. In our dogs, cats, and livestock, we have conducted selection under our own direction. We have chosen animals with traits we like and bred them together, while not breeding those with variants we do not like. Through such *selective breeding*, we have been able to augment particular traits we prefer.

Consider the great diversity of dog breeds (**FIGURE 3.4A**). People generated every type of dog alive today by starting with a single ancestral species and selecting for particular desired traits as individuals were bred together. From Great Dane to Chihuahua, all dogs are able to interbreed and produce viable offspring, yet breeders can maintain striking differences among them by allowing only like individuals to breed with like. This process of selection conducted under human direction is termed **artificial selection**.

Artificial selection has also given us the many crop plants we depend on for food, all of which people domesticated from wild ancestors and carefully bred over years, centuries, or millennia (**FIGURE 3.4B**). Through selective breeding, we have created corn with larger, sweeter kernels; wheat and rice with larger and more numerous grains; and apples, pears, and oranges with better taste. We have diversified single types into many—for instance, breeding variants of the plant *Brassica oleracea* to create broccoli, cauliflower, cabbage, and brussels sprouts. Our entire agricultural system is based on artificial selection; we depend on a working understanding of evolution for the very food we eat.

Evolution generates biological diversity

When Charles Darwin wrote about the wonders of a world full of diverse animals and plants, he conjured up the vision of a "tangled bank" of vegetation harboring all kinds of creatures. Such a vision fits well with the arching vines, dripping leaves, and mossy slopes of Monteverde's tropical cloud forest. Indeed, tropical forests worldwide teem with life and harbor immense biological diversity (see Figure 3.1).

Biological diversity, or **biodiversity** for short, refers to the variety of life across all levels of biological organization, including the diversity of species, their genes, their populations, and their communities (we will introduce communities shortly: p. 62 and Chapter 4). Scientists have described about 1.8 million species, but many more remain undiscovered or unnamed. Estimates for the total number of species in the world range up to 100 million, with many of them thought to occur in tropical forests. In this light, the discovery of a

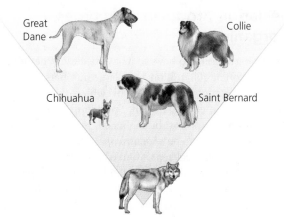

(a) Ancestral wolf and derived dog breeds

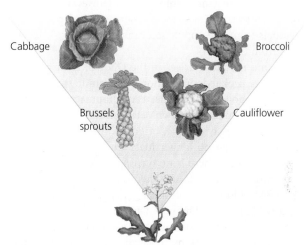

(b) Ancestral *Brassica oleracea* and derived crops

FIGURE 3.4 ▲ Selection imposed by people (selective breeding, or artificial selection) has resulted in the numerous breeds of dogs **(a)**. By starting with the gray wolf (*Canis lupus*) as the ancestral wild species, and by breeding like with like and selecting for the traits we prefer, we have produced breeds as different as Great Danes and Chihuahuas. By this same process we have created the immense variety of crop plants **(b)** we depend on for sustenance. Cabbage, brussels sprouts, broccoli, and cauliflower were all generated from a single ancestral species, *Brassica oleracea*.

new toad species in Costa Rica in 1964 seems far less surprising. Although Costa Rica covers a tiny fraction (0.01%) of Earth's surface area, it is home to 5–6% of all species known to science. Of the 500,000 species scientists estimate exist in the country, only 87,000 (17%) have been inventoried and described.

Tropical rainforests such as Costa Rica's, however, are by no means the only places rich in biodiversity. Step outside anywhere on Earth, even in a major city, and you will find numerous species within easy reach. They may not always be large and conspicuous like Yellowstone's bears or Africa's elephants, but they will be there. Plants poke up from cracks in asphalt in every city in the world, and even Antarctic ice harbors microbes. A handful of backyard soil may contain an entire miniature world of life, including insects, mites, millipedes, nematode worms, plant seeds, fungi, and millions upon millions of bacteria. We will examine Earth's biodiversity in detail in Chapter 11.

Speciation produces new types of organisms

How did Earth come to have so many species? Whether there are 1.8 million or 100 million, such large numbers require scientific explanation. The process by which new species are generated is termed **speciation**. Speciation can occur in a number of ways, but most biologists consider the main mode of species formation to be *allopatric speciation*, species formation due to the physical separation of populations over some geographic distance. To understand allopatric speciation, begin by picturing a population of organisms. Individuals within the population possess many similarities that unify them as a species because they are able to reproduce with one another and share genetic information. However, if the population is broken up into two or more isolated populations, individuals from one population cannot reproduce with individuals from the others.

When a mutation arises in the DNA of an organism in one of these isolated populations, it cannot spread to the other populations. Over time, each population will independently accumulate its own set of mutations. Eventually, the populations may diverge, or grow different enough, that their members can no longer mate with one another. Once this has happened, there is no turning back; the two populations cannot interbreed, and they have embarked on their own independent evolutionary trajectories as separate species (**FIGURE 3.5**). The populations will continue diverging in their characteristics as chance mutations accumulate that cause the populations to become different in random ways. If environmental conditions happen to be different for the two populations, then natural selection may accelerate the divergence.

Populations can be separated in many ways

The long-term geographic isolation of populations that can lead to allopatric speciation can occur in various ways. Glacial ice sheets may move across continents during ice ages and split populations in two. Major rivers may change course and do the same. Mountain ranges may rise and divide regions and their organisms. Drying climate may partially evaporate lakes, subdividing them into smaller bodies of water. Warming or cooling temperatures may cause whole plant communities to move northward or southward, or upslope or downslope, creating new patterns of plant and animal distribution. Regardless of the mechanism of separation, for speciation to occur, populations must remain isolated for a long time, generally thousands of generations.

If the geologic or climatic process that has isolated populations reverses itself—if the glacier recedes, or the river returns to its old course, or warm temperatures turn cool again—then the populations can come back together. This is the moment of truth for speciation. If the populations have not diverged enough, their members will begin interbreeding and reestablish gene flow, mixing the mutations that each population accrued

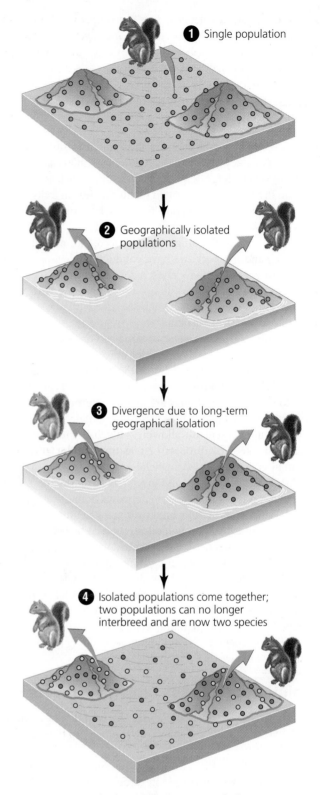

1 Single population

2 Geographically isolated populations

3 Divergence due to long-term geographical isolation

4 Isolated populations come together; two populations can no longer interbreed and are now two species

FIGURE 3.5 ▲ In the long, slow process of allopatric speciation, some geographical barrier splits a population. In this diagram, two mountaintops 1 are turned into islands by rising sea level 2, isolating populations of squirrels. Each isolated population accumulates its own independent set of genetic changes over time, until individuals become genetically distinct and unable to breed with individuals from the other population 3. The two populations now represent separate species and will remain so even if the geographical barrier is removed and the new species intermix 4.

while isolated. However, if the populations have diverged sufficiently, they will not interbreed, and two species will have been formed, each destined to continue on its own evolutionary path.

Although allopatric speciation has long been considered the main mode of species formation, speciation appears to occur in other ways as well. *Sympatric speciation* can occur when populations become reproductively isolated within the same geographic area. For example, populations of some insects may become isolated if they feed and mate exclusively on different types of plants. Or they may mate during different seasons, becoming isolated in time rather than space. In some plants, speciation apparently has occurred as a result of hybridization between species. In others, it has resulted from mutations that changed the numbers of chromosomes—causing speciation to occur in just one generation! Garnering solid evidence for mechanisms of speciation is difficult, so biologists still actively debate the relative prevalence of each of these modes of speciation.

We can infer the history of life's diversification by comparing organisms

Innumerable speciation events have generated complex patterns of diversity at levels above the species level. Evolutionary biologists study such patterns, examining how groups of organisms arose and how they evolved the characteristics they show. For instance, how did we end up with plants as different as mosses, palm trees, daisies, and redwoods? Why do fish swim, snakes slither, and sparrows sing? How and why did birds, bats, and insects each independently evolve the ability to fly? To address such questions, we need to know how the major groups diverged from one another over the course of evolutionary time.

Scientists represent this history of divergence by using branching, treelike diagrams called *cladograms* or **phylogenetic trees**. Similar to family genealogies, these diagrams illustrate scientists' hypotheses as to how divergence took place (**FIGURE 3.6**). Phylogenetic trees can show relationships among species, major groups of species, populations, or genes. Scientists construct these trees by analyzing patterns of similarity among the genes or external characteristics of present-day organisms and inferring which groups share similarities because they are related.

By mapping traits onto a tree according to which organisms possess them, one can trace how the traits may have evolved. For instance, phylogenetic research shows that birds, bats, and insects are distantly related, with many other flightless groups between them. So, it is far simpler to conclude that these three groups evolved flight independently than to conclude that the many flightless groups all lost an ancestral ability to fly. Because phylogenetic trees help biologists make such inferences about so many traits, they have become one of the modern biologist's most powerful tools.

The fossil record teaches us about life's long history

Scientists also decipher life's history by studying fossils. As organisms die, some are buried by sediment. Under certain conditions, the hard parts of their bodies—such as bones, shells, and teeth—may be preserved, as sediments are compressed into rock (pp. 37–39). Minerals replace the organic material, leaving

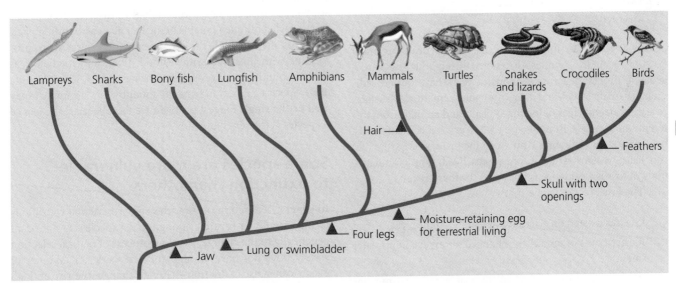

FIGURE 3.6 ▲ Phylogenetic trees show the history of life's divergence. Similar to family genealogies, these diagrams illustrate relationships among groups of organisms, as inferred from comparisons among present-day creatures. The tree shown here is a greatly simplified representation of relationships among groups of vertebrates—just one small portion of the huge and complex "tree of life." Each branch results from a speciation event, and as you follow the tree upward from its trunk to the tips of its branches you proceed forward in time, tracing life's history. By mapping traits onto phylogenetic trees, biologists can study how traits have evolved over time. In this diagram, major traits are mapped using arrows to indicate when they originated. For instance, all vertebrates "above" the point at which jaws are indicated have jaws, whereas lampreys diverged before jaws originated and thus lack them.

FIGURE 3.7 ▲ The fossil record helps reveal the history of life on Earth. The numerous fossils of trilobites indicate that these animals, now extinct, were abundant in the oceans from roughly 540 million to 250 million years ago.

behind a **fossil**, an imprint in stone of the dead organism (**FIGURE 3.7**). In countless locations throughout the world, geologic processes across millions of years have buried sedimentary rock layers and later brought them to the surface, revealing assemblages of fossilized plants and animals from different time periods. By aging the rock layers that contain fossils, paleontologists (scientists who study the history of Earth's life) can learn when particular organisms lived. The cumulative body of fossils worldwide is known as the **fossil record**.

The fossil record shows that:

▶ Life has existed on Earth for at least 3.5 billion years.

▶ Earlier types of organisms changed, or evolved, into later ones.

▶ The number of species existing at any one time has generally increased through time.

▶ The species living today are a tiny fraction of all species that ever lived; the vast majority are long extinct.

▶ There have been several episodes of *mass extinction*, or simultaneous loss of great numbers of species (pp. 59–61, 288–290).

Across life's 3.5 billion years on Earth, complex structures have evolved from simple ones, and large sizes from small ones. However, simplicity and small size have also evolved

when favored by natural selection, and it is easy to argue that Earth still belongs to the bacteria and other microbes, some of them little changed over eons. Even fans of microbes, however, must marvel at some of the exquisite adaptations of animals, plants, and fungi: The heart that beats so reliably for an animal's entire lifetime that we take it for granted. The complex organ system to which the heart belongs. The stunning plumage of a peacock in full display. The ability of each and every plant on the planet to lift water and nutrients from the soil, gather light from the sun, and turn it into food. The staggering diversity of beetles and other insects. The human brain and its ability to reason. All these and more have resulted as the process of evolution has generated new species and whole new branches on the tree of life.

Speciation and extinction together determine Earth's biodiversity

Although speciation generates Earth's biodiversity, it is only one part of the equation, because the fossil record teaches us that the vast majority of species that once lived are now gone—from creatures of the deep past such as the trilobite in Figure 3.7 to recent ones such as Monteverde's golden toad. The disappearance of a species from Earth is called **extinction**. From studying the fossil record, paleontologists calculate that the average time a species spends on Earth is 1–10 million years. The number of species in existence at any one time is equal to the number added through speciation minus the number removed by extinction.

Extinction is a natural process, but human impact can profoundly affect the rate at which it occurs (**FIGURE 3.8**). As we will see in Chapter 11, the biological diversity that makes Earth such a unique planet is being lost at an astounding pace. This loss affects people directly, because other organisms provide us with life's necessities—food, fiber, and medicine, as well as countless vital ecosystem services (pp. 3, 121–122, 148, 160). Species extinction brought about by human impact may well be the single biggest problem we face, because the loss of a species is irreversible.

Some species are more vulnerable to extinction than others

In general, extinction occurs when environmental conditions change rapidly or severely enough that a species cannot adapt genetically to the change; the slow process of natural selection simply does not have enough time to work. All manner of events can cause extinction—climate change, the rise and fall of sea level, the arrival of new species, severe weather events, and more. In general, small populations are vulnerable to extinction because fluctuations in their size could easily, by chance, bring the population size to zero. Species narrowly specialized to some particular resource or way of life are also vulnerable, because environmental changes that make that resource or way of life unavailable can doom them.

The golden toad was a prime example of a vulnerable species. It was **endemic** to the Monteverde cloud forest, meaning that it occurred nowhere else on the planet. Endemic species face relatively high risks of extinction because all their

FIGURE 3.8 ▲ Until 10,000 years ago, the North American continent teemed with mammoths, camels, giant ground sloths, lions, saber-toothed cats, and various types of horses, antelope, bears, and other large mammals. Nearly all of this megafauna went extinct suddenly about the time that people first arrived on the continent. Similar extinctions occurred in other areas simultaneously with human arrival, suggesting to many scientists that overhunting or other human impacts were responsible. See also Figure 11.10, p. 289.

members belong to a single and sometimes small population. At the time of its discovery, the golden toad was known from only a 4-km² (988-acre) area of Monteverde. It also required very specific conditions to breed successfully. During the spring at Monteverde, water collects in shallow pools within the network of roots that span the cloud forest's floor. The golden toad gathered to breed in these root-bound reservoirs, and it was here that Jay Savage and his companions collected their specimens in 1964. Monteverde provided ideal habitat for the golden toad, but the extent of that habitat was minuscule—any environmental stresses that deprived the toad of the resources it needed to survive might doom the entire world population of the species.

In the United States, a number of amphibians are limited to very small ranges and thus are vulnerable to extinction. The Yosemite toad is restricted to a small region of the Sierra Nevada in California, the Houston toad occupies just a few areas of Texas woodland, and the Florida bog frog lives in a tiny region of Florida wetland. Fully 40 salamander

species in the United States are restricted to areas the size of a typical county, and some of these live atop single mountains (**FIGURE 3.9**).

Earth has seen several episodes of mass extinction

Most extinction occurs gradually, one species at a time. The rate at which this type of extinction occurs is referred to as the *background extinction rate*. However, Earth has seen five events of staggering proportions that killed off massive numbers of species at once. These episodes, called **mass extinction events**, have occurred at widely spaced intervals in Earth history and have wiped out 50–95% of our planet's species each time.

The best-known mass extinction occurred 65 million years ago and brought an end to the dinosaurs (although birds are modern representatives of dinosaurs). Evidence suggests that the impact of a gigantic asteroid caused this event, called the

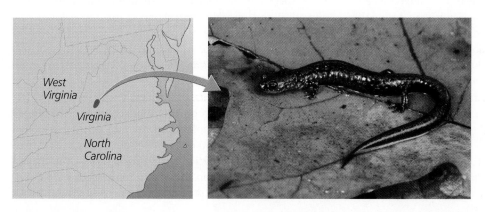

FIGURE 3.9 ◄ The Peaks of Otter salamander (*Plethodon hubrichti*) lives on only a few peaks in Virginia's Blue Ridge Mountains. About 40 other salamander species in the United States are restricted to similarly small ranges. Small range sizes leave these creatures vulnerable to extinction if severe changes occur in their local environment.

The SCIENCE behind the Story

The K-T Mass Extinction

When he first started working at Bottaccione Gorge in Italy, American geologist Walter Alvarez had no idea he might soon help discover what killed off the dinosaurs.

Scientists knew that 65 million years ago, at the transition from the Cretaceous to the Tertiary periods (see Appendix D for the geologic timescale), about 70% of the species then living, including the dinosaurs, disappeared. But they didn't know why.

Alvarez was developing a new method to determine the age of sedimentary rocks, and he'd chosen Bottaccione Gorge because it formed an ideal geologic archive. Its 400-m (1,300-ft) walls are stacked like layer cake with beds of rose-colored limestone that formed between 100 million and 50 million years ago from material that had settled to the bottom of an ancient sea.

While analyzing these layers, Alvarez noticed a band of reddish clay 1 cm (0.4 in.) thick sandwiched between two layers of limestone. The

Dr. Luis Alvarez (left) and Dr. Walter Alvarez (right) with a sample of the iridium layer

older layer just below it was packed with fossils of globotruncana, a tiny animal that lived in the late Cretaceous period. The newer layer just above it contained just a few scattered fossils of a cousin of globotruncana, typical of sedimentary rock formed in the early Tertiary period. The intermediate clay layer, which had formed just as the dinosaurs went extinct, had no fossils at all.

Walter and his father, physicist Luis Alvarez, analyzed the Cretaceous-Tertiary (K-T) clay layer to determine how long it had taken to form, using the rare metal iridium as a kind of clock. Almost all the iridium on

Earth's surface comes from dust from meteorites that burn up in the atmosphere. Because the same amount of meteorite dust rains down each year, the Alvarezes measured iridium levels in the clay layer to determine how many years' worth of iridium had accumulated.

Iridium levels in the limestone were typical for sedimentary rocks, about 0.3 parts per billion. In the clay layer, however, the Alvarezes found levels 30 times higher. To make sure the finding was not unique to Bottaccione Gorge, they checked the K-T clay layer at a Danish sea cliff where the same sequence of rock layers was exposed. It had a concentration of iridium 160 times greater than surrounding rock! The Alvarezes hypothesized that the excess iridium had come from a massive asteroid that smashed into Earth, causing a global environmental catastrophe that resulted in a mass extinction.

To convince themselves and the scientific community that an asteroid

---

Cretaceous-Tertiary, or K-T, event (see **THE SCIENCE BEHIND THE STORY**, above). As massive as this extinction event was, however, it was moderate compared to the mass extinction at the end of the Permian period 250 million years ago (see Appendix D for Earth's geologic periods). Paleontologists estimate that 75–95% of all species may have perished during this event, described by one researcher as the "mother of all mass extinctions." Scientists do not yet know what caused the end-Permian extinction event, although hypotheses include an asteroid impact, massive volcanism, methane releases and global warming, or some combination of factors.

The sixth mass extinction is upon us

Many biologists have concluded that Earth is currently entering its sixth mass extinction event—and that we are the cause. Indeed, the Millennium Ecosystem Assessment (pp. 16, 18)

estimated that today's extinction rate is 100–1,000 times higher than the background rate, and rising. Changes to Earth's natural systems set in motion by human population growth, development, and resource depletion have driven many species extinct and are threatening countless more. The alteration and outright destruction of natural habitats, the hunting and harvesting of species, and the introduction of species from one place to another where they can harm native species—these processes and more have combined to threaten Earth's biodiversity (pp. 290–296).

Amphibians such as the golden toad are disappearing faster than just about any other type of organism. According to the most recent scientific assessments, 40% of frog, toad, and salamander species are in decline, 30% are in danger of extinction, and nearly 170 species have vanished just within the last few years or decades (pp. 294–295). Some species are disappearing in remote and pristine areas, suggesting that chytrid

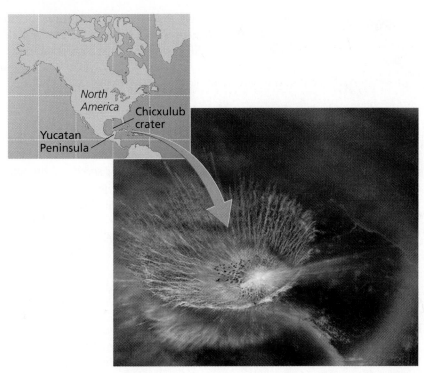

A colossal asteroid impact 65 million years ago is thought to have caused the Cretaceous-Tertiary mass extinction.

impact caused the K-T event, the Alvarezes had to rule out other possible explanations. For example, the extra iridium could have come from seawater. Calculations, however, proved that seawater could not contain enough iridium to account for the high levels in the clay layers. An asteroid 10 km (6.2 mi) wide strikes Earth, on average, every 100 million years. Such an impact would unleash an explosion 1,000 times more forceful than the 1883 eruption of the Indonesian volcano Krakatau (p. 43), which scattered so much dust around the world that global temperatures cooled and sunsets were intense for two years afterward.

An asteroid impact 65 million years ago, they suggested, kicked up enough soot to blot out the sun for several years. This would have inhibited photosynthesis, causing plants to die, animals to starve, and food webs to collapse. Only a few smaller animals survived, feeding on rotting vegetation. When sunlight finally returned, plants sprouted from dormant seeds, and a long recovery began.

Published in the journal *Science* in 1980, the Alvarezes' explanation was immediately challenged by other geologists. Some argued that spectacular and prolonged volcanic eruptions more likely explained the high iridium levels in the K-T layer. Others suggested the extinction could have taken place over thousands of years, which still would appear as an instant in the geologic record.

But throughout the 1980s, scientists kept finding evidence supporting the asteroid-impact hypothesis. Iridium-enriched clay turned up at K-T layers around the world, as did bits of minerals called shocked quartz and stishovite, which form only under the extreme pressure of thermonuclear explosions and asteroid impacts.

The Alvarezes' hypothesis became widely accepted after 1991, once scientists pinpointed a 65-million-year-old crater of the predicted size in ocean sediments off the coast of Mexico. Today, debate continues, but scientists broadly agree that an asteroid impact 65 million years ago caused or contributed to our planet's most recent mass extinction.

fungus could be responsible. But researchers think a variety of causes are affecting amphibians in a "perfect storm" of impacts.

When we look around us, it may not appear as though a human version of an asteroid impact is taking place, but we cannot judge such things on our own timescale. On the geologic timescale, extinction over 100 years or even 10,000 years appears instantaneous. Moreover, speciation is a slow enough process that it will take life millions of years to recover—by which time our own species will most likely not be around.

LEVELS OF ECOLOGICAL ORGANIZATION

The extinction of species, their generation through speciation, and other evolutionary mechanisms and patterns have substantial influence on ecology. **Ecology** is the study of interactions among organisms and between organisms and their environments. It is often said that ecology provides the stage on which the play of evolution unfolds. The two are tightly intertwined in many ways.

We study ecology at several levels

Life occurs in a hierarchy of levels, from atoms, molecules, and cells (pp. 26–29) up through the **biosphere**, which is the cumulative total of living things on Earth and the areas they inhabit. Ecologists study relationships on the higher levels of this hierarchy (**FIGURE 3.10**), namely on the levels of the organism, population, community, ecosystem, and biosphere.

At the level of the organism, the science of ecology describes relationships between an organism and its physical environment. Organismal ecology helps us understand, for example, what aspects of the golden toad's environment were

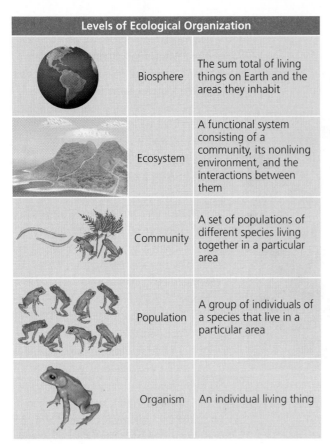

Levels of Ecological Organization

	Biosphere	The sum total of living things on Earth and the areas they inhabit
	Ecosystem	A functional system consisting of a community, its nonliving environment, and the interactions between them
	Community	A set of populations of different species living together in a particular area
	Population	A group of individuals of a species that live in a particular area
	Organism	An individual living thing

FIGURE 3.10 ▲ Life exists in a hierarchy of levels. Ecology includes the study of the organismal, population, community, and ecosystem levels and, increasingly, the level of the biosphere.

important to it, and why. **Population ecology**, in contrast, examines the dynamics of population change and the factors that affect the distribution and abundance of members of a population. It helps us understand why populations of some species (such as the golden toad) decline while populations of others (such as ourselves) increase.

In ecology, **communities** are made up of multiple interacting species that live in the same area. A population of golden toads, a population of resplendent quetzals, and a population of ferns, together with all the other interacting plant, animal, fungal, and microbial populations in the Monteverde cloud forest, would be considered a community. **Community ecology** focuses on patterns of species diversity and on interactions among species, from one-to-one interactions to complex interrelationships involving entire communities. In the case of Monteverde, it allows us to study how the golden toad and many other species of its cloud-forest community interacted.

Ecosystems encompass communities and the abiotic (nonliving) material and forces with which their members interact. Monteverde's cloud-forest ecosystem consists of the community plus the air, water, soil, nutrients, and energy that the community's organisms use. **Ecosystem ecology** reveals patterns, such as the flow of energy and nutrients, by studying living and nonliving components of systems in conjunction. As we will see, changing climate has had a strong influence on the organisms of Monteverde's cloud-forest ecosystem.

As new technologies allow scientists to learn more about the complex dynamics of natural systems on a global scale,

ecologists are increasingly expanding their horizons beyond ecosystems to the biosphere as a whole. In the remainder of this chapter we explore ecology up through the population level. In Chapter 4 we examine community ecology, and in Chapter 5 we consider ecology at the levels of the ecosystem and biosphere.

Each organism has habitat needs

At the level of the organism, each individual relates to its environment in ways that tend to maximize its survival and reproduction. One key relationship involves the specific environment in which an organism lives, its **habitat**. A species' habitat consists of the living and nonliving elements around it, including rock, soil, leaf litter, humidity, plant life, and more. The golden toad lived in a habitat of cloud forest—specifically, on the moist forest floor, using seasonal pools for breeding and burrows for shelter.

Each organism thrives in certain habitats and not in others, leading to nonrandom patterns of **habitat use**. Mobile organisms actively select habitats in which to live from among the range of options they encounter, a process called **habitat selection**. In the case of plants and rooted animals (such as sea anemones), whose young disperse and settle passively, patterns of habitat use result from success in some habitats and failure in others.

Plants known as epiphytes rely on other plants in their habitat; they grow on trees for physical support, obtaining water from the air and nutrients from organic debris that collects among their leaves. Epiphytes thrive in cloud forests because they require a habitat with high humidity, and Monteverde hosts more than 330 species of epiphytes, mostly ferns, orchids, and bromeliads (pineapple relatives). By collecting pools of rainwater and pockets of leaf litter, epiphytes in turn create habitat for many other organisms, including many invertebrates and even frogs that place their eggs or tadpoles in the rainwater pools (**FIGURE 3.11**).

FIGURE 3.11 ▼ The strawberry poison dart frog (*Dendrobates pumilio*) places her tadpoles in pools of rainwater that collect in the leaves of epiphytic plants such as bromeliads. The pools also supply breeding habitat for mosquitoes, whose larvae are eaten by the frog's tadpoles. Some birds specialize on feeding on the insects and other animals that live in epiphytes. And just as epiphytes provide habitat for many small animals, trees provide habitat for epiphytes. The abundance of habitats and niches in just this one small system are an indicator of the rich complexity of tropical cloud forests.

Habitats vary with the body size and needs of the species. A tiny soil mite may use less than a square meter of soil in its lifetime. A vulture, elephant, or whale, in contrast, may traverse miles upon miles of air, land, or water in just a day. Species may also have different habitat needs at different times of year; many migratory birds use distinct breeding habitats, wintering habitats, and migratory habitats.

The criteria by which organisms favor some habitats over others can vary greatly. The soil mite may assess available habitats in terms of the chemistry, moisture, and compactness of the soil and the percentage and type of organic matter. The vulture may ignore not only soil but also topography and vegetation, focusing solely on the abundance of dead animals in the area that it scavenges for food. For a whale, water temperature and salinity and the abundance of marine microorganisms might be the critical characteristics. Every species assesses habitats differently because every species has different needs.

Habitat use is important in environmental science because the availability and quality of habitat are crucial to an organism's well-being. Indeed, because habitats provide everything an organism needs, including nutrition, shelter, breeding sites, and mates, the organism's very survival depends on the availability of suitable habitats. Often this need creates conflict with people who want to alter or develop a habitat for their own purposes.

Niche and specialization are key concepts in ecology

Another way in which an organism relates to its environment is through its niche. A species' **niche** reflects its use of resources and its functional role in a community. This includes its habitat use, its consumption of certain foods, its role in the flow of energy and matter, and its interactions with other organisms. The niche is a multidimensional concept, a kind of summary of everything an organism does. The pioneering ecologist Eugene Odum once wrote that "habitat is the organism's address, and the niche is its profession."

Organisms vary in the breadth of their niches. Species with narrow breadth, and thus very specific requirements, are said to be **specialists**. Those with broad tolerances, able to use a wide array of resources, are **generalists**. For example, in a study of eight Costa Rican bird species that feed from epiphytes, ornithologist T. Scott Sillett found that four were generalists. The other four were specialists on the insect resources the epiphytes provided and spent more than 75% of their foraging efforts feeding from ephiphytes.

Specialists succeed over evolutionary time by being extremely good at the things they do, but they are vulnerable when conditions change and threaten the habitat or resource on which they have specialized. Generalists succeed by being able to live in many different places and weather variable conditions, but they may not thrive in any one situation as much as a specialist does. An organism's habitat preferences, niche, and degree of specialization each reflect adaptations of the species and are products of natural selection. In Chapter 4 (p. 79) we will see how interactions among species can sometimes alter these attributes.

POPULATION ECOLOGY

Individuals of the same species inhabiting a particular area make up a population. Species may consist of multiple populations that are geographically isolated from one another. This is the case with a species characteristic of Monteverde—the resplendent quetzal (*Pharomachrus mocinno*), considered one of the world's most spectacular birds (see Figure 3.1a). Although it ranges from southernmost Mexico to Panama, the resplendent quetzal is adapted to high-elevation tropical forest and is absent from low-elevation areas that surround these natural islands of montane cloud-forest habitat. Moreover, human development has destroyed much of this habitat. Thus, the species today exists in many isolated populations scattered across Central America.

In contrast, the human species is a consummate generalist, and we have spread into nearly every corner of the planet. As a result, it is difficult to define a distinct human population on anything less than the global scale. Some would maintain that in the ecological sense of the word, all 6.9 billion of us comprise one population.

Populations show characteristics that help predict their dynamics

All populations—from humans to quetzals—exhibit characteristics that help population ecologists predict the future dynamics of the population. Attributes such as density, distribution, sex ratio, age structure, and birth and death rates all help the ecologist understand how a population may grow or decline. The ability to predict growth or decline is useful in monitoring and managing threatened and endangered species (pp. 302–307) and in studying human populations (Chapter 8). Understanding human population dynamics, their causes, and their consequences is a central element of environmental science and one of the prime challenges for our society today.

Population size Expressed as the number of individual organisms present at a given time, **population size** may increase, decrease, undergo cyclical change, or remain the same over time. Extinctions are generally preceded by population declines. As late as 1987, scientists documented a golden toad population at Monteverde in excess of 1,500 individuals, but in 1988 scientists sighted only 10 toads, and in 1989 they found just a single individual. By 1990, the species had disappeared.

The passenger pigeon (*Ectopistes migratorius*), also now extinct, illustrates the extremes of population size (**FIGURE 3.12**). Not long ago it was the most abundant bird in North America; flocks of passenger pigeons literally darkened the skies. In the early 1800s, ornithologist Alexander Wilson watched a flock of 2 billion birds 390 km (240 mi) long that took 5 hours to fly over and sounded like a tornado. Passenger pigeons nested in gigantic colonies in the forests of the upper Midwest and southern Canada. Once people began cutting the forests, however, the birds made easy targets for market hunters, who gunned down thousands at a time and shipped them to market by the wagonload. By the end of the 19th century, the passenger pigeon population had declined to such a low number that the birds could not form the large colonies they apparently needed in

(a) Passenger pigeon

(b) 19th-century lithograph of pigeon hunting in Iowa

FIGURE 3.12 ▲ The passenger pigeon **(a)** was once North America's most numerous bird, and its flocks literally darkened the skies when millions of birds passed overhead **(b)**. However, human cutting of forests and hunting drove the species to extinction within just a few decades.

order to breed. In 1914, the last passenger pigeon on Earth died in the Cincinnati Zoo, bringing the continent's most numerous bird species to extinction within just a few decades.

Population density   The flocks and breeding colonies of passenger pigeons showed high population density, another attribute that ecologists assess to understand populations. **Population density** describes the number of individuals in a population per unit area. For instance, the 1,500 golden toads counted in 1987 within 4 km² (988 acres) indicated a density of 375 toads/km². In general, large organisms have low population densities because each individual requires many resources—and thus a great deal of area—to survive.

High population density makes it easier for organisms to group together and find mates, but it can also lead to competition and conflict if space, food, or mates are in limited supply. Overcrowded organisms may become vulnerable to the predators that feed on them, and close contact among individuals can increase the transmission of infectious disease. For these reasons, organisms sometimes leave an area when densities become too high. In contrast, at low population densities, organisms benefit from more space and resources but may find it harder to locate mates and companions.

Overcrowding at high population densities in small remnants of habitat is thought to have doomed Monteverde's harlequin frog (*Atelopus varius*; see Figure 3.1c), an amphibian that disappeared at the same time as the golden toad. The harlequin frog is a specialist, favoring "splash zones," areas alongside rivers and streams that receive spray from waterfalls and rapids. As Monteverde's climate grew warmer and drier in the 1980s and 1990s, water flow decreased and many streams dried up (see **THE SCIENCE BEHIND THE STORY**, pp. 68–69). Splash zones grew smaller and fewer, and harlequin frogs were forced to cluster together in what remained of the habitat. Researchers J. Alan Pounds and Martha Crump recorded frog population densities up to 4.4 times higher than normal, with more than 2 frogs per meter (3.3 ft) of stream. Such overcrowding made the frogs

vulnerable to predator attack, assault from parasitic flies, and transmission of disease pathogens such as chytrid fungus.

A second population of this species of harlequin frog was discovered in 2003 on a private reserve elsewhere in Costa Rica, so there is still hope that the species may continue to persist. The frog was rediscovered by University of Delaware student Justin Yeager, who was doing field research during his study abroad trip that summer.

Population distribution   It was not simply the harlequin frog's density but also its distribution in space that led to its demise at Monteverde. **Population distribution**, or **population dispersion**, describes the spatial arrangement of organisms in an area. Ecologists define three distribution types: random, uniform, and clumped (**FIGURE 3.13**). In a *random distribution*, individuals are located haphazardly in space in no particular pattern. This type of distribution can occur when the resources an organism needs are plentiful and found throughout an area and other organisms do not strongly influence where members of a population settle.

A *uniform distribution* is one in which individuals are evenly spaced. This can occur when individuals hold territories or compete for space. In a desert where water is scarce, each plant needs a certain amount of space for its roots to gather moisture. Plants may even poison one another's roots as a means of competing for space. As a result, plants may end up growing equidistant from one another.

In a *clumped distribution*, the pattern most common in nature, organisms arrange themselves according to the availability of the resources they need to survive. Many desert plants grow in patches around isolated springs or along arroyos that flow with water after rainstorms. During their mating season, golden toads were found clumped at seasonal breeding pools. Humans, too, exhibit clumped distribution; people frequently aggregate in urban centers. Clumped distributions often indicate that species are seeking certain habitats or resources that are themselves clumped.

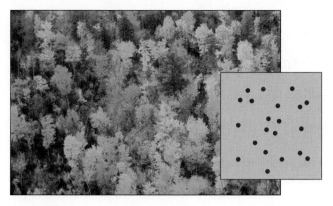

(a) Random

(b) Uniform

(c) Clumped

FIGURE 3.13 ▲ Individuals in a population can spatially distribute themselves in three fundamental ways. In a random distribution **(a)**, organisms are dispersed at random through the environment. In a uniform distribution **(b)**, individuals are spaced evenly, at equal distances from one another. Territoriality or competition can result in such a pattern. In a clumped distribution **(c)**, individuals occur in patches, concentrated more heavily in some areas than in others. Habitat selection, clumped resource patterns, or flocking to avoid predators can result in such a pattern.

Distributions can depend on the scale at which one measures them. On small scales a population may be distributed uniformly, yet this may occur within one patch of a larger, clumped distribution. On very large scales, all organisms show clumped or patchy distributions, because some parts of the total area they inhabit are bound to be more hospitable than others.

Sex ratio A population's **sex ratio** is its proportion of males to females, and this can influence whether the population will increase or decrease in size over time. In monogamous species (in which each sex takes a single mate), a 1:1 sex ratio maximizes population growth, whereas an unbalanced ratio leaves many individuals of one sex without mates. Most species are not monogamous, however, so sex ratios may vary from one species to another.

Age structure Populations generally consist of individuals of different ages. **Age distribution**, or **age structure**, describes the relative numbers of organisms of each age within a population. By combining this information with data on the reproductive potential of individuals in different age classes, a population ecologist can predict how the population may grow or shrink.

For many plants and animals that continue growing in size as they age, older individuals reproduce more; a tree that is large because it is old can produce more seeds, and a fish that is large because it is old may produce more eggs. In some animals, such as birds, the experience they gain with age often makes older individuals better breeders.

Human beings are unusual because we often survive past our reproductive years. As a result, a human population made up largely of older (post-reproductive) individuals will tend to decline over time, whereas one with many young people (of reproductive or pre-reproductive age) will tend to increase. We will use diagrams to explore these ideas further in Chapter 8 (pp. 206–208) as we study human population growth.

Birth and death rates All the preceding factors can influence the rates at which individuals within a population are born and die. Just as individuals of differing ages have different reproductive capacities, individuals of differing ages show different probabilities of dying. For instance, people are more likely to die at old ages than young ages; if you were to follow 1,000 10-year-olds and 1,000 80-year-olds for a year, you would find that at year's end more 80-year-olds had died than 10-year-olds. However, this pattern does not hold for all organisms. Amphibians such as the golden toad produce large numbers of young, which suffer high death rates. For a toad, death is less likely (and survival more likely) at an older age than at a very young age.

To show how the likelihood of survival varies with age, ecologists use graphs called **survivorship curves** (**FIGURE 3.14**). There are three fundamental types of survivorship curves. Humans, with higher death rates at older ages, show a type I survivorship curve. Toads, with highest death rates at young ages, show a type III survivorship curve. A type II survivorship curve is intermediate and indicates equal rates of death at all ages. Many birds are thought to show type II curves.

Populations may grow, shrink, or remain stable

Now that we have outlined some key attributes of populations, we are ready to take a quantitative view of population change by examining some simple mathematical concepts used by population ecologists and by **demographers**

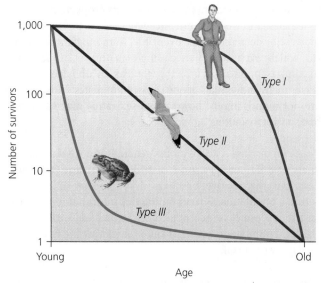

FIGURE 3.14 ▲ In a type I survivorship curve, survival rates are high when organisms are young and decrease sharply when organisms are old. In a type II survivorship curve, survival rates are equivalent regardless of an organism's age. In a type III survivorship curve, most mortality takes place at young ages, and survival rates are greater at older ages. Examples include humans (type I), birds (type II), and amphibians (type III).

(scientists who study human populations). Population growth, or decline, is determined by four factors:

▶ Births within the population (*natality*)

▶ Deaths within the population (*mortality*)

▶ **Immigration** (arrival of individuals from outside the population)

▶ **Emigration** (departure of individuals from the population)

Births and immigration add individuals to a population, whereas deaths and emigration remove individuals. A convenient way to express rates of birth and death is to measure the number of births and deaths per 1,000 individuals per year. These rates are termed the *crude birth rate* and the *crude death rate*.

If we are not interested in the effects of migration, we can measure the **natural rate of population growth** by subtracting the crude death rate from the crude birth rate:

$$\text{(crude birth rate)} - \text{(crude death rate)} = \text{natural rate of population growth}$$

The natural rate of population growth reflects the degree to which a population is growing or shrinking as a result of its own internal factors.

To obtain an overall **population growth rate**, the total rate of change in a population's size per unit time, we must also take into account the effects of migration. Thus, we include terms for immigration and emigration (each expressed per 1,000 individuals per year) in the formula for the natural rate of population growth as follows:

$$\text{(crude birth rate} - \text{crude death rate)} + \text{(immigration rate} - \text{emigration rate)} = \text{population growth rate}$$

The resulting number tells us the net change in a population's size per 1,000 individuals per year. For example, a population with a crude birth rate of 18 per 1,000/yr, a crude death rate of 10 per 1,000/yr, an immigration rate of 5 per 1,000/yr, and an emigration rate of 7 per 1,000/yr would have a population growth rate of 6 per 1,000/yr:

$$(18/1,000 - 10/1,000) + (5/1,000 - 7/1,000) = 6/1,000$$

Thus, a population of 1,000 in one year will reach 1,006 in the next. If the population is 1,000,000, it will reach 1,006,000 the next year. Such population increases are often expressed as percentages, which we can calculate using the following formula:

$$\text{population growth rate} \times 100\%$$

Thus, a growth rate of 6/1,000 would be expressed as:

$$6/1,000 \times 100\% = 0.6\%$$

By measuring population growth in terms of percentages, scientists can compare increases and decreases in species that have far different population sizes. They can also project changes that will occur in the population over longer periods, much like you might calculate the amount of interest your savings account will earn over time.

Unregulated populations increase by exponential growth

When a population increases by a fixed percentage each year, it is said to undergo **exponential growth**. Imagine you put money in a savings account at a fixed interest rate and leave it untouched for years. As the principal accrues interest and grows larger, you earn still more interest, and the sum grows by escalating amounts each year. The reason is that a fixed percentage of a small number makes for a small increase, but that same percentage of a large number produces a large increase. Thus, as savings accounts (or populations) become larger, each incremental increase likewise gets larger. Such acceleration is a characteristic of exponential growth.

We can visualize changes in population size by using population growth curves. The J-shaped curve in **FIGURE 3.15** shows exponential growth. Populations of organisms increase exponentially unless they meet constraints. Each organism reproduces by a certain amount, and as populations get larger, there are more individuals reproducing by that amount. If there are adequate resources and no external limits, ecologists theoretically expect exponential growth.

Normally, exponential growth occurs in nature only when a population is small, competition is minimal, and environmental conditions are ideal for the organism in question. Most often, these conditions occur when the organism is introduced to a new environment that contains abundant resources to exploit. Mold growing on a piece of fruit, or bacteria colonizing a recently dead animal, are cases in point. Plants colonizing regions during primary succession (p. 92) after glaciers recede or volcanoes erupt may also grow exponentially. A current example of exponential growth in the United States is the Eurasian collared dove (see Figure 3.15). Unlike its extinct relative the

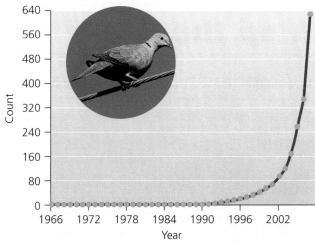

FIGURE 3.15 ▲ Although no species can maintain exponential growth indefinitely, some may grow exponentially for a time when colonizing an unoccupied environment or exploiting an unused resource. The Eurasian collared dove (*Streptopelia decaocto*) is currently spreading across the North American continent, propelled by exponential growth. Data from Sauer, J.R., et al., 2008. The North American Breeding Bird Survey, Results and Analysis 1966–2007. v. 5.15.2008. USGS Patuxent Wildlife Research Center, Laurel, MD.

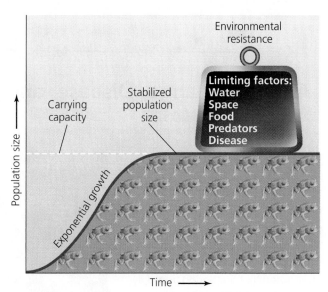

FIGURE 3.16 ▲ The logistic growth curve shows how population size may increase rapidly at first, then grow more slowly, and finally stabilize at a carrying capacity. Carrying capacity is determined both by the *biotic potential* of the organism and by various external limiting factors, collectively termed *environmental resistance*.

passenger pigeon, this species arrived here from Europe and has spread across North America in a matter of years. The Eurasian collared dove thrives in human-disturbed areas and apparently has not (yet) encountered anything to limit its population growth.

Limiting factors restrain population growth

Exponential growth rarely lasts long. If even one single species in Earth's history had increased exponentially for very many generations, it would have blanketed the planet's surface, and nothing else could have survived. Instead, every population eventually is constrained by **limiting factors**—physical, chemical, and biological attributes of the environment that restrain population growth. The interaction of these factors determines the **carrying capacity**, the maximum population size of a species that a given environment can sustain.

Ecologists use the S-shaped curve in **FIGURE 3.16** to show how an initial exponential increase is slowed and eventually brought to a standstill by limiting factors. Called the **logistic growth curve**, it rises sharply at first but then begins to level off as the effects of limiting factors become stronger. Eventually the force of these factors—collectively termed *environmental resistance*—stabilizes the population size at its carrying capacity. One day soon, the Eurasian collared dove population in North America will slow in its growth and then reach carrying capacity. Populations of other European birds that spread across North America in the past, such as the house sparrow and European starling, have peaked and are today declining slowly.

Many factors contribute to environmental resistance and influence a population's growth rate and carrying capacity. Space is one factor that limits the number of individuals a given environment can support; if there is no physical room

for additional individuals, they are unlikely to survive. Other limiting factors for animals in a terrestrial environment include the availability of food, water, mates, shelter, and suitable breeding sites; temperature extremes; prevalence of disease; and abundance of predators. Plants are often limited by amounts of sunlight and moisture and the type of soil chemistry, in addition to disease and attack from plant-eating animals. In aquatic systems, limiting factors include salinity, sunlight, temperature, dissolved oxygen, fertilizers, and pollutants. To determine limiting factors, ecologists may conduct experiments in which they increase or decrease a hypothesized limiting factor and observe its effects on population size.

The influence of some factors depends on population density

A population's density can enhance or diminish the impact of certain limiting factors. Recall that high population density can help organisms find mates but can also increase competition and the risk of predation and disease. Such factors are said to be **density-dependent** factors, because their influence rises and falls with population density. The logistic growth curve in Figure 3.16 represents the effects of density dependence. The larger the population size, the stronger the effects of environmental resistance.

Density-independent factors are limiting factors whose influence is not affected by population density. Temperature extremes and catastrophic events such as floods, fires, and landslides are examples of density-independent factors, because they can eliminate large numbers of individuals without regard to their density.

The logistic curve is a simplified model, and real populations in nature can behave differently. Some may cycle indefinitely above and below the carrying capacity. Some may show

The SCIENCE behind the Story

Climate Change, Disease, and the Amphibians of Monteverde

Soon after the golden toad's disappearance, scientists began to investigate the potential role of global climate change (Chapter 18) in driving cloud-forest species toward extinction. They noted that the period from July 1986 to June 1987 was the driest on record at Monteverde, with unusually high temperatures and record-low stream flows. These conditions caused the golden toad's breeding pools to dry up in the spring of 1987, likely killing nearly all of the eggs and tadpoles in the pools.

By reviewing reams of weather data, scientists found that the number of dry days and dry periods each winter in the Monteverde region had increased between 1973 and 1998. Because amphibians breathe and absorb moisture through their skin, they are susceptible to dry conditions. Based on these facts, herpetologists J. Alan Pounds and Martha Crump in 1994 hypothesized that hot, dry conditions were to blame for high adult mortality and breeding problems among golden toads and other amphibians.

Throughout this period, scientists worldwide were realizing that the atmosphere and oceans were warming because of human release of carbon dioxide and other greenhouse gases (➤ Chapter 18). With this in mind, Pounds and others reviewed the scientific literature on ocean and atmospheric science to analyze the impacts on Monteverde's local climate of warming patterns in the ocean regions around Costa Rica.

Warmer oceans, the researchers found, caused clouds to pass over at

Dr. J. Alan Pounds (left) with Dr. Luis Coloma, looking for harlequin frogs

higher elevations, where they were no longer in contact with the trees. Once the cloud forest's moisture supply was pushed upward, out of reach of the mountaintops, the forest began to dry out (**see first figure**).

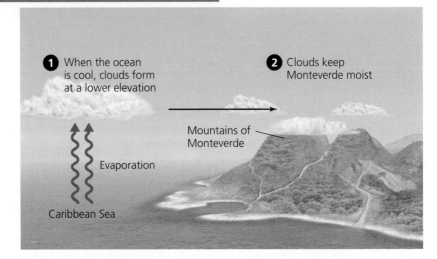

(a) Cool ocean conditions

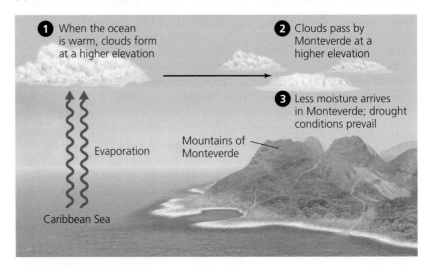

(b) Warm ocean conditions

Monteverde's cloud forest gets its name and life-giving moisture from clouds that sweep inland from the oceans. When ocean temperatures are cool **(a)**, the clouds keep Monteverde moist. Warmer ocean conditions **(b)** resulting from global climate change cause clouds to form at higher elevations and pass over the mountains, drying the cloud forest.

68

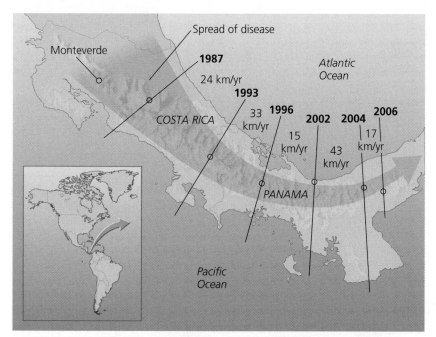

Dr. Karen Lips's research team used known dates of decline (years in figure) in harlequin frog populations at sites in Costa Rica and Panama to infer the spread of a wave of infection (arrow) by chytrid fungus across the region. By their analysis, chytrid reached Monteverde before 1987 and was well into Panama by 2000. Adapted from Lips, K.R., et al., 2008. Riding the wave: Reconciling the roles of disease and climate change in amphibian declines. *PLoS Biology* 6: 441–454.

In a 1999 paper in the journal *Nature,* Pounds and two colleagues reported that climate modification was causing local changes at the species, population, and community levels. They argued that higher clouds and decreasing moisture in the forest could explain the disappearance of the golden toad and harlequin frog, and also the concurrent population crashes and subsequent disappearance of 20 other species of frogs and toads from the Monteverde region.

Moreover, whole communities were being altered. As the montane forests dried out, drought-tolerant species of birds and reptiles shifted upslope, and moisture-dependent species were stranded at the mountaintops by a rising tide of aridity. If a species has nowhere to go, then extinction may result.

Pounds and his colleagues expanded the story further in *Nature* in 2006. Although clouds had risen higher in the sky, the extra moisture evaporating from warming oceans was increasing cloud cover overall, blocking sunlight during the day and trapping heat at night. As a result, at Monteverde and other tropical locations, daytime and nighttime temperatures were becoming more similar.

Such conditions are optimal for chytrid fungi, pathogens that can lethally infect amphibians. In recent years the chytrid fungus *Batrachochytrium dendrobatidis* is thought to have contributed to the likely extinction of 67 of the world's 113 species of harlequin frogs. At Monteverde and elsewhere, Pounds's team argued, climate change is promoting disease epidemics that are driving extinct many of the world's amphibians.

Other researchers agree that chytrid fungus is a major threat but dispute a connection to climate change. In 2008, one team led by biologist Karen Lips of Southern Illinois University–Carbondale reanalyzed the Pounds team's data and also mapped amphibian declines and inferred how the non-native and invasive chytrid fungus has spread rapidly in waves across Central America and South America in recent years (**see second figure**). The analysis by Lips and her team suggested that the dramatic amphibian die-offs at Monteverde and elsewhere are directly due to the arrival of this devastating pathogen and do not necessarily involve climate change.

Later that year, a team led by Jason Rohr of the University of South Florida set out to test the Pounds hypothesis (that climate change was promoting chytrid infection) versus the Lips hypothesis (that chytrid was spreading regardless of climate change). Rohr's group tested for statistical correlations (pp. 12–13) among climate variables, chytrid spread, and the timing of the 67 suspected extinctions of harlequin frog species. This team found a clear spatial pattern in the sequence of extinctions of harlequin frogs that was consistent with the spread of disease, but the researchers could not attribute this beyond doubt to chytrid.

Rohr's team found that harlequin frog extinctions were correlated with rising tropical air temperatures, but the researchers questioned whether this was causal, given that increased extinctions were also correlated with just about every other variable that rose between 1970 and 1990, from human population growth to regional beer and banana production. The researchers concluded that because climate change has been shown to influence many biological systems, it is likely a factor in amphibian declines, but that there is not yet solid evidence for this causal link.

Clearly more research into the effects of both disease and climate change on frogs is needed—and fast. Biologists today are racing to find the answers, hoping to save many of the world's amphibian species from extinction.

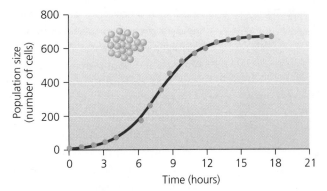

(a) Yeast cells, *Saccharomyces cerevisiae*

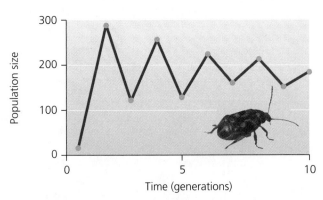

(c) Stored-product beetle, *Callosobruchus maculatus*

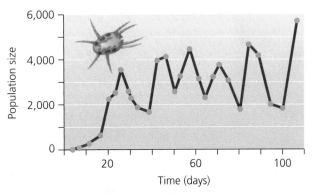

(b) Mite, *Eotetranychus sexmaculatus*

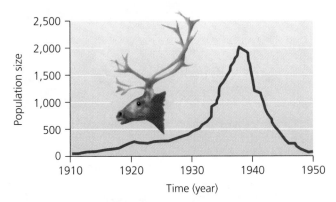

(d) St. Paul reindeer, *Rangifer tarandus*

FIGURE 3.17 ▲ Population growth in nature often departs from the stereotypical logistic growth curve, and it can do so in several fundamental ways. Yeast cells from an early lab experiment show logistic growth **(a)** that closely matches the theoretical model. Some organisms, such as the mite shown in **(b)**, show cycles in which population fluctuates indefinitely above and below the carrying capacity. Population oscillations can also dampen, lessening in intensity and eventually stabilizing at carrying capacity **(c)**, as in a lab experiment with the stored-product beetle. Populations that rise too fast and deplete resources may crash just as suddenly **(d)**, such as the population of reindeer introduced to the Bering Sea island of St. Paul. Data from Pearl, R., 1927. The growth of populations. *Quarterly Review of Biology* 2: 532–548, (a); Huffaker, C.B., 1958. Experimental studies on predation: Dispersion factors and predator-prey oscillations. *Hilgardia* 27: 343–383, Figure 7. Copyright 1958 by the Regents of the University of California. (b); Utida, S., 1967. Damped oscillation of population density at equilibrium. *Researches on Population Ecology* 9: 1–9, (c); Adapted from Scheffer, Victor B., 1951. The rise and fall of a reindeer herd. *The Scientific Monthly* 73: 356–362, Fig. 1. Reprinted with permission from AAAS. (d).

cycles that become less pronounced with time and converge on the carrying capacity. Others may overshoot the carrying capacity and then crash, destined either for extinction or recovery (**FIGURE 3.17**).

Carrying capacities can change

Because environments are complex and ever-changing, carrying capacity can vary. If a fire destroys a forest, for example, the carrying capacities for most forest animals will decline, whereas those for certain species that benefit from fire will increase. Our own species has proved capable of intentionally altering our environment so as to reduce environmental resistance and raise our carrying capacity. When our ancestors began to build shelters and use fire for heating and cooking, they reduced the environmental resistance of areas with cold climates and were able to expand into new territory. As limiting factors are overcome (through development of new technologies or through natural environmental change), the carrying capacity for a species may increase. People have managed so far to increase the planet's carrying capacity for our species, but we have done so by appropriating immense proportions of the planet's resources. In the process, we have reduced the carrying capacities for countless other organisms and have called into question our own long-term survival.

WEIGHING THE ISSUES

Carrying Capacity and Human Population Growth As we have seen (pp. 3–4), the global human population has risen to 6.9 billion, and we have far exceeded our planet's historic carrying capacity for people. What factors increased Earth's carrying capacity for us? Are there limiting factors for the human population? What might they be? Do you think we can keep raising our carrying capacity in the future? Might Earth's carrying capacity for us decrease?

Reproductive strategies vary among species

Limiting factors from an organism's environment provide only half the story of population regulation. The other half comes from the attributes of the organism itself. For example, organisms differ in their *biotic potential*, or capacity to produce offspring. A fish with a short gestation period that lays thousands of eggs at a time has high biotic potential, whereas a whale with a long gestation period that gives birth to a single calf at a time has low biotic potential.

Giraffes, elephants, humans, and other large animals with low biotic potential produce relatively few offspring during their lifetimes. Species that take this approach to reproduction require a long time to gestate and raise each offspring, but the considerable energy and resources they devote to caring for and protecting them helps give these few offspring a high likelihood of survival. Such species are said to be **K-selected** (so named because their populations tend to stabilize over time near carrying capacity, commonly abbreviated as *K*). Because their populations stay close to carrying capacity, these organisms must compete to hold their own in a crowded world. In these species, natural selection favors individuals that invest in producing offspring of high quality that can be good competitors.

In contrast, species that are **r-selected** have high biotic potential and devote their energy and resources to producing as many offspring as possible in a relatively short time. Their offspring do not require parental care after birth, so r-strategists simply leave their survival to chance. The abbreviation *r* denotes the per capita rate at which a population increases in the absence of limiting factors. Population sizes of r-selected species fluctuate greatly, such that they are often well below carrying capacity. This is why natural selection in these species favors traits that lead to rapid population growth. Many fish, plants, frogs, insects, and others are r-selected. The golden toad was one example. Each adult female laid 200–400 eggs, and its tadpoles spent just 5 weeks unsupervised in the breeding pools metamorphosing into adults.

However, it is important to note that *these are two extremes on a continuum* and that most species fall somewhere between the extremes of r-selected and K-selected species. Moreover, many organisms show combinations of traits that do not clearly correspond to a place on the continuum. A redwood tree (*Sequoia sempervirens*), for instance, is large and long-lived, yet it produces many small seeds and offers no parental care.

Changes in populations influence the composition of communities

In the late 1980s, the golden toad and the harlequin frog were the most diligently studied species affected by changing environmental conditions in the Costa Rican cloud forest. However, once scientists began looking at populations of other species at Monteverde, they began to notice further troubling changes. By the early 1990s, not only had golden toads, harlequin frogs, and other organisms been pushed from their cloud-forest habitat into apparent extinction, but also many species from lower, drier habitats had begun to appear at Monteverde. These immigrants included species tolerant of drier conditions, such as blue-crowned motmots (*Momotus momota*) and brown jays (*Cyanocorax morio*). By the year 2000, 15 dry-forest species had moved into the cloud forest and begun to breed.

Meanwhile, population sizes of several cloud-forest bird species declined. After 1987, 20 of 50 frog species vanished from one part of Monteverde, and ecologists later reported more disappearances, including those of two lizards native to the cloud forest. Scientists hypothesized that the climatic trends that researchers were documenting (see **The Science behind the Story**, pp. 68–69) were causing population fluctuations and disease outbreaks and were unleashing changes in the composition of the community.

CONSERVING BIODIVERSITY

Changes in populations and communities have been taking place naturally as long as life has existed, but today human development, resource extraction, and population pressure are speeding the rate of change and altering the types of change. Science is crucial to helping us understand the ways we modify our environment. However, the impacts that threaten biodiversity have complex social, economic, and political roots, and environmental scientists appreciate that we must understand these aspects if we are to develop solutions.

Fortunately, millions of people around the world are already taking action to safeguard biodiversity and the ecological and evolutionary processes that make Earth such a unique place. We will explore these efforts more fully in our discussion of biodiversity and conservation biology in Chapter 11. For now, let us see how Costa Ricans have been confronting the challenges to their nation's biodiversity—because their actions so far show what even a small country of modest means can do.

Social and economic factors affect species and communities

Many of the threats to Costa Rica's species and ecological communities result from past economic and social forces whose influences are still evident. European immigrants and their descendants viewed Costa Rica's lush forests as an obstacle to agricultural development, and timber companies saw them simply as a source of wood products. Costa Rica's leading agricultural products have long included beef and bananas, whose production and cultivation require extensive environmental modification.

In the half-century after 1945, the country's population grew from 860,000 to 3.34 million, and the percentage of land devoted to pasture increased from 12% to 33%. With much of the formerly forested land converted to agriculture, the proportion of the country covered by forest decreased from 80% to 25%. In 1991, Costa Rica was losing its forests faster than any other nation in the world—nearly 140 ha (350 acres) per day. As a result, populations of most species were declining, and some were becoming endangered (**FIGURE 3.18**). As was the case in the United States, few people foresaw the need to conserve biological resources until it became clear that they were being rapidly lost.

(a) Green sea turtle

(b) Red-backed squirrel monkey

FIGURE 3.18 ▲ Costa Rica is home to a number of species classified as globally threatened or endangered. The green sea turtle, *Chelonia mydas* (a), occurs throughout the world's oceans and lays eggs on beaches in Costa Rica and elsewhere, but it has undergone steep population declines. The red-backed squirrel monkey, *Saimiri oerstedii* (b), is endemic to a tiny area in Costa Rica and Panama; its small geographic range makes it vulnerable to forest loss.

Costa Rica took steps to protect its environment

During the 1950s, a group of Quakers, Christian pacifists who opposed the U.S. military draft, emigrated from Alabama to Costa Rica and founded the village of Monteverde. The Quakers ran dairy farms to produce milk and cheese, but they also set aside one-third of their land to remain as forest. The Quakers' efforts, along with contributions from international conservation organizations, provided the beginnings of what is today the Monteverde Cloud Forest Reserve. This privately managed 10,500-ha (26,000-acre) reserve was established in 1972 to protect the forest and its populations of 2,500 plant species, 400 bird species, 500 butterfly species, 100 mammal species, and 120 reptile and amphibian species, including the golden toad.

In 1970, the Costa Rican government and international representatives came together to create the country's first national parks and protected areas. The first parks centered on areas of spectacular scenery, such as Poas Volcano National Park. Santa Rosa National Park encompassed valuable tropical dry forest, Tortuguero National Park protected essential nesting beaches for the green sea turtle (see Figure 3.18a), and Cahuita National Park was meant to safeguard a prominent coral reef system. Initially the government gave the parks little real support. According to Costa Rican conservationist Mario Boza, in their early years the parks were granted only five guards, one vehicle, and no funding.

Today government support is stronger. More than 12% of the nation's area lies within national parks, and an additional 13% is in other types of protected reserves. No other nation on Earth protects over a quarter of its land area in this way. Costa Ricans, along with international biologists, are working to protect endangered species and recover their populations.

Costa Rica and its citizens are now reaping the benefits of their conservation efforts—not only ecological benefits, but economic ones as well. Because of its parks and its reputation for conservation, tourists from around the world visit Costa Rica, a phenomenon called **ecotourism** (**FIGURE 3.19**). Ecotourism draws more than 1 million visitors to Costa Rica each year, provides thousands of jobs to Costa Ricans, and pumps a billion dollars per year into the country's economy.

FIGURE 3.19 ◀ Costa Rica has protected a wide array of its diverse natural areas. This protection has stimulated the nation's economy through ecotourism. Here, visitors experience a walkway through the forest canopy in one of the nation's parks.

How Best to Conserve Biodiversity?

Most people view national parks and ecotourism as excellent ways to help keep ecological systems intact. Yet the golden toad went extinct despite living within a reserve established to protect it, and climate change and disease pay no heed to park boundaries. What lessons can we learn from this about the conservation of biodiversity? Are parks and preserves sufficient? What other approaches might we pursue to save declining species?

It remains to be seen how effectively ecotourism can help preserve natural systems in Costa Rica in the long term. As forests outside the parks disappear, the parks are beginning to suffer from illegal hunting and timber extraction. Conservationists say the parks are still underprotected and underfunded. Ecotourism will likely need to generate still more money to preserve habitat, protect endangered species, and restore altered communities to their former condition. Restoration is a step beyond preservation, and it is being carried out in Costa Rica's Guanacaste Province, where scientists are restoring dry tropical forest from grazed pasture. The restoration of ecological communities is one phenomenon we will examine in our next chapter, as we shift from populations to communities.

➤ CONCLUSION

The golden toad and other organisms of the Monteverde cloud forest have helped illuminate the fundamentals of evolution and population ecology that are integral to environmental science. The evolutionary processes of natural selection, speciation, and extinction help determine Earth's biodiversity. Understanding how ecological processes function at the population level is crucial to protecting biodiversity threatened by the mass extinction event that many biologists maintain is now underway. Population ecology also informs the study of human populations, another key endeavor in environmental science, and one that we take up in Chapter 8.

REVIEWING OBJECTIVES

You should now be able to:

EXPLAIN THE PROCESS OF NATURAL SELECTION AND CITE EVIDENCE FOR THIS PROCESS

- Because organisms produce excess young, individuals vary in their traits, and many traits are inherited, some individuals will prove better at surviving and reproducing. Their genes will be passed on and become more prominent in future generations. (pp. 52–55)
- Mutations and recombination provide the genetic variation for natural selection. (p. 53)
- We have produced our pets, farm animals, and crop plants through artificial selection. (p. 55)

DESCRIBE THE WAYS IN WHICH EVOLUTION RESULTS IN BIODIVERSITY

- Natural selection can act as a diversifying force as species adapt to their environments in myriad ways. (pp. 54–55)
- Speciation by geographic isolation (and other means) produces new species. (pp. 56–57)
- The branching patterns of phylogenetic trees reflect the historical pattern by which lineages of organisms have diverged. (p. 57)
- The fossil record informs us about life's history (pp. 57–58)

DISCUSS REASONS FOR SPECIES EXTINCTION AND MASS EXTINCTION EVENTS

- Extinction may occur when species that are highly specialized or that have small populations encounter rapid environmental change. (pp. 58–59)

- Earth's life has experienced five known episodes of mass extinction, which were due to asteroid impact and possibly volcanism and other factors. (pp. 59–61)
- Today, human impact may be initiating a sixth mass extinction. (pp. 60–61)

LIST THE LEVELS OF ECOLOGICAL ORGANIZATION

- Ecologists study phenomena on the organismal, population, community, and ecosystem levels—and, increasingly, at the level of the biosphere. (pp. 61–62)
- Habitat, niche, and specialization are important ecological concepts. (pp. 62–63)

OUTLINE THE CHARACTERISTICS OF POPULATIONS THAT HELP PREDICT POPULATION GROWTH

- Populations are characterized by population size, population density, population distribution, sex ratio, and age structure. (pp. 63–65)
- Birth and death rates, as well as immigration and emigration, determine how a population will grow or decline. (pp. 65–66)

ASSESS LOGISTIC GROWTH, CARRYING CAPACITY, LIMITING FACTORS, AND OTHER FUNDAMENTAL CONCEPTS IN POPULATION ECOLOGY

- Populations unrestrained by limiting factors will undergo exponential growth until they meet environmental resistance. (pp. 66–67)

- Logistic growth describes the effects of density dependence; growth slows as population size increases, and population size levels off at a carrying capacity. (p. 67)
- Carrying capacity is the maximum size a population can attain in a given environment. (pp. 67, 70)
- K-selection and r-selection describe theoretical extremes in how organisms can allocate growth and reproduction. (p. 71)

IDENTIFY EFFORTS AND CHALLENGES INVOLVED IN THE CONSERVATION OF BIODIVERSITY

- Social and economic factors influence our impacts on natural systems in complex ways. (pp. 71–72)
- Extensive efforts to protect and restore species and habitats are needed to prevent further erosion of biodiversity. (pp. 72–73)

TESTING YOUR COMPREHENSION

1. Explain the premises and logic that support the concept of natural selection.
2. Describe two examples of evidence for natural selection.
3. How does allopatric speciation occur?
4. Name three organisms that have become extinct, and give a probable reason for each extinction.
5. What is the difference between a species and a population? Between a population and a community?
6. Contrast the concepts of habitat and niche.
7. List and describe each of the five major population characteristics discussed in this chapter. Explain how each shapes population dynamics.
8. Can any species undergo exponential growth forever? Explain your answer.
9. Describe how limiting factors relate to carrying capacity.
10. Explain the difference between K-selected species and r-selected species. Can you think of examples of each that were not mentioned in the chapter?

SEEKING SOLUTIONS

1. In what ways has artificial selection changed people's quality of life? Give examples. Can you imagine a way in which artificial selection could be used to improve our quality of life further? Can you imagine a way it could be used to reduce our environmental impact?
2. What types of species are most vulnerable to extinction, and what kinds of factors threaten them? What species in your region are threatened with extinction? What reasons lie behind their endangerment?
3. Do you think the human species can continue raising its global carrying capacity? How so, or why not? Do you think we *should* try to keep raising our carrying capacity? Why or why not?
4. Describe the evidence suggesting that changes in temperature and precipitation led to the extinction of the golden toad and to population crashes for other amphibians at Monteverde. Why do scientists also think that disease played a role? What might be done to help make future such declines in other species less likely?
5. What are the advantages of ecotourism for a country like Costa Rica? Can you think of any disadvantages?
6. **THINK IT THROUGH** You are a population ecologist studying animals in a national park, and your government is asking for advice on how to focus its limited conservation funds. How would you rate the following three species, from most vulnerable (and thus most in need of conservation attention) to least vulnerable? Give reasons for your choices.
   ▶ A bird with an even (1:1) sex ratio that is a habitat generalist
   ▶ A salamander endemic to the park that lives in high-elevation forest
   ▶ A fish that specializes on a few types of invertebrate prey and has a large population size

CALCULATING ECOLOGICAL FOOTPRINTS

Americans love their coffee. In 2009, coffee consumption in the United States topped 2.8 billion pounds (out of nearly 16 billion pounds produced globally). Next to petroleum, coffee is the most valuable (legal) commodity on the world market, and the United States is its leading importer. Given this information, estimate the coffee consumption rates in the table.

Most coffee is produced in large tropical plantations, where coffee is the only tree species and is grown in full sun where natural forests have been cut. However, approximately 2% of coffee is produced in smaller groves where coffee trees are intermingled with other species under a partial rainforest canopy. These *shade-grown* coffee forests maintain greater habitat diversity for tropical rainforest wildlife.

	Population	Pounds of coffee per day	Pounds of coffee per year
You (or the average American)	1	0.025	9.0
Your class			
Your hometown			
Your state			
United States			

Data from O'Brien, T.G., and M.F. Kinnaird, 2003. Caffeine and conservation. *Science* 300:587; and International Coffee Organization.

1. What percentage of global coffee production is consumed in the United States? If U.S. coffee drinkers consumed only shade-grown coffee, how much would shade-grown production need to increase to meet demand?

2. How much extra would you be willing to pay per pound for shade-grown coffee as opposed to standard coffee, if you knew that your money would help to prevent habitat loss or extinction for animals such as resplendent quetzals, endangered Costa Rican squirrel monkeys, and the many songbirds that migrate between Latin America and North America?

3. If everyone in the United States were willing to pay as much extra per pound for shade-grown coffee as you are, how much additional money would that provide for conservation of biodiversity in the tropics each year?

Mastering**ENVIRONMENTALSCIENCE**™

Go to **www.masteringenvironmentalscience.com** for practice quizzes, Pearson eText, videos, current events, and more.

Marsh community along the shore of Lake Ontario

4 SPECIES INTERACTIONS AND COMMUNITY ECOLOGY

UPON COMPLETING THIS CHAPTER, YOU WILL BE ABLE TO:

- Compare and contrast the major types of species interactions
- Characterize feeding relationships and energy flow, using them to construct trophic levels and food webs
- Distinguish characteristics of a keystone species
- Characterize succession, community change, and the debate over the nature of communities

- Perceive and predict the potential impacts of invasive species in communities
- Explain the goals and methods of restoration ecology
- Describe and illustrate the terrestrial biomes of the world

CENTRAL **CASE STUDY**

Black and White, and Spread All Over: Zebra Mussels Invade the Great Lakes

"We are seeing changes in the Great Lakes that are more rapid and more destructive than any time in [their] history."

—Andy Buchsbaum, National Wildlife Federation

"When you tear away the bottom of the food chain, everything that is above it is going to be disrupted."

—Tom Nalepa, National Oceanic and Atmospheric Administration

CANADA

UNITED STATES

Great Lakes

Atlantic Ocean

Pacific Ocean

SOUTH AMERICA

Things had been looking up for the Great Lakes. The pollution-fouled waters of Lake Erie and the other Great Lakes shared by Canada and the United States had become gradually cleaner in the years following the Clean Water Act of 1970 and a binational agreement in 1972. As government regulation brought industrial discharges under control, people once again began to use the lakes for recreation, and populations of fish rebounded.

Then the zebra mussel arrived. Black-and-white-striped shellfish the size of a dime, zebra mussels attach to hard surfaces and feed on algae by filtering water through their gills. This mollusk, given the scientific name *Dreissena polymorpha*, is native to the Caspian Sea, Black Sea, and Azov Sea in western Asia and eastern Europe. In 1988, it was discovered in North American waters at Lake St. Clair, which connects Lake Erie with Lake Huron. People had brought it to this continent by accident when ships arriving from Europe discharged ballast water containing the mussels or their larvae into the Great Lakes.

Within just two years of their discovery in Lake St. Clair, zebra mussels had multiplied and reached all five of the Great Lakes. The next year, these invaders entered New York's Hudson River to the east, and the Illinois River at Chicago to the west. From the Illinois

Aggregation of zebra mussels

River and its canals, they soon reached the Mississippi River, giving them access to a vast watershed covering 40% of the United States. In just three more years, they spread to 19 U.S. states and two Canadian provinces. By 2010, they had colonized waters in 30 U.S. states.

How could a mussel spread so quickly? The zebra mussel's larval stage is well adapted for long-distance dispersal. Its tiny larvae drift freely for several weeks, traveling as far as the currents take them. Adults that attach themselves to boats and ships may be transported from one place to another, even to isolated lakes and ponds well away from major rivers. Moreover, in North America the mussels encountered none of the particular species of predators, competitors, and parasites that had evolved to limit their population growth in the Old World.

Why all the fuss? Zebra mussels clog up water intake pipes at factories, power plants, municipal water supplies, and wastewater treatment facilities (**FIGURE 4.1A**). At one Michigan power plant, workers counted 700,000 mussels per square meter of pipe surface. Great densities of these organisms can damage boat engines, degrade docks, foul fishing gear, and sink buoys that ships use for navigation. Through such impacts, zebra mussels cost Great Lakes economies an estimated $5 billion in the first decade

CHAPTER 4 Species Interactions and Community Ecology

77

(a) Clogging a pipe

(b) Suffocating native clams

FIGURE 4.1 ▲ Zebra mussels clog water intake pipes **(a)** of power plants and industrial facilities. They also foul, starve, and suffocate native clams **(b)** by adhering to their shells and sealing them shut.

of the invasion, and they continue to impose costs of hundreds of millions of dollars each year.

Zebra mussels also have severe impacts on the ecological systems they invade. They eat primarily **phytoplankton**: microscopic photosynthetic algae, protists, and cyanobacteria that drift in open water. Because each mussel filters a liter or more of water every day, they consume so much phytoplankton that they can deplete populations. Phytoplankton is the foundation of the Great Lakes food web, so its depletion is bad news for **zooplankton**, the tiny aquatic animals and protists that eat phytoplankton—and for the fish that eat both. Researchers are finding that water bodies with zebra mussels contain fewer zooplankton and open-water fish than water bodies without them. Zebra mussels also suffocate native mollusks by attaching to their shells (**FIGURE 4.1B**).

However, zebra mussels also benefit some bottom-feeding invertebrates and fish—at least in the short term. By filtering algae and organic matter from open water and depositing nutrients in feces that sink, they shift the community's nutrient balance to the bottom and benefit species that feed there. Once they clear the water, sunlight penetrates more deeply, spurring the growth of large-leafed underwater plants and algae. In the long term, however, eutrophication (pp. 114, 422) may ensue, bringing harm to the system.

In the past several years, scientists have documented a surprising new twist: One invader is being displaced by another. The quagga mussel (*Dreissena bugensis*), a close relative of the zebra mussel from Ukraine, is spreading through the Great Lakes and beyond. This species, named after an extinct zebra-like animal, appears to be replacing the zebra mussel in many locations. Scientists are only beginning to understand what consequences this shift may have for ecological communities.

SPECIES INTERACTIONS

Interactions among species are the threads in the fabric of communities. By interacting with many species in a variety of ways, zebra mussels have set in motion an array of changes in the ecological communities they have invaded. To understand why invasive species cause so much disruption, we must first look at how species naturally interact. Ecologists organize species interactions into several fundamental categories (**TABLE 4.1**).

Competition can occur when resources are limited

When multiple organisms seek the same limited resource, their relationship is said to be one of **competition**. Competing organisms do not usually fight with one another directly and

TABLE 4.1 Effects of Species Interactions on Their Participants

Type of interaction	Effect on species 1	Effect on species 2
Mutualism	+	+
Commensalism	+	0
Predation, parasitism, herbivory	+	−
Neutralism	0	0
Amensalism	−	0
Competition	−	−

"+" denotes a positive effect; "−" denotes a negative effect; "0" denotes no effect.

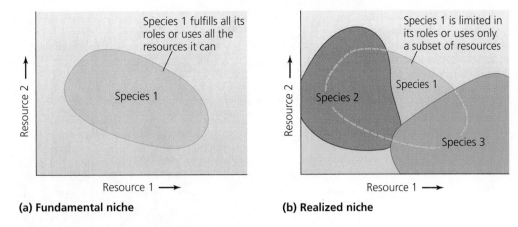

(a) Fundamental niche   **(b) Realized niche**

FIGURE 4.2 ▲ An organism facing competition may be forced to play a more limited ecological role or use fewer resources than it would in the absence of its competitor. With no competitors, an organism can exploit its full fundamental niche **(a)**. But when competitors restrict what an organism can do or what resources it can use, the organism is limited to a realized niche **(b)**, which covers only a subset of its fundamental niche. In considering niches, ecologists have traditionally focused on competition, but they now recognize that other species interactions also are influential.

physically. Competition is generally more subtle and indirect, taking place as organisms vie with one another to procure resources. The resources for which organisms compete can include food, water, space, shelter, mates, sunlight, and more. Competitive interactions can take place among members of the same species (**intraspecific competition**) or among members of two or more different species (**interspecific competition**).

We have already discussed intraspecific competition in Chapter 3, without naming it as such. Recall that density dependence (p. 67) limits the growth of a population; individuals of the same species compete for limited resources, such that competition is more intense when there are more individuals per unit area (denser populations). Thus, intraspecific competition is really a population-level phenomenon.

In contrast, interspecific competition can strongly affect the composition of communities. Interspecific competition can give rise to different types of outcomes. If one species is a very effective competitor, it may exclude another species from resource use entirely. This outcome, called **competitive exclusion**, occurred in parts of the Great Lakes as zebra mussels displaced native mussels—and it is happening now as quagga mussels begin to displace zebra mussels.

Alternatively, if neither competitor fully excludes the other, the species may continue to live side by side. This result, called **species coexistence**, may produce a stable point of equilibrium, in which the relative population sizes of each remain fairly constant through time.

Coexisting species that use the same resources tend to adjust to their competitors to minimize competition with them. Individuals can do this by changing their behavior so as to use only a portion of the total array of resources they are capable of using. In such cases, individuals do not fulfill their entire *niche*. Recall from Chapter 3 (p. 63) that a species' niche reflects its use of resources and its functional role in a community, including its habitat use, food consumption, and other attributes—a kind of multidimensional summary of everything an organism does. The full niche of a species is

called its **fundamental niche** (**FIGURE 4.2A**). An individual that plays only part of its role or uses only part of its resources because of competition or other types of species interactions is said to display a **realized niche** (**FIGURE 4.2B**), the portion of its fundamental niche that is actually "realized," or fulfilled. The quagga mussel can occupy a wider range of water conditions and substrates than the zebra mussel, so it is thought to have a larger fundamental niche. It also appears to be reducing the zebra mussel's realized niche, as it displaces the zebra mussel in many areas.

Species experience similar adjustments over evolutionary time. Over many generations, the process of natural selection (pp. 52–55) may respond to competition by favoring individuals that use slightly different resources or that use shared resources in different ways. If two bird species eat the same type of seeds, individuals that prefer eating larger or smaller seeds might minimize competition and thereby survive and reproduce more effectively. If the seed-eating tendencies are inherited, then these preferences may be passed on to offspring, and over time natural selection might drive one species to specialize on larger seeds and the other to specialize on smaller seeds. Natural selection might also favor one bird species becoming more active in the morning and the other more active in the evening, thus minimizing interference. This process is called **resource partitioning** because the species partition, or divide, the resources they use in common by specializing in different ways (**FIGURE 4.3**).

Resource partitioning can lead to **character displacement**, in which competing species come to diverge in their physical characteristics because of the evolution of traits best suited to the range of resources they use. For birds that specialize on larger seeds, natural selection may favor the evolution of larger bills that enable them to make best use of this resource, whereas for birds specializing on smaller seeds, smaller bills may be favored. This is precisely what extensive recent research has revealed about the finches from the Galápagos Islands that were first described by Charles Darwin (p. 52).

FIGURE 4.3 ▶ When species compete, they tend to partition resources, each specializing on a slightly different resource or way of attaining a shared resource. Various types of birds—including the woodpeckers, creeper, and nuthatch shown here—feed on insects from tree trunks, but they use different portions of the trunk, seeking different foods in different ways.

White-breasted nuthatch climbs down trunk looking for insects

Yellow-bellied sapsucker drills rows of holes and consumes sap and insects stuck in sap

Pileated woodpecker digs deeply into wood to find large insects

Brown creeper climbs up trunk looking for tiny insects

Several types of interactions are exploitative

In competitive interactions, each participant exerts a negative effect on other participants, because each takes resources the others could have used. This is reflected in the two minus signs shown for competition in Table 4.1. In other types of interactions, some participants benefit while others are harmed (note the +/– interactions in Table 4.1). We can think of interactions in which one member exploits another for its own gain as *exploitative* interactions. Such interactions include predation, parasitism, and herbivory.

Predators kill and consume prey

Every living thing needs to procure food, and for most animals, that means eating other living organisms. **Predation** is the process by which individuals of one species—the *predator*—hunt, capture, kill, and consume individuals of another species, the *prey* (**FIGURE 4.4**). Along with competition, predation has traditionally been viewed as one of the primary organizing forces in community ecology. Interactions among predators and prey structure the food webs that we will examine shortly, and they influence community composition by helping to determine the relative abundance of predators and prey.

Zebra mussel predation on phytoplankton has reduced phytoplankton populations by up to 90%, according to studies in the Great Lakes and Hudson River. Zebra mussels also consume the smaller types of zooplankton, and this predation

has caused zooplankton populations to decline by up to 70% in Lake Erie and the Hudson River. (Also contributing to this decline is the fact that the mussels, by eating phytoplankton, compete with zooplankton for food.) Meanwhile, zebra mussels do not eat certain cyanobacteria (p. 31), so concentrations of these cyanobacteria rise in lakes with zebra mussels. Most predators are also prey, however, and zebra mussels have become a food source for a number of North American species since their introduction. These include

FIGURE 4.4 ▼ A fire-bellied snake (*Liophis epinephalus*) devours a frog in the Monteverde cloud forest we studied in Chapter 3. Predation has ecological and evolutionary consequences for both prey and predator.

diving ducks, muskrats, crayfish, flounder, sturgeon, eels, and several types of fish with grinding teeth, such as carp and freshwater drum.

Predation can sometimes drive cyclical population dynamics. An increase in the population size of prey creates more food for predators—and the predators may survive and reproduce more effectively as a result. As the predator population rises, their intensified predation drives down the population of prey. Decreased numbers of prey in turn cause some predators to starve, so that the predator population then declines. This allows the prey population to begin rising again, starting the cycle anew (see Figure 5.2a, p. 110). Most natural systems involve so many complex factors that such simple cycles do not last long, but in some cases we see extended cycles (**FIGURE 4.5**).

Predation also has evolutionary ramifications. Individual predators that are more adept at capturing prey will likely live longer, healthier lives and be better able to provide for their offspring than will less adept individuals. Thus, natural selection on individuals within a predator species leads to the evolution of adaptations (p. 53) that make them better hunters. Prey face an even stronger selective pressure—the risk of immediate death. For this reason, predation pressure has driven the evolution of an elaborate array of defenses against being eaten (**FIGURE 4.6**).

Parasites exploit living hosts

Organisms can exploit other organisms without killing them. **Parasitism** is a relationship in which one organism, the *parasite*, depends on another, the *host*, for nourishment or some other benefit while simultaneously doing the host harm. Unlike predation, parasitism usually does not result in an organism's immediate death, although it sometimes contributes to the host's eventual death.

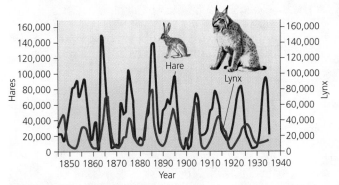

FIGURE 4.5 ▲ Predator–prey systems sometimes show paired cycles, in which increases and decreases in the population of one organism apparently drive increases and decreases in the population of the other. A classic case is that of hares and the lynx that prey on them in Canada. The data come from fur-trapping records of the Hudson Bay Company and represent numbers of each animal trapped. Data from MacLulich, D.A., 1937. Fluctuation in the numbers of varying hare *(Lepus americanus). University of Toronto Studies in Biology Series 43*, Toronto: University of Toronto Press.

Some types of parasites are free-living "social parasites" and come into contact with their hosts infrequently. For example, the cuckoos of Eurasia and the cowbirds of the Americas lay their eggs in other birds' nests and let the host species raise the parasite's young. Other parasites live inside their hosts. These parasites include disease pathogens, such as the protists that cause malaria and dysentery, as well as animals such as tapeworms that live in their hosts' digestive tracts. Still other parasites live on the exterior of their hosts, such as the ticks that attach themselves to their hosts' skin. The sea lamprey *(Petromyzon marinus)* is a parasitic vertebrate that grasps the bodies of fish with a suction-cup mouth and a rasping tongue, sucking blood from the fish for days

FIGURE 4.6 ▼ Natural selection to avoid predation has resulted in many fabulous adaptations. Some prey hide from predators by *crypsis*, or camouflage, such as this gecko **(a)** on tree bark. Other prey are brightly colored to warn predators that they are toxic, distasteful, or dangerous, such as this yellow jacket **(b)**. Still others fool predators with mimicry. This caterpillar **(c)**, when disturbed, swells and curves its tail end and shows false eyespots to look like a snake's head.

(a) Cryptic coloration

(b) Warning coloration

(c) Mimicry

FIGURE 4.7 ▲ With its suction-cup mouth and rasping tongue, the parasitic sea lamprey (*Petromyzon marinus*) attaches itself to fish and sucks their blood for days or weeks, sometimes killing the fish. Sea lampreys wreaked havoc on Great Lakes fisheries after entering the lakes through human-built canals.

or weeks (**FIGURE 4.7**). Sea lampreys invaded the Great Lakes from the Atlantic Ocean after people dug canals to connect the lakes for shipping, and the lampreys soon devastated economically important fisheries of chubs, lake herring, whitefish, and lake trout. Since the 1950s, U.S. and Canadian fisheries managers have reduced lamprey populations by applying chemicals that selectively kill lamprey larvae.

Many insects parasitize other insects, often killing them in the process, and are called *parasitoids*. Various species of parasitoid wasps lay eggs on caterpillars. When the eggs hatch, the wasp larvae burrow into the caterpillar's tissues and slowly consume them. The wasp larvae metamorphose into adults and fly from the body of the dying caterpillar.

Just as predators and prey evolve in response to one another, so do parasites and hosts, in a process termed *coevolution*. **Coevolution** describes a long-term reciprocal process in which two (or more) types of organisms repeatedly respond through natural selection to the other's adaptations. This seesawing process can potentially occur with any type of species interaction. Hosts and parasites often become locked in a duel of escalating adaptations, an *evolutionary arms race*. Like rival nations racing to stay ahead of one another in military technology, host and parasite may repeatedly evolve new responses to the other's latest advance. In the long run, though, it may not be in a parasite's best interest to become too harmful to its host. A parasite might leave more offspring in the next generation—and thus be favored by natural selection—if it allows its host to live longer.

Herbivores exploit plants

One of the most common exploitative interactions is **herbivory**, which occurs when animals feed on the tissues of plants. Insects that feed on plants are the most widespread type of herbivore; just about every plant in the world is attacked by some type of insect (**FIGURE 4.8**). In most cases, herbivory does not kill a plant outright but may affect its growth and reproduction.

FIGURE 4.8 ▲ Herbivory is a common way to make a living. The world holds many thousands, and perhaps millions, of species of plant-eating insects, such as this larva (caterpillar) of the death's head hawk moth (*Acherontia atropos*) of western Europe.

Like animal prey, plants have evolved an impressive arsenal of defenses against the animals that feed on them. Many plants produce chemicals that are toxic or distasteful to herbivores. Others arm themselves with thorns, spines, or irritating hairs. In response, herbivores evolve ways to overcome these defenses, and the plant and the animal may embark on an evolutionary arms race.

Some plants go a step further and recruit certain animals as allies to assist in their defense. Such plants may encourage ants to take up residence by providing swelled stems for the ants to nest in, or nectar-bearing structures for them to feed from. The ants protect the plant in return by attacking other insects that land or crawl on it. Other plants respond to herbivory by releasing volatile chemicals when they are bitten or pierced. The airborne chemicals attract predatory insects that may attack the herbivore. Such cooperative strategies—essentially trading food for protection—are examples of our next type of species interaction, mutualism.

Mutualists help one another

Unlike exploitative interactions, **mutualism** is a relationship in which two or more species benefit from interaction with one another. Generally each partner provides some resource or service that the other needs.

Many mutualistic relationships—like many parasitic relationships—occur between organisms that live in close physical contact. Such physically close association, whether

mutualistic or parasitic, is called **symbiosis**. (Indeed, biologists hypothesize that many mutualistic associations evolved from parasitic ones.) Thousands of terrestrial plant species depend on mutualisms with fungi; plant roots and some fungi together can form symbiotic associations called mycorrhizae. In these relationships, the plant provides energy and protection to the fungus, while the fungus helps the plant absorb nutrients from the soil. In the ocean, coral polyps, the tiny animals that build coral reefs, share beneficial arrangements with algae known as zooxanthellae (p. 439). The coral provide housing and nutrients for the algae in exchange for a steady supply of food that the algae produce through photosynthesis (pp. 31–32).

You, too, are engaged in a symbiotic mutualism. Your digestive tract is filled with hundreds of species of bacteria and other microbes that help you digest food and carry out other bodily functions—microbes that you are providing a place to live. Without these mutualistic microbes, none of us would likely survive for long.

Not all mutualists live in close proximity. **Pollination** (**FIGURE 4.9**), an interaction of key significance to agriculture and our food supply (pp. 260–261), involves free-living organisms that may encounter each other only once. Bees, birds, bats, and other creatures transfer pollen (male sex cells) from one flower to the ova (female cells) of another, fertilizing the female egg, which subsequently grows into a fruit. Most pollinating animals visit flowers for their nectar, a reward the plant uses to entice them. The pollinators receive food, and the plants are pollinated and reproduce. Various types of bees pollinate 73% of our crops, one expert has estimated—from soybeans to potatoes to tomatoes to beans to cabbage to oranges. This is why mysterious die-offs of honeybees today (p. 261) are causing so much concern.

Some interactions have no effect on some participants

Amensalism is a relationship in which one organism is harmed and the other is unaffected. Amensalism is difficult to confirm because it is hard to prove that the organism doing the harm is not in fact besting a competitor for a resource. Some experts suggest that allelopathy, a phenomenon whereby certain plants release poisonous chemicals that harm others nearby, is an example of amensalism. However, allelopathy can also be viewed as one plant deploying chemicals to outcompete others for space.

In **commensalism**, one species benefits and the other is unaffected. For example, one plant may create conditions that happen to make it easier for another plant to establish and grow. Such is the case with palo verde trees in the Sonoran Desert, which provide shade and leaf litter that allow the soil beneath them to hold moisture longer, creating an area that is cooler and moister than the surrounding sun-baked ground. Young plants find it easier to germinate and grow in these conditions, so seedling cacti and other desert plants generally grow up directly beneath "nurse" trees such as palo verde. This phenomenon influences the structure and composition of ecological communities and affects how they change through time.

ECOLOGICAL COMMUNITIES

As we saw in Chapter 3 (Figure 3.10, p. 62), a **community** is an assemblage of populations of organisms living in the same area at the same time. The members of a community interact with one another in the ways described above, and the direct interactions among species often have indirect effects that ripple outward to affect other community members. The strength of interactions also varies, and together species' interactions determine the structure, function, and species composition of communities. Community ecologists are interested in which species coexist, how they relate to one another, how communities change through time, and why these patterns exist.

Energy passes among trophic levels

The interactions among members of a community are many and varied, but some of the most important involve who eats whom. As organisms feed on one another, matter and energy move through the community, from one **trophic level**, or rank in the feeding hierarchy, to another (**FIGURE 4.10**).

Producers *Producers*, or *autotrophs* ("self-feeders"), comprise the first trophic level, as we saw in Chapter 2. Terrestrial green plants, cyanobacteria, and algae capture solar energy and use

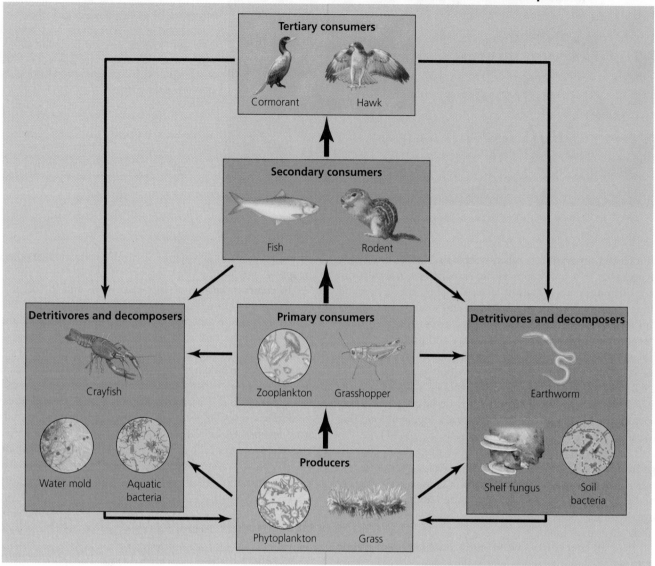

FIGURE 4.10 ▲ Ecologists organize species hierarchically by their feeding rank, or trophic level. The diagram shows aquatic (left) and terrestrial (right) examples at each level. Arrows indicate the direction of energy flow. Producers produce food by photosynthesis, primary consumers (herbivores) feed on producers, secondary consumers eat primary consumers, and tertiary consumers eat secondary consumers. Communities can have more or fewer trophic levels than in this example. Detritivores and decomposers feed on nonliving organic matter and the remains of dead organisms from all trophic levels, and they "close the loop" by returning nutrients to the soil or the water column for use by producers.

photosynthesis to produce sugars (pp. 31–32). The chemosynthetic bacteria of hot springs and deep-sea hydrothermal vents use geothermal energy in a similar way to produce food (p. 33).

Consumers Organisms that consume producers are known as *primary consumers* and comprise the second trophic level. Herbivorous grazing animals, such as deer and grasshoppers, are primary consumers. The third trophic level consists of *secondary consumers*, which prey on primary consumers. Wolves that prey on deer are considered secondary consumers, as are rodents and birds that prey on grasshoppers. Predators that feed at still higher trophic levels are known as *tertiary consumers*. Examples of tertiary consumers include hawks and owls that eat rodents that have eaten grasshoppers. Note that

most primary consumers are *herbivores* because they consume plants, whereas secondary and tertiary consumers are *carnivores* because they eat animals. Animals that eat both plant and animal food are referred to as *omnivores*.

Detritivores and decomposers Detritivores and decomposers consume nonliving organic matter. *Detritivores*, such as millipedes and soil insects, scavenge the waste products or the dead bodies of other community members. *Decomposers*, such as fungi and bacteria, break down leaf litter and other nonliving matter further into simpler constituents that can then be taken up and used by plants. These organisms enhance the topmost soil layers (pp. 227–229) and play essential roles as the community's recyclers, making nutrients from organic matter available for reuse by living members of the community.

In Great Lakes communities, phytoplankton are the main producers, floating freely and photosynthesizing with sunlight that penetrates the upper layer of the water. Zooplankton are primary consumers, feeding on the phytoplankton. Phytoplankton-eating fish are primary consumers, and zooplankton-eating fish are secondary consumers. Higher trophic levels include tertiary consumers such as larger fish and birds that feed on plankton-eating fish. (The left side of Figure 4.10 shows these relationships in a very generalized form.) Zebra mussels and quagga mussels, by eating both phytoplankton and zooplankton, function on multiple trophic levels. When an organism dies and sinks to the bottom, detritivores scavenge its tissues and decomposers recycle its nutrients.

Energy, biomass, and numbers decrease at higher trophic levels

At each trophic level, organisms use energy in cellular respiration (p. 32), and most energy ends up being given off as heat. Only a small amount of the energy is transferred to the next trophic level through predation, herbivory, or parasitism. The first trophic level (producers) contains a large amount of energy, but the second (primary consumers) contains less energy—only that amount gained from consuming producers. The third trophic level (secondary consumers) contains still less energy, and higher trophic levels (tertiary consumers) contain the least. A general rule of thumb is that each trophic level contains just 10% of the energy of the trophic level below it, although the actual proportion can vary greatly.

This pattern can be visualized as a pyramid (**FIGURE 4.11**). The pattern also tends to hold for the numbers of organisms at each trophic level; generally, fewer organisms exist at higher trophic levels than at lower ones. A grasshopper eats many plants in its lifetime, a rodent eats many grasshoppers, and a hawk eats many rodents. Thus, for every hawk in a community there must be many rodents, still more grasshoppers, and an immense number of plants. And because the difference in numbers of organisms among trophic levels tends to be large,

the same pyramid-like relationship also often holds true for **biomass**, the collective mass of living matter in a given place and time.

This pyramid-like pattern illustrates why eating at lower trophic levels—being a vegetarian rather than a meat-eater, for instance—decreases a person's ecological footprint. Each amount of meat or other animal product we eat requires the input of a considerably greater amount of plant material (see Figure 10.19, p. 268). Thus, when we eat animal products, we consume far more energy per calorie gained than when we eat plant products.

WEIGHING **THE ISSUES**

The Footprints of Our Diets What proportion of your diet would you estimate consists of meat, milk, eggs, or other animal products? Would you be willing to decrease this proportion in order to reduce your ecological footprint? Describe some other ways in which you could reduce your footprint through your food choices.

Food webs show feeding relationships and energy flow

As energy is transferred from species on lower trophic levels to species on higher trophic levels, it is said to pass up a **food chain**, a linear series of feeding relationships. Plant, grasshopper, rodent, and hawk make up a food chain—as do phytoplankton, zooplankton, fish, and fish-eating birds. Thinking in terms of food chains is conceptually useful, but in reality ecological systems are far more complex than simple linear chains. A more accurate representation of the feeding relationships in a community is a **food web**, a visual map of energy flow that uses arrows to show the many paths by which energy passes among organisms as they consume one another.

FIGURE 4.12 shows a food web from a temperate deciduous forest of eastern North America. Like virtually all diagrams of ecological systems, it is greatly simplified and leaves out the vast majority of species and interactions that occur. Note, however, that even within this simplified diagram we can pick out a number of different food chains involving different sets of species.

A Great Lakes food web would involve the phytoplankton that photosynthesize near the water's surface; zooplankton, the primary consumers that eat them; fish that eat phytoplankton and zooplankton; larger fish that eat the smaller fish; and lampreys that parasitize the fish. It would include a number of native mussels and clams and, since 1988, the zebra mussels and quagga mussels that are displacing them. It would include diving ducks that used to feed on native bivalves and now prey on the mussels.

This food web would also show that crayfish and other *benthic* (bottom-dwelling) invertebrate animals feed from the refuse of the non-native mussels. Although the mussels' waste promotes bacterial growth and disease pathogens that harm native bivalves, it also provides nutrients that nourish many

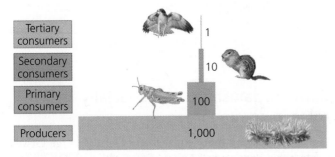

FIGURE 4.11 ▲ Organisms at lower trophic levels generally exist in far greater numbers, with greater energy content and greater biomass, than organisms at higher trophic levels. The reason is that when one organism consumes another, most energy is used up in respiration rather than in building new tissue. The example shown here is generalized; the actual shape of any given pyramid may vary greatly.

Tertiary consumers — 1
Secondary consumers — 10
Primary consumers — 100
Producers — 1,000

Caterpillar and other insects on leaves

Spider

White oak

Eastern chipmunk

Cedar waxwing

Blackberry

Beetles and other insects

Ticks

Red-bellied woodpecker

White-tailed deer

Eastern cottontail

Rat snake

Shelf fungus

Deer mouse

Grasses

American toad

Soil bacteria

Earthworm

FIGURE 4.12 ▲ Food webs represent feeding relationships in a community. This food web pertains to eastern North America's temperate deciduous forest and includes organisms on several trophic levels. In a food web diagram, arrows lead from one organism to another to indicate the direction of energy flow as a result of predation, parasitism, or herbivory. For example, an arrow leads from the grass to the cottontail rabbit to indicate that cottontails consume grasses. The arrow from the cottontail to the tick indicates that parasitic ticks derive nourishment from cottontails. Like most food web diagrams, this one is a simplification, because the actual community contains many more species and interactions than can be shown.

benthic invertebrates. Finally, the food web would include underwater plants and macroscopic algae, whose growth is enhanced as the non-native mussels clarify the water by filtering out phytoplankton, allowing sunlight to penetrate more deeply into the water column. (Jump ahead to Figure 4.16a, p. 94, for an illustration of some of these effects.)

Overall, zebra and quagga mussels alter this food web by shifting productivity from open-water regions to benthic and *littoral* (nearshore) regions. In so doing, the mussels help benthic and littoral fishes and make life harder for open-water fishes (see **THE SCIENCE BEHIND THE STORY**, pp. 88–89).

Some organisms play especially large roles in communities

"Some animals are more equal than others," George Orwell wrote in his classic novel *Animal Farm*. Although Orwell was making wry sociopolitical commentary, his remark hints at a truth in ecology. In communities, ecologists have found, some species exert greater influence than do others. A species that has strong or wide-reaching impact far out of proportion to its abundance is often called a **keystone species**. A keystone is the wedge-shaped stone at the top of an arch that is vital for holding the structure together; remove the keystone, and the arch will collapse (**FIGURE 4.13A**). In an ecological community, removal of a keystone species will have ripple effects that substantially alter the food web.

Often, large-bodied secondary or tertiary consumers near the tops of food chains are considered keystone species. Top predators control populations of herbivores, which otherwise would multiply and could greatly modify the plant community (**FIGURE 4.13B**). Thus, predators at high trophic levels can indirectly promote populations of organisms at low trophic levels by keeping species at intermediate trophic levels in check, a phenomenon ecologists refer to as a **trophic cascade**. In the United States, for example, government bounties promoted the hunting of wolves and mountain lions, which were largely exterminated by the middle of the 20th century. In the absence of these predators, unnaturally dense deer populations have overgrazed forest-floor vegetation and eliminated tree seedlings, causing major changes in forest structure.

The removal of top predators in the United States was an uncontrolled large-scale experiment with unintended consequences, but ecologists have verified the keystone species concept in controlled experiments. Classic research by marine biologist Robert Paine established that the predatory sea star *Pisaster ochraceus* has great influence on the community composition of intertidal organisms (pp. 442–443) on the Pacific coast of North America. When *Pisaster* is present in this community, species diversity is high, with various types of barnacles, mussels, and algae. When *Pisaster* is removed, the mussels it preys on become numerous and displace other species, suppressing species diversity.

More recent work off the U.S. Atlantic coast published in 2007 suggests that the reduction of shark populations by commercial fishing has allowed populations of certain skates and rays to increase, which has depressed numbers of bay scallops and other bivalves they eat.

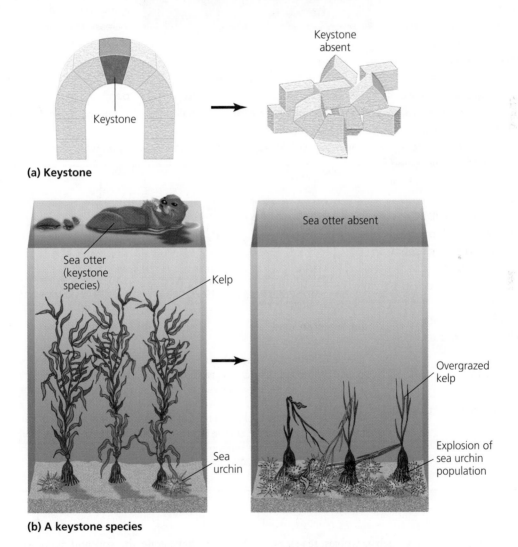

(a) Keystone

(b) A keystone species

FIGURE 4.13 ▲ A keystone is the wedge-shaped stone at the top of an arch that holds its structure together **(a)**. A keystone species, such as the sea otter, is one that exerts great influence on a community's composition and structure **(b)**. Sea otters consume sea urchins that eat kelp in marine nearshore environments of the Pacific. When otters are present, they keep urchin numbers down, allowing lush underwater forests of kelp to grow and provide habitat for many other species. When otters are absent, urchin populations increase and devour the kelp, destroying habitat and depressing species diversity. See **The Science behind the Story**, pp. 90–91.

The SCIENCE behind the Story

Determining Zebra Mussels' Impacts on Fish Communities

When zebra mussels appeared in the Great Lakes, people feared for sport fisheries and estimated that fish population declines could cost billions of dollars. The mussels would deplete the phytoplankton and zooplankton that fish depended on for food, people reasoned.

However, food webs are complicated systems, and disentangling them to infer the impacts of any one species is fraught with difficulty. Thus, even after 15 years, there was no solid evidence of widespread harm to fish populations.

So, aquatic ecologist David Strayer of the Institute of Ecosystem Studies in Millbrook, New York, joined Kathryn Hattala and Andrew Kahnle of New York State's Department of Environmental Conservation (DEC). They mined data sets on fish populations in the Hudson River, which zebra mussels had invaded in 1991.

Strayer and other scientists had been studying aspects of the community for years. Their data showed that after zebra mussels invaded the Hudson:

▸ Biomass of phytoplankton had fallen by 80%.

▸ Biomass of small zooplankton had fallen by 76%.

▸ Biomass of large zooplankton had fallen by 52%.

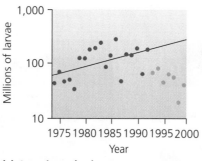

Dr. David Strayer, Institute of Ecosystem Studies, sampling aquatic invertebrates

Zebra mussels increased filter-feeding in the community 30-fold, depleting phytoplankton and small zooplankton and leaving larger zooplankton with less phytoplankton to eat. Overall, the zooplankton and invertebrate animals of the open water (which are eaten by open-water fish) declined by 70%.

However, Strayer had also found that *benthic*, or bottom-dwelling, invertebrates in shallow water (especially in the nearshore, or *littoral*, zone) had increased notably, because the mussels' shells provide habitat structure and their feces provide nutrients.

These contrasting trends in the benthic shallows and the open deep water led Strayer's team to hypothesize that zebra mussels would harm open-water fish that ate plankton but

(a) American shad

(b) Tessellated darter

Larvae of American shad **(a)**, an open-water fish, had been increasing in abundance before zebra mussels invaded (red points and trend line). After zebra mussels invaded, shad larvae decreased (orange points). Juveniles of the tessellated darter **(b)**, a littoral fish, had been decreasing in abundance before zebra mussels invaded (red points and trend line). After zebra mussels invaded, they increased (orange points). SOURCE: Strayer, D., et al., 2004. Effects of an invasive bivalve (*Dreissena polymorpha*) on fish in the Hudson River estuary. *Canadian Journal of Fisheries and Aquatic Sciences* 61: 924–941. © 2004. Reprinted by permission of NRC Research Press.

Animals at high trophic levels, such as wolves, sea stars, sharks, and sea otters (see Figure 4.13), are often viewed as keystone species that can trigger trophic cascades. However, other types of organisms also can exert strong community-wide effects. "Ecosystem engineers" physically modify the environment shared by community members. Beavers build dams and turn streams into ponds, flooding large areas of dry land and turning them to swamp. Prairie dogs dig burrows that aerate the soil and serve as homes for other animals. Ants disperse seeds, redistribute nutrients, and selectively protect or destroy different insects and plants within the radius of their colonies. And zebra and quagga mussels alter the communities they invade by filtering plankton out of the water.

Less conspicuous organisms and those toward the bottoms of food chains may exert still more impact. Remove the

would help littoral-feeding fish. They predicted that following the zebra mussel invasion, larvae and juveniles of six common open-water fish species would decline in number, decline in growth rate, and shift downriver toward saltier water, where mussels are absent. Conversely, they predicted that larvae and juveniles of 10 littoral fish species would increase in number, increase in growth rate, and shift upriver to regions of greatest zebra mussel density.

To test their predictions, the researchers analyzed data from fish surveys carried out by DEC scientists and consultants over 26 years, spanning periods before and after the zebra mussel's arrival. Strayer's team compared data on abundance, growth, and distribution of young fish before and after 1991.

The results supported their predictions. Larvae and juveniles of open-water fish, such as American shad, blueback herring, and alewife, tended to decline in abundance in the years after zebra mussels were introduced (**see first figure, part (a)**). Those of littoral fish, such as tessellated darter, bluegill, and largemouth bass, tended to increase (**see first figure, part (b)**).

Growth rates showed the same trend: Open-water fish grew more slowly after zebra mussels invaded, whereas littoral fish grew more quickly. In terms of distribution in the 248-km (154-mi) stretch of river studied, open-water fish shifted downstream toward areas with fewer zebra mussels, whereas littoral fish shifted upstream toward areas with more zebra mussels (**see second figure**). Overall, the data

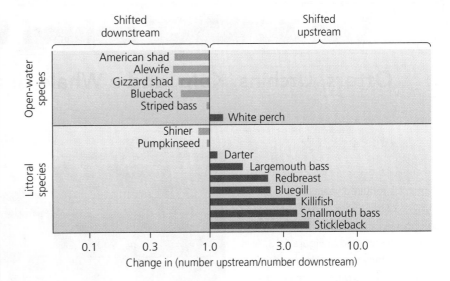

Young of open-water fish, such as American shad, blueback herring, and alewife, tended to shift downstream toward areas with fewer zebra mussels in the years following zebra mussel arrival. Young of littoral fish, such as killifish, bluegill, and largemouth bass, tended to shift upstream toward areas with more zebra mussels. SOURCE: Strayer, D., et al., 2004. Effects of an invasive bivalve (*Dreissena polymorpha*) on fish in the Hudson River estuary. *Canadian Journal of Fisheries and Aquatic Sciences* 61: 924–941. © 2004. Reprinted by permission of NRC Research Press.

supported the hypothesis that the fish community would respond to changes in food resources caused by zebra mussels. The results were published in 2004 in the *Canadian Journal of Fisheries and Aquatic Sciences*.

As Strayer continues his research, he and his colleagues learn more and more. In 2008 they published a broader analysis of the Hudson's food web showing that although littoral species benefited from zebra mussels, the fact that the mussels clarified the water made littoral species more susceptible to variation in clarity due to sediment input at times of high water flow.

Strayer and others also recently showed that populations of native mussels and clams in the Hudson had

crashed after the zebra mussel invaded (likely as a result of competition for food), but that starting in 2000, these native bivalves, instead of going extinct, suddenly stabilized and persisted at about 4–22% of their pre-invasion population sizes. Strayer's team has not yet determined the reason for this turnaround, but they suggested that perhaps the native species could continue to play a role in the community.

Research such as this can help illuminate the often obscure impacts that particular species interactions have on communities as a whole. In this case, the research may also help fisheries biologists to manage commercially and recreationally important fish populations in the Hudson River and other areas invaded by zebra mussels. ■

fungi that decompose dead matter, or the insects that control plant growth, or the phytoplankton that are the base of the marine food chain, and a community may change very rapidly indeed. However, because there are usually more species at lower trophic levels, it is less likely that any single one of them alone has wide influence. Often if one species is removed, other species that remain may be able to perform many of its functions.

Identifying keystone species is no simple task, and there is no universally accepted definition of the term to help us. Community dynamics are complex, species interactions differ in their strength, and the strength of species interactions can vary in time and space. **THE SCIENCE BEHIND THE STORY** (pp. 90–91) gives an idea of the surprises that sometimes lie in store for ecologists studying these interactions.

The SCIENCE behind the Story

Otters, Urchins, Kelp, and a Whale of a Chain Reaction

Ecologists required years of careful study to comprehend the relationships among sea otters, sea urchins, and kelp forests diagrammed in Figure 4.13. And even once they thought they understood it, more surprises were in store.

Sea otters live in coastal waters of the Pacific Ocean. These mammals float on their backs amid the waves, feasting on sea urchins that they pry from the bottom and bring to the surface. Once abundant, sea otters were hunted nearly to extinction for their fur. Protection by international treaty in 1911 allowed their numbers to rebound. Otters returned to high densities in some regions (such as off Alaska and the Aleutian Islands) but failed to return in others.

Biologists had long noted that regions with otters tend to host dense "forests" of kelp. Kelp is a brown alga (seaweed) that anchors to the seafloor, growing up to 60 m (200 ft) high toward the sunlit surface. Kelp forests provide complex physical structure in which diverse communities of fish and invertebrates find shelter and food.

In regions without sea otters, scientists found kelp forests absent. Urchins in such areas become so numerous that they may eat every last bit of kelp, creating empty seafloors called "urchin barrens" that are relatively devoid of life.

Ecologists observed that once areas with urchin barrens were

Dr. James Estes, University of California–Santa Cruz, at sea in Alaska

recolonized by sea otters, urchin numbers declined and kelp forests returned. Through comparative research of this kind, ecologists determined that otters were largely responsible for the presence of the kelp forest community, because they kept urchin numbers in check by preying on them. This research—mostly by James Estes of the University of California–Santa Cruz and his colleagues—established sea otters as a prime example of a keystone species. This single species of predator strongly influences its community through a trophic cascade (p. 87).

But the story does not end there. In the 1990s, otter populations dropped precipitously in Alaska and the Aleutians. No one knew why. So, Estes and his co-workers placed radio tags on Aleutian otters and studied them at sea. They hypothesized that fertility rates had dropped, but the radio-tracking observations showed that females were raising pups without problem.

They then tested a second hypothesis: that otters were simply moving to other locations. But the radio tracking showed no unusual dispersal, and the population declines were taking place evenly over large areas.

They were left with one viable hypothesis: increased mortality. Yet biologists were finding no evidence of disease or starvation. Then one day in 1991, Estes's team witnessed something never seen before. They watched as a sea otter was killed and eaten by an orca, or killer whale. These striking black-and-white predators grow up to 10 m (32 ft) long, hunt in groups, and had always attacked larger prey. A sea otter to them is a mere snack. Yet over the following years, Estes's team saw nine more cases of orca predation on otters. Could killer whales be killing off the otters?

Using data on otter birth rates, death rates, and population age structure (p. 65), the researchers estimated that to account for the otter decline, 6,788 orca attacks per year would have had to occur in their study area. This produced an *expected* rate of observed attacks that matched their *actual* number of observed attacks well, once they calculated their likelihood of actually observing attacks in the limited time and space they could cover on the water.

They then compared a bay where otters were vulnerable to orcas with a lagoon where they were protected.

Communities respond to disturbance in different ways

The removal of a keystone species is just one of many types of disturbance that can modify the composition, structure, or function of an ecological community. Over time, a community may experience natural disturbances ranging from gradual phenomena such as climate change to sudden events such as fires, floods, landslides, or hurricanes. Today, human impacts are major sources of disturbance for ecological communities

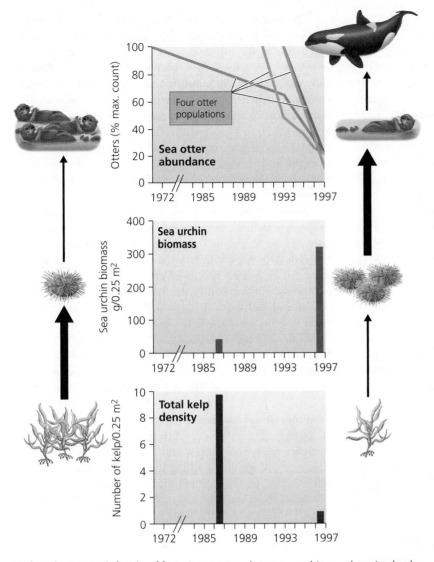

Before the 1990s (left side of figure), sea otters kept sea urchin numbers in check, allowing kelp forests to grow. By the late 1990s (right side of figure), orcas (killer whales), deprived of their usual food sources, were eating otters. This set off a chain reaction across trophic levels: Otters decreased, urchins increased, and kelp decreased. The lines in the top graph indicate trends in otter populations from four Aleutian islands. Width of arrows indicates strength of interaction. SOURCE: Estes, J., et al., 1998. Killer whale predation on sea otters linking oceanic and nearshore ecosystems. *Science* 282: 473–476, Fig. 1. Reprinted with permission from AAAS.

Otter numbers in the lagoon remained stable over four years, whereas those in the bay dropped by 76%. Radio tracking showed no movement between these locations. Together, these lines of evidence led the researchers to propose that predation by orcas was eliminating otters.

As otters declined in the Aleutians, urchins increased, and kelp density fell dramatically (**see figure**). This trophic cascade supported the idea that otters were keystone species—but now it seemed that one keystone species was being controlled by another!

Why had orcas suddenly started eating otters? A possible answer came in 2003, after Alan Springer, Estes, and others determined that sea otters showed only the latest in a series of population crashes in the northern Pacific. Harbor seals had declined by more than 90% in the late 1970s and 1980s. Fur seals had fallen by 60% since the 1970s. Sea lions crashed by 80% in the 1980s. And preceding these declines were the population collapses of the great whales—gray whales, blue whales, humpback whales, and others.

Historically, orcas ate great whales, which they would kill in groups, like a wolf pack taking down an elk. But industrial whaling by Japan, Russia, and other nations since the 1950s caused populations of great whales in the northern Pacific to plummet by 99% between 1965 and 1973. Once human hunting decimated the great whales, the orcas had to turn elsewhere. Springer's team argued that they shifted their diet to smaller, less favored seals and sea lions. When their predation had depleted those populations, the chain reaction continued, and the orcas turned to smaller prey still—sea otters.

Not all scientists are convinced by this bold hypothesis, and the debate has been vigorous. In the meantime, Estes and his colleagues continue to learn more. In 2005, Shauna Reisewitz, Estes, and Charles Simenstad reported that when kelp forests vanish, so do rock greenling, the community's most abundant fish species. The few greenling left also change their diet. Thus, the whale-orca-otter-urchin-kelp system also influences fish and their prey. The ongoing discoveries highlight the intriguing complexity of the species interactions that affect ecological communities. ∎

worldwide—from habitat alteration to pollution to the introduction of non-native species such as the zebra mussel.

Communities are dynamic systems and may respond to disturbance in several ways. A community that resists change and remains stable despite disturbance is said to show **resistance** to the disturbance. Alternatively, a community may show **resilience**, meaning that it changes in response to disturbance but later returns to its original state. Or, a community may be modified by disturbance permanently and may never return to its original state.

Succession follows severe disturbance

If a disturbance is severe enough to eliminate all or most of the species in a community, the affected site may undergo considerable change. To varying extents, this may involve a somewhat predictable series of changes that ecologists have traditionally called **succession**. In the conventional view of this process, there are two types of succession. **Primary succession** follows a disturbance so severe that no vegetation or soil life remains from the community that occupied the site. In primary succession, a biotic community is built essentially from scratch. In contrast, **secondary succession** begins when a disturbance dramatically alters an existing community but does not destroy all living things or all organic matter in the soil. In secondary succession, vestiges of the previous community remain, and these building blocks help shape the process.

At terrestrial sites, primary succession takes place after a bare expanse of rock, sand, or sediment becomes newly exposed to the atmosphere. This can occur when glaciers retreat, lakes dry up, or volcanic lava or ash spreads across the landscape (**FIGURE 4.14**). Species that arrive first and colonize the new substrate are referred to as **pioneer species**. Pioneer species are well adapted for colonization, having traits such as spores or seeds that can travel long distances.

The pioneers best suited to colonizing bare rock are the mutualistic aggregates of fungi and algae known as lichens. In lichens, the algal component provides food and energy via photosynthesis while the fungal component takes a firm hold on rock and captures the moisture that both organisms need to survive. As lichens grow, they secrete acids that break down the rock surface. This begins the formation of soil, and eventually small plants, insects, and worms find the rocky outcrops more hospitable. As new organisms arrive, they provide more nutrients and habitat. As time passes, larger plants establish themselves, the amount of vegetation increases, and species diversity rises.

Secondary succession begins when a fire, a hurricane, logging, or farming removes much of the biotic community. Consider an abandoned agricultural field in eastern North America (**FIGURE 4.15**). After farming ends, the site will be colonized by pioneer species of grasses, herbs, and forbs that were already in the vicinity and that disperse effectively. Soon, shrubs and fast-growing trees such as aspens and poplars begin to grow from the field. As time passes, pine trees rise above these trees and shrubs, forming a pine-dominated forest. This pine forest develops an understory of hardwood trees, because pine seedlings do not grow well under a canopy and some hardwood seedlings do. Eventually the hardwoods outgrow the pines, creating a hardwood forest.

Processes of succession occur in a diversity of ecological systems. For instance, some ponds undergo succession as algae, microbes, plants, and zooplankton grow, reproduce, and die, gradually filling the water body with organic matter. The pond acquires further organic matter and sediments from streams and surface runoff, and eventually it can fill in, becoming a bog (p. 405) or even a terrestrial system.

In the traditional view of succession described here, the transitions between stages lead to a *climax community*, which remains in place with little modification until some disturbance restarts succession. Early ecologists felt that each region had its own characteristic climax community, determined by climate.

Communities may undergo shifts

Today, ecologists recognize that the dynamics of community change are far more variable and less predictable than originally thought. Conditions at one stage may promote progression to another stage, or organisms may, through competition, inhibit a community's progression to another stage. The trajectory of change can vary greatly according to chance factors, such as which particular species happen to gain an early foothold. In addition, climax communities are not determined solely by climate, but rather may vary with soil conditions and other factors from one time or place to another.

Moreover, once a community is disturbed and changes are set in motion, there is no guarantee that the community will ever return to its original state. Sometimes communities may undergo a **phase shift**, or **regime shift**, in which the overall character of the community fundamentally changes. This can happen if some crucial climatic threshold is passed, a keystone species is lost, or an exotic species invades. For instance, in recent years many coral reef communities have

FIGURE 4.14 ▲ As this Norwegian glacier retreats, small plants (foreground) begin the process of primary succession.

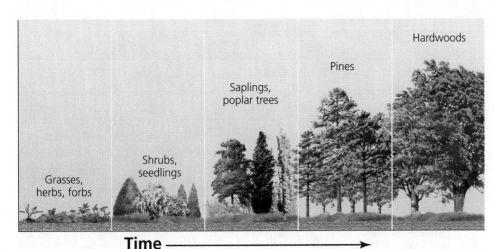

Time

FIGURE 4.15 ◀ Secondary succession occurs after a disturbance, such as a fire, landslide, or farming, removes most vegetation from an area. Here is shown a typical series of changes in a plant community of eastern North America following the abandonment of a farmed field.

undergone a phase shift and become dominated by algae, often as a result of overfishing and population declines of certain fish or turtles that eat algae. Once algae overgrows a reef, the community may never shift back to its original coral-dominated state. Phase shifts make it clear that we cannot count on being able to reverse damage caused by human disturbance; instead, some of the changes we set in motion may become permanent.

How cohesive are communities?

Ecologists who have studied how communities change in response to disturbance have conceptualized communities in different ways. Early in the 20th century, botanist Frederick Clements promoted the view that communities are cohesive entities whose members remain associated over time and space. Communities, he argued, are discrete units with integrated parts, much like organisms. Clements's view implied that the many varied members of a community share similar limiting factors and evolutionary histories.

Henry Gleason disagreed. Gleason, also a botanist, maintained that each species responds independently to its own limiting factors and that a species can join or leave a community without greatly altering its composition. Communities, Gleason argued, are not cohesive units, but temporary associations of individual species that can reassemble into different combinations.

Years of ecological research have led ecologists today to side largely with Gleason, although most see validity in aspects of both men's ideas. Indeed, many ecologists still find it useful to refer to communities by names that highlight certain key plants (such as "oak-hickory forest," "tallgrass prairie," and "pine-bluestem community"). Ecologists find labeling communities as though they were cohesive units to be a pragmatic tool, even though they know the associations could change radically decades or centuries hence.

Invasive species pose new threats to community stability

Traditional concepts of community cohesion and of succession involve species understood to be native to an area. But what if a new species arrives from elsewhere? And what if this non-native (also called *alien* or *exotic*) organism turns *invasive*, spreading widely and becoming dominant in a community? Such **invasive species** can alter a community substantially and have become one of the central ecological forces in today's world.

Most exotic species do not turn invasive, but those that do are generally species that people have *introduced*, intentionally or by accident, from elsewhere in the world. Global trade helped spread zebra and quagga mussels, which were inadvertently transported in the ballast water of cargo ships. To maintain stability at sea, ships take water into their hulls as they begin their voyage and then discharge that water at their destination. Decades of unregulated exchange of ballast water have ferried countless species across the oceans.

Introduced species may become invasive when limiting factors (p. 67) that regulate their population growth are absent. Plants and animals brought to a new area may leave behind the predators, parasites, and competitors that had evolved to exploit them in their native land. This frees them from the natural constraints on their population growth, and the introduced species may thrive and spread if few or no organisms in the new environment prey on it, parasitize it, or compete with it. As the species proliferates, it may exert diverse influences on its fellow community members (**FIGURE 4.16**).

Examples abound of introduced species that have turned invasive and have had major ecological impacts. The chestnut blight, an Asian fungus, killed nearly every mature American chestnut (*Castanea dentata*), the dominant tree species of many eastern North America forests, in the quarter-century preceding 1930. Asian trees had evolved defenses against the fungus over long millennia of coevolution, but the American chestnut had not. A different fungus caused Dutch elm disease, destroying most of the American elms (*Ulmus americana*) that once gracefully lined the streets of many U.S. cities. Fish introduced into streams for sport compete with and exclude native fish. Various grasses introduced in the American West have overrun entire regions, pushing out native vegetation. Hundreds of island-dwelling animals and plants worldwide have been driven extinct by the goats, pigs, and rats introduced by human colonists. We will examine more

FIGURE 4.16 ▶ The zebra mussel and the quagga mussel are biological invaders that are modifying ecological communities. By filtering phytoplankton and small zooplankton from open water, they exert impacts on other species **(a)**, both negative (red downward arrows) and positive (green upward arrows). The map in **(b)** shows known occurrences of the zebra mussel (red dots) and the quagga mussel (green dots) in North America as of 2010. In just two decades they spread from a small area of the Great Lakes to waterways across the continent. People have transported them inadvertently on boat hulls (yellow stars) as far west as California. *Source* (b): U.S. Geological Survey.

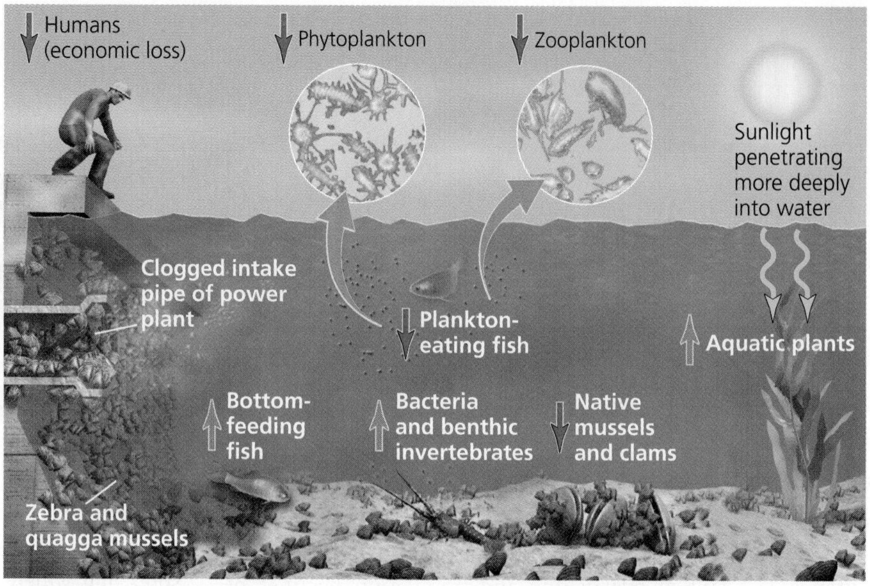

Humans (economic loss)
Phytoplankton
Zooplankton
Sunlight penetrating more deeply into water
Clogged intake pipe of power plant
Plankton-eating fish
Aquatic plants
Bottom-feeding fish
Bacteria and benthic invertebrates
Native mussels and clams
Zebra and quagga mussels

(a) Impacts of zebra and quagga mussels on members of a Great Lakes nearshore community

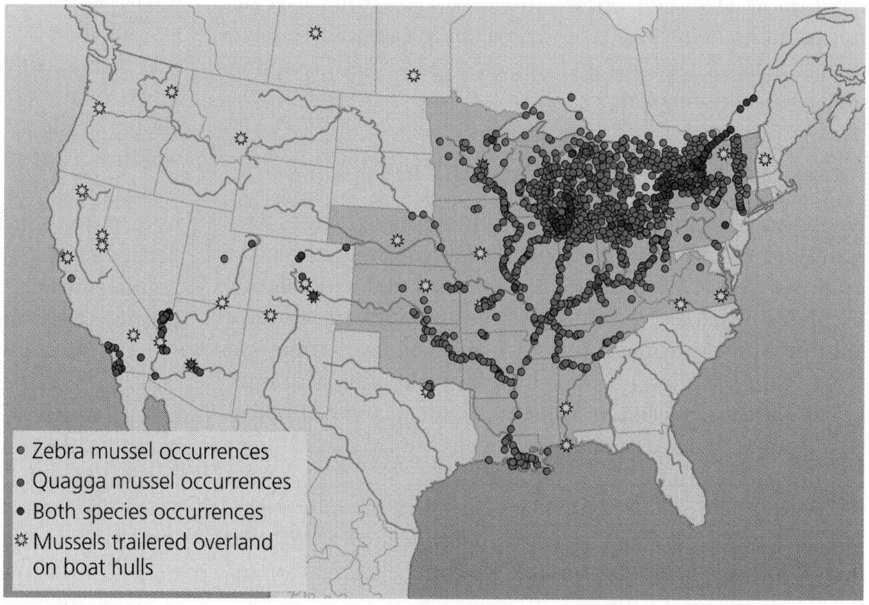

Zebra mussel occurrences
Quagga mussel occurrences
Both species occurrences
Mussels trailered overland on boat hulls

(b) Occurrence of zebra and quagga mussels in North America, as of 2010

examples in our discussion of invasive species and biodiversity in Chapter 11 (pp. 293–294).

Ecologists generally view the impacts of invasive species—and introduced species in general—as overwhelmingly negative. Yet many people enjoy the beauty of introduced ornamental plants in their gardens. Some organisms are introduced intentionally to control pest populations through biocontrol (p. 258). And some introduced species that have turned invasive provide benefits to our economy, such as the European honeybee, which pollinates many of our crops (p. 261). Whatever view one takes, the impacts of invasive species on native species and ecological communities are significant, and they grow year by year with the increasing mobility of people and the globalization of our society.

WEIGHING THE ISSUES

Are Invasive Species All Bad? Some ethicists question the notion that all invasive species should automatically be considered bad. If we introduce a non-native species to a community and it greatly modifies the community, do you think that is a bad thing? What if it drives another species extinct? What if the invasive species arrived on its own, rather than through human intervention? What if it provides economic services to our society, as the European honeybee does? What ethical standard(s) (p. 140) would you apply to determine whether we should battle or accept an invasive species?

In North America, zebra and quagga mussels—and the media attention they continue to generate—helped put invasive species on the map as a major environmental and economic issue. Scientific research proliferated, and many ecologists have come to view invasive species as the second-greatest threat to species and natural systems, behind only habitat destruction. In 1990 the U.S. Congress passed legislation that led to the National Invasive Species Act of 1996. Among other things, this law directed the Coast Guard to ensure that ships dump their freshwater ballast at sea and exchange it with salt water before entering the Great Lakes.

Since then, funding has become widely available for the control and eradication of invasive species. Managers have tried to control the spread of zebra mussels by removing them manually; applying toxic chemicals; drying them out; depriving them of oxygen; introducing predators and diseases; and stressing them with heat, sound, electricity, carbon dioxide, and ultraviolet light. However, most of these are localized and short-term fixes that are not capable of making a dent in the immense populations at large in the environment. In case after case, managers are finding that control and eradication measures are so difficult and expensive that preventive strategies (such as ballast water regulations) represent a much better investment.

If one seeks to prevent invasions before they occur, it helps to be able to predict where a given species might spread. Scientific research can assist by analyzing the biology and needs of the organism and using that knowledge to predict the environmental conditions in which it will thrive. In 2007, researchers used their knowledge of how zebra and quagga mussels use calcium in water to create their shells to predict the geographic areas in which the mussels might do best. The researchers mapped low-risk and high-risk regions across North America, and these mostly conformed to the areas of actual spread. A remaining question is how quagga mussels may differ from zebra mussels. As Figure 4.16 shows, quagga mussels have leapfrogged zebra mussels by spreading into some western states, and no one knows why they have so far been more successful in the West.

Altered communities can be restored

Invasive species are adding to the tremendous transformations that humans have already forced on ecological systems through habitat alteration, deforestation, pollution, climate change, the hunting of keystone species, and other activities. With so much of Earth's landscape altered by human impact—no spot on the planet today is unaffected—many communities and ecosystems are severely degraded. Because ecological systems support our civilization and all of life, when degraded systems cease to function, it threatens our very existence.

This realization has given rise to the science of **restoration ecology**. Restoration ecologists research the historical conditions of ecological communities as they existed before our industrialized civilization altered them. They then try to devise ways to restore altered areas to an earlier condition. In some cases, the intent is primarily to restore the functionality of the system—to reestablish a wetland's ability to filter pollutants and recharge groundwater, for example, or a forest's ability to cleanse the air, build soil, and provide habitat for wildlife. In other cases, the aim is to return a community

all the way to its natural "pre-settlement" condition. Either way, the science of restoration ecology informs the practice of **ecological restoration**, the actual on-the-ground efforts to carry out these visions and restore communities.

Many ecological restoration efforts are underway today. For instance, nearly every last scrap of tallgrass prairie that once covered the eastern Great Plains and parts of the Midwestern United States was converted to agriculture in the 19th century. Now people are restoring patches of prairie by planting native prairie vegetation, weeding out invaders and competitors, and introducing controlled fire to mimic the fires that historically maintained this community (**FIGURE 4.17**). The region outside Chicago, Illinois, boasts several of the largest prairie restoration projects so far, including a 184-ha (455-acre) area inside the massive ring of the Fermilab nuclear accelerator in Batavia.

The world's largest restoration project is the ongoing effort to restore parts of the Florida Everglades, a 7,500-km² (4,700-mi²) ecosystem of marshes and seasonally flooded grasslands. This formerly vast wetland system has been drying out for decades because the water that feeds it has been managed for flood control and overdrawn for irrigation and development. Economically important fisheries have suffered greatly as a result, and the region's famed populations of wading birds have dropped by 90–95%. The 30-year, $7.8-billion restoration project intends to restore natural water flow by undoing the damming and diversions of 1,600 km (1,000 mi) of canals, 1,150 km (720 mi) of levees, and 200 water control structures. Because the Everglades provides drinking water for millions of Florida citizens, as well as considerable tourism revenue, restoring its ecosystem services (pp. 121–122, 148, 160) should prove economically beneficial as well as ecologically valuable. We will explore ecological restoration projects further in Chapter 11 (pp. 308–309) in the context of seeking avenues toward sustainability.

As our population grows and development spreads, ecological restoration is becoming an increasingly vital conservation strategy. However, restoration is difficult, time-consuming, and expensive, and it is not always successful. It is therefore best, whenever possible, to protect natural systems from degradation in the first place, so that restoration does not become necessary.

FIGURE 4.17 ▼ USDA Forest Service ecologists from the Midewin National Tallgrass Prairie inspect native grasses in a prairie restoration area on the site of the former Joliet Arsenal in Illinois.

Restoring "Natural" Communities

Practitioners of ecological restoration in North America often aim to restore communities to their natural state. But what is meant by "natural"? Does it mean the state of the community before industrialization? Before Europeans came to the New World? Before any people laid eyes on the community? Let's say Native Americans altered a forest community 8,000 years ago by burning the underbrush regularly to improve hunting and continued doing so until Europeans arrived 400 years ago and cut down the forest for farming. Today the area's inhabitants want to restore the land to its "natural" forested state. Should restorationists try to recreate the forest of the Native Americans or the forest that existed before Native Americans arrived? What values do you think underlie the desire for restoration?

EARTH'S BIOMES

Across the world, each portion of each continent is home to different sets of species, leading to endless variety in community composition. However, communities in far-flung places often share strong similarities in their structure and function. This allows us to classify communities into broad types. A **biome** is a major regional complex of similar communities—a large-scale ecological unit recognized primarily by its dominant plant type and vegetation structure. The world contains a number of biomes, each covering large geographic areas (**FIGURE 4.18**).

Each biome encompasses a variety of communities that share similarities. For example, the eastern United States supports part of the temperate deciduous forest biome. From New Hampshire to the Great Lakes to eastern Texas, precipitation and temperature are similar enough that most of the region's natural plant cover consists of broad-leafed trees that lose their leaves in winter. Within this region, however, there exist many different types of temperate deciduous forest, such as oak-hickory, beech-maple, and aspen-birch forests, each sufficiently different to be designated a separate community.

Climate influences the locations of biomes

Which biome covers any particular portion of the planet depends on a variety of abiotic factors, including temperature, precipitation, atmospheric circulation, and soil characteristics. Among these factors, temperature and precipitation exert

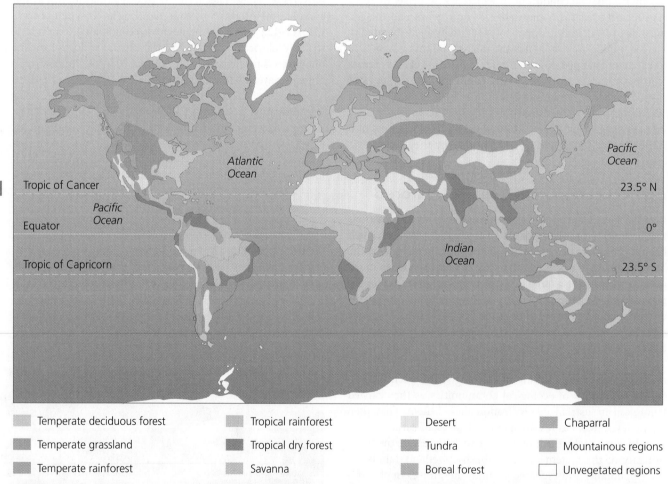

Temperate deciduous forest	Tropical rainforest	Desert	Chaparral
Temperate grassland	Tropical dry forest	Tundra	Mountainous regions
Temperate rainforest	Savanna	Boreal forest	Unvegetated regions

FIGURE 4.18 ▲ Biomes are distributed around the world according to temperature, precipitation, atmospheric and oceanic circulation patterns, and other factors.

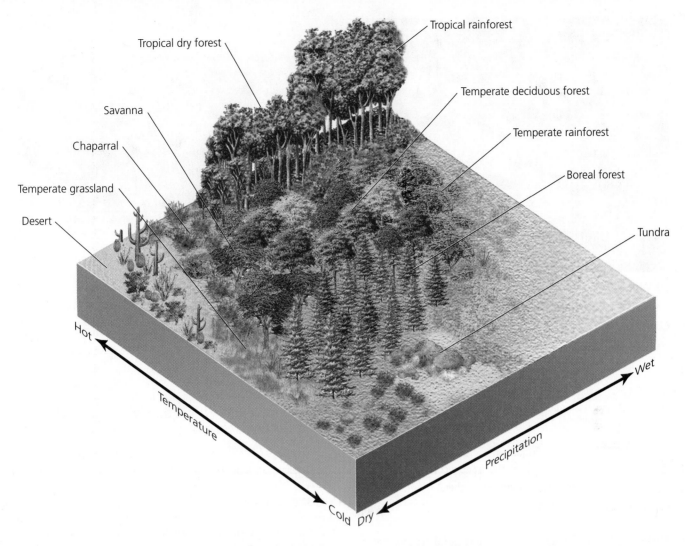

Tropical dry forest

Tropical rainforest

Savanna

Temperate deciduous forest

Chaparral

Temperate rainforest

Temperate grassland

Boreal forest

Desert

Tundra

Hot

Wet

Temperature

Precipitation

Cold Dry

FIGURE 4.19 ▲ As precipitation increases, vegetation generally becomes taller and more luxuriant. As temperature increases, types of plant communities change. Together, temperature and precipitation are the main factors determining which biome occurs in a given area. For instance, deserts occur in dry regions; tropical rainforests occur in warm, wet regions; and tundra occurs in the coldest regions.

the greatest influence (**FIGURE 4.19**). Because biome type is largely a function of climate, and because average monthly temperature and precipitation are among the best indicators of an area's climate, scientists often use climate diagrams, or **climatographs**, to depict such information.

Global climate patterns cause biomes to occur in large patches in different parts of the world. For instance, temperate deciduous forest occurs in eastern North America, Europe, and eastern China. Note in Figure 4.18 how patches representing the same biome tend to occur at similar latitudes. This is due to the north–south gradient in temperature and to atmospheric circulation patterns (p. 467).

Aquatic and coastal systems also show biome-like patterns

In our discussion of biomes, we will focus exclusively on terrestrial systems because the biome concept, as traditionally developed and applied, has been limited to terrestrial systems. Areas equivalent to biomes also exist in the oceans, along coasts, and in freshwater systems, but their geographic distributions would look very different from those of terrestrial biomes if plotted on a world map. One might consider the shallows along the world's coastlines to represent one aquatic system, the continental shelves another, and the open ocean, the deep sea, coral reefs, and kelp forests as still other distinct sets of communities. Many coastal systems—such as salt marshes, rocky intertidal communities, mangrove forests, and estuaries—share both terrestrial and aquatic components. And freshwater systems such as those of the Great Lakes are widely distributed throughout the world.

Unlike terrestrial biomes, aquatic systems are shaped not by air temperature and precipitation, but by factors such as water temperature, salinity, dissolved nutrients, wave action, currents, depth, light levels, and type of substrate (e.g., sandy, muddy, or rocky bottom). Marine communities are also more clearly delineated by their animal life than by their plant life. We will examine freshwater, marine, and coastal systems in the greater detail they deserve in Chapters 15 and 16.

(a) Temperate deciduous forest

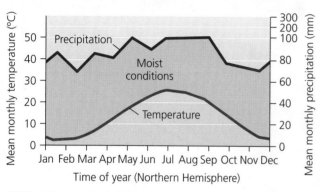

(b) Washington, D.C., USA

FIGURE 4.20 ▲ Temperate deciduous forests **(a)** experience relatively stable seasonal precipitation but more varied seasonal temperatures. Scientists use climate diagrams **(b)** to illustrate an area's average monthly precipitation and temperature. In these diagrams, the *x* axis marks months of the year, and paired *y* axes denote temperature and precipitation. The curves indicate precipitation (blue) and temperature (red) from month to month. When the precipitation curve lies above the temperature curve, as is the case year-round in the temperate deciduous forest around Washington, D.C., the region experiences relatively "moist" conditions, which we indicate with green coloration. Climatograph here and in the following figures adapted from Breckle, S.W., and H. Walter, trans. by G. Lawlor. 2002. *Walter's vegetation of the Earth: The ecological systems of the geo-biosphere*, 4th ed. Originally published by Eugen Ulmer KG, 1999, used by permission.

We can divide the world into roughly ten terrestrial biomes

Temperate deciduous forest The **temperate deciduous forest** (**FIGURE 4.20**) that dominates the landscape around the central and southern Great Lakes is characterized by broad-leafed trees that are *deciduous*, meaning that they lose their leaves each fall and remain dormant during winter, when hard freezes would endanger leaves. These midlatitude forests occur in much of Europe and eastern China as well as in eastern North America—all areas where precipitation is spread relatively evenly throughout the year.

(a) Temperate grassland

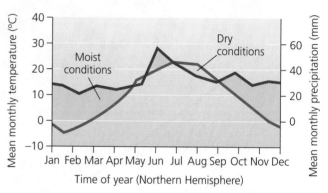

(b) Odessa, Ukraine

FIGURE 4.21 ▲ Temperate grasslands experience temperature variations throughout the year and too little precipitation for many trees to grow. Constructed for Odessa, Ukraine, this climatograph indicates "moist" climate conditions (in green) as well as "dry" climate conditions (in yellow, when the temperature curve is above the precipitation curve). Climatograph adapted from Breckle, S.W., 2002.

Soils of the temperate deciduous forest are relatively fertile, but this biome generally consists of far fewer tree species than are found in tropical rainforests. Oaks, beeches, and maples are a few of the most abundant types of trees in these forests. A sampling of typical animals of the temperate deciduous forest of eastern North America is shown in Figure 4.12 (p. 86).

Temperate grassland Traveling westward from the Great Lakes, temperature differences between winter and summer become more extreme, rainfall diminishes, and we find **temperate grasslands** (**FIGURE 4.21**). This is because the limited amount of precipitation in the Great Plains region west of the Mississippi River can support grasses more easily than trees. Also known as *steppe* or *prairie*, temperate grasslands were once widespread throughout parts of North and South America and much of central Asia.

Today people have converted most of the world's grasslands for agriculture, and almost no undisturbed native grasslands exist in North America. Characteristic vertebrate animals of the native North American grasslands include American

(a) Temperate rainforest

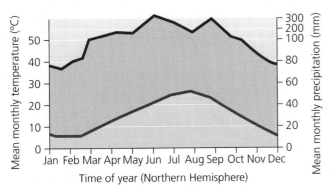

(b) Nagasaki, Japan

FIGURE 4.22 ▲ Temperate rainforests receive a great deal of precipitation and feature moist, mossy interiors. Climatograph adapted from Breckle, S.W., 2002.

(a) Tropical rainforest

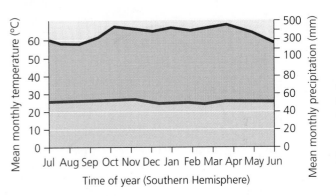

(b) Bogor, Java, Indonesia

FIGURE 4.23 ▲ Tropical rainforests, famed for their biodiversity, grow under constant, warm temperatures and a great deal of rain. Climatograph adapted from Breckle, S.W., 2002.

bison *(Bison bison)*, prairie dogs, pronghorn antelope *(Antilocapra americana)*, and ground-nesting birds such as meadowlarks and prairie chickens. Because we have converted so many grasslands for farming and ranching, most of these animals exist today at only a small fraction of their historic population sizes.

Temperate rainforest Moving further west in North America, the topography becomes varied, and biome types intermix in the Rocky Mountain region. Still further west, the coastal Pacific Northwest region, with its heavy rainfall, features **temperate rainforest** (**FIGURE 4.22**). Coniferous trees, such as cedars, spruces, hemlocks, and Douglas fir *(Pseudotsuga menziesii)* grow very tall in the temperate rainforest, and the forest interior is shaded and damp. Moisture-loving animals such as the bright yellow banana slug *(Ariolimax columbianus)* are common, and old-growth stands hold the endangered spotted owl *(Strix occidentalis)*. The soils of temperate rainforests are usually quite fertile but are susceptible to landslides and erosion if forests are cleared.

Temperate rainforests can produce large volumes of commercially important forest products such as lumber and paper. Controversy has surrounded these forests in the Pacific Northwest, where logging and clear-cutting have eliminated most old growth trees and have driven some species toward extinction. Local people often support further timber extraction, but they also suffer the consequences of overharvesting.

Tropical rainforest In tropical regions we see the same pattern found in temperate regions: Areas of high rainfall grow rainforests, areas of intermediate rainfall host dry or deciduous forests, and areas of lower rainfall are dominated by grasses. However, tropical biomes differ from their temperate counterparts in other ways because they are closer to the equator and therefore warmer on average year-round. For one thing, they hold far greater biodiversity.

Tropical rainforest (**FIGURE 4.23**) is found in Central America, South America, Southeast Asia, west Africa, and other tropical regions and is characterized by year-round rain and uniformly warm temperatures. Tropical rainforests have dark, damp interiors, lush vegetation, and highly diverse biotic communities, with greater numbers of species of insects, birds, amphibians, and various other animals than any other biome. These forests are not dominated by single species of trees, as are

(a) Tropical dry forest

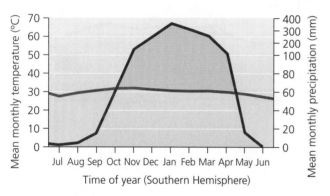

(b) Darwin, Australia

FIGURE 4.24 ▲ Tropical dry forests experience significant seasonal variations in precipitation and relatively stable, warm temperatures. Climatograph adapted from Breckle, S.W., 2002.

(a) Savanna

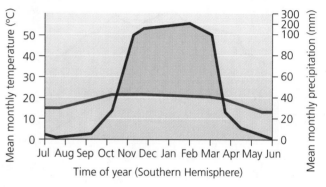

(b) Harare, Zimbabwe

FIGURE 4.25 ▲ Savannas are grasslands with clusters of trees. They experience slight seasonal variation in temperature but significant variation in rainfall. Climatograph adapted from Breckle, S.W., 2002.

forests closer to the poles, but instead consist of very high numbers of tree species intermixed, each at a low density. Any given tree may be draped with vines, enveloped by strangler figs, and loaded with epiphytes (orchids and other plants that grow in trees), such that trees occasionally collapse under the weight of all the life they support.

Despite this profusion of life, tropical rainforests have very poor, acidic soils that are low in organic matter. Nearly all nutrients present in this biome are contained in the plants, not in the soil. An unfortunate consequence is that once tropical rainforests are cleared, the nutrient-poor soil can support agriculture for only a short time (pp. 230–231). As a result, farmed areas are abandoned quickly, farmers move on and clear more forest, and the soil and vegetation recover slowly.

Tropical dry forest   Tropical areas that are warm year-round but where rainfall is lower overall and highly seasonal give rise to **tropical dry forest**, or *tropical deciduous forest* (**FIGURE 4.24**), a biome widespread in India, Africa, South America, and northern Australia. Wet and dry seasons each span about half a year in tropical dry forest. Rains during the wet season can be extremely heavy and, coupled with erosion-prone soils, can lead to severe soil loss where people have cleared forest.

Across the globe, we have converted a great deal of tropical dry forest to agriculture. Clearing for farming or ranching is made easier by the fact that vegetation heights are much lower and canopies less dense than in tropical rainforest. Organisms that inhabit tropical dry forest have adapted to seasonal fluctuations in precipitation and temperature. For instance, plants are deciduous and often leaf out and grow profusely with the rains, then drop their leaves during the driest times of year.

Savanna   Drier tropical regions give rise to **savanna** (**FIGURE 4.25**), tropical grassland interspersed with clusters of acacias or other trees. The savanna biome is found today across stretches of Africa, South America, Australia, India, and other dry tropical regions. Precipitation in savannas usually arrives during distinct rainy seasons and concentrates grazing animals near widely spaced water holes. Common herbivores on the African savanna include zebras, gazelles, and giraffes, and the predators of these grazers include lions, hyenas, and other highly mobile carnivores.

Desert   Where rainfall is very sparse, **desert** (**FIGURE 4.26**) forms. This is the driest biome on Earth; most deserts receive well under 25 cm (9.8 in.) of precipitation per year, much of it

(a) Desert

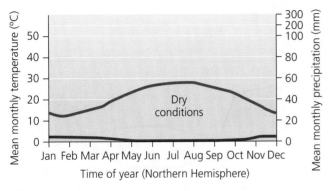

(b) Cairo, Egypt

FIGURE 4.26 ▲ Deserts are dry year-round, but they are not always hot. Precipitation can arrive in intense, widely spaced storm events. The temperature curve is consistently above the precipitation curve in this climatograph of Cairo, Egypt, indicating that the region experiences "dry" conditions all year. Climatograph adapted from Breckle, S.W., 2002.

(a) Tundra

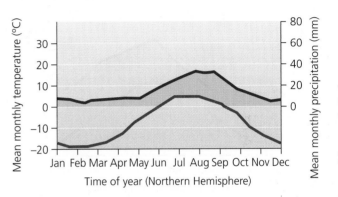

(b) Vaigach, Russia

FIGURE 4.27 ▲ Tundra is a cold, dry biome found near the poles. Alpine tundra occurs atop high mountains at lower latitudes. Climatograph adapted from Breckle, S.W., 2002.

during isolated storms months or years apart. Depending on rainfall, deserts vary in the amount of vegetation they support. Some, like the Sahara and Namib deserts of Africa, are mostly bare sand dunes; others, like the Sonoran Desert of Arizona and northwest Mexico, are quite vegetated.

Deserts are not always hot; the high desert of the western United States is one example. Because deserts have low humidity and relatively little vegetation to insulate them from temperature extremes, sunlight readily heats them in the daytime, but daytime heat is quickly lost at night. As a result, temperatures vary greatly from day to night and across seasons of the year. Desert soils can often be quite saline and are sometimes known as lithosols, or stone soils, for their high mineral and low organic-matter content.

Through convergent evolution (p. 54), desert animals and plants have evolved many adaptations to deal with a harsh climate. Most reptiles and mammals, such as rattlesnakes and kangaroo mice, are active in the cool of night, and many Australian desert birds are nomadic, wandering long distances to find areas of recent rainfall and plant growth. Many desert plants have thick, leathery leaves to reduce water loss, or green trunks so that the plant can photosynthesize without leaves,

minimizing the surface area prone to water loss. The spines of cacti and other desert plants guard those plants from being eaten by herbivores desperate for the precious water they hold.

Tundra   Nearly as dry as desert, **tundra** (**FIGURE 4.27**) occurs at very high latitudes along the northern edges of Russia, Canada, and Scandinavia. Extremely cold winters with little daylight and moderately cool summers with lengthy days characterize this landscape of lichens and low, scrubby vegetation without trees. The great seasonal variation in temperature and day length results from this biome's high-latitude location, angled toward the sun in summer and away from the sun in winter.

Because of the cold climate, underground soil remains more or less permanently frozen and is called **permafrost**. During the long, cold winters, surface soils freeze as well; then, when the weather warms, they melt and produce seasonal accumulations of surface water that make ideal habitat for mosquitoes and other biting insects. The swarms of insects benefit bird species that migrate long distances to breed during the brief but productive summer. Caribou also migrate to the tundra to breed, after which they leave for the winter. Only a few

(a) Boreal forest

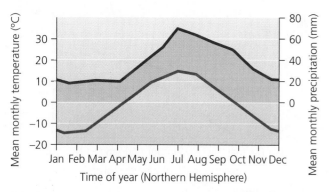

(b) Archangelsk, Russia

FIGURE 4.28 ▲ Boreal forest is characterized by long, cold winters, relatively cool summers, and moderate precipitation. Climatograph adapted from Breckle, S.W., 2002.

(a) Chaparral

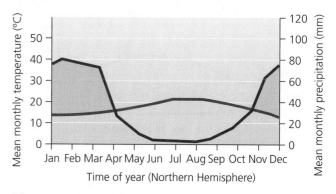

(b) Los Angeles, California, USA

FIGURE 4.29 ▲ Chaparral is a highly seasonal biome dominated by shrubs, influenced by marine weather, and dependent on fire. Climatograph adapted from Breckle, S.W., 2002.

large animals, such as polar bears (*Ursus maritimus*) and musk oxen (*Ovibos moschatus*), can survive year-round here.

Because of its extreme climate, much of the tundra remains intact and relatively unaltered by human occupation and development. However, atmospheric circulation patterns (p. 467) bring our airborne pollutants to this biome, and global climate change is heating high-latitude regions more intensely than other areas. Climate change is melting sea ice, altering seasonal cycles to which animals have adapted over millennia, and melting permafrost, releasing methane gas that further worsens climate change.

Tundra also occurs as *alpine tundra* at the tops of tall mountains in temperate and tropical regions, where high elevation creates conditions similar to those of high latitude.

Boreal forest   The northern coniferous forest, or **boreal forest**, often called *taiga* (**FIGURE 4.28**), stretches in a broad band across much of Canada, Alaska, Russia, and Scandinavia. It consists of a few species of evergreen trees, such as black spruce (*Picea mariana*), that dominate large stretches of forest, interspersed with occasional bogs and lakes. The boreal forest's uniformity over huge areas reflects the climate common to this latitudinal band of the globe: These forests develop in cooler, drier regions than do temperate forests, and they experience long, cold winters and short, cool summers.

Soils are typically nutrient-poor and somewhat acidic. As a result of the strong seasonal variation in day length, temperature, and precipitation, many organisms compress a year's worth of feeding, breeding, and rearing of young into a few warm, wet months. Year-round residents of boreal forest include mammals such as moose (*Alces alces*), wolves (*Canis lupus*), bears, lynx (*Felis lynx*), and many burrowing rodents. This biome also hosts many insect-eating birds that migrate from the tropics to breed during the brief, intensely productive, summer season.

Chaparral   In contrast to the boreal forest's broad, continuous distribution, **chaparral** (**FIGURE 4.29**) is limited to fairly small patches widely flung around the globe. Chaparral consists mostly of evergreen shrubs and is densely thicketed. This biome is also highly seasonal, with mild, wet winters and warm, dry summers. This type of climate is induced by oceanic influences and is often termed "Mediterranean." In addition to ringing the Mediterranean Sea, chaparral occurs along the coasts of California, Chile, and southern Australia. Chaparral communities experience frequent fire, and their plant species are adapted to resist fire or even to depend on it for germination of their seeds.

Altitude creates patterns analogous to latitude

As any hiker or skier knows, climbing in elevation causes a much more rapid change in climate than moving the same distance toward the poles. Vegetative communities change along mountain slopes in correspondence with this altitude-induced climate variation (**FIGURE 4.30**). It is often said that hiking up a mountain in the southwestern United States is like walking from Mexico to Canada. A hiker ascending one of southern Arizona's higher mountains would begin in Sonoran Desert or desert grassland and proceed through oak woodland, pine forest, and finally spruce-fir forest—the equivalent of passing through several biomes. A hiker scaling one of the great peaks of the Andes in Ecuador, near the equator, could begin in tropical rainforest and end amid glaciers in alpine tundra.

Not only does temperature drop at progressively higher altitudes, but atmospheric pressure (and oxygen) also declines, whereas ultraviolet radiation increases. Moreover, mountains can alter local climate in other ways, such as through the **rainshadow** effect. When moisture-laden air ascends a steep slope, it releases precipitation as it cools. By the time it flows over the top of the mountain and down the other side, it can be very dry, creating a relatively arid region in the rainshadow region. All these phenomena affect the nature and distribution of ecological communities in mountainous areas.

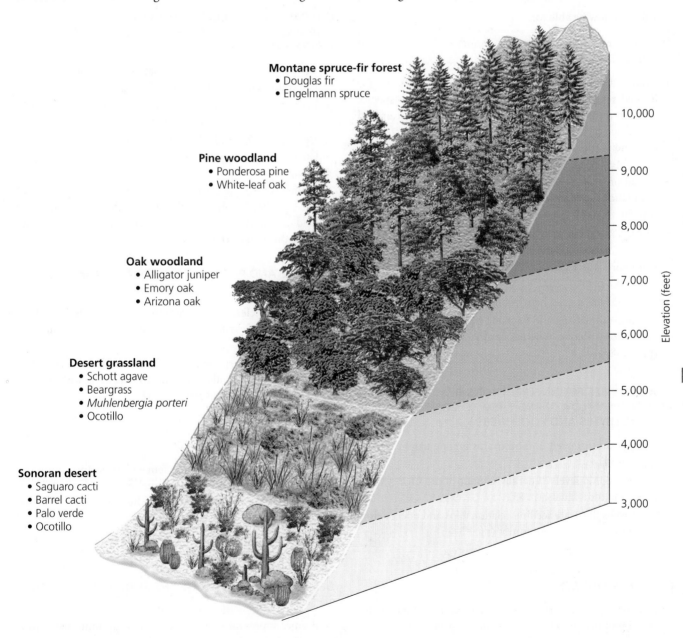

Montane spruce-fir forest
- Douglas fir
- Engelmann spruce

Pine woodland
- Ponderosa pine
- White-leaf oak

Oak woodland
- Alligator juniper
- Emory oak
- Arizona oak

Desert grassland
- Schott agave
- Beargrass
- *Muhlenbergia porteri*
- Ocotillo

Sonoran desert
- Saguaro cacti
- Barrel cacti
- Palo verde
- Ocotillo

Elevation (feet): 10,000 / 9,000 / 8,000 / 7,000 / 6,000 / 5,000 / 4,000 / 3,000

FIGURE 4.30 ▲ As altitude increases, plant communities change in ways similar to the ways they change with latitude as one moves toward the poles. Climbing a mountain in southern Arizona, as pictured here, can be likened to traveling from Mexico to Canada, taking the hiker through the local equivalent of several biomes.

➤ CONCLUSION

The natural world is so complex that we can visualize it in many ways and at various scales. Dividing the world's communities into major types, or biomes, is informative at the broadest geographic scales. Understanding how communities function at more local scales requires understanding how species interact with one another. Species interactions such as predation, parasitism, competition, and mutualism give rise to effects that are both weak and strong, direct and indirect.

Feeding relationships can be represented by the concepts of trophic levels and food webs, and particularly influential species are sometimes called keystone species. Increasingly, people are altering communities, in part by introducing non-native species that sometimes turn invasive. But more and more, through ecological restoration, we are also attempting to undo the changes we have caused.

REVIEWING OBJECTIVES

You should now be able to:

COMPARE AND CONTRAST THE MAJOR TYPES OF SPECIES INTERACTIONS

- Competition results when individuals or species vie for limited resources. It can occur within or among species and can result in coexistence or exclusion. It also can lead to realized niches, resource partitioning, and character displacement. (pp. 78–80)
- In predation, one species kills and consumes another. Predation is the basis of food webs and can influence population dynamics and community composition. (pp. 80–81)
- In parasitism, one species derives benefit by harming (but usually not killing) another. (pp. 81–82)
- Herbivory is an exploitative interaction in which an animal feeds on a plant. (p. 82)
- In mutualism, species benefit from one another. (pp. 82–83)
- In some mutualistic and parasitic interactions, the participants are symbiotic, whereas in others they are free-living. (pp. 82–83)

CHARACTERIZE FEEDING RELATIONSHIPS AND ENERGY FLOW, USING THEM TO CONSTRUCT TROPHIC LEVELS AND FOOD WEBS

- Energy is transferred in food chains among trophic levels. (pp. 83–84)
- Lower trophic levels generally contain more energy, biomass, and individuals. (p. 85)
- Food webs illustrate feeding relationships and energy flow among species in a community. (pp. 85–86)

DISTINGUISH CHARACTERISTICS OF A KEYSTONE SPECIES

- Keystone species exert impacts on communities that are far out of proportion to their abundance. (pp. 86–87)
- Top predators are frequently considered keystone species, but other types of organisms also exert strong effects on communities. (pp. 87–89)

CHARACTERIZE SUCCESSION, COMMUNITY CHANGE, AND THE DEBATE OVER THE NATURE OF COMMUNITIES

- Succession is a stereotypical pattern of change within a community through time. (pp. 92–93)
- Primary succession begins with an area devoid of life. Secondary succession begins with an area that has been severely disturbed. (pp. 92–93)
- Communities may undergo phase shifts involving irreversible change if disturbance is severe enough. (pp. 92–93)
- Clements held that communities are discrete, cohesive units. His view has largely been replaced by that of Gleason, who held that species may be added to and deleted from communities through time. (p. 93)

PERCEIVE AND PREDICT THE POTENTIAL IMPACTS OF INVASIVE SPECIES IN COMMUNITIES

- Invasive species such as the zebra mussel have altered the composition, structure, and function of communities. (pp. 93–95)
- Human beings are the cause of most modern species invasions, but we can also respond to invasions with prevention and control measures. (pp. 94–95)

EXPLAIN THE GOALS AND METHODS OF RESTORATION ECOLOGY

- Restoration ecology is the science of restoring communities to a previous, more functional or more "natural" condition, variously defined as before human impact or before recent industrial impact. (p. 95)
- Restoration ecology informs the growing practice of ecological restoration. (p. 95)

DESCRIBE AND ILLUSTRATE THE TERRESTRIAL BIOMES OF THE WORLD

- Biomes represent major classes of communities spanning large geographic areas. (p. 96)
- The distribution of biomes is determined by temperature, precipitation, and other factors. (pp. 96–97)

- Aquatic and coastal systems can be classified in similar ways, determined by different factors. (p. 97)
- Biomes include temperate and deciduous forest, temperate grassland, temperate rainforest, tropical rainforest, tropical dry forest, savanna, desert, tundra, boreal forest, and chaparral. (pp. 98-103)
- Mountains create mixtures of ecological communities. (p. 103)

TESTING YOUR COMPREHENSION

1. How does competition lead to a realized niche? How does it promote resource partitioning?

2. Contrast the several types of exploitative species interactions. How do predation, parasitism, and herbivory differ?

3. Give examples of symbiotic and nonsymbiotic mutualisms. Describe at least one way in which mutualisms affect your daily life.

4. Compare and contrast trophic levels, food chains, and food webs. How are these concepts related, and how do they differ?

5. What is meant by a keystone species, and what types of organisms are most often considered keystone species?

6. Explain primary succession. How does it differ from secondary succession? Give an example of each.

7. Name five changes to Great Lakes communities that have occurred since the invasion of the zebra mussel.

8. What is restoration ecology? Why is it an important scientific pursuit in today's world?

9. What factors most strongly influence the type of biome that forms in a particular place on land? What factors determine the type of aquatic system that may form in a given location?

10. Draw climate diagrams for a tropical rainforest and for a desert, and label each part of the diagram. How do the two diagrams differ? Describe all of the types of information an ecologist could glean from such diagrams.

SEEKING SOLUTIONS

1. Imagine that you spot two species of birds feeding side by side, eating seeds from the same plant. You begin to wonder whether competition is at work. Describe how you might design scientific research to address this question. What observations would you try to make at the outset? Would you try to manipulate the system to test your hypothesis that the two birds are competing? If so, how?

2. Spend some time outside on your campus or in your yard or in the nearest park or natural area. Find at least 10 species of organisms (plants, animals, or others), and observe each one long enough to watch it feed or to make an educated guess about how it derives its nutrition. Now, using Figure 4.12 as a model, draw a simple food web involving all the organisms you observed.

3. Can you think of one organism not mentioned in this chapter as a keystone species that you believe may be a keystone species? For what reasons do you suspect this? How could you experimentally test whether an organism is a keystone species?

4. Why do scientists consider invasive species to be a problem? What makes a species "invasive," and what ecological effects can invasive species have?

5. From year to year, biomes are stable entities, and our map of world biomes appears to be a permanent record of patterns across the planet. But are the locations and identities of biomes permanent, or could they change over time? Provide reasons for your answers.

6. **THINK IT THROUGH** A federal agency has put you in charge of devising responses to the zebra mussel invasion. Based on what you know from this chapter, how would you seek to control this species' spread and reduce its impacts? What strategies would you consider pursuing immediately, and for which strategies would you commission further scientific research? For each of your ideas, name one benefit or advantage, and identify one obstacle it might face in being implemented. What additional steps might you suggest to deal with the unfolding quagga mussel invasion?

CALCULATING ECOLOGICAL FOOTPRINTS

In 2005, environmental scientists David Pimentel, Rodolfo Zuniga, and Doug Morrison of Cornell University reviewed estimates for the economic and ecological costs inflicted by introduced and invasive species in the United States. They found that approximately 50,000 species have been introduced in the United States and that these account for over $120 billion in economic costs each year. These costs include direct losses and damage as well as costs required for control of the invasive species. (Pimentel's group did not try to quantify monetary estimates for losses of biodiversity, ecosystem services, and aesthetics, which they say would drive total costs several times higher.) Calculate values missing from the table to determine the number of introduced species of each type of organism and the annual cost that each inflicts on our economy.

Group of organism	Percentage of total introduced	Number of species introduced	Percentage of total annual costs	Annual economic costs
Plants	50.0	25,000	27.2	
Microbes	40.0		20.2	
Arthropods	9.0		15.7	
Fish	0.28		4.2	
Birds	0.19		1.5	$1.9 billion
Mollusks	0.18		1.7	
Reptiles and amphibians	0.11		0.009	
Mammals	0.04	20	29.4	$37.5 billion
TOTAL	100	50,000	100	$127.4 billion

Source: Pimentel, D., R. Zuniga, and D. Morrison, 2005. Update on the environmental and economic costs associated with alien-invasive species in the United States. *Ecological Economics* 52: 273–288.

1. Of the 50,000 species introduced into the United States, half are plants. Describe two ways in which exotic plants might be brought to a new environment. How might we help prevent non-native plants from establishing in new areas and posing threats to native communities?

2. Organisms that damage crop plants are the most costly of introduced species; weeds, pathogenic microbes, and arthropods that attack crops together account for half of the costs documented by Pimentel and his colleagues. What steps can we—farmers, governments, and all of us as a society—take to minimize the impacts of invasive species on crops?

3. How might your own behavior influence the influx and ecological impacts of non-native species like those listed above? Name three things you could personally do to help reduce the impacts of invasive species.

Mastering ENVIRONMENTALSCIENCE™

Go to **www.masteringenvironmentalscience.com** for practice quizzes, Pearson eText, videos, current events, and more.

The Mississippi River as it enters the Gulf of Mexico

5 ENVIRONMENTAL SYSTEMS AND ECOSYSTEM ECOLOGY

UPON COMPLETING THIS CHAPTER, YOU WILL BE ABLE TO:

- Describe the nature of environmental systems
- Define ecosystems and evaluate how living and nonliving entities interact in ecosystem-level ecology
- Outline the fundamentals of landscape ecology, GIS, and ecological modeling

- Assess ecosystem services and how they benefit our lives
- Compare and contrast how water, carbon, phosphorus, and nitrogen cycle through the environment
- Explain how human impact is affecting biogeochemical cycles

CENTRAL CASE STUDY

The Gulf of Mexico's "Dead Zone"

"In nature there is no 'above' or 'below,' and there are no hierarchies. There are only networks nesting within other networks."

—Fritjof Capra, theoretical physicist

"Let's say you put Saran Wrap over south Louisiana and suck the oxygen out. Where would all the people go?"

—Nancy Rabalais, biologist for the Louisiana Universities Marine Consortium

NORTH AMERICA

Atlantic Ocean

Mississippi River
Dead zone
Gulf of Mexico

Pacific Ocean

SOUTH AMERICA

Each year, Louisiana's fishermen ply the rich waters of the Gulf of Mexico and send a billion pounds of shrimp, fish, and shellfish to our dinner tables.

But in recent years, fishing in the Gulf has become difficult. Catches of shrimp, menhaden, and other common species are just half of what they were in the 1980s. These fisheries began declining years before oil gushed from the Deepwater Horizon drilling platform and fouled the region in 2010 (pp. 445, 447, 547–548). And they began before Hurricane Katrina and Hurricane Rita pummeled the region in 2005 and left boats, docks, and marinas in ruins. These disasters were like adding insult to injury, worsening a long decline already underway.

The reason for the decline? Each year billions of marine organisms have been suffocating in the Gulf's "dead zone," a region of water so depleted of oxygen that organisms are killed or driven away.

The low concentrations of dissolved oxygen in the bottom waters of this region represent a condition called **hypoxia** (see Figure 5.4, p. 114, and **THE SCIENCE BEHIND THE STORY**, pp. 112–113). Aquatic animals obtain oxygen by respiring through their gills, and, like us, these animals will asphyxiate if deprived of oxygen. Fully oxygenated water contains up to 10 parts per million (ppm) of oxygen. When concentrations drop below 2 ppm, creatures will leave, or die.

Fisherman in the Gulf of Mexico's "dead zone"

The Gulf's dead zone appears each spring and grows through the summer and fall, beginning in Louisiana waters offshore from the mouths of the Mississippi and Atchafalaya rivers. The dead zone reached its largest size in 2002 when it covered 22,000 km^2 (8,500 mi^2)—an area larger than New Jersey. Shrimp boats came up with nets nearly empty. One shrimper derided his meager catch as "cat food." Others, ironically, said they hoped a hurricane would strike and stir some oxygen into the Gulf's stagnant waters.

What's starving these waters of oxygen? Scientists studying the dead zone have identified the culprit: modern Midwestern farm practices and other human impacts hundreds of miles away. Farmers throughout the Mississippi River's vast watershed use fertilizers to nourish their crops. Rain and runoff carry excess nitrogen and phosphorus from these fertilizers off the fields and into streams and rivers, eventually flushing these nutrients down the Mississippi River. As this nutrient pollution from farms heads downriver toward Louisiana and the Gulf, other sources—urban runoff, industrial discharges, fossil fuel emissions, and municipal sewage outflow—add additional nitrogen and phosphorus pollution to the river (**FIGURE 5.1**).

Once the excess nutrients reach the Gulf, they trigger blooms of plankton in the surface waters. As

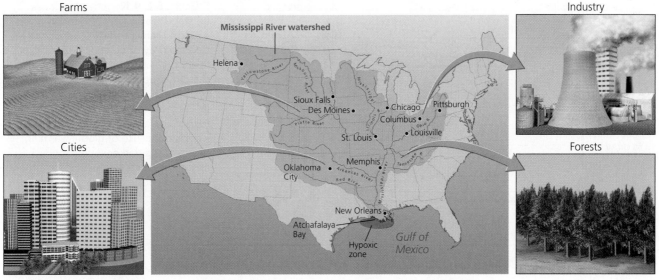

FIGURE 5.1 ▲ The Mississippi River system is the largest in North America. The river's watershed encompasses 3.2 million km² (1.2 million mi²), or 41% of the area of the lower 48 U.S. states. The river carries water, sediment, and pollutants from a variety of sources downriver to the Gulf of Mexico, where nutrient pollution has given rise to a hypoxic zone.

the masses of plankton begin to die and drift toward the bottom, they nourish bacteria, which also become overabundant. Bacteria need oxygen, and as they decompose the masses of dead plankton, they deplete dissolved oxygen from the bottom waters.

In 2010, scientists debated whether the Deepwater Horizon oil spill was worsening the dead zone. An explosion of oil-eating bacteria would worsen the hypoxia—but if oil was killing phytoplankton, that could lessen hypoxia. As of late summer when this book went to press, the dead zone was larger than expected, leading many researchers to suspect that the oil spill had worsened the dead zone.

Responding to the links that scientists have found between the dead zone and nutrient pollution from far upriver, U.S. government regulators have proposed that farmers in states such as Ohio, Iowa, and Illinois cut down on fertilizer use. Farmers' advocates protest that farmers are being singled out while urban pollution sources are ignored. Meanwhile, scientists have documented coastal dead zones in 400 other areas throughout the world, from Chesapeake Bay to Oregon to Denmark to the Black Sea. The story of how researchers determine the causes of these dead zones—and how we as a society respond—involves understanding environmental systems and the complex behavior they exhibit.

EARTH'S ENVIRONMENTAL SYSTEMS

Our planet's environment consists of complex networks of interlinked systems. These include physical systems ranging

from matter and molecules up to magma and mountains (Chapter 2). They include biological systems ranging from organisms and populations (Chapter 3) to communities of interacting species (Chapter 4). At the level of the ecosystem, they involve the interaction of living creatures with the nonliving entities around them. Earth's systems encompass cycles involving rock, air, and water that shape our landscapes and guide the flow of chemical elements and compounds that support life and regulate climate. We depend on these systems for our very survival.

Assessing questions holistically by taking a "systems approach" is helpful in environmental science, in which so many issues are multifaceted and complex. Such a broad and integrative approach poses challenges, because systems often show behavior that is difficult to predict. The scientific method (pp. 10–12) is easiest when researchers can isolate and manipulate small parts of complex systems, focusing on manageable components one at a time. However, environmental scientists are rising to the challenge of studying systems holistically, helping us to develop solutions to problems such as the Gulf of Mexico's hypoxic zone.

Systems show several defining properties

A **system** is a network of relationships among parts, elements, or components that interact with and influence one another through the exchange of energy, matter, or information. Earth's environmental systems receive inputs of energy, matter, or information, process these inputs, and produce outputs. As a system, the Gulf of Mexico receives inputs of fresh water, sediments, nutrients, and pollutants from the Mississippi and other rivers. Shrimpers and fishermen harvest some of the Gulf system's output: matter and energy in the form of shrimp and of fish. This output subsequently becomes input to the nation's economic system and to the digestive systems of all of us who consume the seafood.

Abundant prey

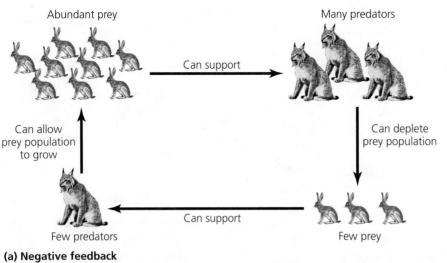

Many predators

Can support →

Can allow
prey population
to grow ↑

Can deplete
prey population ↓

Few predators

← Can support

Few prey

(a) Negative feedback

FIGURE 5.2 ◄ Negative feedback **(a)** stabilizes systems. Abundant prey support many predators, which in turn deplete prey populations—but because fewer prey support fewer predators, the prey population rises again, and the cycle continues. Positive feedback **(b)** destabilizes systems, pushing them toward extremes. As glaciers and sea ice melt because of global warming, this exposes darker surfaces, which absorb more sunlight, causing further warming and melting.

1 In cool climate, sunlight reflects off white surfaces

2 As climate warms, sunlight is absorbed where dark surfaces are exposed

3 Light absorption speeds warming, exposing more dark surfaces

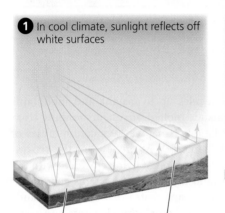

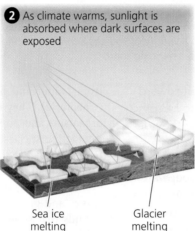

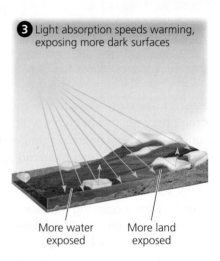

Solid surface of sea ice Glacier completely covers land

Sea ice melting Glacier melting

More water exposed More land exposed

(b) Positive feedback

Sometimes a system's output can serve as input to that same system, a circular process described as a **feedback loop**. Feedback loops are of two types, negative and positive. These terms do *not* respectively mean bad and good. In fact, it's the opposite: just as you'd rather receive a negative test result at the doctor's office, a negative feedback loop is generally better for the stability of an environmental system than a positive one.

In a **negative feedback loop**, output that results from a system acts as input that moves the system in the other direction. Input and output essentially neutralize one another's effects, stabilizing the system. For instance, a thermostat stabilizes a room's temperature by turning the furnace on when the room gets cold and shutting it off when the room gets hot. We encountered one example of a negative feedback loop in Figure 4.5 (p. 81) in Chapter 4. As a population of prey increases, this expands the food supply for predators—but as the predators increase in number, they catch more prey and drive the prey population down. Fewer prey support fewer predators, so the predator population declines, and this allows the prey population to rise again, beginning the cycle anew (**FIGURE 5.2A**). Most systems in nature involve negative feedback loops. Negative feedback enhances stability, and in the long run, only those systems that are stable will persist.

Positive feedback loops have the opposite effect. Rather than stabilizing a system, they drive it further toward an extreme. In positive feedback, increased output leads to increased input, leading to further increased output. Exponential growth in a population (pp. 66–67) provides an example. The more individuals there are, the more offspring can be produced. Another example is the spread of cancer; as cells multiply out of control, the process is self-accelerating.

One positive feedback cycle of great concern to environmental scientists today involves the melting of glaciers and sea ice in the Arctic as a result of global warming (p. 506). Ice and snow, being white, reflect sunlight and keep surfaces cool. But if the climate warms enough to melt the ice and snow, darker surfaces of land and water are exposed, and these darker surfaces absorb sunlight. This warms the surface, causing further melting, which in turn exposes more dark surface area, leading to further warming (**FIGURE 5.2B**). Runaway cycles of positive feedback are rare in nature, but they are common in natural systems altered by human impact, and they can destabilize those systems.

In a system stabilized by negative feedback, when processes move in opposing directions at equivalent rates so that their effects balance out, they are said to be in **dynamic equilibrium**. Processes in dynamic equilibrium can contribute to **homeostasis**, the tendency of a system to maintain constant or stable internal conditions. A system (such as an organism) in homeostasis keeps its internal conditions within a range that allows it to function. However, the steady state of a homeostatic system may itself change slowly over time. For instance, organisms grow and mature. Similarly, Earth has experienced gradual changes in atmospheric composition and ocean chemistry over its long history, yet life persists and our planet remains, by most definitions, a homeostatic system.

It is difficult to understand systems fully by focusing on their individual components because systems can show **emergent properties**, characteristics not evident in the components alone. Stating that systems possess emergent properties is a lot like saying, "The whole is more than the sum of its parts." For example, if you were to reduce a tree to its component parts (leaves, branches, trunk, bark, roots, fruit, and so on) you would not be able to predict the whole tree's emergent properties, which include the role the tree plays as habitat for birds, insects, fungi, and other organisms (**FIGURE 5.3**). You could analyze the tree's chloroplasts (photosynthetic cell organelles), diagram its branch structure, and evaluate the nutritional content of its fruit, but you would still be unable to understand the tree as habitat, as part of a forest landscape, or as a reservoir for carbon storage.

Systems seldom have well-defined boundaries, so deciding where one system ends and another begins can be difficult. Consider a desktop computer. It is certainly a system—a network of parts that interact and exchange energy and information—but where are its boundaries? Is the system merely what sits atop your desk? Or does it include the network you connect it to at school, home, or work? What about the energy grid you plug it into, with its transmission lines and distant power plants? And what of the Internet? Browsing the Web, you draw in digitized text, light, and sound from around the world.

No matter how we attempt to isolate or define a system, we soon see that it has connections to systems larger and smaller than itself. Systems may exchange energy, matter, and information with other systems, and they may contain or be contained within other systems. Thus, where we draw boundaries may depend on the spatial or temporal scale at which we choose to focus.

Environmental systems interact

The Gulf of Mexico and the Mississippi River are systems that interact. On a map, the Mississippi River appears as a branched and braided network of water channels. But where are this system's boundaries? You might argue that the Mississippi consists primarily of water, originates in Minnesota, and ends in the Gulf of Mexico near New Orleans. But what about the tributary rivers that feed into it? And what about the farms, cities, and forests that line its banks (see Figure 5.1)? Major rivers such as the Missouri, Arkansas, and Ohio flow into the Mississippi. Hundreds of smaller tributaries drain vast expanses of farmland, woodland, fields, cities, towns, and industrial areas before their water reaches the Mississippi. These waterways carry with them millions of tons of sediment, hundreds of species of plants and animals, and numerous pollutants.

For an environmental scientist interested in runoff and the flow of water, sediment, or pollutants, it may make the most

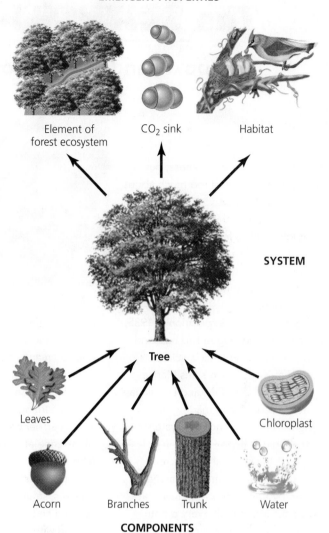

EMERGENT PROPERTIES

Element of forest ecosystem

CO_2 sink

Habitat

SYSTEM

Tree

Leaves

Chloroplast

Acorn

Branches

Trunk

Water

COMPONENTS

FIGURE 5.3 ▲ A system's emergent properties are not evident when we break the system down into its component parts. For example, a tree serves as wildlife habitat and plays roles in forest ecology and global climate regulation, but you would not know that from considering the tree only as a collection of leaves, branches, and chloroplasts. If we try to understand systems solely by breaking them into component parts, we will miss much of what makes them important.

sense to view the Mississippi River's **watershed**—the entire area of land that a river drains (pp. 167, 403)—as a system. However, a scientist interested in the Gulf of Mexico's dead zone may choose to view the Mississippi River watershed together with the Gulf as the system of interest, because their interaction is central to the problem. In environmental science, one's delineation of a system depends on the questions one is addressing.

Researchers studying the Gulf of Mexico's low levels of oxygen attribute this hypoxia to an ecological chain of events triggered by excess nutrients (particularly nitrogen and phosphorus from fertilizers applied to crops) that flow down the Mississippi River into the Gulf. Nutrients such as nitrogen and phosphorus are limiting factors (p. 67) on the growth of *phytoplankton* (p. 78)—photosynthetic algae, protists, and cyanobacteria that drift near the surface. For these organisms, more nutrients mean more growth, so the enhanced input of

The SCIENCE behind the Story

Hypoxia and the Gulf of Mexico's "Dead Zone"

Dr. Nancy Rabalais, LUMCON

She was prone to seasickness, but Nancy Rabalais cared too much about the Gulf of Mexico to let that stop her. Leaning over the side of an open boat idling miles from shore, she hauled a water sample aboard—and helped launch the long effort to breathe life back into the Gulf's "dead zone."

Since that first expedition in 1985, Rabalais, her colleague and husband Eugene Turner, and fellow scientists at the Louisiana Universities Marine Consortium (LUMCON) and Louisiana State University have made great progress in unraveling the mysteries of the region's hypoxia—and in getting it on the political radar screen.

Rabalais and other researchers began by tracking oxygen levels at nine sites in the Gulf every month and continued those measurements for five years. At dozens of other spots near the shore and in deep water they took less frequent oxygen readings. Sensors, as they are lowered into the water, measure oxygen levels and send continuous readings back to a shipboard computer. Further data come from fixed, submerged oxygen meters that continuously measure dissolved oxygen and store the data.

The team also collected hundreds of water samples, using lab tests to measure levels of nitrogen, salt, bacteria, and phytoplankton. LUMCON scientists logged hundreds of miles in their ships, regularly monitoring more than 70 sites in the Gulf. They also donned scuba gear to view firsthand the condition of shrimp, fish, and

other sea life. Such a range of long-term data allowed the researchers to build a "map" of the dead zone, tracking its location and its consequences.

In 1991, Rabalais made that map public, earning immediate headlines. That year, her group mapped the size of the zone at more than 10,000 km^2 (about 4,000 mi^2). Bottom-dwelling shrimp were stretching out of their burrows, straining for oxygen. Many fish had fled. The bottom waters, infused with sulfur from bacterial decomposition, smelled of rotten eggs.

The group's years of monitoring also enabled them to explain and predict the dead zone's emergence. As rivers rose each spring (and as fertilizers were applied in the Midwestern farm states), oxygen would start to disappear in the northern Gulf. The hypoxia would last through the summer or fall, until seasonal storms mixed oxygen into hypoxic areas.

Over time, monitoring linked the dead zone's size to the volume of river flow and its nutrient load. The 1993 flooding of the Mississippi created a zone much larger than the year before, whereas a drought in 2000 brought low river flows, low nutrient loads, and a small dead zone (**see figure**). In 2005, the dead zone was predicted to

be large, but Hurricanes Katrina and Rita stirred oxygenated surface water into the depths, decreasing the dead zone that year.

The source of the problem, Rabalais said, lay back on land. The Mississippi and Atchafalaya rivers draining into the Gulf were polluted with agricultural runoff, and the nutrient pollution from fertilizers spurred algal blooms whose decomposition by bacteria snuffed out oxygen in wide stretches of ocean water. Figure 5.4 (p. 114) illustrates how the process works.

Many Midwestern farming advocates and some scientists, such as Derek Winstanley, chief of the Illinois State Water Survey, challenged the findings. They argued that the Mississippi naturally carries high loads of nitrogen from the rich prairie soil and that Rabalais's team had not ruled out upwelling in the Gulf as a source of nutrients.

But sediment analyses showed that Mississippi River mud contained many fewer nitrates early in the century, and Rabalais and Turner found that silica residue from phytoplankton blooms increased in Gulf sediments between 1970 and 1989, paralleling rising nitrogen levels. In 2000, a federal integrative assessment team of dozens of scientists laid the blame for the dead zone on nutrients from fertilizers and other sources.

Then in 2004, while representatives of farmers and fishermen bickered over political fixes, Environmental Protection Agency water quality scientist Howard

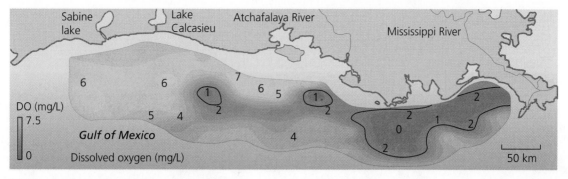

(a) Dissolved oxygen at bottom, July 2009

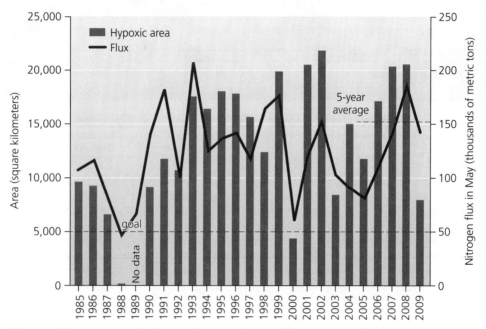

(b) Area of hypoxic zone in the northern Gulf of Mexico

The map in **(a)** shows dissolved oxygen concentrations in bottom waters of the Gulf of Mexico off the Louisiana coast from the July 2009 survey. Areas in red indicate the lowest oxygen levels. Regions considered hypoxic (< 2 mg/L) are encircled with a black line. The graph in **(b)** shows that the size of the hypoxic zone (shown by bars) is correlated with the amount of nitrogen pollution entering from the Mississippi River (shown by line). In addition, floods increase its size by bringing additional runoff (as in 1993), whereas tropical storms decrease its size by mixing oxygen-rich water into the dead zone (as in 2003). Between 2005 and 2009, the hypoxic zone averaged 15,670 km^2 (6,000 mi^2) in size. Scientists and policymakers aim to reduce this to 5,000 km^2 (1,930 mi^2). Data from Nancy Rabalais, Louisiana Universities Marine Consortium (LUMCON).

Marshall suggested that to alleviate the dead zone we'd be best off reducing phosphorus pollution from industry and sewage treatment. His reasoning: Phytoplankton need both nitrogen and phosphorus, but there is now so much nitrogen in the Gulf that phosphorus has become the limiting factor on phytoplankton growth.

Since then, research has supported this contention, and scientists now propose that nitrogen and phosphorus should be managed jointly. Moreover, recent research indicates that a federally mandated 30% reduction in nitrogen in the river will not be adequate to eliminate the dead zone. Scientists also maintain that large-scale restoration of wetlands along the

river and at the river's delta would best filter pollutants before they reach the Gulf.

All this research is guiding a federal plan to reduce farm runoff, clean up the Mississippi, restore wetlands, and shrink the Gulf's dead zone. It has also led to a better understanding of hypoxic zones around the world. ∎

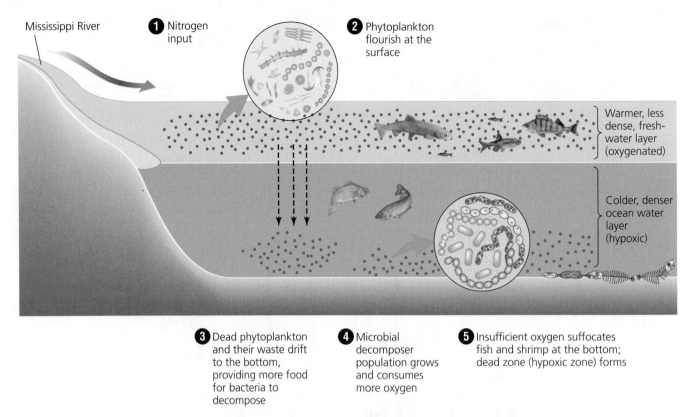

Mississippi River

1 Nitrogen input

2 Phytoplankton flourish at the surface

Warmer, less dense, fresh-water layer (oxygenated)

Colder, denser ocean water layer (hypoxic)

3 Dead phytoplankton and their waste drift to the bottom, providing more food for bacteria to decompose

4 Microbial decomposer population grows and consumes more oxygen

5 Insufficient oxygen suffocates fish and shrimp at the bottom; dead zone (hypoxic zone) forms

FIGURE 5.4 ▲ Excess nitrogen causes eutrophication in coastal marine systems such as the Gulf of Mexico. Coupled with stratification (layering) of water, eutrophication can severely deplete dissolved oxygen. Nitrogen from river water **1** boosts growth of phytoplankton **2**, which die and are decomposed at the bottom by bacteria **3**. Stability of the surface layer prevents deeper water from absorbing oxygen to replace oxygen consumed by decomposers **4**, and the oxygen depletion suffocates or drives away bottom-dwelling marine life **5**. This process gives rise to hypoxic zones like that of the Gulf of Mexico.

nitrogen and phosphorus to the Gulf boosts the growth of phytoplankton. As phytoplankton flourish at the surface and as the tiny animals and protists called zooplankton (p. 78) consume them, more and more dead plankton and their waste products drift to the bottom, and these provide food for organisms (mainly bacteria) that decompose them. The result is a population explosion of bacteria. These decomposers consume enough oxygen though cellular respiration (pp. 32–33) to cause oxygen concentrations in bottom waters to plummet. The resulting pulse of hypoxia suffocates shrimp and fish that live at the bottom.

This process of nutrient overenrichment, blooms of algae, increased production of organic matter, decomposition, and hypoxia is known as **eutrophication** (**FIGURE 5.4**). You are familiar with eutrophication if you have ever noticed a stagnant freshwater pond covered with a thick green layer of algae. As we will see in Chapters 15 and 16 (pp. 422, 448), eutrophication can take place in both freshwater and saltwater systems.

In the Gulf of Mexico, the hypoxic dead zone persists because incoming fresh water from the Mississippi River remains stratified in a layer at the surface and mixes only slowly with the denser salty ocean water beneath it, so that oxygenated surface water does not soon make its way down to the bottom-dwelling life that needs it. Bottom waters receive oxygen only after storms stir up the waters, or in winter as natural mixing occurs while nutrient inputs drop (with less fertilizer application to crops upriver).

Although nitrogen- and phosphorus-rich fertilizers are the main source of excess nutrients in the Mississippi River watershed, there are other sources as well, including livestock manure, nitrogen-fixing crops, sewage treatment facilities, street runoff, and industrial and automobile emissions. These sources all add to the nutrients already entering the river naturally from soil decomposition. In total, agricultural sources are thought to contribute 74% of the nitrate and 65% of the total nitrogen carried in the Mississippi River. Much of the nitrate originates in upper portions of the watershed from corn and soybean fields in Iowa, Illinois, Indiana, Minnesota, and Ohio. Nitrogen fertilizer input to farmland in the Mississippi River watershed has increased dramatically in recent decades.

We may perceive Earth's systems in various ways

There are many ways to delineate natural systems. Categorizing environmental systems can help make Earth's dazzling complexity comprehensible to the human brain and accessible to problem solving. For instance, scientists sometimes divide Earth's components into structural spheres. The **lithosphere** (p. 34) is the rock and sediment beneath our feet, in the planet's uppermost layers. The **atmosphere** (p. 462) is composed of the air surrounding our planet. The **hydrosphere** (p. 403) encompasses all water—salt or fresh, liquid, ice, or vapor—in surface bodies, underground, and in the atmosphere. The **biosphere** (pp. 61–62) consists of all the planet's organisms and the abiotic (nonliving) portions of the environment with which they interact.

Picture a robin plucking an earthworm from the ground after a rain. You are witnessing an organism (the robin) consuming another organism (the earthworm) by removing it from part of the lithosphere (the soil) that the earthworm had been modifying, after rain (from the hydrosphere) moistened the ground. The robin might then fly through the air (the atmosphere) to a tree (an organism), in the process respiring (combining oxygen from the atmosphere with glucose from the organism, and adding water to the hydrosphere and carbon dioxide and heat to the atmosphere). Finally, the bird might defecate, adding nutrients from the organism to the lithosphere below. The study of such interactions among living and nonliving things is a key part of ecology at the ecosystem level.

ECOSYSTEMS

An **ecosystem** consists of all organisms and nonliving entities that occur and interact in a particular area at the same time. The ecosystem concept builds on the idea of the biological community (Chapter 4), but ecosystems include abiotic components as well as biotic ones. In ecosystems, energy flows and matter cycles among these components.

Ecosystems are systems of interacting living and nonliving entities

The ecosystem concept originated early last century with scientists such as British ecologist Arthur Tansley, who recognized that biological entities are tightly intertwined with chemical and physical entities. Tansley felt that there was so much interaction between organisms and their abiotic environments that it made the most sense to view living and nonliving elements together. For instance, in the nearshore ecosystem of the northern Gulf of Mexico, the flow of water, sediment, and nutrients from the Mississippi River intimately affects coastal aquatic organisms. In turn, the photosynthesis, respiration, and decomposition that these organisms carry out influence the chemical and physical conditions of the coastal waters.

Ecologists soon began analyzing ecosystems as an engineer might analyze the operation of a machine. In this view, ecosystems are systems that receive inputs of energy, process and transform that energy while cycling matter internally, and produce outputs (such as heat, water flow, and animal waste products) that enter other ecosystems.

Energy flows in one direction through ecosystems. Most arrives as radiation from the sun, powers the system, and exits in the form of heat (**FIGURE 5.5A**). Matter, in contrast, is generally recycled within ecosystems (**FIGURE 5.5B**). We saw in

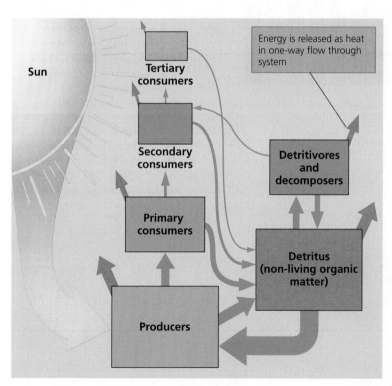

(a) Energy flowing through an ecosystem

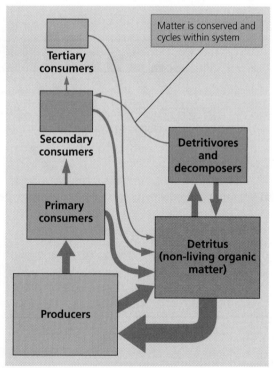

(b) Matter cycling within an ecosystem

FIGURE 5.5 ▲ Energy enters, flows through, and exits an ecosystem. In (a), light energy from the sun (yellow arrow) drives photosynthesis in producers, which begins the transfer of chemical energy (green arrows) among trophic levels and detritus. Energy exits the system through respiration in the form of heat (red arrows). In contrast, matter cycles within an ecosystem. In (b), blue arrows show the movement of nutrients among trophic levels and detritus. In both diagrams, box sizes represent relative magnitudes of energy or matter content, and arrow widths represent relative magnitudes of energy or matter transfer. Such magnitudes may vary tremendously from one ecosystem to another. For simplicity, various abiotic components (such as water, air, and inorganic soil content) of ecosystems are omitted from these schematic diagrams.

Chapter 4 (pp. 85–86) how energy and matter pass among producers, consumers, and decomposers through food-web relationships. Matter is recycled because when organisms die and decay their nutrients remain in the system. In contrast, most energy that organisms take in drives cellular respiration and is released as heat.

Energy is converted to biomass

As autotrophs, such as green plants and phytoplankton, convert solar energy to the energy of chemical bonds in sugars through photosynthesis (pp. 31–32), they perform *primary production*. Specifically, the assimilation of energy by autotrophs is termed **gross primary production**. Autotrophs use a portion of this production to power their own metabolism by cellular respiration (p. 32). The energy that remains after respiration and that is used to generate biomass (p. 85) ecologists call **net primary production**. Thus, net primary production equals gross primary production minus respiration. Net primary production can be measured by the energy or the organic matter stored by autotrophs after they have metabolized enough for their own maintenance.

Another way to think of net primary production is that it represents the energy or biomass available for consumption by heterotrophs. Plant matter not eaten by herbivores becomes fodder for detritivores and decomposers once the plant dies or drops its leaves. Heterotrophs use the energy they gain from plants for their own metabolism, growth, and reproduction. The total biomass that heterotrophs generate by consuming autotrophs is termed *secondary production*.

Ecosystems vary in the rate at which autotrophs convert energy to biomass. The rate at which production occurs is termed **productivity**, and ecosystems whose plants convert solar energy to biomass rapidly are said to have high **net primary productivity**. Freshwater wetlands, tropical forests, coral reefs, and algal beds tend to have the highest net primary productivities, whereas deserts, tundra, and open ocean tend to have the lowest (**FIGURE 5.6A**). Variation

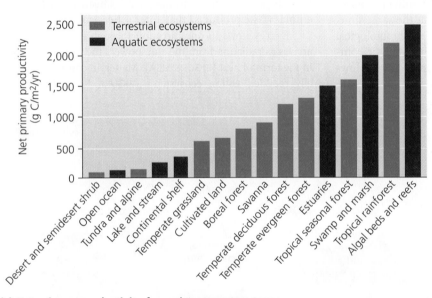

(a) Net primary productivity for major ecosystem types

FIGURE 5.6 ◀ Freshwater wetlands, tropical forests, coral reefs, and algal beds tend to show high net primary productivities **(a)** whereas deserts, tundra, and the open ocean show low values. A world map of net primary production created from satellite data **(b)** shows that on land, net primary production varies geographically with temperature and precipitation. In the world's oceans, net primary production is highest around the margins of continents, where nutrients (of both natural and human origin) run off from land. Data in (a) from Whittaker, R.H., 1975. *Communities and ecosystems*, 2nd ed. New York: MacMillan. Map in (b) from satellite data presented by Field, C.B., et al., 1998. Primary production of the biosphere: Integrating terrestrial and oceanic components. *Science* 281: 237–240. Reprinted with permission from AAAS.

116

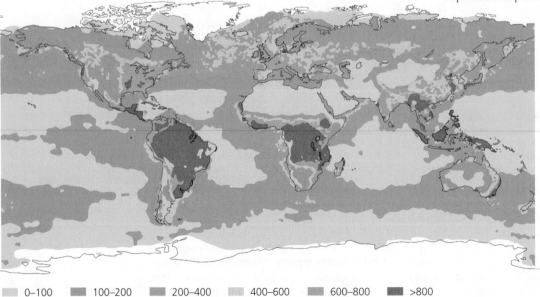

0–100   100–200   200–400   400–600   600–800   >800

(b) Global map of net primary productivity

among ecosystems and among biomes in net primary productivity results in geographic patterns across the globe (**FIGURE 5.6B**). In terrestrial ecosystems, net primary productivity tends to increase with temperature and precipitation. In aquatic ecosystems, net primary productivity tends to rise with light and the availability of nutrients.

Nutrients influence productivity

Nutrients are elements and compounds (pp. 25–27) that organisms consume and require for survival. Organisms need several dozen naturally occurring nutrients to survive. Elements and compounds required in relatively large amounts (such as nitrogen, carbon, and phosphorus) are called *macronutrients*. Nutrients needed in small amounts are called *micronutrients*.

Nutrients stimulate production by plants, and lack of nutrients can limit production. As mentioned earlier, the availability of nitrogen or phosphorus frequently is a limiting factor (p. 67) for plant or algal growth. When these nutrients are added to a system, producers show the greatest response to whichever nutrient has been in shortest supply. Nitrogen tends to be limiting in marine systems, and phosphorus in freshwater systems. Thus the Gulf of Mexico's hypoxic zone is thought to result primarily from excess nitrogen, whereas freshwater ponds and lakes in the Mississippi River Valley tend to experience eutrophication when they receive too much phosphorus. However, organisms also require the right balance of nutrients, and phytoplankton in the Gulf now receive so much nitrogen that it is not always limiting, and instead phosphorus concentrations often control growth.

Canadian ecologist David Schindler and others demonstrated the effects of phosphorus on freshwater systems in the 1970s by experimentally manipulating entire lakes. In one experiment, his team bisected a 16-ha (40-acre) lake in Ontario with a plastic barrier. To one half the researchers added carbon, nitrate, and phosphate; to the other they added only carbon and nitrate. Soon after the experiment began, they witnessed a dramatic increase in algae in the half of the lake that received phosphate, whereas the other half (the control for the experiment, p. 11) continued to host algal levels typical for lakes in the region (**FIGURE 5.7**). This difference held until shortly after they stopped fertilizing seven years later. At that point, algae decreased to normal levels in the half that had previously received phosphate. Such experiments showed clearly that phosphorus addition can markedly increase primary productivity in lakes.

Similar experiments in coastal ocean waters show nitrogen to be the more important limiting factor for primary productivity. In experiments in the 1980s and 1990s, Swedish ecologist Edna Granéli took samples of ocean water from the Baltic Sea and added phosphate, nitrate, or nothing. Chlorophyll and phytoplankton increased greatly in the flasks with nitrate, whereas those with phosphate did not differ from the controls. Experiments in Long Island Sound by other researchers show similar results. For open ocean waters far from shore, research indicates that iron is a highly effective nutrient.

Because nutrients run off from land into the Baltic Sea, Long Island Sound, the Gulf of Mexico, and worldwide, primary productivity in the oceans tends to be greatest in coastal

FIGURE 5.7 ▲ A portion of this lake in Ontario was experimentally treated with the addition of phosphate. This treated portion experienced an immediate, dramatic, and prolonged algal bloom, visible in the opaque water in the topmost part of this photo.

waters and lowest in open ocean areas far from land (see Figure 5.6B). Satellite imaging technology that reveals phytoplankton densities has given scientists an improved view of productivity at regional and global scales, and this has helped them track blooms of algae that contribute to coastal hypoxic zones (**FIGURE 5.8**).

The number of known dead zones is rising globally, with over 400 documented so far. Most are located off the coasts of

FIGURE 5.8 ▼ Satellite images like this one have helped scientists track runoff and phytoplankton blooms. Shown is the Louisiana coast on December 30, 1997. Greenish brown at the top is land, and deep blue at the bottom right is water of the Gulf of Mexico. Along the coast, two light brown plumes of sediment can be seen expanding into the Gulf's waters from the mouths of the Mississippi and Atchafalaya rivers. The shades of turquoise represent phytoplankton blooms that result from nutrient inputs and that foster the dead zone. Image from the SeaWiFS Project, NASA/Goddard Space Flight Center, and ORBIMAGE.

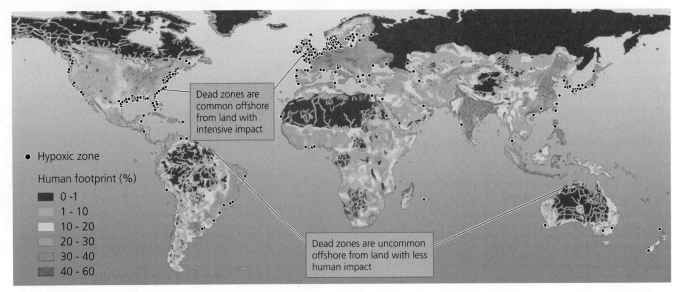

- Hypoxic zone

Human footprint (%)
- 0 -1
- 1 - 10
- 10 - 20
- 20 - 30
- 30 - 40
- 40 - 60

Dead zones are common offshore from land with intensive impact

Dead zones are uncommon offshore from land with less human impact

FIGURE 5.9 ▲ Over 400 marine dead zones have been recorded across the world. These dead zones (shown by dots in the map) occur mostly offshore from areas of land with the greatest human ecological footprints (here, expressed as percentages). Data from Diaz, R. and R. Rosenberg, 2008. Spreading dead zones and consequences for marine ecosystems. *Science* 321: 926–929, Fig. 1. Reprinted with permission from AAAS.

Europe and the eastern United States (**FIGURE 5.9**), and most result from increasing nutrient pollution from farms, cities, and industry. Some are seasonal (like the Gulf of Mexico's), some occur irregularly, and others are permanent. In North America, Chesapeake Bay may be the most severely affected area besides the Gulf of Mexico. Decades of pollution and human impact here have devastated fisheries and greatly altered the bay's ecology.

If one were to add up all the world's marine and coastal dead zones, they would cover an area the size of the state of Michigan. Scientists calculate that the amount of marine life missing from the oceans as a result of dead zones likely exceeds the total amount of shellfish harvested each year from the entire United States—a harvest worth over $2 billion.

The good news is that in locations where people have reduced nutrient runoff, dead zones have begun to disappear. In New York City, hypoxic zones at the mouths of the Hudson and East rivers were nearly eliminated once sewage treatment was improved. The Black Sea, which borders Ukraine, Russia, Turkey, and eastern Europe, had long suffered one of the world's worst hypoxic zones. Then in the 1990s, after the Soviet Union collapsed, industrial agriculture in the region declined drastically. With fewer fertilizers draining into it, the Black Sea began to recover, and today fisheries are reviving. However, agricultural collapse is not a strategy anyone would choose to alleviate hypoxia. Rather, scientists are proposing a variety of innovative and economically acceptable ways to reduce nutrient runoff.

Ecosystems interact spatially

Whether we stand at a river's mouth or peruse a satellite image, we can conceptualize ecosystems at different scales. An ecosystem can be as small as a puddle of water where brine shrimp and tadpoles feed on algae and detritus with mad abandon as the pool dries up. Or an ecosystem might be as large as a bay, lake, or forest. For some purposes, scientists even view the entire biosphere as a single all-encompassing ecosystem. The term is most often used, however, to refer to systems of moderate geographic extent that are somewhat self-contained. For example, the salt marshes that line the outer delta of the Mississippi where its waters mix with those of the Gulf of Mexico may be classified as an ecosystem.

Adjacent ecosystems may share components and interact extensively. For instance, a pond ecosystem is very different from a forest ecosystem that surrounds it, but salamanders that develop in the pond live their adult lives under logs on the forest floor until returning to the pond to breed. Rainwater that nourishes forest plants may eventually make its way to the pond, carrying with it nutrients from the forest's leaf litter. Likewise, coastal dunes, the ocean, and a lagoon or salt marsh all may interact, as do forests and prairie where they converge. Areas where ecosystems meet may consist of transitional zones called **ecotones**, in which elements of each ecosystem mix.

WEIGHING THE ISSUES

Ecosystems Where You Live Think about the area where you live, and briefly describe this region's ecosystems. How do these systems interact? For instance, does any water pass from one to another? Describe the boundaries of watersheds in your region. If one ecosystem were greatly modified (say, if a shopping mall were built atop a wetland or amid a forest), what impacts might that have on nearby systems? (Note: If you live in a city, realize that urban areas can be thought of as ecosystems, too!)

Landscape ecologists study geographic patterns

Because components of different ecosystems may intermix, ecologists often find it useful to view these systems on a larger geographic scale that encompasses multiple ecosystems. For instance, if you are studying large mammals such as black bears, which move seasonally from mountains to valleys or between mountain ranges, you had better consider the overall landscape that includes all these areas. If you study fish such as salmon, which migrate between marine and freshwater ecosystems, you need to know how these systems interact.

In such a broad-scale approach, called **landscape ecology**, scientists study how landscape structure affects the abundance, distribution, and interaction of organisms. Landscape-level approaches are also helping scientists, citizens, planners, and policymakers to plan for sustainable regional development (pp. 353–356).

For a landscape ecologist, a landscape is made up of a spatial array of **patches**. Depending on the researcher's perspective, patches may consist of ecosystems or may simply be areas of habitat for a particular organism. Patches are spread spatially over a landscape in a **mosaic**. This metaphor reflects how natural systems often are arrayed across landscapes in complex patterns, like an intricate work of art. Thus, a forest ecologist may refer to a mosaic of forested patches left standing in an agricultural landscape, or a butterfly biologist might speak of a mosaic of patches of grassland habitat for a particular species of butterfly.

FIGURE 5.10 illustrates a landscape consisting of four ecosystem types, with ecotones along their borders indicated by thick red lines. At this scale, we perceive a mosaic consisting of four patches and a river. However, we can view a landscape at different scales. The figure's inset shows a magnified view of an ecotone. At this finer resolution, we see that the ecotone consists of patches of forest and grassland in a complex arrangement. The scale at which an ecologist focuses will depend on the questions he or she is interested in, or on the organisms he or she is studying.

Every organism has specific habitat needs, so when its habitat is distributed in patches across a landscape, individuals may need to expend energy and risk predation traveling from one to another. If the patches are far apart, the organism's population may become divided into subpopulations, each occupying a different patch in the mosaic. Such a network of subpopulations, most of whose members stay within their respective patches but some of whom move among patches or mate with members of other patches, is called a **metapopulation**. When patches are still more isolated from one another, individuals may not be able to travel between them at all. In such a case, smaller subpopulations may be at risk of extinction.

Because of this extinction risk, metapopulations and landscape ecology are of great interest to **conservation biologists** (pp. 300–302), scientists who study the loss, protection, and restoration of biodiversity. Of particular concern is the

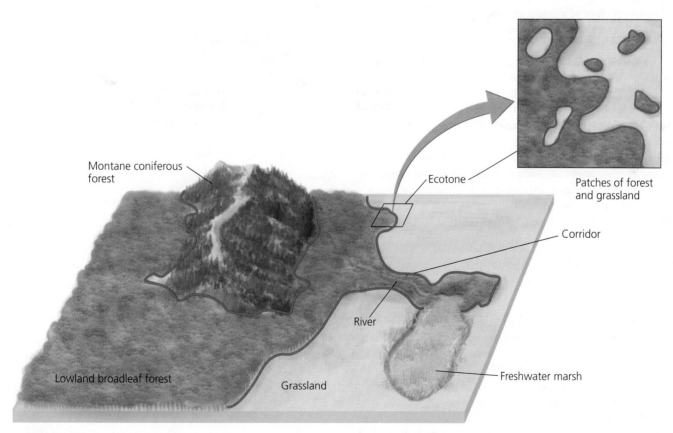

Montane coniferous forest

Ecotone

Patches of forest and grassland

Corridor

River

Lowland broadleaf forest

Grassland

Freshwater marsh

FIGURE 5.10 ▲ Landscape ecology deals with spatial patterns above the ecosystem level. This generalized diagram of a landscape shows a mosaic of patches of five ecosystem types (three terrestrial types, a marsh, and a river). Thick red lines indicate ecotones. A stretch of lowland broadleaf forest running along the river serves as a corridor connecting the large region of forest on the left to the smaller patch of forest alongside the marsh (allowing forest animals to move between patches). The inset shows a magnified view of the forest-grassland ecotone and how it consists of patches on a smaller scale.

fragmentation of habitat into small and isolated patches (pp. 291, 335–340)—something that often results from human development pressures. Establishing corridors of habitat (see Figure 5.10) to link patches and allow animals to move among them is one approach that conservation biologists pursue as they attempt to maintain biodiversity in the face of human impact. We will return to these issues when we study habitat fragmentation and conservation biology in more detail in Chapters 11 and 12.

Remote sensing helps us apply landscape ecology

As more scientists take a landscape perspective, they are benefiting from better and better remote-sensing technologies. Satellites orbiting Earth are sending us more and better data than ever before on how the surface of our planet looks (see Figure 5.8). By helping us monitor our planet from above, satellite imagery is making vital contributions to modern environmental science.

A common tool for research in landscape ecology is the **geographic information system (GIS)**. A GIS consists of computer software that takes multiple types of data (for instance, on geology, hydrology, vegetation, animal populations, and human infrastructure) and combines them on a common set of geographic coordinates. The idea is to create a complete picture of a landscape and to analyze how elements of the different data sets are arrayed spatially and how they may be correlated.

FIGURE 5.11 illustrates in a simplified way how different datasets of a GIS are combined, layer upon layer, to form a composite map. GIS has become a valuable tool used by geographers, landscape ecologists, resource managers, and conservation biologists. GIS technology also brings insights that affect planning and land use decisions. Some conservation groups, such as the Nature Conservancy, now apply the landscape ecology approach widely in their land acquisition and management strategies. Principles of landscape ecology, and tools such as GIS, are increasingly used in regional planning processes (pp. 353–356).

Modeling helps ecologists understand systems

Another way in which ecologists seek to make sense of the complex systems they study is by working with models. In science, a **model** is a simplified representation of a complex natural process, designed to help us understand how the process occurs and to make predictions. **Ecological modeling** is the practice of constructing and testing models that aim to explain and predict how ecological systems function.

Because ecological processes (for ecosystems, communities, or populations) involve so many factors, ecological models can be mathematically complicated. However, the general approach of ecological modeling is easy to understand (**FIGURE 5.12**). Researchers gather data from nature on relationships that interest them and then form a hypothesis about what those relationships are. They construct a model that attempts to explain the relationships in a generalized

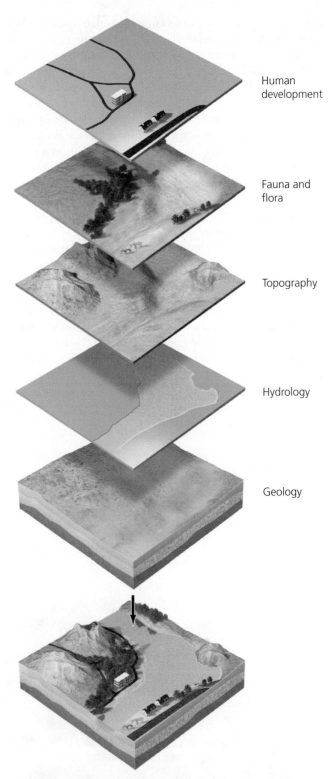

Human development

Fauna and flora

Topography

Hydrology

Geology

FIGURE 5.11 ▲ Geographic information systems (GIS) allow us to layer different types of data on natural landscape features and human land uses so as to produce maps integrating this information. GIS can be used to explore correlations among these data sets and to help in regional planning.

way so that people can use the model to make predictions about how the system will behave. Modelers test their predictions by gathering new data from natural systems, and they use this new data to refine the model, making it increasingly accurate.

120

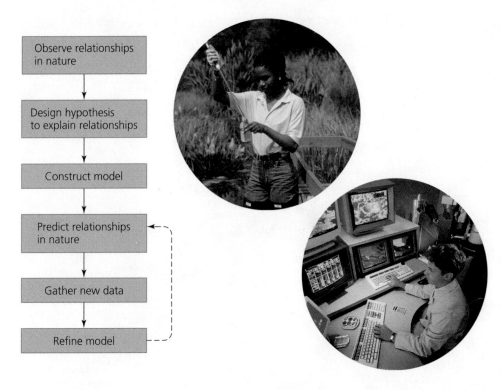

Observe relationships in nature
↓
Design hypothesis to explain relationships
↓
Construct model
↓
Predict relationships in nature
↓
Gather new data
↓
Refine model

FIGURE 5.12 ◄ Ecological modelers observe relationships among variables in nature and then construct models to explain those relationships and make predictions. They test and refine the models by gathering new data from nature and seeing how well the models predict those data.

Note that the process illustrated in Figure 5.12 resembles the scientific method in general, as shown in Figure 1.10 (p. 11). This is because models are essentially hypotheses about how systems function. Accordingly, the use of models is a key part of ecological research today. As just one example, University of Michigan researcher Donald Scavia and his colleagues have modeled nitrogen and phosphorus transport by the Mississippi River along with hypoxia in the Gulf. By applying relationships evident in available data, they were able to reconstruct what past conditions were like before monitoring for the dead zone had begun. They concluded that hypoxic events in the Gulf had started in the mid-1970s. They were also able to forecast future trends. As a result of these forecasts, they advised that the government's recommendation of reducing nutrient loads by 30% would not be enough to prevent hypoxia, but rather that reductions of 40–45% would be needed.

Ecosystems provide vital services

When scientists try to understand how ecosystems function, it is not simply out of curiosity about the world. They also know that human society depends on healthy, functioning ecosystems. When Earth's ecosystems function normally and undisturbed, they provide goods and services that we could not survive without. As we've seen, we rely not just on natural resources (which can be thought of as goods from nature; p. 3), but also on the **ecosystem services** (pp. 3, 148–149, 157, 160–161) that our planet's systems provide. (**TABLE 5.1**).

Ecological processes form the soil that nourishes our crops, purify the water we drink and the air that we breathe, store and stabilize supplies of water that we use, pollinate the food plants we eat, and receive and break down (some of) the waste we dump and the pollution we emit. The negative feedback cycles that are typical of ecosystems regulate and stabilize the climate and help to dampen the impacts of the disturbances we create in natural systems. On top of all these services that are vital for our very existence, ecosystems also provide services that enhance the quality of our lives, ranging from recreational opportunities to scenery for aesthetic enjoyment to inspiration and spiritual renewal. Ecosystem goods and ecosystem services (**FIGURE 5.13**) support our lives and society in profound and innumerable ways.

TABLE 5.1 Ecosystem Services

Ecological processes do many things that benefit us:

▶ Regulate oxygen, carbon dioxide, stratospheric ozone, and other atmospheric gases
▶ Regulate temperature and precipitation with ocean currents, cloud formation, and so on
▶ Protect against storms, floods, and droughts, mainly with vegetation
▶ Store and regulate water supplies in watersheds and aquifers
▶ Prevent soil erosion
▶ Form soil by weathering rock and accumulating organic material
▶ Cycle carbon, nitrogen, phosphorus, sulfur, and other nutrients
▶ Filter waste, remove toxins, recover nutrients, and control pollution
▶ Pollinate plant crops and wild plants so they reproduce
▶ Control crop pests with predators and parasites
▶ Provide habitat for organisms to breed, feed, rest, migrate, and winter
▶ Produce fish, game, crops, nuts, and fruits that people eat
▶ Supply lumber, fuel, metals, fodder, and fiber
▶ Furnish medicines, pets, ornamental plants, and genes for resistance to pathogens and crop pests
▶ Provide recreation such as ecotourism, fishing, hiking, birding, hunting, and kayaking
▶ Provide aesthetic, artistic, educational, spiritual, and scientific amenities

FIGURE 5.13 ▲ Ecological processes naturally provide countless services that we call ecosystem services. Our society, indeed our very survival, depends on these services.

One of the most important ecosystem services is the cycling of nutrients. Through the processes that take place within and among ecosystems, the chemical elements and compounds that we need—carbon, nitrogen, phosphorus, water, and many more—cycle through our environment in complex ways.

BIOGEOCHEMICAL CYCLES

Just as nitrogen and phosphorus from Minnesota soybean fields end up in Louisiana shrimp on our dinner plates, all nutrients move through the environment in intricate ways. Whereas energy enters an ecosystem from the sun, flows from organism to organism, and dissipates to the atmosphere as heat, the physical matter of an ecosystem is circulated over and over again.

Nutrients circulate through ecosystems in biogeochemical cycles

Nutrients move through ecosystems in **nutrient cycles**, also known as **biogeochemical cycles**. In these pathways, chemical elements or molecules travel through the atmosphere, hydrosphere, and lithosphere, and from one organism to another, in

dynamic equilibrium. A carbon atom in your fingernail today might have helped compose the muscle of a cow a year ago, may have resided in a blade of grass a month before that, and may have been part of a dinosaur's tooth 100 million years ago. After we die, the nutrients in our bodies will spread widely through the environment, eventually being incorporated by an untold number of organisms far into the future.

Nutrients and other materials move from one **pool**, or **reservoir**, to another, remaining for varying amounts of time (the **residence time**) in each. The dinosaur, the grass, the cow, and you are each reservoirs for carbon atoms. The rate at which materials move between reservoirs is termed a **flux**, and the flux between any given reservoirs can change over time. When a reservoir releases more materials than it accepts, it is called a **source**, and when a reservoir accepts more materials than it releases, it is called a **sink. FIGURE 5.14** illustrates these concepts in a simple manner.

Human activity has influenced certain fluxes. We have increased the flux of nitrogen from the atmosphere to reservoirs on the Earth's surface, and we have shifted the flux of carbon in the opposite direction. As we discuss biogeochemical cycles, think about how they involve negative feedback loops that promote dynamic equilibrium, and also consider how some human actions can generate destabilizing positive feedback loops.

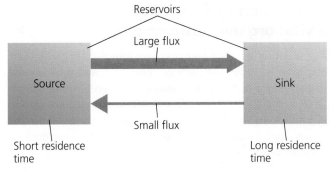

FIGURE 5.14 ▲ The main components of a biogeochemical cycle are reservoirs (places where materials are stored) and fluxes (rates at which materials move among reservoirs). A source releases more materials than it accepts, and a sink accepts more materials than it releases.

The water cycle affects all other cycles

Water is so integral to life and to Earth's fundamental processes that we frequently take it for granted. The essential medium for all manner of biochemical reactions (p. 27), water plays key roles in nearly every environmental system, including each of the nutrient cycles we are about to discuss. Water carries nutrients and sediments from the continents to the oceans via surface runoff, streams, and rivers, and it distributes sediments onward in ocean currents. Increasingly, water also distributes artificial pollutants.

The **water cycle**, or **hydrologic cycle** (**FIGURE 5.15**), summarizes how water—in liquid, gaseous, and solid forms—flows through our environment. Our brief introduction to the water cycle here sets the stage for our more in-depth discussion of freshwater and marine systems in Chapters 15 and 16.

The oceans are the main reservoir in the water cycle, holding 97% of all water on Earth. The fresh water we depend on for our survival accounts for less than 3%, and two-thirds of this small amount is tied up in glaciers, snowfields, and ice caps (p. 403). Thus, considerably less than 1% of the planet's water is in forms that we can readily use—groundwater, surface fresh water, and rain from atmospheric water vapor.

Evaporation and transpiration Water moves from oceans, lakes, ponds, rivers, and moist soil into the atmosphere by **evaporation**, the conversion of a liquid to gaseous form. Warm temperatures and strong winds speed rates of evaporation. A greater degree of exposure has the same effect;

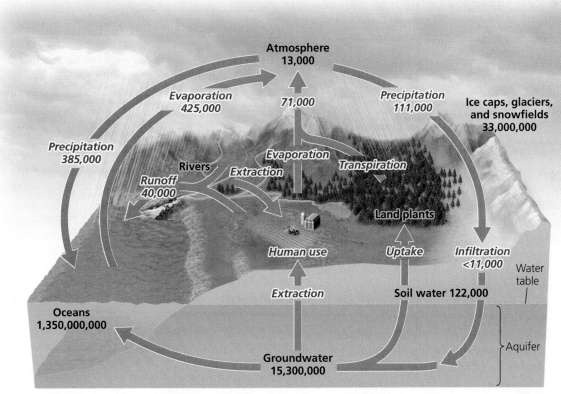

FIGURE 5.15 ▲ The water cycle, or hydrologic cycle, summarizes the many routes that water molecules take as they move through the environment. Gray arrows represent fluxes among reservoirs, or pools, for water. Oceans hold 97% of our planet's water, whereas most fresh water resides in groundwater and icecaps. Water vapor in the atmosphere condenses and falls to the surface as precipitation, then evaporates from land and transpires from plants to return to the atmosphere. Water flows downhill into rivers, eventually reaching the oceans. In the figure, pool names are printed in black type, and numbers in black type represent pool sizes expressed in units of cubic kilometers (km^3). Processes, printed in italic red type, give rise to fluxes, printed in italic red type and expressed in km^3 per year. Data from Schlesinger, W.H., 1997. *Biogeochemistry: An analysis of global change,* © 1997, with permission from Elsevier.

an area logged of its forest or converted to agriculture or residential use will lose water more readily than a comparable area that remains vegetated. Water also enters the atmosphere by **transpiration**, the release of water vapor by plants through their leaves. Transpiration and evaporation act as natural processes of distillation, effectively creating pure water by filtering out minerals carried in solution.

Precipitation, runoff, and surface water   Water returns from the atmosphere to Earth's surface as **precipitation** when water vapor condenses and falls as rain or snow. This moisture may be taken up by plants and used by animals, but much of it flows as **runoff** into streams, rivers, lakes, ponds, and oceans. Amounts of precipitation vary greatly from region to region, helping give rise to our planet's variety of biomes (pp. 96–103).

Groundwater   Some precipitation and surface water soaks down through soil and rock to recharge underground reservoirs known as **aquifers**. Aquifers are spongelike regions of rock and soil that hold **groundwater**, water found underground beneath layers of soil. The upper limit of groundwater held in an aquifer is referred to as the **water table**. (You can jump ahead to Figure 15.6, p. 406, for an illustration of these features.) Aquifers can hold groundwater for long periods of time, so the water may be quite ancient. In some cases groundwater can take hundreds or even thousands of years to recharge fully after being depleted. Groundwater becomes exposed to the air where the water table reaches the surface, and the exposed water can run off toward the ocean or evaporate into the atmosphere.

Our impacts on the water cycle are extensive

Human activity affects every aspect of the water cycle. By clearing forests and other vegetation, we intensify surface runoff and erosion, reduce transpiration, and can lower water tables. By spreading irrigation water on agricultural fields, we can deplete rivers, lakes, and streams, and we increase evaporation. By damming rivers to create reservoirs, we augment evaporation and sometimes increase infiltration of surface water into aquifers. And by emitting into the atmosphere pollutants that dissolve in water droplets, we change the chemical nature of precipitation, in effect sabotaging the natural distillation process that evaporation and transpiration provide. Perhaps most threatening to our future, we are overdrawing groundwater for drinking, irrigation, and industrial use and have thereby begun to deplete groundwater resources. Water shortages have already given rise to conflicts worldwide, from the Middle East to the American West (pp. 401–402, 418).

WEIGHING THE ISSUES

Your Water   Has your region faced any water shortages or conflicts over water use? If not, can you describe how such problems affect some other region? What is the quality of your region's water, and what pollution threats does it face? Given your knowledge of the water cycle, what solutions would you propose for water shortages and/or water pollution in your region?

The carbon cycle circulates a vital organic nutrient

As the definitive component of organic molecules (p. 28), carbon is an ingredient in carbohydrates, fats, and proteins and in the bones, cartilage, and shells of all living things. From DNA to fossil fuels, from plastics to pharmaceuticals, carbon (C) atoms are everywhere. The **carbon cycle** describes the routes that carbon atoms take through the environment (**FIGURE 5.16**).

Photosynthesis, respiration, and food webs   Producers, including plants, algae, and cyanobacteria, pull carbon dioxide out of the atmosphere and out of surface water to use in photosynthesis. Photosynthesis (pp. 31–32) breaks the bonds in carbon dioxide (CO_2) and water (H_2O) to produce oxygen (O_2) and carbohydrates (e.g., glucose, $C_6H_{12}O_6$). Autotrophs use some of the carbohydrates to fuel cellular respiration, thereby releasing some of the carbon back into the atmosphere and oceans as CO_2. When producers are eaten by primary consumers, which in turn are eaten by secondary and tertiary consumers, more carbohydrates are broken down in cellular respiration, producing carbon dioxide and water. The same process occurs as decomposers consume waste and dead organic matter. Cellular respiration from all these organisms releases carbon back into the atmosphere and oceans.

Organisms use carbon for structural growth, so a portion of the carbon an organism takes in becomes incorporated into its tissues. The abundance of plants and the fact that they take in so much carbon dioxide for photosynthesis makes plants a major reservoir for carbon. Because CO_2 is a greenhouse gas of primary concern (pp. 495–497), much research on global climate change is directed toward measuring the amount of CO_2 that plants store. Scientists are working hard to better understand exactly how this portion of the carbon cycle influences Earth's climate (see **THE SCIENCE BEHIND THE STORY**, pp. 126–127).

Sediment storage of carbon   As organisms die, their remains may settle in sediments in ocean basins or freshwater wetlands. As sediment accumulates, older layers are buried more deeply, experiencing high pressure over long periods of time. These conditions can convert soft tissues into fossil fuels—coal, oil, and natural gas (p. 533)—and can turn shells and skeletons into sedimentary rock, such as limestone. Sedimentary rock (pp. 37–40) comprises the largest reservoir in the carbon cycle. Although any given carbon atom spends a relatively short time in the atmosphere, carbon trapped in sedimentary rock may reside there for hundreds of millions of years.

Carbon trapped in sediments and fossil fuel deposits may eventually be released into the oceans or atmosphere by geologic processes such as uplift, erosion, and volcanic eruptions. It also reenters the atmosphere when we extract and burn fossil fuels.

The oceans   The world's oceans are the second-largest reservoir in the carbon cycle. They absorb carbon-containing compounds from the atmosphere, from terrestrial runoff,

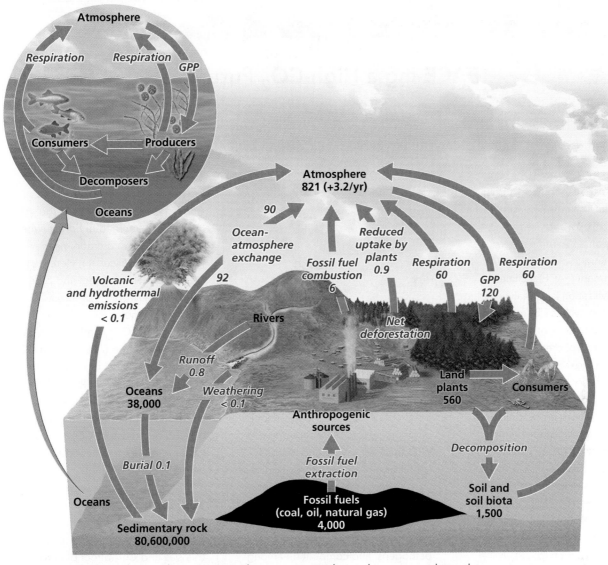

FIGURE 5.16 ▲ The carbon cycle summarizes the many routes that carbon atoms take as they move through the environment. Gray arrows represent fluxes among reservoirs, or pools, for carbon. In the carbon cycle, plants use carbon dioxide from the atmosphere for photosynthesis (gross primary production, or "GPP" in the figure). Carbon dioxide is returned to the atmosphere through cellular respiration by plants, their consumers, and decomposers. The oceans sequester carbon in their water and in deep sediments. The vast majority of the planet's carbon is stored in sedimentary rock. In the figure, pool names are printed in black type, and numbers in black type represent pool sizes expressed in petagrams (units of 10^{15} g) of carbon. Processes, printed in italic red type, give rise to fluxes, printed in italic red type and expressed in petagrams of carbon per year. Data from Schlesinger, W.H., 1997. *Biogeochemistry: An analysis of global change*, © 1997, with permission from Elsevier.

from undersea volcanoes, and from the waste products and detritus of marine organisms. Some carbon atoms absorbed by the oceans—in the form of carbon dioxide, carbonate ions (CO_3^{2-}), and bicarbonate ions (HCO_3^-)—combine with calcium ions (Ca^{2+}) to form calcium carbonate ($CaCO_3$), an essential ingredient in the skeletons and shells of microscopic marine organisms. As these organisms die, their calcium carbonate shells sink to the ocean floor and begin to form sedimentary rock. The rates at which the oceans absorb and release carbon depend on many factors, including temperature and the numbers of marine organisms converting CO_2 into carbohydrates and carbonates.

We are shifting carbon from the lithosphere to the atmosphere

By mining fossil fuel deposits, we are essentially removing carbon from an underground reservoir with a residence time of millions of years. By combusting fossil fuels in our automobiles, homes, and industries, we release carbon dioxide and greatly increase the flux of carbon from the ground to the air. Since the mid-18th century, our fossil fuel combustion has added over 250 billion metric tons (276 billion tons) of carbon to the atmosphere. The movement of CO_2 from the atmosphere back to the hydrosphere, lithosphere, and biosphere has not kept pace.

The SCIENCE behind the Story

FACE-ing a High-CO$_2$ Future

Can fumigating trees with carbon dioxide tell us what to expect from global climate change? Hundreds of scientists think so, and are testing plants' responses to atmospheric change at unique outdoor Free-Air CO$_2$ Enrichment (FACE) facilities.

Our civilization is radically altering Earth's carbon cycle by burning fossil fuels and deforesting landscapes. Today the atmosphere contains 35% more carbon dioxide (CO$_2$) than it did just two centuries ago, and the amount is rapidly increasing. Rising CO$_2$ concentrations are warming our planet, and global climate change brings many unwelcome consequences (Chapter 18).

Plants and other autotrophs remove carbon dioxide from the atmosphere to use in photosynthesis, and all organisms add CO$_2$ to the atmosphere by cellular respiration (p. 32). Will more CO$_2$ mean more plant growth, and will more plants be able to absorb and store much of the extra CO$_2$? Perhaps, but before we rely on forests and phytoplankton to save us from our own emissions, we'd better be sure they can do so.

Historically, if a researcher wanted to measure how plants respond to increased carbon dioxide, he or she would alter gas levels in a small enclosure like a lab or a greenhouse. But can we really scale up results from such small indoor experiments and trust that they will show how entire forests will behave? Many scientists thought not, and so they pioneered Free-Air CO$_2$

Aspen FACE site researcher Dr. Mark Kubiske of the U.S. Forest Service

Enrichment. In FACE experiments, ambient levels of CO$_2$ encompassing areas of forest (or other vegetation) outdoors are precisely controlled. With their large scale and open-air conditions, FACE experiments include most factors that influence a plant community in the wild, such as variation in temperature, sunlight, precipitation, herbivorous insects, disease pathogens, and competition among plants. By measuring how plants respond to changing gas compositions in such real-world conditions, we can better learn how ecosystems may change in the carbon-dioxide-soaked world that awaits us.

Dozens of organizations have sponsored FACE facilities—36 sites in 17 nations so far, including U.S. sites in Arizona, California, Illinois, Minnesota, Nevada, North Carolina, Tennessee, Wisconsin, and Wyoming. The sites cover a variety of ecosystems, from forests to grasslands to rice paddies, and the plots range in size from 1 m to 30 m (3–98 ft) in diameter.

To understand how a typical FACE study works, let's visit the Aspen FACE Experiment at the Harshaw Experimental Forest (where aspen trees are common) near Rhinelander, Wisconsin. Here, tall steel and plastic towers and pipes ring 12 circular plots of forest 30 m (98 ft) in diameter (**see photo**). The pipes release CO$_2$, bathing the plants in an atmosphere 50% richer in CO$_2$ than today's (equal to what is expected for the year 2050). Sensors monitor wind conditions, and computers control for the influence of wind by adjusting CO$_2$ releases, keeping ambient concentrations stable within each plot.

The pipes at the Aspen plots also release tropospheric ozone (O$_3$, a major pollutant in urban smog; p. 477), and researchers study how this gas and CO$_2$ affect plant growth, leaf and root conditions, soil carbon content, and much more. Pipes at some plots release normal air, serving as controls for the treatment plots.

Researchers using the Aspen FACE facility have been asking whether forest trees will sequester more carbon as CO$_2$ levels rise, whether this will change as trees grow, how CO$_2$ interacts with ozone, and how these gases affect trees' interactions with insects and diseases. They have learned a number of things so far, among these:

▶ Insects and diseases that attack aspen and birch trees increase as atmospheric levels of ozone and CO$_2$ rise.

At the Aspen FACE facility in Wisconsin, tall towers and pipes control the atmospheric composition around selected patches of trees.

▶ High CO_2 concentrations delay aspen leaf aging, which makes some aspens vulnerable to frost damage in winter.

▶ Elevated CO_2 levels increase photosynthesis and tree growth—but moderate levels of ozone offset this increased growth (**see graph**).

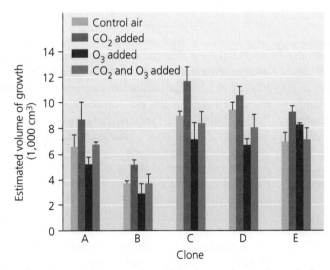

Data from five clones of aspens show that during the study period trees supplied with carbon dioxide grew more than did the control trees, while those supplied with ozone grew less. Trees supplied with both gases did not grow differently from the controls. SOURCE: *Environmental Pollution* 115 (3), Isebrands, J.G., et al, Growth responses of *Populus tremuloides* clones to interacting elevated carbon dioxide and tropospheric ozone, 359–371, Fig 2, © 2001, with permission from Elsevier. http://www.sicnedirect.com/science/journal/02697491.

Because many modelers have not taken ozone into account when estimating how much carbon trees can sequester, the Aspen FACE data suggest that their models may overestimate the amount of CO_2 that trees will pull out of the air.

Together, such results indicate that rising carbon dioxide levels could have a variety of negative impacts on trees and forests. Thus, the old expectation that more CO_2 makes for happier plants appears to be an oversimplification. Indeed, research from other FACE sites is showing that increased growth from enhanced CO_2 is often temporary and that growth rates later flatten out or decline. Recent work on crop plants has even shown that high CO_2 makes some crops less nutritious.

Obtaining solid answers to questions like these often takes years or decades, and FACE experiments are designed to monitor plots for the long term as the plants mature. Some FACE sites have been operating for 20 years and are beginning to produce data that could not be gathered in any other way.

Thus, researchers were shocked in 2008 when the U.S. Department of Energy (DOE), which funds Aspen and other major sites, announced it would cease funding. The DOE advised scientists to cut the trees down and dig up the soil to analyze carbon content. This would provide urgently needed data on carbon sequestration, the DOE said, and then millions of dollars could be shifted toward a new and improved generation of FACE experiments.

Many researchers were aghast, however, and argued that the precious and unique long-term sites still had much to teach us. How else can we know how forests and climate will interact after 25 years, or 50, they asked? Except, of course, to wait and let Earth show us—by which time it may be too late to do anything about it. ■

In addition, cutting down forests removes carbon from the pool of vegetation and releases it to the air. And if less vegetation is left on the surface, there are fewer plants to draw CO_2 back out of the atmosphere.

As a result, scientists estimate that today's atmospheric carbon dioxide reservoir is the largest that Earth has experienced in the past 800,000 years, and likely in the past 20 million years. The ongoing flux of carbon into the atmosphere is the driving force behind today's anthropogenic global climate change (Chapter 18).

Some of the excess CO_2 in the atmosphere is now being absorbed by ocean water. This is causing ocean water to become more acidic, leading to problems that threaten many marine organisms (pp. 440–441, 512).

Our understanding of the carbon cycle is not yet complete. Scientists remain baffled by the so-called missing carbon sink. Of the carbon dioxide we emit by fossil fuel combustion and deforestation, researchers have measured how much goes into the atmosphere and oceans, but there remain roughly 1–2 billion metric tons unaccounted for. Many scientists think this must be taken up by plants or soils of the temperate and boreal forests (pp. 98, 102). They'd like to know for sure, though, because if certain forests are acting as a major sink for carbon (and thus restraining global climate change), we'd like to be able to keep it that way. For if forests that today are sinks were to turn into sources and begin releasing the "missing" carbon, climate change could accelerate drastically.

The nitrogen cycle involves specialized bacteria

Nitrogen (N) makes up 78% of our atmosphere by mass, and is the sixth most abundant element on Earth. It is an essential ingredient in the proteins, DNA, and RNA that build our bodies. Despite its abundance in the air, nitrogen gas (N_2) is chemically inert and cannot cycle out of the atmosphere and into living organisms without assistance from lightning, highly specialized bacteria, or human intervention. For this reason, the element is relatively scarce in the lithosphere and hydrosphere and in organisms. However, once nitrogen undergoes the right kind of chemical change, it becomes biologically active and available to organisms, and it can act as a potent fertilizer. Its scarcity makes biologically active nitrogen a limiting factor for plant growth. For all these reasons the **nitrogen cycle** (**FIGURE 5.17**) is of vital importance to us and to all other organisms.

Nitrogen fixation   To become biologically available, inert nitrogen gas (N_2) must be "fixed," or combined with hydrogen in nature to form ammonia (NH_3), whose water-soluble ions of ammonium (NH_4^+) can be taken up by plants. **Nitrogen fixation** can be accomplished in two ways: by the intense energy of lightning strikes, or when air in the top layer of soil comes in contact with particular types of **nitrogen-fixing bacteria**. These bacteria live in a mutualistic relationship (pp. 82–83) with many types of plants, including soybeans and other legumes, providing them nutrients by converting nitrogen to a usable form. As we will see in Chapter 9, farmers nourish soils by planting crops that host nitrogen-fixing bacteria among their roots (**FIGURE 5.18**).

Nitrification and denitrification   Other types of specialized bacteria then perform a process known as **nitrification**. In this process, ammonium ions are first converted into nitrite ions (NO_2^-), then into nitrate ions (NO_3^-). Plants can take up these ions, which also become available after atmospheric deposition on soils or in water or after application of nitrate-based fertilizer.

Animals obtain the nitrogen they need by consuming plants or other animals. Decomposers obtain nitrogen from dead and decaying plant and animal matter and from animal urine and feces. Once decomposers process nitrogen-rich compounds, they release ammonium ions, making these available to nitrifying bacteria to convert again to nitrates and nitrites.

The next step in the nitrogen cycle occurs when **denitrifying bacteria** convert nitrates in soil or water to gaseous nitrogen via a multistep process. Denitrification thereby completes the cycle by releasing nitrogen back into the atmosphere as a gas.

We have greatly influenced the nitrogen cycle

Historically, nitrogen fixation was a *bottleneck*, a step that limited the flux of nitrogen out of the atmosphere. This changed with the research of two German chemists early in the 20th century. Fritz Haber found a way to combine nitrogen and hydrogen gases to synthesize ammonia, a key ingredient in modern explosives and agricultural fertilizers, and Carl Bosch devised methods to produce ammonia on an industrial scale. The **Haber-Bosch process** enabled people to overcome the limits on productivity long imposed by nitrogen scarcity in nature. By enhancing agriculture, the new fertilizers contributed to the past century's enormous increase in human population. Farmers, homeowners, and golf course managers alike all took advantage of fertilizers, dramatically altering the nitrogen cycle. Today, using the Haber-Bosch process, our species is fixing at least as much nitrogen as is being fixed naturally. We have effectively doubled the rate of nitrogen fixation on Earth (**FIGURE 5.19**).

By fixing atmospheric nitrogen with fertilizers, we increase nitrogen's flux from the atmosphere to Earth's surface. We also enhance this flux by cultivating legume crops whose roots host nitrogen-fixing bacteria. Moreover, we reduce nitrogen's return to the air when we destroy wetlands that filter nutrients; wetland plants host denitrifying bacteria that convert nitrates to nitrogen gas, so wetlands can mop up a great deal of nitrogen pollution.

When our farming practices speed runoff and allow soil erosion, nitrogen flows from farms into terrestrial and aquatic ecosystems, leading to nutrient pollution, eutrophication, and hypoxia. These impacts have become painfully evident to shrimpers and scientists in the Gulf of Mexico, but hypoxia in coastal waters is by no means the only human impact on the nitrogen cycle. When we burn forests and fields, we force nitrogen out of soils and vegetation and into the atmosphere.

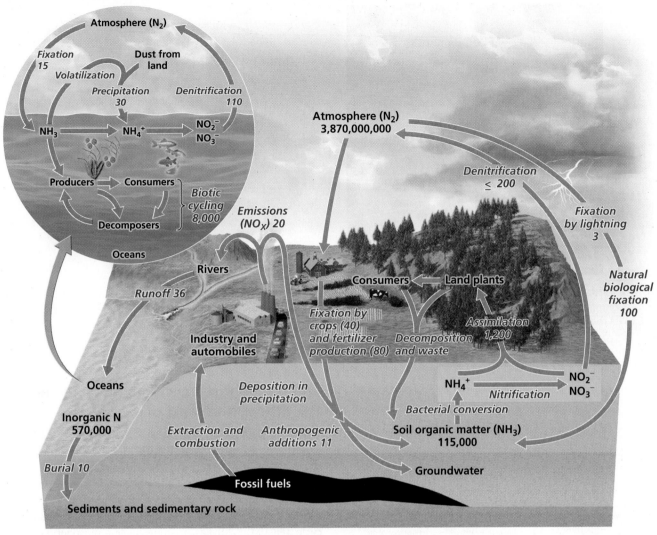

FIGURE 5.17 ▲ The nitrogen cycle summarizes the many routes that nitrogen atoms take as they move through the environment. Gray arrows represent fluxes among reservoirs, or pools, for nitrogen. In the nitrogen cycle, specialized bacteria play key roles in "fixing" atmospheric nitrogen and converting it to chemical forms that plants can use. Other types of bacteria convert nitrogen compounds back to the atmospheric gas N_2. In the oceans, inorganic nitrogen is buried in sediments, whereas nitrogen compounds are cycled through food webs as they are on land. In the figure, pool names are printed in black type, and numbers in black type represent pool sizes expressed in teragrams (units of 10^{12} g) of nitrogen. Processes, printed in italic red type, give rise to fluxes, printed in italic red type and expressed in teragrams of nitrogen per year. Data from Schlesinger, W.H., 1997. *Biogeochemistry: An analysis of global change,* © 1997, with permission from Elsevier.

When we burn fossil fuels, we release nitric oxide (NO) into the atmosphere, where it reacts to form nitrogen dioxide (NO_2). This compound is a precursor to nitric acid (HNO_3), a key component of acid precipitation (pp. 482–486). We introduce another nitrogen-containing gas, nitrous oxide (N_2O), when anaerobic bacteria break down the tremendous volume of animal waste produced in agricultural feedlots (pp. 268–270).

In 1997, a team of scientists led by Peter Vitousek of Stanford University summarized the dramatic changes people have caused in the global nitrogen cycle. According to the Vitousek team's report, we have:

▸ Doubled the rate at which fixed nitrogen enters terrestrial ecosystems (and the rate is still increasing).

▸ Increased atmospheric concentrations of the greenhouse gas N_2O and other nitrogen oxides that produce smog.

▸ Depleted essential nutrients, such as calcium and potassium, from soils, because fertilizer flushes them out.

▸ Acidified surface water and soils.

▸ Greatly increased transfer of nitrogen from rivers to oceans.

▸ Encouraged plant growth, causing terrestrial ecosystems to store more carbon.

▸ Reduced biodiversity, especially plants adapted to low-nitrogen soils.

▸ Altered the composition and function of estuaries and coastal ecosystems.

▸ Harmed many coastal marine fisheries.

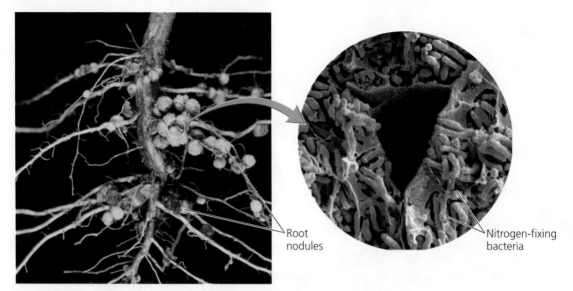

Root
nodules

Nitrogen-fixing
bacteria

FIGURE 5.18 ▲ Specialized bacteria live in nodules on the roots of this legume plant. In the process of nitrogen fixation, the bacteria convert nitrogen to a form that the plant can take up into its roots.

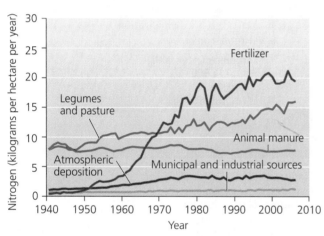

FIGURE 5.19 ▲ In the past several decades, human inputs of nitrogen into the environment have greatly increased, and today fully half of the nitrogen entering the environment is of human origin. These data for the Mississippi River basin show that agricultural fertilizer has for the past 40 years been the leading source of all nitrogen inputs, natural and artificial. *Adapted from Mississippi River Gulf of Mexico Watershed Nutrient Task Force, 2009. Moving forward on Gulf hypoxia: Annual report 2009, using data from Mark David, University of Illinois.*

WEIGHING THE ISSUES

Nitrogen Pollution and Its Financial Impacts Most nitrate that enters the Gulf of Mexico originates from farms and other sources in the upper Midwest, yet many of its negative impacts are borne by downstream users, such as Gulf Coast fishermen. Who do you believe should be responsible for addressing this problem? Should environmental policies on this issue be developed and enforced by state governments, the federal government, both, or neither? Explain the reasons for your answer.

The phosphorus cycle involves mainly the lithosphere and ocean

The element phosphorus (P) is a key component of cell membranes and of several molecules vital for life, including DNA, RNA, ATP, and ADP (pp. 29, 32). Although phosphorus is indispensable for life, the amount of phosphorus in organisms is dwarfed by the vast amounts in rocks, soil, sediments, and the oceans. Unlike the carbon and nitrogen cycles, the **phosphorus cycle** (**FIGURE 5.20**) has no appreciable atmospheric component besides the transport of tiny amounts in windblown dust and sea spray.

Geology and phosphorus availability The vast majority of Earth's phosphorus is contained within rocks and is released only by weathering (pp. 227–228), which releases phosphate ions (PO_4^{3-}) into water. Phosphates dissolved in lakes or in the oceans precipitate into solid form, settle to the bottom, and reenter the lithosphere's phosphorus reservoir in sediments. Because most phosphorus is bound up in rock and only slowly released, environmental concentrations of phosphorus available to organisms tend to be very low. This scarcity explains why phosphorus is frequently a limiting factor for plant growth and why an influx of phosphorus can produce immediate and dramatic effects.

Food webs Plants can take up phosphorus through their roots only when phosphate is dissolved in water. Primary consumers acquire phosphorus from water and plants and pass it on to secondary and tertiary consumers. Consumers also pass phosphorus to the soil through the excretion of waste. Decomposers break down phosphorus-rich organisms and their wastes and, in so doing, return phosphorus to the soil.

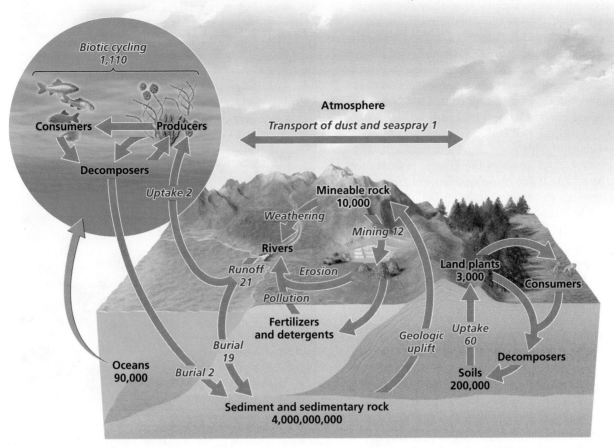

FIGURE 5.20 ▲ The phosphorus cycle summarizes the many routes that phosphorus atoms take as they move through the environment. Gray arrows represent fluxes among reservoirs, or pools, for phosphorus. Most phosphorus resides underground in rock and sediment. Rocks containing phosphorus are uplifted geologically and slowly weathered away. Small amounts of phosphorus cycle through food webs, where this nutrient is often a limiting factor for plant growth. In the figure, pool names are printed in black type, and numbers in black type represent pool sizes expressed in teragrams (units of 10^{12} g) of phosphorus. Processes, printed in italic red type, give rise to fluxes, printed in italic red type and expressed in teragrams of phosphorus per year. Data from Schlesinger, W.H., 1997. *Biogeochemistry: An analysis of global change*, Copyright © 1997, with permission from Elsevier.

We affect the phosphorus cycle

People influence the phosphorus cycle in several ways. We mine rocks containing phosphorus to extract this nutrient for the inorganic fertilizers we use on crops and lawns. Our wastewater discharge also tends to be rich in phosphates. Phosphates that run off into waterways can boost algal growth and cause eutrophication, leading to murkier waters and altering the structure and function of aquatic ecosystems. Phosphates are also present in detergents, so one way each of us can reduce phosphorus input into the environment is to purchase phosphate-free detergents.

People are searching for solutions to the dead zone

In 1998, Congress passed the Harmful Algal Bloom and Hypoxia Research and Control Act. This law called for an "integrated assessment" of the dead zone and its ecological and economic impacts. The assessment report published two

years later proposed that the federal government work with Gulf Coast and Midwestern communities to:

▶ Reduce nitrogen fertilizer use on Midwestern farms.

▶ Time fertilizer applications to minimize rainy-season runoff.

▶ Use alternative crops.

▶ Manage nitrogen-rich livestock manure more effectively.

▶ Restore nutrient-absorbing wetlands in the Mississippi River Basin and near the river's mouth.

▶ Use artificial wetlands to filter farm runoff.

▶ Improve technologies in sewage treatment plants.

▶ Evaluate how these approaches work.

A state-and-federal task force proposed reducing nitrogen flowing down the Mississippi River by 30% by the year 2015 in order to reduce the dead zone's average size to 5,000 km² (1,930 mi²). Farmers' advocates argued that farmers were being unfairly singled out and that many other sources contribute to the problem. They also said restrictions on fertilizer use would hurt farmers economically and decrease crop yields.

In 2004, Congress reauthorized the Harmful Algal Bloom and Hypoxia Research Control Act, encouraging scientists, farmers, and policymakers to search for innovative ways to alleviate pollution while safeguarding agriculture (**FIGURE 5.21**). The law proposed offering farmers insurance and economic incentives for using less fertilizer and testing new farming strategies to maintain yields while decreasing fertilizer use. It promoted planting cover crops in the off-season to reduce runoff from bare fields. It also encouraged farmers to maintain wetlands to serve as natural buffers against pollution. However, Congress never fully funded these programs, so they fell well short of their potential.

Nonetheless, other ongoing efforts helped. From 2000 to 2006, subsidies and incentives from the U.S. Department of Agriculture encouraged farmers to restore, enhance, or create 1.4 million acres of wetlands and 2.3 million acres of conservation buffers (pp. 246–247) in the Mississippi River basin. Similar federal incentives promoted nutrient management and conservation tillage (pp. 239–240) on 43 million acres of additional land. Local efforts helped, too; for instance, Minnesota's Twin Cities were able to reduce phosphorus discharge from their sewage treatment plants by 78%. As a result of all those efforts, nitrogen loads at the river's mouth in the period 2001–2005 were 21% lower than in 1980–1996, and phosphorus loads began declining in the past decade. However, there was no decline in nitrate flow in the spring, which is what contributes the most to the dead zone.

In 2008, the task force issued a new action plan. The plan cited new research confirming that phosphorus was adding to the dead zone, and concluded that reductions in phosphorus as well as nitrogen were needed. It recommended reductions of at least 45% for both nutrients relative to their 1980–1996 averages. The plan also made a plea for federal funding, stating that "resources are insufficient to attain the goals [and] the lack of resources is the primary barrier to successful implementation of the plan." In 2009 the Obama administration responded with $320 million for programs to fight nutrient pollution in the Mississippi River basin.

Planting with the contours of the land reduces erosion

Buffer strips of vegetation reduce erosion

FIGURE 5.21 ▲ On this Iowa farm, farmers plow so as to follow the contours of the land, leave hedgerows within the fields, and leave a wooded buffer along the river bank. Such measures help reduce runoff of soil and nutrients into the river.

The Gulf should also benefit from a separate program to protect and restore the Mississippi River's delta. This state and federal effort aims to reverse some of the effects of river channelization by redistributing water flows back across areas of the delta that had eroded away in recent decades. As sediment rebuilds the Mississippi's shrinking delta, nutrients that had been flowing straight to the Gulf and promoting hypoxia will instead boost the growth of marsh plants, helping to restore coastal wetlands.

➤ CONCLUSION

Thinking in terms of systems is important in understanding Earth's dynamics, so that we may learn how to avoid disrupting its processes and how to mitigate any disruptions we cause. By studying the environment from a systems perspective and by integrating scientific findings with the policy process, people who care about the Mississippi River and the Gulf of Mexico are working today to address the Gulf's dead zone.

Earth hosts many interacting systems, and the way one perceives them depends on the questions in which one is interested. Life interacts with its nonliving environment in ecosystems, systems through which energy flows and matter is recycled. Understanding the biogeochemical cycles that describe the movement of nutrients within and among ecosystems is crucial, because human activities are causing significant changes in the ways those cycles function.

Unperturbed ecosystems use renewable solar energy, recycle nutrients, and are stabilized by negative feedback loops. The environmental systems we see on Earth today are those that have survived the test of time. Our industrialized civilization is young in comparison. Might we not take a few lessons about sustainability from a careful look at the natural systems of our planet?

You should now be able to:

DESCRIBE THE NATURE OF ENVIRONMENTAL SYSTEMS

- Earth's natural systems are complex, so environmental scientists often take a holistic approach to studying environmental systems. (p. 109)
- Systems are networks of interacting components that generally involve feedback loops, show dynamic equilibrium, and result in emergent properties. (pp. 109–111)
- Negative feedback stabilizes systems, whereas positive feedback destabilizes systems. Positive feedback often results from human disturbance of natural systems. (p. 110)
- Because environmental systems interact and overlap, one's delineation of a system depends on the questions in which one is interested. (p. 111)
- Hypoxia in the Gulf of Mexico, resulting from nutrient pollution in the Mississippi River, illustrates how systems are interrelated. (pp. 112–114)

DEFINE ECOSYSTEMS AND EVALUATE HOW LIVING AND NONLIVING ENTITIES INTERACT IN ECOSYSTEM-LEVEL ECOLOGY

- Ecosystems consist of all organisms and nonliving entities that occur and interact in a particular area at the same time. (p. 115)
- Energy flows in one direction through ecosystems, whereas matter is recycled. (pp. 115–116)
- Energy is converted to biomass, and ecosystems vary in their productivity. (p. 116)
- Input of nutrients can boost productivity, but an excess of nutrients can alter ecosystems and cause severe ecological and economic consequences. (pp. 117–118)

OUTLINE THE FUNDAMENTALS OF LANDSCAPE ECOLOGY, GIS, AND ECOLOGICAL MODELING

- Landscape ecology studies how landscape structure influences organisms. (pp. 118–119)
- Landscapes consist of patches spatially arrayed in a mosaic. Organisms dependent on certain types of patches may occur in metapopulations. (p. 119)
- Remote sensing technology and GIS are assisting the use of landscape ecology in conservation and regional planning. (p. 120)

- Ecological modeling helps ecologists make sense of the complex systems they study. (pp. 120–121)

ASSESS ECOSYSTEM SERVICES AND HOW THEY BENEFIT OUR LIVES

- Ecosystems provide "goods" we know as natural resources. (p. 121)
- Ecological processes naturally provide services that we depend on for everyday living. (pp. 121–122)

COMPARE AND CONTRAST HOW WATER, CARBON, NITROGEN, AND PHOSPHORUS CYCLE THROUGH THE ENVIRONMENT

- A source is a reservoir that contributes more of a material than it receives, and a sink is one that receives more than it provides. (pp. 122–123)
- Water moves widely through the environment in the water cycle. (pp. 123–124)
- Most carbon is contained in sedimentary rock. Substantial amounts also occur in the oceans and in soil. Carbon flux between organisms and the atmosphere occurs via photosynthesis and respiration. (pp. 124–128)
- Nitrogen is a vital nutrient for plant growth. Most nitrogen is in the atmosphere, so it must be "fixed" by specialized bacteria or lightning before plants can use it. (pp. 128–130)
- Phosphorus is most abundant in sedimentary rock, with substantial amounts in soil and the oceans. Phosphorus has no appreciable atmospheric reservoir. It is a key nutrient for plant growth. (pp. 130–131)

EXPLAIN HOW HUMAN IMPACT IS AFFECTING BIOGEOCHEMICAL CYCLES

- People are affecting Earth's biogeochemical cycles by shifting carbon from fossil fuel reservoirs into the atmosphere, shifting nitrogen from the atmosphere to the planet's surface, and depleting groundwater supplies, among other impacts. (pp. 124–131)
- Policy can help us address problems with nutrient pollution. (pp. 131–132)

TESTING YOUR COMPREHENSION

1. Which type of feedback loop is most common in nature, and which more commonly results from human action? How might the emergence of a positive feedback loop affect a system in homeostasis?

2. Describe how hypoxic conditions can develop in coastal marine ecosystems such as the northern Gulf of Mexico.

3. What is the difference between an ecosystem and a community?

4. Describe the typical movement of energy through an ecosystem. Now describe the typical movement of matter through an ecosystem.

5. Explain net primary productivity. Name one ecosystem with high net primary productivity and one with low net primary productivity.

6. Why are patches in a landscape mosaic often important to people who are interested in conserving populations of rare animals?

7. What is the difference between evaporation and transpiration? Give examples of how the water cycle interacts with the carbon, phosphorus, and nitrogen cycles.

8. What role does each of the following play in the carbon cycle?
   ▶ Cars
   ▶ Photosynthesis
   ▶ The oceans
   ▶ Earth's crust

9. Distinguish the function performed by nitrogen-fixing bacteria from that performed by denitrifying bacteria.

10. How has human activity altered the carbon cycle? The phosphorus cycle? The nitrogen cycle? What environmental problems have arisen from these changes?

SEEKING SOLUTIONS

1. Once vegetation is cleared from a riverbank, water begins to erode the bank away. This erosion may dislodge more vegetation. Would you expect this to result in a feedback process? If so, which type—negative or positive? Explain your answer. How might we halt or reverse this process?

2. Consider the ecosystem(s) that surround(s) your campus. Describe one way in which energy flows through and matter is recycled. Now pick one type of nutrient and briefly describe how it moves through your ecosystem(s). Does the landscape contain patches? Can you describe any ecotones?

3. For a conservation biologist interested in sustaining populations of the organisms below, why would it be helpful to take a landscape ecology perspective? Explain your answer in each case.
   ▶ A forest-breeding warbler that suffers poor nesting success in small, fragmented forest patches
   ▶ A bighorn sheep that must move seasonally between mountains and lowlands
   ▶ A toad that lives in upland areas but travels cross-country to breed in localized pools each spring

4. A simple change in the flux between just two reservoirs in a single nutrient cycle can potentially have major consequences for ecosystems and, indeed, for the globe. Explain how this can be, using one example from the carbon cycle and one example from the nitrogen cycle.

5. How do you think we might solve the problem of eutrophication in the Gulf of Mexico? Assess several possible solutions, your reasons for believing they might work, and the likely hurdles we might face. Explain who should be responsible for implementing these solutions, and why.

6. **THINK IT THROUGH** You are a shrimper on the Louisiana coast and your income is decreasing because the dead zone is making it harder to catch shrimp. One day your senator comes to town, and you have a one-minute audience with her. What steps would you urge her to take in Washington, D.C., to try to help alleviate the dead zone and bring back the shrimp fishery?

   Now suppose you are an Iowa farmer who has learned that the government is offering incentives to farmers to help reduce fertilizer runoff into the Mississippi River. What types of approaches described in the text and in Figure 5.21 might you be willing to try, and why?

CALCULATING ECOLOGICAL FOOTPRINTS

In the United States, a common dream is to own a suburban home with a weed-free green lawn. Nationwide, Americans tend about 40.5 million acres of lawn grass. But conventional lawn care involves applying fertilizers, pesticides, and irrigation water, and using gasoline or electricity for mowing and other care—all of which raise environmental and health concerns. Using the figures for a typical lawn in the table, calculate the total amount of fertilizer, water, and gasoline used in lawn care across the nation each year.

	Acreage of lawn	Fertilizer used (lbs)	Water used (gal)	Gasoline used (gal)
For the typical 1/4-acre lawn	0.25	37	16,000	4.9
For all lawns in your hometown				
For all lawns in the United States	40,500,000			

Data from Chameides, B., 2008. http://www.nicholas.duke.edu/thegreengrok/lawns.

1. How much fertilizer is applied each year on lawns throughout the United States? Where does the nitrogen for this fertilizer come from? What becomes of the nitrogen and phosphorus applied to a suburban lawn that is not taken up by grass?

2. Leaving grass clippings on a lawn decreases the need for fertilizer by 50%. What else might a homeowner do to decrease fertilizer use in a yard and the environmental impacts of nutrient pollution?

3. How much gasoline could Americans save each year if they did not take care of lawns? At today's gas prices, how much money would this save?

Mastering ENVIRONMENTALSCIENCE™

Go to **www.masteringenvironmentalscience.com** for practice quizzes, Pearson eText, videos, current events, and more.

Ecotourists enjoy scenery at sunset at Kakadu National Park, Australia

6 ENVIRONMENTAL ETHICS AND ECONOMICS: VALUES AND CHOICES

UPON COMPLETING THIS CHAPTER, YOU WILL BE ABLE TO:

- Characterize the influences of culture and worldview on the choices people make
- Outline the nature and historical expansion of ethics in Western cultures
- Compare the major approaches in environmental ethics
- Explain how our economies exist within the environment and rely on ecosystem services
- Describe principles of classical and neoclassical economics and summarize their implications for the environment
- Compare the concepts of economic growth, well-being, and sustainability
- Explain aspects of environmental economics and ecological economics
- Describe how individuals and businesses can help move our economic system in a sustainable direction

CENTRAL CASE STUDY

The Mirarr Clan Confronts the Jabiluka Uranium Mine

"For some people, what they are is not finished at the skin, but continues with the reach of the senses out into the land. If the land is summarily disfigured or reorganized, it causes them psychological pain."

—Barry Lopez, American nature writer

"We say no to uranium mining now and for the future. Our right to say no comes from our ancestors, our heritage, our law and culture, our Native Title."

—Jacqui Katona, speaking for the Mirarr

The remote outback of Australia's Northern Territory is a region of subtropical eucalyptus forests, picturesque cliffs, and vast marshes. It is home to native people (Aboriginal people, or Aborigines) who lived here long before the British colonized the Australian continent. Australia established its largest national park here: Kakadu National Park, a World Heritage Site recognized by the United Nations for its irreplaceable natural and cultural resources.

Protestors against the proposed Jabiluka uranium mine

The region also holds uranium, the naturally occurring radioactive metal valued for its use in nuclear power plants, nuclear weapons, and various medical and industrial tools. Uranium mining is a key contributor to Australia's economy, accounting for 7% of the nation's economic output.

Many of Australia's uranium deposits occur on lands traditionally claimed by Aboriginal groups. This has given rise to conflict between corporations seeking to develop mining operations and Aboriginal people trying to maintain their traditional cultures, which are deeply tied to the landscape. In the Kakadu region, the Mirarr Clan, an extended family of 52 Aboriginal people, lives alongside the Ranger uranium mine. Although Australian law recognizes the Mirarr's traditional claim to the land, the Australian government overruled the clan's objections and approved the Ranger mine's development in 1976.

In recent years, the Mirarr have fought the proposed development of a second mine nearby at the Jabiluka uranium deposit. Like other Aborigines, the Mirarr hold the landscape to be sacred, and the proposed Jabiluka mine would be located among several spiritually important sites. The Mirarr also depend on the land's resources for their daily needs, and Jabiluka is near their traditional hunting and gathering sites, in the floodplain of a river that provides the clan food and water. The Mirarr view Jabiluka as a threat to their health and their environment, particularly given repeated radioactive spills from the Ranger mine. The Mirarr fear that contaminated water would be released into creeks, radioactive radon gas would emanate from stored waste, and dams holding waste could fail in an earthquake.

Environmental activists worldwide joined the Mirarr's struggle against Jabiluka. In 1998, 5,000 people traveled to Kakadu to blockade the mine, and hundreds of protestors were jailed. Mirarr leaders Yvonne Margarula and Jacqui Katona later received the Goldman Prize, the world's foremost award honoring grassroots environmentalism.

Indian Ocean

Pacific Ocean

Kakadu National Park

AUSTRALIA

CHAPTER 6 Environmental Ethics and Economics: Values and Choices

137

The Mirarr's efforts succeeded at last in 2004, when the mine's owner, Energy Resources of Australia, and its international parent company, Rio Tinto, agreed to give the Mirarr veto power over development at Jabiluka. Under the agreement, Jabiluka will not be developed unless the Mirarr agree. The company backfilled and cleaned up the site.

Rio Tinto CEO Sir Robert Wilson cited economic factors (declining uranium prices) as well as ethical factors (concerns about developing the mine without Mirarr consent) as reasons for abandoning Jabiluka's development. Since then, however, the price of uranium has risen on the world market as nations look to revive nuclear power. Today some voices—from the mining industry and nearby Aboriginal groups—are suggesting the mine option be revisited.

Debates pitting economic arguments for uranium mining versus ethical arguments for the preservation of traditional cultures have raged across Australia. At Olympic Dam, a location that holds the world's largest uranium deposit, mining operations are depleting water from a site sacred to the region's Arabunna people, and the mining company has refused to negotiate with the Arabunna.

The Mirarr oppose mining despite the economic benefits the mining company has promised them in jobs, income, development, and a higher material standard of living. In formulating their approaches to the mining proposal, the Mirarr and other Australians weigh economic, social, cultural, spiritual, and philosophical questions as well as scientific ones. Their responses exemplify some of the ways in which values, beliefs, and traditions interact with economic interests to influence the choices all of us make about how to live within our environment.

CULTURE, WORLDVIEW, AND THE ENVIRONMENT

The Mirarr faced difficult choices. They were offered substantial economic benefits, but they felt that mine development ran counter to their ethical respect for their land. Such trade-offs between economic benefits and ethical concerns arise frequently with environmental issues.

Ethics and economics involve values

As we first saw in Chapter 1, environmental science involves a firm understanding of the natural sciences. To address environmental problems, however, we also need to understand how people perceive their environment, how they value it, and how they relate to it philosophically and pragmatically. Ethics and economics are two very different disciplines, but each deals with questions of what we value and how those values influence our decisions and actions. To address any environmental problem, we must aim to understand not only how natural systems work, but also how values shape human behavior.

Culture and worldview influence our perception of the environment

Every action we take affects our environment. Whether we are growing food, building homes, manufacturing products, or fueling vehicles, we meet our needs by extracting resources and altering our surroundings. In deciding how to manipulate our environment to meet our needs, we rely partly on rational assessments of costs and benefits. However, our decisions are also heavily influenced by our culture and our worldview (**FIGURE 6.1**).

(a) Mirarr mother and daughter

(b) Australian uranium miner

FIGURE 6.1 ▲ A mother in the Aboriginal Mirarr group, shown here **(a)** with her daughter and their ancestors' rock art, may have a very different attitude toward the development of a uranium mine than **(b)** a white employee of the Australian mining industry. They may differ in part because of dissimilar worldviews about how to interact with the landscape.

Culture can be defined as the ensemble of knowledge, beliefs, values, and learned ways of life shared by a group of people. Culture, together with personal experience, influences each person's perception of the world and his or her place within it—the person's **worldview**. A worldview reflects beliefs about the meaning, operation, and essence of the world.

People with different worldviews can study the same situation yet draw dramatically different conclusions. For example, many well-meaning people supported the Jabiluka mine while many other well-meaning people opposed it. The officers, employees, and shareholders of the mining company, as well as government officials who supported the mine, view uranium mining as a beneficial source of jobs, income, energy, and economic growth. Mine opponents, in contrast, recognize that uranium mining disturbs the landscape, pollutes air and water, and exposes miners to radiation. Moreover, community disruption, substance abuse, and crime frequently accompany mining booms. Which benefits and drawbacks a given person focuses on may vary according to personal circumstance, culture, values, and worldview.

Many factors shape our worldviews

The traditional culture and worldview of the Mirarr Clan have played large roles in the group's response to the proposed Jabiluka mine. Australian Aborigines view the landscape around them as the physical embodiment of stories that express the beliefs and values central to their culture. The Australian landscape to them is a sacred text, analogous to the Bible in Christianity, the Koran in Islam, or the Torah in Judaism. Aborigines believe that spirit ancestors possessing human and animal features traveled routes called "dreaming tracks," leaving signs and lessons in the landscape. Modern Aborigines still engage in "walkabouts," long walks that retrace the dreaming tracks. By explaining the origins of specific landscape features, dreaming-track stories assign meaning to notable landmarks and help Aborigines construct detailed mental maps of their surroundings. Passed from one generation to the next, the stories also teach lessons concerning family relations, hunting, food gathering, and conflict resolution. The Mirarr feel that open-pit mining desecrates sacred sites and threatens their culture (**FIGURE 6.2**).

WEIGHING **THE ISSUES**

Mining in Mecca? Suppose a mining company discovered uranium near the Sacred Mosque at Mecca—or the site in Bethlehem believed to be the birthplace of Jesus, or the Wailing Wall in Jerusalem. What do you think would happen if the company announced plans to develop a mine close to one of these sacred locations, assuring the public that environmental impacts would be minimal and that the mine would create jobs and stimulate economic growth? How do these unlikely situations resemble the Jabiluka case? Are there differences? Explain your answers.

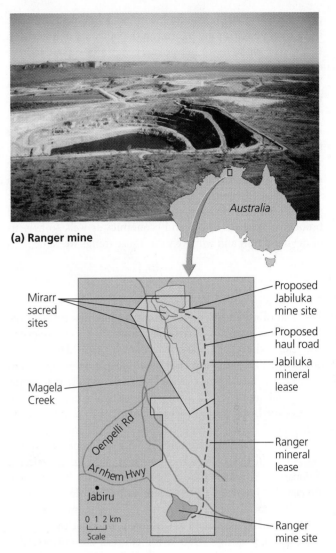

(a) Ranger mine

(b) Map of area

FIGURE 6.2 ▲ The Ranger Mine **(a)**, located on traditional Aboriginal land, has caused enough environmental impact to spark the Mirarr Clan's opposition to the proposed Jabiluka mine, which was planned for development amid several sites **(b)** they consider sacred.

Religion and spiritual beliefs are among many factors that can shape people's worldviews and perception of the environment. A community may also share a particular view of its environment if its members have lived through similar experiences. For example, early European settlers in both Australia and North America viewed their environment as a hostile force because inclement weather and wild animals frequently destroyed crops, killed livestock, and took settlers' lives. Such experiences were shared in stories and songs and helped shape social attitudes in many frontier communities. The view of nature as a hostile adversary to be overcome still influences the way many North Americans and Australians view their surroundings.

A person's political ideology can also shape his or her attitude toward the environment. For instance, one's views on the proper role of government may influence whether one wants government to intervene in a market economy to protect environmental quality. Economic factors also sway how people perceive their environment and make decisions. An individual with a strong interest in the outcome of a decision that may result in his or her private gain or loss is said to have a *vested interest*. Mining company executives and shareholders have a vested interest in opening areas to mining because a new mine can increase profits.

Throughout this book you will encounter scientific data regarding the environmental impacts of our choices (where to make our homes, how to make a living, what to eat, what to wear, how to spend our leisure time, and so on). You will also see that culture, worldviews, and values play critical roles in such choices. Acquiring scientific understanding is only one part of the search for sustainable solutions to environmental problems. We need ethics and economics as well, to help us understand how and why we value those things we value.

ENVIRONMENTAL ETHICS

The field of **ethics** is a branch of philosophy that involves the study of good and bad, of right and wrong. The term *ethics* can also refer to the set of moral principles or values held by a person or a society. Ethicists help clarify how people judge right from wrong by elucidating the criteria, standards, or rules that people use in making these judgments. Such criteria are grounded in values—for instance, promoting human welfare, maximizing individual freedom, or minimizing pain and suffering.

People of different cultures or with different worldviews may differ in their values and thus may differ in the specific actions they consider to be right or wrong. This is why some ethicists are **relativists**, who believe that ethics do and should vary with social context. However, different human societies show a remarkable extent of agreement on what moral standards are appropriate. Thus many ethicists are **universalists**, who maintain that there exist objective notions of right and wrong that hold across cultures and contexts. For both relativists and universalists, ethics is a *prescriptive* pursuit; rather than simply describing behavior, it prescribes how we *ought to* behave.

Ethical standards help us judge right and wrong

Ethical standards are the criteria that help differentiate right from wrong. One classic ethical standard is the *categorical imperative* proposed by German philosopher Immanuel Kant, which advises us to treat others as we would prefer to be treated ourselves. In Christianity this is called the "golden rule," and most of the world's religions teach this same lesson. The universality of the golden rule or categorical imperative makes it a fundamental ethical standard.

Another ethical standard is the *principle of utility*, elaborated by British philosophers Jeremy Bentham and John Stuart Mill. The utilitarian principle holds that something is right when it produces the greatest practical benefits for the most people.

On the question of the Jabiluka mine, defenders of the Mirarr have appealed to the categorical imperative by asking mine proponents how they would feel if outsiders disfigured their sacred homeland. Supporters of mine development, in contrast, appeal to the utilitarian principle by arguing that

uranium from the mine could benefit many thousands of people across Australia and elsewhere by helping to produce electricity in nuclear power plants. We all employ ethical standards as tools for making countless decisions in our everyday lives.

We value things in two ways

People ascribe value to things in two main ways. One way is to value something for the pragmatic benefits it brings us if we put it to use. This is termed **instrumental value** or *utilitarian value*. The other way is to value something for its intrinsic worth, to feel that something has a right to exist and is valuable for its own sake. This notion is termed **intrinsic value** or *inherent value*.

We can perceive instrumental value in a grouse, duck, or rabbit because it can be hunted and provide us food. Alternatively, we can perceive intrinsic value in each of these animals simply because they are living organisms that think and feel and behave and live lives of their own on the planet we share. Likewise, we can ascribe instrumental value to a forest because we can harvest timber from it, hunt game in it, and drink clean water it has captured and filtered. Or we can ascribe intrinsic value to a forest because it provides homes for countless organisms that have intrinsic value. Of course, an animal or a forest or a lake or a mountain can have both intrinsic and instrumental value at the same time. However, different people may emphasize different types of value, and the ways in which people value something have consequences for how they treat it.

Environmental ethics pertains to people and the environment

The application of ethical standards to relationships between people and nonhuman entities is known as **environmental ethics**. This relatively new branch of ethics arose once people began to perceive environmental changes brought about by industrialization. Ethical questions are often difficult to resolve, and those involving our interactions with the environment are no exception. Consider some examples:

1. Does the present generation have an obligation to conserve resources for future generations? If so, how much are we obligated to sacrifice?

2. Are there situations that justify exposing some communities to a disproportionate share of pollution? If not, what actions are warranted in preventing this problem?

3. Are humans justified in driving species to extinction? If destroying a forest would drive extinct an insect species few people have heard of but would create jobs for 10,000 people, would that action be ethically admissible? What if it were an owl species? What if only 100 jobs would be created? What if it were a species harmful to people, such as a mosquito, bacterium, or virus?

The first question above is central to the notion of sustainability (p. 15), which lies at the heart of environmental science today. Sustainability means leaving our descendants a world in which they can meet their needs at least as well as we have met ours. From an ethical perspective, sustainability means treating future generations as we prefer to be treated. The second question goes to the heart of environmental justice, which we will tackle shortly (pp. 144–146). The third set of questions involve intrinsic values (but also instrumental values), and are typical of those that arise in debates over endangered species management, habitat protection, and conservation biology, which we will encounter in Chapter 11 and elsewhere.

We have expanded our ethical consideration over time

Answers to questions like those above depend partly on what ethical standard(s) a person chooses to use. They also depend on the breadth and inclusiveness of the person's domain of ethical concern. A person who ascribes intrinsic value to insects and feels responsibility for their welfare would answer the third set of questions very differently from a person whose domain of ethical concern ends with human beings.

Throughout the history of Western cultures (European and European-derived societies), people have granted intrinsic value and extended ethical consideration to more and more people and things. The enslavement of human beings was common in many societies until recently, for instance. Women in the United States were not allowed to vote until 1920, and many still face lower pay for equal work. Consider, too, how little ethical consideration citizens of one nation extend to those of another on which their government has declared war. In truth, human societies are still struggling to fully embrace the principle that all people be granted equal ethical consideration.

Our expanding domain of ethical concern has begun to include nonhuman entities as well. Concern for the welfare of domesticated animals is evident in the ways many people provide for their pets. Animal rights activists voice concern for animals that are hunted, raised in pens, or used in laboratory testing. Most people now accept that wild animals (at least obviously sentient animals, such as large vertebrates, with which we share similarities) merit ethical consideration. Moreover, increasing numbers of people today see intrinsic value in whole natural communities. Some go further, suggesting that all of nature—living and nonliving things alike—should be ethically recognized.

What is behind this ongoing expansion? Rising economic prosperity in Western cultures, as people became less anxious about their day-to-day survival, has helped to broaden our ethical domain. Science has also played a role, by demonstrating that people do not stand apart from nature, but rather are part of it. Ecology has made clear that all organisms are interconnected and that what affects plants, animals, and ecosystems also affects people. Evolutionary biology has shown that human beings, as one species out of millions, have evolved subject to the same pressures as other organisms.

At the same time, many traditional non-Western cultures have long granted nonhuman entities intrinsic value and ethical standing. Aborigines such as the Mirarr, who view their landscape as sacred and alive, are a case in point. Likewise, many Native American cultures feature ethical systems that encompass both people and aspects of their environment.

We can simplify our continuum of attitudes toward the natural world by dividing it into three ethical perspectives: anthropocentrism, biocentrism, and ecocentrism (**FIGURE 6.3**).

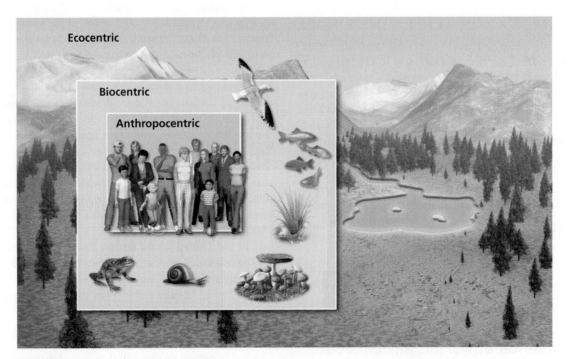

FIGURE 6.3 ▲ We can categorize people's ethical perspectives as anthropocentric, biocentric, or ecocentric. An anthropocentrist extends ethical standing only to human beings and judges actions in terms of their effects on people. A biocentrist values and considers all living things, human and otherwise. An ecocentrist extends ethical consideration to living and nonliving components of the environment and takes a holistic view of the connections among these components, valuing the larger functional systems of which they are a part.

Anthropocentrism People who have a human-centered view of our relationship with the environment show **anthropocentrism**. An anthropocentrist denies or ignores the notion that nonhuman things have intrinsic value. An anthropocentrist also measures the costs and benefits of actions solely according to their impact on people. To evaluate a human action that affects the environment, an anthropocentrist might use criteria such as impacts on human health, economic costs and benefits, and aesthetic concerns. For example, if a proposed mine such as Jabiluka would provide significant economic benefits while doing little harm to aesthetics or human health, the anthropocentrist would conclude it was a worthwhile venture, even if it would destroy habitat for many plants and animals. Conversely, if protecting the area from mining would provide greater economic, spiritual, or other benefits to people, an anthropocentrist would favor its protection. In the anthropocentric perspective, anything not providing benefit to people is considered to be of negligible value.

Biocentrism In contrast to anthropocentrism, **biocentrism** ascribes intrinsic value to certain living things or to the biotic realm in general. In this perspective, nonhuman life has ethical standing, so a biocentrist evaluates actions in terms of their overall impact on living things, human and nonhuman. In the case of a mining proposal, a biocentrist might oppose the mine if it would destroy many plants and animals in the area, even if it would create jobs, generate economic growth, and pose no threat to human health. Some biocentrists advocate equal consideration of all living things, whereas others grant some types of organisms more consideration than others.

Ecocentrism The perspective of **ecocentrism** judges actions in terms of their benefit or harm to the integrity of whole ecological systems, which consist of living and nonliving elements and the relationships among them. An ecocentrist values the well-being of entire species, communities, or ecosystems over the welfare of a given individual. Implicit in this view is that preserving systems generally protects their components, whereas just protecting certain components may not safeguard the entire system. An ecocentrist would respond to the Jabiluka mining proposal by broadly assessing the potential effects of the mine on water quality, air quality, wildlife populations, soil structure, nutrient cycling, and ecosystem services (pp. 3, 121–122). Ecocentrism is a more holistic perspective than biocentrism or anthropocentrism. It encompasses a wider variety of entities and seeks to preserve the connections that tie them together into functional systems.

Environmental ethics has ancient roots

Environmental ethics arose as an academic discipline in the early 1970s, but people have contemplated our ethical relations with nature for thousands of years. The Aboriginal dreaming-track stories are ancient and treat the environment as a source of sacred teachings, endowing boulders, caves, lakes, and other natural features with moral significance worthy of contemplation and protection. In the Western tradition, the ancient Greek philosopher Plato expressed humanity's moral obligation to the environment, writing, "The land is our ancestral home and we must cherish it even more than children cherish their mother."

Some ethicists and theologians have pointed to the religious traditions of Christianity, Judaism, and Islam as sources of anthropocentric hostility toward the environment. They point out biblical passages such as, "Be fruitful and multiply, and fill the earth and subdue it; and have dominion over the fish of the sea and over the birds of the air and over every living thing that moves upon the earth." Such wording, according to many scholars, has justified and encouraged an animosity toward nature that has characterized Western culture over the centuries.

Others interpret sacred texts of these religions to encourage benevolent human stewardship over nature. Consider the directive, "You shall not defile the land in which you live. . . ." If one views the natural world as God's creation, then surely it must be considered a sin to degrade that creation. Indeed, it must be seen as one's duty to act as a responsible steward of the world we have been given.

The industrial revolution inspired philosophers

As industrialization spread in the 19th century from Great Britain and Europe to North America and beyond, it amplified human impacts on the environment. In this period of social and economic transformation, agricultural economies became industrial ones, machines enhanced or replaced human and animal labor, and much of the rural population moved into cities. Consumption of natural resources accelerated, and pollution increased as we burned coal to fuel railroads, steamships, ironworks, and factories.

Some writers of the time drew attention to the drawbacks of industrialization. British critic **John Ruskin** (1819–1900) called cities "little more than laboratories for the distillation into heaven of venomous smokes and smells." Ruskin also worried that people prized the material benefits that nature could provide but no longer appreciated its spiritual and aesthetic benefits. Motivated by such concerns, a number of citizens' groups sprang up in 19th-century England that were forerunners of today's environmental organizations.

In the United States during the 1840s, a philosophical movement called *transcendentalism* flourished, espoused in New England by philosophers **Ralph Waldo Emerson** and **Henry David Thoreau** and by poet **Walt Whitman**. The transcendentalists viewed nature as a direct manifestation of the divine, emphasizing the soul's oneness with nature and God. Like Ruskin, the transcendentalists objected to their fellow citizens' attraction to material things, and through their writing they promoted their holistic view of nature. In some respects, the transcendentalist worldview resembled that of the Mirarr. Both traditions identify a need to experience wild nature, and both view natural entities as symbols or messengers of deeper truths.

Although Thoreau viewed nature as divine, he also observed the natural world closely and came to understand it in the manner of a scientist; indeed, he can be considered one of the first ecologists. His book *Walden*, in which he recorded his observations and thoughts while living at Walden Pond away from the bustle of urban Massachusetts, remains a classic of American literature.

Conservation and preservation arose with the 20th century

One admirer of Emerson and Thoreau was **John Muir** (1838–1914), a Scottish immigrant to the United States who eventually settled in California and made the Yosemite Valley his wilderness home. Although Muir chose to live in isolation in his beloved Sierra Nevada for long stretches of time, he also became politically active and won fame as a tireless advocate for the preservation of wilderness (**FIGURE 6.4**).

Muir was motivated by the rapid deforestation and environmental degradation he witnessed throughout North America and by his belief that the natural world should be treated with the same respect that cathedrals receive. Today he is associated with the **preservation** ethic, which holds that we should protect the natural environment in a pristine, unaltered state. Muir argued that nature deserved protection for its own sake (an ecocentrist argument resting on the notion of intrinsic value), but he also maintained that nature promoted human happiness (an anthropocentrist argument based on instrumental value). "Everybody needs beauty as well as bread," he wrote, "Places to play in and pray in, where nature may heal and give strength to body and soul alike."

FIGURE 6.4 ▲ A pioneering advocate of the preservation ethic, John Muir is also remembered for his efforts to protect the Sierra Nevada from development and for his role in founding the Sierra Club, a leading environmental organization. Here Muir (right) is shown with President Theodore Roosevelt in Yosemite National Park. After his 1903 wilderness camping trip with Muir, the president instructed his interior secretary to increase protected areas in the Sierra Nevada.

FIGURE 6.5 ▲ Gifford Pinchot, the first chief of what would become the U.S. Forest Service, was a leading proponent of the conservation ethic. The conservation ethic holds that people should use natural resources, but should strive to ensure the greatest good for the greatest number for the longest time.

Some of the developments that motivated Muir also inspired the first professionally trained American forester, **Gifford Pinchot** (1865–1946; **FIGURE 6.5**). Pinchot founded what would become the U.S. Forest Service and served as its chief in President Theodore Roosevelt's administration. Like Muir, Pinchot opposed the deforestation and unregulated economic development of North American lands that occurred during their lifetimes. However, Pinchot took a more anthropocentric view of how and why we should value nature. He is today the person most closely associated with the **conservation** ethic, which holds that people should put natural resources to use but that we have a responsibility to manage them wisely. The conservation ethic uses a utilitarian standard stating that in using resources, we should attempt to provide the greatest good to the greatest number of people for the longest time. Whereas preservation aims to preserve nature for its own sake and for our aesthetic and spiritual benefit, conservation promotes the prudent, efficient, and sustainable extraction and use of natural resources for the benefit of present and future generations.

Pinchot and Muir came to represent different branches of the American environmental movement, and their contrasting ethical approaches often pitted them against one another on policy issues of the day. For instance, Pinchot supported damming Hetch Hetchy Valley in Yosemite National Park to provide drinking water for San Francisco, considering this "the highest possible use which could be made of" the valley. Muir was aghast, and proclaimed, "Dam Hetch Hetchy! As well dam for water-tanks the people's cathedrals and churches, for no holier temple has ever been consecrated by the heart of man."

Yet despite their differences, both Muir and Pinchot represented reactions against a prevailing "development ethic," which holds that people should be masters of nature and which promotes economic development without regard to its negative consequences. Both Pinchot and Muir left legacies that reverberate today in the various ethical approaches to environmentalism.

Aldo Leopold's land ethic arose from the conservation and preservation ethics

As a young forester and wildlife manager, **Aldo Leopold** (1887–1949; **FIGURE 6.6**) began his career fully in the conservationist camp, having graduated from Yale Forestry School, which Pinchot had helped found just as Roosevelt and Pinchot were advancing conservation on the national stage. As a forest manager in Arizona and New Mexico, Leopold embraced the government policy of shooting predators, such as wolves, to increase populations of deer and other game animals.

At the same time, Leopold followed the development of ecological science. He eventually ceased to view certain species as "good" or "bad" and instead came to see that healthy ecological systems depend on the protection of all their interacting parts. Drawing an analogy to mechanical maintenance, he wrote, "to keep every cog and wheel is the first precaution of intelligent tinkering."

FIGURE 6.6 ▲ Aldo Leopold, a wildlife manager and pioneering environmental philosopher, articulated a new relationship between people and the environment. In his essay "The Land Ethic" he called on people to include the environment in their ethical framework.

It was not just science, however, that pulled Leopold from an anthropocentric perspective toward a more holistic one. One day he shot a wolf, and when he reached the animal, Leopold was transfixed by "a fierce green fire dying in her eyes." He perceived intrinsic value in the wolf, and the experience remained with him for the rest of his life, helping to lead him to an ecocentric ethical outlook. Years later, as a University of Wisconsin professor, Leopold argued that humans should view themselves and "the land" as members of the same community and that people are obligated to treat the land in an ethical manner. In his 1949 essay "The Land Ethic," he wrote:

> All ethics so far evolved rest upon a single premise: that the individual is a member of a community of interdependent parts. . . . The land ethic simply enlarges the boundaries of the community to include soils, waters, plants, and animals, or collectively: the land. . . . A land ethic changes the role of *Homo sapiens* from conqueror of the land-community to plain member and citizen of it. . . . It implies respect for his fellow-members, and also respect for the community as such.

Leopold intended that the land ethic would help guide decision making. "A thing is right," he wrote, "when it tends to preserve the integrity, stability, and beauty of the biotic community. It is wrong when it tends otherwise." Leopold died before seeing "The Land Ethic" and his best-known book, *A Sand County Almanac*, in print, but today many view him as the most eloquent and important philosopher of environmental ethics.

Ecofeminism critiques male attitudes toward nature and women

As Leopold's influence continued to help expand people's ethical horizons, major social movements such as the civil rights movement and the feminist movement came into full swing during the 1960s and 1970s. A number of feminist scholars saw parallels in human behavior toward nature and men's behavior toward women. The degradation of nature and the social oppression of women shared common roots, these scholars asserted.

Ecological feminism, or **ecofeminism**, argues that the patriarchal (male-dominated) structure of society—which traditionally grants more power and prestige to men than to women—is a root cause of both social and environmental problems. Ecofeminists hold that a traditionally female worldview that interprets the world in terms of interrelationships and cooperation is more compatible with nature than a traditionally male worldview that interprets the world in terms of hierarchies and competition. Ecofeminists maintain that a male tendency to try to dominate and conquer what men hate, fear, or do not understand has historically been exercised against both women and the natural environment.

Environmental justice seeks equal treatment for all races and classes

As our society has slowly granted more equality to women, its domain of ethical concern has also been expanding from rich to poor, and from majority races and ethnic groups to minority ones. This ethical expansion involves applying a standard of fairness and equality, and it has given rise to the environmental justice movement. **Environmental justice** involves the fair and equitable treatment of all people with respect to environmental policy and practice, regardless of their income, race, or ethnicity.

The environmental justice movement has been fueled by the perception that poor people and minorities tend to be exposed to a greater share of pollution, hazards, and environmental degradation than are richer people and whites. Indeed, studies across North America repeatedly document that poor and nonwhite communities bear greater than average burdens of air pollution, lead poisoning, pesticide exposure, toxic waste exposure, and workplace hazards. This is thought to occur because lower-income and minority communities often have less access to information on environmental problems and health risks, less political power with which to protect their interests, and less money to spend on avoiding or alleviating risks. Environmental justice proponents also sometimes blame institutionalized racism and inadequate government policies.

A protest in the early 1980s by residents of Warren County, North Carolina, against a toxic waste dump in their community is widely viewed as the beginning of the environmental justice movement (**FIGURE 6.7**). The state had chosen to site the dump in the county with the highest percentage of African Americans, prompting Warren County residents to suspect "environmental racism."

In Australia, critics have characterized the attempts of the (predominantly white) Australian government and uranium mining companies to open mines on traditional lands of Aboriginal people as violations of environmental justice. Like the African American residents of Warren County, the Mirarr and other Australian Aborigines are minority-race people who have suffered a history of unequal treatment.

FIGURE 6.8 ▲ Native Americans employed as uranium miners in Canada and the United States, such as the Navajo miner shown here, have suffered from adverse effects of mining.

Uranium mining on the lands of Native people has raised environmental justice issues in the United States and Canada as well. From 1948 through the late 1960s, uranium mines employed many Native Americans of the U.S. Southwest, including those of the Navajo nation (**FIGURE 6.8**). Although uranium mining had already been linked to health problems and premature death, the miners were not made aware of radiation and its risks. The Navajo language did not even have a word for *radiation*, and for nearly two decades neither the mining industry nor the U.S. government provided the miners information or safeguards. Many Navajo families built homes and bread-baking ovens consisting of waste rock from the mining process, not realizing it was radioactive.

Cases of lung cancer began to appear among Navajo miners in the early 1960s, but scientific studies of radiation's effects on miners at the time excluded Native Americans. The decision to include only white miners in the studies was attributed to the researchers' desire to study a "homogeneous population." A later generation of Americans perceived this as negligence and discrimination, and their desire for justice led to the Radiation Exposure Compensation Act of 1990, a federal law that compensated Navajo miners who suffered health effects from unprotected work in the mines.

Today, despite much progress toward racial equality, significant inequities remain. Economic gaps between rich and poor have widened. Despite stronger enforcement of environmental laws, minorities and the poor still suffer substandard environmental conditions (**FIGURE 6.9**). Yet today more people are fighting environmental hazards in their communities and winning.

One success story is in California's San Joaquin Valley. The poor, mostly Latino, farm workers in this region who help harvest much of the U.S. food supply also suffer from some of the nation's worst air pollution. Industrial agriculture produces pesticide emissions, dairy feedlot emissions, and wind-blown dust from eroding farmland, yet this pollution was not being regulated. Valley residents enlisted the help of several organizations, including the Center on Race, Poverty, and the

FIGURE 6.7 ▲ Communities of poor people and people of color have suffered more than their share of environmental problems, a situation that has given rise to the environmental justice movement. The movement gained prominence with this protest of a toxic waste dump in Warren County, North Carolina.

FIGURE 6.9 ▲ Hurricane Katrina revealed the need for environmental justice, as the people affected most by the storm and its aftermath were poor and nonwhite. These girls are playing in the Lower Ninth Ward of New Orleans, where many homes were destroyed and water remained unsafe to drink long afterwards. Their mother had moved back here after Katrina destroyed her home, but poverty forced her to accept donated gutted housing once the Federal Emergency Management Agency cut off payments.

Environment, a San Francisco–based environmental justice law firm, and succeeded in convincing California regulators to enforce Clean Air Act provisions and California legislators to pass new legislation regulating agricultural emissions.

WEIGHING THE ISSUES

Environmental Justice Consider the place where you grew up. Where were the factories, waste dumps, and polluting facilities located, and who lived closest to them? Who lives nearest them in the town or city that hosts your campus? Do you think the concerns of environmental justice advocates are justified? If so, what could be done to ensure that poor communities are no more polluted than wealthy ones?

Environmental justice is an international issue

Just as wealthy people often impose their pollution on poorer people, wealthy nations do the same to poorer nations. One common source of environmental injustice among nations concerns the dumping of hazardous waste (pp. 637–638). The millions of tons of hazardous waste that we in developed nations produce in our factories, power plants, and incinerators must go somewhere. Proper disposal is expensive, so companies often find it cheaper to pay cash-strapped nations to take the waste—or cheaper still to dump it illegally. Sometimes hazardous waste is falsely labeled as harmless material, such as fertilizer, fill dirt, or plastics for recycling.

In nations with lax environmental and health regulations, workers and residents are often uninformed of or unprotected against the dangers from this waste. Exposed individuals may

fall ill and can die prematurely. An international treaty, the Basel Convention, prohibits the international export of waste, but trade and illegal dumping have continued. Although 169 nations have ratified the treaty, the United States (the world's largest exporter) has not.

Environmental justice is a key component in pursuing a triple bottom line of environmental, economic, and social sustainability (p. 19). As we explore environmental issues from a scientific standpoint throughout this book, we will also encounter the social, political, ethical, and economic aspects of these issues, and the concept of environmental justice will arise again and again.

ECONOMICS: APPROACHES AND ENVIRONMENTAL IMPLICATIONS

Questions of environmental justice intertwine ethical issues with economic ones, and friction often develops between people's ethical impulses and their economic desires. In the case of the Jabiluka mine proposal, people who oppose the mine do so largely on the basis of ethical and environmental justice concerns. Those who support the mine proposal do so largely on economic grounds, as uranium is a lucrative resource that generates jobs, income, and electricity (**FIGURE 6.10**). Such conflict between ethical inclinations and economic motivations is a recurrent theme in environmental issues worldwide.

Is there a trade-off between the economy and the environment?

Measures to safeguard environmental quality frequently mesh well with ethical considerations, but we often hear it said that environmental protection runs counter to economic interests.

FIGURE 6.10 ▲ Uranium mining provides jobs and income to Australian workers. This Aboriginal man from a group that lives near the Mirarr chose to take a job at the Ranger uranium mine. Unemployment is above 16% among Aborigines in the region, and one-fifth of the mine's employees are Aboriginal.

People argue that environmental protection costs too much money, interferes with progress, or leads to job loss. However, growing numbers of economists assert that there need be no such trade-off—and that in fact, environmental protection is generally *good* for our economy.

The view one takes depends in part on whether one thinks in the short term or the long term. The economic judgments we make often pertain to short time scales—but in the longer term, the degradation of ecosystems feeds back to harm economies. The view one takes also depends on whether one stands to benefit. Often when development results in long-term environmental degradation, private parties benefit economically while the broader public is harmed.

Traditional economic schools of thought have underestimated or ignored the contributions of the environment to our economies, and people viewing the world in this way have equated environmental protection with economic sacrifice. Newer economic schools of thought (such as environmental economics and ecological economics, which we will examine shortly (p. 155), recognize human economies as coupled to the environment and reliant on its goods and services. For people holding this worldview, our economic health depends on some degree of environmental protection. Today, concern over climate change, diminishing fossil fuel supplies, and dependence on foreign oil have led many mainstream economists and policymakers to recognize the immense opportunities in building a green-energy economy based on clean and renewable energy technologies. This presents a clear case of how economic advancement and environmental protection go hand-in-hand.

Economics studies the allocation of resources

Like ethics, economics examines factors that guide human behavior. **Economics** is the study of how people decide to use potentially scarce resources to provide goods and services in the face of demand for them. By this definition, environmental problems are also economic problems that can intensify as population and per capita resource consumption increase. For example, pollution may be viewed as depletion of the scarce resources of clean air, water, or soil. Indeed, the word *economics* and the word *ecology* come from the same Greek root, *oikos,* meaning "household." Economists traditionally have studied the household of human society, and ecologists the broader household of all life.

Several types of economies exist

An **economy** is a social system that converts resources into **goods**, material commodities manufactured for and bought by individuals and businesses; and **services**, work done for others as a form of business. The oldest type of economy is the **subsistence economy**. People in subsistence economies meet most or all of their daily needs directly from nature and their own production; they subsist on what they can gather or produce, rather than working for wages and then purchasing life's necessities. Until very recently, the Mirarr and most other Australian Aboriginal people lived in subsistence economies.

A second type of economy is the **capitalist market economy**. In this system, interactions among buyers and sellers determine which goods and services are produced, how many are produced, and how these are produced and distributed. Capitalist economies contrast with state socialist economies, or **centrally planned economies**, in which government determines in a top-down manner how to allocate resources. In reality, however, virtually all national economies in today's world are hybrid systems, often termed **mixed economies**. The United States, the world's strongest proponent of capitalism, has in fact borrowed a great deal from socialism, and China, the world's largest socialist state, hosts a robust market system. By combining aspects of both approaches, mixed economies encourage stability and prosperity. When things get out of balance—such as when unregulated financial practices led to the recession in 2009—then prosperity may decline and balance may need to be restored.

In modern mixed economies, as we will see in Chapter 7, governments typically intervene in the market for several reasons:

▶ To eliminate unfair advantages held by single buyers or sellers

▶ To provide social services, such as national defense, medical care, and education

▶ To provide "safety nets" (for the elderly, victims of natural disasters, and so on)

▶ To manage the commons (p. 5)

▶ To mitigate pollution and other threats to health and quality of life

The economy exists within the environment

All human economies exist within the natural environment and depend on it in vital ways. Economies receive inputs (such as natural resources) from the environment, process these inputs in complex ways that enable human society to function, and then discharge outputs (such as waste) into the environment. The material inputs and the waste-absorbing capacity that Earth can provide to economies are ultimately finite.

Although the interactions between human economies and the nonhuman environment are readily apparent, traditional economic schools of thought have long overlooked the importance of these connections. Indeed, most mainstream economists still adhere to a worldview that largely ignores the environment (**FIGURE 6.11A**)—and this worldview continues to drive most policy decisions. This mainstream view essentially holds that environmental resources (the inputs into the economy) are free and limitless and that wastes (outputs) can be endlessly exported and absorbed by the environment, at no cost. In contrast, modern economists belonging to the fast-growing fields of environmental economics and ecological economics (p. 155) explicitly recognize that human economies are subsets of the environment and depend crucially upon it for natural resources and ecosystem services (**FIGURE 6.11B**).

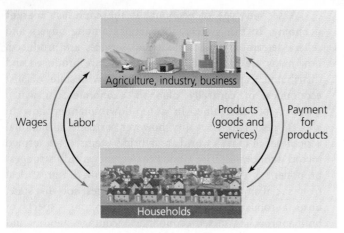

(a) Conventional view of economic activity

148

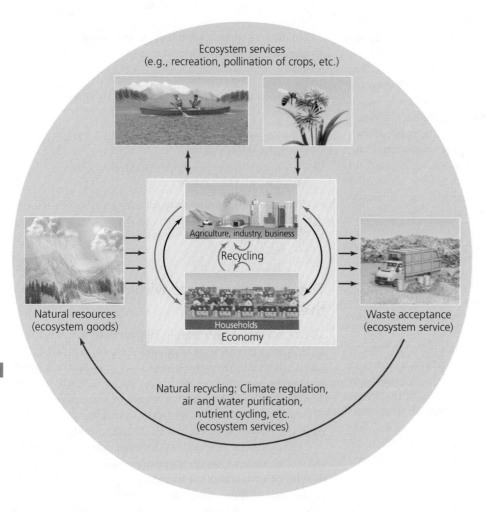

(b) Economic activity as viewed by environmental and ecological economists

FIGURE 6.11 ◀ Standard neoclassical economics focuses on processes of production and consumption between households and businesses **(a)**, viewing the environment as little more than a "factor of production" that helps enable the production of goods. Environmental and ecological economists view the human economy as existing within the natural environment **(b)**, receiving resources from it, discharging waste into it, and interacting with it through various ecosystem services.

Economies rely on ecosystem goods and services

Economic activity uses resources from the environment. Natural resources (pp. 3–4) are the various substances and forces we need to survive: the sun's energy, the fresh water we drink, the trees that provide our lumber, the rocks that provide our metals, and the fossil fuels that power our machines. We can think of natural resources as "goods" produced by nature.

Environmental systems also naturally function in a manner that supports economies. Earth's ecological systems purify air and water, form soil, cycle nutrients, regulate climate, pollinate plants, and recycle the waste generated by our economic activity. Such essential services, called **ecosystem services** (pp. 3, 121–122, 160–161), support the life that makes our economic activity possible.

While our environment enables economic activity by providing ecosystem goods and services, economic activity can also affect the environment. When we deplete natural resources and generate pollution, we can degrade the capacity of ecological systems to function. In fact, the Millennium Ecosystem Assessment (pp. 16, 18) concluded in 2005 that 15 of 24 ecosystem services its scientists surveyed globally were being degraded or used unsustainably. The degradation of ecosystem services will in turn negatively affect economies. Thus, the Mirarr's opposition to mine development on their land is grounded not simply in spirituality and aesthetics, but in an ecologically sound understanding of what the land provides them. They see that the Ranger mine has degraded the ecosystem services they rely on, and they conclude that the Jabiluka mine would do the same. Indeed, across the world, ecological degradation is harming poor and marginalized people more than wealthy ones, the Millennium Ecosystem Assessment found. Thus, restoring ecosystem services stands as a prime avenue for alleviating poverty.

The relationships among economic and environmental conditions are only now becoming widely recognized. Let's briefly examine how economic thought has evolved, tracing the path that is finally beginning to lead economies to become more compatible with the natural systems on which they depend.

Adam Smith proposed an "invisible hand"

Economics shares a common intellectual heritage with ethics, and practitioners of both have long been interested in the relationship between individual action and societal well-being. Some philosophers argued that individuals acting in their own self-interest would harm society (as in the tragedy of the commons, p. 5). Scottish philosopher **Adam Smith** (1723–1790) believed that self-interested behavior could benefit society, as long as the behavior was constrained by the rule of law and private property rights and operated within fairly competitive markets. Known today as a founder of **classical economics**, Smith felt that when people are free to pursue their own economic self-interest in a competitive marketplace, the marketplace will behave as if guided by "an invisible hand" that leads their actions to benefit society as a whole. In his 1776 book *Inquiry into the Nature and Causes of the Wealth of Nations,* Smith wrote:

> It is not from the benevolence of the butcher, the brewer, or the baker that we expect our dinner, but from their regard to their own self-interest. [Each individual] intends only his own security, only his own gain. And he is led in this by an invisible hand to promote an end which was no part of his intention. By pursuing his own interests he frequently promotes that of society more effectually than when he really intends to.

Smith's philosophy remains a pillar of free-market thought today, and many credit it for the tremendous gains in material wealth that industrialized nations have experienced in the past two centuries. Others point out that policies spawned by free-market philosophy tend to worsen inequalities between rich and poor. Still others assert that free-market capitalism promotes environmental degradation.

Neoclassical economics incorporates psychology

Economists subsequently adopted more quantitative approaches as they aimed to explain human behavior. **Neoclassical economics** examines the psychological factors underlying consumer choices, explaining market prices in terms of consumer preferences for units of particular commodities. Standard neoclassical models start from an assumption that people behave rationally and have access to full information.

In neoclassical economic theory, buyers desire the lowest possible price, whereas sellers desire the highest possible price. This conflict between buyers and sellers results in a compromise price being reached and the "right" quantity of commodities being bought and sold. This is often phrased in terms of **supply**, the amount of a product offered for sale at a given price, and **demand**, the amount of a product people will buy at a given price if free to do so. Theoretically, the market automatically moves toward an equilibrium point, a price at which supply equals demand (**FIGURE 6.12A**). Similar reasoning can be applied to environmental issues, such that economists can determine "optimal" levels of resource use or pollution control (**FIGURE 6.12B**). For instance, reducing pollution emitted by a car or factory is often cost-effective at first, but as pollution is reduced, it becomes more and more costly to eliminate each remaining amount. At some point the cost per unit of reduction rises to match the benefits, and this is the optimal level of pollution reduction.

Cost-benefit analysis is a widespread tool

Neoclassical economists commonly use a method referred to as **cost-benefit analysis**. In this approach, estimated costs for a proposed action are totaled up and compared to the sum of benefits estimated to result from the action. If benefits exceed costs, the action should be pursued; if costs exceed benefits, it should not. Given a choice of alternative actions, the one with the greatest excess of benefits over costs should be chosen.

This reasoning seems eminently logical; however, problems often arise because not all costs and benefits can be easily quantified—or even identified or defined. It may be simple to quantify wages paid to uranium miners, the market value of uranium extracted from a mine, or the price of measures to minimize health risks for miners. But it is difficult to assess the intangible cost of a landscape being scarred by mine development, or the cost of radioactive contamination of a stream.

Because some costs and benefits cannot easily be assigned monetary values—and because it is difficult to identify and agree on all costs and benefits—cost-benefit analysis is often controversial. Moreover, because monetary benefits are usually more easily quantified than environmental costs, benefits tend to be overrepresented in traditional cost-benefit analyses. As a result, environmental advocates often feel that

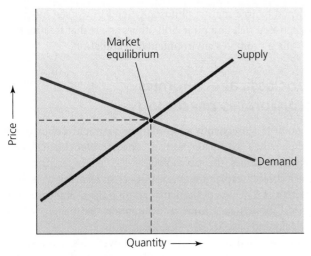

(a) Classic supply–demand curve

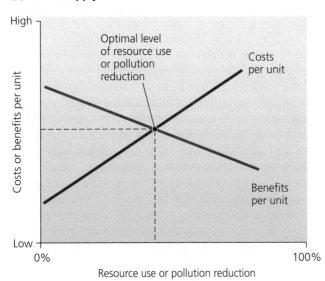

(b) Marginal benefit and cost curves

FIGURE 6.12 ▲ In a basic supply-and-demand graph **(a)**, the demand curve indicates the quantity of a given good (or service) that consumers desire at each price, and the supply curve indicates the quantity produced at each price. The market automatically moves toward an equilibrium point at which supply equals demand. We can use a similar graph **(b)** to determine an "optimal" level of resource use or pollution mitigation. In this graph, the cost per unit of resource use or pollution cleanup (blue line) rises as the resource use or pollution cleanup proceeds and it becomes expensive to extract or clean up the remaining amounts. Meanwhile, the benefits per unit of resource use or pollution cleanup (red line) decrease. The point where the lines intersect gives the optimal level.

these analyses are biased in favor of economic development and against environmental protection.

Aspects of neoclassical economics have profound implications for the environment

Today's capitalist market systems operate largely in accord with the principles of neoclassical economics. These systems have generated unprecedented material wealth, but they have also contributed to environmental degradation. Four fundamental assumptions of neoclassical economics have implications for the environment:

▶ Resources are infinite or substitutable.
▶ Costs and benefits are internal.
▶ Future effects should be discounted.
▶ Growth is good.

Are resources infinite or substitutable? Neoclassical economic models generally treat workers and other resources as being either infinite or largely "substitutable and interchangeable." This implies that once we have depleted a resource—natural, human, or otherwise—we will be able to find a replacement for it. Human resources can substitute for financial resources, for instance, or manufactured resources can substitute for natural resources. Theory allows that the new resource may be less efficient or more costly, but it is assumed that we will always be able to find one. Thus, ecosystem goods and services are treated as "free gifts of nature," infinitely abundant and resilient, and ultimately substitutable by human technological ingenuity.

Certainly it is true that many resources can be replaced. Our societies have made the transition from manual labor to animal labor to steam-driven power to fossil fuel power, and they may yet transition to renewable power sources, such as solar energy. However, Earth's material resources are ultimately limited. Nonrenewable resources such as fossil fuels can be depleted, and many renewable resources can be used up as well if we exploit them faster than they can be replenished. This is what happened to the Easter Islanders (pp. 6–7), who harvested their forests faster than the forests could regrow.

Are costs and benefits internal? A second assumption of neoclassical economics is that all costs and benefits associated with a particular exchange of goods or services are borne by individuals engaging directly in the transaction. In other words, it is assumed that the costs and benefits are "internal" to the transaction, experienced by the buyer and seller alone.

However, many transactions affect other members of society. Pollution from a factory, power plant, or mine can harm people living nearby. The Mirarr have suffered from the contamination of streams by radioactive material discharged from the Ranger mine. On occasion, uranium concentrations in the water have risen 4,000 times higher than allowed by law. When pollution from an industry or activity affects people not involved in the industry or activity, then they suffer—and often taxpayers end up paying the costs of alleviating the problem. Market prices do not take the social, environmental, or economic costs of this pollution into account.

Costs or benefits that affect people other than the buyer or seller are known as **externalities**. A positive externality is a benefit enjoyed by someone not involved in a transaction, and a negative externality, or **external cost**, is a cost borne by someone not involved in a transaction (**FIGURE 6.13**). External costs often harm whole communities of people while allowing certain individuals private gain. External costs commonly include the following:

- ▶ Chronic human health problems
- ▶ Property damage
- ▶ Declines in desirable elements of the environment, such as fewer fish in a stream
- ▶ Aesthetic damage, such as that resulting from air pollution or clear-cutting
- ▶ Stress and anxiety experienced by people downstream or downwind from a pollution source
- ▶ Declining real estate values resulting from these problems

By ignoring negative externalities, economies create a false idea of the true and complete costs of particular choices and unjustly subject people to the consequences of activities in which they did not participate. External costs comprise one reason governments develop environmental legislation and regulations (p. 170). The best way to compensate for external costs is to build the costs into market prices, but it can be challenging to do this. It is difficult to assign a monetary value to such things as illness, premature death, or degradation of an aesthetically or spiritually significant site.

Should future effects be discounted? A third assumption of the neoclassical economic approach is that an event in the future should count less than one in the present; in economic terminology, future effects are *discounted*. Discounting is meant to reflect how human beings—and by extension, markets and their prices—tend to grant more importance to present conditions than to future conditions. Just as you'd rather have an ice cream cone today than being promised one next month, market demand is greater for goods and services that are received sooner rather than later. Economists quantify this by assigning numerical discount rates in their calculations of costs and benefits. As a simple example, applying a 5% annual discount rate to forestry decisions means that a stand of trees whose timber is worth $500,000 to the market today would drop in perceived value by 5% each year. From the perspective of today's market, having the timber in 10 years would be worth only $299,368. By this logic, the more quickly the trees are cut, the more they are worth.

Discounting encourages policymakers to play down the long-term consequences of decisions we make today. Many environmental problems unfold gradually over long time periods, and discounting reduces our incentive to address resource depletion, pollution buildup, and other cumulative forms of environmental degradation. Instead, discounting essentially shunts the costs of dealing with such problems on to future generations. The choice of what discount rate to use is subjective and involves ethical judgment. This has become apparent as economists debate how discount rates influence our estimates of the costs of global climate change to our society (see **THE SCIENCE BEHIND THE STORY**, pp. 152–153).

FIGURE 6.13 ▲ An Indonesian boy wading in a polluted river suffers external costs: costs that are not borne by a buyer or seller. External costs may include water pollution, health problems, property damage, harm to aquatic life, aesthetic degradation, declining real estate values, and other impacts.

Is growth good? **Economic growth** can be defined as an increase in an economy's production and consumption of goods and services. Neoclassical economics views economic growth as essential for maintaining social order, because a growing economy can alleviate the discontent of the poor by creating opportunities for the poor to become wealthier. A rising tide raises all boats, as the saying goes; if we make the overall economic pie larger, then each person's slice can become larger, even if some people still have much smaller slices than others.

Note, however, that economic activity and true wealth are not the same. To the extent that rising consumption and economic growth are means to an end—pathways toward greater human happiness—they are good things. However, many observers today worry that growth is no longer the best path to happiness. In fact, sociologists have coined a word for the way that consumption and material affluence often fail to bring people contentment: **affluenza**. In addition, people in poverty may feel poorer and less happy as the gap between rich and poor widens, even if they gain more wealth themselves.

Critics of the growth paradigm note that runaway growth resembles the multiplication of cancer cells, which eventually overwhelm and destroy the organism in which they grow. These critics fear that runaway growth could destroy our economic system. Resources are ultimately limited, they argue, so nonstop growth is not sustainable and will fail as a long-term strategy.

We live in a growth-oriented economy

Under the mainstream assumption that all economic growth is good, growth has become the quantitative yardstick by which we measure progress. Among policymakers, economists, and the news media today, increases in an industry's output or

The SCIENCE behind the Story

Ethics in Economics: Discounting and Global Climate Change

How much will it cost our society if we do nothing about global climate change? To address this question and to help decide how to respond to climate change, the British government commissioned economist Nicholas Stern to lead a team to assess the economic costs that a changing climate may impose on society. But once Stern's report was published in 2006, a debate ensued that had as much to do with ethics as with economics. The dispute centered on discounting (p. 151), an economic practice heavy with ethical implications.

To produce the *Stern Review on the Economics of Climate Change*, Stern's research team surveyed the burgeoning scientific literature on the impacts of rising temperatures, changing rainfall patterns, and increasing storminess (see Chapter 18). It then estimated the economic consequences of these climatic changes and tried to put a price tag on the global cost. The report concluded that without action to forestall it, climate change would cause losses in annual global gross

Sir Nicholas Stern, head of the Government Economic Service, United Kingdom

domestic product (GDP) of 5–20% by the year 2200 (**see figure**).

Stern's team also calculated that by paying just 1% of GDP annually starting now, our society could stabilize atmospheric greenhouse gas concentrations and prevent most of these future economic losses. The bottom line: Spending a relatively small amount of money now will save us much larger expenses in the future. This conclusion caught the attention of governments worldwide; for the first time, leading economists were advancing a strong economic argument for tackling climate change immediately.

However, the *Stern Review*'s numbers depend partly on how one

chooses to weigh impacts in the future versus impacts happening now. Stern used two discount factors. One accounted for the likelihood that people in the future will be richer than we are today and thus better able to handle economic costs. The other considered whether the future should be discounted simply because it is the future. This latter discount factor (called a *pure time discount factor*) is essentially an ethical issue because it assigns explicit values to the welfare of future versus current generations.

The *Stern Review* used a pure time discount rate of 0.1%. This means that an impact occurring next year is judged 99.9% as important as one occurring this year. It means that the welfare of a person born 100 years from now is valued at 90% of one born today. This discount rate treats current and future generations *nearly* equally. Future generations are slightly downweighted only because of the (very small) possibility that our species could go extinct (in which case, there would be no future generations to be concerned about).

growth in a country's economy are always touted as good news, and decreases, stability, or even a minor drop in the *rate* of growth are presented as bad news.

The amount by which our global economy has grown in recent decades is unprecedented in human history, and the world economy is seven times the size it was just half a century ago. All measures of economic activity—trade, rates of production, amount and value of goods manufactured—are higher than they have ever been. This has brought many people much greater material wealth (although not equitably, and gaps between rich and poor are wide and growing).

The modern-day United States exemplifies the view that "more and bigger" is always better. Spurred on by advertising (which has doubled in just the past 20 years) and the increased availability of goods due to technological advances and expanded global trade, Americans have embarked on a frenzy of consumption unparalleled in history. The dramatic rise in per capita consumption of goods and services has numerous consequences (**FIGURE 6.14**). The many environmental and social impacts of rising per capita consumption will become clearer as we progress through this book.

Can growth go on forever?

Today's mainstream economic theory assumes that growth can continue forever. This view rests on the assumption that resources are infinite or substitutable and on the view that the environment is merely a factor of production, rather than an all-encompassing context and home for the economy.

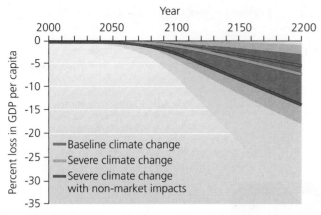

According to the *Stern Review*, baseline climate change as forecast by the IPCC *Third Assessment Report* (p. 504) could bring global losses of 5.3% of per capita GDP annually by the year 2200. Severe climate change, as forecast by research since that report was issued, could bring annual losses of 7.3%—and adding nonmarket values (pp. 157, 160) raises this figure to 13.8%. Gray-shaded areas show ranges of future values judged statistically to be 95% likely; darkest gray indicates where ranges overlap for all three lines, and lightest gray for just one line. SOURCE: Data from HM Treasury, 2007. *Stern review on the economics of climate change.* London, U.K. By permission.

Some economists viewed this discount rate as too low. Yale University economist William Nordhaus proposed a discount rate starting at 3% and falling to 1% in 300 years. He maintained that such numbers are more objective because they reflect people's values as revealed by the prices they pay in the market. Indeed, when economists use discounting to assess capital investments or construction projects (say, development of a railroad, dam, or highway) they typically choose discount rates close to those Nordhaus suggests. Nordhaus argued that the *Stern Review*'s near-zero discount rate overweights the future, forcing people today to pay too much to address hypothetical future impacts.

In their response to Nordhaus and other critics, Stern and his team argued that discount rates of 3% or 1% may be useful for assessing development projects but are too high for long-term environmental problems that directly affect human welfare. A 3% rate, Stern pointed out, means that a person born in 1990 is valued only half as much as a person born in 1965. It means a grandchild is judged to be worth far less than a grandparent simply because of the dates they were born. For various reasons, Stern argued, the market should not be used to guide ethical decisions.

Stern's group also published sensitivity analyses that examined how their conclusions would vary under different discount rates. These showed that the report's main message—that it's cheaper to deal with climate change now than later—was robust to a fair amount of fluctuation in the discount rate, up to at least a rate of 1.5%.

At the end of the day, the choice of a discount rate is an ethical decision on which well-intentioned people may differ. As governments, businesses, and individuals begin to invest in addressing climate change, the debate over the *Stern Review* reveals how ethics and economics remain intertwined.

Addressing global climate change is becoming urgent as changes are happening faster than scientists had predicted just a few years ago (e.g., p. 510). In 2008, Stern announced that the new scientific forecasts of accelerated change meant that it would be more expensive to mitigate global warming than the *Stern Review* had calculated. Instead of 1% of GDP annually, the world would need to spend 2% of GDP each year in order to control climate change, Stern said. This unfortunate development helped drive home the message: The quicker we get going, the better off we'll be. ∎

Economic growth stems from two sources: (1) an increase in inputs to the economy (e.g., greater inputs of labor and natural resources) and (2) improvements in the efficiency of production due to better technologies and approaches (i.e., ideas and equipment that enable us to produce more goods with fewer inputs). This second approach—whereby we produce more with less—is often termed "economic development."

As our population and consumption grow, it is becoming clearer that we cannot sustain growth forever using only the first approach. Nonrenewable resources on Earth are finite in quantity, and there are limits on the rates that we can harvest many renewable resources.

As for the second approach, we have indeed been able to use technological innovation to push back the limits on growth time and again. More-efficient technologies for extracting minerals, fossil fuels, and groundwater allow us to exploit these resources more fully with less waste. Automated farm machinery, fertilizers, and chemical pesticides let us grow more food per unit area of land (pp. 254–255). Better machinery in our factories speeds our manufacturing. In these ways, we can produce more goods and services while using fewer resources. A prime example of doing more with less is the extreme miniaturization of computer technology (**FIGURE 6.15**). We have been making chips more powerful while also making them smaller and using fewer resources. This leap in productivity has helped propel our economy into the Information Age.

Can we conclude, then, that improvements in technology will help us continue economic growth indefinitely? Answering "yes" are the Cornucopians (p. 18), primarily economists, businesspeople, and policymakers. Responding "no" are the

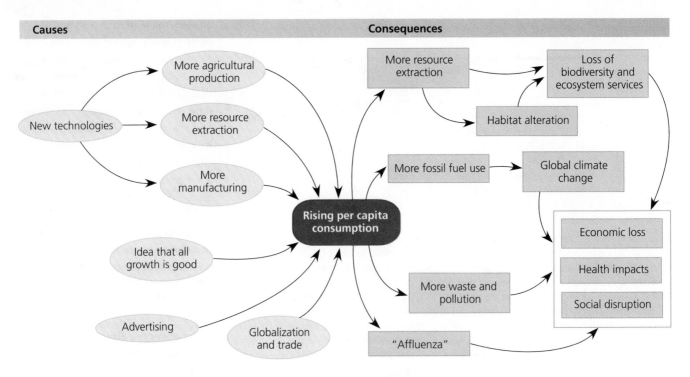

Causes

Consequences

New technologies → More agricultural production

New technologies → More resource extraction

New technologies → More manufacturing

More agricultural production → Rising per capita consumption

More resource extraction → Rising per capita consumption

More manufacturing → Rising per capita consumption

Idea that all growth is good → Rising per capita consumption

Advertising → Rising per capita consumption

Globalization and trade → Rising per capita consumption

Rising per capita consumption

More resource extraction → Loss of biodiversity and ecosystem services

More resource extraction → Habitat alteration

Habitat alteration → Loss of biodiversity and ecosystem services

More fossil fuel use → Global climate change

Global climate change → Economic loss / Health impacts / Social disruption

More waste and pollution → Economic loss / Health impacts / Social disruption

"Affluenza" → Social disruption

FIGURE 6.14 ▲ The rise in per capita consumption of goods and services stems from multiple causes (ovals on left) and results in a diversity of environmental, social, and economic consequences (boxes on right). Arrows in this concept map lead from causes to consequences. Note that items grouped within an outlined box do not necessarily share any special relationship; the outlined box is intended merely to streamline the figure.

Solutions

As you progress through this chapter, try to identify as many solutions to rising per capita consumption as you can. What could you personally do to help address this issue? Consider how each action or solution might affect items in the concept map above.

Cassandras (p. 18), who include a growing number of scientists and others.

The Cassandra view has been articulated most famously by a group of researchers who published a series of books including *The Limits to Growth* (1972), *Beyond the Limits* (1992), and *Limits to Growth: The Thirty-year*

FIGURE 6.15 ▲ Advances in technology have enabled us to push back the natural limits on growth and continue expanding our economies. The miniaturization of computer chips is a striking advance in efficiency, allowing us to store and use far more information with far less material input.

Update (2004). Starting from the premise that Earth's resources are finite, and using calculations of resource availability and consumption, they ran simulation models to predict how our economies would fare in the future. Model runs using data matching our society's current consumption patterns consistently predicted economic collapse as resources become scarce (**FIGURE 6.16A**). The team experimented to see what parameters would produce model runs predicting a sustainable civilization (**FIGURE 6.16B**), and they used these results to make recommendations for achieving sustainability.

Cornucopians, who include the economist Julian Simon and, more recently, the statistician Bjorn Lomborg, counter that Cassandras such as the *Limits to Growth* team underestimate the human capacity to innovate and the degree to which more powerful technologies can increase our access to resources. They also argue that market forces help us avoid resource depletion. As resources become scarce, prices rise, and individuals and firms have incentive to turn to different resources, shift to different products, or reuse and recycle. All of these actions will alleviate pressure on the resource. As Simon put it:

> The natural world allows, and the developed world promotes through the marketplace, responses to human needs and shortages. . . . The main fuel to speed our progress is our stock of knowledge, and the brake is our lack of imagination. The ultimate resource is people—skilled, spirited, and hopeful people who will exert their wills and imaginations for their own benefit, and so, inevitably, for the benefit of us all.

Why does this divergence in views exist between Cornucopians and Cassandras? It may be because we are living in

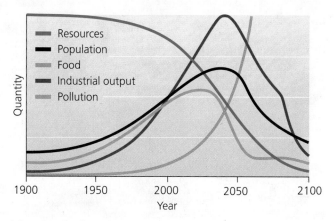

(a) Projection based on status quo policies

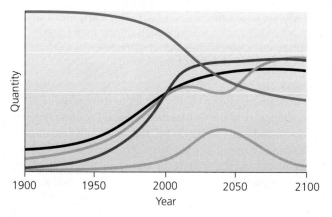

(b) Projection based on policies for sustainability

FIGURE 6.16 ▲ Environmental scientists Donella Meadows, Jorgen Randers, and Dennis Meadows used computer simulations to project future trends in human population, resource availability, food production, industrial output, and pollution. Their projections are based on past data and current scientific understanding of the environment's biophysical limits. Shown in **(a)** is their projection for a world in which "society proceeds in a traditional manner without any major deviation from the policies pursued during most of the twentieth century," and allowing for twice as many resources to be discovered and extracted as are now known to be accessible. In this projection, population and production increase until declining nonrenewable resources make further growth impossible, causing population and production to decline rather suddenly while pollution continues upward. In contrast, **(b)** shows a scenario with policies aimed at sustainability. In this projection, population levels off below 8 billion, production and resource availability stabilize at medium-high levels, and pollution declines to low levels. Data from Meadows, D., et al., 2004. *Limits to growth: The 30-year update.* White River Junction, VT: Chelsea Green Publishing. Used by permission.

a unique time of transition. Throughout our history, we have lived in a world with low numbers of people. In such an "empty world," we could always rely on being able to exploit more resources. Today, however, the human population is so large that it is beginning to strain Earth's systems and deplete its resources. In this new "full world," we will encounter limits. Because we are only now starting to learn what those limits are, people have a wide diversity of viewpoints.

As we will see with many issues in the pages ahead, both Cornucopians and Cassandras make good points. Cassandras have indeed often underestimated our ability to innovate and find ways to adapt—but ultimately, all nonrenewable resources are finite and many renewable resources can be exploited only at limited rates. If our population and consumption continue to grow and we do not shift to full reuse and recycling, we will deplete more and more resources and put still greater demands on our capacity to innovate.

WEIGHING THE ISSUES

Cornucopian or Cassandra? Would you consider yourself a Cornucopian or a Cassandra? Which aspect(s) of each point of view do you share? Do you think our society would be better off if everyone were a Cornucopian or if everyone were a Cassandra—or do we need both?

Environmental economists address shortcomings of mainstream economics

The Cornucopian view has held great sway over mainstream economists, but economists in the newer field of **environmental economics** maintain that economic growth may be unsustainable if we do not reduce population growth and make resource use far more efficient. Most environmental economists feel that, with effort, we can make changes needed to attain sustainability within our current economic systems. By modifying the principles of neoclassical economics to address environmental challenges, environmental economists argue that we can keep our economies growing and developing in positive ways.

Environmental economists were the first to develop methods to tackle the problems of external costs and discounting, some of which we will examine shortly. In doing so, they blazed a trail. Ecological economists then went further, proposing that sustainability requires more far-reaching changes. Thus, whereas environmental economists implemented reform, ecological economists call for revolution.

Ecological economists propose a steady-state economy

Ecological economics, which has emerged as a discipline in the past two decades, applies principles of ecology and systems science (Chapters 3–5) to the analysis of economic systems. In nature, every population has a carrying capacity (p. 67), and systems generally operate in self-renewing cycles, not in a linear or progressive manner. Ecological economists maintain that civilizations, like natural populations, cannot permanently overcome their environmental limitations and that we should not expect endless economic growth. Most ecological economists argue that if population growth and resource consumption are not reined in, depleted natural systems could plunge our economies into ruin. Many of these economists advocate economies that neither grow nor shrink, but are stable. Such **steady-state economies** are intended to mirror natural ecological systems.

One idea of a steady-state economy was articulated back in the 19th century, when British economist **John Stuart Mill** (1806–1873) hypothesized that as resources became harder to find and extract, economic growth would slow and eventually stabilize. Individuals and society would subsist sustainably on steady flows of natural resources and on savings accrued during occasional productive periods of growth. Such a model appears to describe the economies of Aboriginal Australians and many other traditional societies worldwide before the global expansion of European culture.

Modern proponents of a global steady-state economy, such as American economist **Herman Daly**, are not optimistic that a steady state will evolve on its own from a capitalist market system. Instead, most steady-state proponents believe we will need to rethink our assumptions and fundamentally change the way we conduct economic transactions.

Critics of a steady-state economy often assume that an end to growth will bring an end to a rising quality of life, but ecological economists respond that quality of life should continue to rise. Technological advances will not cease just because growth stabilizes, they argue, and neither will behavioral changes (such as greater use of recycling) that enhance sustainability. Instead, wealth and happiness can continue to rise after economic growth has leveled off.

Attaining sustainability will certainly require the reforms that environmental economists advocate, and may well require the fundamental shifts in thinking, values, and behavior that ecological economists say are needed. Environmental and ecological economists are actively developing strategies for sustainability, and we will now survey a few that they offer.

We can measure economic progress differently

For decades, policymakers and the public have assessed each national economy by calculating its **Gross Domestic Product (GDP)**, the total monetary value of final goods and services it produces each year. Governments use GDP to make financial policy decisions that affect billions of people. However, GDP is a poor measure of economic well-being, because it does not account for nonmarket values (pp. 157–160) and because it is not an expression solely of *desirable* economic activity. Instead, GDP includes *all* economic activity, so GDP can rise even when the economic activities that drive it harm society.

For example, crime boosts GDP because crime forces people to invest in security measures and to replace stolen items. The *Exxon Valdez* oil spill (**FIGURE 6.17**) increased the U.S. GDP because oil spills require cleanups, which cost money and, as a result, increase the production of goods and services. Natural disasters such as Hurricane Katrina boost GDP because of all the costs incurred in emergency measures and cleanup and rebuilding operations. A radiation leak at a uranium mine on Aboriginal homelands in Australia would add to the Australian GDP because of the monetary transactions required for cleanup and medical care. In fact, pollution generally causes GDP to increase twice—once when the polluting substance is manufactured and once when society pays to clean up the pollution.

Some economists have tried to develop indicators that differentiate between desirable and undesirable economic activity

FIGURE 6.17 ▲ The *Exxon Valdez* oil spill, which devastated Alaskan beaches and fisheries in 1989, actually increased the nation's GDP. The many monetary transactions resulting from cleanup efforts mean that pollution increases GDP—one of many reasons GDP is not a reliable indicator of well-being.

and that can function as better guides to citizens' welfare. One such alternative to the GDP is the **Genuine Progress Indicator (GPI)**, introduced in 1995 by the nonprofit organization Redefining Progress. To calculate GPI, economists begin with conventional economic activity and then add to it positive contributions that are not paid for with money, such as volunteer work and parenting. They then subtract negative impacts, such as crime, pollution, gaps between rich and poor, and other detrimental factors (**FIGURE 6.18A**). The GPI thereby summarizes more forms of economic activity than does the GDP, and it also differentiates between activity that increases societal well-being and activity that decreases it.

The two indices can differ strikingly; **FIGURE 6.18B** compares changes in per capita GPI and per capita GDP in the United States from 1950 to 2004. On a per capita basis, the nation's GDP has risen greatly, but its GPI has remained largely flat for the past 35 years.

Other "green accounting" indicators include Net Economic Welfare (NEW), which adjusts GDP by adding the value of leisure time and personal transactions while deducting costs of environmental degradation; the United Nations' Human Development Index, which assesses a nation's standard of living, life expectancy, and education; and the Index of Sustainable Economic Welfare (ISEW), which gave rise to the GPI. Any of these indices, ecological economists maintain, provide a more accurate portrait of a nation's well-being than does GDP.

Adopting green accounting can prove difficult in practice, however. China introduced a green accounting system in 2006 that subtracted costs of pollution and resource depletion from its GDP, but it proved so controversial that the government suppressed the 2007 report.

We can give ecosystem goods and services monetary values

Earth's ecosystems provide us essential life-support services, including arable soil, waste treatment, clean water, and clean air. Yet any survey of modern trends—deforestation, biodiversity loss, pollution, collapsed fisheries, climate change, and so

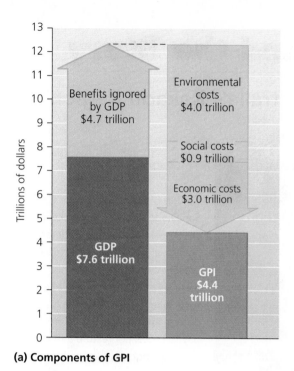

(a) Components of GPI

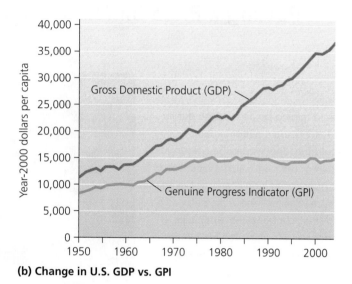

(b) Change in U.S. GDP vs. GPI

FIGURE 6.18 ▲ The Genuine Progress Indicator (GPI; orange bar at right) adds to the Gross Domestic Product (GDP; red bar at left) benefits such as volunteering and parenting (upward pointing gold arrow). The GPI then subtracts external environmental costs such as pollution, social costs such as divorce and crime, and economic costs such as borrowing and the gap between rich and poor (downward pointing gold arrow). Shown in **(a)** are values for GDP and GPI for the United States in 2004. The graph in **(b)** shows that, although per capita U.S. GDP has increased dramatically since 1950, per capita U.S. GPI has leveled off since 1975. GPI advocates say this indicates that we are spending more money but that our lives are not much better. Data from Talberth, J., C. Cobb, and N. Slattery, 2007. *The Genuine Progress Indicator 2006: A tool for sustainable development*. Redefining Progress, Oakland, CA. By permission of John Talberth, Ph.D. All data are adjusted for inflation by using year-2000 dollars.

on—makes clear that our society often abuses the very systems that sustain it. Why is this? From the economist's perspective, people exploit natural resources and systems largely because the market assigns these entities no quantitative monetary value or, at best, assigns values that underestimate their true worth. Economists have historically failed to account for how much ecosystem goods and services contribute economically to human welfare, and the values of these vital contributions are routinely underrepresented in cost-benefit analyses.

Think for a minute about some of these services. The aesthetic and recreational pleasure we obtain from natural landscapes, whether wildernesses or city parks, is something of real value. Yet because we do not generally pay money for this, its value is hard to quantify and appears in no traditional measures of economic worth. Or consider Earth's water cycle (p. 123), by which rain fills our reservoirs with drinking water, rivers give us hydropower and flush away our waste, and water evaporates, purifying itself of contaminants and readying itself to fall again as rain. This natural cycle is vital to our very existence, yet because we do not quantify its value, markets impose no financial penalties when we interfere with it.

Ecosystem services are said to have **nonmarket values**, values not usually included in the price of a good or service (**TABLE 6.1** and **FIGURE 6.19**, p. 160). Because markets do not assign value to ecosystem services, debates such as that over the Jabiluka mine often involve comparing apples and oranges—in this case, the intangible ecological, cultural, and

spiritual arguments of the Mirarr versus the financial numbers of mine proponents.

To resolve this dilemma, environmental and ecological economists have sought ways to assign values to ecosystem services. One technique, **contingent valuation**, uses surveys to determine how much people are willing to pay to protect or restore a resource. Such an approach was used to assess a mining proposal in the Kakadu region in the early 1990s that preceded the Jabiluka proposal (see **THE SCIENCE BEHIND THE STORY**, pp. 158–159).

Whereas contingent valuation measures people's *expressed preferences*, other methods aim to measure people's *revealed preferences*—preferences as revealed by data on actual behavior. For example, researchers may use the amount of money, time, or effort people expend to travel to parks for recreation to measure the value people place on parks. Economists may compare housing prices for similar homes in different environmental settings to infer the dollar value of landscapes, views, and peace and quiet. They may also assign environmental amenities value by calculating the cost required to restore natural systems that have been damaged, to replace those systems' functions with technology, or to mitigate harm from pollution.

In 1997 a research team headed by environmental economist Robert Costanza set out to calculate the overall global economic value of all the services that oceans, forests, wetlands, and other ecosystems provide. Costanza's team

Dr. Richard Carson, University of California–San Diego

Should mining be allowed on public land alongside a national park, or should the land instead be preserved as part of the park? Australians faced this question at world-famous Kakadu National Park several years before the Mirarr confronted the Jabiluka proposal in the same region. When environmental economists set out to address the question, their work became a prime example of how to apply monetary figures to the nonmarket values inherent in natural features and ecosystem services.

The Kakadu Conservation Zone was a government-owned 50-km^2 (19-mi^2) plot of land surrounded by Kakadu National Park in northern Australia. Kakadu National Park is a United Nations World Heritage Site consisting of tropical dry forest, seasonal wetlands, geological wonders, and spectacular concentrations of wildlife (**see photo**). The region is home to many Aboriginal groups, so the park and its surroundings also boast unique cultural resources such as world-renowned Aboriginal rock art (see Figures 6.1a and 6.19g, pp. 138, 160).

However, the region is also rich in mineral resources. Besides uranium, the Kakadu Conservation Zone (KCZ) holds deposits of gold, platinum, and palladium. In 1990, the Australian government aimed to decide whether the KCZ should be opened to mining or preserved and added to the park. The government had calculated that mining would produce about $102 million in economic benefits each year. The question was: How much

was the land "worth" economically if preserved undeveloped?

To address this question, the Australian government's Resource Assessment Commission sponsored environmental economist Richard

Carson of the University of California–San Diego and Australians Leanne Wilks and David Imber to determine how much Australian citizens valued preservation of the KCZ. This research team selected the method of contingent valuation, which surveys people to determine how much they are willing to pay to protect or restore a resource.

Contingent valuation is the only way to measure both "use values" such as recreation (see Figure 6.19a, p. 160) and "nonuse values" such as existence value (see Figure 6.19b, p. 160, which is the value people place on something's existence, whether they experience it directly or not. Because many Australians might never visit Kakadu, it was expected

An ecotourist inspects a giant termite mound at Kakadu National Park. Scenery and wildlife at the Kakadu Conservation Zone inspired Australian survey respondents to place a high value on preserving the land by preventing mining there.

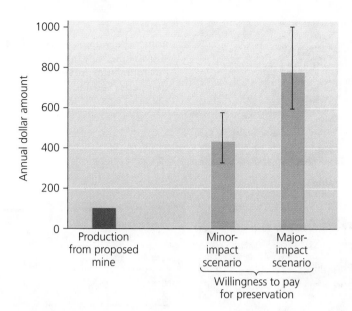

Survey data indicated that mine development (red bar at left) was worth less than land preservation, under both minor-impact (middle bar) and major-impact (right bar) scenarios. Amounts are annual dollar estimates of mine production (left bar) and of Australian citizens' willingness to pay (right two bars). Error bars denote ranges of values judged statistically to be 95% likely. SOURCE: Data from Carson, R., L. Wilks, and D. Imber, 1994. Valuing the preservation of Australia's Kakadu Conservation Zone. *Oxford Economic Papers* 46: 727–749.

that some of what they might be willing to pay would be based on existence value.

Carson's team hired and trained interviewers, who went door-to-door throughout Australia and interviewed 2,034 citizens. The interviewers first described the Kakadu region, showing each person maps and photos of the landscape and of animals living there, and sharing information on tourism and Aboriginal claims. They also showed photos of sites where pits had been dug for uranium mining in the 1950s. The interviewers then described the proposed mining operations, with each mine planned to be 1 square kilometer, mostly underground, and employing 150 workers for 10 years.

When the interviewers turned to describing potential impacts of the mining, they had to tread carefully. The mining industry was predicting very little environmental and social impact, whereas environmentalists were predicting extensive damage. Thus, to cover all possibilities, the interviewers presented two scenarios:

(1) a "minor-impact" scenario based on predictions of the mining industry, and (2) a "major-impact" scenario based on predictions of environmentalists. The interviewers presented both scenarios in detail, complete with photographs and artists' renderings. Each scenario involved information on mine-related traffic, chemicals used to extract minerals, water use, waste rock, and possible effects on wildlife and the landscape.

Finally, the interviewers asked the respondents how much their households would be willing to pay to prevent mine development based on whether each scenario, in turn, was correct.

On average, respondents said their households would pay $80 per year to prevent the minor-impact scenario and $143 per year to prevent the major-impact scenario. Multiplying these figures by the number of households in Australia (5.4 million at the time), the researchers found that preservation was "worth" $435 million annually to the Australian population under the

minor-impact scenario, and $777 million under the major-impact scenario.

Both of these numbers exceeded the $102 million in annual economic benefits expected from mine development (**see graph**). Thus, the researchers concluded that preserving the land undeveloped was worth more than opening it to mining.

The Carson team's study came under attack from mine proponents and also from some economists who criticized the contingent valuation approach. Because contingent valuation relies on survey questions and not actual expenditures, critics complained that people may volunteer idealistic (inflated) values rather than realistic ones, knowing that they will not actually have to pay the price they name.

In part because of such concerns, the Resource Assessment Commission decided not to use the Carson team's results. In the end, the government chose to preserve the land and not open it to mining, but it stated that the decision was based on Aboriginal opposition rather than economic valuation.

Publishing their results in the *Oxford Economic Papers* series in 1994, Carson's team acknowledged some shortcomings of their method but stressed the precautions they had taken to dissuade people from reporting inflated values. They also referenced a 1993 government panel report in the United States that responded to similar critiques in the aftermath of the *Exxon Valdez* oil spill. Americans were debating the same issues as Australians at this time, and the panel report found that contingent valuation, when done right, can be accurate and informative.

Clearly, contingent valuation is one tool in the economist's arsenal to ascribe monetary figures to nonmarket values that are difficult to quantify—but are undeniably important. ■

(a) Use value

(b) Existence value

(c) Option value

(d) Aesthetic value

(e) Scientific value

(f) Educational value

(g) Cultural value

FIGURE 6.19 ◀ Accounting for nonmarket values such as those shown here may help us make better environmental and economic decisions. See Table 6.1 for further detail.

TABLE 6.1 Values That Modern Market Economies Generally Do Not Address

Nonmarket value	Is the worth we ascribe to things . . .
Use value	that we use directly
Existence value	simply because they exist, even though we may never experience them directly (e.g., an endangered species in a far-off place)
Option value	that we do not use now but might use later
Aesthetic value	for their beauty or emotional appeal
Scientific value	that may be the subject of scientific research
Educational value	that teach us about ourselves or the world
Cultural value	that sustain or help define our culture

combed the scientific literature and evaluated over 100 studies that used various valuation methods to estimate dollar values for 17 major ecosystem services such as water purification, climate regulation, plant pollination, and pollution cleanup (**FIGURE 6.20**). To improve the accuracy of estimates, the researchers reevaluated the data using multiple valuation techniques. They then multiplied average estimates for each ecosystem by the global area occupied by each, and calculated that the biosphere in total provides at least $33 trillion ($46 trillion in 2011 dollars) worth of ecosystem services each year—more than the GDP of all nations combined!

Published in the journal *Nature* in 1997, this research sparked excitement and controversy. Some economists critiqued the methods, and some ethicists argued that we should not put dollar figures on services such as clean air and water, because they are priceless and we would perish without them.

In a follow-up study in 2002, Costanza joined Andrew Balmford and 17 other colleagues to compare the benefits and costs of preserving natural systems intact versus converting wild lands for agriculture, logging, or fish farming. After reviewing many studies, they reported in the journal *Science* that a global network of nature reserves covering 15% of Earth's land surface and 30% of the ocean would be worth between $4.4 and $5.2 trillion. This amount is 100 times greater than the value of those areas were they to be converted for direct exploitative human use—demonstrating, in their words, that "conservation in reserves represents a strikingly good bargain."

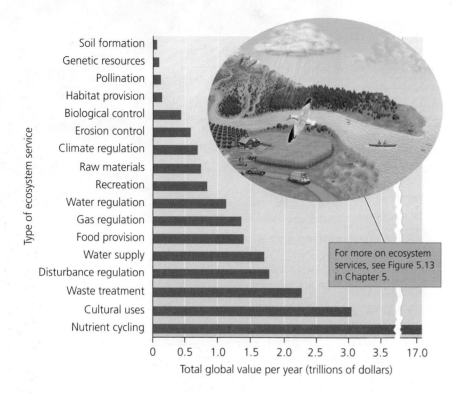

For more on ecosystem services, see Figure 5.13 in Chapter 5.

FIGURE 6.20 ◄ In 1997, environmental economist Robert Costanza and colleagues estimated the value of the world's ecosystem services at more than $33 trillion ($46 trillion in 2011 dollars). Shown are subtotals for each major type of ecosystem service. The $33 trillion figure is an underestimate because it does not include values from ecosystems for which adequate data were unavailable. Data from Costanza, R., et al., 1997. The value of the world's ecosystem services and natural capital. *Nature* 387: 253–260.

Markets can fail

When markets do not take into account the environment's positive effects on economies (such as ecosystem services) or when they do not reflect the negative impacts of economic activity on people or the environment (external costs), economists call this **market failure**. Market failure occurs when markets do not reflect the full costs and benefits of actions.

Traditionally, market failure has been countered by government intervention. Governments can restrain corporate behavior through laws and regulations. They can institute *green taxes,* which penalize environmentally harmful activities. Alternatively, they can design economic incentives that use market mechanisms to promote fairness, resource conservation, and economic sustainability. One technique is to create markets in permits for activities such as the emission of gases that pollute and warm the atmosphere. We will examine legislation, regulation, green taxation, subsidies, and market incentives such as cap-and-trade permit trading in Chapter 7 in our discussion of environmental policy.

Ecolabeling helps address market failure

Another way of mitigating market failure is to provide better information to consumers so that they can use the information in making purchasing decisions. Increasingly, manufacturers are designating on their labels how their products were grown, harvested, or manufactured. This method, called **ecolabeling**, serves to inform consumers which brands use environmentally benign processes (**FIGURE 6.21**).

By preferentially buying ecolabeled products, consumers can provide businesses a powerful incentive to switch to more sustainable processes. One early example of this was labeling cans of tuna as "dolphin-safe," indicating that the methods used to catch the tuna avoid the accidental capture of dolphins. Other examples include labeling recycled paper (p. 626), organically grown foods (pp. 271–272), genetically modified foods (labeled in Europe but not North America; p. 267), fair trade and shade-grown coffee (p. 273), and lumber harvested through sustainable forestry (pp. 331–332).

FIGURE 6.21 ▼ Ecolabeling gives consumers information on products with low environmental impact and enables consumers to encourage more sustainable business practices through their purchasing decisions. Organic juices are just one example of ecolabeled products that have become widely available in the mass market.

In a similar vein, individuals who invest their money in the stock market can choose to pursue *socially responsible investing*, which entails investing in companies that have met criteria for environmental or social sustainability. A fast-growing trend, in 2007 U.S. investors sank $2.7 trillion into socially responsible investing, representing 11% of all investment.

Corporations are responding to sustainability concerns

As more consumers and investors express preferences for sustainable products and services, more industries, businesses, and corporations are finding that they can make money by "greening" their operations.

Some companies have cultivated an eco-conscious image from the start, such as Ben & Jerry's (ice cream), Patagonia (outdoor apparel), and Seventh Generation (household products). The phone company Credo (formerly Working Assets) donates a portion of its proceeds to environmental and progressive nonprofit groups according to how its customers vote to distribute the funds. Some newer businesses are trying to go even further than these pioneers. The outdoor apparel startup Nau manufactures stylish items from materials made of corn biomass and recycled bottles while helping to fund environmental nonprofits. At the local level, entrepreneurs are starting thousands of sustainably oriented businesses all across the world.

Today corporate sustainability is going mainstream, and some of the world's largest corporations have joined in. They are seeking to ride what Daniel Esty and Andrew Winston, authors of the 2006 book *Green to Gold*, call a "green wave" of consumer preference for sustainable products and business practices. These "waveriders" include corporations as varied as McDonald's, Starbucks, Intel, Microsoft, Ford Motor Company, Toyota, IKEA, Dow, DuPont, BASF, and British Petroleum.

Carpeting company Interface, Inc., provides one of the best-known cases of a company changing its practices to promote sustainability (p. 633), but there are many more. Hewlett-Packard was judged the greenest large company in America in a survey run by *Newsweek* magazine and three research agencies in 2009. Hewlett-Packard runs ambitious programs to reuse and recycle used products and components such as computers and other electronics, toner cartridges, and plastics.

Nike, Inc., collects millions of used sneakers each year and recycles their materials to create synthetic surfaces for basketball courts, tennis courts, and running tracks. Nike also uses recycled materials in its shoes, uses a high proportion of organic cotton, and has developed less-toxic rubber and adhesives. The Gap, Inc., has cut energy use in its stores and distribution centers, runs alternative transportation programs for its employees, and has a sustainably designed headquarters building (**FIGURE 6.22**). Today many corporations are finding ways to increase energy efficiency, reduce toxic substances, increase the use of recycled materials, and minimize greenhouse gas emissions. In doing so, they often find that they reduce costs and increase profit.

FIGURE 6.22 ▲ A green roof and solar panels top the corporate headquarters of The Gap, Inc., in San Bruno, California. Many large corporations and small businesses alike are taking steps toward more sustainable operations. Many are saving money by doing so.

Of course, corporations exist to make money for their shareholders, so they cannot be expected to pursue goals that do not turn a profit. Moreover, many corporate greening efforts are more rhetoric than reality, and corporate **greenwashing** may mislead consumers into thinking the companies are acting more sustainably than they are. The bottled water industry presents a stark example of greenwashing. Advertising with words such as "pure" and "natural" and images of forests and alpine springs leads us to believe that bottled water is cleaner and healthier for us to drink—when in reality the water is often less safe, the plastic bottles create a major source of waste, immense amounts of oil are burned to transport the bottles, and the industry depletes aquifers in local communities (pp. 416–418).

Wal-Mart provides the most celebrated recent example of corporate greening efforts. Advocates of sustainability had long criticized the world's largest retailer for its environmental and social impacts. The company responded by launching a quest to sell organic and sustainable products, reduce packaging and use recycled materials, enhance fuel efficiency in its truck fleet, reduce energy use in its stores, power itself with renewable energy, cut carbon dioxide emissions, and preserve one acre of natural land for every acre developed. It is also developing a "sustainability index" that will rate the products it carries to help inform eco-conscious consumers.

Many observers remain skeptical of Wal-Mart's commitment to sustainability. Yet, if it achieves even some of its stated goals, the social and environmental benefits could be substantial. The corporation's vast global reach and its ability to persuade suppliers to alter their ways in order to retain its business mean that any change Wal-Mart enacts could have far-reaching impacts.

In the end, corporate actions hinge on consumer behavior. It is up to all of us in our roles as consumers to encourage trends in sustainability by rewarding those businesses that truly promote sustainable solutions. This is one way that each of us can express and promote the ethical values we cherish through the economic system in which we live.

➤ CONCLUSION

Corporate sustainability, ecolabeling, alternative ways of measuring growth, and valuation of ecosystem services are a few of the recent developments that have brought economic approaches to bear on environmental protection and resource conservation. As economics becomes more oriented toward sustainability, it renews some of its historic ties to ethics.

Environmental ethics has expanded people's sphere of ethical consideration outward to encompass more societies, cultures, creatures, and even nonliving entities. This ethical expansion involves the concept of distributional equity, or equal treatment, which is the aim of environmental justice. One type of distributional equity is equity among generations. Concern by today's generations for the welfare of tomorrow's generations is the basis of sustainability.

Is sustainability a pragmatic pursuit for us? The answer largely depends on whether we believe that economic well-being and environmental well-being are incompatible goals or whether we accept that they can work in tandem. Equating economic well-being with economic growth, as economists traditionally have, suggests that economic welfare entails a trade-off with environmental quality. However, if we can enhance economic welfare in the absence of growth, then economic health and environmental quality will be mutually reinforcing.

REVIEWING OBJECTIVES

You should now be able to:

CHARACTERIZE THE INFLUENCES OF CULTURE AND WORLDVIEW ON THE CHOICES PEOPLE MAKE

- A person's culture influences his or her worldview. Factors such as religion and political ideology are especially influential. (pp. 138–139)

OUTLINE THE NATURE AND HISTORICAL EXPANSION OF ETHICS IN WESTERN CULTURES

- Environmental ethics applies ethical standards to relationships between people and aspects of their environments. (p. 140)
- Our society's domain of ethical concern has been expanding, such that we have granted more and more entities ethical consideration. (p. 141)
- Anthropocentrism values people above all else, whereas biocentrism values all life and ecocentrism values ecological systems. (pp. 141–142)

COMPARE THE MAJOR APPROACHES IN ENVIRONMENTAL ETHICS

- The preservation ethic (preserving natural systems intact) and the conservation ethic (promoting responsible long-term use of resources) have guided branches of the environmental movement. (p. 143)
- The philosophy of Aldo Leopold is viewed as centrally important to modern environmental ethics. (p. 144)
- Environmental justice seeks equal treatment for people of all races and income levels. (pp. 144–146)
- Environmental justice is also an international issue between developed and developing nations. (p. 146)

EXPLAIN HOW OUR ECONOMIES EXIST WITHIN THE ENVIRONMENT AND RELY ON ECOSYSTEM SERVICES

- Human economies are a subset of the environment and depend on ecological systems for natural resources and for ecosystem services. (pp. 147–149)
- Ecosystem services, provided for free by nature, enable our economies to prosper. To degrade ecosystem services is to threaten our economic well-being. (pp. 148–149)

DESCRIBE PRINCIPLES OF CLASSICAL AND NEOCLASSICAL ECONOMICS AND SUMMARIZE THEIR IMPLICATIONS FOR THE ENVIRONMENT

- Classical economic theory proposes that individuals acting for their own economic good can benefit society as a whole. This view has provided a philosophical basis for free-market capitalism. (p. 149)
- Neoclassical economics focuses on supply and demand as forces that drive economic activity, and quantifies costs and benefits. (pp. 149–150)
- Four assumptions of neoclassical economic theory promote negative environmental impacts. (pp. 150–151)

COMPARE THE CONCEPTS OF ECONOMIC GROWTH, WELL-BEING, AND SUSTAINABILITY

- Conventional economic theory has promoted endless economic growth, with little regard to possible environmental impact. (pp. 151–155)
- Economic growth does not necessarily lead to social well-being. (pp. 151–155)
- Some economists believe that developing a steady-state economy will be needed to achieve sustainability. (pp. 155–156)

EXPLAIN ASPECTS OF ENVIRONMENTAL ECONOMICS AND ECOLOGICAL ECONOMICS

- Environmental economists advocate reforming economic practices to promote sustainability. Key approaches are to identify external costs, assign value to nonmonetary items, and try to make market prices reflect real costs and benefits. (pp. 155–161)
- Ecological economists support these efforts but advocate going further and pursuing a steady-state economy. (pp. 155–156)

DESCRIBE HOW INDIVIDUALS AND BUSINESSES CAN HELP MOVE OUR ECONOMIC SYSTEM IN A SUSTAINABLE DIRECTION

- Consumer choice in the marketplace, facilitated by ecolabeling, can encourage businesses to pursue sustainable practices. (p. 161)
- Many corporations are modifying their operations to become more sustainable, and this often is financially profitable. (p. 162)

TESTING YOUR COMPREHENSION

1. What does the study of ethics encompass? Describe and differentiate two classic ethical standards, the categorical imperative and the principle of utility. What is environmental ethics?

2. Compare and contrast the perspectives of anthropocentrism, biocentrism, and ecocentrism. How would you characterize the perspective of the Mirarr Clan?

3. Differentiate the preservation ethic from the conservation ethic. Explain the contributions of John Muir and Gifford Pinchot in the history of environmental ethics.

4. Describe Aldo Leopold's "land ethic." How did Leopold define the "community" to which ethical standards should be applied?

5. Define the concept of environmental justice. Give an example of an inequity relevant to environment justice that exists in your city, state, or country.

6. Name four key contributions the environment makes to the economy.

7. Describe Adam Smith's metaphor of the "invisible hand." How did neoclassical economists refine classical economics?

8. Describe four ways in which critics hold that neoclassical economic approaches can worsen environmental problems.

9. Compare and contrast the views of neoclassical economists, environmental economists, and ecological economists, particularly regarding the issue of economic growth.

10. What is contingent valuation, and what is one of its weaknesses? Describe an alternative method that avoids this weakness.

SEEKING SOLUTIONS

1. Describe your worldview as it pertains to your relationship with your environment. How do you think your culture has influenced your worldview? How do you think your personal experience has affected it? How do you think your worldview shapes your decisions? Do you feel that you fit into any particular category discussed in this chapter? Why or why not?

2. How would you analyze the case of the Mirarr Clan and the proposed Jabiluka uranium mine from each of the following perspectives? In your description, list at least one question that a person of each perspective would likely ask when attempting to decide whether the mine should be developed. Be as specific as possible, and be sure to identify similarities and differences in approaches:

   ▶ Preservationist
   ▶ Conservationist
   ▶ Environmental justice advocate
   ▶ Neoclassical economist
   ▶ Ecological economist

3. Visit the U.S. EPA's "Environmental Justice Geographic Assessment" website at www.epa.gov/compliance/whereyoulive/ejtool.html, and enter your city, county, or zip code. On the map page, check the six types of "Regulated sites" in the list of "Map Features" on the right, and redraw the map. Then at the top of the new map, click on the button for "Add demographics layer." Apply various layers in turn, including at least "per capita income," "percent minority," and "percent below poverty." The site will map these parameters and show how the EPA-regulated facilities are located relative to them.

   What patterns do you see? Where are most facilities located, relative to wealth, race, and other factors? What do you think accounts for the patterns you find? What could be done to alleviate any problems you may perceive?

4. Do you think that a steady-state economy is a practical alternative to our current approach that prioritizes economic growth? Why or why not?

5. Do you think we should attempt to quantify and assign market values to ecosystem services and other entities that have only nonmarket values? Why or why not?

6. THINK IT THROUGH A manufacturing facility on a river near your home provides jobs for 200 people in your community of 10,000 people, and it pays $2 million in taxes to the local government each year. Sales taxes from purchases made by plant employees and their families contribute an additional $1 million to local government coffers. However, newspaper reports and recent peer-reviewed studies in well-respected scientific journals reveal that the plant has been discharging large amounts of waste into the river, causing a 25% increase in cancer rates, a 30% reduction in riverfront property values, and a 75% decrease in native fish populations.

The plant owner says the facility can stay in business only because there are no regulations mandating expensive treatment of waste from the plant. If such regulations were imposed, he says he would close the plant, lay off its employees, and relocate to a more business-friendly community.

How would you recommend resolving this situation? What further information would you want to know before making a recommendation? In arriving at your recommendation, how did you weigh the costs and benefits associated with each of the plant's impacts?

CALCULATING ECOLOGICAL FOOTPRINTS

The population of the United States grew from 152,271,000 to 292,892,000 between 1950 and 2004, but the nation's Gross Domestic Product (GDP) grew still faster. Taking these two trends together, Figure 6.18b (p. 157) shows how per capita GDP rose over this half century. During this period, the Genuine Progress Indicator (GPI) and the per capita GPI grew as well, although more slowly than GDP.

Given the values of the major components of the GPI for the United States in 1950 and 2004, and the population figures above, calculate and enter the per capita rates for these components, as well as the overall GPI values. Refer to Figure 6.18a (p. 157) to check how to calculate the GPI.

Components of GPI	U.S. total in 1950 (trillions of dollars)	Per capita in 1950 (thousands of dollars)	U.S. total in 2004 (trillions of dollars)	Per capita in 2004 (thousands of dollars)
GDP	1.153		7.589	
Benefits	1.041		4.746	
Environmental costs	0.407		3.990	
Social and economic costs	0.476		3.926	
GPI				

Data from Talberth, J., C. Cobb, and N. Slattery, 2007. *The Genuine Progress Indicator 2006: A tool for sustainable development.* Redefining Progress, Oakland, CA.

1. How many times greater was the GDP in 2004 than in 1950? By how many times did the GPI increase between 1950 and 2004? What does this comparison tell you?

2. By how many times, respectively, did benefits, environmental costs, and social and economic costs increase between 1950 and 2004? How are trends in each of these components driving the overall trend between the GPI and the GDP? Which component has gotten worst over the years? How would you account for these trends?

3. There are many ways to define and measure the various types of benefits and costs that go into the GPI. How do

you think a person with a biocentric worldview would measure these differently from a person with an anthropocentric worldview? Whose GPI for the year 2005 would likely be higher?

4. Now consider your own life. Very roughly, what would you estimate are the values of the benefits, environmental costs, and social and economic costs you experience? What could you do to help improve these trends in your own personal accounting?

Mastering ENVIRONMENTALSCIENCE™

Go to **www.masteringenvironmentalscience.com** for practice quizzes, Pearson eText, videos, current events, and more.

Brown plume of wastewater from Tijuana River entering Pacific Ocean near Tijuana

7 ENVIRONMENTAL POLICY: DECISION MAKING AND PROBLEM SOLVING

UPON COMPLETING THIS CHAPTER, YOU WILL BE ABLE TO:

- Describe environmental policy and assess its societal context
- Identify the institutions important to U.S. environmental policy and recognize major U.S. environmental laws
- Categorize the different approaches to environmental policy

- Delineate the steps of the environmental policy process and evaluate its effectiveness
- Discuss the role of science in the policy process
- List the institutions involved with international environmental policy and describe how nations handle transboundary issues

CENTRAL CASE STUDY

San Diego and Tijuana: Sewage Pollution Problems and Policy Solutions

"Never doubt that a small group of thoughtful, committed citizens can change the world—indeed it is the only thing that ever has."

—Anthropologist Margaret Mead

"It is the continuing policy of the Federal Government . . . to create and maintain conditions under which man and nature can exist in productive harmony and fulfill the social, economic, and other requirements of present and future generations of Americans."

—National Environmental Policy Act, 1969

The beaches south of San Diego boast some of the world's best waves for surfing. These days, however, most surfers avoid the temptation. For it is here that the heavily polluted Tijuana River flows across the international border from Mexico and empties into the sea, disgorging millions of gallons of untreated wastewater.

"When it rains, I call it the sewage tsunami," says surfer and environmentalist Serge Dedina. "For 40 square miles, from Imperial Beach to Coronado, there is a brown plume as far as the eye can see."

Such incidents occur when heavy rains overwhelm the ability of sewage treatment plants to process wastewater. San Diego's coastal waters receive stormwater runoff when rains wash pollutants into local rivers. Across the border in the Mexican city of Tijuana, the city's aging, leaky sewer system becomes clogged with debris, causing raw sewage to overflow into the streets and, eventually, into the Tijuana River.

In 2007, the most recent year with full data, San Diego County officials closed beaches 27 times, issued 668 days' worth of health advisories, and kept one area off-limits to swimming the entire year. Of these

Contaminated beach near Tijuana River mouth

beach closures, 76% were due to pollution from the Tijuana River.

The Tijuana River winds northwestward through the arid landscape of northern Baja California, Mexico, crossing the U.S. border south of San Diego. A river's **watershed** consists of all the land from which water drains into the river, and the Tijuana River's watershed covers 4,500 km² (1,750 mi²) and is home to 2 million people of two nations. The Tijuana River watershed is a *transboundary* watershed (so named because it crosses a political boundary—in this case a national border), with approximately 70% of its area in Mexico (**FIGURE 7.1**). On the Mexican side of the border, the river and creeks that flow into it are lined with farms, apartments, shanties, and factories, as well as leaky sewage treatment plants and toxic dump sites. Rains wash pollutants from all these sources into the Tijuana River and eventually onto U.S. and Mexican beaches.

Although pollution has flowed in the Tijuana River for decades, the problem has grown worse in recent years as the region's population has boomed, outstripping the capacity of aging sewage treatment facilities. Beach closures and pollution advisories have become commonplace, and garbage carried by

CANADA

UNITED STATES

San Diego

Tijuana

MEXICO

Pacific Ocean

Atlantic Ocean

SOUTH AMERICA

DANGER CONTAMINATED WATER KEEP OUT

PELIGRO AGUA CONTAMINADA ALEJESE

CHAPTER 7 Environmental Policy: Decision Making and Problem Solving

167

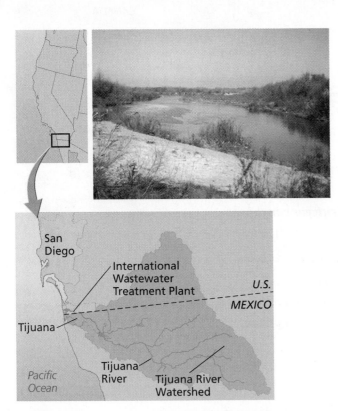

FIGURE 7.1 ▲ The Tijuana River winds northwestward from Mexico into California just south of San Diego, draining 4,500 km² (1,750 mi²) of land in its watershed (colored green in map). Pollution entering the river affects Mexican residents of the watershed and U.S. citizens on San Diego County beaches. During high-water episodes that overwhelm treatment facilities, millions of gallons of raw sewage can flush down the river and into the Pacific Ocean.

the flows litters the beaches. "Every day I find broken glass, balloons, or can pop-tops. I've even found hypodermic needles. It's really sad," one resident of Imperial Beach told her local newspaper.

Things are worse on the Mexican side because most Mexican residents of the Tijuana River watershed live in poverty relative to their U.S. neighbors. Close to one-third of Tijuana's homes are not connected to a sewer system, and in poor neighborhoods such as Loma Taurina, river pollution directly affects people's day-to-day lives. The rise of U.S.-owned factories, or *maquiladoras*, on the Mexican side of the border has contributed to the river's pollution, both through direct disposal of industrial waste and by attracting thousands of new workers to the already crowded region.

As impacts have intensified, people in the San Diego and Tijuana areas, from coastal residents to grassroots activists to businesspeople, have pressed policymakers to take action to address these problems. As one result, Mexico and the United States worked together to construct a wastewater treatment plant to handle excess sewage from Tijuana. The South Bay International Wastewater Treatment Plant

(IWTP) began operating just north of the border in 1997 and treats up to 95 million L (25 million gal) of wastewater each day. Unfortunately, the facility reached its capacity within three years because Tijuana's population grew so quickly, and excess sewage simply flowed downriver. We will see during the course of this chapter how people are now making progress by using science and policy to help address the region's pollution challenges.

ENVIRONMENTAL POLICY: AN OVERVIEW

When a society reaches broad agreement that a problem exists, it may persuade its leaders to try to resolve the problem through the making of policy. **Policy** consists of a formal set of general plans and principles intended to address problems and guide decision making in specific instances. **Public policy** is policy made by governments, including those at the local, state, federal, and international levels. Public policy consists of laws, regulations, orders, incentives, and practices intended to advance societal welfare. **Environmental policy** is policy that pertains to human interactions with the environment. It generally aims to regulate resource use or reduce pollution to promote human welfare and/or protect natural systems.

Forging effective policy requires input from science, ethics, and economics. Science (Chapter 1) provides the information and analysis needed to identify and understand problems and devise potential solutions to them. Ethics and economics (Chapter 6) offer criteria to assess problems and to help clarify how society might address them. Government interacts with individual citizens, organizations, and the private sector in a variety of ways to formulate policy (**FIGURE 7.2**).

The pollution problems of the Tijuana River watershed illustrate how science, economics, and ethics each inform and motivate the making of policy. Scientific research tells us that raw sewage carries pathogens, organisms that can cause illness in humans and other animals. Science also reveals how untreated sewage can alter conditions for aquatic and marine life by lowering concentrations of dissolved oxygen, increasing mortality for many species. In terms of economic impact, pollution and beach closures reduce recreation, tourism, and other economic activity associated with clean coastal areas. This is a major consideration both in Mexico and in southern California, whose beaches each year host 175 million visitors who spend over $1.5 billion. Ethically, water pollution poses problems because in most cases pollution from upstream users degrades water quality for downstream users.

Many environmental problems share this combination of impacts—harming human health, altering ecological systems, inflicting economic damage, and creating inequities among people. In the Tijuana River watershed, as in many other environmental systems, all of this takes place in a transboundary context, with political pressures from both domestic and international sources.

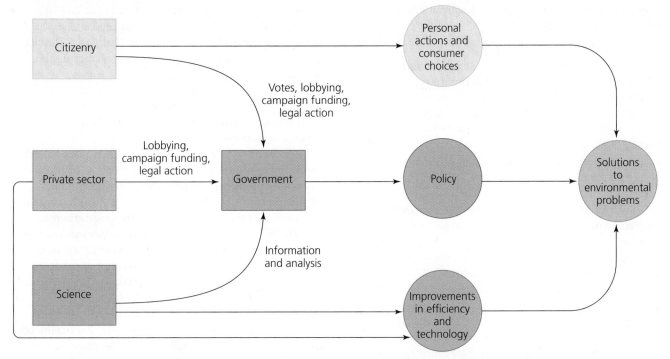

FIGURE 7.2 ▲ Policy plays a central role in how we as a society address environmental problems. Voters, the private sector, and lobbying groups representing various interests influence government representatives. Scientific research also informs government decisions. Governmental representatives and agencies formulate policy that aims to address social problems, including environmental problems. Public policy—along with improvements in technology and efficiency from the private sector and personal actions and consumer purchasing choices exercised by citizens—can produce lasting solutions to environmental problems.

Environmental policy addresses issues of fairness and resource use

The capitalist market system is driven by short-term economic gain rather than long-term social and environmental stability. Market capitalism provides little incentive for businesses or individuals to behave in ways that minimize environmental impact or equalize costs and benefits among parties. As we saw in Chapter 6, market prices often do not reflect the full value of environmental contributions to economies or the full costs imposed on the public by private parties when their actions degrade the environment. Such *market failure* (p. 161) is traditionally viewed as justification for government intervention. Environmental policy aims to protect environmental quality and the natural resources people use and to promote equity or fairness in people's use of resources.

The tragedy of the commons Policy to protect resources held and used in common by the public is intended to safeguard these resources from depletion or degradation. As Garrett Hardin explained in his essay "The Tragedy of the Commons" (p. 5), a resource held in common that is accessible to all will eventually become overused and degraded. Therefore, he argued, it is in our best interest to develop guidelines for the use of such resources. In Hardin's example of a common pasture, guidelines might limit the number of animals each individual can graze or might require pasture users to pay to restore and manage the shared resource. These two

concepts—restriction of use and management—are central to environmental policy today.

Although public oversight through government can alleviate the tragedy of the commons, this dilemma can also be addressed in other ways. One is a cooperative approach: Resource users band together and cooperate to prevent overexploitation. Another is privatization: The resource is subdivided and each allotment is owned privately, such that the owners each have incentive to conserve the resource in ways that maximize its productivity. The cooperative approach may work if the resource is fairly localized and its use is easily enforced, but often these conditions do not hold. Privatization may work if property rights can be clearly assigned (as with land), but it may not work with resources such as air or water. Thus in many cases public oversight and regulation are the best ways to avoid the tragedy of the commons.

Free riders Another reason to develop policy for publicly held resources is the **free rider** predicament. Let's say a community on a river suffers from water pollution that emanates from 10 different factories. The problem could in theory be solved if every factory voluntarily agreed to reduce its own pollution. However, once they all begin reducing their pollution, it becomes tempting for any one of them to stop doing so. Such a factory, by avoiding the sacrifices others are making, would in essence get a "free ride." If enough factories take a free ride, the whole effort will collapse. Because of the free rider problem, private voluntary efforts are often less effective than efforts mandated by public policy, which can ensure that all parties sacrifice equitably.

External costs   Environmental policy also aims to promote fairness by dealing with *external costs* (pp. 150–151), harmful impacts borne by people not involved in the market transactions that created them. For example, a factory may reap greater profits by discharging waste freely into a river and avoiding paying for waste disposal or recycling. Its actions, however, impose external costs (water pollution, decreased fish populations, aesthetic degradation, or other problems) on downstream users of the river. U.S.-owned *maquiladoras* in the Tijuana River watershed dump waste that contaminates the river, affecting Mexican families downstream (**FIGURE 7.3**). Likewise, wastewater from the growing number of people living in the watershed further pollutes the river, imposing external costs on families farther downstream and on beachgoers in Mexico and California.

These goals of environmental policy—to protect resources against the tragedy of the commons and to promote fairness by eliminating free riders and addressing external costs—are reflected in today's diversity of approaches to environmental policy. As just one example, the **polluter-pays principle** specifies that the party responsible for pollution should also be held responsible for covering the costs of its impacts. This principle helps protect resources such as clean water and air, promotes just treatment of all parties, and helps shift external costs into the actual market prices of goods and services.

Many factors hinder environmental policy

If the goals of environmental policy are seemingly so noble, why is it that environmental laws and regulations are so often challenged and that citizens and policymakers repeatedly ignore or reject the ideas of environmental advocates?

In the United States, most environmental policy has come in the form of regulations handed down from government. Businesses and individuals sometimes view these regulations as overly restrictive, bureaucratic, or unresponsive to human needs. For instance, many landowners fear that zoning regulations (p. 355) or protections for endangered species (p. 303) will impose restrictions on the use of their land. Developers complain of time and money lost to bureaucracy in obtaining permits; reviews by government agencies; surveys for endangered species; and required environmental controls, monitoring, and mitigation. In the eyes of such property owners and businesspeople, environmental regulation often means inconvenience and economic loss.

Another reason people sometimes do not see a need for environmental policy stems from the nature of most environmental problems, which often develop gradually. The degradation of ecosystems and public health caused by human impact on the environment are long-term processes. Human behavior, however, is geared toward addressing short-term needs, and this tendency is reflected in our social institutions. Businesses usually opt for short-term economic gain over long-term considerations. The news media have a short attention span based on the daily news cycle, whereby new and sudden events are given more coverage than slowly developing long-term trends. Politicians often act out of short-term interest because they depend on reelection every few years. For all these reasons, many environmental policy goals that seem admirable and that attract wide public support may be obstructed in their practical implementation.

U.S. ENVIRONMENTAL LAW AND POLICY

The United States provides a good focus for understanding environmental policy in constitutional democracies worldwide, for several reasons. First, the United States has pioneered innovative environmental policy. Second, U.S. policies have served as models—of both success and failure—for many other nations and international government bodies. Third, the United States exerts a great deal of influence on the affairs of other nations. Finally, understanding U.S. environmental policy on the federal level enables us to better understand it at local, state, and international levels.

170

FIGURE 7.3 ▶ River pollution raises many issues that have been viewed as justification for environmental policy, including the notion of external costs. This woman washing clothes in the river may suffer upstream pollution from factories, and her use of detergents may cause further pollution for people living downstream.

Federal policy arises from the three branches of government

Like all federal U.S. policy, environmental policy results from actions of the three branches of government—legislative, executive, and judicial—established under the U.S. Constitution. Statutory law, or **legislation**, is created by Congress, which consists of the Senate and the House of Representatives. For instance, the House of Representatives passed the Tijuana River Valley Estuary and Beach Sewage Cleanup Act in 2000 to fund the treatment of sewage flowing into the United States in the Tijuana River. When the International Wastewater Treatment Plant was built, its planners had to heed the Clean Water Act (which requires that harmful toxins and bacteria be removed from wastewater discharged into U.S. waterways) and the Endangered Species Act (which protects species that inhabit the Tijuana River Valley, such as the Pacific pocket mouse).

Legislation is enacted (approved) or vetoed (rejected) by the president, who may also issue *executive orders,* specific legal instructions for government agencies. Once laws are enacted, their implementation and enforcement is assigned to the appropriate administrative agency within the executive branch. Administrative agencies (**FIGURE 7.4**), which may be established by Congress or by presidential order, are sometimes nicknamed the "fourth branch" of government because they are the source of a great deal of policy, in the form of regulations. **Regulations** are specific rules or requirements intended to help achieve the objectives of the more broadly written statutory law. Besides issuing regulations, administrative agencies monitor compliance with laws and regulations and enforce them when they are violated.

The judiciary, consisting of the Supreme Court and various lower courts, is charged with interpreting legislation. This is necessary because social norms, societal conditions, and technologies change over time, and because Congress must write laws broadly to ensure that they apply to varied circumstances throughout the nation. Decisions rendered by the courts make up a body of law known as *case law.* Previous rulings serve as *precedents,* or legal guides, for later cases, steering judicial decisions through time. The judiciary has been an important arena for environmental policy. Grassroots environmental advocates and nongovernmental organizations use lawsuits to help level the playing field with large corporations and government agencies. Conversely, the courts hear complaints from businesses and individuals challenging the constitutional validity of environmental laws they feel to be infringing on their rights.

Federal Administrative Agencies that Influence Environmental Policy	
Executive Office of the President	**Department of the Interior**
Council on Environmental Quality	Bureau of Indian Affairs
	Bureau of Land Management (BLM)
Department of Agriculture (USDA)	Bureau of Reclamation
	Minerals Management Service
Natural Resources Conservation Service	National Park Service (NPS)
U.S. Forest Service (USFS)	Office of Surface Mining
	U.S. Fish and Wildlife Service (USFWS)
Department of Commerce	U.S. Geological Survey (USGS)
Bureau of the Census	**Department of Justice**
National Marine Fisheries Service	
National Oceanic and Atmospheric Administration (NOAA)	Environmental and Natural Resources Division
Department of Defense	**Department of Labor**
Army Corps of Engineers	Occupational Safety and Health Administration (OSHA)
Department of Energy (DOE)	**Department of State**
Energy Efficiency and Renewable Energy	Bureau of Oceans and International Environmental and Scientific Affairs
Energy Information Administration (EIA)	Bureau of Population, Refugees, and Migration
Federal Energy Regulatory Commission (FERC)	
National Laboratories and Technology Centers	**Department of Transportation**
Office of Environmental Management	
Office of Fossil Energy	Federal Transit Administration
Department of Health and Human Services	**Independent Agencies**
Agency for Toxic Substances and Disease Registry	Consumer Product Safety Commission
Centers for Disease Control and Prevention (CDC)	Environmental Protection Agency (EPA)
Food and Drug Administration (FDA)	National Aeronautics and Space Administration (NASA)
	National Transportation Research Center
	Nuclear Regulatory Commission (NRC)
	Tennessee Valley Authority (TVA)
	U.S. Agency for International Development (USAID)

FIGURE 7.4 ▲ Administrative agencies of the executive branch are the source of most U.S. environmental policy. This chart lists those agencies that deal with environmental policy, and the Departments within which they fall. *Source:* U.S. General Services Administration, Washington, D.C.

State and local governments also make environmental policy

The structure of the federal government is mirrored at the state level with governors, legislatures, judiciaries, and agencies. States, counties, and municipalities all generate environmental policy of their own and interact to address environmental problems. For instance, state and local agencies helped regulate the impact of the International Wastewater Treatment Plant on nearby communities and on the quality of California's coastal waters (**FIGURE 7.5**). Ideally, states and localities can act as laboratories for experimenting with novel policy concepts, so that ideas that succeed may be adopted elsewhere. In general, densely populated states, such as California, New York, and Massachusetts, have strong environmental laws and well-funded environmental agencies. Citizens of such states put a premium on safeguarding environmental quality because they have already witnessed extensive environmental impacts.

State laws cannot violate principles of the U.S. Constitution, and if state and federal laws conflict, federal laws take precedence. Federal policymakers may influence environmental policy at the level of the states by:

▶ Supplanting state law to force change. (This is uncommon.)

▶ Providing financial incentives to encourage change. (This can be effective if funds are adequate.)

▶ Following an approach of "cooperative federalism," whereby a federal agency sets national standards and then works with state agencies to achieve them in each state. (This is most common.)

In recent years pressure to weaken federal oversight and hand over power to the states has been growing, but political scientists argue that retaining strong federal control over environmental policy is a good idea because:

▶ Citizens of all states should have equitable protection against environmental and health impacts.

▶ Dealing with environmental problems often requires an "economy of scale" in resources, such that one strong national effort is far more efficient than 50 state efforts.

▶ Many environmental issues involve transboundary disputes, and nationwide efforts minimize disputes among states.

To safeguard public health in locations such as San Diego's beaches, California legislators in 1997 required state environmental health officials to set standards for testing waters for bacterial contamination. Officials issue an advisory, or warning, when bacterial concentrations in nearshore waters exceed health limits established by California law. In 1999, further legislation passed by the California State Assembly mandated another state agency to develop methods for locating sources of contamination. As we proceed through our discussion of federal policy, keep in mind that important environmental policy is also created at the state and local levels.

Some constitutional amendments bear on environmental law

The U.S. Constitution lays out several principles that have come to be especially relevant to environmental policy. One of these is the clause from the Fourteenth Amendment prohibiting a state from denying "equal protection of its laws" to any person. This amendment provides the constitutional basis for environmental justice concerns (pp. 144–146).

The Fifth Amendment ensures, in part, that private property shall not "be taken for public use without just compensation." Courts have interpreted this clause, known as the *takings clause*, to ban not only the literal taking of private property but also what is known as regulatory taking. A **regulatory taking** occurs when the government, by means of a law or regulation, deprives a property owner of all or some economic uses of his or her property. Many people cite the takings clause in opposing environmental regulations that restrict development on privately-owned land. For example, some would contend that zoning regulations (p. 355) that prohibit a landowner from opening a hazardous waste dump in a residential neighborhood deprive the landowner of an economically valuable use of the land and, therefore, violate the Fifth Amendment.

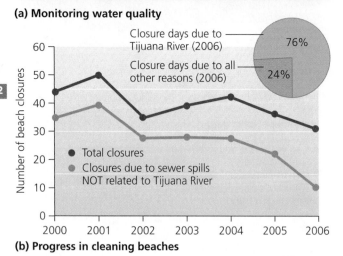

(a) **Monitoring water quality**

(b) **Progress in cleaning beaches**

FIGURE 7.5 ▲ A water quality technician with the State of California **(a)** gathers data on the quality of ocean water along a San Diego County beach. With the help of such data, policy efforts by state, local, and national governments are beginning to cleanse the region's waters. Numbers of beach closures have been declining **(b)**, and pollution from the Tijuana River is now the bulk of the remaining problem. *Source:* County of San Diego Department of Environmental Health. *San Diego County 2006 Beach Closure & Advisory Report*, p. 15, Fig 13. Used by permission.

172

FIGURE 7.6 ▲ Beaches subject to erosion and hurricane damage like this one in South Carolina have given rise to environmental policy debates. Should government be able to restrict development in areas where erosion, storms, and flooding pose risks to life and property? If so, does this constitute a taking of private property rights, and should the property owner be compensated? These were the questions addressed in *Lucas v. South Carolina Coastal Council.*

In a landmark case in 1992, the Supreme Court ruled that a state land use law intended to "prevent serious public harm" violated the takings clause. The case, *Lucas v. South Carolina Coastal Council,* involved a developer named Lucas who in 1986 purchased beachfront property in South Carolina for $975,000. In 1988, before Lucas began to build, South Carolina's legislature passed a law banning construction on eroding beaches (**FIGURE 7.6**). A state agency classified the Lucas property as an eroding beach and prohibited residential construction there.

Believing that this constituted a regulatory taking, Lucas asked a state court to overrule the new law and allow him to build homes on his property. The state court agreed and ruled that Lucas was entitled to $1.2 million in compensation. South Carolina appealed the case to the state supreme court, which overturned the lower court's decision. Lucas then appealed to the U.S. Supreme Court, which overturned the state supreme court's ruling and sided with the lower court, declaring that the state law deprived Lucas of all economically beneficial uses of his land. Today, homes stand on the land, and regulatory taking remains a contentious area of law—a key issue in the sensitive balance between private rights and the public good.

WEIGHING THE ISSUES

Regulatory Takings Suppose you have purchased land and plan to make money by clearing its forest and building condominiums on it. A local environmental group finds an endangered plant species on the property and petitions the government to prevent development of the land. Use the takings clause to argue why you should be allowed to build or should be financially compensated if you are not allowed to build.

Now suppose that you are a member of the environmental group and also a neighbor of the landowner in question. The landowner's land holds the last stand of forest in this region, which has experienced extensive forest loss over the past decade. You feel your quality of life and that of your neighbors will be compromised if the development goes ahead. Make a case why the landowner should not be allowed to build.

Which argument do you feel has greater merit? Would it make a difference to you if the land in question had already been zoned, either for development or for preservation, when the purchase was made? What do you think is the best way to balance private property rights with protection of the public good in cases like this?

Early U.S. environmental policy promoted development

The laws that comprise U.S. environmental policy were created largely in three periods. Laws enacted during the first period, from the 1780s to the late 1800s, accompanied the westward expansion of the nation and were intended mainly to promote settlement and the extraction and use of the continent's abundant natural resources. Among these early laws were the *General Land Ordinances of 1785 and 1787,* which gave the federal government the right to manage unsettled lands and created a grid system for surveying them and readying them for private ownership. Between 1785 and the 1870s, the federal government promoted settlement in the Midwest and West on lands it had appropriated from Native Americans by doling out as many of these lands as possible to its citizens.

Western settlement provided these citizens with means to achieve prosperity and also served to relieve crowding in eastern cities. It expanded the geographical reach of the United States at a time when the young nation was still jostling with European powers for control of the continent. It also wholly displaced the millions of Native Americans who had inhabited these lands. U.S. environmental policy of this era reflected the public perception that Western lands were practically infinite and inexhaustible in natural resources. Following are a few laws typical of this era:

▶ The *Homestead Act of 1862* allowed any citizen to claim, for a $16 fee, 65 ha (160 acres) of public land by living there for 5 years and cultivating the land or building a home (**FIGURE 7.7A**). A waiting period of only 14 months was available to those who could pay $176 for the land.

▶ The *General Mining Act of 1872* legalized and promoted mining by private individuals on public lands for just $5 per acre, subject to local customs, with no government oversight (**FIGURE 7.7B**). This law is still on the books today (pp. 182, 655).

▶ The *Timber Culture Act of 1873* granted 65 ha (160 acres) to any citizen promising to cultivate trees on one-quarter of that area (**FIGURE 7.7C**).

In addition, many millions of acres of land were doled out to railroad companies, which built rail lines to transport

(a) Settlers in Custer County, Nebraska, circa 1860

N.P. White's Derrick on Lynx Creek, Alaska.

(b) Nineteenth-century mining operation, Lynx Creek, Alaska

(c) Loggers felling an old-growth tree, Washington

FIGURE 7.7 ▲ Settlers **(a)** took advantage of the federal government's early land policies, including the Homestead Act of 1862. Early mining activities on public lands **(b)** were largely unregulated under laws such as the General Mining Act of 1872. Although the Timber Culture Act of 1873 promoted tree planting on settled agricultural lands, elsewhere the timber industry was allowed to clear-cut the nation's ancient trees **(c)** with little government policy to limit logging or encourage replanting or conservation.

people, resources, and goods across the continent. All these policies encouraged settlers, entrepreneurs, and land speculators to move west.

The second wave of U.S. environmental policy encouraged conservation

In the late 1800s, as the continent became more populated and its resources were increasingly exploited, public perception and government policy toward natural resources began to shift. Reflecting the emerging conservation and preservation ethics (p. 143) in American society, laws of this period aimed to mitigate some of the environmental impacts associated with westward expansion.

In 1872, Congress designated Yellowstone as the world's first national park. In 1891, Congress, to prevent overharvesting and protect forested watersheds, passed a law authorizing the president to create "forest reserves" off-limits to logging. In 1903, President Theodore Roosevelt created the first national wildlife refuge. These acts enabled the creation, over the next few decades, of a national park system, national forest system, and national wildlife refuge system that still stand as global models (pp. 324, 334). These developments reflected a new understanding that the West's resources were in fact exhaustible and required legal protection.

Land management policies continued through the 20th century, targeting soil conservation in the Dust Bowl years (p. 236) and extending through the Wilderness Act of 1964 (pp. 333–334), which sought to preserve pristine lands "untrammeled by man, where man himself is a visitor who does not remain."

The third wave of U.S. environmental policy responded to pollution and public outcry

Further social changes in the mid- to late-20th century gave rise to the third major period of U.S. environmental policy. In a more densely populated nation driven by technology, heavy industry, and intensive resource consumption, Americans found themselves better off economically but living amid dirtier air, dirtier water, and more waste and toxic chemicals. During the

FIGURE 7.8 ▲ Scientist, writer, and citizen activist Rachel Carson illuminated the problem of pollution from DDT and other pesticides in her 1962 book, *Silent Spring*.

1960s and 1970s, several events triggered increased awareness of environmental problems and brought about a shift in public priorities and important changes in public policy.

A landmark event was the 1962 publication of *Silent Spring*, a book by American scientist and writer Rachel Carson (**FIGURE 7.8**). *Silent Spring* awakened the public to the negative ecological and health effects of pesticides and industrial chemicals (pp. 379–380). The book's title refers to Carson's warning that pesticides might kill so many birds that few would be left to sing in springtime.

Ohio's Cuyahoga River (**FIGURE 7.9**) also drew attention to the hazards of pollution. The Cuyahoga was so polluted with oil and industrial waste that the river actually caught fire

FIGURE 7.9 ▶ In a spectacular display of the need for better control over water pollution, Ohio's Cuyahoga River caught fire several times in the 1950s and 1960s. The Cuyahoga was so polluted with oil and industrial waste that the river would burn for days at a time.

(a) Denis Hayes (on phone) organizing the first Earth Day, 1970

(b) Schoolchildren celebrating Earth Day, Kathmandu, Nepal, 2002

FIGURE 7.10 ▲ April 22, 1970, marked the first Earth Day celebration **(a)**. This public outpouring of support for environmental protection helped spark the third wave of environmental policy in the United States. Decades later **(b)**, Earth Day is celebrated by millions of people across the globe.

near Cleveland more than half a dozen times during the 1950s and 1960s. This spectacle, coupled with an enormous oil spill off the Pacific coast near Santa Barbara, California, in 1969, moved the public to prompt Congress and the president to do more to protect the environment.

Today, largely because of grassroots activism and environmental policies enacted since the 1960s, pesticides are more strictly regulated, and the nation's air and water are considerably cleaner. Public enthusiasm for environmental protection remains strong today, with polls repeatedly showing that an overwhelming majority of Americans favor environmental protection. Such support is evident each year in April, when millions of people worldwide celebrate Earth Day in thousands of events featuring speeches, demonstrations, hikes, bird-walks, cleanup parties, and more. Since the first Earth Day, on April 22, 1970, participation in this event has grown and spread to nearly every country in the world (**FIGURE 7.10**).

NEPA gives citizens input into environmental policy decisions

Besides Earth Day, two federal actions marked 1970 as the dawn of the modern era of environmental policy in the United States. On January 1, 1970, President Richard Nixon signed the **National Environmental Policy Act (NEPA)** into law. NEPA created an agency called the Council on Environmental Quality and required that an **environmental impact statement (EIS)** be prepared for any major federal action that might significantly affect environmental quality. An EIS is a report of results from detailed studies that assess the potential impacts on the environment that would likely result from development projects undertaken or funded by the federal government.

NEPA's effects have been far-reaching. The EIS process forces government agencies and businesses that contract with them to evaluate impacts on the environment before proceeding with a new dam, highway, or building project. The EIS process uses a cost-benefit approach typical of neoclassical economics (p. 149) and generally does not halt development projects. However, it does provide incentives to lessen the environmental damage resulting from a development or activity. NEPA also grants ordinary citizens input in the policy process by requiring that environmental impact statements be made publicly available and that public comment on them be solicited and considered.

The South Bay International Wastewater Treatment Plant provides a good example of the EIS process. Before building the plant, the U.S. government was legally required by NEPA to assess its environmental impact. In 1991 the government released for public comment a draft EIS that assessed likely impact on biological and cultural resources, public health and safety, scenic and recreational values, water quality, and other factors. After three years of research and public discussion, a final EIS was released, in which the government endorsed opening the plant in phases. In the first phase, large particles would be filtered out of wastewater using an advanced primary treatment process, but other pollutants—including toxic metals and bacteria—would remain untreated until facilities were constructed for a secondary treatment process. Until those facilities were completed, the IWTP would release polluted water into the ocean, in violation of the Clean Water Act.

In response, two environmental groups sued the government, arguing that it had failed to consider the best type of secondary treatment. The suit was settled out of court, but it spurred the government to conduct a supplemental EIS that assessed seven different secondary treatment alternatives. Based on this analysis, it decided to endorse the alternative urged by the environmental groups.

Creation of the EPA marked a shift in environmental policy

Six months after signing NEPA into law, Nixon issued an executive order calling for a new, integrated approach to environmental policy based on the understanding that environmental problems are interrelated. "The Government's environmentally-related activities have grown up piecemeal over the years," the order stated. "The time has come to organize them rationally and systematically." Nixon's order moved elements of agencies regulating water quality, air pollution, solid waste, and other issues into the newly created **Environmental Protection Agency (EPA)**. The order charged the EPA with conducting and evaluating research, monitoring environmental quality, setting and enforcing standards for pollution levels, assisting the states in meeting standards and goals, and educating the public. The EPA has been the main U.S. agency working to develop solutions to pollution issues in the Tijuana River watershed.

Other prominent laws followed

Ongoing public demand for a cleaner environment during this period resulted in a number of key laws that remain fundamental to U.S. environmental policy (**FIGURE 7.11**). These laws established ways of cleaning up air and water, protecting rare and endangered species, and controlling hazardous waste and toxic substances.

For problems such as the Tijuana River's pollution, a crucial law has been the Clean Water Act of 1977. Prior to passage of federal laws such as the Clean Water Act, water pollution problems were left largely to local and state government or were addressed through lawsuits. The flaming waters of the Cuyahoga, however, indicated to many people that tough legislation was needed. Thanks to restrictions on pollutants by the Federal Water Pollution Control Acts of 1965 and 1972, and then the Clean Water Act, U.S. waterways finally began to recover. These laws regulated the discharge of wastes, especially from industry, into rivers and streams. The Clean Water Act also aimed to protect wildlife and establish a system for granting permits for the discharge of pollutants.

Historians suggest that major advances in environmental policy occurred in the 1960s and 1970s because several factors converged. First, evidence of environmental problems became

Key Environmental Protection Laws, 1963–1980

1963	Clean Air Act
1964	Wilderness Act
1965	Federal Water Pollution Control Act, Solid Waste Disposal Act
1968	Wild and Scenic Rivers Act
1970	National Environmental Policy Act
1972	Marine Mammal Protection Act, Federal Pesticide Act
1973	Endangered Species Act
1974	Safe Drinking Water Act
1976	Toxic Substances Control Act
1977	Clean Water Act, Soil and Water Conservation Act
1980	CERCLA ("Superfund")

FIGURE 7.11 ▲ Most major laws in modern U.S. environmental policy were enacted in the 1960s and 1970s.

widely and readily apparent. Second, people could visualize policies to deal with the problems. Third, the political climate was ripe, with a supportive public and leaders who were willing to act. Fourth, people's economic wealth had improved to the point at which life was reasonably comfortable for many. With the basic necessities for survival ensured, people found themselves willing to make financial sacrifices and behavioral changes to obtain a cleaner, healthier environment for themselves and their children. In addition, photographs from the space program allowed humanity to see, for the first time ever, images of the Earth from space (see photos on pp. 2 and 687). It is hard for us today to understand the full power of those photos, but they changed many people's worldview by making us aware of the finite nature of our planet.

In the 1980s, Congress strengthened, broadened, and elaborated upon the major laws of the 1970s. For example, major amendments were made to the Clean Water Act in 1987 and to the Clean Air Act in 1990. Today thousands of federal, state, and local laws and regulations help protect health and environmental quality in the United States and abroad.

Many people reacted against regulation

By the 1990s, the political climate in the United States had changed. Although public support for the goals of environmental protection remained high, many citizens and policy experts began to feel that the legislative and regulatory means

FIGURE 7.12 ◄ College students and activists at the 2009 Power Shift event in Washington D.C. urge U.S. leaders to enact policies to help bring the atmosphere's carbon dioxide concentration back down to 350 parts per million. This was one of several major events that year that expressed grassroots support for addressing global climate change through the political process.

used to achieve environmental policy goals too often imposed economic burdens on businesses and personal burdens on individuals. Increasingly, attempts were made at the federal level to roll back or weaken environmental laws. These began with the Reagan administration and culminated in an array of efforts by the George W. Bush administration and by the Republican-controlled Congresses in power from 1994 through 2006.

As advocates of environmental protection watched their hard-won gains eroding, many began to feel that new perspectives and strategies were needed. In a provocative 2004 essay titled "The Death of Environmentalism," political consultants Michael Shellenberger and Ted Nordhaus argued that the modern American environmental movement needed to reinvent its approach. Environmental advocates needed to appeal to people's core values and not simply offer technical policy fixes, Shellenberger and Nordhaus maintained. They needed to stop labeling problems as "environmental" and start showing people why these problems are actually human issues that lie at the very heart of our quality of life. They needed to be more responsive to people's needs and to build broad and meaningful alliances with labor unions, advocates for the poor, and other progressive causes in order to establish a political majority that could govern. Most of all, they needed to articulate a positive, inspiring vision for the future.

Many environmental advocates reacted defensively to Shellenberger and Nordhaus's suggestions, but their views opened a productive discussion. The resulting new perspectives contributed to the election of Barack Obama as president in 2008.

Today's environmental policy focuses on sustainability and climate change

With Obama's election and enhanced Democratic majorities in Congress came a widespread expectation that the United States might reestablish its international leadership in environmental policy. During the country's years of inactivity, other nations had increased their attention to environmental

issues. The 1992 Earth Summit at Río de Janeiro, Brazil, and then the 2002 World Summit on Sustainable Development in Johannesburg, South Africa, were the largest international diplomatic conferences ever held, unifying leaders from 200 nations around the idea of sustainable development (pp. 19, 677).

Indeed, we may now be embarking on a fourth wave of environmental policy, one focused on sustainability and sustainable development. This new policy approach tries to find ways to safeguard the functionality of natural systems while raising living standards for the world's poorer people. As part of this trend, U.S. environmental policy is becoming more integrated with that of other nations.

Moreover, the pressing issue of global climate change (Chapter 18) has come to dominate much of the discussion over environmental policy in the United States and across the world (**FIGURE 7.12**). As we continue to feel the social, economic, and ecological effects of climate change and other environmental impacts, environmental policy and the search for sustainable solutions will become central parts of governance and everyday life for all of us in the years ahead.

APPROACHES TO ENVIRONMENTAL POLICY

When most people think of environmental policy, what comes to mind are major laws, such as the Clean Water Act, or government regulations, such as those specifying what an industry can and cannot dump into a river. However, environmental policy is far more diverse and can take a variety of approaches (**FIGURE 7.13**).

Conflicts can be addressed in court

Prior to the legislative push of recent decades, most environmental policy questions in the United States were addressed with lawsuits in the courts through **tort law**, which is law that deals with harm caused to one entity by another. (The word *tort*

PROBLEM: Pollution from factory harms people's health

FIGURE 7.13 ▲ For any given environmental problem, such as pollution from a factory, we may consider three major types of policy approaches: ❶ seeking damages through tort law, ❷ limiting pollution through command-and-control legislation and regulation, and ❸ reducing pollution through market-based or other economic strategies.

SOLUTIONS: Three policy approaches

❶ Sue factory in court seeking damages and/or injunction.

EPA

❷ Government regulation restricts emissions allowed.

❸ Market-based approaches: factories that pollute less (on right) outcompete polluting factory (on left) through permit-trading, avoiding green taxes, collecting subsidies, or selling ecolabeled products. Polluting factory must find ways to cut emissions to survive in marketplace.

is French for a wrong or an injustice.) Most pollution issues were subject primarily to *nuisance law*, one form of tort law. Individuals suffering external costs from pollution would seek redress through lawsuits against polluters, one case at a time. In many cases the courts punished polluters by issuing injunctions to stop their operations or by demanding that damages be paid to the affected individuals. However, as industrialization proceeded and population grew denser, pollution became harder to avoid, and judges were reluctant to hinder industry, which was viewed to be promoting society's economic development.

In 1970, the U.S. Supreme Court heard the case *Boomer v. Atlantic Cement Company*. The court ruled that residents of Albany, New York, who were suffering pollution from a neighborhood cement plant were entitled to financial compensation. However, the court refused to shut down the plant. Instead, it allowed the plant to continue operating once it paid the residents. The court had calculated that the economic costs to the company of controlling its pollution were greater than the economic costs of the pollution to the residents, and the court based its decision on an attempt to minimize overall

costs. In handing down this ruling, the justices were essentially letting the market decide between right and wrong. For people concerned about the pervasive spread of pollution throughout society, rulings like these showed that tort law was no longer a viable avenue for preventing pollution. This helped spur the drive for legislation and regulation as the primary means of protecting public health and safety through environmental policy.

Command-and-control policy has improved our lives

Most environmental laws of recent decades, and most regulations enforced by agencies today, use a **command-and-control** approach. In the command-and-control approach, a regulating agency prohibits certain actions, or sets rules, standards, or limits, and threatens punishment for those who violate these terms. This simple and direct approach to policymaking has brought citizens of the United States and many other nations cleaner air, cleaner water, safer workplaces, healthier

The SCIENCE behind the Story

Comparing Costs and Benefits of Environmental Regulations

Federal regulations to protect environmental quality often result in clearer air or cleaner water, but are such regulations worth the costs to industry, businesses, and consumers? In 2003, a federal study determined that the answer to that question is a resounding yes.

The study, conducted by the White House Office of Management and Budget (OMB) under the George W. Bush administration, weighed the economic costs and benefits of 107 federal regulations enacted in the United States from 1992 to 2002, many of which dealt with environmental issues. The researchers found that these regulations cost $36–42 billion annually. However, their estimated value in terms of public good totaled $146–230 billion. Thus, the economic benefits of these regulations far exceeded their costs. Moreover, of all the regulations, those dealing with environmental protection were found to be especially cost-effective (**see figure**).

Most regulatory benefits were reflected in health and social gains resulting from clean air. Over the studied

Truck with cleaner diesel exhaust thanks to federal regulations

10-year period, the benefits of clean-air regulations were found to be 6–10 times greater than the costs of complying with the regulations. Reductions in hospitalization and emergency room visits, premature deaths, and lost workdays together accounted for savings of $118–177 billion.

The authors of the OMB review relied on cost-benefit analyses (p. 149) previously made by experts in the EPA and other agencies and scrutinized them for appropriateness and accuracy before using them. To see how the EPA performed its economic analysis and how the OMB double-checked the EPA's findings, we can examine one EPA regulation included in the OMB report—an air pollution regulation called the Heavy Duty

Engine/Diesel Fuel Rule, which today is bringing much cleaner diesel fuel to the nation's highways.

The diesel regulation aimed to address diesel engines that emit fine soot, sulfur compounds, and nitrogen oxides (NO_x). In the 1990s, the EPA regulation was intended to make diesel trucks and buses run more cleanly by lowering the fuel's sulfur content. However, achieving cleaner diesel fuel would mean potentially costly changes to the way that fuel is made and handled.

To determine potential costs, EPA analysts examined how fuel processors, engine manufacturers, and vehicle operators would need to change their processes or equipment to clean up the fuel. For example, a heavy-duty engine redesigned to run more efficiently on cleaner fuel would cost approximately $4,600 more to operate, the analysts calculated. Total research and development costs for emission control were predicted to exceed $600 million. The EPA asked for input from affected parties, scientists, and the public, finally determining that removing sulfur from diesel fuel would be neither cheap nor

neighborhoods, and many other improvements in quality of life. The relatively safe, healthy, comfortable lives most of us enjoy today owe much to the command-and-control environmental policy of the past few decades.

Even in plain financial terms, putting health and quality of life aside, command-and-control policy has largely been effective. In 2003, the White House Office of Management and Budget undertook an extensive analysis of U.S. regulatory policy to determine whether regulations resulted in more economic costs or more economic benefits. The analysis revealed that the benefits outweighed the costs by a great deal and that environmental regulations were the most beneficial of all (see THE SCIENCE BEHIND THE STORY, above).

Despite these successes, many people have grown disenchanted with the top-down, sometimes heavy-handed nature of the command-and-control approach. Sometimes government actions are well intentioned but not well enough informed, so they can lead to unforeseen consequences. Regulatory policy can also fail if a government does not live up to its responsibilities to protect its citizens or treat them equitably. This may occur when leaders allow themselves to be unduly influenced by *interest groups*, small groups of people seeking private gain that work against the larger public interest. In addition, the command-and-control approach can fail when it generates opposition to government policy. If citizens view laws and regulations primarily as restrictions on their freedom, those policies will not last long in a constitutional democracy.

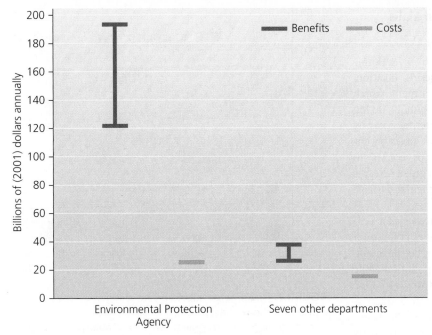

The OMB study showed that annual economic benefits of regulations exceeded costs considerably and that EPA regulations were even more cost-efficient than those of agencies in seven other departments. Red lines indicate range of values of benefits. Orange ranges of costs are so narrow that they appear as bars on the graph. SOURCE: U.S. Office of Management and Budget, 2003. *Informing regulatory decisions: 2003 report to Congress on the costs and benefits of federal regulations and unfunded mandates on state, local, and tribal entities.* Washington, D.C.

quick, with total annualized costs of about $4.2 billion by 2030.

EPA analysts then considered the benefits of reduced diesel pollution. Using air quality data from more than 940 smog monitors around the country, the EPA estimated diesel-related air pollution levels with and without the proposed diesel regulation. The analysts examined public health statistics, especially data related to respiratory illnesses such as asthma and chronic bronchitis. They then estimated how many cases of disease or premature death might be prevented if people breathed fewer diesel-related pollutants.

They determined the economic impacts of such health problems by tallying up medical, workforce, and social costs. Each premature death was assigned an impact of $6 million, and each case of chronic bronchitis was pegged at $331,000. Estimates were regionalized to account for different economic conditions in various parts of the country.

In the end, the EPA analysis determined that the benefits of cleaner diesel fuel far exceeded the costs. If sulfur content of diesel were cut from 500 to 15 parts per million, then 2.6 million tons of smog-causing NO_x emissions would be eliminated each year. An estimated 8,300 premature deaths, 5,500 cases of chronic bronchitis, over 360,000 asthma attacks, and about 1.5 million lost workdays would be prevented each year. By 2030, the analysis determined, benefits would reach about $70 billion.

Following this analysis, the Heavy Duty Engine/Diesel Fuel Rule took effect in 2001 just before President Bill Clinton left office. The George W. Bush administration stalled it for three years and then let it proceed. The program has been phased in gradually through 2010, and today trucks are using much cleaner diesel fuel on the nation's highways, for only 3–5 cents per gallon more.

The OMB, in compiling its own review of federal rules, scrutinized the EPA study and concluded that most of the estimates did indeed make economic sense. The OMB analysts made some minor changes when calculating their own figures, but for the most part they confirmed the validity of the EPA estimates and used most of the EPA's economic data in the OMB report. The final OMB report is now being used to evaluate and support efforts at environmental regulation around the country. ■

Economic policy tools can help achieve environmental goals

The most common critique of command-and-control policy is that it achieves its goals in a more costly and less efficient manner than the marketplace can. Whereas command-and-control policy mandates particular solutions to problems, private entities competing in a free market can innovate and may produce new or better solutions at lower cost. As a result, political scientists, economists, and policymakers today are exploring alternative policy approaches to channel the innovation and economic efficiency of the market in directions that benefit the public. Such approaches use economic incentives to encourage desired outcomes, discourage undesired outcomes, and set market dynamics in motion to achieve goals in an economically efficient manner.

Like regulation and like the tort law approach, economic policy tools aim to "internalize" external costs suffered by the public by building them into the prices paid in the marketplace. Each of these three major approaches has strengths and weaknesses, and each is best suited to different conditions. These three approaches may, however, be used together. For instance, government regulation is often needed to frame market-based efforts, and citizens can use the courts to ensure the enforcement of regulations. Let's now explore several main economic policy tools.

Green taxes discourage undesirable activities

The most straightforward economic policy tool—taxation—can be used to discourage undesirable activities. In taxation, money passes from private parties to the government, which then reapportions it in services to benefit the public. Taxing undesirable activities helps to "internalize" external costs by making them part of the cost of doing business. Taxes on environmentally harmful activities and products have come to be called **green taxes**. When a business pays a green tax, it is essentially reimbursing the public for environmental damage it causes that affects the public.

Under green taxation, a firm owning a factory that pollutes a waterway would pay taxes on the amount of pollution it discharges—the more pollution, the higher the tax payment. This gives firms a financial incentive to reduce pollution while allowing them the freedom to decide how best to do so. One polluter might choose to invest in technologies to reduce its pollution if doing so is less costly than paying the taxes. Another polluter might find abating its pollution more costly and could choose to pay the taxes instead—funds the government might then apply toward mitigating pollution in some other way.

Green taxes have yet to gain widespread support in the United States, although similar "sin taxes" on cigarettes and alcohol are long-accepted tools of U.S. social policy. Taxes on pollution have been widely instituted in Europe, where many nations have adopted the polluter-pays principle (p. 170). Today there is debate worldwide about whether "carbon taxes"—taxing gasoline, coal-based electricity, and fossil-fuel-intensive products according to the carbon emissions they produce—should be instituted to fight global climate change (p. 524).

Green taxation provides incentive for industry to lower emissions not merely to a level specified in a regulation, but to still-lower levels. However, green taxes do have drawbacks: Businesses will most likely pass on their tax expenses to consumers, and these increased costs may affect low-income consumers disproportionately more than high-income ones.

Subsidies promote certain activities

Another type of economic policy tool is the **subsidy**, a government giveaway of money or publicly owned resources that is intended to encourage a particular industry or activity. With a subsidy, money flows in the opposite direction than in taxation: It flows from the public to private parties. A government provides subsidies to certain types of businesses or individuals to support and promote industries or activities that it deems desirable in some way. Subsidies take many forms. A *tax break* is a common form of subsidy. Relieving the tax burden on an industry, firm, or individual assists it by reducing its expenses. Because a tax break deprives the government's treasury of funds it would otherwise collect, it has the same financial effect as a direct giveaway of money.

Subsidies can be used to promote environmentally sustainable activities, but all too often they are used to prop up unsustainable ones. For instance, from 2002 to 2008, the U.S. government gave $72 billion of its citizens' money to fossil fuel

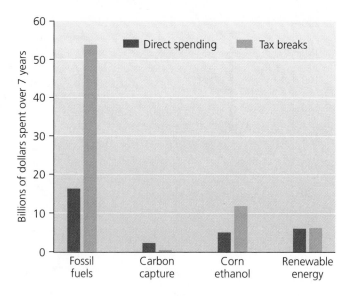

FIGURE 7.14 ▲ Energy subsidies are still geared toward supporting the well-established fossil fuel industries more than the young and struggling renewable energy industries. In the United States from 2002 to 2008, fossil fuels received two-and-a-half times more subsidies than renewable energy sources, and most of the money for renewables went toward corn ethanol. Data from Internal Revenue Service, U.S. Department of Energy, Congressional Joint Committee on Taxation, Office of Management and Budget, and U.S. Department of Agriculture.

corporations while spending only $29 billion on renewable energy efforts (**FIGURE 7.14**). Thus, although the nation's leaders spoke of supporting renewable energy and moving away from fossil fuels to ensure energy security and fight climate change, they were still sending taxpayers' money—$240 for each man, woman, and child—predominantly toward fossil fuels. About $54 billion of the fossil fuel subsidies were in the form of tax breaks. Moreover, most of the subsidies for renewable energy went toward corn ethanol, which, as we will see in Chapter 20, is not widely viewed as a sustainable fuel.

In September 2009, President Obama and other leaders of the Group of 20 (G-20) nations resolved to gradually phase out their collective $300 billion of annual fossil fuel subsidies. If all these nations actually keep this promise, this could help hasten the shift to cleaner renewable energy sources, and thereby reduce global greenhouse gas emissions an estimated 10–12% by 2050.

Many other environmentally harmful subsidies exist. Under the General Mining Act of 1872 (pp. 173, 655), mining companies extract $500 million to $1 billion in minerals from U.S. public lands each year without paying a penny in royalties to the taxpayers who own these lands. Since this law was enacted, the U.S. government has given away nearly $250 billion of mineral resources, and mining activities have polluted more than 40% of watersheds in the West. The 140-year-old law still allows mining companies to buy public lands for $5 or less per acre.

On the national forests, the U.S. Forest Service spends $35 million of taxpayer money each year building roads to allow private timber corporations access to cut trees in the forests, which the companies then sell at a profit (**FIGURE 7.15**).

In total, the world's governments spend roughly $1.45 *trillion* of their citizens' money each year on subsidies judged

FIGURE 7.15 ▲ When companies cut timber in U.S. national forests, taxpayers pay the costs of building and maintaining access roads. "By absorbing these costs, the federal government shields the timber industry from the true cost of doing business," the nonprofit group Taxpayers for Common Sense concludes.

to be harmful to the environment and to the economy, according to environmental scientists Norman Myers and Jennifer Kent. This amount is larger than the economies of all but five nations. Advocates of sustainable resource use have long urged governments to subsidize environmentally sustainable activities instead. This is beginning to happen, but the shift could be much faster.

We can harness market dynamics to promote sustainability

With subsidies and green taxes, the government decides to employ financial incentives in direct and selective ways. However, we may also pursue policy goals by establishing financial incentives and then letting marketplace dynamics run their course. In Chapter 6 we discussed *ecolabeling* (p. 161), the practice whereby sellers who use sustainable practices in growing, harvesting, or manufacturing their products advertise this fact on their labels, hoping to win approval from buyers. In many cases, ecolabeling has developed following initial steps taken by government to require the disclosure of information to consumers. Yet, once established, ecolabeling can spread in a free market as more and more businesses seek this path to winning consumer confidence and outcompeting less sustainably produced brands.

Permit trading can save money and produce results

In the innovative market-based approach known as **permit trading** the government creates a market in permits for an environmentally harmful activity, and then companies and utilities are allowed to buy, sell, or trade rights to conduct the activity. For instance, to decrease emissions of air pollutants, a government might grant emissions permits and set up an **emissions trading** system. In a **cap-and-trade** emissions trading system the government first determines the overall amount of pollution it will accept (i.e., it caps the amount, at a level below what it would be in the absence of the program), and then issues permits to polluters that allow them each to emit a certain fraction of that amount. Polluters may buy, sell, and trade these permits with other polluters.

Suppose, for example, you are an industrial plant owner with permits to release 10 units of pollution, but you find that you can become more efficient and release only 5 units instead. You then have a surplus of permits, which might be very valuable to some other plant owner who is having trouble reducing pollution or who wants to expand production. In such a case, you can sell your extra permits to the other plant owner. Doing so generates income for you and meets the needs of the other plant, while the total amount of pollution does not rise. By providing companies an economic incentive to find ways to reduce emissions, permit trading can reduce expenses for both industry and the public relative to a conventional regulatory system.

To lower emissions further, the government may reduce the amount of overall emissions allowed year by year. Moreover, environmental organizations may buy up permits and "retire" them, reducing the overall amount of pollution still further. To see an illustration of how a cap-and-trade system works, jump ahead to Figure 18.31, p. 523.

A cap-and-trade system has been in place in the United States, established by the 1990 amendments to the Clean Air Act (p. 471) that mandated lower emissions of sulfur dioxide, a major contributor to acidic deposition (pp. 482–486). Since then, sulfur dioxide emissions from sources in the program have declined by 52% (**FIGURE 7.16**), sulfate deposition has been reduced, and air quality and visibility have improved. The 52% reduction in pollution was greater than the amount actually required by the legislation, offering evidence that cap-and-trade systems can sometimes cut pollution more effectively than command-and-control regulation. Moreover, the cuts were attained at much less cost than was predicted, and with no apparent effect on electricity supply or economic growth. Savings from the permit trading system have been estimated at billions of dollars per year, and the EPA calculates that the program's benefits outweigh its costs by about 40 to 1. Other similar programs have also shown success, including one in the Los Angeles basin to reduce smog (p. 471) and one among northeastern states aimed at nitrogen oxides.

Although cap-and-trade programs can reduce pollution overall, they do allow hotspots of pollution to occur around plants that buy permits to pollute more. Moreover, large firms can hoard permits deterring smaller new firms from entering the market, and thereby suppressing competition. Nonetheless,

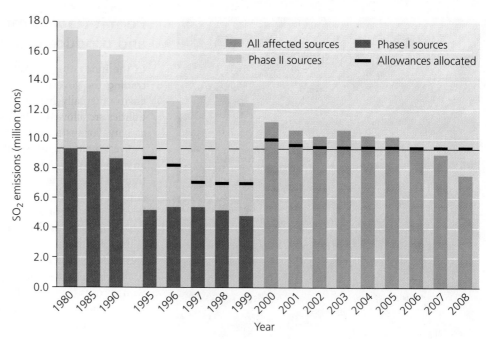

FIGURE 7.16 ▶ Emissions of sulfur dioxide from sources participating in the emissions trading program mandated by the 1990 Clean Air Act amendments have fallen 52% since 1990. Emissions dropped steeply in 1995, the first year of the program, as a result of reductions among large sources included in Phase I of the program (red portions of bars). Emissions fell again in 2000, when Phase II began, which required reductions from smaller sources (yellow portions of bars). As of 2008, emissions from both types of sources (orange bars) had dropped below the amounts allocated in permits (black bands). *Source:* Data from U.S. Environmental Protection Agency.

permit trading shows promise for safeguarding environmental quality while granting industries the flexibility to lessen their impacts in ways that are economically palatable.

In recent years, several markets in carbon emissions have begun to address the greenhouse gas emissions that drive global climate change. In the European Union Emission Trading Scheme (p. 524), each participating nation takes the emissions permits it is allowed and allocates them to its industries according to their emissions at the start of the program. The industries then can trade permits freely, establishing a market whereby the price of a carbon emissions permit fluctuates according to supply and demand. In the United States, the Chicago Climate Exchange and the Regional Greenhouse Gas Initiative (pp. 522–524) are involved in carbon trading, and other such markets are on the horizon. Federal legislation to establish a nationwide carbon trading program is currently being debated.

Emissions trading programs need to be properly set up, however, to be effective. For instance, in the first phase of the European Union Emission Trading Scheme, nations allocated too many permits, causing the price to fall precipitously as the permits became nearly worthless. The overallocation was partly addressed in 2008 as the program entered its second phase, and prices have since risen to moderate levels that restore incentives for industries to reduce emissions.

WEIGHING THE ISSUES

A License to Pollute? Some environmental advocates oppose emissions trading because they view it as giving polluters "a license to pollute." How do you feel about emissions trading as a means of reducing air pollution? Would you favor command-and-control regulation instead? What advantages and disadvantages do you see in each approach?

Market incentives also operate at the local level

You may have already taken part in transactions involving financial incentives as policy tools. Many municipalities charge residents for waste disposal according to the amount of waste they generate. Other cities place taxes or disposal fees on items whose safe disposal is costly, such as tires and motor oil. Still others give rebates to residents who buy water-efficient toilets and appliances, because the rebates can cost the city less than upgrading its sewage treatment system. Likewise, power utilities sometimes offer discounts to customers who buy high-efficiency lightbulbs and appliances, because doing so is cheaper for the utilities than expanding the generating capacity of their plants. At all levels, from the local to the international, market-based incentives that are well planned and implemented can reduce environmental impact while minimizing overall costs to industry and easing concerns about the intrusiveness of government regulation.

The public and private sectors can work as partners

In responding to environmental challenges, our society is increasingly trying to take advantage of the most effective aspects of government regulation and of private enterprise. One approach is to combine them in an arrangement called a **public-private partnership**. In most public-private partnerships, a for-profit entity takes charge of performing the work while operating within confines agreed upon with a public entity that acts as an overseer. The hope is that the public policy goal will be achieved in a timely and cost-efficient manner as the private entity tries to find ways to be efficient so that it can profit from its work.

In the Tijuana River valley, Bajagua LLC, a private consortium of San Diego businesspeople, sought to create a public-private partnership with U.S. and Mexican federal and local governments to address the region's sewage treatment problems. Bajagua LLC proposed to build on the Mexican side of the border a plant with a capacity to treat 223 million L (59 million gallons) of wastewater per day. This plant would give secondary treatment to wastewater from the existing and inadequate South Bay International Wastewater Treatment Plant (IWTP) on the U.S. side, as well as full treatment to additional wastewater from the Tijuana River. Investors in the Bajagua Project hoped to make money from selling reclaimed water for landscaping, industry, and agriculture. Supporters of the project hoped that some of this money could offset costs to U.S. taxpayers and that the new treatment plant would end wastewater pollution in Tijuana and on San Diego's beaches for good.

In 2008, a federal commission considering the Bajagua Project decided instead to pursue a publicly funded $88 million upgrade of the IWTP to add secondary treatment on site. The commission concluded that the publicly funded option was less expensive and faced fewer logistical and regulatory hurdles. The failure of the Bajagua Project to win approval illustrates the general challenge of designing workable public-private partnerships that solve problems while serving both private and public interests.

SCIENCE AND THE ENVIRONMENTAL POLICY PROCESS

In constitutional democracies such as the United States, each and every person has a political voice and can make a difference. However, money wields influence, and some people and organizations are far more influential than others. We will explore some of these dynamics as we examine the main steps of the policymaking process and examine the role that science plays in policy.

Environmental policy results from a stepwise process

Creating environmental policy results from a multiple-step process that requires initiative, dedication, and the support of many people (**FIGURE 7.17**). Our discussion of this process pertains both to citizens at the grassroots level and to large organizations.

❶ Identify a problem The first step in the policy process is to identify an environmental problem. This requires curiosity, observation, record keeping, and an awareness of our relationship with the environment. For example, assessing the contamination of San Diego- and Tijuana-area beaches required understanding the ecological and health impacts of untreated wastewater. It also required being able to detect contamination on beaches and understanding water flow dynamics among the beaches, the Pacific Ocean, and the Tijuana River watershed.

❷ Pinpoint causes of the problem The next step in the policy process is to discover specific causes of the problem.

❶ Identify a problem

❷ Pinpoint causes of the problem

❸ Envision solution

❹ Get organized

❺ Cultivate access and influence

❻ Shepherd the solution into law

❼ Implement, assess, and interpret policy

FIGURE 7.17 ▲ Understanding the steps of the policy process is an essential element of solving environmental problems.

A person seeking causes for the Tijuana River's pollution might notice that transboundary pollution began to worsen in the 1960s, when U.S. companies started opening *maquiladoras* on the Mexican side of the border. Advocates of the *maquiladora* system argue that these factories provide much-needed jobs south of the border while keeping companies' costs low by paying Mexican workers far less than U.S. workers. Critics argue that the factories are waste-generating, water-guzzling polluters whose transboundary nature makes them difficult to regulate.

Identifying problems and their causes requires scientific research. Much of this work takes place in the arena of *risk assessment* (pp. 392–394), in which scientists evaluate the extent and nature of problems and judge the risks that they pose to public health or environmental quality.

❸ Envision a solution The better one can pinpoint causes of a problem, the more effectively one can envision solutions to it. Science plays a vital role here too, through the process of *risk management* (pp. 393–394), in which scientists help develop strategies to minimize risk. Frequently, though, solutions involve primarily social or political action. In San Diego, citizen activists wanted Tijuana to enforce its own pollution laws more effectively—something that, once visualized, began to happen when San Diego city employees started training and working with their Mexican counterparts to keep hazardous waste out of the sewage treatment system.

❹ Get organized When it comes to influencing policy, organizations are generally more effective than individuals. Yet as effective as large organizations can be, small coalitions and even individual citizens who are motivated, informed, and organized can help solve environmental problems.

San Diego-area resident Lori Saldaña provides an example. Concerned about the Tijuana River's pollution, Saldaña reviewed plans for the international wastewater treatment plant

the U.S. government had proposed to build (**FIGURE 7.18**). She concluded that it would merely shift pollution from the river to the ocean, where sewage would be released through a massive underwater pipe 5.6 km (3.5 mi) offshore. Working with her local Sierra Club chapter, Saldaña protested the plant's design and participated in the lawsuit that forced the EPA to conduct further studies and eventually implement design changes.

Later, Saldaña and oceanographer Tim Baumgartner boated out above the pipe's opening with scientific instruments to measure water quality. Their results showed a plume of pollution emanating from the pipe. They posted these data on the Internet and urged policymakers to speed construction of secondary treatment facilities. For her efforts, Saldaña received awards and was appointed to a commission on border environmental issues by President Bill Clinton. After a decade of activism, Saldaña ran for the California State Assembly in 2004 and won, becoming the representative from California's 76th district.

❺ Cultivate access and influence The fifth step in the policy process entails gaining access to policymakers who have the clout to enact change. People gain access and influence through lobbying, campaign contributions, and the revolving door. Anyone who spends time or money trying to influence an elected official's decisions is engaged in **lobbying**. Although anyone can lobby, it is much more difficult for an ordinary citizen than for the full-time professional lobbyists employed by the many businesses and organizations seeking a voice in politics. Environmental advocacy organizations are not the most influential of lobbying groups. Indeed, the American Petroleum Institute spends nearly as much lobbying as the entire budgets of the top five U.S. environmental advocacy groups combined.

Supporting a candidate's election efforts with money is another way to make one's voice heard. Environmental policy often regulates the activities of corporations and industries, so they have a strong interest in shaping it. Corporations and

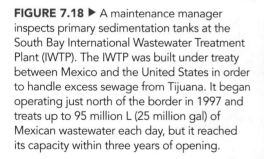

FIGURE 7.18 ▶ A maintenance manager inspects primary sedimentation tanks at the South Bay International Wastewater Treatment Plant (IWTP). The IWTP was built under treaty between Mexico and the United States in order to handle excess sewage from Tijuana. It began operating just north of the border in 1997 and treats up to 95 million L (25 million gal) of Mexican wastewater each day, but it reached its capacity within three years of opening.

industries may not legally make direct campaign contributions, but they are allowed to establish political action committees (PACs), which raise money to help candidates win elections, in hope of gaining access to those individuals once they are elected. In the wake of a controversial 2010 U.S. Supreme Court decision, corporations (and unions) are now also allowed to purchase political ads supporting or opposing candidates.

Some individuals employed by government-regulated industries gain political influence when they take jobs with the very government agencies responsible for regulating their industry. Businesses also often hire former government bureaucrats who regulated their industries. This movement of individuals between the private sector and government agencies is known as the **revolving door**. Defenders of the revolving door system assert that corporate executives who take government jobs regulating their own industry bring with them an intimate knowledge of the industry that makes them highly qualified and likely to benefit society with well-informed policy. Critics of the system contend that taking a job regulating your former employer is a clear conflict of interest that undermines the effectiveness of the regulatory process.

All three of these methods—lobbying, campaign contributions, and the revolving door—were employed by the Bajagua LLC consortium that sought to contract with the U.S. government to build a wastewater treatment plant to supplement the IWTP. Backers of the Bajagua Project lobbied government officials, contributed to California congressmen, and hired former government officials, all of whom they hoped would support their cause. Following years of such efforts, in 2006 the International Boundary and Water Commission (which owns and operates the IWTP) agreed to support the Bajagua Project—although the commission changed its mind two years later.

6 Shepherd the solution into law   The next step is to prepare a bill, or draft law, that embodies the desired solutions. Anyone can draft a bill, but one must find legislators to sponsor it. At the federal level, this means finding members of the House and Senate willing to introduce the bill and shepherd it from subcommittee through full committee and on to passage by the full Congress (**FIGURE 7.19**). If it passes through all of these steps, the bill may become law with the president's signature, but it can die in countless fashions along the way.

7 Implement, assess, and interpret policy   Following a law's enactment, administrative agencies design specific regulations and take charge of implementing and enforcing them. Over time, regulators, policymakers, and interested citizens may evaluate the policy's successes and failures and may revise the policy as necessary. Moreover, the judicial branch interprets law in response to suits in the courts, and much environmental policy has lived and died by judicial interpretation. The policy process is cyclical, so as all these developments proceed, and as societal and environmental conditions change, the policy process may circle back to its first step as fresh problems are identified, and the process may begin anew.

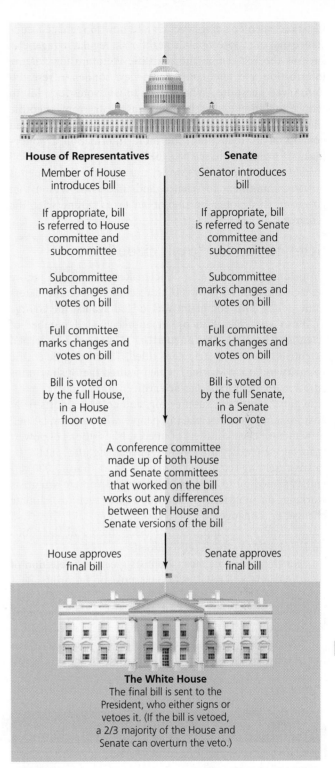

FIGURE 7.19 ▲ Before a bill becomes U.S. law, it must clear a number of hurdles in both legislative bodies. If the bill passes the House and Senate, a conference committee must work out any differences between the House and Senate versions before the bill is sent to the president. The president may then sign or veto the bill.

Science plays a role in policy

Economic interests, ethical values, and political ideology all influence the policy process, yet effective environmental policy is generally also informed by scientific research. For instance, when deciding whether and how to regulate a substance that may pose a public health risk, regulatory agencies such as the EPA comb the scientific literature for existing information and may commission new studies to research unresolved questions. When trying to win votes for a bill to reduce pollution, a legislator may use data from scientific studies to quantify the cost of the pollution or the predicted benefits of its reduction. The more information a policymaker can glean from science, the better policy he or she will be able to create. In today's world, a nation's strength depends on its commitment to science, and this is the reason that governments devote a portion of our taxes to fund scientific research.

Science can be "politicized"

Unfortunately, sometimes policymakers choose to ignore science and instead allow political ideology alone to determine policy. This complaint was lodged against the George W. Bush administration by an unprecedented number of scientists. In 2004, the nonpartisan Union of Concerned Scientists released a statement titled, "Restoring Scientific Integrity in Policy Making," which faulted the Bush administration for manipulating scientific information for political ends; censoring, suppressing, and editing reports from government scientists; placing people who are unqualified or who have clear conflicts of interest in positions of power; ignoring scientific advice; and misleading the public by misrepresenting scientific knowledge. More than 12,000 scientists signed on to this letter. Many government scientists working on politically sensitive issues such as climate change or endangered species protection said they had found their work suppressed or discredited and their jobs threatened. Many chose self-censorship.

For these reasons, most scientists greeted the election of Barack Obama with relief, as Obama spoke of "restoring scientific integrity to government" and ensuring "that scientific data [are] never distorted or concealed to serve a political agenda." Yet, of course, either political party can politicize science.

Whenever taxpayer-funded science is suppressed or distorted for political ends, by either the right or the left, we all lose. Abuses of power generally come to light only when brave government scientists risk their careers to alert the public and when journalists work hard to uncover and publicize these issues. We cannot simply take for granted that science will play a role in policy. As scientifically literate citizens of a democracy, we all need to stay vigilant and help make sure our government representatives are making proper use of the tremendous scientific assets we have at our disposal.

INTERNATIONAL ENVIRONMENTAL POLICY

Environmental systems pay no heed to political boundaries, so environmental problems often are not restricted to the confines of particular nations. Most of the world's major rivers straddle or cross international borders, so problems like those along the Tijuana River are frequently international in scope. Because U.S. law has no authority in Mexico or any other nation outside the United States, international policy is vital to solving transboundary problems.

International law includes conventional and customary law

Whereas U.S. law arises from the Constitution and the Bill of Rights, international environmental law is more nebulous in its origins and authority. International law known as **customary law** arises from long-standing practices, or customs, held in common by most cultures. International law known as **conventional law** arises from **conventions**, or **treaties**, into which nations enter. One example is the Montreal Protocol, a 1987 accord among more than 160 nations to reduce the emission of airborne chemicals that deplete the ozone layer (pp. 481–482). Another example is the Kyoto Protocol to reduce greenhouse gas emissions that contribute to global climate change (pp. 520–521). **TABLE 7.1** shows a selection of major environmental treaties.

The South Bay International Wastewater Treatment Plant that treats wastewater from the Tijuana River watershed was built as a result of a 1990 treaty between Mexico and the

TABLE 7.1 Some Major International Environmental Treaties

Convention or protocol	Year it came into force	Nations that have ratified it	U.S. status
Convention on International Trade in Endangered Species of Wild Fauna and Flora (CITES) (p. 305)	1975	175	Ratified
Ramsar Convention on Wetlands of International Importance (p. 414)	1975	159	Ratified
Protocol on Substances that Deplete the Ozone Layer (Montreal Protocol), of the Vienna Convention for the Protection of the Ozone Layer (p. 482)	1989	196	Ratified
Basel Convention on the Control of Transboundary Movements of Hazardous Wastes and Their Disposal (p. 637)	1992	172	Signed but has not ratified
Convention on Biological Diversity (p. 305)	1993	168	Signed but has not ratified
Stockholm Convention on Persistent Organic Pollutants (p. 396)	2004	152	Signed but has not ratified
Kyoto Protocol, of the UN Framework Convention on Climate Change (p. 520)	2005	184	Signed but has not ratified

United States. Many social, economic, and environmental issues along the U.S.–Mexican border are influenced by the 1994 **North American Free Trade Agreement (NAFTA)** among Mexico, Canada, and the United States (see **THE SCIENCE BEHIND THE STORY**, pp. 190–191).

Cross-border cooperation helps address environmental problems

Often nations make progress on international issues not through legislation or treaties, but through creative multilateral agreements hammered out after a lot of hard work and diplomacy. Such was the case with the Tijuana Master Plan for Water and Wastewater Infrastructure, an effort begun in 2002 to manage drinking water and wastewater in the Tijuana metropolitan area. In 2002, the U.S. Congress also funded the EPA to work with Mexican officials to upgrade Tijuana's sewer system. The Tijuana Sewer Rehabilitation Project, known as *Tijuana Sana* ("Healthy Tijuana"), is repairing leaky sewer pipes in Tijuana. The $43 million project aims to replace 131 km (81 mi), or 7.5%, of Tijuana's sewer pipes to reduce or eliminate the most severe sewage spills into the Tijuana River.

Collaborative efforts today continue with the Border 2012 program, a U.S.–Mexican effort to improve quality of life for people living on both sides of the border. Of the 12 million people living near the border, 90% live in 14 paired cities, among them San Diego and Tijuana. The program aims to improve environmental health and reduce pollution of water, air, and land. So far in the Tijuana region it has improved wastewater treatment and air quality monitoring, launched a $1.7 million ecological restoration project at the Tijuana River estuary, and cleaned up a lead-smelting *maquiladora* that was abandoned by its U.S. owner.

Several organizations shape international environmental policy

A number of international organizations regularly act to shape environmental policy and influence the behavior of nations by providing funding, applying peer pressure, and/or directing media attention.

The United Nations   In 1945, representatives of 50 countries founded the **United Nations (UN),** which today serves every nation in the world. Headquartered in New York City, this organization's purpose is "to maintain international peace and security; to develop friendly relations among nations; to cooperate in solving international economic, social, cultural and humanitarian problems and in promoting respect for human rights and fundamental freedoms; and to be a centre for harmonizing the actions of nations in attaining these ends."

The United Nations has taken an active role in shaping international environmental policy (**FIGURE 7.20**). Of several agencies within it that influence environmental policy, most notable is the *United Nations Environment Programme* (*UNEP*), created in 1972, which helps nations understand and solve environmental problems. Based in Nairobi, Kenya, its mission is sustainability, enabling countries and their citizens "to improve their quality of life without compromising that of

FIGURE 7.20 ▲ The United Nations is active in international environmental policymaking. Here, the executive director of UNEP (left) and the president of the UN Governing Council (right) converse with Indonesian President Susilo Bambang Yudhoyono at a 2010 meeting of national environment ministers in Bali, Indonesia.

future generations." UNEP's extensive research and outreach activities provide a wealth of information useful to policymakers and scientists throughout the world.

The World Bank   Established in 1944 and based in Washington, D.C., the **World Bank** is one of the globe's largest sources of funding for economic development. This institution shapes environmental policy through its funding of dams, irrigation infrastructure, and other major development projects. In fiscal year 2009, the World Bank provided $47 billion in loans and support for projects designed to benefit the poorest people in the poorest countries around the world.

Despite its admirable mission, the World Bank has frequently been criticized for funding unsustainable projects that cause more environmental problems than they solve. Providing for the needs of growing human populations in poor nations while minimizing damage to the environmental systems on which people depend can be a tough balancing act. Environmental scientists today agree that the concept of sustainable development must be the guiding principle for such efforts.

The European Union   The **European Union (EU)** seeks to promote Europe's unity and its economic and social progress (including environmental protection) and to "assert Europe's role in the world." The EU can sign binding treaties on behalf of its 27 member nations and can enact regulations that have the same authority as national laws in each member nation. It can also issue *directives*, which are more advisory in nature. The EU's European Environment Agency works to address waste management, noise pollution, water pollution, air pollution, habitat degradation, and natural hazards. The EU also seeks to remove trade barriers among member nations. It has classified some nations' environmental regulations as barriers to trade because some northern European nations have traditionally had more stringent environmental laws that prevent the import and sale of environmentally harmful products from other member nations.

The SCIENCE behind the Story

Assessing the Environmental Impacts of NAFTA

A number of international treaties address environmental issues (see Table 7.1, p. 188). But treaties not aimed specifically at environmental concerns may still have major environmental consequences. The North American Free Trade Agreement (NAFTA) is one such treaty.

Mexico, the United States, and Canada signed NAFTA to promote free trade among them. NAFTA, which came into force in 1994, eliminated trade barriers such as tariffs on imports and exports. Nations erect tariffs in order to raise prices on foreign goods so that competition from foreign industries won't drive domestic industries out of business. But if nations can agree to mutually eliminate tariffs, it makes goods cheaper for everyone.

Yet many people worried that NAFTA also threatened to undermine protections for workers and the environment. For instance, if a nation's regulations to protect environmental quality or worker safety are viewed as a barrier to trade or investment, under NAFTA those regulations can potentially be overturned.

Workers assemble television circuit boards for Panasonic in a maquiladora in Tijuana, Mexico.

Moreover, many people felt that industries, motivated to decrease costs and increase profits, would move their factories (and jobs) to the nation with the weakest regulations. This could create "pollution havens." Many people thought that Mexico would be overrun by *maquiladoras* seeking to profit from lax regulation and that Mexico would thereby suffer intensive pollution (**see figure**). Once such a migration began, people predicted, this could lead to a "race to the bottom," whereby all three nations would begin gutting their regulations in an attempt to lure business. This was also the reason that NAFTA caused so many blue-collar U.S. workers to fear that their jobs would migrate to Mexico.

In response to these fears, two side agreements for labor and environmental concerns were negotiated. The environmental agreement, the North American Agreement on Environmental Cooperation, set up a tri-national Commission on Environmental Cooperation (CEC). The CEC has monitored NAFTA's effects on the environment over the years, testing several hypotheses about impacts NAFTA was predicted to have. The CEC has held four symposia, for which dozens of researchers have published over 50 research papers analyzing different aspects of the topic.

This research suggests that, for the most part, the feared consequences have not occurred or have not been due to NAFTA. In 2006, Chantal Line Carpentier, head of the CEC's Environment, Economy, and Trade Program, summarized 11 years of studies as follows: "Environmental effects have been neither very bad nor very good, and are policy-dependent."

Researchers testing for a "race to the bottom" did not find evidence for it. Instead, they found that several

The World Trade Organization Based in Geneva, Switzerland, the **World Trade Organization (WTO)** represents multinational corporations and promotes free trade by reducing obstacles to international commerce and enforcing fairness among nations in trading practices. Whereas the United Nations and the European Union have limited influence over nations' internal affairs, the WTO has real authority to impose financial penalties on nations that do not comply with its directives. These penalties can sometimes play major roles in shaping environmental policy.

Like the EU, the WTO has interpreted some national environmental laws as unfair barriers to trade. For instance, in 1995 the U.S. EPA issued regulations requiring cleaner-burning gasoline in U.S. cities, following Congress's amendments of the Clean Air Act. Brazil and Venezuela filed a complaint with the WTO, saying the new rules unfairly discriminated against the petroleum they exported to the United States, which did not burn as cleanly. The WTO agreed, ruling that even though the South American gasoline posed a threat to human health in the United States, the EPA rules represented an illegal trade barrier. The ruling forced the United States to alter its approach to regulating gasoline. Not surprisingly, critics have frequently charged that the WTO aggravates environmental problems.

Maquiladoras in the border areas of Mexico flourished in the years after NAFTA, employing many people but creating substantial pollution. Perhaps the worst pollution occurred at the site shown here. At this abandoned lead recycling plant, *Metales y Derivados*, the U.S. owner left 6,600 tons of hazardous waste amid a working-class Tijuana neighborhood.

reducing hazards. The three nations began conferring on how to protect threatened species and habitats.

Moreover, consumer demand in the United States and Canada for sustainably produced products helped encourage improvement in Mexico. *Maquiladoras* selling products north of the border made more environmental improvements than those selling only in Mexico, researchers found. Many Mexican coffee farmers converted to sustainable plantations (p. 273).

However, despite all these improvements in efficiency, and despite the fact that NAFTA spurred Mexico's government to strengthen its environmental regulation, environmental impacts in Mexico grew worse after NAFTA. For instance, air and water pollution from *maquiladoras* in border areas such as Tijuana increased greatly. Yet, researchers determined that this was due not to a pollution haven or race to the bottom but to the fast increase in economic growth in Mexico. Economic growth simply outpaced the country's ability to enhance regulation and environmental protection.

Overall, CEC-sponsored research in the years since NAFTA has shown that trade liberalization can lead to environmental improvements, but only if policymakers pay close attention to trends and are ready to make adjustments. Opportunities for creating win-win policies that benefit both trade and environmental quality exist, researchers say, and are ready to be seized. ∎

measures of environmental quality improved during the 1990s and that NAFTA did not seem to influence these measures.

Researchers also have not found strong evidence for creation of a "pollution haven" in Mexico. One reason is that businesses have many factors to weigh when considering whether to relocate, and environmental regulations are merely one of these. Ironically, the one clear case of a pollution haven occurred not in Mexico, but in Canada! Exports of hazardous waste (mostly from steel and chemical factories) from the United States to Canada more than quadrupled soon after NAFTA went into effect. Disposal was cheaper in Canada, with fewer regulations and liability concerns. Canada responded by tightening its regulations.

CEC-sponsored researchers also tested hypotheses that NAFTA might enhance sustainability by facilitating the spread of environmentally superior technology, products, and approaches among nations. Researchers found plenty of examples. Canada began using the EPA's Energy Star program (pp. 518, 557), and Mexico banned the pesticide DDT. Chemical manufacturers set up a transnational program for

WEIGHING THE ISSUES

Trade Barriers and Environmental Protection If Nation A has stricter laws for environmental protection than Nation B, and if these laws restrict the ability of Nation B to export its goods to Nation A, then by the policy of the WTO and the EU, Nation A's environmental protection laws could be overruled in the name of free trade. Do you think this is right? What if Nation A is a wealthy industrialized country and Nation B is a poor developing country that needs every economic boost it can get?

Nongovernmental organizations A number of nongovernmental organizations (NGOs) have become international in scope and exert influence over international environmental policy (**FIGURE 7.21**). These groups are diverse in their size and mission; those that advocate for aspects of environmental protection are known as environmental NGOs ("ENGOs"). Groups such as the Nature Conservancy focus on accomplishing conservation objectives on the ground (in its case, purchasing and managing land and habitat for rare species) without becoming politically involved. Other groups, such as Conservation

FIGURE 7.21 ▲ Pursuing different visions of what makes for international environmental progress, nongovernmental organizations, such as the environmental advocacy group Greenpeace, sometimes clash with international institutions, such as the World Bank and the World Trade Organization.

International, the World Wide Fund for Nature, Greenpeace, and Population Connection, attempt to shape policy through research, education, lobbying, or protest. NGOs apply more funding and expertise to environmental problems—and conduct more research intended to solve them—than do many national governments.

International institutions wield influence in a globalizing world

As globalization proceeds, our world becomes ever more interconnected. As a result, human societies and Earth's ecological systems are being altered at unprecedented rates. Trade and technology have expanded the global reach of all societies, especially those such as the United States that consume resources from across the world. Highly consumptive nations that import goods and resources from far and wide exert extensive impacts on the planet's environmental systems. Multinational corporations operate outside the reach of national laws and rarely have incentive to conserve resources or conduct their business sustainably in the nations where they operate. For all these reasons, in today's globalizing world the organizations and institutions that shape international policy are becoming increasingly influential.

➤ CONCLUSION

Environmental policy is a problem-solving pursuit that makes use of science, ethics, and economics and that requires an astute understanding of the political process. Conventional command-and-control approaches of legislation and regulation remain the most common approaches to policymaking, but tort law retains influence, and innovative market-based policy tools are increasingly being developed. The United States has historically often led the way with environmental policy, but as we have seen in the case of the Tijuana River, environmental issues often span political boundaries and require international cooperation. Through the hard work of concerned citizens interacting with their government representatives, the political process is gradually producing promising solutions in the Tijuana River valley.

We will draw on the fundamentals of environmental policy introduced in this chapter throughout the remainder of this book. By integrating these fundamentals with your knowledge of the natural sciences, you will be well equipped to develop your own creative solutions to many of the challenging problems we will encounter.

REVIEWING OBJECTIVES

You should now be able to:

DESCRIBE ENVIRONMENTAL POLICY AND ASSESS ITS SOCIETAL CONTEXT

- Policy is a tool for decision making and problem solving that makes use of information from science and values from ethics and economics. (p. 168)
- Environmental policy aims to protect natural resources and environmental amenities from degradation or depletion and to promote equitable treatment of people. It addresses the tragedy of the commons, free riders, and external costs. (pp. 169–170)

IDENTIFY THE INSTITUTIONS IMPORTANT TO U.S. ENVIRONMENTAL POLICY AND RECOGNIZE MAJOR U.S. ENVIRONMENTAL LAWS

- The legislative, executive, and judicial branches, together with administrative agencies, all play roles in U.S. environmental policy. (p. 171)
- State and local governments also implement environmental policy. (p. 172)
- The concept of a regulatory taking arises from the Fifth Amendment. (pp. 172–173)
- U.S. environmental policy came in three waves. The first promoted frontier expansion and resource extraction.

The second aimed to mitigate impacts of the first through conservation. The third targeted pollution and gave us many of today's major environmental laws. (pp. 173–176)

- Some major U.S. laws include the National Environmental Policy Act, the Clean Air Act, and the Clean Water Act. (pp. 176–177)
- Currently, a fourth wave of environmental policy, occurring internationally, may be building around sustainable development and addressing global climate change. (p. 178)

CATEGORIZE THE DIFFERENT APPROACHES TO ENVIRONMENTAL POLICY

- Tort law has been a traditional approach to resolving environmental disputes. (pp. 178–179)
- Legislation from Congress and regulations from administrative agencies make up most federal policy. These top-down approaches are referred to as command-and-control. (pp. 179–180)
- Shortcomings of the command-and-control approach have led many economists to advocate economic policy tools. (p. 181)
- Economic policy tools include green taxes, subsidies, and market-based approaches such as ecolabeling and permit trading. (pp. 182–184)
- Public-private partnerships are challenging to make work, but they can offer a promising approach. (pp. 184–185)

DELINEATE THE STEPS OF THE ENVIRONMENTAL POLICY PROCESS AND EVALUATE ITS EFFECTIVENESS

- The policy process entails (1) identifying a problem, (2) pinpointing causes of the problem, (3) envisioning solutions, (4) getting organized, (5) gaining access to power, (6) guiding a solution into law, and (7) implementing, assessing, and interpreting the policy. (pp. 185–187)
- In a democracy, anyone can use the policy process, but corporations and organizations with money and resources tend to have the most clout. (pp. 186–187)

DISCUSS THE ROLE OF SCIENCE IN THE POLICY PROCESS

- Data from scientific research are vital for informing policy. (pp. 187–188)
- Policymakers may sometimes ignore or distort science for political ends, so we in the public need to remain vigilant. (p. 188)

LIST THE INSTITUTIONS INVOLVED WITH INTERNATIONAL ENVIRONMENTAL POLICY AND DESCRIBE HOW NATIONS HANDLE TRANSBOUNDARY ISSUES

- Many environmental problems cross political boundaries and must be addressed internationally. (pp. 188–189)
- International policy includes customary law (law by shared traditional custom) and conventional law (law by treaty). (p. 188)
- Institutions such as the United Nations, World Bank, European Union, World Trade Organization, and nongovernmental organizations all shape international policy. (pp. 189–192)

TESTING YOUR COMPREHENSION

1. Describe two common justifications for environmental policy, and discuss three problems that environmental policy commonly seeks to address.
2. Outline the primary responsibilities of the legislative, executive, and judicial branches of the U.S. government. What is the "fourth branch" of the U.S. government?
3. What is meant by a *regulatory taking*?
4. Summarize the differences among the first, second, and third waves of environmental policy in U.S. history. Describe what now appears to be the fourth wave.
5. What did the National Environmental Policy Act accomplish? Briefly describe the origin and mission of the U.S. Environmental Protection Agency.

6. Compare and contrast the three major approaches to environmental policy: tort law, command-and-control, and economic policy tools.
7. Explain how each of the following work: a subsidy, a green tax, and marketable emissions permits.
8. Describe the environmental policy process, from identification of a problem through the implementation, assessment, and interpretation of policy.
9. What is the difference between customary law and conventional law? What special difficulties do transboundary environmental problems present?
10. Why are environmental regulations sometimes considered to be unfair barriers to trade?

SEEKING SOLUTIONS

1. Many free-market advocates maintain that environmental laws and regulations are an unnecessary government intrusion into private affairs. As you may recall from Chapter 6 (p. 149), Adam Smith argued that individuals can benefit society by pursuing their own self-interest. Do you agree? Can you describe a situation in which an individual acting in his or her self-interest could harm society by causing an environmental problem? Can you describe how environmental policy might rectify the situation? What are some advantages and disadvantages of instituting environmental laws and regulations, versus allowing unfettered exchange of materials and services?

2. Reflect on the causes for the transitions in U.S. history from one type of environmental policy to another. Now peer into the future, and think about how life might be different in 25, 50, or 100 years. What would you speculate about the environmental policy of the future? What issues might it address? Do you predict we will have more or less environmental policy?

3. Consider the main approaches to environmental policy—tort law, command-and-control, and economic policy tools. Describe an advantage and a disadvantage of each. Do you think any one approach is most effective? Could we do with just one approach, or does it help to have more than one?

4. Think of one environmental problem that you would like to see solved. From what you've learned about the policy-making process, describe how you think you could best shepherd your ideas through the process to address this problem.

5. Compare the roles of the United Nations, the World Bank, the European Union, the World Trade Organization, and nongovernmental organizations. If you could gain the support of just one of these institutions for a policy you favored, which would you choose? Why?

6. **THINK IT THROUGH** You have just been elected to the U.S. House of Representatives. People in your district suffer an unusual amount of water pollution from untreated municipal wastewater, chemical discharges from factories, and oil spillage from commercial and recreational boats. What policy approaches would you choose to pursue to search for solutions to these problems? Give reasons for your choices.

CALCULATING ECOLOGICAL FOOTPRINTS

The League of Conservation Voters (LCV) is a nongovernmental organization that tracks the voting records of U.S. policymakers on environmental issues and provides information to citizens. Go to the LCV's website at www.lcv.org. Under "Scorecard," click on the link for "Real-Time Vote Tracker", and find the representatives for your state. The Ys and Ns show how they voted on key bills. To read descriptions of the bills, click on the links for those bills. Use this information to address Question 1 and Question 2 below.

Next, click on the link under "Scorecard" for the previous year, and then click on the link to "Party Leaders' Scores." This should enable you to explore the voting records of the Congressional Leadership, the five most powerful members of the House and the Senate. Pick one of them to use for Question 3.

Voting Records of Your Congressional Representatives		
	Name of Congressperson	How he/she voted on your bill of interest
A House representative from your state		
A U.S. senator from your state		
A member of the Congressional Leadership		

1. After exploring the voting records of your state's representatives in the House, find one vote that you disagree with. In the first row of the table, write the representative's name, and describe his or her vote on your bill of interest. Now outline a letter that you would write to this representative explaining why you feel he or she should support your position.

2. Next, explore the voting records of your state's two U.S. senators, and find one vote that you disagree with. In the second row of the table, write the senator's name, and describe his or her vote on your bill of interest. Outline a letter that you would write to this senator explaining why you feel he or she should support your position.

3. Finally, explore the voting records of the five Congressional leaders, and find one vote that you disagree with. In the third row of the table, write the congressperson's name, and describe his or her vote on your bill of interest. Outline a letter that you would write to this Congressional leader explaining why you feel he or she should support your position.

4. Select one of the bills you see described, and explain how you think passage of the bill might influence the ecological footprint of people in your state.

Mastering ENVIRONMENTALSCIENCE™

Go to **www.masteringenvironmentalscience.com** for practice quizzes, Pearson eText, videos, current events, and more.

ENVIRONMENTAL ISSUES AND THE SEARCH FOR SOLUTIONS

Workers attach sensors to a wind turbine blade at a National Renewable Energy Laboratory research center in Boulder, Colorado.

Crowded street in Guangzhou, one of China's largest cities

8 HUMAN POPULATION

UPON COMPLETING THIS CHAPTER, YOU WILL BE ABLE TO:

- Perceive the scope of human population growth
- Assess divergent views on population growth
- Evaluate how human population, affluence, and technology affect the environment
- Explain and apply the fundamentals of demography
- Outline and assess the concept of demographic transition

- Describe how wealth and poverty, the status of women, and family planning affect population growth
- Characterize the dimensions of the HIV/AIDS epidemic
- Link population goals to sustainable development goals

China's One-Child Policy

"Population growth is analogous to a plague of locusts. What we have on this earth today is a plague of people."

—Ted Turner, media magnate and supporter of the United Nations Population Fund

"As you improve health in a society, population growth goes down. . . . Before I learned about it, I thought it was paradoxical."

—Bill Gates, Chair, Microsoft Corporation

ASIA

EUROPE

China

AFRICA

Indian
Ocean

AUSTRALIA

The People's Republic of China is the world's most populous nation, home to one-fifth of the 6.9 billion people living on Earth at the start of 2011.

When Mao Zedong founded the country's current regime 62 years earlier, roughly 540 million people lived in a mostly rural, war-torn, impoverished nation. Mao believed population growth was desirable, and under his rule China grew and changed. By 1970, improvements in food production, food distribution, and public health allowed China's population to swell to 790 million people. At that time, the average Chinese woman gave birth to 5.8 children in her lifetime.

However, the country's burgeoning population and its industrial and agricultural development were eroding the nation's soils, depleting its water, leveling its forests, and polluting its air. Chinese leaders realized that the nation might not be able to feed its people if their numbers grew much larger. They saw that continued population growth could exhaust resources and threaten the stability and progress of Chinese society. The government decided to institute a population-control program that prohibited most Chinese couples from having more than one child.

The program began with education and outreach efforts encouraging people to marry later and have fewer children (**FIGURE 8.1**). Along with these efforts, the Chinese government increased the availability of

Mother with one son in northern China

contraceptives and abortion. By 1975, China's annual population growth rate had dropped from 2.8% to 1.8%.

To further decrease the birth rate, in 1979 the government took the more drastic step of instituting a system of rewards and punishments to enforce a one-child limit. One-child families received better access to schools, medical care, housing, and government jobs, and mothers with only one child were given longer maternity leaves. In contrast, families with more than one child were subjected to monetary fines, employment discrimination, and social scorn and ridicule. In some cases, the fines exceeded half of a couple's annual income.

Population growth rates dropped still further, but public resistance to the policy was simmering. Beginning in 1984, the one-child policy was loosened, strengthened, and then loosened again as government leaders explored ways to maximize population control while minimizing public opposition. Today the one-child program is less strict than in past years and applies mostly to families in urban areas. Many farmers and ethnic minorities in rural areas are exempted, because success on the farm often depends on having multiple children.

In enforcing its policies, China has been conducting one of the largest and most controversial social experiments in history. In purely quantitative terms, the experiment has been a major success: The nation's

CHAPTER 8 Human Population

FIGURE 8.1 ▲ Billboards and murals like this one in the Chinese city of Chengdu promote the national one-child policy. Although the mural shows a baby girl, the policy has resulted in a preference for baby boys, like this one being toted by his grandfather.

growth rate is now down to 0.5%, making it easier for the country to deal with its many social, economic, and environmental challenges.

However, the one-child policy has also produced unintended consequences. Traditionally, Chinese culture has valued sons because they carry on the family name, assist with farm labor in rural areas, and care for aging parents. Daughters, in contrast, will most likely marry and leave their parents, as the culture dictates. As a result, they cannot provide the same benefits to their parents as will sons. Thus, faced with being limited to just one child, many Chinese couples prefer a son to a daughter. Tragically, this has led to selective abortion, killing of female infants, an unbalanced sex ratio, and a black-market trade in teenaged girls for young men who cannot find wives.

Further problems are expected in the near future, including an aging population and a shrinking workforce. Moreover, China's policies have elicited criticism worldwide from people who oppose government intrusion into personal reproductive choices.

As other nations become more and more crowded, might their governments also feel forced to turn to drastic policies that restrict individual freedoms? In this chapter, we examine human population dynamics worldwide, consider their causes, and assess their consequences for the environment and our society.

OUR WORLD AT SEVEN BILLION

While China works to slow its population growth, numbers of people continue to rise in most nations of the world. Most population growth is occurring in poverty-stricken developing nations that are ill-equipped to handle it. India is on course to surpass China as the world's most populous nation in coming decades (**FIGURE 8.2**). Although the *rate* of human population growth is slowing, we are still increasing in absolute numbers, and sometime during the year 2011 our global population will surpass 7 billion.

Just how much is 7 billion? We often have trouble conceptualizing huge numbers. Keep in mind that 1 billion is 1,000 times greater than 1 million. If you were to count once each second without ever sleeping, it would take over 30 years to reach a billion. To put a billion miles on your car, you

FIGURE 8.2 ▲ Population growth is driving our global numbers to 7 billion and beyond. India is on course to surpass China soon as the world's most populous nation. In this photo, Indian women wait in line for immunizations for their babies.

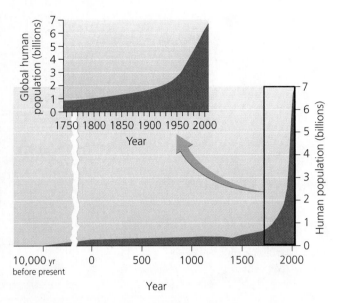

FIGURE 8.3 ▲ Viewing global human population size over a long time scale (bottom graph) and since the industrial revolution (inset top graph) shows how nearly all growth has occurred in just the past 200 years. We have risen from fewer than 1 billion in 1800 to nearly 7 billion today. Data from U.S. Bureau of Census.

would need to drive from New York City to Los Angeles more than 350,000 times. To travel 7 billion miles you'd need to make 29,000 trips to the Moon.

The human population is growing rapidly

Our global population grows by over 80 million people each year. This is the equivalent of adding all the people of California, Texas, and New York to the world annually—and it means that we add 2.6 people to the planet *every second*. Take a look at **FIGURE 8.3** and note just how recent and sudden our rapid increase has been. It took until after 1800, virtually all of human history, for our population to reach 1 billion. Yet we reached 2 billion by 1930, and 3 billion in just 30 more years. Our population added its next billion in just 15 years, and it has taken only 12 years to add each of the next three installments of a billion people. Consider when you were born and estimate the number of people added to the planet just since that time. No previous generations have *ever* lived amid so many other people.

You may not feel particularly crowded, but consider that we in the United States are not experiencing as much population growth and density as people in most other nations. In today's world, rates of annual growth vary greatly from region to region (**FIGURE 8.4**) and are highest in developing nations.

What accounts for our unprecedented growth? We saw in Chapter 3 (pp. 66–67) how exponential growth—the increase in a quantity by a fixed percentage per unit time—accelerates increase in population size, just as compound interest accrues in a savings account. The reason, you will recall, is that a fixed percentage of a small number makes for a small increase, but that the same percentage of a large number produces a large increase. Thus, even if the growth *rate* remains steady, population *size* will increase by greater increments with each successive generation.

For much of the 20th century, the growth rate of the human population rose from year to year. This rate peaked at 2.1% during the 1960s and has declined to 1.2% since then. Although 1.2% may sound small, exponential growth endows

small numbers with large consequences. A hypothetical population starting with one man and one woman that grows at 1.2% gives rise to a population of 2,939 after 40 generations and 112,695 after 60 generations.

At a 1.2% annual growth rate, a population doubles in size in just 58 years. We can roughly estimate doubling times with a handy rule of thumb. Just take the number 70 (which is 100 times 0.7, the natural logarithm of 2) and divide it by the annual percentage growth rate: 70/1.2 = 58.3. Had China not instituted its one-child policy, and had its growth rate remained at 2.8%, it would have taken only 25 years to double in size.

To help you visualize just how fast our world population has grown—and will continue to grow in your lifetime—take a look at the **ENVISIONIT** feature on p. 204. In each chapter from now on, you will find an *EnvisionIt* page that uses visual images to bring to life important concepts in environmental science.

Is population growth a problem?

Our spectacular growth in numbers has resulted largely from technological innovations, improved sanitation, better medical care, increased agricultural output, and other factors that have brought down death rates and infant mortality rates. Birth rates have not declined as much, so births have outpaced deaths for many years now. Thus, our population explosion has arisen from a very good thing—our ability to keep more of our fellow human beings alive longer! Why then do so many people view population growth as a problem?

Let's start with a bit of history. At the outset of the industrial revolution (p. 4), population growth was universally regarded as a good thing. For parents, high birth rates meant more children to support them in old age. For society, it meant a greater pool of labor for factory work. However,

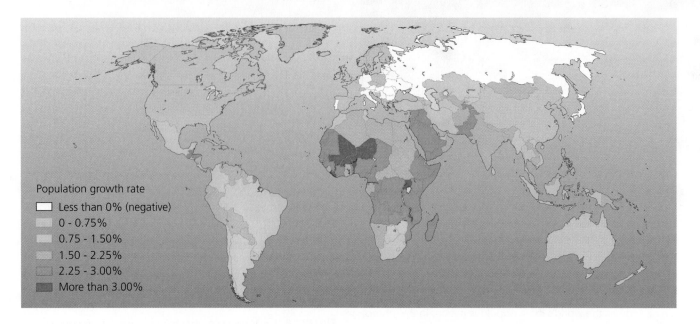

Population growth rate

- [ ] Less than 0% (negative)
- 0 - 0.75%
- 0.75 - 1.50%
- 1.50 - 2.25%
- 2.25 - 3.00%
- More than 3.00%

FIGURE 8.4 ▲ Population growth rates vary greatly from place to place. Population is growing fastest in poorer nations of the tropics and subtropics but is now beginning to decrease in some northern industrialized nations. Shown are natural rates of population change (p. 66) as of 2010. Data from Population Reference Bureau, 2010. *2010 World population data sheet.* By permission.

British economist **Thomas Malthus** (1766–1834) had a different view. Malthus (**FIGURE 8.5A**) argued that unless population growth were controlled by laws or other social strictures, the number of people would eventually outgrow the available food supply. Malthus's most influential work, *An Essay on the Principle of Population*, published in 1798, argued that if society did not limit births (through abstinence and contraception, for instance), then rising death rates would reduce the population through war, disease, and starvation.

In our day, biologists Paul and Anne Ehrlich of Stanford University have been called "neo-Malthusians" because they too have warned that our population may grow faster than our ability to produce and distribute food. In his best-selling 1968 book, *The Population Bomb*, Paul Ehrlich (**FIGURE 8.5B**) predicted that population growth would unleash famine and conflict that would consume civilization by the end of the 20th century.

Although human population quadrupled in the past 100 years—the fastest it has ever grown (see Figure 8.3)—Ehrlich's forecasts have not fully materialized. This is due, in part, to the way we have intensified food production in recent decades (pp. 254–255). Population growth has indeed contributed to famine, disease, and conflict—but as we shall see, enhanced prosperity, education, and gender equality have also helped to reduce birth rates.

Does this mean we can disregard the concerns of Malthus and Ehrlich? Some Cornucopians (p. 18) say yes. Under the Cornucopian view that many economists hold, population growth poses no problem if new resources can be found or created to replace depleted ones (p. 150). In contrast, environmental scientists recognize that not all resources can be replaced. Once a species has gone extinct, for example, we cannot replicate its exact functions in an ecosystem or know what benefits it might have provided us. Land, too, is irreplaceable; we cannot expand Earth like a balloon to increase the space we have in which to live.

Even if resource substitution could hypothetically enable our population to grow indefinitely, could we maintain the *quality* of life that we desire for ourselves and our descendants? Unless the availability and quality of all resources keeps pace forever with population growth, the average person in the future will have less space in which to live, less food to eat, and less material wealth than the average person does today. Thus, population growth is indeed a problem if it depletes resources, stresses social systems, and degrades the natural environment, such that our quality of life declines (**FIGURE 8.6**).

Some national governments now fear falling populations

Despite these considerations, many policymakers find it difficult to let go of the notion that population growth increases a nation's economic, political, and military strength. This notion

(a) Thomas Malthus   **(b) Paul Ehrlich**

FIGURE 8.5 ▲ Thomas Malthus **(a)** and Paul Ehrlich **(b)** each argued that runaway population growth would surpass food supply and lead to disaster.

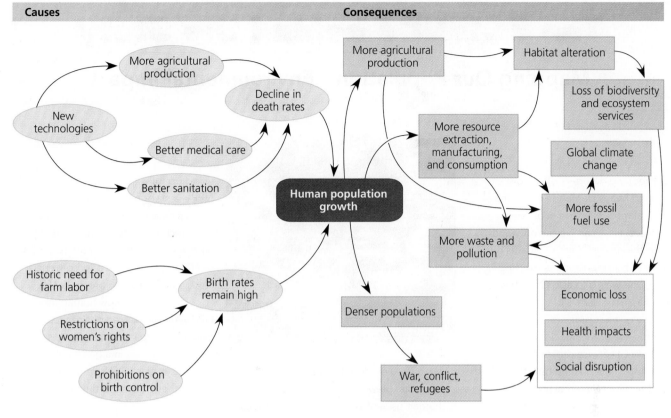

FIGURE 8.6 ▲ Human population growth has diverse causes (ovals on left) and consequences (boxes on right), as we shall see throughout this chapter. The consequences of rapid population growth are generally negative for society and the environment. Arrows in this concept map lead from causes to consequences. Note that items grouped within outlined boxes do not necessarily share any special relationship; the outlined boxes are merely to streamline the figure.

Solutions

As you progress through this chapter, try to identify as many solutions to human population growth as you can. What could you personally do to help address this issue? Consider how each action or solution might affect items in the concept map above.

has held sway despite the clear fact that in today's world, population growth is correlated with poverty, not wealth: Strong and wealthy nations tend to have slow population growth, whereas nations with fast population growth tend to be weak and poor. While China and India struggle to get their population growth under control, many other national governments are offering financial and social incentives that encourage their own citizens to produce more children.

Much of the concern is that when birth rates decline, a population grows older. In aging populations, larger numbers of elderly people will need social services, but fewer workers will be available to pay taxes to fund these services. Moreover, in nations with declining birth rates that are also accepting large numbers of foreign immigrants, some native-born citizens may worry that their country's culture is at risk of being diluted or lost. For both these reasons, governments of nations now experiencing population declines (such as many in Europe) feel uneasy. According to the Population Reference Bureau, two of every three European governments now feel their birth rates are too low, and none states that its country's birth rate is too high. However, outside Europe, 49% of national governments still feel their birth rates are too high, and only 12% feel they are too low.

Population is one of several factors that affect the environment

One widely used formula gives us a handy way to think about population and other factors that affect environmental quality. Nicknamed the **IPAT model**, it is a variation of a formula proposed in 1974 by Paul Ehrlich and John Holdren, a Harvard University environmental scientist who today is President Obama's science advisor. The IPAT model represents how our total impact (I) on the environment results from the interaction among population (P), affluence (A), and technology (T):

$$I = P \times A \times T$$

Increased population intensifies impact on the environment as more individuals take up space, use resources, and generate waste. Increased affluence magnifies environmental impact through the greater per capita resource consumption that generally has accompanied enhanced wealth. Technology that enhances our abilities to exploit minerals, fossil fuels, old-growth forests, or fisheries generally increases impact, but technology to reduce smokestack emissions, harness renewable energy, or improve manufacturing efficiency can decrease impact.

Mapping Our Population's Environmental Impact

Dr. Helmut Haberl of Austria's Institute for Social Ecology

Burgeoning numbers of people are making heavy demands on Earth's natural resources and ecosystem services. How can we quantify and map the environmental impacts our expanding population is exerting?

One way is to ask. Of all the biomass that Earth's plants can produce, what proportion do human beings use (for food, clothing, shelter, etc.), or otherwise prevent from growing? This was the question asked by nine environmental scientists led by Helmut Haberl of the Institute of Social Ecology in Austria. They teamed up to measure our consumption of *net primary production* (NPP; p. 116), the net amount of energy stored in plant matter as a result of photosynthesis. Human overuse of NPP diminishes resources for other species; alters habitats, communities, and ecosystems; and threatens our future ability to derive ecosystem services.

Haberl's team began with a well-established model that maps how vegetation varies with climate across the globe and used it to produce a detailed world map of "potential NPP"—vegetation that would exist if there were no human influence. The team then gathered data for the

year 2000 on crop harvests, timber harvests, grazing pressure, and other human uses of vegetation from various global databases from the United Nations Food and Agriculture Organization (FAO) and other sources. They also gathered data on how people affect vegetation indirectly, such as through fires, erosion and soil degradation, and other changes due to land use. To calculate the proportion of NPP that people appropriate, the researchers divided the amounts used up in these impacts by the total "potential" amount.

When all the data crunching was done, Haberl's group concluded that people harvest 12.5% of global NPP and that land use reduces it 9.6% further and fires 1.7% further (see **first figure**). This makes us responsible for

using up fully 23.8% of the planet's NPP—a staggeringly large amount for just a single species! Half of this use occurred on cropland, where 83.5% of NPP was used. In urban areas, 73.0% of NPP was consumed; on grazing land, 19.4%; and in forests, 6.6%.

To determine how human use of NPP varies across regions of the world, Haberl's group layered the data sets atop one another in a geographic information systems (GIS) approach (p. 120). Again, they calculated the proportion of

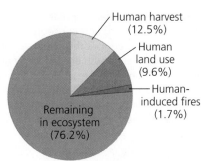

Humanity uses or causes Earth to lose 23.8% of the planet's net primary production. Direct harvesting (of crops, timber, etc.) accounts for most of this, and land use impacts and fire also contribute. Data from Haberl, H., et al., 2007. Quantifying and mapping the human appropriation of net primary production in Earth's terrestrial ecosystems. *Proc. Natl. Acad. Sci.* 104:12942–12947.

We might also add a sensitivity factor (S) to the equation to denote how sensitive a given environment is to human pressures:

$$I = P \times A \times T \times S$$

For instance, the arid lands of western China are more sensitive to human disturbance than the moist regions of southeastern China. Plants grow more slowly in the arid west, making the land more vulnerable to deforestation and soil degradation. Thus, adding an additional person to western

China has more environmental impact than adding one to southeastern China.

We could refine the IPAT equation further by adding terms for the effects of social institutions such as education, laws and their enforcement, stable and cohesive societies, and ethical standards that promote environmental well-being. Such factors all affect how population, affluence, and technology translate into environmental impact.

Impact can be thought of in various ways, but we can generally boil it down either to pollution or resource consumption.

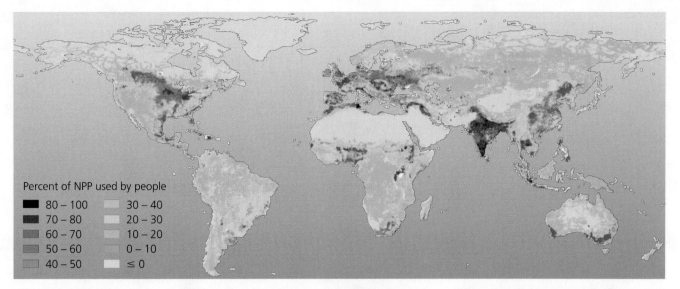

The proportion of Earth's net primary production that people appropriate varies from region to region. Regions that are densely populated or intensively farmed exert the heaviest impact. *Source:* Haberl, H., et al., 2007. Quantifying and mapping the human appropriation of net primary production in Earth's terrestrial ecosystems. *Proc. Natl. Acad. Sci.* 104:12942–12947, Fig 1b. © 2007 National Academy of Sciences, U.S.A. By permission.

NPP that we appropriate and produced a global map (see **second figure**). The researchers published their results in 2007 in the *Proceedings of the National Academy of Sciences of the USA*.

In their global map of NPP consumption, densely populated and heavily farmed regions such as India, eastern China, and Europe show the greatest proportional use of NPP. The influence of population is clear. For instance, although people in southern Asia consume very little per capita, dense populations here result in a 63% use of NPP. In contrast, in sparsely inhabited regions of the world (such as the boreal forest, Arctic tundra, Himalayas, and Sahara Desert), humans consume almost no NPP. In North America, NPP use is heaviest in the East, Midwest, and Great Plains. In general, the map shows heavy appropriation of NPP in areas where population is dense relative to the area's vegetative production.

The map does not fully show the effects of resource consumption due to affluence. Wealthy societies tend to import food, fiber, energy, and products from other places, and this consumption can drive environmental degradation in poorer regions. For instance, North Americans and Europeans import timber logged from the Amazon basin, as well as soybeans and beef grown in areas where Amazonian forest was cleared. Through global trade, we redistribute the products we gain from the planet's NPP. As a result, the environmental impacts of our consumption are often felt far from where we consume products.

By showing areas of high and low impact, maps like the one produced in this project can help us to make better decisions and minimize our impacts on ecosystems and ecosystem services. Haberl's team also says its data show that we are using a great deal of Earth's plant matter already, so that schemes to expand biomass energy production (pp. 576–583) may not be wise.

Environmental scientists who have commented on the paper agree and say the team's data should give us pause. As Jonathan Foley of the University of Wisconsin and his colleagues put it, "Ultimately, we need to question how much of the biosphere's productivity we can appropriate before planetary systems begin to break down. 30%? 40%? 50%? More? . . . Or have we already crossed that threshold?" ■

The depletion of resources by larger and hungrier populations has been a focus of scientists and philosophers since Malthus's time. Today, researchers calculate that humanity is appropriating for its own use nearly one-quarter of Earth's terrestrial net primary production (see **THE SCIENCE BEHIND THE STORY**, above).

One reason our population has kept growing, despite limited resources, is that we have developed technology—the T in the IPAT equation—time and again to increase efficiency, alleviate our strain on resources, and allow us to expand further. For instance, we have employed technological advances to increase global agricultural production faster than our population has risen (p. 253).

Modern-day China shows how all elements of the IPAT formula can combine to cause tremendous environmental impact in little time. The world's fastest-growing economy over the past two decades, China is "demonstrating what happens when large numbers of poor people rapidly become more affluent," in the words of Earth Policy Institute president Lester Brown. While millions of Chinese are increasing their

ENVISION IT

1900

Earth has become more crowded as our global population has grown.

1950

From 1900 to 1950 our numbers grew by 50%.

2000

Then our numbers exploded: In 2000 our population was 2.4 times greater than in 1950.

2050

For every two people alive in 2000, there will be three in 2050. Will enough natural resources remain to provide our children the standard of living we enjoy today?

YOU CAN MAKE A DIFFERENCE

➤ Support women's rights and education in developing nations.
➤ Even as population grows, reducing your resource consumption will lessen your ecological footprint.

204

material wealth and their resource consumption, the country is battling unprecedented environmental challenges brought about by its pell-mell economic development. Intensive agriculture has expanded westward out of the country's moist rice-growing areas, causing farmland to erode and literally blow away, much like the Dust Bowl tragedy that befell the U.S. heartland in the 1930s (p. 235). China has overpumped aquifers and has drawn so much water for irrigation from the Yellow River that the once-mighty waterway now dries up in many stretches. Although China is reducing its air pollution from industry and charcoal-burning homes, the country faces new urban pollution and congestion threats from rapidly rising numbers of automobiles. As the world's industrializing countries try to attain the material prosperity that industrialized nations enjoy, China is a window on what much of the rest of the world could soon become.

DEMOGRAPHY

It is a fallacy to think of people as being somehow outside nature. We exist within our environment as one species out of many. As such, all the principles of population ecology outlined in Chapter 3 that apply to toads, frogs, and passenger pigeons apply to humans as well. The application of principles from population ecology to the study of statistical change in human populations is the focus of **demography**.

Earth has a carrying capacity for us

Environmental factors set limits on our population growth, and as we saw in Chapter 3, the environment has a carrying capacity (p. 67) for our species, just as it does for every other. We happen to be a particularly successful species, however, and we have repeatedly raised this carrying capacity by developing technology to overcome the natural limits on our population growth.

Environmental scientists who have tried to pin a number to the human carrying capacity have come up with wildly differing estimates. The most rigorous estimates range from 1–2 billion people living prosperously in a healthy environment to 33 billion living in extreme poverty in a degraded world of intensive cultivation without natural areas. As our population climbs beyond 7 billion, we may yet continue to find ways to raise our carrying capacity. Given our knowledge of population ecology and logistic growth (p. 67), however, we have no reason to presume that human numbers can go on growing indefinitely.

Demography is the study of human population

Demographers study population size, density, distribution, age structure, sex ratio, and rates of birth, death, immigration, and emigration of people, just as population ecologists study these characteristics in other organisms. Each of these characteristics is useful for predicting population dynamics and environmental impacts.

Population size Our global human population of nearly 7 billion is spread among 200 nations with populations ranging up to China's 1.34 billion, India's 1.21 billion, and the 312 million of the United States (**FIGURE 8.7**). The United Nations Population Division estimates that by the year 2050 the global population

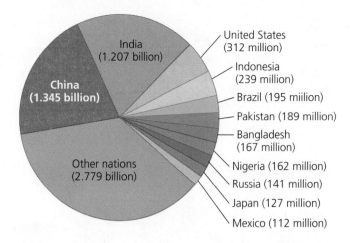

FIGURE 8.7 ▲ Almost one in five people in the world are Chinese, and more than one of every six live in India. Three of every five people live in one of the 11 nations that have populations above 100 million. Data are projected for mid-2011, based on Population Reference Bureau, 2010. *2010 world population data sheet.*

will surpass 9 billion (**FIGURE 8.8**). However, population size alone—the absolute number of individuals—doesn't tell the whole story. Rather, a population's environmental impact depends on its density, distribution, and composition (as well as on affluence, technology, and other factors outlined earlier).

Population density and distribution People are distributed unevenly over our planet. In ecological terms, our distribution is clumped (p. 64) at all spatial scales. At the

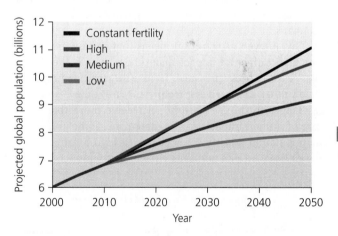

FIGURE 8.8 ▲ The United Nations predicts world population growth. In the latest projection (made in 2008), population is estimated to reach 11.0 billion in the year 2050 if fertility rates remain constant at 2008 levels (top line in graph). However, U.N. demographers expect fertility rates to continue falling, so they arrived at a best guess (*medium* scenario) of 9.15 billion for 2050. In the *high* scenario, if women on average have 0.5 child more than in the medium scenario, population will reach 10.5 billion in 2050. In the *low* scenario, if women have 0.5 child fewer than in the medium scenario, the world will contain 8.0 billion people in 2050. Adapted by permission from Population Division of the Department of Economic and Social Affairs of the United Nations Secretariat, 2009. *World population prospects: The 2008 revision,* http://esa.un .org/unpp, Fig 1. (c) United Nations, 2009.

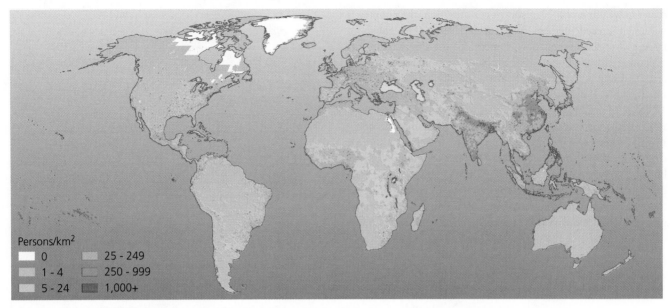

FIGURE 8.9 ▲ Human population density varies tremendously from one region to another. Arctic and desert regions have the lowest population densities, whereas areas of India, Bangladesh, and eastern China have the densest populations. The World: Population Density, 2000. Center for International Earth Science Information Network (CIESIN), Columbia University; and Centro Internacional de Agricultura Tropical (CIAT). 2005. Gridded Population of the World Version 3 (GPWv3). Palisdes, NY: Socioeconomic Data and Applications Center (SEDAC), Columbia University. By permission.

Persons/km²
- 0
- 1 - 4
- 5 - 24
- 25 - 249
- 250 - 999
- 1,000+

global scale (**FIGURE 8.9**), population density is highest in regions with temperate, subtropical, and tropical climates, such as China, Europe, Mexico, southern Africa, and India. Population density is lowest in regions with extreme-climate biomes, such as desert, rainforest, and tundra. Dense along seacoasts and rivers, human population is less dense away from water. At more local scales, we cluster together in cities and towns.

This uneven distribution means that certain areas bear more environmental impact than others. Just as the Yellow River experiences pressure from Chinese farmers, the world's other major rivers, from the Nile to the Danube to the Ganges to the Mississippi, all receive more than their share of human impact. At the same time, some areas with low population density are sensitive (a high S value in our revised IPAT model) and thus vulnerable to impact. Deserts

and arid grasslands, for instance, are easily degraded by development that commandeers too much water.

Age structure As we learned in Chapter 3 (p. 65), age structure describes the relative numbers of individuals of each age class within a population. Data on age structure are especially valuable to demographers trying to predict future dynamics of human populations. A population made up mostly of individuals past reproductive age will tend to decline over time. In contrast, a population with many individuals of reproductive age or pre-reproductive age is likely to increase. A population with an even age distribution will likely remain stable as births keep pace with deaths.

Age structure diagrams, often called *population pyramids*, are visual tools scientists use to illustrate age structure (**FIGURE 8.10**). The width of each horizontal bar represents

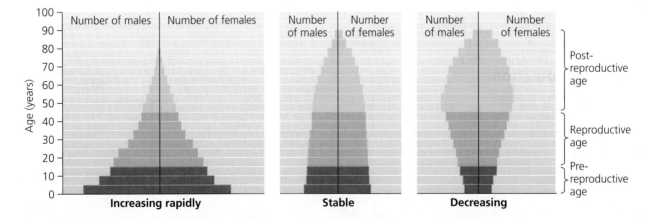

FIGURE 8.10 ▲ Age structure diagrams show numbers of individuals of different age classes in a population. A diagram like that on the left is weighted toward young age classes, indicating a population that will grow quickly. A diagram like that on the right is weighted toward old age classes, indicating a population that will decline. Populations with balanced age structures, like the one shown in the middle diagram, will remain stable.

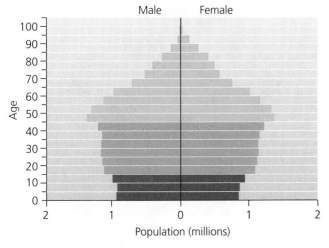

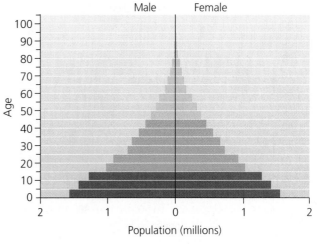

(a) Age pyramid of Canada in 2010

(b) Age pyramid of Madagascar in 2010

FIGURE 8.11 ▲ Canada **(a)** shows a fairly balanced age structure, with relatively even numbers of individuals in various age classes. Madagascar **(b)** shows an age distribution heavily weighted toward young people. Madagascar's population growth rate (2.9%) is over seven times greater than Canada's (0.4%). Data from Population Division of the Department of Economic and Social Affairs of the United Nations Secretariat, 2009. *World population prospects: The 2008 revision*, http://esa.un.org/unpp. © United Nations, 2009. By permission.

the number of people in each age class. A pyramid with a wide base denotes a large proportion of people who have not yet reached reproductive age—and this indicates a population soon capable of rapid growth. In this respect, a wide base of a population pyramid is like an oversized engine on a rocket—the bigger the booster, the faster the increase.

As an example, compare age structures for the nations of Canada and Madagascar (**FIGURE 8.11**). Madagascar's large concentration of individuals in young age groups portends a great deal of reproduction. Not surprisingly, Madagascar has a much greater population growth rate than does Canada.

Today, populations are aging in many nations (**FIGURE 8.12**). The global median age today is 28, but it will be 38 in the year 2050. Population aging is pronounced in the United States, where the "baby-boom" generation is now approaching retirement age. In coming years you will likely be asked to pay more taxes to support Social Security and Medicare benefits for your elders.

By causing dramatic reductions in the number of children born since 1970, China's one-child policy virtually guaranteed that the nation's population age structure would change. Indeed, in 1970 the median age in China was 20; by 2050 it will be 45. In 1970 there were more children under age 5 than people over 60 in China, but by 2050 there will be 12 times more people over 60 than under 5! Today there are 111 million Chinese people older than 65, but that number will triple by 2050 (**FIGURE 8.13**). This dramatic shift in age structure will challenge China's economy, health care system, families, and military forces because fewer working-age people will be available to support social programs to assist the rising number of older people.

On the one hand, older populations will present new challenges for many nations as increasing numbers of elderly will require the care and assistance of relatively fewer working-age citizens.

On the other hand, a shift in age structure toward an older population reduces the proportion of dependent children. Fewer young adults may mean a decrease in crime rates. Moreover, older people are often productive members of society, contributing volunteer activities and services to their children

and grandchildren. In many ways both good and bad, "graying" populations will affect societies in China, the United States, and elsewhere throughout our lifetimes.

Sex ratios   The ratio of males to females also can affect population dynamics. Note that population pyramids give data on sex ratio by representing numbers of males and females on opposite sides of each diagram. To understand the consequences of sex ratio variation, imagine two islands, one populated by 99 men and 1 woman and the other by 50 men and 50 women. Where would we be likely to see the greatest population increase over time? Of course, the island with an equal number of

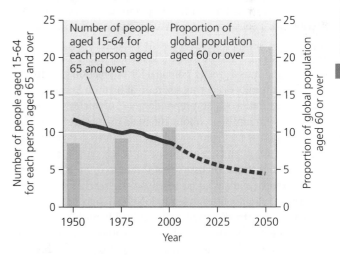

FIGURE 8.12 ▲ Populations are aging worldwide as people of "baby boom" generations grow older and as birth rates decline. Gray bars in this graph show the proportion of the world's population over age 60 at three past years (darker gray bars) and two future years (lighter gray bars). The line shows the number of people aged 15–64 for each person over age 65. As this ratio declines, fewer working-age people are available to support the elderly. Data from Popluation Divsion of the Department of Economic and Social Affairs of the United Nations Secretariat, 2009. *World population prospects: The 2008 revision*, http://esa.un.org/unpp. © United Nations, 2009. By permission.

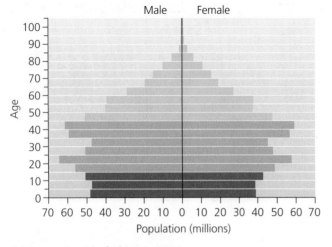

(a) Age pyramid of China in 2010

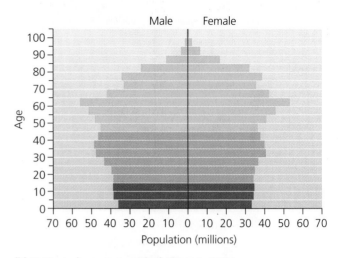

(b) Projected age pyramid of China in 2050

(c) Young female factory workers in Hong Kong

(d) Elderly Chinese

FIGURE 8.13 ▲ As China's population ages, older people will outnumber the young. Population pyramids show the predicted graying of the Chinese population from 2010 **(a)** to 2050 **(b)**. Today's children may, as working-age adults **(c)**, face pressures to support greater numbers of elderly citizens **(d)** than has any previous generation. Data from United Nations Department of Economic and Social Affairs, Population Division: World Population Aging, 2009.

men and women would have a greater number of potential mothers and thus a greater potential for population growth.

The naturally occurring sex ratio at birth in human populations features a slight preponderance of males; for every 100 female infants born, about 106 male infants are born. This phenomenon is an evolutionary adaptation (p. 53) to the fact that males are slightly more prone to death during any given year of life. It tends to ensure that the ratio of men to women will be approximately equal when people reach reproductive age. Thus, a slightly uneven sex ratio at birth may be beneficial. However, a greatly distorted ratio can lead to problems.

In recent years, demographers have witnessed an unsettling trend in China: The ratio of newborn boys to girls has become strongly skewed. In the 2000 census, 120 boys were reported born for every 100 girls. Some provinces reported sex ratios as high as 138 boys for every 100 girls. The leading hypothesis for these unusual sex ratios is that many parents, having learned the sex of their fetuses by ultrasound, are selectively aborting female fetuses.

Recall that Chinese culture has traditionally valued sons over daughters. Sociologists maintain that this cultural gender preference, combined with the government's one-child policy, has led some couples to selectively abort female fetuses or to abandon or kill female infants. The Chinese government reinforced this gender discrimination when in 1984 it exempted rural peasants from the one-child policy if their first child was a girl, but not if the first child was a boy.

China's skewed sex ratio may further lower population growth rates. However, it has proved tragic for the "missing girls." It is also having the undesirable social consequence of leaving many Chinese men single. This, in turn, has resulted in a grim new phenomenon. In parts of rural China, teenaged girls are being kidnapped and sold to families in other parts of the country as brides for single men.

WEIGHING THE ISSUES

China's Reproductive Policy Consider the benefits as well as the problems associated with a reproductive policy such as China's. Do you think a government should be able to enforce strict penalties for citizens who fail to abide by such a policy? If you disagree with China's policy, what alternatives can you suggest for dealing with the resource demands of a rapidly growing population?

Population change results from birth, death, immigration, and emigration

Rates of birth, death, immigration, and emigration determine whether a population grows, shrinks, or remains stable. The formula for measuring population growth that we used in Chapter 3 (p. 66) pertains to people: Birth and immigration add individuals to a population, whereas death and emigration remove individuals. Technological advances have led to a dramatic decline in human death rates, widening the gap between birth rates and death rates and resulting in the global human population expansion.

In today's ever-more-crowded world, immigration and emigration play increasing roles. Refugees, people forced to flee their home country or region, have become more numerous as a result of war, civil strife, and environmental degradation. The United Nations puts the number of refugees who flee to escape poor environmental conditions at 25 million per year and possibly many more. Often the movement of refugees causes environmental problems in the receiving region as these desperate victims try to eke out an existence with no livelihood and no cultural or economic attachment to the land or incentive to conserve its resources. The millions who fled Rwanda following the genocide there in the 1990s, for example, inadvertently destroyed large areas of forest while trying to obtain fuelwood, food, and shelter to stay alive in the Democratic Republic of Congo (**FIGURE 8.14**).

For most of the past 2,000 years, China's population was relatively stable. The first significant increases resulted from enhanced agricultural production and a powerful government during the Qing Dynasty in the 1800s. Population growth began to outstrip food supplies by the 1850s, and quality of life for the average Chinese peasant began to decline. Over the next 100 years, China's population grew slowly, at about 0.3% per year, amid food shortages and political instability. Population growth rates rose again following

TABLE 8.1 Trends in China's Population Growth

Measure	1950	1970	1990	2010
Total fertility rate	5.8	5.8	2.2	1.5
Rate of natural population increase (% per year)	1.9	2.6	1.4	0.5
Doubling time (years)	37	27	49	140
Population (billions)	0.56	0.83	1.15	1.34

Data from China Population Information and Research Center; and Population Reference Bureau, 2010. *2010 World population data sheet.*

Mao's establishment of the People's Republic in 1949, and they have declined since the advent of the one-child policy (**TABLE 8.1**).

In recent decades, falling growth rates in many countries have led to an overall decline in the global growth rate (**FIGURE 8.15**). This decline has come about, in part, from a steep drop in birth rates. Note, however, that although the *rate of growth* is slowing, the *absolute size* of the population continues to increase. Even though our percentage increases are getting smaller year by year, these are percentages of

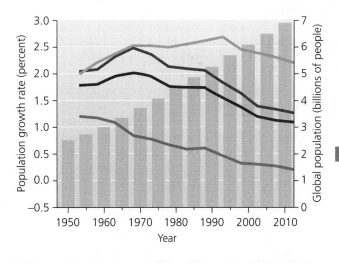

FIGURE 8.14 ▼ The flight of refugees from Rwanda into the Democratic Republic of Congo in 1994 following the Rwandan genocide caused unimaginable hardship for the refugees and tremendous stress on the environment into which they moved.

FIGURE 8.15 ▲ The annual growth rate of the global human population (dark blue line) peaked in the late 1960s and has declined since then. Growth rates of developed nations (green line) have fallen since 1950, while those of developing nations (red line) have fallen since the global peak in the late 1960s. For the world's least developed nations (orange line), growth rates began to fall in the 1990s. Although growth *rates* are declining, global population size (gray bars) is still growing about the same amount each year, because smaller percentage increases of ever-larger numbers produce roughly equivalent additional amounts. Data from Population Division of the Department of Economic and Social Affairs of the United Nations Secretariat, 2009. *World population prospects: The 2008 revision,* 2008. *World population prospects: The 2008 revision,* http://esa.un.org/unpp. © United Nations, 2009. By permission.

ever-larger numbers, so we continue to add over 80 million people to the planet each year.

Total fertility rate influences population growth

One key statistic demographers calculate to examine a population's potential for growth is the **total fertility rate (TFR)**, the average number of children born per woman during her lifetime. **Replacement fertility** is the TFR that keeps the size of a population stable. For humans, replacement fertility roughly equals a TFR of 2.1. (Two children replace the mother and father, and the extra 0.1 accounts for the risk of a child dying before reaching reproductive age.) If the TFR drops below 2.1, population size (in the absence of immigration) will shrink.

Various factors have driven TFR downward in many nations in recent years. Historically, people tended to conceive many children, which helped ensure that at least some would survive. Today's improved medical care has reduced infant mortality rates, making it less necessary to bear multiple children. Increasing urbanization has also driven TFR down; whereas rural families need children to contribute to farm labor, in urban areas children are usually excluded from the labor market, are required to go to school, and impose economic costs on their families. Moreover, if a government provides some form of social security, as most do these days, parents need fewer children to support them in their old age. Finally, with greater educational opportunities and changing roles in society, women tend to shift into the labor force, putting less emphasis on child rearing.

All these factors have come together in Europe, where TFR has dropped from 2.6 to 1.6 in the past half-century. Nearly every European nation now has a fertility rate below the replacement level, and populations are declining in 17 of 45 European nations. In 2010, Europe's overall annual **natural rate of population change** (change due to birth and death rates alone, excluding migration) was between 0.0% and 0.1%. Worldwide by 2010, 72 countries had fallen below the replacement fertility of 2.1. These countries make up roughly 45% of the world's population and include China (with a TFR of 1.5). **TABLE 8.2** shows TFRs of major continental regions.

TABLE 8.2 Total Fertility Rates for Major Continental Regions	
Region	**Total fertility rate (TFR)**
Africa	4.7
Australia and South Pacific	2.5
Latin America and Caribbean	2.3
Asia	2.2
North America	2.0
Europe	1.6

Data from Population Reference Bureau, 2010. *2010 World population data sheet.*

Consequences of Low Fertility? In the United States, Canada, and almost every European nation, the total fertility rate is now at or below the replacement fertility rate (although some of these nations are still growing because of immigration). What economic or social consequences do you think might result from below-replacement fertility rates? Would you rather live in a society with a growing population, a shrinking population, or a stable population? Why?

Many nations are experiencing the demographic transition

Many nations with lowered birth rates and TFRs are experiencing a common set of interrelated changes. In countries with good sanitation, effective health care, and reliable food supplies, more people than ever before are living long lives. As a result, over the past 50 years the life expectancy for the average person has increased from 46 to 69 years as the global death rate has dropped from 20 deaths per 1,000 people to 8 deaths per 1,000 people. Strictly speaking, **life expectancy** is the average number of years that an individual in a particular age group is likely to continue to live, but often people use this term to refer to the average number of years a person can expect to live from birth. Much of the increase in life expectancy is due to reduced rates of infant mortality. Societies going through these changes are generally those that have undergone urbanization and industrialization and have generated personal wealth for their citizens.

To make sense of these trends, demographers developed a concept called the **demographic transition**. This is a model of economic and cultural change first proposed in the 1940s and 1950s by demographer Frank Notestein to explain the declining death rates and birth rates that have occurred in Western nations as they industrialized. Notestein believed nations move from a stable pre-industrial state of high birth and death rates to a stable post-industrial state of low birth and death rates (**FIGURE 8.16**). Industrialization, he proposed, causes these rates to fall by first decreasing mortality and then lessening the need for large families. Parents thereafter choose to invest in quality of life rather than quantity of children. Because death rates fall before birth rates fall, a period of net population growth results. Thus, under the demographic transition model, population growth is seen as a temporary phenomenon that occurs as societies move from one stage of development to another.

The pre-industrial stage The first stage of the demographic transition model is the **pre-industrial stage**, characterized by conditions that have defined most of human history. In pre-industrial societies, both death rates and birth rates are high. Death rates are high because disease is widespread, medical care rudimentary, and food supplies unreliable and difficult to obtain. Birth rates are high because people must compensate for infant mortality by having several children. In this stage, children are valuable as workers who can help meet a family's basic needs. Populations within the pre-industrial

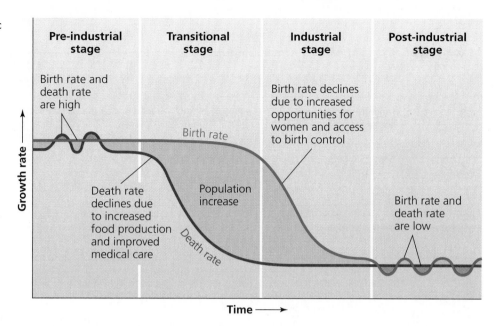

FIGURE 8.16 ▶ The demographic transition models a process that has taken some populations from a pre-industrial stage of high birth rates and high death rates to a post-industrial stage of low birth rates and low death rates. In this diagram, the wide green area between the two curves illustrates the gap between birth and death rates that causes rapid population growth during the middle portion of this process. Adapted from Kent, M. and K. Crews, 1990. *World population: Fundamentals of growth.* By permission of the Population Reference Bureau.

In the figure:

Pre-industrial stage — Birth rate and death rate are high

Transitional stage — Death rate declines due to increased food production and improved medical care; Population increase

Industrial stage — Birth rate declines due to increased opportunities for women and access to birth control

Post-industrial stage — Birth rate and death rate are low

Axes: Growth rate (vertical), Time (horizontal). Curves labeled Birth rate and Death rate.

stage are not likely to experience much growth, which is why the human population was relatively stable until the industrial revolution.

Industrialization and falling death rates Industrialization initiates the second stage of the demographic transition, known as the **transitional stage**. This transition from the pre-industrial stage to the industrial stage is generally characterized by declining death rates due to increased food production and improved medical care. Birth rates in the transitional stage remain high, however, because people have not yet grown used to the new economic and social conditions. As a result, population growth surges.

The industrial stage and falling birth rates The third stage in the demographic transition is the **industrial stage**. Industrialization increases opportunities for employment outside the home, particularly for women. Children become less valuable, in economic terms, because they do not help meet family food needs as they did in the pre-industrial stage. If couples are aware of this, and if they have access to birth control, they may choose to have fewer children. Birth rates fall, closing the gap with death rates and reducing population growth.

The post-industrial stage In the final stage, the **post-industrial stage**, both birth and death rates have fallen to low and stable levels. Population sizes stabilize or decline slightly. The society enjoys the fruits of industrialization without the threat of runaway population growth. The United States is an example of a nation in this stage, although the U.S. population is growing faster than most other post-industrial nations because of a high immigration rate.

Is the demographic transition a universal process?

The demographic transition has occurred in many European countries, the United States, Canada, Japan, and several other developed nations over the past 200 to 300 years. It is a model that may or may not apply to all developing nations as they industrialize now and in the future. On the one hand, note in Figure 8.15 (p. 209) how growth rates fell first for industrialized nations, then for less developed nations, and finally for least developed nations, suggesting that it may merely be a matter of time before all nations experience the transition. On the other hand, population dynamics may differ for developing nations that try to adopt the Western world's industrial model rather than devising their own. And some demographers assert that the transition will fail in cultures that place greater value on childbirth or grant women fewer freedoms.

Moreover, natural scientists estimate that for people of all nations to attain the material standard of living that North Americans now enjoy, we would need the natural resources of four-and-a-half more planet Earths. Whether developing nations (which include the vast majority of the planet's people) pass through the demographic transition is one of the most important and far-reaching questions for the future of our civilization and Earth's environment.

POPULATION AND SOCIETY

Demographic transition theory links the quantitative study of how populations change with the societal factors that influence (and are influenced by) population dynamics. Let's now examine a few of these societal factors more closely.

Birth control is a key approach for controlling population growth

Perhaps the greatest single factor enabling a society to slow its population growth is the ability of women and couples to practice birth control. **Birth control** is the effort to control the number of children one bears, particularly by reducing the frequency of pregnancy. This relies on **contraception**, the deliberate attempt to prevent pregnancy despite sexual intercourse. There is a wide variety of contraceptive methods for

Some Common Methods of Contraception

Method	How it works	Percent use by U.S. women who use contraception	Effectiveness (pregnancies per 100 women per year[1])
IUD	The intrauterine device (IUD) is a flexible plastic T-shaped device that a doctor inserts into a woman's uterus. Different types prevent pregnancy in different ways. IUDs remain effective for 5-12 years.	<5% in U.S., but is most popular method internationally	Very effective (<1)
Hormone injection	A woman gets a shot from a doctor, providing hormones that last 3 months.	5%	Effective (<1-3)
Birth control pill	A woman takes 1 pill daily. The pill provides hormones that prevent ovaries from releasing eggs and that thicken cervical mucus to block sperm.	30%	Effective (<1-8)
Condom	A man places this thin latex or plastic sheath on the penis before sex. By blocking semen, it prevents pregnancy and also stops the spread of sexually transmitted diseases. Can be used with other methods.	18%	Moderately effective (2-15)
Withdrawal	In the world's oldest birth control method, the man pulls out before ejaculating.	4%	Low effectiveness (4-27)
Spermicide	A couple uses a gel or cream containing chemicals that stop sperm from moving. Can be used with other methods.		Low effectiveness if used alone (15-29)
Emergency contraception	A woman may take a so-called "morning-after pill" up to 5 days after having intercourse.		Decreases risk of pregnancy by up to 89% if used within 72 hours
Nonreversible (permanent) methods			
Tubal ligation	A woman who is sure she never wants to have children can undergo this surgical operation to become permanently sterile. A surgeon closes or blocks the fallopian tubes so eggs cannot be released.	27%	Nearly 100% effective
Vasectomy	A man who is sure he never wants to have children can undergo this surgical operation to become permanently sterile. A doctor blocks the vasa deferentia so sperm cannot be released.	10%	Nearly 100% effective

[1]The first number is the number of expected pregnancies per 100 women per year if the method is *always* used as directed. Where there is a range of numbers, the second number is the number expected if the method is *not always* used as directed.

FIGURE 8.17 ▲ A wide variety of contraceptive methods for birth control are available today. This chart shows some of the more common methods. For further information, the website of Planned Parenthood is an excellent and reliable source. Data from Planned Parenthood (R) website, www.plannedparenthood.org.

reducing the chance of pregnancy (**FIGURE 8.17**). Some methods physically prevent sperm and egg from coming into contact. Others function by altering a woman's hormone balance. Some, like the "morning-after pill," can even be effective following intercourse. And some methods are permanent; in fact, the most common method of contraception in the United States is voluntary sterilization; fully 37% of U.S. women who practice contraception do so through tubal ligation or their partner's vasectomy. For further information on various methods, a reliable source is the website of Planned Parenthood.

Planned Parenthood is one example of a family-planning organization. **Family planning** is the effort to plan the number and spacing of one's children, so as to offer children and parents the best quality of life possible. Family-planning programs and clinics offer information and counseling to potential mothers and fathers about reproductive health and reproductive decisions. They also frequently offer free or discounted contraceptives.

In many societies, religious doctrine or cultural influences hinder pregnancy planning, so contraceptives are denied to people who might otherwise use them. Yet in 2010, 55% of married women worldwide (aged 15–49) reported using some modern method of contraception to plan or prevent pregnancy. China, at 86%, had the highest rate of contraceptive use of any nation. Six European nations showed rates of contraceptive use of 70% or more, as did Australia, Brazil, Canada, Costa Rica, Cuba, Micronesia, New Zealand, Nicaragua, Paraguay, Puerto Rico, Thailand, the United States, and Uruguay. At the other end of the spectrum, 17 African nations had rates below 10%. These low rates of contraceptive use contribute to high fertility rates in sub-Saharan Africa, where the region's TFR is 5.2 children per woman. By comparison, the TFR in Asia, which in 1950 was 5.9, today is 2.2—in part a legacy of programs in China, India, Thailand, and other nations that make contraception widely available.

Empowering women reduces fertility rates

Fertility rates have dropped most noticeably in nations where women have gained improved access to contraceptives and to family planning (see **THE SCIENCE BEHIND THE STORY**, pp. 214–215).

The data showing that fertility rates decline as family planning and birth control become available clearly indicate that in societies where women have little power, substantial numbers of pregnancies are unintended. Tragically, millions of women still lack the information and personal freedom of choice to allow them to make their own decisions about when to have children and how many to have. Today, many social scientists and policymakers recognize that for population growth to slow and stabilize, women should be granted equal power with men in societies worldwide. This would have many benefits; studies show that where women are freer to decide whether and when to have children, fertility rates fall, and the resulting children are better cared for, healthier, and better educated.

Equality for women involves expanding educational opportunities for them, especially in Muslim nations and regions in the developing world where girls are kept out of school or discouraged from pursuing an education. Over two-thirds of the world's people who cannot read are women. And data clearly show that as women receive educational opportunities, fertility rates decline (**FIGURE 8.18**). Education helps

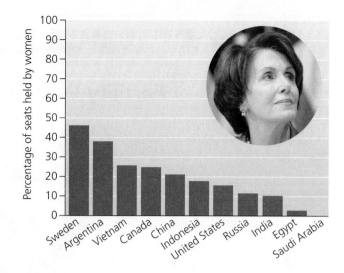

FIGURE 8.19 ▲ Although women make up half the world's population, they are vastly underrepresented in positions of political power. Measured by the proportion of seats held by women in national legislatures, women in some nations fare better than those in others. Although the United States finally gained a female Speaker of the House (Nancy Pelosi, inset), the United States lags behind the world average in the proportion of women in its legislative branch. Data are for 2010 from Inter-Parliamentary Union, Women in National Parliaments.

more women pursue careers, delay childbirth, and have greater say in reproductive decisions.

Sadly, we are still a long way from achieving gender equality. Over 60% of the world's people living in poverty are women. Violence against women remains shockingly common. In many societies, men restrict women's decision-making abilities, including decisions as to how many children they will bear. Even in progressive Western societies where women have made great strides, they still often face institutionalized gender discrimination, including lower pay for equal work.

The gap between the power held by men and by women is obvious at the highest levels of government. Worldwide, only 18.9% of elected government officials in national legislatures are women (**FIGURE 8.19**). The United States lags behind not only Europe but also many developing nations in the proportion of female representatives in its government. As more women win positions of power, perhaps we will advance toward gender equality. Such equality would have environmental consequences, for when women have economic and political power and access to education, they gain the option, and often the motivation, to limit the number of children they bear.

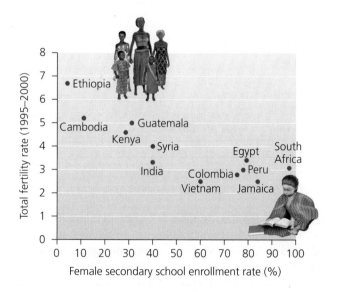

FIGURE 8.18 ▲ Increasing female literacy is strongly associated with reduced birth rates in many nations. Data from McDonald, M., and D. Nierenberg, 2003. Linking population, women, and biodiversity. *State of the world 2003*. Washington, D.C.: Worldwatch Institute.

Population policies and family-planning programs are working around the globe

Data show that funding and policies that encourage family planning can lower population growth rates in all types of nations, even those that are least industrialized. No nation has pursued a sustained population control program as intrusive as China's, but other rapidly growing nations have implemented programs that are less restrictive.

The SCIENCE behind the Story

Fertility Decline in Bangladesh

Research in developing nations indicates that poverty and overpopulation can create a vicious cycle, in which poverty promotes high fertility and high fertility obstructs economic development. Are there policy steps that such countries can take to bring down fertility rates? Scientific analysis of family-planning programs in the South Asian nation of Bangladesh suggests that there are.

Bangladesh is one of the poorest and most densely populated countries on the planet. Its 166 million people live in an area about the size of Wisconsin, and 45% of them live below the poverty line. With few natural resources and 1,150 people per km^2 (nearly 3,000 per mi^2—2.5 times the population density of New Jersey), limiting population growth is critically important. Back in 1976, Bangladeshi president Ziaur Rahman declared, "If we cannot do something about population, nothing else that we accomplish will matter much."

Since then, Bangladesh has made striking progress in slowing its population growth. Despite stagnant economic development, low literacy rates, poor health care, and limited rights for women, the nation's total fertility rate (TFR) has dropped markedly.

Dr. James Phillips of Columbia University

In the 1970s, the average woman in Bangladesh gave birth to nearly 7 children over the course of her life. Today, the TFR is 2.4 (see **first figure**).

Researchers hypothesized that family-planning programs were responsible for Bangladesh's rapid reduction in TFR. Because conducting an experiment to test such a hypothesis is difficult, some researchers took advantage of a natural experiment (pp. 12–13). By comparing Bangladesh to countries that are socioeconomically similar

but have had less success in lowering TFR, such as Pakistan (see Figure 8.20, p. 216), researchers concluded that Bangladesh succeeded because of aggressive, well-funded outreach efforts that were sensitive to the values of its traditional society.

However, no two nations are identical, so it is difficult to draw firm conclusions from such broad-scale comparisons. This is why the Matlab

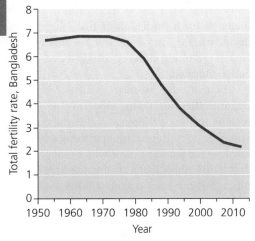

Total fertility rate has declined markedly in Bangladesh in the past 40 years, in part because of family-planning programs. Data from Population Division of the Department of Economic and Social Affairs of the united Nations Secretariat, 2009. *World population prospects: The 2008 revision*, http://esa.un.org/unpp. © United Nations, 2009. By permission.

India was the first nation to implement population control policies. However, when some policymakers introduced forced sterilization in the 1970s, the resulting outcry brought down the government. Since then, India's efforts have been more modest and far less coercive, focusing on family planning and reproductive health care. However, today a number of Indian states also run programs of incentives and disincentives promoting a "two-child norm," and current debate centers on whether this is a just and effective

approach. Regardless, unless India strengthens its efforts to slow population growth, it seems set to overtake China and become the world's most populous nation about the year 2030.

The government of Thailand has reduced birth rates and slowed population growth. In the 1960s, Thailand's growth rate was 2.3%, but today it stands at 0.6%. This decline was achieved without a one-child policy, resulting instead from an education-based approach to family planning and the increased

Family Planning and Health Services Project, in the isolated rural area of Matlab, Bangladesh, has become one of the best-known experiments in family planning in developing countries.

The Matlab Project was an intensive outreach program run collaboratively by the Bangladeshi government and international aid organizations. Each household in the project area received biweekly visits from local women offering counseling, education, and free contraceptives. Compared to a similar government-run program in a nearby area, the Matlab Project featured more training, more services, and more frequent visits. In both areas, a highly organized health surveillance system gave researchers detailed information about births, deaths, and health-related behaviors such as contraceptive use. The result was an experiment comparing the Matlab Project with the government-run project.

When Matlab Project director James Phillips and his colleagues reviewed a decade's worth of data in 1988, they found that fertility rates had declined in both areas. The decline appeared to be due almost entirely to a rise in contraceptive use, because other factors—such as the average age of marriage—remained the same. Phillips and his colleagues also found the declines to be significantly greater in the Matlab area than in the government-run area. These findings suggested that high-intensity outreach efforts could affect TFR even in the absence of significant improvements in women's

In the Matlab Project, Bangladeshi households received visits from local women offering counseling, education, and free contraceptives.

status, education, or economic development.

Why was the outreach program successful? One hypothesis was that visits from health care workers helped convince local women that small families are desirable. However, in 1999, Mary Arends-Kuenning, a University of Michigan graduate student, and her colleagues reported that there was no relationship between the number of visits made by outreach workers and women's perception of the ideal family size, either in Matlab or nearby areas. Ideal family size declined equally in all areas. Instead of creating new demand for birth control, the Matlab Project appears to have helped

women convert an already-existing desire for fewer children into behaviors, such as contraceptive use, that reduce fertility.

Bangladesh's ability to rein in fertility rates despite unfavorable social and economic conditions bodes well for impoverished nations facing explosive population growth. However, further reductions may require political and socioeconomic changes that are difficult to implement in traditional, resource-strapped countries such as Bangladesh. Nonetheless, scientific research from Matlab has illuminated the impacts of family-planning programs on fertility, helping to inform population control efforts in Bangladesh and elsewhere. ■

availability of contraceptives. Brazil, Mexico, Iran, Cuba, and many other developing countries have instituted active programs to reduce their population growth that entail setting targets and providing incentives, education, contraception, and reproductive health care. These programs are working: Studies show that nations with such programs have lower fertility rates than similar nations without them (**FIGURE 8.20**).

In 1994, the United Nations hosted a milestone conference on population and development in Cairo, Egypt, at which 179 nations endorsed a platform calling on all governments to offer universal access to reproductive health care within 20 years. The conference marked a turn away from older notions of top-down command-and-control (p. 179) population policy geared toward pushing contraception and lowering population to preset targets. Instead, it urged governments to offer better education and health care and to address social needs that affect population from the bottom up (such as alleviating poverty, disease, and

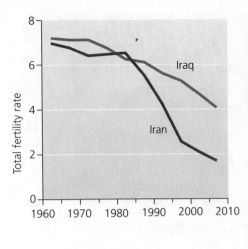

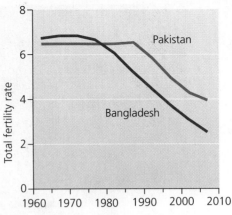

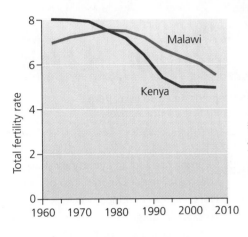

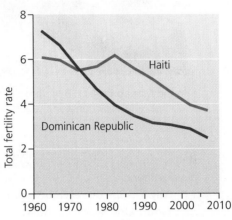

FIGURE 8.20 ◀ Data from four pairs of neighboring countries demonstrate the effectiveness of family planning in reducing fertility rates. In each case, the nation that invested in family planning and (in some cases) made other reproductive rights, education, and health care more available to women (blue lines) reduced its total fertility rate (TFR) more than its neighbor (red lines). Data from Harrison, P. and F. Pearce, 2000. *AAAS Atlas of population and environment*, edited by the American Association for the Advancement of Science, © 2000 by the American Association for the Advancement of Science. Used by permission of the publisher, University of California Press.

sexism). However, worldwide funding for family planning fell by well over a third in the decade following the Cairo conference.

WEIGHING THE ISSUES

Abstaining from International Family Planning? Over the years, the United States has joined 180 other nations in providing millions of dollars to the United Nations Population Fund (UNFPA), which advises governments on family planning, sustainable development, poverty reduction, reproductive health, and AIDS prevention in many nations, including China. Starting in 2001, the Bush administration withheld funds, saying that U.S. law prohibits funding any organization that "supports or participates in the management of a program of coercive abortion or involuntary sterilization" and claiming that the Chinese government has been implicated in both. Many nations criticized the U.S. decision, and the European Union offered UNFPA additional funding to offset the loss of U.S. contributions. Once President Obama came to office, he reinstated funding, asking Congress to allocate $60 million in 2009. What do you think U.S. policy should be? Should the United States fund family-planning efforts in other nations? What conditions, if any, should it place on the use of such funds?

Poverty is correlated with population growth

The alleviation of poverty was a prime target of the Cairo conference because poorer societies tend to show higher population growth rates than do wealthier societies (**FIGURE 8.21**). This correlation is consistent with demographic transition theory. Note that the causality is thought to operate in both directions: Poverty exacerbates population growth, and rapid population growth worsens poverty.

These trends also influence the distribution of people on the planet. In 1960, 70% of the world's population lived in developing nations. As of 2010, 82% of all people live in these countries. Moreover, fully 99% of the next billion people to be added to the global population will be born into these poor, less developed regions (**FIGURE 8.22**).

This is unfortunate from a social standpoint, because these people will be added to the nations that are least able to provide for them. It is also unfortunate from an environmental standpoint, because poverty often results in environmental degradation. People who depend on agriculture in areas of poor farmland, for instance, may need to try to farm even if doing so degrades the soil and is not sustainable. This is largely why Africa's Sahel region—like many arid areas of western China—turns to desert during climatically dry periods (**FIGURE 8.23**). Poverty also drives people to cut forests and to deplete biodiversity. For example, impoverished settlers and miners hunt large mammals for "bush meat" in Africa's forests, including the great apes that are now heading toward extinction.

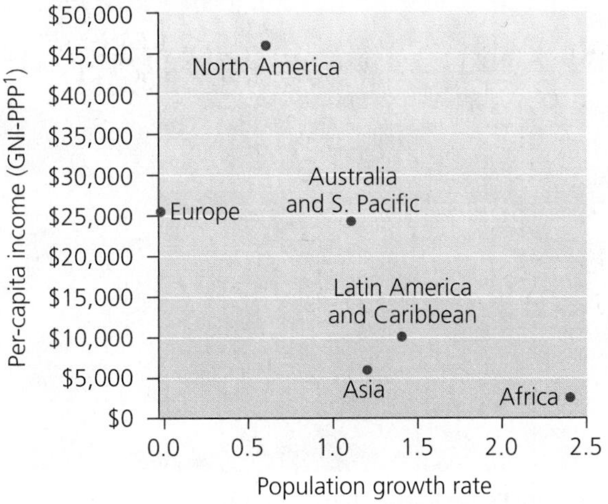

FIGURE 8.21 ▲ Poverty and population growth show a fairly strong correlation, despite the influence of many other factors. Regions with the most rapid population growth tend to show lower per capita incomes. In a feedback effect, the causality likely works in both directions. Per capita income is here measured in GNI PPP, or "gross national income in purchasing power parity," a measure that standardizes income among nations by converting it to "international" dollars, which indicate the amount of goods and services one could buy in the United States with a given amount of money. Data from Population Reference Bureau, 2010. *2010 World population data sheet.*

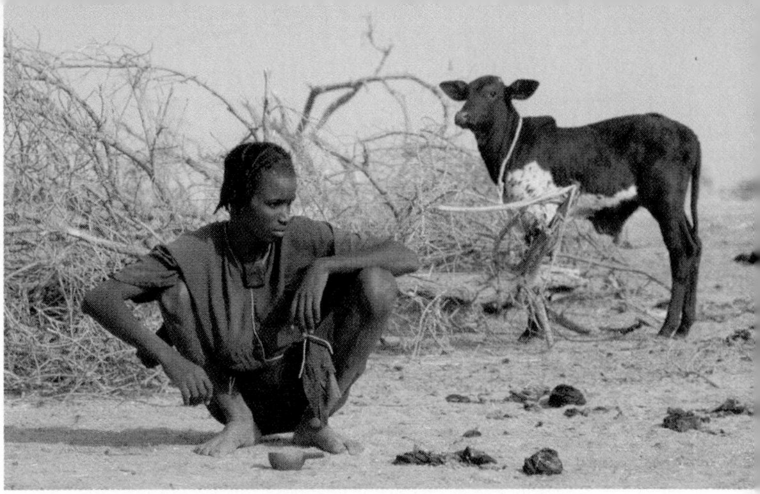

FIGURE 8.23 ▲ In the semi-arid Sahel region of Africa, population may be increasing beyond the land's ability to handle it. Here, drought and dependence on grazing agriculture have led to environmental degradation.

Consumption from affluence creates environmental impact

Poverty can lead people into environmentally destructive behavior, but wealth can produce even more severe and far-reaching environmental impacts. The affluence of a society such as the United States, Japan, or the Netherlands is built on levels of resource consumption unprecedented in human history. Much of this chapter has dealt with numbers of people rather than on the amount of resources each member of the population consumes or the amount of waste each member produces. The environmental impact of human activities, however, depends not only on the number of people involved but also on the way those people live. Recall the A (for affluence) in the IPAT equation. Affluence and consumption are spread unevenly across the world, and wealthy societies generally consume resources from regions far beyond their own.

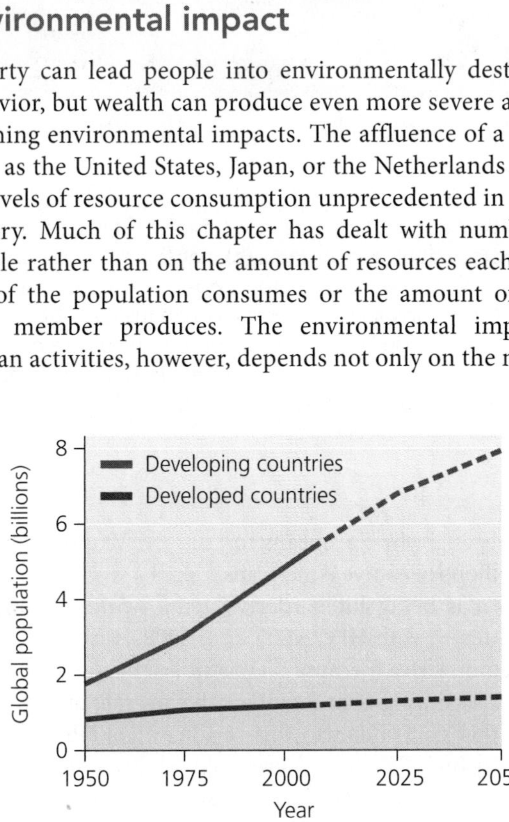

FIGURE 8.22 ▲ Over 99% of the next 1 billion people added to Earth's population will be born into the less developed, poorer parts of the world. Dashed portions of the lines indicate projected future trends. Data from U.N. Population Division, 2008. *World population prospects: The 2008 revision.*

In Chapter 1 (pp. 5, 16), we explored the concept of the *ecological footprint*, the cumulative amount of Earth's surface area required to provide the raw materials a person or population consumes and to dispose of or recycle the waste produced. Individuals from affluent societies leave considerably larger per capita ecological footprints (see Figure 1.14, p. 16). In this sense, the addition of one American to the world has as much environmental impact as the addition of 4.5 Chinese, 10 Indians, or 19 Afghans. This fact reminds us that the "population problem" does not lie solely with the developing world!

Indeed, just as population is rising, so is consumption, and some environmental scientists have calculated that we are already living beyond the planet's means to support us sustainably. One recent analysis concludes that humanity's global ecological footprint surpassed Earth's capacity to support us in 1987 and that our species is now living 30% beyond its means (**FIGURE 8.24**). This is the *overshoot* we learned of in Chapter 1 (see Figure 1.5, p. 5). In this analysis, our ecological footprint can be compared to the amount of biologically productive land and sea available to us—an amount termed *biocapacity*. For any given area, if the footprint is greater than the biocapacity, there is an "ecological deficit." If the footprint is less than the biocapacity, there is an "ecological reserve." Because our footprint exceeds our biocapacity by 30% worldwide, we are running a global ecological deficit, gradually draining our planet of its natural capital and its long-term ability to support our civilization. The rising consumption that is accompanying the rapid industrialization of China, India, and other populous nations makes it all the more urgent for us to find a path to global sustainability.

The wealth gap and population growth contribute to conflict

The stark contrast between affluent and poor societies in today's world is the cause of social as well as environmental stress. Over

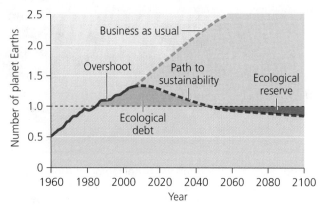

(a) A family living in the United States

Average per U.S. resident:
• $46,970 annual income
• 31,000 m² of land
• 19.0 metric tons of CO_2 emitted per year

FIGURE 8.24 ▲ As we saw in Chapter 1 (Figure 1.5, p. 5), the global ecological footprint of the human population is estimated to be 30% greater than what Earth can bear in the long run. If population and consumption continue to rise (orange dashed line), we will increase our ecological deficit, or degree of overshoot, until systems give out and populations crash. If, instead, we pursue a path to sustainability (red dashed line), we can eventually repay our ecological debt and sustain our civilization. Adapted from WWF International. 2008. Living Planet Report 2008. Published in October 2008 by WWF-World Wide Fund for Nature. © 2008 WWF. (panda.org). Zoological Society of London, and Global Footprint Network.

Average per resident of India:
• $2,960 annual income
• 2,800 m² of land
• 1.1 metric tons of CO_2 emitted per year

(b) A family living in India

FIGURE 8.25 ▲ A typical U.S. family **(a)** may own a large house with a wealth of material possessions. A typical family in a developing nation such as India **(b)** may live in a small, sparsely furnished dwelling with few material possessions and little money or time for luxuries. Compared with the average resident of India, the average U.S. resident enjoys 16 times more income, uses 11 times more land, and emits 17 times more carbon dioxide emissions. Data from Population Reference Bureau, 2009. 2009 World Population Data Sheet.

half the world's people live below the internationally defined poverty line of U.S. $2 per day. The richest one-fifth of the world's people possesses over 80 times the income of the poorest one-fifth (**FIGURE 8.25**). The richest one-fifth also uses 86% of the world's resources. That leaves only 14% of global resources—energy, food, water, and other essentials—for the remaining four-fifths of the world's people to share.

As the gap between rich and poor grows wider and as the sheer numbers of those living in poverty continue to increase, it seems reasonable to predict increasing tensions between "haves" and "have-nots." This is why the inequitable distribution of wealth is a key factor the U.S. Departments of Defense and State take into account when assessing the potential for armed conflict, whether it be conventional warfare or terrorism.

HIV/AIDS is exerting major impacts on African populations

The rising material wealth and falling fertility rates of many developed nations today is slowing population growth in accord with the demographic transition model. However, nations where the human immunodeficiency virus (HIV) and acquired immunodeficiency syndrome (AIDS) has taken hold are not following Notestein's script. Instead, in these countries death rates have increased, presenting a scenario more akin to Malthus's fears.

The AIDS epidemic (**FIGURE 8.26**) is having the greatest impact on human populations of any communicable disease since the Black Death killed roughly one-third of Europe's population in the 14th century and since smallpox and other

diseases brought by Europeans to the New World wiped out likely millions of Native Americans.

Africa is being hit hardest. Of the world's 33 million people infected with HIV/AIDS as of 2008, two-thirds live in sub-Saharan Africa. Because HIV is spread by the exchange of bodily fluids during sexual contact, the low rate of contraceptive use that contributes to this region's high TFR also fuels the spread of HIV. One in every 20 adults in sub-Saharan Africa is infected with HIV, and for southern African nations, the figure is 1 in 5. As AIDS takes roughly 3,800 lives in sub-Saharan Africa every day, the epidemic unleashes a variety of demographic changes. Infant mortality here has risen to 8 deaths out of 100 live births—13 times the rate in the developed world. Infant mortality and the premature deaths of young adults has caused life expectancy in parts of southern

— Ecological footprint
▪▪▪ Projected ecological footprint
---- Biocapacity

Graph axes: Number of planet Earths (0 to 2.5); Year (1960–2100)
Labels: Business as usual; Overshoot; Path to sustainability; Ecological reserve; Ecological debt

218

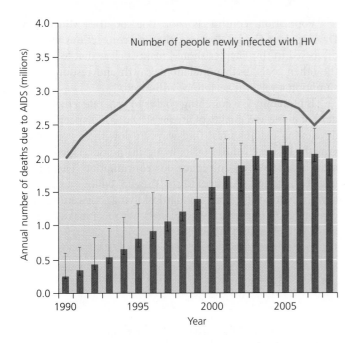

FIGURE 8.26 ▲ Two million people die from AIDS worldwide each year (red bars), and even more are newly infected with HIV each year (green line). However, extensive efforts to combat the disease are beginning to bear fruit; infections have been decreasing since 1998, and deaths began decreasing in 2006. Data from UNAIDS and WHO, 2009. *AIDS epidemic update 2009*, Geneva Switzerland.

Africa to fall from almost 60 years in the early 1990s back down to 40–50 years, where it stood decades earlier. AIDS also leaves behind millions of orphans.

Africa is not the only region with reason to worry. HIV is well established in the Caribbean and in Southeast Asia, and it is spreading in eastern Europe and central Asia. In China, the government has been reluctant to monitor or publicize the status of the disease because of the extreme social stigma that the culture attaches to it. The government has admitted to fewer than 1 million cases, although some international experts fear there could be 10 times more and that vigilance is needed to keep the number from rising.

Demographic change has social and economic repercussions

Because it removes young and productive members of society, AIDS undermines the ability of poorer nations to develop. For example, in 1999 Zambia lost 600 teachers to AIDS, and only 300 new teachers graduated to replace them. South Africa loses an estimated $7 billion per year to declines in its labor force as AIDS patients fill the nation's hospitals. The loss of productive household members causes families and communities to break down as income and food production decline while medical expenses and debt skyrocket.

These problems are hitting many African countries at a time when their governments are already experiencing *demographic fatigue*. Demographically fatigued governments face overwhelming challenges related to population growth, including educating and finding jobs for their swelling ranks of young people. With the added stress of HIV/AIDS, these

governments face so many demands that they are stretched beyond their capacity to address problems. As a result, the problems grow worse, and citizens lose faith in their governments' abilities to help them. This constellation of challenges is making it difficult for some African nations to advance through the demographic transition.

There is good news, however. Improved public health efforts (including sex education, contraceptives, and intravenous drug abuse policies) across the world have slowed HIV transmission rates, and improved medical treatments are lengthening the lives of people who are infected. Note in Figure 8.26 that the number of AIDS deaths began to decrease in 2006, following a drop in new HIV infections since the late 1990s. We may finally be turning the corner on this challenge, thanks to government policy, international collaboration, medical research, nonprofit aid groups, and the grassroots efforts of patients and their advocates (**FIGURE 8.27**). With continued work, we can venture to hope that the multipronged effort to combat HIV/AIDS might eventually become a success story in humanity's timeless battle against disease.

Sustainable development and population goals go hand-in-hand

In 2000, world leaders came together to adopt the *United Nations Millennium Declaration*, which set forth eight **Millennium Development Goals** for humanity (**TABLE 8.3**). Each broad goal for sustainable development has several specific underlying targets that may be met through implementation of concrete strategies. For instance, strategies to "develop a global partnership for development" include working with governments and corporations of wealthy nations to help provide poor nations with more financial aid, more debt relief, freer access to global markets, and better access to inexpensive drugs and to technologies such as cell phones and Internet access.

FIGURE 8.27 ▼ Africans are fighting back against AIDS. Rolake Odetoyinbo Nwagwu is HIV-positive, but directs an organization to support people with AIDS, called Positive Action for Treatment Access. Here she speaks to journalists in Nigeria.

TABLE 8.3 United Nations Millennium Development Goals for 2015

▶ Eradicate extreme poverty and hunger
▶ Achieve universal primary education
▶ Promote gender equality and empower women
▶ Reduce child mortality
▶ Improve maternal health
▶ Combat HIV/AIDS, malaria, and other diseases
▶ Ensure environmental sustainability
▶ Develop a global partnership for development

Source: United Nations. *End poverty 2015: Millennium Development Goals.* © United Nations. Reproduced with permission.

Population control is not one of the Millennium Development Goals. However, the interconnections we have discussed in this chapter should make it clear that to achieve the other goals, both population growth and resource consumption levels will need to be addressed.

If humanity's overarching goal is to generate a high standard of living and quality of life for all the world's people, then developing nations must find ways to slow their population growth. However, those of us living in the developed world must also be willing to reduce our consumption. Earth does not hold enough resources to sustain all 7 billion of us at the current North American standard of living, nor can we venture out and bring home extra planets. We must make the best of the one place that supports us all.

➤ CONCLUSION

Today's human population is larger than at any time in the past. Our growing population and our growing consumption affect the environment and our ability to meet the needs of all the world's people. Sadly, 90% of children born today are likely to live their lives in conditions far less healthy and prosperous than most of us in the industrialized world are accustomed to.

However, there are at least two major reasons to be encouraged. First, although global population is still rising, the *rate* of growth has decreased nearly everywhere, and some countries are even seeing population declines. Most developed nations have passed through the demographic transition, showing that it is possible to lower death rates while stabilizing population and creating more prosperous societies. Second, progress has been made in expanding rights for women worldwide. Although there is still a long way to go, women are obtaining better education, more economic independence, and more ability to control their reproductive decisions. Aside

from the clear ethical progress these developments entail, they are helping to slow population growth.

Human population cannot continue to rise forever. The question is *how* it will stop rising: Will it be through the gentle and benign process of the demographic transition, through restrictive governmental intervention such as China's one-child policy, or through the miserable Malthusian checks of disease and social conflict caused by overcrowding and competition for scarce resources? Moreover, sustainability demands a further challenge—that we stabilize our population size in time to avoid destroying the natural systems that support our economies and societies. We are indeed a special species. We are the only one to achieve such dominance and to alter much of Earth's landscape and even its climate system. We are also the only species with the intelligence needed to turn around an increase in our own numbers before we destroy the very systems on which we depend.

REVIEWING OBJECTIVES

You should now be able to:

PERCEIVE THE SCOPE OF HUMAN POPULATION GROWTH

- Our global population of nearly 7 billion people adds over 80 million people per year (2.6 people every second). (p. 199)
- The global population growth rate peaked at 2.1% in the 1960s and now stands at 1.2%. Growth rates vary among regions. (p. 199)

ASSESS DIVERGENT VIEWS ON POPULATION GROWTH

- Thomas Malthus and Paul Ehrlich warned that overpopulation would deplete resources and harm humanity, whereas Cornucopians see little to fear in population growth (p. 200)
- Rising population can deplete resources, intensify pollution, stress social systems, and degrade ecosystems, such that environmental quality and our quality of life decline. (p. 200)

- Population decline and population aging in some nations have given rise to fears of economic decline. (pp. 200–201)

EVALUATE HOW HUMAN POPULATION, AFFLUENCE, AND TECHNOLOGY AFFECT THE ENVIRONMENT

- The IPAT model summarizes how environmental impact (I) results from interactions among population size (P), affluence (A), and technology (T). (p. 201)
- Rising population and rising affluence may each increase consumption and environmental impact. Technology has frequently worsened environmental degradation, but it can also help mitigate our impacts. (pp. 201–205)

EXPLAIN AND APPLY THE FUNDAMENTALS OF DEMOGRAPHY

- Demography applies principles of population ecology to the statistical study of human populations. (p. 205)

- Demographers study size, density, distribution, age structure, and sex ratios of populations, as well as rates of birth, death, immigration, and emigration. (pp. 205–210)
- Total fertility rate (TFR) contributes greatly to change in population size. (p. 210)

OUTLINE AND ASSESS THE CONCEPT OF DEMOGRAPHIC TRANSITION

- The demographic transition model explains why population growth has slowed in industrialized nations. Industrialization and urbanization reduce the economic need for children, while education and the empowerment of women decrease unwanted pregnancies. Parents in developed nations choose to invest in quality of life rather than quantity of children. (pp. 210–211)
- The demographic transition may or may not proceed to completion in all of today's developing nations. Whether it does is of immense importance for the quest for sustainability. (p. 211)

DESCRIBE HOW WEALTH AND POVERTY, THE STATUS OF WOMEN, AND FAMILY PLANNING AFFECT POPULATION GROWTH

- Many birth control methods serve to reduce unwanted pregnancies. (pp. 211–212)
- When women are empowered and achieve equality with men, fertility rates fall, and children tend to be better cared for, healthier, and better educated. (p. 213)

- Family-planning programs and reproductive education have reduced population growth in many nations. (pp. 214–216)
- Poorer societies tend to show faster population growth than do wealthier societies. (pp. 216–217)
- The intensive consumption of affluent societies often makes their ecological impact greater than that of poorer nations with larger populations. (pp. 217–218)
- Wealth gaps spawn conflict. (pp. 217–218)

CHARACTERIZE THE DIMENSIONS OF THE HIV/AIDS EPIDEMIC

- About 33 million people worldwide are infected with HIV, and 2 million die from AIDS each year. Most live in sub-Saharan Africa. (pp. 218–219)
- Epidemics that claim many young and productive members of society influence population dynamics and can have severe social and economic ramifications. (p. 219)
- We may at last be turning the corner on HIV/AIDS, thanks to a multifaceted effort. (p. 219)

LINK POPULATION GOALS TO SUSTAINABLE DEVELOPMENT GOALS

- Achieving the U.N. Millennium Development Goals will require attention to population growth and resource consumption. (pp. 219–220)

TESTING YOUR COMPREHENSION

1. What is the approximate current human global population? How many people are being added to the population each day?

2. Why has the human population continued to grow despite environmental limitations? Do you think this growth is sustainable? Why or why not?

3. Contrast the views of environmental scientists with those of Cornucopian economists and policymakers regarding whether population growth is a problem. Name several reasons why population growth is commonly viewed as a problem.

4. Explain the IPAT model. How can technology either increase or decrease environmental impact? Provide at least two examples.

5. Describe how demographers use size, density, distribution, age structure, and sex ratio of a population to estimate how it may change. How do each of these factors help determine the impact of human populations on the environment?

6. What is the total fertility rate (TFR)? Why is the replacement fertility for humans approximately 2.1? How is Europe's TFR affecting its natural rate of population change?

7. Why have fertility rates fallen in many countries?

8. How does the demographic transition model explain the increase in population growth rates in recent centuries? How does it explain the recent decrease in population growth rates in many countries?

9. Why are the empowerment of women and the pursuit of gender equality viewed as important to controlling population growth? Describe the aim of family-planning programs.

10. Why do poorer societies have higher population growth rates than wealthier societies? How does poverty affect the environment? How does affluence affect the environment?

SEEKING SOLUTIONS

1. China's reduction in birth rates is leading to significant change in the nation's age structure. Review Figure 8.13, which shows that the population is growing older, leading to the top-heavy population pyramid for the year

2050. What effects might this ultimately have on Chinese society? What steps could be taken in response?

2. The World Bank estimates that half the world's people survive on less than $2 per day. How do you think this

situation affects the political stability of the world? Explain your answer.

3. Apply the IPAT model to the example of China provided in the chapter. How do population, affluence, technology, and ecological sensitivity affect China's environment? Now consider your own country or your own state. How do population, affluence, technology, and ecological sensitivity affect your environment? How can we minimize the environmental impacts of growth in the human population?

4. Do you think that all of today's developing nations will complete the demographic transition and come to enjoy a permanent state of low birth and death rates? Why or why not? What steps might we as a global society take to help ensure that they do? Now think about developed nations such as the United States and Canada. Do you think these nations will continue to lower and stabilize their birth and death rates in a state of prosperity? What factors might affect whether they do so?

5. **THINK IT THROUGH** India's prime minister puts you in charge of that nation's population policy. India has a population growth rate of 1.6% per year, a TFR of 2.7, a 49% rate of contraceptive use, and a population that is 71% rural. What policy steps would you recommend, and why?

6. **THINK IT THROUGH** Now suppose that you have been tapped to design population policy for Germany. Germany is losing population at an annual rate of 0.2%, has a TFR of 1.3, a 66% rate of contraceptive use, and a population that is 73% urban. What policy steps would you recommend, and why?

CALCULATING ECOLOGICAL FOOTPRINTS

A nation's population size and the affluence of its citizens each influence its resource consumption and environmental impact. As of 2010, the world's population passed 6.9 billion, and average per capita income was $10,030 per year, while the latest estimate for the world's average ecological footprint was 2.7 hectares (ha) per person. The sampling of data in the table will allow you to explore patterns in how population, affluence, and environmental impact are related.

Nation	Population (millions of people)	Affluence (per capita income, in GNI PPP)[1]	Personal impact (per capita footprint, in ha/person)	Total impact (national footprint, in millions of ha)
Belgium	10.8	$34,760	5.1	55
Brazil	193.3	$10,070	2.4	
China	1,338.1	$6,020	2.1	
Ethiopia	85.0	$870	1.4	
India	1,188.8	$2,960	0.9	
Japan	127.4	$35,220	4.9	
Mexico	110.6	$14,270	3.4	
Russia	141.9	$15,630	3.7	
United States	309.6	$46,970	9.4	2,884

[1]GNI PPP is "gross national income in purchasing power parity," a measure that standardizes income among nations by converting it to "international" dollars, which indicate the amount of goods and services one could buy in the United States with a given amount of money.

DATA SOURCES: Population and affluence data are from Population Reference Bureau, 2010. *World population data sheet 2010.* Footprint data are for 2005, from WWF International, Zoological Society of London, and Global Footprint Network. Living Planet Report 2008.

1. Calculate the total impact (national ecological footprint) for each country.

2. Draw a graph illustrating per capita impact (on the y axis) vs. affluence (on the x axis). What do the results show? Explain why the data look the way they do.

3. Draw a graph illustrating total impact (on the y axis) in relation to population (on the x axis). What do the results suggest to you?

4. Draw a graph illustrating total impact (on the y axis) in relation to affluence (on the x axis). What do the results suggest to you?

5. You have just used three of the four variables in the IPAT equation. Now give one example of how the T (technology) variable could potentially increase the total impact of the United States, and one example of how it could potentially decrease the U.S. impact.

Farmland in Iowa near the Mississippi River

9 SOIL AND AGRICULTURE

UPON COMPLETING THIS CHAPTER, YOU WILL BE ABLE TO:

- Explain the importance of soils to agriculture
- Describe the impacts of agriculture on soils
- Outline major developments in the history of agriculture
- Delineate the fundamentals of soil science, including soil formation and soil properties

- Analyze the types and causes of soil erosion and land degradation
- Explain the principles of soil conservation and provide solutions to soil erosion and land degradation
- Summarize major policy approaches for pursuing soil conservation

Iowa's Farmers Practice No-Till Agriculture

"The nation that destroys its soil destroys itself."

—U.S. President Franklin D. Roosevelt

"There are two spiritual dangers in not owning a farm. One is the danger of supposing that breakfast comes from the grocery, and the other that heat comes from the furnace."

—Conservationist and Philosopher Aldo Leopold

Iowa farmers Nate Ronsiek (left) and Paul "Butch" Schroeder (right)

Iowa farmers Paul "Butch" Schroeder and his brother David know that their livelihoods depend on keeping their soil healthy and productive. The Schroeder brothers farm 1,200 ha (3,000 acres) in western Iowa, where rich prairie soils have historically made for bountiful grain harvests.

Yet the Schroeders also know that repeated cycles of plowing and planting since farmers first settled the region have diminished the soil's productivity. They know that much of the topsoil—the valuable surface layer richest in organic matter and nutrients—has been lost to erosion: washed away by water and blown away by wind. Turning the earth by tilling (plowing, disking, harrowing, or chiseling) aerates the soil and works weeds and old crop residue into the soil to nourish it, but tilling also leaves the surface bare, allowing wind and water to erode away precious topsoil.

As a result, the Schroeder brothers abandoned the conventional practice of tilling the soil after harvests and instead turned to **no-till** farming. Rather than plowing after each harvest, they began leaving crop residues atop their fields, keeping the soil covered with plant material at all times. To plant the next crop, they cut a thin, shallow groove into the soil surface, dropped in seeds, and covered them. By planting seeds of the new crop through the residue of the old, less soil erodes away, organic material accumulates,

and the soil soaks up more water—all of which encourages better plant growth. On portions of their land where some tilling is required, they practice **conservation tillage,** an approach involving limited tilling—only as much as is needed.

The Schroeders have found that no-till farming saves time and money as well. By foregoing tilling, they reduce the number of passes they need to make on the tractor, which saves fuel, time, effort, and wear and tear on equipment.

Butch and David Schroeder practice other conservation measures on their land as well. They take soil samples to determine how much fertilizer different areas need, so as not to overapply it. They employ planting methods designed to reduce erosion. They forego farming on lands that are vulnerable or have poor soil and retire them as part of the U.S. government's Conservation Reserve Program, which pays farmers to take highly erodible land out of cultivation. All these measures save farmers money while protecting environmental quality and nurturing soil as an investment for sustainable yields in the future.

Future investments are important when a family farm is passed down to the next generation. In Sioux County in northwest Iowa, 26-year-old Nate Ronsiek has taken charge of the farm his late father Vince left to him. Nate and his wife Rachel are putting into practice the strategies for conservation that Nate's father taught him, as well as those he learned as a student at Kansas State University.

"Nate has been farming a short time, but he's doing a lot of things right on his farm that are saving him money and improving the environment," says Greg Marek, a conservationist with the local Natural Resource Conservation Service office.

To raise grain for their 65 stock cows, Ronsiek practices no-till farming. He wanted to test the approach for himself, so he worked with agricultural extension agents from Iowa State University to conduct experiments on his own land and compare results from no-till fields and conventional fields. After three years, his no-till fields produced as much corn as the conventional fields while requiring less time and money. Ronsiek says he can clearly see how water infiltrates better in the no-till fields and how those fields suffer less erosion.

"I'm excited," Ronsiek says. "In future growing seasons yields are expected to increase—profit potential, too. What I don't spend on fuel and time for conventional tillage field trips I can invest elsewhere on the farm."

By enhancing soil conditions and reducing erosion, no-till techniques are benefiting Iowa's people and environment as well, cutting down on pollution in its air, waterways, and ecosystems. Similar effects are being felt elsewhere in the world where no-till methods are being applied.

No-till farming is not a panacea everywhere. But in suitable regions, proponents say it can help make agriculture sustainable. We will need sustainable agriculture if we are to feed the world's human population while protecting the natural environment, including the soils that vitally support our production of food.

SOIL: THE FOUNDATION FOR AGRICULTURE

As the human population has grown, so have the amounts of land and resources we devote to agriculture, which currently covers 38% of Earth's land surface. We can define **agriculture** as the practice of raising crops and livestock for human use and consumption. We obtain most of our food and fiber from **cropland**, land used to raise plants for human use, and from **rangeland**, or pasture, land used for grazing livestock.

Healthy soil is vital for agriculture, for forests (Chapter 12), and for the functioning of Earth's natural systems. **Soil** is not merely lifeless dirt; it is a complex system consisting of disintegrated rock, organic matter, water, gases, nutrients, and microorganisms. Productive soil is a renewable resource. Once depleted, soil may renew itself over time, but this renewal generally occurs very slowly. If we abuse soil through careless or uninformed practices, we can greatly reduce its ability to support life for long time periods.

As population and consumption increase, soils are being degraded

If we are to feed the world's rising human population, we will need to change our diet patterns or increase agricultural production—and do so sustainably, without degrading the environment and its ability to support agriculture. However, we cannot simply keep expanding agriculture into new areas, because land suitable and available for farming is running out. Instead, we must find ways to improve the efficiency of food production in areas already under cultivation, and we must pursue agricultural methods that cause less impact on natural systems.

Today many lands unsuitable for agriculture are being farmed or grazed, causing considerable environmental damage. Mismanaged agriculture can turn grasslands into deserts; remove ecologically precious forests; diminish biodiversity; encourage invasive species; pollute soil, air, and water with toxic chemicals; and allow fertile soil to be blown and washed away.

Each year, our planet gains 80 million people yet loses 5–7 million ha (12–17 million acres, about the size of West Virginia) of productive cropland. Throughout the world, especially in drier regions, it has gotten more difficult to raise crops and graze livestock as soils have deteriorated in quality and declined in productivity—a process termed **soil degradation** (FIGURE 9.1A). Soil degradation has resulted primarily from forest removal, cropland agriculture, and overgrazing of livestock (FIGURE 9.1B).

Over the past 50 years, scientists estimate that soil degradation has reduced potential rates of global grain production by 13% on cropland and 4% on rangeland. By the middle of this century, there will likely be 2.2 billion more mouths to feed. For these reasons, it is imperative that we learn to practice agriculture in sustainable ways that maintain the integrity of our soil.

Agriculture arose starting 10,000 years ago

During most of the human species' 160,000-year existence, we were hunter-gatherers, depending on wild plants and animals for our food and fiber. Then about 10,000 years ago, as glaciers retreated and the climate warmed, people in some cultures began to raise plants from seed and to domesticate animals.

Agriculture most likely began as hunter-gatherers brought wild fruits, grains, and nuts back to their encampments. Some of these foods fell to the ground, were thrown away, or were eaten but survived passage through the digestive system. The plants that grew from these seeds likely produced fruits larger and tastier than those in the wild, since they sprang from seeds of fruits that people had selected because they were especially large and delicious. As these plants bred with others nearby that shared their characteristics, they gave rise to subsequent generations of plants with large and flavorful fruits.

(a) Farmer with degraded soil

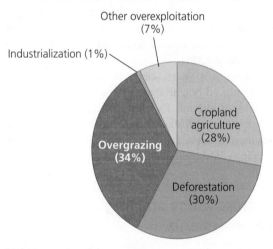

(b) Causes of soil degradation

FIGURE 9.1 ▲ Intensive agriculture has degraded many of the world's soils. Here **(a)** a farmer shows degraded soil in China's Yunnan Province. Most of the world's soil degradation **(b)** results from cropland agriculture, overgrazing by livestock, and deforestation. Data (b) from Wali, M.K., et al., 1999. Assessing terrestrial ecosystem sustainability: Usefulness of regional carbon and nitrogen models. *Nature and Resources* 35: 21–33.

Eventually, people realized they could guide this process, and our ancestors began intentionally planting seeds from plants whose produce was most desirable. This practice of selective breeding (p. 55) continues to the present day and has produced the many hundreds of crops we enjoy, all of which are artificially selected versions of wild plants. People

followed the same process of selective breeding with animals, creating livestock from wild species.

Evidence from archaeology and paleoecology suggests that agriculture was invented independently by different cultures in at least five areas of the world, and possibly 10 or more (**FIGURE 9.2**). The earliest widely accepted evidence for plant and animal domestication is from the "Fertile Crescent" region of the Middle East about 10,500 years ago. By studying radioisotopes (p. 26) of carbon from crop remains, scientists have determined that wheat and barley originated here, as did rye, peas, lentils, onions, garlic, carrots, grapes, and other food plants familiar to us today. The people of the Fertile Crescent also domesticated goats and sheep. In China, domestication began 9,500 years ago, leading to the rice, millet, and pigs we know today. Agriculture in Africa (coffee, yams, sorghum, and more) and the Americas (corn, beans, squash, potatoes, llamas, and more) developed in several regions 4,500–7,000 years ago, and likely as much as 10,000 years ago.

Once our ancestors learned to cultivate crops and raise animals, they began to settle in more permanent camps and villages, often near water sources. Agriculture and a sedentary lifestyle likely reinforced one another in a positive feedback cycle (p. 110): The need to harvest crops kept people sedentary, and once they were sedentary, it made sense to plant more crops. As food supplies became more abundant, carrying capacities (p. 67) increased, and populations rose. Population increase, in turn, further promoted the intensification of agriculture. Moreover, the ability to grow excess farm produce enabled some people to live off the food that others produced, leading to the development of professional specialties, commerce, technology, densely populated urban centers, social stratification, and politically powerful elites. For better or worse, the advent of agriculture eventually brought us the civilization we know today.

For thousands of years, the work of cultivating, harvesting, storing, and distributing crops was performed by human and animal muscle power, along with hand tools and simple machines—an approach known as **traditional agriculture**. In the oldest form of traditional agriculture, known as *subsistence agriculture*, farming families produce only enough food for themselves. As farmers began integrating into market economies and producing excess food to sell, they started using teams of animals for labor and significant quantities of irrigation water and fertilizer.

Industrial agriculture dominates today

The industrial revolution (p. 4) introduced large-scale mechanization and fossil fuel combustion to agriculture just as it did to industry, enabling farmers to replace horses and oxen with faster and more powerful means of cultivating, harvesting, transporting, and processing crops. Such **industrial agriculture** also boosted yields by intensifying irrigation and introducing synthetic fertilizers, while the advent of chemical pesticides reduced competition from weeds and herbivory by crop pests. To be efficient, however, industrial agriculture demands that vast areas be planted with single types of crops. The uniform planting of a single crop, termed **monoculture**, is distinct from

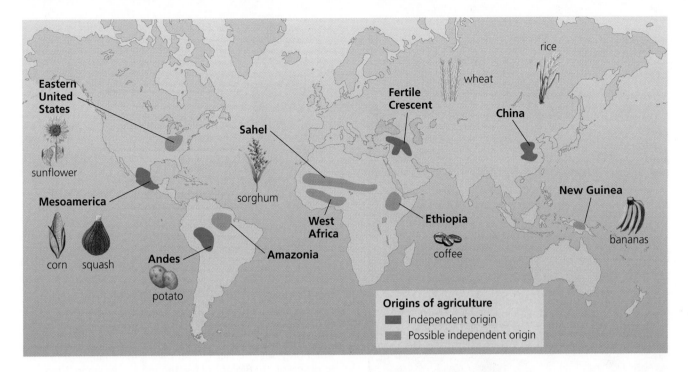

FIGURE 9.2 ▲ Agriculture originated independently in multiple locations throughout the world as different cultures domesticated certain plants and animals from wild species living in their environments. This depiction summarizes conclusions from diverse sources of research on early agriculture. Areas where people are thought to have invented agriculture independently are colored green. (China may represent two independent origins.) In areas colored blue, people either invented agriculture independently or obtained the idea from cultures of other regions. A few of the many crop plants domesticated in each region are shown. Data from syntheses in Diamond, J., 1997. *Guns, germs, and steel*. New York: W.W. Norton; and Goudie, A., 2000. *The human impact*, 5th ed. Cambridge, MA: MIT Press.

the *polyculture* approach of much traditional agriculture, such as Native American farming systems that mixed maize, beans, squash, and peppers in the same fields. Today, industrial agriculture occupies over 25% of the world's cropland and dominates areas such as Iowa.

Industrial agriculture spread from developed nations to developing nations with the advent of the **Green Revolution**, a phenomenon we will explore in Chapter 10 (pp. 254–256). Beginning around 1950, the Green Revolution introduced new technology, crop varieties, and farming practices to the developing world. These advances dramatically increased yields and helped millions avoid starvation. But despite its successes, the Green Revolution is exacting a price. The intensive cultivation of farmland with pesticides, fertilizers, and monocultures creates new problems and worsens old ones—many of which pertain to the integrity of soil, the very foundation of our terrestrial food supply.

SOIL AS A SYSTEM

We generally overlook the startling complexity of soil. Derived from rock, soil also contains a large biotic component, supports plant growth, and is molded by living organisms (**FIGURE 9.3**). By volume, soil consists very roughly of 50% mineral matter and up to 5% organic matter. The rest consists of pore space taken up by air or water. The organic matter in soil includes living and dead microorganisms as well as decaying material derived from plants and animals. A single teaspoon of soil can contain millions of bacteria and thousands of fungi, algae, and protists. Soil also provides habitat for earthworms, insects, mites, millipedes, centipedes, nematodes, sow bugs, and other invertebrates, as well as burrowing mammals, amphibians, and reptiles. The composition of a region's soil can have as much influence on its ecosystems as do climate, latitude, and elevation. In fact, because soil is composed of living and nonliving components that interact in complex ways, soil itself meets the definition of an ecosystem (p. 115).

Soil formation is a slow process

The formation of soil plays a key role in terrestrial primary succession (p. 92), which begins when the lithosphere's parent material is exposed to the effects of the atmosphere, hydrosphere, and biosphere (p. 115). **Parent material** is the base geologic material in a particular location. It can be hardened lava or volcanic ash; rock or sediment deposited by glaciers; wind-blown dunes; sediments deposited by rivers, in lakes, or in the ocean; or **bedrock**, the continuous mass of solid rock that makes up Earth's crust. Parent material is broken down by **weathering**, the physical, chemical, and biological processes that convert large rock particles into smaller particles (**FIGURE 9.4**).

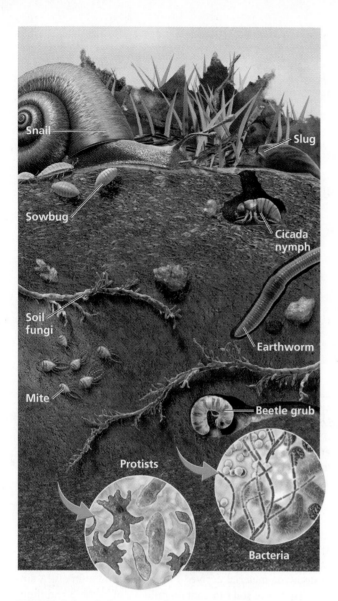

FIGURE 9.3 ▲ Soil is a complex mixture of organic and inorganic components and is full of living organisms whose actions help keep it fertile. In fact, entire ecosystems exist in soil. Most soil organisms, from bacteria to fungi to insects to earthworms, decompose organic matter. Many, such as earthworms, also help to aerate the soil.

Once weathering has produced fine particles, biological activity next contributes to soil formation through the deposition, decomposition, and accumulation of organic matter. As plants, animals, and microbes die or deposit waste, this material is incorporated amid the weathered rock particles, mixing with minerals. The deciduous trees of temperate forests, for example, drop their leaves each fall, making leaf litter available to the detritivores and decomposers (p. 84) that break it down and incorporate its nutrients into the soil. In decomposition, complex organic molecules are broken down into simpler ones, including those that plants can take up through their roots. Partial decomposition of organic matter creates *humus*, a dark, spongy, crumbly mass of material made up of complex organic compounds. Soils with high humus content hold moisture well and are productive for plant life.

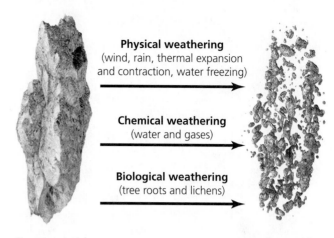

Physical weathering
(wind, rain, thermal expansion
and contraction, water freezing)

Chemical weathering
(water and gases)

Biological weathering
(tree roots and lichens)

Parent material
(rock)

Smaller particles
of parent material

FIGURE 9.4 ▲ *Physical* or *mechanical weathering* breaks rocks down without triggering chemical changes in the parent material. Wind, rain, freezing, and thawing all contribute to physical weathering. In *chemical weathering*, water or gases chemically alter parent material, causing rock to disintegrate. *Biological weathering* occurs when living things break down parent material (by either physical or chemical means). Lichens initiate primary succession by producing acid, which chemically weathers rock, whereas trees may speed physical weathering as their growing roots rub against rock.

Weathering and the accumulation and transformation of organic matter are the key processes of soil formation, but these are influenced by five main factors:

▶ Climate: Soil forms faster in warm, wet climates, because heat and moisture speed most physical, chemical, and biological processes.

▶ Organisms: Plants and decomposers add organic matter to soil over time.

▶ Topography: Hills and valleys affect exposure to sun, wind, and water, and they influence how soil moves.

▶ Parent material: Its attributes influence properties of the resulting soil.

▶ Time: Soil formation can take decades, centuries, or millennia.

WEIGHING THE ISSUES

Earth's Soil Resources It can take hundreds to thousands of years to produce just 1 inch of topsoil. Do you think such a resource should be considered "renewable"? How would you define a "renewable resource"? How might your definition change according to whether you take an anthropocentric, biocentric, or ecocentric ethical perspective (pp. 141–142)? How do you think soil's long renewal time should influence its management? What types of practices encourage the formation of new topsoil?

A soil profile consists of layers known as horizons

As wind, water, and organisms move and sort the fine particles that weathering creates, distinct layers eventually develop. Each layer of soil is termed a **horizon**, and the cross-section as a whole, from surface to bedrock, is known as a **soil profile**.

The simplest way to categorize soil horizons is to recognize A, B, and C horizons corresponding respectively to topsoil, subsoil, and parent material. However, soil scientists often recognize at least three additional horizons (**FIGURE 9.5**). Soils from different locations vary, and few soil profiles contain all six of these horizons, but any given soil contains at least some of them.

Generally, the degree of weathering and the concentration of organic matter decrease as one moves downward in a soil profile. Minerals are generally transported downward as a result of **leaching**, the process whereby solid particles suspended or dissolved in liquid are transported to another location. Soil that undergoes leaching is a bit like coffee grounds in a drip filter. When it rains, water infiltrates the soil, dissolves some of its components, and carries them downward into the lower horizons. Minerals commonly leached from the E horizon include iron, aluminum, and silicate clay. In some soils, minerals may be leached so rapidly that plants are deprived of nutrients. Minerals that leach rapidly from soils may be carried into groundwater and can pose human health threats when the water is extracted for drinking.

A crucial horizon for agriculture and ecosystems is the A horizon, or **topsoil**. Topsoil consists mostly of inorganic mineral components such as weathered substrate, with organic matter and humus from above mixed in. Topsoil is the portion of the soil that is most nutritive for plants, and it takes its loose texture, dark coloration, and strong water-holding capacity from its humus content. The O and A horizons are home to most of the countless organisms that give life to soil. Topsoil is vital for agriculture, but agriculture practiced unsustainably over time will deplete organic matter, reducing the soil's fertility and ability to hold water. When a farmer practices no-till farming, he or she essentially creates an O horizon of crop residue to cover the topsoil and then plants seeds of the new crop through this O horizon into the protected topsoil layer.

Soils differ in color, texture, structure, and pH

The six horizons presented in Figure 9.5 depict an idealized soil, but soils display great variety. Soil scientists classify soils—and farmers judge their quality for farming—based largely on properties such as color, texture, structure, and pH.

Soil color   To a scientist or a farmer, the color of soil can indicate its composition and sometimes its fertility. Black or dark brown soils are usually rich in organic matter, whereas a pale gray to white color often indicates leaching or low organic content.

Soil texture   Soil texture is determined by the size of particles (**FIGURE 9.6**). **Clay** consists of particles less than 0.002 mm in diameter; **silt**, of particles 0.002–0.05 mm; and **sand**, of particles 0.05–2 mm. Sand grains, as any beachgoer knows, are large enough to see individually and do not adhere to one another. Clay particles, in contrast, readily adhere to one another and give clay a sticky feeling when moist. Silt is intermediate, feeling somewhat powdery when dry, and smooth when wet. Soil with an even mixture of the three particle sizes is known as **loam**.

Soils with large particles are porous and allow water to pass through (and beyond the reach of plants' roots) quickly—so crops planted in sandy soils require frequent irrigation. Conversely, soils with very fine particles have much smaller pore spaces because particles pack more closely together, making it difficult for water and air to pass through. Thus clay soils are characterized by slower infiltration of water and lower amounts of oxygen available to soil life. For these reasons, silty soils with medium-sized pores, or loamy soils with mixtures of pore sizes, are generally best for plant growth and agriculture.

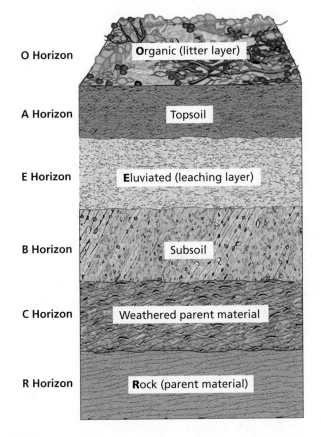

O Horizon — **O**rganic (litter layer)

A Horizon — Topsoil

E Horizon — **E**luviated (leaching layer)

B Horizon — Subsoil

C Horizon — Weathered parent material

R Horizon — **R**ock (parent material)

FIGURE 9.5 ▲ Mature soil consists of layers, or horizons, that have different compositions and characteristics. Uppermost is the **O horizon**, or litter layer (O = organic), consisting mostly of organic matter deposited by organisms. Below it lies the **A horizon**, or topsoil, consisting of some organic material mixed with mineral components. Minerals and organic matter tend to leach out of the **E horizon** (E = eluviation, or leaching) into the **B horizon**, or subsoil, where they accumulate. The **C horizon** consists largely of weathered parent material and overlies an **R horizon** (R = rock) of pure parent material.

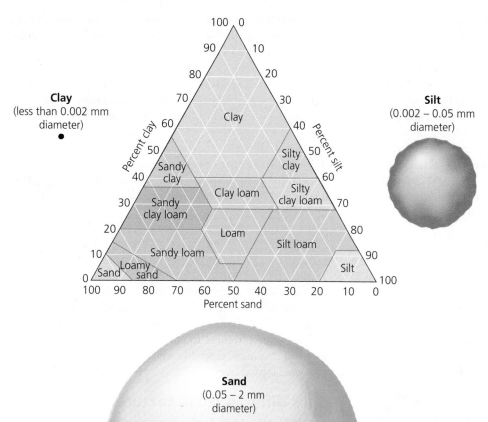

FIGURE 9.6 ◄ The texture of soil depends on its mix of particle sizes. Using the triangular diagram shown, scientists classify soil texture according to the relative proportions of sand, silt, and clay. After measuring the percentage of each particle size in a soil sample, a scientist can trace the appropriate white lines extending inward from each side of the triangle to determine what type of soil texture that particular combination of values creates. Loam is generally the best for plant growth, although some types of plants grow better in other textures of soil.

Soil structure   Soil structure is a measure of the "clumpiness" of soil. An intermediate degree of clumpiness is generally best for plant growth. Repeated tilling can compact soil, reducing its ability to absorb water and inhibiting the penetration of plants' roots.

Soil pH   Plants can die in soils that are too acidic or alkaline (p. 28), whereas moderate variation influences the availability of nutrients for plants' roots. During leaching, for instance, acids from organic matter may remove some nutrients from the sites of exchange between plant roots and soil particles, and water carries these nutrients deeper.

Cation exchange is vital for plant growth

Plants gain many nutrients through a process called *cation exchange*. Soil particle surfaces that are negatively charged hold cations, or positively charged ions (p. 26), such as those of calcium, magnesium, and potassium. In cation exchange, plant roots donate hydrogen ions to the soil in exchange for these nutrient ions, which the soil particles then replenish by exchange with soil water.

Cation exchange capacity expresses a soil's ability to hold cations (preventing them from leaching and thus making them available to plants) and is a useful measure of soil fertility. Soils with fine texture (e.g., clay) and soils rich in organic matter have high cation exchange capacity. As soil pH becomes lower (more acidic), cation exchange capacity

diminishes, nutrients leach away, and soil instead may supply plants with harmful aluminum ions. This is one way in which acid precipitation (pp. 482–486) can damage soils and plant communities.

Regional differences in soil traits can affect agriculture

Soil characteristics vary from place to place. For example, although tropical rainforests have high primary productivity (p. 116), most of their nutrients are tied up in plant tissues and not in the soil. As a result, the soil of the Amazonian rainforest is much less productive than the soil of temperate grassland in Iowa. The enormous amount of rain that falls in the Amazon readily leaches minerals and nutrients out of the topsoil and E horizon. Those not captured by plants are taken quickly down to the water table, below the reach of plants' roots. Warm temperatures speed the decomposition of leaf litter and the uptake of nutrients by plants, so amounts of humus remain small in the thin topsoil layer.

As a result, when forest is cleared for farming, cultivation quickly depletes the soil's fertility. This is why the traditional form of agriculture in tropical forested areas is *swidden* agriculture, in which the farmer cultivates a plot for one to a few years and then moves on to clear another plot, leaving the first to grow back to forest (**FIGURE 9.7A**). At low population densities this can be sustainable, but with today's dense human populations, soils may not be allowed enough time to

(a) Tropical swidden agriculture on nutrient-poor soil

(b) Industrial agriculture on Iowa's rich topsoil

FIGURE 9.7 ▲ In tropical forested areas such as Surinam **(a)**, the traditional form of farming is *swidden* agriculture. In this practice, forest is cut, the plot is farmed for one to a few years, and the farmer then moves on to clear another plot. Frequent movement is necessary because tropical rainforest soils (see inset) are nutrient-poor and easily depleted. Burning the cut vegetation adds nutrients to the soil, which is why this practice is often called *slash-and-burn* agriculture. On the Iowa prairie **(b)**, less rainfall means that nutrients are not leached from the topsoil, and organic matter accumulates, forming rich soil that can be sustainably farmed. Note the thick dark layer of topsoil in the inset for the Iowa farm.

regenerate. As a result, intensive agriculture has degraded the soils and forests of many tropical areas.

On the Iowa prairie, in contrast (**FIGURE 9.7B**), lower rainfall means leaching is reduced and nutrients remain within reach of plants' roots. Plants return nutrients to the topsoil when they die, maintaining its fertility. The thick, rich topsoil of temperate grasslands can be farmed repeatedly with minimal loss of fertility if farming techniques such as no-till and conservation tillage are used.

LAND DEGRADATION AND SOIL CONSERVATION

Some soils are naturally more productive than others, but today human impact is limiting productivity by degrading soils in many areas. The degradation of soil quality due to aspects of agriculture is central to the broader problem known as land degradation. **Land degradation** refers to a general deterioration of land that diminishes its productivity and biodiversity, impairs the functioning of its ecosystems, and reduces the ecosystem services the land can offer us.

Land degradation is caused by the cumulative impacts of intensive and unsustainable agriculture, as well as by deforestation and urban development. It is a global phenomenon that affects up to one-third of the world's people (**FIGURE 9.8**). Land degradation is manifested in processes such as soil erosion, nutrient depletion, water scarcity, salinization (p. 242), waterlogging (p. 242), chemical pollution, changes in soil structure and pH, and loss of organic matter from the soil. Scientists, farmers, and ranchers are working hard to discover and implement solutions for all of these problems to help conserve the productivity of soil, one of our most valuable renewable resources. (see **ENVISIONIT**, p. 234).

Erosion can degrade ecosystems and agriculture

Erosion is the removal of material from one place and its transport toward another by the action of wind or water (**FIGURE 9.9**). *Deposition* occurs when eroded material arrives at a new location and is deposited. Erosion and deposition are natural processes that in the long run can help create soil. Flowing water can deposit freshly eroded sediment rich in nutrients across river valleys and deltas, producing rich and productive soils. This is why floodplains are excellent for farming—and why flood control measures can decrease long-term farming productivity. However, erosion often is a problem locally for ecosystems and agriculture because it generally occurs much more quickly than soil is formed. Erosion also tends to remove topsoil, the most valuable soil layer for living things. In general, steeper slopes, greater precipitation intensities, and sparser vegetative cover all lead to greater water erosion.

People have made land more vulnerable to erosion through three widespread practices:

▶ Overcultivating fields through poor planning or excessive tilling

▶ Overgrazing rangeland with more livestock than the land can support

▶ Clearing forests on steep slopes or with large clear-cuts (pp. 325–326)

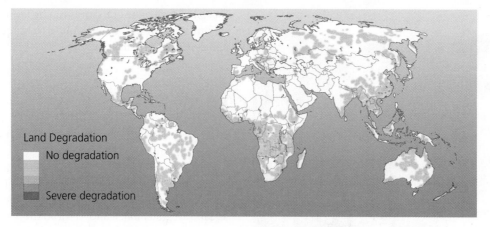

FIGURE 9.8 ▲ Land degradation is a problem worldwide, manifested in soil erosion, depletion of nutrients and organic matter from the soil, water scarcity, pollution, and other problems. Dark areas on this map show regions of the world suffering land degradation, principally as a result of agriculture and forest removal. Degradation is worsening most quickly in developing nations in Africa and Asia. Data from Bai, Z.G., et al., 2008. *Global assessment of land degradation and improvement.* *1. Identification by remote sensing.* Adapted by permission of ISRIC–World Soil Information.

One study determined that at erosion rates typical for the United States, U.S. croplands lose about 2.5 cm (1 in.) of topsoil every 15–30 years, reducing corn yields by 4.7–8.7% and wheat yields by 2.2–9.5%. Erosion can be hard to detect and difficult to measure, even when it may be having substantial consequences. For example, a farmer could lose 12 tons/ha (5 tons/acre) of valuable topsoil, yet this loss represents only a penny's thickness of soil. In many parts of the world, scientists and agriculturalists are carefully measuring erosion rates in hopes of identifying vulnerable areas before they become too badly eroded (see **THE SCIENCE BEHIND THE STORY,** pp. 236–237).

To prevent erosion in vulnerable locations, we can erect a variety of physical barriers that capture soil. In the long-term and across large areas, however, the growth of vegetation is what prevents soil loss. Vegetation slows wind and water flow while plant roots hold soil in place and take up water. This is why no-till agriculture has been so successful—leaving residue on fields after harvest physically shields the topsoil from erosion. Many no-till farmers go a step further and plant *cover crops* during periods between their main crops. Cover crops serve to cover and anchor the soil during a time when conventional farmers would leave it tilled and bare, susceptible to the ravages of wind and water.

According to U.S. government figures, erosion rates in the United States declined from 9.1 tons/ha (3.7 tons/acre) in 1982 to 5.9 tons/ha (2.4 tons/acre) in 2003, thanks to conservation tillage and other soil conservation measures we will discuss soon. Yet in spite of these measures, the erosive impacts of industrial agriculture cause U.S. farmlands to lose 5 tons of soil for every ton of grain harvested.

Soil erosion is a global problem

In today's world, humans are the primary cause of erosion, and we have accelerated it to unnaturally high rates. In a 2004 study, geologist Bruce Wilkinson analyzed prehistoric erosion rates from the geologic record and compared these with modern rates. He concluded that human activities move over 10 times more soil than all other natural processes on the surface of the planet combined. A 2007 study by soil scientist David Montgomery found an even greater degree of impact (**FIGURE 9.10**). Montgomery's data also hinted at a solution, revealing that land farmed under conservation approaches erodes at slower and more natural rates than land under conventional farming.

More than 19 billion ha (47 billion acres) of the world's croplands suffer from erosion and other forms of soil degradation resulting from human activities. In the past decade, China lost an area of arable farmland the size of Indiana. In Kazakhstan, industrial cropland agriculture was imposed on land better suited for grazing, and wind erosion then degraded tens of millions of hectares. For Africa, soil degradation over the next 40 years could reduce crop yields

FIGURE 9.9 ▼ Water erosion can readily remove soil from areas where soil is exposed, such as farmland.

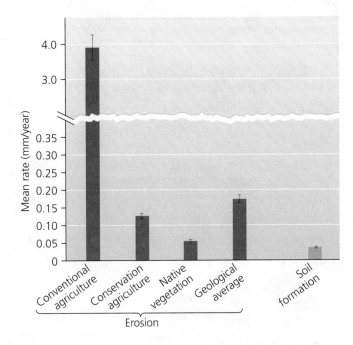

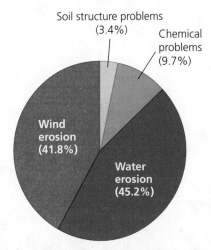

FIGURE 9.11 ▲ Soil degradation on drylands is due primarily to erosion by wind and water. Data from U.N. Environment Programme. 2002. *Tackling land degradation and desertification.* Washington and Rome: Global Environment Facility and International Fund for Agricultural Development.

FIGURE 9.10 ▲ Erosion rates from conventional agriculture (leftmost bar) are much higher than rates in areas covered by native vegetation and than rates averaged over the geologic record. They also greatly exceed the average global rate of soil formation (orange bar). Encouragingly, agriculture under conservation tillage, terracing, and other conservation approaches shows erosion rates much closer to natural rates than to the rates of conventional agriculture. (Estimates for geological rates exceed those for soil production because most erosion estimates are from steeper terrain, where erosion is faster.) Data from Montgomery, David R., 2007. Soil erosion and agricultural sustainability. *Proc. Natl. Acad. Sci.* 104: 13268–13272.

by half. Couple these declines in soil quality and crop yields with the rapid population growth occurring in many of these areas, and we begin to see why some observers describe the future of agriculture as a crisis situation.

Desertification reduces productivity of arid lands

Much of the world's population lives and farms in *drylands*, arid and semi-arid environments that cover about 40% of Earth's land surface. Precipitation in these regions is too meager to meet the demand for water from growing human populations, so drylands are prone to desertification. **Desertification** describes a form of land degradation in which more than 10% of productivity is lost as a result of erosion, soil compaction, forest removal, overgrazing, drought, salinization, climate change, water depletion, and other factors. Most such degradation results from wind and water erosion (**FIGURE 9.11**). Severe desertification can expand existing desert areas and create new ones. This process has occurred in areas of the Middle East that have been inhabited, farmed, and grazed for long periods—including the Fertile Crescent region, where agriculture first originated (pp. 226–227).

These arid lands—in present-day Iraq, Syria, Turkey, Lebanon, and Israel—are not so fertile anymore.

By some estimates, desertification endangers the food supply or well-being of more than 1 billion people and costs people in over 100 countries tens of billions of dollars in income each year. China alone loses $6.5 billion annually from desertification. In its western reaches, desert areas are expanding and joining one another because of overgrazing from over 400 million goats, sheep, and cattle. In the Sistan Basin along the border of Iran and Afghanistan, an oasis that supported a million livestock recently turned barren in just 5 years, and windblown sand buried more than 100 villages. In Kenya, overgrazing and deforestation fueled by rapid population growth has left 80% of its land vulnerable to desertification. In a positive feedback cycle (p. 110), soil degradation forces ranchers to crowd onto poorer land and farmers to reduce the fallow periods during which land lies unplanted and can regain nutrients. Both of these actions further worsen soil degradation.

A 2007 United Nations report estimated that desertification could worsen as climate change alters rainfall patterns, making some areas drier, and could displace 50 million people in 10 years. The report suggested that industrialized nations fund reforestation projects in drylands of the developing world to slow desertification while gaining carbon credits in emissions trading programs (pp. 183, 523–524).

As a result of desertification, in recent years gigantic dust storms from denuded land in China have blown across the Pacific Ocean to North America, and dust storms from Africa's Sahara Desert have blown across the Atlantic Ocean to the Caribbean Sea (see Figure 17.11C, p. 469). Such massive dust storms occurred in the United States during the Dust Bowl days of the early 20th century, when desertification shook American agriculture and society to their very roots.

Each year the world loses 12 million hectares of land to desertification—an area larger than Pennsylvania, or 2,000 times the size of Manhattan.

Dunes encroaching on farms in China

Agriculture in arid areas contributes to land degradation, and droughts from climate change may worsen the situation.

Dust storm

Revegetation helps prevent erosion and desertification.

Planting trees in China

Drought in India

YOU CAN MAKE A DIFFERENCE

➤ Ask your federal legislators to support the Conservation Reserve Program and other measures to fight erosion in agriculture.

➤ Use water in moderation to help conserve your local supply.

➤ Volunteer for tree-planting and land stewardship projects in your region.

The Dust Bowl was a monumental event in the United States

Prior to large-scale cultivation of North America's Great Plains, native prairie grasses of this temperate grassland region held soils in place. In the late 19th and early 20th centuries, many homesteading settlers arrived in Oklahoma, Texas, Kansas, New Mexico, and Colorado with hopes of making a living there as farmers. Between 1879 and 1929, cultivated area in the region soared from around 5 million ha (12 million acres) to 40 million ha (100 million acres). Farmers grew abundant wheat, and ranchers grazed many thousands of cattle, sometimes expanding onto unsuitable land. Both types of agriculture contributed to erosion by removing native grasses and altering soil structure.

In the early 1930s, a drought exacerbated the ongoing human impacts, and the region's strong winds began to erode millions of tons of topsoil (**FIGURE 9.12**). Dust storms traveled up to 2,000 km (1,250 mi), blackening rain and snow as far away as New York and Vermont. Some areas lost as much as 10 cm (4 in.) of topsoil in a few years. The most-affected region in the southern Great Plains became known as the **Dust Bowl**, a term now also used for the historical event itself. The "black blizzards" of the Dust Bowl forced thousands of farmers off their land, and many who remained had to rely on government assistance programs to survive.

The Soil Conservation Service pioneered measures to address soil degradation

In response to the devastation in the Dust Bowl, the U.S. government, along with state and local governments, increased support for research into soil conservation measures. The U.S. Congress passed the Soil Conservation Act of 1935, establishing the Soil Conservation Service (SCS). This new agency worked closely with farmers to develop conservation plans for individual farms, following several aims and principles:

▶ Assess the land's resources, problems, and opportunities for conservation.

▶ Draw on science to prepare an integrated plan for each property.

▶ Work closely with land users to ensure that conservation plans harmonize with users' objectives.

▶ Implement conservation measures on individual properties to contribute to the overall quality of life in the watershed or region.

The early teams that the SCS formed to combat erosion typically included soil scientists, forestry experts, engineers, economists, and biologists. These teams were among the earliest examples of interdisciplinary approaches to environmental problem solving. The first director of the SCS, Hugh Hammond Bennett, was an innovator and evangelist for soil conservation. Under his leadership, the agency promoted soil conservation practices through county-based **conservation districts**. These districts operate with federal direction, authorization, and funding, but they are organized by the states. The districts implement soil conservation programs locally and aim to empower local residents to plan and set priorities in their home areas. In 1994 the SCS was renamed the **Natural Resources Conservation Service (NRCS)**, and its responsibilities were expanded to include water quality protection and pollution control.

In addition, most state universities employ *agricultural extension agents*, experts who assist farmers by providing information on new research and helping them apply this knowledge with new techniques. Today in Iowa and elsewhere across the country, extension agents from both universities and government agencies are advising farmers on countless matters and helping them switch to methods such as no-till farming (**FIGURE 9.13A**). Nate Ronsiek of Sioux County, Iowa, is one of many farmers who take advantage of the expertise and resources that these experts have to offer. "If there is a way to do it, they know it," he says. "NRCS keeps me up-to-date with the latest in conservation techniques, practices, and

(a) Kansas dust storm, 1930s

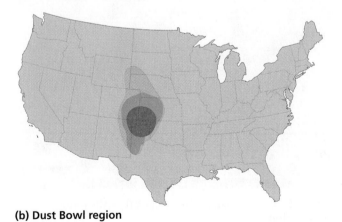

(b) Dust Bowl region

FIGURE 9.12 ▲ Drought combined with poor agricultural practices brought devastation and despair to millions of U.S. farmers in the 1930s in the Dust Bowl region of the southern Great Plains. The photo **(a)** shows towering clouds of dust approaching houses near Dodge City, Kansas, in a 1930s dust storm. The map **(b)** shows the Dust Bowl region, with darker colors indicating the areas most affected.

Measuring Erosion with Pins and . . . Nuclear Fallout?

Erosion continually shapes the landscapes around us, but it is fiendishly difficult to measure with accuracy. Soil scientists such as the late Jerry Ritchie, a noted U.S. Department of Agriculture researcher who died in 2009, have adopted both low-tech and high-tech methods—from simple measuring pins to complex radiation detectors—to address this vexing challenge.

In the 1990s, Ritchie and his team of researchers studied grass hedges—widely used to trap eroding soil by slowing runoff from rain—to measure how well they actually work. Ritchie's team began by measuring erosion around hedges planted near gullies in Maryland. They relied on simple, inexpensive tools known as erosion pins (**see figure**), which were developed in

Dr. Jerry Ritchie, measuring soil movement with Carole Ritchie

the 1960s and 1970s by scientists working for the United Nations Food and Agriculture Organization. Erosion pins are spikes that can be made from almost anything, including bamboo stakes or pieces of plastic pipe. The pins, each cut to a uniform length, are driven into the soil until their tops are level with the ground's surface. Over time, if soil in the area is eroding, the soil surface will recede, and the erosion pins will be increasingly exposed.

By using many pins over a wide area and averaging their readings, scientists can determine an overall erosion rate for the area.

Researchers often dig expansive holes, called catchpits, nearby. These pits are lined with plastic and serve to collect eroding soil. Researchers measure the volume of soil that accumulates in the catchpits and compare that data with the extent of exposure on erosion pins. In the Maryland experiment, Ritchie used such techniques and found that 1–2 cm (0.4–0.8 in.) of soil was accumulating upslope from the hedges per year, indicating that the grass was indeed trapping soil.

Erosion pins and catchpits work well in one spot but are impractical over a large region. To get evidence

ways to save money." Ronsiek was just one of 27 farmers who worked with extension agents from Iowa State University to test the effectiveness of no-till farming versus several types of tilling.

Soil conservation efforts are thriving internationally

The SCS and NRCS have served as models for efforts elsewhere in the world (**FIGURE 9.13B**). In South America, no-till agriculture has exploded in popularity and now covers a large proportion of farmland in Argentina, Brazil, and Paraguay. The shift to no-till farming across this vast region came about largely through local grassroots organization by farmers, with the help of agronomists and government extension agents who provided them information and resources. Southernmost Brazil alone boasts thousands of "Friends of the Land" clubs in which local farmers collaborate with trained experts.

From Argentina to Iowa, no-till agriculture is one of many approaches to soil conservation. Hugh Hammond

(a) NRCS extension agent Greg Marek (L) visits farm of Nate Ronsiek (R).

FIGURE 9.13 ▲ Agricultural extension agents assist farmers by providing information on new research and techniques that can help them farm productively while minimizing damage to the land. In Iowa **(a)**, NRCS extension agent Greg Marek (left) helps farmer Nate Ronsiek check a terrace between his no-till fields. In South America **(b)**, an extension agent from Colombia's Instituto Colombiano Agropecuario inspects yuca plants grown by farmer Pedro Gomez.

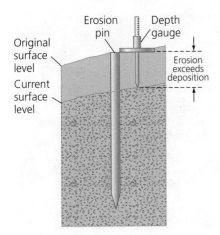

Erosion pin   Depth gauge

Original surface level
Current surface level

Erosion exceeds deposition

As soil erodes around an erosion pin, more of the pin is exposed, enabling the soil scientist to measure the amount of soil loss.

of erosion on a wide scale, scientists have turned to measuring a modern-day leftover rarely considered an environmental benefit—nuclear fallout from atomic weapons testing. In 1945, the United States exploded the world's first nuclear bomb, and since then more than 2,000 nuclear devices have been tested by the United States, the former Soviet Union, and other nations. Nuclear weapons testing has spread radioactive material through the atmosphere worldwide, and fallout

has covered Earth's surface with minuscule but measurable amounts of radioactive debris.

Fallout includes cesium-137, a radioactive isotope (p. 26) of the element cesium. Cesium-137, a product of nuclear fuel and weapons reactions, has a half-life of 30 years—a duration that enables soil scientists to use the isotope as a universal environmental tracer for erosion and sediment deposits. Soil tends to absorb cesium-137 quickly and evenly, so if soil in an area hasn't moved or been heavily disturbed, testing will show fairly uniform levels of the isotope. If some areas show low concentrations of cesium-137 and others show high concentrations, then erosion may be at work. Ritchie decided to sample for cesium-137 in the area around the hedges and compare the results against his physical soil measurements.

In studies involving cesium-137, soil samples are tested using a gamma spectrometer. This device measures gamma rays, which serve as unique signatures of energy emitted by chemical elements in soil or rock. As they are emitted, gamma rays show up

as sharp emission lines on a spectrum. The energy represented in these emissions reflects which elements are present, and the intensity of the lines reveals the concentration of each element. Thus researchers can calculate the amount of an isotope such as cesium-137 in a test sample. In erosion tests, each sample from the study area is measured for cesium-137, and those levels are compared to baseline levels for the region. By pinpointing places in the study area with lower or higher levels of accumulated cesium, scientists can detect where soil has moved and how much has shifted.

In Maryland, the radioactive testing helped Ritchie determine that hedges may offer only partial help against erosion. Although his team's physical measurements of soil accumulation showed soil being deposited near the hedges, the cesium-137 tests revealed that the area around the hedges had nonetheless undergone a net loss of soil over four decades. Grass hedges can help, Ritchie wrote when releasing his findings for the USDA's Agricultural Research Service in 2000, but they "should not be seen as a panacea." ∎

(b) Colombian extension agent and farmer

Bennett advocated a complex approach, combining techniques such as crop rotation, contour farming, strip cropping, terracing, grazing management, and reforestation, as well as wildlife management. Such measures are now widely applied in many places around the world.

Farmers can protect soil against degradation in various ways

A number of farming techniques can reduce the impacts of conventional cultivation on soils (**FIGURE 9.14**). Some of these have been promoted by the SCS since the Dust Bowl. Some, like no-till farming in Iowa, are finding popularity more recently. Others have been practiced by certain cultures for centuries.

Crop rotation   In **crop rotation**, farmers alternate the type of crop grown in a given field from one season or year to the next (**FIGURE 9.14A**). Rotating crops can return nutrients to the soil, break cycles of disease associated with continuous cropping, and minimize the erosion that can come from letting fields lie fallow. Many U.S. farmers rotate their

(a) Crop rotation

(b) Contour farming

(c) Terracing

(d) Intercropping

(e) Shelterbelts

(f) No-till farming

FIGURE 9.14 ▲ The world's farmers have adopted various strategies to conserve soil. Rotating crops such as soybeans and corn **(a)** helps restore soil nutrients and reduce impacts of crop pests. Contour farming **(b)** reduces erosion on hillsides. Terracing **(c)** minimizes erosion in steep mountainous areas. Intercropping **(d)** can reduce soil loss while maintaining soil fertility. Shelterbelts **(e)** protect against wind erosion. In **(f)**, corn grows up from amid the remnants of a "cover crop" used in no-till agriculture.

fields between wheat or corn and soybeans from one year to the next. Soybeans are legumes, plants that have specialized bacteria on their roots that fix nitrogen (p. 130), revitalizing soil that the previous crop had partially depleted of nutrients. Crop rotation also reduces insect pests; if an insect is

adapted to feed and lay eggs on one crop, planting a different type of crop will leave its offspring with nothing to eat.

In a practice similar to crop rotation, many no-till farmers such as those in Iowa plant cover crops designed to prevent erosion and nitrogen loss during times of year when the main crops are not growing. Using legumes (such as clover) as cover crops will restore nitrogen to the soil.

Contour farming   Water running down a hillside with little vegetative cover can easily carry soil away, so farmers have developed several methods for cultivating slopes. **Contour farming** (**FIGURE 9.14B**) consists of plowing furrows sideways across a hillside, perpendicular to its slope and following the natural contours of the land, to help prevent formation of rills and gullies. In contour farming, the side of each furrow acts as a small dam that slows runoff and captures soil. Contour farming is most effective on gradually sloping land with crops that grow well in rows. Iowa farmers Butch and David Schroeder practice contour farming on all their fields. They also plant buffer strips of vegetation along the borders of their fields and along nearby streams (see Figure 5.21, p. 132), which further protects against erosion and water pollution.

Terracing   On extremely steep terrain, terracing (**FIGURE 9.14C**) is the most effective method for preventing erosion. Terraces are level platforms, sometimes with raised edges, that are cut into steep hillsides to contain water from irrigation and precipitation. **Terracing** transforms slopes into series of steps like a staircase, enabling farmers to cultivate hilly land without losing huge amounts of soil to water erosion. Farmers have used terracing for centuries in ruggedly mountainous regions, such as the foothills of the Himalayas and the Andes. Terracing is labor-intensive to establish but in the long term is likely the only sustainable way to farm in mountainous terrain.

Intercropping   Farmers may also minimize erosion by **intercropping**, planting different types of crops in alternating bands or other spatially mixed arrangements (**FIGURE 9.14D**). Intercropping helps slow erosion by providing more ground cover than does a single crop. Like crop rotation, intercropping reduces vulnerability to insects and disease and, when a nitrogen-fixing legume is used, replenishes the soil. In southern Brazil, some no-till farmers intercrop food crops with cover crops. The cover crops are physically mixed with primary crops, which include maize, soybeans, wheat, onions, cassava, grapes, tomatoes, tobacco, and orchard fruit.

Shelterbelts   A widespread technique to reduce erosion from wind is to establish **shelterbelts**, or *windbreaks* (**FIGURE 9.14E**). These are rows of trees or other tall plants that are planted along the edges of fields to slow the wind. On the Great Plains, fast-growing species such as poplars are often used. Shelterbelts can be combined with intercropping by planting mixed crops in rows surrounded by or interspersed with rows of trees that provide fruit, wood, or protection from wind.

Conservation tillage   *Conservation tillage* describes an array of approaches that reduce the amount of tilling relative to conventional farming; one common definition is any method of limited tilling that leaves more than 30% of crop residue covering the soil after harvest. No-till farming is the ultimate form of conservation tillage. To plant using the no-till method (**FIGURE 9.14F**), a tractor pulls a "no-till drill" (**FIGURE 9.15**) that cuts furrows through the O horizon of dead weeds and crop residue and the upper levels of the A horizon. The device drops seeds into the furrow and closes

FIGURE 9.15 ▲ Farmers practice no-till agriculture with a no-till drill like the one shown here. The drill ❶ cuts a furrow through the soil surface, ❷ drops in a seed, and ❸ closes the furrow over the seed.

the furrow over the seeds. Often a localized dose of fertilizer is added to the soil along with the seed.

By increasing organic matter and soil biota while reducing erosion, no-till farming and conservation tillage can restore and improve soil quality. Based on results in Iowa and elsewhere, proponents of no-till farming credit the practice with a number of benefits (**TABLE 9.1**). One benefit that looms larger and larger is that of carbon storage (pp. 519, 546–547). To mitigate global climate change (Chapter 18), we must find ways to reduce the atmospheric concentration of carbon dioxide. By adding organic matter to the soil, no-till farming captures carbon that otherwise would make its way to the atmosphere and instead stores it in the soil. In addition, because no-till farming requires fewer rounds on the tractor, the farmer burns less gasoline. Some farmers are now even receiving money from carbon offsets through the Chicago Climate Exchange (p. 523) for using no-till methods.

In the United States today, nearly one-quarter of farmland is under no-till cultivation, and over 40% is farmed

TABLE 9.1   Benefits of No-Till Farming
Benefits of no-till farming
▶ Requires less labor
▶ Saves time
▶ Reduces wear on machinery
▶ Decreases fossil fuel use
▶ Improves soil productivity in long term
▶ Enhances surface water quality
▶ Reduces soil erosion
▶ Enhances moisture retention by soil
▶ Improves water infiltration into soil
▶ Lessens soil compaction
▶ Enhances conditions for wildlife
▶ Reduces release of carbon to atmosphere
▶ Decreases air pollution

Adapted from International Soil Tillage Research Organization, (ISTRO), Info-extra, Vol. 3, No. 1, January 1997.

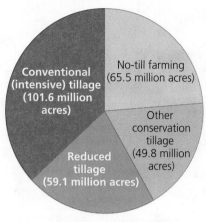

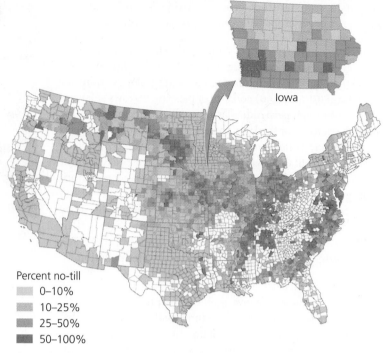

(a) Types of tillage in United States

Percent no-till
- 0–10%
- 10–25%
- 25–50%
- 50–100%

(b) Distribution of no-till farming in United States, and in Iowa

FIGURE 9.16 ▲ No-till farming is practiced on 23.7% of U.S. farmland **(a)**, and other forms of conservation tillage are implemented on a further 18.0%. "Reduced tillage" (15–30% crop residue left on ground after harvest) takes place on 21.4% of land, and conventional farming (0–15% residue left on ground) occurs on 36.8% of land. Across the United States **(b)**, red and orange colors on this county-by-county map show that no-till methods are practiced most in the Midwest and along the central and southern Atlantic seaboard. In Iowa, no-till methods are most common in the southern half of the state. Data are for 2007 (a) and 2004 (b), from Conservation Technology Information Center, National Crop Residue Management Survey.

using conservation tillage (**FIGURE 9.16A**). Forty percent of soybeans, 21% of corn, and 18% of cotton receive no-till treatment. No-till farming in the United States is most prevalent in the Midwest, the Great Plains, the Ohio and Mississippi River valleys, and along the central and southern Atlantic seaboard (**FIGURE 9.16B**). Iowa ranks second only to Illinois in total area under no-till farming, but its rate of no-till farming (23%) is close to the national average. Four states use no-till methods on over half their farmland: Tennessee, Virginia, Maryland, and Kentucky.

No-till and conservation tillage methods were pioneered in the United States and the United Kingdom, but they have become most widespread in subtropical and temperate South America. In Brazil, Argentina, and Paraguay, over half of all cropland is now under no-till cultivation. In this part of the world, heavy rainfall promotes erosion, causing tilled soils to lose organic matter and nutrients, and hot weather can overheat tilled soil. Thus, the no-till approach is especially helpful here, and results have exceeded those in the United States: Crop yields increased (in some cases they nearly doubled), erosion was reduced, soil quality was enhanced, and pollution declined, all while costs to farmers dropped by roughly 50%.

Critics of no-till farming in the United States have noted that this approach often requires substantial use of chemical herbicides (because weeds are not physically removed from fields) and synthetic fertilizer (because other plants take up a significant portion of the soil's nutrients).

In many industrialized countries, this has indeed been the case. Proponents, however, point out that in developing regions of South America, farmers have departed from the industrialized model by relying more heavily on *green manures* (dead plants as fertilizer) and by rotating fields with cover crops, including nitrogen-fixing legumes. The manures and legumes nourish the soil, and cover crops reduce weeds by taking up space the weeds might occupy.

Critics maintain, however, that green manures are generally not practical for large-scale intensive agriculture. Certainly, conservation tillage methods work well in some areas but not in others, and they work better with some crops than with others. Farmers will do best by educating themselves on the options and doing what is best for their particular crops on their own land.

WEIGHING THE ISSUES

How Would You Farm? You are a farmer owning land on both sides of a steep ridge. You want to plant a sun-loving crop on the sunny, but very windy, south slope of the ridge and a crop that needs a great deal of irrigation on the north slope. What type of farming techniques might maximize conservation of your soil? What other factors might you want to know about before you decide to commit to one or more methods?

FIGURE 9.17 ▲ Vast swaths of countryside in northern and western China have been planted with fast-growing poplar trees. These "reforestation" efforts do not create ecologically functional forests—the plantations are too biologically simple—but they do reduce soil erosion.

Erosion control practices protect and restore plant cover

Farming methods to control erosion make use of the general principle that maximizing vegetative cover will protect soils, and this principle has been applied widely beyond farming. When grazing livestock, people try to move animals from place to place before the plant cover in any one area is reduced too much. In the cultivation of trees through the practice of forestry (Chapter 12), methods such as clear-cutting—the removal of all trees from an area at once—can lead to severe erosion, particularly on steep slopes, but alternative methods that remove fewer trees over longer periods of time help to minimize soil loss. When banks along creeks and roadsides erode, we plant plants to anchor the soil. In areas with severe and widespread erosion, some nations have planted vast plantations of fast-growing trees. China has embarked on the world's largest tree-planting program to slow its soil loss (**FIGURE 9.17**). Although such "reforestation" efforts help slow erosion, they do not at the same time produce ecologically functioning forests, because tree species are selected only for their fast growth and are planted in monocultures.

Irrigation boosts productivity but can damage soil

Erosion is not the only threat to the health and integrity of soils. Soil can also be degraded when we apply water to crops. The artificial provision of water to support agriculture is known as **irrigation**. Some crops, such as rice and cotton, require large amounts of water, whereas others, such as beans and wheat, require relatively little. Other factors influencing the amount of water required for growth include the rate of evaporation and the soil's ability to hold water

and make it available to plant roots. If the climate is too dry or too much water evaporates or runs off before it can be absorbed into the soil, crops may require irrigation. By irrigating crops, people have managed to turn previously dry and unproductive regions into fertile farmland.

Seventy percent of all fresh water that people withdraw is used for irrigation. Irrigated acreage has increased dramatically worldwide, reaching almost 400 million ha (nearly 1 billion acres) in 2007, twice the area of Mexico. In some cases, withdrawing water for irrigation has depleted aquifers and dried up rivers and lakes. Currently, irrigation efficiency worldwide is low; plants end up using only 43% of the water that we apply. Drip irrigation systems (**FIGURE 9.18**) that target water directly to plants are one solution. These systems allow more control over where water is aimed and waste far less water. Once considered expensive to install, they are becoming cheaper, such that more farmers in developing countries will be able to afford them. We will examine irrigation further in Chapter 15 (p. 408).

(a) Conventional irrigation

(b) Drip irrigation

FIGURE 9.18 ▲ Currently, plants take up less than half the water we apply in irrigation. Conventional methods **(a)** lose a great deal of water to evaporation. In more efficient irrigation approaches, water is precisely targeted to plants. In drip irrigation systems, such as this one watering grape vines in California **(b)**, hoses are arranged so that water drips directly onto the plants.

Salinization is easier to prevent than to correct

If some water is good for plants and soil, it might seem that more must be better. But we can supply too much water. Overirrigated soils saturated with water may experience **waterlogging** when the water table rises to the point that water bathes plant roots, depriving them of access to gases and essentially suffocating them.

A more frequent problem is **salinization**, the buildup of salts in surface soil layers. In dryland areas where precipitation is minimal and evaporation rates are high, the evaporation of water from the soil's A horizon may pull water from lower horizons upward by capillary action. As this water rises through the soil, it carries dissolved salts. When the water evaporates at the surface, those salts precipitate, often turning the soil surface white. Irrigation in arid areas generally hastens salinization, and irrigation water often contains some dissolved salt in the first place, introducing new salt to the soil. Salinization now inhibits production on one-fifth of all irrigated cropland globally, costing more than $11 billion each year.

Remedying salinization once it has occurred is more expensive and difficult than preventing it in the first place. The best way to prevent salinization is to avoid planting crops that require a great deal of water in areas that are prone to salinization. A second way is to irrigate with water low in salt content. A third way is to irrigate efficiently, supplying no more water than the crop requires. This minimizes the amount of water that evaporates and hence the amount of salt that accumulates in the topsoil.

If salinization has occurred, one potential way to mitigate it is to stop irrigating and wait for rain to flush salts from the soil. However, this solution is unrealistic because salinization generally becomes a problem in dryland areas, where precipitation is rarely adequate to flush soils. A better option may be to plant salt-tolerant plants, such as barley, that can be used as food or pasture. A third option is to bring in large quantities of less-saline water with which to flush the soil. However, using too much water may cause waterlogging.

(a) Inorganic fertilizer

(b) Organic fertilizer

FIGURE 9.19 ▲ Farmers apply fertilizers to soils to add nitrogen, phosphorus, and other nutrients that encourage plant growth. In **(a)**, a farmer applies synthetically manufactured granules of inorganic fertilizer to a wheat crop. In **(b)**, earthworms crawl through rich compost used as organic fertilizer on an organic farm.

Fertilizers boost crop yields but can be overapplied

Aside from salinization, we can also chemically damage soil by overapplying fertilizers. Plants require nitrogen, phosphorus, and potassium to grow, as well as smaller amounts of over a dozen other nutrients. Plants remove these nutrients from soil as they grow, and leaching likewise removes nutrients. If agricultural soils come to contain too few nutrients, crop yields decline. Therefore, we go to great lengths to enhance nutrient-limited soils by adding **fertilizer**, any of various substances that contain essential nutrients (**FIGURE 9.19**).

There are two main types of fertilizers. **Inorganic fertilizers** are mined or synthetically manufactured mineral supplements. **Organic fertilizers** consist of the remains or wastes of organisms and include animal manure; crop residues; fresh vegetation (*green manure*); and *compost*, a mixture produced when decomposers break down organic matter, including food and crop waste, in a controlled environment. Organic fertilizers can provide some benefits that inorganic fertilizers cannot. The proper use of compost improves soil structure, nutrient retention, and water-retaining capacity, helping to prevent erosion. As a form of recycling, composting reduces the amount of waste consigned to landfills and incinerators (pp. 625–626). However, organic fertilizers are no panacea. For instance, when manure is applied in amounts needed to supply sufficient nitrogen for a crop, it may introduce excess phosphorus that can run off into waterways.

Inorganic fertilizers are generally more susceptible to leaching and runoff, and they are somewhat more likely to

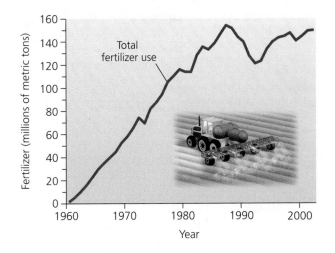

FIGURE 9.20 ▲ Use of synthetic, inorganic fertilizers has risen sharply over the past half century and now stands at over 147 million metric tons annually. (The temporary drop during the early 1990s was due to economic decline in countries of the former Soviet Union following that nation's dissolution.) Data from Food and Agriculture Organization of the United Nations (FAO).

cause unintended off-site impacts. The use of inorganic fertilizer worldwide has grown tremendously in recent decades (**FIGURE 9.20**). Its use has greatly boosted our global food production, but its overapplication is causing increasingly severe pollution problems.

Applying fertilizer to croplands can have impacts far beyond the boundaries of the fields (**FIGURE 9.21**). We saw in Chapter 5 how nitrogen and phosphorus runoff from farms and other sources in the Mississippi River basin each year spurs phytoplankton blooms in the Gulf of Mexico and creates an oxygen-depleted "dead zone" that kills fish and shrimp. Such eutrophication (pp. 114, 421–422) occurs at countless river mouths, lakes, and ponds throughout the world. Moreover, nitrates readily leach through soil and contaminate groundwater, and components of some nitrogen fertilizers can even volatilize (evaporate) into the air. Through these processes, unnatural amounts of nitrates and phosphates spread through ecosystems and pose human health risks, including cancer and methemoglobinemia, or blue-baby syndrome, which can asphyxiate and kill infants. The U.S. Environmental Protection Agency (EPA) has determined that nitrate concentrations in excess of 10mg/L

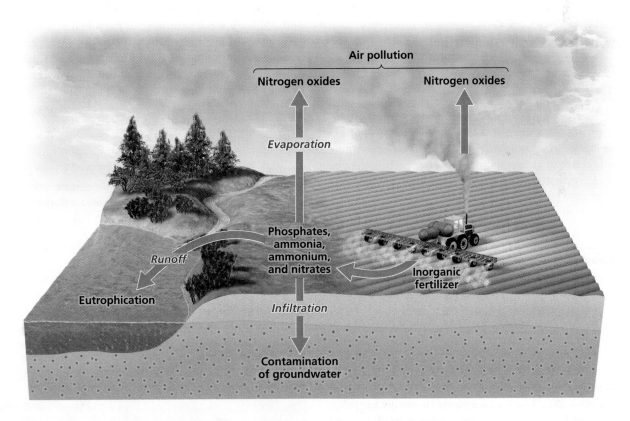

FIGURE 9.21 ▲ The overapplication of inorganic (or organic) fertilizers can have effects beyond the farm field, because nutrients not taken up by plants end up elsewhere. Nitrates can leach into groundwater, where they can pose a threat to human health in drinking water. Phosphates and some nitrogen compounds can run off into surface waterways and alter the ecology of streams, rivers, ponds, and lakes through eutrophication. Some compounds, such as nitrogen oxides, can even enter and pollute the air. Human inputs of nitrogen have greatly modified the nitrogen cycle (pp. 128–130) and now account for one-half the total nitrogen flux on Earth.

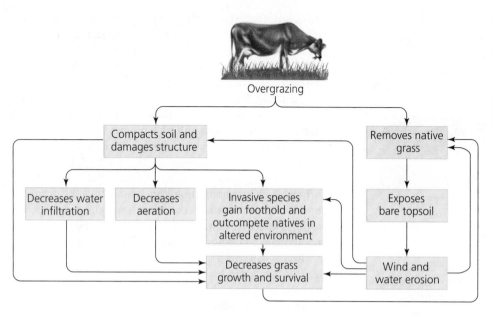

Overgrazing

Compacts soil and damages structure

Removes native grass

Decreases water infiltration

Decreases aeration

Invasive species gain foothold and outcompete natives in altered environment

Exposes bare topsoil

Decreases grass growth and survival

Wind and water erosion

FIGURE 9.22 ◄ When grazing by livestock exceeds the carrying capacity of rangelands and their soil, overgrazing can set in motion a series of consequences and positive feedback loops that degrade soils and grassland ecosystems.

for adults and 5 mg/L for infants in drinking water are unsafe, yet many sources around the world exceed even the looser standard of 50 mg/L set by the World Health Organization.

Grazing practices can contribute to soil degradation

We have focused in this chapter largely on the cultivation of crops as a source of impacts on soils and ecosystems, but raising livestock also has impacts. When sheep, goats, cattle, or other livestock graze on the open range, they feed primarily on grasses. As long as livestock populations do not exceed a range's carrying capacity (p. 67) and do not consume grasses faster than grasses can be replaced, grazing may be sustainable. Moreover, human use of rangeland does not necessarily exclude its use by wildlife or its continued functioning as a grassland ecosystem. However, when too many livestock eat too much of the plant cover, impeding plant regrowth and preventing the replacement of biomass, the result is **overgrazing**.

Rangeland scientists have shown that overgrazing causes a number of impacts that damage soils, natural communities, and the land's productivity for grazing (**FIGURE 9.22**). When livestock remove too much plant cover, soil is exposed and made vulnerable to erosion. In a positive feedback cycle, soil erosion makes it difficult for vegetation to regrow, a problem that perpetuates the lack of cover and gives rise to more erosion. Moreover, non-native weedy plants may invade denuded soils (**FIGURE 9.23**). These invasive plants are usually less palatable to livestock and can outcompete native vegetation in the new, modified environment, further decreasing native plant cover.

Too many livestock trampling the ground compacts soils and alters their structure. Soil compaction makes it more difficult for water to infiltrate, for soils to be aerated, for plants' roots to expand, and for roots to conduct cellular respiration (p. 32). All of these effects further decrease the growth and survival of native plants.

As a cause of soil degradation worldwide, overgrazing is equal to cropland agriculture, and it is a greater cause of desertification. Humans keep a total of 3.4 billion cattle, sheep, and goats. Rangeland classified as degraded now adds up to 680 million ha (1.7 billion acres), five times the area of U.S. cropland, although some estimates put the number as high as 2.4 billion ha (5.9 billion acres), fully 70% of the world's rangeland area. Rangeland degradation is estimated to cost $23.3 billion per year. Grazing exceeds the sustainable supply of grass in India by 30% and in parts of China by up to 50%. To relieve pressure on rangelands, both nations are now beginning to feed crop residues to livestock.

Range managers in the United States do their best to assess the carrying capacity of rangelands and inform livestock owners of these limits, so that herds are rotated from site to site as needed to conserve grass cover and soil integrity. Managers also can establish and enforce limits on grazing on publicly owned land. Yet U.S. ranchers have

FIGURE 9.23 ▼ The effects of overgrazing can be striking, as shown in this photo along a fence line separating a grazed field (left) from an ungrazed field (right). The overgrazed field has lost its grass cover and has been invaded by weedy plants that outcompete native grasses in degraded soils.

traditionally had little incentive to conserve rangelands because most grazing has taken place on public lands leased from the government and because the U.S. government has heavily subsidized grazing. As a result of this classic "tragedy of the commons" situation (p. 5), extensive overgrazing has caused many environmental problems in the American West.

Today increasing numbers of ranchers are working cooperatively with government agencies, environmental scientists, and even environmental advocates to find ways to ranch more sustainably and safeguard the health of the land (see **THE SCIENCE BEHIND THE STORY**, pp. 246–247).

AGRICULTURAL POLICY

Agriculture is the most widespread type of human land use. The 38% of Earth's terrestrial surface that it covers is more than the area of North America and Africa combined. Of this land, 26% is rangeland that supports grazing, and 12% is cropland. Approaches such as no-till farming, organic farming (pp. 271–275), and responsible grazing can be sustainable, but today's industrial agriculture exerts tremendous impacts on land and ecosystems, including soil erosion, pollution, and other impacts we have examined above and will further explore in Chapter 10. Because these impacts exert external costs (pp. 150–151), and because agriculture is such a central part of our economy, public policy plays a role in managing agriculture and its impacts.

Some policies worsen land degradation

In theory, the marketplace should discourage people from farming and grazing using intensive methods that degrade land they own if such practices are not profitable. But land degradation often unfolds gradually, whereas farmers and ranchers generally cannot afford to go without profits in the short term, even if they know conservation is in their long-term interests.

Moreover, many nations spend billions of dollars in government subsidies to support agriculture. Roughly one-fifth of the income of the average U.S. farmer comes from subsidies. Proponents of such subsidies stress that the uncertainties of weather make profits and losses from farming unpredictable from year to year. To persist, these proponents say, an agricultural system needs some way to compensate farmers for bad years. This may be the case, but subsidies can encourage people to cultivate land that would otherwise not be farmed; to produce more food than is needed, driving down prices for other producers; and to practice methods that further degrade the land. Thus, opponents of subsidies argue that subsidizing environmentally destructive agricultural practices is unsustainable. They suggest that a better model is for farmers to buy insurance to protect themselves against short-term production failures.

Ranchers in the United States also benefit from government subsidies. Most U.S. rangeland is federally owned and managed by the Bureau of Land Management (BLM). The BLM is the nation's single largest landowner, with 106 million ha (261 million acres) across 12 western states (see Figure 12.10, p. 324). Ranchers are allowed to graze livestock on BLM lands for inexpensive fees; a grazing permit for federal lands in 2009 was just $1.35 per month per "animal unit" (one horse, one cow plus calf, five sheep, or five goats). Such low fees can encourage overgrazing and lead to degradation of grazing land through the tragedy of the commons scenario.

For this reason, ranchers and environmentalists have traditionally been at loggerheads. In the recent years, however, some ranchers and environmentalists have been finding common ground, teaming up to preserve ranchland against what each of them views as a common threat—the encroaching housing developments of suburban sprawl (pp. 348–352). Although developers often pay high prices for ranchland, many ranchers do not want to see the loss of the wide-open spaces and the ranching lifestyle that they cherish.

Wetlands have been drained for farming

Many of our crops grow on the sites of former *wetlands* (p. 405)—swamps, marshes, bogs, and river floodplains—that people drained and filled in (**FIGURE 9.24**). Today, less than half the original wetlands of the lower 48 U.S. states and southern Canada remain.

This loss is the result of decades of laborious efforts, encouraged by government policy, to drain wetlands for agriculture. To promote settlement and farming, the United States passed a series of laws known as the Swamp Lands Acts in 1849, 1850, and 1860. The government transferred over 24 million ha (60 million acres) of wetlands to state ownership (and eventually to private hands) to stimulate drainage, conversion, and flood control. In the Mississippi River valley, the Midwest, and a handful of states from Florida to Oregon, these transfers eradicated malaria (because the mosquitoes that transmit the disease breed in wetlands) and created over 10 million ha (25 million acres) of new farmland. A U.S. Department of Agriculture (USDA) program in 1940 provided monetary aid and technical assistance to farmers draining wetlands on their property, resulting in the conversion of almost 23 million ha (57 million acres).

Today, we have a new view of wetlands. Rather than viewing them as worthless swamps, science has made clear that they are vital ecosystems that provide valuable wildlife habitat, improve water quality, control flooding, and recharge water supplies (Chapter 15). This scientific knowledge, along with a preservation ethic, has persuaded policymakers to develop regulations intended to safeguard our remaining wetlands. However, because of loopholes, differing state laws, development pressures, and debate over the legal definition of wetlands, many of these vital ecosystems are still being lost.

In current U.S. policy, financial incentives are also being used as a means to protect wetlands and influence agricultural land use. Under the *Wetlands Reserve Program*, the U.S. government offers payments to landowners who protect, restore, or enhance wetland areas on their property. Over 2 million

The SCIENCE behind the Story

Restoring the Malpai Borderlands

In the high desert of southern Arizona and New Mexico, scientists and cattle ranchers trying to heal the scars left by decades of overgrazing found they had to contend with a creature even more damaging than a hungry steer: Smokey Bear.

Wildfires might seem a natural enemy of grasslands, but researchers in the Malpai Borderlands realized that people's efforts to suppress fire had done far more harm. Before large numbers of settlers and ranchers arrived more than a century ago, this semi-arid Western landscape thrived under an ecological cycle common to many grasslands. Trees such as mesquite grew near creeks. Hardy grasses such as black grama covered the drier plains. Periodic wildfires, usually sparked by lightning, burned back shrubs and trees and kept grasslands open. Deer, rabbits, and bighorn sheep grazed on the grasses but rarely ate enough to deplete the range. Fed by seasonal rains, new grasses sprouted without being overeaten or crowded out by larger plants.

By the 1990s, however, those grasslands were increasingly scarce.

Burning brush to re-establish grass cover in the Malpai Borderlands

Ranchers had brought large cattle herds to the area in the late 1800s. The cows chewed through vast expanses of grass, trampled the soil, and scattered mesquite seed into areas where grasses had dominated. Ranchers fought wildfires to keep their herds safe, and the federal government joined in the firefighting efforts.

The Malpai's ranching families found themselves struggling to make a living. Decades of photos taken by ranchers and by botanist Raymond Turner showed how soil had eroded and how trees and brush had overgrown the grass. The ranchers knew their cattle were part of the problem, but they also suspected that firefighting efforts were to blame.

In 1993, a group of ranchers launched an innovative plan. They formed the Malpai Borderlands Group, designating about 325,000 ha (800,000 acres) of land for protection and study. Ranchers joined government agencies, environmentalists, and scientists to study the region's ecology, bring back grasses, and restore native animal species. They sought to return periodic fire to the landscape by conducting prescribed burns (p. 329) and by allowing natural fires to run their course (**see first figure**).

The group's research efforts have centered on the Gray Ranch, a 121,000-ha (300,000-acre) parcel in the heart of the borderlands. At McKinney Flats within the Gray Ranch, scientists led by researcher Charles Curtin divided rangeland into four study areas of about 890 ha (2,200 acres) each. Each area is further divided into four "treatments," or areas with varying land management techniques:

▶ In Treatment 1, land is burned and grazed.
▶ In Treatment 2, land is burned but not grazed.

---

acres are currently enrolled in this program nationwide. The Wetlands Reserve Program was one of 15 conservation programs passed by the U.S. Congress in the latest farm bill legislation.

A number of U.S. and international programs promote soil conservation

The U.S. Congress has enacted a number of provisions promoting soil conservation through the farm bills it passes every five to six years. Many of these provisions require farmers to adopt soil conservation plans and practices before they can receive government subsidies.

The **Conservation Reserve Program**, established in the 1985 farm bill, pays farmers to stop cultivating highly erodible cropland and instead place it in conservation reserves planted with grasses and trees. Lands under the Conservation Reserve Program now cover an area nearly the size of Iowa, and the United States Department of Agriculture (USDA) estimates that each dollar invested in this program saves nearly 1 ton of topsoil. Besides reducing erosion, the Conservation Reserve Program generates income for farmers, improves water quality, and provides habitat for wildlife. Butch and David Schroeder are among the many farmers in Iowa and across the nation who have taken advantage of this program, taking land with damaged and erodible soil out of production.

To restore native grasslands, the Malpai Borderlands Group reinstated fire as a natural landscape process, conducting controlled burns. Monitoring indicates that restoring fire has improved ecological conditions in the region.

▶ In Treatment 3, land is grazed but not burned.
▶ In Treatment 4, land is neither grazed nor burned.

Treatments 1 and 3, which allow grazing, also feature small fenced-off areas that prevent animals from eating grass. These "exclosures" allow scientists to make precise side-by-side comparisons of how grazing affects grasses. Scientists measure rainfall in each area and monitor soils for degradation and erosion. Teams of wildlife and vegetation specialists monitor each treatment for the distribution and abundance of birds, insects, animals, and vegetation (**see second figure**).

By comparing areas where fire is suppressed to those where fire can

Researchers measure vegetation on research plots in the Malpai Borderlands.

burn, researchers have documented how the suppression of fire leads to more brush and trees and less grass. Conversely, when fire burns an area, woody plants such as mesquite trees are damaged and their seed production is disrupted, and grass subsequently returns to replace the

trees. Such changes follow both natural fires and carefully monitored, deliberately set controlled burns.

All told, the landscape-level (p. 119) McKinney Flats project may be the largest replicated terrestrial ecological experiment in North America. It has helped spur a number of other research initiatives, and the Malpai Borderlands Group now hosts a scientific conference each year featuring research from the region.

The scientists have also helped ranchers develop a cycle of "grass-banking," in which cattle are allowed to graze on shared plots of land while other areas recover. Ranchers must also work with the weather. Scientists have found that controlled burns or grassbanking should track with cycles of rain and drought to bring back grass.

Because of such research, controlled burns are now a regular part of the Malpai landscape. Ranchers have burned more than 100,000 ha (250,000 acres) since 1994, and natural fires are often allowed to run their course with little or no intervention. Damaged areas have been reseeded with native grasses.

Scientists increasingly believe that sustainable ranching, if managed properly, can help restore grasslands damaged by overgrazing. "We cannot assume rangelands will recover on their own," Curtin wrote in a recent study on the Malpai Borderlands. "Conservation of grazed lands requires restoring and sustaining natural processes." ■

"There is no reason to farm the land if it's in such poor condition that you can't make money," Butch Schroeder says.

Farmers like the Schroeder brothers apply for the program, and the U.S. Farm Service Agency selects the farmers to whom it will award contracts based on an "environmental benefits index" that includes the predicted benefits to wildlife, water quality, air quality, and the farm through reduced erosion, as well as benefits likely to endure beyond the contract period, with all these balanced against the cost of payments. Contracts generally run for 10–15 years.

Of all U.S. states, Iowa is most heavily invested in the program; the state has 107,000 contracts with 53,000 farms worth a total of $200 million per year. Nationwide, the government pays farmers about $1.8 billion per year for the conservation of lands totaling between 12–16 million ha (30–40 million acres). The area varies with market prices for crops; for instance, conservation reserves decreased from 2007 to 2009 in response to higher food prices as farmers withdrew some lands from the program and planted them with crops.

Congress reauthorized and expanded the Conservation Reserve Program in the farm bills of 1996, 2002, and 2008. The 2008 bill limited total conservation reserve lands to 32 million acres but also funded 14 other similar programs (including the Wetlands Reserve Program) for the conservation of grasslands, wetlands, wildlife habitat, and other resources.

FIGURE 9.24 ◀ Most of North America's wetlands have been drained and filled, and the land converted to agricultural use. The northern Great Plains region of Canada and the United States was once pockmarked with thousands of "prairie potholes," water-filled depressions that served as nesting sites for most of the continent's waterfowl. Today, many of these wetlands have been lost; shown are farmlands encroaching on prairie potholes in North Dakota.

WEIGHING THE ISSUES

Soil, Subsidies, and Sustainability Do you think that financial incentive programs such as the Conservation Reserve Program and the Wetlands Reserve Program are a good use of taxpayers' money? Are financial incentives more effective than government regulation for promoting certain land use goals? Do you think they can help lead us toward agriculture that is truly sustainable?

Internationally, the United Nations promotes soil conservation and sustainable agriculture through a variety of programs led by the Food and Agriculture Organization (FAO). As just one example, FAO's Farmer-Centered Agricultural Resource Management Program (FARM) supports innovative approaches to resource management and sustainable agriculture in eight Asian nations. This program studies agricultural success stories and tries to help farmers elsewhere duplicate successful efforts. Rather than following a top-down, government-controlled approach, the FARM program calls on the creativity of local communities to educate and encourage farmers to conserve soils and secure their food supply. Indeed, we will need improved policies and practices in all areas of the world if we are to sustainably provide for our planet's growing human population.

➤ CONCLUSION

Our species has enjoyed a 10,000-year history with agriculture, yet despite all we have learned about land degradation and soil conservation, challenges remain. Many of the policies enacted and the practices developed to combat soil degradation in the United States and worldwide have been quite successful, particularly in reducing topsoil erosion. It is clear, however, that even the best-conceived soil conservation programs require research, education, funding, and commitment from both farmers and governments if they are to fulfill their potential. In light of continued population growth, we will likely need better technology and wider adoption of soil conservation techniques to avoid an eventual food crisis. Increasingly, it seems relevant to consider whether the universal adoption of Aldo Leopold's land ethic (p. 144) will also be required if we are to feed the 9 billion people expected to crowd our planet by midcentury.

REVIEWING OBJECTIVES

You should now be able to:

EXPLAIN THE IMPORTANCE OF SOILS TO AGRICULTURE

- Successful agriculture requires healthy soil. (p. 225)
- Soil is a renewable resource, but its renewal occurs slowly. (p. 225)

DESCRIBE THE IMPACTS OF AGRICULTURE ON SOILS

- As the human population grows and consumption increases, pressures from agriculture are degrading Earth's soil, and we lose 5–7 million ha (12–17 million acres) of productive cropland annually. (p. 225)

OUTLINE MAJOR DEVELOPMENTS IN THE HISTORY OF AGRICULTURE

- Beginning about 10,000 years ago, people began breeding crop plants and domesticating animals through the process of selective breeding, or artificial selection. (pp. 225–226)
- Agriculture originated multiple times independently in different cultures across the world. (pp. 226–227)
- Industrial agriculture is replacing traditional agriculture, which largely replaced hunting and gathering. (pp. 226–227)

DELINEATE THE FUNDAMENTALS OF SOIL SCIENCE, INCLUDING SOIL FORMATION AND SOIL PROPERTIES

- Soil includes diverse biotic communities that decompose organic matter. (pp. 227–228)
- Weathering and biological activity help form soil, influenced by climate, organisms, topography, parent material, and time. (pp. 227–228)
- Soil profiles consist of distinct horizons with characteristic properties. (p. 229)
- Soil can be characterized by color, texture, structure, and pH. (pp. 229–230)
- Soil properties differ among regions and affect plant growth and agriculture. (pp. 230–231)

ANALYZE THE TYPES AND CAUSES OF SOIL EROSION AND LAND DEGRADATION

- Soil deterioration is the major component of land degradation globally. (pp. 231–235)
- Some agricultural practices have resulted in high rates of erosion, lowering crop yields. (pp. 233–235)
- Desertification affects many of the world's soils in arid regions. (pp. 233–235)
- Overirrigation can cause salinization and waterlogging, which lower crop yields and are difficult to mitigate. (pp. 241–242)
- Overapplying fertilizers can cause pollution problems that affect ecosystems and human health. (pp. 242–243)

- Overgrazing can cause soil degradation on grasslands and diverse impacts to native ecosystems. (pp. 244–247)

EXPLAIN THE PRINCIPLES OF SOIL CONSERVATION AND PROVIDE SOLUTIONS TO SOIL EROSION AND LAND DEGRADATION

- The Dust Bowl in the United States and similar events elsewhere have inspired scientists and farmers to develop better ways to conserve topsoil. (p. 235)
- Extension agents and agencies such as the National Resources Conservation Service (NRCS) educate and assist farmers. (pp. 235–236)
- Farming techniques such as crop rotation, contour farming, intercropping, terracing, shelterbelts, and reduced tillage enable farmers to reduce soil erosion and boost crop yields. (pp. 237–240)
- As a rule, vegetation helps anchor soil and prevent erosion. (p. 241)

SUMMARIZE MAJOR POLICY APPROACHES FOR PURSUING SOIL CONSERVATION

- Some existing policies and subsidies worsen land degradation. (p. 245)
- In the United States and across the world, governments are devising innovative policies and programs for agricultural conservation, such as the Conservation Reserve Program and the Wetlands Reserve Program (pp. 245–248)

TESTING YOUR COMPREHENSION

1. On what basis do scientists hypothesize that the practices of selective breeding and human agriculture began roughly 10,000 years ago? Summarize the influence of agriculture on the development and organization of human communities.

2. Describe the methods used in traditional agriculture. How does industrial agriculture differ from traditional agriculture?

3. What processes are most responsible for the formation of soil? Describe the three types of weathering that contribute to soil formation. Name the five primary factors thought to influence soil formation.

4. How are soil horizons created? What is the general pattern of distribution of organic matter in a typical soil profile?

5. Why is erosion generally considered a destructive process? Name three human activities that can promote soil erosion. Describe four kinds of soil erosion by water. What factors affect the intensity of water erosion?

6. List the innovations in soil conservation introduced by Hugh Hammond Bennett, first director of the Soil Conservation Service (SCS). What other farming techniques can help reduce the risk of erosion due to conventional cultivation methods?

7. Explain the method of no-till farming. Why does this method reduce soil erosion?

8. How do fertilizers boost crop growth? How can large amounts of fertilizer added to soil also end up in water supplies and the atmosphere?

9. Describe the effects of overgrazing on soil. What policies can be linked to the practice of overgrazing? What conditions characterize sustainable grazing practices?

10. Explain the goals and methods of the Conservation Reserve Program.

SEEKING SOLUTIONS

1. How do you think a farmer can best help to conserve soil? How do you think a scientist can best help to conserve soil? How do you think a national government can best help to conserve soil?

2. How and why might actual soils differ from the idealized six-horizon soil profile presented in the chapter? How might departures from the idealized profile indicate the impact of human activities? Provide at least three examples.

CHAPTER 9 Soil and Agriculture

249

3. Discuss how the methods of no-till or conservation tillage farming, as described in this chapter, can enhance soil quality. What other benefits can these methods have? What drawbacks or negative effects might no-till or conservation tillage practices have on soil and water quality, and how might these be prevented?

4. Besides no-till farming, select two other methods or approaches described in this chapter that you feel are promoting the science or practice of soil conservation. In reference to Aldo Leopold's land ethic (p. 144), how are humans and the soil members of the same community?

5. **THINK IT THROUGH** You are a land manager with the U.S. Bureau of Land Management (BLM, p. 324) and have just been put in charge of 500,000 acres of public lands that have been degraded by decades of overgrazing. Soil is eroding, creating large gullies. Invasive weeds are replacing native grasses. Shrubs are encroaching on grassland areas because fire was suppressed. Environmentalists want an end to ranching on the land. Ranchers want continued grazing, but they are concerned about the land's condition and are willing to entertain new ideas. What steps would you take to assess the land's condition and begin restoring its soil and vegetation? Would you allow grazing, and if so, would you set limits on it?

6. **THINK IT THROUGH** You are the head of an international granting agency that assists farmers with soil conservation and sustainable agriculture. You have $10 million to disburse. Your agency's staff has decided that the funding should go to (1) farmers in an arid area of Africa prone to salinization, (2) farmers in a fast-growing area of Indonesia where swidden agriculture is practiced, (3) farmers in Argentina practicing no-till agriculture, and (4) farmers in a dryland area of Mongolia undergoing desertification. What types of projects would you recommend funding in each of these areas, how would you apportion your funding among them, and why?

CALCULATING ECOLOGICAL FOOTPRINTS

In the United States, approximately 5 tons of topsoil are lost for every ton of grain harvested. Erosion rates vary greatly with soil type, topography, tillage method, and crop type. For simplicity, let us assume that the 5:1 ratio applies to all plant crops and that a typical diet includes 1 pound of plant material or its derived products (sugar, for example) per day. In the first two columns of the table, calculate the annual topsoil losses associated with growing this food for you and for other groups, assuming the same diet.

	Plant products consumed (lb)	Soil loss at 5:1 ratio (lb)	Soil loss at 3.25:1 ratio (lb)	Reduced soil loss at 3.25:1 relative to 5:1 ratio (lb)
You	365	1,825	1,186	639
Your class				
Your state				
United States				

1. Improved soil conservation measures reduced erosion in the United States by 35% from 1982 to 2003. If additional measures were again able to reduce the current rate of soil loss by this percentage, the ratio of soil lost to grain harvested would fall from 5:1 to 3.25:1. Calculate the soil losses associated with food production at a 3.25:1 ratio, and record your answers in the third column of the table.

2. Calculate the amount of topsoil hypothetically saved by the additional conservation measures in Question 1, and record your answers in the fourth column of the table.

3. Define a "sustainable" rate of soil loss. Describe how you might determine whether a given farm were practicing sustainable use of soil.

Mastering**ENVIRONMENTALSCIENCE**™

Organic vegetable farm in Whatcom County, Washington

10 AGRICULTURE, BIOTECHNOLOGY, AND THE FUTURE OF FOOD

UPON COMPLETING THIS CHAPTER, YOU WILL BE ABLE TO:

- Explain the challenge of feeding a growing human population
- Identify the goals, methods, and consequences of the Green Revolution
- Describe approaches for preserving crop diversity
- Categorize strategies for pest management

- Discuss the importance of pollination
- Describe the science behind genetically modified food
- Evaluate the debate over genetically modified food
- Assess feedlot agriculture for livestock and poultry
- Weigh approaches in aquaculture
- Evaluate sustainable agriculture

Possible Transgenic Maize in Southern Mexico

"Worrying about starving future generations won't feed them. Food biotechnology will."

—Advertising Campaign of the Monsanto Company

"Industrial agriculture has not produced more food. It has destroyed diverse sources of food, and it has stolen food from other species . . . using huge quantities of fossil fuels and water and toxic chemicals in the process."

—Vandana Shiva, Director of the Research Foundation for Science, Technology, and Natural Resource Policy, India

Corn is a staple grain of the world's food supply. We can trace its ancestry back roughly 9,000 years, when people in the highland valleys of southern Mexico first domesticated that region's wild maize plants. The corn we eat today arose from some of the many varieties that evolved from the early selective crop breeding conducted by the people of this region.

Today southern Mexico remains a world center of biodiversity for maize, with many locally adapted domesticated varieties (called **landraces**) growing in the rich, well-watered soil. Preserving such traditional varieties of crops in their ancestral homelands is important for securing the future of our food supply, scientists maintain, because these varieties serve as reservoirs of genetic diversity—reservoirs we may need to draw upon to sustain or advance our agriculture.

Thus, many food experts around the world expressed alarm in 2001 when Mexican government scientists conducting routine genetic tests of farmers' maize from the state of Oaxaca announced that they had turned up DNA that matched genes from genetically modified (GM) corn. GM corn was widely grown in the United States, but Mexico had banned its cultivation in 1998. Corn is one of many crops that scientists have genetically engineered to express desirable traits such as large size, fast growth, and resistance to insect pests. To genetically engineer crops, scientists extract genes from the DNA of one organism and transfer them into the DNA

A Mexican farmer inspects his maize

of another. The aim is to improve crop performance and feed the world's hungry, but many people worry that **transgenes**, the genes engineered and moved into these **transgenic** plants, may have unintended consequences. One concern is that transgenic crops might crossbreed with local landraces and thereby "contaminate" the genetic makeup of native crops.

Two researchers at the University of California at Berkeley, Ignacio Chapela and his postdoctoral associate David Quist, shared these concerns, and they ventured to Oaxaca to test samples of local maize landraces. Their analyses seemed to confirm the government scientists' findings, revealing what they argued were traces of DNA from genetically engineered corn in the genes of native maize plants. They also suggested that the invading genes had split up and spread throughout the maize genome. Quist and Chapela published their findings in the scientific journal *Nature* in November 2001.

Activists opposed to GM food trumpeted the news and urged a ban on imports of transgenic crops from producer countries such as the United States into developing nations. The agrobiotech industry defended the safety of its crops and questioned the validity of the research—as did many of Quist and Chapela's peers. Responding to criticisms from researchers, *Nature* took the unprecedented step of stating that Quist and Chapela's paper should not have been published—a decision that ignited a firestorm of controversy (see **THE SCIENCE BEHIND THE STORY**, pp. 264–265).

Subsequent research by Mexican government scientists reported that in 15 localities, 3–60% of maize contained transgenes and that 37% of maize grains distributed by the government to farmers were transgenic. These results were never published in a peer-reviewed scientific journal, but a commission of experts convened under the North American Free Trade Agreement (NAFTA; pp. 190–191) to study the issue reasoned that corn shipments from the United States contain a mix of GM and non-GM grain, that small farmers plant some of this seed, and that transgenes entering Mexico could spread by wind pollination and by interbreeding with native landraces of maize.

Meanwhile, teams of Mexican and American scientists were conducting their own surveys of Oaxacan maize. In 2005, one team reported finding no transgenes at all, despite analyzing 154,000 maize seeds. Then in 2007 and 2009, other teams reported evidence of transgenes in native maize at low frequency from scattered locations (see **The Science behind the Story**, pp. 264–265). Scientists are still debating the issue, and the entire episode reveals the difficulty researchers sometimes face in achieving certainty in their data and consensus on its interpretation, especially when the stakes are high and commercial interests are involved. However, most scientists today agree that if transgenes are not yet present in Mexican maize, they probably soon will be, for there appears to be little to stop them.

A harder question to answer is how the genetic modification of crops may affect people and the environment (potentially in both positive and negative ways), and whether it constitutes progress toward a goal of sustainable agriculture. In this chapter, we explore the quest for sustainable agriculture as we take a wide-ranging view of the ways we increase our agricultural output and how we might manage the environmental and social consequences of these efforts.

THE RACE TO FEED THE WORLD

Although human population growth has slowed, we can still expect our numbers to swell to 9 billion by the middle of this century. For every three people living today, there will be four in 2050. Feeding 2 billion more mouths while protecting the integrity of soil, water, and ecosystems will require sustainable agriculture. Progressing toward a sustainable model for agriculture could involve a diversity of approaches, ranging from organic farming to the genetically modified crops that have elicited so much controversy in Mexico.

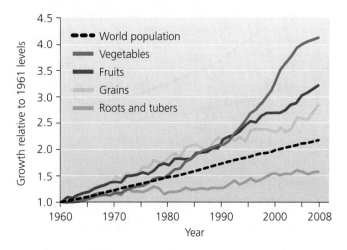

FIGURE 10.1 ▲ Global production of most foods has risen more quickly than world population over the past half-century. This means that we have produced more food per person each year. Trend lines show cumulative increases relative to 1961 levels (for example, a value of 2.0 means twice the 1961 amount). Population is measured in number of people, and food is measured by weight. Data from U.N. Food and Agriculture Organization (FAO).

We are producing more food per person

Over the past half-century, our ability to produce food has grown even faster than global population (**FIGURE 10.1**). We have increased food production by devoting more fossil fuel energy to agriculture; intensifying our use of irrigation, fertilizers, and pesticides; planting and harvesting more frequently; cultivating more land; and developing (through crossbreeding and genetic engineering) more productive crop and livestock varieties.

Improving people's quality of life by producing more food per person is a monumental achievement of which humanity can be proud. However, making our food supply sustainable depends on maintaining healthy soil, water, and biodiversity. As we saw in Chapter 9, careless agricultural practices can damage environmental systems and diminish the capacity of soils to continue supporting crops and livestock. Today many of the world's soils are in decline, and most of the planet's arable land has already been claimed. Even though agricultural production has outpaced population growth so far, there is no guarantee that it will continue to do so.

We face undernutrition, overnutrition, and malnutrition

Despite our rising food production, 1 billion people worldwide do not have enough to eat. These people suffer from **undernutrition**, receiving fewer calories than the minimum dietary energy requirement. As a result, every 5 seconds, somewhere in the world, a child starves to death. In most cases, the reasons for undernutrition are economic. One-fifth of the world's people live on less than $1 per day, and over half live on less than $2 per day, the World Bank estimates. Political obstacles, conflict, and inefficiencies in distribution contribute significantly to hunger as well.

Most people who are undernourished live in the developing world. However, hunger is a problem even in the United States,

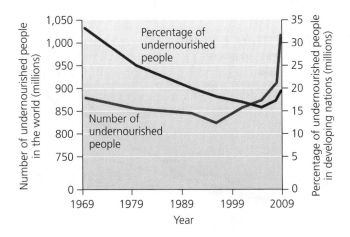

FIGURE 10.2 ▲ The number of people suffering undernutrition in the world declined from 1970 through the late 1990s, while the percentage of people in developing nations who were undernourished fell still more. However, both of these indicators of hunger rose recently in response to economic conditions. Data from Food and Agriculture Organization of the United Nations, Rome, 2009. *The state of food insecurity in the world, 2009.* Rome.

FIGURE 10.3 ▲ Millions of children, including this child with his mother in Somalia, suffer from forms of malnutrition, such as kwashiorkor and marasmus.

where the U.S. Department of Agriculture (USDA) has classified 36 million Americans as "food insecure," lacking the income required to reliably procure sufficient food. Agricultural scientists and policymakers worldwide pursue a goal of **food security**, the guarantee of an adequate, safe, nutritious, and reliable food supply available to all people at all times.

Globally, the number of people suffering from undernutrition fell steadily from 1970 through the late 1990s. The percentage of undernourished people fell even more steeply. These encouraging trends were reversed as a result of a global spike in food prices in 2006–2008 and the economic slump of 2008–2009, and both the number and percentage of hungry people increased (**FIGURE 10.2**). However, in percentage terms, we still have managed to reduce hunger substantially, from 26% of the population in 1970 to 15% in 2009.

Although 1 billion people lack access to adequate food, many others consume more than is healthy and suffer from **overnutrition**, receiving too many calories each day. Overnutrition is a problem in developed nations such as the United States, where food is abundant, junk food is cheap, and people tend to lead sedentary lives with little exercise. As a result, more than three in five U.S. adults are technically overweight, and more than one in four are obese. Obesity leads to cardiovascular disease, diabetes, and other health problems, impairing people's quality of life, shortening their life spans, and imposing greater health care costs on society.

Lower-income people in wealthy societies often suffer overnutrition the most, because many rely on inexpensive mass-marketed foods that are calorie-rich but nutrient-poor. However, people of all income levels are eating more processed energy-rich foods that are high in fat and sugar yet low in nutrition, and they are getting less exercise. Worldwide, the World Health Organization estimates that in 2005, 1.6 billion adults were overweight and that at least 400 million were obese. It predicts that by 2015 those numbers will rise to 2.3 billion and 700 million.

Just as the *quantity* of food a person eats is important for health, so is the *quality* of food. **Malnutrition**, a shortage of nutrients the body needs, occurs when a person fails to obtain a complete complement of vitamins and minerals. Malnutrition can easily lead to disease (**FIGURE 10.3**). For example, people who eat a diet that is high in starch but deficient in protein or essential amino acids (p. 28) can develop *kwashiorkor*. Children who have recently stopped breast-feeding are most at risk for developing kwashiorkor, which causes bloating of the abdomen, deterioration and discoloration of hair, mental disability, immune suppression, developmental delays, anemia, and reduced growth. Protein deficiency together with a lack of calories can lead to *marasmus*, which causes wasting or shriveling among millions of children in the developing world.

The Green Revolution boosted agricultural production

The desire for greater quantity and quality of food for our growing population led in the mid- and late-20th century to the **Green Revolution** (first introduced in Chapter 9, p. 227). Realizing that farmers could not go on forever cultivating additional land to increase crop output, agricultural scientists devised methods and technologies to increase crop output per

FIGURE 10.4 ▲ Norman Borlaug holds examples of the wheat variety he bred that helped launch the Green Revolution. The high-yielding, disease-resistant wheat helped boost agricultural productivity in many developing countries.

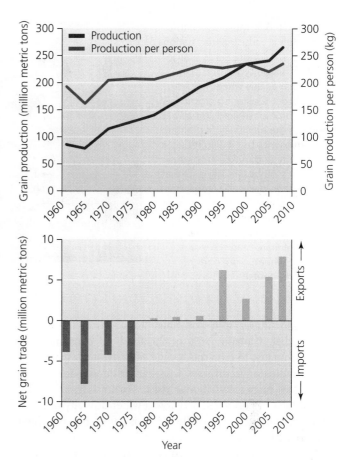

FIGURE 10.5 ▲ Thanks to Green Revolution technology, India increased its grain production even faster than its population grew, so that grain per person also increased (upper graph). As production rose, India was able to stop importing grain and begin exporting it to other nations instead (lower graph). Data from U.N. Food and Agriculture Organization (FAO).

unit area of existing cultivated land. As a result, industrialized nations dramatically increased their per-area yields. The average hectare of U.S. cornfield raised its corn output fivefold during the 20th century, for instance. Many people saw such growth in production and efficiency as key to ending starvation in developing nations.

The transfer of technology to the developing world that marked the Green Revolution began in the 1940s, when the late U.S. agricultural scientist **Norman Borlaug** introduced Mexico's farmers to a specially bred type of wheat (**FIGURE 10.4**). This strain of wheat produced large seed heads, was resistant to diseases, was short in stature to resist wind, and produced high yields. Within two decades of planting this new crop, Mexico tripled its wheat production and began exporting wheat. The stunning success of this program inspired others. Borlaug—who won the Nobel Peace Prize for his work—took his wheat to India and Pakistan and helped transform agriculture there.

Soon many developing countries were doubling, tripling, or quadrupling their yields using selectively bred strains of wheat, rice, corn, and other crops from industrialized nations. When Borlaug died in 2009 at age 95, he was widely celebrated as having "saved more lives than anyone in history"—perhaps as many as a billion.

The Green Revolution brought mixed consequences

Along with the new grains, developing nations imported the methods of industrial agriculture. They began applying large amounts of synthetic fertilizers and chemical pesticides on their fields, irrigating crops generously with water, and using more machinery powered by fossil fuels. From 1900 to 2000, humans increased energy inputs into agriculture by 80 times while expanding the world's cultivated area by just 33%.

This high-input agriculture succeeded dramatically in producing more corn, wheat, rice, and soybeans from each hectare of land. Intensified agriculture saved millions in India

from starvation in the 1970s and eventually turned that nation into a net exporter of grain (**FIGURE 10.5**).

The effects of these developments on the environment have been mixed. On the positive side, the intensified use of already-cultivated land reduced pressures to convert additional natural lands for new cultivation. Between 1961 and 2008, food production rose 150% and population rose 100%, while area converted for agriculture increased only 10%. In this way, the Green Revolution prevented some degree of deforestation and habitat conversion and thus helped preserve biodiversity and natural ecosystems. On the negative side, the intensive application of water, fossil fuels, inorganic fertilizers, and synthetic pesticides worsened pollution, erosion, salinization, and desertification (Chapter 9).

The planting of crops in **monocultures**, large expanses of single crop types (p. 226; **FIGURE 10.6**), makes planting and harvesting more efficient and thereby increases output. However, monocultures also reduce biodiversity over huge areas, because many fewer wild organisms are able to live in monocultures than in native habitats or in traditional small-scale polycultures. Moreover, when all plants in a field are genetically similar, as in monocultures, all are equally susceptible to viral diseases, fungal pathogens, or insect pests that can spread quickly from plant to plant (see Figure 10.6). For this reason, monocultures bring risks of catastrophic failure.

FIGURE 10.6 ▶ Most agricultural production in industrialized nations comes from monocultures—large stands of single types of crop plant, such as this wheat field in Washington. Clustering crops uniformly in immense fields greatly improves the efficiency of planting and harvesting, but it also reduces biodiversity and makes crops susceptible to outbreaks of pests that specialize on particular crops. For example, outbreaks of armyworms **(inset)** can substantially reduce yields. The caterpillars of these moths defoliate many crops, including wheat, corn, cotton, alfalfa, and beets.

Monocultures also contribute to a narrowing of the human diet. Globally, 90% of the food we consume now comes from just 15 crop species and eight livestock species—a drastic reduction in diversity from earlier times. Such dietary restriction carries nutritional risks. Fortunately, expanded global trade has provided access to a wider diversity of foods from around the world, although this effect has benefited wealthy people more than poor people. One reason farmers and scientists are so concerned about transgenic contamination of southern Mexico's native maize is that Mexican maize varieties serve as valuable sources of genetic variation in a world where so much variation is being lost.

Today, yields are declining in some Green Revolution regions, likely as a result of soil degradation from the heavy use of fertilizers, pesticides, and irrigation. Moreover, wealthier farmers with larger plots of land were best positioned to invest in Green Revolution technologies. As a result, many low-income farmers who could not invest in these technologies were driven out of business and moved to cities, adding to the immense migration of poor rural people to urban areas of the developing world (pp. 345–346).

WEIGHING THE ISSUES

The Green Revolution and Population
In the 1960s, India's population was skyrocketing, and its traditional agriculture was not producing enough food to support the growth. By adopting Green Revolution agriculture, India sidestepped mass starvation. However, Norman Borlaug called his Green Revolution methods "a temporary success in man's war against hunger and deprivation," something to give us breathing room in which to deal with what he called the "Population Monster." Indeed, in the years since intensifying its agriculture, India has added several hundred million more people and continues to suffer widespread poverty and hunger.

Do you think the Green Revolution has solved problems, deferred problems, or created new ones? Which aspects of the Green Revolution do you think help in the quest for sustainability, which do not, and why? Consider our discussion of demographic transition theory from Chapter 8 (pp. 210–211). Have we been dealing with the "Population Monster" during the breathing room the Green Revolution has bought us?

Biofuels affect food supplies

Just as the Green Revolution's noble intentions and significant successes gave rise to some problematic side effects, in recent years some well-intentioned efforts to promote renewable energy have had unintended consequences. **Biofuels** (pp. 579–582) are fuels derived from organic materials and used in internal combustion engines as replacements for petroleum. In an effort to shift from fossil fuels to renewable energy sources, policymakers have encouraged the production of biofuels from crops. In the United States, **ethanol** made from corn is the primary biofuel (pp. 579–581). Following expanded subsidies in 2007, U.S. ethanol production nearly doubled as new ethanol facilities opened and farmers began selling their corn for ethanol instead of for food.

This caused a scarcity of corn worldwide, and prices for basic foods (such as tortillas in Mexico) skyrocketed. Prices for other staple grains also rose because farmers shifted fields formerly devoted to other food crops into biofuel production. For low-income people, the steep rise in food prices was frightening. Thousands staged protests, and riots erupted in Mexico and many other nations. This food crisis contributed to the increase in global hunger shown in Figure 10.2. The world realized belatedly that growing crops for biofuels could compete directly with growing food for people to eat. We will examine the complicated issues surrounding biofuels in Chapter 20.

PRESERVING CROP DIVERSITY

The monocultures of modern industrial agriculture essentially place all our eggs in one basket, such that any single catastrophe could potentially wipe out entire crops. Our wide adoption of industrial agriculture has also reduced the diversity of crops, and today we as a global society rely on a much smaller number of plant types than in decades and centuries past. This is partly why so many people grew concerned over the possibility that genes from transgenic corn might be moving into local Mexican landraces of maize.

Crop diversity provides insurance against failure

Preserving the integrity of diverse native crop variants gives us a bulwark against the potential failure of our homogenized commercial crops. The wild relatives of crop plants and their locally adapted landraces contain a diversity of genes that we may someday need to introduce into our commercial crops (through crossbreeding or genetic engineering) to confer resistance to disease or pests or to meet other unforeseen challenges.

Because accidental interbreeding can diminish the diversity of local variants, many scientists argue that we need to protect landraces in areas like southern Mexico that remain important repositories of crop biodiversity. For this reason, the Mexican government helped create the Sierra de Manantlan Biosphere Reserve around an area harboring the localized plant thought to be the direct ancestor of maize. For this reason, too, it imposed a national moratorium in 1998 on the planting of transgenic corn. However, this ban was lifted in 2009 as multinational agribusiness corporations were allowed to begin experimental plantings in northern Mexico. A Mexican government agency also disburses grain to farmers that includes millions of tons of U.S. corn (one-third of it transgenic), and farmers can also acquire U.S. seed on their own. Given global trade and the increasing use of genetically modified corn worldwide, gene flow between transgenic corn and Mexico's native landraces would seem inevitable at some point.

Worldwide, we have already lost a great deal of genetic diversity in crops in the past century. Only 30% of the maize varieties that grew in Mexico in the 1930s exist today. The number of wheat varieties in China dropped from 10,000 in 1949 to 1,000 by the 1970s. In the United States, many fruit and vegetable crops have decreased in diversity by 90% in less than a century. Market forces have discouraged diversity in the appearance of fruits and vegetables: Commercial food processors prefer items to be uniform in size and shape for convenience, and consumers are often wary of unusual-looking food products. However, now that local and organic agriculture (pp. 271–277) is growing in affluent societies, consumer preferences for diversity are increasing.

Seed banks are living museums

Protecting areas and cultures that maintain a wealth of crop diversity is one way to preserve genetic assets for our agriculture. Another is to collect and store seeds from diverse crop varieties. This is the work of **seed banks**, institutions that preserve seed types as a kind of living museum of genetic diversity (**FIGURE 10.7**). These facilities keep seed samples in cold, dry conditions to encourage long-term viability, and they plant and harvest them periodically to renew the stocks.

Major seed banks include the Royal Botanic Garden's Millennium Seed Bank in Britain, the U.S. National Seed

(a) Traditional food plants of the Desert Southwest

(b) Pollination by hand

FIGURE 10.7 ▲ Seed banks preserve genetic diversity of traditional crop plants. Native Seeds/SEARCH of Tucson, Arizona, preserves seeds of food plants important in traditional diets of Native Americans of the Southwest **(a)**. Beans, chiles, squashes, gourds, maize, cotton, and lentils are all in its collections, as well as traditional foods such as mesquite flour, prickly pear pads, chia seeds, tepary beans, and cholla cactus buds that help fight the diabetes that Native Americans frequently suffer after adopting a Western diet. At the farm where seeds are grown, care is taken to pollinate varieties by hand **(b)** to protect their genetic distinctiveness.

Storage Laboratory at Colorado State University, Seed Savers Exchange in Iowa, and the Wheat and Maize Improvement Center (CIMMYT) in Mexico. In total, 1,400 such facilities house 1–2 million distinct types of seeds worldwide.

The most renowned seed bank is the so-called "doomsday seed vault" established in 2008 on the island of Spitsbergen in Arctic Norway. The internationally funded Svalbard Global Seed Vault (**FIGURE 10.8**) is storing millions of seeds from around the world (spare sets from other seed banks) as a safeguard against global agricultural calamity—"an insurance policy for the world's food supply." The doomsday seed vault is an admirable effort, but we would be well advised not to rely on it to save us. Far better to manage our agriculture wisely and sustainably so that we never need to break into the vault!

FIGURE 10.8 ▲ The Svalbard Global Seed Vault in Arctic Norway (nicknamed the "doomsday seed vault") will store millions of seed samples in frozen conditions as insurance against global agricultural catastrophe. This secured facility is built deep into a mountain in an area of permanently frozen ground, with refrigeration to keep seeds still colder to ensure their longevity. There is no tectonic activity here, and little natural radiation or humidity; moreover, the site is high enough above sea level to stay dry even if climate change melts all the planet's ice.

PESTS AND POLLINATORS

Throughout the history of agriculture, the insects, fungi, viruses, rats, and weeds that eat or compete with our crop plants have taken advantage of the ways we cluster food plants into agricultural fields. As just one example, various species of moth caterpillars known as armyworms (see Figure 10.6) reduce yields of everything from beets to sorghum to millet to canola to pasture grasses. Pests and weeds pose an especially great threat to monocultures, where a pest adapted to specialize on the crop can move easily from plant to plant.

What people term a *pest* is any organism that damages crops that are valuable to us. What we term a *weed* is any plant that competes with our crops. These are subjective categories that we define entirely by our own economic interests. There is nothing inherently malevolent in the behavior of a pest or a weed. These organisms are simply trying to survive and reproduce. From the perspective of an insect that feeds on corn, grapes, or apples, encountering a grain field, vineyard, or orchard is like discovering an endless buffet.

We have developed thousands of chemical pesticides

To suppress pests and weeds, people have developed thousands of chemicals to kill insects (*insecticides*), plants (*herbicides*), and fungi (*fungicides*). Such poisons are collectively termed **pesticides**. All told, roughly 400 million kg (900 million lb) of active ingredients from conventional pesticides are applied in the United States each year. Three-quarters of this total is applied on agricultural land. Since 1960, pesticide use has risen fourfold worldwide. Usage in industrialized nations has leveled off in the past two decades, but it continues to rise in the developing world. Today more than $32 billion is expended annually

on pesticides, with one-third of that total spent in the United States. We will address the health consequences of synthetic pesticides for people and other organisms in Chapter 14.

Pests evolve resistance to pesticides

Despite the toxicity of these chemicals, their effectiveness tends to decline with time as pests evolve resistance to them. Recall from our discussion of natural selection (pp. 52–55) that organisms within populations vary in their traits. Because most insects, weeds, and microbes can occur in huge numbers, it is likely that a small fraction of individuals may by chance have genes that confer immunity to a given pesticide. For instance, even if an insecticide application kills 99.99% of the insects in a field, 1 in 10,000 survives. If an insect that is genetically resistant to an insecticide survives and mates with other resistant individuals, the insect population may recover. This new population will consist of individuals that are resistant to the insecticide. As a result, insecticide applications will cease to be effective (**FIGURE 10.9**).

In many cases, industrial chemists are caught up in an evolutionary arms race (p. 82) with the pests they battle, racing to increase or retarget the toxicity of their chemicals while the armies of pests evolve ever-stronger resistance to their efforts. Because we seem to be stuck in this cyclical process, it has been nicknamed the "pesticide treadmill." As of 2008, among arthropods (insects and their relatives) alone, there were more than 8,000 known cases of resistance by 556 species to over 300 insecticides. Hundreds more weed species and plant diseases have evolved resistance to herbicides and other pesticides. Many species, including insects such as the green peach aphid, Colorado potato beetle, and diamondback moth, have evolved resistance to multiple chemicals.

An additional problem is that pesticides often kill nontarget organisms, including the predators and parasites of the pests. When these valuable natural enemies are eliminated, pest populations become that much harder to control.

Biological control pits one organism against another

Because of pesticide resistance, toxicity to nontarget organisms, and human health risks from some synthetic chemicals, agricultural scientists increasingly battle pests and weeds with organisms that eat or infect them. This strategy, called **biological control** or **biocontrol**, operates on the principle that "the enemy of one's enemy is one's friend." For example, parasitoid wasps (p. 82) are natural enemies of many caterpillars. These wasps lay eggs on a caterpillar, and the larvae that hatch from the eggs feed on the caterpillar, eventually killing it. Parasitoid wasps are frequently used as biocontrol agents and have often succeeded in controlling pests and reducing chemical pesticide use.

One classic case of successful biological control is the introduction of the cactus moth, *Cactoblastis cactorum*, from Argentina to Australia in the 1920s to control invasive prickly pear cactus that was overrunning rangeland. Within just a few years, the moth managed to free millions of hectares of rangeland from the cactus (**FIGURE 10.10**).

A widespread modern biocontrol effort has been the use of *Bacillus thuringiensis* (Bt), a naturally occurring soil bacterium that produces a protein that kills many caterpillars and some

1 Pests attack crops

2 Pesticide is applied

3 Most pests are killed. A few with innate resistance survive

4 Survivors breed and produce a pesticide-resistant population

5 Pesticide is applied again

6 Pesticide has little effect. New, more toxic, pesticides are developed

FIGURE 10.9 ▲ Through the process of natural selection (pp. 52–55), crop pests may evolve resistance to the poisons we apply to kill them. When a pesticide is applied to an outbreak of insect pests, it may kill all individuals except those few with an innate immunity, or resistance, to the poison (resistant individuals are colored red in the diagram). Those surviving individuals may establish a population with genes for resistance to the poison. Future applications of the pesticide may then be ineffective, forcing us to develop a more potent poison or an alternative means of pest control.

fly and beetle larvae. Farmers spray Bt spores on their crops to protect against insect attack. (In addition, as we will see (p. 263), scientists have managed to isolate the gene responsible for the bacterium's poison and engineer it into crop plants.)

Biological control agents themselves can become pests

When a pest is not native to the region where it is damaging crops, scientists may consider introducing a natural enemy (a predator, parasite, or pathogen) of the pest from its native range, in the expectation that the enemy will attack it. Alternatively, scientists occasionally consider importing a biocontrol agent from abroad that the pest has not encountered, reasoning that the pest has not evolved ways to avoid the biocontrol agent. In either case, this approach involves introducing an animal or microbe from a foreign ecosystem into a new ecological context. This is risky, because no one can know for certain what effects the biocontrol agent might have.

Following the cactus moth's success in Australia, for example, it was introduced in other countries to control nonnative prickly pear. Moths introduced to Caribbean islands spread to Florida on their own and are now eating their way through rare native cacti in the southeastern United States. If these moths reach Mexico and the southwestern United

States, they could decimate many native and economically important species of prickly pear there.

In Hawaii, wasps and flies have been introduced to control pests at least 122 times over the past century, and biologists Laurie Henneman and Jane Memmott suspected that some of these might be harming native Hawaiian caterpillars that were not pests. They sampled parasitoid wasp larvae from 2,000 caterpillars of various species in a remote mountain swamp far from farmland. In this wilderness preserve, they found that fully 83% of the parasitoids were biocontrol agents that had been intended to combat lowland agricultural pests.

Because of concerns about unintended impacts, researchers study biocontrol proposals carefully before putting them into action, and government regulators must approve these efforts. If biological control works as planned, it can be a permanent, effective, and environmentally benign solution. Yet there will never be a sure-fire way of knowing in advance whether a given biocontrol program will work as planned.

Integrated pest management combines biocontrol and chemical methods

As it became clear that both chemical and biocontrol approaches pose risks, agricultural scientists and farmers began developing more sophisticated strategies, trying to combine the best attributes of each approach. **Integrated pest management**

(a) Before cactus moth introduction

(b) After cactus moth introduction

FIGURE 10.10 ▲ In a classic case of biocontrol, larvae of the cactus moth, *Cactoblastis cactorum*, were used to clear non-native prickly pear cactus from millions of hectares of rangeland in Queensland, Australia. These photos from the 1920s show an Australian ranch before **(a)** and after **(b)** introduction of the moth.

(IPM) incorporates numerous techniques, including biocontrol, use of chemicals when needed, close monitoring of populations, habitat alteration, crop rotation, transgenic crops, alternative tillage methods, and mechanical pest removal.

IPM has become popular in many parts of the world. Indonesia stands as an exemplary case (**FIGURE 10.11**). This nation had subsidized pesticide use heavily for years, but its scientists came to understand that pesticides were actually making pest problems worse. They were killing the natural enemies of the brown planthopper, which began to devastate rice fields as its populations exploded. Concluding that pesticide subsidies were costing money, causing pollution, and apparently decreasing yields, the Indonesian government in 1986 banned the import of 57 pesticides, slashed pesticide subsidies, and promoted IPM. International experts helped teach Indonesian rice farmers about IPM in "Farmer Field Schools," collaborative groups of farmers who traded information and experimented with new approaches. Within just four years, pesticide production fell by half, imports fell by two-thirds, and subsidies were phased out (saving $179 million annually). Rice yields rose 13%.

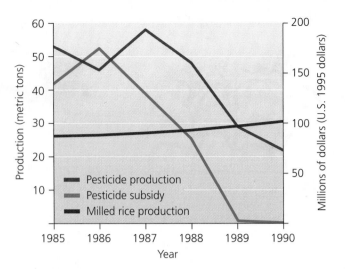

FIGURE 10.11 ▲ The Indonesian government threw its weight behind integrated pest management starting in 1986. Within just a few years, pesticide production and pesticide imports were down drastically, pesticide subsidies were phased out, and yields of rice increased slightly.

Since that time, over 1 million Indonesian farmers have been educated in IPM in Farmer Field Schools, and the approach has spread to dozens of other nations. Across Asia, studies show that farmers in these programs are generally able to increase yields of rice and other crops while greatly reducing pesticide use.

We depend on insects to pollinate crops

Managing insect pests is such a major issue in agriculture that it is easy to fall into a habit of thinking of all insects as somehow bad or threatening. But in fact, most insects are harmless to agriculture, and some are absolutely essential. The insects that pollinate crops are among the most vital, yet least understood and appreciated, factors in our food production. Pollinators are the unsung heroes of agriculture.

Pollination (p. 83) is the process by which male sex cells of a plant (pollen) fertilize female sex cells of a plant; it is the botanical version of sexual intercourse. Without pollination, no plant could reproduce sexually, and no plant species would persist for long. Plants such as conifer trees and grasses achieve pollination by the wind. Millions of minuscule pollen grains are blown long distances, and by chance a small number land on the female parts of other plants of their species. In contrast, the many kinds of plants that sport showy flowers are typically pollinated by animals, such as hummingbirds, bats, and insects (see Figure 4.9, p. 83). Flowers are, in fact, evolutionary adaptations that function to attract pollinators. The sugary nectar and protein-rich pollen in flowers serve as rewards to lure these sexual intermediaries, and the sweet smells and bright colors of flowers are signals to advertise these rewards.

Our staple grain crops are derived from grasses and are wind-pollinated, but many other crops depend on insects for pollination. The most complete survey to date, by tropical bee biologist Dave Roubik, documented 800 types of cultivated plants that rely on bees and other insects for pollination. An estimated 73% of these types are pollinated, at least in part, by bees; 19% by flies; 5% by wasps; 5% by beetles; and 4% by moths and butterflies. Bats pollinate 6.5%, and birds 4%. Overall, native species of bees in the United States alone are estimated to provide $3 billion of pollination services each year to crop agriculture.

Populations of native pollinators have declined precipitously, however. As one example of many, the U.S. Great Basin states are a world center for the production of alfalfa seed, and alfalfa flowers are pollinated mostly by native alkali bees that live in the soil as larvae. In the 1940s to 1960s, farmers began plowing the land and increasing pesticide use in an effort to boost yields. These measures killed vast numbers of the soil-dwelling bees, and alfalfa seed production plummeted.

Conservation of pollinators is vital

Preserving the biodiversity of native pollinators is especially important today because the domesticated workhorse of pollination, the honeybee (*Apis mellifera*), is also declining. American farmers regularly hire beekeepers to bring colonies of this introduced Old World honeybee to their fields when it is time to pollinate crops (**FIGURE 10.12**). Honeybees pollinate over 100 crops that comprise one-third of the U.S. diet, contributing an estimated $15 billion in services.

In recent years, two accidentally introduced parasitic mites have swept through honeybee populations, decimating hives and pushing many beekeepers toward financial ruin. On top of this, starting in 2006, entire hives inexplicably began dying off. In each of the last several years, up to one-third of all honeybees in the United States have vanished from what is being called *colony collapse disorder*. Scientists are racing to discover the cause of this mysterious syndrome. Leading hypotheses are insecticide exposure, an unknown new parasite, or a combination of stresses that weaken bees' immune systems and destroy social communication within the hive.

Farmers and homeowners can help maintain populations of pollinators by reducing or eliminating pesticide use. All insect pollinators are vulnerable to the vast arsenal of insecticides we apply to crops, lawns, and gardens. When people try to control the "bad" bugs that threaten the plants they value, they all too often kill the "good" insects as well. Homeowners can help pollinating insects by planting gardens of flowering plants and by providing nesting sites for bees. Farmers who allow flowering plants (such as clover) to grow around the edges of their fields can maintain a diverse community of insects, some of which will pollinate their crops.

GENETICALLY MODIFIED FOOD

The Green Revolution enabled us to feed a greater number and proportion of the world's people, but relentless population growth is demanding still more innovation. A new set of potential solutions began to arise in the 1980s and 1990s as advances in genetics enabled scientists to directly alter the genes of organisms, including crop plants and livestock. The genetic modification of organisms that provide us food holds promise to enhance nutrition and the efficiency of agriculture while lessening impacts on the planet's environmental systems. However, genetic modification may also pose risks that are not yet well understood. This possibility has given rise to anxiety and protest by consumer advocates, small farmers, environmental activists, and critics of big business.

Genetic modification of organisms depends on recombinant DNA

The genetic modification of crops and livestock is one type of genetic engineering. **Genetic engineering** is any process whereby scientists directly manipulate an organism's genetic material in the laboratory by adding, deleting, or changing segments of its DNA (p. 29). **Genetically modified (GM) organisms** are organisms that have been genetically engineered using **recombinant DNA**, which is DNA that has been patched together from the DNA of multiple organisms. The goal is to place genes that produce certain proteins and code for certain desirable traits (such as rapid growth, disease resistance, or high nutritional content) into the genomes of organisms lacking those traits.

Recombinant DNA technology was developed by scientists studying the bacterium *Escherichia coli*. As shown in **FIGURE 10.13**, scientists first isolate *plasmids*, which are small, circular DNA molecules, from a bacterial culture. At the same time, DNA containing a gene of interest is removed from the cells of another organism. Scientists insert the gene of interest into the plasmid to form recombinant DNA. This recombinant DNA enters new bacteria, which then reproduce, generating many copies of the desired gene. Scientists then introduce the gene into cells of the organism they want to transform.

For instance, Quist and Chapela and other researchers examining Mexican maize were looking for a stretch of DNA called the 35S promoter. The 35S promoter was derived originally from a common natural plant virus, the cauliflower mosaic virus. The 35S promoter directs the expression of many plant genes in nature, and today scientists use it in most GM crops that are engineered for resistance to herbicides.

FIGURE 10.12 ▼ North American farmers hire beekeepers to bring hives (in white boxes in photo) of European honeybees (**inset**, on apple blossom) to their crops when it is time for flowers to be pollinated. Recently, honeybees have suffered devastating epidemics, making it increasingly important to conserve native species of pollinators.

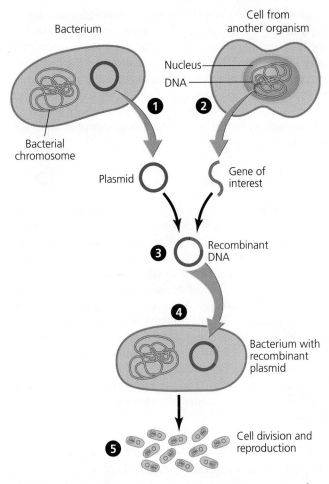

FIGURE 10.13 ▲ To create recombinant DNA, scientists first isolate *plasmids* ❶, small circular DNA molecules, from a bacterial culture. DNA containing a gene of interest ❷ is then removed from another organism. Scientists insert the gene of interest into the plasmid to form recombinant DNA ❸. This recombinant DNA enters new bacteria ❹, which then reproduce ❺, generating many copies of the desired gene. If all goes as planned, the new gene will be expressed in the genetically modified organism as a desirable trait, such as rapid growth or high nutritional content in a food crop.

As we learned at the outset of our case study, an organism that contains DNA from another species is called a **transgenic** organism, and the genes that have moved between them are called **transgenes**. The creation of transgenic organisms is one type of **biotechnology**, the material application of biological science to create products derived from organisms. Biotechnology has helped us develop medicines, clean up pollution, understand the causes of cancer and other diseases, dissolve blood clots after heart attacks, and make better beer and cheese. **FIGURE 10.14** shows several notable developments in GM foods. The stories behind them illustrate both the promises and pitfalls of food biotechnology.

Genetic engineering is like, and unlike, traditional breeding

The genetic alteration of plants and animals by people is nothing new; through artificial selection (p. 55), we have influenced the genetic makeup of our livestock and crop plants for thousands of years. As we saw in Chapter 9 (pp. 225–226), our ancestors altered the gene pools of our domesticated plants and animals through selective breeding by preferentially mating individuals with favored traits so that offspring would inherit those traits. Early farmers selected plants and animals that grew faster, were more resistant to disease and drought, and produced large amounts of fruit, grain, or meat.

Proponents of GM crops often stress this continuity with our past and say today's GM food will be just as safe as selectively bred food. Dan Glickman, head of the USDA from 1995 to 2001, remarked:

> Biotechnology's been around almost since the beginning of time. It's cavemen saving seeds of a high-yielding plant. It's Gregor Mendel, the father of genetics, cross-pollinating his garden peas. It's a diabetic's insulin, and the enzymes in your yogurt. . . . Without exception, the biotech products on our shelves have proven safe.

However, as critics are quick to point out, the techniques geneticists use to create GM organisms differ from traditional selective breeding in several ways. For one, selective breeding mixes genes from individuals of the same or similar species, whereas scientists creating recombinant DNA routinely mix genes of organisms as different as viruses and crops, or spiders and goats. For another, selective breeding deals with whole organisms living in the field, whereas genetic engineering works with genetic material in the lab. Third, traditional breeding selects from combinations of genes that come together on their own, whereas genetic engineering creates the novel combinations directly. Thus, traditional breeding changes organisms through the process of selection (pp. 52–55), whereas genetic engineering is more akin to the process of mutation (p. 53).

Biotechnology is transforming the products around us

In just three decades, GM foods have gone from science fiction to big business. As recombinant DNA technology first developed in the 1970s, scientists debated whether the new methods were safe. They collectively regulated and monitored their own research until most scientists were satisfied that reassembling genes in bacteria did not create dangerous superbacteria. Once the scientific community declared itself confident in the 1980s that the technique was safe, industry leaped at the chance to develop hundreds of applications, from improved medicines (such as hepatitis B vaccine and insulin for diabetes) to designer plants and animals.

Most GM crops today are engineered to resist herbicides, so that farmers can apply herbicides to kill weeds without having to worry about killing their crops. Other crops are engineered to resist insect attack. Some are modified for both types of resistance. Resistance to herbicides and pests enables large-scale commercial farmers to grow crops more efficiently. As a result, sales of GM seeds to these farmers in the United States and other countries have risen quickly.

Today 85% of the U.S. corn harvest and 90% of U.S. soybeans, cotton, and canola consist of genetically modified strains, and 41% of these crops are engineered for more than one trait. Worldwide, three of every four soybean plants are now transgenic, as is one of every four corn plants, one of

Food	Development	Food	Development
Bt crops	By equipping plants with the ability to produce their own pesticides, scientists hoped to reduce crop losses from insects. Scientists working with *Bacillus thuringiensis* (Bt) pinpointed the genes responsible for producing that bacterium's toxic effects on insects, and inserted the genes into the DNA of crops. The USDA and EPA approved Bt versions of 18 crops for field testing, from apples to broccoli to cranberries. Corn and cotton are the most widely planted Bt crops today. Proponents say Bt crops reduce the need for chemical pesticides. Critics worry that they induce insects to evolve resistance to the toxins, cause allergic reactions in people, and harm nontarget species.	Roundup Ready crops	The Monsanto Company's widely used herbicide, Roundup, kills weeds, but it kills crops too. So, Monsanto engineered soybeans, corn, cotton, and canola to withstand the effects of its herbicide. With these "Roundup Ready crops," farmers can spray Roundup without killing their crops. Of course, this also creates an incentive for farmers to use Roundup rather than a competing brand. Unfortunately, Roundup's active ingredient, glyphosate, is a leading cause of illness for California farm workers, and weeds are starting to evolve resistance to glyphosate.
Golden rice	Millions of people in the developing world get too little vitamin A in their diets, causing diarrhea, blindness, immune suppression, and even death. The problem is worst with children in east Asia, where the staple grain, white rice, contains no vitamin A. Researchers took genes from plants that produce vitamin A and spliced the genes into rice DNA to create more-nutritious "golden rice" (the vitamin precursor gives it a golden color). Critics charged that biotech companies over-hyped their product.	Sunflowers and superweeds	Research on Bt sunflowers suggests that transgenes might spread to other plants and turn them into vigorous weeds that compete with the crop. This is most likely to happen with crops like squash, canola, and sunflowers that can breed with their wild relatives. Researchers bred wild sunflowers with Bt sunflowers and found that hybrids with the Bt gene produced more seeds and suffered less herbivory than hybrids without it. They concluded that if Bt sunflowers were planted commercially, the Bt gene might spread and turn wild sunflowers into superweeds.
Ice-minus strawberries	Researchers removed a gene that facilitated the formation of ice crystals from the DNA of a bacterium, *Pseudomonas syringae*. The modified, frost-resistant bacteria could then serve as a kind of antifreeze when sprayed on the surface of crops such as strawberries, protecting them from frost damage. However, news coverage of scientists spraying plants while wearing face masks and protective clothing caused public alarm.	StarLink corn	StarLink corn, a variety of Bt corn, had been approved and used in the United States for animal feed but not for human consumption. In 2000, StarLink corn DNA was discovered in taco shells and other corn products. These products were recalled amid fears of allergic reactions. No such health effects were confirmed, but the corn's manufacturer chose to withdraw the product from the market.

FIGURE 10.14 ▲ As genetically modified foods were developed, a number of products ran into public opposition or trouble in the marketplace. A selection of such cases serves to illustrate some of the issues that proponents and opponents of genetically modified foods have been debating.

every five canola plants, and half of all cotton plants. Globally in 2009, 14 million farmers grew GM crops on 134 million ha (331 million acres) of farmland—nearly 9% of all cropland in the world, and an area the size of Kansas, Nebraska, Colorado, Minnesota, and California combined.

Soybeans account for most of the world's GM crops (**FIGURE 10.15A**, p. 266). Of the 25 nations growing GM crops in 2009, six (the United States, Brazil, Argentina, India, Canada, and China) accounted for 95% of production, with the United States alone growing nearly half the global total (**FIGURE 10.15B**, p. 266). The market value of GM crops in 2009 was estimated at $10.5 billion.

What are the benefits and impacts of GM crops?

As GM crops were adopted and as biotech business expanded, many citizens, scientists, and policymakers became concerned. Some feared the new foods might be dangerous for people to eat. Others were concerned that transgenes might escape and

harm nontarget organisms. Still others worried that pests would evolve resistance to the supercrops and become "superpests" or that transgenes would be transferred from crops to other plants and turn them into "superweeds." Some, like Quist and Chapela, worried that transgenes might ruin the integrity of native ancestral landraces of crops.

Many conventional crops can interbreed with their wild relatives (rice can breed with wild rice, for instance), so there seems little reason to believe that transgenic crops would not do the same. In the first confirmed case, GM oilseed rape was found hybridizing with wild mustard. In another case, creeping bentgrass engineered for use on golf courses—a GM plant not yet approved by the USDA—pollinated wild grass up to 21 km (13 mi) away from its test growing site in Oregon. Most scientists think transgenes will inevitably make their way from GM crops into wild plants, but the consequences of this are open to debate.

Supporters of GM crops maintain that transgenic crops are beneficial for the environment. Indeed, adoption of insect-resistant Bt crops appears to reduce the use of chemical

The SCIENCE behind the Story

Transgenic Contamination of Native Maize?

Dr. Ignacio Chapela (left) and Dr. David Quist (right), University of California, Berkeley

David Quist and Ignacio Chapela's *Nature* paper reporting the presence of DNA from genetically engineered corn in native Mexican maize (p. 252) was controversial from the moment it was published.

A number of geneticists soon questioned their bold conclusion that transgenes had entered the genomes of native maize landraces in remote areas of Oaxaca. Some of these critics argued that the low levels of transgenic DNA Quist and Chapela detected indicated, at most, first-generation hybrids between local landraces and transgenic varieties. Other critics noted that the technique they relied on, the polymerase chain reaction (PCR), is highly sensitive to contamination and careless laboratory practices.

Quist and Chapela's claim that the transgenes also had been jumping throughout the genome came under particular fire. They had used a technique called inverse PCR (iPCR) and had reported that the transgenes were surrounded by essentially random sequences of DNA. Several geneticists argued that iPCR was unreliable, that the pair had used insufficient controls, and that their results could have arisen from similarities between the transgenes and stretches of maize DNA.

In April 2002, *Nature* published two letters from scientists critical of the study and stated that "the evidence available is not sufficient to justify the publication of the original paper."

Quist and Chapela responded by acknowledging that some of their initial findings based on iPCR were probably invalid. But they also presented new analyses to support their fundamental claim and pointed to a Mexican government study that also found transgenes in Mexican maize.

The correspondence published in *Nature* set off a broader debate on the editorial pages of scientific journals, in newspapers and magazines, on the Internet, and in the halls of academe. The dispute sometimes turned personal. GM supporters pointed out that Chapela and Quist had long opposed transgenic crops

and biotechnology corporations. Chapela had spoken out against his university's plan to enter into a $25 million partnership with the biotechnology firm Novartis (now Syngenta), sparking a debate that divided the UC Berkeley faculty. GM opponents countered that Quist and Chapela's most vocal critics received funding from biotechnology corporations. Even *Nature*, whose impartiality is critical to its reputation as a first-tier science journal, was engaged in commercial partnerships with biotechnology corporations.

Chapela was denied tenure by Berkeley's administration, even though his department and dean had recommended him. Alleging unfair treatment, Chapela waged a three-year fight to win tenure and gained a following that included 230 academics who signed a letter of support. In 2005, a new chancellor awarded him tenure.

Meanwhile, in hopes of resolving whether transgenes had in fact entered Mexican maize, dozens of other researchers began work on the issue. In 2005, a research team led by Sol Ortiz-García of Mexico's Instituto Nacional de Ecologia and Allison Snow of Ohio State University published results from an extensive survey

---

insecticides, because farmers use fewer insecticides if their crops do not need them. However, herbicide-tolerant crops tend to result in more herbicide use because farmers may apply more herbicide if their crops can withstand it. In addition, weeds may evolve resistance to herbicides, making heavier applications necessary. For instance, by some estimates, 68 species of weeds have already evolved resistance to Monsanto's Roundup herbicide, and populations of these weeds nearly doubled between 2007 and 2009. A 2010 study by the Institute of Science in Society used USDA data to

calculate that as GM crops expanded in the United States between 1996 and 2008, insecticide use declined by 64 million pounds, but herbicide use increased by 383 million pounds.

Although supporters of GM crops maintain that GM foods have led to no demonstrated ill health effects, the overall increase in pesticide application does pose health risks, as well as ecological impacts. On the other hand, herbicide-tolerant crops help enable no-till farming (pp. 224–225, 239–240) by freeing farmers from having to till to control weeds—and no-till farming can bring various environmental benefits, as we

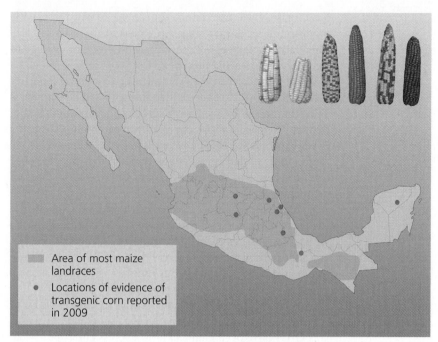

No report of transgenic contamination of Mexican maize has been without controversy, but the number of reports is growing. Mapped here are incidences (red dots) reported in a peer-reviewed 2009 research paper. The green-shaded areas are regions of greatest density of native maize landraces. Adapted from Dyer, G.A., et al., 2009. Dispersal of transgenes through maize seed systems in Mexico. *Fig 1. PLoS ONE* 4(5) e5734:1–9.

Legend:
- Area of most maize landraces
- Locations of evidence of transgenic corn reported in 2009

localities. Intensive sampling at two of these localities showed transgenes from 11 of 60 fields, and in roughly 1% of the samples.

In explaining why their team found transgenes whereas Ortiz-García's had not, Piñeyro-Nelson's group mentioned that a company that had run genetic tests for both research teams used methods liable to overlook evidence for transgenes. The company, they maintained, was likely producing "false negatives" by being too conservative.

Their company's reputation on the line, the vice president and the CEO of the firm, Genetic ID, penned a letter to the editor of *Molecular Ecology*, the journal that had published the paper. Nothing was wrong with the firm's techniques, they argued; instead, Piñeyro-Nelson's group had likely contaminated the samples and was claiming "false positives." Piñeyro-Nelson's team responded with a thorough defense and presented new data to support its conclusions. This included tests applying herbicide to plants, revealing that some showed resistance presumably induced by the transgene.

By 2010, two other papers published by still more researchers presented additional evidence of transgenes in Mexican maize (**see figure**). The debate goes on, but it increasingly appears that transgenes have in fact moved into native Mexican maize, at low frequencies in scattered locations. Whether this will have substantial effects on agriculture, health, or the environment, however, remains unknown. ■

of maize from across Oaxaca, in the *Proceedings of the National Academy of Sciences of the USA*. They analyzed 154,000 seeds from 870 maize plants in 125 fields at 18 localities in 2003 and 2004, yet they detected no evidence of transgenes from GM corn. Their statistical analysis indicated with 95% certainty that if transgenes were present at all, they likely occurred in fewer than 1 out of 10,000 seeds.

To explain how their results could differ so greatly from Quist and Chapela's findings, Ortiz-García's team suggested that transgenes could have become rarer since 2000–2001 either

(1) by chance, (2) by hybridization with local landraces, (3) by being weeded out by natural selection, or (4) if farmers stopped planting GM seed once extension agents began warning them about it.

Then in 2009, a new research team announced that they had found transgenes in Mexican maize. This team, led by Alma Piñeyro-Nelson and Elena Álvarez-Buylla of the Universidad Nacional Autónoma de México, had collected maize from Oaxaca in 2001, 2002, and 2004, analyzing it with multiple methods. They found evidence of transgenes in three

saw in Chapter 9. In addition, crops engineered for drought tolerance can save on water needed for irrigation, and GM crops that increase yields help reduce the clearing of forests and the conversion of natural lands for agriculture.

Because GM technology is rapidly changing, and because its large-scale introduction into our environment is recent, there remains much that we don't yet know about the impacts of transgenic crops. Millions of Americans eat GM foods every day without outwardly obvious signs of harm, and direct evidence for negative ecological effects is limited so far.

However, it is still too early to dismiss concerns about environmental impacts without further scientific research. Many experts feel we should proceed with caution, adopting the **precautionary principle**, the idea that one should not undertake a new action until the ramifications of that action are well understood.

The best efforts so far to scientifically test the ramifications of GM crops were made in Great Britain. The British government, in considering whether to allow the planting of GM crops, commissioned three large-scale studies. The first study

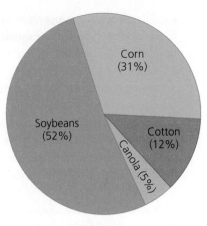

(a) GM crops by type

Corn (31%)
Soybeans (52%)
Cotton (12%)
Canola (5%)

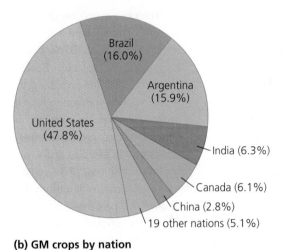

(b) GM crops by nation

Brazil (16.0%)
Argentina (15.9%)
United States (47.8%)
India (6.3%)
Canada (6.1%)
China (2.8%)
19 other nations (5.1%)

FIGURE 10.15 ▲ Of the world's genetically modified crops (a), soybeans constitute the majority so far. Of the world's nations (b), the United States devotes the most land area to GM crops. Data are for 2009, from the International Service for the Acquisition of Agri-Biotech Applications.

found that GM crops could produce long-term economic benefits for Britain, although short-term benefits would be minor. The second study found little to no evidence of harm to human health, but noted that effects on wildlife and ecosystems should be tested before crops are approved. The third study (involving 19 researchers, 200 sites, and $8 million in funding) tested effects on bird and invertebrate populations from four GM crops modified for herbicide resistance. Results showed that fields of GM beets and GM spring oilseed rape supported less biodiversity than fields of their non-GM counterparts. Fields of GM maize supported more, however, and fields of winter oilseed rape showed mixed results. Policymakers had hoped that this study would end all debate, but the science showed that the impacts of GM crops are complex.

Debate over GM foods involves more than science

Far more than science is involved in society's debate over GM foods. Ethical issues play a large role. For many people, the idea of "tinkering" with the food supply seems dangerous or morally wrong. Furthermore, because every person relies on food for survival and cannot choose *not* to eat, the genetic modification of dietary staples such as corn, wheat, and rice essentially forces people to consume GM products or to go to unusual effort to avoid them.

The perceived lack of control over one's own food has driven widespread concern that the global food supply is being dominated by a few large agrobiotech corporations that develop GM technologies, among them Monsanto, Syngenta, Bayer CropScience, Dow, DuPont, and BASF. Many critics say these multinational corporations threaten the independence and well-being of the small farmer. This perceived loss of democratic local control is a driving force in the opposition to GM foods, especially in Europe and the developing world. Critics of biotechnology also voice concern that much of the research into the safety of GM organisms is funded, overseen, or conducted by the corporations that stand to profit if their transgenic crops are approved for human consumption, animal feed, or ingredients in other products.

So far, GM crops have not lived up to their promise of feeding the world's hungry. Nearly all commercially available GM crops have been engineered to express either pesticidal properties (e.g., Bt crops) or herbicide tolerance. Often, a company's GM crops are tolerant to herbicides that the same company manufactures and profits from (e.g., Monsanto's Roundup Ready crops). Crops with traits that might benefit poor, small-scale farmers of developing countries (such as increased nutrition, drought tolerance, and salinity tolerance) have not been widely commercialized, perhaps because corporations have less economic incentive to do so. Whereas the Green Revolution was a largely public venture, the "gene revolution" promised by GM crops is largely driven by the financial interests of corporations selling proprietary products.

Indeed, corporations patent the transgenes they develop, and some go to great lengths to protect their investments. For instance, thousands of North American farmers have found themselves at the mercy of the St. Louis–based Monsanto Company. This first came to light in Canada, when Monsanto's private investigators secretly took seed samples from canola plants grown by Saskatchewan farmer Percy Schmeiser (**FIGURE 10.16**) and then charged him with violating Canada's law that makes it illegal for farmers to reuse patented seed or grow the seed without a contract with the company. Schmeiser maintained that pollen from Monsanto's Roundup Ready canola (see Figure 10.14) used by his neighbors blew onto his land and pollinated his non-GM canola. When Schmeiser harvested his seed and replanted it the next year, he unwittingly grew some plants containing Monsanto's patented herbicide-resistance gene. Schmeiser never purchased Monsanto's patented seed and said he did not want the crossbreeding. Still, Monsanto sued Schmeiser, and the court sided with Monsanto, ordering the farmer to pay the corporation roughly $238,000. Schmeiser appealed the case to Canada's Supreme Court, which also ruled, 5–4, that Schmeiser had violated Monsanto's patent.

In 2008, Schmeiser and Monsanto reached an out-of-court settlement whereby Monsanto agreed to pay all clean-up costs of the contamination. But the company continued suing many other farmers. As of late 2007, Monsanto had launched 112 such lawsuits against 372 farmers and 49 farm companies

FIGURE 10.16 ▲ Saskatchewan farmer Percy Schmeiser was accused by the Monsanto Company of planting its patented Roundup Ready canola in his field without a contract with the company. Schmeiser said his non–genetically modified plants were contaminated with Monsanto's transgenes from neighboring farms. Monsanto won the David-and-Goliath case in Canada's Supreme Court, but Schmeiser won an out-of-court settlement and became a hero to small farmers and activists opposed to genetically modified foods worldwide.

in 27 U.S. states, winning recorded judgments averaging $385,000, according to the Center for Food Safety (CFS), a watchdog organization. In addition, Monsanto has forced 2,400–4,500 other farmers into out-of-court settlements resulting in payments of $85–160 million, the CFS estimates. Monsanto says it is merely demanding that farmers heed the patent laws. North Dakota farmer Tom Wiley sees it differently, saying, "Farmers are being sued for having GMOs on their property that they did not buy, do not want, will not use, and cannot sell."

Given such developments, the future of GM foods seems likely to hinge on social, economic, legal, and political factors as well as scientific ones—and these factors vary in different nations. European consumers have expressed widespread unease about GM technologies. In contrast, U.S. consumers have largely accepted the GM crops approved by U.S. agencies, generally without even realizing that the majority of their food contains GM products. Opposition to GM foods in Europe blocked the import of hundreds of millions of dollars in U.S. agricultural products from 1998 to 2003. The United States complained that the European Union was hindering free trade and brought a successful case before the World Trade Organization (p. 190). Europeans now demand that GM foods be labeled and criticize the United States for not joining 100 other nations in signing the *Cartagena Protocol on Biosafety*, a treaty that lays out guidelines for open information about exported crops.

Transnational spats between Europe and the United States will surely affect the future direction of agriculture. So will policies of the world's large rapidly industrializing nations. Brazil, India, and China are now aggressively pursuing GM crops, even as ethical, economic, and political debates over the costs and benefits of these foods continue.

RAISING ANIMALS FOR FOOD: LIVESTOCK, POULTRY, AND AQUACULTURE

Food from cropland agriculture makes up a large portion of the human diet, but most of us also eat animal products. Doing so has significant environmental, social, agricultural, and economic impacts.

Consumption of animal products is growing

As wealth and global commerce has increased, so has humanity's consumption of meat, milk, eggs, and other animal products (**FIGURE 10.17**). The world population of domesticated animals raised for food rose from 7.2 billion animals to 24.9 billion animals between 1961 and 2008. Most of these animals are chickens. Global meat production has increased fivefold since 1950, and per capita meat consumption has doubled. The United Nations Food and Agriculture Organization (FAO) estimates that as more developing nations go through the demographic transition (pp. 210–211) and become wealthier, total meat consumption will nearly double by the year 2050.

Our food choices are also energy choices

What we choose to eat has ramifications for how we use energy and the land that supports agriculture (**FIGURE 10.18**). Recall our discussions of trophic levels and pyramids of energy (pp. 83–85). Every time energy moves from one trophic level to the next, as much as 90% is used up in cellular

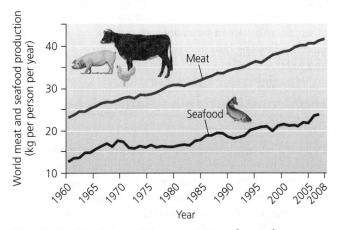

FIGURE 10.17 ▲ Per capita consumption of meat from farmed animals has risen steadily worldwide, as has per capita consumption of seafood (marine and freshwater, harvested and farmed). Data from U.N. Food and Agriculture Organization (FAO).

FIGURE 10.18 ▲ What we choose to eat has environmental consequences. Consuming meat intensifies demands and pressures on natural resources relative to a vegetarian diet.

respiration. For example, if we feed grain to a cow and then eat beef from the cow, we lose most of the grain's energy to the cow's metabolism. Energy is used up as the cow converts the grain to tissue as it grows, and as the cow conducts cellular respiration on a daily basis to maintain itself. For this reason, eating meat is far less energy-efficient than relying on a vegetarian diet and leaves a far greater ecological footprint. In contrast, if we eat lower in the food chain (a more vegetarian diet), we put a greater proportion of the sun's energy to use as food. The lower in the food chain we eat, the greater percentage of solar energy we put to use, and the more people Earth can support.

Some animals convert grain feed into milk, eggs, or meat more efficiently than others (**FIGURE 10.19**). Scientists have calculated relative energy-conversion efficiencies for different types of animals. Such energy efficiencies have ramifications for land use because land and water are required to raise food for the animals, and some animals require more than others. **FIGURE 10.20** shows the area of land and weight of water required to produce 1 kg (2.2 lb) of food protein for milk, eggs, chicken, pork, and beef. Producing eggs and chicken meat requires the least space and water, whereas producing beef requires the most. Such differences make clear that when we choose what to eat, we are also indirectly choosing how to make use of resources such as land and water.

High consumption has led to feedlot agriculture

In traditional agriculture, livestock are kept by farming families near their homes or are grazed on open grasslands by nomadic herders or sedentary ranchers. These traditions survive, but the advent of industrial agriculture has brought a new method. **Feedlots**, also known as *factory farms* or *concentrated animal feeding operations (CAFOs)*, are essentially huge warehouses or pens designed to deliver energy-rich food to animals living at extremely high densities (**FIGURE 10.21**). Today nearly half the world's pork and most of its poultry come from feedlots.

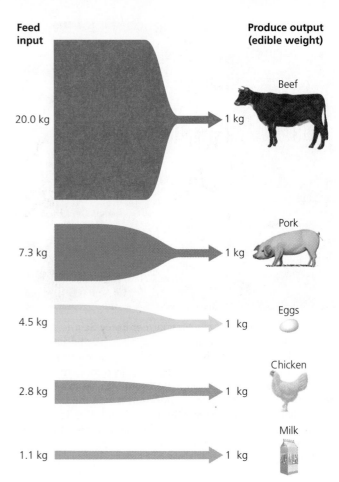

Feed input — Produce output (edible weight)

20.0 kg — Beef 1 kg

7.3 kg — Pork 1 kg

4.5 kg — Eggs 1 kg

2.8 kg — Chicken 1 kg

1.1 kg — Milk 1 kg

FIGURE 10.19 ▲ Different animal food products require different amounts of input of animal feed. Chickens must be fed 2.8 kg of feed for each 1 kg of resulting chicken meat, for instance, whereas 20 kg of feed must be provided to cattle to produce 1 kg of beef. Data from Smil, V., 2001. *Feeding the world: A challenge for the twenty-first century.* Cambridge, MA: MIT Press.

Feedlot animals are generally fed grain grown on cropland. One-third of the world's cropland is devoted to growing feed for animals, and 45% of our global grain production goes to livestock and poultry.

Feedlot operations allow for economic efficiency and greater food production and are probably necessary for a nation with a level of meat consumption as high as that of the United States. Feedlots offer one overarching benefit for environmental quality: Taking cattle and other livestock off rangeland and concentrating them in feedlots reduces the grazing impacts they would otherwise exert across large portions of the landscape. In Chapter 9 (pp. 244–247) we saw how overgrazing degrades soils and vegetation. Animals that are densely concentrated in feedlots will not contribute to overgrazing. However, intensified animal production through the industrial feedlot model exerts other environmental impacts.

Livestock agriculture pollutes water and air

Localized concentrations of pollution are the most noticeable environmental impacts of feedlots. Livestock produce prodigious amounts of manure and urine, and their waste can

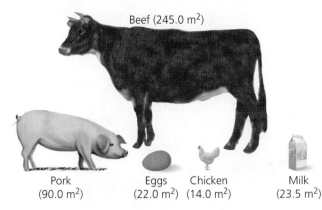

Beef (245.0 m²)

Pork
(90.0 m²)

Eggs
(22.0 m²)

Chicken
(14.0 m²)

Milk
(23.5 m²)

(a) Land required to produce 1 kg of protein

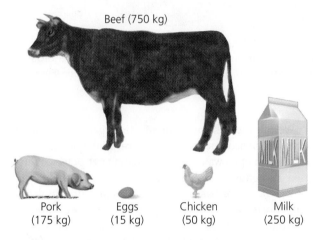

Beef (750 kg)

Pork
(175 kg)

Eggs
(15 kg)

Chicken
(50 kg)

Milk
(250 kg)

(b) Water required to produce 1 kg of protein

FIGURE 10.20 ▲ Producing different types of animal products requires different amounts of land and water. Raising cattle for beef requires by far the most land and water of all animal products. Data from Smil, V., 2001. *Feeding the world: A challenge for the twenty-first century.* Cambridge, MA: MIT Press.

pollute surface water and groundwater. Rich in nitrogen and phosphorus, livestock waste is a common cause of eutrophication (pp. 114, 422) in freshwater systems. It can also release a wide array of bacterial and viral pathogens that can sicken people, including *Salmonella, E. coli, Giardia, Microsporidia, Pfiesteria,* and pathogens that cause diarrhea, botulism, and parasitic infections.

The crowded and dirty conditions under which animals are often kept necessitate heavy use of antibiotics to control disease. Hormones are administered to livestock as well, and feed is spiked with heavy metals that spur growth. Livestock excrete most of these chemicals, which end up in wastewater and may be transferred up the food chain in downstream ecosystems. Some of the chemicals that remain in livestock meat are transferred to us when we eat the meat. The overuse of antibiotics can also cause microbes to evolve resistance to the antibiotics (just as pests evolve resistance to pesticides; pp. 258–259), making these drugs less effective. All in all, the FAO estimates that livestock (including both feedlot and grazed animals) in the United States account for 55% of soil erosion, 37% of pesticide applications, 50% of antibiotics consumed, and one-third of the nitrogen and phosphorus pollution in our waterways.

Feedlot impacts can be minimized when properly managed, and both the EPA and the states regulate U.S. feedlots. Wastewater and manure may be stored in lagoons, where it undergoes a degree of treatment somewhat similar to that of municipal wastewater (pp. 426–427). The resulting sludge may then be applied to farm fields as fertilizer (or injected into the ground where plants need it), reducing the need for chemical fertilizers.

Raising animals for food also results in air pollution. Besides the strong odors that emanate from feedlots and wastewater lagoons, livestock are a major source of greenhouse gases that lead to climate change (pp. 495–497). A comprehensive FAO report in 2006 concluded that livestock

FIGURE 10.21 ▼ Most meat eaten in the United States comes from animals raised in feedlots or factory farms. These locations house thousands of chickens **(a)** or cattle **(b)** at high densities. The animals are dosed liberally with antibiotics to control disease.

(a) Chicken factory farm in Arkansas

(b) Cattle feedlot in Nebraska

agriculture contributes 9% of our carbon dioxide emissions, 37% of our methane emissions, and 65% of our nitrous oxide emissions—altogether, 18% of the emissions driving climate change, a larger share than automobile transportation! Livestock release methane and nitrous oxide in their metabolism and waste. Nitrous oxide is also released from certain feed crops and from fertilizers applied to feed crops. Carbon dioxide is released when forests are cleared for ranching or for growing feed, and when fossil fuels are burned to grow feed, transport animals, and more. In addition to all these greenhouse emissions, livestock waste also emits ammonia—68% of our total emissions—which contributes to acid rain (pp. 482–486).

WEIGHING THE ISSUES

Feedlots and Animal Rights Animal rights activists decry factory farming because they say it mistreats animals. Chickens, pigs, and cattle are kept crowded together in small pens their entire lives, fattened up, and slaughtered. Should we concern ourselves with the quality of life—and death—of the animals that constitute part of our diet? Do you think animal rights concerns are as important as the environmental issues? Are conditions at feedlots a good reason for being vegetarian?

We also raise fish on "farms"

Besides growing plants as crops and raising animals on rangelands and in feedlots, we rely on aquatic organisms for food. Wild fish populations are plummeting throughout the world's oceans as increased demand and new technologies have led us to overharvest most marine fisheries (pp. 448–453). This means that raising fish and shellfish on "fish farms" may be the only way to meet our growing demand for these foods (**FIGURE 10.22**).

FIGURE 10.22 ▼ People practice many types of aquaculture. Here, a French oyster farmer tends his oyster beds on an island along the Atlantic coast of France.

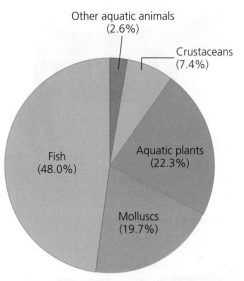

FIGURE 10.23 ▲ Aquaculture involves many types of fish, but also a wide diversity of other marine and freshwater organisms. Data from U.N. Food and Agriculture Organization (FAO).

We call the cultivation of aquatic organisms for food in controlled environments **aquaculture**. Many aquatic species are grown in open water in large, floating net-pens. Others are raised in ponds or holding tanks. Small-scale community-based aquaculture has been a focus of sustainable development efforts in the developing world, whereas large-scale industrialized aquaculture produces large amounts of food while exerting environmental impacts. People pursue aquaculture with over 220 freshwater and marine species ranging from fish to shrimp to clams to seaweeds (**FIGURE 10.23**). Some species, such as carp, are grown for local consumption, whereas others, such as salmon and shrimp, are exported to affluent nations.

Aquaculture is the fastest-growing type of food production; in the past 20 years, global output has increased fivefold. Most widespread in Asia, aquaculture today produces $80 billion worth of food and provides three-quarters of the freshwater fish and over half of the shellfish that we eat.

Aquaculture brings benefits

When conducted on a small scale by families or villages, as in China and much of the developing world, aquaculture helps ensure people a reliable protein source. Such small-scale aquaculture can be sustainable, and it is compatible with other activities. For instance, uneaten fish scraps make excellent fertilizers for crops, and food waste can be fed to fish. Aquaculture on larger scales can help improve a region's or nation's food security by increasing supplies of fish available.

Aquaculture on any scale helps reduce fishing pressure on overharvested and declining wild stocks. Reducing fishing pressure also lessens the *by-catch* (p. 450; the unintended catch of nontarget organisms) that results from commercial fishing. Furthermore, aquaculture consumes fewer fossil fuels and provides a safer work environment than does commercial fishing. Fish farming can also be remarkably energy-efficient, producing as much as 10 times more fish per unit area than is harvested from waters of the continental shelf and up to 1,000 times more than is harvested from the open ocean.

Aquaculture has negative impacts

Along with its benefits, aquaculture has disadvantages. Dense concentrations of farmed animals can increase disease, which reduces food security, necessitates antibiotic treatment, and results in expense. A virus outbreak wiped out half a billion dollars in shrimp in Ecuador in 1999, for instance. Aquaculture can also produce prodigious amounts of waste, both from the farmed organisms and from the feed that goes uneaten and decomposes in the water. Commercially farmed fish often are fed grain, and as we have discussed, growing grain to feed animals that we then eat is energy-inefficient. In other cases, farmed fish are fed fish meal made from wild ocean fish such as herring and anchovies, whose harvest may place additional stress on wild fish populations. For all these reasons, industrial-scale aquaculture can leave a large ecological footprint.

If farmed aquatic organisms escape into ecosystems where they are not native (as several carp species have done in U.S. waters), they may spread disease to native stocks or may outcompete native organisms for food or habitat. These possibilities raise special concern when the farmed animals are genetically modified. Genetic engineering of Atlantic and Pacific salmon may produce transgenic fish that grow to several times the normal size for their species (**FIGURE 10.24**). Such GM fish may outcompete their wild cousins while also spreading disease to them. They may also interbreed with native and hatchery-raised fish and weaken already troubled stocks. Researchers have concluded that under certain circumstances, escaped transgenic salmon may increase the extinction risk for native populations of their species, in part because the larger male fish (such as those carrying a gene for rapid and excessive growth) have better odds of mating successfully.

FIGURE 10.24 ▼ Transgenic salmon (top) can be considerably larger than wild salmon of the same species.

SUSTAINABLE AGRICULTURE

Industrial agriculture has allowed food production to keep pace with our growing population, but it involves many adverse environmental and social impacts. These range from the degradation of soils (Chapter 9) to reliance on fossil fuels (Chapter 19) to problems arising from pesticide use, genetic modification, and intensive feedlot and aquaculture operations. Although intensive commercial agriculture may help alleviate certain environmental pressures, it often worsens others. Industrial agriculture in some form seems necessary to feed our planet's 6.9 billion people, but many experts feel we will be better off in the long run by raising animals and crops in ways that are less polluting and less resource-intensive.

We are moving toward sustainable agriculture

Sustainable agriculture is agriculture that does not deplete soils faster than they form. It is farming and ranching that do not reduce the amount of healthy soil, clean water, and genetic diversity essential to long-term crop and livestock production. Simply put, sustainable agriculture is agriculture that can be practiced in the same way far into the future. Approaches such as no-till farming and the soil conservation methods we examined in Chapter 9 help move us toward sustainable agriculture. So do more traditional approaches, such as the Chinese practice of carp aquaculture in small ponds and the cultivation of diverse landraces of Mexican maize.

One key component of making agriculture sustainable is reducing the fossil-fuel-intensive inputs we invest in it and decreasing the pollution these inputs cause. *Low-input agriculture* describes agriculture that uses lesser amounts of pesticides, fertilizers, growth hormones, antibiotics, water, and fossil fuel energy than are used in industrial agriculture. Food-growing practices that aim to use no synthetic fertilizers, insecticides, fungicides, or herbicides—but instead rely on biological approaches such as composting (pp. 625–626) and biocontrol—are termed **organic agriculture**.

Organic approaches reduce inputs and pollution

The bounty of organic agriculture is increasingly available to us (**FIGURE 10.25**). But what exactly is meant by the term *organic*? In 1990, the U.S. Congress passed the Organic Food Production Act to establish national standards for organic products and facilitate their sale. Under this law, the USDA in 2000 issued criteria by which crops and livestock could be officially certified as organic (**TABLE 10.1**), and these standards went into effect in 2002 as part of the National Organic Program. In crafting a formal definition for the word *organic*, the National Organic Standards Board wrote:

> Organic agriculture . . . is based on minimal use of off-farm inputs and on management practices that restore, maintain and enhance ecological harmony. . . . Methods are used to minimize pollution from air, soil, and water. . . . The primary goal of organic agriculture is to optimize the health and productivity of interdependent communities of soil life, plants, animals, and people.

FIGURE 10.25 ▲ You have likely encountered organic produce at your local grocery. Look for the USDA organic logo (inset) to ascertain whether a product is certified organic under the National Organic Program.

California, Washington, and Texas established stricter state guidelines for labeling foods organic, and today many U.S. states and over 70 nations have laws spelling out organic standards.

For farmers, organic farming can bring a number of benefits: lower input costs, enhanced income from higher-value produce, and reduced chemical pollution and soil degradation (see **THE SCIENCE BEHIND THE STORY**, pp. 274–275). In many cases more pests attack organic crops because of the lack of chemical pesticides, but biocontrol methods (p. 258) can help keep pests in check. Moreover, the lack of synthetic chemicals maintains soil quality and encourages pollinating insects. Surveys reveal that farmers who adopt organic techniques do so primarily because they want to practice stewardship toward the land and to safeguard their family's health. Farmers face obstacles to adopting organic methods, however. Foremost among these are the risks and costs of shifting to new methods, particularly during the transition period, because U.S. farmers need to meet standards for three years before their products can be certified.

Many consumers favor organic food out of concern about potential health risks posed by the pesticides, hormones, and antibiotics used in conventional agriculture. Consumers also buy organic products out of a desire to improve environmental quality. Recent surveys show that about 55% of U.S. and Canadian consumers who buy organic products do so for health reasons and about one-third do so for environmental reasons. The main obstacle for consumers is price. Organic products tend to be 10–30% more expensive than conventional ones, and some (such as milk) can cost twice as much. However, enough consumers are willing to pay more for organic products that grocers and other businesses are making them more widely available.

Organic agriculture is booming

Just a decade or two ago, few farmers grew organic food, few consumers wanted it, and the only place to buy it was in specialty stores. Today that has changed; three of four Americans buy organic food at least occasionally, and more than four of five retail groceries offer it. In fact, consumer demand for organic food is now so great that farmers cannot keep up with demand, and shortages regularly occur. U.S. consumers spent $22.9 billion on organic food in 2008, amounting to 3.5% of all food sales (**FIGURE 10.26A**). Worldwide, sales of organic food tripled between 2000 and 2008, when sales surpassed $50 billion.

Production is increasing along with demand (**FIGURE 10.26B**). Although organic agriculture takes up less than 1% of agricultural land worldwide (35 million ha, or 86.5 million acres, in 2008), this area is rapidly expanding. Two-thirds of this area is in developed nations, and two-thirds is grazing land. In addition, 430,000 ha (1.1 million acres) of aquaculture is now certified organic. As of 2008 in the United States, nearly

TABLE 10.1 USDA Criteria for Certifying Crops and Livestock as Organic

For crops to be considered organic . . .
▶ The land must be free of prohibited substances for 3 years.
▶ The crops must not be genetically engineered.
▶ The crops must not be irradiated to kill bacteria.
▶ Sewage sludge cannot be used.
▶ Use of organic seeds and planting stock is preferred.
▶ Farmers must not use synthetic fertilizers. Only crop rotation, cover crops, animal or crop wastes, or approved synthetic materials are allowed.
▶ Use of most conventional pesticides is prohibited. Crop pests, weeds, and diseases should be managed with biocontrol, mechanical practices, or approved synthetic substances.

For livestock to be considered organic . . .
▶ Mammals must be raised organically from the last third of gestation; poultry, from the second day of life.
▶ Livestock feed must be 100% organic; however, vitamin and mineral supplements are allowed.
▶ Dairy cows must receive 80% organic feed for 9 months, followed by 3 months of 100% organic feed.
▶ Hormones and antibiotics are prohibited; vaccines are permitted.
▶ Animals must have access to the outdoors.

Adapted from the National Organic Program. 2002. *Organic production and handling standards.* Washington, DC: U.S. Department of Agriculture.

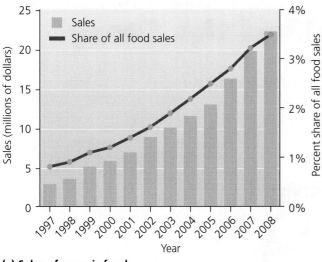

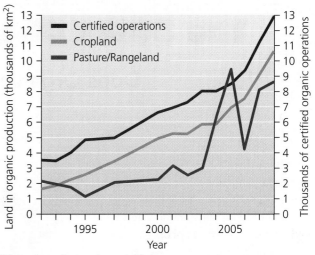

(a) Sales of organic food

(b) Extent of organic agriculture

FIGURE 10.26 ▲ Sales of organic food in the United States **(a)** have increased rapidly in the past decade, both in total dollar amounts (bars) and as a percentage of the overall food market (line). Since the mid-1990s in the United States **(b)**, acreage devoted to organic crops and livestock has each quadrupled, and the number of certified operations has more than tripled. (a) Adapted by permission from Willer, Helga and Lukas Kilcher (Eds.) (2009) *The world of organic agriculture, statistics and emerging trends* 2009. FiBL-IFOAM Report. IFOAM, Bonn; FiBL, Frick; ITC, Geneva. Data used with permission from Organic Trade Association OTA: Manufacturer Survey 2007. Additional, updated data provided by Organic Trade Association. (b) Data from USDA Economic Research Service.

2 million ha (4.8 million acres) were under organic management. Europe boasts still more: 8.2 million ha (20.3 million acres) in 2008, most in nations of the European Union, where 4% of the agricultural area is in organic production. Today farmers in all 50 U.S. states and 1.4 million farmers in more than 150 nations practice organic farming commercially to some extent.

Mexico is a leader in organic agriculture among developing nations. Overall, 2.9% of Mexico's agricultural land is organic, one of the highest percentages for developing nations, and 130,000 Mexican farmers farm organically, the third-highest number in the world. Fully 30% of Mexico's coffee crop is now organic, and Mexico produces more organic coffee than any other nation (**FIGURE 10.27**). Besides coffee, Mexico grows cocoa and a variety of fruits and vegetables organically. Most of this produce is destined for export to the United States and Europe.

Government initiatives have also spurred the growth of organic farming. The European Union adopted a policy in 1993 to support farmers financially during conversion. This is an example of a subsidy (p. 182) aimed at reducing the external costs (pp. 150–151) to society of industrial agriculture. The United States offers no such subsidies, which may be why U.S. organic production lags behind that of Europe. However, the 2008 Farm Bill (p. 247) did set aside $112 million over five years for organic agriculture, and the government helps to defray certification expenses. Government support is helpful, because conversion often means a temporary loss in income for farmers. Once conversion is complete, though, studies suggest that reduced inputs and higher market prices can make organic farming at least as profitable for the farmer as conventional methods.

Locally supported agriculture is growing

Apart from organic methods, another component of the move toward sustainable agriculture is an attempt to reduce fossil fuel use from the long-distance transport of food (see **ENVISIONIT**, p. 276). The average food product sold in a

U.S. supermarket travels at least 2,300 km (1,400 mi) between the farm and the grocery. The extensive transportation and storage infrastructure that gives industrial agriculture its long-distance reach distributes a wide and reliable variety of foods to all areas, but it does so by burning immense amounts of petroleum, with all the environmental impacts that entails. In addition, because of the travel time, supermarket produce is often chemically treated to preserve freshness and color.

In response, increasing numbers of farmers and consumers in developed nations are supporting local small-scale agriculture. Farmers' markets (**FIGURE 10.28**) are springing up throughout North America as people rediscover the joys of fresh, locally grown produce. At **farmers' markets**, consumers buy meats and fresh fruits and vegetables in season

FIGURE 10.27 ▼ These farmers are harvesting coffee beans from an organic plantation of shade-grown coffee in Chiapas, Mexico. Mexico is a world leader in the production of organic coffee.

Does Organic Farming Work?

Organic farming puts fewer synthetic chemicals into the soil, air, and water, but can it produce large enough crop yields to feed the world's people?

To search for the answer we can turn to Switzerland, a leader in organic farming. One in every nine hectares of agricultural land here is managed organically—the fourth-highest rate in the world—and Swiss scientists have taken the lead in examining the benefits and costs of organic farming. The longest-running and most-cited studies of organic agriculture come from experimental farms at Therwil, Switzerland, established back in 1977. Long-term studies are rare and valuable in agriculture and in ecology, because they can reveal slow processes or subtle effects that get swamped out by variation in shorter-term studies.

At the Swiss research site, wheat, potatoes, and other crops are grown in plots cultivated in different treatments:

- Conventional farming using large amounts of chemical pesticides, herbicides, and inorganic fertilizer
- Conventional farming that also uses organic fertilizer (cattle manure)

Dr. Paul Mäder, Research Institute of Organic Agriculture, Switzerland

- Organic farming using only manure, mechanical weeding, and plant extracts to control pests
- Organic farming that also adds natural boosts, such as herbal extracts in compost

Researchers record crop yields at harvest each year. They analyze the soil regularly, measuring nutrient content, pH, structure, and other variables. They also measure the biological diversity and activity of microbes and invertebrates in the soil.

Such indicators of soil quality help researchers assess the potential for long-term farm productivity.

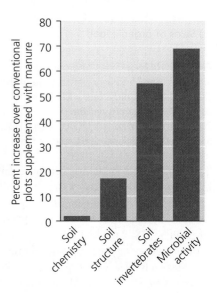

Organic fields outperformed conventional fields supplemented with manure in their soil quality, including averaged values for soil chemistry (six variables), soil structure (three variables), soil invertebrates (five variables), and microbial activity (six variables). They outperformed conventional fields without manure (not shown) still more. Data from Mäder, P., et al., 2002. Soil fertility and biodiversity in organic farming. *Science* 296: 1694–1697.

from local producers. These markets generally offer a wide choice of organic items and unique local varieties not found in supermarkets.

Some consumers are even partnering with local farmers in a phenomenon called **community-supported agriculture (CSA)**. In a CSA program, consumers pay farmers in advance for a share of their yield, usually a weekly delivery of produce. Consumers get fresh seasonal produce, while farmers get a guaranteed income stream up front to invest in their crops—a welcome alternative to taking out loans and being at the mercy of the weather. As of 2007, over 12,500 U.S. farms were supplying hundreds of thousands of families per week in community-supported agriculture programs.

Sustainable agriculture mimics natural ecosystems

The best approach for making an agricultural system sustainable is to mimic the way a natural ecosystem functions. Ecosystems are sustainable because they operate in cycles and are internally stabilized with negative feedback loops. In this way they provide a useful model for agriculture.

One example comes from Japan, where some small-scale rice farmers are reviving ancient traditions and finding them superior to modern industrial methods. Takao Furuno is one such farmer. Starting 20 years ago, he and his wife added a crucial element to their rice paddies: the crossbred

In 2002, Paul Mäder and colleagues from two Swiss research institutes reported in the journal *Science* results from 21 years of data. Over this time, the organic fields yielded 80% of what the conventional fields produced. Organic crops of winter wheat yielded 90% of the conventional yield. Organic potato crops averaged 58–66% of conventional yields because of nutrient deficiency and disease.

Although the organic plots produced 20% less, they did so while receiving 35–50% less fertilizer than the conventional fields and 97% fewer pesticides. Thus, Mäder's team concluded, the organic plots were highly efficient and represent "a realistic alternative to conventional farming systems."

How can organic fields produce decent yields without synthetic agricultural chemicals? The answer lies in the soil. Mäder's team found that soil in the organic plots had better structure, better supplies of some nutrients, and much more microbial activity and invertebrate biodiversity (**first figure**). Differences were visible just by looking at the fields (**second figure**).

The Swiss data reflect those found by scientists elsewhere. For instance, U.S. researchers comparing organic and conventional farms in North Dakota and Nebraska found that organic farming produced soils with more microbial life, earthworm activity, water-holding capacity, topsoil depth, and naturally occurring nutrients. Given such differences in soil quality, researchers expect that organic fields should perform better

(a) Organic plot

(b) Conventional plot

Organic plots **(a)** showed more weeds but more earthworm castings and healthier soil structure than conventional plots **(b)** in the Swiss experiment.

and better relative to conventional fields as time goes by—in other words, that they are more sustainable.

Studies are continuing at the Swiss plots to determine whether this is true. As one example, in 2009, Jens Leifeld and two colleagues at a Zurich research institute analyzed soil carbon content after 27 years. They found that soil carbon had decreased in all the treatments, but that it had declined most in conventional plots

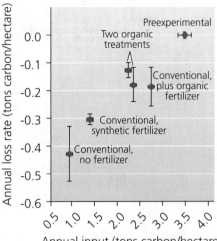

Organic plots lost less soil carbon over 27 years of farming than did conventional plots with no fertilizer or only synthetic fertilizer. Conventional plots with organic fertilizer supplements retained just as much soil carbon as the organic plots. Data from Leifeld, J., et al., 2009. Consequences of conventional versus organic farming on soil carbon: Results from a 27-year field experiment. *Agronomy Journal* 101: 1204–1218.

using inorganic fertilizer or no fertilizer. Conventional plots that supplemented synthetic fertilizer with cow manure, however, did just as well as the organic plots in retaining soil carbon (**third figure**).

As more long-term data are published and new studies begin, we are learning more and more about the benefits of organic agriculture for soil quality and about how we might improve conventional methods to maximize crop yields while protecting the long-term sustainability of agriculture. ∎

---

aigamo duck. Each spring after they plant rice seeds, the Furunos release hundreds of *aigamo* ducklings into their paddies (**FIGURE 10.29**). The ducklings eat weeds that compete with the rice, as well as insects and snails that attack the rice. The ducklings also fertilize the rice plants with their waste and oxygenate the water by paddling. Furuno and the scientists and extension agents who have worked with him have found that rice plants grow larger and yield far more rice in paddies that have ducks. Once the rice grains form, the ducks are taken out of the paddies (because they would eat the rice grains) and kept in sheds, where they are fed waste grain. They mature, lay eggs, and can be sold at market.

Besides the ducks, Furuno raises fish in the paddies, and these provide food and fertilizer as well. He also lets the aquatic fern *Azolla* cover the water surface. This plant fixes nitrogen, feeds the ducks, hides the fish, and provides habitat for insects, plankton, and aquatic invertebrates, which provide further food for the fish and ducks. Because fast-growing *Azolla* can double its biomass in 3 days, surplus plant matter is harvested and used as cattle feed. The end result is a productive, functioning ecosystem in which pests and weeds are transformed into resources and which yields organic rice, eggs, and meat from ducks and fish. From 2 hectares of paddies and 1 hectare of organic vegetables, the Furunos annually produce 7 tons of rice, 300 ducks, 4,000 ducklings, and

The average grocery store item travels 1,400 miles from its origin to your shopping cart.

This long-distance transport consumes oil and contributes to pollution and climate change.

YOU CAN MAKE A DIFFERENCE

➤ Ask your campus's food services director to buy locally grown produce and locally made products.

➤ Shop at farmers' markets, or help to start one on campus.

➤ Partner with farmers in community-supported agriculture programs.

Locally grown produce and locally made products cut down on the carbon footprint of our food.

FIGURE 10.28 ▲ Farmers' markets like this one in San Francisco are beginning to flourish as consumers rediscover the benefits of buying fresh, locally grown produce.

FIGURE 10.29 ▲ Ducks help rice crops grow in a sustainable agricultural system developed in Japan, known as the "*aigamo* method."

enough vegetables to feed 100 people. At this rate—twice the productivity of the region's conventional farmers—just 2% of Japan's people could supply the nation's food needs.

Takao Furuno wrote a book to popularize the "*aigamo* method," and today over 10,000 Japanese farmers are using it, increasing their yields 20–50% and regaining huge amounts of time they used to spend manually weeding. The approach is now spreading to other Asian nations. Furuno's approach shows how working with nature rather than against it can produce success in agriculture.

Treating agricultural systems as ecosystems is a key aspect of sustainable agriculture, and this general lesson applies regardless of location, scale, or the crop involved. Southern Mexico's organic coffee farmers promote an ecosystem approach by letting insect pests be eaten by birds, spiders, and other insects, rather than chemically poisoning all of them and their soil as well. Those coffee-growers practicing shade-grown techniques enhance this sustainable ecosystem model further. Likewise, Mexico's traditional maize farmers who cultivate native landraces are preserving and promoting genetic diversity characteristic of natural systems. Efforts like these, together with other approaches farmers around the world are taking to help make agriculture sustainable, will be crucial for all of us as we progress through the coming century.

➤ CONCLUSION

Many practices of intensive commercial agriculture exert substantial negative environmental impacts. At the same time, it is important to realize that aspects of industrial agriculture have helped to relieve certain pressures on land or resources. Whether Earth's natural systems would be under more pressure from 6.9 billion people practicing traditional agriculture or from 6.9 billion people living under our modern industrialized model is a very complex question.

What is certain is that if our planet is to support 9 billion people by midcentury without further degrading the soil, water, pollinators, and other resources and ecosystem services that support our food production, we must find ways to shift to sustainable agriculture. Approaches such as biological pest control, organic agriculture, pollinator conservation, preservation of native crop diversity, sustainable aquaculture, and likely some degree of careful and responsible genetic modification of food may all be parts of the game plan we will need to set in motion to work together toward a sustainable future.

R E V I E W I N G   O B J E C T I V E S

You should now be able to:

EXPLAIN THE CHALLENGE OF FEEDING A GROWING HUMAN POPULATION

- Our food production has outpaced our population growth, yet 1 billion people still go hungry. (pp. 253–254)
- Undernutrition, overnutrition, and malnutrition are all challenges. (pp. 253–254)

- Growing crops for biofuels can result in food shortages. (p. 256)

IDENTIFY THE GOALS, METHODS, AND CONSEQUENCES OF THE GREEN REVOLUTION

- The Green Revolution aimed to increase agricultural enhanced productivity per unit area of land in developing nations. (pp. 254–255)

- Scientists used selective breeding to develop crop strains that grew quickly, were more nutritious, or were resistant to disease or drought. (p. 255)
- The expanded use of fossil fuels, chemical fertilizers, and pesticides has increased pollution. (pp. 255–256)
- The increased efficiency of production has fed more people while reducing the amount of natural land converted for farming. (pp. 255–256)

DESCRIBE APPROACHES FOR PRESERVING CROP DIVERSITY

- Protecting diversity of native crop varieties provides insurance against failure of major crops. (pp. 256–257)
- Seed banks preserve rare and local varieties of seed, acting as storehouses for genetic diversity. (pp. 257–258)

CATEGORIZE STRATEGIES FOR PEST MANAGEMENT

- We kill "pests" and "weeds" with synthetic chemicals that can pollute the environment and pose health hazards. (p. 258)
- Pests can evolve resistance to chemical pesticides, leading us to design more toxic poisons. (pp. 258–259)
- We employ natural enemies of pests against them in the practice of biological control. (pp. 258–259)
- Integrated pest management combines various techniques and minimizes the use of synthetic chemicals. (pp. 259–260)

DISCUSS THE IMPORTANCE OF POLLINATION

- Insects and other organisms are essential for the reproduction of many crop plants. (pp. 260–261)
- Conservation of pollinating insects is vitally important to our food security. (p. 261)

DESCRIBE THE SCIENCE BEHIND GENETICALLY MODIFIED FOOD

- Genetic modification uses recombinant DNA technology to move genes for desirable traits from one type of organism to another. (pp. 261–262)
- Genetic engineering is both like and unlike traditional selective breeding. (p. 262)

- GM crops may have ecological impacts, including the spread of transgenes, an increase in chemical pollution, and indirect impacts on biodiversity. (pp. 263–266)

EVALUATE THE DEBATE OVER GENETICALLY MODIFIED FOOD

- Many people have ethical qualms about altering food through genetic engineering. (p. 266)
- Opponents of GM foods view multinational biotechnology corporations as a threat to the independence of small farmers. (pp. 266–267)
- Nations have adopted differing stances on GM foods. (p. 267)

ASSESS FEEDLOT AGRICULTURE FOR LIVESTOCK AND POULTRY

- As wealth has increased, so has the consumption of animal products. (p. 267)
- Eating animal products leaves a greater ecological footprint than eating plant products. (pp. 267–269)
- Feedlots create waste and other environmental impacts, but they also relieve pressure on lands that could otherwise be overgrazed. (pp. 268–270)

WEIGH APPROACHES IN AQUACULTURE

- Aquaculture provides economic benefits and food security, relieves pressures on wild fish stocks, and can be sustainable. (pp. 270–271)
- Aquaculture also gives rise to pollution, habitat loss, and other environmental impacts. (p. 271)

EVALUATE SUSTAINABLE AGRICULTURE

- Organic agriculture exerts fewer environmental impacts than industrial agriculture. It comprises a small part of the market but is growing rapidly. (pp. 271–275)
- Locally supported agriculture, as shown by farmers' markets and community-supported agriculture, is also growing. (pp. 273–277)
- Mimicking natural ecosystems is a key approach to making agriculture sustainable. (pp. 274–277)

TESTING YOUR COMPREHENSION

1. What kinds of techniques have people employed to increase agricultural production? How did Norman Borlaug help inaugurate the Green Revolution?

2. Explain how pesticide resistance occurs.

3. Explain the concept of biocontrol. List several components of a system of integrated pest management (IPM).

4. About how many and what types of cultivated plants are known to rely on insects for pollination? Why is it important to preserve the biodiversity of native pollinators?

5. What is recombinant DNA? How is a transgenic organism created? How is genetic engineering different from traditional agricultural breeding? How is it similar?

6. Describe several reasons why many people support the development of genetically modified organisms, and name several uses of such organisms that have been developed so far.

7. Describe the scientific concerns held by opponents of GM crops. Describe some of their other concerns.

8. Name several positive and negative environmental effects of feedlot operations. Why is beef an inefficient food from the perspective of energy consumption?

9. What are some economic benefits of aquaculture? What are some negative environmental impacts?

10. What are the objectives of sustainable agriculture? What factors are driving the growth of organic agriculture?

SEEKING SOLUTIONS

1. Assess several ways in which high-input industrial agriculture can be beneficial for the environment and several ways in which it can be detrimental. Now suggest several ways in which we might modify industrial agriculture to mitigate its environmental impacts.

2. Describe two fundamental approaches for preserving the genetic diversity of crop variants. Give a specific example of each approach. What advantages and disadvantages, what benefits and risks do you see in each approach?

3. What factors make for an effective biological control strategy of pest management? What risks are involved in biocontrol? If you had to decide whether to use biocontrol against a particular pest, what questions would you want to have answered before you decide?

4. People who view GM foods as solutions to world hunger and pesticide overuse often want to speed their development and approval. Others adhere to the precautionary principle and want extensive testing for health and environmental safety. How much caution do you think is warranted before a new GM crop is introduced?

5. **THINK IT THROUGH** You are a USDA official and must decide whether to allow the planting of a new genetically modified strain of cabbage that produces its own pesticide and has twice the vitamin content of regular cabbage. What questions would you ask of scientists before deciding whether to approve the new crop? What scientific data would you want to see? Would you also consult nonscientists or consider ethical, economic, and social factors?

6. **THINK IT THROUGH** You are a farmer in the U.S. Midwest, own 1,000 acres, and intend to farm corn for a living. Would you choose to grow genetically modified corn? How would you manage pests and weeds? Would you grow corn for people, livestock, or both? Think of the various ways corn is grown, purchased, and valued in different places—such as the United States, Europe, and southern Mexico—as you formulate your answer.

CALCULATING ECOLOGICAL FOOTPRINTS

As food production became more industrialized during the 20th century, more and more energy has been expended to store food and ship it to market. In the United States today, food travels an average of 1,400 miles from the field to your table. The price you pay for the food covers the cost of this long-distance transportation, which in 2004 was approximately $1 per ton per mile. Assuming that the average person eats 2 pounds of food per day, calculate the food transportation costs for each category in the table below.

Consumer	Daily cost	Annual cost
You	$1.40	$511
Your class		
Your town		
Your state		
United States		

Pirog, R. and A. Benjamin, 2003. Checking the Food Odometer: comparing Food Miles for Local Versus Conventional Produces Sales to Iowa Institutions. Leopold Center for Sustainable Agriculture, Iowa State University, Ames.

1. What specific challenges to environmental sustainability are imposed by a food production and distribution system that relies on long-range transportation to bring food to market?

2. A 2003 study noted that locally produced food traveled only 50 miles or so to market, thus saving 96% of the transportation costs. Locally grown foods may be fresher and cause less environmental impact as they are brought to market, but what are the disadvantages to you as a consumer in relying on local food production? Do you think the advantages outweigh those disadvantages?

3. What has happened to gasoline prices recently? How would future increases in the price of gas affect your answers to the preceding questions?

A diversity of fruits from Panamanian rainforest trees

11 BIODIVERSITY AND CONSERVATION BIOLOGY

UPON COMPLETING THIS CHAPTER, YOU WILL BE ABLE TO:

- Characterize the scope of biodiversity on Earth
- Contrast the background extinction rate with periods of mass extinction
- Evaluate the primary causes of biodiversity loss
- Specify the benefits of biodiversity

- Assess the science and practice of conservation biology
- Analyze efforts to conserve threatened and endangered species
- Compare and contrast conservation efforts above the species level

CENTRAL CASE STUDY

Saving the Siberian Tiger

"Future generations would be truly saddened that this century had so little foresight, so little compassion, such lack of generosity of spirit for the future that it would eliminate one of the most dramatic and beautiful animals this world has ever seen."

—George Schaller, wildlife biologist, on the tiger

"If you kill a tiger, you can buy a motorbike."

—Anonymous poacher, on selling tiger parts

Historically, tigers roamed widely across Asia from Turkey to northeast Russia to Indonesia. Within the past 200 years, however, people have driven the majestic striped cats from most of their historic range. Today, tigers are exceedingly rare and are sliding toward extinction.

Of the tigers that survive, those of the subspecies known as the Siberian tiger are the largest cats in the world. Males can reach 3.7 m (12 ft) in length and weigh up to 360 kg (800 lb). These regal animals today find their last refuge in the forests of the remote Sikhote-Alin Mountains of the Russian Far East.

A Siberian tiger in the Sikhote-Alin Mountains

For thousands of years the Siberian tiger coexisted with the region's native people and held a prominent place in their lore. These people referred to the tiger as "Old Man" or "Grandfather" and viewed it as a guardian of the mountains and forests. They rarely killed a tiger unless it had preyed on a person.

The Russians who moved into the region and exerted control in the early 20th century had no such cultural traditions. They hunted tigers for sport and hides, and some Russians reported killing as many as 10 tigers in a single hunt. Later, poachers began killing tigers illegally to sell their body parts to China and other Asian countries, where they are used in traditional medicine and as alleged aphrodisiacs. Meanwhile, road building, logging, and agriculture began to fragment tiger habitat and provide easy access for poachers. The tiger population dipped to perhaps 20–30 animals.

International conservation groups got involved just in time, working with Russian biologists to save the dwindling tiger population. One such group was the Hornocker Wildlife Institute, now part of the Wildlife Conservation Society (WCS). In 1992 the group helped launch the Siberian Tiger Project, devoted to studying the tiger and its habitat. The team put together a plan to protect the tiger, began educating people regarding the tiger's importance and value, and worked closely with those who live near the big cats.

Today, WCS biologists track tigers with radio-collars, monitor their movements and health, determine causes of death when they die, study aspects of the tiger's ecosystem, and help fund local wildlife officials to deter and capture poachers.

Thanks to such efforts, the Siberian tiger population has been holding steady, even while the world's other tiger populations have declined. The last range-wide survey, in 2005, found 428–502 Siberian tigers in the wild, while 1,500 more survive in zoos and captive breeding programs around the world. Poaching remains rampant, however, and partial surveys since 2005 have shown decreasing numbers, so the outlook for the cat's survival still is in question. Fortunately, many people are trying to save these endangered animals. It is one of numerous efforts around the world today to stem the loss of our planet's priceless biological diversity.

OUR PLANET OF LIFE

Our rising human population and resource consumption are putting ever-greater pressure on the flora and fauna of our planet, from tigers to tiger beetles. We are diminishing Earth's diversity of life, the very quality that makes our planet so special. In Chapter 3 (p. 55) we introduced the concept of **biological diversity**, or **biodiversity**, as the variety of life across all levels of biological organization, including the diversity of species, their genes, their populations, and their communities. In this chapter we will probe further, examining current biodiversity trends and their relevance to our lives. We will then explore science-based solutions to biodiversity loss.

Biodiversity encompasses multiple levels

Biodiversity is a concept as multifaceted as life itself, and biologists employ different working definitions according to their own aims and philosophies. Nonetheless, there is broad agreement that the concept applies across the major levels in the organization of life, encompassing diversity in genes, species, populations, communities, and ecosystems (**FIGURE 11.1**). The level that is easiest to visualize and most commonly used is species diversity.

Species diversity   As you will recall (p. 52), a **species** is a distinct type of organism, a set of individuals that uniquely share certain characteristics and can breed with one another and produce fertile offspring. Biologists use differing criteria to delineate species boundaries. Some biologists emphasize characteristics that species share because of common ancestry, whereas others emphasize the ability to interbreed. The spirited debate on this topic could easily fill a textbook in itself! In practice, however, scientists broadly agree on species identities.

We can express **species diversity** in terms of the number or variety of species in a particular region. One component of species diversity is *species richness*, the number of species. Another is *evenness* or *relative abundance*, the extent to which different species are similar in their numbers of individuals (**FIGURE 11.2**).

As we have seen, speciation (pp. 56–57) generates new species, adding to species richness, whereas extinction (pp. 58–59) diminishes species richness. Although immigration, emigration, and local extinction may change species richness locally, only speciation and extinction change it globally.

Taxonomists, the scientists who classify species, use an organism's physical appearance and genetic makeup to determine its species. Taxonomists also group species by their similarity into a hierarchy of categories meant to reflect evolutionary relationships. Related species are grouped together into *genera* (singular, *genus*), related genera are grouped into families, and so on (**FIGURE 11.3**). Every species is given a two-part Latin or Latinized scientific name denoting its genus and species. The tiger, *Panthera tigris*, is similar to the world's other species of large cats, such as the jaguar (*Panthera onca*), the leopard (*Panthera pardus*), and the African lion (*Panthera leo*). These four species are closely related in evolutionary terms, a

Ecosystem diversity

Species diversity

Genetic diversity

FIGURE 11.1 ▲ The concept of biodiversity encompasses several levels in the hierarchy of life. Species diversity (middle frame of the figure) refers to the number or variety of species. Genetic diversity (bottom frame) refers to variation in DNA composition among individuals within a species. Ecosystem diversity (top frame) and related concepts refer to variety at levels above the species level, such as ecosystems, communities, habitats, or landscapes.

relationship that is indicated by the genus name they share, *Panthera*. They are more distantly related to cats in other genera such as the cheetah (*Acinonyx jubatus*) and the bobcat (*Felis rufus*), although all cats are classified together in the family Felidae.

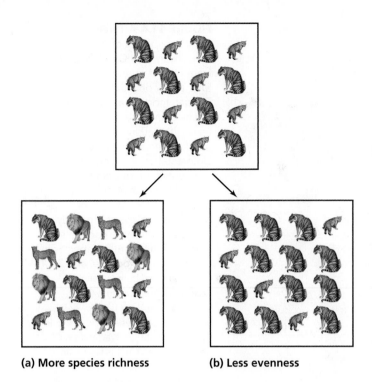

(a) More species richness   **(b) Less evenness**

FIGURE 11.2 ▲ Compared with the boxed area at the top, **(a)** has greater *species richness* because it contains four species rather than two. Compared with the boxed area at the top, **(b)** has reduced *evenness* because the proportions of the two species are less even; that is, tigers show a greater *relative abundance* than bobcats.

Biodiversity exists below the species level in the form of *subspecies*, populations of a species that occur in different geographic areas and differ from one another in some characteristics. Subspecies arise by the same processes that drive speciation (pp. 56–57) but result when divergence stops short of forming separate species. Scientists denote subspecies with a third part of the scientific name. The Siberian tiger, *Panthera tigris altaica*, is one of five subspecies of tiger still surviving (**FIGURE 11.4**). Tiger subspecies differ in color, coat thickness, stripe patterns, and size. For example, *Panthera tigris altaica* is taller at the shoulder than the Bengal tiger (*Panthera tigris tigris*) of India and Nepal, and it has a thicker coat and larger paws.

Genetic diversity Scientists designate subspecies when they recognize substantial, genetically based differences among individuals from different populations of a species. However, all species consist of individuals that vary genetically from one another to some degree, and this genetic diversity is an important component of biodiversity. **Genetic diversity** encompasses the differences in DNA composition (p. 29) among individuals.

Genetic diversity provides the raw material for adaptation to local conditions. A diversity of genes for coat thickness in tigers allowed natural selection (pp. 52–55) to favor genes for thin fur in Bengal tigers living in warm regions, and genes for thick fur for Siberian tigers living in cold regions. In the long term, populations with more genetic diversity may stand better chances of persisting, because their variation provides

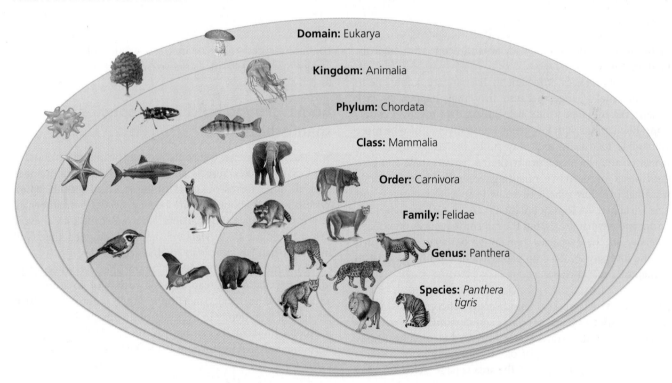

FIGURE 11.3 ▲ Taxonomists classify organisms using a hierarchical system meant to reflect evolutionary relationships. Species that are similar in their appearance, behavior, and genetics (because they share recent common ancestry) are placed in the same genus. Organisms of similar genera are placed within the same family. Families are placed within orders, orders within classes, classes within phyla, phyla within kingdoms, and kingdoms within domains. For instance, tigers belong to the class Mammalia, along with elephants, kangaroos, and bats. However, the differences between these species, which have evolved and diverged over millions of years, are great enough that they are placed in different orders, families, and genera.

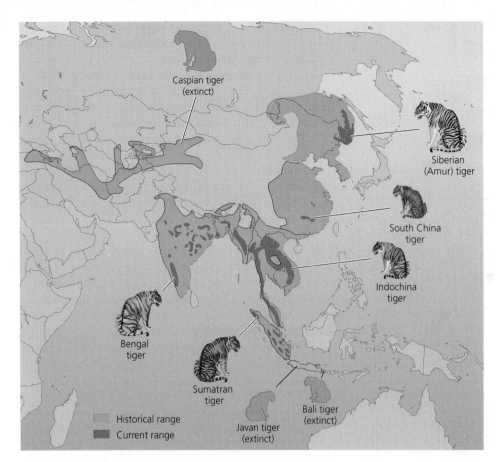

FIGURE 11.4 ◀ Three of the eight subspecies of tiger—the Bali, Javan, and Caspian tigers—went extinct during the 20th century. Today only the Siberian, Bengal, Indochina, Sumatran, and South China tigers persist, and fewer than 30 individuals of the South China tiger survive. Deforestation, hunting, and other pressures from people have caused tigers of all subspecies to disappear from 93% of the geographic range they historically occupied. This map contrasts the ranges of the eight subspecies in the years 1800 (orange) and 2000 (red). Adapted from Millennium Ecosystem Assessment, 2005. *Ecosystems and human well-being: biodiversity synthesis*, Fig 4.4. World Resources Institute, Washington DC. Categories used in this figure are modified from the source so as to match biome categorization in this book.

them more genetic options with which to cope with environmental change.

Populations with little genetic diversity are vulnerable to environmental change, because they may happen to lack genetic variants that would help them adapt to novel conditions. Populations with low genetic diversity may also be more vulnerable to disease and may suffer *inbreeding depression*, which occurs when genetically similar parents mate and produce weak or defective offspring. Scientists have sounded warnings over low genetic diversity in species that in the past have dropped to low population sizes, including cheetahs, bison, and elephant seals, but the full consequences of reduced diversity in these species remain to be seen. Diminished genetic diversity in our crop plants also is a prime concern to humanity (pp. 256–258).

Ecosystem diversity Biodiversity encompasses levels above the species level, as well. *Ecosystem diversity* refers to the number and variety of ecosystems, but biologists may also refer to the diversity of biotic communities or habitats within some specified area. If the area is large, scientists may also consider the geographic arrangement of habitats, communities, or ecosystems at the landscape level, including the sizes, shapes, and connections among patches. Under any of these concepts, a seashore of rocky and sandy beaches, forested cliffs, offshore coral reefs, and ocean waters would hold far more biodiversity than the same acreage of a monocultural cornfield. A mountain slope whose vegetation changes with elevation from desert to hardwood forest to coniferous forest to alpine meadow would hold more biodiversity than an area the same size consisting of only desert, forest, or meadow.

Some groups hold more species than others

Species are not evenly distributed among taxonomic groups. In number of species, insects show a staggering predominance over all other forms of life (**FIGURE 11.5** and **FIGURE 11.6**). Within insects, about 40% are beetles. Beetles outnumber all non-insect animals and all plants. No wonder the 20th-century British biologist J.B.S. Haldane famously quipped that God must have had "an inordinate fondness for beetles."

In some taxonomic groups, many species evolved rapidly as populations spread across a variety of environments and adapted to local conditions. Other groups diversified because of a tendency to become subdivided by barriers that promote allopatric speciation (pp. 56–57). Still other groups accumulated species through time because of low rates of extinction.

Measuring biodiversity is not easy

Scientists often express biodiversity in terms of its most easily measured component, species diversity, and in particular, species richness. Species richness is a good gauge for overall biodiversity, but we still are profoundly ignorant of the number of species that exist. So far, scientists have identified and described about 1.8 million species of plants, animals, and

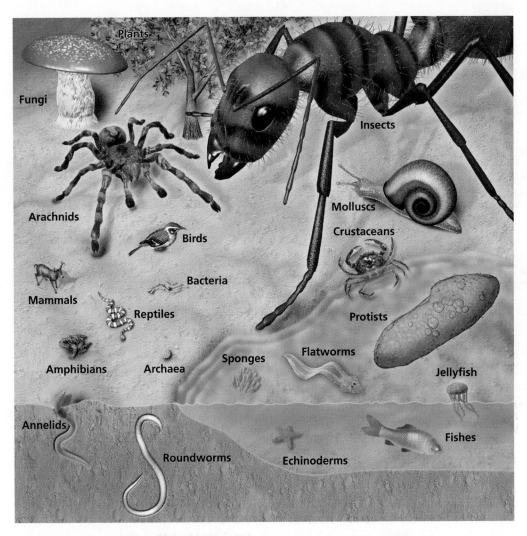

FIGURE 11.5 ▲ This illustration shows organisms scaled in size to the number of species known from each major taxonomic group. This gives a visual sense of the disparity in species richness among groups. However, because most species are not yet discovered or described, some groups (such as bacteria, archaea, insects, nematodes, protists, fungi, and others) may contain far more species than we now know of. Data from Groombridge, B., and M.D. Jenkins, 2002. *Global biodiversity: Earth's living resources in the 21st century*. UNEP-World Conservation Monitoring Centre. Cambridge, U.K.: Hoechst Foundation.

microorganisms. However, most of Earth's species remain undiscovered. Estimates for the total number that actually exist range from 3 million to 100 million, with the most widely accepted estimates in the neighborhood of 14 million.

Our knowledge of species numbers is incomplete for several reasons. First, many species are tiny and easily overlooked, including bacteria, nematodes (roundworms), fungi, protists, and soil-dwelling arthropods. Second, many organisms are so difficult to identify that ones thought to be identical sometimes turn out, once biologists look more closely, to be multiple species. This is frequently the case with microbes, fungi, and small insects, but also sometimes with organisms as large as birds, trees, and whales. Third, some areas of Earth remain little explored. We have barely sampled the ocean depths, hydrothermal vents (p. 33), or the tree canopies and soils of tropical forests. As one example, a 2005 expedition to the remote Foja Mountains of New Guinea discovered over 40 new species of vertebrates, plants, and butterflies in less

than a month, while research in marine waters nearby turned up another 50 species.

Smithsonian Institution entomologist Terry Erwin pioneered one method of estimating species numbers. In 1982, Erwin's crews fogged rainforest trees in Central America with clouds of insecticide and then collected insects, spiders, and other arthropods as they died and fell from the treetops. Erwin concluded that 163 beetle species specialized on the tree species *Luehea seemannii*. If this were typical, he figured, then the world's 50,000 tropical tree species would hold 8,150,000 beetle species and—because beetles represent 40% of all arthropods—20 million arthropod species. If canopies hold two-thirds of all arthropods, then arthropod species in tropical forests alone would number 30 million. Many assumptions were involved in this calculation, and several follow-up studies have revised Erwin's estimate downward, but it remains one of the better efforts at estimating species numbers.

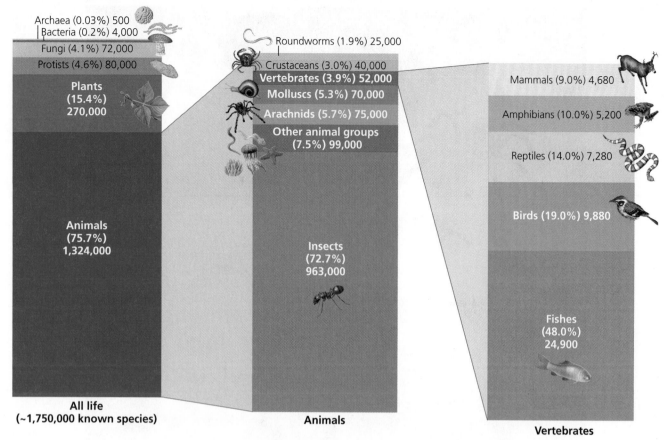

All life
(~1,750,000 known species)

Archaea (0.03%) 500
Bacteria (0.2%) 4,000
Fungi (4.1%) 72,000
Protists (4.6%) 80,000
Plants
(15.4%)
270,000
Animals
(75.7%)
1,324,000

Animals

Roundworms (1.9%) 25,000
Crustaceans (3.0%) 40,000
Vertebrates (3.9%) 52,000
Molluscs (5.3%) 70,000
Arachnids (5.7%) 75,000
Other animal groups
(7.5%) 99,000
Insects
(72.7%)
963,000

Vertebrates

Mammals (9.0%) 4,680
Amphibians (10.0%) 5,200
Reptiles (14.0%) 7,280
Birds (19.0%) 9,880
Fishes
(48.0%)
24,900

FIGURE 11.6 ▲ In the left portion of the figure, we see that three-quarters of known species are animals. The central portion subdivides animals, revealing that nearly three-quarters of animals are insects and that vertebrates comprise only 3.9% of animals. Among vertebrates (right portion of figure), nearly half are fishes, and mammals comprise only 9%. As noted, most species are not yet discovered or described, so some groups may contain far more species than we now know of. Data from Groombridge, B., and M.D. Jenkins, 2002. *Global biodiversity: Earth's living resources in the 21st century.* Cambridge, UK: Hoechst Foundation.

Biodiversity is unevenly distributed

Numbers of species tell only part of the story of Earth's biodiversity. Living things are distributed across our planet unevenly, and scientists have long sought to explain the distributional patterns they see. For instance, species richness generally increases as one approaches the equator (**FIGURE 11.7A**). This pattern of variation with latitude, called the *latitudinal gradient*, is one of the most obvious patterns in ecology, yet one of the most difficult for scientists to explain.

Hypotheses abound for the cause of the latitudinal gradient in species richness, but it seems likely that plant productivity and climate stability play key roles (**FIGURE 11.7B**). Greater amounts of solar energy, heat, and humidity at tropical latitudes lead to more plant growth, making areas nearer the equator more productive and able to support larger numbers of animals. The relatively stable climates of equatorial regions—their similar temperatures and rainfall from day to day and season to season—discourage single species from dominating ecosystems, and instead allow numerous species to coexist. Whereas variable environmental conditions favor generalists—species that can tolerate a wide range of circumstances but that do no single thing especially well—stable conditions favor organisms with specialized niches that do particular things very well.

Additionally, polar and temperate regions may be relatively species-poor because glaciation events repeatedly forced organisms from these regions toward more tropical latitudes.

The latitudinal gradient influences the species diversity of Earth's biomes (pp. 96–103). Tropical dry forests and rainforests tend to support far more species than tundra and boreal forests, for instance. Tropical biomes typically show more evenness as well, whereas in high-latitude biomes with low species richness, particular species greatly outnumber others. For example, Canada's boreal forest is dominated by immense expanses of black spruce, whereas Panama's tropical forest contains hundreds of tree species, no one of which greatly outnumbers the others.

At smaller scales, diversity patterns vary with habitat type. Generally, habitats that are structurally diverse allow for more ecological niches (pp. 63, 79) and support greater species richness and evenness. For instance, forests generally support more diversity than grasslands.

For any given area, species diversity tends to increase with diversity of habitats, because each habitat supports a somewhat different set of organisms. Thus, ecotones (pp. 118–119), transition zones where two or more habitats intermix, often support high biodiversity. Human disturbance (such as when we clear open patches within a forested landscape) can sometimes create

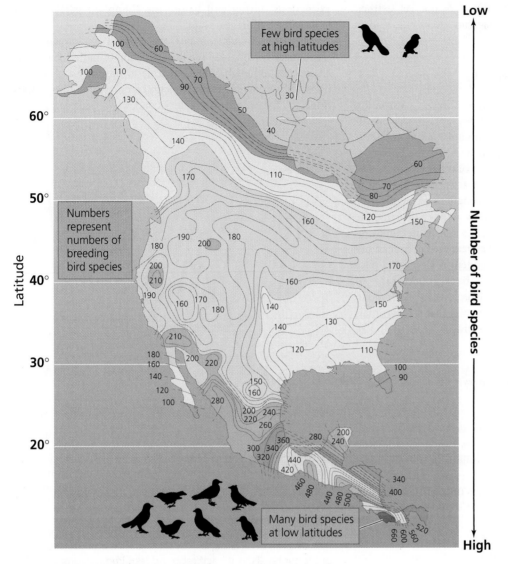

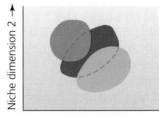

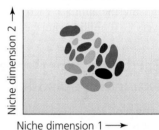

(a) Latitudinal gradient in species richness for birds in North America

(b) One hypothesis to explain the latitudinal gradient

Temperate and polar latitudes: Variable climate favors fewer species, and species that are widespread generalists.

Tropical latitudes: Greater solar energy, heat, and humidity promote more plant growth to support more organisms. Stable climate favors specialist species. Together these encourage greater diversity of species.

Few bird species at high latitudes

Numbers represent numbers of breeding bird species

Many bird species at low latitudes

Latitude

Number of bird species

Low

High

Niche dimension 2

Niche dimension 1 ⟶

Niche dimension 2

Niche dimension 1 ⟶

FIGURE 11.7 ▲ Birds in North and Central America **(a)** provide an example of the latitudinal gradient in species richness: Regions of arctic Canada and Alaska are home to 30–100 breeding species, whereas in areas of Costa Rica and Panama the number is over 600. In one hypothesis for the latitudinal gradient **(b)**, the variable climates of polar and temperate latitudes favor generalist organisms that can survive a wide range of conditions. In tropical latitudes, intense solar energy, heat, and humidity induce greater plant growth (which supports more organisms), and the stable climate favors specialist species. Together these factors are thought to promote greater species richness in the tropics. (a) From Cook, R.E. et al. 1969. Latitudinal gradient in species richness adapted from Variation in species density in North American Birds. *Systematic biology* 18: 63–84. (Originally published as *Systematic zoology*). By permission of Oxford University Press.

ecotones or patchwork combinations of habitats. Because this increases habitat diversity, species diversity may often rise in disturbed areas. However, this is true only at local scales. At larger scales, human disturbance decreases diversity because species that rely on large, unbroken expanses of single habitat will disappear.

Understanding patterns of biodiversity is vital for landscape ecology (pp. 118–120), regional planning (p. 353), and forest management (pp. 322–332). We will discuss geographic patterns in Chapter 12 (pp. 335–340) and later in this chapter as we explore solutions to the ongoing loss of biodiversity that our planet is currently experiencing.

BIODIVERSITY LOSS AND SPECIES EXTINCTION

Biodiversity at all levels is being lost to human impact, most irretrievably in the extinction of species. Once vanished, a species can never return. **Extinction** (pp. 58–59) occurs when the last member of a species dies and the species ceases to exist, as was the case with Monteverde's golden toad (Chapter 3). The disappearance of a particular population from a given area, but not the entire species globally, is referred to as **extirpation**. The tiger has been extirpated from most of its historic range (see Figure 11.4), but it is not yet extinct. Extirpation is an erosive process that can, over time, lead to extinction.

Extinction occurs naturally

Human impact is responsible for most cases of extirpation and extinction today, but these processes do also occur naturally, albeit at a much slower rate. If organisms did not naturally go extinct, we would be up to our ears in dinosaurs, trilobites, ammonites, and the millions of other creatures that vanished from Earth during the immense span of time before humans appeared. Paleontologists estimate that roughly 99% of all species that ever lived are now extinct, leaving the remaining 1% as the wealth of species on our planet today.

Most extinctions preceding the appearance of human beings occurred one by one for independent reasons, at a pace that paleontologists refer to as the **background rate of extinction**. By studying traces of these organisms preserved in the fossil record (pp. 57–58), scientists infer that for mammals and marine animals, each year, on average, 1 species out of every 1–10 million vanished.

Earth has experienced five mass extinction episodes

Extinction rates rose far above this background rate at several points in Earth's history, when episodes of **mass extinction** (pp. 59–61) occurred. In the past 440 million years, our planet experienced five major mass extinction events (**FIGURE 11.8**). Each event eliminated more than one-fifth of life's families and at least half its species (**TABLE 11.1**). The most severe episode occurred at the end of the Permian period, 248 million years ago, when close to 90% of all species went extinct. The best-known episode occurred at the end of the Cretaceous period, 65 million years ago, when an apparent asteroid impact brought an end to the dinosaurs and many other groups (pp. 60–61). Evidence exists for earlier mass

extinctions during and before the Cambrian period, more than half a billion years ago.

If current trends continue, the modern era, known as the Quaternary period, may see the extinction of more than half of all species. Although similar in scale to previous mass extinctions, today's ongoing mass extinction is different in two primary respects. First, we are causing it. Second, we will suffer as a result.

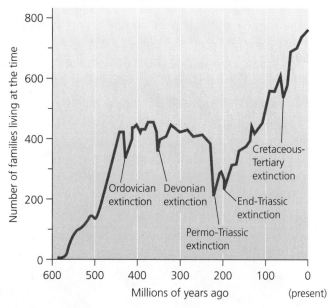

FIGURE 11.8 ▲ The fossil record shows evidence for five episodes of mass extinction during the past half-billion years of Earth history. At the end of the Ordovician, Devonian, Permian, Triassic, and Cretaceous periods, 50–95% of the world's species appear to have gone extinct. (This graph shows families, not species, which is why the drops appear less severe.) Each time, biodiversity later rebounded to equal or higher levels, but each rebound required millions of years. Data from Raup, D.M., and J.J. Sepkoski, 1982. Mass extinctions in the marine fossil record. *Science* 215: 1501–1503. Fig 2. Reprinted with permission from AAAS and the author.

TABLE 11.1	Mass Extinctions			
Event	Date (millions of years ago [mya])	Cause	Types of life most affected	Percentage of life depleted
Ordovician	440 mya	Unknown	Marine organisms; terrestrial record is unknown	>20% of families
Devonian	370 mya	Unknown	Marine organisms; terrestrial record is unknown	>20% of families
Permo-Triassic	250 mya	Possibly volcanism	Marine organisms; terrestrial record is less known	>50% of families; 80–95% of species
End-Triassic	202 mya	Unknown	Marine organisms; terrestrial record is less known	20% of families; 50% of genera
Cretaceous-Tertiary	65 mya	Asteroid impact	Marine and terrestrial organisms, including dinosaurs	5% of families; >50% of species
Current	Beginning 0.01 mya	Human impacts	Large animals, specialized organisms, island organisms, organisms harvested by people	Ongoing

Humans are setting the sixth mass extinction in motion

Over just the past few centuries, we have recorded hundreds of instances of species extinction caused by people. Sailors documented the extinction of the dodo on the Indian Ocean island of Mauritius in the 17th century, for example, and today only a few body parts of this unique bird remain in museums. Among North American birds in the past two centuries alone, we have driven into extinction the Carolina parakeet, great auk, Labrador duck, passenger pigeon (pp. 63–64), and probably the Bachman's warbler and Eskimo curlew. Several more species, including the whooping crane, Kirtland's warbler, and California condor (p. 306), teeter on the brink of extinction. Recent possible sightings of the ivory-billed woodpecker (**FIGURE 11.9**) in swampy woodlands of Arkansas, Florida, and Louisiana provide some hope that this charismatic species may still persist.

However, species extinctions caused by humans precede written history. Indeed, people may have been hunting species to extinction for thousands of years. Archaeological evidence shows that in case after case, a wave of extinctions followed close on the heels of human arrival on islands and continents (**FIGURE 11.10**). After Polynesians reached Hawaii, half its birds went extinct. Birds, mammals, and reptiles vanished following human arrival on many other oceanic islands, including large

FIGURE 11.9 ▲ The ivory-billed woodpecker (*Campephilus principalis*) was one of North America's most majestic birds and lived in old-growth forests throughout the southeastern United States. Forest clearing and timber harvesting eliminated the mature trees it needed for food, shelter, and nesting, and this symbol of the South appeared to go extinct. In recent years, fleeting and controversial observations in Arkansas, Louisiana, and Florida have raised hopes that the species persists, but proof has been elusive.

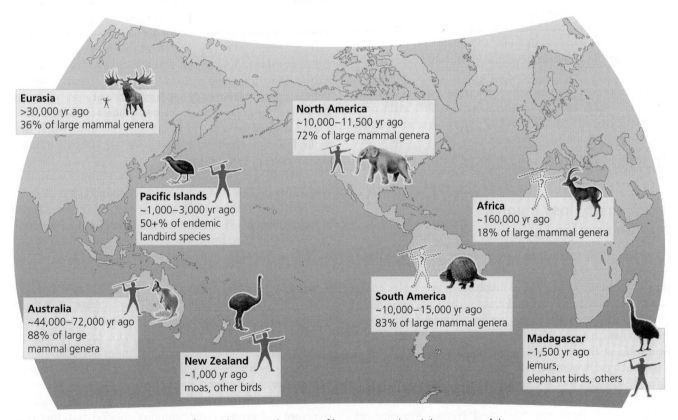

Eurasia
>30,000 yr ago
36% of large mammal genera

North America
~10,000–11,500 yr ago
72% of large mammal genera

Pacific Islands
~1,000–3,000 yr ago
50+% of endemic landbird species

Africa
~160,000 yr ago
18% of large mammal genera

Australia
~44,000–72,000 yr ago
88% of large mammal genera

South America
~10,000–15,000 yr ago
83% of large mammal genera

Madagascar
~1,500 yr ago
lemurs,
elephant birds, others

New Zealand
~1,000 yr ago
moas, other birds

FIGURE 11.10 ▲ This map shows for each region the time of human arrival and the extent of the recent extinction wave. Illustrated are representative extinct megafauna from each region. The human hunter icons are sized according to the degree of evidence that human hunting was a cause of extinctions; larger icons indicate more certainty that humans (as opposed to climate change or other factors) were the cause. Data for South America and Africa are so far too sparse to be conclusive, and future archaeological and paleontological research could well alter these interpretations. Adapted from Barnosky, A.D., et al., 2004. Assessing the causes of late Pleistocene extinctions on the continents. *Science* 306: 70–75; and Wilson, E.O., 1992. *The diversity of life.* Cambridge, MA: Belknap Press.

island masses such as New Zealand and Madagascar. The pattern appears to hold for at least two continents, as well. Dozens of species of large vertebrates died off in Australia after people arrived roughly 50,000 years ago, and North America lost 33 genera of large mammals after people arrived on the continent more than 10,000 years ago (see Figure 3.8, p. 59).

Current extinction rates are much higher than normal

Today, species loss is accelerating as our population growth and resource consumption put increasing strain on habitats and wildlife. In 2005, scientists with the Millennium Ecosystem Assessment (pp. 16, 18) calculated that the current global extinction rate is 100 to 1,000 times greater than the background rate. They projected that the rate would increase tenfold or more in future decades.

To keep track of the current status of endangered species, the International Union for Conservation of Nature (IUCN) maintains the **Red List**, an updated list of species facing high risks of extinction. The 2010 Red List reported that 21% (1,143) of mammal species, 12% (1,223) of bird species, and 30% (1,895) of amphibian species are threatened with extinction. Among other major groups (for which assessments are not fully complete), 17% to 74% of species are judged to be at high risk of extinction. In the United States alone over the past 500 years, 237 animals and 30 plants are confirmed to have gone extinct. For all these figures, the actual numbers of species are without doubt greater than the known numbers.

Among the 1,143 mammals facing possible extinction is the tiger, which despite—or perhaps because of—its tremendous size and reputation as a fierce predator, is one of the most endangered large animals on the planet. In 1950, eight tiger subspecies existed (see Figure 11.4). Today, three are extinct. The Bali tiger, *Panthera tigris balica*, went extinct in the 1940s; the Caspian tiger, *Panthera tigris virgata*, during the 1970s; and the Javan tiger, *Panthera tigris sondaica*, during the 1980s.

Biodiversity loss involves more than extinction

Extinction is only part of the story of biodiversity loss. The larger part of the story is the decline in population sizes of many organisms. As a species declines in its numbers, its geographic range often shrinks as it is extirpated from various portions of the range. Thus, many species today are less numerous and occupy less area than they once did. Tigers numbered well over 100,000 worldwide in the 19th century but number only 3,400–5,100 today. Such declines mean that genetic diversity and ecosystem diversity, as well as species diversity, are being lost.

To measure and quantify this degradation, scientists at the World Wildlife Fund and the United Nations Environment Programme (UNEP) developed a metric called the *Living Planet Index*. This index summarizes trends in the populations of 887 terrestrial species, 458 freshwater species, and 341 marine species that are sufficiently monitored to provide reliable data. Between 1970 and 2005, the Living Planet Index fell by roughly 28% (**FIGURE 11.11**), driven primarily by biodiversity losses in tropical regions.

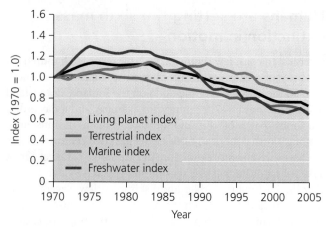

FIGURE 11.11 ▲ The Living Planet Index serves as an indicator of the state of global biodiversity. Index values summarize trends for 4,642 populations of 1,686 vertebrate species. Between 1970 and 2005, the Living Planet Index fell by roughly 28%. The index for terrestrial species fell by 33%; for freshwater species, 35%; and for marine species, 14%. Data from World Wide Fund for Nature, 2008. *The Living Planet Report, 2008*. Gland, Switzerland.

Several major causes of biodiversity loss stand out

Reasons for the decline of a given population or species are often complex and difficult to determine. The current collapse of amphibian populations throughout the world provides an example. Frogs, toads, and salamanders worldwide are decreasing drastically in abundance, and as we saw in Chapter 3 with the golden toad, some have already gone extinct. As these creatures disappear before our eyes, scientists are struggling to explain why. Recent studies have implicated a wide array of factors, and most biologists now suspect that multiple factors may be interacting (see **THE SCIENCE BEHIND THE STORY**, pp. 294–295).

Overall, scientists have identified four primary causes of population decline and species extinction: habitat loss, invasive species, pollution, and overharvesting. Global climate change (Chapter 18) now is becoming the fifth. Each of these causes is intensified by human population growth and by our increase in per capita consumption of resources.

Habitat loss Habitat loss is the single greatest cause of biodiversity loss today. Species lose their habitats when those habitats are destroyed outright, but habitat loss also occurs when habitats are altered through more subtle processes, including fragmentation and other forms of degradation.

Nearly every human activity can alter or destroy habitat. Farming replaces diverse natural communities with simplified ones of only a few plant species. Grazing modifies the structure and species composition of grasslands. Either type of agriculture can lead to desertification. Clearing forests removes the food, shelter, and other resources that forest-dwelling organisms need to survive. Hydroelectric dams turn rivers into reservoirs upstream and affect water conditions and floodplain communities downstream. Urban sprawl supplants diverse natural ecosystems with simplified human-made ones, driving many species from their homes

FIGURE 11.12 ▲ Development of land for new housing is one of many ways in which habitat is altered or destroyed.

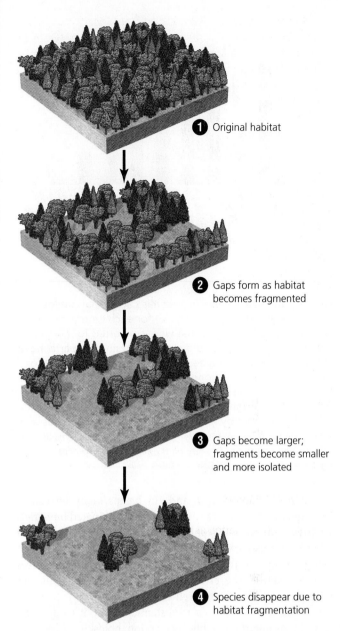

1. Original habitat

2. Gaps form as habitat becomes fragmented

3. Gaps become larger; fragments become smaller and more isolated

4. Species disappear due to habitat fragmentation

FIGURE 11.13 ▲ Fragmentation of habitat ❶ begins when gaps are created ❷ within a natural habitat. As development proceeds, these gaps expand ❸, join together, and eventually come to dominate the landscape ❹, stranding islands of habitat in their midst. As habitat becomes fragmented, fewer populations can persist, and numbers of species in the fragments decline.

(**FIGURE 11.12**). Because organisms have adapted over thousands or millions of years to the habitats in which they live, any major change in their habitat is likely to render it less suitable for them.

The sudden and complete elimination of habitat is uncommon at large scales; instead, what we usually witness is gradual, piecemeal degradation, such as **habitat fragmentation** (**FIGURE 11.13**). Farming, logging, road building, and other types of human land use and development often intrude into forested habitats, for example, breaking up a continuous expanse of forest into an array of fragments, or patches. Animals and plants adapted to unbroken forest will die out as a forest becomes fragmented. As habitat fragmentation proceeds across a landscape, species requiring the habitat gradually disappear, winking out from one fragment after another. In response to habitat fragmentation, conservation biologists have designed landscape-level strategies to try to optimize the arrangement of areas to be preserved. We will examine some of these strategies in our discussion of parks and protected areas in Chapter 12 (pp. 332–340).

Human habitat alteration benefits some species. Animals such as house sparrows, pigeons, gray squirrels, and cockroaches thrive in urban and suburban environments and benefit from our modification of natural habitats. However, the species that benefit are relatively few; for every species that wins, more lose. Furthermore, the species that do well in our midst tend to be weedy generalists that are in little danger of disappearing any time soon.

Habitat loss is the primary source of population declines for 83% of threatened mammals and 85% of threatened birds, according to UNEP data. For example, the prairies native to North America's Great Plains are today almost entirely converted to agriculture. Less than 1% of original prairie habitat remains. As a result, grassland bird populations have declined by an estimated 82–99%. Many grassland species have been extirpated from large areas, and the two species of prairie chickens persisting in pockets of the Great Plains could soon go extinct.

Habitat destruction has occurred widely in nearly every biome (**FIGURE 11.14**). Over half of temperate forests, grasslands, and shrublands had been converted by the year 1950 (mostly for agriculture). Across Asia, scientists estimate that 40% of the tiger's remaining habitat has disappeared just in the last decade. Today habitat is being lost most rapidly in tropical rainforests, tropical dry forests, and savannas.

Pollution Pollution can harm organisms in many ways. Air pollution (Chapter 17) degrades forest ecosystems. Water pollution (Chapter 15) impairs fish and amphibians. Agricultural runoff (including fertilizers, pesticides, and

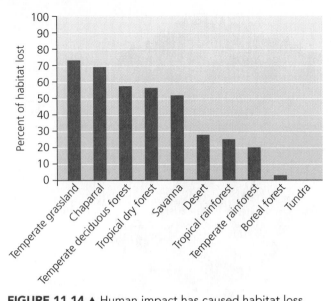

FIGURE 11.14 ▲ Human impact has caused habitat loss across the world's biomes. Shown for each biome are percentages of original area that were converted for human use through 1990. Temperate grassland and chaparral have lost over 70% of their area, whereas tundra and boreal forest have lost virtually none. In recent decades, tropical dry forest and savanna have lost the greatest fraction. These data are for outright conversion and do not include areas indirectly affected by human activity in other ways. Adapted from Millennium Ecosystem Assessment, 2005. *Ecosystems and human well-being: Biodiversity synthesis.* World Resources Institute, Washington, D.C.

sediments; Chapters 5, 9, and 10) harms many terrestrial and aquatic species. Heavy metals, polychlorinated biphenyls (PCBs), endocrine-disrupting compounds, and other toxic chemicals poison people and wildlife (Chapter 14). Garbage in the ocean can strangle, drown, or choke marine creatures (pp. 445–446). The effects of oil and chemical spills on wildlife are dramatic and well known.

Although pollution is a substantial threat, it tends to be less significant than public perception holds it to be. The damage to wildlife and ecosystems caused by pollution can be severe, but this tends to be surpassed by the harm caused by habitat alteration or invasive species.

Overharvesting For most species, a high intensity of hunting or harvesting by people will not in itself pose a threat of extinction, but for some species it can. The Siberian tiger is one such species. Large in size, few in number, long-lived, and raising few young in its lifetime—a classic K-selected species (p. 71)—the Siberian tiger is just the type of animal to be vulnerable to hunting. The advent of Russian hunting nearly drove the animal extinct, whereas decreased hunting during and after World War II allowed a population increase. By the mid-1980s, the Siberian tiger population was likely up to 250 individuals.

The political freedom that came with the Soviet Union's breakup in 1989, however, brought with it a freedom to harvest Siberia's natural resources, including the tiger, without regulations or rules, and poachers illegally killed at least 180 Siberian tigers between 1991 and 1996. This coincided with an economic expansion in many Asian countries, where tiger penises are believed to boost human sexual performance and where tiger bones, claws, whiskers, and other body parts are used to try to treat a variety of maladies (**FIGURE 11.15**). Although no proof of their effectiveness has been demonstrated, sale of the parts from one tiger fetches at least $15,000 in today's black market—a powerful economic incentive for poachers in poor regions.

Over the past century, hunting has led to steep declines in the populations of many K-selected animals. The Atlantic gray whale went extinct, and several other whales remain threatened or endangered. Gorillas and other primates that are killed for their meat may face extinction soon. Thousands of sharks are killed each year simply for their fins, which are used in soup. Today the oceans contain only 10% of the large animals they once did (p. 451).

To combat overharvesting, governments have passed laws, signed treaties, and strengthened anti-poaching efforts. Scientists have begun using genetic analyses to expose illegal hunting and wildlife trade. For instance, DNA testing can reveal the geographic origins of elephant

FIGURE 11.15 ◄ Body parts from tigers are used as traditional medicines and aphrodisiacs in some Asian cultures. Poachers have illegally killed countless tigers through the years to satisfy the surging market demand for these items. Here a street vendor in northern China displays tiger body parts for sale.

Invasive Species		
Species	Invasive in...	Effects
Gypsy moth (*Lymantria dispar*) 	Northeastern United States (Native to Eurasia)	In the 1860s, a scientist introduced the gypsy moth to Massachusetts in the belief that it might help produce a commercial-quality silk. The moth failed to start a silk industry, and instead spread through the northeastern United States, where its outbreaks defoliate trees over large regions every few years.
European starling (*Sturnus vulgaris*) 	North America (Native to Europe)	The bird was first introduced to New York City in the late 19th century by Shakespeare devotees intent on bringing every bird mentioned in Shakespeare's plays to America. It only took 75 years for starlings to spread to all corners of North America, becoming one of the continent's most abundant birds. Starlings are thought to outcompete native birds for nest holes.
Cheatgrass (*Bromus tectorum*) 	Western United States (Native to Eurasia)	In just 30 years after its introduction to Washington state in the 1890s, cheatgrass has spread across much of the western United States. It crowds out other plants, uses up the soil's nitrogen, and burns readily. Fire kills many of the native plants, but not cheatgrass, which grows back even stronger amid the lack of competition.
Brown tree snake (*Boiga irregularis*) 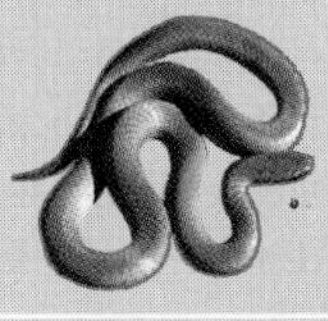	Guam (Native to Southeast Asia)	Nearly all native forest bird species on the South Pacific island of Guam have disappeared. The culprit is the brown tree snake, brought to the island inadvertently as stowaways in cargo bays of military planes in World War II. Guam's birds had not evolved with tree snakes, and so had no defenses against the snake's nighttime predation. The snakes have spread to other islands where they are repeating their ecological devastation. The arrival of this snake is the greatest fear of conservation biologists in Hawaii.
Kudzu (*Pueraria montana*) 	Southeastern United States (Native to Japan)	Kudzu is a vine that can grow 30 m (100 ft) in a single season. The U.S. Soil Conservation Service introduced kudzu in the 1930s to help control erosion. Adaptable and extraordinarily fast-growing, kudzu has taken over thousands of hectares of forests, fields, and roadsides.
Asian long-horned beetles (*Anoplophora glabripennis*) 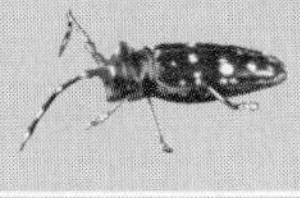	United States (Native to Asia)	Having arrived in imported lumber in the 1990s, these beetles burrow into trees and interfere with the trees' ability to absorb and process water and nutrients. They may wipe out the majority of hardwood trees in an area. Several U.S. cities, including Chicago and Seattle, have cleared thousands of trees after detecting these invaders.
Rosy wolfsnail (*Euglandina rosea*) 	Hawaii (Native to Southeastern United States and Latin America)	In the 1950s, well-meaning scientists introduced the rosy wolfsnail to Hawaii to prey upon and reduce the population of another invasive species, the giant African land snail. Within a few decades, however, the carnivorous rosy wolfsnail had instead driven more than half of Hawaii's native species of banded tree snails to extinction.

FIGURE 11.16 ▲ Invasive species are species that thrive in areas where they are introduced, outcompeting, preying on, or otherwise harming native species. This chart shows a few of the many thousands of invasive species.

ivory and whether whale meat sold in markets is from animals caught illegally (see **THE SCIENCE BEHIND THE STORY**, pp. 304–305).

Invasive species Our introduction of non-native species to new environments, where some may become invasive (pp. 93–95), also pushes native species toward extinction (**FIGURE 11.16**). Some introductions are accidental. Examples include aquatic organisms transported in the ballast water of ships (such as zebra mussels; Chapter 4), animals that escape from the pet trade, and weeds whose seeds cling to our socks as we travel from place to place. Other introductions are intentional. People have long brought food crops and animals with them as they colonized new places, and today we continue international trade in exotic pets and ornamental plants, often heedless of the ecological consequences.

Species native to islands are especially vulnerable to disruption from introduced species because the native island species have existed in isolation for millennia with relatively few parasites, predators, and competitors. As a result, they have not evolved the defenses necessary to resist invaders that are better adapted to these pressures.

Most organisms introduced to new areas perish, but the few types that survive may do very well, especially if they find themselves without the predators and parasites that attacked them back home, or without the competitors that had limited their access to resources. Once released from the limiting factors (p. 67) of predation, parasitism, and competition, an

The SCIENCE behind the Story

Amphibian Diversity and Decline

Amphibians illustrate the two most salient aspects of Earth's biodiversity today. Scientists are discovering more and more species—but at the same time, more and more populations and species are vanishing.

With most classes of vertebrates, new species are discovered at a rate of just one or a few per year, but the number of known species of amphibians (which include frogs, toads, salamanders, and others)—over 6,300 as of 2010—has jumped by 38% since 1985.

At the same time, nearly 2,500 of these species are in decline. Researchers have named some species just before they went extinct (**see photo**). We are doubtless losing others before they are even discovered. Nearly 170 species of frogs, toads, and salamanders studied just years or decades ago, including the golden toad (Chapter 3), are now thought to be gone.

Amphibians rely on both aquatic and terrestrial environments and can breathe and absorb water through their skin, so they are sensitive to pollution and other stresses. For this reason, amphibians are widely regarded

Dr. Madhava Meegaskumbura searching for frogs at night in Sri Lanka

as "biological indicators" that tell us whether ecosystems are in good shape. Thus, studying the reasons for amphibian declines can tell us much about the state of our environment.

From the American tropics to Africa to southern Asia, scientific scrutiny and improved technology have revealed amphibian biodiversity "hotspots." In the 1990s, an international team of scientists set out to determine whether Sri Lanka, a large tropical island off the coast of India, held more than the 40 frog species that were already known. Researcher Madhava Meegaskumbura and his

team combed through trees, rivers, ponds, and leaf litter for 8 years, collecting more than 1,400 frogs at 300 study sites. The scientists analyzed the frogs' physical appearance, habitat use, and vocalizations. They also examined the frogs' genes by analyzing several regions of their DNA (p. 29). They then compared these genetic, physical, and behavioral characteristics to those of known species of frogs.

The team found that the DNA from many of their frogs didn't match that of

A baby southern gastric brooding frog peers out from its mother's mouth. This Australian species, one of only two animals to raise its young in its stomach, likely produced chemicals that could help treat stomach ulcers for millions of people. Sadly, this frog went extinct soon after its discovery, taking its secrets with it.

---

introduced species may increase rapidly, spread, and displace native species. Invasive species cause billions of dollars in economic damage each year.

Climate change The preceding four types of human impacts affect biodiversity in discrete places and times. In contrast, our manipulation of Earth's climate system (Chapter 18) is having global impacts on biodiversity. As our emissions of greenhouse gases from fossil fuel combustion cause temperatures to warm worldwide, we are modifying global weather patterns and increasing the frequency of extreme weather events.

Extreme weather events such as droughts increase stress on populations, and warming temperatures are forcing organisms to shift their geographic ranges toward the poles and higher in altitude. Some species will not be able to adapt. Like the cloud-forest fauna at Monteverde (Chapter 3), mountaintop organisms cannot move further upslope to escape warming temperatures, so they will likely perish. Trees may not be able to move poleward fast enough. Animals and plants may find themselves among different communities of prey, predators, and parasites to which they are not adapted. In the Arctic, where warming has been greatest, thawing ice hinders the ability of the polar bear (**FIGURE 11.17**) to hunt seals

known species. And they found that many of their frogs looked different, sounded different, or behaved differently from known species. Clearly, they had discovered new species unknown to science. Some of these novel species live on rocks and sport leg fringes and markings that help disguise them as clumps of moss. Others are tree frogs that lay their eggs in baskets they construct. In all, more than 100 new species were discovered—all on an island only slightly larger than West Virginia! Reported in the journal *Science* in 2002, the study caught the attention of biologists worldwide.

Such promising discoveries come against a backdrop of distressing declines. Observed numbers of amphibians are down around the globe. Scientists are racing to discover the reasons and have found evidence for causes as varied as habitat destruction, chemical pollution, disease, invasive species, and climate change (**see graph**).

In some cases, researchers surmise that a combination of factors may be at work, multiplying one another's effects. For example, one set of field and lab experiments revealed that wood frogs are more vulnerable to parasitic infections when exposed to pesticides, while another study showed that tadpoles of the gray tree frog suffer unusually high mortality when exposed to pesticides while being stressed with the presence of predators.

Most worrisome is that many amphibian populations are vanishing in remote and pristine environments

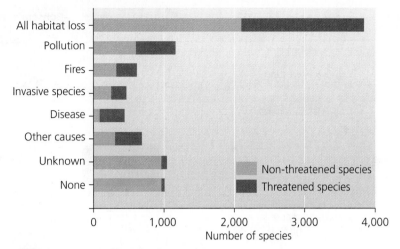

Habitat loss is the main cause of population decline for the world's amphibian species. Pollution is second in importance, but many declines are not yet attributable to cause. Data from IUCN, 2008. Global amphibian assessment.

when no direct damage is apparent. In many of these cases, the culprit seems to be a fungal disease called chytridiomycosis, caused by the pathogen *Batrachochytrium dendrobatidis*. We encountered this devastating disease in our **Science behind the Story** in Chapter 3 on Costa Rica's golden toad (pp. 68-69).

Researchers do not know whether the fatal disease has recently spread around the world from some unknown locale, whether it has always been around but has recently evolved greater strength, or whether amphibian immune systems have recently weakened. What is clear is that the disease is sweeping through populations worldwide. Human influence may or may not be responsible for its introduction and/or transmission.

As scientists learn more about the causes of amphibian declines, they are designing responses. A conservation action plan published by the IUCN set out recommended steps, including protecting and restoring habitat, cracking down on illegal harvesting, enhancing disease monitoring, and establishing captive breeding programs.

In the meantime, researchers are warning that the fate of amphibians may foreshadow the future for other organisms. "Amphibians have been around for 300 million years. They're tough, and yet they're checking out all around us," says David Wake, a biologist at the University of California at Berkeley, who was among the first to note the creatures' decline. "We really do see amphibians as biodiversity bellwethers." ■

FIGURE 11.17 ▶ As Arctic warming melts the sea ice from which it hunts seals, polar bears (*Ursus maritimus*) must swim farther for food. A lawsuit from environmental groups forced the U.S. Fish and Wildlife Service in 2008 to list the polar bear as threatened under the Endangered Species Act as a result of climate change.

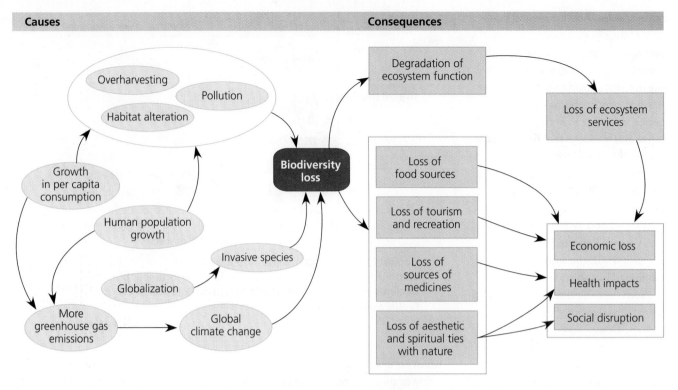

FIGURE 11.18 ▲ The loss of biodiversity stems from a variety of causes (ovals on left) and results in a number of consequences (boxes on right) for ecological systems and human well-being. Arrows in this concept map lead from causes to consequences. Note that items grouped within outlined boxes do not necessarily share any special relationship; the outlined boxes are intended merely to streamline the figure.

Solutions

As you progress through this chapter, try to identify as many solutions to biodiversity loss as you can. What could you personally do to help address this issue? Consider how each action or solution might affect items in the concept map above.

(p. 514). For this reason, the polar bear was added in 2008 to the U.S. endangered species list (p. 302). All in all, scientists predict that a 1.5–2.5 °C (2.7–4.5 °F) global temperature increase could put 20–30% of the world's plants and animals at increased risk of extinction.

All five of these causes of biodiversity loss are intensified by human population growth and rising per capita consumption, the ultimate reasons behind the proximate threats to biodiversity. Just as we now have a solid scientific understanding of the causes of biodiversity loss, we are also coming to appreciate its consequences (**FIGURE 11.18**), as our actions erode the many benefits that biodiversity brings us.

BENEFITS OF BIODIVERSITY

Why does biodiversity loss matter? There are several ways to answer this question. One is to consider the many tangible, pragmatic ways that biodiversity benefits people and supports human society. Another is to focus on the ethical dimensions to biodiversity preservation, because many people feel that organisms have an intrinsic right to exist. Finally, we might consider a metaphor first offered by Paul and Anne Ehrlich (p. 200): The loss of one rivet from an airplane's wing—or two, or three—may not cause the plane to crash. But at some point as rivets are removed the structure will be

compromised, and eventually the loss of just one more rivet will cause it to fail. Keeping this metaphor in mind, we would be wise to preserve as many functional components of our ecosystems as possible to make sure our environmental systems continue to function.

Biodiversity provides ecosystem services free of charge

Contrary to popular opinion, some things in life can indeed be free—as long as we protect the ecological systems that provide them. Intact forests provide clean air and water, and they buffer hydrologic systems against flooding and drought. Native crop varieties provide insurance against disease and drought. Abundant wildlife can attract tourists and boost the economies of developing nations. Intact ecosystems provide these and other valuable processes, known as **ecosystem services** (pp. 3, 121–122, 148), for all of us, free of charge.

Maintaining these ecosystem services is one clear benefit of protecting biodiversity. According to UNEP, biodiversity

▶ Provides food, fuel, fiber, and shelter

▶ Purifies air and water

▶ Detoxifies and decomposes wastes

▶ Stabilizes Earth's climate

- Moderates floods, droughts, and temperature extremes
- Cycles nutrients and renews soil fertility
- Pollinates plants, including many crops
- Controls pests and diseases
- Maintains genetic resources for crop varieties, livestock breeds, and medicines
- Provides cultural and aesthetic benefits
- Gives us the means to adapt to change

Organisms and ecosystems support a vast number of vital processes that people cannot replicate or would need to pay for if nature did not provide them. As we have seen (pp. 160–161), the annual value of just 17 of these ecosystem services may be in the neighborhood of $46 trillion per year—more than the gross domestic product (GDP) of all national economies combined.

Biodiversity helps maintain ecosystem function

Ecological research demonstrates that biodiversity tends to enhance the *stability* of communities and ecosystems. Research has also found that biodiversity tends to increase the *resilience* (p. 91) of ecological systems—their ability to weather disturbance, bounce back from stress, or adapt to change. Thus, when we lose biodiversity, it can diminish a natural system's ability to function and to provide services to our society.

Will the loss of a few species really make much difference in an ecosystem's ability to function? Ecological research suggests that it depends on which species are lost. Removing a species that can be functionally replaced by others may make little difference. Recall, however, our discussion of keystone species (pp. 86–91). Like the keystone that holds together an arch, removing a keystone species can significantly alter an ecological system. If a keystone species is lost, other species may disappear in response.

Top predators such as tigers are often considered keystone species. A single top predator may prey on many other carnivores, each of which may prey on many herbivores, each of which may consume many plants. Thus the removal of a single individual at the top of a food chain can have impacts that multiply as they cascade down the food chain. Moreover, top predators such as tigers, wolves, and grizzly bears are among the species most vulnerable to human impact. Large animals are frequently hunted, and they also need large areas of habitat, making them susceptible to habitat loss and fragmentation. Top predators are also vulnerable to the buildup of toxic pollutants in their tissues through the process of biomagnification (pp. 385–386).

"Ecosystem engineers" (p. 88) such as ants and earthworms can be every bit as influential as keystone species, so the loss of an ecosystem engineer from a system can likewise set major changes in motion. Ecosystems are complex, and it is difficult to predict which particular species may be important. Thus, many people prefer to apply the precautionary principle (p. 265) in the spirit of Aldo Leopold (p. 144), who advised, "To keep every cog and wheel is the first precaution of intelligent tinkering."

Biodiversity enhances food security

Biodiversity provides the food we eat. Throughout our history, human beings have used 7,000 plant species and several thousand animal species for food. Today many nutritional experts are worried because industrial agriculture has narrowed our diet. Globally, we get 90% of our food from just 30 crops and 14 livestock species, and this lack of diversity leaves us vulnerable to failures of particular crops. In a world where 1 billion people go hungry and more are malnourished, we could improve food security (the guarantee of an adequate, safe, nutritious, and reliable food supply to all people at all times; p. 254) by finding sustainable ways to capitalize on the nutritional opportunities offered by wild species and rare crop varieties.

Many new or underutilized food sources could be harvested or farmed in sustainable ways (**FIGURE 11.19**). The

Food Security and Biodiversity: Potential new food sources		
Species	**Native to...**	**Potential uses and benefits**
Amaranths (three species of *Amaranthus*) 	Tropical and Andean America	Grain and leafy vegetable; livestock feed; rapid growth, drought resistant
Buriti palm (*Mauritia flexuosa*) 	Amazon lowlands	"Tree of life" to Amerindians; vitamin-rich fruit; pith as source for bread; palm heart from shoots
Maca (*Lepidium meyenii*) 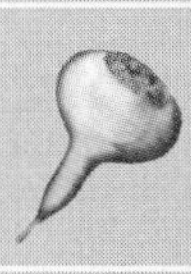	Andes Mountains	Cold-resistant root vegetable resembling radish, with distinctive flavor; near extinction
Babirusa (*Babyrousa babyrussa*) 	Indonesia: Moluccas and Sulawesi	A deep-forest pig; thrives on vegetation high in cellulose and hence less dependent on grain
Capybara (*Hydrochoeris hydrochoeris*) 	South America	World's largest rodent; meat esteemed; easily ranched in open habitats near water
Vicuna (*Lama vicugna*) 	Central Andes	Threatened species related to llama; source of meat, fur, and hides; can be profitably ranched
Chachalacas (*Ortalis*, many species) 	South and Central America	Tropical birds; adaptable to human habitations; fast-growing

FIGURE 11.19 ▲ By protecting biodiversity, we enhance food security. The wild species shown here are a fraction of the many plants and animals that could supplement our food supply. Adapted from Wilson, E.O., 1992. *The diversity of life.* Cambridge, MA: Belknap Press.

babassu palm (*Orbignya phalerata*) of the Amazon produces more vegetable oil than any other plant. The serendipity berry (*Dioscoreophyllum cumminsii*) generates a sweetener 3,000 times sweeter than table sugar. Several species of salt-tolerant grasses and trees are so hardy that farmers can irrigate them with saltwater. These plants produce animal feed, a vegetable oil substitute, and other products.

For our existing crops, having genetic diversity available in crop relatives and wild ancestors is enormously valuable. Our discussion of native landraces of corn in Mexico in Chapter 10 provides just one example. In 1995, Turkey's wheat crops received $50 billion worth of disease resistance from wild wheat strains. California's barley crops annually receive $160 million in disease resistance benefits from Ethiopian strains of barley. In the 1970s a researcher discovered a maize species in the mountains of Jalisco, Mexico, known as *Zea diploperennis*. This maize is highly resistant to disease, and it is a perennial, able to grow back year after year without being replanted. Yet we had almost lost this valuable plant; at the time of its discovery, its entire range was limited to a single 10-ha (25-acre) plot of land.

Organisms provide drugs and medicines

People have made medicines from plants for centuries, and many of today's drugs are derived from chemical compounds from wild plants (**FIGURE 11.20**). The rosy periwinkle

(*Catharanthus roseus*) produces compounds that treat Hodgkin's disease and a deadly form of leukemia. Had this plant from Madagascar become extinct, these two fatal diseases would have claimed far more victims. In Australia, a rare species of cork, *Duboisia leichhardtii*, provides hyoscine, a compound that physicians use to treat cancer, stomach disorders, and motion sickness. The Pacific yew (*Taxus brevifolia*) of North America's Pacific Northwest produces a compound that forms the basis for the anti-cancer drug taxol. In all, each year across the world, pharmaceutical products owing their origin to wild species generate up to $150 billion in sales and save thousands of human lives.

The world's biodiversity holds a still-greater treasure chest of medicines yet to be discovered. It can truly be said that every species that goes extinct represents one lost opportunity to find a cure for cancer or AIDS. A recent international survey highlighted animals that show particular promise, yet may be lost to extinction before we can profit from what they have to offer (**FIGURE 11.21**). We lost such an opportunity in the two species of gastric brooding frogs (*Rheobatrachus* spp.) recently discovered in the rainforests of Queensland, Australia (see photo in **The Science behind the Story,** p. 294). Females of these bizarre frogs raised their young inside their stomachs, where in any other animal stomach acids would soon destroy them! Apparently the young frogs exuded substances that neutralized their mother's acid production. Any such substance could be of immense use for treating human stomach ulcers, which affect 25 million U.S. citizens. Sadly, both frog species went extinct in the 1980s, taking their medical secrets with them forever.

Medicines and Biodiversity: Natural sources of pharmaceuticals		
Plant	Drug	Medical application
Pineapple (*Ananas comosus*)	Bromelain	Controls tissue inflammation
Autumn crocus (*Colchicum autumnale*)	Colchicine	Anticancer agent
Yellow cinchona (*Cinchona ledgeriana*)	Quinine	Antimalarial
Common thyme (*Thymus vulgaris*)	Thymol	Cures fungal infection
Pacific yew (*Taxus brevifolia*)	Taxol	Anticancer (especially ovarian cancer)
Velvet bean (*Mucuna deeringiana*)	L-Dopa	Parkinson's disease suppressant
Common foxglove (*Digitalis purpurea*)	Digitoxin	Cardiac stimulant

FIGURE 11.20 ▲ By protecting biodiversity, we enhance our ability to treat illness. Shown are just a few of the plants found to provide chemical compounds of medical benefit. Adapted from Wilson, E.O., 1992. *The diversity of life.* Cambridge, MA: Belknap Press.

WEIGHING THE ISSUES

Bioprospecting in Costa Rica

Bioprospectors working for pharmaceutical companies scour biodiversity-rich countries, searching for organisms that can provide new drugs, foods, medicines, or other valuable products. Many have been criticized for "biopiracy"—harvesting indigenous species to create commercial products without compensating the country of origin. To make sure it would not lose the benefits of its own biodiversity, the nation of Costa Rica reached an agreement with the Merck pharmaceutical company in 1991. The nonprofit National Biodiversity Institute of Costa Rica (INBio) allowed Merck to evaluate a number of Costa Rica's species for their commercial potential in return for $1.1 million, a small royalty rate on any products developed, and training for Costa Rican scientists.

Do you think both sides win in this agreement? What if Merck discovers a compound that it turns into a multi-billion-dollar drug? Does this provide a good model for other countries? For other companies?

Major types of animals that offer potential medical benefits but that are threatened with extinction of key species		
Animal	Potential medical uses	Extinction risks
Amphibians 	• Antibiotics, alkaloids for pain-killers, chemicals for treating heart disease and high blood pressure • Natural adhesives for treating tissue damage • Study of newts and salamanders that can regenerate organs and tissues could suggest how we might, too. • "Antifreeze" compounds that allow frogs to survive freezing might help us preserve organs for transplants.	30% of all species are threatened with extinction (see The Science behind the Story, pp. 294–295).
Bears 	• An acid from bears' gall bladders already treats gallstones and liver disease, and prevents bile buildup during pregnancy. • While hibernating, bears build bone mass. If we learn how they do it, we could treat osteoporosis and hip fractures, which lead to 740,000 deaths per year. • Hibernating bears excrete no waste for months. If we can learn how, this could help treat renal disease, which kills 80,000 people in the U.S. each year.	Nine species are at risk of extinction.
Cone snails 	• Compounds found from these snails include one that may prevent death of brain cells from head injuries or strokes, and a pain-killer 1,000 times more potent than morphine. But so far just a few hundred of the 70,000–140,000 compounds these snails produce have been studied.	Most live in coral reefs, which are threatened ecosystems (pp. 439–442).
Sharks 	• Squalamine from sharks' livers could lead to novel antibiotics, appetite-suppressants, drugs to shrink tumors, and drugs to fight vision loss. • Study of sharks' salt glands is helping address kidney diseases. • Study of sharks' immune systems may shed light on our own.	Overfishing has reduced populations of most species, and some now risk extinction.
Horseshoe crabs 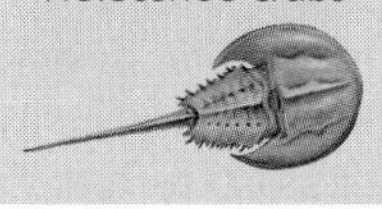	• A number of antibiotics are being developed. • The compound T140 may be more effective than AZT in treating AIDS, and could also help treat arthritis and several cancers. • Cells from blood can help detect cerebral meningitis in people.	Overfishing is sharply diminishing populations.

FIGURE 11.21 ▲ The major types of animals shown above offer significant potential medical benefits that researchers are just beginning to study, but we could lose these benefits if these animals become extinct. Adapted from Chivian, E., and A. Bernstein, 2008. *Sustaining life: How human health depends on biodiversity.* Oxford Univ. Press.

Biodiversity generates economic benefits through tourism and recreation

Besides providing for our food and health, biodiversity can generate income through tourism, particularly for developing countries in the tropics that boast impressive species diversity. Many people like to travel to experience protected natural areas, and in so doing they create economic opportunities for residents living near those areas. Visitors spend money at local businesses, hire local people as guides, and support parks that employ local residents. Ecotourism (p. 72) thus can bring jobs and income to areas that otherwise might be poverty-stricken.

Ecotourism has become a vital source of income for nations such as Costa Rica, with its rainforests; Australia, with its Great Barrier Reef; Belize, with its reefs, caves, and rainforests; and Kenya and Tanzania, with their savanna wildlife. The United States, too, benefits from ecotourism; its national parks draw millions of visitors from around the world. Ecotourism can serve as a powerful financial incentive for nations, states, and local communities to preserve natural areas and reduce impacts on the landscape and on native species.

As ecotourism increases, however, too many visitors to natural areas can degrade the outdoor experience and disturb wildlife. Anyone who has been to Yosemite, the Grand Canyon, or the Great Smokies on a crowded summer weekend can attest to this. As ecotourism continues to grow, so will debate over its costs and benefits for local communities and for biodiversity.

People value and seek out connections with nature

Not all of the benefits of biodiversity to people can be expressed in the hard numbers of economics or the day-to-day practicalities of food and medicine. Some scientists and philosophers argue that there is a deeper value in biodiversity. Edward O. Wilson (**FIGURE 11.22**) has popularized the notion of **biophilia**, asserting that human beings have an instinctive love for nature and feel an emotional bond with other living things (**FIGURE 11.23**). Wilson and others cite as evidence of biophilia our affinity for parks and wildlife, our love for pets, the high value of real estate with a view of natural landscapes, and our interest—despite being far removed from a hunter-gatherer lifestyle—in hiking, bird-watching, fishing, hunting, backpacking, and similar outdoor pursuits.

In a 2005 book, writer Richard Louv adds that as today's children are increasingly deprived of outdoor experiences and direct contact with wild organisms, they suffer what he calls "nature-deficit disorder." Although it is not a recognized medical condition, Louv argues that this alienation from biodiversity and nature may damage childhood development and lie behind many of the emotional and physical problems young people in developed nations face today.

FIGURE 11.22 ▲ Edward O. Wilson is the world's most recognized authority on biodiversity and its conservation, and he has inspired many people who study our planet's life. A Harvard professor and world-renowned expert on ants, Wilson has written over 20 books and has won two Pulitzer prizes. His books *The Diversity of Life*, *The Future of Life*, and *The Creation* address the value of biodiversity and its outlook for the future.

WEIGHING THE ISSUES

Biophilia and Nature-Deficit Disorder What do you think of the concepts of biophilia and "nature-deficit disorder"? Have you ever felt a connection to other living things that you couldn't explain in scientific or economic terms? Conversely, is there such a thing as "biophobia"—a human fear of other organisms? Do you think that our affinity for (or fear of) other living things is innate, or is it learned from our culture and upbringing? Explain your answer.

FIGURE 11.23 ▼ An Indonesian girl peers into a flower of *Rafflesia arnoldii*, the largest flower in the world. The concept of biophilia holds that human beings have an instinctive love and fascination for nature and a deep-seated desire to affiliate with other living things.

Do we have ethical obligations toward other species?

Aside from all of biodiversity's pragmatic benefits, many people feel that living organisms have an inherent right to exist. In this view, biodiversity conservation is justified on ethical grounds alone. When Maurice Hornocker and his associates first established the Siberian Tiger Project, he wrote: "Saving the most magnificent of all the cat species and one of the most endangered should be a global responsibility. . . . If they aren't worthy of saving, then what are we all about? What is worth saving?"

We human beings are part of nature, and like any other animal we need to use resources and consume other organisms to survive. In that sense, there is nothing immoral about our doing so. However, we have conscious reasoning ability and are able to control our actions and make deliberate decisions. Our ethical sense has developed from this intelligence and ability to choose. As our society's sphere of ethical consideration has widened over time, and as more of us take up biocentric or ecocentric worldviews (pp. 141–142), more people have come to feel that other organisms have intrinsic value and an inherent right to exist.

Yet despite our expanding ethical convictions and despite biodiversity's many benefits, the future of biodiversity remains far from secure. The search for solutions to today's biodiversity crisis is dynamic and exciting, and scientists are helping to develop innovative approaches to maintaining the diversity of life on Earth.

CONSERVATION BIOLOGY: THE SEARCH FOR SOLUTIONS

Today, more and more scientists and citizens perceive a need to do something to stem the loss of biodiversity. In his 1994 autobiography, *Naturalist*, E.O. Wilson wrote:

> When the [20th] century began, people still thought of the planet as infinite in its bounty. [Yet] in one lifetime, exploding human populations have reduced wildernesses to threatened nature reserves. Ecosystems and species are vanishing at the fastest rate in 65 million years. Troubled by what we have wrought, we have begun to turn in our role from local conqueror to global steward.

Conservation biology arose in response to biodiversity loss

The urge to act as responsible stewards of natural systems, and to use science as a tool in that endeavor, helped spark the rise of conservation biology. **Conservation biology** is a scientific discipline devoted to understanding the factors, forces, and processes that influence the loss, protection, and restoration of biological diversity. It arose as biologists became increasingly alarmed at the degradation of the natural systems they had spent their lives studying. Conservation biologists choose questions and pursue research with the aim of developing solutions to such problems as habitat degradation and species loss (See **ENVISIONIT**, p. 301). Conservation biology

Sampling insects in Madagascar

Conservation biologists are racing to save species from decline and extinction.

Tracking rhinos in Africa

Blood sample from Seychelles Magpie Robin

They work in the field, in the lab, at zoos, and with local people in areas that need protecting ...

... striving to recover populations of plants and animals threatened by habitat loss and other causes.

Sea turtle with transmitter, Hong Kong

Sampling plants in Florida

Radiotracking birds in Spain

YOU CAN MAKE A DIFFERENCE

➤ Preserve or restore wildlife habitat in your yard, in your community, or on your campus.
➤ Volunteer time or money to organizations working to save species and habitat.
➤ Buy shade-grown coffee, organic produce, and other wildlife-friendly products.

301

is thus an applied and goal-oriented science, with implicit values and ethical standards.

Conservation biologists work at multiple levels

Conservation biologists integrate an understanding of evolution and extinction with ecology and the dynamic nature of environmental systems. They use field data, lab data, theory, and experiments to study our impacts on other organisms. They also attempt to design, test, and implement ways to mitigate human impact. These researchers address the challenges facing biological diversity at all levels, from genes to species to ecosystems.

At the genetic level, *conservation geneticists* study genetic attributes of organisms to infer the status of their populations. If two populations of a species are found to be genetically distinct, they may have different ecological needs and may require different types of management. Moreover, as a population dwindles, genetic variation is lost from the gene pool. Conservation geneticists investigate how small a population can become and how much genetic variation it can lose before running into problems such as inbreeding depression (p. 284), whereby genetic similarity causes parents to produce weak or defective offspring. By determining a *minimum viable population size* for a population, conservation geneticists and population biologists help wildlife managers decide how vital it may be to increase the population. Problems for populations spell problems for species, because declines and local extirpation can lead to range-wide endangerment and extinction.

Other efforts in conservation biology revolve around habitats, communities, ecosystems, and landscapes. Changes at these levels affect populations and species, and conservation efforts at these levels are often informed by studies of genes, populations, and species. As we saw in our discussion of landscape ecology (pp. 118–120), organisms are sometimes distributed across a landscape as a *metapopulation*, or a network of subpopulations. Because small and isolated subpopulations are most vulnerable to extirpation, conservation biologists pay special attention to them. By examining how organisms disperse from one habitat patch to another, and how their genes flow among subpopulations, conservation biologists try to learn how likely a population is to persist or succumb in the face of habitat change or other threats.

Endangered species are a focus of conservation efforts

A great deal of effort in conservation biology is focused at the species level, and this has implications for policy. The primary legislation for protecting biodiversity in the United States is the **Endangered Species Act (ESA)**. Passed in 1973, the Endangered Species Act forbids the government and private citizens from taking actions that destroy endangered species or their habitats. The ESA also forbids trade in products made from endangered species. The aim is to prevent extinctions, stabilize declining populations, and enable populations to recover. As of 2010, there were 1,010 species in the United States listed as "endangered" and 314 more listed as "threatened," the

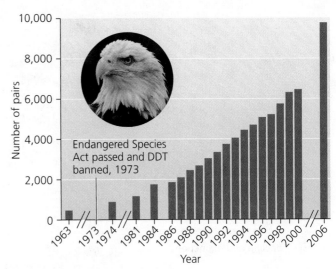

FIGURE 11.24 ▲ The recovery of the bald eagle (*Haliaeetus leucocephalus*) is a success story of the U.S. Endangered Species Act (ESA). The national symbol of the United States was close to extinction in the Lower-48 states in the 1960s. Following its protection under the ESA and a ban on the pesticide DDT in 1973, the eagle population began a long rebound. With its Lower-48 population around 10,000 pairs in 2007, the bald eagle was declared recovered and was removed from the Endangered Species List. Data from U.S. Fish and Wildlife Service, based on annual volunteer eagle surveys. The paucity of data after 2000 is because surveys began to be discontinued once it became clear that the eagle was recovering.

status considered one notch less severe than endangered. Nearly 90% of these species have active recovery plans whereby government agencies are taking steps to protect them and stabilize or increase their populations.

The ESA has had a number of notable successes. Following the ban on the pesticide DDT (pp. 379–380) and years of intensive effort by wildlife managers, the peregrine falcon, brown pelican, bald eagle, and other birds have recovered and are no longer listed as endangered (**FIGURE 11.24**). Intensive management programs with other species, such as the red-cockaded woodpecker (Figure 12.15, p. 327), have held formerly declining populations steady in the face of continued pressure on habitat. In fact, roughly 40% of declining populations have been stabilized.

This success comes despite the fact that the U.S. Fish and Wildlife Service and the National Marine Fisheries Service, the agencies responsible for upholding the ESA, are perennially underfunded for the job. Reauthorization of the ESA faced stiff opposition from the Republican Congresses in power from 1994 to 2006. Efforts to weaken the ESA by stripping it of its ability to safeguard habitat were narrowly averted in 2006 after 5,700 scientists sent Congress a letter of protest.

Polls repeatedly show that most Americans support the idea of protecting endangered species. Yet some opponents feel that the ESA places more value on the life of an endangered organism than it does on the livelihood of a person. This was a common perception in the Pacific Northwest in the 1990s, when protection for the northern spotted owl (**FIGURE 11.25**) slowed logging in old-growth rainforest and

FIGURE 11.25 ▲ The northern spotted owl (*Strix occidentalis occidentalis*), a bird of the Pacific Northwest rainforest, is disappearing because the mature forests it depends on have been logged (and now because the barred owl, a competitor, is invading its logged forests). Proponents of logging argue that protecting the spotted owl under the Endangered Species Act has caused economic harm and job loss. Advocates of its protection maintain that unsustainable logging practices pose a larger risk of job loss.

many loggers began to fear for their jobs. In addition, many landowners worry that federal officials will restrict the use of private land on which threatened or endangered species are found. This has led in many cases to a practice described as "shoot, shovel, and shut up," among landowners who want to conceal the presence of such species on their land.

ESA supporters maintain that such fears are overblown, pointing out that the ESA has stopped few development projects. Moreover, a number of provisions of the ESA and its amendments promote cooperation with landowners. *Habitat conservation plans* and *safe harbor agreements* are two types of arrangements that the government may make with private landowners that allow the landowners to harm species in some ways if they voluntarily improve habitat for the species in others.

Today a number of nations have laws protecting species, although they are not always effective. When Canada enacted its long-awaited endangered species law in 2002, the *Species at Risk Act (SARA)*, the Canadian government was careful to stress cooperation with landowners and provincial governments, rather than presenting the law as a decree from the national government. Environmental advocates and many scientists, however, protested that SARA was too weak and failed to protect habitat adequately.

In Russia, the government issued Decree 795 in 1995, creating a Siberian tiger conservation program and declaring the tiger a natural and national treasure. In 2007 it established a national park in the Sikhote-Alin Mountains to help protect the tiger. However, funding from the state for tiger conservation is so meager that the Wildlife Conservation Society feels it necessary to help pay for Russians to enforce their own anti-poaching laws.

Conservation efforts include international treaties

The United Nations has facilitated several international treaties to protect biodiversity. The 1973 **Convention on International Trade in Endangered Species of Wild Fauna and Flora (CITES)** protects endangered species by banning the international transport of their body parts. When nations enforce it, CITES can protect tigers and other rare species whose body parts are traded internationally.

In 1992, leaders of many nations agreed to the **Convention on Biological Diversity**. This treaty embodies three goals: to conserve biodiversity, to use biodiversity in a sustainable manner, and to ensure the fair distribution of biodiversity's benefits. The Convention aims to help

▶ Provide incentives for biodiversity conservation

▶ Manage access to and use of genetic resources

▶ Transfer technology, including biotechnology

▶ Promote scientific cooperation

▶ Assess the effects of human actions on biodiversity

▶ Promote biodiversity education and awareness

▶ Provide funding for critical activities

▶ Encourage each nation to report regularly on its biodiversity conservation efforts

Among its many accomplishments so far, the treaty has prompted nations to increase their area in protected reserves, has enhanced global markets for shade-grown coffee and other crops grown without removing forests, has ensured that African nations share in the economic benefits of ecotourism with their wildlife preserves, and has replaced pesticide-intensive farming practices with sustainable ones in some rice-producing Asian nations. Yet the treaty's overall goal—"to achieve, by 2010, a significant reduction of the current rate of biodiversity loss at the global, regional and national level"—has not been met. As of 2010, 193 nations had become parties to the Convention on Biological Diversity. The only ones choosing *not* to do so are tiny Andorra, the Vatican, and the United States.

Captive breeding, reintroduction, and cloning are being used to save species

In the effort to save threatened and endangered species, zoos and botanical gardens have become centers for **captive breeding**, in which individuals are bred and raised in controlled conditions with the intent of reintroducing them into the wild. The IUCN counts 65 plant and animal species that now exist *only* in captivity or cultivation.

One example of captive breeding and reintroduction is the program to save the California condor, North America's largest

The SCIENCE behind the Story

Using Forensics to Uncover Illegal Whaling

As any television buff knows, forensic science is a crucial tool in solving mysteries and fighting crime. In recent years conservation biologists have been using forensics to unearth secrets and catch bad guys in the multi-billion-dollar illegal global wildlife trade. One such detective story comes from the Pacific Ocean and Japan.

The meat from whales has long been a delicacy in Japan and other nations. Whaling ships decimated populations of most species of whales in the 20th century through overhunting, and the International Whaling Commission (IWC) outlawed commercial whaling worldwide beginning in 1986. Yet whale meat continues to be sold at market to wealthy consumers today (**see photo**). This meat comes legally from several sources:

1. From scientific hunts. Japan and several other nations negotiated with the IWC to continue to hunt limited numbers of whales for research purposes, and this meat may be sold afterwards.
2. From whales killed accidentally when caught in fishing nets meant for other animals (bycatch; p. 450).
3. Possibly from stockpiles frozen before the IWC's moratorium.

However, conservation biologists long suspected that much of the whale meat on the market was actually caught illegally for the purpose of selling for food and that fleets from Japan and other nations were killing more whales than international law allowed. Once DNA sequencing technology

Dr. C. Scott Baker of Oregon State University

was developed, scientists could use this tool to find out.

The detectives in this story are conservation geneticists C. Scott Baker, Stephen Palumbi, Frank Cipriano, and their colleagues. For close to two decades they have been traveling to Asia on what has amounted to top-secret grocery shopping trips.

Minke whales from the western Pacific Ocean may end up as meat in Japanese or Korean markets.

It began in 1993, when Baker and Palumbi bought samples of whale meat—all labeled simply as *kujira*, the generic Japanese term for whale meat—from a number of markets in Japan and sequenced DNA from these samples. Law forbids the export of whale meat, so the researchers had to run their analyses in their hotel rooms with portable genetic kits. Once they were back home in the United States, they compared their data with sequences from known whale species.

By analyzing which samples matched which, they concluded that they had sampled meat from nine minke whales, four fin whales, one humpback whale, and two dolphins. Moreover, because subspecies of whales from different oceans differ genetically, the researchers were able to analyze the genetic variation in their samples and learn that one fin whale came from the Atlantic whereas the other three were from the Pacific, and that eight of the nine minke whales came from the Southern Hemisphere.

Because several of these species and/or subspecies were off-limits to hunting, the data suggested that some of the meat had been hunted, processed, or traded illegally. Baker and Palumbi concluded in a 1994 paper in *Science* that "the existence of legal whaling serves as a cover for the sale of illegal whale products." They urged that the international community monitor catches more closely.

Two years later, Baker, Palumbi, and Cipriano presented results from markets in South Korea and Japan. Again their genetic sleuthing revealed a diversity of whale species, and they stated that their data were "difficult to reconcile" with records of legal catches (scientific whaling by Japan and fishing by-catch by South Korea) reported by these nations to the IWC. Among the whales they detected were two specimens of what seemed to be a subspecies or species of whale new to science.

In 2000, the team analyzed 655 samples labeled as whale meat from Japanese and South Korean markets and found evidence for 12 species or subspecies of whales, along with orcas, porpoises, and dolphins—and even sheep and horses! Seven of the whale species were internationally protected, and together these constituted 10% of the whale meat for sale in Japanese markets.

Genetic analyses of minke whale samples from Japan's markets also indicated that a surprisingly large percentage came from animals from the Sea of Japan, where Korea and Japan harvested them as fishing by-catch. One-third of the meat on the market was coming from the Sea of Japan, meaning that four times as many whales were being killed there as Japan was reporting (**see top figure**). The research team calculated that Japan and Korea together were taking so many minke whales from the Sea of Japan that they would eventually wipe out the population (**see bottom figure**).

In 2007, Baker led a team that combined genetic forensics with ecological methods to estimate numbers of individual whales whose meat was passing through Korean markets. They inferred that meat from 827 minke whales had passed through South Korea's market in five years. The nation had reported catching only 458 minke whales as fishing by-catch, leading the researchers to conclude

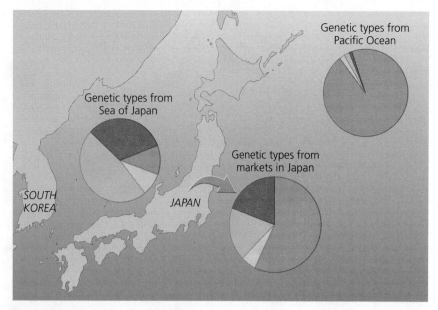

The distribution of genetic types from minke whale meat in Japanese markets (center pie chart) shows evidence of whales from the Sea of Japan (left pie chart) as well as of whales from the open Pacific Ocean (right pie chart). Note how proportions of the types from the market are intermediate between those of each geographic area of ocean. Adapted from Lukoschek, V., et al., 2009. High proportion of protected minke whales sold on Japanese markets due to illegal, unreported, or unregulated exploitation. *Animal Conservation* 12: 385–395, Fig 2. By permission of John Wiley and Sons. www.interscience.wiley.com.

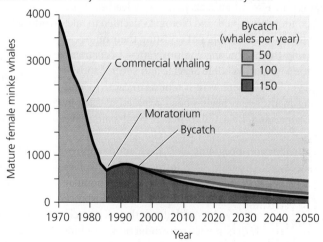

Minke whales in the Sea of Japan declined sharply (orange color in graph) until the 1986 moratorium on their capture. Population models forecast that by-catch of 150, 100, or even 50 minke whales per year would be enough to prevent the population's recovery. Data suggest that actual by-catch from the Sea of Japan has been close to 150 per year. Adapted from Baker, C.S., et al., 2000. Predicted decline of protected whales based on molecular genetic monitoring of Japanese and Korean markets. *Proc. Roy. Soc. Lond. B* 267: 1191–1199, Fig 3. By permission of The Royal Society and the author.

that the remainder had been taken illegally.

The governments of Japan and South Korea have tried to rebuff these findings. Yet the technology and approaches that turn scientists into forensic detectives are now influencing the debate and the negotiation over whaling policy at the international level.

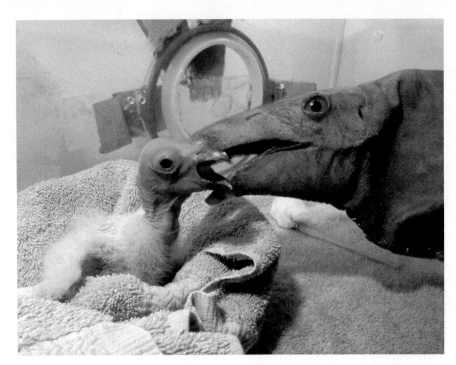

FIGURE 11.26 ◄ To save the California condor (*Gymnogyps californianus*) from extinction, biologists have raised hundreds of chicks in captivity with the help of hand puppets designed to look and feel like the heads of adult condors. Using these puppets, biologists feed the growing chicks in an enclosure and shield them from all contact with humans, so that when the chick is grown it does not feel an attachment to people.

bird (**FIGURE 11.26**). Condors were persecuted by people in the early 20th century and also frequently collided with electrical wires. In addition, many condors succumbed to lead poisoning after scavenging carcasses of animals killed with lead shot. By 1982, only 22 condors remained, and biologists decided to take all the birds into captivity, in hopes of boosting their numbers and then releasing them. The ongoing program—a collaboration between the Fish and Wildlife Service and several zoos—is succeeding. As of 2010 there were 176 birds in captivity and 180 birds living in the wild, having been released at sites in California, Arizona, and Baja California. Several pairs have begun nesting.

Other reintroduction programs have been more controversial. The successful program to reintroduce gray wolves (*Canis lupus*) to Yellowstone National Park has proven popular with the American public but continues to meet stiff resistance from ranchers, who fear the wolves will attack their livestock. In Arizona and New Mexico, a wolf reintroduction program is making slow headway, but a number of wolves there have been shot.

For the Siberian tiger, China is considering a similar reintroduction program. The Chinese government says it is preparing 600 captive Siberian tigers for release into its forests in the far northeastern portion of the country. Critics note that the forests are so fragmented that efforts would be better focused on improving habitat first.

The newest idea for saving species from extinction is to create individuals by cloning them. In this technique, DNA from an endangered species is inserted into a cultured egg without a nucleus, and the egg is implanted into a closely related species that acts as a surrogate mother. So far several mammals have been cloned in this way, with mixed results. Some scientists even talk of recreating extinct species from DNA recovered from preserved body parts. Indeed, in 2009 a subspecies of Pyrenean ibex (*Capra pyrenaica*, a type of mountain goat) was cloned from cells taken from the last surviving individual, which had died in 2000. The cloned baby ibex died shortly after birth. However, even if cloning can succeed from a technical standpoint, such efforts are not an adequate response to biodiversity loss. Without ample habitat and protection in the wild, having cloned animals in a zoo does little good.

Forensics is being used to protect threatened species

To counter poaching and other illegal harvesting, scientists now have a new tool at their disposal. *Forensic science*, or *forensics*, involves the scientific analysis of evidence to make an identification or answer a question, most often relating to a crime or accident. Conservation biologists are now employing forensics to protect species at risk. By analyzing DNA (p. 29) from organisms or their body parts sold at market, researchers can often determine the species or subspecies of organism, and sometimes its geographic origin. This can help detect illegal activity, enhancing the enforcement of laws protecting wildlife. One prime example is the analysis of whale meat sold in Asian markets (see **The Science behind the Story**, pp. 304–305).

Another example is the effort to track the geographic origin of tusks of African elephants killed for ivory. Trade in ivory has been banned in an effort to stop the slaughter of elephants. After customs agents seized 6.5 tons of tusks in Singapore in 2002, researchers led by Samuel Wasser of the University of Washington analyzed DNA from the tusks to determine the geographic origin of the elephants that were killed. The researchers sought to find out whether the tusks belonged to savanna elephants killed in Zambia (the origin of the shipment), or whether they came from forest elephants from other locations. The DNA matched known samples from Zambian elephants, indicating that many more elephants were being killed there than Zambia's government had claimed. As a result of the findings, the Zambian government replaced its wildlife director and imposed harsher sentences on poachers and ivory smugglers.

Some species act as "umbrellas" to protect habitat and communities

The Endangered Species Act and CITES provide legal means and resources for protecting species, but no such law or treaty exists for communities or ecosystems. Scientists know that protecting habitat and conserving communities and ecosystems are vital goals and that protecting species does little good if the larger systems they rely on are not also protected. For these reasons, conservation biologists often use particular species as tools to try to conserve communities and ecosystems. Such species are frequently called *umbrella species* because they serve as a kind of umbrella to protect many other species. Umbrella species often are large species that roam great distances, such as the Siberian tiger. Because such species require large areas of habitat, meeting their habitat needs automatically helps meet those of thousands of less charismatic animals, plants, and fungi that might never elicit as much public interest.

Environmental advocacy organizations have found that using large and chari~~smatic~~ vertebrates as spearheads for biodiversity conse~~rvation is an~~ effective strategy. This approach of promo~~ting flagship species~~ is evident in the long-ti~~me logo of the World W~~ide Fund for Nature (World ~~Wildlife Fund in Americ~~a), the panda. The panda (see ~~p. 318) is an endan~~gered animal requiring size-~~able areas of bamboo~~ forest. Its lovable appear-~~ance makes it dear to the~~ public—and an effective ~~means to raise funds for conserv~~ation efforts that protect ~~its habitat.

Many conservatio~~n organizations today ~~use this umbrella-species a~~pproach. The Nature ~~Conservancy buys up com~~munities and land-~~scapes, and a novel idea may~~ be the Wildlands ~~Project, which envisions large~~ amounts of North ~~America being reconnected.

Parks and protected areas help ~~preserve biodiversit~~y at the ~~landscape level

Althoug~~h most legislation, funding, and resources for biodiversity conservation go toward single-species approaches, the practice of setting aside areas of undeveloped land to be preserved in parks and protected areas helps to conserve habitats, communities, ecosystems, and landscapes. Currently we have set aside 12% of the world's land area in various types of preserves and protected areas, including national parks, state parks, provincial parks, wilderness areas, biosphere reserves, and many others. Many of these lands are managed for human recreation, water quality protection, or other purposes, rather than for biodiversity, and many suffer from illegal logging and resource extraction because enforcement is lacking. Yet these areas do offer animals and plants some degree of protection from human persecution, and many are large enough to preserve whole natural systems that otherwise would be fragmented, degraded, or destroyed. We will explore parks and protected areas in full in Chapter 12 (pp. 332–340).

Biodiversity hotspots pinpoint regions of high diversity

One international approach oriented around geographic regions, rather than single species, is the effort to map **biodiversity hotspots**. The concept of biodiversity hotspots was introduced in 1988 by British ecologist Norman Myers as a way to prioritize regions that are most important globally for biodiversity conservation. A hotspot supports an especially great number of species that are **endemic** (p. 58) to the region, that is, found nowhere else in the world (**FIGURE 11.27**). To qualify as a hotspot, a location must harbor at least 1,500 endemic plant species (0.5% of the world's total plant species). In addition, a hotspot must have already lost 70% of its habitat as a result of human impact and be in danger of losing more.

FIGURE 11.27 ▲ The ring-tailed lemur *(Lemur catta)*, like all lemur species, is a primate that is endemic to the island of Madagascar. Over 2,000 individuals survive in zoos worldwide thanks to captive breeding, but in Madagascar the lemur's natural habitat is fast disappearing. Madagascar has lost over 90% of its forests because of human population growth, poverty, and resource extraction. One recent president encouraged conservation and ecotourism, but when his government fell, illegal logging resumed, destroying large areas of protected forest. Many lemurs were killed and sold for their meat, and timber was exported to meet demand from wealthy nations.

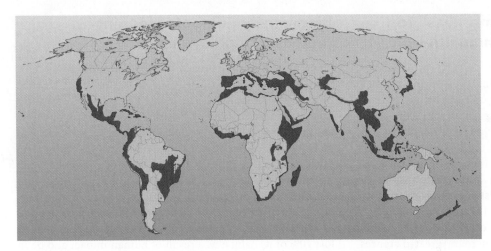

FIGURE 11.28 ◀ Some areas of the world possess exceptionally high numbers of species found nowhere else. Many conservation biologists have supported prioritizing habitat preservation in these areas, dubbed *biodiversity hotspots*. Shown in red are the 34 biodiversity hotspots mapped by Conservation International. Only about 15% of the area in red is actually habitat for these species; most is developed. Data from Conservation International, 2008.

The nonprofit group Conservation International maintains a list of 34 biodiversity hotspots (**FIGURE 11.28**). The ecosystems of these areas together once covered 15.7% of the planet's land surface but today, because of habitat loss, cover only 2.3%. This small amount of land is the exclusive home for half the world's plant species and 42% of all terrestrial vertebrate species. The hotspot concept gives incentive to focus on these areas, where the greatest number of unique species can be protected per unit effort.

Innovative economic strategies are being employed

Economic incentives can help to facilitate international conservation efforts. One strategy is the *debt-for-nature swap*. In such a swap, a conservation organization raises money and offers to pay off a portion of a developing nation's international debt in exchange for a promise by the nation to set aside reserves, fund environmental education, and better manage protected areas.

The U.S. government committed itself to a program of debt-for-nature swaps through its 1998 Tropical Forest Conservation Act. As of 2009, 15 deals had been struck with 13 developing nations, enabling $218 million in their funds intended for debt payments to go to conservation efforts instead. In the largest such deal yet, the U.S. government forgave Indonesia $30 million of debt while two conservation groups paid Indonesia $2 million. In return, Indonesia will preserve forested areas on Sumatra that are home to the Sumatran tiger and other species.

A newer strategy that Conservation International has pioneered is the *conservation concession*. Developing nations often sell "concessions" to foreign multinational corporations, allowing them to extract resources from the nation's land. For instance, a poor nation may earn money by selling to foreign timber companies the right to log its forests. Conservation International has stepped in and purchased concessions for conservation rather than for resource extraction. In these deals, the nation gets the money *and* keeps its natural resources intact. The South American country of Surinam, which still has extensive areas of pristine rainforest, entered into such an agreement and has virtually halted logging while pulling in $15 million.

We can restore degraded ecosystems

Protecting natural areas before they become degraded is the best way to safeguard biodiversity and ecological systems. However, in some cases we can restore degraded natural systems to some semblance of their former condition, through the practice of **ecological restoration**, which we first explored in Chapter 4 (pp. 95–96). Informed by the science of *restoration ecology*, efforts to restore damaged ecosystems aim not just to bring back populations of animals and plants, but to reestablish the processes—the cycling of matter and the flow of energy—that make an ecosystem function. By restoring complex natural systems such as the Illinois prairies (p. 95), the Florida Everglades (p. 95), or the southeastern longleaf pine forest (p. 327), restoration ecologists aim to recreate systems that filter pollutants, cleanse water and air, build soil, and recharge groundwater, providing both habitat for wildlife and services for people.

One of today's highest-profile restoration projects is the ongoing effort to restore the vast marshes of southern Iraq, an area historically so lush and productive that it is often claimed to have been the setting for the biblical Garden of Eden. For several thousand years, tens of thousands of people (nicknamed "Marsh Arabs") lived sustainably among the wetlands and channels of this region in the floodplain of the Tigris and Euphrates rivers, thriving on its plentiful fish and shellfish (**FIGURE 11.29**). However, these wetlands, the largest in the Middle East, began to be drained for oil exploration in the 1970s and during the Iran-Iraq War in the 1980s. Iraqi ruler Saddam Hussein, a Sunni Muslim, viewed the Shiite people living in the marshes to be disloyal and worried that the marshes offered refuge to Shiite rebels who sought his overthrow. In the 1990s following the Persian Gulf War, Saddam ordered a huge system of dikes and canals built to drain the marshes, devastating the wetlands and the communities of people living there.

After the U.S. invasion of Iraq in 2003, restoration ecologists from many nations joined the people of the marshes in a multimillion-dollar ecological restoration effort. The international project, led by Iraqi scientist Azzam Allwash, was able to restore natural water flow to about 75% of the region, so that vegetation grew back and wildlife and people

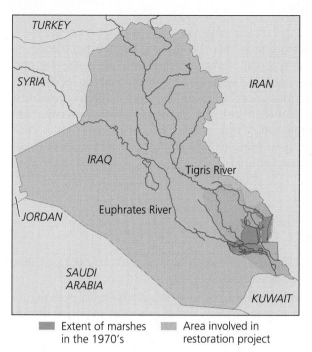

(a) Extent of Iraq's marshes

■ Extent of marshes in the 1970's   ■ Area involved in restoration project

(b) Communities of "Marsh Arabs" before destruction

FIGURE 11.29 ▲ Saddam Hussein's intentional destruction of the vast marshlands in southern Iraq **(a)** devastated the largest wetlands in the Middle East and disrupted the communities **(b)** of tens of thousands of people. Since his removal from power, scientists and engineers have been trying to restore the marshlands to their former condition.

began returning. The effort was on the verge of success, but drought descended on the region and two of Iraq's neighbors, Turkey and Syria, began diverting water from the rivers for their own purposes. As of 2010, the marsh restoration project was in limbo, poised for success but held back by the drought and the diversions. Today ecologists and human rights supporters alike are searching for ways to complete the restoration project.

WEIGHING THE ISSUES

Single-Species Conservation? What would you say are some advantages of focusing on conserving single species, versus trying to conserve broader communities, ecosystems, or landscapes? What might be some of the disadvantages? Which do you think is the better approach, or should we use both?

Community-based conservation is growing

Aiming to help people, wildlife, and ecosystems all at the same time, as the Iraqi marsh restoration project intends, is the focus of many current efforts in conservation biology. In past decades, conservationists from developed nations, in their zeal to preserve ecosystems in other nations, too often neglected the needs of people in the areas they wanted to protect. Developing nations came to view this international

environmentalism as a kind of neocolonialism. Today this has largely changed, and many conservation biologists actively engage local people in efforts to protect land and wildlife.

This approach, called **community-based conservation**, is being pursued to help protect tigers in many places across Asia. In India, multiple projects offer education, health care, and development assistance to communities living amid the shrinking habitat of the Bengal tiger. In Cambodia, people who used to hunt Indochinese tigers are being retrained and paid salaries as forest guards to protect the animals from poachers or as wildlife technicians to help with science and monitoring. And in the Russian Far East, the Wildlife Conservation Society is working with local hunters to reduce poaching of Siberian tigers and to increase the populations of deer and other animals that are prey for both tigers and hunters. The WCS is also establishing a market in the West for sustainably harvested products from the region that are certified "tiger-friendly." Proceeds from sales of the products aim to supplement the incomes of up to 1,000 local people by 12–25%.

Making conservation beneficial for local people requires hard work, investment, and trust on all sides. While setting aside land for preservation may deprive local people of access to natural resources, it also helps ensure that these resources will not be used up or sold to foreign corporations, but can instead be sustainably managed. Moreover, parks and reserves draw ecotourism that supports local economies. Community-based conservation has not always been successful, but in a world of increasing human population, we will require locally based management that sustainably meets people's needs.

➤ CONCLUSION

Data from scientists worldwide confirm what any naturalist who has watched the habitat change in his or her hometown already knows: From amphibians to tigers, biological diversity is being lost rapidly and visibly within our lifetimes. This erosion of biodiversity threatens to result in a mass extinction event equivalent to those of the geologic past. Habitat alteration, invasive species, pollution, overharvesting of biotic resources, and climate change are the primary causes of biodiversity loss. This loss matters, because human society cannot function without biodiversity's pragmatic benefits. Conservation biologists are rising to the challenge of conducting science aimed at saving endangered species, protecting their habitats, recovering populations, and preserving and restoring natural ecosystems. The innovative strategies these scientists are pursuing hold promise to slow the loss of biodiversity on Earth.

REVIEWING OBJECTIVES

You should now be able to:

CHARACTERIZE THE SCOPE OF BIODIVERSITY ON EARTH

- Biodiversity can be thought of at three levels: species diversity, genetic diversity, and ecosystem diversity. (pp. 282–284)
- Roughly 1.8 million species have been described so far, but scientists agree that the world holds millions more. (pp. 284–285)
- Some taxonomic groups (such as insects) hold far more diversity than others. (pp. 284–286)
- Researchers make global estimates of biodiversity by extrapolating from local areas and certain taxonomic groups. (pp. 284–285)
- Diversity is unevenly spread across different habitats, biomes, and regions of the world. (pp. 286–287)

CONTRAST THE BACKGROUND EXTINCTION RATE WITH PERIODS OF MASS EXTINCTION

- Species have gone extinct at a background rate of roughly one species per 1–10 million species each year. Most of the species that have ever lived are now extinct. (p. 288)
- Earth has experienced five mass extinction events in the past 440 million years. (p. 288)
- Human impact is now initiating a sixth mass extinction. (pp. 289–290)

EVALUATE THE PRIMARY CAUSES OF BIODIVERSITY LOSS

- Habitat alteration is the main cause of current biodiversity loss. (pp. 290–291)
- Pollution, overharvesting, and invasive species are also important causes. (pp. 291–294)
- Climate change is becoming a major cause. (pp. 294–296)
- Amphibians are facing a global crisis, probably from a mix of factors (pp. 294–295)

SPECIFY THE BENEFITS OF BIODIVERSITY

- Biodiversity supports functioning ecosystems and the services they provide us. (pp. 296–297)

- Wild species are sources of food, medicine, and economic development. (pp. 297–299)
- Many people feel that we have a psychological need to connect with the natural world, as well as an ethical duty to preserve nature. (pp. 299–300)

ASSESS THE SCIENCE AND PRACTICE OF CONSERVATION BIOLOGY

- Conservation biology studies biodiversity loss and seeks ways to protect and restore biodiversity. (pp. 300–301)
- Conservation biologists integrate research at the genetic, population, species, ecosystem, and landscape levels. (pp. 301–302)

ANALYZE EFFORTS TO CONSERVE THREATENED AND ENDANGERED SPECIES

- The U.S. Endangered Species Act has been controversial but effective. (pp. 302–303)
- CITES and the Convention on Biological Diversity are major international treaties to safeguard biodiversity. (p. 303)
- Modern recovery strategies include captive breeding and reintroduction programs. (pp. 303, 306)
- Forensics can help us trace products from illegally poached animals. (pp. 304–306)

COMPARE AND CONTRAST CONSERVATION EFFORTS ABOVE THE SPECIES LEVEL

- Charismatic species are often used in popular appeals to conserve habitats and ecosystems. (p. 307)
- Parks and protected areas conserve biodiversity at the landscape level. (p. 307)
- Biodiversity hotspots help prioritize regions globally for conservation. (pp. 307–308)
- Debt-for-nature swaps and conservation concessions provide economic incentives for conservation. (p. 308)
- Ecological restoration efforts are restoring degraded ecosystems. (pp. 308–309)
- Community-based conservation empowers people to invest in conserving their local species and ecosystems. (p. 309)

TESTING YOUR COMPREHENSION

1. What is biodiversity? Describe three levels of biodiversity.

2. What are the five primary causes of biodiversity loss? Give one specific example of each.

3. List three invasive species, and describe their impacts.

4. Define the term *ecosystem services*. Give three examples of ecosystem services that people would have a hard time replacing if their natural sources were eliminated.

5. What is the relationship between biodiversity and food security? Between biodiversity and pharmaceuticals? Give three examples of potential benefits of biodiversity conservation for food supplies and medicine.

6. Describe three reasons why people suggest biodiversity conservation is important.

7. Name two successful accomplishments of the U.S. Endangered Species Act. Now name two reasons some people have criticized it.

8. Describe how captive breeding can help with endangered species recovery, and give an example. Explain why cloning would never be, in itself, an effective response to species loss.

9. What is the difference between an umbrella species and a keystone species? Could one species be both an umbrella species and a keystone species?

10. What is a debt-for-nature swap? How does it compare to a conservation concession?

SEEKING SOLUTIONS

1. Many arguments have been advanced for the importance of preserving biodiversity. Which argument do you think is most compelling, and why? Which argument do you think is least compelling, and why?

2. Some people declare that we shouldn't worry about endangered species because extinction has always occurred. How would you respond to this view?

3. Advocates of biodiversity preservation from developed nations have long pushed to set aside land in biodiversity-rich regions of developing nations. Leaders of developing nations have responded by accusing these advocates of neocolonialism. "Your nations attained prosperity and power by overexploiting their environments decades or centuries ago," these leaders ask, "so why should we now sacrifice our development by setting aside our land and resources?" What would you say to these leaders? What would you say to the environmental advocates? Do you see ways that both preservation and development goals might be reached?

4. Compare the biodiversity hotspot approach to the approach of community-based conservation. What are the advantages and disadvantages of each? Can we—and should we—follow both approaches?

5. **THINK IT THROUGH** You are an influential legislator in a nation that has no endangered species act, and you want to introduce legislation to protect your nation's vanishing biodiversity. Consider the U.S. Endangered Species Act and the Canadian Species At Risk Act, as well as international efforts such as CITES and the Convention on Biological Diversity. What strategies would you write into your legislation? How would your law be similar to and different from each of these efforts?

6. **THINK IT THROUGH** As a citizen and resident of your community, and a parent of two young children, you attend a town meeting called to discuss the proposed development of a shopping mall and condominium complex. The development would eliminate a 100-acre stand of forest, the last sizeable forest stand in your town. The developers say the forest loss will not matter because plenty of 1-acre stands still exist scattered throughout the area. Consider the development's possible impacts on the community's biodiversity, children, and quality of life. What will you choose to tell your fellow citizens and the town's decision-makers at this meeting, and why?

CALCULATING ECOLOGICAL FOOTPRINTS

Of the five major causes of biodiversity loss discussed in this chapter, habitat alteration arguably has the greatest impact. In their 1996 book introducing the ecological footprint concept, authors Mathis Wackernagel and William Rees present a consumption/land use matrix for an average North American. Each cell in the matrix lists the number of hectares of land of that type required to provide for the different categories of a person's consumption (food, housing, transportation, consumer goods, and services). Of the 4.27 hectares required to support this average person, 0.59 hectares are forest, with most (0.40 hectares) being used to meet the housing demand. Using this information, calculate the missing values in the table.

	Hectares of forest used for housing	Total forest hectares used
You	0.40	0.59
Your class		
Your state		
United States		

Data from Wackernagel, M., and W. Rees, 1996. *Our ecological footprint: Reducing human impact on the earth*. British Columbia, Canada: New Society Publishers.

1. Approximately two-thirds of the forests' productivity is consumed for housing. To what use(s) would you speculate that most of the other third is put?

2. If the harvesting of forest products exceeds the sustainable harvest rate, what will be the likely consequence for the forest? For communities surrounding the forest?

3. What impacts would you expect on biodiversity in each of the following cases, and why?

   a. The cutting of small plots of forest within a large forest

   b. The clear-cutting (pp. 325–326) of an entire forest

   c. The clear-cutting of an entire forest followed by planting of a monocultural plantation of young trees

   Does the spatial scale at which you judge the effects on biodiversity affect your answer? In what ways?

Mastering**ENVIRONMENTALSCIENCE**™

Go to **www.masteringenvironmentalscience.com** for practice quizzes, Pearson eText, videos, current events, and more.

Hiawatha National Forest, in Michigan's Upper Peninsula

12 FORESTS, FOREST MANAGEMENT, AND PROTECTED AREAS

- Summarize the ecological and economic contributions of forests
- Outline the history and current scale of deforestation
- Assess aspects of forest management and describe methods of harvesting timber
- Identify federal land management agencies and the lands they manage
- Recognize types of parks and protected areas and evaluate issues involved in their design

Certified Sustainable Paper in Your Textbook

"As a company and as individuals, we must be able to look ourselves in the mirror and know that we are truly doing what is right and good for the environment."

—Rick Willett, president and CEO, NewPage Corporation

"FSC is the high bar in forest certification, because its standards protect forests of significant conservation value and consistently deliver meaningful improvements in forest management on the ground."

—Kerry Cesareo, deputy director, WWF-US Forests Program

CANADA

Escanaba, Michigan

UNITED STATES

Great Lakes

As you turn the pages of this textbook, you are handling paper made from trees that were grown, managed, harvested, and processed using certified sustainable practices.

If you were to trace the paper in this book back to its origin, you would find yourself standing in a diverse mixed forest of aspen, birch, beech, maple, spruce, and pine in the Upper Peninsula of Michigan. This sparsely populated region near the Canadian border, flanked by Wisconsin, Lake Michigan, and Lake Superior, remains heavily forested, despite supplying timber to our society for nearly 200 years.

The trees cut to make this book's paper were selected for harvest based on a sustainable management plan designed to avoid depleting the forest of its mature trees or degrading the ecological functions the forest performs. The logs were then transported to a nearby pulp and paper mill at Escanaba, a community of 13,000 people on the shore of Lake Michigan. The mill is Escanaba's largest employer, providing jobs to about 1,100 residents.

A worker examines paper being produced at the mill in Escanaba, Michigan

At the Escanaba mill, the wood is chipped and then fed into a digester, where the chips are cooked with chemicals to break down the wood's molecular bonds. Lignin in the wood (the polymer that gives wood its sturdiness) is separated out and discarded when white paper is made, leaving only the wood's cellulose fibers. Millworkers then bleach, wash, and screen the cellulose fibers and mix them with a lot of water and a little dye. This smooth, watery mix is poured onto a moving mat, where the water drains off and heavy rollers press the cellulose mix into thin sheets. The sheets of newly formed paper are then dried on heated rollers and made ready to receive any of a variety of coatings. The paper is wound into immense reels, later to be cut into sheets.

In this way the Escanaba mill produces about 770,000 tons of paper each year. It does its best to recycle chemicals and water used in the process, and it combusts discarded lignin to help power the mill.

At every stage in this process, independent third-party inspectors from the Forest Stewardship Council (FSC) examine the practices being used, to

314

ensure that they meet the FSC's strict criteria for sustainable forest management and paper production. The Forest Stewardship Council is the most demanding of a number of organizations that officially certify forests, companies, and products that meet sustainability standards. For instance, FSC-certified timber harvesting operations in Michigan's Upper Peninsula are required to protect rare species and sensitive habitats, safeguard water sources, control erosion, minimize pesticide use, and maintain the diversity of the forest and its ability to regenerate after harvesting, among many other considerations. FSC certification is the best way for consumers of forest products to know that they are supporting sustainable practices that protect the welfare of the world's forests.

The Escanaba mill is one of 11 mills run by the NewPage Corporation, based in Miamisburg, Ohio. NewPage's mills (in Kentucky, Maine, Maryland, Michigan, Minnesota, Wisconsin, and Nova Scotia) together produce 4.4 million tons of coated paper each year for books, magazines, newspapers, food packaging, and more. Only some of this paper is FSC certified, but NewPage seeks out suppliers of wood from private land who are certified at least to the less-rigorous standards of the Sustainable Forestry Initiative®. It states that it does not use wood from old-growth forests, rainforests, or forests of exceptional conservation value.

The paper in this book is FSC-certified from sources that strive to follow sustainable practices. As you flip through this book, you can feel satisfied that you are doing a small part to help safeguard the world's forests by supporting sustainable forestry practices.

FOREST ECOSYSTEMS AND FOREST RESOURCES

Forests currently cover 31% of Earth's land surface (**FIGURE 12.1**). They provide habitat for countless organisms; help maintain soil, air, and water quality; and play key roles in our planet's biogeochemical cycles (pp. 122–131). Forests have also long provided humanity with wood for fuel, construction, paper production, and more.

Many kinds of forests exist

A **forest** is any ecosystem with a high density of trees. Many kinds of forest exist, as we saw in our survey of the major forest biomes in Chapter 4 (pp. 96–103). Most of the world's forests occur as *boreal forest* (p. 102), a biome that stretches across much of Canada, Scandinavia, and Russia; or as *tropical rainforest* (p. 99), a biome that occurs in South and Central America, equatorial Africa, Indonesia, and Southeast Asia. *Temperate deciduous forests* (p. 98), *temperate rainforests* (p. 99), and *tropical dry forests* (p. 100) also cover large regions of the globe. Ecosystems with lower densities of trees, consisting of trees with open areas among them, are called *woodlands*.

Within each forest biome, the nature of the plant community varies from region to region because of differences in soil and climate. As a result, ecologists and forest managers find it useful to classify forests into *forest types*, categories

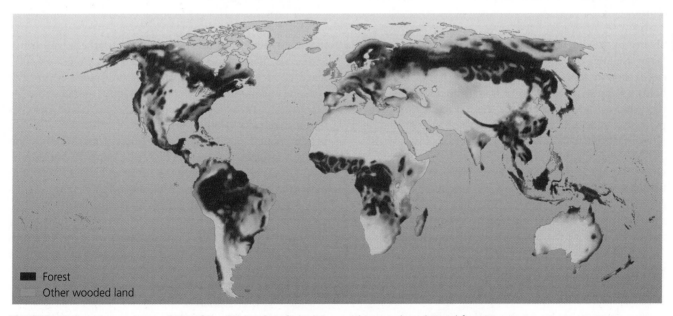

FIGURE 12.1 ▲ Forests cover 31% of Earth's land surface. Most widespread are boreal forests in the north and tropical forests in South America and Africa. Additional areas are classified as "wooded land," which supports trees at sparser densities. Data from Food and Agriculture Organization of the United Nations. 2010. *Global forest resources assessment* 2010. By permission.

Forest
Other wooded land

(a) Maple-beech-birch forest, Michigan's Upper Peninsula

(b) Oak-hickory forest, West Virginia

(c) Ponderosa pine forest, northern Arizona

(d) Redwood forest, coastal northern California

FIGURE 12.2 ▲ A few of the 23 forest types found in the continental United States include **(a)** maple-beech-birch forest, **(b)** oak-hickory forest, **(c)** ponderosa pine forest, and **(d)** redwood forest.

defined by their predominant tree species. The eastern United States contains 10 forest types, ranging from spruce-fir to oak-hickory to longleaf-slash pine. The western United States holds 13 forest types, ranging from Douglas fir and hemlock–sitka spruce forests of the moist Pacific Northwest to ponderosa pine and pinyon-juniper woodlands of the drier interior. **FIGURE 12.2** shows a sampling of U.S. forest types.

Michigan's Upper Peninsula holds a diverse mix of forest types, because it lies where the temperate deciduous forest biome of the eastern United States merges into the boreal forest of Canada. In the region surrounding Escanaba, areas of spruce-fir forest intermix with forests of aspen, birch, maple, and beech, with some areas of white, red, and jack pine.

Forests are ecologically complex

Because of their structural complexity and their ability to provide many niches for organisms, forests comprise some of the richest ecosystems for biodiversity (**FIGURE 12.3**). Trees, shrubs, and forest-floor plants furnish food and shelter for an immense diversity of animals. Countless insects, birds,

mammals, and other organisms subsist on the leaves, fruits, and seeds these plants produce. Moreover, plants are colonized by an extensive array of fungi and microbes, in both parasitic and mutualistic relationships (pp. 81–83).

In a forest's *canopy* (the upper level of leaves and branches in the treetops), beetles, caterpillars, and other leaf-eating insects abound, providing food for birds such as warblers and tanagers (**FIGURE 12.4**), while arboreal mammals from squirrels to sloths to monkeys consume fruit and leaves. Other animals live and feed amid the middle and lower portions of trees in the *subcanopy*, and still others utilize the bark, branches, and trunks. Cavities in trunks provide nest and shelter sites for a wide variety of animals. Dead and dying trees (called *snags*) are also valuable; as insects break down the wood, they provide food for animals, including woodpeckers that create cavities that other animals may later use.

Shrubs and small trees in the *understory* and ground-cover plants on the forest floor provide food and habitat for still more organisms. In fact, much of a forest's biodiversity resides on the forest floor, where the fallen leaves and branches of the leaf litter nourish the soil. As we saw in Chapter 9

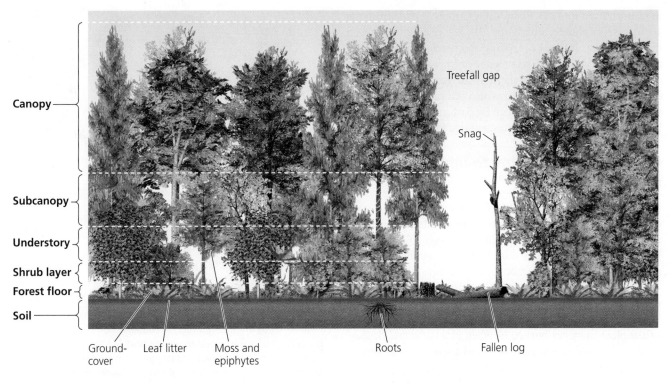

Treefall gap

Snag

Subcanopy

Understory

Shrub layer

Forest floor

Soil

Ground-cover   Leaf litter   Moss and epiphytes   Roots   Fallen log

FIGURE 12.3 ▲ Mature forests are complex ecosystems. In this cross-section of a mature forest, the crowns of the largest trees form the canopy, and smaller trees beneath them form the shaded subcanopy and understory. Shrubs and groundcover grow just above the forest floor, which may be covered in leaf litter rich with invertebrate animals. Vines, mosses, lichens, and epiphytes cover portions of trees and the forest floor. Snags (standing dead trees), whose wood can be easily hollowed out, provide food and nesting sites for woodpeckers and other animals. Fallen logs nourish the soil and provide habitat for countless invertebrates. Treefall gaps caused by fallen trees let light through the canopy and create small openings in the forest, allowing early successional plants to grow in patches within the mature forest.

(pp. 227–228), a multitude of soil organisms helps decompose plant material and cycle nutrients.

Rainforests feature additional complexity not shown in Figure 12.3. In tropical rainforests, especially tall individual trees called *emergent trees* protrude here and there above the rainforest canopy. Epiphytes (p. 62), plants specialized to grow atop other plants, add to the biomass and species diversity at all levels of a rainforest. Epiphytes include many ferns, mosses, lichens, orchids, and bromeliads.

Forests with a greater diversity of plants tend to host a greater diversity of organisms overall. As forests change over time through the process of succession (pp. 92–93), their species composition changes along with their structure. In general, old-growth forests host more biodiversity than younger forests, because older forests contain more structural diversity and thus more microhabitats and resources for more species.

Forests provide ecosystem services

Besides promoting biodiversity, forests provide people with many cultural, aesthetic, health, and recreation values (pp. 157–160). Forests also supply us with many vital ecosystem services (p. 3, 121–122, 148, 160–161). As plants grow (whether in forest, grassland, or other biomes), their roots stabilize the soil and help to prevent erosion. When rain falls, leaves and leaf litter slow runoff by intercepting water. By slowing runoff, forest vegetation prevents flooding, helps water soak into the ground to nourish roots and recharge aquifers, reduces soil erosion, and helps keep streams and rivers clean.

Forest plants also filter pollutants and purify water as they take it up from the soil and release it to the atmosphere

FIGURE 12.4 ▲ The scarlet tanager (*Piranga olivacea*) is one of many birds that lives in the canopy of eastern North America's temperate deciduous forest during the spring and summer, provisioning its young with insects that eat the leaves of oaks, hickories, beeches, and maples.

in transpiration (p. 124). Plants draw carbon dioxide from the air for use in photosynthesis (pp. 31–32), release the oxygen that we breathe, influence moisture and precipitation patterns, and moderate climate. Trees' roots draw minerals up from deep soil layers and deliver them to surface soil layers where other plants can use them. Plants also return organic material to the topsoil in the form of litter. By performing all these ecological functions, forests are indispensable for our survival.

Carbon storage helps to limit climate change

Of all the ecosystem services that forests provide, their storage of carbon has become of great international interest as nations debate how to control global climate change (Chapter 18). Because trees absorb carbon dioxide from the air during photosynthesis and then store carbon in their tissues, forests serve as a major reservoir for carbon. Scientists estimate that the world's forests store over 280 billion metric tons of carbon in living tissue, which is more than the atmosphere contains. When plant matter is burned or when plants die and decompose, carbon dioxide is released—and thereafter less vegetation remains to soak it up. Carbon dioxide is the primary greenhouse gas contributing to global climate change (p. 496). Therefore, when we cut forests, we worsen climate change. The more forests we preserve or restore, the more carbon we keep out of the atmosphere, and the better we can address climate change.

Forests provide us valuable resources

Besides carbon storage and other ecosystem services, which alone make them priceless to our society, forests also provide many economically valuable resources. Among these are plants for medicines, dyes, and fibers; animals, plants, and mushrooms for food; and, of course, wood from trees. For millennia, wood from forests has fueled our fires, keeping us warm and well fed. It has built the houses that keep us sheltered. It built the ships that carried people and cultures between continents. And it gave us paper, the medium of the first information revolution.

In recent decades, industrial harvesting has allowed us to extract more timber than ever before, supplying all these needs of a rapidly growing human population and its expanding economy. The exploitation of forest resources has been instrumental in helping our society achieve the standard of living we now enjoy.

Most commercial logging today takes place in Canada, Russia, and other nations with large expanses of boreal forest, and in tropical nations with large areas of rainforest, such as Brazil and Indonesia. In the United States, most logging takes place in pine plantations of the South and conifer forests of the West.

Nations maintain and use forests for all these economic and ecological reasons. An international survey in 2010 found that globally, 30% of forests were designated primarily for timber production (**FIGURE 12.5**). Lesser amounts were designated for a variety of functions, including conservation of biodiversity, protection of soil and water quality, and "social services" such as recreation, tourism, education, and conservation of culturally important sites.

318

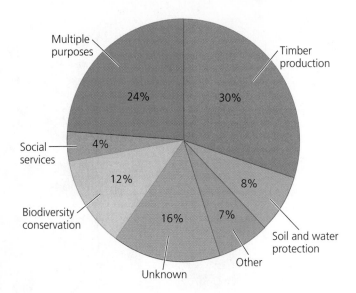

FIGURE 12.5 ▲ Worldwide, nations designate 30% of forests primarily for production of timber and other forest products. Smaller areas are designated for conservation of biodiversity, protection of soil and water quality, and "social services" such as recreation, tourism, education, and conservation of culturally important sites. One-quarter of forests are designated for combinations of these functions. Data from Food and Agriculture Organization of the United States, *Global forest resources assessment* 2010. By permission.

FOREST LOSS

Our demand for wood and paper products and our need for open land for agriculture have led people to clear forested land. **Deforestation**, the clearing and loss of forests, has altered landscapes and ecosystems across much of our planet. As we eliminate some forests and alter others, we lose biodiversity, worsen climate change, and disrupt the ecosystem services that support our societies. The alteration, fragmentation, and outright loss of forested land represent one of our society's primary challenges (see **ENVISIONIT**, p. 320).

Agriculture and demand for wood has led to deforestation

From the slash-and-burn farmer cutting tropical rainforest to the American suburbanite shopping at a grocery store, we all depend on food and fiber grown on cropland and rangeland—much of which occupies lands where forests once stood. All of us also depend in some way on wood, from the subsistence herder in Nepal cutting trees for firewood to the American student consuming reams of paper in the course of getting a degree. For such reasons, people have cleared forests for millennia.

Deforestation has negative impacts, however, and as we saw with Easter Island (pp. 6–7), it has sometimes brought civilizations to ruin. In tropical areas, deforestation causes major losses of biodiversity, and in arid regions it promotes desertification (p. 233). Moreover, forest loss adds carbon dioxide to the atmosphere, contributing to global climate change. In the time it takes you to read this sentence, 2 hectares (5 acres) of tropical forest will have been cleared.

How can we determine how much forest we are losing worldwide? In 2010, the U.N. Food and Agriculture

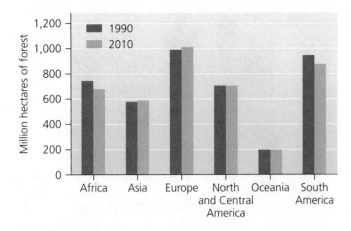

FIGURE 12.6 ▲ Africa and South America are experiencing deforestation as they attempt to develop, extract resources, and provide new agricultural land for their growing populations. In Europe, forested area is slowly increasing as some formerly farmed areas are abandoned and allowed to grow back into forest. Data for North and Central America reflect a balance of forest regrowth in North America and forest loss in Central America. In Asia, natural forests are being lost, but China's extensive planting of tree plantations (here counted as forests) to combat desertification has increased forest cover for Asia since 1990. Data from Food and Agriculture Organization of the United States, *Global forest resources assessment* 2010. By permission.

Organization (FAO) released its latest *Global Forest Resources Assessment*, for which researchers combined remote sensing data from satellites, analysis from forest experts, questionnaire responses, and statistical modeling to form a comprehensive picture of the world's forests. The assessment concluded that we are deforesting 13 million hectares (32 million acres) each year. Subtracting annual regrowth from this amount makes for an annual net loss of 5.2 million hectares (12.8 million acres)—an area about half the size of Kentucky or twice the size of Massachusetts. The good news is that this rate (for the decade 2000–2010) is lower than the deforestation rate for the 1990s, when 8.3 million ha (20.5 million acres) were lost worldwide each year.

Forests are being felled most quickly in the tropical rainforests of Latin America and Africa (**FIGURE 12.6**). Developing nations in these regions are striving to expand areas of settlement for their burgeoning populations and to boost their economies by extracting natural resources and selling them abroad. Moreover, many people in these societies harvest fuelwood for their daily cooking and heating needs (p. 577). In contrast, some areas of Europe and North America are slowly gaining forest cover as they recover from past deforestation.

Deforestation fed the growth of the United States

Deforestation for timber and farmland propelled the phenomenal expansion of the United States and Canada westward across the North American continent over the past 400 years. The vast deciduous forests of the East were virtually stripped of their trees by the mid-19th century, making way for countless small farms. Timber from these forests built the cities of the Atlantic seaboard. Later, cities such as Chicago were constructed with timber felled in the vast pine and hardwood forests of Wisconsin and Michigan.

As a farming economy shifted to an industrial one, wood was used to stoke the furnaces of industry. Logging operations moved south to the Ozarks of Missouri and Arkansas, and then the pine woodlands and bottomland hardwood forests of the South were logged and converted to pine plantations. Once most mature trees were removed from these areas, timber companies moved west, cutting the continent's biggest trees in the Rocky Mountains, the Sierra Nevada, the Cascade Mountains, and the Pacific Coast ranges.

By the 20th century, very little **primary forest**—natural forest uncut by people—remained in the lower 48 U.S. states, and today even less is left (**FIGURE 12.7**). Nearly all

(a) 1620: Areas of primary (uncut) forest

(b) Today: Areas of primary (uncut) forest

FIGURE 12.7 ▲ When Europeans first were colonizing North America **(a)**, the entire eastern half of the continent and substantial portions of the western half were covered in primary forest, shown in green. Today, nearly all this primary forest is gone **(b)**, having been cut to make way for agriculture and to provide timber. Much of the landscape has become reforested with secondary forest, which generally contains smaller trees of different species compositions. *Sources:* (a) adapted from Greeley, W.B., 1925. The relation of geography to timber supply, *Economic Geography* 1:1–11; and (b) Map by George Draffin, www.endgame.org.

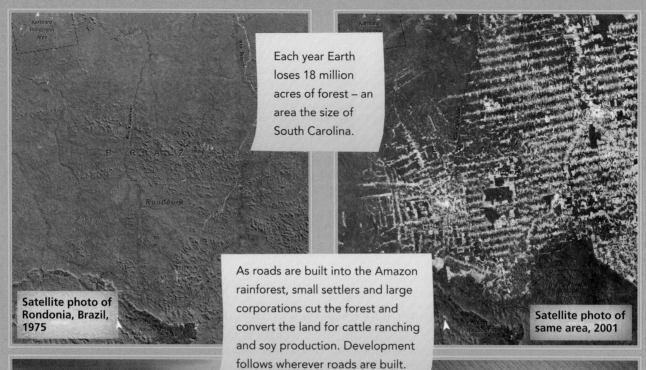

Each year Earth loses 18 million acres of forest – an area the size of South Carolina.

Satellite photo of Rondonia, Brazil, 1975

Satellite photo of same area, 2001

As roads are built into the Amazon rainforest, small settlers and large corporations cut the forest and convert the land for cattle ranching and soy production. Development follows wherever roads are built.

Cattle on burned and cleared land

A soy farm leaves a tiny forest fragment

Forest loss drives biodiversity loss, climate change, erosion, and other problems.

Settlers in the Amazon

Often people can farm the land for only a few years before the soil gives out and they have to cut more forest.

YOU CAN MAKE A DIFFERENCE

➤ Buy certified sustainable wood products.

➤ Eat less meat (raising livestock requires more land than raising crops).

➤ Contribute to forest protection efforts in your region and around the world.

of the largest oaks and maples found in eastern North America today, and even most redwoods of the California coast, are merely *second-growth* trees: trees that have sprouted and grown to partial maturity after old-growth timber was cut. Such second-growth trees characterize **secondary forest**. Secondary forest generally contains smaller trees than does primary forest, and the species composition, structure, and nutrient balance of a secondary forest may differ markedly from the primary forest that it replaced.

The fortunes of loggers have risen and fallen with the availability of big trees. As each region was denuded of trees, the timber industry declined, and the timber companies moved on while local loggers lost their jobs. If the remaining ancient trees of North America—most in British Columbia and Alaska—are cut, many loggers will be out of jobs once again. Their employers will move on to nations of the developing world, as many already have.

Forests are being lost rapidly in developing nations

Uncut tropical forests still remain in many developing countries, and these nations are in the position the United States and Canada enjoyed a century or two ago: having a resource-rich frontier that they can develop. Today's advanced technology, however, allows these countries to exploit their resources and push back their frontiers even faster than occurred in North America. As a result, deforestation is rapid in places such as Brazil, Indonesia, and West Africa.

Developing nations are often desperate enough for economic development and foreign capital that they impose few or no restrictions on logging. Often their timber is extracted by foreign multinational corporations, which pay fees to the developing nation's government for a **concession**, or right to extract the resource. Once a concession is granted, the corporation has little or no incentive to manage forest resources sustainably. Local people may receive temporary employment from the corporation, but once the timber is gone they no longer have the forest and the ecosystem services it provided. Thus, most economic benefits are short term, reaped not by local residents but by the foreign corporation. Moreover, much of the wood extracted from forests in developing nations is not used in those nations, but rather is exported to Europe and North America. Thus, consumption by those of us in wealthy nations is fueling much of the forest destruction in poorer nations.

In Sarawak, in the Malaysian portion of the island of Borneo, foreign corporations granted logging concessions have deforested several million hectares of tropical rainforest since 1963. The clearing of this forest—one of the world's richest, the home of orangutans and countless other organisms—has directly affected the 22 tribes of people who live as hunter-gatherers in Sarawak's rainforest. The Malaysian government did not consult the tribes about the logging, which diminished the wild game on which these people depended. Oil palm agriculture was established afterward, leading to pesticide and fertilizer runoff that killed fish in local streams.

The tribes protested peacefully and eventually began blockading logging roads. The government jailed them at first, then negotiated, but insists on converting the tribes to a farming way of life.

Throughout Southeast Asia and Indonesia today, vast swaths of tropical rainforest are being clear-cut to establish plantations of oil palms (**FIGURE 12.8**). Oil palm fruit produces palm oil, which we use in our snack foods, soaps, cosmetics, and now as a biofuel. In Indonesia, the world's largest palm oil producer, oil palm plantations have displaced over 15 million acres of rainforest. Clearing for plantations encourages further development and eases access for people to enter the forest and conduct logging illegally.

FIGURE 12.8 ▲ Oil palm plantations like this one in Borneo are replacing primary forest across large areas of Southeast Asia and Indonesia. Forest clearing for plantations promotes further development, illegal logging, and forest degradation. Since 1950, the immense island of Borneo (maps at bottom) has lost most of its forest cover. Data from Radday, M., WWF-Germany, 2007. Designed by Hugo Ahlenius, UNEP/GRID-Arendal. Extent of deforestation in Borneo 1950–2001, and projection towards 2020. http://maps.grida.no/go/graphic/extent-of-deforestation-in-borneo-1950–2005-and-projection-towards-2020.

Solutions to deforestation are emerging

A number of solutions are now being proposed to address deforestation in developing nations. For example, we discussed conservation concessions in Chapter 11, whereby a conservation organization buys a concession and uses it to preserve forest rather than cut it down (p. 308).

In Indonesia, NewPage Corporation is teaming up with the nonprofit World Resources Institute (WRI) and is funding a three-year project called POTICO (Palm Oil, Timber and Carbon Offsets) that aims to reduce deforestation and illegal logging. In the POTICO project, WRI will work with palm oil companies that own concessions to clear virgin rainforest, steering them instead to land that is already logged and degraded. It will then protect the forests that were slated for conversion or else allow FSC-certified sustainable forestry in them. WRI and NewPage hope this will encourage oil palm plantations to become ecologically and economically sustainable while preserving primary rainforest. And because primary forest stores far more carbon than oil palm plantations, these land swaps can reduce Indonesia's greenhouse gas emissions and qualify for credit via carbon offsets (p. 524).

Carbon offsets are a key to some emerging international plans to curb deforestation and climate change together. Forest loss accounts for an estimated 12–25% of the world's greenhouse gas emissions—as much as all the world's vehicles emit—yet the Kyoto Protocol (pp. 520–521) did not address it. Thus, for the 2009 Copenhagen climate conference (pp. 521–522), negotiators outlined a program called *Reducing Emissions from Deforestation and Forest Degradation (REDD)*, whereby wealthy industrialized nations would pay poorer developing nations to conserve forest. Under this plan, the poor nations would gain much-needed income while the rich nations would receive carbon credits to offset their emissions in an international cap-and-trade system (pp. 183, 523–524).

The Copenhagen conference ended without a binding agreement, so the REDD plan fell through. However, leaders of rich nations did agree to transfer $100 billion per year to poor nations by 2020, and much of this could end up going toward REDD.

The small South American nation of Guyana—poor financially but rich in forests—is taking a leading role in international discussions of REDD. Guyana's president Bharrat Jagdeo in 2008 commissioned the consulting firm McKinsey and Company to calculate the amount of money his nation would make if it cut down its forests for agriculture. The figure came to $580 million per year over 25 years. In a free market under purely financial considerations, this is the amount that wealthy nations "should" pay Guyana to forego cutting its forest. Guyana has already forged a deal with Norway in which Norway will pay Guyana $30 million in 2010 for conserving its forest, and up to $250 million in 2015, depending on Guyana's progress in reducing forest loss.

FOREST MANAGEMENT

Because our demand for forest resources is rising even as we clear and alter our remaining forests, we need to manage the forests from which we take resources. *Foresters* are professionals who manage forests through the practice of **forestry** (also called **silviculture**), and they must balance our society's demand for forest products against the central importance of forests as ecosystems. As a society, we must weigh the long-term value of preserving forests against the short-term benefits of clearing them.

The good news is that sustainable forest management practices are spreading as informed consumers demand sustainably produced products. Just as your textbook uses FSC-certified paper from sustainably managed forests, more and more paper, lumber, and other forest products are now made using certified sustainable practices. In this way, consumer choice is influencing the ways forests are managed.

Forest management is one type of resource management

Debates over how to manage forest resources reflect broader questions about how to manage natural resources in general. We need to manage the resources we take from the natural world because many of them are limited. We have seen (pp. 3–4) how resources such as fossil fuels and many minerals are nonrenewable on human time scales, whereas resources such as the sun's energy are perpetually renewable.

Between these extremes lie resources that are renewable if they are not exploited too rapidly or carelessly. Besides timber, these resources include soils, fresh water, rangeland, wildlife, and fisheries.

Elsewhere in this book we examine how we can conserve soil resources with farming practices that fight erosion (pp. 237–241); safeguard the supply and quality of surface water and groundwater (pp. 418–428); encourage low-impact grazing to manage rangeland sustainably (pp. 244–247); and protect fisheries and wildlife from overharvesting (pp. 300–310, 452–456). We also examine how all these resources serve functions in ecosystems.

Resource management describes our use of strategies to manage and regulate the harvest of renewable resources. Sustainable resource management is the practice of harvesting renewable resources in ways that do not deplete them. Resource managers are guided by research in the natural sciences, but social, political, and economic factors also influence their decisions.

Resource managers follow several strategies

A key question in managing resources is whether to focus narrowly on the resource of interest or to look more broadly at the environmental system of which it is a part. Taking a broader view often helps avoid damaging the system and thereby helps sustain the resource in the long term.

Maximum sustainable yield Traditionally, a guiding principle in resource management has been **maximum sustainable yield**. Its aim is to achieve the maximum amount of resource extraction without depleting the resource from one harvest to the next. Recall the logistic growth curve (see Figure 3.16, p. 67), which reflects how the limiting factors in the environment slow exponential population growth and then cap it at a carrying capacity. The logistic curve indicates that a population grows most quickly when it is at an intermediate size—specifically, at one-half of carrying capacity. A fisheries manager aiming for maximum sustainable yield will therefore prefer to keep fish populations at intermediate levels so that they rebound quickly after each harvest. Doing so should result in the greatest amount of fish harvested over time while sustaining the population indefinitely (**FIGURE 12.9**).

This management approach, however, keeps the fish population at only half its carrying capacity—well below the size it would attain in the absence of fishing. Reducing one population in this way will likely affect other species and alter the food web dynamics of the community. From an ecological point of view, management for maximum sustainable yield may set in motion significant ecological changes.

In forestry, maximum sustainable yield argues for cutting trees shortly after they go through their fastest stage of growth, and trees often grow most quickly at intermediate ages. Thus, trees are generally cut long before they have grown as large as they would in the absence of harvesting. This practice maximizes timber production over time, but it also alters forest ecology and eliminates habitat for species that depend on mature trees.

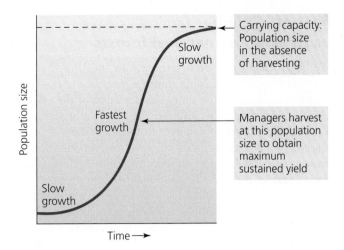

FIGURE 12.9 ▲ Using the concept of maximum sustainable yield, resource managers attempt to maximize the amount of resource harvested while keeping the harvest sustainable in perpetuity. For a wildlife population or fisheries stock that grows according to a logistic growth curve, managers aim to keep the population at roughly half the carrying capacity, because populations grow fastest at intermediate sizes.

Ecosystem-based management Because of these dilemmas, increasing numbers of managers today espouse **ecosystem-based management**, which attempts to manage resource harvesting so as to minimize impact on the ecosystems and ecological processes that provide the resource. Many sustainably certified forestry plans protect certain forested areas, restore ecologically important habitats, and consider patterns at the landscape level (p. 119), allowing timber harvesting while preserving the functional integrity of the forest ecosystem. This means fostering the area's ecological processes, including succession (pp. 92–93), in which the forest community naturally changes over time.

It can be challenging, however, to determine how best to implement this type of management. Ecosystems are complex, and our understanding of how they operate is limited. Thus, ecosystem-based management has come to mean different things to different people.

Adaptive management Some management actions will succeed, and some will fail. A wise manager will try new approaches if old ones are not effective. **Adaptive management** involves systematically testing different approaches and aiming to improve methods through time. For managers, it entails monitoring the results of one's practices and adjusting them as needed, based on what is learned. Adaptive management is intended as a true fusion of science and management, because hypotheses about how best to manage resources are explicitly tested. This process can be time-consuming and complicated, but if followed as intended, adaptive management can be highly effective.

Adaptive management was featured in the *Northwest Forest Plan*, a 1994 plan crafted by the administration of President Bill Clinton to resolve disputes between loggers and preservationists over the last remaining old-growth temperate rainforests in the continental United States. This plan sought to allow limited logging to continue in the Pacific Northwest, with adequate protections for species such as the spotted owl (p. 303), and to let science guide management.

Fear of a "timber famine" drove us to establish national forests

We began managing forest resources in the United States around the turn of the 20th century in response to the rampant deforestation of the time. The depletion of the eastern U.S. forests prompted widespread fear of a "timber famine." This led the federal government to form a system of forest reserves: public lands set aside to grow trees, produce timber, protect water quality, and serve as insurance against scarcities of lumber. Today the U.S. **national forest** system consists of 77 million ha (191 million acres) spread across all but a few states (**FIGURE 12.10**). The system, managed by the U.S. Forest Service, covers over 8% of the nation's land area.

The U.S. Forest Service was established in 1905 under the leadership of Gifford Pinchot (p. 143). Pinchot and others developed the concepts of resource management, maximum sustainable yield, and conservation during the Progressive Era, a time of social reform when people applied science to public policy to improve society. In line with Pinchot's conservation ethic (p. 143), the Forest Service aimed to manage the forests for "the greatest good of the greatest number in the long run." Pinchot believed the nation should extract and use resources from its public lands, so timber harvesting was, from the start, a goal of the national forests. But conservation meant planting trees as well as harvesting them, and the Forest Service set out to manage its timber resources wisely.

Timber is extracted from public and private lands

Private timber companies extract timber from the U.S. national forests, as well as from publicly held state forests. U.S. Forest Service employees manage timber sales and build roads to provide access for logging companies, which sell the timber they harvest for profit. Thus, taxpayers subsidize private timber harvesting on public land, as examined in our discussion of subsidies in Chapter 7 (pp. 182–183).

Timber companies extracted 10.7 million m³ (378 million ft³) of live timber from national forests in 2006, the most recent year for which comprehensive data are available. However, this is less than the amount harvested from other public lands, and it is much less than the amount cut on private lands (**FIGURE 12.11**). The vast majority of timber harvesting in the United States today takes place on private land owned by the timber industry and private land owned by small landowners. Logging has declined on national forests since the 1980s, and in 2006 tree regrowth outpaced tree removal on these lands by 11 to 1. Overall, timber harvesting in the United States and other developed nations has remained stable for the past 40 years, while it has more than doubled in developing countries.

On timber industry land, companies manage their resources in accordance with maximum sustainable yield, so as to obtain maximal profits over many years. On public lands, rates of tree removal and growth reflect social and political factors as well as economic ones, and these evolve over time. On the U.S. national forests, private timber extraction began to increase in the 1950s as the nation experienced a postwar economic boom, paper consumption rose, and the population expanded into newly built suburban homes. More recently, harvests from national forests decreased as economic trends shifted, public concern over clear-cutting grew, and forest management philosophy evolved.

Note, however, that even when regrowth outpaces removal, the character of forests may change. Once primary forest is lost and is replaced by younger secondary forest or single-species plantations, the resulting forest may be very different from the original forest and generally is less ecologically valuable.

FIGURE 12.10 ▶ Federal agencies own and manage well over 250 million ha (600 million acres) of land in the United States, particularly in the western states. These include national forests, national parks, national wildlife refuges, Native American reservations, and Bureau of Land Management lands. Data from United States Geological Survey.

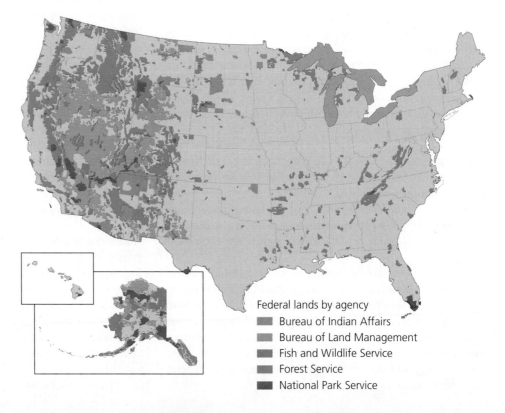

Federal lands by agency
- Bureau of Indian Affairs
- Bureau of Land Management
- Fish and Wildlife Service
- Forest Service
- National Park Service

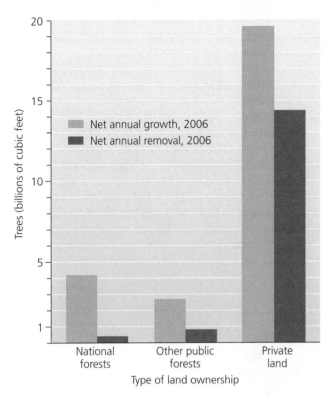

FIGURE 12.11 ▲ As the United States recovers from deforestation, trees (measured in cubic feet of wood biomass) are growing at a faster rate than they are being removed, particularly on national forests and other public lands. "Private land" in this figure includes land owned by the timber industry and by small landholders. However, it is important to remember that the forests that regrow after logging often differ substantially from the forests that were removed. Data from USDA Forest Service, 2008. *Forest resources of the United States, 2007.* U.S. Department of Agriculture, Washington, D.C.

Plantation forestry has grown

Today the U.S. timber industry focuses on production in the South (and to a lesser degree the Pacific coast states and the northern states) from plantations of fast-growing tree species planted in single-species monocultures (pp. 255–256). Because all trees in a given stand are planted at the same time, the stands are **even-aged**, with all trees the same age (**FIGURE 12.12**). Stands are cut after a certain number of years (called the *rotation time*), and the land is replanted with seedlings. Such plantation forestry is growing quickly worldwide, and today fully 7% of the world's forests are plantations. One-quarter of these feature non-native tree species.

Ecologists and foresters view plantations more as crop agriculture than as ecologically functional forests. Because there are few tree species and little variation in tree age, plantations do not offer many forest organisms the habitat they need. For instance, stands of red pine planted near Escanaba, Michigan, host far less biodiversity than the more-diverse forests of multiple tree species that surround them. Even-aged single-species plantations managed for timber production lack the structural complexity that characterizes a healthy natural forest as seen in Figure 12.3 (p. 317). Plantations are also vulnerable to outbreaks of pest species such as the pine beetle, as we shall soon see. For all these reasons, some harvesting

FIGURE 12.12 ▲ Even-aged stand management is practiced on tree plantations where all trees are of equal age, as seen in the stand in the foreground that is regrowing after clear-cutting. In uneven-aged tree stand management, harvests are designed to maintain a mix of tree ages, as seen in the more mature forest in the background. The increased structural diversity of uneven-aged stands provides superior habitat for most wild species and, if diverse tree species also are intermixed, makes these stands more akin to ecologically functional forests.

methods aim to maintain **uneven-aged** stands, where a mix of ages (and often a mix of tree species) makes the stand more similar to a natural forest.

We harvest timber by several methods

When timber companies harvest trees, they may choose from several methods. In the simplest method, **clear-cutting**, all trees in an area are cut, leaving only stumps (**FIGURE 12.13**). Clear-cutting is the most cost-efficient method in the short term, but it has the greatest ecological impacts. In the best-case scenario, clear-cutting may mimic natural disturbance events

FIGURE 12.13 ▼ Clear-cutting is cost-efficient for timber companies but can have severe ecological consequences, including soil erosion, water pollution, and altered community composition. Some species use clear-cuts as they regrow, but most people find these areas aesthetically unappealing, and public reaction to clear-cutting has driven changes in forestry methods.

such as fires, tornadoes, or windstorms that knock down trees across large areas. In the worst-case scenario, entire ecological communities are destroyed, soil erodes, and the penetration of sunlight to ground level changes microclimatic conditions such that new types of plants replace those of the original forest. Clear-cutting essentially sets in motion an artificially driven process of succession (pp. 92–93) in which the resulting climax community may be quite different from the original climax community.

Clear-cutting occurred widely across North America in the mid- and late-20th century as public awareness of environmental issues was blossoming. The combination produced public outrage toward the timber industry and forest managers. Eventually the industry integrated other harvesting methods (**FIGURE 12.14**). Clear-cutting (**FIGURE 12.14A**) is still widely practiced, but alternative methods involve cutting some trees and leaving others standing. In the *seed-tree* approach (**FIGURE 12.14B**), small numbers of mature and vigorous seed-producing trees are left standing so that they can reseed the logged area. In the *shelterwood* approach (also represented by Figure 12.14b), small numbers of mature trees are left in place to provide shelter for seedlings as they grow. These three methods all lead to even-aged stands of trees.

In contrast, *selection systems* (**FIGURE 12.14C**) allow uneven-aged stand management, because only some trees are cut at any one time. In single-tree selection, widely spaced trees are cut one at a time, whereas in group selection, small patches of trees are cut. The stand's overall rotation time may be the same as in an even-aged approach, because multiple harvests are made, but the stand remains mostly intact between harvests. Selection systems maintain uneven-aged stands, but they are not ecologically harmless. Moving

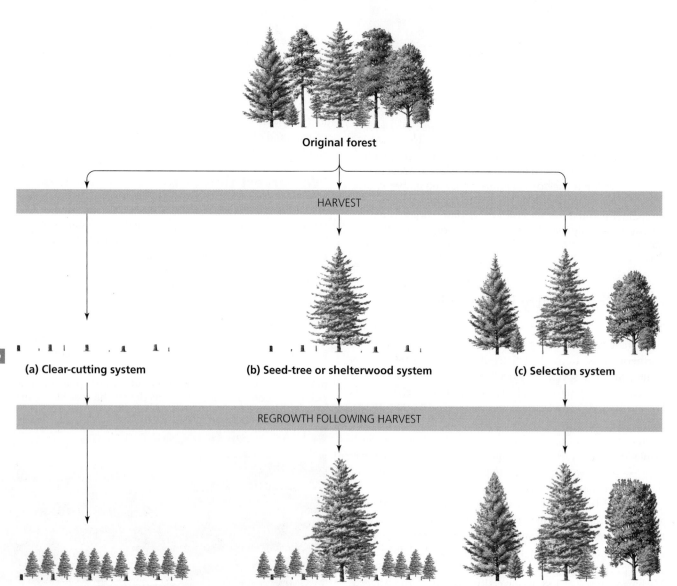

FIGURE 12.14 ▲ Foresters have devised various methods to harvest timber from forests. In clear-cutting **(a)**, all trees in an area are cut, extracting a great deal of timber inexpensively but leaving a vastly altered landscape. In seed-tree systems and shelterwood systems **(b)**, small numbers of large trees are left in clear-cuts to help reseed the area or provide shelter for growing seedlings. In selection systems **(c)**, a minority of trees is removed at any one time, while most are left standing. These latter methods involve less environmental impact than clear-cutting, but all methods can cause significant changes to the structure and function of natural forest communities.

trucks and machinery over a network of roads and trails to access individual trees compacts the soil and disturbs the forest floor. Selection methods are also unpopular with timber companies because they are expensive, and with loggers because they are more dangerous than clear-cutting.

All methods of logging disturb habitat and affect an area's plants and animals. All methods alter forest structure and composition. For example, when we log an eastern U.S. oak-hickory forest, the forest that regrows is often dominated by maples, beech, and tulip trees. Most logging methods speed runoff, cause flooding, and increase soil erosion, leading to the siltation of waterways, which degrades habitat and affects water quality. When steep hillsides are clear-cut, landslides can result.

Public forests may be managed for recreation and ecosystems

In recent decades, public awareness of the impacts of logging has grown, and citizens have urged that national and state forests be managed for recreation, wildlife, and ecosystem integrity, rather than only for timber. Many citizens also do not want taxpayers subsidizing the extraction of publicly held resources by private corporations. Scientists analyzing government subsidies have concluded that the U.S. Forest Service loses at least $100 million of taxpayers' money each year by selling timber below the costs it incurs for marketing and administering the harvest and for building access roads. Subsidies also inflate harvest levels beyond what would occur in a free market.

On paper, the U.S. Forest Service has long had a policy of attending to interests besides timber production. For the past half-century, forest management has nominally been guided by the policy of **multiple use**, meaning that the national forests were to be managed for recreation, wildlife habitat, mineral extraction, and various other uses. In reality, however, timber production was most often the primary use.

In 1976 the U.S. Congress passed the **National Forest Management Act**, which mandated that every national forest draw up plans for renewable resource management. These plans were to be explicitly based on the concepts of multiple use and sustained yield and were to be subject to public input under the National Environmental Policy Act (p. 176). Guidelines specified that these plans:

▶ Consider both economic and environmental factors.

▶ Provide for diverse ecological communities and preserve the regional diversity of tree species.

▶ Ensure research and monitoring of management practices.

▶ Permit increases in harvest levels only if sustainable.

▶ Ensure that timber is harvested only where soils and wetlands will not be irreversibly damaged, lands can be restocked quickly, and profit alone does not guide the choice of harvest method.

▶ Ensure that logging is conducted only where impacts have been assessed; cuts are shaped to the terrain; maximum size limits are established; and cuts do not threaten timber regeneration or soil, watershed, fish, wildlife, recreation, or aesthetic resources.

Following passage of the National Forest Management Act, the U.S. Forest Service developed new programs to manage

FIGURE 12.15 ▲ Ecosystem-based management is practiced in this longleaf pine forest, an ecosystem of the southeastern United States. Foresters and biologists restore and nurture mature longleaf pine trees while burning and removing brush from the understory. This habitat is home to a number of specialized species, including the endangered red-cockaded woodpecker (inset).

wildlife, non-game animals, and endangered species. It pushed for ecosystem-based management and ran programs of ecological restoration, attempting to recover plant and animal communities that had been lost or degraded (**FIGURE 12.15**). Timber harvesting methods were brought more in line with ecosystem-based management goals. A set of approaches dubbed **new forestry** called for timber cuts that mimicked natural disturbances. For instance, "sloppy clear-cuts" that leave a variety of trees standing were intended to mimic the changes a forest might experience if hit by a severe windstorm.

The Hiawatha National Forest provides a typical example of management under the National Forest Management Act. Headquartered in Escanaba, the Hiawatha National Forest covers 360,000 ha (895,000 acres) in Michigan's Upper Peninsula. The 2006 revision of its forest management plan included a wide range of goals and guidelines that together seek to achieve a balance of multiple uses in the forest (**TABLE 12.1**).

Forest management is subject to political influence, however. In 2004, President George W. Bush's administration freed forest managers from many requirements of the National Forest Management Act, granting them more flexibility in managing forests, but loosening environmental protections and restricting public oversight.

Then in 2005, the Bush administration repealed the *roadless rule* of President Bill Clinton's administration, by which 23.7 million ha (58.5 million acres)—31% of national forest land and 2% of total U.S. land—were in 2001 put off-limits to further road construction or maintenance (and thus off-limits to logging). The roadless rule had been supported by a record 4.2 million public comments, but the Bush administration overturned it and required state governors to petition the federal government if they wanted to keep areas in their states roadless. Several states sued, asking that the roadless rule be reinstated, and after a series of court rulings, most of the roadless policy was reinstated in 2009.

TABLE 12.1 Goals and Guidelines from the Management Plan of the Hiawatha National Forest in Michigan's Upper Peninsula

Timber harvesting

▶ Harvest timber for mills in region, and replant seedlings.

▶ Use clear-cutting, shelterwood, and selection systems, according to location and tree type.

▶ Harvest trees according to specified rotation ages (e.g., 35–70 years for aspen but 60–160 years for white pine).

▶ Manage a mix of even-aged and uneven-aged stands of maple, beech, red pine, and other trees for timber and wildlife habitat.

▶ Harvest timber in ways that simulate natural disturbances.

▶ Protect 52,000 acres of old-growth forest.

Ecosystem management

▶ Restore 300 acres of wetlands and 9–13 miles of streams.

▶ Conduct prescribed fire and remove brush on 1,000 acres per year.

▶ Minimize soil damage from logging and other activities.

▶ Do not allow livestock grazing.

▶ Protect historic and archaeological sites, caves, ponds, rivers, snags, water quality, and wilderness areas.

Fish and wildlife

▶ Monitor fish and wildlife populations.

▶ Maintain 6,700 acres of young jack pine stands for endangered Kirtland's warbler.

▶ Maintain habitats for threatened plants and animals, including lakeside daisy, Canada lynx, yellow rail, and others.

▶ Control invasive species, especially forest pests.

Recreation and other

▶ Allow diverse recreation—hiking, fishing, boating, snowmobiling, dog mushing, and more—but specify areas for recreation (e.g., off-road vehicles in some places, no vehicles in others).

▶ Build no new roads; reconstruct 10 miles per year; decommission 5 miles per year.

▶ Acquire private land (inholdings) within wilderness areas when possible.

Adapted from USDA Forest Service, 2006. Hiawatha National Forest 2006 Forest Plan.

Fire policy also stirs controversy

Another area of policy debate involves how to handle wildfire. We all know that Smokey Bear, the Forest Service's beloved cartoon bear in a ranger's hat, advises us to fight forest fires. Alas, many scientists assert that Smokey's message has harmed American forests.

For over a century, the Forest Service and other agencies suppressed fire whenever and wherever it broke out. Yet ecological research now clearly shows that many species and communities depend on fire. Certain plants have seeds that germinate only in response to fire, and researchers studying tree rings have documented frequent fires in North America's grasslands and pine woodlands. (Burn marks in a tree's rings reveal past fires, giving scientists an accurate history of fire events extending back hundreds or even thousands of years.) Ecosystems dependent on fire are adversely affected by its suppression. Pine woodlands become cluttered with hardwood

FIGURE 12.16 ▲ The suppression of fire by people over the past century has led to a buildup of leaf litter, woody debris, and young trees, which serve as fuel to increase the severity of fires when they do occur. As a result, catastrophic wildfires that damage ecosystems and threaten homes in the wildland-urban interface (such as this fire in southern California in 2007) have become more common in recent years. To avoid these unnaturally severe fires, most fire ecologists suggest allowing natural fires to burn when possible and instituting controlled burns to reduce fuel loads and restore forest ecosystems.

understory that ordinarily would be cleared away by fire, for instance, and animal diversity and abundance decline.

In the long term, fire suppression can lead to catastrophic fires that truly do damage forests, destroy human property, and threaten human lives (**FIGURE 12.16**). Fire suppression allows limbs, logs, sticks, and leaf litter to accumulate on the forest floor, producing kindling for a catastrophic fire. Such fuel buildup worsened the 1988 fires in Yellowstone National Park, the 2009 fires in southern California, and thousands of other wildfire episodes. Catastrophic fires have become more numerous in recent years (**FIGURE 12.17**). At the same time,

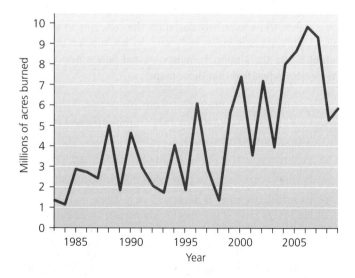

FIGURE 12.17 ▲ Wildfires have become more numerous and have consumed more acreage in the United States in recent years, as shown by this graph of acres burned each year throughout the nation. Decades of fire suppression and resulting fuel buildup have contributed to this trend. Data from National Interagency Fire Center.

increased residential development along the edges of forested land—in the so-called **wildland-urban interface**—is placing more homes in fire-prone situations.

To reduce fuel load and improve the condition of forests, the Forest Service and other agencies burn areas of forest under carefully controlled conditions. These **prescribed burns**, or **controlled burns**, clear away fuel loads, nourish the soil with ash, and encourage the vigorous growth of new vegetation. Because they are time-intensive, however, prescribed burns are conducted on only a small proportion of land (about 2 million acres per year). All too often, these worthy efforts have been impeded by public misunderstanding and by interference from politicians who have not taken time to understand the science behind the approach.

In the wake of major fires in California in 2003, the U.S. Congress passed the Bush Administration's *Healthy Forests Restoration Act*. Although this legislation encouraged some prescribed burning, it primarily promoted the physical removal of small trees, underbrush, and dead trees by timber companies. The removal of dead trees, or snags, following a natural disturbance (such as a fire, windstorm, insect damage, or disease) is called **salvage logging**. From an economic standpoint, salvage logging may seem to make good sense. However, ecologically, snags have immense value; the insects that decay them provide food for wildlife, and many animals depend on holes in snags for nesting and roosting. Removing timber from recently burned land can also cause erosion and soil damage. Moreover, it may impede forest regeneration and promote further wildfires (see **THE SCIENCE BEHIND THE STORY**, pp. 330–331).

Climate change is altering forests

Global climate change (Chapter 18) is now bringing warmer weather to most of North America and drier weather to much of the American West. Scientific climate models predict further warming and drying (pp. 407, 502). This is expected to worsen wildfire risk, especially in the western states. Indeed, research already suggests that the recent increase in fires in North America has been driven in part by warmer, drier climate (**FIGURE 12.18A**).

Climate change also appears to be driving the dynamics of certain pest insects that destroy forest trees. Pine beetles are small black beetles that feed within the bark of conifer trees. Milder winters and hotter, drier summers have resulted in major outbreaks of the mountain pine beetle. These infestations can rapidly kill huge areas of trees (**FIGURE 12.18B**).

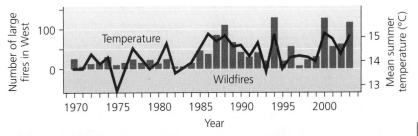

(a) Wildfires correlate with hotter summer temperatures

FIGURE 12.18 ▶ Global climate change is affecting North American forests by increasing fires due to heat and drought and by promoting outbreaks of insect pests. The number of large wildfires (>400 ha [1,000 acres]) in the western United States **(a)** has increased along with the region's summer temperatures. Throughout the West, immense areas of trees have been killed **(b)** by the mountain pine beetle (inset), which thrives with warm summers and mild winters. *Source* (a): Westerling, A.L., et al., 2006. Warming and earlier spring increase western U.S. forest wildfire activity. *Science* 313: 940–943, Fig 1A. Reprinted with permission from AAAS.

(b) Mountain pine beetles kill more trees in a warmer climate

The SCIENCE behind the Story

Fighting over Fire and Forests

It's not often that a single scientific paper throws an entire college into turmoil and lands a graduate student in a Congressional hearing to face hostile questioning from federal lawmakers. But such is the political sensitivity of salvage logging.

When a fire burns a forest, should the killed trees be cut and sold for timber? Proponents of salvage logging say yes: we should not let economically valuable wood go to waste. Opponents of post-fire logging say the burned wood is more valuable left in place—for erosion control, wildlife habitat (snags provide holes for cavity-dwelling animals and food for insects and birds), and organic material to enhance the soil and nurse the growth of future trees.

Proponents of salvage logging argue that forests regenerate best after a fire if they are logged and replanted with seedlings. Moreover, they maintain, salvage logging reduces fire risk by removing woody debris that could serve as fuel for the next fire. When the Biscuit Fire consumed 200,000 ha (500,000 acres) in Oregon in 2002, foresters from the College of Forestry at Oregon State University (OSU) made these arguments in support of plans to log portions of the burned area.

OSU graduate student Daniel Donato testifies in a Congressional hearing

Meanwhile, OSU forestry graduate student Daniel Donato and five other OSU researchers were setting up research plots in areas burned by the Biscuit Fire to test whether salvage logging really does reduce fire risk and help seedlings regenerate. They measured seedling growth and survival and the amount of woody debris in a number of study plots on burned land before (2004) and after (2005) salvage logging took place, and on burned land that was not logged.

The researchers found that conifer seedlings sprouted naturally in the burned areas at densities exceeding what foresters aim for when they replant sites manually. This suggested that artificial planting of seedlings may be unnecessary. In contrast, natural seedling densities in logged areas were only 29% as high (**first figure**). This indicated that salvage logging was hindering seedling survival,

presumably because logging disturbs the soil and crushes many seedlings.

Donato's team also found that salvage logging more than tripled the amount of woody debris on the ground relative to unlogged sites (**second figure**). The research team suggested that the best strategy after fire may be to leave the site alone and leave dead trees standing, so that seedlings regenerate safely.

These conclusions directly contradicted what some OSU forestry professors had argued following the

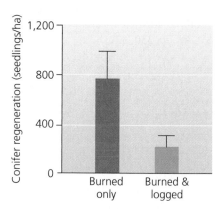

In the wake of the Biscuit Fire, natural growth of conifer seedlings was lower in areas (orange bar) that underwent salvage logging. Data from Donato, D.C., et al., 2006. Post-wildfire logging hinders regeneration and increases fire risk. *Science* 311: 352, Fig 1A. Reprinted with permission from AAAS.

Most at risk are plantation forests dominated by single species that the beetles prefer. Mountain pine beetle outbreaks devastated more than 11 million ha (27 million acres) in the western United States and Canada between 1990 and 2010. Pine beetle outbreaks also create a positive feedback loop (p. 110); by killing trees, they reduce the amount of carbon dioxide pulled from the air, and thereby intensify climate change.

Together, the factors driven by climate change could alter the nature of forest ecosystems in profound ways. Some dense, moist

forests could eventually be replaced by drier woodlands, shrublands, grasslands, or novel types of ecosystems not seen today.

Sustainable forestry is gaining ground

All these challenges can be addressed through the practices of sustainable forestry. Although the world overall is losing forested land, a number of advances are being made toward sustainable forestry practices (**FIGURE 12.19**).

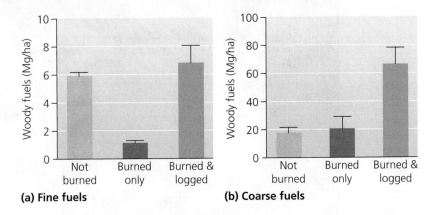

(a) Fine fuels **(b) Coarse fuels**

Burned sites that were salvage-logged (orange bar) contained more fine woody debris **(a)** and coarse woody debris **(b)** than unlogged burned sites (red bar) or than sites that did not burn (yellow bar). Data from Donato, D.C., et al., 2006. Post-wildfire logging hinders regeneration and increases fire risk. *Science* 311: 352, Fig 1B. Reprinted with permission from AAAS.

2002 fire. When they learned that the prestigious journal *Science* had accepted Donato's team's paper for publication, they took the unusual step of asking the journal's editors to reconsider their decision. Claiming that *Science*'s peer review process had failed, professor John Sessions and others tried through back channels to stop publication of the paper—actions that were widely condemned as an attempt at censorship.

The paper's publication in 2006 unleashed a torrent of bizarre events. U.S. Congressmen Greg Walden of Oregon and Brian Baird of Washington felt that the paper threatened legislation they had sponsored to accelerate salvage logging. Walden and Baird called Donato and others before a hearing of the House of Representatives' Committee on Resources and grilled them before a packed crowd in Medford, Oregon. There the 29-year-old graduate student stood up for his team's findings under harsh questioning.

The Bureau of Land Management (BLM) suspended the team's research grant, in what many viewed as a response to pressure from congressmen or the Bush administration. This highly unusual action drew media attention, and the BLM quickly reinstated the funding.

A heated debate roiled for months in the OSU College of Forestry. The college receives 12% of its funding from taxes on timber sales, leading many to suggest that the college is open to influence from industry. E-mail correspondence was subpoenaed and showed the college's dean, Hal Salwasser, collaborating with timber industry representatives to refute the paper. As publicity built, the college's reputation suffered, graduate admissions declined, and a faculty committee on academic freedom criticized Salwasser for "significant failures of leadership." The dean admitted mistakes, survived a no-confidence vote of the faculty, and pledged to make reforms.

In the pages of *Science* and elsewhere, scientific criticisms of the study's methods were largely rebutted, yet many felt that the study's conclusions stretched beyond what its short-term data could support. Scientists on both sides of the debate agreed that long-term research was needed to truly assess the effects of salvage logging on forest regeneration and fire risk.

Two studies by different OSU forestry scientists soon provided the first such long-term data. Jeffrey Shatford and colleagues documented widespread natural regrowth of conifers across areas of Oregon and northern California that had burned 9–19 years earlier. And Jonathan Thompson and colleagues examined satellite data, aerial photography, and government records for regions within the Biscuit Fire area that had burned in a previous fire 15 years earlier. They found that of the regions burned in that 1987 fire, those that were salvage logged burned more severely in 2002 than regions that were not logged. In a paper in the *Proceedings of the National Academy of Sciences*, Thompson's team concluded that salvage logging increases the risk of severe fires, even when debris is removed and seedlings are manually planted.

More long-term studies on the impacts of salvage logging are forthcoming. In the end, these should help us to better understand and manage our forests amid increasingly frequent wildfire. ∎

Any company can claim that its timber harvesting practices are sustainable, but how is the purchaser of wood products to know whether they really are? Organizations such as the Forest Stewardship Council (FSC) examine the practices of firms and rate them against criteria for sustainability. They then grant **sustainable forest certification** to forests, companies, and products produced using methods they consider sustainable.

A variety of certification organizations exist, but the Forest Stewardship Council is considered to have the strictest standards. Thus, all of the products produced by NewPage's Escanaba mill attain certification by the Sustainable Forestry Initiative® program, but only some achieve the stricter FSC certification. The paper for this textbook is "chain-of-custody certified" by the FSC, meaning that all steps in the life cycle of the paper's production—from timber harvest to transport

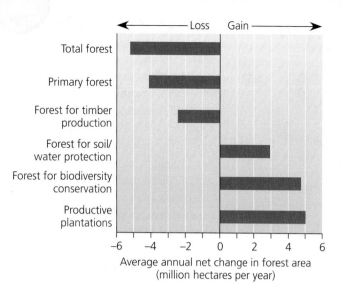

FIGURE 12.19 ▲ Progress worldwide toward sustainable forestry is mixed, as shown by six leading indicators. On the one hand, forest uses are shifting toward conservation, and timber production is increasingly coming from plantations. On the other hand, global forest area is still shrinking. Data from U.N. Food and Agriculture Organization, 2010. *Global forest resources assessment 2010.* Data reflect averages as reported by FAO for most-recent years, ending in 2010 but extending back variably to 2005, 2000, or 1990.

FIGURE 12.20 ▲ An inspector marks wood with the FSC logo after confirming that it was harvested in accord with FSC criteria for sustainable harvesting. A consumer movement centered on independent certification of sustainable wood products is allowing consumer choice to promote sustainable forestry practices.

to pulping to production—have met strict standards. FSC certification rests on 10 general principles (**TABLE 12.2**) and 56 more-detailed criteria.

The number of FSC-certified forests, companies, and products is growing quickly. Five percent of the world's forests managed for timber production are now FSC certified. This totaled over 128 million ha (320 million acres) in 80 nations as of spring 2010. FSC-labeled products accounted for over $20 billion in sales in 2008. About 1,000 operations are certified, and over 16,000 companies are chain-of-custody certified. Some of the

growth of certification is tied to the increasing construction of sustainable, or green, buildings (pp. 363, 366).

This expanding participation means that FSC-certified products are becoming easier for consumers to find. Consumer demand for sustainable wood has been great enough that many major retail businesses such as Home Depot now sell sustainable wood. The purchasing decisions of such retailers are influencing the practices of timber companies. You can look for the logos of certifying organizations (**FIGURE 12.20**) on forest products where they are sold. By asking retailers whether they carry certified wood, you make businesses aware of your preferences and help drive demand for these products. Through our requests and our purchases, we as consumers can influence practices in forests from Michigan to Malaysia to Manaus. Sustainable forestry is more costly for the timber industry, but if certification standards are kept adequately strong, then consumer choice in the marketplace can be a powerful driver of sustainable forestry practices.

PARKS AND PROTECTED AREAS

As our world fills with more people consuming more goods, the conservation and sustainable management of resources from forests and other ecosystems becomes ever more important. So does our need to preserve land and functional ecosystems by setting aside tracts of undisturbed land intended to remain forever undeveloped.

Preservation has been part of the American psyche ever since John Muir rallied support for saving scenic lands in the Sierras (p. 143). For ethical reasons as well as pragmatic ecological and economic ones, Americans and people worldwide have chosen to set aside areas of land in perpetuity to be protected from development. Today 12% of the world's land area is designated for preservation in various types of parks, reserves, and protected areas.

TABLE 12.2 Ten Principles of Forest Stewardship Council (FSC) Certification

To receive FSC certification, forest product companies must:

▶ Comply with all laws and treaties.

▶ Show uncontested, clearly defined, long-term land rights.

▶ Recognize and respect indigenous peoples' rights.

▶ Maintain or enhance long-term social and economic well-being of forest workers and local communities, and respect workers' rights.

▶ Use and share benefits derived from the forest equitably.

▶ Reduce environmental impacts of logging and maintain the forest's ecological functions.

▶ Continuously update an appropriate management plan.

▶ Monitor and assess forest condition, management activities, and social and environmental impacts.

▶ Maintain forests of high conservation value.

▶ Promote restoration and conservation of natural forests.

Why create parks and reserves?

We can cite a number of reasons why people establish parks and protected areas:

▶ Enormous or unusual scenic features such as the Grand Canyon, Mount Rainier, or Yosemite Valley inspire people to preserve them (**FIGURE 12.21**).

▶ Protected areas offer recreational value for hiking, fishing, hunting, kayaking, and other pursuits.

▶ Parks generate revenue from ecotourism (p. 72).

▶ Undeveloped land offers us peace of mind, health, exploration, wonder, and spiritual solace. Children especially benefit from healthy exposure to the outdoors (p. 299).

▶ Protected areas offer utilitarian benefits through ecosystem services. For example, undeveloped watersheds provide cities with clean drinking water and a buffer against floods.

▶ Reserves protect biodiversity. A park or reserve is a kind of Noah's Ark, an island of habitat that can maintain species (and communities and ecosystems) that human impact might otherwise cause to disappear.

FIGURE 12.21 ▲ The awe-inspiring beauty of places like Yosemite National Park inspires millions of people from North America and abroad to visit America's national parks.

Federal parks and reserves began in the United States

The striking scenery of the American West impelled the U.S. government to create the world's first **national parks**, publicly held lands protected from resource extraction and development but open to nature appreciation and various forms of recreation. Yellowstone National Park was established in 1872, and Sequoia, General Grant (now Kings Canyon), Yosemite, Mount Rainier, and Crater Lake National Parks followed after 1890. The Antiquities Act of 1906 gave the U.S. president authority to declare selected public lands as *national monuments*, which may later become national parks.

The National Park Service (NPS) was created in 1916 to administer the growing system of parks and monuments, which today numbers 392 sites totaling 34 million ha (84 million acres) and includes national historic sites, national recreation areas, national wild and scenic rivers, and other types of areas (see Figure 12.10). This most widely used park system in the world received 285 million reported recreation visits in 2009—almost one per U.S. resident.

Because America's national parks are open to everyone and showcase the nation's natural beauty in a democratic way, writer Wallace Stegner famously called them "the best idea we ever had." An Escanaba mill worker's family can camp along the sandstone cliffs of Pictured Rocks National Lakeshore on Lake Superior, head to remote Isle Royale National Park to search for moose and wolves, or take a trip across Lake Michigan to the immense sand dunes of Sleeping Bear Dunes National Lakeshore. On their way they would pass Mackinac Island State Park, the first officially designated state park in the nation and now one of 3,700 state parks across the United States.

Another type of federal protected area in the United States is the **national wildlife refuge**. The system of national wildlife refuges, begun in 1903 by President Theodore Roosevelt, now totals over 550 sites comprising 39 million ha (96 million acres; see Figure 12.10), plus an additional 22 million ha (55 million acres) of ocean and islands within the immense new Pacific Remote Islands Marine National Monument designated in 2009 that stretches northwest from Hawaii. There are wildlife refuges within an easy drive of nearly every major American city.

The U.S. Fish and Wildlife Service (USFWS) administers the national wildlife refuges, which not only serve as havens for wildlife, but also in many cases encourage hunting, fishing, wildlife observation, photography, environmental education, and other public uses. Some wildlife advocates find it ironic that hunting is allowed at many refuges, but hunters have long been in the forefront of the conservation movement and have traditionally supplied the bulk of funding for land acquisition and habitat management. Many refuges are managed for waterfowl, but the USFWS increasingly considers non-game species, manages at the habitat and ecosystem levels, and restores marshes and grasslands.

Wilderness areas are established on federal lands

In response to the public's desire for undeveloped areas of land, in 1964 the U.S. Congress passed the Wilderness Act,

FIGURE 12.22 ▲ Wilderness areas were designated on various federally managed lands in the United States following the 1964 Wilderness Act. These include areas that are little disturbed by human activities, such as this remote shoreline where Lake Superior meets Michigan's Upper Peninsula.

which allowed some areas of existing federal lands to be designated as **wilderness areas**. These areas are off-limits to development but are open to hiking, nature study, and other low-impact public recreation (**FIGURE 12.22**).

Congress declared that wilderness areas were necessary "to assure that an increasing population, accompanied by expanding settlement and growing mechanization, does not occupy and modify all areas . . . leaving no lands designated for preservation and protection in their natural condition." Despite these stirring words, however, some preexisting extractive land uses, such as grazing and mining, were allowed to continue within wilderness areas as a political compromise so the act could be passed. Wilderness areas are established within national forests, national parks, national wildlife refuges, and land managed by the Bureau of Land Management (BLM). They are overseen by the agencies that administer those areas. In Michigan's Upper Peninsula, several wilderness areas are designated in the Hiawatha and Ottawa National Forests, one in Pictured Rocks National Lakeshore, and one in Seney National Wildlife Refuge. Overall there are 756 wilderness areas totaling 44 million ha (109 million acres). These cover 5% of U.S. land area (2.7% if Alaska is excluded).

Not everyone supports land set-asides

The restriction of activities in wilderness areas has generated some opposition to U.S. land protection policies, especially among citizens and policymakers in western states. When these states came into existence, the federal government retained ownership of much of their acreage. Idaho, Oregon, and Utah own less than half the land within their borders, and in Nevada 80% of the land is federally owned. Some western state governments have sought to obtain land from the federal government and encourage resource extraction and development.

The drive to secure local control of lands, extract resources, and expand recreational access is epitomized by the *wise-use movement*, a loose confederation that coalesced in the 1980s and 1990s in response to the successes of environmental advocacy. Wise-use advocates are dedicated to protecting private property rights; opposing government regulation; transferring federal lands to state, local, or private hands; and promoting motorized recreation on public lands. They include farmers, ranchers, trappers, and mineral prospectors at the grassroots level, as well as groups representing the industries that extract timber, mineral, and fossil fuel resources. Wise-use advocates have driven debates over national park policy, such as whether recreational activities that disturb wildlife, such as snowmobiles and jet-skis, should be allowed. They found a sympathetic ear in the George W. Bush administration, which shifted policies and enforcement away from conservation and wilderness protection and toward recreation and extractive uses.

For very different reasons, groups of indigenous people also frequently oppose government actions to set aside land. In the case of Native Americans, the U.S. government forced them from the land in the first place and imposed rules of property ownership that were foreign to most Native American cultures. Some sites used today for recreation in the national parks are sites long sacred to Native cultures. For instance, the huge basalt rock formation at Devil's Tower National Monument in Wyoming is sacred to many Plains Indian tribes, yet it is also popular with technical rock climbers. Native Americans view rock climbing as a desecration of the site. As a compromise, the National Park Service asks climbers not to climb the tower in June, when Native Americans travel there for religious ceremonies—an accommodation the U.S. Supreme Court upheld after a lengthy court battle.

On the other hand, protected areas sometimes serve the interests of indigenous people. In Brazil and some other Latin American nations, rainforest parks and reserves have sometimes been established to encompass regions occupied by indigenous tribes. The tribes are thus protected from conflict with miners, farmers, and settlers, and they can continue their traditional way of life.

Nonfederal entities also protect land

Efforts to set aside land—and the debates over such efforts—at the federal level are paralleled at regional and local levels. In the United States, each state has agencies that manage resources on public lands, as do many counties and municipalities. In the 19th century, New York State created Adirondack State Park in a mountainous area where streams converge to form the Hudson River, which thereafter flows south past Albany to New York City. Seeing the need for river water to power industries, keep canals filled, and provide drinking water, the state made a far-sighted decision that has paid dividends through the years.

Private nonprofit groups also preserve land. **Land trusts** are local or regional organizations that purchase land to preserve it in its natural condition. The Nature Conservancy, the world's largest land trust, uses science to select areas and ecosystems in greatest need of protection. Smaller land trusts are plentiful in number and diverse in their missions and methods. Nearly 1,700 local and state land trusts in the United States together own 690,000 ha (1.7 million acres) and have helped preserve an additional 4.1 million ha (10.2 million acres), including scenic areas such as Big Sur on the California coast, Jackson Hole in Wyoming, and Maine's Mount Desert Island.

Parks and reserves are increasing internationally

Many nations have established national park systems and other protected areas and are benefiting from ecotourism as a result—from Costa Rica (Chapter 3) to Ecuador to Thailand to Tanzania. The total worldwide area in protected parks and reserves has increased sixfold since 1970, and today the world's 114,000 protected areas cover 12% of the planet's land area. However, parks in developing nations do not always receive the funding they need to manage resources, provide for recreation, and protect wildlife from poaching and timber from logging. Thus many of the world's protected areas are merely *paper parks*—protected on paper but not in reality.

Some types of protected areas fall under national sovereignty but are designated or partly managed internationally by the United Nations. **Biosphere reserves** are tracts of land with exceptional biodiversity that couple preservation with sustainable development to benefit local people. Biosphere reserves are designated by UNESCO (the United Nations Educational, Scientific, and Cultural Organization) following application by local stakeholders. Each biosphere reserve consists of (1) a core area that preserves biodiversity, (2) a buffer zone that allows local activities and limited development, and (3) an outer transition zone where agriculture, human settlement, and other land uses can be pursued in a sustainable way (**FIGURE 12.23A**).

The Maya Biosphere Reserve in Guatemala (**FIGURE 12.23B**) is an example. Rainforest here is protected in core areas, and in the transition zone timber harvesting takes place in concessions, some of which are FSC-certified. A 2008 study by the nonprofit Rainforest Alliance found that FSC certification gave people economic incentives to conserve the forest. The study found that rates of deforestation and wildfire were much lower in the FSC-certified areas where sustainable logging occurred than in the core area that was supposed to be fully protected.

World heritage sites are another type of international protected area. Over 890 such sites in 148 countries are listed for their natural or cultural value. One such site is Australia's Kakadu National Park, the setting for our central case study in Chapter 6. Another is a mountain gorilla reserve shared by three African countries. This reserve, which integrates national parklands of Rwanda, Uganda, and the Democratic Republic of Congo, is also an example of a *transboundary park*, an area of protected land overlapping national borders. Transboundary

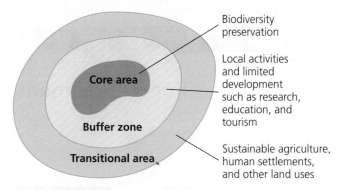

(a) The three zones of a biosphere reserve

(b) Mayan temple ruins at the Maya Biosphere Reserve

FIGURE 12.23 ▲ Biosphere reserves are international protected areas that couple preservation with sustainable development to benefit local residents. Each reserve **(a)** includes a core area that preserves biodiversity, a buffer zone that allows limited development, and a transition zone that permits various uses. At the Maya Biosphere Reserve in Guatemala **(b)**, Forest Stewardship Council (FSC) certification in the transition zone helped prevent illegal deforestation that was rampant in the core area.

parks account for 10% of protected areas worldwide, involving over 100 nations. An example is Waterton–Glacier National Parks on the Canadian–U.S. border.

Some transboundary reserves function as *peace parks*, helping to ease tensions by acting as buffers between nations that quarrel over boundary disputes. This is the case with Peru and Ecuador as well as Costa Rica and Panama, and many people hope that peace parks might also help resolve conflicts between Israel and its neighbors.

Habitat fragmentation makes preserves still more vital

Protecting large areas of land has taken on new urgency now that scientists understand the risks posed by habitat fragmentation (p. 291). Often, it is not the outright destruction of forests and other habitats that threatens species and ecosystems, but rather their fragmentation (see **THE SCIENCE BEHIND THE STORY**, pp. 336–337). Expanding agriculture, spreading cities, highways, logging, and other

Forest Fragmentation in the Amazon

What happens to animals, plants, and ecosystems when we fragment a forest? A massive experiment smack in the middle of the Amazon rainforest is helping scientists learn the answers to this question.

Running for over 30 years and stretching across 1,000 km^2 (386 mi^2), the Biological Dynamics of Forest Fragments Project (BDFFP) is the world's largest and longest-running experiment on forest fragmentation. Hundreds of researchers have muddied their boots in the rainforest here, publishing over 650 scientific research papers, graduate theses, and books. Ironically, this megaproject is today threatened by the very phenomenon it studies: forest fragmentation.

The story starts in the 1970s as conservation biologists debated how to apply island biogeography theory (pp. 338–339) to forested landscapes being fragmented by development. Biologist Thomas Lovejoy decided some good hard data were needed. He conceived a huge experimental project to test ideas about forest fragmentation and established it in the heart of the biggest primary rainforest on the planet—South America's Amazon rainforest.

Farmers, ranchers, loggers, and miners were streaming into the Amazon then, and deforestation was rife. If scientists could learn how large fragments had to be to retain their species, it would help them work with policymakers to preserve forests in the face of development pressures.

Lovejoy's team of Brazilians and Americans worked out a deal with Brazil's government: Ranchers could clear some forest within the study area if they left square plots of forest standing as fragments within those

Dr. Thomas Lovejoy, founder of the BDFFP

clearings. By this process, 11 fragments of three sizes (1 ha [2.5 acres], 10 ha [25 acres], and 100 ha [250 acres]) were left standing, isolated as "islands" of forest, surrounded by "seas" of cattle pasture (see **photo**). Each fragment was fenced to keep cattle out. Then, 12

study plots (of 1, 10, 100 and 1,000 ha) were established within the large expanses of continuous forest still surrounding the pastures, to serve as control plots against which the fragments could be compared.

Besides comparing treatments (fragments) and controls (continuous forest), the project also surveyed populations in fragments before and after they were isolated. With these data on trees, birds, mammals, amphibians, and invertebrates, they documented declines in the diversity of most groups (see **second figure**).

As researchers studied the fragments over the years, they found that small fragments lost more species, and lost them faster, than large fragments— just as island biogeography theory

336

Experimental forest fragments of 1, 10, and 100 ha were created in the BDFFP.

predicts. To slow down species loss by 10 times, researchers found that a fragment needs to be 1,000 times bigger. Even 100-ha fragments were not large enough for some animals and lost half their species in less than 15 years. Monkeys died out because they need large ranges. So did colonies of army ants and the birds that follow them to eat insects scared up as the ants swarm across the forest floor.

Fragments distant from continuous forest lost more species, as predicted by island biogeography theory, but data revealed that even very small openings can stop organisms adapted to deep interior forest from dispersing to recolonize fragments. Many understory birds would not traverse cleared areas of only 30–80 m (100–260 ft). Distances of just 15–100 m (50–330 ft) were insurmountable for some bees, beetles, and tree-dwelling mammals.

Soon, a complication ensued: Ranchers abandoned many of the pastures because the soil was unproductive, and these areas began filling in with secondary forest. As this young secondary forest grew, it made the fragments less like islands—but it led to new insights. Researchers learned that this habitat can act as a corridor for some species, allowing them to disperse from continuous forest and recolonize fragments from which

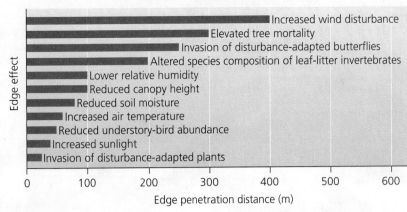

This sampling of 11 of the many edge effects documented by the BDFFP shows how far each impact extends into the interior of forest fragments. Data from various studies summarized in Laurance, W.F., et al., 2002. Ecosystem decay of Amazonian forest fragments: A 22-year investigation. *Conservation Biology* 16: 605–618, Fig 3, adapted. Reprinted by permission of John Wiley & Sons, Inc.

they'd disappeared. By documenting which species did this, scientists learned which might be more resilient to fragmentation. For the Amazon as a whole, it suggested that regrowing forest (as opposed to pasture) can buffer fragments against some species loss.

The secondary forest habitat also introduced new species—generalists adapted to disturbed areas. Frogs, leaf-cutter ants, and small mammals and birds that thrive in second-growth soon became common in adjacent fragments. Open-country butterfly species moved in, displacing interior-forest butterflies.

The invasion of open-country species into the fragments illustrated one type of edge effect. There were more: Edges receive more sunlight, heat, and wind than interior forest, and these conditions can kill trees adapted to the dark, moist interior. Tree death results in the flux of carbon dioxide to the atmosphere, which worsens climate change. Sunlight also promotes growth of vines and shrubs that create a thick, tangled understory along edges. As BDFFP researchers documented these impacts, they found that many edge effects extended far into the forest (see **third figure**). Tree mortality (from wind, sun and water stress, and parasitic vines) reached in 300 m (nearly 1,000 ft). Small fragments essentially became "all edge."

The results on edge effects are relevant for all of Amazonia, because forest clearance and road construction create an immense amount of edge.

A satellite image study in the 1990s estimated that for every 2 acres of land deforested, 3 acres were brought within 1 km of a road or pasture edge.

BDFFP scientists emphasize that impacts across the Amazon will be more severe than the impacts revealed by their experiments. This is because most real-life fragments, unlike the experimental ones, (1) are not protected from hunting, logging, mining, and fires; (2) do not have secondary forest to provide connectivity; (3) are not near large tracts of continuous forest that provide recolonizing species and maintain humidity and rainfall; and (4) are not square in shape, and thus feature more edge. The years of data argue, scientists say, for preserving numerous tracts of Amazonian forest that are as large as possible.

Ironically, today the BDFFP study site is itself threatened by forest fragmentation. A Brazilian government agency has settled colonists just outside the site and has proposed settling 180 families inside it. Development is proceeding up a new highway from Manaus, a city of 1.7 million people, to Venezuela, and wherever roads are built and people settle, logging, hunting, mining, and burning follow. BDFFP researchers can now hear chainsaws and shotgun blasts from their study plots. "It would be tragic," says leading BDFFP scientist William Laurance, "to see a site that's given us so much information be lost so easily."

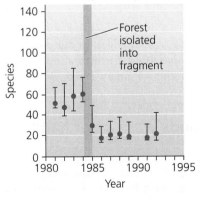

Species richness of understory birds declined in this 1-ha forest plot after the plot was isolated as a fragment (at the time shown by the shaded bar). Data from Ferraz, G., et al., 2003. Rates of species loss from Amazonian forest fragments. *Proc. Natl. Acad. Sci.* 100: 14069–14073. © 2003 National Academy of Sciences. By permission.

(a) Fragmentation from clear-cuts in Mount Hood National Forest, Oregon

1831

1882

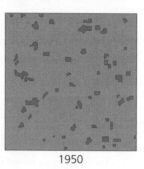
1950

(b) Fragmentation of wooded area (green) in Cadiz Township, Wisconsin

(c) Wood thrush

FIGURE 12.24 ▲ As human populations have grown and human impacts have intensified, most large expanses of natural habitat have been fragmented into smaller, disconnected areas. Forest fragmentation from clear-cutting **(a)** is evident on the Mount Hood National Forest, Oregon. Forest fragmentation has been extreme in the eastern and Midwestern United States; shown in **(b)** are historical changes in forested area in Cadiz Township, Wisconsin, between 1831 and 1950. Fragmentation affects forest-dwelling species such as the wood thrush, *Hylocichla mustelina* **(c)**, a migrant songbird of eastern North America. In forest fragments, wood thrush nests are parasitized by cowbirds that thrive in open country and edge habitats. *Source* (b): Curtis, J.T., 1956. The modification of mid-latitude grasslands and forests by man. In Thomas, W.L. Jr., Ed., *Man's role in changing the face of the earth*. Copyright 1956. Used by permission of the publisher, University of Chicago Press.

338

impacts routinely chop up large contiguous expanses of habitat into small, disconnected ones, as we saw in Chapter 11 (see Figure 11.13, p. 291). Forests are being fragmented everywhere these days, as a result of logging (**FIGURE 12.24A**) and other factors, including agriculture and residential development (**FIGURE 12.24B**). Even in areas where forest cover is increasing (such as the northeastern United States), regrowing forests are generally becoming fragmented into ever-smaller parcels.

When forests are fragmented, many species suffer. For some, a fragment may simply not contain enough area; bears, mountain lions, and other animals that need large ranges in which to roam may disappear. For other species, the problem may lie with **edge effects**, impacts that result because the conditions along a fragment's edge are different from conditions in the interior. Bird species that thrive in the interior of forests may fail to reproduce when forced near the edge of a fragment (**FIGURE 12.24C**). Their nests often are attacked by predators and parasites that favor open

habitats or that travel along habitat edges. Because of edge effects, avian ecologists judge forest fragmentation to be a main reason why populations of many North American songbirds are declining.

Insights from islands warn us of habitat fragmentation

In assessing the impacts of habitat fragmentation on populations, ecologists and conservation biologists have leaned on concepts from **island biogeography theory**. Introduced by E.O. Wilson (p. 300) and ecologist Robert MacArthur in 1963, this theory explains how species come to be distributed among oceanic islands. Since then, researchers have applied it to "habitat islands"—patches of one habitat type isolated within "seas" of others.

Island biogeography theory explains how the number of species on an island results from a balance between the number added by immigration and the number lost through

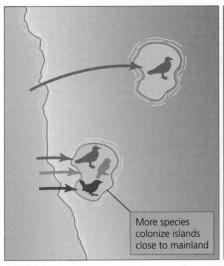
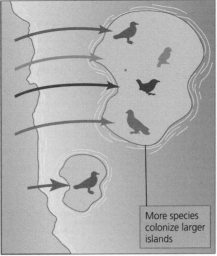
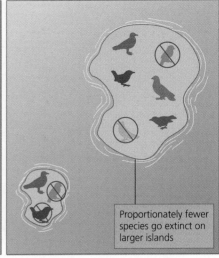

| (a) Distance effect | (b) Target size | (c) Differential extinction |

FIGURE 12.25 ▲ According to island biogeography theory, islands located close to a continent receive more immigrants than islands that are distant **(a)**, and thus they end up with more species. Larger islands present fatter targets for dispersing organisms to encounter **(b)**, so that more species immigrate to large islands than to small islands. Large islands also experience lower extinction rates **(c)**, because larger area allows for larger populations.

extirpation. It predicts an island's species richness based on the island's size and its distance from the mainland:

▶ The farther an island lies from a continent, the fewer species tend to find and colonize it. Thus, remote islands host few species because of low immigration rates (**FIGURE 12.25A**). This is the *distance effect*.

▶ Larger islands have higher immigration rates because they present fatter targets for dispersing organisms to encounter (**FIGURE 12.25B**).

▶ Larger islands have lower extinction rates because more space allows for larger populations, which are less likely to drop to zero by chance (**FIGURE 12.25C**).

Together, the latter two trends give large islands more species than small islands—a phenomenon called the *area effect*. Large islands also contain more species because they tend to possess more habitats than smaller islands, providing suitable environments for a wider variety of species. Very roughly, the number of species on an island is expected to double as island size increases tenfold. This is illustrated by *species-area curves* (**FIGURE 12.26**).

These patterns hold up for terrestrial habitat islands as well, such as forests fragmented by logging and road building. Small "islands" of forest lose diversity fastest, starting with large species that were few in number to begin with. One of the first researchers to show this experimentally was University of Michigan graduate student William Newmark, who in 1983 examined historical records of mammal sightings in North American national parks. The parks, increasingly surrounded by development, were islands of natural habitat isolated by farms, roads, and cities (**FIGURE 12.27**).

Newmark found that many parks were missing a few species and that 42 species had disappeared in all. The red fox and river otter had vanished from Sequoia and Kings Canyon National Parks, for example, and the white-tailed jackrabbit and spotted skunk no longer lived in Bryce Canyon National Park. As island biogeography theory predicted, the largest parks retained the most species, and the smallest parks lost the most species. Species disappeared because the parks were too small to sustain their populations, Newmark concluded, and because the parks had become too isolated to be recolonized by new arrivals.

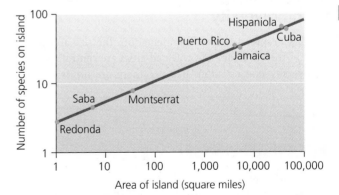

FIGURE 12.26 ▲ Larger islands hold more species—a prediction confirmed by data from around the world. By plotting the number of amphibian and reptile species on Caribbean islands against the islands' areas, this species-area curve shows that species richness increases with area. The increase is not linear, but logarithmic; note the scales of the axes. Data from MacArthur, R.H. *The theory of island biogeography.* © 1967 Princeton University Press, 1995 renewed PUP. Reprinted by permission of Princeton University Press.

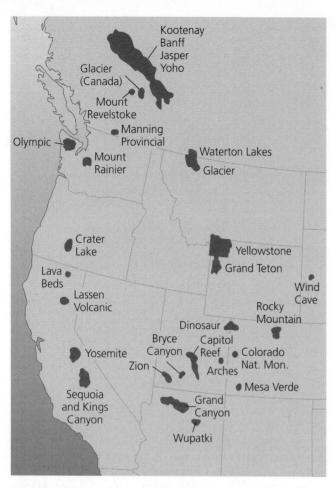

FIGURE 12.27 ▲ William Newmark visualized national parks of western North America as islands of natural habitat in a sea of development. His data showed that mammal species were disappearing from these terrestrial "islands" in accordance with island biogeography theory. Adapted from Quammen, D.; Maps by Kris Ellingsen. 1996. *The song of the dodo: Island biogeography in an age of extinctions.* © 1996. Reprinted by permission.

Reserve design has consequences for biodiversity

Amid widespread habitat fragmentation, the size and placement in the landscape of protected areas are central issues in biodiversity conservation. Because there are limits on how much land can feasibly be set aside, conservation biologists have argued heatedly about whether it is better to make reserves large in size and few in number, or many in number but small in size. Nicknamed the **SLOSS dilemma**, for "single large or several small," this debate is complex, but it seems clear that large species that roam great distances, such as the Siberian tiger (Chapter 11), benefit more from the "single large" approach to reserve design. In contrast, creatures such as insects that live as larvae in small areas may do just fine in a number of small isolated reserves, if they can disperse as adults by flying from one reserve to another. The SLOSS debate was one motivation for establishing the Amazonian forest fragmentation project described in our **Science behind the Story** (pp. 336–337).

A related issue is how effectively **corridors** of protected land allow animals to travel between islands of habitat. In theory, connections between fragments provide animals access to more habitat and encourage gene flow to maintain populations in the long term. Many land management agencies and environmental groups try to join new reserves to existing reserves for these reasons.

Climate change threatens our investment in protected areas

As if we did not face enough challenges in designing, establishing, and guarding protected areas to preserve species, communities, and ecosystems, global climate change (Chapter 18) now threatens to undo our efforts. As temperatures become warmer, species ranges shift toward naturally cooler climes: toward the poles and upward in elevation (p. 512). In a landscape of fragmented habitat, it may sometimes be impossible for individuals of a species to move if they cannot fly or disperse long distances. Species we had hoped to protect in parks may, in a warming world, become trapped in them. We saw this with the golden toad in Monteverde in Chapter 3, and indeed, high-elevation species are most at risk from climate change, because there is nowhere for them to go once a mountaintop becomes too warm or dry. For this reason, corridors to allow movement become still more important, and conservation biologists are now looking beyond parks and protected areas as they explore strategies for saving biodiversity.

➤ CONCLUSION

Forests are ecologically vital and economically valuable, yet we are losing them around the world. Forest management in North America reflects trends in land and resource management in general. Early emphasis on resource extraction evolved into policies on sustainable yield and multiple use as land and resource availability declined and as the public became more aware of environmental degradation. Public forests today are managed not only for timber production, but also for recreation, wildlife habitat, and ecosystem integrity. Sustainable forest certification provides economic incentives for responsible conservation on all forested lands.

Meanwhile, public support for the preservation of natural lands has led to the establishment of a variety of parks and protected areas in the United States and abroad. As development spreads across the landscape, fragmenting habitats and subdividing populations, scientists trying to conserve species, communities, and ecosystems are thinking and working at the landscape level to design protected areas.

REVIEWING OBJECTIVES

You should now be able to:

SUMMARIZE THE ECOLOGICAL AND ECONOMIC CONTRIBUTIONS OF FORESTS

- Many kinds of forests exist. (pp. 315–316)
- Forests are ecologically complex and support a wealth of biodiversity (pp. 316–317)
- Forests contribute ecosystem services, including carbon storage. (pp. 317–318)
- Forests provide us timber and other economically important products and resources. (p. 318)

OUTLINE THE HISTORY AND CURRENT SCALE OF DEFORESTATION

- We have lost forests to clearance for agriculture and over-harvesting for wood. (pp. 318–320)
- Developed nations deforested much of their land as settlement, farming, and industrialization proceeded. (pp. 319, 321)
- Today deforestation is taking place most rapidly in developing nations. (p. 320–321)
- Carbon offsets are one new potential solution to deforestation. (p. 322)

ASSESS ASPECTS OF FOREST MANAGEMENT AND DESCRIBE METHODS OF HARVESTING TIMBER

- Forestry is one type of resource management. (p. 322)
- Resource managers have long managed for maximum sustainable yield and are beginning to implement ecosystem-based management and adaptive management. (p. 323)
- The U.S. national forests were established to conserve timber and allow its sustainable extraction. (p. 324)
- Most U.S. timber today comes from private lands. (pp. 324–325)
- Plantation forestry, featuring single-species, even-aged stands, is widespread and growing. (p. 325)

- Harvesting methods include clear-cutting and other even-aged techniques, as well as selection strategies that maintain uneven-aged stands that more closely resemble natural forest. (pp. 325–327)
- Foresters are beginning to manage for recreation, wildlife habitat, and ecosystem integrity. (pp. 327–328)
- Fire policy has been controversial, but scientists agree that we need to address the impacts of a century of fire suppression. (pp. 328–331)
- Climate change is affecting forests. (pp. 329–330)
- Certification of sustainable forest products allows consumer choice in the marketplace to influence forestry practices. (pp. 330–332)

IDENTIFY FEDERAL LAND MANAGEMENT AGENCIES AND THE LANDS THEY MANAGE

- The U.S. Forest Service, National Park Service, Fish and Wildlife Service, and Bureau of Land Management manage U.S. national forests, national parks, national wildlife refuges, and BLM land. (pp. 324, 333)

RECOGNIZE TYPES OF PARKS AND PROTECTED AREAS AND EVALUATE ISSUES INVOLVED IN THEIR DESIGN

- Public demand for preservation and recreation led to the creation of parks, reserves, and wilderness areas. (pp. 332–334)
- Biosphere reserves are one of several types of international protected lands. (p. 335)
- Because habitat fragmentation affects wildlife, conservation biologists are working on how best to design parks and reserves. (pp. 335–340)
- Island biogeography helps inform conservation planning, but climate change is posing threats to protected areas. (pp. 338–340)

TESTING YOUR COMPREHENSION

1. Name at least two reasons why natural primary forests contain more biodiversity than single-species forestry plantations.

2. Describe three ecosystem services that forests provide.

3. Name several major causes of deforestation. Where is deforestation most severe today?

4. Compare and contrast maximum sustainable yield, ecosystem-based management, and adaptive management. How may pursuing maximum sustainable yield sometimes conflict with what is ecologically desirable?

5. Compare and contrast the major methods of timber harvesting.

6. Describe several ecological impacts of logging. How has the U.S. Forest Service responded to public concern over these impacts?

7. Are forest fires a bad thing? Explain your answer.

8. Name at least four reasons why people have created parks and reserves. Why did the U.S. Congress determine in 1964 that wilderness areas were necessary? How do these areas differ from national parks?

9. What percentage of Earth's land is protected? Describe one type of protected area that has been established outside North America.

10. Give two examples of how forest fragmentation affects animals. How does island biogeography theory help us design reserves?

SEEKING SOLUTIONS

1. People in developed nations are fond of warning people in developing nations to stop destroying rainforest. People of developing nations often respond that this is hypocritical, because the developed nations became wealthy by deforesting their land and exploiting its resources in the past. What would you say to the president of a developing nation, such as Brazil, that is seeking to clear much of its forest?

2. Do you think maximum sustainable yield represents an appropriate policy for resource managers to follow? Why or why not?

3. What might you tell a wise-use advocate to help him or her understand why a wilderness hiker wants scenic land in Utah federally protected? What might you tell a wilderness hiker to help him or her understand why a wise-use advocate opposes the protection? How might you help them find common ground?

4. Given the effects that climate change may have on species' ranges, if you were trying to preserve an endangered mammal that occurs in a small area, and if you had unlimited funds to acquire land to help restore its population, how would you design a protected area for it? Would you use corridors? Would you include a diversity of elevations? Would you design few large reserves or many small ones? Explain your answers.

5. **THINK IT THROUGH** You have just become the supervisor of a national forest. Timber companies are requesting to cut as many trees as you will let them, and environmentalists want no logging at all. Ten percent of your forest is old-growth primary forest, and the remaining 90% is secondary forest. Your forest managers are split among preferring maximum sustainable yield, ecosystem-based management, and adaptive management. What management approach(es) will you take? Will you allow logging of all old-growth trees, some, or none? Will you allow logging of secondary forest? If so, what harvesting strategies will you encourage? What would you ask your scientists before deciding on policies on fire management and salvage logging?

6. **THINK IT THROUGH** You run a major nonprofit environmental advocacy organization and are trying to save an ecologically priceless tract of tropical forest in a poor developing nation. You have worked in this region for years and know and care for the local people, who want to save the forest and its animals but also need to make a living and use the forest's resources. The nation's government plans to sell a concession to a foreign multinational timber corporation to log the entire forest unless your group can work out some other solution. Describe what solution(s) you would you try to arrange. Consider the range of issues and options discussed in this chapter, including government protected areas, private protected areas, biosphere reserves, forest management techniques, carbon offsets, FSC-certified sustainable forestry, and more. Explain reasons for your choice(s).

CALCULATING ECOLOGICAL FOOTPRINTS

We all rely on forest resources. The average North American consumes nearly 300 kg (660 lb) of paper and paperboard per year. Using the estimates of paper and paperboard consumption for each region of the world, calculate the per capita consumption for each region using the population data in the table. Note: 1 metric ton = 2,205 pounds.

	Population (millions)	Total paper consumed (millions of metric tons)	Per capita paper consumed (pounds)
Africa	944	7	16
Asia	4,010	148	
Europe	733	105	
Latin America	569	22	
North America	335	96	
Oceania	35	5	
World	6,625		127

All data are for 2007, from Population Reference Bureau and U.N. Food and Agriculture Organization (FAOSTAT).

1. How much paper would North Americans save each year if we consumed paper at the rate of Europeans?

2. How much paper would be consumed if everyone in the world used as much paper as the average European? As the average North American?

3. Why do you think people in other regions consume less paper, per capita, than North Americans? Name three things you could do to reduce your paper consumption.

4. Describe three ways in which consuming FSC-certified paper rather than conventional paper can reduce the environmental impacts of paper consumption.

Mastering ENVIRONMENTAL SCIENCE ™

Intersection near Shibuya station, Tokyo, Japan

13 THE URBAN ENVIRONMENT: CREATING LIVABLE AND SUSTAINABLE CITIES

UPON COMPLETING THIS CHAPTER, YOU WILL BE ABLE TO:

- Describe the scale of urbanization
- Assess urban and suburban sprawl
- Outline city and regional planning and land use strategies
- Evaluate transportation options

- Describe the roles of urban parks
- Analyze environmental impacts and advantages of urban centers
- Assess urban ecology, green building efforts, and the pursuit of sustainable cities

343

Managing Growth in Portland, Oregon

"Sagebrush subdivisions, coastal condomania, and the ravenous rampage of suburbia in the Willamette Valley all threaten to mock Oregon's status as the environmental model for the nation."

—Oregon Governor Tom McCall, 1973

"We have planning boards. We have zoning regulations. We have urban growth boundaries and 'smart growth' and sprawl conferences. And we still have sprawl."

—Environmental scientist Donella Meadows, 1999

CANADA

Portland, Oregon

UNITED STATES

Atlantic Ocean

Pacific Ocean

With the fighting words above, Oregon governor Tom McCall challenged his state's legislature in 1973 to take action against runaway sprawling development, which many Oregon residents feared would ruin the communities and landscapes they loved. McCall echoed the growing concerns of state residents that farms, forests, and open space were being gobbled up for development, including housing for people moving in from California and elsewhere.

Foreseeing a future of subdivisions, strip malls, and traffic jams engulfing

People enjoying Tom McCall Waterfront Park in Portland, Oregon

the pastoral Willamette Valley, Oregon acted. The state legislature passed Senate Bill 100, a sweeping land use law that would become the focus of acclaim, criticism, and careful study for years afterward by other states and communities trying to manage their own urban and suburban growth.

Oregon's law required every city and county to draw up a comprehensive land use plan in line with statewide guidelines that had gained popular support from the state's electorate. As part of each land use plan, each metropolitan area had to establish an **urban growth boundary (UGB)**, a line on a map intended to separate areas desired to be urban from areas

desired to remain rural. Development for housing, commerce, and industry would be encouraged within these urban growth boundaries but severely restricted beyond them. The intent was to revitalize city centers, prevent suburban sprawl, and protect farmland, forests, and open landscapes around the edges of urbanized areas.

Residents of the area around Portland, the state's largest city, established a new regional entity to plan and to apportion land in their region. The Metropolitan Service District, or Metro, represents 25 municipalities and three counties. Metro adopted the Portland-area urban growth boundary in 1979 and has tried to focus growth on existing urban centers and to build communities where people can walk or take mass transit between home, work, and shopping. These policies have largely worked as intended. Portland's downtown and older neighborhoods have thrived, regional urban centers are becoming denser and more community oriented, mass transit continues to expand, and development has been limited on land beyond the UGB. Portland began attracting international attention for its "livability."

To many Portlanders today, the UGB remains the key to maintaining quality of life in city and countryside alike. In the view of its critics, however, the "Great Wall of Portland" is an elitist and intrusive government regulatory tool. In 2004, Oregon voters approved a ballot measure that threatened to eviscerate the land use rules that most citizens had backed for three decades. Ballot Measure 37 required the state to compensate certain landowners if government regulation had decreased the value of their land. For example, regulations prevent landowners outside UGBs from subdividing their lots and selling them for housing development. Under Measure 37, the state had to pay these landowners to make up for theoretically lost income or else allow them to ignore the regulations. Because state and local governments did not have enough money to pay such claims, the measure was on track to gut Oregon's zoning, planning, and land use rules.

Landowners filed over 7,500 claims for payments or waivers affecting 295,000 ha (730,000 acres). Although the measure had been promoted to voters as a way to protect the rights of small family landowners, most claims were filed by large developers. Neighbors suddenly found themselves confronting the prospect of massive housing subdivisions, gravel mines, strip malls, or industrial facilities being developed next to their homes—and many who had voted for Measure 37 now had misgivings.

The state legislature, under pressure from opponents and supporters alike, settled on a compromise: to introduce a new ballot measure. Oregon's voters passed Ballot Measure 49 in 2007. It protects the rights of small landowners to gain income from their property by developing small numbers of homes, but it also restricts large-scale development and development in sensitive natural areas.

In 2010, Metro finalized a historic agreement with representatives and citizens of its region's three counties to determine where urban growth will and will not be allowed over the next 50 years. Metro and the counties apportioned over 121,000 ha (300,000 acres) of undeveloped land into "urban reserves" open for development and "rural reserves" where farmland and forests would be preserved, precisely mapping the boundaries. The hope is that this will give clarity and direction for landowners and governments alike for half a century.

People are confronting similar issues in communities throughout North America, and debates and negotiations like those in Oregon will determine how our cities and landscapes will change in the future.

OUR URBANIZING WORLD

We live at a turning point in human history. Beginning in 2009, for the first time ever, more people were living in urban areas (cities and suburbs) than in rural areas. This shift from the countryside into towns and cities, called **urbanization**, may be the single greatest change our society has undergone since its transition from a nomadic hunter-gatherer lifestyle to a sedentary agricultural one.

As we undergo this shift to urban areas, two pursuits become ever more important. One is to make our urban areas more livable by meeting residents' needs for a safe, clean, healthy urban environment and a high quality of life. Another is to make our urban systems sustainable by creating cities that can continue to function effectively and prosperously for the long-term future while minimizing our ecological footprint and working with natural systems (rather than against them).

Industrialization has driven the move to urban centers

Since 1950, the world's urban population has more than quadrupled. Urban populations are growing for two reasons: (1) the human population overall is growing (Chapter 8), and (2) more people are moving from farms to cities than are moving from cities to farms.

The shift from country to city began long ago. Agricultural harvests that produced surplus food freed a proportion of citizens from farm life and allowed the rise of specialized manufacturing professions, class structure, political hierarchies, and urban centers (p. 226). The industrial revolution (p. 4) spawned technological innovations that created jobs and opportunities in urban centers for people who were no longer needed on farms. Industrialization and urbanization bred further technological advances that increased production efficiencies, both on the farm and in the city. This process of positive feedback continues today.

Worldwide, the proportion of population that is urban rose from 30% half a century ago to just over 50% today. Since 1950, the global urban population has grown by 4.7 times, whereas the rural population has not quite doubled. The United Nations projects that the urban population will nearly double between now and 2050, whereas the rural population will decline by 16%.

Trends differ between developed and developing nations, however (**FIGURE 13.1**). In developed nations such as the United States, urbanization has slowed, because three of every four people already live in cities, towns, and **suburbs**, the smaller communities that ring cities. Back in 1850, the U.S. Census Bureau classified only 15% of U.S. citizens as urban dwellers. That percentage passed 50% shortly before 1920 and now stands at 80%. Most U.S. urban dwellers reside in suburbs; fully 50% of the U.S. population today is suburban.

In contrast, today's developing nations, where most people still reside on farms, are urbanizing rapidly (**FIGURE 13.2**). As industrialization diminishes the need for farm labor and increases urban commerce and jobs, rural people are streaming from farms to cities. Sadly, wars, conflict, and ecological degradation are also driving millions of people out of the countryside and into urban centers. For all these reasons, most fast-growing cities today are in the developing world. In cities such as Delhi, India; Lagos, Nigeria; and Karachi, Pakistan, population growth often exceeds economic growth, and the result is overcrowding, pollution, and poverty. United Nations demographers estimate

FIGURE 13.1 ▶ In developing nations today, urban populations are growing quickly, whereas rural populations are leveling off and are expected to begin declining. Developed nations are already largely urbanized, so in these nations urban populations are growing more slowly, and rural populations are falling. Solid lines in the graph indicate past data, and dashed lines indicate projections of future trends. Data from Population Division of the Department of Economic and Social Affairs of the United Nations Secretariat, 2010. *World urbanization prospects: The 2009 revision*, Fig 3. By permission.

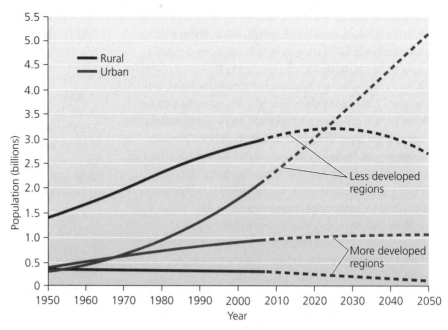

that urban areas of developing nations will absorb nearly all of the world's population growth from now on.

Across the world today, 21 "megacities" are each home to 10 million residents or more (**TABLE 13.1**). Still, such megacities are the exception. The majority of urban dwellers live in smaller cities, such as Portland, Omaha, Winnipeg, Raleigh, Austin, and their still-smaller suburbs.

Environmental factors influence the location of urban areas

Real estate agents have long used the saying, "Location, location, location," to stress how much a home's whereabouts determines its value. Location is vitally important for urban centers as well. Environmental variables such as climate, topography, and the configuration of waterways influence whether a small settlement will become a large city—and successful cities tend to be located in places that give them

economic advantages. Think of any major city, and chances are it's situated along a major river, seacoast, railroad, or highway—some corridor for trade that has driven economic growth (**FIGURE 13.3**).

Many well-located cities have served as linchpins in trading networks, funneling in resources from agricultural regions, processing them and manufacturing products, and shipping those products to other markets. Portland got its start in the mid-19th century as pioneers arriving by the Oregon Trail settled where the Willamette River flowed into the Columbia River. Situated at the juncture of these two major rivers, and just upriver from where the Columbia flows into the Pacific Ocean, Portland had a strategic advantage in trade. The city grew as it received, processed, and shipped overseas the produce from the farms of Oregon's Willamette River valley and the upper Columbia River valley, and as it imported products shipped in from other North American ports and from Asia.

FIGURE 13.2 ▶ The population shift from country to city is happening most quickly today in developing nations. Urban lifestyles are alluring and exciting to those with the economic means to take advantage of the cultural amenities they have to offer. Yet sadly, many new immigrants to the cities end up in poverty, as seen here in this sprawling *favela* (slum neighborhood) in Rio de Janeiro, Brazil.

(a) St. Louis, Missouri

(b) Fort Worth, Texas

FIGURE 13.3 ▲ St. Louis **(a)** is situated on the Mississippi River near its confluence with the Missouri River—and this strategic location for river trade drove its growth in the 19th and early 20th centuries. In contrast, Fort Worth, Texas **(b)**, grew most in the late 20th century as a result of linkages by the interstate highway system and a major international airport.

We can see such geographic patterns over and over again with major cities. A prime example is Chicago, which grew with extraordinary speed in the 19th and early 20th centuries as railroads funneled through it the resources from the vast lands of the Midwest and West on their way to consumers and businesses in the populous cities of the East. Chicago became a center for grain processing, livestock slaughtering, meatpacking, and much else.

Today, however, powerful technologies and cheap transportation enabled by fossil fuels have allowed cities to thrive even in resource-poor regions. The Dallas–Fort Worth area prospers from—and relies on—oil-fueled transportation by interstate highways and a major airport. Los Angeles, Las Vegas, Phoenix, and other cities in the Southwest have flourished in desert regions by appropriating water from distant sources. Whether such cities can sustain themselves as oil and water become increasingly scarce in the future is an important question.

TABLE 13.1 Metropolitan Areas with 10 Million Inhabitants or More

City, Country	Millions of people
Tokyo, Japan	36.5
Delhi, India	21.7
Sao Paulo, Brazil	20.0
Mumbai (Bombay), India	19.7
Mexico City, Mexico	19.3
New York–Newark, United States	19.3
Shanghai, China	16.3
Calcutta, India	15.3
Dhaka, Bangladesh	14.3
Buenos Aires, Argentina	13.0
Karachi, Pakistan	12.8
Los Angeles–Long Beach–Santa Ana, United States	12.7
Beijing, China	12.2
Rio de Janeiro, Brazil	11.8
Manila, Philippines	11.4
Osaka-Kobe, Japan	11.3
Cairo, Egypt	10.9
Moscow, Russia	10.5
Paris, France	10.4
Istanbul, Turkey	10.4
Lagos, Nigeria	10.2

Source: United Nations Population Division, 2010. *World urbanization prospects: The 2009 revision.* New York: UNPD.

In recent years, many cities in the southern and western United States have undergone growth spurts as people (particularly retirees) from northern and eastern states have moved south and west in search of warmer weather or more space. Between 1990 and 2008, the population of the Dallas–Fort Worth and Houston metropolitan areas each grew by over 50%, that of the Atlanta area grew by 82%; that of the Phoenix region grew by 91%; and that of the Las Vegas metropolitan area grew by a whopping 119%.

WEIGHING THE ISSUES

What Made Your City? Consider the town or city in which you live, or the major urban center located nearest you. Why do you think it developed where it did? What physical, social, or environmental factors may have aided its growth? Do you think it will prosper in the future? Why or why not?

Cities grew, and then suburbs grew

American cities grew rapidly throughout the 19th and early 20th centuries, as a result of immigration from abroad and increased trade as the nation expanded westward. The bustling economic activity of downtown districts held people in cities despite growing crowding, poverty, and crime. However, by the mid-20th century, many affluent city dwellers were choosing to move outward to the cleaner, less crowded, and more park-like suburban communities beginning to

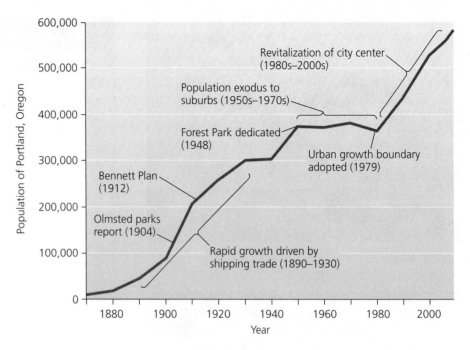

FIGURE 13.4 ▶ Once Portland established its position as a strategically located port, international shipping trade provided jobs and boosted its economy, and the city's population grew rapidly. City residents began leaving for the suburbs in the 1950s to 1970s, but policies designed to enhance the city center then revitalized the city's growth.

surround the cities. These people were pursuing more space, better economic opportunities, cheaper real estate, less crime, and better schools for their children.

As affluent people moved outward into the expanding suburbs, jobs followed. This exodus to the suburbs hastened the economic decline of downtown districts throughout the nation, and American cities stagnated. Chicago's population declined to 80% of its peak because so many residents moved to its suburbs. Philadelphia's population fell to 76% of its peak, Washington, D.C.'s to 71%, and Detroit's to just 55%.

Portland followed this trajectory but also is an example of a city that bounced back. After growing rapidly for many decades, Portland's population growth stalled in the 1950s to 1970s as crowding and deteriorating economic conditions drove city dwellers to the suburbs. However, subsequent policies to revitalize the city center helped restart Portland's growth (**FIGURE 13.4**).

Several factors enabled people to move to suburbs in the mid- and late 20th century. The rise of the automobile was one, together with an expanding road network and inexpensive and abundant oil. Cars, roads, and cheap oil allowed millions of people to commute by car to their downtown workplaces from new homes in suburban "bedroom communities." By facilitating long-distance transport, fossil fuels and highway networks also made it easier for businesses to import and export resources, goods, and waste. The U.S. government's development of the interstate highway system was pivotal in encouraging these trends.

Since then, technology has reinforced the spread of urban and suburban areas. In our age of jet travel, television, cell phones, and the Internet, being located in a city's downtown on a river or seacoast is no longer so vital to success. As globalization continues to connect distant societies, businesses and individuals can more easily communicate from locations away from major city centers.

In most ways, suburbs have delivered the qualities people sought in them. The wide spacing of houses, with each house on its own plot of land, gives families room and privacy. However, by allotting more space to each person, suburban growth has spread human impact across the landscape. Natural areas have disappeared as housing developments are constructed. We have built extensive road networks to ease travel, but suburbanites now find themselves needing to drive everywhere. They commute longer distances to work and spend more time in congested traffic. The expanding rings of suburbs surrounding cities have grown larger than the cities themselves, and towns are running into one another. These aspects of suburban growth have inspired a new term: *sprawl*.

SPRAWL

The term *sprawl* has become laden with meanings and suggests different things to different people. To some, sprawl is aesthetically ugly, environmentally harmful, and economically inefficient. To others, it is the collective outgrowth of reasonable individual desires and decisions in a world of growing human population. We can begin our discussion by giving **sprawl** a simple, nonjudgmental definition: the spread of low-density urban or suburban development outward from an urban center.

Urban areas spread outward

As urban and suburban areas grow in population, they also grow spatially. This is clear from maps and satellite images of rapidly spreading cities such as Las Vegas (**FIGURE 13.5**). Another example is Chicago, whose metropolitan area now consists of more than 9.5 million people spread across 23,000 km² (9,000 mi²)—an area 40 times the size of the city. All in all, houses and roads supplant over 1 million ha (2.5 million acres) of U.S. land each year—over 2,700 ha (6,700 acres) every day.

(a) Las Vegas, Nevada, 1984

(b) Las Vegas, Nevada, 2009

FIGURE 13.5 ▲ Satellite images show the type of rapid urban and suburban expansion that many people have dubbed *sprawl*. Las Vegas, Nevada, is currently one of the fastest-growing cities in North America. Between 1984 **(a)** and 2009 **(b)**, its population and its developed area each tripled.

Several types of development approaches can lead to sprawl (**FIGURE 13.6**, p. 352). These approaches allot each person more space than do cities. For example, the average resident of Chicago's suburbs takes up 11 times more space than a resident of the city. As a result, the outward spatial growth of suburbs across the landscape generally outpaces growth in numbers of people. In fact, many researchers define *sprawl* as the physical spread of development at a rate that exceeds the rate of population growth. For instance, the population of Phoenix grew 12 times larger between 1950 and 2000, yet its land area grew 27 times larger. Between 1950 and 1990, the population of 58 major U.S. metropolitan areas rose by 80%, but the land area they covered rose by 305%. Even in 11 metro areas where population declined between 1970 and 1990 (for instance, Rust Belt cities such as Detroit, Cleveland, and Pittsburgh), the amount of land covered increased.

Sprawl has several causes

There are two main components of sprawl. One is human population growth—there are simply more people alive each year. The other is per capita land consumption—each person takes up more land. The amount of sprawl is a function of the number of people added to a region times the amount of land the average person occupies.

A study of U.S. metropolitan areas between 1970 and 1990 found that each of these two factors contributes about equally to sprawl but that cities varied in which was more important. The Los Angeles metro area increased in population density by 9% between 1970 and 1990, becoming the nation's most densely populated metro area. Increasing density should

be a good recipe for preventing sprawl. Yet L.A. grew in size by a whopping 1,021 km² (394 mi²). This spatial growth, despite the increase in density, resulted from an overwhelming influx of new people. In contrast, the Detroit metro area lost 7% of its population between 1970 and 1990, yet it expanded in area by 28%. In this case, sprawl clearly was caused solely by increased per capita land consumption.

We discussed reasons for human population growth in Chapter 8. As for the rise in per capita land consumption, factors just mentioned—better highways, inexpensive gasoline, and technologies such as telecommunications and the Internet—fostered movement away from city centers by freeing businesses from dependence on the centralized infrastructure a major city provides and by giving workers greater flexibility to live where they desire. The primary reason for greater per capita land consumption, however, is that most people simply like having space and privacy and dislike congestion. Furthermore, the consumption-oriented American lifestyle promotes bigger houses, bigger cars, and bigger TVs, and having more space to house one's possessions becomes important. Unless there are overriding economic or social disadvantages, most people prefer living in less congested, more spacious, more affluent communities.

Economists, politicians, and city boosters have encouraged the unbridled spatial expansion of cities and suburbs. The conventional assumption has been that growth is always good and that attracting business, industry, and residents will enhance a community's economic well-being, political power, and cultural influence. Today, however, this assumption is being challenged. As growing numbers of people feel the negative effects of sprawl on their lifestyles, they have begun to question the mantra that all growth is good.

The SCIENCE behind the Story

Measuring the Impacts of Sprawl

Dr. Reid Ewing, Rutgers University

Critics of sprawl have blamed it for so many societal ills that a person can be made to feel guilty just for being born in the suburbs or shopping at a mall. But what does scientific research tell us are the actual consequences of sprawl?

When Reid Ewing of Rutgers University and his team set out to measure the impacts of sprawl, they discovered that it was difficult even to agree on a definition of the term or on how to measure it. Surveying the literature, Ewing's team found that researchers using different criteria ranked cities in very different ways. For instance, in most studies Los Angeles was deemed more sprawling than Portland, but in some, Portland was judged more sprawling than L.A.

So Ewing, Rolf Pendall of Cornell University, and Don Chen of the nonprofit group Smart Growth America tried to define *sprawl* as simply as possible, without mixing sprawl's consequences into the definition. They decided that sprawl occurs when the spread of development across the landscape far outpaces population growth. They then devised four criteria by which to rank 83 of the largest U.S. metropolitan

areas in terms of sprawl. Sprawling cities would show:

▶ Low residential density
▶ Distant separation of homes, employment, shopping, and schools
▶ Lack of focused activity in community centers and downtown areas
▶ Street networks that make many streets hard to access

For each criterion, Ewing's team measured multiple factors (22 variables in all), analyzed them, and arrived at a cumulative index of sprawl. They obtained data from municipalities throughout the country to assess with their index.

The research team's rankings showed that the nation's most sprawling area was the quickly expanding Riverside–San Bernardino, California, region. The area with the least sprawl was New York City, whose historically dense population and vibrant neighborhoods kept it geographically compact relative to its number of inhabitants.

The **two tables** list the 10 most- and least-sprawling areas.

The researchers next correlated their sprawl scores with a number of transportation variables. They found that people in the 10 most-sprawling metros owned more cars (1.8 per household) than people in the 10 least-sprawling metros (1.6 per household). They determined that residents of the most-sprawling metros drove an average of 43 km (27 mi) per day, whereas those of the least-sprawling metros drove only 34 km (21 mi). Moreover, people in the most-sprawling metros

Traffic congestion is frequently blamed on sprawl.

What is wrong with sprawl?

Sprawl means different things to different people. To some, the word evokes strip malls, homogenous commercial development, and tracts of cookie-cutter houses encroaching on farmland or ranchland. It may suggest traffic jams, destruction of wildlife habitat, and loss of natural land around cities.

For other people, sprawl represents the collective result of choices made by millions of well-meaning individuals trying to make a better life for themselves and their families. In this

view, those who decry sprawl are being elitist and fail to appreciate the good things about suburban life. Let us try, then, to leave the emotional debate aside and assess what research can tell us about the impacts of sprawl (see **THE SCIENCE BEHIND THE STORY**, above).

Transportation   Most studies show that sprawl constrains transportation options, essentially forcing people to drive cars. These constraints include the need to own a vehicle and to drive it most places, the need to drive greater distances or

used public transit far less and suffered 67% more traffic fatalities.

Strikingly, the study found no significant difference in commute time between home and work for people of sprawling versus less-sprawling metro areas. Critics of sprawl have long blamed sprawl for traffic congestion and commute delays (**photo**). Advocates of suburban spread have argued that more streets ease commutes and that regions can sprawl their way out of congestion. The Ewing team's results seem to suggest that each side may have a point and that the effects may cancel one another out.

Because vehicle emissions pollute the air, the researchers also measured levels of tropospheric ozone (p. 472). Sprawling areas had worse air pollution, they found; ozone levels were 40% higher in the most-sprawling metros.

None of these results prove that sprawl *causes* these impacts, because statistical correlation alone does not imply causation. However, taken together, the findings are certainly suggestive. The results were published in 2003 in the *Transportation Research Record* and in a 2002 report published by Smart Growth America.

Since then, Ewing and his colleagues have examined other impacts of sprawl, and their recent studies find that sprawl correlates with high per capita energy use and with several measures of health. Because research had shown that residents of sprawling areas depend more on cars and walk less, Ewing's team hypothesized that they would find more obesity and poorer health in people living in sprawling areas.

The 10 Most-Sprawling American Urban Areas

Rank	Metropolitan Region
1	Riverside–San Bernardino, CA
2	Greensboro–Winston-Salem–High Point, NC
3	Raleigh–Durham, NC
4	Atlanta, GA
5	Greenville–Spartanburg, SC
6	West Palm Beach–Boca Raton–Delray Beach, FL
7	Bridgeport–Stamford–Norwalk–Danbury, CT
8	Knoxville, TN
9	Oxnard–Ventura, CA
10	Fort Worth–Arlington, TX

1 = most sprawling; 10 = less sprawling.

Source: Ewing, R., et al. 2002. *Measuring sprawl and its impact.* Washington, D.C.: Smart Growth America.

The 10 Least-Sprawling American Urban Areas

Rank	Metropolitan Region
83	New York, NY
82	Jersey City, NJ
81	Providence–Pawtucket–Woonsocket, RI
80	San Francisco, CA
79	Honolulu, HI
78	Omaha, NE-IA
77	Boston–Lawrence–Salem–Lowell–Brockton, MA
76	Portland, OR
75	Miami–Hialeah, FL
74	New Orleans, LA

83 = least sprawling; 74 = more sprawling.

Source: Ewing, R., et al., 2002. *Measuring sprawl and its impact.* Washington, D.C.: Smart Growth America.

Indeed, they found that people from sprawling metros are on average heavier for their height and show increased instances of high blood pressure (although not more instances of diabetes or cardiovascular disease).

Recent research from other groups has produced a mix of findings on sprawl, obesity, and health; some support Ewing's conclusions, some contradict them, and some find correlations but are reluctant to conclude that sprawl causes the health effects. As more studies accrue, we will gain a better idea of the benefits and costs of our choices in urban design. ∎

to spend more time in vehicles, lack of mass transit options, and more traffic accidents. Across the United States, during the 1980s and 1990s the average length of work trips rose by 36%, and total vehicle miles driven rose three times faster than population growth. An automobile-oriented culture also increases dependence on nonrenewable petroleum, with its economic and environmental consequences (pp. 545–554).

Pollution   By promoting automobile transportation, sprawl increases pollution. Carbon dioxide emissions from vehicles contribute to global climate change (Chapter 18) while nitrogen- and sulfur-containing air pollutants lead to tropospheric ozone, urban smog, and acid precipitation (pp. 482–486). Motor oil and road salt from roads and parking lots pollute waterways, posing risks to ecosystems and human health. Runoff of polluted water from paved areas is estimated to be about 16 times greater than from naturally vegetated areas.

Health   Aside from the health impacts of pollution, some research suggests that sprawl promotes physical inactivity because

(a) Uncentered commercial strip development

(b) Low-density single-use development

(c) Scattered, or leapfrog, development

(d) Sparse street network

FIGURE 13.6 ▲ Several standard approaches to development can result in sprawl. In uncentered commercial strip development **(a)**, businesses are arrayed in a long strip along a roadway, and no attempt is made to create a centralized community with easy access for consumers. In low-density, single-use residential development **(b)**, homes are located on large lots in residential tracts far away from commercial amenities. In scattered, or leapfrog, development **(c)**, developments are created at great distances from a city center and are not integrated. In developments with a sparse street network **(d)**, roads are far enough apart that moderate-sized areas go undeveloped, but not far enough apart for these areas to function as natural areas or sites for recreation. All these development approaches necessitate frequent automobile use.

driving cars largely takes the place of walking during daily errands. Physical inactivity increases obesity and high blood pressure, which can in turn lead to other ailments. A 2003 study found that people from the most-sprawling U.S. counties weigh 2.7 kg (6 lb) more for their height than people from the least-sprawling U.S. counties and that slightly more people from the most-sprawling counties show high blood pressure.

Land use The spread of low-density development means that more land is developed while less is left as forests, fields, farmland, or ranchland. Of the estimated 1 million ha (2.5 million acres) of U.S. land converted each year, roughly 60% is agricultural land and 40% is forest. These lands provide vital resources, recreation, aesthetic beauty, wildlife habitat, air and water purification, and other ecosystem services (pp. 3, 121–122, 160). Sprawl generally diminishes all these amenities.

Economics Sprawl drains tax dollars from existing communities and funnels them into infrastructure for new development on the fringes of those communities. Money that could be spent maintaining and improving downtown centers is instead spent on extending the road system, water and sewer system, electricity grid, telephone lines, police and fire service, schools, and libraries. These costs of extending infrastructure are generally paid by taxpayers of the community and are not charged to developers. For instance, one study calculated that sprawling development at Virginia Beach, Virginia, would require 81% more in infrastructure costs and would drain 3.7 times more from the community's general fund each year than compact urban development. Advocates for sprawling development argue that as owners of newly developed homes and businesses pay property taxes, this revenue eventually reimburses the community's investment in extending infrastructure. However, studies have found that in most cases taxpayers continue to subsidize new development unless municipalities specifically require that developers pay new infrastructure costs.

WEIGHING THE ISSUES

Sprawl Near You Is there sprawl in the area where you live? Does it bother you, or not? Has development in your area had any of the impacts described above? Do you think your city or town should use its resources to encourage outward growth? Why or why not?

CREATING LIVABLE CITIES

To respond to the challenges that sprawl presents, architects, planners, developers, and policymakers across North America today are trying to restore the vitality of city centers and to plan and manage how urbanizing areas develop. (see **ENVISIONIT**, p. 354)

City and regional planning help to create livable urban areas

How can we design cities so as to maximize their efficiency, functionality, and beauty? These are the questions central to **city planning** (also known as **urban planning**). City planners advise policymakers on development options, transportation needs, public parks, and other matters.

Washington, D.C., is the first—and still finest and grandest—example of city planning in the United States. President George Washington hired French architect Pierre Charles L'Enfant in 1791 to design a capital city for the new nation from scratch on undeveloped land along the Potomac River. L'Enfant laid out a Baroque-style plan of diagonal avenues cutting across a grid of streets, with plenty of space allotted for majestic public monuments, and the city was built according to his plan (**FIGURE 13.7**). A century later, as the city became crowded and dirty, a special commission in 1901 undertook new planning efforts to beautify the city while staying true to the intentions of L'Enfant's original plan. These planners imposed a height restriction on new buildings, which kept the magnificent government edifices and monuments from being crowded and dwarfed by modern skyscrapers, and preserved the spacious, airy, and stately feel of the city.

At the turn of the 20th century, city planning was coming into its own across North America as urban leaders sought to beautify and impose order on fast-growing, unruly cities. Landscape architect Daniel Burnham's 1909 *Plan of Chicago* was perhaps the grandest effort of this time, and it was largely implemented over the following years and decades. Burnham's plan expanded Chicago's city parks and playgrounds, improved neighborhood living conditions, streamlined traffic systems, and cleared industry and railroads from the shore of Lake Michigan to provide public access to the lake.

Portland gained its own comprehensive plan just three years later, in 1912. Edward Bennett's *Greater Portland Plan* recommended rebuilding the harbor; dredging the river channel; constructing new docks, bridges, tunnels, and a waterfront railroad; superimposing wide radial boulevards on the old city

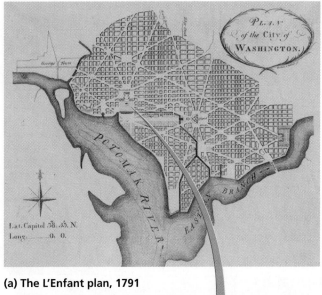

(a) The L'Enfant plan, 1791

(b) Washington, D.C., today

FIGURE 13.7 ▲ Pierre Charles L'Enfant's 1791 plan for the new capital city of Washington, D.C., **(a)** laid out a series of splendid diagonal avenues cutting across gridded streets, allowing plenty of space for the magnificent public monuments that still grace the city today **(b)**.

street grid; establishing civic centers downtown; and greatly expanding the number of parks. Voters approved the plan by a two-to-one margin, but they defeated a bond issue that would have paid for park development. As the century progressed, several other major planning efforts were conducted, and some ideas, such as establishing a downtown public square, came to fruition.

WEIGHING THE ISSUES

Your Urban Area Think of your favorite parts of the city you know best. What aspects do you like about them? What do you dislike about some of your least favorite parts of the city? What could this city do to improve the quality of life for its inhabitants?

In today's world of sprawling metropolitan areas, **regional planning** has become at least as important as city planning. Regional planners deal with the same issues as city planners, but

Do we really need to get stuck in endless traffic jams?

Or can we deal with urban sprawl and design cities to make our lives healthier and happier?

Years ago, cities were more like college campuses, letting us move around by foot, bike, and public transport.

Toronto park and skyline

What if we could recreate that style of city today?

Biker and bus in London

San Antonio Riverwalk

YOU CAN MAKE A DIFFERENCE

➤ Walk or ride your bike to class.
➤ Choose a place to live that has easy access to amenities and a short commute to work.
➤ Join with groups working to make your community a better place to live.

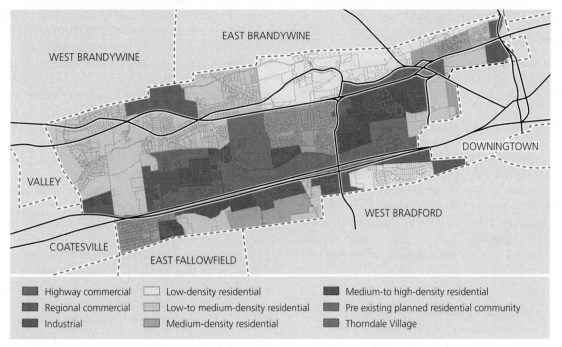

FIGURE 13.8 ▲ By zoning areas for different uses, planners guide how a community develops. This map for Caln Township, Pennsylvania, shows several common zoning patterns. Public and institutional uses are clustered together in a downtown area. Industrial uses are clustered together, away from most residential areas. Commercial uses are clustered along major roadways, and residential zones generally are higher in density toward the center of town.

they work on broader geographic scales and must coordinate their work with multiple municipal governments. In some places, regional planning has been institutionalized in formal government bodies; the Portland area's Metro is the epitome of such a regional planning entity. When Metro and its region's three counties in 2010 announced their collaborative plan on how to apportion their remaining undeveloped land into "urban reserves" and "rural reserves," it marked a historic accomplishment in regional planning. The agreement, hammered out after two years of complex and intensive negotiations, designates 28,000 acres for urban use and 272,000 acres for rural use. It means that homeowners, farmers, developers, and governments alike can feel more secure knowing what kinds of land uses lie in store on and near their land over the next half-century.

Zoning is a key tool for planning

One tool that planners use is **zoning**, the practice of classifying areas for different types of development and land use (**FIGURE 13.8**). For instance, to preserve the cleanliness and tranquility of residential neighborhoods, industrial facilities may be kept out of districts zoned for residential use. The specification of zones for different types of development gives planners a powerful means of guiding what gets built where. Zoning also gives home buyers and business owners security, because they know in advance what types of development can and cannot be located nearby.

Zoning involves government restriction on the use of private land and represents a top-down constraint on personal property rights. For this reason, some people consider zoning a regulatory taking (p. 172) that violates individual freedoms. Many others defend zoning, saying that government has a

proper role in setting limitations on property rights for the good of the community. Similar debates arise with endangered species management (p. 303) and other environmental issues.

Oregon voters sided with private property rights when in 2004 they passed Ballot Measure 37, which shackled government's ability to enforce zoning regulations with landowners who bought their land before the regulations were enacted. Subsequently, many Oregonians began witnessing new development they did not condone, and in 2007 they passed Ballot Measure 49 in order to restore public oversight over development. The passage of Oregon's Measure 37 spawned similar ballot measures in other U.S. states, although voters have defeated most of these. For the most part, people have supported zoning over the years because the common good it produces for communities is widely felt to outweigh the restrictions on private use.

WEIGHING THE ISSUES

Zoning and Development Imagine you own a 10-acre parcel of land that you want to sell for housing development—but the local zoning board rezones the land so as to prohibit the development. How would you respond?

Now imagine that you live next to someone else's undeveloped 10-acre parcel. You enjoy the privacy it provides—but the local zoning board rezones the land so that it can be developed into a dense housing subdivision. How would you respond?

What factors do you think members of a zoning board should take into consideration when deciding how to zone or rezone land in a community?

Urban growth boundaries are now widely used

Planners intended Oregon's urban growth boundaries to limit sprawl by containing growth largely within existing urbanized areas (**FIGURE 13.9**). The UGBs aimed to revitalize downtowns; protect farms, forests, and their industries; and ensure urban dwellers some access to open space near cities.

Since Oregon began its experiment, a number of other states, regions, and cities have adopted UGBs—from Boulder, Colorado, to Lancaster, Pennsylvania, to many California communities. In their own ways, all the UGBs aim to concentrate development, prevent sprawl, and preserve working farms, orchards, ranches, and forests. UGBs also appear to reduce the amounts that municipalities have to pay for infrastructure, compared to sprawl. The best estimate nationally is that UGBs save taxpayers about 20% on infrastructure costs. However, UGBs also seem to increase housing prices within their boundaries. In the Portland area, housing has become less affordable, but in most other ways its UGB (see Figure 13.9) is working as intended. It has lowered prices for land outside the UGB while raising prices within it. It has restricted development outside the UGB. It has increased the density of new housing inside the UGB by over 50% as homes are built on smaller lots and as multistory apartments fulfill a vision of "building up, not out." Downtown employment rose by 73% between 1970 and 1995 as businesses and residents alike invested in the central city. And Portland has been able to absorb considerable immigration while avoiding rampant sprawl.

However, urbanized area still increased by 101 km² (39 mi²) in the decade after Portland's UGB was established, because 146,000 people were added to the population. This fact suggests that relentless population growth may thwart even the best anti-sprawl efforts and that livable cities may fall victim to their own success if they are in high demand as places to live. Indeed, Metro has enlarged the Portland-area UGB three dozen times and plans to expand it more in the future, into the newly designated urban reserves.

"Smart growth" aims to counter sprawl

As more people feel negative effects of sprawl on their everyday lives, efforts to control growth are springing up throughout North America. Oregon's Senate Bill 100 was one of the first, and since then dozens of states, regions, and cities have adopted similar land use policies. Urban growth boundaries and many other ideas from these policies have coalesced under the concept of **smart growth** (**TABLE 13.2**).

Proponents of smart growth want municipalities to manage the rate, placement, and style of development so as to promote healthy neighborhoods and communities, jobs and economic development, transportation options, and environmental quality. They aim to rejuvenate the older existing communities that so often are drained and impoverished by sprawl. Smart growth means "building up, not out"—focusing development and economic investment in existing urban centers and favoring multistory shop-houses and high-rises.

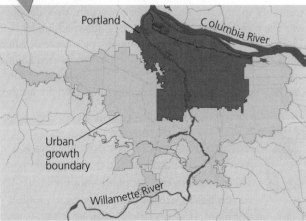

FIGURE 13.9 ▲ Oregon's urban growth boundaries (UGBs) were a response to fears that suburban sprawl might one day stretch in a great megalopolis from Eugene, Oregon, to Seattle, Washington. The Portland area's 956-km² (369-mi²) UGB encompasses Portland (dark gray) and portions of 24 other communities and three counties (light gray). Its jagged boundary separates areas planners have earmarked for urban development from areas they have chosen to protect from urban development.

TABLE 13.2 Ten Principles of "Smart Growth"
▶ Mix land uses
▶ Take advantage of compact building design
▶ Create a range of housing opportunities and choices
▶ Create walkable neighborhoods
▶ Foster distinctive, attractive communities with a strong sense of place
▶ Preserve open space, farmland, natural beauty, and critical environmental areas
▶ Strengthen existing communities, and direct development toward them
▶ Provide a variety of transportation choices
▶ Make development decisions predictable, fair, and cost-effective
▶ Encourage community and stakeholder collaboration in development decisions

Source: U.S. Environmental Protection Agency.

FIGURE 13.10 ◀ This plaza at Mizner Park in Boca Raton, Florida, is part of a planned community built in the new urbanist style. Homes, schools, and businesses are mixed close together in a centered neighborhood so that most amenities are within walking distance.

"New urbanism" is now in vogue

A related movement among architects, planners, and developers is **new urbanism**. This approach seeks to design neighborhoods on a walkable scale, with homes, businesses, schools, and other amenities all close together for convenience. The aim is to create functional neighborhoods in which most of a family's needs can be met close to home without the use of a car. Green spaces, trees, a mix of architectural styles, and creative street layouts add to the visual interest and pleasantness of new-urbanist developments (**FIGURE 13.10**). These developments mimic the traditional urban neighborhoods that existed until the advent of suburbs.

New-urbanist neighborhoods are generally connected to public transit systems. In **transit-oriented development**, compact communities in the new-urbanist style are arrayed around stops on a major rail line, enabling people to travel most places they need to go by train and foot alone. Several lines of the Washington, D.C., Metro system are developed in this manner.

Over 600 communities in the new-urbanist style across North America are in planning or construction, and many are now complete. Among them are Seaside, Florida; Kentlands in Gaithersburg, Maryland; Addison Circle in Addison, Texas; Mashpee Commons in Mashpee, Massachusetts; Harbor Town in Memphis, Tennessee; Celebration in Orlando, Florida; and Orenco Station, west of Portland.

To develop urban centers in these ways, zoning rules must allow it. Traditionally, zoning rules have often limited the density of development. Although well-intentioned, such rules encourage sprawl, so some are now being rethought in communities that desire new urbanism and smart growth.

Transportation options are vital to livable cities

Traffic jams on roadways cause air pollution, stress, and countless hours of lost personal time, and they cost the U.S. economy an estimated $74 billion each year in fuel and lost productivity. So, a key ingredient in any planner's recipe for improving the quality of urban life is making alternative transportation options available to citizens. These options include public buses, trains, subways, and *light rail* (smaller rail systems powered by electricity).

Mass transit rail systems ease traffic congestion, take up less space than road networks, and emit less pollution than cars. A 2005 study calculated that each year rail systems in U.S. metropolitan areas save taxpayers a total of $67.7 billion in costs for congestion, consumer transportation, parking, road maintenance, and accidents—far more than the $12.5 billion that governments spend to subsidize rail systems each year. As long as an urban center is large enough to support the infrastructure necessary, both train and bus systems are cheaper, more energy-efficient, and cleaner than roadways choked with cars (**FIGURE 13.11**).

In Portland, the bus system carries 66 million riders per year, and an average Portland bus is estimated to keep about 250 cars off the road each day. Portland also boasts downtown streetcars and one of the nation's leading light rail systems (**FIGURE 13.12**). Metro policy encourages the development of self-sufficient neighborhood communities in the new-urbanist style along the rail lines, and light rail ridership is steadily increasing as the system expands.

The nation's most-used train systems are the extensive heavy rail systems in America's largest cities, such as New York City's subways, Washington, D.C.'s Metro, the T in Boston, and the San Francisco Bay area's BART. Each of these carries more than one-fourth of its city's daily commuters.

In general, however, the United States lags behind most nations when it comes to mass transit. Many countries, rich and poor alike, have extensive and accessible bus systems that ferry citizens within and between towns and cities cheaply and effectively. And while Japan and many European nations have developed modern high-speed "bullet" trains (**FIGURE 13.13**), the United States has been starving its only national passenger rail network, Amtrak, of funding.

The United States has neglected mass transit for the past several decades largely because in a nation where population density was low and gasoline was cheap, it chose instead to

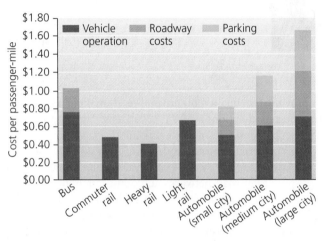

(a) Energy consumption for different modes of transit

(a) MAX light rail train

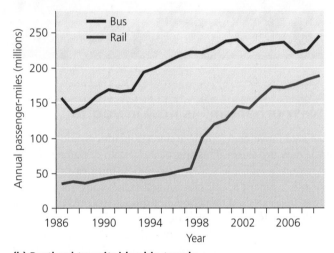

(b) Operating costs for different modes of transit

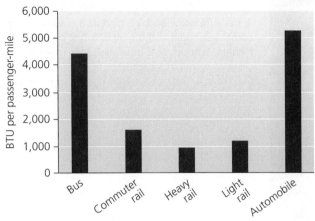

(b) Portland transit ridership trends

FIGURE 13.11 ▲ Rail transit consumes far less energy per passenger mile **(a)** than bus or automobile transit. Rail transit involves fewer costs per passenger mile **(b)** than bus or auto-mobile transit. Data from Litman, T., 2005. *Rail transit in America: A comprehensive evaluation of benefits.* © 2005 T. Litman. Used with permission.

FIGURE 13.12 ▲ Its light rail system **(a)** is one component of an urban planning strategy that has helped make Portland one of North America's most livable cities. In Portland, bus ridership has been increasing slowly, whereas ridership on the MAX light rail system is growing quickly **(b)** as the system expands.

358

invest in road networks for cars and trucks. As energy costs and population each rise, however, mass transit becomes increasingly appealing, and citizens clamor for the establishment or expansion of train and bus systems in their communities. The nation may even see its first bullet train before long; the 2009 stimulus bill passed by Congress set aside $8 billion for developing high-speed rail, and the Obama administration has identified 10 potential corridors for development of such trains.

Establishing mass transit is not always easy, however. Once a road system is in place, lined with businesses and homes, it can be difficult and expensive to replace or complement it with a mass transit system. Strong and visionary political leadership may be required. Such was the case in Curitiba, Brazil. Faced with an influx of immigrants from outlying farms in the 1970s, city leaders led by Mayor Jaime Lerner decided to pursue an aggressive planning process so that they could direct growth rather than being overwhelmed by it. They established a large fleet of public buses and reconfigured Curitiba's road system to maximize its efficiency. Today this metropolis of 2.5 million people has an outstanding bus system that is used each day by three-quarters of the population. The 340 bus routes, 250 terminals,

FIGURE 13.13 ▲ The "bullet trains" of the high-speed rail systems of Japan and Europe can travel at 150–220 mph. China, too, is investing in high-speed trains, such as this one speeding through the city of Qingdao. Plans to establish such lines in the United States are just getting underway.

and 1,900 buses accompany measures to encourage bicycles and pedestrians. All of this has resulted in a steep drop in car use, despite the city's rapidly growing population.

To make urban transportation more efficient, governments can also raise fuel taxes, tax inefficient modes of transport, reward carpoolers with carpool lanes, encourage bicycle use and bus ridership, and charge trucks for road damage. They can choose to minimize investment in infrastructure that encourages sprawl and to stimulate investment in renewed urban centers.

Parks and open space are key elements of livable cities

City dwellers often desire some sense of escape from the noise, commotion, and stress of urban life. Natural lands, public parks, and open space provide greenery, scenic beauty, freedom of movement, and places for recreation. These lands also keep ecological processes functioning by regulating climate, producing oxygen, filtering air and water pollutants, and providing habitat for wildlife. The animals and plants of urban parks and natural lands also serve to satisfy biophilia (pp. 299–300), our natural affinity for contact with other organisms. In the wake of urbanization and sprawl, protecting natural lands and establishing public parks has become more important, as many urban and suburban dwellers come to feel increasingly disconnected from nature.

America's city parks began to arise in the late 19th century as politicians and citizens yearning to make their crowded and dirty cities more livable established public spaces using aesthetic ideals borrowed from European parks, gardens, and royal hunting grounds. The lawns, shaded groves, curved pathways, and pastoral vistas we see today in many American city parks originated with these European ideals, as interpreted by the leading American landscape architect, Frederick Law Olmsted. Olmsted designed New York's Central Park (**FIGURE 13.14**) and many other urban park systems.

East Coast cities such as New York, Boston, and Philadelphia developed parks early on, but cities further west were not far behind. In San Francisco, William Hammond Hall transformed 2,500 ha (1,000 acres) of the peninsula's natural landscape of dunes into a verdant playground of lawns, trees, gardens, and sports fields. Golden Gate Park remains today one of the world's foremost city parks. Portland's quest for urban parks began in 1900, when city leaders created a parks commission and then hired Frederick Law Olmsted's son, John Olmsted, to design a citywide park system. His 1904 plan recommended acquiring land to ring the city generously with parks, but no action was taken. A full 44 years later, citizens pressured city leaders to create Forest Park along a large forested ridge on the northwest side of the city. At 11 km (7 mi) long, it is today the largest city park in the United States.

Parklands come in various types

Large city parks are a key component of a healthy urban environment, but even small spaces can make a big difference. Playgrounds provide places where children can be active outdoors and interact with their peers. Community gardens allow people to grow their own vegetables and flowers in a neighborhood setting (**FIGURE 13.15**).

FIGURE 13.14 ▲ Central Park in Manhattan was one of the first American city parks to be developed, and it remains one of the largest and finest.

FIGURE 13.15 ▲ Urban community gardens like this one in Seattle provide city residents with a place to grow vegetables and also serve as green spaces that beautify cities.

Greenways are strips of land that connect parks or neighborhoods, often run along rivers, streams, or canals, and provide access to networks of walking trails. They can protect water quality, boost property values, and serve as corridors for the movement of wildlife and people alike. The Rails-to-Trails Conservancy has spearheaded the conversion of abandoned railroad rights-of-way into greenways for walking, jogging, and biking. To date, 24,000 km (15,000 mi) of 1,500 rail lines have been converted across North America.

The concept of the corridor is sometimes implemented on a large scale. *Greenbelts* are long and wide corridors of parklands, often encircling an entire urban area. One example is the system of forest preserves that stretches through Chicago's suburbs like a necklace (**FIGURE 13.16**). In Canada, cities including Toronto, Ottawa, and Vancouver employ greenbelts as urban growth boundaries, containing sprawl and preserving open space for city residents.

Within urban parklands, many cities are working to enhance the "naturalness" of these areas through ecological restoration (pp. 95–96, 308–309), the practice of restoring native communities. In Portland's parks, volunteer teams plant native vegetation and remove English ivy, an invasive plant that covers trees and smothers native plants on the forest floor. At some Chicago-area forest preserves, scientists and volunteers use prescribed burns (p. 329) to restore prairie native to the region. In San Francisco's Presidio, areas are being restored to the native dune communities that were displaced by urban development.

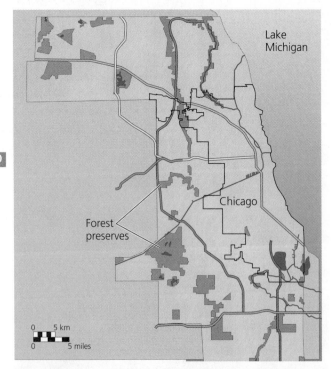

FIGURE 13.16 ▲ The forest preserves of Cook County, DuPage County, Lake County, and Will County, Illinois, wind through the suburbs surrounding the city of Chicago. This regional greenbelt system features 40,000 ha (100,000 acres) of woodlands, fields, marshes, and prairies, comprising the largest holding of locally owned public conservation land in the United States.

URBAN SUSTAINABILITY

Many of the approaches that planners, political leaders, and ordinary citizens are taking to make their cities more safe, clean, healthy, and pleasant are also helping to make these cities more sustainable. A sustainable city is one that can function effectively and prosperously over the long term, providing generations of residents a good quality of life far into the future. In part, this entails minimizing the city's impacts on the natural systems and resources that nourish it. Urban centers exert both positive and negative environmental impacts. The extent and nature of these impacts depend strongly on how we utilize resources, produce goods, transport materials, and deal with waste.

Urban resource consumption brings a mix of environmental effects

You might guess that urban living has a greater environmental impact than rural living. However, the picture is not so simple; instead, urbanization brings a complex mix of consequences.

Resource sinks Cities and towns are sinks (pp. 122-123) for resources, having to import from source areas beyond their borders nearly everything they need to feed, clothe, and house their inhabitants. Urban and suburban areas rely on large expanses of land elsewhere to supply food, fiber, water, timber, metal ores, and mined fuels. Urban centers also need areas of natural land to provide ecosystem services, including purification of water and air, nutrient cycling, and waste treatment (pp. 3, 121–122, 160). Indeed, for their day-to-day survival, major cities such as New York, Boston, San Francisco, and Los Angeles depend on water they pump in from faraway watersheds (**FIGURE 13.17**).

The long-distance transportation of resources and goods from countryside to city requires a great deal of fossil fuel use and thus has considerable environmental impacts. However, imagine that all the world's 3.5 billion urban residents were instead spread evenly across the landscape. What would the transportation requirements be, then, to move all those resources and goods around to all those people? A world without cities would likely require *more* transportation to provide people the same level of access to resources and goods.

Efficiency Once resources have arrived at an urban center where people are densely concentrated, cities minimize per capita consumption through the efficient delivery of goods and services. For instance, providing electricity from a power plant for densely packed urban houses and apartments is more efficient than providing electricity to far-flung homes in the countryside. The density of cities facilitates the provision of many social services that improve quality of life, including medical services, education, water and sewer systems, waste disposal, and public transportation.

More consumption Because cities draw resources from afar, their ecological footprints are much greater than their actual land areas. For instance, urban scholar Herbert Girardet

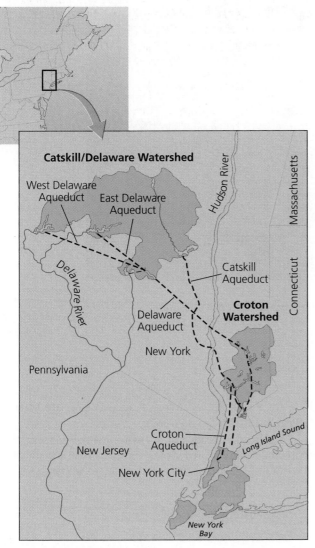

FIGURE 13.17 ▲ New York City pipes in its drinking water from upstate reservoirs in the Croton and Catskill/Delaware watersheds. The city has gone to lengths to acquire, protect, and manage watershed land to minimize pollution of these water sources. When New York City was confronted in 1989 with an order by the Environmental Protection Agency to build a $6 billion filtration plant to protect its citizens against waterborne disease, the city opted instead to purchase and better protect watershed land—for a fraction of the cost.

calculated that the ecological footprint of London, England, extends 125 times larger than the city's actual area. By another estimate, cities take up only 2% of the world's land surface but consume over 75% of its resources.

However, the ecological footprint concept is most meaningful when used on a per capita basis. So, in asking whether urbanization causes increased resource consumption, we must ask whether the average urban dweller has a larger footprint than the average rural dweller. The answer is yes, but urban and suburban residents also tend to be wealthier than rural residents, and wealth correlates with resource consumption. Thus, although urban citizens tend to consume more than rural ones, the reason could be simply that they tend to be wealthier.

Urbanization preserves land

Because people pack densely together in cities, more land outside cities is left undeveloped. Indeed, this is the very idea behind urban growth boundaries. If cities did not exist, and if instead all 6.9 billion of us were evenly spread across the planet's land area, no large blocks of land would be left uninhabited, and we would have much less room for agriculture, wilderness, biodiversity, or privacy. The fact that half the human population is concentrated in discrete locations helps allow room for natural ecosystems to continue functioning and provide the ecosystem services on which all of us, urban and rural, depend.

Urban centers suffer and export pollution

Just as cities import resources, they export wastes, either passively through pollution or actively through trade. In so doing, urban centers transfer the costs of their activities to other regions—and mask the costs from their own residents. Citizens of Toronto may not recognize that pollution from coal-fired power plants in their region worsens acid precipitation hundreds of miles to the east. Citizens of New York City may not realize how much garbage their city produces if it is shipped elsewhere for disposal.

However, not all waste and pollution leaves the city. Urban residents are exposed to heavy metals, industrial compounds, and chemicals from manufactured products that accumulate in soil and water. Airborne pollutants cause photochemical smog, industrial smog, and acid precipitation. Fossil fuel combustion releases greenhouse gases and pollutants that pose health risks.

Urban residents also suffer noise pollution and light pollution. **Noise pollution** consists of undesired ambient sound. Excess noise degrades one's surroundings aesthetically, can cause stress, and at intense levels (such as with prolonged exposure to the sounds of leaf blowers, lawn mowers, and jackhammers) can harm hearing. The glow of **light pollution** from city lights obscures the night sky, impeding the visibility of stars.

City residents suffer thermal pollution as well, because cities often have ambient temperatures that are several degrees higher than those of surrounding areas. This **urban heat island effect** results from the concentration of heat-generating buildings, vehicles, factories, and people, and from the way that buildings and dark paved surfaces absorb heat and then release it slowly, warming the air and interfering with patterns of convective circulation that would otherwise cool the city (**FIGURE 13.18**).

These various forms of pollution and the health risks they pose are not evenly shared among urban residents. Those who receive the brunt of the pollution are often those who are too poor to live in cleaner areas. Environmental justice concerns (pp. 144–146) center on the fact that a disproportionate number of people living near, downstream from, or downwind from factories, power plants, and other polluting facilities are people who are poor and, often, people of racial minorities.

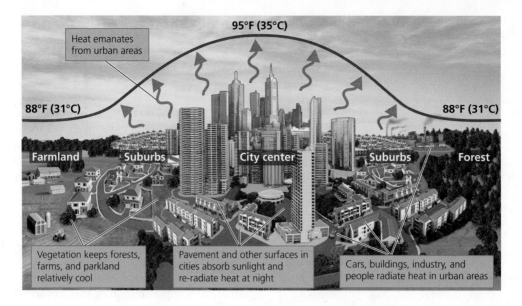

95°F (35°C)

Heat emanates from urban areas

88°F (31°C)

88°F (31°C)

Farmland   Suburbs   City center   Suburbs   Forest

Vegetation keeps forests, farms, and parkland relatively cool

Pavement and other surfaces in cities absorb sunlight and re-radiate heat at night

Cars, buildings, industry, and people radiate heat in urban areas

FIGURE 13.18 ▲ Cities produce urban heat islands, creating temperatures warmer than surrounding areas. The high concentration of people and objects generates heat, and pavement, buildings, and other surfaces absorb daytime heat and then release it slowly at night.

Urban centers foster innovation

Cities promote a flourishing cultural life and, by mixing together diverse people and influences, spark innovation and creativity. The urban environment can promote education and scientific research, and cities have long been viewed as engines of technological and artistic inventiveness. This inventiveness can lead to solutions to societal problems, including ways to reduce environmental impacts.

For instance, research into renewable energy sources is helping us develop ways to replace fossil fuels. Technological advances have helped us reduce pollution. Wealthy and educated urban populations provide markets for low-impact goods, such as organic produce. Recycling programs help reduce the solid waste stream. Environmental education is helping people choose their own ways to live cleaner, healthier, lower-impact lives. All these phenomena arise from the education, innovation, science, and technology that are part of urban culture.

Urban ecology helps cities take steps toward sustainability

Modern cities that import all their resources and export all their wastes have a linear, one-way metabolism. Linear models of production and consumption tend to destabilize environmental systems and are not sustainable. Proponents of sustainability for cities stress the need to develop circular systems, akin to systems found in nature, which recycle materials and use renewable sources of energy.

Researchers in the field of **urban ecology** hold that cities can be viewed explicitly as ecosystems and that the fundamentals of ecosystem ecology and systems science (Chapter 5) apply to the urban environment. Major urban ecology projects are ongoing in Baltimore and Phoenix, where researchers are studying

these cities explicitly as ecological systems (see **THE SCIENCE BEHIND THE STORY**, pp. 364–365).

To help cities reduce their environmental impacts while improving quality of life for their citizens, urban sustainability advocates suggest that cities follow an ecosystem-centered model by striving to:

▶ Maximize efficient use of resources.

▶ Recycle as much as possible (pp. 626–627).

▶ Develop environmentally friendly technologies.

▶ Account fully for external costs (pp. 150–151).

▶ Offer tax incentives to encourage sustainable practices.

▶ Use locally produced resources.

▶ Use organic waste and wastewater to restore soil fertility.

▶ Encourage urban agriculture.

Planners and visionary leaders have come up with designs for entire cities—"eco-cities"—built from scratch that follow cyclical and fully sustainable patterns of resource use and waste recycling. So far, however, none of these efforts has panned out. In recent years, several high-profile attempts in China collapsed amid faulty implementation, corruption, and misunderstandings between Western architects and Chinese leaders. New efforts are ongoing and give some cause for optimism, however.

In practice, the pursuit of urban sustainability is happening piecemeal—but at a rapid pace now—across the world as more and more cities are adopting sustainable strategies. For instance, urban agriculture is a growing pursuit in many urban areas, from Portland to Cuba to Japan. Singapore produces all its meat and 25% of its vegetable needs within its city limits, while in Berlin, Germany, 80,000 people grow food in community gardens. Municipal recycling programs continue to grow across the United States and the world. Curitiba,

Brazil, shows the kind of success that can result when a city invests in well-planned infrastructure. Besides the highly effective bus transportation network described earlier, the city provides recycling, environmental education, job training for the poor, and free health care. Surveys show that its citizens are unusually happy and better off economically than people living in other Brazilian cities.

In 2007, New York City unveiled an extensive plan that Mayor Michael Bloomberg hoped would make it "the first environmentally sustainable 21st-century city." Bloomberg's "PlaNYC" is a 127-item program to reduce greenhouse gas emissions, improve mass transit, plant trees, clean up polluted land and rivers, and enhance access to parkland. The aim of PlaNYC is to make New York City a better place to live as it accommodates 1 million more people by 2030 (**TABLE 13.3**). The most controversial aspect of the program, a proposal to charge drivers for driving into downtown Manhattan, has not gained approval from the state government, but other aspects of the plan are succeeding. As of 2010, just three years into the program, accomplishments included planting 322,000 trees;

opening or renovating 113 school playgrounds; acquiring 29,000 acres to protect the upstate water supply; installing 200 miles of bike lanes and 5,000 bike racks; retrofitting the Staten Island Ferry to reduce pollution; converting 25% of the taxi fleet to hybrid vehicles; and reducing greenhouse gas emissions by 9%.

Green buildings are a key step toward sustainable cities

Another accomplishment of PlaNYC so far is its initiation of 224 energy efficiency projects in city buildings. Constructing or renovating buildings using new efficient technologies is probably the most effective way that cities can reduce energy consumption and the greenhouse gas emissions that lead to climate change. Buildings consume 40% of the energy and 70% of the electricity used in the United States, so any efforts to improve the efficiency of buildings will go a long way.

Today there is a thriving movement in architecture and construction to design and build **green buildings**, structures that incorporate various means of reducing the ecological footprint of a building's construction and operation. Green buildings are built from sustainable materials, minimize their use of energy and water, minimize health impacts on their occupants, limit their pollution, and recycle their waste (**FIGURE 13.19**). The U.S. Green Building Council promotes these efforts by running a certification program called the **Leadership in Energy and Environmental Design (LEED)** program. Buildings (either new buildings or renovation projects) apply for certification and, depending on their performance, may be granted silver, gold, or platinum status (**TABLE 13.4**, p. 366).

Green building techniques add expense to construction, but the added cost is generally less than 3% for a LEED-silver building and 10% for a LEED-platinum building. LEED certification is booming throughout the United States. Portland now features several dozen LEED-certified buildings, including nine LEED-platinum structures. As one example, the Rosa Parks Elementary School, completed in 2007 and certified LEED-gold, was built with locally sourced and nontoxic materials, uses

---

TABLE 13.3 Goals of New York City's PlaNYC Sustainability Program

▶ Create affordable, sustainable housing for 1 million more people

▶ Ensure that all New Yorkers live within a 10-minute walk of a park

▶ Clean up all contaminated land in the city

▶ Reduce water pollution, and open 90% of waterways for recreation

▶ Repair and maintain water delivery network

▶ Add substantial new mass transit capacity

▶ Bring subways and roads up to full state of good repair

▶ Upgrade energy infrastructure to provide clean power

▶ Achieve the cleanest air quality of any big city in America

▶ Reduce greenhouse gas emissions by 30%

Source: City of New York, Office of the Mayor *plaNYC Progress Report 2010: A Greener, Greater New York.*

FIGURE 13.19 ◀ Green buildings incorporate many features to reduce the building's energy use, water use, and ecological footprint generally. Made from sustainable materials, they are built to be healthy for their occupants, to limit pollution, and to recycle waste. The features may even extend outside the building: Chicago's City Hall boasts a green roof, or ecoroof.

The SCIENCE behind the Story

Baltimore and Phoenix Showcase Urban Ecology

Researchers in urban ecology examine how ecosystems function in cities and suburbs, how species respond to urbanization, and how people interact with the urban environment. Today, Baltimore and Phoenix are centers for urban ecology.

These two cities are very different: Baltimore is an Atlantic port city on Chesapeake Bay with a long history, whereas Phoenix is a young and fast-growing Southwestern metropolis sprawling across the desert. Each was picked by the U.S. National Science Foundation to serve as a research site in its prestigious Long Term Ecological Research (LTER) program, which funds multi-decade ecological research. Since 1997, hundreds of researchers have studied Baltimore and Phoenix explicitly as ecosystems, examining nutrient cycling, biodiversity, air and water quality, how people react to environmental health threats, and more.

Research teams in both cities are combining old maps, aerial photos, and new remote sensing satellite data to reconstruct the history of landscape change. In Phoenix, one group showed how urban development spread across the desert in a "wave of advance," affecting soils, vegetation, and microclimate as it went. In Baltimore, mapping efforts showed that the amount of forest remained the same over the past 100 years but that

An urban ecologist takes a water sample under an overpass in Baltimore

residential and commercial development fragmented the forest into smaller patches.

The study regions designated for each city encompass both heavily urbanized central city areas and rural and natural areas on the urban fringe. To detect the impacts of urbanization, many research projects compare conditions in these two types of areas.

The Baltimore project's scientists can see ecological effects of urbanization just by comparing the urban lower end of their site's watershed with its less-developed upper end. In the lower end, pavement, rooftops, and compacted soil prevent rainfall from infiltrating the soil, so water runs off quickly into streams. The rapid flow cuts streambeds deeply into the earth while leaving the surrounding soil

drier. As a result, wetland-adapted trees and shrubs are vanishing, replaced by dry-adapted upland trees and shrubs.

The fast flow of water also worsens pollution. In natural areas, streams and wetlands break down nitrogen compounds, effectively filtering pollution. But in urban areas, where wetlands dry up and runoff from pavement creates flash floods, streams lose their filtering ability. In Baltimore, the resulting pollution ends up in Chesapeake Bay, which suffers eutrophication and a large hypoxic dead zone (pp. 108–109, 112–114). Baltimore-project scientists studying nutrient cycling (pp. 122–131) found that urban and suburban watersheds have far more nitrate pollution than natural forests (**first figure**).

Baltimore research also revealed that applying salt to icy roads in winter has major environmental impacts. The salt makes its way into streams, which become up to 100 times saltier than natural streams, even in summer when road salt is not applied. Such high salinity kills organisms (**second figure**), degrades habitat and water quality, and impairs streams' ability to remove nitrate.

To study contamination of groundwater and drinking water, researchers are using isotopes (p. 26) in a forensics approach to trace where salts in the most polluted streams are coming

---

24% less energy and water than comparable buildings, and diverted nearly all of its construction waste from the landfill. The schoolchildren learn about renewable energy in their own building by watching a display of the electricity produced by its solar photovoltaic (PV) system. We will explore green buildings at colleges and universities in Chapter 24 (pp. 670–671) in our discussion of campus sustainability efforts.

Steps toward livability enhance sustainability

The best news of all is that most of the steps being taken to make cities more livable—more pleasant, safer, cleaner, and healthier for residents—are also helping to make cities more sustainable. Planning and zoning are pursuits that specifically

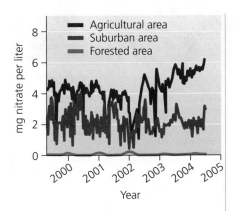

Streams in suburban areas in Baltimore had nitrate levels much higher than streams in nearby forested areas, but lower than streams in nearby agricultural areas, where fertilizers are used liberally. With kind permission from Springer Science and Business Media and the author, Groffman, P.M., et al., 2004. Nitrogen fluxes and retention in urban watershed ecosystems. *Ecosystems* 7: 393–403; Fig 4. Also from Baltimore Ecosystem Study.

from. On the bright side, Baltimore is improving water quality substantially by spending $900 million upgrading its sewer system.

Urbanization also affects species and ecological communities. Cities and suburbs facilitate the spread of non-native species, because people introduce exotic ornamental plants and because urbanization's impacts on the soil, climate, and landscape favor weedy generalist species over more specialized native ones. In Baltimore, non-native plant species are most abundant in urban areas. In Phoenix's dry climate, pollen from some non-native plants causes allergy problems for city residents.

Community ecologists studying the wild animals and plants that persist within Phoenix are finding that urbanization alters the relationships among them. Compared with natural landscapes, cities offer steady and reliable food resources—think of people's bird feeders, or food scraps from dumpsters.

Growing seasons are extended, and seasonal variation is buffered in cities, as well. The urban heat island effect (pp. 361–362) raises nighttime temperatures closer to daytime ones and makes temperatures more similar year-round. Buildings and ornamental vegetation shelter animals from extreme conditions, and irrigation in yards and gardens makes water available. In a desert city like Phoenix, watering boosts primary productivity and lowers daytime temperatures. Together, all these changes shift patterns of predation and competition among species. This tends to lead to higher population densities of animals but lower species diversity as generalists thrive and displace specialists.

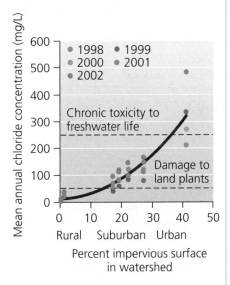

Salt concentrations (from runoff of road salt) in Baltimore-area streams were high enough to damage plants in the suburbs and to kill aquatic animals in urban areas. Adapted from Kaushal, S.S., et al., 2005. Increased salinization of fresh water in the northeastern United States. *Proc. Natl. Acad. Sci.* USA 102: 13517–13520, Fig 2. Copyright 2005 National Academy of Sciences, U.S.A. By permission.

Urban ecologists in Phoenix and Baltimore are studying social and demographic aspects of the urban environment as well. Some studies measure how natural amenities affect property values. For example, one study found that proximity to a park increases a home's property values—unless crime is pervasive. If the robbery rate surpasses 6.5 times the national average (as it does in Baltimore) then proximity to a park begins to depress property values.

Environmental justice concerns (pp. 144–146) are the focus of other studies, and these have repeatedly found that sources of industrial pollution tend to be located in neighborhoods that are less affluent and that are home to people of racial and ethnic minorities. Phoenix researchers mapped patterns of air pollution and toxic chemical releases and found that minorities and the poor are exposed to a greater share of these hazards. As a result, they also suffer from higher rates of childhood asthma.

In Baltimore, researchers found a more complex pattern. Toxic release sites were more likely to be in working-class white neighborhoods than in African American neighborhoods. This, the researchers concluded, was a result of historical inertia. In the past, living close to one's workplace—the factories that release toxic chemicals—was something people preferred, and white workers claimed the privilege of living near them.

Whether addressing the people, natural communities, or changing ecosystems of the urban environment, studies on urban ecology like those in Phoenix and Baltimore will be vitally informative in our ever-more-urban world. ∎

entail a long-term vision. By projecting farther into the future than political leaders or businesses generally do, planning and zoning are powerful forces for sustaining urban communities. The principles and practices of smart growth and new urbanism cut down on energy consumption, thus helping us address the looming societal problems of peak oil (pp. 539–544) and climate change (Chapter 18). Moving transportation away from cars and toward mass transit systems reduces gasoline consumption and carbon emissions. And parks offer ecosystem services in the city as well as promoting mental health for residents.

In general, to pursue urban sustainability, experts advise that developed countries should invest in resource-efficient technologies to reduce their impacts and enhance their

TABLE 13.4 Green Building Approaches for LEED (Leadership in Energy and Environmental Design) Certification

Measure	Techniques that are rewarded	Maximum points
Sustainable sites	Build on previously developed land; minimize erosion, pollution, runoff, and impacts on waterways; use regionally appropriate landscaping; integrate with transportation options	21
Water efficiency	Use efficient appliances inside; landscape for water conservation outside	11
Energy and atmosphere	Monitor energy use; use efficient design, construction, appliances, systems, and lighting; use clean, renewable sources of energy (often generated on-site)	37
Materials and resources	Use local or sustainably grown, harvested, and produced products and materials; reduce, reuse, and recycle waste	14
Indoor environmental quality	Use strategies that improve indoor air, provide access to natural daylight and views; improve acoustics	17
Points above may total up to 100. Then, up to 10 bonus points may be awarded for:		
Innovation in design	Use new and innovative technologies and strategies to go beyond what is required by other LEED credits	6
Regional priority	Address environmental concerns identified as locally most important for the specific region of the country	4

Out of 110 possible points, 40 are required for LEED certification, 50 for silver, 60 for gold, and 80 for platinum levels. Point allocation shown is for new construction. Allocation varies slightly for renovations and specific types of buildings.
Source: U.S. Green Building Council.

economies, whereas developing countries should invest in basic infrastructure to improve health and living conditions. Successes from Portland to Curitiba to New York City suggest that cities may one day make urban sustainability a reality.

Indeed, because they affect the environment in some positive ways and have the potential for efficient resource use, urban centers can and should be a key element in achieving progress toward global sustainability.

➤ CONCLUSION

As half the human population has shifted from rural to urban lifestyles, the nature of our environmental impact has changed. As urban and suburban dwellers, our impacts are less direct but often more far-reaching. Resources must be delivered to us over long distances, requiring the use of still more resources. Limiting the waste of those resources by making urban and suburban areas more sustainable will be vital for the future. Fortunately, the innovative cultural environment that cities foster has helped us develop solutions to alleviate impact and promote sustainability.

Part of seeking urban sustainability lies in making urban areas better places to live. A key component of these efforts

involves expanding transportation options to relieve congestion and make cities run more efficiently. Another lies in ensuring access to adequate parklands to keep us from becoming wholly isolated from nature. Accomplishments in city and regional planning have improved many American cities, and we should be encouraged about such progress. Proponents of smart growth and new urbanism believe they have solutions to the challenges posed by urban and suburban sprawl. Continuing experimentation in cities everywhere will help us determine how best to ensure that urban growth improves our quality of life while not degrading the quality of our environment.

REVIEWING OBJECTIVES

You should now be able to:

DESCRIBE THE SCALE OF URBANIZATION

- The world's population is becoming predominantly urban. (p. 345)
- The shift from rural to urban living is driven largely by industrialization and is proceeding fastest now in the developing world. (pp. 345–346)
- Nearly all future population growth will be in cities of the developing world. (p. 346)

- Environmental factors influence the location and growth of cities. (pp. 346–347)
- The geography of urban areas is changing as cities decentralize and suburbs grow and expand. (pp. 347–348)

ASSESS URBAN AND SUBURBAN SPRAWL

- Sprawl covers large areas of land with low-density development. Both population growth and increased per capita land use contribute to sprawl. (pp. 348–349)

- Sprawl results from the home-buying choices of individuals who prefer suburbs to cities, and it has been facilitated by government policy and technological developments. (p. 349)
- Sprawl may lead to negative impacts involving transportation, pollution, health, land use, natural habitat, and economics. (pp. 350–352)

OUTLINE CITY AND REGIONAL PLANNING AND LAND USE STRATEGIES

- City and regional planning and zoning are key tools for improving the quality of urban life. (pp. 353, 355)
- Urban growth boundaries, "smart growth," and "new urbanism" attempt to re-create compact and vibrant urban spaces. (pp. 356–357)

EVALUATE TRANSPORTATION OPTIONS

- Mass transit systems can enhance the efficiency and sustainability of urban areas. (pp. 357–358)
- The United States lags behind other nations in mass transit, but as population and demand increase, new efforts are being made. (pp. 357–358)

DESCRIBE THE ROLES OF URBAN PARKS

- Urban parklands provide recreation, soothe the stress of urban life, and keep people in touch with natural areas. (p. 359)

- A variety of types of parklands exist, including playgrounds, community gardens, greenways, and greenbelts. (pp. 359–360)

ANALYZE ENVIRONMENTAL IMPACTS AND ADVANTAGES OF URBAN CENTERS

- Cities are resource sinks with high per capita resource consumption, and they create substantial waste and pollution. (pp. 360–362)
- Cities also can maximize efficiency, allow natural lands to be preserved, and foster innovation that can lead to solutions for environmental problems. (pp. 360–362)

ASSESS URBAN ECOLOGY, GREEN BUILDING EFFORTS, AND THE PURSUIT OF SUSTAINABLE CITIES

- The linear mode of consumption and production is unsustainable, and more circular modes will be needed to create sustainable cities. (p. 362)
- Although a true "eco-city" has yet to be built, many cities worldwide are taking steps to decrease their ecological footprints. (pp. 362–363)
- The burgeoning green building movement is a key part of urban sustainability efforts. (pp. 363–364, 366)
- Most steps taken for urban livability also enhance sustainability. (pp. 364–366)

TESTING YOUR COMPREHENSION

1. What factors lie behind the shift of population from rural areas to urban areas? What types of cities and countries are experiencing the fastest urban growth today, and why?

2. Why have so many city dwellers in the United States, Canada, and other nations moved into suburbs?

3. Give two definitions of *sprawl*. Describe five negative impacts that have been suggested to result from sprawl.

4. What are city planning and regional planning? Contrast planning with zoning. Give examples of some of the suggestions made by early planners, such as Daniel Burnham and Edward Bennett.

5. How are some people trying to prevent or slow sprawl? Describe some key elements of "smart growth." What

effects, positive and negative, do urban growth boundaries tend to have?

6. Describe several apparent benefits of rail transit systems. What is a potential drawback?

7. How are city parks thought to make urban areas more livable? Give three examples of types of parks or public spaces.

8. Why do urban dwellers tend to consume more resources per capita than rural dwellers?

9. Describe the connection between urban ecology and sustainable cities. List three actions a city can take to enhance its sustainability.

10. Name two positive effects of urban centers on the natural environment.

SEEKING SOLUTIONS

1. Evaluate the causes of the spread of suburbs and of the environmental, social, and economic impacts of sprawl. Overall, do you think the spread of urban and suburban development that many people label *sprawl* is predominantly a good thing or a bad thing? Do you think it is inevitable? Give reasons for your answers.

2. Would you personally want to live in a neighborhood developed in the style of the new urbanism? Would you like

to live in a city or region with an urban growth boundary? Why or why not?

3. Consider the variety of approaches used to construct or renovate a building using LEED-certified green building techniques. Are there any LEED-certified buildings on your campus? If so, how do they differ from conventional buildings? Think about a building on your campus that you think is unhealthy or environmentally wasteful.

Name three to five specific ways in which green building techniques could be used to improve this building.

4. All things considered, do you feel that cities are a positive thing or a negative thing for environmental quality? How much do you feel we may be able to improve the sustainability of our urban areas?

5. **THINK IT THROUGH** You are the president of your college or university, and students are clamoring for you to help make your campus into the world's first fully sustainable campus. Considering what you have learned about enhancing livability and sustainability in cities, what lessons might you try to apply to your college or university? You are scheduled to give a speech to the campus community about your plans and will need to name five specific actions you plan to take to pursue a sustainable campus. What will they be, and what will you say about each choice to describe its importance?

6. **THINK IT THROUGH** After you earn your college degree, you are offered three equally desirable jobs, in three very different locations. If you take the first, you will live in the midst of a densely populated city. If you accept the second, you will live in a suburb where you have more space but where sprawl may soon surround you for many miles. If you select the third, you will live in a rural area with plenty of space but few cultural amenities. You are a person who aims to live in the most ecologically sustainable way you can. Where would you choose to live? Why? What considerations will you factor into your decision?

CALCULATING ECOLOGICAL FOOTPRINTS

One way of altering your ecological footprint is to consider transportation alternatives. Each gallon of gasoline is converted to approximately 20 lb of carbon dioxide (CO_2) during combustion, and this CO_2 is then released into the atmosphere. The table below lists typical amounts of CO_2 released for each person per mile, through various forms of transportation, assuming typical fuel efficiencies.

For an average North American person who travels 12,000 miles per year, calculate and record in the table the CO_2 emitted yearly for each transportation option, and the reduction in CO_2 emission that one could achieve by relying solely on each option.

	CO_2 per person per mile	CO_2 per person per year	CO_2 emission reduction	Your estimated mileage per year	Your CO_2 emissions per year
Automobile (driver only)	0.825 lb	9,900 lb	0		
Automobile (2 persons)	0.413 lb				
Automobile (4 persons)	0.206 lb				
Vanpool (8 persons)	0.103 lb				
Bus	0.261 lb				
Walking	0.082 lb				
Bicycle	0.049 lb				
				Total = 12,000	

1. Which transportation option will give you the most miles traveled per unit of carbon dioxide emitted?

2. Clearly, it is unlikely that any of us will walk or bicycle 12,000 miles per year or travel only in vanpools of eight people. In the last two columns, estimate what proportion of the 12,000 annual miles you think that you actually travel by each method, and then calculate the CO_2 emissions that you are responsible for generating over the course of a year. Which transportation option accounts for the most emissions for you?

3. How could you reduce your CO_2 emissions? How many pounds of emissions do you think you could realistically eliminate over the course of the next year by making changes in your transportation decisions?

MasteringENVIRONMENTALSCIENCE™

A health care worker in Kenya demonstrates a mosquito net to help villagers combat malaria

14 ENVIRONMENTAL HEALTH AND TOXICOLOGY

UPON COMPLETING THIS CHAPTER, YOU WILL BE ABLE TO:

- Identify major environmental health hazards and explain the goals of environmental health
- Describe the types, abundance, distribution, and movement of toxic substances in the environment
- Discuss the study of hazards and their effects, including wildlife toxicology, case histories, epidemiology, animal testing, and dose-response analysis

- Evaluate risk assessment and risk management
- Compare philosophical approaches to risk
- Describe regulatory policy in the United States and internationally

Poison in the Bottle: Is Bisphenol A Safe?

"Babies in the U.S. are born pre-polluted with BPA. What more evidence do we need to act?"

—Dr. Janet Gray, Director of the Environmental Risks and Breast Cancer Project, Vassar College

"There is no basis for human health concerns from exposure to BPA."

—The American Chemistry Council

Bisphenol A: Worldwide

Is this baby ingesting toxic substances?

How is it that a chemical found to alter reproductive development in animals gets used in baby bottles? How can it be that a substance linked to breast cancer, prostate cancer, and heart disease is routinely used in food and drink containers? The chemical **bisphenol A** (**BPA** for short) has been associated with everything from neurological effects to miscarriages. Yet it's in hundreds of products we use every day, and there's a better than 9 in 10 chance that it is coursing through your body right now.

To understand how chemicals that may pose health risks come to be widespread in our society, we need to explore how scientists and policymakers study toxic substances and other environmental health risks—and the vexing challenges these pursuits entail.

Chemists first synthesized bisphenol A, an organic compound (p. 28) with the chemical formula $C_{15}H_{16}O_2$, in 1891. As they began producing plastics in the 1950s, chemists found bisphenol A to be useful in creating epoxy resins used in lacquers and coatings. Epoxy resins containing BPA were soon being used to line the insides of metal food and drink cans and the insides of pipes for our water supply, as well as in enamels, varnishes, adhesives, and even dental sealants for our teeth.

Chemists also found that linking BPA molecules into polymers (p. 28) helped create polycarbonate plastic, a hard, clear type of plastic that soon found use in water bottles, food containers, eating utensils, eyeglass lenses, CDs and DVDs, laptops and other electronics, auto parts, refrigerator shelving, sports equipment, baby bottles, and children's toys. With so many uses, bisphenol A has become one of the world's most-produced chemicals; each year we make 1 pound of BPA for each person on the planet, and over 6 pounds per person in the United States!

Unfortunately, bisphenol A leaches out of its many products and into our food, water, air, and bodies. Fully 93% of Americans carry detectable concentrations in their urine, according to the latest National Health and Nutrition Examination Survey conducted by the Centers for Disease Control and Prevention (CDC). Because most BPA passes through the body within hours, these data suggest that we are receiving almost continuous exposure. Babies and children accumulate the most BPA because they eat more for their body weight and metabolize the chemical less effectively.

What, if anything, is BPA doing to us? To address such questions, scientists run experiments on laboratory animals, administering known doses of the substance and measuring the health impacts that result. Over 200 studies with rats, mice, and other animals have shown many apparent effects of BPA, including a wide range of reproductive abnormalities. A few recent studies suggest human health impacts as well (see **THE SCIENCE BEHIND THE STORY**, pp. 382–383).

Many of these effects occur at extremely low doses—much lower than the exposure levels set so far by regulatory agencies for human safety. Scientists say this is because BPA mimics the female sex hormone estrogen; that is, it is structurally similar to estrogen and can induce some of its effects in animals. Hormones such as estrogen function at minute concentrations, so when a synthetic chemical similar to estrogen reaches the body in a similarly low concentration, it can fool the body into responding.

In reaction to the burgeoning research, a growing number of researchers, doctors, and consumer advocates are calling on governments to regulate bisphenol A and for manufacturers to stop using it. The chemical industry insists that BPA is safe, pointing to industry-sponsored research that finds no health impacts. To sort through the debate, several expert panels have convened to assess the fast-growing body of scientific studies. Some panels have found nothing to worry about, others have expressed concern, and most have struggled with the fact that traditional research methods are not geared to test hormone-mimicking substances that exert effects at low doses. Indeed, dealing with substances like BPA is forcing us toward a challenging paradigm shift in the way we assess environmental health risks.

For instance, the U.S. Food and Drug Administration (FDA) insisted in 2008 that it saw no reason to regulate BPA, but its own science advisory committee disagreed, and in 2009 the FDA decided to start a testing program. Soon afterwards, the U.S. Environmental Protection Agency (EPA) announced it would begin its own assessment.

In 2008 Canada became the first nation to declare bisphenol A toxic. It banned the sale, import, and advertising of baby products using BPA. Since then, Denmark has issued a similar ban, and several U.S. states and municipalities have also taken action. Parents filed a class-action lawsuit against baby-bottle manufacturers, and legislators introduced bills in Congress to outlaw BPA's use in food and drink containers.

In the face of mounting press coverage and public concern, many companies are choosing to voluntarily remove BPA from their products. The six major U.S. manufacturers of plastic baby bottles promised in 2008 to stop using BPA and to find alternatives. Nalgene phased out its BPA-containing polycarbonate water bottles. Wal-Mart and Toys "R" Us decided to stop carrying children's products with BPA. The manufacturer Sunoco stopped selling BPA to companies that use it in children's products.

As of 2010, concerned parents can now more easily find BPA-free items for their infants and children. The rest of us remain exposed through most food

FIGURE 14.1 ▲ Researchers for *Consumer Reports* magazine tested these common packaged foods and more in 2009; they found that nearly all of them contained bisphenol A that had leached from the linings of their containers.

cans, many drink containers, and thousands of other products (**FIGURE 14.1**).

Bisphenol A is by no means one of our greatest environmental health threats. However, it provides a timely example of how we as a society assess health risks and decide how to manage them. As scientists and government regulators assess BPA's potential risks, their efforts give us a window on how hormone-disrupting chemicals are challenging the way we appraise and control the environmental health risks we face.

ENVIRONMENTAL HEALTH

Examining the impacts of human-made chemicals such as bisphenol A is just one aspect of the broad field of environmental health. The study and practice of **environmental health** assesses environmental factors that influence our health and quality of life. These factors include wholly natural aspects of the environment over which we have little or no control, as well as anthropogenic (human-caused) factors. Practitioners of environmental health seek to prevent adverse effects on human health and on the ecological systems that are essential to our well-being.

We face four types of environmental hazards

Many environmental health hazards exist in the world around us. We can categorize them into four main types: physical, biological, chemical, and cultural. For each type of hazard, there is some amount of risk that we cannot avoid—but there is also some amount of risk that we *can* avoid by taking precautions. Much of environmental health consists of taking steps to minimize the risks of encountering hazards and to mitigate the impacts of the hazards we do encounter.

Physical hazards *Physical hazards* arise from processes that occur naturally in our environment and pose risks to human life or health. Some such physical processes are ongoing natural phenomena, such as ultraviolet (UV) radiation from sunlight (**FIGURE 14.2A**). Excessive exposure to UV radiation damages

DNA and has been tied to skin cancer, cataracts, and immune suppression. We can reduce our exposure and risk to UV light by using clothing and sunscreen to shield our skin from intense sunlight.

Other physical hazards include discrete events such as earthquakes, volcanic eruptions, fires, floods, blizzards, landslides, hurricanes, and droughts. We can do little to predict the timing of a natural disaster such as an earthquake, and nothing to prevent one. However, we can minimize risk by preparing ourselves. Scientists can map geologic faults to determine areas at risk of earthquakes, engineers can design buildings to resist damage, and governments and individuals can take steps to prepare for a quake's aftermath.

Some common practices make us vulnerable to certain physical hazards. Clearing forests from slopes makes landslides more likely, for instance, and channelizing rivers promotes flooding in some areas while preventing it in others. We can reduce risk from such hazards by improving our forestry and flood control practices and by regulating development in areas prone to floods, landslides, fires, or coastal waves.

Chemical hazards *Chemical hazards* include many of the synthetic chemicals that our society manufactures, such as pharmaceuticals, disinfectants, and pesticides (**FIGURE 14.2B**). Some substances produced naturally by organisms (such as venoms) also can be hazardous, as can many substances that we find in nature and then process for our use (such as hydrocarbons, lead, and asbestos). Following our overview of environmental health,

much of this chapter will focus on chemical health hazards and the ways we study and regulate them.

Biological hazards *Biological hazards* result from ecological interactions among organisms (**FIGURE 14.2C**). When we become sick from a virus, bacterial infection, or other pathogen, we are suffering parasitism (pp. 81–82) by other species that are simply fulfilling their ecological roles. This is what we call **infectious disease**. Some infectious diseases are spread when pathogenic microbes attack us directly. With others, infection occurs through a *vector*, an organism (generally an arthropod, such as a mosquito) that transfers the pathogen to the host. Infectious diseases such as malaria, cholera, tuberculosis, and influenza (flu) are major environmental health hazards, especially in developing nations with widespread poverty and few resources for health care. As with physical and chemical hazards, it is impossible for us to avoid risk from biological agents completely, but through monitoring, sanitation, and medical treatment we can reduce the likelihood and impacts of infection.

Cultural hazards Hazards that result from our place of residence, our socioeconomic status, our occupation, or our behavioral choices can be thought of as *cultural hazards* or *lifestyle hazards*. We can minimize or prevent some of these cultural or lifestyle hazards, whereas others may be beyond our control. For instance, choosing to smoke cigarettes, or living or working with people who smoke, greatly increases our risk of lung cancer

FIGURE 14.2 ▶ Environmental health hazards come in four types. The sun's ultraviolet radiation is an example of a physical hazard **(a)**. Excessive exposure increases the risk of skin cancer. Chemical hazards **(b)** include both synthetic and natural chemicals. Much of our exposure comes from pesticides and household chemical products. Biological hazards **(c)** include diseases and the organisms that transmit them. Some mosquitoes are vectors for pathogenic microbes, including those that cause malaria. Cultural or lifestyle hazards **(d)** include the behavioral decisions we make, as well as the socioeconomic constraints forced on us. Smoking is a lifestyle choice that raises one's risk of lung cancer and other diseases.

(a) Physical hazard

(b) Chemical hazard

(c) Biological hazard

(d) Cultural hazard

(**FIGURE 14.2D**). Choosing to smoke is a personal behavioral decision, but exposure to secondhand smoke in the home or workplace may be beyond one's control. Much the same might be said for drug use, diet and nutrition, crime, and mode of transportation. As advocates of environmental justice (pp. 144–146) argue, health factors such as living near toxic waste sites or working unprotected with pesticides are often correlated with socioeconomic deprivation. The fewer economic resources or political clout one has, the harder it is to avoid cultural hazards and environmental health risks in general.

Many environmental health hazards exist indoors

We are exposed to environmental health hazards both outdoors and indoors (**TABLE 14.1**). We spend roughly 90% of our lives indoors, however, and the spaces inside our homes and workplaces can be rife with hazards, from dust to cigarette smoke to emissions from plastics, household chemicals, and cleaning products. We will briefly survey a few indoor health hazards here, before taking a closer look at indoor pollutants in Chapter 17 (pp. 486–490).

Cigarette smoke and radon are leading indoor hazards (pp. 487–488) and are the top two causes of lung cancer in developed nations. *Radon* is a highly toxic radioactive gas that is colorless and undetectable without specialized kits. Radon seeps up from the ground in areas with certain types of bedrock and can accumulate in basements and homes with poor circulation. The EPA estimates that slightly less than 1 person in 1,000 may contract lung cancer as a result of a lifetime of radon exposure at average levels for U.S. homes.

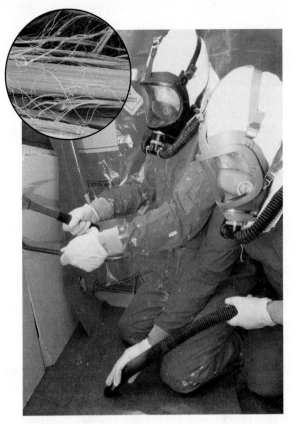

FIGURE 14.3 ▲ Asbestos was once widely used in insulation and other products. A cause of lung cancer and asbestosis, the fibrous substance (inset) has now been removed from many buildings in which it was used. Its removal poses risks as well, however, and removal workers must wear protective clothing and respirators.

Another indoor hazard is asbestos. There are several types of *asbestos*, each of which is a mineral that forms long, thin, microscopic fibers. This structure allows asbestos to trap heat, muffle sound, and resist fire. Because of these qualities, asbestos was used widely as insulation in buildings, as well as in many products. Unfortunately, its fibrous structure also makes asbestos dangerous when inhaled. When asbestos gets lodged in lung tissue, the body produces acid in an attempt to eliminate it. The acid scars the lung tissue but does little to dislodge or dissolve the asbestos. Within a few decades, the scarred lungs may cease to function, a disorder called *asbestosis*. Asbestos can also cause types of lung cancer. Because of these risks, asbestos has been removed from many schools and offices (**FIGURE 14.3**). However, the removal process can release some asbestos into the air, increasing people's exposure, so an alternative approach is to encase the material in place.

Lead poisoning is another indoor health hazard. When ingested, lead, a heavy metal, can cause damage to the brain, liver, kidney, and stomach; learning problems and behavioral abnormalities; anemia; hearing loss; and even death. It has been suggested that the downfall of ancient Rome was caused in part by chronic lead poisoning, because Romans drank wine sweetened with mixtures prepared in leaden vessels. The composer Beethoven suffered maladies that some historians attribute to lead poisoning. Today lead poisoning can result from drinking water that has passed through the lead pipes common in older homes. Even in newer pipes, lead solder was widely used into the 1980s and is still sold in stores.

TABLE 14.1 Selected Environmental Hazards
Outdoor Air
▸ Chemicals from automotive exhaust
▸ Chemicals from industrial pollution
▸ Photochemical smog (pp. 477–478)
▸ Pesticide drift
▸ Dust and particulate matter
Water
▸ Pesticide and herbicide runoff
▸ Nitrates and fertilizer runoff
▸ Mercury, arsenic, and other heavy metals in groundwater and surface water
Food
▸ Natural toxins
▸ Pesticide and herbicide residues
Indoors
▸ Smoking and secondhand smoke
▸ Radon
▸ Asbestos
▸ Lead in paint and pipes
▸ Toxicants in plastics and consumer products (bisphenol A, PBDEs, phthalates, etc.)
▸ Dust and particulate matter

Lead may be most dangerous through its presence in paint. Until 1978, most paints contained lead, and most home interiors were painted with lead-based paint. If you strip layers of paint from woodwork in an older home, you will likely be exposed to airborne lead and should take precautions. Babies and young children often peel paint from walls and may ingest or inhale it. Lead poisoning among U.S. children has greatly declined in recent years (**FIGURE 14.4**) as a result of education campaigns and the fact that lead-based paints and leaded gasoline have been phased out (p. 8), thanks to government regulation to protect public health. Not all nations have acted accordingly, however. In 2007, millions of children's toys and other items exported from China were found to contain lead-based paint. The resulting consumer outcry in the United States forced recalls of these items, and China eventually agreed to limit and monitor the use of lead-based paint in its manufacturing.

One recently recognized hazard is a group of chemicals known as *polybrominated diphenyl ethers (PBDEs)*. These compounds provide fire-retardant properties and are used in a diverse array of consumer products, including computers, televisions, plastics, and furniture. They are released during production and disposal of products and may also evaporate at very slow rates throughout the lifetime of products. These chemicals persist and accumulate in living tissue, and their abundance in the environment and in people in the United States is doubling every few years.

Like bisphenol A, PBDEs appear to act as hormone disruptors; lab testing with animals shows them to affect thyroid hormones. Animal testing also suggests that PBDEs affect the development of the brain and nervous system and may cause cancer. Concern about PBDEs rose after a study showed that concentrations in the breast milk of Swedish mothers had increased exponentially from 1972 to 1997. U.S. studies also show rising concentrations in breast milk. The European Union decided in 2003 to ban PBDEs, and industries in Europe phased them out. As a result, concentrations in breast milk of European mothers have fallen substantially. In the United States, however, there has so far been little movement to address the issue.

Disease is a major focus of environmental health

Among the hazards people face, disease stands preeminent. Despite all our technological advances, we still find ourselves battling disease, which causes the vast majority of human deaths worldwide (**FIGURE 14.5A**). Major killers such as cancer, heart disease, and respiratory disorders have some genetic basis, but they are influenced by environmental factors. For instance, whether a person develops asthma depends not only on his or her genes, but also on environmental conditions. Studies have shown that pollutants from fossil fuel combustion worsen asthma, and children raised on farms suffer less asthma than children raised in cities. Malnutrition (p. 254), poverty, and poor hygiene can each foster various illnesses. Lifestyle choices also matter: Smoking can lead to lung cancer, and lack of exercise to heart disease.

Over half the world's deaths result from noninfectious diseases, such as cancer and heart disease, and 1 death in 11 is due to injuries. Infectious diseases account for 1 of every 4 deaths that occur each year—nearly 15 million people worldwide (**FIGURE 14.5B**). Infectious disease is a greater problem in developing countries, where it accounts for close to half of all deaths. Infectious disease causes many fewer deaths in developed nations because their wealth allows their citizens better hygiene and access to medicine.

In the United States and other developed nations, lifestyle trends are altering the prevalence of noninfectious disease in ways both good and bad. In the last two decades, the percentage of Americans who smoke cigarettes has dropped by 38%. But as we exercise less and eat fattier diets, the percentage of Americans who suffer obesity has more than doubled (**FIGURE 14.6**).

Although infectious disease accounts for fewer deaths than noninfectious disease, infectious disease robs society of more years of human life, because it tends to strike people at all ages, including the very young. The World Health Organization (WHO) estimates that of all the years of life lost to deaths each year, infectious disease takes 51%, whereas noninfectious disease takes 34% and injuries take 14%. Decades of public health efforts have lessened the impacts of infectious disease and even have eradicated some diseases—yet other diseases are posing new challenges. Some, such as acquired immunodeficiency syndrome (AIDS), continue to spread globally despite concerted efforts to stop them (pp. 218–219). Others, such as tuberculosis and strains of malaria, are evolving resistance to our antibiotics, in the same way that pests evolve resistance to our pesticides (pp. 258–259).

Infectious disease interacts with social and environmental influences

Many diseases are spreading because we are so mobile in our modern era of globalization. A pathogen can now hop continents in a matter of hours by airplane in its human host. As a

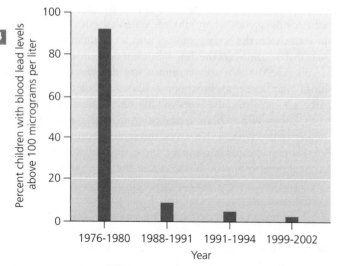

FIGURE 14.4 ▲ The percentage of U.S. children aged 1–5 with high levels of lead in the blood has declined steeply as a result of public health education and the government-mandated removal of lead from paint and gasoline. Data from National Center for Environmental Health, Centers for Disease Control and Prevention.

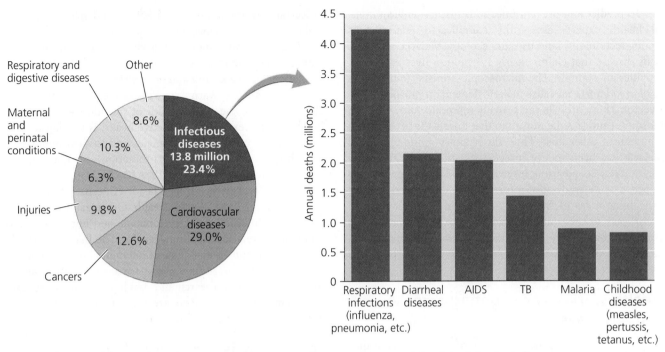

Respiratory and digestive diseases

Other

8.6%

Maternal and perinatal conditions

10.3%

6.3%

Injuries

9.8%

12.6%

Cancers

Infectious diseases
13.8 million
23.4%

Cardiovascular diseases
29.0%

(a) Leading causes of death across the world

(y-axis) Annual deaths (millions)

Respiratory infections (influenza, pneumonia, etc.) | Diarrheal diseases | AIDS | TB | Malaria | Childhood diseases (measles, pertussis, tetanus, etc.)

(b) Leading causes of death by infectious diseases

FIGURE 14.5 ▲ Infectious diseases are the second-leading cause of death worldwide **(a)**, accounting for over one-quarter of all deaths. Six types of diseases **(b)**—respiratory infections, diarrhea, AIDS, tuberculosis (TB), malaria, and childhood diseases—account for 80% of all deaths from infectious disease. Data from World Health Organization, 2009. *World health statistics 2009*. WHO, Geneva, Switzerland.

result, some diseases have moved into new areas. For example, West Nile virus, native to the Old World, was detected near New York City in 1999, and within just five years had spread to all lower 48 U.S. states! Although birds are most affected by this mosquito-transmitted virus, more than 27,000 human cases, including 1,000 deaths, had been recorded in the United States through the start of 2010.

In our world of global mobility and dense human populations, novel diseases (or new strains of old diseases) that emerge in one location are more likely to spread to other locations, even worldwide. Recent examples include severe acute respiratory syndrome (SARS) in 2003, the H5N1 avian flu starting in 2004, and the H1N1 swine flu that spread across the globe in 2009–2010 (**FIGURE 14.7**). Diseases like influenza, whose pathogens evolve rapidly, give rise to a variety of strains, making it more likely that one may turn exceedingly dangerous and threaten a global pandemic.

The changes we cause to our environment can also cause diseases to spread. Human-induced global warming of the climate (Chapter 18) is causing tropical diseases such as malaria, dengue, cholera, and yellow fever to begin expanding into the temperate zones. And habitat alteration can affect the abundance, distribution, and movement of certain disease vectors.

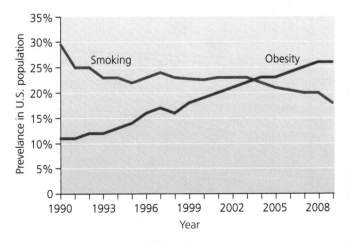

FIGURE 14.6 ▲ The prevalence of smoking has decreased in the United States in the past 20 years, but obesity is on the rise. Data from United Health Foundation, 2009. *America's health rankings, 2009 edition*. Minnetonka, MN.

FIGURE 14.7 ▼ The H1N1 swine flu outbreak in 2009–2010 showed how a strain of a fast-evolving infectious disease can spread globally in a densely populated world where people are highly mobile. Over 18,000 people had died of this flu as of July 2010.

To predict and prevent infectious disease, environmental health experts assess the complicated relationships among technology, land use, and ecology. Malaria, an infectious disease that claims nearly 1.3 million lives each year, provides an example. The microscopic protists (four species of *Plasmodium*) that cause malaria depend on mosquitoes as a vector. These protists can sexually reproduce only within a mosquito, and it is the mosquito that injects the protists into a human or other host. Thus, the primary mode to control malaria has been to use insecticides such as DDT to kill mosquitoes. People using insecticides and draining wetlands in broad-scale eradication projects have eliminated malaria from large areas of the temperate world where it used to occur, such as the southern United States. However, human land disturbance that creates pools of standing water in formerly well-drained areas can boost mosquito populations and allow malaria to reinvade.

Health workers are fighting disease in many ways

We have many ways in which to fight disease, and thousands of people—from doctors and nurses to policymakers to philanthropists—are dedicating their lives and careers to improving human health (see **ENVISIONIT**, p. 377). Perhaps the best way to reduce disease is to improve the basic living conditions of the world's poor. Besides providing food security (p. 254), this means ensuring access to safe drinking water and improving sanitation by minimizing exposure to human waste, garbage, and wastewater. In recent years, we have made slow but steady progress in providing adequate drinking water and sanitation to the world's people (**FIGURE 14.8**).

Another important pursuit is to expand access to health care. In developing nations, this includes opening clinics, immunizing children against diseases, providing prenatal and postnatal care for mothers and babies, and making generic and inexpensive pharmaceuticals available.

Education campaigns play a vital role in rich and poor nations alike. Public service announcements and government regulations on packaging and advertising have helped to decrease cigarette use, warn the public of the risks of drugs and alcohol, and advise us on issues of nutrition and exercise. Education on sex and reproductive health is helping to save women from unwanted pregnancies and to slow population growth (pp. 211–216), and the promotion of condom use is helping to slow the spread of HIV/AIDS (p. 219).

Such efforts are being spearheaded internationally by the United Nations, the World Health Organization, the U.S. Agency for International Development, and nongovernmental organizations and funding agencies. The Bill and Melinda Gates Foundation alone donates $800 million to global health programs each year. Meanwhile, pharmaceutical corporations invest vast sums to research and develop new medicines and treatments. All these efforts are continuing to enhance the quality of people's health throughout the world.

Toxicology is the study of poisonous substances

Although most indicators of human health are improving as the world's wealth increases, our modern society is exposing us to more and more synthetic chemicals. Some of these substances pose threats to human health, but figuring out which of them do—and how, and to what degree—is a complicated scientific endeavor. **Toxicology** is the science that examines the effects of poisonous substances on humans and other organisms. Toxicologists assess and compare substances to determine their **toxicity**, the degree of harm a chemical substance can inflict. A toxic substance, or poison, is called a **toxicant**, but any chemical substance may exert negative impacts if we ingest or expose

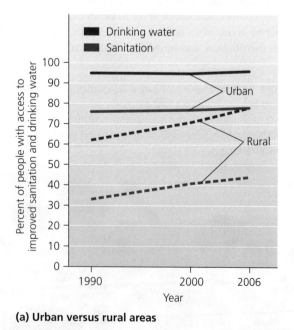

(a) **Urban versus rural areas**

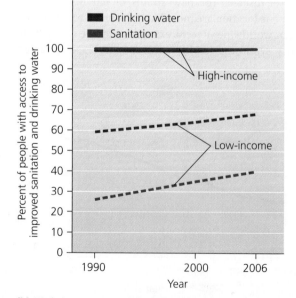

(b) **High-income versus low-income regions of the world**

FIGURE 14.8 ▲ We are gradually improving sanitation and drinking water for the world's people. Urban areas **(a)** are better off than rural areas, and high-income regions **(b)** are better off than low-income regions. Data from World Health Organization, 2009. *World health statistics 2009*, Geneva.

Environmental health efforts around the world are helping people cope with natural and human-created health risks.

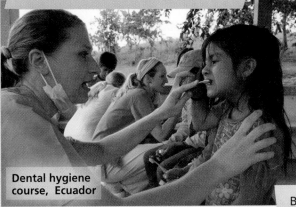

Dental hygiene course, Ecuador

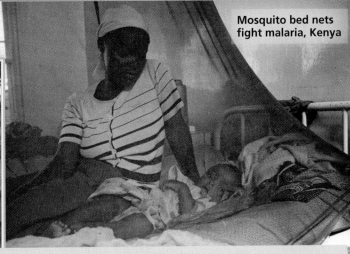

Mosquito bed nets fight malaria, Kenya

By improving sanitation and providing clean drinking water and better medical care, we can help fight infectious diseases like malaria, which threaten lives and sap the vigor of whole societies.

Water purification in a Ugandan village

X-ray shows tumor from lung cancer

In developed nations, cigarette smoke and radon exposure (which both lead to lung cancer) are major environmental health risks.

Smoke-Free Air Is A HUMAN RIGHT

YOU CAN MAKE A DIFFERENCE

➤ Keep your loved ones healthy by minimizing environmental hazards in your home, lifestyle, and community.

➤ Support public health initiatives to reduce obesity, smoking, radon exposure, and other hazards.

➤ Joining organizations like the Peace Corps allows U.S. students to help improve health in developing nations.

ourselves to enough of it. Conversely, a toxicant in a small-enough quantity may pose no health risk at all. These facts are often summarized in the catchphrase, "The dose makes the poison." In other words, a substance's toxicity depends not only on its chemical identity, but also on its quantity.

In recent decades, our ability to produce new chemicals has expanded, concentrations of chemical contaminants in the environment have increased, and public concern for health and the environment have grown. These trends have driven the rise of **environmental toxicology**, which deals specifically with toxic substances that come from or are discharged into the environment. Environmental toxicology studies health effects on humans, other animals, and ecosystems, and it represents one approach within the broader scope of environmental health. Toxicologists generally focus on human health, using other organisms as models and test subjects. In environmental toxicology, animals are also studied out of concern for their welfare and because—like canaries in a coal mine—animals can serve as indicators of health threats that could soon affect people.

Risks must be balanced against rewards

The job of toxicologists and other scientists who study environmental health hazards is to learn as much as they can about the hazards, but the rest of us need to take this information and weigh it against any benefits we obtain from exposing ourselves to the hazards. With most hazards, there is some tradeoff between risk and reward, and we must judge as best we can how these compare. In regard to bisphenol A, its usefulness for many purposes means that despite its health risks, we may as a society choose to continue using it. The availability of safer and affordable alternatives is important in such decisions. Industry is finding replacements for BPA polymers in baby bottles and water containers, but until a replacement is found for it as an epoxy liner for food cans, it will likely continue to serve this function.

As we review the impacts of toxic substances throughout this chapter, it is important to keep in mind that artificially produced chemicals have played a crucial role in giving us the standard of living we enjoy today. These chemicals have helped create the industrial agriculture that produces our food, the medical advances that protect our health and prolong our lives, and many of the modern materials and conveniences we use every day. It is appropriate to remember these benefits as we examine some of the unfortunate side effects of these advances and as we search for better alternatives.

TOXIC SUBSTANCES IN THE ENVIRONMENT

Our environment contains countless natural substances that may pose health risks. These include oil oozing naturally from the ground; radon gas seeping up from bedrock; and **toxins**, toxic chemicals manufactured in the tissues of living organisms—for example, chemicals that plants use to ward off herbivores or that insects use to defend themselves from predators. In addition, we are exposed to many synthetic (artificial, or human-made) chemicals.

TABLE 14.2 Estimated Numbers of Chemicals in Commercial Substances

Type of chemical	Estimated number
Chemicals in commerce	100,000
Industrial chemicals	72,000
New chemicals introduced per year	2,000
Pesticides (21,000 products)	600
Food additives	8,700
Cosmetic ingredients (40,000 products)	7,500
Human pharmaceuticals	3,300

Data are for the 1990s, from Harrison, P., and F. Pearce, 2000. *AAAS atlas of population and environment.* Berkeley, CA: University of California Press.

Synthetic chemicals are all around us

Tens of thousands of synthetic chemicals have been manufactured (**TABLE 14.2**), and synthetic chemicals surround us in our daily lives (**FIGURE 14.9**). Each year in the United States, we manufacture or import 113 kg (250 lb) of chemical substances for every man, woman, and child. Many of these substances find their way into soil, air, and water, as revealed by researchers who monitor environmental quality. For instance, scientists at the U.S. Geological Survey's National Water-Quality Assessment Program (NAWQA) have carried out systematic surveys for synthetic chemicals in U.S waterways and aquifers since the 1980s. A 2002 study found that 80% of U.S. streams contain at least trace amounts of 82 wastewater contaminants, including antibiotics, detergents, drugs, steroids,

FIGURE 14.9 ▲ Synthetic chemicals, such as those in household products, are everywhere around us in our everyday lives. Some of these compounds may potentially pose environmental or human health risks.

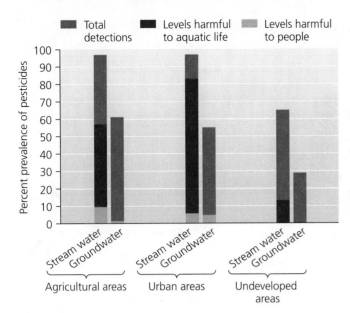

FIGURE 14.10 ▲ Nearly all U.S. streams and most aquifers in agricultural and urban areas contain pesticides throughout the year. Fewer than 10% of tested samples violate human health standards, but most violate standards for aquatic life. Data from Gilliom, Robert J., et al., 2006. *Pesticides in the nation's streams and ground water, 1992–2001.* Circular 1291, National Water-Quality Assessment Program, U.S. Geological Survey.

plasticizers, disinfectants, solvents, perfumes, and other substances. A 2006 study of groundwater detected 42 volatile organic compounds (VOCs, p. 473) in 18% of wells and 92% of aquifers tested throughout the nation, although fewer than 2% of samples violated federal health standards for drinking water. (VOCs are emitted from products such as gasoline, paints, and plastics, and they come from many sources, including urban runoff; engine exhaust; industrial emissions; wastewater; and leaky storage tanks, landfills, and septic systems.)

The pesticides we use to kill insects and weeds (p. 258) on farms, lawns, and golf courses are some of the most widespread synthetic chemicals. A 2006 NAWQA study concluded that pesticides are regularly present in streams and groundwater nationwide, finding traces of at least one pesticide in every stream that was tested. The data showed that concentrations were seldom high enough to pose health risks to people, but they were often high enough to affect aquatic life or fish-eating animals (**FIGURE 14.10**). Pesticide contamination is most severe in the farming states of the Midwest and Great Plains.

Synthetic chemicals are in all of our bodies

As a result of all this exposure, every one of us carries traces of hundreds of industrial chemicals in our bodies. The U.S. government's latest National Health and Nutrition Examination Survey (the one that found 93% of Americans showing traces of BPA in their urine; p. 370) gathered data on 148 foreign compounds in Americans' bodies. Among these were several toxic persistent organic pollutants restricted by international treaty (p. 396). Depending on the pollutant, these were detected in 41% to 100% of the people tested. Smaller-scale surveys have found similar results.

In 2009, science writer Arianne Cohen decided to get her own body surveyed to find out what chemicals were present inside her. She worked with researchers, shelled out over $4,000 to undergo a battery of tests, and then wrote up the results in the December 2009 issue of *Popular Science* magazine. The verdict: Her body contained BPA, persistent pollutants like dioxins, nitrates from food, chemicals from plastics, and plenty more. And why not?, she asked. After all, we encounter countless chemicals all day, from shampoo in our morning shower to packaging and nonstick pans at meals to pesticides on our lawns in the afternoon to flame retardants on our sheets at bedtime.

Our exposure to synthetic chemicals begins in the womb, as substances our mothers ingested while pregnant were transferred to us. A 2009 study by the nonprofit Environmental Working Group found 232 chemicals in the umbilical cords of 10 newborn babies it tested. Nine of the 10 umbilical cords contained BPA, leading researchers to note that we are born "pre-polluted."

All this should not necessarily be cause for alarm. Not all synthetic chemicals pose health risks, and relatively few are known with certainty to be toxic. However, of the roughly 100,000 synthetic chemicals on the market today, very few have been thoroughly tested. For the vast majority, we simply do not know what effects, if any, they may have.

Why are there so many synthetic chemicals around us? Let's consider herbicides and insecticides, made widespread by advances in chemistry and production capacity during and after World War II. As material prosperity grew in Westernized nations following the war, people began using pesticides for agriculture, for lawns and golf courses, and to kill insects in their homes and offices. Pesticides were viewed as a means toward a better quality of life. It was not until the 1960s that people began to learn about the risks of exposure to pesticides. The key event was the publication of Rachel Carson's 1962 book *Silent Spring* (p. 175), which brought the insecticide dichloro-diphenyl-trichloroethane (DDT) to the public's attention.

Silent Spring began the public debate over synthetic chemicals

Rachel Carson was a naturalist, author, and government scientist. In *Silent Spring*, she brought together a diverse collection of scientific studies, medical case histories, and other data that no one had previously synthesized and presented to the general public. Her message was that DDT in particular, and artificial pesticides in general, were hazardous to people, wildlife, and ecosystems. Carson wrote at a time when large amounts of pesticides virtually untested for health effects were indiscriminately sprayed over residential neighborhoods and public areas, on an assumption that the chemicals would do no harm to people (**FIGURE 14.11**). Most consumers had no idea that the store-bought chemicals they used in their houses, gardens, and crops might be toxic.

The chemical industry challenged Carson's book vigorously, attempting to discredit the author's science and personal reputation. Carson suffered from cancer as she finished *Silent Spring*, and she lived only briefly after its publication. However, the book was a best-seller and helped generate significant social change in views and actions toward the environment. The use of DDT was banned in the United States in 1973 and is now illegal in a number of nations.

FIGURE 14.11 ▶ Children on a Long Island, New York, beach are fogged with DDT from a pesticide spray machine being tested in 1945. Before the 1960s, the environmental and health effects of potent pesticides such as DDT were not widely known. Public parks and neighborhoods were regularly sprayed for insect control without safeguards against excessive human exposure.

U.S. chemical companies still manufacture and export DDT, because developing countries with tropical climates use DDT to control disease vectors, such as mosquitoes that transmit malaria. In these countries, malaria represents a greater health threat than do the toxic effects of the pesticide.

WEIGHING THE ISSUES

A Circle of Poison? Although the United States has banned the use of DDT, U.S. companies still manufacture and export the compound to developing nations. Thus, it is possible that pesticide-laden food can be imported back into the United States in what has been called a "circle of poison." How do you feel about this? Is it unethical for one country to sell to others a substance that it has deemed toxic? Or would it be unethical for the United States *not* to sell DDT to African nations if they desire it for controlling malaria?

Toxicants come in different types

Toxicants can be classified based on their particular effects on health. The best-known are **carcinogens**, which are substances or types of radiation that cause cancer. In cancer, malignant cells grow uncontrollably, creating tumors, damaging the body, and often leading to death. Cancer frequently has a genetic component, but a wide variety of environmental factors are thought to raise the risk of cancer. Indeed, in 2010 the President's Cancer Panel concluded that the prevalence of environmentally induced cancer has been "grossly underestimated." In our society today, the greatest number of cancer cases is thought to result from carcinogens contained in cigarette smoke. Polycyclic aromatic hydrocarbons (PAHs; p. 28) comprise some of the carcinogens found in cigarette smoke. PAHs also occur in charred meats and are released from the combustion of coal, oil , and natural gas.

Carcinogens can be difficult to identify because there may be a long lag time between exposure to the agent and the detectable onset of cancer—up to 15–30 years in the case of cigarette smoke. Moreover, as with all risks, only a portion of people

exposed to a carcinogen will eventually get cancer. Cancer is a leading cause of death that kills millions and leaves few families untouched. Two of every five Americans are diagnosed with cancer at some time in their lives, and one of every five die from it. Thus, the study of carcinogens has played a large role in shaping the way that toxicologists pursue their work.

Mutagens are substances that cause genetic mutations in the DNA of organisms (p. 29). Although most mutations have little or no effect, some can lead to severe problems, including cancer and other disorders. If mutations occur in an individual's sperm or egg cells, then the individual's offspring suffer the effects.

Chemicals that cause harm to the unborn are called **teratogens**. Teratogens that affect development of human embryos in the womb can cause birth defects. One example involves the drug thalidomide, developed in the 1950s as a sleeping pill and to prevent nausea during pregnancy. Tragically, the drug turned out to be a powerful teratogen, and its use caused birth defects in thousands of babies. Even a single dose during pregnancy could result in limb deformities and organ defects. Thalidomide was banned in the 1960s once scientists recognized its connection with birth defects. Ironically, today the drug shows promise in treating a wide range of diseases, including Alzheimer's disease, AIDS, and various types of cancer.

Other chemical toxicants, known as **neurotoxins**, assault the nervous system. Neurotoxins include venoms produced by animals such as snakes and stinging insects, heavy metals such as lead and mercury, pesticides, and some chemical weapons developed for use in war. A famous case of neurotoxin poisoning occurred in Japan, where a chemical factory dumped mercury waste into Minamata Bay between the 1930s and 1960s. Thousands of people there ate fish contaminated with the mercury and soon began suffering from slurred speech, loss of muscle control, sudden fits of laughter, and in some cases death. The company and the government eventually paid about $5,000 in compensation to each poisoned resident.

The human immune system protects our bodies from disease. Some toxicants weaken the immune system, reducing the body's ability to defend itself against bacteria, viruses, allergy-causing agents, and other attackers. Others, called

allergens, overactivate the immune system, causing an immune response when one is not necessary. One hypothesis for the increase in asthma in recent years is that allergenic synthetic chemicals are more prevalent in our environment. Allergens are not universally considered toxicants, however, because they affect some people but not others and because one's response does not necessarily correlate with the degree of exposure.

Most recently, scientists have recognized **endocrine disruptors**, toxicants that interfere with the *endocrine system*, or hormone system. The endocrine system consists of chemical messengers (*hormones*) that travel through the bloodstream at extremely low concentrations and have many vital functions. They stimulate growth, development, and sexual maturity, and they regulate brain function, appetite, sex drive, and many other aspects of our physiology and behavior. Some hormone-disrupting toxicants affect an animal's endocrine system by blocking the action of hormones or accelerating their breakdown. Others are so similar to certain hormones in their molecular structure and chemistry that they "mimic" the hormone by interacting with receptor molecules just as the actual hormone would (**FIGURE 14.12**).

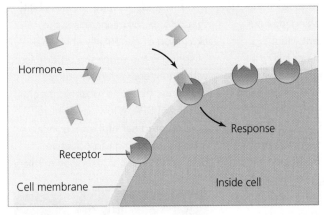

(a) Normal hormone binding

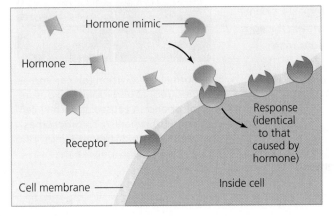

(b) Hormone mimicry

FIGURE 14.12 ▲ Many endocrine-disrupting substances mimic the structure of hormone molecules. Like a key similar enough to fit into another key's lock, the hormone mimic binds to a cellular receptor for the hormone, causing the cell to react as though it had encountered the hormone.

(a) Exposure through toys

(b) Exposure through cosmetics

FIGURE 14.13 ▲ Many soft plastic children's toys **(a)** and many cosmetics **(b)** contain phthalates, a hormone-disrupting chemical. Banned in Europe, phthalates remain widespread in the United States.

Bisphenol A is one of many chemicals that appear to mimic the female sex hormone estrogen and bind to estrogen receptors. Many plastic products also contain another class of hormone-disrupting chemical, called *phthalates*. Used to soften plastics and enhance fragrances, phthalates are used widely in children's toys (**FIGURE 14.13A**), perfumes and cosmetics (**FIGURE 14.13B**), and other items. As with BPA, nearly every American carries phthalates within his or her body as a result of environmental exposure. Health research on phthalates has linked them to birth defects, breast cancer, reduced sperm counts, and other reproductive effects. The European Union and nine other nations have banned phthalates, California and Washington enacted bans for children's toys, and the United States in 2008 banned six types of phthalates in toys. Still, across North America many routes of exposure remain.

It is important to note that any given substance may have multiple effects. For example, bisphenol A is primarily known as an endocrine disruptor, but research shows it to have mutagenic and teratogenic effects as well, and because studies link it to breast cancer and prostate cancer in lab animals, it also appears to act as a carcinogen.

Toxicants may concentrate in water

Toxic substances are not evenly distributed in the environment, and they move about in specific ways (**FIGURE 14.14**).

The SCIENCE behind the Story

Testing the Safety of Bisphenol A

Dr. Patricia Hunt, Case Western Reserve University

Of the many studies documenting health impacts of bisphenol A on lab animals, one of the first came about because a lab assistant reached for the wrong soap.

At a laboratory at Case Western Reserve University in Ohio in 1998, geneticist Patricia Hunt was making a routine check of her female lab mice. As she extracted and examined developing eggs from the ovaries, she began to wonder what had gone wrong. About 40% of the eggs showed problems with their chromosomes, and 12% had irregular amounts of genetic material, a dangerous condition called aneuploidy, which can lead to miscarriages or birth defects in mice and people alike.

A bit of sleuthing revealed that a lab assistant had mistakenly washed the lab's plastic mouse cages and water bottles with an especially harsh soap. The soap damaged the cages so badly that parts of them seemed to have melted.

The cages were made from polycarbonate plastic, which contains bisphenol A (BPA). Hunt knew at the time that BPA mimics estrogen and that some studies had linked the chemical to reproductive abnormalities in mice, such as low sperm counts and early sexual development. Other research indicated that BPA leaches out of plastic into water and food when the plastic is treated with heat, acidity, or harsh soap.

Hunt wondered whether the chemical might be adversely affecting the mice in her lab. Deciding to re-create the accidental cage-washing incident in a controlled experiment, Hunt instructed researchers in her lab to wash polycarbonate cages and water bottles using varying levels of the harsh soap. They then compared mice kept in damaged cages with plastic water bottles to mice kept in undamaged cages with glass water bottles.

The developing eggs of mice exposed to BPA through the deliberately damaged plastic showed significant problems during meiosis, the division of chromosomes during egg formation—just as they had in the original incident (see **photograph**). In contrast, the eggs of mice in the control cages were normal.

In another round of tests, Hunt's team gave sets of female mice daily oral doses of BPA over 3, 5, and 7 days. They observed the same meiotic abnormalities in these mice, although at lower levels (**first graph**). The mice given BPA for 7 days were most severely affected.

Published in 2003 in the journal *Current Biology*, Hunt's findings set off a new wave of concern over the safety of bisphenol A. The findings were disturbing because sex cells of mice and of people divide and function in similar ways. "We have observed meiotic defects in mice at exposure levels close to or even below those considered 'safe' for humans," the research paper stated. "Clearly, the possibility that BPA exposure increases the likelihood of genetically abnormal offspring is too serious to be dismissed without extensive further study."

Since that time, dozens of other studies of BPA at low doses have documented harmful effects in lab animals, including reproductive disorders related to estrogen mimicry

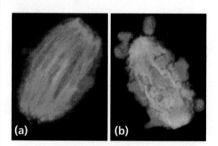

In normal cell division **(a)**, chromosomes (red) align properly. Exposure to bisphenol A causes abnormal cell division **(b)**, whereby chromosomes scatter and are distributed improperly and unevenly between daughter cells.

Water running off from land often transports toxicants from large areas and concentrates them in small volumes of surface water. The NAWQA findings on water quality reflect this concentrating effect. If chemicals persist in soil, they can leach into groundwater and contaminate drinking water supplies.

Many chemicals are soluble in water and enter organisms' tissues through drinking or absorption. For this reason, aquatic animals such as fish, frogs, and stream invertebrates are effective indicators of pollution. When aquatic organisms become sick, we can take it as an early warning that something

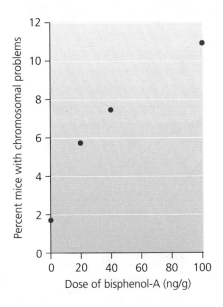

In this dose-response experiment, the percentage of mice showing chromosomal problems during cell division rose with increasing dose of bisphenol A. In the United States and Europe, regulators have set safe intake levels for people at doses of 50 ng/g of body weight per day. Data from Hunt, P.A., et al., 2003. Bisphenol A exposure causes meiotic aneuploidy in the female mouse. *Current Biology* 13: 546–553.

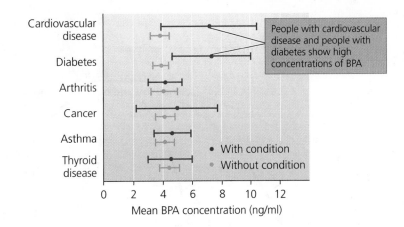

In a 2008 epidemiology study, average bisphenol A concentrations were significantly higher for Americans with diabetes and cardiovascular disease, but not with various other conditions. Error bars are 95% confidence intervals. Adapted from Lang, I.A., et al., 2008. Association of urinary bisphenol A concentration with medical disorders and laboratory abnormalities in adults. *JAMA* 300: 1303–1310. Used by permission of the American Medical Association.

but also other maladies ranging from thyroid problems to liver damage to obesity. Scientist Frederick vom Saal, whose research in 1997 had shown the first evidence for BPA's effects, said in 2007, "This chemical is harming snails, insects, lobsters, fish, frogs, reptiles, birds, and rats, and the chemical industry is telling people that because you're human, unless there's human data, you can feel completely safe."

Vom Saal did not have to wait long for the first human study to appear. In 2008, the *Journal of the American Medical Association* published research led by Iain Lang and David Melzer of the Peninsula Medical School, Exeter, U.K. Lang and Melzer's team took an epidemiological approach (p. 387) to assess BPA's possible effects on people, by using data from the U.S. government's latest National Health and Nutrition Examination Survey. Using data from 1,455 survey participants, they got a representative sample of adults in the U.S. population. After controlling the data for race/ethnicity, education, income, smoking, body mass, and other variables, they tested for statistical correlations between a series of major health disorders and the concentration of BPA in people's urine.

These researchers' analyses showed that Americans with high BPA concentrations showed high rates of diabetes and cardiovascular disease (**second graph**), as well as abnormal concentrations of three liver enzymes. The team found no association with a number of other conditions such as cancer, stroke, arthritis, thyroid disease, and respiratory diseases. The researchers also explored correlations with other estrogenic compounds and found that these did not show the associations that BPA showed.

Previous studies of the mechanisms by which BPA acts in cell cultures and in rodents' bodies helped explain how and why BPA might affect liver enzymes and diabetes. However, the reasons for cardiovascular effects remain unclear.

This first direct indication of human health impacts from BPA was a correlative study that does not establish causation. To demonstrate that BPA actually *causes* the observed effects, researchers would need to track people with low and high BPA levels for years, predict who would most likely get sick, and test these predictions with future data. It will take many years to complete such long-term studies. In the meantime, more and more scientists are urging regulators to restrict BPA based on the evidence already at hand. ■

is amiss. When scientists find low concentrations of pesticides harming frogs, fish, and invertebrates, they view this as a warning that people could be next. The contaminants that wash into streams and rivers also flow and seep into the water we drink and drift through the air we breathe.

Airborne substances can travel widely

Because many chemical substances can be transported by air (Chapter 17), chemicals can exert impacts far from the site of their origin and use. Airborne transport of pesticides is

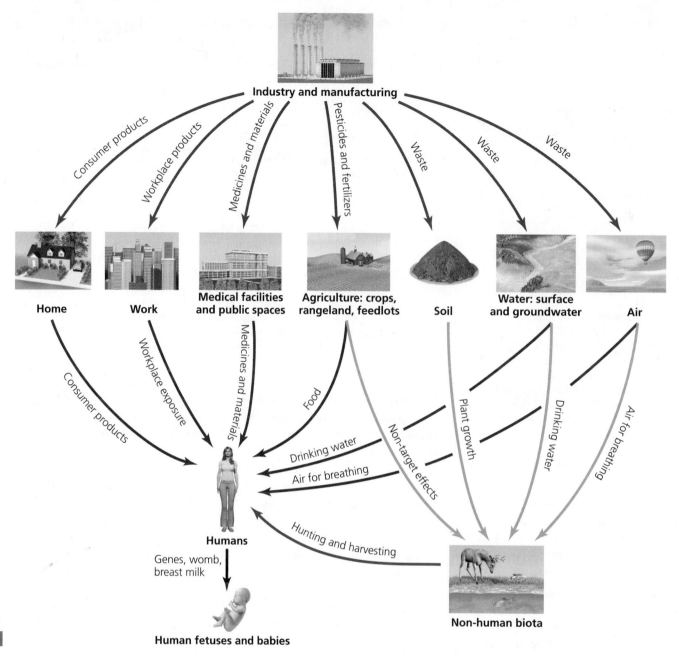

FIGURE 14.14 ▲ Synthetic chemicals take many routes in traveling through the environment. People take in only a tiny proportion of these compounds, and many compounds are harmless. However, people receive small amounts of toxicants from many sources, and developing fetuses and babies are particularly sensitive.

sometimes termed *pesticide drift*. The Central Valley of California is the world's most productive agricultural region, but because it is naturally arid, food production depends on the intensive use of irrigation, fertilizers, and pesticides. The region's frequent winds often blow airborne pesticide spray—and dust particles containing pesticide residue—for long distances. In the mountains of the Sierra Nevada, research has associated pesticide drift from the Central Valley with population declines in four species of frogs. Families living in towns in the Central Valley suffer health impacts, and advocates for farm workers maintain that hundreds of thousands of the state's residents are at risk.

Because so many substances are carried by the wind, synthetic chemicals are ubiquitous worldwide, even in

seemingly pristine areas. Scientists who travel to the most remote alpine lakes in the wilderness of British Columbia find them contaminated with industrial toxicants, such as polychlorinated biphenyls (PCBs), which are by-products of chemicals used in transformers and other electrical equipment. Earth's polar regions are particularly contaminated, because natural patterns of global atmospheric circulation (p. 467) tend to move airborne chemicals toward the poles (**FIGURE 14.15**). Thus, although we manufacture and apply synthetic substances mainly in temperate and tropical regions, contaminants are strikingly concentrated in the tissues of Arctic polar bears, Antarctic penguins, and people living in Greenland.

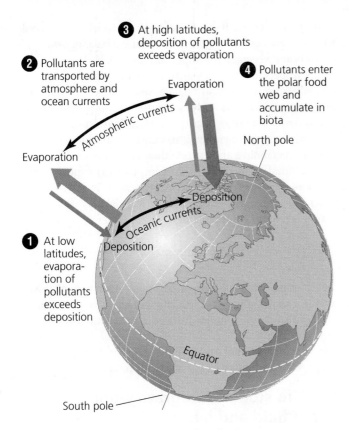

② Pollutants are transported by atmosphere and ocean currents

③ At high latitudes, deposition of pollutants exceeds evaporation

④ Pollutants enter the polar food web and accumulate in biota

Evaporation

Atmospheric currents

Evaporation

North pole

Deposition

Oceanic currents

Deposition

① At low latitudes, evaporation of pollutants exceeds deposition

Equator

South pole

FIGURE 14.15 ▲ In a process called *global distillation*, pollutants that evaporate and rise high into the atmosphere at lower latitudes **①**, or are deposited in the ocean, are carried toward the poles **②** by atmospheric currents of air and oceanic currents of water. This process concentrates pollutants near the poles **③**. Thus, polar organisms take in more than their share of toxicants **④**, despite the fact that relatively few synthetic chemicals are manufactured or used near the poles.

Some toxicants persist

Once a toxic substance arrives somewhere, it may degrade quickly and become harmless, or it may remain unaltered and persist for many months, years, or decades. The rate at which a given substance degrades depends on its chemistry and on factors such as temperature, moisture, and sun exposure. The Bt toxin (pp. 258–259, 263) used in biocontrol and genetically modified crops has a very short persistence time, whereas chemicals such as DDT and PCBs persist for decades. Atrazine, our most widely used herbicide, is highly variable in its persistence, depending on environmental conditions.

Persistent synthetic chemicals exist in our environment today because we have designed them to persist. The synthetic chemicals used in plastics, for instance, are used precisely because they resist breakdown. Sooner or later, however, most toxicants degrade into simpler compounds called *breakdown products*. Often these are less harmful than the original substance, but sometimes they are just as toxic as the original chemical, or more so. For instance, DDT breaks down into DDE, a highly persistent and toxic compound in its own right. Atrazine produces a large number of breakdown products whose effects have not been fully studied.

Toxic substances may accumulate and move up the food chain

Of the toxic substances that organisms absorb, breathe, or consume, some are quickly excreted, and some are degraded into harmless breakdown products. Others persist intact in the body. Substances that are fat-soluble or oil-soluble (including organic compounds such as DDT and DDE) are absorbed and stored in fatty tissues. Substances such as methylmercury (CH_3Hg^+) may be stored in muscle tissue. Such persistent toxicants accumulate in an animal's body in a process termed **bioaccumulation**, which results in the animal's having a greater concentration of the substance than exists in the surrounding environment.

Toxic substances that bioaccumulate in an organism's tissues may be transferred to other organisms as predators consume prey. When one organism consumes another, the predator takes in any stored toxicants and stores them itself. Thus bioaccumulation takes place on all trophic levels. Moreover, each individual consumes many individuals from the trophic level beneath it, so with each step up the food chain, concentrations of toxicants become magnified. This process, called **biomagnification**, occurred throughout North America with DDT. Top predators, such as birds of prey, ended up with high concentrations of the pesticide because concentrations became magnified as DDT moved from water to algae to plankton to small fish to larger fish and finally to fish-eating birds (**FIGURE 14.16**).

Biomagnification caused populations of many North American birds of prey to decline precipitously from the 1950s to the 1970s. The peregrine falcon was almost totally wiped out in the eastern United States, and the bald eagle, the U.S. national bird, was virtually eliminated from the lower 48 states. The brown pelican vanished from its Atlantic Coast range, remaining only in Florida, and the osprey and other hawks saw substantial population declines. Eventually scientists determined that DDT was causing these birds' eggshells to grow thinner, so that eggs were breaking while in the nest.

In a remarkable environmental success story, populations of all these birds have rebounded (p. 302) since the United States banned DDT. However, biomagnification is by no means a thing of the past. DDT continues to impair wildlife in parts of the world where it is still used, and mercury buildup in fish poses risks to human health in North America (p. 448). In the Arctic, polar bears at the top of the food chain feed on seals that contain biomagnified toxicants. Despite their remote location, the polar bears of Svalbard Island in Arctic Norway show extremely high levels of PCB contamination, as a result of biomagnification and the concentrating effect of the process of global distillation shown in Figure 14.15. Polar bear cubs suffer immune suppression, hormone disruption, and high mortality—and because the cubs receive PCBs in their mothers' milk, contamination persists and accumulates across generations.

Not all toxicants are synthetic, and not all synthetic chemicals are toxic

Although many toxicologists focus on synthetic chemicals, toxic substances also exist naturally in the environment around us and in the foods we eat. Thus, it would be a mistake to assume

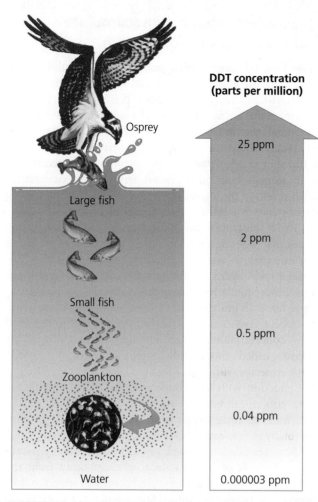

DDT concentration (parts per million)

25 ppm

2 ppm

0.5 ppm

0.04 ppm

0.000003 ppm

Osprey

Large fish

Small fish

Zooplankton

Water

FIGURE 14.16 ▲ Organisms at the lowest trophic level take in fat-soluble compounds such as DDT from water. As animals at higher trophic levels eat organisms lower on the food chain, each organism passes its load of toxicants up to its consumer, such that organisms on all trophic levels bioaccumulate the substance in their tissues. Concentrations increase at higher trophic levels by the process of biomagnification, because consumers each consume many individuals from lower trophic levels. In a classic case of biomagnification, DDT moves from zooplankton through various types of fish, becoming highly concentrated in fish-eating birds such as ospreys.

that all artificial substances are unhealthy and that all natural substances are healthy. In fact, the plants and animals we eat contain many chemicals that can cause us harm. Recall that plants produce toxins to ward off animals that eat them. In domesticating crop plants, we have selected (p. 55) for strains with reduced toxin content, but we have not eliminated these dangers. Furthermore, when we consume animal meat, we ingest toxins the animals obtained from plants or animals they ate.

Scientists are actively debating just how much risk natural toxicants pose. Biochemist Bruce Ames of the University of California at Berkeley maintains that the amounts of synthetic chemicals in our food from pesticide residues are dwarfed by the quantities of natural toxicants. Ames also holds that the natural defenses against toxins that our bodies have evolved are largely effective against synthetic toxicants. Therefore, he reasons, synthetic chemicals are a relatively minor worry. Moreover, he contends, if fears of pesticide residues, or price increases resulting from their removal,

cause people to eat fewer fruits and vegetables, then people will be at greater risk for disease, because a balanced diet rich in fruits and vegetables is important for health.

A respected senior scientist, Ames was a hero of environmentalists for his early work. The *Ames test*, a bacterial assay for mutagens, allows easy screening of suspected toxins. Today Ames is a target of criticism from environmental advocates who say that synthetic toxicants (1) are harder for the body to metabolize and excrete than natural ones, (2) persist and accumulate in the environment, and (3) expose people (such as farm workers and factory workers) to risks in ways other than the ingestion of food. What is clear is that more research is required on these questions.

STUDYING EFFECTS OF HAZARDS

Determining health effects of particular environmental hazards is a challenging job, especially because any given person or organism has a complex history of exposure to many hazards throughout life. Scientists rely on several different methods with people and with wildlife, ranging from correlative surveys to manipulative experiments (pp. 12–13).

Wildlife studies integrate work in the field and lab

Scientists study the impacts of environmental hazards on wild animals to help conserve animal populations and also to understand potential risks to people. Just as placing the proverbial canary in a coal mine helped miners determine whether the air was safe for them to breathe, studying how wild animals respond to pollution and other hazards can help us detect environmental health threats before they do us too much harm.

Wildlife toxicologists use a variety of approaches in their research. When scientists were zeroing in on the impacts of DDT, one key piece of evidence came from museum collections of wild birds' eggs from the decades before synthetic pesticides were manufactured. Old eggs from museum collections had measurably thicker shells than the eggs scientists were studying in the field from present-day birds, indicating that something was thinning the present-day shells.

Often wildlife toxicologists work in the field with animals to take measurements, document patterns, and generate hypotheses, before heading to the laboratory to run controlled manipulative experiments to test their hypotheses. For instance, as biologist Louis Guillette studied alligators in Florida (**FIGURE 14.17A**), he discovered that many showed bizarre reproductive problems. Females had trouble producing viable eggs, young alligators had abnormal gonads, and male hatchlings had too little of the male sex hormone testosterone while female hatchlings had too much of the female sex hormone estrogen. Because certain lakes received agricultural runoff that included insecticides such as DDT and dicofol and herbicides such as atrazine, he hypothesized that chemical contaminants were disrupting the endocrine systems of alligators during their development in the egg. Indeed, when Guillette and his coworkers compared alligators in polluted lakes with those in cleaner lakes, they found the ones in polluted lakes to be suffering far more problems. Moving into the lab, the researchers

(a) Louis Guillette taking blood sample from alligator

FIGURE 14.17 ▲ Researchers Louis Guillette **(a)** and Tyrone Hayes **(b)** found that alligators and frogs, respectively, show reproductive abnormalities that they attribute to endocrine disruption by pesticides.

(b) Tyrone Hayes in lab with frog

found that several contaminants detected in alligator eggs and young could bind to receptors for estrogen and reverse the sex of male embryos. Their experiments showed that atrazine appeared to disrupt hormones by inducing production of aromatase, an enzyme that converts testosterone to estrogen.

Following Guillette's work, researcher Tyrone Hayes (**FIGURE 14.17B**) found similar reproductive problems in frogs and attributed them to atrazine. In lab experiments, male frogs raised in water containing very low doses of the herbicide became feminized and hermaphroditic, developing both testes and ovaries. Hayes then moved to the field to look for correlations between herbicide use and reproductive impacts in the wild. His field surveys showed that leopard frogs across North America experienced hormonal problems in areas of heavy atrazine usage.

Human studies rely on case histories, epidemiology, and animal testing

In studies of human health, we gain much knowledge by studying sickened individuals directly. Medical professionals have long treated victims of poisonings, so the effects of common poisons are well known. Autopsies help us understand what constitutes a lethal dose. This process of observation and analysis of individual patients is known as a **case history** approach. Case histories have advanced our understanding of human illness, but they do not always help us infer the effects of rare hazards, new hazards, or chemicals that exist at low environmental concentrations and exert minor long-term effects. Case histories also tell us little about probability and risk, such as how many extra deaths we might expect in a population due to a particular cause.

For such questions, which are common in environmental toxicology, we need **epidemiological studies**, large-scale comparisons among groups of people, usually contrasting a group

known to have been exposed to some hazard and a group that has not. Epidemiologists track the fate of all people in the study for a long period of time (often years or decades) and measure the rate at which deaths, cancers, or other health problems occur in each group. The epidemiologist then analyzes the data, looking for observable differences between the groups, and statistically tests hypotheses accounting for differences. When a group exposed to a hazard shows a significantly greater degree of harm, it suggests that the hazard may be responsible. For example, epidemiologists have tracked asbestos miners for evidence of asbestosis, lung cancer, and mesothelioma. Survivors of the Chernobyl nuclear disaster have been monitored for thyroid cancer and other illnesses (pp. 572–573). Canadian epidemiologists are now tracking people for impacts of bisphenol A exposure.

The epidemiological process is akin to a natural experiment (pp. 12–13), in which the experimenter studies groups of subjects made possible by some event that has occurred. A similar approach was followed by anthropologist Elizabeth Guillette, who happens to be married to alligator biologist Louis Guillette (see **THE SCIENCE BEHIND THE STORY**, pp. 388–389). The advantages of epidemiological studies are their realism and their ability to yield relatively accurate predictions about risk. Drawbacks include the need to wait a long time for results and an inability to address future effects of new hazards, such as products just coming to market. In addition, participants in epidemiological studies encounter many factors that affect their health besides the one under study. Epidemiological studies measure a statistical association between a health hazard and an effect, but they do not confirm that the hazard *causes* the effect.

To establish causation, manipulative experiments are needed. However, subjecting people to massive doses of toxic substances in a lab experiment would clearly be unethical. This is why researchers have traditionally used nonhuman animals as

The SCIENCE behind the Story

Pesticides and Child Development in Mexico's Yaqui Valley

With spindly arms and big, round eyes, one set of pictures shows the sorts of stick figures drawn by young children everywhere. Next to them is another group of drawings, mostly disconnected squiggles and lines, resembling nothing. Both sets of pictures are intended to depict people. The main difference identified between the two groups of young artists: long-term pesticide exposure.

Children's drawings are not a typical tool of toxicology, but anthropologist Elizabeth Guillette was interested in the effects of pesticides on children and wanted to try new methods. She devised tests to measure childhood development based on techniques from anthropology and medicine. Searching for a study site, Guillette found the Yaqui valley region of northwestern Mexico.

The Yaqui valley is farming country, worked for generations by the indigenous group that gives the region its name. Synthetic pesticides arrived in the area in the 1940s. Some Yaqui embraced the agricultural innovations, spraying their farms in the valley to increase their yields. Yaqui farmers in the surrounding foothills, however, generally chose to bypass the chemicals and to continue

A Yaqui family

following more traditional farming practices. Although differing in farming techniques, Yaqui in the valley and foothills continued to share the same culture, diet, education system, income levels, and family structure.

At the time of the study, in 1994, valley farmers planted crops twice a year, applying pesticides up to 45 times from planting to harvest. A previous study conducted in the valley in 1990, focusing on areas with the largest farms, had indicated high levels of multiple pesticides in the breast milk of mothers and in the umbilical cord blood of newborn babies. In contrast, foothill families avoided chemical pesticides in their gardens and homes.

To understand how pesticide exposure affects childhood development, Guillette and fellow researchers studied 50 preschoolers aged 4 to 5, of whom 33 were from the valley and 17 from the foothills.

Each child underwent a half-hour exam, during which researchers showed a red balloon, promising to give the balloon later as a gift, and using the promise to evaluate long-term memory. Each child was then put through a series of physical and mental tests:

- Catching a ball from distances of up to 3 m (10 ft) away, to test coordination
- Jumping in place for as long as possible, to assess endurance
- Drawing a picture of a person, as a measure of perception
- Repeating a short string of numbers, to test short-term memory
- Dropping raisins into a bottle cap from a height of about 13 cm (5 in.), to gauge fine-motor skills

The researchers also measured each child's height and weight. When all tests were completed, each child was asked what he or she had been promised and received a red balloon.

Although the two groups of children did not differ in height and weight, they differed markedly in other measures of development. Valley children lagged far behind the foothill children in coordination, physical

test subjects. Foremost among these animal models have been laboratory strains of rats, mice, and other mammals (**FIGURE 14.18**). Because of shared evolutionary history, the bodies of all mammals function similarly, so substances that harm mice and rats are reasonably likely to harm us. Some people feel the use of animals for testing is unethical, but animal testing enables scientific and medical advances that would be impossible or far more difficult otherwise. Still, new techniques (with

human cell cultures, bacteria, or tissue from chicken eggs) are being devised that may soon replace some live-animal testing.

Dose-response analysis is a mainstay of toxicology

The standard method of testing with lab animals in toxicology is **dose-response analysis**. Scientists quantify the toxicity of

endurance, long-term memory, and fine motor skills:

- Valley children had great difficulty catching the ball.
- The average valley child could jump for 52 seconds, compared to 88 seconds for foothill children.
- Most valley children missed the bottle cap when dropping raisins, whereas foothill children dropped them into the caps more often.
- Each group did fairly well repeating numbers, but valley children showed poor long-term memory. At the end of the test, all but one of the foothill children remembered that they had been promised a balloon, and 59% remembered it was red. However, of the valley children only 27% remembered the color of the balloon, only 55% remembered they'd be getting a balloon, and 18% were unable to remember anything about a balloon.

The children's drawings exhibited the most dramatic difference between valley and foothill children (**see figure**). The researchers determined each drawing could earn 5 points, with 1 point each for a recognizable feature: head, body, arms, legs, and facial features. Foothill children drew pictures that looked like people, averaging about 4.5 points per drawing. Valley children, in contrast, averaged 1.6 points per drawing; their scribbles resembled little that looked like a person. By the standards of developmental medicine, the 4- and 5-year-old valley children drew at the level of 2-year-olds.

4-year-olds   5-year-olds

Drawings by children in the foothills

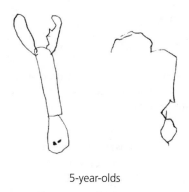

4-year-olds   5-year-olds

Drawings by children in the valley

Data from children in Mexico's Yaqui valley offer a startling example of apparent neurological effects of pesticide poisoning. Young children from foothills areas where pesticides were not commonly used drew recognizable figures of people. Children the same age from valley areas where pesticides were used heavily in industrial agriculture could draw only scribbles. Adapted from Guillette, E.A., et al., 1998. *Environmental Health Perspectives* 106: 347–353.

Some scientists greeted Guillette's study skeptically, pointing out that its sample size was too small to be meaningful. Others said that factors the researchers missed, such as different parenting styles or unknown health problems, could be to blame. Prominent toxicologists argued that because the researchers lacked time and money to take blood or tissue samples to check for pesticides or other toxic substances, the study results couldn't be tied to agricultural chemicals.

Regardless of these criticisms, Guillette maintains that her findings show that nontraditional study methods are a valid way to track the effects of environmental toxicants and that pesticides present complex, long-term risks to human growth and health. ∎

a substance by measuring the strength of its effects or the number of animals affected at different doses. The **dose** is the amount of substance the test animal receives, and the **response** is the type or magnitude of negative effects the animal exhibits as a result. The response is generally quantified by measuring the proportion of animals exhibiting negative effects. The data are plotted on a graph, with dose on the *x* axis and response on the *y* axis (**FIGURE 14.19A**). The resulting curve is called a **dose-response curve.**

Once they have plotted a dose-response curve, toxicologists can calculate a convenient shorthand gauge of a substance's toxicity: the amount of the substance it takes to kill half the population of study animals used. This lethal dose for 50% of individuals is termed the **LD_{50}**. A high LD_{50} indicates low toxicity, and a low LD_{50} indicates high toxicity.

If the experimenter is interested in nonlethal health effects, he or she may want to document the level of toxicant at which 50% of a population of test animals is affected in

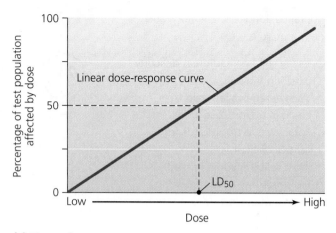

(a) Linear dose-response curve

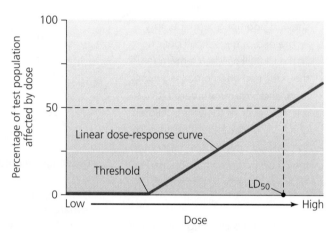

(b) Dose-response curve with threshold

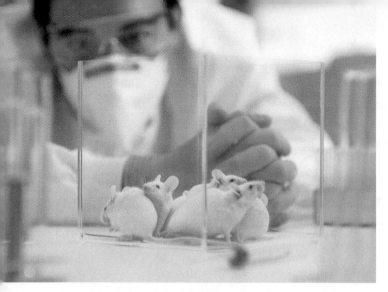

FIGURE 14.18 ▲ Laboratory testing with specially bred strains of mice, rats, and other animals allows researchers to study the toxicity of substances, develop safety guidelines, and make medical advances in ways they simply could not achieve without these animals.

some other way (for instance, what level of toxicant causes 50% of lab mice to lose their hair?). Such a level is called the effective-dose-50%, or **ED$_{50}$**.

Some substances can elicit effects at any concentration, but for others, responses may occur only above a certain dose, or threshold. Such a **threshold** dose (**FIGURE 14.19B**) might be expected if the body's organs can fully metabolize or excrete a toxicant at low doses but become overwhelmed at high concentrations. It might also occur if cells can repair damage to their DNA only up to a certain point.

Researchers generally give lab animals much higher doses relative to body mass than people would receive in the environment. This is so that the response is great enough to be measured and so that differences between the effects of small and large doses are evident. Data from a range of doses give shape to the dose-response curve. Once the data from animal tests are plotted, researchers can extrapolate downward to estimate responses to still-lower doses from a hypothetically large population of animals. This way, they can come up with an estimate of, say, what dose causes cancer in 1 mouse in 1 million. A second extrapolation is required to estimate the effect on humans, with our greater body mass. Because these two extrapolations stretch beyond the actual data obtained, they introduce uncertainty into the interpretation of what doses are safe for people.

Sometimes a response may *decrease* as a dose increases. Toxicologists are finding that some dose-response curves are U-shaped, J-shaped, or shaped like an inverted U (**FIGURE 14.19C**). Such counterintuitive curves contradict toxicology's traditional assumption that "the dose makes the poison." These unconventional dose-response curves often occur with endocrine disruptors, likely because the hormone system is geared to respond to minute concentrations of substances (normally, hormones in the bloodstream). Because the endocrine system responds to minuscule amounts of chemicals, it may be vulnerable to disruption by contaminants that are dispersed through the environment

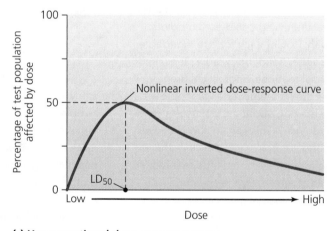

(c) Unconventional dose-response curve

FIGURE 14.19 ▲ In a classic linear dose-response curve **(a)**, the percentage of animals killed or otherwise affected by a substance rises with the dose. The point at which 50% of the animals are killed is labeled the *lethal-dose-50*, or *LD$_{50}$*. For some toxic substances, a threshold dose **(b)** exists, below which doses have no measurable effect. Some substances, in particular endocrine disruptors, show unconventional, nonlinear dose-response curves **(c)** that are U-shaped, J-shaped, or inverted. These curves indicate that some low doses can elicit stronger responses than higher doses.

and that reach our bodies in very low concentrations. In research with bisphenol A, a number of studies with lab animals have found unconventional dose-response curves.

Endocrine disruption poses challenges for toxicology

As today's emerging understanding of endocrine disruption leads toxicologists to question their assumptions, unconventional dose-response curves are presenting challenges for scientists studying toxic substances and for policymakers trying to set safety standards for them. Knowing the shape of a dose-response curve is crucial if one is using it to predict responses at doses below those that have been tested. Because so many novel synthetic chemicals exist in very low concentrations over wide areas, many scientists suspect that we may have underestimated the dangers of compounds that exert impacts at low concentrations.

Scientists first noted endocrine-disrupting effects decades ago, but the idea that synthetic chemicals might be altering the hormones of animals was not widely appreciated until the 1996 publication of the book *Our Stolen Future*, by Theo Colburn, Dianne Dumanoski, and J.P. Myers. Colburn, a scientist with the World Wildlife Fund, had spent years bringing toxicologists, medical doctors, and wildlife biologists together to share their findings. She and her co-authors then presented these data and stories in *Our Stolen Future*. Like *Silent Spring*, this book integrated scientific work from various fields and presented a unified picture that shocked many readers—and brought criticism from some scientists and from the chemical industry.

Today, thousands of studies have linked hundreds of substances to effects on reproduction, development, immune function, brain and nervous system function, and other hormone-driven processes. Evidence is strongest so far in nonhuman animals, but many studies suggest impacts on people. Some researchers suggest that endocrine disruptors may account for rising rates of testicular cancer, undescended testicles, and genital birth defects in men. Others argue that the sharp rise in breast cancer rates (one in eight U.S. women today develops breast cancer) may be due to hormone disruption, because an excess of estrogen appears to feed tumor development in older women. Still other scientists attribute an apparent drop in sperm counts among men worldwide to endocrine disruptors. Danish researchers reported in 1992 that the number and motility of sperm in men's semen had declined by 50% since 1938 (**FIGURE 14.20**). Subsequent studies by other researchers have largely supported these findings (although tremendous geographic variation remains unexplained). As Louis Guillette put it when he testified to Congress at a hearing on this research, "Every man in this room is half the man his grandfather was."

Much of the research into hormone disruption has brought about strident debate. This is partly because a great deal of scientific uncertainty is inherent in any young and developing field. Another reason is that negative findings about chemicals pose an economic threat to the manufacturers of those chemicals. For instance, Tyrone Hayes's work has met with fierce criticism from scientists associated with atrazine's manufacturer, which stands

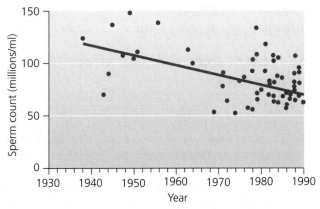

FIGURE 14.20 ▲ Research in 1992 synthesized the results of 61 studies that had reported sperm counts in men since 1938. Together the studies included 15,000 men from 20 nations on six continents. The data were highly variable but showed a significant decrease in sperm counts over time. Scientists hypothesize that this decline may result from environmental exposure to endocrine-disrupting chemicals. Data from Carlsen, E., et al. 1992. Evidence for decreasing quality of semen during the last 50 years. *British Medical Journal* 305: 609–613, as adapted by Toppari, J., et al. 1996. Male reproductive health and environmental xenoestrogens. *Environmental Health Perspectives* 104 (Suppl 4): 741–803.

to lose many millions of dollars if its top-selling herbicide were to be banned or restricted in the United States.

With bisphenol A, the chemical industry has consistently sworn that BPA is safe, and it has generated research showing a lack of health effects. Regulatory agencies such as the FDA have relied primarily on industry-sponsored research in vouching for the chemical's safety. However, independent academic scientists unaffiliated with industry are reaching different conclusions. By one count through the end of 2006, 151 of the 178 published studies with lab animals reported harm from low doses of bisphenol A. Almost without exception, the studies reporting harm received public government funding, whereas those reporting no harm were funded by industry. The difference in results between industry and academic research is so stark that U.S. legislators convened Congressional hearings in 2008 and chastised the FDA for cherry-picking certain papers and ignoring others.

Individuals vary in their responses to hazards

Different individuals may respond quite differently to identical exposures to hazards. These differences can be genetically based or can be due to a person's current condition. People in poorer health are often more sensitive to biological and chemical hazards. Sensitivity also can vary with sex, age, and weight. Because of their smaller size and rapidly developing organ systems, fetuses, infants, and young children tend to be much more sensitive to toxicants than are adults. Regulatory agencies such as the U.S. Environmental Protection Agency (EPA) typically set standards for adults and extrapolate downward for infants and children. However, many scientists contend that these linear extrapolations often do not offer adequate protection to fetuses, infants, and children.

The type of exposure can affect the response

The risk posed by a hazard often varies according to whether a person experiences high exposure for short periods of time, known as **acute exposure**, or low exposure over long periods of time, known as **chronic exposure**. Incidences of acute exposure are easier to recognize, because they often stem from discrete events, such as accidental ingestion, an oil spill, a chemical spill, or a nuclear accident. Lab tests and LD_{50} values generally reflect acute toxicity effects. However, chronic exposure is more common—and more difficult to detect and diagnose. Chronic exposure often affects organs gradually, as when smoking causes lung cancer, or when alcohol abuse leads to liver or kidney damage. Pesticide residues on food or arsenic in drinking water (pp. 424–425) also pose chronic risk. Because of the long time periods involved, relationships between cause and effect may not be readily apparent.

Mixes may be more than the sum of their parts

It is difficult enough to determine the impact of a single hazard, but the task becomes astronomically more difficult when multiple hazards interact. Chemical substances, when mixed, may act together in ways that cannot be predicted from the effects of each in isolation. Mixed toxicants may sum each other's effects, cancel out each other's effects, or multiply each other's effects. Whole new types of impacts may arise when toxicants are mixed together. Interactive impacts that are greater than the simple sum of their constituent effects are called **synergistic effects**.

With Florida's alligators, lab experiments have indicated that the DDT breakdown product DDE can either help cause or inhibit sex reversal, depending on the presence of other chemicals. Mice exposed to a mixture of nitrate, atrazine, and aldicarb have been found to show immune, hormone, and nervous system effects that were not evident from exposure to each of these chemicals alone.

Traditionally, environmental health has tackled effects of single hazards one at a time. In toxicology, the complex experimental designs required to test interactions, and the sheer number of chemical combinations, have meant that single-substance tests have received priority. This approach is changing, but scientists in environmental health and toxicology will never be able to test all possible combinations.

RISK ASSESSMENT AND RISK MANAGEMENT

Policy decisions on whether to ban chemicals or restrict their use generally follow years of rigorous testing for toxicity. Likewise, strategies for combating disease and other health threats are based on extensive scientific research. However, policy and management decisions also incorporate economics and ethics. And all too often, they are influenced by political pressure from powerful interests. The steps between the collection and interpretation of scientific data and the formulation of policy involve assessing and managing risk.

We express risk in terms of probability

Exposure to an environmental health threat does not invariably produce some given effect. Rather, it causes some probability of harm, some statistical chance that damage will result. To understand a health threat, a scientist must know more than just its identity and strength. He or she must also know the chance that one will encounter it, the frequency with which one may encounter it, the amount of substance or degree of threat to which one is exposed, and one's sensitivity to the threat. Such factors help determine the overall risk posed.

Risk can be measured in terms of *probability*, a quantitative description of the likelihood of a certain outcome. The probability that some harmful outcome (for instance, injury, death, environmental damage, or economic loss) will result from a given action, event, or substance expresses the **risk** posed by that phenomenon.

Our perception of risk may not match reality

Every action we take and every decision we make involves some element of risk, some (generally small) probability that things will go wrong. We try in everyday life to behave in ways that minimize risk, but our perceptions of risk do not always match statistical reality (**FIGURE 14.21**). People often worry unduly about negligibly small risks yet happily engage in other activities that pose high risks. For instance, most people perceive flying in an airplane as a riskier activity than driving a car, but driving a car is statistically far more dangerous.

Psychologists agree that this difference between perception and reality stems from the fact that we feel more at risk when we are not controlling a situation and more safe when we are "at the wheel"—regardless of the actual risk involved. When we drive a car, we feel we are in control, even though statistics show we are at greater risk than as a passenger in an airplane.

This psychology may help account for people's anxiety over nuclear power, toxic waste, and pesticide residues on foods—environmental hazards that are invisible or little understood and whose presence in our lives is largely outside our personal control. In contrast, people are more ready to accept and ignore the risks of smoking cigarettes, overeating, and not exercising—voluntary activities statistically shown to pose far greater risks to health.

Risk assessment analyzes risk quantitatively

The quantitative measurement of risk and the comparison of risks involved in different activities or substances together are termed **risk assessment**. Risk assessment is a way to identify and outline problems. In environmental health, it helps ascertain which substances and activities pose health threats to people or wildlife and which are largely safe.

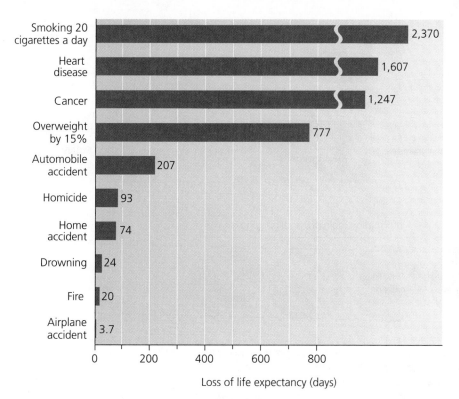

FIGURE 14.21 ◄ Our perceptions of risk do not always match the reality of risk. Listed here are several leading causes of death in the United States, along with a measure of the risk each poses. Risk is measured in days of lost life expectancy, that is, the number of days of life lost by people suffering the hazard, spread across the entire population—a measure commonly used by insurance companies. By this measure, one common source of anxiety, airplane accidents, poses 20 times less risk than home accidents, over 50 times less risk than auto accidents, and over 200 times less risk than being overweight. Data from Cohen, B., 1991. Catalog of risks extended and updated. *Health Physics* 61: 317–335.

Assessing risk for a chemical substance involves several steps. The first steps involve the scientific study of toxicity we examined above—determining whether a substance has toxic effects and, through dose-response analysis, measuring how effects vary with the degree of exposure. Subsequent steps involve assessing the individual's or population's likely extent of exposure to the substance, including the frequency of contact, the concentrations likely encountered, and the length of encounter.

To assess risk from a widely used substance such as bisphenol A, teams of scientific experts may be convened to review hundreds of studies so that regulators and the public can benefit from informed summaries. Between 2001 and 2008, the U.S. government sponsored three such panels on BPA, and the American Plastics Council, an industry group, sponsored two. The 2008 panel, sponsored by the government's National Toxicology Program, assessed BPA as being of intermediate concern for human health, on the third level of their five-level scale. The panel voiced "some concern for effects on brain, behavior, and prostate gland in fetuses, infants, and children at current human exposures," and wrote that "the possibility that bisphenol A may alter human development cannot be dismissed."

Risk management combines science and other social factors

Accurate risk assessment is a vital step toward effective **risk management**, which consists of decisions and strategies to minimize risk (**FIGURE 14.22**). In most nations, risk management is handled largely by federal agencies. In the United States, these include agencies such as the EPA, the Centers for Disease Control and Prevention (CDC), and the Food and Drug Administration (FDA). In risk management,

scientific assessments of risk are considered in light of economic, social, and political needs and values. Risk managers assess costs and benefits of addressing risk in various ways with regard to both scientific and nonscientific concerns before making decisions on whether and how to reduce or eliminate risk.

In environmental health and toxicology, comparing costs and benefits (p. 149) can be difficult because the benefits are often economic, whereas the costs often pertain to health. Moreover, economic benefits are generally known, easily quantified, and of a discrete and stable amount, whereas health risks are hard-to-measure probabilities, often involving a small percentage of people likely to suffer greatly and a large majority likely to experience little effect. When a government agency bans a pesticide, it may mean considerable economic loss for the manufacturer and potential economic loss for the farmer, whereas the benefits accrue less predictably over the long term to some percentage of factory workers, farmers, and the general public. Because of the lack of equivalence in the way costs and benefits are measured, risk management frequently tends to stir up debate.

In the case of bisphenol A, eliminating plastic linings in our food and drink cans could do more harm than good, because the linings help prevent metal corrosion and the contamination of food by pathogens. Alternative substances exist for most of BPA's uses, but replacing BPA with alternatives will entail economic costs to industry, and these costs get passed on to consumers in the prices of products. Such complex considerations can make risk management decisions difficult even if the science of risk assessment is fairly clear. This difficulty may help account for the hesitancy of U.S. regulatory agencies to issue restrictions on BPA so far. As of 2010, both the FDA and the EPA were continuing to review options for managing risk from BPA.

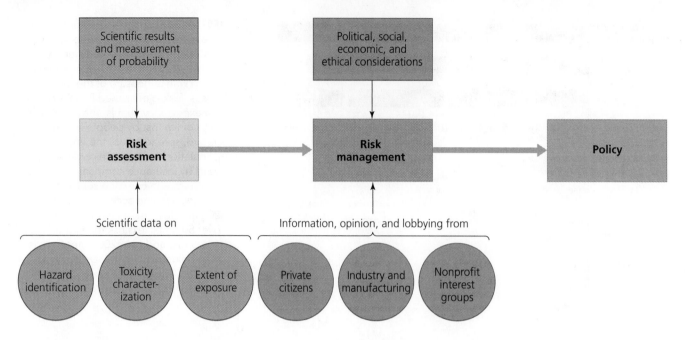

FIGURE 14.22 ▲ The first step in addressing the risk of an environmental hazard is *risk assessment*, a process of quantifying the risk of the hazard and comparing it to other risks. Once science identifies and measures risks, then *risk management* can proceed. In this process, economic, political, social, and ethical issues are considered in light of the scientific data from risk assessment. The consideration of all these types of information is intended to result in policy decisions that minimize the risk of the environmental hazard.

PHILOSOPHICAL AND POLICY APPROACHES

Because we cannot know a substance's toxicity until we measure and test it, and because there are so many untested chemicals and combinations, science will never eliminate the many uncertainties that accompany risk assessment. In such a world of uncertainty, there are two basic philosophical approaches to categorizing substances as safe or dangerous (**FIGURE 14.23**).

Two approaches exist for determining safety

One approach is to assume that substances are harmless until shown to be harmful. We might nickname this the *innocent-until-proven-guilty approach*. Because thoroughly testing every existing substance (and combination of substances) for its effects is a hopelessly long, complicated, and expensive pursuit, the innocent-until-proven-guilty approach has the virtue of facilitating technological innovation and economic activity. However, it has the disadvantage of putting into wide use some substances that may later turn out to be dangerous.

The other approach is to assume that substances are harmful until shown to be harmless. This approach follows the *precautionary principle* (p. 265). This more cautious approach should enable us to identify troublesome toxicants before they are released into the environment, but it may also impede the pace of technological and economic advance.

These two approaches are actually two ends of a continuum of possible approaches. The two endpoints differ mainly

in where they lay the burden of proof—specifically, whether product manufacturers are required to prove safety or whether government, scientists, or citizens are required to prove danger.

Philosophical approaches are reflected in policy

This choice of philosophical approach has direct implications for policy, and nations vary in how they blend the two approaches when it comes to regulating synthetic substances. European nations have recently embarked on a policy course that incorporates the precautionary principle, whereas the United States largely follows an innocent-until-proven-guilty approach. For instance, compounds in cosmetics require no FDA review or approval before being sold to the public.

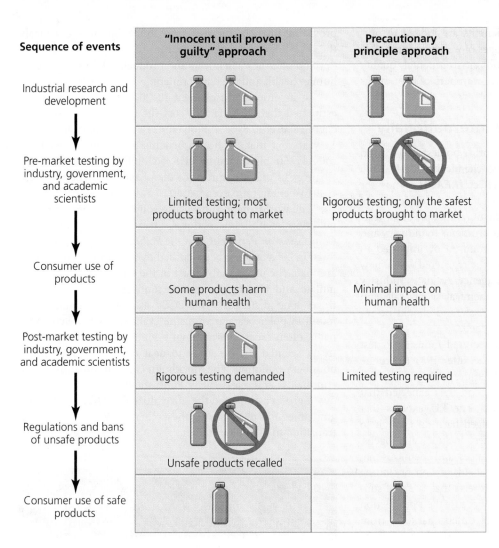

Sequence of events	"Innocent until proven guilty" approach	Precautionary principle approach
Industrial research and development		
Pre-market testing by industry, government, and academic scientists	Limited testing; most products brought to market	Rigorous testing; only the safest products brought to market
Consumer use of products	Some products harm human health	Minimal impact on human health
Post-market testing by industry, government, and academic scientists	Rigorous testing demanded	Limited testing required
Regulations and bans of unsafe products	Unsafe products recalled	
Consumer use of safe products		

FIGURE 14.23 ◄ Two main approaches can be taken to introduce new substances to the market. In one approach, substances are "innocent until proven guilty"; they are brought to market relatively quickly after limited testing. Products reach consumers more quickly, but some fraction of them (blue bottle in diagram) may cause harm to some fraction of people. The other approach is to adopt the precautionary principle, bringing substances to market cautiously, only after extensive testing. Products that reach the market should be safe, but many perfectly safe products (purple bottle in diagram) will be delayed in reaching consumers.

In the United States, several federal agencies apportion responsibility for tracking and regulating synthetic chemicals. The FDA, under the Food, Drug, and Cosmetic Act of 1938 and its subsequent amendments, monitors foods and food additives, cosmetics, drugs, and medical devices. The EPA regulates pesticides under the Federal Insecticide, Fungicide, and Rodenticide Act of 1947 (FIFRA) and its amendments. The Occupational Safety and Health Administration (OSHA) regulates workplace hazards under a 1970 act. Several other agencies regulate other substances. Synthetic chemicals not covered by other laws are regulated by the EPA under the 1976 Toxic Substances Control Act (TSCA).

EPA regulation is only partly effective

The **Toxic Substances Control Act** directs the EPA to monitor thousands of industrial chemicals manufactured in or imported into the United States, ranging from PCBs to lead to asbestos, and including bisphenol A and many other compounds in plastics. The act gives the agency power to regulate these substances and ban them if they are found to pose excessive risk.

However, many public health advocates view TSCA as being far too weak. The 62,000 chemicals existing when TSCA became law were grandfathered in, and the number has since swollen to 83,000, while just five have been restricted. Health advocates note that the screening required of industry is minimal

and that to mandate more extensive and meaningful testing, the EPA must show proof of the chemical's toxicity. In other words, the agency is trapped in a Catch-22: To push for studies looking for toxicity, it must have proof of toxicity already. The result is that most synthetic chemicals are not thoroughly tested before being brought to market. Of those that fall under TSCA, only 10% have been thoroughly tested for toxicity; only 2% have been screened for carcinogenicity, mutagenicity, or teratogenicity; fewer than 1% are regulated; and almost none have been tested for endocrine, nervous, or immune system damage, according to the U.S. National Academy of Sciences.

In its regulation of pesticides under FIFRA, the EPA is charged with "registering" each new pesticide that manufacturers propose to bring to market. The registration process involves risk assessment and risk management. The EPA first asks the manufacturer to provide information, including results of safety assessments the company has performed according to EPA guidelines. The EPA examines the company's research and all other relevant scientific research. It examines the product's ingredients and how the product will be used and tries to evaluate whether the chemical poses risks to people, other organisms, or water or air quality. The EPA then approves, denies, or sets limits on the chemical's sale and use. It also must approve language used on the product's label.

Because the registration process takes economic considerations into account, critics say it allows hazardous chemicals to be approved if the economic benefits are judged to outweigh the hazards. Here the challenges of weighing intangible risks involving human health and environmental quality against the tangible and quantitative numbers of economics become apparent.

Toxicants are regulated internationally

The European Union is taking the world's boldest step toward testing and regulating manufactured chemicals. In 2007, the EU's **REACH** program went into effect (*REACH* stands for Registration, Evaluation, Authorization, and Restriction of Chemicals). REACH largely shifts the burden of proof for testing chemical safety from national governments to industry and requires that chemical substances produced or imported in amounts of over 1 metric ton per year be registered with a new European Chemicals Agency. This agency evaluates industry research and decides whether the chemical seems safe and should be approved, whether it is unsafe and should be restricted, or whether more testing is needed.

Previous European policy had required industry to test only those chemicals brought to market after 1981 (there are 4,300 such chemicals). REACH will require testing for chemicals already on the market in 1981, and it is expected that 30,000 substances will be registered. The EU expects that roughly 1,500 substances may be judged harmful enough to be replaced with safer substances. These include carcinogens, mutagens, teratogens, persistent substances, and chemicals that bioaccumulate. (EU commissioners held off on including endocrine disruptors.) The EU also expects that 1–2% of substances may cease to be manufactured because their production will no longer be profitable. REACH differs markedly from TSCA (**TABLE 14.3**), illustrating how the approach that Europe is pursuing is different from the approach being pursued in the United States.

The REACH policy also aims to help industry, by giving it a single streamlined regulatory system and by exempting it from having to file paperwork on substances under 1 metric ton. By requiring stricter review of major chemicals already in use, exempting chemicals made only in small amounts, and

providing financial incentives for innovating new chemicals, the EU hopes to help European industries research and develop safer new chemicals and products while safeguarding human health and the environment.

In an impacts assessment in 2003, EU commissioners estimated that REACH will cost the chemical industry and chemical users 2.8–5.2 billion euros (US $3.8–7.0 billion) over 11 years but that the health benefits to the public would be roughly 50 billion euros (US $67 billion) over 30 years. Changes in the program since then have made the predicted cost-to-benefit ratio even better.

The world's nations have also sought to address chemical pollution with international treaties. The *Stockholm Convention on Persistent Organic Pollutants (POPs)* came into force in 2004 and has been ratified by over 150 nations. POPs are toxic chemicals that persist in the environment, bioaccumulate and biomagnify up the food chain, and often can travel long distances. The PCBs and other contaminants found in polar bears are a prime example. Because contaminants often cross international boundaries, an international treaty seemed the best way to deal fairly with such transboundary pollution. The Stockholm Convention aims first to end the use and release of 12 POPs shown to be most dangerous, a group nicknamed the "dirty dozen" (**TABLE 14.4**). It sets guidelines for phasing out these chemicals and encourages transition to safer alternatives.

TABLE 14.3 American versus European Approaches to Chemical Regulation

TSCA (United States)	REACH (European Union)
▶ Government bears burden of proof to show harm	▶ Industry bears burden of proof to show safety
▶ Little data on new chemicals are required from industry	▶ More data on new chemicals are required from industry
▶ Chemicals in use before 1976 are not regulated	▶ Chemicals in use before 1981 bear scrutiny like that directed toward newer chemicals
▶ Prioritizing problems is hampered by lack of data	▶ Problems are prioritized using data on risk
▶ Industry is allowed to keep trade secrets from the public	▶ Database will allow public access to chemical information

Adapted from Schwarzman, M.R., and M.P. Wilson, 2009. New science for chemicals policy. *Science* 326: 1065–1066.

TABLE 14.4 The "Dirty Dozen" Persistent Organic Pollutants (POPs) Targeted by the Stockholm Convention

Toxicant	Description
Aldrin	Insecticide to kill termites and crop pests
Chlordane	Insecticide to kill termites and crop pests
DDT	Insecticide to protect against insect-spread disease; still applied in some countries to control malaria
Dieldrin	Insecticide to kill termites, textile pests, crop pests, and disease vectors
Dioxins	By-product of incomplete combustion and chemical manufacturing; released in metal recycling, pulp and paper bleaching, auto exhaust, tobacco smoke, and wood and coal smoke
Endrin	Pesticide to kill rodents and crop insects
Furans	By-product of processes that release dioxins; also present in commercial mixtures of PCBs
Heptachlor	Broad-spectrum insecticide
Hexachlorobenzene	Fungicide for crops; released by chemical manufacture and processes that release dioxins and furans
Mirex	Household insecticide; fire retardant in plastics, rubber, and electronics
PCBs	Industrial chemical used in heat-exchange fluids, electrical transformers and capacitors, paints, sealants, and plastics
Toxaphene	Insecticide to kill crop insects and livestock parasites

Data from United Nations Environment Programme (UNEP), 2001.

➤ CONCLUSION

International agreements such as REACH and the Stockholm Convention represent a sign that governments may act to protect the world's people, wildlife, and ecosystems from toxic substances and other environmental hazards. At the same time, solutions often come more easily when they do not arise from government regulation alone. Consumer choice exercised through the market can often be an effective way to influence industry's decision making. Consumers of products, from plastics to pesticides to cosmetics to kids' toys, can make decisions that influence industry when they have full information from scientific research regarding the risks involved. Once scientific results are in, a society's philosophical approach to risk management will determine what policy decisions are made.

Whether the burden of proof is laid at the door of industry or of government, we will never attain complete scientific knowledge of any risk. Rather, we must make choices based on the information available. Synthetic chemicals have brought us innumerable modern conveniences, a larger food supply, and medical advances that save and extend human lives. Human society would be very different without them. Yet a safer and happier future, one that safeguards the well-being of both people and the environment, depends on knowing the risks that some hazards pose and on having means in place to phase out harmful substances and replace them with safer ones.

R E V I E W I N G   O B J E C T I V E S

You should now be able to:

IDENTIFY MAJOR ENVIRONMENTAL HEALTH HAZARDS AND EXPLAIN THE GOALS OF ENVIRONMENTAL HEALTH

- Environmental health seeks to assess and mitigate environmental factors that adversely affect human health and ecological systems. (p. 371)
- Environmental health threats include physical, chemical, biological, and cultural hazards. (pp. 371–372)
- Several major environmental hazards (cigarette smoke, radon, asbestos, lead, and PBDEs) exist indoors. (pp. 373–374)
- Disease is a major focus of environmental health. We have successfully fought some infectious diseases, but others are spreading. (pp. 374–376)
- Sanitation, clean water, food security, education, and access to medical care are strategies to enhance environmental health. (pp. 376–377)
- Toxicology is the study of poisonous substances. (pp. 376, 378)

DESCRIBE THE TYPES, ABUNDANCE, DISTRIBUTION, AND MOVEMENT OF TOXIC SUBSTANCES IN THE ENVIRONMENT

- Thousands of synthetic substances exist in our environment, and many are present in our bodies. (pp. 378–379)
- Toxicants include carcinogens, mutagens, teratogens, allergens, neurotoxins, and endocrine disruptors. (pp. 380–381)
- Toxic substances may concentrate in and move through surface water and groundwater, or may travel long distances through the atmosphere. (pp. 381–385)
- Some chemicals break down very slowly and thus persist in the environment. (p. 385)
- Some organic poisons bioaccumulate and move up the food chain, poisoning consumers at high trophic levels through the process of biomagnification. (pp. 385–386)
- Toxicants may be natural as well as synthetic. (pp. 385–386)

DISCUSS THE STUDY OF HAZARDS AND THEIR EFFECTS, INCLUDING WILDLIFE TOXICOLOGY, CASE HISTORIES, EPIDEMIOLOGY, ANIMAL TESTING, AND DOSE-RESPONSE ANALYSIS

- Studies of wildlife can inform human health efforts. (pp. 386–387)
- In case histories, researchers study health problems in individual people. (p. 387)
- Epidemiology involves gathering data from many people over long time periods and comparing groups with and without exposure to the factor being assessed. (p. 387)
- In dose-response analysis, scientists measure the response of test animals to various doses of the suspected toxicant. (pp. 388–390)
- Unconventional dose-response curves from endocrine disrupting substances are causing toxicologists to reassess some of their assumptions and methods. (pp. 390–391)
- Toxicity or strength of response may be influenced by the dose or amount of exposure, the nature of exposure (acute or chronic), individual variation, and synergistic interactions with other hazards. (pp. 391–392)

EVALUATE RISK ASSESSMENT AND RISK MANAGEMENT

- Risk assessment involves quantifying and comparing risks involved in different activities or substances. (pp. 392–393)
- Risk management integrates science with political, social, and economic concerns, in order to design strategies to minimize risk. (pp. 393–394)

COMPARE PHILOSOPHICAL APPROACHES TO RISK

- An innocent-until-proven-guilty approach assumes that a substance is safe unless shown to be harmful. This puts the burden of proof on the public to prove that a substance is harmful. (pp. 394–395)
- A precautionary approach assumes that a substance may be harmful unless proven safe. This puts the burden of proof on the manufacturer to prove that a substance is safe. (pp. 394–395)

DESCRIBE REGULATORY POLICY IN THE UNITED STATES AND INTERNATIONALLY

- The EPA, CDC, FDA, and OSHA are responsible for regulating environmental health hazards under U.S. policy. (p. 395)
- European nations take a more precautionary approach than does the United States when it comes to testing chemical products, as shown by a comparison of REACH and TSCA. (p. 396)
- The Stockholm Convention aims to ban 12 persistent organic pollutants. (p. 396)

TESTING YOUR COMPREHENSION

1. What four major types of health hazards does research in the field of environmental health encompass?

2. In what way is disease the greatest hazard that people face? What kinds of interrelationships must environmental health experts study to learn about how diseases affect human health?

3. Where does most exposure to lead, asbestos, radon, and PBDEs occur? How has each been addressed?

4. When did concern over the effects of pesticides start to grow in the United States? Describe the argument presented by Rachel Carson in *Silent Spring*. What policy resulted from the book's publication? Where and how is DDT still used?

5. List and describe the six general categories of toxicants described in this chapter.

6. How do toxic substances travel through the environment, and where are they most likely to be found? Describe and contrast the processes of bioaccumulation and biomagnification.

7. What are epidemiological studies, and how are they most often conducted?

8. Why are animals used in laboratory experiments in toxicology? Explain the dose-response curve. Why is a substance with a high LD_{50} considered safer than one with a low LD_{50}?

9. What factors may affect an individual's response to a toxic substance? Why is chronic exposure to toxic agents often more difficult to measure and diagnose than acute exposure? What are synergistic effects, and why are they difficult to measure and diagnose?

10. How do scientists identify and assess risks from substances or activities that may pose health threats?

SEEKING SOLUTIONS

1. Describe some environmental health hazards that you think you may be living with indoors. How do you think you may have been affected by indoor or outdoor hazards in the past? How could you best deal with these hazards in the future?

2. Do you feel that laboratory animals should be used in experiments in toxicology? Why or why not?

3. Why has research on endocrine disruption spurred so much debate? What steps do you think could be taken to help establish greater consensus among scientists, industry, regulators, policymakers, and the public?

4. Describe differences in the policies of the United States and the European Union toward the study and management of the risks of synthetic chemicals. Which do you believe is better, the policies of the United States or those of the European Union, and why?

5. **THINK IT THROUGH** You are the parent of two young children, and you want to minimize the environmental health risks your kids are exposed to. Name five steps that you could take in your household and in your daily life that would minimize your children's exposure to environmental health hazards.

6. **THINK IT THROUGH** You have just been hired as the office manager for a high-tech startup company that employs bright and motivated young people but is located in an old, dilapidated building. Despite their youth and vigor, the company's employees seem perpetually sick with colds, headaches, respiratory ailments, and other unexplained illnesses. Looking into the building's history, you discover that the water pipes and ventilation system are many decades old, that there have been repeated termite infestations, and that part of the building was remodeled just before your company moved in but there are no records of what was done in the remodel. Your company has all the latest furniture, computers, and other electronics. Most windows are sealed shut. There is an indoor smoking lounge.

    You want to figure out what is making the employees sick, and you want to convince your boss to hire professionals to examine the building for hazards. What will you tell your boss to interest her in funding inspections? Name five specific hazards you will ask the inspectors to look for. What questions will you ask employees in order to help focus the inspections?

CALCULATING ECOLOGICAL FOOTPRINTS

In 2001, the last year the U.S. EPA gathered and reported data on pesticide use, Americans used 1.20 billion pounds of pesticide active ingredients, and world pesticide use totaled 5.05 billion pounds of active ingredients. In that same year, the U.S. population was 285 million, and the world's population totaled 6.16 billion. Pesticides include hundreds of chemicals used as insecticides, fungicides, herbicides, rodenticides, repellants, and disinfectants. They are used by farmers, governments, industries, and individuals. In the table, calculate your share of pesticide use as a U.S. citizen in 2001 and the amount used by (or on behalf of) the average citizen of the world.

	Annual pesticide use (pounds of active ingredient)
You	
Your class	
Your state	
United States	1.20 billion
World (total)	5.05 billion
World (per capita)	

1. What is the ratio of your annual pesticide use to the world's per capita average?

2. Refer to the "Calculating Ecological Footprints" section in Chapter 1 (pp. 21–22), and find the ecological footprints of the average U.S. citizen and the average world citizen. Compare the ratio of pesticide usage with the ratio of the overall ecological footprints. How do these differ, and how would you account for this?

3. Does the figure for per capita pesticide use for you as a U.S. citizen seem reasonable for you personally? Why or why not? Do you find this figure alarming or of little concern? What else would you like to know to assess the risk associated with this level of pesticide use?

MasteringENVIRONMENTALSCIENCE™

Go to **www.masteringenvironmentalscience.com** for practice quizzes, Pearson eText, videos, current events, and more.

Hoover Dam on the Colorado River

15 FRESHWATER SYSTEMS AND RESOURCES

UPON COMPLETING THIS CHAPTER, YOU WILL BE ABLE TO:

- Explain water's importance to people and ecosystems, and describe the distribution of fresh water on Earth
- Describe major types of freshwater systems
- Discuss how we use water and alter freshwater systems

- Assess problems of water supply and propose solutions to address depletion of fresh water
- Assess problems of water quality and propose solutions to address water pollution
- Explain how we treat drinking water and wastewater

Gambling with Water in the Colorado River Basin

"Water promises to be to the 21st century what oil was to the 20th century: the precious commodity that determines the wealth of nations."

—Fortune Magazine, May 2000

"The wars of the 21st century will be fought over water."

—World Water Commission Chairman Ismail Serageldin

UNITED STATES

Las Vegas, Nevada

Atlantic Ocean

Pacific Ocean

We don't generally look to a desert when we're searching for water, but Las Vegas is placing a high-stakes bet by doing exactly that. Leaders of this fast-growing metropolis are intent on pushing through a controversial plan to mine water from beneath the ground in a distant area of the Great Basin desert. The $3.5-billion project would pipe the water 450 km (280 mi) out of rural Nevada and straight to Vegas. Nevada's largest city is desperate for water because it is outgrowing its allotment from the Colorado River, whose declining supply Nevada is legally obligated to share with neighboring states.

The Colorado River originates in the high peaks of the Rocky Mountains, charges through arid canyonlands, crosses into Mexico, and empties into the Gulf of California, draining 637,000 km² (246,000 mi²) of southwestern North America (**FIGURE 15.1**). Its raging waters chiseled through thousands of feet of bedrock, creating the Grand Canyon and leaving extraordinary scenery along its 2,330-km (1,450-mi) length.

Today, however, only a trickle reaches the river's mouth. Massive dams impound the Colorado's waters to provide flood control, recreation, and hydroelectric power. Water is siphoned off for irrigation to make agriculture possible in this dry region. It is piped to cities in canals and aqueducts to quench the thirst of

Water piped in from the Colorado River surges from a fountain outside a Las Vegas casino.

30 million people. It keeps golf courses green in the desert, fills swimming pools in southern California backyards, and gushes through fountains at Las Vegas casinos. Without the river, none of this could exist.

Since 1922, the seven states along the Colorado have divided the river's water among themselves, guided by the Colorado River Compact they signed that year. In this treaty, Arizona, California, Colorado, Nevada, New Mexico, Utah, and Wyoming apportioned water according to the needs and negotiating power each state had at the time. Mexico won an allotment in a later treaty.

Today the populations of these states are booming, even as the region has suffered severe drought. The Colorado River's flow fell below average for the entire decade following 2000, leaving reservoirs at half capacity. Worse still, scientific data paint a grim picture for the future: Tree-ring studies show that megadroughts were common in the past, and climate modelers predict that global climate change will bring still more drought.

Las Vegas in particular faces disaster if drought conditions continue. If water levels drop much further at Lake Mead, not only will the city need to stop withdrawing water, but Hoover Dam could shut down the electric power production that keeps the city

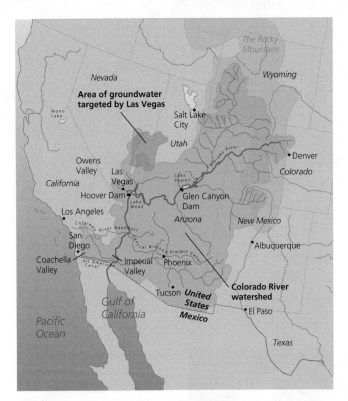

FIGURE 15.1 ▲ Water from portions of seven states drains into the Colorado River as it meanders through the southwestern United States and into Mexico before disgorging into the Gulf of California. The Colorado River's watershed is shown in orange. Diversions along the river's length sometimes cause the river to run dry at its mouth.

alight. Vegas is feeling the heat not only because it gets just 11 cm (4.5 in.) of rain a year, but also because in 1922 Nevada was barely populated, and so it won only 4% of the river's water in the Colorado River Compact. Today, rapidly growing Las Vegas needs more water—fast. The city already returns treated wastewater to the river so it can legally take more than its share, and it pays its residents to rip up their lawns to reduce water use. Even some of its casinos—legendary for their extravagance—are adopting water conservation measures. But in a city adding tens of thousands of people each year, such steps may not be enough.

Facing growing demand for water amid dwindling supplies, the seven states in the Colorado River Compact faced each other anew at the negotiating table. In 2007, they worked out a plan that allowed upper-basin states to withhold more water when needed and encouraged lower-basin states to develop supplies elsewhere. The plan backed Las Vegas's controversial proposal to mine groundwater from rural areas of eastern Nevada. The other states feared that if Las Vegas cannot get adequate water from somewhere, it might back out of the Colorado River Compact or challenge it in court, which might jeopardize their own shares of Colorado River water.

Ranchers and other local residents of the scenic area of Great Basin desert from which Vegas proposes to take water oppose the plan and have sued to stop it. So have environmental advocates. These opponents protest that groundwater withdrawals will wreck the ecology and economy of the entire region. A federal environmental review of the vast project was due in 2010.

In the meantime, Las Vegas's contentious plan has bounced around between courts and government agencies like a pinball in a pinball machine. The first phase of the plan won approval from Nevada's state engineer, but a district judge overturned this approval and blocked the project. The Southern Nevada Water Authority appealed the judge's ruling. In 2010, Nevada's Supreme Court gave the whole process a clean slate, preserving Las Vegas's applications for water rights but allowing new protests from the desert regions to be heard. It is just the latest episode in the colorful history of water politics among the states and regions jockeying for the Southwest's increasingly scarce water.

FRESHWATER SYSTEMS

"Water, water, everywhere, nor any drop to drink." The well-known line from the poem *The Rime of the Ancient Mariner* describes the situation on our planet well. Water may seem abundant, but water that we can drink is quite rare and limited (**FIGURE 15.2**). About 97.5% of Earth's water resides in the oceans and is too salty to drink or to use to water crops. Only 2.5% is considered **fresh water**, water that is relatively pure, with few dissolved salts. Because most fresh water is tied up in glaciers, icecaps, and underground aquifers, just over 1 part in 10,000 of Earth's water is easily accessible for human use.

Although water is a limited resource, it is also a renewable resource, as long as we manage our use sustainably. Water is renewed and recycled as it moves through the *hydrologic cycle* (see Figure 5.15, pp. 123–124), sometimes changing among liquid, solid (ice), and gaseous (water vapor) states. As water moves and changes, it redistributes heat, erodes mountain ranges, builds river deltas, maintains organisms and ecosystems, shapes civilizations, and gives rise to political conflicts. Let's first examine the portions of the hydrologic cycle that are most conspicuous to us—surface water bodies—and take stock of the ecological systems they support.

Surface water converges within watersheds

Liquid fresh water occurs either as surface water or groundwater. **Surface water**, water located atop Earth's surface, accounts for just 1% of fresh water, but it is vital for our survival and for the planet's ecological systems. Once water falls from the sky as rain, emerges from springs, or melts from snow or a glacier, it may soak into the ground or

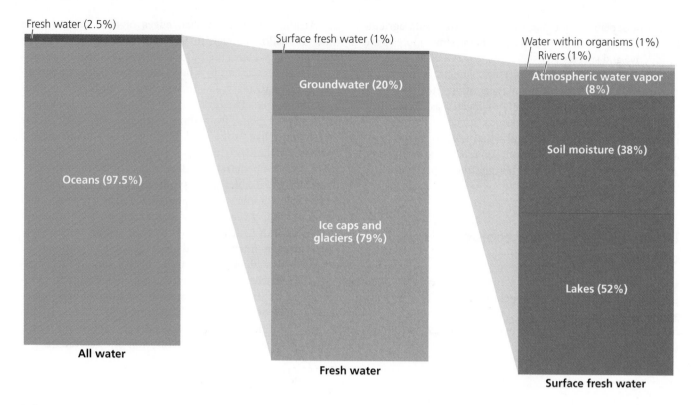

Fresh water (2.5%)

Oceans (97.5%)

All water

Surface fresh water (1%)

Groundwater (20%)

Ice caps and glaciers (79%)

Fresh water

Water within organisms (1%)
Rivers (1%)

Atmospheric water vapor (8%)

Soil moisture (38%)

Lakes (52%)

Surface fresh water

FIGURE 15.2 ▲ Only 2.5% of Earth's water is fresh water. Of that 2.5%, most is tied up in glaciers and ice caps. Of the 1% that is surface water, most is in lakes and soil moisture. Data from United Nations Environment Programme (UNEP) and World Resources Institute.

may flow downhill over land. Water that flows over land is called **runoff**.

As it flows downhill, runoff converges where the land dips lowest, forming streams, creeks, or brooks. These small watercourses may merge into rivers, whose water eventually reaches a lake or ocean. A smaller river flowing into a larger one is a *tributary*. The area of land drained by a *river system*— a river and all its tributaries—is that river's **watershed** (pp. 111, 167), also called a *drainage basin*. If you could stand at the mouth of the Colorado River and trace every drop of water in it back to the spot it first fell as precipitation, then you will have delineated the Colorado River's watershed, the very area shaded in Figure 15.1.

Rivers and streams wind through landscapes

Landscapes determine where rivers flow, but rivers shape the landscapes through which they run. A river that runs through a steeply sloped region and carries a great deal of sediment may flow as an interconnected series of watercourses called a *braided river*. In flatter regions, most rivers are *meandering rivers*. In a meandering river, the force of water rounding a bend gradually eats away at the outer shore, eroding soil from the bank. Meanwhile, sediment is deposited along the inside of the bend, where water currents are weaker. Over time, river bends become exaggerated in shape, forming oxbows (**FIGURE 15.3**). If water erodes a shortcut from one end of the loop to the other, pursuing a

direct course, the oxbow is cut off and remains as an isolated, U-shaped water body called an *oxbow lake*.

Over thousands or millions of years, a meandering river may shift from one course to another, back and forth over a large area, carving out a flat valley. Areas nearest a river's course that are flooded periodically are said to be within the river's **floodplain**. Frequent deposition of silt from flooding

FIGURE 15.3 ▼ Rivers and streams shape landscapes. Horseshoe Bend is an oxbow of the Colorado River, formed over millennia as the river cut through the underlying rock.

makes floodplain soils especially fertile. As a result, agriculture thrives in floodplains, and *riparian* (riverside) forests are productive and species-rich.

Rivers and streams host diverse ecological communities. Algae and detritus support many types of invertebrates, from water beetles to crayfish. Insects as diverse as dragonflies, mayflies, and mosquitoes develop as larvae in streams and rivers before maturing into adults that take to the air. Fish and amphibians consume aquatic invertebrates and plants, and birds such as kingfishers, herons, and ospreys dine on fish and amphibians.

Lakes and ponds are ecologically diverse systems

Lakes and ponds are bodies of standing surface water. Their physical conditions and the types of life within them vary with depth and the distance from shore. As a result, scientists have described several zones typical of lakes and ponds (**FIGURE 15.4**).

Around the nutrient-rich edges of a water body, the water is shallow enough that aquatic plants grow from the mud and reach above the water's surface. This region, named the *littoral zone*, abounds in invertebrates—such as insect larvae, snails, and crayfish—on which fish, birds, turtles, and amphibians feed. The *benthic zone* extends along the bottom of the lake or pond, from shore to the deepest point. Many invertebrates live in the mud, feeding on detritus or on one another. In the open portion of a lake or pond, far from shore, sunlight penetrates shallow waters of the *limnetic zone*. Because light enables photosynthesis (pp. 31–32), the limnetic zone supports phytoplankton (algae, protists, and cyanobacteria; p. 78), which in turn support zooplankton (p. 78), both of which are eaten by fish. Within the limnetic zone, sunlight intensity (and therefore water temperature) decreases with depth. Clear water allows sunlight to penetrate deeply, whereas turbid water does not. Below the limnetic zone lies the *profundal zone*, the volume of open water that sunlight does not reach. This zone lacks photosynthetic life and thus is lower in the dissolved oxygen on which aquatic animals rely. The concentration of dissolved oxygen

404

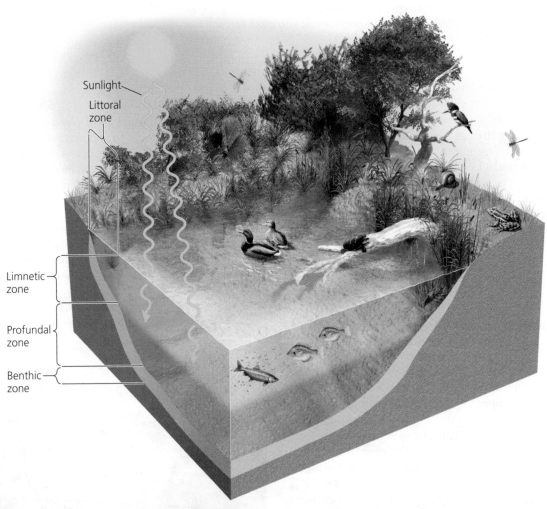

FIGURE 15.4 ▲ In lakes and ponds, emergent plants grow along the shoreline in the littoral zone. The limnetic zone is the layer of open, sunlit water, where photosynthesis takes place. Sunlight does not reach the deeper profundal zone. The benthic zone, at the bottom of the water body, often is muddy, rich in detritus and nutrients, and low in oxygen.

depends on the amount released by photosynthesis and the amount removed by animal and microbial respiration, among other factors.

Ponds and lakes change over time as streams and runoff bring them sediment and nutrients. *Oligotrophic* lakes and ponds, which are low in nutrients and high in oxygen, may slowly give way to the high-nutrient, low-oxygen conditions of *eutrophic* water bodies (jump ahead to see Figure 15.23, p. 422). Eventually, water bodies may fill in completely by the process of aquatic succession (p. 92). As lakes or ponds change over time, species of fish, plants, and invertebrates adapted to oligotrophic conditions may give way to those that thrive in eutrophic conditions. These changes occur naturally, but eutrophication can also result from human-caused nutrient pollution, as we saw with the Gulf of Mexico's dead zone in Chapter 5.

The largest lakes are sometimes known as inland seas. North America's Great Lakes are prime examples. Lake Baikal in Asia is the world's deepest lake, at 1,637 m (just over 1 mile) deep. The Caspian Sea is the world's largest body of fresh water, covering nearly as much area as Montana or California.

Wetlands include marshes, swamps, bogs, and seasonal pools

Wetlands are systems in which the soil is saturated with water and which generally feature shallow standing water with ample vegetation. There are many types of wetlands, and most are enormously rich and productive. In *freshwater marshes*, shallow water allows plants such as cattails and bulrushes to grow above the water surface. *Swamps* also consist of shallow water rich in vegetation, but they occur in forested areas. In cypress swamps of the southeastern United States, cypress trees grow in standing water. In northern forests, swamps are created when beavers dam streams with limbs from trees they have cut, flooding wooded areas upstream. *Bogs* are ponds covered with thick floating mats of vegetation and can represent a stage in aquatic succession.

Some wetlands are seasonal, being wet only at some times of year. *Vernal pools* are an example. In many parts of North America, these pools form in early spring from rain and snowmelt, and then dry up once weather becomes warmer. Numerous animals have evolved to take advantage of the ephemeral time windows in which they exist each year (**FIGURE 15.5**).

Wetlands are extremely valuable as habitat for wildlife. They also provide important ecosystem services by slowing runoff, reducing flooding, recharging aquifers, and filtering pollutants. Despite these vital roles, people have drained and filled wetlands extensively for agriculture (pp. 245, 248). Southern Canada and the United States have lost well over half their wetlands since European colonization. In 2006, the U.S Supreme Court was asked to decide whether the Clean Water Act (p. 177) protected small isolated wetlands (in the case *Rapanos v. United States*, dealing with a Michigan developer who filled wetlands on his property). The Court's split decision required the Army Corps of

FIGURE 15.5 ▲ This vernal pool in Pennsylvania is a type of seasonal wetland. The spotted salamander, *Ambystoma maculatum* (inset), breeds in temporary pools like these in the early spring, and its larvae mature here before metamorphosing and heading into the forest.

Engineers to devise guidelines to determine when such wetlands are ecologically valuable enough to protect by law.

Groundwater plays key roles in the hydrologic cycle

Any precipitation reaching Earth's land surface that does not evaporate, flow into waterways, or get taken up by organisms infiltrates the surface. Most percolates downward through the soil to become **groundwater**, water beneath the surface held within pores in soil or rock. Groundwater makes up one-fifth of Earth's fresh water supply and plays a key role in meeting human water needs.

Groundwater flows downward and from areas of high pressure to areas of low pressure. However, a typical rate of flow might be only about 1 m (3 ft) per day, so groundwater can remain underground for a long time. When we pump groundwater through wells, we are drawing up ancient water. The average age of groundwater has been estimated at 1,400 years, and some is tens of thousands of years old.

Groundwater and surface water interact, and water can flow from one type of system to the other. Surface water becomes groundwater by infiltration and percolation. Groundwater becomes surface water through springs (and human-drilled wells), often keeping streams flowing or wetlands moist when surface conditions are otherwise dry. Each day in the United States, 1.9 trillion L (492 billion gal) of groundwater are released into bodies of surface water—nearly as much as the daily flow of the Mississippi River.

As we first saw in Chapter 5 (p. 124), groundwater is contained within **aquifers**: porous, spongelike formations of rock,

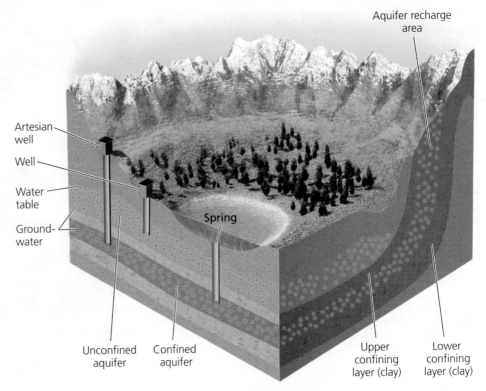

FIGURE 15.6 ▶ Groundwater may occur in unconfined aquifers above impermeable layers or in confined aquifers under pressure between impermeable layers. Water may rise naturally to the surface at springs and through the wells we dig. Artesian wells tap into confined aquifers to mine water under pressure.

Labels: Aquifer recharge area; Artesian well; Well; Water table; Ground-water; Spring; Unconfined aquifer; Confined aquifer; Upper confining layer (clay); Lower confining layer (clay)

sand, or gravel that hold water (**FIGURE 15.6**). An aquifer's upper layer, or *zone of aeration*, contains pore spaces partly filled with water. In the lower layer, or *zone of saturation*, the spaces are completely filled with water. The boundary between these two zones is the **water table**. Picture a sponge resting partly submerged in a tray of water; the lower part of the sponge is saturated, whereas the upper portion contains plenty of air in its pores. Any area where water infiltrates Earth's surface and reaches an aquifer below is known as a *recharge zone*.

Like a tray of lasagna, the earth underground consists of layers of materials with different textures and densities. When a porous, water-bearing layer of rock, sand, or gravel is trapped between upper and lower layers of less permeable substrate (often clay), we have a **confined aquifer**, or **artesian aquifer**. In such a situation, the water is under great pressure. In contrast, an **unconfined aquifer** has no impermeable upper layer to confine it, so its water is under less pressure and can be readily recharged by surface water.

The largest known aquifer is the Ogallala Aquifer, which underlies the Great Plains of the United States (**FIGURE 15.7**). Water from this massive aquifer has enabled American farmers to create the most bountiful grain-producing region in the world.

Water is unevenly distributed across Earth

Although Great Plains farmers benefit from the Ogallala Aquifer as a source of fresh water, most other areas are not so endowed. Different regions possess vastly different amounts of groundwater, surface water, and precipitation. Precipitation ranges from about 1,200 cm (470 in.) per year at Mount Waialeale on the Hawaiian island of Kauai to virtually zero in Chile's Atacama Desert.

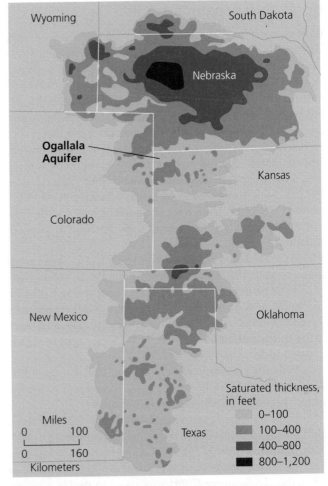

Labels: Wyoming; South Dakota; Nebraska; Ogallala Aquifer; Kansas; Colorado; New Mexico; Oklahoma; Texas

Saturated thickness, in feet: 0–100; 100–400; 400–800; 800–1,200

Miles 0 100
Kilometers 0 160

FIGURE 15.7 ▲ The Ogallala Aquifer is the world's largest aquifer, and it held 3,700 km³ (881 mi³) of water before pumping began. This aquifer underlies 453,000 km² (175,000 mi²) of the Great Plains beneath eight U.S. states. Overpumping for irrigation is currently reducing the volume and extent of this aquifer.

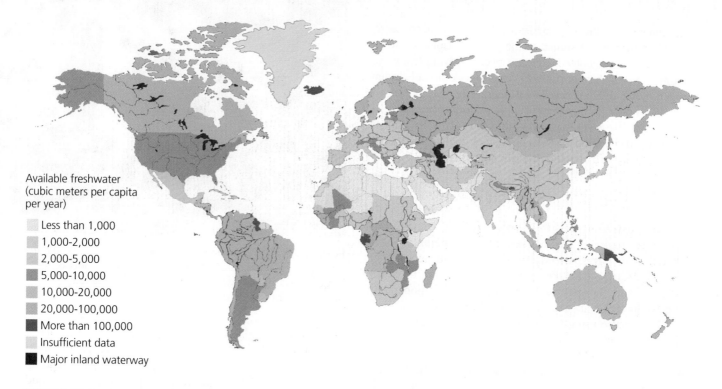

Available freshwater
(cubic meters per capita
per year)

- Less than 1,000
- 1,000-2,000
- 2,000-5,000
- 5,000-10,000
- 10,000-20,000
- 20,000-100,000
- More than 100,000
- Insufficient data
- Major inland waterway

FIGURE 15.8 ▲ Nations vary tremendously in the amount of fresh water per capita available to their citizens. For example, Iceland, Papua New Guinea, Gabon, and Guyana (dark blue in this map) each have over 100 times more water per person than do many Middle Eastern and North African countries. Data from Harrison, P., and F. Pearce, 2000. *AAAS atlas of population and the environment*, edited by the American Association for the Advancement of Science, © 2000 by the American Association for the Advancement of Science. Used by permission of the publisher, University of California Press.

People are not distributed across the globe in accordance with water availability (**FIGURE 15.8**). For example, Canada boasts 20 times more water per citizen than does China. The Amazon River carries 15% of the world's runoff, but its watershed holds less than half a percent of the world's human population. Many densely populated nations, such as Pakistan, Iran, and Egypt, face serious water shortages. Asia possesses the most water of any continent but has the least water available per person, whereas Australia, with the least amount of water, boasts the most water available per person. Because of the mismatched distribution of water and population, human societies have always struggled to transport fresh water from its source to where people need it.

Fresh water is distributed unevenly in time as well as space. India's monsoon storms can dump half of a region's annual rain in just a few hours, for example. The uneven distribution of water over time is one reason people have erected dams to store water, so that we may distribute it when needed.

Climate change may bring shortages

As if the existing mismatches between water availability and human need were not enough, global climate change (Chapter 18) will worsen conditions in many regions by altering precipitation patterns, melting glaciers, causing early-season runoff, and intensifying droughts and flooding. A drought has afflicted the Colorado River basin for most of the past decade (**FIGURE 15.9**), and climate models (pp. 502–503) predict a drier future for the American Southwest.

For the Colorado River basin, the risks from climate change come on top of the recent realization that we'd already been overestimating the amount of rainfall to expect in the region. Tree-ring studies by scientists have recently found that

FIGURE 15.9 ▼ Water has dropped to abnormally low levels at Lake Mead in recent years, as shown by the "bathtub ring" of exposed shoreline in this photograph. As of 2010, the reservoir was at only 41% of capacity. This is the result of a multiyear drought, but if global climate change makes the region more arid, then water levels could keep dropping. In 2008, scientists predicted it was 50% likely that Lake Mead would be completely dry by 2021.

precipitation in the region historically has averaged nearly 20% less than was assumed at the time the Colorado River Compact was negotiated. The prospect of having less water for more people has spurred Las Vegas to seek to mine distant groundwater and has caused four other western states to commit $2.5 billion to new water supply projects. Nationwide, a 2009 report by water managers estimated that water utilities will need to cough up $448–944 billion by 2050 to adapt their infrastructure and operations to climate change.

HOW WE USE WATER

In our efforts to harness fresh water, we have achieved impressive engineering accomplishments. In so doing, we have altered many environmental systems. An estimated 60% of the world's largest 227 rivers (and 77% of those in North America and Europe) have been strongly or moderately affected by artificial dams, dikes, and diversions.

We are also using too much water. Data indicate that at present our consumption of fresh water in much of the world is unsustainable, and we are depleting many sources of surface water and groundwater (see **ENVISIONIT**, p. 410). Already, one-third of the world's people are affected by water shortages, according to a recent global assessment.

Water supplies households, industry, and especially agriculture

We all use water at home for drinking, cooking, and cleaning. Most mining, industrial, and manufacturing processes require water. Farmers and ranchers use water to irrigate crops and water livestock. Globally, we allot about 70% of our annual fresh water use to agriculture. Industry accounts for roughly 20%, and residential and municipal uses for only 10%. Nations with arid climates tend to use a greater share for agriculture, whereas heavily industrialized nations use a great deal for industry.

When we remove water from an aquifer or surface water body and do not return it, this is called **consumptive use**. Our primary consumptive use of water is for **irrigation**, which as we saw in Chapter 9 is the provision of water to crops (p. 241). In contrast, **nonconsumptive use** of water does not remove, or only temporarily removes, water from an aquifer or surface water body. Using water to generate electricity at hydroelectric dams is an example of nonconsumptive use; water is taken in, passed through dam machinery to turn turbines, and released downstream.

Why do we allocate 70% of our water use to agriculture? Our rapid population growth requires us to feed and clothe more people each year, and the intensification of agriculture during the Green Revolution (pp. 227, 254–256) required significant increases in irrigation. As a result, we withdraw 70% more water for irrigation today than we did 50 years ago and have doubled the amount of land under irrigation. So far, expansion of irrigated agriculture has kept pace with population growth; irrigated area per person has remained for four decades at around 460 m^2 (4,950 ft^2)—one-tenth of a football field for each of us.

Irrigation can more than double crop yields by allowing farmers to apply water when and where it is needed. The

FIGURE 15.10 ▲ Farming cotton in desert regions of southern California, as seen here, requires immense amounts of irrigation water. The water in these irrigation canals is diverted from the Colorado River.

18% of world farmland that we irrigate yields fully 40% of our produce, including 60% of the global grain crop. Still, most irrigation remains highly inefficient, and crop plants end up using only 40% of the water that we apply. Inefficient "flood and furrow" irrigation, in which fields are liberally flooded with water that may evaporate from standing pools, accounts for 90% of irrigation worldwide. Overirrigation leads to waterlogging and salinization (p. 242), which affect one-fifth of farmland today and reduce farming income worldwide by $11 billion.

Many national governments subsidize irrigation to promote agricultural self-sufficiency. Unfortunately, inefficient irrigation methods in arid regions such as the Middle East are using up huge amounts of groundwater for little gain. In the Colorado River Valley, large amounts of water are diverted from the river to irrigate cotton and other crops being grown in desert regions of Arizona and southern California (**FIGURE 15.10**). Farmers in California's arid Imperial Valley, the nation's largest irrigation district, soak over 200,000 ha (500,000 acres) of land with subsidized water for which they pay just one penny per 835 L (220 gal).

We divert surface water to suit our needs

People have long diverted water from rivers and lakes to farm fields, homes, and cities. The Colorado River's water is heavily diverted and utilized (**FIGURE 15.11**). Early in its course, some Colorado River water is piped through a mountain tunnel and down the Rockies' eastern slope to supply the city of Denver. More is removed for Las Vegas and other cities and for farmland as the water proceeds downriver. When it reaches Parker Dam on the California–Arizona state line, large amounts are diverted into the Colorado River Aqueduct, which brings water to millions of people in the Los Angeles and San Diego areas via a long, open-air canal. From Parker Dam, Arizona also draws water, transporting it in canals of the Central Arizona Project. Further south at Imperial Dam, water is diverted into the Coachella and All-American Canals (see photo in Figure 15.11), destined for agriculture, mostly in California's Imperial Valley.

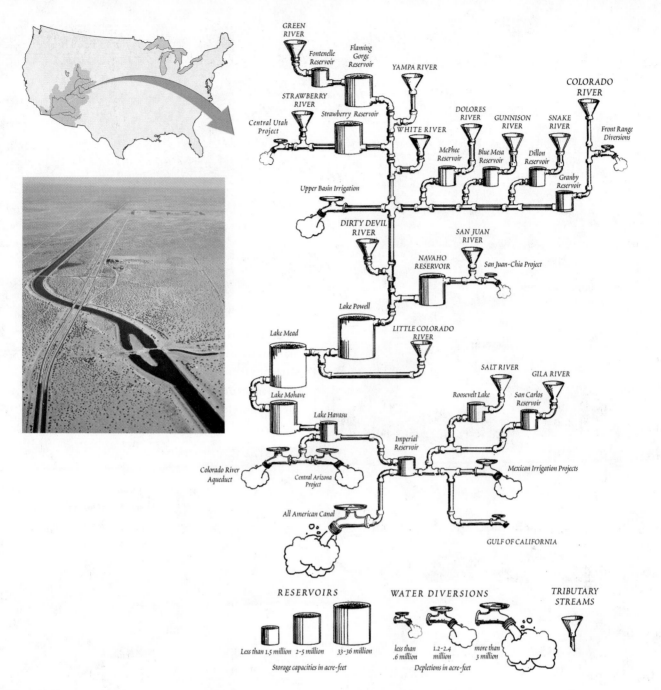

RESERVOIRS

Less than 1.5 million   2-5 million   33-36 million

Storage capacities in acre-feet

WATER DIVERSIONS

less than .6 million   1.2-2.4 million   more than 3 million

Depletions in acre-feet

TRIBUTARY STREAMS

FIGURE 15.11 ▲ The degree to which we have engineered the once-wild Colorado River led cartoonist Lester Dore to portray the Colorado and its tributaries as an immense plumbing system. Thirteen major dams pool water in enormous reservoirs along the lengths of the Colorado River and its tributaries. Several major canals (such as the All American Canal, the world's largest irrigation canal, shown winding through the California desert in the photo) divert water from the river, mostly to irrigate crops in desert regions. *Source: High Country News,* 10 November 1997.

The world's largest diversion project is underway in China. There, government leaders are pushing through an ambitious plan to pipe water from the Yangtze River in southern China (where water is plentiful) to northern China's Yellow River, which routinely dries up at its mouth because the climate is drier and people withdraw its water. Three sets of massive aqueducts, totaling 2,500 km (1,550 mi) in length, are being built to move billions of tons of water northward. China's leaders hope the diversions will solve water shortages for northern farms and cities.

Many scientists say the project won't transfer enough to make a difference, yet would cause extensive environmental impacts and displace hundreds of thousands of people.

In the past, large-scale diversion projects have enabled politically strong yet water-poor regions to forcibly appropriate water from communities too weak to keep it for themselves. The city of Los Angeles grew by commandeering water from the rural Owens Valley 350 km (220 mi) away. In so doing, it turned the environment of this region to desert, creating dustbowls and destroying its economy. Then in 1941,

Walking miles for
water in India

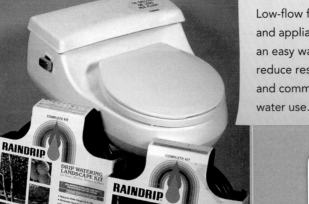

Desperation for
water in Ethiopia
has led to conflict

In developed
nations we use over
10 times more water
per person per day
than in developing
nations.

Agriculture consumes two
of every three gallons we
consume, and inefficient
irrigation wastes water.

Low-flow faucets
and appliances are
an easy way to
reduce residential
and commercial
water use.

Desert irrigation
in Saudi Arabia

RAINDRIP
DRIP WATERING
LANDSCAPE KIT

RAINDRIP
LOW VOLUME
SPRINKLER KIT

SHOWER MASSAGE

SuperSaver

Low-flow
products

YOU CAN MAKE A DIFFERENCE

➤ Persuade your college or university to
install low-flow, water-saving appliances
campus-wide.

➤ Turn the water off while you're brushing
your teeth.

➤ Eat less meat—raising livestock and their
feed requires lots of water.

L.A. decided to divert streams feeding into Mono Lake, over 565 km (350 mi) away in northern California. As the lake level fell 14 m (45 ft) over 40 years, salt concentrations doubled, and aquatic communities suffered. Today history seems to be repeating itself as Las Vegas tries to win approval for its 450-km (280-mi) pipeline to import groundwater from sparsely populated eastern Nevada, where local residents adamantly oppose the diversion plan and where scientists fear for the impacts on ecologically sensitive areas.

Reaching for Water The burgeoning metropolis of Los Angeles went to great lengths to obtain water. Today, other cities in the arid Southwest—such as Las Vegas, Phoenix, and Denver—are expected to double in population in coming decades, and Las Vegas is now reaching toward a distant aquifer for water. Do you think such diversions are ethically justified? How will people living in the source areas of the water be affected? If rural communities and wetland ecosystems of eastern Nevada are destroyed by Las Vegas's diversion project, is such destruction for the greater good if the water will benefit more people in the city? How else might these cities meet their future water needs? Or should we encourage an end to development in water-scarce areas and require residents to adopt a more sustainable lifestyle with regard to water consumption?

We build dikes and levees to control floods

Among the reasons we control the movement of fresh water, flood prevention ranks high. People have always been attracted to riverbanks for their water supply and for the flat topography and fertile soil of floodplains. But if one lives in a floodplain, one must be prepared to face flooding. **Flooding** is a normal, natural process that occurs when snowmelt or heavy rain swells the volume of water in a river so that water spills over the river's banks. In the long term, floods are immensely beneficial to both natural systems and human agriculture, because floodwaters build and enrich soil by spreading nutrient-rich sediments over large areas.

In the short term, however, floods can do tremendous damage to the farms, homes, and property of people who choose to live in floodplains. To protect against floods, communities and governments have built *dikes* and *levees* (long raised mounds of earth) along riverbanks to hold water in main channels. Many dikes are small and locally built, but the U.S. Army Corps of Engineers has constructed thousands of miles of massive levees along major waterways. These structures prevent flooding at most times and places but can sometimes worsen flooding because they force water to stay in channels and accumulate, building up enormous energy and leading to occasional catastrophic overflow events (**FIGURE 15.12**). In one tragic example, a major levee along the Mississippi River failed after Hurricane Katrina, allowing large portions of New Orleans to be flooded.

FIGURE 15.12 ▲ Channelizing rivers with levees and dikes can prevent flooding in some areas but may make it more likely in others. When levees give way, as shown here in California's Central Valley, catastrophic flooding can result.

When we engineer rivers to stay in their channels, we often end up increasing the frequency of floods in downriver areas. Human development also tends to worsen flooding because pavement and compacted soils speed runoff, sending intense pulses of water into rivers. In contrast, forests and wetlands make flooding less likely because they allow water to spread out while vegetation slows its flow and porous soil soaks it up. Thus, major flood events may become more frequent in developed areas. Indeed, a "100-year flood" is not something that can occur only once every 100 years. Rather, it is a statistical probability: in any given year, there is a 1% chance of a 100-year flood. Like spinning a roulette wheel, one occasionally gets the same unlikely result twice in a row. Scientists calculate the likelihood of major floods based on natural historical conditions, but if conditions change (through increased development, habitat loss, engineering and channelization, or global climate change), then the frequency of flooding can change.

On the Colorado River, immense dams have controlled water levels so as to virtually eliminate floods in some areas. Ecologists realized that the lack of flooding was altering the nature of the river and its plant and animal communities, and they eventually convinced water managers to create artificial floods in Grand Canyon National Park as a means of ecological restoration (pp. 95, 308–309). Today occasional, carefully timed releases through Glen Canyon Dam from Lake Powell send water coursing through the canyon to simulate natural flooding.

We have erected thousands of dams

A **dam** is any obstruction placed in a river or stream to block its flow. Dams create **reservoirs**, artificial lakes that store water for human use. We build dams to prevent floods, provide drinking water, facilitate irrigation, and generate electricity. Power generation with hydroelectric dams is discussed in Chapter 20 (pp. 583–586).

Worldwide, we have erected more than 45,000 large dams (greater than 15 m, or 49 ft, high) across rivers in over 140 nations. We have built tens of thousands of smaller dams. Only a few major rivers in the world remain undammed and free-flowing. These run through the tundra and taiga of Canada, Alaska, and Russia and in remote regions of Latin America and Africa.

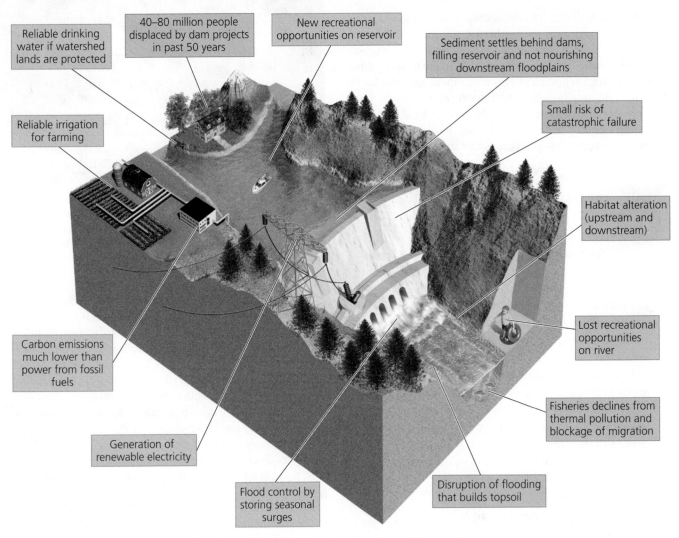

Reliable drinking water if watershed lands are protected

40–80 million people displaced by dam projects in past 50 years

New recreational opportunities on reservoir

Sediment settles behind dams, filling reservoir and not nourishing downstream floodplains

Reliable irrigation for farming

Small risk of catastrophic failure

Habitat alteration (upstream and downstream)

Carbon emissions much lower than power from fossil fuels

Lost recreational opportunities on river

Generation of renewable electricity

Fisheries declines from thermal pollution and blockage of migration

Flood control by storing seasonal surges

Disruption of flooding that builds topsoil

FIGURE 15.13 ▲ Damming rivers has diverse consequences for people and the environment. The generation of clean and renewable electricity is one of several major benefits (green boxes) of hydroelectric dams. Habitat alteration is one of several negative impacts (red boxes).

Our largest dams comprise some of humanity's greatest engineering feats. The two behemoths of the Colorado River, Hoover Dam and Glen Canyon Dam, each stand over 200 m (700 ft) high and consist of 7 million tons of concrete and steel. Hoover Dam holds back a reservoir (Lake Mead) 177 km (110 mi) long and 152 m (500 ft) deep. Glen Canyon Dam holds back a reservoir (Lake Powell) 300 km (186 mi) long and 171 m (560 ft) deep. Together these reservoirs store four times more water than flows down the river in an entire year.

Dams produce a mix of benefits and costs, as illustrated in **FIGURE 15.13**. As an example of this complex mix, we can consider the world's largest dam project. The Three Gorges Dam on China's Yangtze River, 186 m (610 ft) high and 2.3 km (1.4 mi) wide, was completed in 2008 (**FIGURE 15.14A**). Its reservoir stretches for 616 km (385 mi; as long as Lake Superior). This project is now beginning to provide flood control, enable boats and barges to travel farther upstream, and generate enough hydroelectric power to replace dozens of large coal or nuclear plants.

However, the Three Gorges Dam cost $39 billion to build, and its reservoir flooded 22 cities and the homes of 1.24 million people, requiring the largest resettlement project in China's history (**FIGURE 15.14B**). The rising water submerged 10,000-year-old archaeological sites, productive farmlands, and wildlife habitat. The reservoir slows the river's flow so that suspended sediment settles behind the dam. Because the river downstream is deprived of sediment, the tidal marshes at the Yangtze's mouth are eroding away, leaving the city of Shanghai with a degraded coastal environment and less coastal land to develop. Many scientists worry that the Yangtze's many pollutants will also be trapped in the reservoir, making the water undrinkable. The Chinese government plans to sink $5 billion into building hundreds of sewage treatment and waste disposal facilities. On top of all these worries, the 2008 earthquake in southern China (p. 41) raised fears that a future quake could damage the dam, perhaps even leading to a catastrophic collapse.

(a) The Three Gorges Dam in Yichang, China

(b) Displaced people in Sichuan Province, China

FIGURE 15.14 ▲ China's Three Gorges Dam **(a)**, completed in 2008, is the world's largest dam. Over 1.2 million people were displaced and whole cities were leveled for its construction, as shown here **(b)** in Sichuan Province.

Some dams are being removed

People who feel that the costs of some dams outweigh their benefits are pushing for such dams to be dismantled. By removing dams and letting rivers flow free, they say, we can restore riparian ecosystems, reestablish economically valuable fisheries, and revive river recreation such as fly-fishing and rafting. Increasingly, private dam owners and the Federal Energy Regulatory Commission (FERC), the U.S. government agency charged with renewing licenses

for dams, have agreed. Many aging dams are in need of costly repairs or have outlived their economic usefulness, and roughly 400 dams have been removed in the United States in the past decade.

The drive to remove dams first gathered steam in 1999 with the dismantling of the Edwards Dam on Maine's Kennebec River. FERC had determined that the environmental benefits of removing the dam outweighed the economic benefits of relicensing it. Within a year after the 7.3-m (24-ft) high, 279-m (917-ft) long dam was removed, large numbers of 10 species of migratory fish, including salmon, sturgeon, shad, herring, alewife, and bass, ventured upstream and began using the 27-km (17-mi) stretch of river above the dam site. Some property owners along the former reservoir who had opposed the dam's removal had a change of heart once they saw the healthy and vibrant river that now ran past their property. More dams will come down as over 500 FERC licenses come up for renewal in the next decade.

We are depleting surface water

As a result of our diversions and our consumption, in many places we are withdrawing surface water at unsustainable rates. What water is left in the Colorado River after all the diversions comprises just a trickle. On some days, water does not reach the Gulf of California at all. This reduction in flow (**FIGURE 15.15**) threatens the future of the cities and farms that depend on the river. And it has drastically altered the ecology of the river and its delta, changing plant communities, wiping out populations of fish and invertebrates, and devastating fisheries.

The Colorado's plight is not unique. Several hundred miles to the east, the Rio Grande also frequently runs dry, the victim of overextraction by both Mexican and U.S. farmers in times of drought. China's Yellow River, as we have seen, often fails to reach the sea. Even the river that has nurtured human

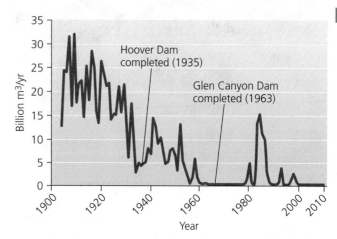

FIGURE 15.15 ▲ Flow at the mouth of the Colorado River has greatly decreased over the past century as a result of withdrawals, mostly for agriculture. The river now often runs dry at its mouth. Data from Postel, S., 2005. Worldwatch Institute, *Liquid assets: The critical need to safeguard freshwater ecosystems*, Fig 3. www.worldwatch.org. By permission.

civilization as long as any other, the Nile in Egypt, now peters out before reaching its mouth.

Nowhere are the effects of surface water depletion so evident as at the Aral Sea. Once the fourth-largest lake on Earth, just larger than Lake Huron, it lost over four-fifths of its volume in just 45 years (**FIGURE 15.16**). This dying

(a) Satellite image of Aral Sea, 1987

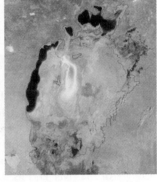

(b) Satellite image of Aral Sea, 2009

(c) Ships stranded by the Aral Sea's fast-receding waters

FIGURE 15.16 ▲ The Aral Sea in central Asia was once the world's fourth largest lake. However, it has been shrinking **(a, b)** because so much water was withdrawn to irrigate cotton crops. Along its former shorelines, ships lie stranded in the sand **(c)** because the waters receded so far and so quickly. Today restoration efforts are beginning to reverse the decline in the northern portion of the sea, and waters there are slowly rising.

inland sea, on the border of present-day Uzbekistan and Kazakhstan, is the victim of irrigation practices. The former Soviet Union instituted industrial cotton farming in this dry region by flooding the land with water from the two rivers leading into the Aral Sea. For a few decades this boosted Soviet cotton production, but it shrunk the Aral Sea, and the irrigated soil became salty and waterlogged. Today 60,000 fishing jobs are gone, winds blow pesticide-laden dust up from the dry lakebed, and little cotton grows on the blighted soil. However, all may not be lost: Scientists, engineers, and local people struggling to save the northern portion of the Aral Sea and its ecosystems may now have begun reversing its decline.

Worldwide, roughly 15–35% of water withdrawals for irrigation are thought to be unsustainable. In areas where agriculture is demanding more fresh water than can be sustainably supplied, *water mining*—withdrawing water faster than it can be replenished—is taking place (**FIGURE 15.17**). In these areas, aquifers are being depleted or surface water is being piped in from other regions.

The world is losing wetlands

From the Colorado River Delta to the Aral Sea, wetlands are being lost as we divert and withdraw water, channelize rivers, build dams, and otherwise engineer natural waterways. These actions add to the extensive draining of wetlands for agriculture discussed in Chapter 9 (pp. 245, 248). As wetlands disappear, we lose the many ecosystem services (pp. 3, 121–122, 160–161) they provide us, such as filtering pollutants, harboring wildlife, controlling floods, and helping to maintain drinking water supplies.

Fortunately, many people today see the value in wetlands and are trying to protect and restore them. In 1971 an international agreement was reached in Ramsar, Iran, to document and protect wetlands around the world. The *Ramsar Convention on Wetlands of International Importance* seeks the "conservation and wise use of all wetlands" through the "maintenance of their ecological character . . . within the context of sustainable development." Today the Colorado River Delta and nearly 1,900 other sites covering 185 million ha (714,000 mi²) across the globe are granted a degree of protection as Ramsar Wetlands, wetlands noted for their ecological, social, and economic importance.

We are depleting groundwater

Groundwater is more easily depleted than surface water because most aquifers recharge very slowly. If we compare an aquifer to a bank account, we are making more withdrawals than deposits, and the balance is shrinking. Today we are mining groundwater, extracting 160 km³ (5.65 trillion ft³) more water each year than returns to the ground. This is a problem because one-third of Earth's human population—including 99% of the rural population of the United States—relies on groundwater for its needs.

As aquifers are mined, water tables drop. Groundwater becomes more difficult and expensive to extract, and

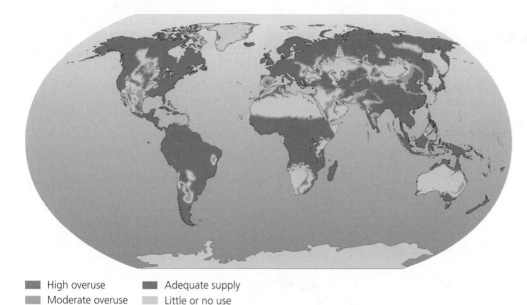

FIGURE 15.17 ◀ Irrigation for agriculture is the main contributor to unsustainable water use. Mapped are regions where overall use of fresh water (for agriculture, industry, and domestic use) exceeds the available supply, requiring groundwater depletion or diversion of water from other regions. The map understates the problem, because it does not reflect seasonal shortages. Data from UNESCO, 2006. *Water: A shared responsibility*. World Water Development Report 2. UNESCO and Berghahn Books.

- ▓ High overuse
- ▓ Moderate overuse
- ▓ Low overuse
- ▓ Adequate supply
- ▓ Little or no use

eventually it may run out. In parts of Mexico, India, China, and multiple Asian and Middle Eastern nations, water tables are falling 1–3 m (3–10 ft) per year. In the United States, by the late 1990s overpumping had drawn the Ogallala Aquifer down by 85 trillion gallons, a volume equal to the yearly flow of 18 Colorado Rivers.

When groundwater is overpumped in coastal areas, salt water can intrude into aquifers, making water undrinkable. This has occurred in Florida, the Middle East, and other locations. Moreover, as aquifers lose water, they can become less capable of supporting overlying strata, and the land surface above may subside. For this reason, cities from Venice to Bangkok to Beijing are slowly sinking. Mexico City's downtown has sunk over 10 m (33 ft) since the time of Spanish arrival; streets are buckled, old buildings lean at angles, and underground pipes break so often that 30% of the system's water is lost to leaks (**FIGURE 15.18A**).

Sometimes land subsides suddenly, creating **sinkholes**, areas where the ground gives way with little warning, occasionally swallowing homes and businesses (**FIGURE 15.18B**). Once the ground subsides, soil and rock becomes compacted, losing the porosity that enabled it to hold water. Recharging a depleted aquifer thereafter becomes more difficult. Estimates suggest that compacted aquifers under California's Central Valley have lost storage capacity equal to that of 40% of the state's surface reservoirs.

Falling water tables also do vast ecological harm. Permanent wetlands exist where water tables reach the surface, so when water tables drop, wetland ecosystems dry up. In Jordan, the Azraq Oasis covered 7,500 ha (18,500 acres) and enabled hundreds of thousands of migratory birds and other animals to find water in the desert. The water table dropped 4.5 m (14.7 ft) in the 1980s because of increased well use by the nearby city of Amman, and the oasis dried up. Today

CHAPTER 15 Freshwater Systems and Resources

415

FIGURE 15.18 ▼ When too much groundwater is withdrawn, the land may weaken and subside. This can cause buildings to lean, as seen **(a)** in Mexico City. Large areas of land may sometimes collapse suddenly in sinkholes, as seen **(b)** in Florida.

(a) 16th-century chapel in Mexico City

(b) Sinkhole in Florida

The SCIENCE behind the Story

Is It Better in a Bottle?

Which is safer and healthier for you to drink, tap water or bottled water?

If you said bottled water, you're not alone. Most people think bottled water is safer and healthier, which is why sales have doubled each decade for the past 20 years. But is bottled water really as pure and clean as its marketers want us to think?

It's hard to know the answer, because companies are not required to tell us anything about the quality of the water in their bottles, or even where the water comes from.

Municipalities that provide tap water to their residents need to submit regular reports to the Environmental Protection Agency describing their sources, treatment methods, and contaminants. In contrast, bottled water is regulated much more lightly as a "food" by the Food and Drug Administration (FDA). Bottling companies do not have to inform the public or the government where their water comes from or how it is treated, and they are not required to test samples with certified laboratories or notify the FDA of contamination problems.

In 2009, Congress held a hearing on this discrepancy after the U.S. Government Accountability Office and the nonprofit Environmental Working Group each published

30–40 billion plastic water bottles are thrown away in the U.S. each year.

reports detailing how consumers cannot get basic information about the bottled water they purchase. The Environmental Working Group (EWG), based in Washington, D.C., surveyed the labels and websites of 188 brands of bottled water. Only 2 of these brands disclosed information comparable to that required of municipal tap water providers. The FDA told Congress it would step up inspections, but the agency is overworked and understaffed as it is.

So, to find out what's in bottled water, scientists have had to do some detective work. In 2008, research scientists at the EWG sent samples of 10 major brands of bottled water to the University of Iowa's Hygienic Laboratory for analysis. The lab's chemists ran a battery of tests and detected 38 chemical pollutants, including traces of heavy metals, radioactive isotopes, caffeine and pharmaceuticals from wastewater pollution, nitrate and ammonia from

fertilizer, and various industrial compounds such as solvents and plasticizers (**first figure**). Each brand contained 8 contaminants on average, and two brands had levels of chemicals that exceeded legal limits in California and industry safety guidelines.

Two brands showed the chemical composition of standard municipal water treatment—including chlorine, fluoride, and other by-products of disinfection—indicating that these were identical to tap water. Indeed, an estimated 25–44% of bottled water *is* tap water. Corporations can essentially just turn on the faucet, filter the water, bottle it, and sell it to us at marked-up prices.

EWG also sent water samples to a lab at the University of Missouri, where researchers tested for cancer-causing effects in cell cultures. Their assays measure whether a substance causes an increase in the growth of breast cancer cells. One bottled water brand caused a 78% increase in cancer cell growth relative to the control sample, suggesting that something in the water was carcinogenic.

Other scientists have compared bottled water to tap water in different ways. In 2009, researchers Martin Wagner and Jörg Oehlmann of Johann Wolfgang Goethe University in Frankfurt, Germany, tested

---

international donors are helping Jordan's government to try to find alternative sources of water and restore this oasis.

For all these reasons, residents of eastern Nevada are anxious and angry about Las Vegas's plans to mine groundwater from the aquifer beneath them. Like the residents of Owens Valley before them, they fear that once their water is taken by a large city, their wells will run dry, their farming will fail, their wetlands will vanish, and their communities will disintegrate.

Can we quench our thirst for bottled water?

These days, our groundwater is being withdrawn for a new purpose: to be packaged in plastic bottles and sold on supermarket shelves (**FIGURE 15.19**). Bottled water is booming business. In 2008 the average American drank 29 gallons of bottled water, and sales topped $11 billion in the United States

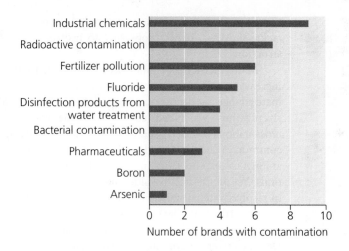

Of 10 leading brands of bottled water tested, most contained industrial chemicals, radioactive isotopes, and fertilizer pollution, as well as other contaminants. *Source:* Naidenko, O., et al., 2008. *Bottled water quality investigation: 10 major brands, 38 pollutants.* Environmental Working Group, Washington, D.C.

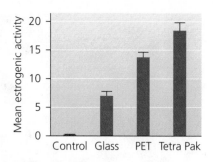

Estrogenic activity (as determined by a yeast cell culture test) was highest in bottled water from Tetra Pak containers, followed by PET plastic containers, and then glass containers. A negative control showed no appreciable estrogenic potency. With kind permission from Springer Science and Business Media and the author, from Wagner, M., and J. Oehlmann, 2009. Endocrine disruptors in bottled mineral water: Total estrogenic burden and migration from plastic bottles. *Environmental Science and Pollution Research* 16: 278–286, Fig 3a.

20 brands of bottled water for the presence of hormone-disrupting chemicals that mimic estrogen. As we saw in Chapter 14, endocrine disruptors such as bisphenol A, phthalates, and other compounds found in plastics can exert a wide array of health impacts, even at very low doses.

Wagner and Oehlmann compared nine brands packaged in glass bottles, nine brands packaged in plastic bottles (PET, or polyethylene terephthalate, the type of plastic with a #1 symbol on the bottom), and two brands packaged in "Tetra Pak" paperboard boxes with an inner plastic coating. They placed samples (as well as tap-water samples as controls) in a "yeast estrogen screen," a standard test-tube screening procedure that uses yeast cells engineered with genes to change color when exposed to estrogen-mimicking compounds.

The researchers detected estrogenic contamination in 60% of the samples (**second figure**). Both Tetra Pak brands and seven of nine plastic brands contained hormone-mimicking substances that apparently leached from the packaging. So did three of the glass-bottled brands, presumably from contamination at the bottling plant.

Wagner and Oehlmann then tested whether chemicals in the water would affect a living animal, a type of snail that is known to increase its reproduction when confronted with an estrogenic substance. They raised some snails in PET plastic containers and others in glass containers. After 56 days, the snails in the PET containers had produced 39–122% more embryos than snails in control conditions, whereas snails in glass containers showed no difference. This suggested that estrogenic compounds were leaching from the

plastic. Their results were published in the journal *Environmental Science and Pollution Research*.

Research on the quality and safety of bottled water is just getting started, but already these studies and others like them have indicated that bottled water can contain a range of contaminants, some of which may pose health risks. Tap water may also contain plenty of pollutants, but municipalities are required to test and report on their water quality, so it is much easier for consumers to get information.

For maximum safety and health, the EWG recommends drinking filtered tap water (using a filter on your faucet) instead of bottled water. Depending on the quality of your local tap water, it may be as good or better than bottled water, even when unfiltered. You can consult water quality information from your utility to find out what you are drinking. ■

and $60 billion worldwide. Americans now drink more bottled water than beer or milk and pay more per gallon for it than for gasoline.

Most people who buy bottled water do so for portability and convenience, or because they believe it will taste better or be safer and healthier than tap water. However, in blind taste tests people think tap water tastes just as good, and chemical

analyses show that bottled water is no safer or healthier than tap water (see **THE SCIENCE BEHIND THE STORY**, above). Much of it is municipal tap water that corporations have simply purified, packaged, and marked up in price to sell. In fact, when you buy a bottle of water, you have no idea where it came from or what its quality really is. The U.S. government strictly regulates the tap water that municipal governments provide us, but

FIGURE 15.19 ▲ Bottled water is more popular than ever, but it is not demonstrably safer or healthier than tap water, it requires large amounts of energy to produce and distribute, and some communities are opposing the corporations that pump, bottle, and sell their groundwater.

it requires almost nothing of the corporations that sell us water in bottles at prices up to 1,900 times more expensive than water from the tap.

Bottled water also exerts substantial ecological impact because it is heavily packaged and because we transport it long distances using fossil fuels. A 2009 study calculated the energy costs of bottled water to be 1,000–2,000 times greater than the energy costs of tap water. Most energy was used in manufacturing the bottle and transporting the product. Other studies indicate that producing a liter of bottled water requires a quarter-liter of oil and 3–5 liters of additional water.

After use, at least three out of four bottles in the United States are thrown away, and not recycled. That's 30–40 billion containers per year (5 containers for every human being on Earth), and close to 1.5 million tons of plastic waste.

Bottled water is also engendering resentment from people who live where the bottling firms are removing water. Most of the world's bottled water is sold by a handful of multinational corporations—Nestle, Pepsi-Cola, Coca-Cola, and a few others. These corporations may move into a community, extract its groundwater, package and sell it at a profit, and then move on, leaving the community with a degraded resource base. Just as the people of eastern Nevada don't want Las Vegas mining their groundwater, people in small communities from Poland, Maine (Poland Springs) to Adams County, Wisconsin (Perrier) to Salida, Colorado (Arrowhead) are fighting the corporations that are mining their groundwater resources.

Will we see a future of water wars?

Depletion of fresh water leads to shortages, and resource scarcity can lead to conflict. On the Colorado River in 1933, the governor of Arizona foresaw that California's water diversion might endanger Arizona's future allotment, so he sent the state's National Guard to threaten the construction of Parker Dam. After a long standoff, the U.S. interior secretary halted the project to avoid hostilities while the issue was mediated in court. Arizona won the court case, but California persuaded

Congress to authorize the dam, and Arizona chose not to tackle the U.S. Army troops sent to protect the dam's construction.

Many predict that water's role in regional conflicts will increase as human population continues to grow and as climate change alters precipitation patterns. A total of 261 major rivers (whose watersheds cover 45% of the world's land area) cross national borders, and transboundary disagreements are common. Water is already a key element in the hostilities among Israel, the Palestinian people, and neighboring nations. Yet on the positive side, many nations have cooperated to resolve water disputes. India has struck agreements to co-manage transboundary rivers with Pakistan, Bangladesh, Bhutan, and Nepal. In Europe, nations along the Rhine and Danube rivers have signed water-sharing treaties. Such progress gives reason to hope that water wars will be few and far between.

SOLUTIONS TO DEPLETION OF FRESH WATER

Population growth, expansion of irrigated agriculture, and industrial development doubled our annual fresh water use between 1960 and 2000. We now use an amount equal to 10% of total global runoff. The hydrologic cycle makes fresh water a renewable resource, but if we take more than a lake, river, or aquifer can provide, we must either reduce our use, find another water source, or be prepared to run out of water.

Solutions can address supply or demand

To address shortages of fresh water, we can aim either to increase supply or to reduce demand. We can increase supply temporarily through more intensive extraction, but this is generally not sustainable. Diversions may solve supply problems in one area while causing shortages in others. In contrast, strategies for reducing demand include conservation and efficiency measures. Lowering demand is more difficult politically in the short term but may be necessary in the long term. In the developing world, international aid agencies are increasingly funding demand-based solutions over supply-based solutions, because demand-based solutions offer better economic returns and cause less ecological and social damage.

Desalination "makes" more water

A supply strategy with some potential for sustainability is to generate fresh water by **desalination**, or *desalinization*, the removal of salt from seawater or other water of marginal quality. One method of desalination mimics the hydrologic cycle by evaporating allotments of ocean water with heat and then condensing the vapor—essentially *distilling* fresh water. Another method forces water through membranes to filter out salts; the most common such process is *reverse osmosis*.

Over 7,500 desalination facilities are operating worldwide. However, desalination is expensive, requires large inputs of fossil fuel energy, kills aquatic life at water intakes, and generates concentrated salty waste. As a result, desalination is pursued mostly in wealthy oil-rich nations where water is scarce. In Saudi Arabia, desalination produces half the nation's drinking

FIGURE 15.20 ▲ The Yuma Desalting Plant along the Colorado River in Arizona was shut down years ago because it was not cost-effective. Today, amid the region's dwindling supply and rising demand, engineers are testing the plant to evaluate whether to restart it. Here, U.S. Bureau of Reclamation area manager Jim Cherry fills a bottle with desalinized water from a faucet at the plant during a trial run.

water. The largest facility in the United States is in Tampa, Florida, whose groundwater suffers from saltwater intrusion.

In 1992, the world's largest reverse osmosis plant was completed along the Colorado River near Yuma, Arizona (**FIGURE 15.20**). It was intended to remove salt from irrigation runoff reentering the river, but it proved too expensive to operate and closed after only eight months. In 2007 and again in 2009 it was reopened experimentally, allowing engineers to evaluate whether they can restart the plant for good in a cost-effective way.

Agricultural demand can be reduced

Because most water is used for agriculture, it makes sense to look first to agriculture for ways to decrease demand. Farmers can improve efficiency by lining irrigation canals to prevent leaks, leveling fields to minimize runoff, and adopting efficient irrigation methods. Low-pressure spray irrigation squirts water downward toward plants, and drip irrigation systems target individual plants and introduce water directly onto the soil (see Figure 9.18b, p. 241). Both methods reduce water lost to evaporation and runoff. Experts estimate that drip irrigation (in which as little as 10% of water is wasted) could cut water use in half while raising yields by 20–90% and

producing $3 billion in extra annual income for farmers of the developing world.

Choosing crops to match the land and climate in which they are farmed can save huge amounts of water. Currently, crops that require a great deal of water, such as cotton, rice, and alfalfa, are often planted in arid areas with government-subsidized irrigation, such as occurs in California's Imperial valley. As a result of the subsidies, the true cost of water is not part of the costs of growing the crop. Eliminating subsidies and growing crops in climates with adequate rainfall could greatly reduce water use.

In addition, selective breeding (p. 55) and genetic modification (pp. 261–267) can produce crop varieties that require less water. Finally, as individuals we can all help reduce agricultural water use by eating less meat, because producing meat requires far greater water inputs than producing grain or vegetables (see Figure 10.20b, p. 269).

We can lower residential and industrial water use

In our households, we can reduce water use by installing low-flow faucets, showerheads, washing machines, and toilets. Automatic dishwashers, studies show, use less water than does washing dishes by hand. Catching rain runoff from your roof in a barrel—*rainwater harvesting*—will reduce the amount you need to use from the hose. And if your city allows it, you can use *gray water*—the wastewater from showers and sinks—to water your yard. If you have a lawn, it is best to water it at night, when water loss from evaporation is minimal. Better yet, you can replace a water-intensive lawn with native plants adapted to your region's natural precipitation patterns. *Xeriscaping*, landscaping using plants adapted to arid conditions, has become popular in the U.S. Southwest (**FIGURE 15.21**). Each of us can cut our daily water use dozens, hundreds, or even thousands of gallons a day by reexamining aspects of our daily lives.

FIGURE 15.21 ▼ Homeowners and businesses can reduce water consumption by xeriscaping their grounds—landscaping them with drought-tolerant plants. This xeriscaped yard is in Boulder City, Nevada, near Lake Mead.

Industry and municipalities can take water-saving steps as well. Manufacturers are shifting to processes that use less water and in doing so are reducing their costs. Las Vegas is one of many cities that are recycling treated municipal wastewater for irrigation and industrial uses. Governments in Arizona and in England are capturing excess runoff and pumping it into aquifers. Finding and patching leaks in pipes has saved some cities and companies large amounts of water—and money. Boston and its suburbs reduced water demand by 31% over 17 years by patching leaks, retrofitting homes with efficient plumbing, auditing industry, and promoting conservation to the public. This program enabled Massachusetts to avoid an unpopular $500 million river diversion scheme.

Market-based approaches to water conservation are being debated

Economists who want to use market-based strategies to achieve sustainable water use have suggested ending government subsidies of inefficient practices and instead letting water become a commodity whose price reflects the true costs of its extraction. Others worry that making water a fully priced commodity would make it less available to the world's poor and increase the gap between rich and poor. Because industrial use of water can be 70 times more profitable than agricultural use, market forces alone might favor uses that would benefit wealthy and industrialized people, companies, and nations at the expense of the rural poor.

Similar concerns surround another potential solution, the privatization of water supplies. During the 1990s, many public water systems were partially or wholly privatized, as governments transferred construction, maintenance, management, or ownership to private companies. This was done to enhance efficiency, but firms have little incentive to allow equitable access to water for rich and poor alike. Already in some developing countries, rural residents without access to public water supplies find themselves forced to buy water from private vendors and pay 12 times more than those connected to public supplies.

Other experiences indicate that decentralization of control over water, from the national level to the local level, can help conserve water. In Mexico, the effectiveness of irrigation systems improved dramatically once they were transferred from public ownership to the control of 386 local water user associations.

Regardless of how demand is addressed, the shift from supply-side to demand-side solutions is paying dividends. In Europe, a new focus on demand (through government mandates and public education) has decreased public water consumption, and industries are becoming more water-efficient. The United States decreased its water consumption by 5% from 1980 to 2005 thanks to conservation measures, even while its population grew 31%.

FRESH WATER POLLUTION AND ITS CONTROL

The quantity and distribution of fresh water poses one set of environmental and social challenges. Safeguarding the *quality* of water involves another array of environmental and human health dilemmas. To be safe for human consumption and for other organisms, water must be relatively free of disease-causing organisms and toxic substances.

Developed nations have made admirable advances in cleaning up water pollution over the past few decades. Still, the World Commission on Water recently concluded that over half the world's major rivers remain "seriously depleted and polluted, degrading and poisoning the surrounding ecosystems, threatening the health and livelihood of people who depend on them." The largely invisible pollution of groundwater, meanwhile, has been termed a "covert crisis."

Water pollution comes from point and non-point sources

Pollution is the release into the environment of matter or energy that causes undesirable impacts on the health or well-being of humans or other organisms. Pollution can be physical, chemical, or biological and can affect water, air, or soil. **Water pollution** comes in many forms and can cause diverse impacts on aquatic ecosystems and human health.

Some water pollution is emitted from **point sources**—discrete locations, such as a factory or sewer pipe. In contrast, **non-point-source** pollution is cumulative, arising from multiple inputs over larger areas, such as farms, city streets, and residential neighborhoods (**FIGURE 15.22**). The U.S. Clean Water Act (p. 177) addressed point-source pollution with some success by targeting industrial discharges. As a result, water quality in the United States today suffers most from non-point-source pollution, resulting from countless common activities such as applying fertilizers and pesticides to lawns, applying salt to roads in winter, and changing automobile oil. To minimize non-point-source pollution of drinking water, governments limit development on watershed land surrounding reservoirs.

Water pollution takes many forms

We can categorize water pollution into several types, including toxic chemical pollution, physical pollution by sediment, thermal pollution, nutrient pollution, and biological pollution by disease-causing organisms.

Toxic chemicals Our waterways have become polluted with toxic organic substances of our own making, including pesticides, petroleum products, and other synthetic chemicals (pp. 378–379, 381–383). Many of these can poison animals and plants, alter aquatic ecosystems, and cause an array of human health problems, including cancer. In addition, toxic metals such as arsenic, lead, and mercury damage human health and the environment, as do acids from acid precipitation (pp. 482–484) and from acid drainage from mining sites (p. 649). Health impacts of toxic substances are discussed in Chapter 14.

Issuing and enforcing more stringent regulations of industry can help reduce releases of many toxic chemicals. We can also modify our industrial processes and our purchasing decisions to rely less on these substances.

Sediment Although flooding helps build fertile farmland, sediment that rivers transport can also impair aquatic ecosystems. Clear-cutting, mining, clearing land for housing

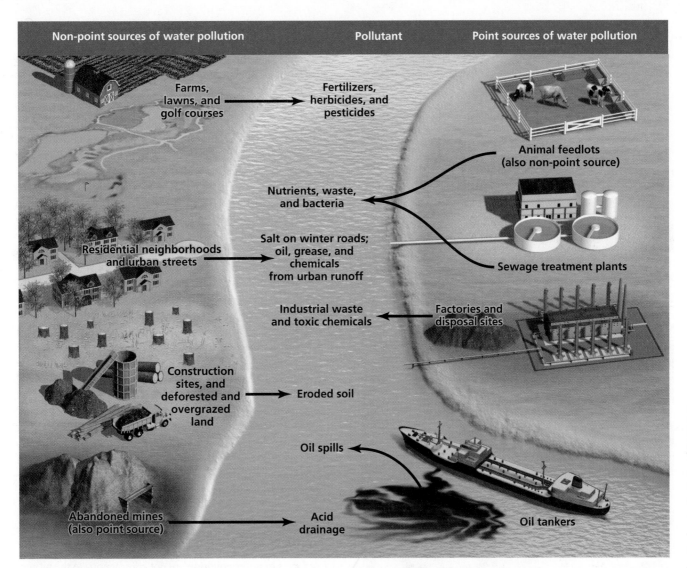

Non-point sources of water pollution	Pollutant	Point sources of water pollution

Farms, lawns, and golf courses → Fertilizers, herbicides, and pesticides

Nutrients, waste, and bacteria ← Animal feedlots (also non-point source)

Sewage treatment plants

Residential neighborhoods and urban streets → Salt on winter roads; oil, grease, and chemicals from urban runoff

Industrial waste and toxic chemicals ← Factories and disposal sites

Construction sites, and deforested and overgrazed land → Eroded soil

Oil spills ← Oil tankers

Abandoned mines (also point source) → Acid drainage

FIGURE 15.22 ▲ Point-source pollution (on right) comes from discrete facilities or locations, usually from single outflow pipes. Non-point-source pollution (such as runoff from streets, residential neighborhoods, lawns, and farms; on left) originates from numerous sources spread over large areas.

development, and cultivating farm fields all expose soil to wind and water erosion (pp. 231–233). Some water bodies, such as the Colorado River and China's Yellow River, are naturally sediment-rich, but many others are not. When a clear-water river receives a heavy influx of eroded sediment, aquatic habitat changes dramatically, and fish adapted to clear water may be killed. We can reduce sediment pollution by better managing farms and forests and avoiding large-scale disturbance of vegetation.

Thermal pollution   Water's ability to hold dissolved oxygen decreases as temperature rises, so some aquatic organisms may not survive when human activities raise water temperatures. When we withdraw water from a river and use it to cool an industrial facility, we transfer heat from the facility back into the river where the water is returned. People also raise water temperatures by removing streamside vegetation that shades water.

Too little heat can also cause problems. On the Colorado and many other dammed rivers, water at the bottoms of reservoirs is colder than water at the surface. When dam

operators release water from the depths of a reservoir, downstream water temperatures drop suddenly. In the Colorado River system, these pulses of cold water have favored cold-loving invasive trout over an endangered native species of suckerfish.

Nutrient pollution   We saw in Chapter 5 with the Gulf of Mexico's dead zone how nutrient pollution from fertilizers and other sources can lead to eutrophication and hypoxia in coastal marine waters (see Figure 5.4, p. 114). Eutrophication proceeds in a similar fashion in freshwater systems, where phosphorus is usually the nutrient that spurs growth (p. 117). When excess phosphorus enters a water body, it fertilizes algae and aquatic plants, boosting their growth. Algae then spread and cover the water's surface, depriving underwater plants of sunlight. As algae die off, bacteria consume them. Because this decomposition requires oxygen, the increased bacterial activity drives down levels of dissolved oxygen. These levels can drop too low to support fish and shellfish, leading to dramatic changes in aquatic ecosystems.

(a) Oligotrophic water body

(b) Eutrophic water body

FIGURE 15.23 ▲ An oligotrophic water body **(a)** with clear water and low nutrient content may eventually become a eutrophic water body **(b)** with abundant algae and high nutrient content. Pollution of freshwater bodies by excess nutrients accelerates the process of eutrophication.

Eutrophication (**FIGURE 15.23**) is a natural process, but nutrient input from runoff from farms, golf courses, lawns, and sewage can dramatically increase the rate at which it occurs. We can reduce nutrient pollution by treating wastewater, reducing fertilizer application, using phosphate-free detergents, and planting vegetation and protecting natural areas to increase nutrient uptake.

Pathogens and waterborne diseases   Disease-causing organisms (pathogenic viruses, protists, and bacteria) can enter drinking water supplies when these are contaminated with human waste from inadequately treated sewage or with animal waste from feedlots (pp. 268–269). Specialists monitoring water quality can tell when water has been contaminated by waste when they detect fecal coliform bacteria, which live in the intestinal tracts of people and other

vertebrates. These bacteria are usually not pathogenic themselves, but they serve as indicators of fecal contamination, alerting us that the water may hold other pathogens that can cause ailments such as giardiasis, typhoid, or hepatitis A.

Biological pollution by pathogens causes more human health problems than any other type of water pollution. In the United States, an estimated 20 million people fall ill each year from drinking water contaminated with pathogens. Worldwide, the United Nations estimates that 3,800 children die every day from diseases associated with unsafe drinking water, such as cholera, dysentery, and typhoid fever.

As we saw in our discussion of environmental health in Chapter 14, we are making slow but steady progress worldwide in supplying people with safe drinking water and with sewer or sanitation facilities (see Figure 14.8, p. 376). According to World Health Organization data, 86% of the world's people have access to safe water as a result of improvement in their water supply—up from 76% in 1990. However, nearly a billion people are still without safe water. Similar progress has been made with sanitation, but 2.6 billion people—4 of 10 people in the world—still lack adequate sewer or sanitation facilities. Most of these people live in rural areas in Asia or Africa. These conditions contribute to widespread health impacts and 5 million deaths per year.

We reduce the risks posed by waterborne pathogens by using chemical or other means to disinfect drinking water (p. 426) and by treating wastewater (pp. 426–427). Other measures to lessen health risks include public education to encourage personal hygiene and government enforcement of regulations to ensure the cleanliness of food production, processing, and distribution.

Scientists use several indicators of water quality

Most forms of water pollution are not conspicuous to the human eye, so scientists and technicians measure certain physical, chemical, and biological properties of water to characterize its quality (**FIGURE 15.24**). Biological properties include the

FIGURE 15.24 ▼ Scientists and technicians characterize water quality by measuring certain physical, chemical, and biological properties. Here a researcher collects water samples from a lagoon near a landfill in England.

presence of fecal coliform bacteria and disease-causing organisms, as discussed above. Algae and aquatic invertebrates are also commonly used as biological indicators of water quality.

Chemical properties include nutrient concentrations, pH (p. 28), taste, odor, and hardness. Hard water contains high concentrations of calcium and magnesium ions, prevents soap from lathering, and leaves chalky deposits behind when heated. An important chemical characteristic is dissolved oxygen content. Dissolved oxygen is an indicator of aquatic ecosystem health because water low in dissolved oxygen is less capable of supporting aquatic life.

Among physical characteristics, the color of water can reveal particular substances present in a sample. Some forest streams run the color of iced tea because of chemicals called tannins that occur naturally in decomposing leaf litter. Temperature can also be used to assess water quality. High temperatures can interfere with some biological processes, and warmer water holds less dissolved oxygen. Finally, turbidity measures the density of suspended particles in a water sample. Fast-moving rivers that cut through arid or eroded landscapes, such as the Colorado and Yellow rivers, carry a great deal of sediment and are turbid and muddy-looking as a result. If scientists can measure only one parameter, they will often choose turbidity, because it tends to correlate with many others and is thereby a good indicator of water quality.

Groundwater pollution is a difficult problem

Most pollution control efforts focus on surface water. Yet groundwater sources once assumed to be pristine are regularly polluted by industry and agriculture. Groundwater pollution is hidden from view and difficult to monitor; it can be out-of-sight, out-of-mind for decades until widespread contamination of drinking supplies is discovered.

Groundwater pollution is also more difficult to address than surface water pollution. Rivers flush their pollutants fairly quickly, but groundwater retains its contaminants until they decompose, which in the case of persistent pollutants can be many years or decades. The long-lived pesticide DDT, for instance, is found widely in U.S. aquifers even though it was banned over 35 years ago. Moreover, chemicals break down much more slowly in aquifers than in surface water or soils. Decomposition is slower in groundwater because it is not exposed to sunlight, contains fewer microbes and minerals, and holds less dissolved oxygen and organic matter. For example, concentrations of the herbicide alachlor decline by half after 20 days in soil, but in groundwater this takes almost 4 years.

There are many sources of groundwater pollution

Some chemicals that are toxic at high concentrations, including aluminum, fluoride, nitrates, and sulfates, occur naturally in groundwater. After all, groundwater is in contact with rock for thousands of years, and during that time all kinds of compounds, both toxic and benign, may leach into the water (including the calcium and magnesium that lead to "hardness," mentioned above). The poisoning of wells by arsenic in

the Asian nation of Bangladesh is one case of natural contamination (see **THE SCIENCE BEHIND THE STORY**, pp. 424–425).

However, groundwater pollution resulting from human activity is widespread. Industrial, agricultural, and urban wastes—from heavy metals to petroleum products to solvents to pesticides—can leach through soil and seep into aquifers. Pathogens and other pollutants can enter groundwater through improperly designed wells and from the pumping of liquid hazardous waste below ground (pp. 638–639). A recent 17-year study of volatile organic compounds (VOCs; p. 473) detected 42 types of VOCs from manufactured products and industrial processes in nearly all U.S. aquifers and in 18% of wells sampled (p. 379).

Leakage of carcinogenic pollutants (such as chlorinated solvents and gasoline) from underground tanks of oil and industrial chemicals also poses a threat to groundwater. Across the United States, the Environmental Protection Agency (EPA) has embarked on a nationwide cleanup program to unearth and repair leaky tanks (**FIGURE 15.25**). After 15 years of work, the EPA by March 2010 had confirmed leaks from 492,000 tanks, had initiated cleanups on 467,000 of them, and had completed cleanups of 395,000.

Agriculture contributes to groundwater pollution in several ways. Pesticides were detected in most of the shallow aquifer sites tested in the United States in the 1990s, although levels generally did not violate EPA safety standards for drinking water. Nitrate from fertilizers has leached into aquifers in Canada and in 49 U.S. states. Nitrate in drinking water has been linked to cancers, miscarriages, and "blue-baby" syndrome, which reduces the oxygen-carrying capacity of infants' blood. Agriculture can also contribute pathogens; in 2000, the groundwater supply of Walkerton, Ontario, became contaminated with the bacterium *Escherichia coli*, or *E. coli*. Two thousand people became ill, and seven died.

Manufacturing facilities and military sites have been heavy polluters through the years. Legislation has forced them to reduce discharges, but groundwater can bear a toxic legacy long afterwards. During World War II, the U.S. Army operated the

FIGURE 15.25 ▼ Leaky underground storage tanks are a major source of groundwater pollution. Under an EPA program, hundreds of thousands of these tanks are being unearthed and repaired.

The SCIENCE behind the Story

Arsenic in the Aquifer: Tackling A Tragedy in Bangladesh

In the 1970s, UNICEF, with the help of environmental scientists at the British Geological Survey, launched a campaign to improve access to fresh water in the poverty-stricken south-Asian nation of Bangladesh. By digging thousands of small artesian wells, they hoped to reduce Bangladeshis' dependence on disease-ridden surface waters.

Soon, however, scientists began to suspect that the wells dug to improve Bangladeshis' health were contaminated with arsenic, a naturally occurring heavy metal that can cause serious skin disorders and other illnesses, including cancer. Today, researchers are trying to discover how and why arsenic is poisoning the water, so that solutions can be devised.

A medical doctor had sounded the first alarm. In 1983, dermatologist K.C. Saha of the School of Tropical Medicine in Calcutta, India, saw the first of many patients who showed signs of arsenic poisoning. Through a process of elimination, contaminated well water was identified as the likely

Skin lesions caused by arsenic poisoning in Bangladesh

cause of the poisoning. The hypothesis was confirmed by groundwater testing and by the work of epidemiologists, among them Dipankar Chakraborti of Calcutta's Jadavpur University.

However, it was not until 2001 that the British Geological Survey and Bangladesh's government finished testing 3,524 wells and published their final report. Of the shallow wells, 46% exceeded the World Health Organization's maximum recommended safety level of arsenic. Extrapolating across all of Bangladesh, the scientists estimated that as many as 2.5 million wells serving 57 million people were contaminated. Arsenic contamination is most prevalent in

southern Bangladesh, but localized hotspots occur in northern regions of the country (**figure**).

Scientists agreed that the arsenic is of natural origin, but it remained unclear how the low levels of arsenic naturally present in rock and soil became dissolved in the aquifers in elemental and highly toxic form. One initial explanation, suggested by Chakraborti and his colleagues, blamed agricultural irrigation. By drawing large amounts of water out of aquifers during Bangladesh's dry season, they argued, irrigation permitted oxygen to enter the aquifers, prompting the release of arsenic from pyrite, a common mineral.

Other scientists contended that pyrite oxidation could not explain most cases of arsenic contamination. In a 1998 paper in the journal *Nature*, Ross Nickson of the British Geological Survey and his colleagues suggested that arsenic was being released from iron oxides carried into Bangladesh by the Ganges River. They showed

world's largest facility producing trinitrotoluene (TNT) near St. Louis. Nitroaromatic by-products seeped into the drinking water for miles around, and the site became an early priority for cleanup under the Superfund program (pp. 639–640). At another Superfund site, the Hanford Nuclear Reservation in Washington State, groundwater has been contaminated with radioactive waste, some of it with a half-life of a quarter-million years.

It is better to prevent pollution than to deal with it later

Preventing pollution is easier and more effective than mitigating it later, so the best solutions involve prevention rather than "end-of-pipe" treatment and cleanup. For instance, filtering polluted groundwater is expensive; facilities in the U.S. Midwest by one estimate spend $400 million annually just to

remove the herbicide atrazine from water supplies. Pumping water out of an aquifer, treating it, then injecting it back in, repeatedly, takes an impracticably long time. We are doing this at Superfund sites, but work has begun on just over 1% of all sites with heavily polluted groundwater, and the eventual cleanup bill is estimated at $1 trillion.

There are many things ordinary people can do to help prevent freshwater pollution. One is to exercise the power of consumer choice in the marketplace by purchasing sustainably made products. Another is to become involved in protecting local waterways. Community-based "riverwatch" groups and watershed associations enlist volunteers to collect data, monitor pollution, and help agencies safeguard the health of local water bodies. And of course, it is important to let government representatives know that you value clean water and to urge them to pursue policies to fight pollution.

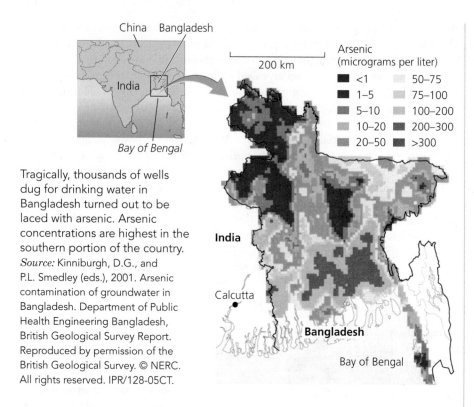

Tragically, thousands of wells dug for drinking water in Bangladesh turned out to be laced with arsenic. Arsenic concentrations are highest in the southern portion of the country. *Source:* Kinniburgh, D.G., and P.L. Smedley (eds.), 2001. Arsenic contamination of groundwater in Bangladesh. Department of Public Health Engineering Bangladesh, British Geological Survey Report. Reproduced by permission of the British Geological Survey. © NERC. All rights reserved. IPR/128-05CT.

how, contrary to the predictions of the pyrite oxidation hypothesis, a hydrochemical survey found that arsenic increased with aquifer depth and was inversely correlated with concentrations of sulfur, a component of pyrite. Nickson's team concluded that chemical conditions created by buried organic matter, such as peat, had probably leached arsenic from iron oxides over thousands of years.

Subsequently, Massachusetts Institute of Technology hydrologist Charles Harvey and colleagues

suggested that irrigation may contribute to the arsenic problem after all, but not because of pyrite oxidation. In a 2002 paper in the journal *Science*, they describe digging more than a dozen experimental wells near the capital city of Dhaka. Contrary to the pyrite oxidation hypothesis, and in agreement with Nickson's team, they found little evidence for a connection between sulfur or oxygen and arsenic. Yet they also found that they could increase arsenic concentrations by injecting organic matter, such as

molasses, into their experimental wells. While being metabolized by microbes, the molasses appeared to be freeing arsenic from iron oxides. A similar process might take place naturally, Harvey's team argued, when runoff from rice paddies, ponds, and rivers recharges aquifers depleted by heavy pumping for irrigation.

Then in 2009, Harvey, his graduate student Rebecca Neumann, and others reported that organic matter appeared to be making its way into aquifers through the numerous ponds that formed in depressions people had dug for fill dirt to build up dry land for villages. Neumann and Harvey's team described in the journal *Nature Geoscience* how they traced chemical isotopes (p. 26) and found that carbon from the ponds was being metabolized underground, causing arsenic to dissolve into shallow groundwater.

In light of these findings, scientists suggest that wells be dug deeper into the aquifers to reach drinking water that is not contaminated by these surface processes. However, they caution that irrigation water should perhaps still be drawn from the shallow wells. They point out that the area's rice farming seems to decrease arsenic levels and that sucking too much water from deep in an aquifer might pull contaminated surface water downward, causing it to mix throughout the aquifer.

Legislative and regulatory efforts have helped to reduce pollution

As numerous as our freshwater pollution problems may seem, it is important to remember that many were worse a few decades ago, when the Cuyahoga River would catch fire (p. 175). Citizen activism and government response during the 1960s and 1970s in the United States resulted in legislation such as the Federal Water Pollution Control Act of 1972 (later amended and renamed the Clean Water Act in 1977). These acts made it illegal to discharge pollution from a point source without a permit, set standards for industrial wastewater, set standards for contaminant levels in surface waters, and funded construction of sewage treatment plants. Thanks to such legislation, point-source pollution in the United States was reduced, and rivers and lakes became notably cleaner.

In the past decade, however, enforcement of water quality laws grew weaker, as underfunded and understaffed state and federal regulatory agencies succumbed to pressure from industry and from politicians who receive money from industry. A comprehensive investigation by *The New York Times* in 2009 revealed that violations of the Clean Water Act have risen and that documented violations now number over 100,000 per year (to say nothing of undocumented instances). The EPA and the states act on only a tiny percentage of these violations, the *Times* found. As a result, 1 in 10 Americans have been exposed to unsafe drinking water—for the most part unknowingly, because many pollutants cannot be detected by smell, taste, or color. In response, new EPA Administrator Lisa Jackson promised to strengthen enforcement.

The Great Lakes of Canada and the United States represent an encouraging success story in fighting water pollution.

In the 1970s these lakes, which hold 18% of the world's surface fresh water, were badly polluted with wastewater, fertilizers, and toxic chemicals. Algal blooms fouled beaches, and Lake Erie was pronounced "dead." Today, efforts of the Canadian and U.S. governments have paid off. According to Environment Canada, releases of seven toxic chemicals are down by 71%, municipal phosphorus has decreased by 80%, and chlorinated pollutants from paper mills are down by 82%. Levels of PCBs and DDE are down by 78% and 91%, respectively. Bird populations are rebounding, and Lake Erie is now home to the world's largest walleye fishery. The Great Lakes' troubles are by no means over—sediment pollution is still heavy, PCBs and mercury still settle from the air, and fish are not always safe to eat. However, the progress so far shows how conditions can improve when citizens push their governments to take action.

We treat our drinking water

Technological advances as well as government regulation have improved our control of pollution. The treatment of drinking water is a widespread and successful practice in developed nations today. Before being sent to your tap, water from a reservoir or aquifer is treated with chemicals to remove particulate matter; passed through filters of sand, gravel, and charcoal; and/or disinfected with small amounts of an agent such as chlorine. The U.S. EPA sets standards for over 90 drinking water contaminants, which local governments and private water suppliers are obligated to meet.

We treat our wastewater

Wastewater treatment is also now a mainstream practice. **Wastewater** refers to water that people have used in some way. It includes water carrying sewage; water from showers, sinks, washing machines, and dishwashers; water used in manufacturing or industrial cleaning processes; and storm water runoff. Natural systems can process moderate amounts of wastewater, but the large and concentrated amounts that our densely populated areas generate can harm ecosystems and pose health threats. Thus, attempts are now widely made to treat wastewater before it is released into the environment.

In rural areas, **septic systems** are the most popular method of wastewater disposal. In a septic system, wastewater runs from the house to an underground septic tank, inside which solids and oils separate from water. The clarified water proceeds downhill to a drain field of perforated pipes laid horizontally in gravel-filled trenches underground. Microbes decompose pollutants in the wastewater these pipes emit. Periodically, solid waste from the septic tank is pumped out and taken to a landfill.

In more densely populated areas, municipal sewer systems carry wastewater from homes and businesses to centralized treatment locations. There, pollutants are removed by physical, chemical, and biological means (**FIGURE 15.26**). At a treatment facility, **primary treatment**, the physical removal of contaminants in settling tanks or clarifiers, removes about 60% of suspended solids. Wastewater then proceeds to **secondary treatment**, in which water is stirred and aerated so that aerobic bacteria degrade organic pollutants. Roughly 90%

of suspended solids may be removed after secondary treatment. Finally, the clarified water is treated with chlorine, and sometimes ultraviolet light, to kill bacteria. Most often, the treated water, called *effluent*, is piped into rivers or the ocean following primary and secondary treatment. However, Las Vegas and many other municipalities are recycling "reclaimed" water for lawns and golf courses, for irrigation, or for industrial purposes such as cooling water in power plants.

As water is purified throughout the treatment process, the solid material removed is termed *sludge*. Sludge is sent to digesting vats, where microorganisms decompose much of the matter. The result, a wet solution of "biosolids," is then dried and either disposed of in a landfill, incinerated, or used as fertilizer on cropland. Methane-rich gas created by the decomposition process is sometimes burned to generate electricity, helping to offset the cost of treatment. Each year about 6 million dry tons of sludge are generated in the United States.

WEIGHING **THE ISSUES**

Sludge on the Farm It is estimated that 38% to over half the biosolids from sewage sludge produced each year is used as fertilizer on farmland. This practice makes productive use of the sludge, increases crop output, and conserves landfill space, but many people have voiced concern over accumulation of toxic metals, proliferation of dangerous pathogens, and odors. Do you feel that this practice represents an efficient use of resources or an unnecessary risk? What further information would you want to know to inform your decision?

Constructed wetlands can aid treatment

Long before people built the first wastewater treatment plants, natural wetlands were filtering and purifying water. Recognizing this, engineers have begun manipulating wetlands and even constructing new wetlands to employ them as tools to cleanse wastewater. Generally in this approach, wastewater that has gone through primary or secondary treatment at a conventional facility is pumped into the wetland, where microbes living amid the algae and aquatic plants decompose the remaining pollutants. Water cleansed in the wetland can then be released into waterways or allowed to percolate underground.

One of the first constructed wetlands was established in Arcata, a town on northern California's scenic Redwood Coast (**FIGURE 15.27**). This 35-ha (86-acre) engineered wetland system was built in the 1980s after residents objected to a $50 million system plan to build a large treatment plant and pump treated wastewater into the ocean. In the wetland system that was built instead, oxidation ponds send partially treated wastewater to the wetland, where plants and microbes continue to perform secondary treatment. The project cost just $7 million, and the site also serves as a haven for wildlife and human recreation. In fact, the Arcata Marsh and Wildlife Sanctuary has brought the town's waterfront back to life; more

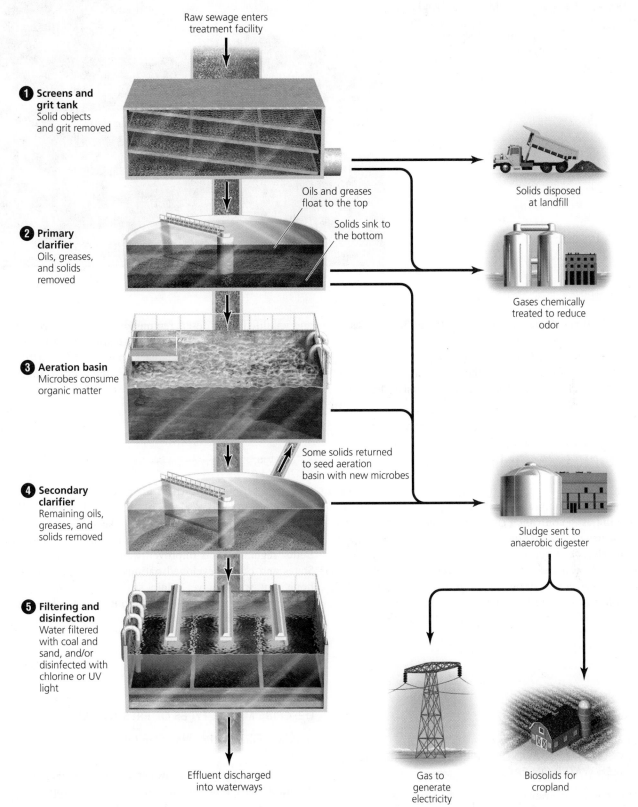

Raw sewage enters
treatment facility

1 Screens and grit tank
Solid objects and grit removed

Oils and greases float to the top

Solids sink to the bottom

2 Primary clarifier
Oils, greases, and solids removed

Solids disposed at landfill

Gases chemically treated to reduce odor

3 Aeration basin
Microbes consume organic matter

Some solids returned to seed aeration basin with new microbes

4 Secondary clarifier
Remaining oils, greases, and solids removed

Sludge sent to anaerobic digester

5 Filtering and disinfection
Water filtered with coal and sand, and/or disinfected with chlorine or UV light

Effluent discharged into waterways

Gas to generate electricity

Biosolids for cropland

FIGURE 15.26 ▲ Shown here is a generalized process from a modern, environmentally sensitive wastewater treatment facility. Wastewater initially passes through screens to remove large debris and into grit tanks to let grit settle **1**. It then enters tanks called primary clarifiers **2**, in which solids settle to the bottom and oils and greases float to the top for removal. Clarified water then proceeds to aeration basins **3** that oxygenate the water to encourage decomposition by aerobic bacteria. Water then passes into secondary clarifier tanks **4** for removal of further solids and oils. Next, the water may be purified **5** by chemical treatment with chlorine, passage through carbon filters, and/or exposure to ultraviolet light. The treated water (called *effluent*) may then be piped into natural water bodies, used for urban irrigation, flowed through a constructed wetland, or used to recharge groundwater. In addition, most treatment facilities control odor in the early steps and use anaerobic bacteria to digest sludge removed from the wastewater. Biosolids from digesters may be sent to farm fields as fertilizer, and gas from digestion may be used to generate electric power.

FIGURE 15.27 ▶ The Arcata Marsh and Wildlife Sanctuary is the site of an artificially engineered, constructed wetland system that helps treat this northern California city's wastewater. Upland areas around the marsh are open to the public for recreation.

than 100,000 people visit each year, and over 300 species of birds have been observed here.

Other cities have looked to Arcata as a model for their own efforts. At Sweetwater Wetlands in Tucson, Arizona, the artificial marshes comprise a wetland oasis for birds and wildlife in this desert region while helping to recharge a depleted aquifer. The practice of treating wastewater with artificial wetlands is growing fast; today over 500 artificially constructed or restored wetlands in the United States are performing this service.

➤ CONCLUSION

Citizen action, government legislation and regulation, new technologies, economic incentives, and public education are all helping us to confront a rising challenge of our new century: ensuring adequate quantity and quality of fresh water for ourselves and for the planet's ecosystems. Accessible fresh water comprises a minuscule percentage of the hydrosphere, but we generally take it for granted. Our expanding population and increasing water use are bringing us toward conditions of widespread scarcity. Water depletion has become a serious concern in many areas of the developing world and in arid regions of developed nations. Water pollution, meanwhile, continues to take a toll on the health, economies, and societies of nations both rich and poor. Better regulation has improved water quality in the United States and other developed nations, and there is reason to hope that we may yet attain sustainability in our water use. Potential solutions are numerous, and the issue is too important to ignore.

R E V I E W I N G   O B J E C T I V E S

You should now be able to:

EXPLAIN WATER'S IMPORTANCE TO PEOPLE AND ECOSYSTEMS, AND DESCRIBE THE DISTRIBUTION OF FRESH WATER ON EARTH

- Water is a renewable but limited resource, so we must manage it sustainably. (p. 402)
- A functioning water cycle is vital to maintaining our civilization and the natural systems that support it. (p. 402)
- Of all the water on Earth, only about 1% is readily available for our use. (pp. 402–403)
- Water availability varies in space and time, and regions vary greatly in how much they possess. (pp. 406–407)
- Climate change may bring water shortages in some regions. (pp. 407–408)

DESCRIBE MAJOR TYPES OF FRESHWATER SYSTEMS

- A watershed is the area of land drained by a river system. (p. 403)
- The main types of freshwater ecosystems include rivers and streams, lakes and ponds, and wetlands. (pp. 403–405)
- Groundwater is contained within aquifers. (pp. 405–406)

DISCUSS HOW WE USE WATER AND ALTER FRESHWATER SYSTEMS

- We use water for agriculture, industry, and residential use. Globally, 70% is used for agriculture. (p. 408)
- We divert water with canals and irrigation ditches to bring water where it is desired. (pp. 408–409, 411)
- We attempt to control floods with dikes and levees. (p. 411)

- We have dammed most of the world's rivers. Dams bring a diversity of benefits and costs. Some dams are now being removed. (pp. 411–413)
- We pump water from aquifers and surface water bodies, sometimes at unsustainable rates. (pp. 413–416)

ASSESS PROBLEMS OF WATER SUPPLY AND PROPOSE SOLUTIONS TO ADDRESS DEPLETION OF FRESH WATER

- Surface water extraction has caused rivers to run dry and water bodies to shrink. (pp. 413–414)
- Many wetlands have been lost, and we are now trying to restore some. (p. 414)
- Water tables are dropping in many areas from unsustainable groundwater extraction. (pp. 414–416)
- Some of our water extraction now goes to bottled water, which is hugely popular despite being no healthier than tap water and creating substantial plastic waste. (pp. 416–418)
- Political tensions over water may heighten in the future. (p. 418)
- Desalination increases water supply, but it is expensive and energy-intensive. (pp. 418–419)
- Solutions to reduce demand include technology, market-based approaches, and consumer products that increase efficiency in agriculture, industry, and the home. (pp. 419–420)
- Privatization of water supplies is a much-debated issue. (p. 420)

ASSESS PROBLEMS OF WATER QUALITY AND PROPOSE SOLUTIONS TO ADDRESS WATER POLLUTION

- Water pollution stems from point sources and non-point sources. (pp. 420–421)
- Water pollutants include toxic chemicals, sediment, thermal pollution, excessive nutrients, and microbial pathogens. (pp. 420–422)
- Scientists who monitor water quality use biological, chemical, and physical indicators. (pp. 422–423)
- Groundwater pollution is more persistent and difficult to address than surface water pollution. (pp. 423–424)
- Preventing water pollution is better than mitigation. (p. 424)
- Legislation and regulation have improved water quality in developed nations in recent decades. (pp. 425–426)

EXPLAIN HOW WE TREAT DRINKING WATER AND WASTEWATER

- Municipalities treat drinking water by filtering and disinfection in a multistep process. (p. 426)
- Septic systems help treat wastewater in rural areas. (p. 426)
- Wastewater is treated physically, biologically, and chemically in a series of steps at municipal wastewater treatment facilities. (pp. 426–427)
- Artificial wetlands enhance wastewater treatment while restoring habitat for wildlife. (pp. 426, 428)

TESTING YOUR COMPREHENSION

1. Compare and contrast the main types of freshwater ecosystems. Name and describe the major zones of a typical pond or lake.

2. Why are sources of fresh water unreliable for some people and plentiful for others?

3. Describe three benefits and three costs of damming rivers. What particular environmental, health, and social concerns has China's Three Gorges Dam and its reservoir raised?

4. Why do the Colorado, Rio Grande, Nile, and Yellow rivers now slow to a trickle or run dry before reaching their deltas?

5. Why are water tables dropping around the world? What are some negative impacts of falling water tables?

6. Name three major types of water pollutants, and provide an example of each. Now list three properties of water that scientists use to determine water quality.

7. Define *groundwater*. Why do many scientists consider groundwater pollution a greater problem than surface water pollution?

8. What are some anthropogenic (human) sources of groundwater pollution?

9. Describe how drinking water is treated. How does a septic system work?

10. Describe and explain the major steps in the process of wastewater treatment. How can artificial wetlands aid such treatment?

SEEKING SOLUTIONS

1. How can we lessen agricultural demand for water? Describe some ways we can reduce household water use. How can industrial uses of water be reduced?

2. How might desalination technology help "make" more water? Describe two methods of desalination. Why is this technology mostly being used in places like Saudi Arabia?

3. Describe three ways in which your own actions contribute to water pollution. Now describe three ways in which you could diminish these impacts.

4. Have the provisions of the Clean Water Act been effective? Discuss some of the methods we can adopt, in addition to "end-of-pipe" solutions, to prevent water pollution.

5. **THINK IT THROUGH** Your state's governor has put you in charge of water policy for the state. The aquifer beneath your state has been overpumped, and many wells have run dry. Agricultural production last year decreased for the first time in a generation, and farmers are clamoring for you to do something. Meanwhile, the state's largest city is

growing so fast that more water is needed for its burgeoning urban population. What policies would you consider to restore your state's water supply? Would you try to take steps to increase supply, decrease demand, or both? Explain why you would choose such policies.

6. **THINK IT THROUGH** Having solved the water depletion problem in your state, your next task is to deal with pollution of the groundwater that provides your state's drinking water supply. Recent studies have shown that one-third of the state's groundwater has levels of pollutants that violate EPA standards for human health, and citizens are fearful for their safety. What steps would you consider taking to safeguard the quality of your state's groundwater supply, and why?

CALCULATING ECOLOGICAL FOOTPRINTS

One of the single greatest personal uses of water is for showering. Old-style showerheads that were standard in homes and apartments built before 1992 dispense at least 5 gallons of water per minute, but low-flow showerheads produced after that year dispense just 2.5 gallons per minute. Given an average daily shower time of 8 minutes, calculate the amounts of water used and saved over the course of a year with old standard versus low-flow showerheads, and record your results in the table.

	Annual water use with standard showerheads (gallons)	Annual water use with low-flow showerheads (gallons)	Annual water savings with low-flow showerheads (gallons)
You			
Your class			
Your state			
United States			

1. Beginning in 2010, the EPA is promoting, under its WaterSense program, showerheads that produce still-lower flows of 2 gallons per minute (gpm). Some cities are already requiring these, and some models today go even lower. How much water would you save per year by using a 2-gpm showerhead instead of a 2.5-gpm showerhead?

2. How much water would you be able to save annually by shortening your average shower time from 8 minutes to 6 minutes? Assume you use a 2.5-gpm showerhead.

3. Compare your answers to questions 1 and 2. Do you save more water by showering 8 minutes with a 2-gpm showerhead or 6 minutes with a 2.5-gpm showerhead?

4. Can you think of any factors that are not being considered in this scenario of water savings? Explain.

Mastering**ENVIRONMENTALSCIENCE**™

Go to **www.masteringenvironmentalscience.com** for practice quizzes, Pearson eText, videos, current events, and more.

Schooling marine fish

16 MARINE AND COASTAL SYSTEMS AND RESOURCES

UPON COMPLETING THIS CHAPTER, YOU WILL BE ABLE TO:

- Identify physical, geographical, chemical, and biological aspects of the marine environment
- Explain how the oceans influence, and are influenced by, climate
- Describe major types of marine ecosystems

- Assess impacts from marine pollution
- Review the state of ocean fisheries and reasons for their decline
- Evaluate marine protected areas and reserves as innovative solutions

Collapse of the Cod Fisheries

"All of a sudden they just crashed."

—Donald Paul, Newfoundland fisherman

"Either we have sustainable fisheries, or we have no fishery."

—Canadian Fisheries Minister David Anderson

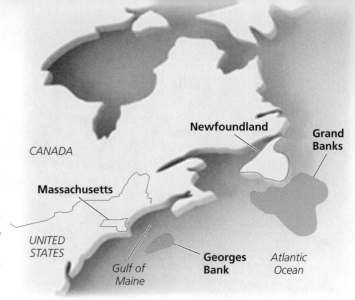

No fish has had more impact on human civilization than the Atlantic cod. Europeans exploring the coasts of North America 500 years ago discovered that they could catch these abundant fish merely by dipping baskets over the railings of their ships. The race that ensued to harvest this resource helped lead to the colonization of the New World. Starting in the early 1500s, schooners captured countless millions of cod, and the fish became a dietary staple in cultures on both sides of the Atlantic.

Massachusetts cod fishermen haul in a dwindling catch

Since then, cod fishing has been the economic engine for hundreds of communities in coastal New England and eastern Canada. Massachusetts honored the fish by naming Cape Cod after it and by erecting a carved wooden cod statue in its statehouse. In many Canadian coastal villages, cod fishing has been a way of life for generations. So it came as a shock when the cod all but disappeared, and governments had to step in and close the fisheries.

The Atlantic cod (*Gadus morhua*) is a type of *groundfish*, a name given to fish that live or feed on the bottom. People have long coveted groundfish such as halibut, pollack, haddock, and flounder. Adult cod eat smaller fish and invertebrates, commonly grow 60–70 cm long, and can live 20 years. A mature female cod can produce several million eggs. Atlantic cod inhabit cool ocean waters on both sides of the North Atlantic and occur in 24 discrete populations, called *stocks*. One stock inhabits the Grand Banks off Newfoundland, and another lives on Georges Bank off Massachusetts (**FIGURE 16.1**).

The Grand Banks provided ample fish for centuries. With advancing technology, however, ships became larger and more effective at finding fish. By the 1960s, massive industrial trawlers from Europe were vacuuming up unprecedented numbers of groundfish. In 1977, Canada exercised its legal right to the waters 200 nautical miles from shore, kicked out foreign fleets, and claimed most of the Grand Banks for itself. Canada developed the same industrial technologies and revved up its fishing industry like never before.

Then came the crash. Catches dwindled in the 1980s because too many fish had been taken and because bottom-trawling (fishing by dragging weighted nets across the seafloor, pp. 449–451) had destroyed so much underwater habitat. By 1992 the situation was dire: Scientists reported that mature cod were at just 10% of their long-term abundance. Canadian Fisheries Minister John Crosbie announced a 2-year ban on commercial cod fishing off Labrador and Newfoundland, where the $700 million fishery supplied income to 16% of the province's workforce. To compensate fishers, the government offered 10 weekly payments of $225, along with training for new job skills and incentives for early retirement. Over the next two years, 40,000 fishers and processing-plant workers lost their jobs, and some coastal communities faced economic ruin.

Cod stocks did not rebound by 1994, so the government extended the moratorium, enacted bans on all other major cod fisheries, and scrambled to offer more compensation, eventually spending over

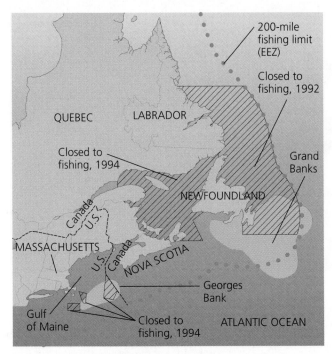

FIGURE 16.1 ▲ Populations of Atlantic cod inhabit areas of the northwestern Atlantic Ocean, including the Grand Banks and Georges Bank, regions of shallow water that are especially productive for groundfish. Portions of these and other areas have been closed to fishing in recent years because cod populations have collapsed after being overfished.

$4 billion. In 1997–1998, Canada partially reopened some fisheries, but data soon confirmed that the stocks were not recovering. In 2003, the cod fisheries were closed indefinitely, to recreational fishing as well as commercial fishing. It became illegal for Canadians in these areas to catch even one cod for their family's dinner. Fishers challenging the ban were arrested, fined, and jailed.

Then in 2009, one area of the Grand Banks was reopened to cod fishing after data showed the stock was recovering slightly. Some dreamed of a comeback for the fishery, but others thought the decision ill-advised. Quotas were set beyond what a scientific review board had recommended, and many feared another hasty reopening would simply decimate the stock yet again.

Across the border in U.S. waters, cod stocks had collapsed in the Gulf of Maine and on Georges Bank. In 1994, the National Marine Fisheries Service (NMFS) closed three prime fishing areas on Georges Bank. Over the next several years, NMFS designed a number of regulations, but these steps were too little, too late. A 2005 report revealed that the cod were not recovering, and further restrictions were enacted. As of 2008, managers announced that the Gulf of Maine stock was 58% of what it needed to be to be sustainable, and the Georges Bank stock was only 12% as large as it needed to be. Today scientists are struggling to explain why cod are not recovering. Research so far suggests that once mature cod were eliminated, the species they preyed upon proliferated, and now those fish compete with and prey on young cod.

There is good news, however: The Georges Bank closures have helped some other species to rebound. Seafloor invertebrates have begun to recover in the absence of trawling. Spawning stock of haddock and yellowtail flounder has risen. Sea scallops have increased in biomass 14-fold. Recoveries like these in no-fishing areas are showing scientists, fishers, and policymakers that protecting areas of ocean can help save dwindling marine populations and restore fisheries.

THE OCEANS

It's been said that our planet Earth should more properly be named "Ocean." After all, ocean water covers most of our planet's surface. The oceans touch and are touched by virtually every environmental system and every human endeavor. They shape our planet's climate, teem with biodiversity, provide us resources, and facilitate our transportation and commerce. Even if you live in a landlocked region far from the coast, the oceans affect you. They provide fish for people to eat in Iowa, they supply oil for cars in Missouri, and they influence the weather in Nebraska.

Oceans cover most of Earth's surface

The world's five oceans—Pacific, Atlantic, Indian, Arctic, and Southern—are all connected, comprising a single vast body of water (**FIGURE 16.2**). This one "world ocean" covers 71% of Earth's surface and contains 97.5% of its water. The oceans take up most of the hydrosphere, influence the atmosphere and lithosphere, and encompass much of the biosphere. Let's first briefly survey the physical and chemical makeup of the oceans—for although they may look homogenous from a beach, boat, or airplane, marine systems are complex and dynamic.

Seafloor topography can be rugged

Most maps depict oceans as smooth swaths of blue, but when we examine what's beneath the waves we see that the geology of the ocean floor can be intricate. Underwater volcanoes shoot forth enough magma to build islands above sea level, such as the Hawaiian Islands (p. 42). Steep canyons as large as Arizona's Grand Canyon lie just offshore of some continents. The lowest spot in the oceans—the Mariana Trench in the South Pacific—is deeper than Mount Everest is high, by over 2.1 km (1.3 mi). Our planet's longest mountain range is under water: the Mid-Atlantic Ridge (p. 36) runs the length of the Atlantic Ocean.

To comprehend underwater geographic features, we can examine a stylized map (**FIGURE 16.3**) that reflects *bathymetry* (the measurement of ocean depths) and *topography* (physical

FIGURE 16.2 ▶ The world's oceans are connected in a single vast body of water but are given different names. The Pacific Ocean is the largest and, like the Atlantic and Indian Oceans, includes both tropical and temperate waters. The Arctic and Southern Oceans include the waters in the polar regions. Many smaller bodies of water are named as seas or gulfs; a selected few are shown here.

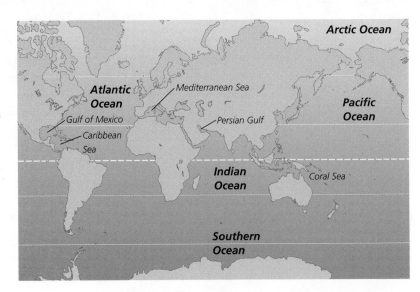

geography, or the shape and arrangement of landforms). In bathymetric profile, gently sloping **continental shelves** underlie the shallow waters bordering the continents. Continental shelves vary tremendously in width but average 80 km (50 mi) wide, with an average slope of just 1.9 m/km (10 ft/mi). These shelves drop off at the *shelf-slope break*, where the *continental slope* angles more steeply downward to the deep ocean basin below.

Some island chains, such as the Florida Keys, are formed by reefs (p. 439) and lie atop the continental shelf. Others, such as the Aleutian Islands, which curve across the North Pacific from Alaska toward Russia, are volcanic in origin. The Aleutians are also the site of a deep trench that, like the Mariana Trench, formed at a convergent tectonic plate boundary, where one slab of crust dives beneath another in the process of subduction (p. 36).

Wherever reefs, volcanism, or other processes create physical structure underwater, life thrives. Marine animals make use of physical structure as habitat, and topographically complex areas often make for productive fishing grounds. Georges Bank and the Grand Banks are examples; they are essentially huge underwater mounds formed during the ice ages when glaciers dumped debris at their southernmost extent. As climate warmed and the glaciers retreated, sea level rose, and these hilly areas were submerged in the ocean's salty water.

Ocean water contains salts

Ocean water contains approximately 96.5% H_2O by mass. Most of the remainder consists of ions from dissolved salts (**FIGURE 16.4**). Ocean water is salty primarily because ocean basins are the final repositories for water that runs off the

FIGURE 16.3 ▶ A stylized bathymetric profile shows key geologic features of the submarine environment. Shallow water exists around the edges of continents over the continental shelf, which drops off at the shelf-slope break. The steep continental slope gives way to the more gradual continental rise, all of which are underlain by sediments from the continents. Vast areas of seafloor are flat abyssal plain. Seafloor spreading occurs at oceanic ridges, and oceanic crust is subducted in trenches (p. 36). Volcanic activity along trenches may give rise to island chains such as the Aleutian Islands. Features on the left side of this diagram are more characteristic of the Atlantic Ocean, and features on the right side of the diagram are more characteristic of the Pacific Ocean. Adapted from Thurman, H.V., 1990. *Essentials of oceanography*, 4th ed. New York: Macmillan.

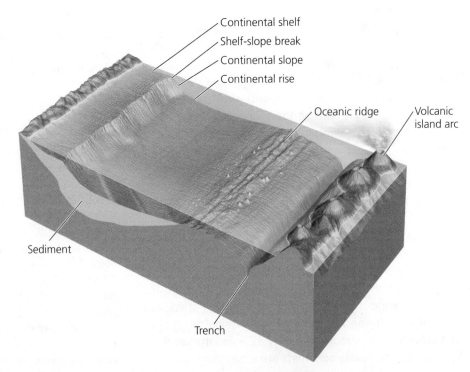

enters the oceans. We shall soon see (pp. 440–441) how this affects ocean pH, turning the water more acidic and posing problems for marine life.

Ocean water is vertically structured

Sunlight warms the ocean's surface but does not penetrate deeply, so ocean water is warmest at the surface and becomes colder with depth. Surface waters in tropical regions receive more solar radiation and therefore are warmer than surface waters in temperate or polar regions (**FIGURE 16.5**). Warmer water is less dense than cooler water, but water also becomes denser as it gets saltier because as salt dissolves in water, it increases mass more than it expands volume. These relationships give rise to different layers of water: heavier (colder and saltier) water sinks, whereas lighter (warmer and less salty) water remains nearer the surface. Waters of the surface zone are heated by sunlight and stirred by wind such that they are of similar density down to a depth of about 150 m (490 ft). Below this zone lies the *pycnocline*, a region in which density increases rapidly with depth. The pycnocline contains about 18% of ocean water by volume, compared to the surface zone's 2%. The remaining 80% lies in the deep zone beneath the pycnocline. The dense water in this zone is sluggish and unaffected by winds, storms, sunlight, and temperature fluctuations.

Despite the daily heating and cooling of surface waters, ocean temperatures are much more stable than temperatures on land. Midlatitude oceans experience yearly temperature variation of only around 10°C (18°F), and tropical and polar oceans are still more stable. The reason for this stability is that water has a high *heat capacity*, a measure of the heat required to increase temperature by a given amount. It takes more

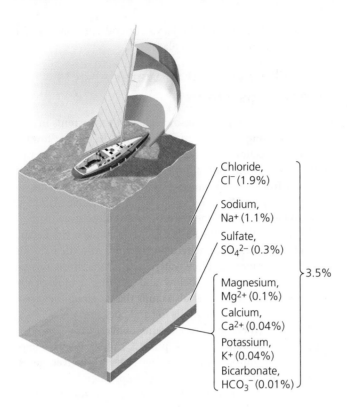

FIGURE 16.4 ▲ Ocean water consists of 3.5% salt, by mass, as shown by the proportionally thin colored slices of the cube in this diagram. Most of this salt is NaCl in solution, so sodium and chloride ions are abundant. A number of other ions and trace elements are also present.

Chloride, Cl^- (1.9%)

Sodium, Na^+ (1.1%)

Sulfate, SO_4^{2-} (0.3%)

} 3.5%

Magnesium, Mg^{2+} (0.1%)

Calcium, Ca^{2+} (0.04%)

Potassium, K^+ (0.04%)

Bicarbonate, HCO_3^- (0.01%)

land. Rivers carry sediment and dissolved salts from the continents into the oceans, as do winds. Evaporation from the ocean surface then removes pure water, leaving a higher concentration of salts. If we were able to evaporate all the water from the oceans, their basins would be left covered with a layer of dried salt 63 m (207 ft) thick.

The salinity of ocean water generally ranges from 33 to 37 parts per thousand, varying from place to place because of differences in evaporation, precipitation, and freshwater runoff from land and glaciers. Coastal waters are often less saline because of the influx of freshwater runoff. Salinity near the equator is low because this region has a great deal of precipitation, which is relatively salt-free. In contrast, surface salinity is high at latitudes roughly 30–35 degrees north and south, where evaporation exceeds precipitation.

Besides the dissolved salts shown in Figure 16.4, nutrients such as nitrogen and phosphorus occur in seawater in trace amounts (well under 1 part per million) and play essential roles in nutrient cycling (pp. 122–131) in marine ecosystems. Another aspect of ocean chemistry is dissolved gas content. Roughly 36% of the gas dissolved in seawater is oxygen, which is produced by photosynthetic plants, bacteria, and phytoplankton (p. 78) and enters by diffusion from the atmosphere. Oxygen concentrations are highest in the upper layer of the ocean, reaching 13 ml/L of water. Marine animals depend on dissolved oxygen, and as we saw in Chapter 5, if oxygen is depleted, a hypoxic "dead zone" may ensue, killing animals or forcing them to leave. Another gas that is soluble in ocean water is carbon dioxide (CO_2). As we pump excess CO_2 into the atmosphere by burning fossil fuels, more CO_2

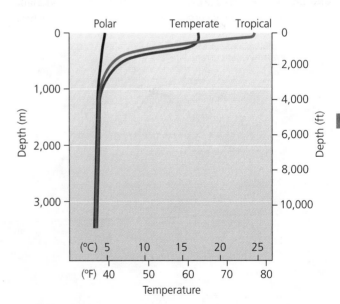

FIGURE 16.5 ▲ Ocean water varies in temperature with depth. Water temperatures near the surface are warmer because of daily heating by the sun, and within the top 1,000 m (3,300 ft) they become rapidly colder with depth. This temperature differential is greatest in the tropics because of intense solar heating and is least in the polar regions. Deep water at all latitudes is equivalent in temperature. *Source:* From Garrison, T., 2005. *Oceanography*, 5th ed., Fig 6.15. Belmont, CA: Brooks/Cole © 2005. By permission of Cengage Learning.

energy to increase the temperature of water than it does to increase the temperature of air. High heat capacity enables ocean water to absorb a tremendous amount of heat from the air. In fact, just the top 2.6 m (8.5 ft) of the oceans holds as much heat as the entire atmosphere! By absorbing heat and releasing it to the atmosphere, the oceans help regulate Earth's climate (Chapter 18). They also influence climate by moving heat from place to place via the ocean's surface circulation, a system of currents that move in the pycnocline and the surface zone.

Surface water flows horizontally in currents

Earth's ocean is composed of vast, riverlike flows driven by density differences, heating and cooling, gravity, and wind. Surface **currents** flow within the upper 400 m (1,300 ft) of water, horizontally and for great distances, in long-lasting patterns across the globe (**FIGURE 16.6**). Warm-water currents carry water heated by the sun from equatorial regions, while cold-water currents carry water cooled in high-latitude regions or from deep below. Some surface currents are very slow. Others, like the Gulf Stream, are rapid and powerful. From the Gulf of Mexico, the Gulf Stream flows up the U.S. Atlantic coast and past the eastern edges of Georges Bank and the Grand Banks at nearly 2 m/sec, or over 4 mph. Averaging 70 km (43 mi) across, the Gulf Stream continues across the North Atlantic, bringing warm water to Europe and moderating that continent's climate (p. 499), which otherwise would be much colder.

Besides influencing climate, ocean currents have aided navigation and shaped human history. Currents helped carry Polynesians to Easter Island, Darwin to the Galapagos, and Europeans to the New World. Currents transport heat, nutrients, pollution, and the larvae of cod and many other marine species from place to place.

Vertical movement of water affects marine ecosystems

Surface winds and heating also create vertical currents in seawater. **Upwelling**, the upward flow of cold, deep water toward the surface, occurs where horizontal surface currents diverge, or flow away from one another. Upwelled water is rich in nutrients from the bottom, so upwellings are often sites of high primary productivity (p. 116) and lucrative fisheries. Upwellings also occur where strong winds blow away from or parallel to coastlines (**FIGURE 16.7**). An example is the Pacific coast of North America, where north winds and the Coriolis effect (p. 467) move surface waters away from shore, raising nutrient-rich water from below and creating a biologically rich region. The cold water also chills the air along the coast, giving San Francisco its famous fog and cool summers.

In areas where surface currents converge, or come together, surface water sinks, a process called **downwelling**. Downwelling transports warm water rich in dissolved gases, providing an influx of oxygen for deep-water life. Vertical currents also occur in the deep zone, where differences in density can lead to rising and falling convection currents, similar to those in molten rock (p. 35) and in air (p. 467).

Currents affect climate

The horizontal and vertical movements of ocean water can have far-reaching effects on climate globally and regionally. The **thermohaline circulation** is a worldwide current system in which warmer, fresher water moves along the surface and

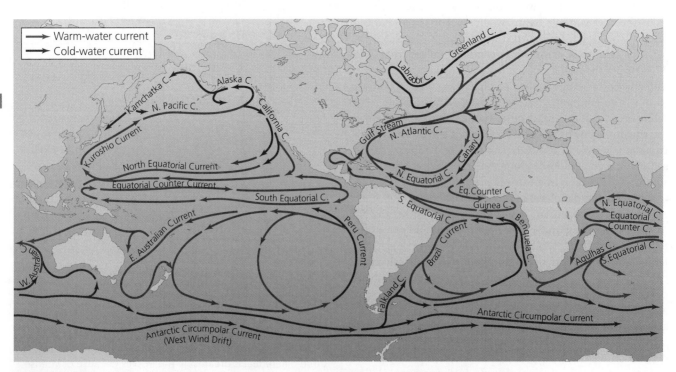

FIGURE 16.6 ▲ The upper waters of the oceans flow in surface currents, long-lasting and predictable global patterns of water movement. Warm- and cold-water currents interact with the planet's climate system, and people have used them for centuries to navigate the oceans. *Source: From Garrison, T., 2005. Oceanography, 5th ed., Fig 9.8(b). Belmont, CA: Brooks/Cole © 2005. By permission of Cengage Learning.*

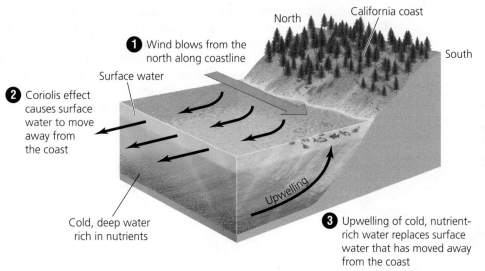

① Wind blows from the north along coastline

North

California coast

South

Surface water

② Coriolis effect causes surface water to move away from the coast

Upwelling

Cold, deep water rich in nutrients

③ Upwelling of cold, nutrient-rich water replaces surface water that has moved away from the coast

FIGURE 16.7 ◀ Upwelling is the movement of bottom waters upward. This often brings nutrients up to the surface, creating rich areas for marine life. For example, north winds blow along the California coastline ①, while the Coriolis effect (p. 467) draws wind and water away from the coast ②. Water is then pulled up from the bottom ③ to replace the water that moves away from shore.

colder, saltier water (which is more dense) moves deep beneath the surface (**FIGURE 16.8**). One segment of this worldwide conveyor-belt system includes the warm surface water in the Gulf Stream that flows across the Atlantic Ocean to Europe. As this water releases heat to the air, keeping Europe warmer than it would otherwise be, the water cools, becomes saltier through evaporation, and thus becomes denser and sinks, creating a region of downwelling known as the *North Atlantic Deep Water (NADW)*.

Scientists hypothesize that interrupting the thermohaline circulation could trigger rapid climate change. If global warming (Chapter 18) causes much of Greenland's ice sheet to melt, the resulting freshwater runoff into the North Atlantic would make surface waters less dense (because fresh water is less dense than salt water). This could potentially stop the NADW formation and shut down the northward flow of warm water, causing Europe to cool rapidly. Some data suggest that the thermohaline circulation in this region

is already slowing, but other researchers maintain that Greenland will not produce enough runoff to cause a shutdown this century.

Another interaction between ocean currents and the atmosphere that influences climate is the **El Niño–Southern Oscillation (ENSO)**, a systematic shift in atmospheric pressure, sea surface temperature, and ocean circulation in the tropical Pacific Ocean. Under normal conditions, prevailing winds blow from east to west along the equator, from a region of high pressure in the eastern Pacific to one of low pressure in the western Pacific, forming a large-scale convective loop in the atmosphere (**FIGURE 16.9A**). The winds push surface waters westward, causing water to "pile up" in the western Pacific. As a result, water near Indonesia can be 50 cm (20 in.) higher and 8 °C warmer than water near South America. The westward-moving surface waters allow cold water to rise up from the deep in a nutrient-rich upwelling along the coast of Peru and Ecuador.

El Niño conditions are triggered when air pressure decreases in the eastern Pacific and increases in the western Pacific, weakening the equatorial winds and allowing the warm water to flow eastward (**FIGURE 16.9B**). This suppresses upwelling along the Pacific coast of the Americas, shutting down the delivery of nutrients that support marine life and fisheries. Coastal industries such as Peru's anchovy fisheries are devastated by El Niño events, and the 1982–1983 El Niño caused over $8 billion in economic losses worldwide. El Niño events alter weather patterns around the world, creating rainstorms and floods in areas that are generally dry (such as southern California) and causing drought and fire in regions that are typically moist (such as Indonesia).

La Niña events are the opposite of El Nino events; in a La Niña event, cold waters rise to the surface and extend westward in the equatorial Pacific when winds blowing to the west strengthen, and weather patterns are affected in opposite ways. ENSO cycles are periodic but irregular, occurring every 2–8 years. Scientists are exploring whether the globally warming air and sea temperatures we examine in Chapter 18 may be increasing the frequency and strength of these cycles.

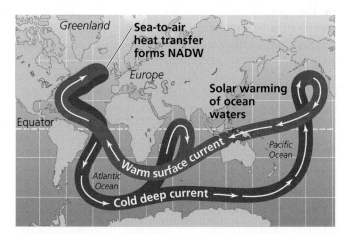

Greenland

Sea-to-air heat transfer forms NADW

Europe

Equator

Solar warming of ocean waters

Warm surface current

Atlantic Ocean

Pacific Ocean

Cold deep current

FIGURE 16.8 ▲ As part of the oceans' thermohaline circulation, warm surface currents carry heat from equatorial waters northward toward Europe, where they warm the atmosphere and then cool and sink, forming the North Atlantic Deep Water (NADW). Scientists debate whether rapid melting of Greenland's ice sheet could interrupt this heat flow and cause Europe to cool dramatically.

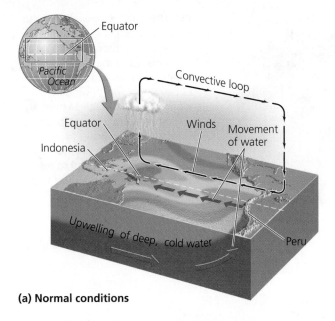

(a) Normal conditions

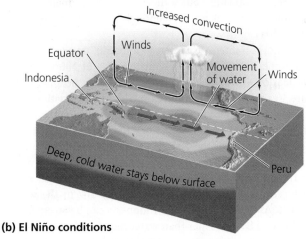

(b) El Niño conditions

FIGURE 16.9 ▲ In these diagrams, red and orange colors denote warmer water, and blue and green colors denote colder water. Under normal conditions **(a)**, prevailing winds push warm surface waters toward the western Pacific. Under El Niño conditions **(b)**, winds weaken and the warm water flows back across the Pacific toward South America, like water sloshing in a bathtub. This shuts down upwelling along the American coast and alters precipitation patterns regionally and globally. Adapted from National Oceanic and Atmospheric Administration, Tropical Atmospheric Ocean Project.

Climate change is altering the oceans

Scientists today are learning a great deal about how global climate change will affect ocean chemistry and biology. The oceans absorb carbon dioxide (CO_2) from the atmosphere, as we first saw in the carbon cycle (see Figure 5.16, p. 125). As our civilization pumps excess carbon dioxide into the atmosphere by burning fossil fuels for energy and removing vegetation from the land, the buildup of atmospheric CO_2 is causing the planet to grow warmer, setting in motion many consequences (Chapter 18).

The oceans have soaked up roughly a third of the excess CO_2 that we've added to the atmosphere so far, and this has slowed the onset of global climate change. However, there are

two concerns. One is that the ocean's surface water may soon become saturated with as much CO_2 as it can hold. Once it reaches this limit, then climate change will accelerate.

The second concern is that as ocean water soaks up CO_2, it becomes more acidic. As **ocean acidification** proceeds, many sea creatures have difficulty forming shells because the chemicals they need are less available. This process is harming corals in particular (see **THE SCIENCE BEHIND THE STORY**, pp. 440–441). Corals build reefs, which are hubs for marine biodiversity and provide billions of dollars' worth of ecosystem services. Because of the decline of coral reefs, scientists are warning that ocean acidification is shaping up to be one of most damaging consequences of global climate change.

MARINE AND COASTAL ECOSYSTEMS

With their variation in topography, temperature, salinity, nutrients, and sunlight, marine and coastal environments feature a variety of ecosystems. Regions of ocean water differ greatly, and some zones support more life than others. The uppermost 10 m (33 ft) of water absorbs 80% of solar energy, so nearly all of the oceans' primary productivity occurs in the top layer, or **photic zone**. Generally, the warm, shallow waters of continental shelves are most biologically productive and support the greatest species diversity. Habitats and ecosystems occurring between the ocean's surface and floor are termed **pelagic**, whereas those that occur on the ocean floor are called **benthic**. Most marine and coastal ecosystems are powered by solar energy, with sunlight driving photosynthesis by phytoplankton in the photic zone. Yet even the darkest ocean depths host life.

Open-ocean ecosystems vary in their biodiversity

Biological diversity in pelagic regions of the open ocean is highly variable in its distribution. Primary production (p. 116) and animal life near the surface are concentrated in regions of nutrient-rich upwelling. Microscopic phytoplankton constitute the base of the marine food chain in the pelagic zone. These photosynthetic algae, protists, and cyanobacteria feed zooplankton (p. 78), which in turn become food for fish, jellyfish, whales, and other free-swimming animals (**FIGURE 16.10**). Predators at higher trophic levels include larger fish, sea turtles, and sharks. Fish-eating birds such as puffins, petrels, and shearwaters feed at the surface of the open ocean, returning periodically to nesting sites on islands and coastlines.

In the little-known deep-water ecosystems, animals have adapted to tolerate extreme water pressures and to live in the dark without food from autotrophs. Many of these often bizarre-looking creatures scavenge carcasses or organic detritus that falls from above. Others are predators, and still others attain food from symbiotic mutualistic (pp. 82–83) bacteria. Some species carry bacteria that produce light chemically by bioluminescence (**FIGURE 16.11**).

As we saw in Chapter 2 (p. 33), some ecosystems form around hydrothermal vents, where heated water spurts from

FIGURE 16.10 ▲ The uppermost reaches of ocean water contain billions upon billions of phytoplankton—tiny photosynthetic algae, protists, and bacteria that form the base of the marine food chain—as well as zooplankton, small animals and protists that dine on phytoplankton and comprise the next trophic level.

the seafloor, carrying minerals that precipitate to form rocky structures. Tubeworms, shrimp, and other creatures in these recently discovered systems use symbiotic bacteria to derive their energy from chemicals in the heated water rather than from sunlight. They manage to thrive within amazingly narrow zones between scalding-hot and icy cold water.

Kelp forests harbor many organisms

Along many temperate coasts, large brown algae, or **kelp**, grow from the floor of continental shelves, reaching up toward the sunlit surface. Some kelp reaches 60 m (200 ft) in height and can grow 45 cm (18 in.) per day. Dense stands of kelp form underwater "forests" (**FIGURE 16.12**). Kelp forests supply shelter and

FIGURE 16.11 ▼ Life is scarce in the dark depths of the deep ocean, but the creatures that do live there can appear bizarre. The anglerfish lures prey toward its mouth with a bioluminescent (glowing) organ that protrudes from its head.

FIGURE 16.12 ▲ "Forests" of tall brown algae known as *kelp* grow from the floor of the continental shelf. Numerous fish and other creatures eat kelp or find refuge among its fronds.

food for invertebrates and fish, which in turn provide food for predators such as seals and sharks. Indeed, kelp forests were the setting for our discussion of keystone species in Chapter 4 (pp. 86–91). Recall that sea otters control sea urchin populations and that when otters disappear, urchins overgraze the kelp, destroying the forests. Kelp forests also absorb wave energy and protect shorelines from erosion. People in Asian cultures eat some types of kelp, and kelp provides compounds known as alginates, which serve as thickeners in consumer products including cosmetics, paints, paper, soaps, and ice cream.

Coral reefs are treasure troves of biodiversity

Shallow subtropical and tropical waters are home to coral reefs. A reef is an underwater outcrop of rock, sand, or other material. A **coral reef** is a mass of calcium carbonate composed of the skeletons of tiny marine animals known as *corals*. A coral reef may occur as an extension of a shoreline, along a *barrier island* paralleling a shoreline, or as an *atoll*, a ring around a submerged island.

Corals are tiny invertebrate animals related to sea anemones and jellyfish. They remain attached to rock or existing reef and capture passing food with stinging tentacles (see photo on p. 440). Corals also derive nourishment from symbiotic algae known as *zooxanthellae*, which inhabit their bodies and produce food through photosynthesis. Most corals are colonial, and the colorful surface of a coral reef consists of millions of densely packed individuals. (The colors come from zooxanthellae.) As corals die, their skeletons remain part of the reef while new corals grow atop them, increasing the reef's size.

Like kelp forests, coral reefs protect shorelines by absorbing wave energy. They also host tremendous biodiversity (**FIGURE 16.13A**, page 442). This is because coral reefs provide complex physical structure (and thus many habitats) in shallow nearshore waters, which are regions of high primary productivity. Besides the staggering diversity of anemones, sponges, hydroids, tubeworms, and other sessile (stationary) invertebrates, innumerable molluscs, flatworms, sea stars, and urchins patrol reefs, while thousands of fish species find food and shelter in reef nooks and crannies.

Coral reefs are experiencing alarming declines worldwide, however. Many have undergone "coral bleaching," a

The SCIENCE behind the Story

Will Climate Change Rob Us of Coral Reefs?

The oceans have helped delay the advance of global climate change by soaking up a portion of the excess carbon dioxide that we have pumped into the atmosphere by burning fossil fuels and clearing forests. But when we alter a flux in a biogeochemical cycle, there are usually consequences. What consequences might there be for this increase in the flux of carbon from the atmosphere to the oceans? Hundreds of scientists in recent years have been working to find out.

Let's start with the chemistry (**first figure**). Carbon dioxide (CO_2) from the air ❶ can react on the ocean's surface with water (H_2O) to form carbonic acid

Tiny corals, the builders of reefs

(H_2CO_3) ❷. This is what happens in soft drinks when carbon dioxide is added to give them fizz:

$$CO_2 + H_2O \rightarrow H_2CO_3$$

This carbonic acid molecule soon dissociates ❸ into a bicarbonate ion (HCO_3^-) and a proton, or hydrogen ion:

$$H_2CO_3 \rightarrow HCO_3^- + H^+$$

The free proton then combines with a carbonate ion (CO_3^{2-}) ❹ to form a second bicarbonate ion:

$$H^+ + CO_3^{2-} \rightarrow HCO_3^-$$

This is bad news for marine organisms such as corals, which depend on carbonate ions in order to form the calcium carbonate ($CaCO_3$) that comprises their hard shells. As the availability of carbonate ions in the water declines, it gets more difficult to build

❶ Human activity increases carbon dioxide in air — CO_2

❷ Carbon dioxide reacts with ocean water, forming carbonic acid — $CO_2 + H_2O \rightarrow H_2CO_3$

❸ Carbonic acid dissociates into a bicarbonate ion and a proton — $H_2CO_3 \rightarrow HCO_3^- + H^+$

❹ Proton combines with carbonate ion to form another bicarbonate ion — $H^+ + CO_3^{2-} \rightarrow HCO_3^-$

❺ As carbonate ions are pulled into solution, acidity dissolves calcium carbonate in coral reefs — $CaCO_3 \rightarrow Ca^{2+} + CO_3^{2-}$ (coral)

Excess carbon dioxide from the atmosphere reacts with seawater to create carbonic acid, eventually diminishing the number of carbonate ions available for corals to use to build their shells. Adapted from Hoegh-Guldberg, O., et al., 2007. Coral reefs under rapid climate change and ocean acidification. *Science* 318: 1737–1742. Fig 1a. Reprinted with permission from AAAS.

calcium carbonate shells by the process known as calcification. In fact, as carbonic acid and bicarbonate ions become more common in the water, the calcium carbonate shells of marine creatures like corals begin dissolving **❺**:

$$CaCO_3 \rightarrow Ca^{2+} + CO_3^{2-}$$

Researchers have been testing how increasing acidity affects corals in the lab and in the field. They've also been studying the current global distribution of coral reefs, the massive calcium carbonate structures built by millions of tiny corals that create habitat for so many other marine animals. Through these research avenues they aim to make predictions about how coral reefs will fare in the future as climate change proceeds. In 2007, a team led by Ove Hoegh-Guldberg of the University of Queensland in Australia reviewed existing knowledge of these questions and published its conclusions in the journal *Science*.

Chemistry tests in the lab show that coral shells begin to erode faster than they are built once the carbonate ion concentration falls below 200 micromoles/kg of seawater. Researchers studying coral reefs in the field are finding the same thing: Reefs are growing only in waters with greater than 200 micromoles/kg of carbonate ion availability. A few calculations reveal that this concentration is what we would expect to result from an atmospheric CO_2 concentration of 480 parts per million (ppm). Earth's natural ("preindustrial") level was 280 ppm, and today we have raised it above 380 ppm. Thus we are a little more than halfway to the point at which coral reefs will begin to dissolve in most of the areas they now exist. At the rate CO_2 is accumulating in the atmosphere, we could easily reach 480 ppm around the middle of this century.

Hoegh-Guldberg's team sought some long-term context, so they used data from the Vostok ice core (p. 500), drilled from Antarctic ice, to look for patterns from the past 420,000 years. They found that in all that time, carbonate ions concentrations in the oceans had

never been as low as they are today. Indeed, current sea temperatures are 0.7°C warmer and pH is 0.1 unit lower than in the 420,000 years previous. These researchers concluded that our oceans are well on the way to losing coral reefs.

Paleontological studies of the fossil record lend support to this idea, showing that coral reefs and other marine organisms using calcification apparently disappeared in the early Triassic period during an episode of extremely high atmospheric CO_2 levels.

What should we expect in the future? A number of scientists have looked to the *Fourth Assessment Report* of the Intergovernmental Panel on Climate Change (IPCC; p. 504), the 2007 consensus document summarizing scientific knowledge on climate change to that point. Using the IPCC's known and predicted data on atmosphere and ocean temperatures and pH, these researchers modeled how oceanic carbonate ion concentrations should change at various values for atmospheric CO_2.

The results predicted that as atmospheric CO_2 levels rise, coral reefs will shrink in distribution, diversity, and density (**second figure**). By the time atmospheric CO_2 levels pass 500 ppm, little area of ocean will be left with conditions to support coral reefs. Most of the world's coral reefs, home to so much vibrant biodiversity, "will become rapidly eroding rubble banks," Hoegh-Guldberg's team wrote.

Without living and growing coral reefs, the rich communities of reef-dependent animals will likely collapse. For human society, fisheries will decline, tourism will wither, and coasts will lose protection against storm surges. Developing nations and island nations in the tropics—those already expected to suffer the most from climate change—will be hit hardest.

Scientists are hoping there may be silver linings. Organisms vary in how they react, and not all will suffer from the ocean's chemistry changes. Corals that are more tolerant to temperature changes (such as a variety

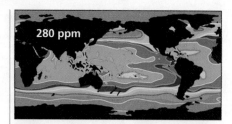

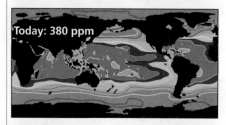

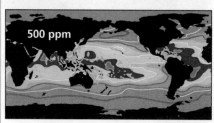

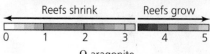

Ω aragonite

Most of Earth's tropical and subtropical oceans were suitable for the growth of coral reefs (blue colors; top panel) before people began emitting carbon dioxide to the atmosphere. Today, at 380 ppm atmospheric CO_2 (middle panel), fewer regions are suitable. When the planet reaches 500 ppm (bottom panel), very few areas of the ocean will have conditions suitable for coral reefs. Adapted from Hoegh-Guldberg, O., et al., 2007. Coral reefs under rapid climate change and ocean acidification. *Science* 318: 1737–1742. Fig 4. Reprinted with permission from AAAS.

called *Porites*) may succeed where others fail. Perhaps certain rapidly colonizing species (such as those in the genus *Acropora*) will be able to colonize areas vacated by dying corals.

What can we do about the threats from ocean acidification and loss of coral reefs? Ultimately the only effective solution is to reduce our carbon emissions, and soon. In the meantime, scientists say, we can deal with more localized threats to reefs, such as nutrient pollution and overfishing, so as to relieve them of many of the stresses they already face. ∎

(a) Coral reef community

(b) Bleached coral

FIGURE 16.13 ▲ Corals reefs provide food and shelter for a tremendous diversity **(a)** of fish and other creatures. Today these reefs face multiple stresses from human impacts. Many corals have died as a result of coral bleaching **(b)**, in which corals lose their zooxanthellae. Bleaching is evident in the whitened portion of this coral.

process that occurs when zooxanthellae die or leave the coral, depriving it of nutrition. Corals lacking zooxanthellae lose color and frequently die, leaving behind ghostly white patches in the reef (**FIGURE 16.13B**). Coral bleaching is thought to result from increased sea surface temperatures associated with global climate change, from the influx of pollutants, from unknown natural causes, or from combinations of these factors.

Nutrient pollution in coastal waters also promotes the growth of algae, which are smothering reefs in the Florida Keys and many other regions. Moreover, coral reefs sustain damage when divers stun fish with cyanide to capture them for food or for the pet trade, a common practice in waters of Indonesia and the Philippines. Finally, as global climate change proceeds, the oceans are becoming more acidic as excess carbon dioxide from the atmosphere reacts with seawater

to form carbonic acid. Acidification threatens to deprive corals of the carbonate ions they need to produce their structural parts (see **The Science behind the Story**, pp. 440–441).

A few coral species thrive in waters outside the tropics and build reefs on the ocean floor at depths of 200–500 m (650–1,650 ft). These little-known reefs, which occur in coldwater areas off the coasts of Norway, Spain, the British Isles, and elsewhere, are only now beginning to be studied by scientists. Already, however, many have been badly damaged by bottom-trawling (pp. 449–451)—the same practice that has so degraded the benthic habitats of groundfish such as the Atlantic cod. Norway and other countries are now beginning to protect some of these deep-water reefs.

Intertidal zones undergo constant change

Where the ocean meets the land, **intertidal**, or **littoral**, ecosystems (**FIGURE 16.14**) spread between the uppermost reach of the high tide and the lowest limit of the low tide. **Tides** are the periodic rising and falling of the ocean's height at a given location, caused by the gravitational pull of the moon and sun (see Figure 21.19, p. 608). High and low tides occur roughly 6 hours apart, so intertidal organisms spend part of each day submerged in water, part of the day exposed to air and sun, and part of the day being lashed by waves. These creatures must also protect themselves from marine predators at high tide and terrestrial predators at low tide.

The intertidal environment is a tough place to make a living, but it is home to a remarkable diversity of organisms. Life abounds in the crevices of rocky shorelines, which provide shelter and pools of water (tide pools) during low tides. Sessile animals such as anemones, mussels, and barnacles live attached to rocks, filter-feeding on plankton in the water that washes over them. Urchins, sea slugs, chitons, and limpets eat intertidal algae or scrape food from the rocks. Sea stars (starfish) creep slowly along, preying on the filter-feeders and herbivores. Crabs clamber around the rocks, scavenging detritus.

The rocky intertidal zone is so diverse because environmental conditions such as temperature, salinity, and moisture change dramatically from the high to the low reaches. This environmental variation gives rise to horizontal bands dominated by different sets of organisms arrayed according to their habitat needs, competitive abilities, and adaptation to exposure. Sandy intertidal areas, such as those of Cape Cod, host less biodiversity, yet plenty of organisms burrow into the sand at low tide to await the return of high tide, when they emerge to feed.

Salt marshes line temperate shorelines

Along many of the world's coasts at temperate latitudes, **salt marshes** occur where the tides wash over gently sloping sandy or silty substrates. Rising and falling tides flow into and out of channels called *tidal creeks* and at highest tide spill over onto elevated marsh flats (**FIGURE 16.15**). Marsh flats grow thick with salt-tolerant grasses, as well as rushes, shrubs, and other herbaceous plants.

Salt marshes boast very high primary productivity and provide critical habitat for shorebirds, waterfowl, and many

(a) Tidal zones

(b) Tide pools at low tide

FIGURE 16.14 ▲ The rocky intertidal zone stretches along rocky shorelines between the lowest and highest reaches of the tides **(a)**, providing niches for a diversity of organisms including sea stars (starfish), barnacles, crabs, sea anemones, corals, bryozoans, snails, limpets, chitons, mussels, nudibranchs (sea slugs), and sea urchins. Fish swim in tidal pools **(b)**, and many types of algae cover the rocks. Areas higher on the shoreline are exposed to the air more frequently and for longer periods, so organisms that tolerate exposure best specialize in the upper intertidal zone. The lower intertidal zone is exposed less frequently and for shorter periods, so organisms less tolerant of exposure thrive in this zone.

commercially important fish and shellfish species. Salt marshes also filter pollution and stabilize shorelines against storm surges. However, because people desire to live and do business along coasts, we have altered or destroyed vast expanses of salt marshes to make way for coastal development. When salt marshes are destroyed, we lose the ecosystem services they provide. When Hurricane Katrina struck the Gulf Coast, for instance, the flooding was made worse because vast areas of salt marshes had vanished during the preceding decades because of development, subsidence from oil

and gas drilling, and dams that held back marsh-building sediment.

Mangrove forests line coasts in the tropics and subtropics

In tropical and subtropical latitudes, mangrove forests replace salt marshes along gently sloping sandy and silty coasts. **Mangroves** are some of the few types of trees that are salt-tolerant, and they have unique types of roots that curve

FIGURE 16.15 ▲ Salt marshes occur in temperate intertidal zones where the substrate is muddy. Tidal waters flow in channels called *tidal creeks* amid flat areas called *benches*, sometimes partially submerging the salt-adapted grasses.

upward like snorkels to attain oxygen lacking in the mud or that curve downward like stilts to support the tree in changing water levels (**FIGURE 16.16**). Fish, shellfish, crabs, snakes, and other organisms thrive among the root networks, and birds feed and nest in the dense foliage of these coastal forests. Besides serving as nurseries for fish and shellfish that people harvest, mangroves also provide materials that people use for food, medicine, tools, and construction.

From Florida to Mexico to the Philippines, half the world's mangrove forests have been destroyed as people have developed coastal areas. Shrimp farming in particular has driven the conversion of large areas of mangroves. When mangroves are removed, coastal areas lose the ability to slow runoff, filter pollutants, and retain soil. As a result, offshore systems such as coral reefs and eelgrass beds are more readily degraded. Moreover, mangrove forests protect coastal communities against storm surges and tsunamis. The 2004 Indian Ocean tsunami (p. 46) devastated areas where mangroves had been removed but caused less damage where mangroves were intact.

Fresh water meets salt water in estuaries

Many salt marshes and mangrove forests occur in or near **estuaries**, water bodies where rivers flow into the ocean, mixing fresh water with salt water. Estuaries are biologically productive ecosystems that experience fluctuations in salinity as tides and freshwater runoff vary daily and seasonally. Sheltered from crashing surf, the shallow water of estuaries nurtures eelgrass beds and other plant life, producing abundant food and resources. For shorebirds and for many commercially important shellfish species, estuaries provide critical habitat. For fishes such as salmon, which spawn in streams and mature in the ocean, estuaries provide a transitional zone where young fish make the passage from fresh water to salt water.

Estuaries everywhere have been affected by coastal development, water pollution, habitat alteration, and overfishing. Think of any major estuary in the United States, from New York Harbor to Chesapeake Bay to Florida Bay to the mouth of the Mississippi to San Francisco Bay. They all have suffered pollution and other impacts from human development. Coastal ecosystems have borne the brunt of human impact because two-thirds of Earth's people choose to live within 160 km (100 mi) of the ocean.

WEIGHING THE ISSUES

Coastal Development A developer wants to build a large marina on an estuary in your coastal town. The marina would boost the town's economy but eliminate its salt marshes. What consequences would you expect for property values? For water quality? For wildlife? As a homeowner living adjacent to the marshes, how would you respond? Do you think that developers or town officials should offer homeowners insurance against damage from storm surges when a protective salt marsh is destroyed?

FIGURE 16.16 ▶ Mangrove forests line tropical and subtropical coastlines. Mangrove trees, with their unique roots, are adapted for growing in salt water and provide habitat for many fish, birds, crabs, and other animals.

MARINE POLLUTION

People have long made the oceans a sink for waste and pollutants. Even into the mid-20th century, it was common for coastal U.S. cities to dump trash and untreated sewage along their shores. The Clean Water Act (p. 177) reduced the discharge of point-source pollution (pp. 420–421) into the oceans. But a great deal of pollution is non-point-source pollution that comes from countless small things we all do and from items we throw away far inland. For the oceans are downstream from everywhere: As long as rivers flow to the sea, much of our pollution on land sooner or later ends up in the oceans.

Oil, plastic, toxic chemicals, and excess nutrients all eventually make their way into the oceans. Sewage and trash from cruise ships and abandoned fishing gear from fishing boats add to the input. The scope of trash in the sea can be gauged by the amount picked up each September by volunteers who trek beaches in the Ocean Conservancy's annual International Coastal Cleanup. In this nonprofit organization's 2009 cleanup, 499,000 people from 108 nations picked up 3.4 million kg (7.4 million lb) of trash from 24,000 km (15,000 mi) of shoreline.

Nets and plastic debris endanger marine life

Discarded fishing nets, plastic bags and bottles, gloves, fishing line, buckets, floats, abandoned cargo, and much other trash can harm marine organisms. Fishing nets that are lost or intentionally discarded can snare and drown marine animals (**FIGURE 16.17**). Marine mammals, seabirds, fish, and sea turtles may mistake floating plastic debris for food and can die as a result of ingesting material they cannot digest or expel. One study of northern fulmars, a type of seabird, found dead on Dutch beaches found that 98% had plastic in their stomachs. Recently a sperm whale was found dead with 200 kg (440 lb) of fishing gear in its stomach. Individuals of all five species of sea turtle in the Gulf of Mexico are known to have died from consuming or contacting marine debris.

Because plastic is designed not to degrade, it can drift for decades before washing up on beaches. Plastic does degrade

FIGURE 16.17 ▼ This grey seal became entangled in a discarded fishing net. Each year many thousands of marine mammals, birds, and sea turtles are killed by plastic debris, abandoned nets, and other trash that people dump in the ocean.

slowly in seawater and sunlight, but in doing so, breaks down into smaller and smaller bits that become more and more numerous. This can actually make things worse, as the oceans are now filled with uncountable trillions of tiny pellets of plastic floating just under the surface like confetti. Small fish ingest these and pass them up the food chain to larger animals.

In recent years scientists have learned that plastic trash is accumulating in certain regions of the oceans where it is brought and trapped by circulating currents. One such area is within the North Pacific Gyre, the region ringed by the clockwise flow of major currents in the northern Pacific Ocean, stretching from California to Hawaii to Japan. An immense area within this region is often referred to as "the Great Pacific Garbage Patch" because of the density of plastic trash in the water. Most items are tiny, but one study documented 3.3 plastic bits per square meter, and the patch is often estimated as being twice the size of Texas. California ship captain Charles Moore discovered this area early on, and founded the Algalita Marine Research Foundation to conduct research and raise awareness. Aspects of plastic pollution in the Pacific are illustrated in **ENVISIONIT** (p. 446).

In 2006, the U.S. Congress responded to ocean pollution by passing the Marine Debris Research, Prevention, and Reduction Act. However, more is needed. We can all help by reducing our use of unnecessary plastic, reusing the plastic items we do use, and recycling the plastic we discard. Industry can reduce packaging and use more biodegradable materials. And volunteers and businesses can support efforts such as the International Coastal Cleanup.

Oil pollution comes from spills of all sizes

For many people, ocean pollution first brings to mind oil pollution. About 30% of our crude oil and nearly half of our natural gas come from seafloor deposits. Most offshore oil and gas is concentrated in petroleum-rich regions such as the North Sea and the Gulf of Mexico, but energy companies extract smaller amounts from diverse locations, among them the Grand Banks and adjacent Canadian waters. Proposals to drill for oil and gas in Georges Bank and the Gulf of Maine have been stalled by both the U.S. and Canadian governments, in large part because any spilled oil could damage the region's valuable fisheries.

The danger of oil spills to fisheries, economies, and ecosystems became clear in 2010 when British Petroleum's *Deepwater Horizon* offshore drilling platform exploded, killing 11 workers and sinking into the ocean off the Louisiana coast. Oil gushed from the underwater well at rates of 1,800 gallons per minute, rose to the surface, and spread (**FIGURE 16.18A**). Eventually the oil polluted hundreds of miles of water and shoreline along the coasts of Louisiana, Mississippi, Alabama, and Florida.

Stopping the flow was immensely challenging because the well was a mile below sea level. A blowout preventer had failed to operate, and for three months, British Petroleum (BP) was unable to stop the leak, despite trying numerous methods. The saga illustrated how inadequate our technology is for dealing with such a deepwater oil spill.

The U.S. Coast Guard, BP workers, out-of-work fishermen that BP hired, and thousands of volunteers joined together to try to prevent oil from reaching beaches, clean up fouled beaches, and treat affected wildlife. These noble efforts

Roadside litter and dumped garbage end up in streams, then rivers, then the ocean.

Booms trap trash headed to the ocean in the Los Angeles River

Countless trillions of pieces of plastic trash are accumulating in the oceans, concentrated by currents.

Charles Moore holds water from the "Great Pacific Garbage Patch"

Dead albatross with stomach full of plastic

Marine debris kills animals that get entangled in nets or swallow plastic.

Endangered Hawaiian Monk Seal caught in fishing tackle

YOU CAN MAKE A DIFFERENCE

➤ Clean a beach with the International Coastal Cleanup.

➤ Pick up litter wherever you are; trash in a stream today could be in the ocean tomorrow.

➤ Reduce consumption, and reuse and recycle items. You will generate far less waste.

were productive, but in the face of such a large spill, they could only do so much.

The economic and ecological impacts of the spill were daunting. The region's mighty fishing industry was brought to a standstill. Beach tourism plummeted. Plants, animals, and ecosystems were devastated. Vegetation was killed along hundreds of miles of shoreline, destabilizing and eroding the protective marshes. Shrimp, crabs, and fish died and washed ashore in great numbers. Sea turtles, dolphins, and birds died when they swallowed oil, when oil interfered with their breathing, or when oil coated birds' feathers, leading to hypothermia. The large and charismatic brown pelicans, Louisiana's state bird, were particularly hard hit. The impacts of this largest accidental oil spill in world history are further illustrated in our *EnvisionIt* feature in Chapter 19 (p. 548).

Such major oil spills from platforms and from the tanker ships that transport oil make headlines and cause severe environmental and economic problems. Yet despite the severity of events like the *Deepwater Horizon* oil spill, most oil pollution in the oceans accumulates from innumerable, widely spread, small, non-point sources, including leakage from small boats and runoff from human activities on land. Moreover, the amount of petroleum spilled into the oceans in recent years is equaled by the amount that seeps up from naturally occurring seafloor deposits (**FIGURE 16.18B**).

In response to headline-grabbing oil spills, governments have begun to implement stricter safety standards for tankers, such as requiring industry to pay for tugboat escorts in coastal waters and to develop prevention and response plans for major spills. The U.S. Oil Pollution Act of 1990 created a $1

FIGURE 16.18 ▼ Pollution was severe **(a)** after BP's *Deepwater Horizon* oil drilling platform exploded and disgorged millions of gallons of crude oil into the Gulf of Mexico in 2010. However, of the 1.3 million metric tons of petroleum entering the world's oceans each year, nearly half is from natural seeps **(b)**. Non-point-source pollution from petroleum consumption by people accounts for 38% of total input, including numerous diffuse sources, especially runoff from rivers and coastal communities and leakage from two-stroke engines. Less oil is being spilled into ocean waters today in large tanker spills **(c)**, thanks in part to regulations on the oil shipping industry and improved spill response techniques. The bar chart shows cumulative quantities of oil spilled worldwide from nonmilitary spills over 7 metric tons. Data from: (b) National Research Council, 2003. *Oil in the sea III. Inputs, fates, and effects.* Washington, DC: National Academies Press; and (c) International Tanker Owners Pollution Federation Ltd.

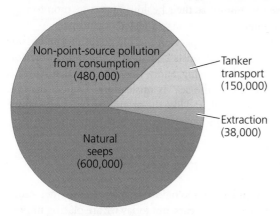

(a) Boat proceeds through *Deepwater Horizon* oil spill

(pie chart)
Non-point-source pollution from consumption (480,000)
Tanker transport (150,000)
Extraction (38,000)
Natural seeps (600,000)

(b) Sources of petroleum input into oceans (metric tons)

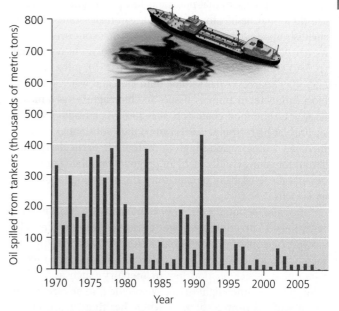

(c) Quantity of petroleum spilled from tankers, 1970–2009

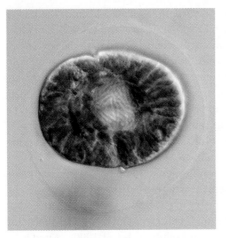

(a) Dinoflagellate (*Gymnodinium*)

FIGURE 16.19 ▲ In a harmful algal bloom, certain types of algae multiply to great densities in surface waters, producing toxins that can harm organisms. Red tides are a type of algal bloom in which the algae, such as dinoflagellates of the genus *Gymnodinium* **(a)**, produce pigment that turns the water red **(b)**.

(b) Red tide, Gulf of Carpentaria, Australia

billion prevention and cleanup fund and required that by 2015 all oil tankers in U.S. waters be equipped with double hulls as a precaution against puncture. The oil industry has resisted such safeguards. However, over the past three decades, the amount of oil spilled by tankers worldwide has decreased (**FIGURE 16.18C**), in part because of an increased emphasis on spill prevention and response. In the wake of the *Deepwater Horizon* spill, the U.S. government is considering tighter regulations on offshore drilling operations.

Toxic pollutants can contaminate seafood

Aside from the harm that pollutants like petroleum and plastic can do to marine life, toxic pollutants can make some fish and shellfish unsafe for people to eat. One prime concern today is mercury contamination. Mercury is a toxic heavy metal (p. 380) emitted from coal combustion (p. 475), mine tailings, and other sources. After settling onto land and water, mercury bioaccumulates in animals' tissues and biomagnifies as it makes its way up the food chain (pp. 385–386). As a result, fish and shellfish at high trophic levels can contain substantial levels of mercury. Eating seafood high in mercury is particularly dangerous for young children and for pregnant or nursing mothers, because the fetus, baby, or child can suffer neurological damage as a result.

Because seafood is an important part of a healthy diet, nutritionists do not advocate avoiding seafood entirely. However, people in at-risk groups should avoid fish high in mercury (such as swordfish, shark, and albacore tuna) while continuing to eat seafood low in mercury (such as catfish, salmon, and canned light tuna). We should also be careful not to eat seafood from local areas where health advisories have been issued.

Excess nutrients cause algal blooms

Pollution from fertilizer runoff or other nutrient inputs can create dead zones in coastal marine ecosystems, as we saw with the Gulf of Mexico in Chapter 5. The release of excess nutrients into surface waters can spur unusually rapid growth of phytoplankton, causing eutrophication (pp. 114, 422) in freshwater and saltwater systems.

Excessive nutrient concentrations sometimes give rise to population explosions among several species of marine algae that produce powerful toxins that attack the nervous systems of vertebrates. Blooms of these algae are known as **harmful algal blooms**. Some algal species produce reddish pigments that discolor surface waters, and blooms of these species are nicknamed **red tides** (**FIGURE 16.19**). Harmful algal blooms can cause illness and death among zooplankton, birds, fish, marine mammals, and people as their toxins are passed up the food chain. They also cause economic loss for communities dependent on fishing or beach tourism. Reducing nutrient runoff into coastal waters can lessen the frequency of these outbreaks. When they occur, we can minimize their health impacts by monitoring to prevent human consumption of affected organisms.

As severe as the impacts of marine pollution can be, however, most marine scientists concur that the more worrisome dilemma is overharvesting. Sadly, the old cliché that "there are always more fish in the sea" is misleading. The oceans today have been overfished, and like the groundfish of the Northwest Atlantic, many stocks have been largely depleted.

EMPTYING THE OCEANS

The oceans and their biological resources have met human needs for thousands of years, but today we are placing unprecedented pressure on marine resources. Over half the world's

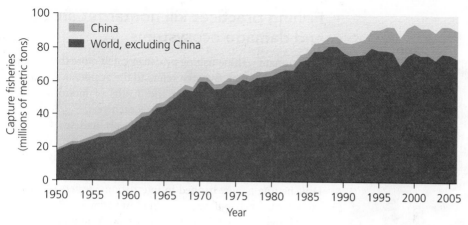

FIGURE 16.20 ◄ After rising for decades, the total global fisheries catch has stalled for the past 20 years, and many fear that a global decline is imminent if conservation measures are not taken. The figure shows trends with and without China's data, because research suggests that China's data may be somewhat inflated. Data from the Food and Agriculture Organization of the United Nations. 2009. *The state of world fisheries and aquaculture* 2008. Fig 3. By permission.

marine fish populations are fully exploited, meaning that we cannot harvest them more intensively without depleting them, according to the U.N. Food and Agriculture Organization (FAO). An additional 28% of marine fish populations are overexploited and already being driven toward extinction. Only one-fifth of the world's marine fish populations can yield more than they are already yielding without being driven into decline.

Total global fisheries catch, after decades of increases, leveled off after about 1988 (**FIGURE 16.20**), despite increased fishing effort. In 2008, the FAO concluded that "the maximum wild capture fisheries potential from the world's oceans has probably been reached." Fishery collapses such as those off Newfoundland and New England are ecologically devastating and take a severe economic toll on human communities that depend on fishing. If current trends continue, a comprehensive 2006 study in the journal *Science* predicted, populations of *all* ocean species that we fish for today will collapse by the year 2048. Aquaculture (raising fish in tanks or pens) is booming and is helping to relieve pressure on wild stocks, but fish farming comes with its own set of environmental dilemmas (pp. 270–271). All this makes it vital, many scientists and fisheries managers say, that we turn immediately to more sustainable fishing practices.

We have long overfished

People have always harvested fish, shellfish, turtles, seals, and other animals from the oceans. Much of this harvesting was sustainable, but scientists are learning that people began depleting some marine populations centuries or millennia ago. Overfishing then accelerated during the colonial period of European expansion and intensified further in the 20th century. At each stage, improved technologies and increasingly global markets intensified our impact.

Recent syntheses of historical evidence by marine biologists and historians reveal that ancient overharvesting likely affected ecosystems in ways we only partially understand today. Large animals, including the Caribbean monk seal, Steller's sea cow, and Atlantic gray whale, were hunted to extinction prior to the 20th century—before scientists were able to study them or the ecological roles they played. Overharvesting of the vast oyster beds of Chesapeake Bay led to the collapse of its oyster fishery in the late 19th century. With few oysters to filter algae and bacteria from the water, eutrophication and hypoxia similar to that of the Gulf of Mexico (Chapter 5) resulted. In the Caribbean, green sea

turtles ate sea grass and likely kept it cropped low, like a lawn. But with today's turtle population a fraction of what it was, sea grass grows thickly, dies, and rots, giving rise to sea grass wasting disease, which ravaged Florida Bay sea grass in the 1980s. The best-known case of historical overharvesting is the near-extinction of many species of whales. This resulted from commercial whaling that began centuries ago and was curtailed only in 1986. Since then, some species (such as the humpback whale) have been recovering, but others have not.

Groundfish in the Northwest Atlantic historically were so abundant that the people who harvested them never imagined they could be depleted. Yet careful historical analysis of fishing records has revealed that even in the 19th century, fishers repeatedly experienced locally dwindling catches, and each time needed to introduce some new approach or technology to extend their reach and restore their catch rate.

Fishing has industrialized

Today's industrialized commercial fishing fleets employ fossil fuels, huge vessels, and powerful new technologies to capture fish in great volumes. *Factory fishing* vessels even process and freeze their catches while at sea. The global reach of today's fleets makes our impacts much more rapid and intensive than in the past.

The modern fishing industry uses several methods to capture fish at sea that are highly efficient but also environmentally damaging. Some vessels set out long *driftnets* that span large expanses of water (**FIGURE 16.21A**). These chains of transparent nylon mesh nets are arrayed to drift with currents so as to capture passing fish, and are held vertical by floats at the top and weights at the bottom. Drift netting usually targets species that traverse open water in immense schools, such as herring, sardines, and mackerel. Specialized forms of drift netting are used for sharks, shrimp, and other animals.

Longline fishing (**FIGURE 16.21B**) involves setting out extremely long lines (up to 80 km (50 mi) long) with up to several thousand baited hooks spaced along their lengths. Tuna and swordfish are among the species targeted by longline fishing.

Trawling entails dragging immense cone-shaped nets through the water, with weights at the bottom and floats at the top. Trawling in open water captures pelagic fish, whereas *bottom-trawling* (**FIGURE 16.21C**) involves dragging weighted nets across the floor of the continental shelf to catch groundfish and other benthic organisms, such as scallops.

(a) Driftnetting

(b) Longlining

(c) Bottom-trawling

FIGURE 16.21 ▲ Commercial fishing fleets use several methods of capture. In drift netting **(a)**, long transparent nylon nets are set out to drift through open water to capture schools of fish. In longlining **(b)**, lines with numerous baited hooks are set out in open water. In bottom-trawling **(c)**, weighted nets are dragged along the floor of the continental shelf. All methods result in large amounts of bycatch, the capture of nontarget animals. The illustrations above are schematic for clarity and do not portray the immense scale that these technologies can attain; for instance, industrial trawling nets can be large enough to engulf multiple Boeing 747 jumbo jets.

Fishing practices kill nontarget animals and damage ecosystems

Unfortunately, these fishing practices catch more than just the species they target. **Bycatch** refers to the accidental capture of animals, and it accounts for the deaths of millions of fish, sharks, marine mammals, and birds each year.

Drift netting captures dolphins, seals, and sea turtles, as well as countless nontarget fish. Most of these end up drowning (mammals and turtles need to surface to breathe) or dying from air exposure on deck (fish breathe through gills in the water). Drift netting is now banned in international waters because of excessive bycatch, but continues in most national waters.

Similar bycatch problems exist with longline fishing, which kills turtles, sharks, and albatrosses, magnificent seabirds with wingspans up to 3.6 m (12 ft). Several methods are being developed to limit bycatch from longline fishing (such as using flagging to scare birds away from the lines), but an estimated 300,000 seabirds of various species die each year when they become caught on hooks while trying to ingest bait.

The scale of bycatch, and solutions to address it, both are illustrated by the story of dolphins and tuna. Several species of dolphins that associate with yellowfin tuna become caught in purse seines set for the tuna in the tropical Pacific. In purse seining, boats surround schools of tuna with a net and drawn the net in, trapping both tuna and dolphins. Hundreds of thousands of dolphins were being needlessly killed each year throughout the 1960s. The U.S. Marine Mammal Protection Act of 1972 forced U.S. fleets to try to free dolphins from seines before they were hauled up and spurred fishing gear to be modified to allow dolphins to escape. Bycatch dropped greatly as a result (**FIGURE 16.22A**).

However, as other nations' ships began catching tuna, dolphin bycatch rose again. Because U.S. fleets were operating under more restrictions, the U.S. government required that tuna imported from foreign fleets also minimize dolphin bycatch, and it supported ecolabeling efforts (p. 161) to label tuna as "dolphin-safe" if its capture used methods designed to avoid bycatch. These measures helped reduce dolphin deaths from 133,000 in 1986 to less than 2,000 per year since 1998. We can celebrate this success story but should also recognize that there is little accountability for the many "dolphin-safe" labels, that sharks and other animals continue to be caught as bycatch, and that dolphin populations have not recovered (**FIGURE 16.22B**).

Bottom-trawling not only results in bycatch but also can destroy entire ecosystems. The weighted nets crush organisms in their path and leave long swaths of damaged sea bottom. Bottom-trawling is especially destructive to structurally complex areas, such as reefs, that provide shelter and habitat for animals. In recent years, underwater photography has begun to reveal the extent of structural and ecological disturbance done by bottom-trawling (**FIGURE 16.23**). Bottom-trawling is often likened to clear-cutting (p. 325) and strip-mining (pp. 648–649), and in heavily fished areas, the bottom may be damaged multiple times. At Georges Bank, it is estimated that the average expanse of bottom has been trawled three times. Bottom-trawling here is known to destroy young cod as bycatch, and this is thought to be a main reason why the Georges Bank cod stock is not recovering.

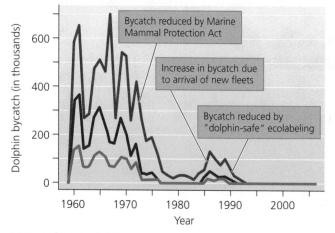

Legend:
— Total, all dolphins
— Northeastern offshore spotted dolphins
— Eastern spinner dolphins

Bycatch reduced by Marine Mammal Protection Act

Increase in bycatch due to arrival of new fleets

Bycatch reduced by "dolphin-safe" ecolabeling

(a) Reduction in dolphin bycatch

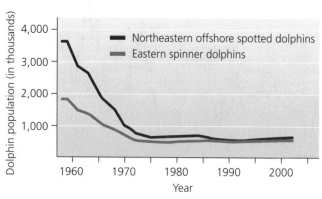

Northeastern offshore spotted dolphins
Eastern spinner dolphins

(b) Lack of recovery by dolphin populations

FIGURE 16.22 ▲ Bycatch of dolphins by tuna fleets **(a)** declined first as a result of regulations following the 1972 U.S. Marine Mammal Protection Act and later, when "dolphin-safe" ecolabeling encouraged fleets to adopt methods that reduced bycatch. However, despite this success, populations of the two dolphin species most affected **(b)** have not re-bounded, possibly because there are fewer fish for them to eat. Data from (a) National Oceanic and Atmospheric Administration, www.noaa.gov and (b) Wade, P.R., et al., 2007. Depletion of spotted and spinner dolphis in the eastern tropical Pacific: Modeling hypothe-ses for their lack of recovery. *Mar. Ecol. Prog. Ser.* 343: 1–14.

Modern fishing fleets deplete marine life rapidly

We can see the effects of large-scale industrialized fishing in the catch records of groundfish from the Northwest Atlantic. Although cod had been harvested since the 1500s on the Grand Banks, catches more than doubled once immense industrial trawlers from Europe, Japan, and the United States appeared in the 1960s (**FIGURE 16.24A**). These record-high catches lasted only a decade; the industrialized approach removed so many fish that the stock has not recovered. Likewise, on Georges Bank, cod catches rose greatly in the 1960s, remained high for 30 years, then collapsed (**FIGURE 16.24B**).

Throughout the world's oceans, today's industrialized fish-ing fleets are depleting marine populations with astonishing

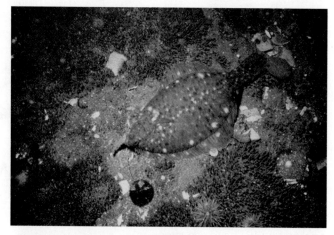

(a) Before trawling at Georges Bank

(b) After trawling at Georges Bank

FIGURE 16.23 ▲ Bottom-trawling causes severe structural dam-age to reefs and benthic habitats, and it can decimate underwater communities and ecosystems. A photo of an untrawled location **(a)** on the seafloor of Georges Bank shows a vibrant and diverse benthic community. A photo of the same site after trawling **(b)** shows a flattened and lifeless expanse of sea bottom.

speed. In a 2003 study, Canadian fisheries biologists Ransom Myers and Boris Worm analyzed data from FAO archives and found the same pattern for region after region: In just a decade after the arrival of industrialized fishing, catch rates dropped pre-cipitously, with 90% of large-bodied fish and sharks eliminated. Populations then stabilized at 10% of their former levels. This means, Myers and Worm concluded, that the oceans today con-tain only one-tenth of the large-bodied animals they once did.

As we have seen (pp. 86–91), when animals at high trophic levels are removed from a food web, the proliferation of their prey can alter the nature of the entire community. Many scientists conclude that most marine communities may have been very different prior to industrial fishing.

Several factors mask declines

Although industrialized fishing has depleted fish stocks in region after region, the overall global catch has remained roughly stable for two decades (see Figure 16.20). This seem-ing stability can be explained by several factors that mask population declines. Fishing fleets are now traveling longer distances to reach less-fished portions of the ocean. They

FIGURE 16.24 ▶ In the North Atlantic off the coast of Newfoundland, commercial catches of Atlantic cod **(a)** increased with intensified fishing by industrial trawlers in the 1960s and 1970s. The fishery subsequently crashed, and moratoria imposed in 1992 and 2003 have not brought it back. A similar pattern is seen in the cod catches at Georges Bank **(b)**; industrial fishing produced 30 years of high catches, followed by a collapse and the closure of some areas to fishing. Note also that in each case, there is one peak before 1977 and one after 1977. The first peak and decline resulted from foreign fishing fleets, whereas the second peak and decline resulted from Canadian and U.S. fleets, respectively, after they laid claim to their 200-mile exclusive economic zones. Data (a) from *Millennium Ecosystem Assessment, 2005. Ecosystems and human well-being: biodiversity synthesis.* World Resources Institute. Washington, DC. Used with permission and (b) O'Brien, L., et al., 2008. Georges Bank Atlantic cod. In *Assessment of 19 Northeast groundfish stocks through 2007.* Northeast Fisheries Science Center, Woods Hole, MA.

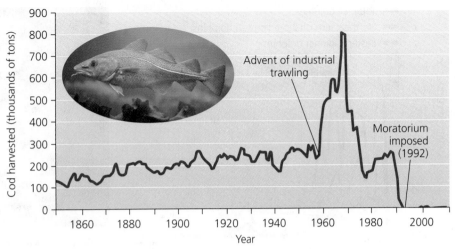

(a) Cod harvested from Grand Banks

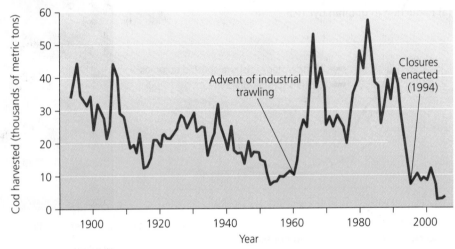

(b) Cod harvested from Georges Bank

also are fishing in deeper waters; average depth of catches was 150 m (495 ft) in 1970 and 250 m (820 ft) in 2000. Moreover, fleets are spending more time fishing and are setting out more nets and lines—expending increasing effort just to catch the same number of fish. For example, a 2010 study showed that British trawlers were working 17 times harder just to catch the same number of cod and other fish as 120 years ago.

More powerful technology also helps explain large catches despite declining stocks. Today's Japanese, European, Canadian, and U.S. fleets can reach almost any spot on the globe with vessels that attain speeds of 80 kph (50 mph). They boast an array of technologies that militaries have developed for spying and for chasing enemy submarines, including advanced sonar mapping equipment, satellite navigation, and thermal sensing systems. Some fleets rely on airplanes to find schools of commercially valuable fish such as bluefin tuna.

We are "fishing down the food chain"

Numbers of fish do not tell the whole story of fisheries depletion. Analyses of fisheries data reveal in case after case that as fishing increases, the size and age of fish caught decline. This is not only because fishers prefer to take large fish, but also because under intense fishing pressure, few fish escape being caught for very many years, so that few have a chance to grow to large size. Cod caught in the Northwest Atlantic today average much smaller than decades ago, and it is now rare to find a cod over 10 years of age, even though cod of this age formerly were common. Because large female fish produce far more young than small ones, their depletion makes it harder for populations to recover.

In addition, as particular species become too rare to fish profitably, fleets begin targeting other species that are more abundant. Generally this means shifting from large, desirable species to smaller, less desirable ones. Time and again, fleets have depleted popular food fish (such as cod) and shifted to species of lower value (such as capelin, smaller fish that cod eat). Because this often entails catching species at lower trophic levels, this phenomenon has been termed "fishing down the food chain."

Our purchasing choices can influence fishing practices

To most of us, marine fishing practices may seem a distant phenomenon over which we have no control. Yet by exercising careful choice when we buy seafood, we as consumers can encourage more sustainable fisheries practices (**FIGURE 16.25**). Purchasing ecolabeled seafood, such as dolphin-safe tuna, is one way to exercise choice, but in most cases consumers have no readily available information about how their seafood

FIGURE 16.25 ▲ You can discourage overfishing and support sustainable fisheries by exercising choice as a consumer when buying seafood. Here, a choosy customer shops for sustainably harvested fish at a fish market on the island of Menorca, Spain.

was caught. Thus, several organizations have devised concise guides to help consumers differentiate fish and shellfish that are overfished or whose capture is ecologically damaging from those that are harvested more sustainably. For instance, the Monterey Bay Aquarium provides a wealth of this information on its website, and **TABLE 16.1** includes a few examples.

WEIGHING THE ISSUES

Eating Seafood After reading this chapter, do you plan to alter your decisions about eating seafood in any way? If so, how? If not, why not? Do you think consumer buying choices can exert an influence on fishing practices? On mercury contamination in seafood?

TABLE 16.1 Seafood Choices for Consumers

Best choices*	Seafood to avoid†
Catfish (U.S.-farmed)	Atlantic cod
Clams (farmed)	Caviar (sturgeon, imported, wild)
Dungeness crab	Chilean seabass/toothfish
Pacific halibut	Atlantic halibut, flounders, soles
Mussels (farmed)	Mahi mahi/dolphinfish (imported)
Oysters (farmed)	Orange roughy
Salmon (wild-caught, Alaska)	Salmon (farmed, Atlantic)
Bay scallops (farmed)	Sharks
Striped bass (farmed)	Shrimp (imported)
Tilapia (U.S.-farmed)	Red snapper
Rainbow trout (farmed)	Swordfish (imported)
Skipjack tuna (troll/pole-caught)	Tuna: bluefin, tongol, yellowtail

*Fish or shellfish that are abundant, well managed, and fished or farmed in environmentally friendly ways.

†Fish or shellfish from sources that are overfished and/or fished or farmed in ways that harm other marine life or the environment.

Adapted from: *Monterey Bay Aquarium. Seafood watch national sustainable seafood guide.* January 2010.

Marine biodiversity loss erodes ecosystem services

Overfishing, pollution, habitat change, and other factors that deplete biodiversity can threaten the ecosystem services we derive from the oceans. In the 2006 study that predicted global fisheries collapse by 2048 (p. 449), the study's 14 authors analyzed all existing scientific literature to summarize the effects of biodiversity loss on ecosystem function and ecosystem services. They found that across 32 different controlled experiments conducted by various researchers, systems with reduced species diversity or genetic diversity showed less primary and secondary production and were less able to withstand disturbance.

The team also found that when biodiversity was reduced, so were habitats that serve as nurseries for fish and shellfish. Moreover, biodiversity loss was correlated with reduced filtering and detoxification (as from wetland vegetation and oyster beds). This can lead to harmful algal blooms, dead zones, fish kills, and beach closures.

MARINE CONSERVATION

Because we bear responsibility and stand to lose a great deal if valuable ecological systems collapse, marine scientists have been working to develop solutions to the problems that threaten the oceans. Many have begun by taking a hard look at the strategies used traditionally in fisheries management.

Fisheries management has been based on maximum sustainable yield

Fisheries managers conduct surveys, study fish population biology, and monitor catches. They then use that knowledge to regulate the timing of harvests, the scale of harvests, and the techniques used to catch fish. The goal is to allow for maximal harvests while keeping fish available for the future—the concept of *maximum sustainable yield* (p. 323). Recall that this means keeping populations at about half the level they would otherwise achieve, because populations growing by logistic growth grow fastest at this size (pp. 67, 323). If data indicate that current yields are unsustainable, managers might limit the number or biomass of fish that can be harvested, or they might restrict the type of gear fishers can use.

Despite such efforts, a number of fish and shellfish stocks have plummeted. Thus, many scientists and managers feel it is time to shift the focus away from individual species and toward viewing marine resources as elements of larger ecological systems. This means considering the impacts of fishing practices on habitat quality, species interactions, and other factors that may have indirect or long-term effects on populations. One key aspect of such *ecosystem-based management* (p. 323) is to set aside areas of ocean where systems can function without human interference.

We can protect areas in the ocean

Hundreds of **marine protected areas (MPAs)** have been established, most of them along the coastlines of developed countries. However, despite their name, nearly all MPAs allow

The SCIENCE behind the Story

Do Marine Reserves Work?

Dr. Callum M. Roberts, University of York, monitoring reef fish in Belize

Marine reserves off the coasts of Florida and the Caribbean island of St. Lucia provided some of the first clear evidence that marine reserves can benefit nearby fisheries. In 2001, a team of fisheries scientists led by York University researcher Callum Roberts published these results in the journal *Science*.

Following the establishment in 1995 of the Soufrière Marine Management Area (SMMA), a network of reserves intended to help restore St. Lucia's severely depleted coral reef fishery, Roberts and his colleague, Julie Hawkins, conducted annual surveys of fish abundance in the reserves and nearby areas. Within three years, they found that the biomass of five commercially important families of fish—surgeonfishes, parrot fishes, groupers, grunts, and snappers—had

tripled inside the reserves and doubled outside them (**first figure**).

Roberts and Hawkins also interviewed local fishers and found that those with large traps were catching 46% more fish per trip in 2000–2001 than they had in 1995–1996, and that fishers with small traps were catching 90% more (**second figure**). Roberts and his colleagues concluded that "in 5 years, reserves have led to improvement in the SMMA fishery, despite the 35% decrease in area of fishing grounds."

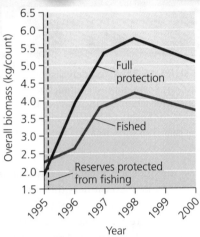

Established in 1995, the Soufrière Marine Management Area (SMMA) along the coast of St. Lucia had a rapid impact. By 1998, fish biomass within the five reserves tripled, and in adjacent fished areas it doubled. Data from Roberts, C., et al., 2001. Effects of marine reserves on adjacent fisheries. *Science* 294: 1920–1923. Fig 2. Reprinted with permission from AAAS and the author.

---

fishing or other extractive activities. As one report from an environmental advocacy group put it, MPAs "are dredged, trawled, mowed for kelp, crisscrossed with oil pipelines and fiber-optic cables, and swept through with fishing nets."

Because of the lack of true refuges from fishing pressure, many scientists want to establish areas where fishing is prohibited. Such "no-take" areas have come to be called **marine reserves**. Designed to preserve ecosystems intact, marine reserves are also intended to improve fisheries. Scientists argue that marine reserves can act as production factories for fish for surrounding areas, because fish larvae produced inside reserves will disperse outside and stock other parts of the ocean (see **THE SCIENCE BEHIND THE STORY**, above). By serving both purposes, proponents maintain, marine reserves are a win-win proposition for environmentalists and fishers alike.

Many commercial and recreational fishers dislike the idea of no-take reserves, however, just as most were opposed to the Canadian groundfish moratoria and the Georges Bank closures. Nearly every marine reserve that has been established or proposed has met with opposition from people and businesses who use the area for fishing or recreation. Some protests have turned violent. To protest fishing restrictions,

fishermen in the Galápagos Islands destroyed offices at Galápagos National Park and threatened researchers and park managers with death.

Clearly, when marine reserves are established, it pays to be sensitive to the concerns of people of the area. In 2006, President Bush established the Papahanaumokuakea Marine National Monument around the northwestern Hawaiian Islands—at 362,000 km² (140,000 mi²), an area larger than all U.S. national parks combined. In this new monument, native Hawaiians were given fishing rights in areas where fishing was otherwise made off-limits.

In contrast, the United Kingdom in 2010 established the world's largest marine reserve around the Chagos Archipelago, a group of 55 islands it controls in the Indian Ocean. The indigenous people of these islands were forcibly evicted from the main island, Diego Garcia, in 1967–1973 so that the United States could build a military base there. The roughly 4,000 Chagossian people have been fighting in courts ever since for the right to return to their islands. They now fear that with fishing banned, they will never be able to return because there would be no way to make a living. The U.K. government has said that if the people return the fishing bans will be reassessed

Roberts and his colleagues also studied the oldest fully protected marine reserve in the United States, the Merritt Island National Wildlife Refuge, established in 1962 as a buffer around what is today the Kennedy Space Center on Cape Canaveral, Florida. In a previous study, fisheries biologists Darlene Johnson, James Bohnsack, and Nicholas Funicelli had found that the reserve contained more and larger fish than did nearby unprotected areas. They also found that some of the reserve's fish appeared to be migrating to nearby fishing areas.

Bohnsack, Roberts, and their colleagues corroborated the evidence for migration by analyzing trophy records from the International Game Fish Association. They found that the proportion of Florida's record-sized fish caught near Merritt Island increased significantly after 1962. The researchers hypothesized that the reserve was allowing fish to grow to trophy size before they migrated to nearby areas, where recreational fishers caught them.

Not everyone saw the St. Lucia and Merritt Island cases as proof that marine reserves could rescue depleted fisheries. In 2002, several alternative interpretations of the evidence were published as

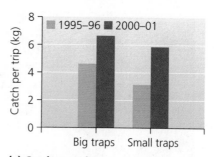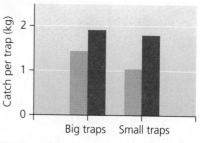

(a) Catch per trip   (b) Catch per trap

Roberts and his colleagues studied biomass of fish caught at the SMMA over two 5-month periods in 1995–1996 and 2000–2001. Per fishing trip (a), the catch for fishers with big traps increased by 46%, and the catch for fishers with small traps increased by 90%. Per trap (b), fishers with big traps caught 36% more fish, and fishers with small traps caught 80% more fish. Data from Roberts, C., et al., 2001. Effects of marine reserves on adjacent fisheries. *Science* 294: 1920–1923. Fig 3. Reprinted with permission from AAAS and the author.

letters in *Science*. Mark Tupper, a fisheries scientist at the University of Guam, suggested that the St. Lucia results were relevant only to coral reef fisheries in developing nations and that Florida's boost in fish populations was due primarily to limits on recreational fishing. Karl Wickstrom, editor-in-chief of *Florida Sportsman* magazine, suggested that the increase in trophy fish near Merritt Island was caused by commercial fishing regulations and changes in how trophies were recorded and promoted. And Ray Hilborn, a fisheries scientist at the University of Washington, pointed out that the St. Lucia study lacked a control condition.

In response, Roberts and his colleagues reaffirmed the validity of their results while acknowledging some limitations. They agreed with Tupper that marine reserves are not always effective and often need to be complemented by other management tools, such as size limits. "We agree that inadequately protected reserves are useless," they wrote, "but our study shows that well-enforced reserves can be extremely effective and can play a critical role in achieving sustainable fisheries."

and that in the meantime the bans will save the region's fish from being depleted by industrial fleets from other countries.

WEIGHING THE ISSUES

Preservation at Sea Almost 4% of U.S. land area is designated as wilderness, yet before the recent actions in Hawaii and the Chagos islands, far less than 1% of coastal waters were protected in reserves. Why do you think it is taking so long for the preservation ethic to make the leap to the oceans? What challenges do you see with preserving areas of coastal ocean? Do you think local people should be given fishing rights in marine reserves? What could be done to better protect marine ecosystems, fisheries, and local fishing cultures?

Reserves can work for both fish and fishers

Over the past two decades, data from marine reserves around the world have been indicating that reserves can work as win-win solutions that benefit ecosystems, fish populations, and fishing economies. A comprehensive review of data from marine reserves as of 2001 revealed that just one to two years after their establishment, marine reserves:

▶ Increased densities of organisms on average by 91%.

▶ Increased biomass of organisms on average by 192%.

▶ Increased average size of organisms by 31%.

▶ Increased species diversity by 23%.

That year, 161 prominent marine scientists signed a "consensus statement" summarizing the effects of marine reserves. Besides boosting fish biomass, total catch, and record-sized fish, the report stated, marine reserves yield several benefits. Within reserve boundaries, they:

▶ Produce rapid and long-term increases in abundance, diversity, and productivity of marine organisms.

▶ Decrease mortality and habitat destruction.

▶ Lessen the likelihood of species extirpation.

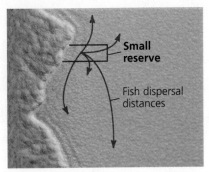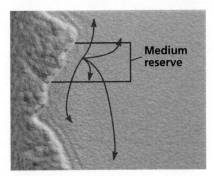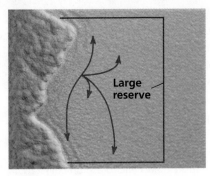

FIGURE 16.26 ▲ Marine reserves of different sizes may have varying effects on ecological communities and fisheries. Young and adult fish and shellfish of different species can disperse different distances, as indicated by the red arrows in the figure. A small reserve (left panel) may fail to protect animals because too many disperse out of the reserve. A large reserve (right panel) may protect fish and shellfish very well but will provide relatively less "spillover" into areas where people can legally fish. Thus medium-sized reserves (middle panel) may offer the best hope of preserving species and ecological communities while also providing adequate fish to people. *Source:* Adapted from Halpern, B.S., and R.R. Warner, 2003. Matching marine reserve design to reserve objectives. *Proceedings of the Royal Society of London B* 270: 1871–1878, Fig 1. Used by permission of The Royal Society and the author.

Outside reserve boundaries, marine reserves:

▶ Can create a "spillover effect" as individuals of protected species spread outside reserves.

▶ Allow larvae of species protected within reserves to "seed the seas" outside reserves.

The consensus statement was backed up by research into reserves worldwide. At Apo Island in the Philippines, biomass of large predators increased eightfold inside a marine reserve, and fishing improved outside the reserve. At two coral reef sites in Kenya, commercially fished and keystone species were up to 10 times more abundant in the protected area as in the fished area. At Leigh Marine Reserve in New Zealand, snapper increased 40-fold, and spiny lobsters were increasing by 5–11% yearly. Spillover from this reserve improved fishing and ecotourism, and local residents who once opposed the reserve now support it. Since that time, further research has shown that reserves create a fourfold increase in catch per unit effort in fished areas surrounding reserves, and that they can greatly increase ecotourism by divers and snorkelers.

On Georges Bank, once commercial trawling was halted in 1994, populations of many organisms began to recover. As benthic invertebrates began to come back, numbers of groundfish such as haddock and yellowtail flounder rose inside the closed areas, and scallops increased by 14 times. Moreover, fish from the closure areas appear to be spilling over into adjacent waters, because fishers have been catching more groundfish from Georges Bank as a whole since the late 1990s. From these and other datasets, increasing numbers of scientists, fishers, and policymakers are advocating the establishment of fully protected marine reserves as a central management tool.

How should reserves be designed?

If marine reserves work in principle, the question becomes how best to design reserves and arrange them into networks. Studies are modeling how to optimize the size and spacing of reserves so that ecosystems are protected, fisheries are sustained, and people are not overly excluded from marine areas (**FIGURE 16.26**). Scientists are asking how large reserves need to be, how many there need to be, and where they need to be placed to take best advantage of ocean currents. Of several dozen studies that have estimated how much area of the ocean should be protected in no-take reserves, estimates range from 10% to 65%, with most falling between 20% and 50%. Most scientists feel that involving fishers directly in the planning process is crucial for coming up with answers to all these questions. If marine reserves can be made to work and to be accepted, then they may well seed the seas and help lead us toward solutions to one of our most pressing environmental problems.

➤ CONCLUSION

Oceans cover most of our planet, and we are still exploring their diverse topography and ecosystems. As we learn more about marine and coastal environments, we continue to intensify our use of their resources and to exert more severe impacts. Ocean acidification, loss of coral reefs, several types of pollution, and the ongoing depletion of many of the world's marine fish stocks are all major challenges that we will need to address. Yet scientists are demonstrating that setting aside protected areas of the ocean can serve to maintain and restore natural systems and also to enhance fisheries. In the meantime, all of us can contribute through consumer choices that can help move us toward sustainable fishing practices.

REVIEWING OBJECTIVES

You should now be able to:

IDENTIFY PHYSICAL, GEOGRAPHICAL, CHEMICAL, AND BIOLOGICAL ASPECTS OF THE MARINE ENVIRONMENT

- Oceans cover 71% of Earth's surface and contain over 97% of its water. (p. 433)
- Seafloor topography can be complex. (pp. 433–434)
- Ocean water contains 96.5% H_2O by mass and various dissolved salts. (pp. 434–435)
- Colder, saltier water is denser and sinks. Water temperatures vary with latitude, and temperature variation is greater in surface layers. (p. 435)
- Surface currents move horizontally through the oceans, driven by wind and other factors. (p. 436)
- Vertical water movement includes upwelling and downwelling, which affect the distribution of nutrients and life. (pp. 436–437)

EXPLAIN HOW THE OCEANS INFLUENCE, AND ARE INFLUENCED BY, CLIMATE

- The thermohaline circulation shapes regional climate, for instance, keeping Europe warm. Global warming could potentially shut down existing circulation patterns. (pp. 436–437)
- El Niño and La Niña events alter climate and affect fisheries. (pp. 437–438)
- The oceans sequester atmospheric carbon and have slowed global climate change, but they could soon become saturated (p. 438)
- Absorption of excess carbon dioxide leads to ocean acidification, which hinders corals in forming reefs. (pp. 438, 440–441).

DESCRIBE MAJOR TYPES OF MARINE ECOSYSTEMS

- Major types of marine and coastal ecosystems include pelagic and deep-water open ocean systems, kelp forests, coral reefs, intertidal zones, salt marshes, mangrove forests, and estuaries. (pp. 438–444)
- Many of these systems are highly productive and rich in biodiversity. Many also suffer heavy impacts from human influence. (pp. 439–444)

ASSESS IMPACTS FROM MARINE POLLUTION

- People pollute ocean waters with trash, including nets and plastic that harm marine life. (pp. 445–446)

- Plastic trash accumulates in ocean regions where it is trapped by currents. (pp. 445–446)
- Marine oil pollution results from non-point sources on land as well as from spills at sea from tankers and drilling platforms. (pp. 445–448)
- Heavy metal contaminants in seafood affect human health. (p. 448)
- Nutrient pollution can lead to dead zones and harmful algal blooms. (p. 448)

REVIEW THE STATE OF OCEAN FISHERIES AND REASONS FOR THEIR DECLINE

- Over half the world's marine fish populations are fully exploited, 28% are overexploited, and only 20% can yield more without declining. (pp. 448–449)
- Global fish catches have stopped growing since the late 1980s, despite increased fishing effort and improved technologies. (p. 449)
- People began depleting marine resources long ago, but impacts have intensified in recent decades. (p. 449)
- Commercial fishing practices include drift netting, longline fishing, and trawling, all of which capture nontarget organisms, called bycatch. (pp. 449–451)
- Today's oceans hold just one-tenth as many large animals as they did before industrialized commercial fishing. (p. 451)
- As fishing intensity increases, fish become smaller. (p. 452)
- Consumers can encourage good fishery practices by shopping for sustainable seafood. (pp. 452–453)
- Marine biodiversity loss affects ecosystem services. (p. 453)
- Traditional fisheries management has not stopped declines, so many scientists feel that ecosystem-based management is needed. (p. 453)

EVALUATE MARINE PROTECTED AREAS AND RESERVES AS INNOVATIVE SOLUTIONS

- We have established fewer protected areas in the oceans than we have on land, and most marine protected areas allow many extractive activities. (pp. 453–454)
- No-take marine reserves can protect ecosystems while also boosting fish populations and making fisheries sustainable. (pp. 454–456)

CHAPTER 16 Marine and Coastal Systems and Resources

457

TESTING YOUR COMPREHENSION

1. What proportion of Earth's surface do oceans cover? What is the average salinity of ocean water? How are density, salinity, and temperature related in each layer of ocean water?

2. What factors drive ocean currents? Give an example of how a surface current affects climate regionally.

3. What is causing ocean acidification? What consequences do scientists expect ocean acidification to bring about?

4. Where in the oceans are productive areas of biological activity likely to be found?

5. Describe three kinds of ecosystems found near coastal areas and the types of life they support.

6. Why are coral reefs biologically valuable? How are they being degraded by human impact? What is causing the disappearance of mangrove forests and salt marshes?

7. What is meant by the "Great Pacific Garbage Patch"? Discuss three ways in which people are fighting pollution in the oceans and on our coasts.

8. Describe an example of how overfishing can lead to ecological damage and fishery collapse.

9. Name three industrial fishing practices, and explain how they create bycatch and harm marine life.

10. How does a marine reserve differ from a marine protected area? Why do many fishers oppose marine reserves? Explain why many scientists say no-take reserves will be good for fishers.

SEEKING SOLUTIONS

1. What benefits do you derive from the oceans? How does your behavior affect the oceans? Give specific examples.

2. We have been able to reduce the amount of oil we spill into the oceans, but petroleum-based products such as plastic continue to litter our oceans and shorelines. Discuss some ways that we can reduce this impact on the marine environment.

3. Describe the trends in global fish capture over the past 55 years and over the past 20 years, and explain several factors that account for these trends.

4. Consider what you know about biological productivity in the oceans, about the scientific data on marine reserves, and about the social and political issues surrounding the establishment of marine reserves. What types of ocean regions do you think it would be particularly appropriate to establish as marine reserves? Why?

5. **THINK IT THROUGH** You make your living fishing on the ocean, just as your father and grandfather did, and as most of your neighbors do in your small coastal village. Your region's fishery has just collapsed, however, and everyone is blaming it on overfishing. The government

has closed the fishery for three years, and scientists are pushing for a permanent marine reserve to be established on your former fishing grounds. You have no desire to move away from your village, so what steps will you take now? Will you protest the closure? What compensation will you ask of the government if it prevents you from fishing? Will you work with scientists to establish a reserve that improves fishing in the future, or will you oppose their attempts to create a reserve? What data and what assurances will you ask of them?

6. **THINK IT THROUGH** You are mayor of a coastal town where some residents are employed as commercial fishers and others make a living serving ecotourists who come to snorkel and scuba dive at the nearby coral reef. In recent years, several fish stocks have crashed, and ecotourism is dropping off as fish disappear from the increasingly degraded reef. Scientists are urging you to help establish a marine reserve around portions of the reef, but most commercial and recreational fishers are opposed to this idea. What steps would you take to restore your community's economy and environment?

CALCULATING ECOLOGICAL FOOTPRINTS

The relationship between the ecological goods and services used by individuals and the amount of *land* area needed to provide those goods and services is relatively well developed. People also use goods and services from Earth's oceans, where the concept of *area* is less useful. It is clear, however, that our removal of fish from the oceans has an impact, or an ecological footprint.

The table shows data on the mean annual per-person consumption from ocean fisheries for North America, China, and the world as a whole. Using the data provided, calculate the amount of fish each consumer group would consume per year, given the annual per capita consumption rates, for each of these three regions. Record your results in the table.

	Annual Consumption		
Consumer group	North America (24.1 kg per person)	China (26.1 kg per person)	World (16.4 kg per person)
You			
Your class			
Your state			
United States			
World			

Data from Food and Agriculture Organization of the United Nations 2009. *The state of world fisheries and aquaculture: 2008.*

1. Calculate the ratio of North America's per capita fish consumption rate to that of the world. Compare this ratio to the ratio of the per capita ecological footprints for the United States, Canada, and Mexico (see Figure 1.14, p. 16) versus the world average footprint of 2.7 ha/person/year. Can you account for similarities and differences between these ratios?

2. The population of China has grown at an annual rate of 1.1% since 1987, while over the same period fish consumption in China has grown at an annual rate of 8.9%. Speculate on the reasons behind China's rapidly increasing consumption of fish.

3. What ecological concerns do the combined trends of human population growth and increasing per capita fish consumption raise for you? What role might you play in contributing to these concerns or to their solutions?

Mastering**ENVIRONMENTALSCIENCE**™

Go to **www.masteringenvironmentalscience.com** for practice quizzes, Pearson eText, videos, current events, and more.

CHAPTER 16 Marine and Coastal Systems and Resources

Earth's atmosphere from space

17 ATMOSPHERIC SCIENCE AND AIR POLLUTION

UPON COMPLETING THIS CHAPTER, YOU WILL BE ABLE TO:

- Describe the composition, structure, and function of Earth's atmosphere
- Relate weather and climate to atmospheric conditions
- Identify major pollutants, outline the scope of outdoor air pollution, and assess potential solutions

- Explain stratospheric ozone depletion and identify steps taken to address it
- Define *acid deposition* and illustrate its consequences
- Characterize the scope of indoor air pollution and assess potential solutions

CENTRAL **CASE STUDY**

L.A. and Its Sister Cities Struggle for a Breath of Clean Air

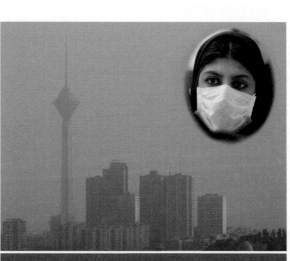

"Poisonous gases emitted from vehicles and industries continue to wreak havoc on Tehran's air."

—Yousef Rashidi, director of Tehran's Air Quality Control Company

"The danger posed by Tehran's air pollution crisis is no less than an earthquake. . . . The destructive effects of an earthquake are instantaneous, whereas air pollution kills innocent people gradually."

—Mohammed Hadi Haidarzadeh, head of the Tehran City Council's clean air office

Los Angeles has long symbolized air pollution in Americans' minds. Smog blanketed the city throughout the 1970s, 1980s, and 1990s, a byproduct of its car culture and the exhaust of millions of automobiles on the freeways. But in recent years, policy efforts and technological advances have improved air quality in Los Angeles. Today L.A. still suffers the nation's worst smog, but its skies are clearer than in some of its "sister cities" elsewhere in the world.

A bad air day in Tehran, Iran

The Sister Cities program was devised half a century ago by President Eisenhower to encourage world peace and understanding by fostering relationships between people of U.S. cities and others around the world. Today hundreds of American cities share government visits and cultural ties with their sister cities across the world. One of L.A.'s sister cities is Tehran, the capital of Iran. As it happens, one thing these two cities share is smog.

When people were gunned down in the streets of Tehran in 2009 while protesting an apparently fraudulent presidential election, the story made international headlines, while cell-phone videos of the deaths circulated around the world. Yet every single day people die in Iran's capital from the more mundane menace of air pollution. Health authorities blame several thousand premature deaths per year in Tehran on lung and respiratory diseases resulting from air pollution. In 2006, fully 3,600 people succumbed in just a month.

Although Tehran's parks and tree-lined streets make it a gorgeous city, the quality of its air is so bad that residents commonly wear face masks when they step outside to go to the bank, the laundromat, or the grocery store. On especially bad days, the government closes schools and advises people to stay indoors.

"On Saturday afternoon after walking for only 15 minutes," says Tehran journalist Alborz Maleki, "I got a terrible headache. People suffer from eye allergies, nose irritation, . . . headaches, burning eyes. Especially children and elderly people are facing many [health] problems."

Mehdi Gharakhanlou, a 38-year-old businessman, says he returns home from the central city each winter day "fatigued, nauseated, with itchy eyes, a sore throat, a bad headache or even with all of these symptoms . . . . I wear a mask when I am out, but even that doesn't help much."

As in Los Angeles, traffic generates most of the pollution in Tehran. An estimated 80% of Tehran's pollutants come from the exhaust of its 3 million vehicles, many of which are over 20 years old and lack basic pollutant-filtering technology such as catalytic converters. Tehran's limited public transportation system forces most people to rely on cars, motorcycles, and taxis to get around. Moreover, Iran's government spends billions of dollars in subsidies to give its citizens some of the cheapest gasoline in the world. With 40-cent-per-gallon gas, people have little financial incentive to conserve, and Iranian vehicles guzzle gas at four times the rate of European vehicles.

As with Los Angeles, topography worsens the problem. Tehran lies in a valley, surrounded by mountains that trap the city's pollution. Wintertime is the worst, when thermal inversions confine pollutants over the city and there is little wind to blow them away.

And as with Los Angeles in recent decades, people are streaming into Tehran from elsewhere. As a result, the government's efforts to rein in pollution are being overwhelmed by population growth.

In short, Tehran today is dangerously polluted for the same reasons that Los Angeles was a few decades ago. L.A. and Tehran today typify a difference between cities of developed nations and cities of developing nations. Nations that are industrializing as they try to build wealth for their citizens are running into the same urban and industrial air pollution problems that plagued the United States and other wealthy nations a generation and more ago.

Los Angeles has 24 other sister cities. Many of them struggle with severe air pollution as well. Athens, Greece; Jakarta, Indonesia; Mumbai, India; Guangzhou, China; Taipei, Taiwan; and San Salvador, El Salvador all experience poor air quality on a regular basis as new emissions from increased auto exhaust and fossil fuel combustion are added to more traditional pollution sources. Best known among these sister cities is Mexico City, Mexico, with perhaps the most polluted air in the world.

Most of these cities are taking steps to improve their air quality, just as Los Angeles and other American cities have done before them. We will examine the solutions sought in Los Angeles, Tehran, and L.A.'s other sister cities throughout this chapter as we learn about Earth's atmosphere and how to reduce the pollutants we release into it.

THE ATMOSPHERE

Every breath we take reaffirms our connection to the **atmosphere**, the thin layer of gases that surrounds Earth. The atmosphere provides us oxygen, absorbs hazardous solar radiation, burns up incoming meteors, transports and recycles water and nutrients, and moderates climate.

Earth's atmosphere consists of 78% nitrogen (N_2) and 21% oxygen (O_2). The remaining 1% is composed of argon (Ar) and minute concentrations of several other gases (**FIGURE 17.1**).

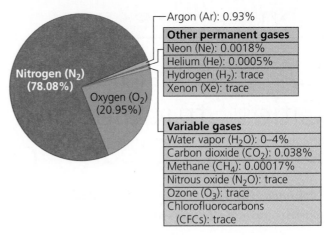

FIGURE 17.1 ▲ Earth's atmosphere consists mostly of nitrogen, secondarily of oxygen, and lastly of argon and a mix of gases at dilute concentrations. Permanent gases are fixed in concentration. Variable gases vary in concentration as a result of either natural processes or human activities. Data from Ahrens, C.D., 2007. *Meteorology today*, 8th ed. Belmont, CA: Brooks/Cole.

These include *permanent gases* that remain at stable concentrations and *variable gases* that vary in concentration from time to time or place to place as a result of natural processes or human activities.

Over our planet's long history, the atmosphere's composition has changed. Long ago our atmosphere was dominated by carbon dioxide (CO_2), nitrogen, carbon monoxide (CO), and hydrogen (H_2), but about 2.7 billion years ago, oxygen began to build up with the emergence of autotrophic microbes that emitted oxygen by photosynthesis (pp. 31–32). Today, human activity is altering the quantities of some atmospheric gases, such as carbon dioxide, methane (CH_4), and ozone (O_3). In this chapter and in Chapter 18, we will explore the atmospheric changes brought about by our pollutants, but we begin with an overview of Earth's atmosphere.

The atmosphere is layered

The atmosphere that stretches so high above us and seems so vast is actually just a thin coating about 1/100 of Earth's diameter, like the fuzzy skin of a peach. This coating consists of four layers that scientists recognize by differences in temperature, density, and composition (**FIGURE 17.2**).

The bottommost layer, the **troposphere**, blankets Earth's surface and provides us the air we need to live. The movement of air within the troposphere is also largely responsible for the planet's weather. Although it is thin (averaging 11 km [7 mi] high) relative to the atmosphere's other layers, the troposphere contains three-quarters of the atmosphere's mass, because air is denser near Earth's surface. Tropospheric air gets colder with altitude, dropping to roughly –52 °C (–62 °F) at the top of the troposphere. At this point, temperatures cease to decline with altitude, marking a boundary called the *tropopause*. The tropopause acts like a cap, limiting mixing between the troposphere and the atmospheric layer above it, the stratosphere.

The **stratosphere** extends 11–50 km (7–31 mi) above sea level. Similar in composition to the troposphere, the stratosphere is 1,000 times drier and less dense. Its gases experience

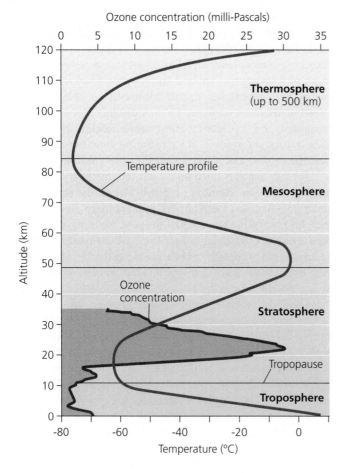

FIGURE 17.2 ▲ Temperature (red line) drops with altitude in the troposphere, rises with altitude in the stratosphere, drops in the mesosphere, then rises again in the thermosphere. The tropopause separates the troposphere from the stratosphere. Ozone (blue shaded area) reaches a peak in a portion of the stratosphere, giving rise to the term *ozone layer*. Adapted from Jacobson, M.Z., 2002. *Atmospheric pollution: History, science, and regulation.* Cambridge: Cambridge University Press; and Parson, E.A., 2003. *Protecting the ozone layer: Science and strategy.* Oxford: Oxford University Press.

little vertical mixing, so once substances (including pollutants) enter it, they tend to remain for a long time. The stratosphere's air becomes warmer with altitude, attaining a maximum temperature of –3 °C (27 °F) at its top. The reason is that ozone and oxygen absorb and scatter the sun's ultraviolet (UV) radiation (p. 31), so that much of the UV radiation penetrating the upper stratosphere fails to reach the lower stratosphere. Most of the atmosphere's minute amount of ozone concentrates in a portion of the stratosphere roughly 17–30 km (10–19 mi) above sea level, a region we have come to call Earth's **ozone layer**. The ozone layer greatly reduces the amount of UV radiation that reaches Earth's surface. Because UV light can damage living tissue and induce mutations in DNA, the ozone layer's protective effects are vital for life on Earth.

Above the stratosphere lies the *mesosphere*, which extends 50–80 km (31–56 mi) above sea level. Air pressure is extremely low here, and temperatures decrease with altitude. The *thermosphere*, our atmosphere's top layer, extends upward to an altitude of 500 km (300 mi).

Atmospheric properties include temperature, pressure, and humidity

Air moves dynamically within the lower atmosphere as a result of differences in the physical properties of air masses. Among these properties are pressure and density, relative humidity, and temperature.

Gravity pulls gas molecules toward Earth's surface, causing air to be most dense near the surface and less so as altitude increases. **Atmospheric pressure**, which measures the force per unit area produced by a column of air, also decreases with altitude, because at higher altitudes fewer molecules are pulled down by gravity (**FIGURE 17.3**). At sea level, atmospheric pressure averages 14.7 lb/in.² or 1,013 millibars (mb). Mountain climbers trekking to Mount Everest, the world's highest mountain, can look up and view their destination from Kala Patthar, a nearby peak, at roughly 5.5 km (18,000 ft). At this altitude, pressure is 500 mb, and half the atmosphere's air molecules are above the climber whereas half are below. A climber who reaches Everest's peak (8.85 km [29,035 ft]), where the "thin air" is just over 300 mb, stands above two-thirds of the molecules in the atmosphere. When we fly on a commercial jet airliner at a typical cruising altitude of 11 km (36,000 ft), we are above roughly 80% of the atmosphere's molecules.

Another property of air is **relative humidity**, the ratio of water vapor a given volume of air contains to the maximum amount it *could* contain at a given temperature. Average daytime relative humidity in June in the desert at Phoenix, Arizona, is only 31% (meaning that the air contains less than a third of the water vapor it possibly can at its temperature),

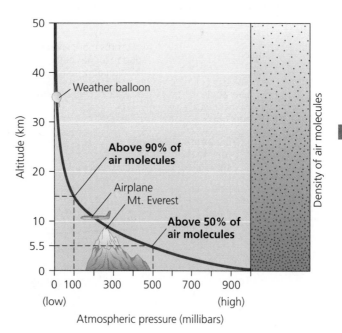

FIGURE 17.3 ▲ As one climbs higher through the atmosphere, gas molecules become less densely packed. As density decreases, so does atmospheric pressure. Because most air molecules lie low in the atmosphere, one needs to be only 5.5 km (3.4 mi) high to be above half the planet's air molecules. Adapted from Ahrens, C.D., 2007. *Meteorology today,* 8th ed. Fig 1.9. © 2007. Belmont, CA: Brooks/Cole. By permission of Cengage Learning.

whereas on the tropical island of Guam, relative humidity rarely drops below 88%. People are sensitive to changes in relative humidity because we perspire to cool our bodies. When humidity is high, the air is already holding nearly as much water vapor as it can, so sweat evaporates slowly and the body cannot cool itself efficiently. This is why high humidity makes it feel hotter than it really is. Low humidity speeds evaporation and makes it feel cooler.

The temperature of air also varies with location and time. At the global scale, temperature varies over Earth's surface because the sun's rays strike some areas more directly than others. At more local scales, temperature varies because of topography, plant cover, proximity of water to land, and many other factors. Sometimes this local variation is striking; a hillside sheltered from wind or sunlight may have a very different *microclimate*, or localized pattern of weather conditions, than the other side of the hill.

Solar energy heats the atmosphere, helps create seasons, and causes air to circulate

Energy from the sun heats air in the atmosphere, drives air movement, helps create seasons, and influences weather and climate. An enormous amount of solar energy continuously bombards the upper atmosphere—over 1,000 watts/m², many thousands of times more than the total output of electricity generated by human society. Of that solar energy, about 70% is absorbed by the atmosphere and planetary surface, while the rest is reflected back into space (see Figure 18.1, p. 496).

The spatial relationship between Earth and the sun determines how much solar radiation strikes each point on Earth's surface. Sunlight is most intense when it shines directly overhead and meets the planet's surface at a perpendicular angle. At this angle, sunlight passes through a minimum of energy-absorbing atmosphere, and Earth's surface receives a maximum of solar energy per unit surface area. Conversely, solar energy that approaches Earth's surface at an oblique angle loses intensity as it traverses a longer distance through the atmosphere, and it is less intense when it reaches the surface. This is why, on average, solar radiation intensity is highest near the equator and weakest near the poles (**FIGURE 17.4**).

Because Earth is tilted on its axis (an imaginary line connecting the poles, running perpendicular to the equator) by about 23.5 degrees, the Northern and Southern Hemispheres each tilt toward the sun for half the year, resulting in the seasons (**FIGURE 17.5**). Regions near the equator experience about 12 hours each of sunlight and darkness per day throughout the year, but near the poles, day length varies greatly and seasonality is pronounced.

Land and surface water absorb solar energy and then radiate heat, causing some water to evaporate. Air near Earth's surface therefore tends to be warmer and moister than air at higher altitudes. These differences set into motion a process of **convective circulation** (**FIGURE 17.6**). Warm air, being less dense, rises and creates vertical currents. As air rises into regions of lower atmospheric pressure, it expands and cools. Once the air cools, it descends and becomes denser, replacing warm air that is rising. The air picks up heat and moisture near ground level and prepares to rise again, continuing the process. Similar convective circulation patterns occur in ocean waters (pp. 437–438), in magma beneath Earth's surface (p. 36), and even in a simmering pot of soup. Convective circulation influences both weather and climate.

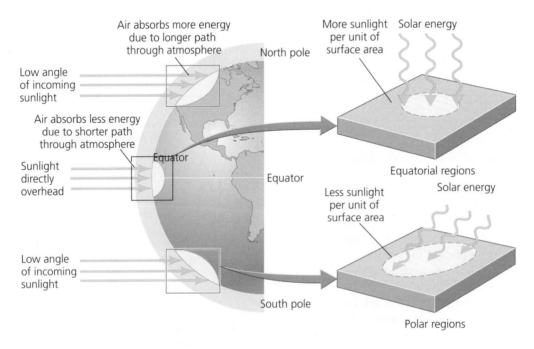

FIGURE 17.4 ▲ Because of Earth's curvature, polar regions receive on average less solar energy than equatorial regions. One reason is that sunlight gets spread over a larger area when striking the surface at an angle. Another reason is that sunlight approaching at a lower angle near the poles must traverse a longer distance through the atmosphere, during which more energy is absorbed or reflected. These patterns represent year-round averages; the latitude at which radiation approaches the surface perpendicularly varies with the seasons (see Figure 17.5).

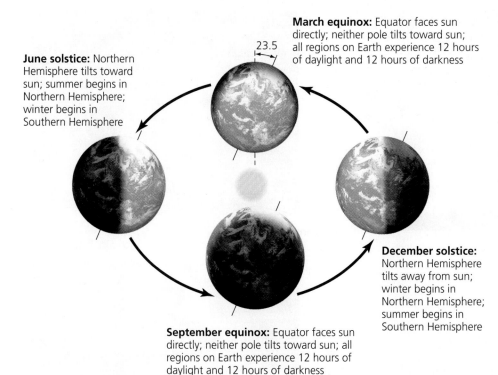

June solstice: Northern Hemisphere tilts toward sun; summer begins in Northern Hemisphere; winter begins in Southern Hemisphere

23.5

March equinox: Equator faces sun directly; neither pole tilts toward sun; all regions on Earth experience 12 hours of daylight and 12 hours of darkness

December solstice: Northern Hemisphere tilts away from sun; winter begins in Northern Hemisphere; summer begins in Southern Hemisphere

September equinox: Equator faces sun directly; neither pole tilts toward sun; all regions on Earth experience 12 hours of daylight and 12 hours of darkness

FIGURE 17.5 ◄ The seasons occur because Earth is tilted on its axis by 23.5 degrees. As Earth revolves around the sun, the Northern Hemisphere tilts toward the sun for one half of the year, and the Southern Hemisphere tilts toward the sun for the other half of the year. In each hemisphere, summer occurs during the period in which the hemisphere receives the most solar energy because of its tilt toward the sun.

The atmosphere drives weather and climate

Weather and climate each involve the physical properties of the troposphere, such as temperature, pressure, humidity, cloudiness, and wind. **Weather** specifies atmospheric conditions over short time periods, typically hours or days, and

Heat radiates to space

Cool, dry air

Condensation and precipitation

Air sinks, compresses, and warms

Air rises, expands, and cools

Warm, dry air

Hot, moist air

Air picks up moisture and heat (moist surface warmed by sun)

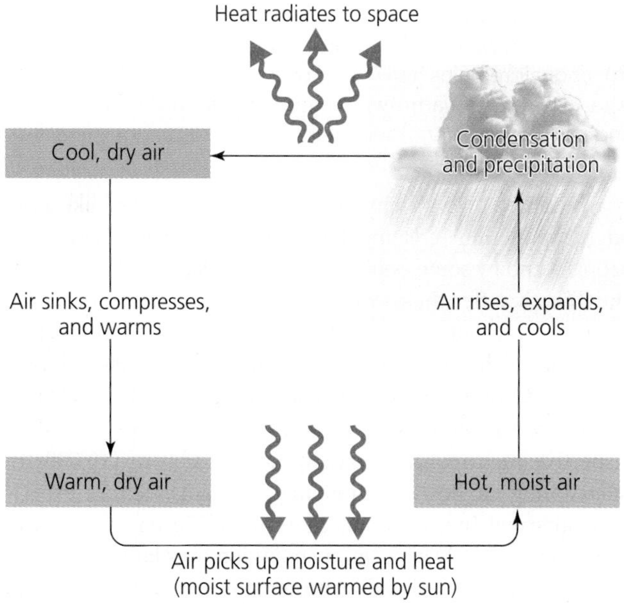

FIGURE 17.6 ▲ Weather is driven in part by the convective circulation of air in the atmosphere. Air being heated near Earth's surface picks up moisture and rises. Once aloft, this air cools, and moisture condenses, forming clouds and precipitation. Cool, drying air begins to descend, compressing and warming in the process. Warm, dry air near the surface begins the cycle anew.

within relatively small geographic areas. **Climate**, in contrast, describes the pattern of atmospheric conditions found across large geographic regions over long periods of time, typically seasons, years, or millennia. Mark Twain once noted the distinction between climate and weather by saying, "Climate is what we expect; weather is what we get." For example, Los Angeles has a "Mediterranean" climate characterized by reliably warm, dry summers and mild, rainy winters, but on occasional days in the fall, dry Santa Ana winds can blow in from the desert and bring extremely hot weather.

Air masses interact to produce weather

Weather can change quickly when air masses with different physical properties meet. The boundary between air masses that differ in temperature and moisture (and therefore density) is called a **front**. The boundary along which a mass of warmer, moister air replaces a mass of colder, drier air is termed a **warm front** (**FIGURE 17.7A**). Some of the warm, moist air along the leading edge of a warm front usually rises over the cooler air mass blocking its progress. As it rises, the warm air cools and the water vapor within condenses, forming clouds and light rain. A **cold front** (**FIGURE 17.7B**) is the boundary along which a colder, drier air mass displaces a warmer, moister air mass. The colder air, being denser, tends to wedge beneath the warmer air. The warmer air rises, expands, then cools to form clouds and thunderstorms. Once a cold front passes through, the sky usually clears, and temperature and humidity drop.

Adjacent air masses may also differ in atmospheric pressure. A **high-pressure system** contains air that descends because it is cool and then spreads outward as it nears the ground. High-pressure systems typically bring fair weather. In a **low-pressure system**, warmer air rises, drawing air inward toward the center of low atmospheric pressure. The rising air expands and cools, and clouds and precipitation often result.

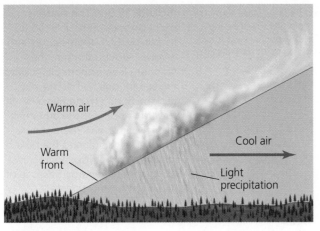

(a) Warm front

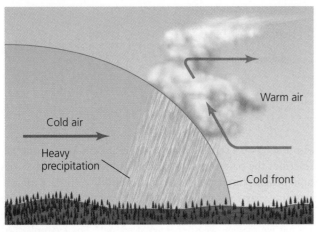

(b) Cold front

FIGURE 17.7 ▲ When a warm front approaches **(a)**, warmer air rises over cooler air, causing light or moderate precipitation as moisture in the warmer air condenses. When a cold front approaches **(b)**, colder air pushes beneath warmer air, and the warmer air rises, resulting in condensation and heavy precipitation.

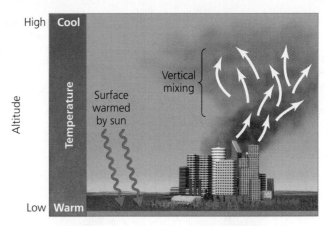

(a) Normal conditions

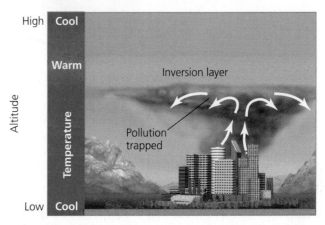

(b) Thermal inversion

FIGURE 17.8 ▲ A thermal inversion is a natural atmospheric occurrence that can worsen air pollution locally. Under normal conditions **(a)**, tropospheric temperature decreases with altitude, and air of different altitudes mixes, dispersing pollutants upward and outward. During a thermal inversion **(b)**, cool air remains near the ground beneath an "inversion layer" of air that warms with altitude. Little mixing occurs, and pollutants are trapped near the surface.

Under most conditions, air in the troposphere decreases in temperature as altitude increases. Because warm air rises, vertical mixing results. Occasionally, however, a layer of cool air occurs beneath a layer of warmer air. This departure from the normal temperature profile is known as a **temperature inversion**, or **thermal inversion** (**FIGURE 17.8**). The band of air in which temperature rises with altitude is called an **inversion layer** (because the normal direction of temperature change is inverted). The cooler air at the bottom of the inversion layer is denser than the warmer air at the top of the inversion layer, so it resists vertical mixing and remains stable. Thermal inversions can occur in different ways, sometimes involving cool air at ground level and sometimes producing an inversion layer higher above the ground (as shown in Figure 17.8b). One common type of inversion occurs in mountain valleys where slopes block morning sunlight, keeping ground-level air within the valley shaded and cool.

Whereas vertical mixing normally allows air pollution to be carried upward and diluted, thermal inversions trap pollutants near the ground. A thermal inversion sparked a "killer smog" crisis in London, England, in 1952. A high-pressure

system settled over the city for several days and acted like a cap on the pollution, creating foul conditions that killed 4,000 people—and by some estimates up to 12,000. Both Los Angeles and Tehran suffer their worst pollution on days when thermal inversions prevent pollutants from being dispersed. Both cities are encircled by mountains, which promote inversion layers. The San Gabriel Mountains north of L.A. and the Alborz Mountains surrounding Tehran on three sides tower over their respective metropolitan areas by up to 3,000 m (10,000 ft), interrupting air flow and trapping pollutants. Tehran experiences thermal inversions on more than 250 days each year. Inversions regularly concentrate pollution over large metropolitan areas in valleys ringed by mountains, from Mexico City to Seoul, Korea, to Rio de Janeiro and São Paulo, Brazil.

Large-scale circulation systems produce global climate patterns

At larger geographic scales, convective air currents contribute to broad climatic patterns (**FIGURE 17.9A**). Near the equator, solar radiation sets in motion a pair of convective cells known

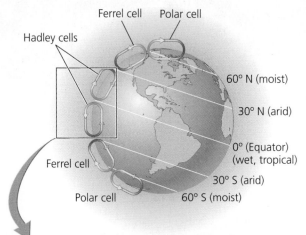

FIGURE 17.9 ◄ A series of large-scale convective cells **(a)** helps determine global patterns of humidity and aridity. Warm air near the equator rises, expands, and cools, and moisture condenses, giving rise to a wet climate in tropical regions. Air travels toward the poles and descends around 30 degrees latitude. This air, which loses its moisture in the tropics, causes regions around 30 degrees latitude to be arid. This convective circulation, a Hadley cell, occurs on both sides of the equator. Between roughly 30 and 60 degrees latitude north and south, Ferrel cells occur; and between 60 and 90 degrees latitude, polar cells occur. As a result, air rises at around 60 degrees latitude, creating a moist climate, and falls at around 90 degrees, creating a dry climate. Global wind currents **(b)** show latitudinal patterns as well. Trade winds between the equator and 30 degrees latitude blow westward, whereas westerlies between 30 and 60 degrees latitude blow eastward.

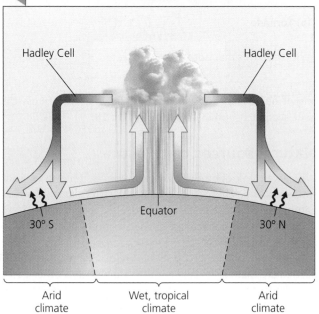

(a) Convection currents

(b) Global wind patterns

467

as **Hadley cells**. Here, where sunlight is most intense, surface air warms, rises, and expands. As it does so, it releases moisture, producing the heavy rainfall that gives rise to tropical rainforests near the equator. After releasing much of its moisture, this air diverges and moves in currents heading northward and southward. The air in these currents cools and descends back to Earth at about 30 degrees latitude north and south. Because the descending air has low relative humidity, the regions around 30 degrees latitude are quite arid, giving rise to deserts. Two pairs of similar but less intense convective cells, called **Ferrel cells** and **polar cells**, lift air and create precipitation around 60 degrees latitude north and south and cause air to descend at around 30 degrees latitude and in the polar regions.

These three pairs of cells account for the latitudinal distribution of moisture across Earth's surface: wet climates near the equator; arid climates near 30 degrees latitude; moist regions near 60 degrees latitude; and dry conditions near the poles. These patterns, combined with temperature variation, help explain why biomes tend to be arrayed in latitudinal bands (Figure 4.18, p. 96).

The Hadley, Ferrel, and polar cells interact with Earth's rotation to produce the global wind patterns shown in **FIGURE 17.9B**. As Earth rotates on its axis, locations on the equator spin faster than locations near the poles. As a result, the north–south air currents of the convective cells appear to be deflected from a straight path as some portions of the globe move beneath them more quickly than others. This apparent deflection, called the **Coriolis effect**, results in the curving global wind patterns evident in Figure 17.9b. Near the equator lies a region with few winds known as the *doldrums*. Between the equator and 30 degrees latitude, the *trade winds* blow from east to west. From 30 to 60 degrees latitude, the *westerlies* blow from west to east. People used these global circulation patterns for centuries to facilitate ocean travel by wind-powered sailing ships.

The atmosphere interacts with the oceans to affect weather, climate, and the distribution of biomes. Winds and convective circulation in ocean water together maintain ocean currents (pp. 436–437). Trade winds weaken periodically, leading to El Niño conditions (pp. 437–438). And oceans and atmosphere sometimes interact to create violent storms.

Storms pose hazards

Atmospheric conditions can sometimes create storms that threaten life and property. **Hurricanes** (**FIGURE 17.10A**) form when winds rush into areas of low pressure where warm,

| (a) Hurricane | (b) Tornado |

FIGURE 17.10 ▲ Hurricanes **(a)** and tornadoes **(b)** are two types of cyclonic storms that pose hazards to our life and property.

moisture-laden air over tropical oceans is rising. In the Northern Hemisphere, the winds turn counterclockwise because of the Coriolis effect; in other regions of Earth, such cyclonic storms are called *cyclones* or *typhoons*. The powerful convective currents of these storms draw up immense amounts of water vapor, which condenses and falls heavily as rain. In North America, the Gulf Coast and Atlantic Coast are most susceptible to hurricanes.

Tornadoes (**FIGURE 17.10B**) form when a mass of warm air meets a mass of cold air and the warm air rises quickly, setting a powerful convective current in motion. The spinning funnel of rising air picks up soil and objects in its path with winds up to 500 km per hour (310 mph). In North America, tornadoes are most apt to form in the Great Plains and the Southeast, where cold air from Canada and warm air from the Gulf of Mexico frequently meet.

Understanding how the atmosphere functions can help us predict violent storms and warn people of their approach. Such knowledge can also help us comprehend how our pollution of the atmosphere affects climate, ecological systems, economies, and human health.

OUTDOOR AIR POLLUTION

Throughout human history, we have made the atmosphere a dumping ground for our airborne wastes. Whether from primitive wood fires or modern coal-burning power plants, people have generated **air pollutants**, gases and particulate material added to the atmosphere that can affect climate or harm people or other organisms. **Air pollution** refers to the release of air pollutants. In recent decades, government policy and improved technologies have helped us reduce most types of **outdoor air pollution** (often called **ambient air pollution**) in industrialized nations. However, outdoor air pollution remains a problem, particularly in developing nations and in urban areas.

The greatest air pollution problem today may well be our emission of greenhouse gases that contribute to global climate change. Addressing our release of excess carbon dioxide, methane, and other gases that warm the atmosphere stands as

one of our civilization's primary challenges. We discuss this issue separately and in depth in Chapter 18.

Natural sources can pollute

When we think of outdoor air pollution, we tend to envision smokestacks belching smoke from industrial plants. However, natural processes produce a great deal of air pollution. Some of these natural impacts are made worse by human activity and land use policies.

Volcanic eruptions (pp. 42–43) release large quantities of particulate matter, as well as sulfur dioxide and other gases, into the troposphere (**FIGURE 17.11A**). Major eruptions may blow matter into the stratosphere, where it can circle the globe for months or years. Sulfur dioxide reacts with water and oxygen and then condenses into fine droplets, called **aerosols**, which reflect sunlight back into space and thereby cool the atmosphere and surface. The 1991 eruption of Mount Pinatubo in the Philippines ejected nearly 20 million tons of ash and aerosols and cooled global temperatures by 0.5 °C (0.9 °F).

Fires from burning vegetation also pollute the atmosphere with soot and gases. Over 60 million ha (150 million acres) of forest and grassland burn in a typical year (**FIGURE 17.11B**). Fires occur naturally, but human influence can make them more severe. In North America, fuel buildup from decades of fire suppression has promoted destructive forest fires in recent years (p. 328). In regions like the Los Angeles basin, where residential development has encroached deeply into chaparral ecosystems (p. 102) that are naturally fire-prone, fires may cause extensive and costly damage. In the tropics, many farmers set fires to clear forest for farming and grazing using a "slash-and-burn" approach (p. 231). In 1997, a severe drought brought on by the 20th century's strongest El Niño event caused unprecedented forest fires in Indonesia, Mexico, Central America, and Africa. Immense swaths of rainforest burned while smoke pollution sickened 20 million Indonesians and caused a plane to crash and ships to collide. Scientists predict that global climate change (Chapter 18) may increase fires due to drought in many regions.

(a) Mount Saint Helens eruption, 1980

(b) Natural fire in California

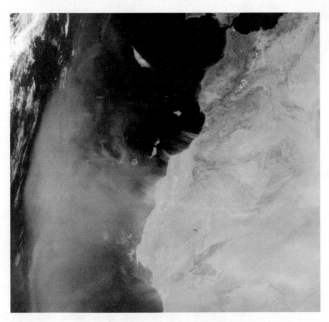

(c) Dust storm blowing dust from Africa to the Americas

FIGURE 17.11 ▲ Volcanoes are one source of natural air pollution, as shown by Mount Saint Helens **(a)**, which erupted in the state of Washington in 1980. Fires **(b)** in forests and grasslands are another source of soot and gaseous pollutants. Dust storms are a third source of natural air pollution. Trade winds blowing soil across the Atlantic Ocean from Africa to the Americas **(c)** carry fungal and bacterial spores linked to die-offs in Caribbean coral reef systems, although they also bring nutrients to the Amazon basin. Strong westerlies sometimes lift soil from deserts in Mongolia and China and blow it across the Pacific Ocean to North America.

Winds sweeping over arid terrain can send huge amounts of dust aloft—sometimes across oceans from one continent to another (**FIGURE 17.11C**). In July 2009, the bustling city of Tehran came to a standstill when windstorms blew sand and dust from drought-stricken Iraq into Iran, enveloping half of the country. Businesses, schools, and government offices in Tehran were closed for several days, airplane flights were cancelled, and people were warned to stay indoors to safeguard their health. Dust storms occur naturally, but they are made far worse by unsustainable farming and grazing practices that strip vegetation from the soil, promote wind erosion, and lead to desertification (p. 233). Continental-scale dust storms took place in the United States in the 1930s, when soil from the drought-plagued Dust Bowl states blew eastward to the Atlantic (p. 235).

We create outdoor air pollution

Since the onset of industrialization, human activity has introduced a variety of sources of air pollution. As with water pollution, anthropogenic (human-caused) air pollution can emanate from mobile or stationary sources, and from *point sources* or *non-point sources* (pp. 420–421). A point source describes a specific location from which large quantities of pollutants are discharged. A coal-fired power plant is an example of a point source (see **ENVISIONIT**, p. 470). Non-point sources are more diffuse, often consisting of many small sources spread across a wide area. In major cities from Los Angeles to Tehran, power plants and factories act as stationary point sources, whereas millions of automobiles on the roadways comprise a mobile non-point source.

Burning coal in power plants gives us the electricity that helps us do so much in our daily lives.

Electricity is everywhere in our lives

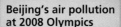

Beijing's air pollution at 2008 Olympics

But coal combustion releases soot that dirties urban air ...

... sulfur dioxide that causes acid rain ...

Statue eroded by acid rain

... and carbon dioxide that drives global climate change.

Home with PV solar, wind turbine, and solar water heater

We can reduce pollution by cleansing power plant exhaust with scrubbers. Engineers are searching for ways to sequester CO_2 below ground. Switching to renewable energy sources will clean our air most effectively.

YOU CAN MAKE A DIFFERENCE

➤ Turn off and power down: Conserve electricity.

➤ Use energy-efficient light bulbs and appliances.

➤ Push your campus to invest in clean renewable energy from wind, solar, or geothermal power.

Once in the atmosphere, pollutants may cause harm directly, or they may induce chemical reactions that produce harmful compounds. **Primary pollutants**, such as soot and carbon monoxide, are pollutants emitted into the troposphere in a form that can cause harm or can react to form harmful substances. Harmful substances produced when primary pollutants interact or react with constituents of the atmosphere are called **secondary pollutants**. Secondary pollutants include tropospheric ozone, sulfuric acid, and other examples we explore below.

Pollutants differ in the amount of time they spend in the atmosphere—called their **residence time**—because substances differ in how readily they react in air and in how quickly they settle to the ground. Pollutants with brief residence times exert localized impacts over short time periods. Most particulate matter and most pollutants from automobile exhaust stay aloft only hours or days, which is why air quality in a city like Tehran or Los Angeles can change substantially from one day to the next. In contrast, pollutants with long residence times can exert impacts regionally or globally for long periods, even centuries. The pollutants that drive global climate change and those that deplete Earth's ozone layer (two separate phenomena) are able to cause these global and long-lasting impacts because they persist in the atmosphere for so long. **FIGURE 17.12** shows this relationship with some examples. As we discuss many of these pollutants below, you may want to refer to this figure to help understand how residence time allows substances to cause the impacts they do.

Clean Air Act legislation addresses pollution in the United States

To address air pollution in the United States, Congress has passed a series of laws, beginning with the Air Pollution Control Act of 1955. The Clean Air Act of 1963 funded research into pollution control and encouraged emissions standards for automobiles and stationary point sources, such as industrial plants. Subsequent amendments expanded the legislation's scope and established a nationwide air quality monitoring system.

In 1970, Congress thoroughly revised the law in what came to be known as the **Clean Air Act of 1970**. This legislation set stricter standards for air quality, imposed limits on emissions from new stationary and mobile sources, provided new funds for pollution-control research, and enabled citizens to sue parties violating the standards. Some of these goals came to be viewed as too ambitious, so amendments in 1977 altered some standards and extended some deadlines for compliance.

The **Clean Air Act of 1990** sought to strengthen regulations pertaining to air quality standards, auto emissions, toxic air pollution, acid deposition, and stratospheric ozone depletion. It also introduced an emissions trading program for sulfur dioxide (pp. 183–184). Beginning in 1995, businesses and utilities were allocated permits for emitting this pollutant and could then buy, sell, or trade these allowances. Each year the overall amount of allowed pollution was decreased. This market-based incentive program has helped reduce sulfur dioxide emissions nationally (see Figure 7.16, p. 184). It has also spawned similar cap-and-trade programs for other pollutants, including greenhouse gases (p. 523). The Los Angeles region adopted its own cap-and-trade program in 1994. The RECLAIM (Regional Clean Air Incentives Market) program helped the L.A. basin decrease sulfur dioxide emissions by 47% and nitrogen oxide emissions by 61% by 2003, and further cuts are being achieved as the program continues.

As a result of Clean Air Act legislation, the U.S. Environmental Protection Agency (EPA) sets nationwide standards for emissions of pollutants and for concentrations of pollutants in ambient air throughout the nation. It is largely up to the states to monitor air quality and develop, implement, and enforce regulations within their borders. States submit implementation plans to the EPA for approval, and if a state's plans are not adequate, the EPA can take over enforcement in that state.

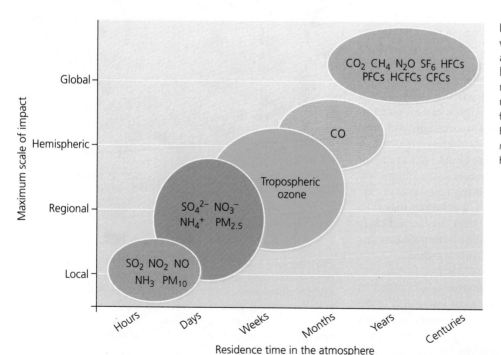

FIGURE 17.12 ◀ Substances with short residence times in the atmosphere affect air quality only locally, whereas those with long residence times affect air quality regionally or globally. Data from United Nations Environment Programme, 2007. *Global environmental outlook (GEO-4)*, Nairobi, Kenya.

The EPA sets standards for "criteria pollutants"

The EPA and the states focus on six **criteria pollutants**, pollutants judged to pose especially great threats to human health—carbon monoxide (CO), sulfur dioxide (SO_2), nitrogen dioxide (NO_2), tropospheric ozone (O_3), particulate matter, and lead (Pb). For these, the EPA has established *national ambient air quality standards (NAAQS)*, which are maximum allowable concentrations of these pollutants in ambient outdoor air.

Carbon monoxide   **Carbon monoxide** is a colorless, odorless gas produced primarily by the incomplete combustion of fuel. Vehicles and engines account for 78% of CO emissions in the United States; other sources include industrial processes, combustion of waste, and residential wood burning. Carbon monoxide poses risk to humans and other animals because it can bind irreversibly to hemoglobin in red blood cells, preventing the hemoglobin from binding with oxygen.

Sulfur dioxide   Like CO, **sulfur dioxide** is a colorless gas, but unlike CO it has a pungent odor. The vast majority of SO_2 pollution results from the combustion of coal for electricity generation and industry. During combustion, elemental sulfur (S) in coal reacts with oxygen (O_2) to form SO_2. Once in the atmosphere, SO_2 may react to form sulfur trioxide (SO_3) and sulfuric acid (H_2SO_4), which may then settle back to Earth in the form of acid deposition (pp. 482–486).

Nitrogen dioxide   **Nitrogen dioxide** is a highly reactive, foul-smelling reddish brown gas that contributes to smog and acid deposition. Along with nitric oxide (NO), NO_2 belongs to a family of compounds called **nitrogen oxides** (NO_x). Nitrogen oxides result when atmospheric nitrogen and oxygen react at the high temperatures created by combustion engines. Most U.S. NO_x emissions result from combustion in vehicle engines. Electrical utility and industrial combustion account for most of the rest.

Tropospheric ozone   Although ozone in the stratosphere shields us from the dangers of UV radiation, ozone from human activity forms and accumulates low in the troposphere and acts as a pollutant. In the troposphere, this colorless gas with an objectionable odor results from the interaction of sunlight, heat, nitrogen oxides, and volatile carbon-containing chemicals. **Tropospheric ozone** (also called **ground-level ozone**) is therefore categorized as a secondary pollutant. A major component of smog, O_3 can pose health risks as a result of its instability as a molecule; this triplet of oxygen atoms will readily release one of its threesome, leaving a molecule of oxygen gas and a free oxygen atom. The oxygen atom may then participate in reactions that can injure living tissues and cause respiratory problems. Tropospheric ozone is the pollutant that most frequently exceeds its EPA air quality standard.

Particulate matter   **Particulate matter** is composed of solid or liquid particles small enough to be suspended in the atmosphere. Particulate matter includes primary pollutants such as dust and soot, as well as secondary pollutants such as sulfates and nitrates. Particulate matter can damage respiratory tissues when inhaled. The EPA classifies particulate pollution by the size of the particles. PM_{10} pollution consists of particles less than 10 microns in diameter (one-seventh the width of a human hair), whereas $PM_{2.5}$ pollution consists of still-finer particles less than 2.5 microns in diameter. Most PM_{10} pollution is dust, whereas most $PM_{2.5}$ pollution results from combustion processes.

Lead   **Lead** is a heavy metal that enters the atmosphere as a particulate pollutant. The lead-containing compounds tetraethyl lead and tetramethyl lead, when added to gasoline, improve engine performance. However, exhaust from the combustion of leaded gasoline emits airborne lead, which can be inhaled or can be deposited on land and water. Lead can enter the food chain, accumulate within body tissues, and cause central nervous system malfunction and many other ailments (p. 373). Since the 1980s, leaded gasoline has been phased out in most industrialized nations (p. 8), and lead pollution has plummeted. The United States led the way, Iran began phasing out leaded gasoline a decade ago, and today more developing nations are following suit. Although auto exhaust still creates significant lead pollution in many developing nations, the main source of atmospheric lead pollution in developed nations today is industrial metal smelting.

EPA monitoring finds that large numbers of Americans live in areas where concentrations of criteria pollutants regularly reach unhealthy levels. For instance, residents of Los Angeles County breathe air that violates safety standards for four of the six criteria pollutants. So do people in four adjacent southern California counties, according to data from 2010 (**FIGURE 17.13**). All together, as of 2008, 127 million Americans lived in counties that violated the national ambient air quality standards for at least one of the six criteria pollutants.

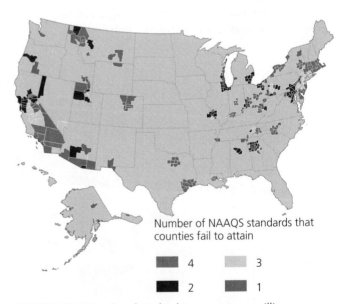

Number of NAAQS standards that counties fail to attain

- 4
- 3
- 2
- 1

FIGURE 17.13 ▲ One hundred twenty-seven million Americans live in counties that in 2008 failed to meet the EPA's national ambient air quality standards (NAAQS) for at least one criteria pollutant. This map shows counties, as of 2010, that have persistently failed to attain the standards for one (green) through four (red) of the six criteria pollutants. Data from U.S. EPA.

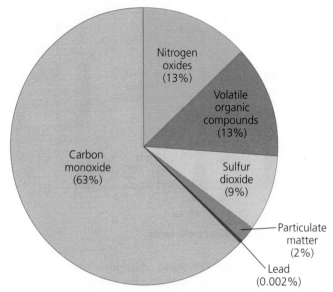

WEIGHING THE ISSUES

Your Region's Air Quality Locate where you live on the map in Figure 17.13. What is the status of your county's air quality, and how does your county compare to the rest of the nation? Now explore one of the EPA websites that lets you browse information on the air you breathe: www.airnow.gov, www.epa.gov/aircompare, or www.epa.gov/air/emissions/where.htm. What factors do you think influence the quality of your region's air? Can you propose three steps for reducing air pollution in your region?

FIGURE 17.14 ▲ In 2008, the United States emitted 123 million tons of the six major pollutants whose emissions are monitored by the EPA and state and local agencies. Carbon monoxide accounted for most of these emissions, by mass. In this graph, particulate matter does not include dust and fires, because distinguishing human from natural emissions of these is difficult. Data from U.S. EPA.

Agencies monitor emissions

Besides measuring concentrations of the six criteria pollutants in ambient air, state and local agencies also monitor, calculate, and report to the EPA emissions of certain major pollutants that affect ambient concentrations of the criteria pollutants. These include the four criteria pollutants that are primary pollutants (carbon monoxide, sulfur dioxide, particulate matter, and lead), as well as all nitrogen oxides (because NO reacts readily in the atmosphere to form NO_2, which is both a primary and secondary pollutant). Tropospheric ozone is a secondary pollutant only; we do not emit it. Instead, agencies monitor emissions of volatile organic compounds, which can react to produce ozone and other secondary pollutants.

Volatile organic compounds (VOCs) are carbon-containing chemicals used in and emitted by vehicle engines and a wide variety of solvents and industrial processes, as well as by many household chemicals and consumer items. One group of VOCs consists of hydrocarbons (p. 28) such as methane (CH_4, the primary component of natural gas), propane (C_3H_8, used as a portable fuel), butane (C_4H_{10}, found in cigarette lighters), and octane (C_8H_{18}, a component of gasoline). Human activities account for about half the VOC emissions in the United States. The remainder comes from natural sources; for example, plants produce isoprene (C_5H_8) and terpenes such as $C_{10}H_{15}$, compounds that produce a bluish haze that has given the Blue Ridge Mountains their name.

In the United States in 2008, human activity polluted the air with 123 million tons of the six monitored pollutants. Carbon monoxide was the most abundant pollutant by mass, followed by NO_X, VOCs, and SO_2 (**FIGURE 17.14**). Each of us contributes to air pollution when we drive a car. Each year that we drive the average amount for a U.S. driver (12,000 miles per year), we release close to 6 metric tons of carbon dioxide, 275 kg (605 lbs) of methane, and 19 kg (41 lb) of nitrous oxide, all of them greenhouse gases (p. 495) that drive climate change.

We have reduced U.S. air pollution

Since the Clean Air Act of 1970, we have reduced emissions of each of the six monitored pollutants (**FIGURE 17.15A**), and total emissions of the six together have declined by 60%. These dramatic reductions in emissions have occurred despite substantial increases in the nation's population, energy consumption, miles traveled by vehicle, and gross domestic product (**FIGURE 17.15B**).

We have achieved these emissions reductions as a result of policy steps and technological developments, both ultimately motivated by grassroots social demand for cleaner air. Cleaner-burning motor vehicle engines and automotive technologies such as catalytic converters (p. 682) have played a large part, decreasing the emissions of carbon monoxide and several other pollutants. The sulfur dioxide permit trading program (pp. 183–184) and clean coal technologies (p. 546) have reduced SO_2 emissions. Technologies such as baghouse filters, electrostatic precipitators, and **scrubbers** (**FIGURE 17.16**) that chemically convert or physically remove airborne pollutants before they are emitted from smokestacks have allowed factories, power plants, and refineries to reduce emissions of several pollutants. And the leaded gasoline phaseout caused U.S. lead emissions to plummet by 93% in the 1980s alone.

Overall, the reduction of outdoor air pollution since 1970 represents one of the greatest accomplishments of the United States in safeguarding human health and environmental quality. Plenty of room for improvement remains, however. Some pollutants are not declining, some new ones are emerging, and our greenhouse gas emissions continue to rise. Indeed, U.S. carbon dioxide emissions rose 44% from 1970 to 2008. The U.S. Supreme Court in 2007 ruled that the EPA has the legal authority to regulate carbon dioxide as a pollutant. It has not yet chosen to do so, because of the logistical challenge and formidable political opposition. Yet if we have been able to reduce emissions of other major pollutants by 60% since 1970 while improving our economy, we can hope that the same or better might soon be achieved with greenhouse gas emissions.

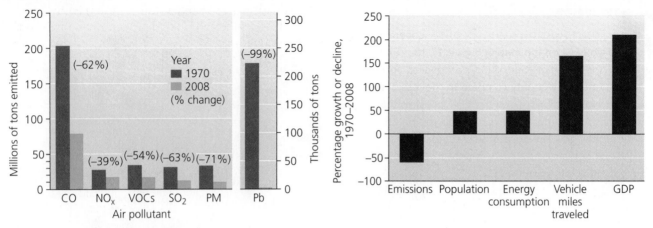

(a) Declines in six major pollutants

(b) Trends in major indicators

FIGURE 17.15 ▲ The EPA tracks emissions of several major pollutants into U.S. air. Each of these pollutants has shown substantial declines since 1970 **(a)**, and emissions from all six together have declined by 60%. We have achieved these reductions in emissions despite increases **(b)** in U.S. population, energy consumption, vehicle miles traveled, and gross domestic product. Data for particulate matter is for PM_{10} since 1985. Data from U.S. EPA.

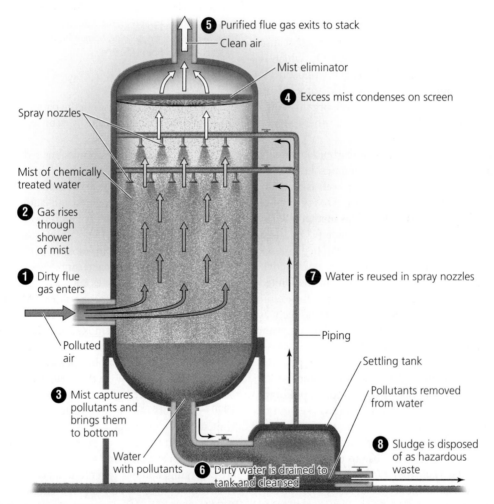

FIGURE 17.16 ▲ In this spray-tower wet scrubber, polluted air ❶ rises through a chamber while arrays of nozzles spray a mist of water mixed with lime or other active chemicals ❷. The falling mist captures pollutants and carries them to the bottom of the chamber ❸, essentially washing them out of the air. Excess mist is captured on a screen ❹, and air emitted from the scrubber has largely been cleansed ❺. Periodically, the dirty water is drained from the chamber ❻, cleansed in a settling tank, and recirculated ❼ through the spray nozzles. The resulting sludge must be disposed of ❽ as hazardous waste (pp. 638–639). Scrubbers and other pollution-control devices come in many designs; the type shown here typically removes at least 90% of particulate matter and gases such as sulfur dioxide.

Toxic pollutants pose health risks

We are also reducing emissions of **toxic air pollutants**, substances known to cause cancer, reproductive defects, or neurological, developmental, immune system, or respiratory problems in people and other organisms. Under the 1990 Clean Air Act, the EPA regulates 188 different toxic air pollutants produced by a variety of human activities, including metal smelting, sewage treatment, and industrial processes. These pollutants range from the heavy metal mercury (from coal-burning power plant emissions and other sources) to VOCs such as benzene (a component of gasoline) and methylene chloride (found in paint stripper). Among the 188 pollutants are 21 from mobile sources (such as diesel exhaust) and 33 "urban hazardous" pollutants judged to pose the greatest health risks in urban areas.

State and federal agencies operate 300 monitoring sites across the United States for toxic air pollutants. Periodic reports assess health impacts as well as emissions, and the latest national assessment (published in 2009 using 2002 data) estimated that toxic air pollutants cause cancer in 1 out of every 28,000 Americans (36 cancer cases per 1 million people). This national risk assessment mapped how risks vary geographically for cancer (**FIGURE 17.17A**) and non-cancerous respiratory ailments (**FIGURE 17.17B**). Although some areas such as the Los Angeles region still experience high health risks, the EPA estimates that Clean Air Act regulations have helped to reduce emissions of toxic air pollutants since 1990 by more than 35%.

Industrializing nations are suffering increasing air pollution

Although the United States and other industrialized nations have improved their air quality, outdoor air pollution is growing worse in many industrializing countries. In these societies, rapidly proliferating factories and power plants are releasing more emissions as governments encourage economic growth while making little effort to control pollution. Meanwhile, more citizens own and drive automobiles (**FIGURE 17.18A**). At the same time, most people continue to burn traditional sources of fuel, such as wood, charcoal, and coal, for cooking and home heating. Iran is typical, and studies find that Tehran's residents inhale 7–9 kg (15–20 lbs) of dust each year and that levels of CO, SO_2, particulate matter, and other pollutants are all well above international safety standards and still increasing.

The people of China suffer some of the world's worst air pollution. China has fueled its rapid industrial development with its abundant reserves of coal, the most-polluting fossil fuel. Power plants and factories have sprung up pell-mell across the nation, often using outdated, inefficient, heavily polluting technology because it is cheaper and quicker to build. Car ownership is skyrocketing; in the capital of Beijing alone 1,500 new cars hit the roads each day. In many cities on a regular basis, the haze is too thick for people to see the sun. The health impacts are enormous yet inadequately documented. Reports by Chinese scientists, the World Bank, and the World Health Organization all estimate that outdoor air pollution causes over 300,000 premature deaths each year.

Winds carry some of China's pollution to neighboring nations such as South Korea and Japan, and some even travels across the Pacific Ocean to Los Angeles and other western U.S. cities.

China's government is now trying hard to decrease pollution. Its well-publicized efforts to clean Beijing's air before hosting the Olympic Games in 2008 (**FIGURE 17.18B**) were just the tip of the iceberg. The government is closing down some heavily polluting factories and mines, phasing out some subsidies for polluting industries, installing pollution controls in power plants, and encouraging the development of wind, solar, and nuclear power. It is subsidizing people to buy efficient electric heaters for their homes, to substitute for dirty, inefficient coal stoves. It is mandating cleaner formulations for gasoline and diesel and raising standards for fuel

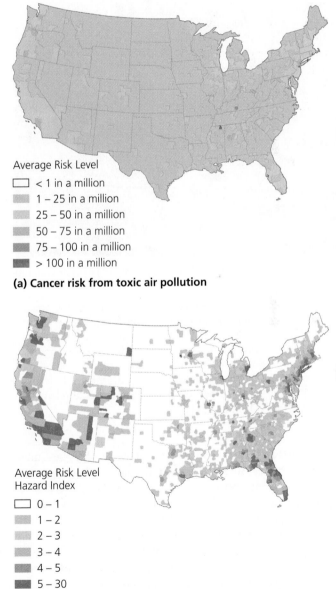

Average Risk Level

- ☐ < 1 in a million
- ▨ 1 – 25 in a million
- ▨ 25 – 50 in a million
- ▨ 50 – 75 in a million
- ▨ 75 – 100 in a million
- ■ > 100 in a million

(a) Cancer risk from toxic air pollution

Average Risk Level
Hazard Index

- ☐ 0 – 1
- ▨ 1 – 2
- ▨ 2 – 3
- ▨ 3 – 4
- ▨ 4 – 5
- ■ 5 – 30

(b) Non-cancer respiratory risk from toxic air pollution

FIGURE 17.17 ▲ The EPA's most recent national-scale assessment of toxic air pollutants mapped health risks county-by-county throughout the United States for cancer **(a)** and non-cancer respiratory ailments **(b)**. Darker colors reflect higher risk. Data from U.S. EPA.

(a) Traffic congestion causes pollution in Tehran

(b) Pollution enshrouds Beijing's "Bird's Nest" Olympic stadium

FIGURE 17.18 ▲ Automobile traffic in Tehran **(a)** illustrates one major cause of air pollution in today's industrializing nations. Officials estimate that Tehran's vehicles emit about 5,000 tons of pollutants each day as traffic creeps along at an average of 18 kph (11 mph). In Beijing **(b)**, Chinese officials tried to clean the air of one of world's dirtiest cities before the 2008 Olympic Games. Here, smog envelops the famed "Bird's Nest" Olympic stadium.

efficiency and emissions for cars well above what the United States requires. In Beijing, mass transit is being expanded, many buses run on natural gas, and heavily polluting vehicles are restricted from operating in the central city. All these efforts are beginning to pay off in Beijing, where the air is starting to get cleaner. However, enforcement and execution are often lacking in the rest of the country.

Pollution from autos, industry, agriculture, and wood-burning stoves in China, India, Iran, and other industrializing nations of Asia has resulted in a persistent 2-mile-thick layer of pollution that hangs over southern Asia throughout the dry season each December through April. Dubbed the *Asian Brown Cloud*, or *Atmospheric Brown Cloud*, this massive layer of brownish haze is estimated to reduce the sunlight reaching Earth's surface in southern Asia by 10–20%; decrease rice productivity by 5–10%; promote flooding in some areas and

drought in others; speed the melting of Himalayan glaciers by depositing dark soot that absorbs sunlight; and contribute to many thousands of deaths each year.

Air quality is a rural issue, too

Although we often focus on pollution in cities, air quality is not only an urban issue. In rural areas, people suffer from drift of airborne pesticides from farms, as well as industrial pollutants that drift far from cities, factories, and power plants. A great deal of rural air pollution emanates from feedlots (pp. 268–270), where cattle, hogs, or chickens are raised in dense concentrations. The huge numbers of animals at feedlots and the voluminous amounts of waste they produce release dust as well as methane, hydrogen sulfide, and ammonia. These gases create objectionable odors, and ammonia contributes to nitrogen deposition. Studies show that people working at and living near feedlots have high rates of respiratory illness.

Smog is our most common air quality problem

Despite the gains in air quality made in many nations in recent years, hundreds of millions of people worldwide suffer from *smog*, an unhealthy mixture of air pollutants that often forms over urban areas as a result of fossil fuel combustion. Since the onset of the industrial revolution, cities have suffered a type of smog we call **industrial smog**, or gray-air smog. When coal or oil is burned, some portion is completely combusted, forming CO_2; some is partially combusted, producing CO; and some remains unburned and is released as soot (particles of carbon). Moreover, coal contains varying amounts of contaminants, including mercury and sulfur. Sulfur reacts with oxygen to form sulfur dioxide, which can undergo a series of reactions to form sulfuric acid and ammonium sulfate (**FIGURE 17.19A**). These chemicals and others produced by further reactions, along with soot, are the main components of industrial smog and create its characteristic gray color.

The thick and blinding "killer smog" of 1952 in London occurred when weather conditions trapped emissions from the coal used to fire the city's industries and heat people's homes. Sulfur dioxide and particulate matter caused most of the 4,000–12,000 deaths from that episode. In the wake of this catastrophe and others, the governments of most developed nations began regulating industrial emissions to minimize the external costs (pp. 150–151) they impose on citizens. As a result, industrial smog is far less common today in developed nations than it was 50–100 years ago. However, many regions that are industrializing today, such as China, India, and eastern Europe, rely heavily on coal burning (by industry and by citizens heating and cooking in their homes). Combined with lax pollution control, this produces industrial smog that poses significant health risks in many areas.

Although coal combustion supplies the chemical constituents for industrial smog, weather also plays a role. Four years before London's killer smog, a similar event occurred in Pennsylvania in a small town named Donora (**FIGURE 17.19B**). Here, air near the ground cooled during the night, and because Donora is located in hilly terrain, too little morning sun reached the valley floor to warm and disperse the cold air. The

dangers to life because they allow more ultraviolet radiation to reach Earth's surface. (pp. 479–481)

- Ozone depletion is most severe over Antarctica, where an "ozone hole" appears each spring. (pp. 479–481)

- The Montreal Protocol and its follow-up agreements have proven remarkably successful in reducing emissions of ozone-depleting substances. (p. 482)

- The long residence time of CFCs in the atmosphere accounts for a time lag between the protocol and the full restoration of stratospheric ozone. (p. 482)

DEFINE *ACID DEPOSITION* AND ILLUSTRATE ITS CONSEQUENCES

- Acid deposition results when pollutants such as SO_2 and NO react in the atmosphere to produce strong acids that are deposited on Earth's surface. (pp. 482–483)

- Acid deposition may be wet (e.g., "acid rain") or dry, and it may occur a long distance from the source of pollution. (pp. 482–485)

- Water bodies, soils, plants, animals, and ecosystems all experience negative impacts from acidic deposition. (pp. 483–486)

- Regulation, cap-and-trade programs, and technology are all helping to reduce acid deposition in North America, yet more needs to be done, and industrializing nations will need to tackle the problem as well. (pp. 485–486)

CHARACTERIZE THE SCOPE OF INDOOR AIR POLLUTION AND ASSESS POTENTIAL SOLUTIONS

- Indoor air pollution causes far more deaths and health problems worldwide than outdoor air pollution. (pp. 486–487)

- Indoor burning of fuelwood is the developing world's primary indoor air pollution risk. (p. 487)

- Tobacco smoke and radon are the deadliest indoor pollutants in the developed world. (pp. 487–488)

- Volatile organic compounds and living organisms can pollute indoor air. (pp. 488–489)

- Using low-toxicity materials, keeping spaces clean, monitoring air quality, and maximizing ventilation all help to reduce indoor air pollution. (pp. 489–490)

TESTING YOUR COMPREHENSION

1. About how thick is Earth's atmosphere? Name one characteristic of each of the four atmospheric layers.

2. Where is the "ozone layer" located? How and why is stratospheric ozone beneficial for people, whereas tropospheric ozone is harmful?

3. How does solar energy influence weather and climate? How do Hadley, Ferrel, and polar cells help to determine long-term climatic patterns and the location of biomes?

4. Describe a thermal inversion. How do inversions contribute to severe smog episodes like the ones in London and in Donora, Pennsylvania?

5. How does a primary pollutant differ from a secondary pollutant? Give an example of each.

6. What has happened with concentrations of "criteria pollutants" in U.S. ambient air in recent decades? What has happened with our emissions of major pollutants? Name one health risk from toxic air pollutants.

7. How does photochemical smog differ from industrial smog? How do the weather and topography influence smog formation?

8. How do chlorofluorocarbons (CFCs) deplete stratospheric ozone? Why is this depletion considered a long-term international problem? What was done to address this problem?

9. Why are the effects of acid deposition often felt in areas far from where the primary pollutants are produced? List three impacts of acid deposition.

10. Name five common sources of indoor pollution. For each, describe one way to reduce one's exposure to this source.

SEEKING SOLUTIONS

1. Consider the photochemical smog pollution that has plagued Los Angeles, Tehran, Mexico City, and other metropolitan areas. Describe several ways in which major cities have tried to improve their air quality.

2. Name one type of natural air pollution, and discuss how human activity can sometimes worsen it. What potential solutions can you think of to minimize this human impact?

3. Describe how and why emissions of major pollutants have been reduced by over 50% in the United States since 1970, despite increases in population and economic activity.

4. International regulatory action has produced reductions in CFCs, but other transboundary pollution issues, including acid deposition, have not yet been addressed as effectively. What types of actions do you feel are appropriate for pollutants that cross political boundaries?

5. **THINK IT THROUGH** You have become the head of your county health department, and the EPA informs you that your county has failed to meet the national ambient air quality standards for ozone, sulfur dioxide, and nitrogen dioxide. Your county is partly rural but is home to a city of 200,000 people and 10 sprawling suburbs. There are several large and aging coal-fired power plants, a number of factories with advanced pollution control technology, and no public transportation system. What steps would you urge the county government to take to meet the air quality standards? Explain how you would prioritize these steps.

6. **THINK IT THROUGH** You have been elected mayor of the largest city in your state. Your city's residents are complaining about photochemical smog and traffic congestion. Traffic engineers and city planners project that population and traffic will grow by 20% in the next decade. Some experts are urging you to restrict traffic into the city, allowing only cars with odd-numbered license plates on odd-numbered days, and those with even-numbered plates on even-numbered days. However, business owners fear losing money should these measures discourage shoppers from visiting. Consider the particulars of your city, and then decide whether you will pursue an odd-day/even-day driving program, and explain why or why not. What other steps would you take to address your city's smog problem?

CALCULATING ECOLOGICAL FOOTPRINTS

"While only some motorists contribute to traffic fatalities, all motorists contribute to air pollution fatalities." So stated a writer for the Earth Policy Institute, in pointing out that air pollution kills far more people than vehicle accidents. According to EPA data, emissions of nitrogen oxides in the United States in 2008 totaled 16.3 million tons. Nitrogen oxides come from fuel combustion in motor vehicles, power plants, and other industrial, commercial, and residential sources, but fully 9.5 million tons of the 2008 total came from vehicles. The U.S. Census Bureau estimates the nation's population to have been 304.1 million in 2008 and projects that it will reach 334.1 million in 2020. Considering these data, calculate the missing values in the table below (1 ton = 2,000 lb).

	Total NO_x emissions (lb)	NO_x emissions from vehicles (lb)
You		
Your class		
Your state		
United States		

Data from U.S. EPA.

1. By what percentage is the U.S. population projected to increase between 2008 and 2020? Do you think that NO_x emissions will increase, decrease, or remain the same over that period of time? Why? (You may want to refer to Figure 17.15.)

2. Assuming you are an average American driver, how many pounds of NO_x emissions are you responsible for creating? How many pounds would you prevent if you were to reduce by half the vehicle miles you travel? What percentage of your total NO_x emissions would that be?

3. How might you reduce your vehicle miles traveled by 50%? What other steps could you take to reduce the NO_x emissions for which you are responsible?

Mastering ENVIRONMENTALSCIENCE™

Go to **www.masteringenvironmentalscience.com** for practice quizzes, Pearson eText, videos, current events, and more.

Malé, capital city of the Maldives

18 GLOBAL CLIMATE CHANGE

UPON COMPLETING THIS CHAPTER, YOU WILL BE ABLE TO:

- Describe Earth's climate system and explain the many factors influencing global climate
- Characterize human influences on the atmosphere and on climate

- Summarize how researchers study climate
- Outline current and future trends and impacts of global climate change
- Suggest ways we may respond to climate change

Rising Seas May Flood the Maldives

"Global warming and climate change can effectively kill us off, make us refugees. . . ."

—Ismail Shafeeu, Minister of Environment, Maldives, 2000

"If we can't save the Maldives today, we can't save London, New York, or Hong Kong tomorrow."

—Mohamed Nasheed, President, Maldives, 2009

EUROPE

ASIA

AFRICA

INDIA

Maldives

Indian
Ocean

On October 17, 2009, President Mohamed Nasheed of the Maldives, a nation of low-lying islands in the Indian Ocean, donned scuba gear and dove into the blue waters of Girifushi Island lagoon. He was followed by his entire cabinet.

These officials were holding a cabinet meeting underwater—no doubt the world's first such meeting ever. Sitting at a table beneath the waves, they signed a declaration reading:

The Maldives' underwater cabinet meeting

> *SOS from the front line:* Climate change is happening and it threatens the rights and security of everyone on Earth. With less than one degree of global warming, the glaciers are melting, the ice sheets collapsing, and low-lying areas are in danger of being swamped. We must unite in a global effort to halt further temperature rises, by slashing carbon dioxide emissions to a safe level of 350 parts per million.

Known for its spectacular tropical setting, colorful coral reefs, and sun-drenched beaches, the Maldives seems a paradise to its many visiting tourists—while for 370,000 Maldives residents, it is home. But residents and tourists alike now fear that the Maldives could soon be submerged by the rising seas brought by global climate change.

Nearly 80% of the Maldives' land area lies less than 1 m (39 in.) above sea level. In a nation of 1,200 islands whose highest point is just 2.4 m (8 ft) above sea level, rising seas are a matter of life or death. The world's oceans rose 10–20 cm (4–8 in.) during the 20th century as warming temperatures expanded ocean water and as melting polar ice discharged water into the ocean. According to current projections, sea level will rise another 18–59 cm (7–23 in.) by the year 2100.

Higher seas are expected to flood large areas of the Maldives and cause salt water to contaminate drinking water supplies. Storms intensified by warmer water temperatures will erode beaches, cause flooding, and damage the coral reefs that are so vital to the tourism and fishing industries that drive the nation's economy. The Maldives government recently evacuated residents from several of the lowest-lying islands, and residents of other islands are considering moving voluntarily.

President Nasheed is making sure the world knows of his country's tribulations. He points out that small island nations like his are not responsible for the carbon emissions driving global climate change, yet they are the ones bearing the brunt of the consequences. "If things go business as usual," he said, "we will not live; we will die. Our country will not exist."

The underwater cabinet meeting was part of a global campaign to draw attention to the impacts of climate change, sponsored by the nonprofit group 350.org. This campaign culminated in an International Day of Climate Action on October 24, 2009, when 5,200 events took place in 181 nations.

Residents of the Maldives are not alone in their predicament. Other island nations, from the Galápagos to Fiji to the Seychelles, also fear a future of encroaching seawater. These island nations have organized themselves to make their position on climate change known to the world through AOSIS, the Alliance of Small Island States.

Mainland coastal areas of the world, from the hurricane-battered coasts of Florida, Louisiana, Texas, the Carolinas, and other states, to coastal cities such as New York and San Francisco, will face similar challenges from sea level rise—and this is just one of the many consequences of global climate change. In one way or another, climate change will affect each and every one of us for the remainder of our lifetimes.

OUR DYNAMIC CLIMATE

Climate influences virtually everything around us, from the day's weather to major storms, from crop success to human health, and from national security to the ecosystems that support our economies. If you are a student in your teens or twenties, the accelerating changes in our climate today may well be *the* major event of your lifetime and the phenomenon that most shapes your future.

Climate change is also the fastest-developing area of environmental science. New scientific papers that refine our understanding of climate are published every week, and policymakers and businesspeople make decisions and announcements just as quickly. By the time you read this chapter, some of its information will already be out of date. We urge you to explore further, with your instructor and on your own, the most recent information on climate change and the impacts it will have on your future.

What is climate change?

Climate describes an area's long-term atmospheric conditions, including temperature, moisture content, wind, precipitation, barometric pressure, solar radiation, and other characteristics. As we learned (p. 465), *climate* differs from *weather* in that weather specifies conditions at localized sites over hours or days, whereas climate describes conditions across broader regions over seasons, years, or centuries. **Global climate change** describes trends in Earth's climate, involving aspects such as temperature, precipitation, and storm frequency and intensity. People often use the term *global warming* synonymously in casual conversation, but **global warming** refers specifically to an increase in Earth's average surface temperature. Global warming is only one aspect of global climate change, although warming does in turn drive other components of climate change.

When people speak of climate change and global warming, they generally are referring to the trends taking place right now. Scientists may also sometimes use these terms to refer to trends in the geologic past. Indeed, our planet's climate varies naturally through time and to some extent is always changing. However, the climatic changes unfolding today are proceeding at an exceedingly rapid rate. Moreover, scientists agree that human activities, notably fossil fuel combustion and deforestation, are largely responsible. Understanding how and why today's climate is changing requires understanding how our planet's

climate functions. Thus, we first will survey the fundamentals of Earth's climate system—a complex and finely tuned system that has nurtured life for billions of years.

The sun and atmosphere keep Earth warm, while other factors regulate climate

Four factors exert the most influence on Earth's climate. The first is the sun. Without it, Earth would be dark and frozen. The second is the atmosphere. Without it, Earth would be as much as 33 °C (59 °F) colder on average, and temperature differences between night and day would be far greater than they are. The sun and the atmosphere, therefore, keep our planet warm enough to sustain life. Two further factors help regulate how climate varies across our planet: The oceans shape climate by storing and transporting heat and moisture, and cycles in the ways our planet spins, tilts, and moves through space influence how climate varies over long periods of time.

The sun supplies most of our planet's energy. Earth's atmosphere, clouds, land, ice, and water together absorb about 70% of incoming solar radiation and reflect the remaining 30% back into space (**FIGURE 18.1**). The 70% that is absorbed powers many of Earth's processes, from winds to waves to evaporation to photosynthesis. We will examine each of the major four factors shortly but will focus first on the atmosphere.

Greenhouse gases warm the lower atmosphere

As Earth's surface absorbs solar radiation, the surface increases in temperature and emits infrared radiation (p. 31). Infrared radiation has longer wavelengths than the visible and ultraviolet light that had arrived from the sun and passed through the atmosphere. Atmospheric gases having three or more atoms in their molecules tend to absorb this infrared radiation very effectively. These include water vapor, ozone (O_3), carbon dioxide (CO_2), nitrous oxide (N_2O), and methane (CH_4), as well as halocarbons, a diverse group of mostly human-made gases that includes chlorofluorocarbons (CFCs; p. 479). Such gases are known as **greenhouse gases**. After absorbing radiation emitted from the surface, greenhouse gases subsequently re-emit infrared radiation in all directions. Some of this re-emitted energy is lost to space, but some travels back downward, warming the atmosphere (specifically the *troposphere*; p. 462) and the planet's surface in a phenomenon known as the **greenhouse effect**.

The greenhouse effect is a natural phenomenon, and greenhouse gases have been present in our atmosphere for all of Earth's history. It's a good thing, too. Without the natural greenhouse effect, our planet would be too cold to support life as we know it. Thus, it is not the natural greenhouse effect that concerns scientists today, but rather the *anthropogenic* (human-generated) intensification of the greenhouse effect. By adding novel greenhouse gases (certain halocarbons) to the atmosphere, and by increasing the concentrations of several natural greenhouse gases over the past 250–300 years, we are intensifying our planet's greenhouse effect beyond what our species has ever experienced.

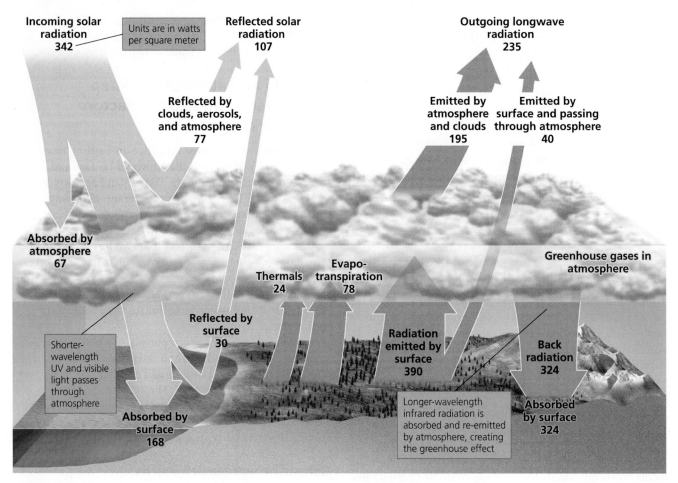

FIGURE 18.1 ▲ Our planet absorbs nearly 70% of the solar radiation it receives from the sun, and it reflects the rest back into space (yellow arrows). Most visible and ultraviolet radiation from the sun readily passes through the atmosphere and reaches the surface, and this radiation is absorbed and then re-emitted (orange arrows) as infrared radiation, which has longer wavelengths. Greenhouse gases in the atmosphere absorb some of this long-wavelength radiation and then re-emit it, sending some downward to warm the atmosphere and the surface by the greenhouse effect. This illustration, showing major pathways of energy flow in watts per square meter, indicates that our planet naturally emits and reflects 342 watts/m², the same amount it receives from the sun. Arrow thicknesses in the diagram are proportional to flows of energy in each pathway. Data from Kiehl, J.T., and K.E. Trenberth, 1997. Earth's annual global mean energy budget. *Bulletin of the American Meteorological Society* 78: 197–208. © American Meteorological Society (AMS). By permission.

Greenhouse gases differ in their ability to warm the troposphere and surface. *Global warming potential* refers to the relative ability of one molecule of a given greenhouse gas to contribute to warming. **TABLE 18.1** shows global warming potentials for several greenhouse gases. Values are expressed in relation to carbon dioxide, which is assigned a global warming potential of 1. Thus, a molecule of methane is 25 times as potent as a molecule of carbon dioxide, and a molecule of nitrous oxide is 298 times as potent as a CO_2 molecule.

Carbon dioxide is the greenhouse gas of primary concern

Although carbon dioxide is less potent on a per-molecule basis than other greenhouse gases such as methane and nitrous oxide, it is far more abundant in the atmosphere, so it contributes more to the natural greenhouse effect. It also contributes more to the anthropogenic greenhouse effect, because our greenhouse gas emissions consist mostly of carbon

TABLE 18.1 Global Warming Potentials of Four Greenhouse Gases

Greenhouse gas	Relative heat-trapping ability (in CO_2 equivalents)
Carbon dioxide	1
Methane	25
Nitrous oxide	298
Hydrochlorofluorocarbon HFC-23	14,800

Data are for a 100-year time horizon, from Intergovernmental Panel on Climate Change, 2007. *Fourth assessment report. Climate change 2007: The physical science basis.*

dioxide. That is, even after accounting for the greater global warming potential of molecules of other gases, carbon dioxide's abundance in our emissions makes it the major contributor to anthropogenic global warming. According to the

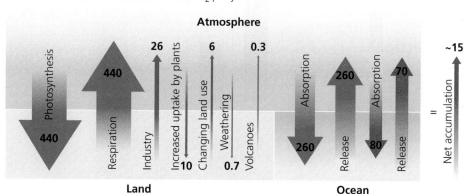

Natural fluxes ■ Anthropogenic fluxes ■
Units are in billions of metric tons of CO_2 per year

Atmosphere

Photosynthesis 440
Respiration 440
Industry 26
Increased uptake by plants 6
Changing land use 10
Weathering 0.7
Volcanoes 0.3

Absorption 260
Release 260
Absorption 70
Release 80

Net accumulation = ~15

Land **Ocean**

FIGURE 18.2 ◀ Human activities since the industrial revolution have sent more carbon dioxide from the Earth's surface to its atmosphere than is moving back from the atmosphere to the surface. Shown here are all current fluxes of carbon dioxide, with arrows sized according to their mass of CO_2. Green arrows indicate natural fluxes, and red arrows indicate anthropogenic fluxes. Adapted from Intergovernmental Panel on Climate Change, 2007. *Fourth assessment report.*

latest data, CO_2 is exerting nearly six times more impact than methane, nitrous oxide, and halocarbons combined.

Fossil fuel use and deforestation release carbon dioxide

What has caused atmospheric carbon dioxide levels to rise so rapidly? As you may recall from our discussion of the carbon cycle in Chapter 5 (pp. 124–125), and as we will see further in Chapter 19 (p. 533), most carbon is stored for long periods in the upper layers of the lithosphere. The deposition, partial decay, and compression of organic matter (mostly plants) that grew in wetland or marine areas hundreds of millions of years ago led to the formation of coal, oil, and natural gas in buried sediments. In the absence of human activity, these carbon reservoirs would remain buried for many millions more years. However, over the past two centuries we have extracted these fossil fuels from the ground and burned them in our homes, factories, and automobiles, transferring large amounts of carbon from one reservoir (the underground deposits that stored the carbon for millions of years) to another (the atmosphere). This sudden flux of carbon from lithospheric reservoirs into the atmosphere is the main reason atmospheric carbon dioxide concentrations have increased so dramatically.

At the same time, people have cleared and burned forests to make room for crops, pastures, villages, and cities. Forests serve as a reservoir for carbon as plants conduct photosynthesis and then store carbon in their tissues. Thus, when we clear forests it reduces the biosphere's ability to remove carbon dioxide from the atmosphere. In this way, deforestation (pp. 318–321) has contributed to rising atmospheric CO_2 concentrations. **FIGURE 18.2** summarizes scientists' current understanding of the fluxes (both natural and anthropogenic) of carbon dioxide between the atmosphere and reservoirs on Earth's surface.

Other greenhouse gases add to warming

Human activities have boosted Earth's atmospheric concentration of carbon dioxide from 280 parts per million (ppm) in the late 1700s to 389 ppm in 2010 (**FIGURE 18.3**). The atmospheric CO_2 concentration is now at its highest level by

far in over 800,000 years, and likely the highest in the last 20 million years.

Carbon dioxide is not the only greenhouse gas becoming more prevalent in our atmosphere. Methane concentrations are also rising—2.5-fold since 1750 (Figure 18.3)—and today's atmospheric concentration is the highest by far in over 800,000 years. We release methane by tapping into fossil fuel deposits, raising livestock that emit methane as a metabolic waste product, disposing of organic matter in landfills, and growing certain crops, such as rice.

Human activities have also enhanced atmospheric concentrations of nitrous oxide. This greenhouse gas, a by-product of feedlots, chemical manufacturing plants, auto emissions, and synthetic nitrogen fertilizers, has risen by 18% since 1750 (Figure 18.3).

Among other greenhouse gases, ozone concentrations in the troposphere have risen roughly 36% since 1750 because of photochemical smog (pp. 477–478). The contribution of halocarbon gases to global warming has begun to slow because of the Montreal Protocol and subsequent controls on their production and use (pp. 481–482).

Water vapor is the most abundant greenhouse gas in our atmosphere and contributes most to the natural greenhouse

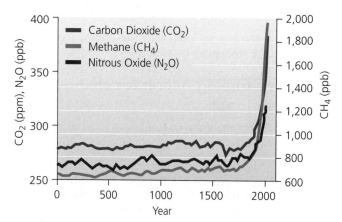

FIGURE 18.3 ▲ Since the start of the industrial revolution, global atmospheric concentrations of carbon dioxide (red line), methane (green line), and nitrous oxide (blue line) have increased markedly. Data from Intergovernmental Panel on Climate Change, 2007. *Fourth assessment report.* FAQ 2.1, Fig 1, in The physical science basis: Contribution of working Group I."

effect. Its concentrations vary locally, but its global concentration has not changed over recent centuries, so it is not viewed as having driven industrial-age climate change.

Feedback complicates our predictions

As tropospheric temperatures increase, Earth's water bodies should transfer more water vapor into the atmosphere, but scientists aren't yet sure how this will affect our climate. On one hand, more atmospheric water vapor could lead to more warming, which could lead to more evaporation and water vapor, in a positive feedback loop (p. 110) that would amplify the greenhouse effect. On the other hand, more water vapor could give rise to increased cloudiness, which might, in a negative feedback loop (p. 110), slow global warming by reflecting more solar radiation back into space. In this second scenario, depending on whether low- or high-elevation clouds result, they might either shade and cool Earth (negative feedback) or else contribute to warming and accelerate evaporation and further cloud formation (positive feedback). We simply don't yet know which effect might outweigh the other. Because of feedback loops (see Figure 5.2, p. 110), minor modifications of components of the atmosphere can potentially lead to major effects on climate. This poses challenges for making accurate predictions of future climate change.

Most aerosols exert a cooling effect

Whereas greenhouse gases exert a warming effect on the atmosphere, **aerosols** (p. 468), microscopic droplets and particles, can have either a warming or cooling effect. Soot particles, or black carbon aerosols, generally cause warming by absorbing solar energy, but most other tropospheric aerosols cool the atmosphere by reflecting the sun's rays. Sulfate aerosols produced by fossil fuel combustion may slow global warming, at least in the short term. When sulfur dioxide enters the atmosphere, it undergoes various reactions, some of which lead to acid precipitation (pp. 482–483). These reactions, along with volcanic eruptions, can form a sulfur-rich aerosol haze in the upper atmosphere that reduces the sunlight reaching Earth's surface. Aerosols released by major volcanic eruptions can exert cooling effects on Earth's climate for up to several years. This occurred as recently as 1991 with the eruption of Mount Pinatubo in the Philippines (p. 468).

Radiative forcing expresses change in energy input

To measure the degree of impact that any given factor exerts on Earth's temperature, scientists calculate its radiative forcing. **Radiative forcing** is the amount of change in thermal energy that a given factor causes. Positive forcing warms the surface, whereas negative forcing cools it. **FIGURE 18.4** shows researchers' best calculations of the radiative forcing that our planet is experiencing today. These quantitative estimates indicate the degree of influence that aerosols, greenhouse gases, and other factors exert over Earth's energy balance.

When scientists sum up the effects of all factors, they find that Earth today is experiencing overall radiative forcing of about 1.6 watts/m². This means that compared with the

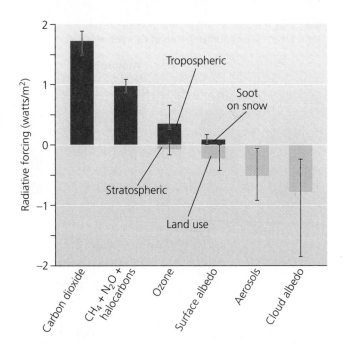

FIGURE 18.4 ▲ For each emitted gas or other human impact on the atmosphere since the industrial revolution, we can estimate the warming or cooling effect it has had on Earth's climate. We express this as *radiative forcing*, which in this graph is shown as the amount of influence on climate today relative to 1750, in watts per square meter. Red bars indicate positive forcing (warming), and blue bars indicate negative forcing (cooling). *Albedo* (p. 506) refers to the reflectivity of a surface. A number of more minor influences are not shown. In total, scientists estimate that human impacts on the atmosphere exert a cumulative radiative forcing of 1.6 watts/m². Data from Intergovernmental Panel on Climate Change, 2007. *Fourth assessment report.*

pre-industrial Earth of 1750, today's planet is receiving and retaining 1.6 watts/m² more of thermal energy than it is emitting into space. For context, look back at Figure 18.1 and note that Earth is estimated naturally to receive and give off 342 watts/m² of energy. Although 1.6 may seem like a small proportion of 342, it is enough to alter climate significantly.

Several factors influence climate

Besides atmospheric composition, our climate is influenced by cyclic changes in Earth's rotation and orbit, variation in energy released by the sun, absorption of carbon dioxide by the oceans, and ocean circulation patterns.

Milankovitch cycles In the 1920s, Serbian mathematician Milutin Milankovitch described three types of periodic changes in Earth's rotation and orbit around the sun. Over thousands of years, our planet wobbles on its axis, varies in the tilt of the axis, and experiences change in the shape of its orbit, all in regular long-term cycles of different lengths. These variations, now known as **Milankovitch cycles**, alter the way solar radiation is distributed over Earth's surface (**FIGURE 18.5**). By modifying patterns of atmospheric heating, these cycles trigger long-term climate variation. This includes periodic episodes of *glaciation* during which global surface temperatures drop and

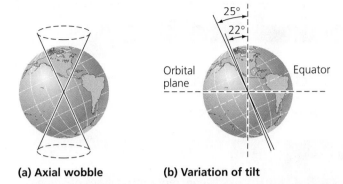

(a) Axial wobble   (b) Variation of tilt

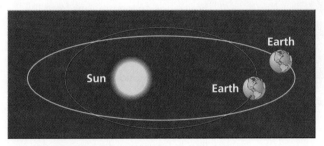

(c) Variation of orbit

FIGURE 18.5 ▲ There are three types of Milankovitch cycles. The first is an axial wobble (a) that occurs on a 19,000- to 23,000-year cycle. The second is a 3-degree shift in the tilt of Earth's axis (b) that occurs on a 41,000-year cycle. The third is a variation in Earth's orbit from almost circular to more elliptical (c), which repeats itself every 100,000 years. These variations affect the intensity of solar radiation that reaches portions of Earth at different times, contributing to long-term changes in global climate.

ice sheets advance from the poles toward the midlatitudes, as well as intervening warm *interglacial* periods.

Solar output   The sun varies in the amount of radiation it emits, over both short and long timescales. For example, at each peak of its 11–year sunspot cycle the sun may emit solar flares, bursts of energy strong enough to disrupt satellite communications. However, scientists are concluding that the variation in solar energy reaching our planet in recent centuries has simply not been great enough to drive significant temperature change on Earth's surface. Estimates place the radiative forcing of natural changes in solar output at only about 0.12 watts/m² —less than any of the anthropogenic causes shown in Figure 18.4.

Ocean absorption   The oceans hold 50 times more carbon than the atmosphere holds. They absorb carbon dioxide from the atmosphere when the gas dissolves directly in water and when marine phytoplankton use it for photosynthesis. However, the oceans are absorbing less CO_2 than we are adding to the atmosphere (see Figure 5.16, p. 125). Thus, carbon absorption by the oceans is slowing global warming but is not preventing it. Moreover, recent evidence indicates that the rate of absorption is now decreasing. As ocean water warms, it absorbs less CO_2 because gases are less soluble in warmer water—a positive feedback effect (p. 110) that accelerates warming of the atmosphere.

Ocean circulation   Ocean water exchanges tremendous amounts of heat with the atmosphere, and ocean currents move energy from place to place. In equatorial regions, such as the area around the Maldives, the oceans receive more heat from the sun and atmosphere than they emit. Near the poles, the oceans emit more heat than they receive. Because cooler water is denser than warmer water, the cooling water at the poles tends to sink, and the warmer surface water from the equator moves to take its place. This is one principle underlying global ocean circulation patterns (pp. 436–437).

As discussed in Chapter 16, the oceans' thermohaline circulation system has influential regional effects. For example, it moves warm tropical water northward toward Europe, providing that continent a far milder climate than it would otherwise have. As we saw (p. 437), scientists are studying how much freshwater input from Greenland's melting ice sheet this warm-water flow can withstand without being shut down—an occurrence that could have devastating impacts on Europe.

Also recall from Chapter 16 that multiyear climate variability results from the El Niño–Southern Oscillation (pp. 437–438), which involves systematic shifts in atmospheric pressure, sea surface temperature, and ocean circulation in the tropical Pacific Ocean. These shifts, in turn, overlie longer-term variability from a phenomenon known as the Pacific Decadal Oscillation. El Niño and La Niña events alter weather patterns from region to region in diverse ways, often leading to rainstorms and floods in dry areas and drought and fire in moist areas. This in turn, as we saw, leads to impacts on wildlife, agriculture, and economically important fisheries.

STUDYING CLIMATE CHANGE

To comprehend any phenomenon that is changing, we must study its past, present, and future. Scientists monitor present-day climate, but they also have devised clever means of inferring past change and have developed sophisticated methods to predict future conditions.

Proxy indicators tell us about the past

Evidence about **paleoclimate**, climate in the geologic past, is vital for giving us a baseline against which to measure changes happening in our climate today. Environmental scientists have developed a number of ingenious methods to decipher clues from the past, taking advantage of the record-keeping capacity of the natural world. **Proxy indicators** are types of indirect evidence that serve as proxies, or substitutes, for direct measurement and that shed light on past climate.

For example, Earth's ice caps, ice sheets, and glaciers hold clues to climate history. Over the ages, these huge expanses of snow and ice have accumulated to great depths, preserving within their layers tiny bubbles of the ancient atmosphere (**FIGURE 18.6**, p. 502). Scientists can examine the trapped air bubbles by drilling into the ice and extracting long columns, or cores. The layered ice, accumulating season after season over thousands of years, provides a timescale. By studying the chemistry of the ice and the bubbles in each layer in these ice cores, scientists can determine atmospheric composition, greenhouse gas concentrations, temperature trends, snowfall,

The SCIENCE behind the Story

Reading History in the World's Longest Ice Core

In the most frigid reaches of our planet, snow falling year after year for millennia compresses into ice and stacks up into immense sheets that scientists can mine for clues to Earth's climate history. The ice sheets of Antarctica and Greenland trap tiny air bubbles, dust particles, and other proxy indicators (p. 499) of past conditions. By drilling boreholes and extracting ice cores, researchers can tap into these valuable archives.

Recently, researchers drilled and analyzed the deepest core ever. At a remote and pristine site in Antarctica named Dome C, they drilled down 3,270 m (10,728 ft) to bedrock and pulled out more than 800,000 years' worth of ice. The longest previous ice core (from Antarctica's Vostok station) had gone back "only" 420,000 years.

Ice near the top of these cores was laid down most recently, and ice at the bottom is oldest, so by analyzing ice at intervals along the core's length, researchers can generate a timeline of environmental change. Researchers date layers of the ice core by analyzing deuterium isotopes (p. 26) to determine the rate of ice accumulation, referencing studies and models of how ice compacts over time, and calibrating the timeline by

An EPICA researcher prepares a Dome C ice core sample for analysis

matching recent events in the chronology (for example, major volcanic eruptions) with independent data sets from previous cores, tree rings, and other sources.

Dome C, a high summit of the Antarctic ice sheet, is one of the coldest spots on the planet, with an annual mean temperature of –54.5 °C (–98.1 °F). The Dome C ice core was drilled by the European Project for Ice Coring in Antarctica (EPICA), a consortium of researchers from 10 European nations.

In 2004, this team of 56 researchers published a paper in the journal *Nature*, reporting data across 740,000 years. The researchers obtained data on surface air temperature by measuring the ratio of deuterium isotopes to normal hydrogen in the ice, because this ratio is temperature-dependent.

From 2005 to 2008, five follow-up papers in the journals *Science* and

Nature reported analyses of greenhouse gas concentrations from the EPICA ice core and extended the gas and temperature data back to cover all 800,000 years. By analyzing air bubbles trapped in the ice, the researchers quantified atmospheric concentrations of carbon dioxide (**red line in figure**), methane (**green line in figure**), and nitrous oxide.

These data demonstrate that by emitting these greenhouse gases since the industrial revolution, we have brought their atmospheric concentrations well above the highest levels they reached naturally any time in the last 800,000 years. Today's carbon dioxide spike is too recent to show up in the ice core, but the current concentration (of 389 ppm in 2010) is far above previous maximum values (of ~300 ppm) shown in the red line of the figure. These data reveal that we as a society have brought ourselves deep into uncharted territory.

The EPICA results also confirm that temperature swings in the past were tightly correlated with concentrations of greenhouse gases (**compare top two datasets in figure with temperature dataset at bottom**). This finding bolsters the scientific consensus that

---

solar activity, and even (from trapped soot particles) frequency of forest fires and volcanic eruptions during each time period. By extracting ice cores from Antarctica, scientists have now been able to go back in time 800,000 years, reading Earth's history across eight glacial cycles (see **THE SCIENCE BEHIND THE STORY**, above.)

Researchers also drill cores into beds of sediment beneath bodies of water. Sediments often preserve pollen grains and other remnants from plants that grew in the past, as we saw with the study of Easter Island (pp. 6–7). Because climate

influences the types of plants that grow in an area, knowing what plants were present can tell us a great deal about the climate at that place and time.

Tree rings provide another proxy indicator. The width of each ring of a tree trunk cut in cross-section reveals how much the tree grew in a particular growing season. A wide ring means more growth, generally indicating a wetter year. Long-lived trees such as bristlecone pines can provide records of precipitation and drought going back several thousand years. Tree rings are also used to study local fire history, since

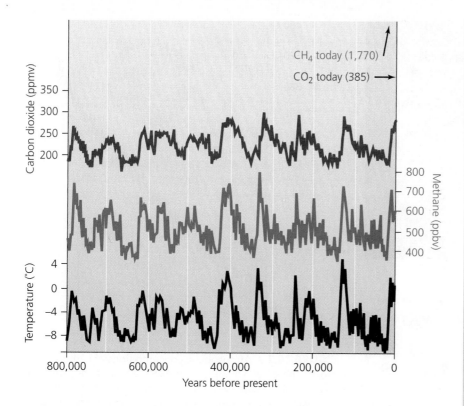

Data from the EPICA ice core reveal changes in surface temperature (black line), atmospheric methane concentration (green line), and atmospheric carbon dioxide concentration (red line), across 800,000 years. High peaks in temperature indicate warm interglacial periods, and low troughs indicate cold glacial periods. Atmospheric concentrations of carbon dioxide and methane rise and fall in tight correlation with temperature. Today's current values are included at the far top right of the graph, for comparison. Adapted by permission of Macmillan Publishers Ltd: Brook, E. 2008. Paleoclimate: Windows on the greenhouse. *Nature* 453: 291-292, Fig 1a.

greenhouse gas emissions are causing our planet to warm today.

Also clear from the data is that temperature has varied with swings in solar radiation due to Milankovitch cycles (pp. 498–499). The complex interplay of these cycles produces periodic temperature fluctuations on Earth resulting in periods of *glaciation* (when temperate regions of the planet are covered in ice) and in warm *interglacial* periods. The Dome C ice core spans eight glacial cycles.

Other findings from the ice core are not easily explained. Intriguingly, the earlier glacial cycles are of a different character from the more recent cycles (**see black line in figure**). For the most recent cycles, the Dome C core indicates that glacial periods are long, whereas interglacial periods are brief, with a rapid rise and fall of temperature. Interglacials thus appear on a graph of temperature through time as tall thin spikes. However, older glacial cycles revealed by the Dome C core look different: The glacial and interglacial periods are of more equal duration, and the warm extremes of interglacials are not as great.

This change in the nature of glacial cycles through time had been noted before by researchers working with oxygen isotope data from the fossils of marine organisms. But why glacial cycles should be so different before and after the 450,000-year mark, no one knows.

Today polar scientists are searching for a site that might provide an ice core stretching back more than 1 million years. For at that time, data from marine isotopes tell us that glacial cycles switched from a periodicity of roughly 41,000 years (conforming to the influence of planetary tilt), to intervals of about 100,000 years (more similar to orbital changes). An ice core that captures cycles on both sides of the 1-million-year divide might help clarify the influence of Milankovitch cycles or perhaps offer other explanations.

The intriguing patterns revealed by the Dome C ice core show that we still have plenty to learn about our complex climate history. However, the clear relationship between greenhouse gases and temperature evident in the EPICA data suggest that, if we want to prevent sudden global warming, we will need to reduce our society's greenhouse emissions. ■

a charred ring indicates that a fire took place in the region in that year.

In arid regions such as the U.S. Southwest, packrat middens are a valuable source of climate data. Packrats are rodents that carry seeds and plant parts back to their middens, or dens, in caves and rock crevices sheltered from rain. In arid locations, plant parts may be preserved for centuries, allowing researchers to study the past flora of the region.

Researchers gather data on past ocean conditions from coral reefs (pp. 439–442). Living corals take in trace elements and isotope ratios (p. 26) from ocean water as they grow, and they incorporate these chemical clues to ocean conditions, layer by layer, into growth bands in the reefs they build.

Proxy indicators often tell us information about local or regional areas. To get a global perspective, scientists need to combine multiple records from various areas. Because the number of available indicators decreases the further back in time we go, estimates of global climate conditions for the recent past tend to be more reliable than those for the distant past.

(a) Ice core

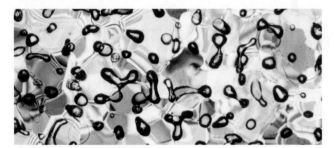

(b) Micrograph of ice core

FIGURE 18.6 ▲ In Greenland and Antarctica, scientists have drilled deep into ancient ice sheets and removed cores of ice like this one **(a)**, held by Dr. Gerald Holdsworth of the University of Calgary, to extract information about past climates. Bubbles (black shapes) trapped in the ice **(b)** contain small samples of the ancient atmosphere.

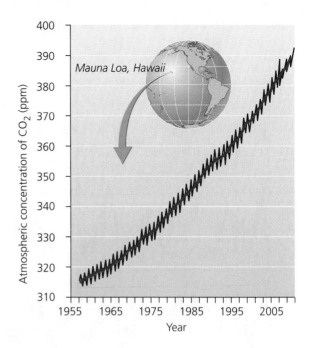

FIGURE 18.7 ▲ Atmospheric concentrations of carbon dioxide have risen steeply since 1958, when Charles Keeling began collecting these data at the Mauna Loa Observatory in Hawaii. The jaggedness of the upward trend reflects seasonal variation; because the Northern Hemisphere has more land area and thus more vegetation than the Southern Hemisphere, more carbon dioxide is absorbed during the northern summer, when Northern Hemisphere plants are more photosynthetically active. Data from National Oceanic and Atmospheric Administration, Earth System Research Laboratory, Global Monitoring Division, 2010.

Direct measurements tell us about the present

Today we measure temperature with thermometers, rainfall with rain gauges, wind speed with anemometers, and air pressure with barometers, using computer programs to integrate and analyze this information in real time. With these technologies and more, we document in detail the fluctuations in weather day-by-day and hour-by-hour across the globe. As a result, we have gained an understanding of present-day climate conditions in every region of our planet.

We also measure the chemistry of the atmosphere and the oceans with a range of equipment. Direct measurements of carbon dioxide concentrations in the atmosphere began in 1958, when scientist Charles Keeling started analyzing hourly air samples from a monitoring station at the Mauna Loa Observatory in Hawaii. These data show that atmospheric CO_2 concentrations have increased from 315 ppm in 1958 to 389 ppm in 2010 (**FIGURE 18.7**). As scientists continue these measurements today, they build upon the best long-term dataset we have of direct atmospheric sampling of a greenhouse gas.

Direct measurements of climate variables such as temperature and precipitation extend back in time somewhat

further. Precise and reliable thermometer measurements cover more than a century. People have also kept records of economically important impacts of climate that allow scientists to infer climate conditions of recent centuries: Fishers have recorded the timing of sea ice formation, and winemakers have kept meticulous records of precipitation and the length of the growing season. However, accurate records of all these types extend back at most only a few hundred years. To truly understand climate and how it behaves over time—and to predict future change—scientists must learn what climate conditions were like thousands and millions of years ago.

Models help us predict the future

To understand how climate systems function and to predict future climate change, scientists simulate climate processes with sophisticated computer programs. **Climate models** are programs that combine what is known about atmospheric circulation, ocean circulation, atmosphere–ocean interactions, and feedback cycles to simulate climate processes (**FIGURE 18.8**). This requires manipulating vast amounts of data with complex mathematical equations—a task not possible until the advent of modern computers.

Climate modelers essentially provide starting information to the model, set up rules for the simulation, and then let it run. Researchers strive for accuracy by building in as much information as they can from what is understood about how

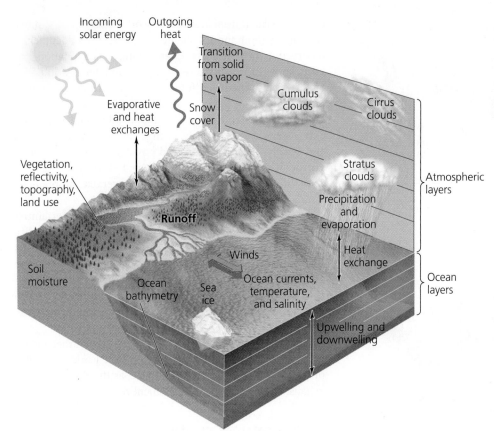

Incoming solar energy

Outgoing heat

Transition from solid to vapor

Evaporative and heat exchanges

Snow cover

Cumulus clouds

Cirrus clouds

Vegetation, reflectivity, topography, land use

Runoff

Stratus clouds

Atmospheric layers

Precipitation and evaporation

Heat exchange

Soil moisture

Ocean bathymetry

Sea ice

Winds

Ocean currents, temperature, and salinity

Ocean layers

Upwelling and downwelling

FIGURE 18.8 ◄ Modern climate models incorporate many factors, including processes involving the atmosphere, land, oceans, ice, and biosphere. Such factors are shown graphically here, but the actual models deal with them as mathematical equations in computer simulations.

the climate system functions. They then test the efficacy of a model by entering past climate data and running the model toward the present. If a model accurately reconstructs current climate, based on well-established data from the past, then we have reason to believe that it simulates climate mechanisms realistically and that it may accurately predict future climate.

FIGURE 18.9 (see next page) shows temperature results from three such simulations. Results in **FIGURE 18.9A** are based on natural climate-changing factors alone (such as volcanic activity and variation in solar energy). Results in **FIGURE 18.9B** are based on anthropogenic factors only (such as human emissions of greenhouse gases and sulfate aerosols). Results in **FIGURE 18.9C** are based on natural and anthropogenic factors combined. As you can see, the results in Figure 18.9c produce the closest match between predictions and actual climate. Such results support the hypothesis that both natural and human factors contribute to climate dynamics. They also indicate that global climate models can produce reliable predictions.

Plenty of challenges remain for climate modelers, because the climate system is so complex and because many uncertainties remain in our understanding of feedback processes (p. 110). Yet as scientific knowledge of climate processes improves, as computing power intensifies, and as we glean enhanced data from proxy indicators, climate models are becoming increasingly reliable. They are improving in resolution and are beginning to predict climate change region by region for various areas of the world.

CURRENT AND FUTURE TRENDS AND IMPACTS

In recent years, it seems that virtually everyone is detecting climatic changes. Fishermen in the Maldives note the seas encroaching on their home island. Ranchers in west Texas suffer a multiyear drought. Homeowners in Florida find it impossible to obtain insurance against the hurricanes and storm surges that increasingly threaten them. New Yorkers, Bostonians, Chicagoans, and Los Angelenos face one unprecedented weather event after another.

Such impressions are indeed part of a real pattern. Scientific evidence that climate has changed worldwide since industrialization is now overwhelming and indisputable. Climate change in recent years has already had numerous impacts on the physical properties of our planet, on organisms and ecosystems, and on human well-being. If we continue to emit greenhouse gases into the atmosphere, the consequences of climate change will only grow more severe.

The IPCC summarizes evidence and predicts impacts

The most thoroughly reviewed and widely accepted synthesis of scientific information concerning climate change is a series of reports issued by the **Intergovernmental Panel on Climate Change (IPCC)**. This international panel consists of many hundreds of scientists and government officials. Established in 1988 by the United Nations Environment Programme

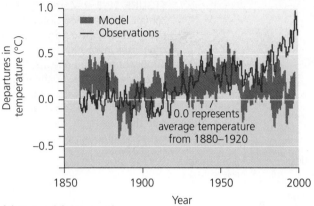

(a) Natural factors only

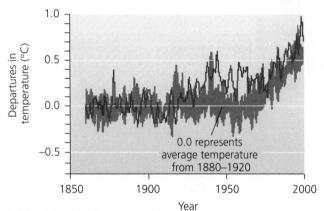

(b) Anthropogenic factors only

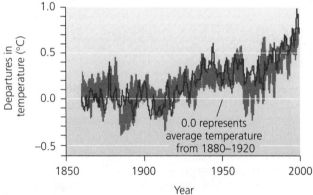

(c) All factors

FIGURE 18.9 ▲ Scientists test climate models by entering climate data from past years and comparing model predictions (blue areas) with actual observed data (red lines). Models that incorporate only natural factors **(a)** or only human-induced factors **(b)** do not predict real climate trends as well as models **(c)** that incorporate both natural and anthropogenic factors. Data from the Intergovernmental Panel on Climate Change, 2001. *Third assessment report.*

(UNEP) and the World Meteorological Organization (WMO), the IPCC was awarded the Nobel Peace Prize in 2007 for its work in informing the world of the trends and impacts of climate change.

In that year, the IPCC released its **Fourth Assessment Report**, which represented the current consensus of scientific climate research from around the world. This report summarized many thousands of scientific studies, and it documented observed trends in surface temperature, precipitation patterns, snow and ice cover, sea levels, storm intensity, and other factors. It also predicted future changes in these phenomena after considering a range of potential scenarios for future greenhouse gas emissions. The report addressed impacts of current and future climate change on wildlife, ecosystems, and society. Finally, it discussed possible strategies we might pursue in response to climate change. **FIGURE 18.10** summarizes some of the IPCC report's major observed and predicted trends and impacts.

Like all science, the IPCC report deals in uncertainties. Its authors therefore took great care to assign statistical probabilities to its conclusions and predictions. In addition, its estimates regarding impacts on society are conservative, because its scientific conclusions had to be approved by representatives of the world's national governments, some of which are reluctant to move away from a fossil-fuel-based economy.

Climate scientists at the WMO and many other institutions, agencies, and universities around the world are continuing to monitor our changing climate, as the IPCC begins its exhaustive work on a fifth assessment report, due out in 2014.

Temperatures continue to rise

The IPCC's 2007 report concluded that average surface temperatures on Earth increased by an estimated 0.74 °C (1.33 °F) in the century from 1906 to 2005 (**FIGURE 18.11**, p. 506), with most of this increase occurring in the last few decades. The numbers of extremely hot days and heat waves have increased globally, whereas the number of cold days has decreased. According to the WMO, the 16 warmest years on record since global measurements began 150 years ago have all been since 1990. The decade from 2000–2009 was the hottest ever, and since the 1960s each decade has been warmer than the last.

In the next 20 years, we can expect average surface temperatures on Earth to rise roughly 0.4 °C (0.7 °F), according to IPCC analysis. Even if we were to cease greenhouse gas emissions today from fossil fuel use and deforestation, temperatures would still rise 0.1 °C (0.2 °F) per decade because of a time lag: Some gases already in the atmosphere have yet to exert their full influence. At the end of the 21st century, the IPCC predicts global temperatures will be 1.8–4.0 °C (3.2–7.2 °F) higher than today's, depending on the emission scenario. Unusually hot days and heat waves will become more frequent. Future changes in temperature are predicted to vary from region to region in ways that parallel regional differences already apparent (**FIGURE 18.12**, p. 506). For example, polar regions will continue to experience the most intense warming.

Sea surface temperatures are also increasing as the oceans absorb heat from the atmosphere. The record number of hurricanes and tropical storms in 2005—Hurricane Katrina and 27 others—left many people wondering whether global warming was to blame. Scientists are not yet sure. Recent analyses of storm data suggest that warmer seas may not be increasing the number of tropical storms but may be increasing their power and, possibly, their duration.

Major Observed and *Predicted* Trends and Impacts of Climate Change, from IPCC Fourth Assessment Report, 2007

Global physical indicators	Social indicators
Earth's average surface temperature increased 0.74 °C (1.33 °F) in the past 100 years, *and will rise 1.8–4.0 °C (3.2–7.2 °F) in the 21st century.*	Farmers and foresters have had to adapt to altered growing seasons and disturbance regimes.
Eleven of the years from 1995 to 2006 were among the 12 warmest on record.	*Temperate-zone crop yields will rise until temperature warms beyond 3 °C (5.4 °F), but in the dry tropics and subtropics, crop productivity will fall and lead to hunger.*[5]
Oceans absorbed >80% of heat added to the climate system, and warmed to depths of at least 3,000 m (9,800 ft).	*Impacts on biodiversity will cause losses of food, water, and other ecosystem goods and services.*[2]
Glaciers, snow cover, ice caps, ice sheets, and sea ice will continue melting, contributing to sea-level rise.	*Sea level rise will displace people from islands and coasts.*[3]
Sea level rose by an average of 17 cm (7 in.) in the 20th century, *and will rise 18–59 cm (7–23 in.) in the 21st century.*	*Melting of mountain glaciers will reduce water supplies to millions of people.*[2]
Ocean water became more acidic by about 0.1 pH unit, *and will decrease in pH by 0.14–0.35 units more by century's end.*	*Economic costs will outweigh benefits as climate change worsens;*[2] *costs could average 1–5% of GDP globally for 4 °C (7.2 °F) of warming.*
Storm surges increased, *and will increase further.*[1]	Poorer nations and communities are suffering more from climate change, because they rely more on climate-sensitive resources and have less capacity to adapt.[2]
Carbon uptake by terrestrial ecosystems will peak by mid-21st century and then weaken or reverse, amplifying climate change.[2]	*Human health will suffer as increased warm-weather health hazards outweigh decreased cold-weather health hazards.*[2]
Regional physical indicators	**Biological indicators**
Arctic areas warmed fastest. *Future warming will be greatest in the Arctic and greater over land than over water.*	Species ranges are shifting toward the poles and upward in elevation, *and will continue to shift.*
Summer Arctic sea ice thinned by 7.4% per decade since 1978.	The timing of seasonal phenomena (such as migration and breeding) is shifting, *and will continue to shift.*
Precipitation will increase at high latitudes and decrease at subtropical latitudes, making wet areas wetter and dry ones drier.[1]	*About 20–30% of species studed so far will face extinction risk if temperature rises more than 1.5–2.5 °C (2.7–4.5 °F).*[5]
Droughts became longer, more intense, and more widespread since the 1970s, especially in the tropics and subtropics.[1]	*Species interactions and ecosystem structure and function could change greatly, resulting in biodiversity loss.*
Droughts and flooding will increase, leading to agricultural losses.[2]	*Corals will experience further mortality from bleaching and ocean acidification.*[5,4]
Hurricanes intensified in the North Atlantic since 1970[1], *and will continue to intensify.*[1]	
The thermohaline circulation will slow, but will not shut down and chill Europe in the 21st century.[3]	

FIGURE 18.10 ▲ Climate change has had numerous consequences already and is predicted to have many more. Listed here are some of the main observed and predicted trends and impacts described in the Intergovernmental Panel on Climate Change's *Fourth Assessment Report.* Observed phenomena are in plain text, whereas predicted future phenomena are in italicized text. For simplicity, this table expresses mean estimates only. The IPCC report provides ranges of estimates as well. [1]Certainty level = 66–90% probability of being correct. [2]Certainty level = ~80% probability of being correct. [3]Certainty level = 90–99% probability of being correct. [4]Certainty level = >99% probability of being correct. [5]Certainty level = ~50% probability of being correct. Data from the Intergovernmental Panel on Climate Change (IPCC), 2007. *Fourth assessment report.*

Precipitation is changing, too

A warmer atmosphere holds more water vapor, but changes in precipitation patterns have been complex, with some regions of the world receiving more rain and snow than usual and others receiving less. In regions such as the southwestern United States, droughts have become more frequent and severe, harming agriculture, worsening soil erosion, reducing water supplies, and triggering wildfire. Meanwhile, in dry and humid regions alike, heavy rain events have increased. Intense rainstorms, combined with land use changes, contribute to flooding, such as the June 2008 floods in Iowa and other parts of the U.S. Midwest that killed 13 people, left thousands homeless, and inflicted over $10 billion in damage.

Future changes in precipitation are predicted to vary among regions in ways that parallel regional differences seen over the past century (**FIGURE 18.13**, p. 507). In general, precipitation will increase at high latitudes and decrease at low and middle latitudes, magnifying differences in rainfall that already exist and worsening water shortages in many developing countries of the arid subtropics. In many areas, heavy precipitation events will become more frequent, increasing the risk of flooding.

Melting ice and snow have far-reaching effects

As the world warms, mountaintop glaciers are disappearing (**FIGURE 18.14**, p. 507). Between 1980 and 2008, the World

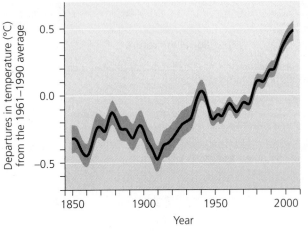

(a) Global temperature measured since 1850

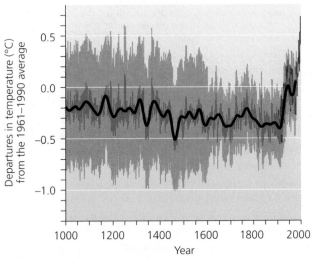

(b) Northern Hemisphere temperature over the past 1,000 years

FIGURE 18.11 ▲ Data from thermometers **(a)** show changes in Earth's average surface temperature since 1850. Gray-shaded area indicates range of uncertainty. In **(b)**, proxy indicators (blue line) and thermometer data (red line) together show average temperature changes in the Northern Hemisphere over the past 1,000 years. The gray-shaded zone represents the 95% confidence range. This record shows that 20th-century warming has eclipsed the magnitude of change during both the "Medieval Warm Period" (10th–14th centuries) and the "Little Ice Age" (15th–19th centuries). Data from the Intergovernmental Panel on Climate Change, 2007. *Fourth assessment report* (a); and Intergovernmental Panel on Climate Change, 2001. *Third assessment report* (b).

Glacier Monitoring Service estimates that the world's glaciers on average have each lost mass equivalent to 13 m (43 ft) vertical thickness of water. Many glaciers on tropical mountaintops have disappeared already. In Glacier National Park in Montana, only 26 of 150 glaciers present at the park's inception remain, and scientists estimate that by 2020 or 2030 even these will be gone.

Mountains accumulate snow in the winter and release meltwater gradually during the summer. Over one-sixth of the world's people live in regions that depend on mountain meltwater. As warming temperatures continue to diminish mountain glaciers, this will reduce summertime water supplies to millions of people, likely forcing whole communities to look elsewhere for water or to move.

Warming temperatures are also melting vast amounts of ice in the Arctic. Recent research reveals that the immense ice sheet that covers Greenland is melting faster and faster (see **THE SCIENCE BEHIND THE STORY**, pp. 510–511). At the other end of the world, in Antarctica, coastal ice shelves the size of Rhode Island have disintegrated as a result of contact with warmer ocean water, although increased precipitation is supplying the continent's interior with extra snow, making its ice sheet thicker even as it loses ice around its edges.

One reason warming is accelerating in the Arctic is that as snow and ice melt, darker, less-reflective surfaces are exposed, and Earth's *albedo*, or capacity to reflect light, decreases. Pools of meltwater are darker than ice or snow, and bare ground is darker still. As a result, more of the sun's rays are absorbed at the surface, fewer reflect back into space, and the surface warms. In a process of positive feedback, this warming causes more ice and snow to melt, which in turn causes more absorption of radiation and more warming (see Figure 5.2b, p. 110).

Scientists predict that snow cover and ice sheets will decrease near the poles and that sea ice will continue to shrink in both the Arctic and Antarctic. Some emission scenarios show Arctic sea ice disappearing completely by the late 21st century, creating new shipping lanes for commerce and a rush to exploit underwater oil and mineral reserves. Already, Russia, Canada, the United States, and other nations are jockeying for position, using new survey data to try to lay claim to regions of the Arctic as the ice melts.

Warmer temperatures in the Arctic are also causing *permafrost* (permanently frozen ground) to thaw. As ice crystals within permafrost melt, the thawing soil settles, destabilizing buildings, pipelines, and other infrastructure. When permafrost

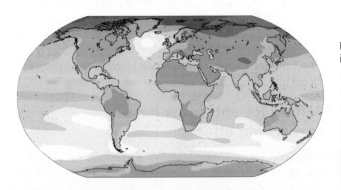

Percent increase in temperature (°C)

- ☐ 0.5–1.5
- 1.5–2.5
- 2.5–3.5
- 3.5–4.5
- 4.5–5.5
- 5.5–6.5
- 6.5–7.5

FIGURE 18.12 ◀ This map shows projected increases in surface temperature for the decade 2090–2099, relative to temperatures in 1980–1999. Land masses will warm more than oceans, and the Arctic will warm the most, up to 7.5 °C (13.5 °F). The IPCC uses multiple climate models when predicting regional variation in how temperature will change, and this map was generated using an emissions scenario that is intermediate in its assumptions, involving an average global temperature rise of 2.8 °C (5.0 °F) by 2100. Data from Intergovernmental Panel on Climate Change, 2007. *Fourth assessment report.* Climate Change 2007: Synthesis Report, Fig SPM.6.

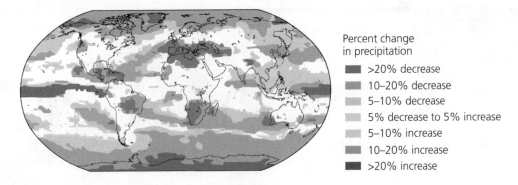

Percent change in precipitation

■ >20% decrease
■ 10–20% decrease
■ 5–10% decrease
■ 5% decrease to 5% increase
■ 5–10% increase
■ 10–20% increase
■ >20% increase

FIGURE 18.13 ▲ This map shows projected changes in June–August precipitation for the decade 2090–2099, relative to precipitation levels in 1980–1999. Browner shades indicate less precipitation, and bluer shades indicate more precipitation. White indicates areas for which models could not agree. This map was generated using an emissions scenario that is intermediate in its assumptions, involving an average global temperature rise of 2.8 °C (5.0 °F) by 2100. Data from Intergovernmental Panel on Climate Change, 2007. *Fourth assessment report.* Climate Change 2007: Synthesis Report, Fig 3.3.

thaws, it also can release methane that has been stored for thousands of years. Because methane is a potent greenhouse gas, this acts as a positive feedback mechanism (p. 110) that intensifies climate change. No one knows how much methane lies beneath Arctic permafrost, but it is a very large amount, and many researchers worry that its release could drive climate change beyond our control.

Rising sea levels will affect hundreds of millions of people

As glaciers and ice sheets melt, increased runoff into the oceans causes sea levels to rise. Sea levels also are rising because ocean water is warming, and water expands in volume as its temperature increases. In fact, recent sea-level rise has resulted primarily from the thermal expansion of seawater. Worldwide, average sea levels rose an estimated 17 cm (6.7 in.) during the 20th century (**FIGURE 18.15**, p. 509). Seas rose by an estimated 1.8 mm/year from 1961–2003 and 3.1 mm/year from 1993–2006. These numbers represent vertical rises in water level, and on most coastlines a vertical rise of a few inches means many feet of horizontal incursion inland.

Higher sea levels lead to beach erosion, coastal flooding, intrusion of salt water into aquifers, and other impacts (see **ENVISIONIT**, p. 508). In 1987, unusually high waves struck the Maldives and triggered a campaign to build a large seawall around Malé, the nation's capital. Known as "The Great Wall of Malé," the seawall is intended to protect buildings and

(a) Grinnell Glacier in 1938

(b) Grinnell Glacier in 2005

FIGURE 18.14 ▲ Glaciers are melting rapidly around the world as global warming proceeds. The Grinnell Glacier in Glacier National Park, Montana, retreated substantially between 1938 **(a)** and 2005 **(b)**. The graph shows declines in mass in 30 of the world's major glaciers monitored since 1980. Data from World Glacier Monitoring Service.

People on ocean islands are feeling the impacts of climate change.

Storm surge on Tuvalu

Flooding on Tuvalu

Salt water seeps into wells and kills crops.

Rising seas erode shorelines, submerge land, inundate cemeteries, and force people from their homes.

Funafuti: 4 meters above sea level

Island nations are begging us to curb our greenhouse gas emissions.

Maldives and Kiribati officials

As climate change proceeds, impacts will intensify along our shores, too.

Storm surge in Massachusetts

YOU CAN MAKE A DIFFERENCE

➤ Lower your carbon footprint by reducing your energy use.
➤ Pressure your campus to invest in energy efficiency and clean renewable energy sources.
➤ Lobby policymakers to take action to curb emissions. Doing this with a group like 350.org makes it fun.

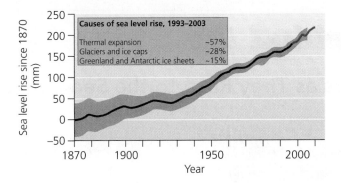

FIGURE 18.15 ▲ Data from tide gauges (black line) and satellite observations (red line) show that global average sea level has risen about 200 mm (7.9 in.) since 1870. Gray-shaded area indicates range of uncertainty. Thermal expansion of water accounts for most sea level rise. Data from Intergovernmental Panel on Climate Change, 2007. *Fourth assessment report.*

roads by dissipating the energy of incoming waves during storm surges. A *storm surge* is a temporary and localized rise in sea level brought on by the high tides and winds associated with storms. The higher the sea level is to begin with, the further inland a destructive storm surge can reach.

On December 26, 2004, the Maldives got a taste of what could be in store in the future when a massive *tsunami* (pp. 44, 46), or tidal wave, devastated coastal areas throughout the Indian Ocean. The tsunami killed 100 Maldives residents and left 20,000 homeless. Schools, boats, tourist resorts,

hospitals, and transportation and communication infrastructure were damaged or destroyed. The World Bank estimated that direct damage in the Maldives totaled $470 million, an astounding 62% of the nation's gross domestic product (GDP). Indirect damage from soil erosion, saltwater contamination of aquifers, and other impacts continues to cause further long-term economic losses. The tsunami was caused *not* by climate change, but by an earthquake. Yet as sea levels rise, the damage that natural events can inflict increases considerably.

The Maldives has fared better than many other island nations. It saw sea levels rise about 2.5 mm per year throughout the 1990s, but most Pacific islands are experiencing greater rises in sea level. Regions experience differing amounts of sea level change because land may be rising or subsiding naturally, depending on local geological conditions.

In the United States, 53% of the population lives in coastal counties (**FIGURE 18.16A**). Vulnerability to storm surges became tragically apparent in 2005 when Hurricane Katrina struck New Orleans and the Gulf Coast, followed shortly thereafter by Hurricane Rita (**FIGURE 18.16B**). Outside New Orleans today, marshes of the Mississippi River delta are being lost rapidly as rising seas eat away at coastal vegetation. These coastal wetlands are also being lost because dams upriver hold back silt that once maintained the delta, land has subsided because of petroleum extraction, and salt water is encroaching up channelized waterways (killing freshwater plants). Pollution from the *Deepwater Horizon* oil spill of 2010 (pp. 445, 447, 547–548) is expected to further eliminate wetlands by killing coastal vegetation.

0 1 2 3 5 8 12 20 35 60 80
Height above sea level (m)

FIGURE 18.16 ▲ Substantial areas of the Atlantic and Gulf Coasts of the United States lie within a few meters of sea level **(a)** and thus are susceptible to long-term sea level rise and storm surges if climate change is not brought under control. In a world of higher sea levels and stronger storms, rescues like this one from the floodwaters of Hurricane Katrina's storm surge in coastal Mississippi in 2005 **(b)** could become more frequent.

The SCIENCE behind the Story

Timing Greenland's Glaciers as They Race to the Sea

Scientists have known for years that the Arctic is bearing the brunt of global warming and that the massive ice sheet covering Greenland is melting around its edges. But data from 1993 to 2003 showed Greenland's ice loss accounting for only 4–12% of global sea level rise, about 0.21 mm/yr. And as authors of the IPCC's *Fourth Assessment Report* used results from climate models to predict Greenland's future contributions to seal level rise, the models told them to expect more of the same.

However, some brand-new research hadn't made it into the models. Scientists studying how ice moves were learning that ice sheets can collapse more quickly than expected and that Greenland's ice loss is accelerating. As a result, they said, the IPCC report underestimated the likely speed and extent of future sea level rise.

Greenland's ice sheet is massive, averaging nearly a mile deep and covering as much area as Texas, California, Michigan, and Minnesota combined. If the entire ice sheet were

Outlet glaciers melting into Scoresby Sund, Greenland's largest fjord

to melt, global sea level would rise by a whopping 7 m (23 ft).

The ice sheet gains mass by accumulating snow during cold weather, which compresses into ice over time. It loses mass as surface ice melts in warm weather, generally at the periphery, where ice is thinnest or contacts seawater. If melting and runoff outpace accumulation, then the ice sheet shrinks.

But researchers are now learning that the internal physical dynamics of how ice moves may be more important. These dynamics can speed the flow of *outlet glaciers* downhill toward the coast, where eroding ice sloughs off and melts into the sea.

The first good indication of this process came in 2002 when a team led

by Jay Zwally of NASA's Goddard Space Flight Center in Maryland noted that ice in outlet glaciers flows more quickly during warm months, when pools of meltwater form on the surface. In a paper published in the journal *Science* in 2002, Zwally's group proposed that meltwater leaks down through crevasses and vertical tunnels called *moulins* to the bottom of the glacier. There, the water runs downhill in a layer between bedrock and ice, lubricating the bedrock surface and enabling the ice to slide downhill, like a car hydroplaning on a wet road. In addition, the meltwater weakens ice on its way down and warms the base of the glacier, melting some of it to create more water (**see figure**).

Other scientists proposed an additional mechanism: Warming ocean water melts ice shelves along the coast, depriving outlet glaciers of the buttressing support that holds them in place. Without a floating ice tongue at its terminus, a glacier slides into the ocean more readily.

These physical dynamics involve positive feedback, researchers say;

All told, more than 2.5 million ha (1 million acres) of Louisiana's coastal wetlands have vanished since 1940. Continued wetland loss will deprive New Orleans (much of which is below sea level, safeguarded only by levees) of protection against future storm surges.

At the end of the 21st century, the IPCC predicts mean sea level will be 18–59 cm (7–23 in.) higher than today's, depending on our level of emissions. However, these estimates do not take into account recent findings on accelerated ice melting in Greenland (see **The Science behind the Story**, above)—and apparently Antarctica as well—because that research was so new that it had not yet been incorporated into climate models at the time of the latest IPCC report. If polar melting continues to accelerate, then sea levels will rise more quickly.

If sea levels rise as predicted, hundreds of millions of people will be displaced or will need to invest in costly efforts to protect against high tides and storm surges. Densely populated regions on low-lying river deltas, such as Bangladesh, would be most affected. So would storm-prone regions such as Florida, coastal cities such as Houston and Charleston, and areas where land is subsiding, such as the U.S. Gulf Coast. Many Pacific islands would need to be evacuated. Already some nations such as Tuvalu and the Maldives fear for their very existence. In the meantime, island nations such as the Maldives are likely to suffer from shortages of fresh water as rising seas bring salt water into aquifers, just as the 2004 tsunami did. The contamination of groundwater and soils by seawater also threatens coastal areas such as Tampa, Florida, which depend on small lenses of fresh water that float atop saline groundwater.

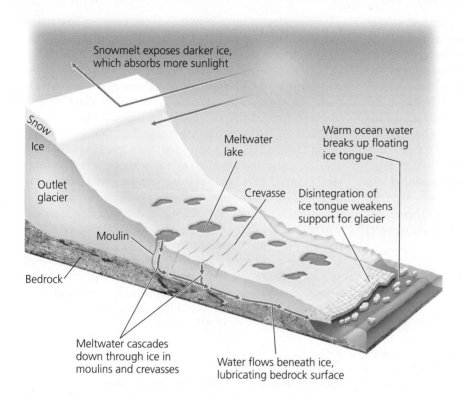

Snowmelt exposes darker ice, which absorbs more sunlight

Snow

Ice

Outlet glacier

Meltwater lake

Warm ocean water breaks up floating ice tongue

Crevasse

Disintegration of ice tongue weakens support for glacier

Moulin

Bedrock

Meltwater cascades down through ice in moulins and crevasses

Water flows beneath ice, lubricating bedrock surface

Outlet glaciers from Greenland's ice sheet are accelerating their slide into the ocean because meltwater is descending through moulins and crevasses, lubricating the bedrock surface. Moreover, melting snow exposes darker ice, which absorbs sunlight and speeds melting. Finally, breakup of floating ice tongues weakens support for the glacier.

once global warming initiates these processes, they encourage further melting. As such, we should expect that Greenland's melting will accelerate.

Several research groups soon attempted to measure the rates of Greenland's ice loss and account for the impact of these physical dynamics. In 2006, Eric Rignot of NASA's Jet Propulsion Laboratory in Pasadena,

California, and Pannir Kanagaratnam of the University of Kansas, Lawrence, published research in *Science* showing that rates of ice loss had more than doubled between 1996 and 2005.

To measure how ice moves over such a scale and in such inhospitable terrain, Rignot and Kanagaratnam analyzed satellite data showing how fast Greenland's ice was moving, along

with aircraft radar measurements giving data on ice thickness. This let them estimate how much ice was lost during this period.

The researchers found that the physical dynamics of ice movement were responsible for the loss of 56 km³ of ice in 1996, 92 km³ in 2000, and 167 km³ in 2005. The 2005 amount is equivalent to 44 trillion gallons of water, enough to supply the New York City metro area for over a century.

They also determined that the physical dynamics of ice flow accounted for two-thirds of total ice loss. The other third was due to runoff outpacing snow accumulation, and this too was accelerating. Including these amounts showed that Greenland lost a grand total of 91 km³ of ice in 1996, 138 km³ in 2000, and 224 km³ in 2005. This last amount exceeds all the water consumed in the United States in an entire year.

Clearly, ice losses were accelerating quickly. Rignot and Kanagaratnam calculated that Greenland's contribution to sea level rise increased from 0.23 mm/yr in 1996 to 0.57 mm/yr in 2005. Scientists are now keeping a close eye on Greenland's ice sheet to see whether it continues to discharge water more quickly. If it does, climate modelers will need to incorporate into their models the new and evolving understanding of the physical dynamics of ice—in time for the next IPCC report scheduled for 2014. ∎

WEIGHING THE ISSUES

Environmental Refugees Citizens of the Maldives see an omen of their future in the Pacific island nation of Tuvalu, which has been losing 9 cm (3.5 in.) of elevation per decade to rising seas. Appeals from Tuvalu's 11,000 citizens were heard by New Zealand, which began accepting these environmental refugees in 2003. Do you think the rest of the world should grant such environmental refugees international status and assume some responsibility for taking care of them? Do you think a national culture can survive if its entire population is relocated? Think of the tens of thousands of refugees from Hurricane Katrina. How did their lives and culture fare in the wake of that tragedy?

Coral reefs are threatened by climate change

Maldives residents also worry about damage to their coral reefs (pp. 439–442), marine ecosystems that are critical for their economy. Coral reefs provide habitat for important food fish that are consumed locally and exported. They offer snorkeling and scuba diving sites for tourism. Reefs also reduce wave intensity, protecting coastlines from erosion. Around the world, rising seas are eating away at the coral reefs, mangrove forests, and salt marshes that serve as barriers protecting our coasts (pp. 439–444).

Climate change poses two additional threats to coral reefs, as we saw in Chapter 16. First, warmer waters contribute to coral bleaching (p. 442), which kills corals. Second, enhanced CO_2 concentrations in the atmosphere are altering ocean chemistry. As ocean water absorbs atmospheric CO_2, it becomes more acidic. This increased acidity impairs the ability of coral and other organisms to build exoskeletons of calcium carbonate. Ocean acidification and the potential loss of coral reefs worldwide (explained in Chapter 16's *Science behind the Story*, pp. 440–441) threaten to become one of the most serious and far-reaching impacts of global climate change. The oceans have already decreased by 0.1 pH unit, and they are predicted to decline in pH by 0.14–0.35 more units over the next 100 years. This could easily be enough to destroy most or all of our planet's living coral reefs. Such destruction would reduce marine biodiversity and fisheries significantly, because so many organisms depend on living coral reefs for food and shelter.

Climate change affects organisms and ecosystems

As illustrated by the coral reef crisis, changes in Earth's physical systems often have direct consequences for living things. Organisms are adapted to their environments, so they are affected when those environments are altered. As global warming proceeds, it is modifying all manner of biological phenomena that are regulated by temperature. In the spring, plants are leafing out earlier, insects are hatching earlier, birds are migrating earlier, and animals are breeding earlier. These shifts can create mismatches in seasonal timing. For example, European birds known as great tits had evolved to time their breeding so as to raise their young at the time of peak caterpillar abundance. Now caterpillars are peaking earlier, but the birds have been unable to adjust, and fewer young birds are surviving.

Biologists are also recording spatial shifts in the ranges of organisms, with plants and animals moving toward the poles or upward in elevation (i.e., toward cooler regions) as temperatures warm. As these trends continue, some organisms will not be able to cope, and the IPCC estimates that as many as 20–30% of all plant and animal species could be threatened with extinction. Trees may not be able to shift their distributions fast enough. Rare species may be forced out of preserves into developed areas, undercutting the effectiveness of refuges as tools for conservation. Animals and plants adapted to montane environments may be forced uphill until there is nowhere left to go (**FIGURE 18.17**). Recall that we saw in Chapter 3 how this is occurring with organisms in the cloud forests of Monteverde.

Effects on plant communities comprise an important component of climate change, because by drawing in CO_2 for photosynthesis, plants act as reservoirs for carbon. If higher CO_2 concentrations enhance vegetative growth, this could help mitigate carbon emissions, in a process of negative feedback. However, if climate change decreases plant growth (through drought, fire, or disease, for instance), then positive feedback could increase carbon flux to the atmosphere. As we saw in Chapter 5's *Science behind the Story* (pp. 126–127), Free-Air CO_2 Enrichment (FACE) experiments are revealing complex answers, showing that extra carbon dioxide can bring both positive and negative results for plant growth.

In regions where precipitation and stream flow increase, erosion and flooding will pollute and alter aquatic systems. In regions where precipitation decreases, lakes, ponds, wetlands, and streams will shrink, affecting aquatic organisms, as well as human health and well-being. The many impacts on ecological systems will diminish the ecosystem goods and services we receive from nature and that our societies depend on, from food to clean air to drinking water.

Climate change affects people

Damage from drought, flooding, storm surges, sea level rise, and other impacts of climate change has already taken a toll on the lives and livelihoods of millions of people. However, climate change will have still more consequences for human populations, including impacts on agriculture, forestry, economics, and health.

Agriculture For farmers, earlier springs require earlier crop planting. For some crops in the temperate zones, moderate warming may slightly increase production because growing seasons become longer. Additional carbon dioxide being available to plants for photosynthesis may also increase yields, but as discussed above, elevated CO_2 can have mixed results. Moreover, some research shows that crops become less nutritious when supplied with more carbon dioxide. If rainfall shifts in space and time, intensified droughts and floods will likely cut into agricultural productivity. Considering all factors together, the IPCC predicts global crop yields to increase somewhat, but beyond a rise of 3 °C (5.4 °F), it expects crop yields to decline. In seasonally dry tropical and subtropical regions, growing seasons may be shortened, and harvests may be more susceptible to drought. Thus, scientists predict that

FIGURE 18.17 ◄ The pika (*Ochotona princeps*) is a unique mammal that lives at high elevations in mountains of western North America. It is apparently highly vulnerable to climate change, as many populations in the Great Basin have been extirpated already as temperatures have warmed, pushing montane organisms upslope.

crop production will fall in these regions even with minor warming. This would worsen hunger in many of the world's developing nations.

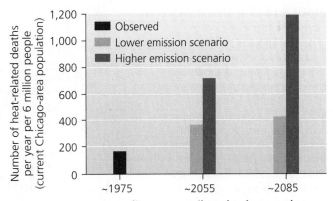

Average summer mortality rates attributed to hot weather episodes

FIGURE 18.18 ▲ Heat stress will increase with climate change. In the United States, the annual number of days over 100 °F is predicted to increase greatly by the end of the century, under either low- or high-emissions scenarios. This will lead to a greater risk of death from heat stress in cities such as Chicago, which experienced a killer heat wave in 1995. Data from Karl, Thomas R., et al. (eds.), 2009. *Global climate change impacts in the United States.*, p. 90. U.S. Global Change Research Program and Cambridge University Press.

WEIGHING THE ISSUES

Agriculture in a Warmer World The IPCC predicts that warming will shift agricultural belts toward the poles. Locate the nations of Russia, Canada, and Argentina on a world map, and hypothesize how such a poleward shift of agriculture might affect these nations. Now locate India, Nigeria, and Ethiopia, and hypothesize how the shift might affect them. Many developing nations near the equator have long suffered from food shortages and soil degradation. Do you think climate change could magnify inequities between developed and developing nations? If so, what might we do to alleviate this problem?

Forestry In the forests that provide our timber and paper products, enriched atmospheric CO_2 may spur greater growth in the near term, but other climatic effects such as drought, fire, and disease may eliminate these gains. For example, droughts brought about by a strong El Niño in 1997–1998 allowed immense forest fires to destroy millions of hectares of rainforest in Indonesia, Brazil, Mexico, and elsewhere. In North America, forest managers increasingly find themselves battling catastrophic fires, invasive species, and insect and disease outbreaks. Catastrophic fires are caused in part by decades of fire suppression (p. 328) but are also promoted by longer, warmer, drier fire seasons (see Figure 12.18a, p. 329). Milder winters and hotter, drier summers are promoting outbreaks of pine beetles, pest insects that are now destroying millions of acres of trees in North American forests (see Figure 12.18b, pp. 329–330).

Health As climate change proceeds, we will face more heat waves—and heat stress can cause death (**FIGURE 18.18**), especially among older adults. A 1995 heat wave in Chicago killed at least 485 people, and a 2003 heat wave in Europe killed 35,000 people. A warmer climate also exposes us to other health problems:

▶ Respiratory ailments from air pollution, as hotter temperatures promote formation of photochemical smog (pp. 477–478)

▶ Expansion of tropical diseases, such as dengue fever, into temperate regions as vectors of infectious disease (such as mosquitoes) move toward the poles

▶ Disease and sanitation problems when floods overcome sewage treatment systems

▶ Injuries and drowning if storms become more frequent or intense

Health hazards from cold weather will decrease, but most researchers feel that the increase in warm-weather hazards will more than offset these gains.

Economics People will experience a variety of economic costs and benefits from the many impacts of climate change, but on the whole researchers predict that costs will outweigh benefits, especially as climate change grows more severe. Climate change is also expected to widen the gap between rich and poor, both within and among nations. Poorer people have less wealth and technology with which to adapt to climate change, and poorer people rely more on resources (such as local food and water) that are sensitive to climatic conditions.

From a wide variety of economic studies, the IPCC estimated that climate change will cost 1–5% of GDP on average globally, with poor nations losing proportionally more than rich nations. Economists trying to quantify damages from climate change by measuring its external costs (pp. 150–151) have proposed costs of anywhere from $10 to $350 per ton of carbon. The highest-profile economic study to date has been the *Stern Review* commissioned by the British government (see *The Science behind the Story*, Chapter 6, pp. 152–153). This exhaustive review maintained that climate change could cost us roughly 5–20% of GDP by the year 2200, but that investing just 1% of GDP starting now could enable us to avoid these future costs. Regardless of the precise numbers, many economists and policymakers are concluding that spending money now to mitigate climate change will save us a great deal more money in the future.

Impacts will vary regionally

Future impacts of climate change will be subject to regional variation, so the way each of us experiences these impacts over the coming decades will vary depending on where we live. Temperature changes have been greatest in the Arctic (**FIGURE 18.19**), and scientists are still debating why. Here, ice sheets are melting, sea ice is thinning, storms are increasing, and altered conditions are posing challenges for people and

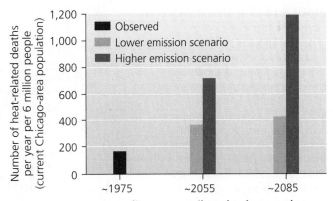

CHAPTER 18 Global Climate Change

513

FIGURE 18.19 ▶ The Arctic has borne the brunt of climate change's impacts so far. As Arctic sea ice melts, it recedes from large areas, as shown by the map indicating the mean minimum summertime extent of sea ice for the recent past, present, and future. Inuit people find it difficult to hunt and travel in their traditional ways, and polar bears starve because they are less able to hunt seals. Human-made structures are damaged as permafrost thaws beneath them: Buildings can lean, buckle, crack, and fall. Map data from National Center for Atmospheric Research and National Snow and Ice Data Center.

wildlife. As sea ice melts earlier, freezes later, and recedes from shore, it becomes harder for Inuit people and for polar bears alike to hunt the seals they each rely on for food. Thin sea ice is dangerous for people to travel and hunt upon, and in recent years, polar bears have been dying of exhaustion and starvation as they try to swim long distances between ice floes. Permafrost is thawing in the Arctic, destabilizing countless buildings. The strong Arctic warming is melting ice caps and ice sheets, contributing to sea level rise.

514

WEIGHING THE ISSUES

Climate Change and Human Rights In 2005, a group representing North America's Inuit people sent a legal petition to the Inter-American Commission on Human Rights, demanding that the United States restrict its greenhouse gas emissions, which the Inuit maintained were destroying their way of life in the Arctic. The Commission dismissed the petition. Do you think Arctic-living people deserve some sort of compensation from industrialized nations whose emissions have caused climate change that disproportionately affects the Arctic? What ethical or human rights issues, if any, do you think climate change presents? How could these best be resolved?

For the United States, potential impacts are analyzed and summarized by the U.S. Global Change Research Program, which Congress created in 1990 to coordinate federal climate research. In 2009, scientists for this program reviewed current research and issued a comprehensive report highlighting the effects of climate change on the United States and issuing predictions of future impacts (**TABLE 18.2**).

The report predicted that some impacts would be felt across the nation. Average temperatures in most of the United

TABLE 18.2 Some Predicted Impacts of Climate Change in the United States

▶ Average temperatures will rise 2.2–6.1 °C (4–11 °F) further by the end of this century.

▶ Droughts and flooding will worsen.

▶ Longer growing seasons and enhanced CO_2 will favor crops, but more drought, heat stress, pests, and diseases will decrease most yields.

▶ Snowpack will decrease in the West; water shortages will worsen.

▶ Cold-weather illness will decline, but health problems due to heat stress, disease, and pollution will rise. Some tropical diseases will spread north.

▶ Sea level rise and storm surges will erode beaches and destroy coastal wetlands and real estate.

▶ Alpine ecosystems and barrier islands will begin to vanish.

▶ Drought, fire, and pest outbreaks will continue to alter forests.

▶ Northeast forests will lose sugar maples; Southeast forests will be invaded by grassland; Southwest ecosystems will turn more desertlike.

▶ Melting permafrost will undermine Alaskan buildings and roads.

Adapted from Karl, Thomas R., et al. (eds.), 2009. *Global climate change impacts in the United States.* U.S. Global Change Research Program and Cambridge University Press.

**Lower Emissions Scenario
Projected Temperature Change (°F)**

End-of-Century (2080-2099 average)

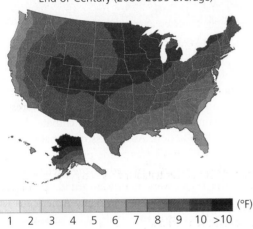

**Higher Emissions Scenario
Projected Temperature Change (°F)**

End-of-Century (2080-2099 average)

(°F)

1   2   3   4   5   6   7   8   9   10 >10

FIGURE 18.20 ◄ Average temperatures across the United States are predicted to rise by 4–6 °F by the end of this century under a low-emissions scenario, and 7–11 °F under a high-emissions scenario. Data from Karl, Thomas R., et al. (eds.), 2009. *Global climate change impacts in the United States.* U.S. Global Change Research Program and Cambridge University Press.

States have already increased by 0.6–1.1 °C (1–2 °F) since the 1960s and 1970s, and they will rise by another 2.2–6.1 °C (4–11 °F) by the end of this century (**FIGURE 18.20**), according to the report. Plant communities will likely change in all areas of the country, in general shifting northward and upward in elevation. Extreme weather events are projected to become more frequent.

Other impacts will likely vary by region, with each region of the United States facing its own challenges (**FIGURE 18.21**). For instance, winter and spring precipitation is projected to decrease across the South but increase across the North. Drought may strike in some regions and flooding in others. Sea level rise may affect the East Coast more than the West Coast. Agriculture may experience a wide array of effects that will vary from one region to another. As climate models improve, scientists become able to present graphical depictions summarizing the predicted impacts of climate change on particular geographic areas (**FIGURE 18.22**).

All these impacts of climate change are projected consequences of the warming effect of our greenhouse gas emissions (**FIGURE 18.23**). We are bound to experience further consequences, but by addressing the root causes of anthropogenic climate change now, we may still be able to prevent the most severe future impacts.

CHAPTER 18 Global Climate Change

515

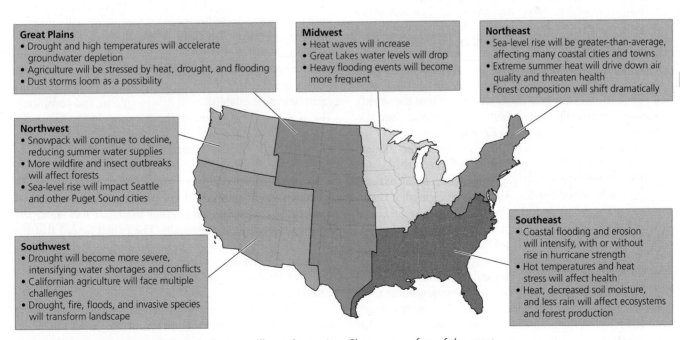

Great Plains
• Drought and high temperatures will accelerate groundwater depletion
• Agriculture will be stressed by heat, drought, and flooding
• Dust storms loom as a possibility

Midwest
• Heat waves will increase
• Great Lakes water levels will drop
• Heavy flooding events will become more frequent

Northeast
• Sea-level rise will be greater-than-average, affecting many coastal cities and towns
• Extreme summer heat will drive down air quality and threaten health
• Forest composition will shift dramatically

Northwest
• Snowpack will continue to decline, reducing summer water supplies
• More wildfire and insect outbreaks will affect forests
• Sea-level rise will impact Seattle and other Puget Sound cities

Southwest
• Drought will become more severe, intensifying water shortages and conflicts
• Californian agriculture will face multiple challenges
• Drought, fire, floods, and invasive species will transform landscape

Southeast
• Coastal flooding and erosion will intensify, with or without rise in hurricane strength
• Hot temperatures and heat stress will affect health
• Heat, decreased soil moisture, and less rain will affect ecosystems and forest production

FIGURE 18.21 ▲ Impacts of climate change will vary by region. Shown are a few of the most important impacts scientists expect for each region by the end of the century. Adapted from Karl, Thomas R., et al. (eds.), 2009. *Global climate change impacts in the United States.* U.S. Global Change Research Program and Cambridge University Press.

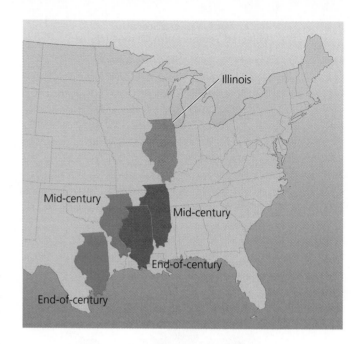

FIGURE 18.22 ▶ Researchers used climate models to predict the future climate of Illinois. Under a low-emissions scenario (purple), scientists predict that Illinois will become warmer, attaining in midcentury a climate similar to northern Mississippi's, and at the end of the century a climate similar to northern Louisiana's. Under a high-emissions scenario (green), Illinois is predicted to become still hotter, but drier, attaining in midcentury a climate similar to that of southwest Arkansas, and at the end of the century a climate similar to that of east Texas. Adapted from Karl, Thomas R., et al. (eds.), 2009. *Global climate change impacts in the United States.* U.S. Global Change Research Program and Cambridge University Press.

Are we responsible for climate change?

Scientists agree that most or all of today's global warming is due to the well-documented recent increase in greenhouse gas concentrations in our atmosphere. They also agree that this rise in greenhouse gases results from our combustion of fossil fuels for energy and secondarily from land use changes, including deforestation and agriculture.

By the time the IPCC's *Fourth Assessment Report* came out, many scientists had already become concerned enough about the consequences of climate change to put themselves on record urging governments to address the issue. In 2005, the national academies of science from 11 nations (Brazil, Canada, China, France, Germany, India, Italy, Japan, Russia, the United Kingdom, and the United States) issued a joint statement urging political leaders to take action. The statement read, in part:

> The scientific understanding of climate change is now sufficiently clear to justify nations taking prompt action [to reduce] global greenhouse gas emissions. . . . A lack of full scientific certainty about some aspects of climate change is not a reason for delaying an immediate response that will, at a reasonable cost, prevent dangerous anthropogenic interference with the climate system.

Such a broad and clear consensus statement from the world's scientists was virtually unprecedented, on any issue—and scientists' concerns have only grown since that time.

Yet despite the overwhelming evidence for climate change and its impacts, many people, especially in the United States, long tried to deny that it was happening. Many of these naysayers now admit that the climate is changing but doubt that we are the cause. Indeed, while most of the world's nations moved forward to confront climate change through international dialogue, in the United States public discussion of climate change remained mired in outdated debates over whether the phenomenon was real and whether humans were to blame. These debates were fanned by spokespeople from conservative think tanks and a handful of scientists, many funded by ExxonMobil and other corporations in the fossil fuel industries. These people aimed to cast doubt on the scientific consensus, and their views were amplified by the American news media, which seeks to present two sides to every issue, even when the sides' arguments are not equally supported by evidence.

For many Americans, former Vice President Al Gore's 2006 movie and book, *An Inconvenient Truth*, presented an eye-opening summary of the science of climate change and a compelling call to action. Awareness of climate change grew as the 2007 IPCC report was made publicly available on the Internet and was widely covered in the media, and later as Gore and the IPCC were jointly awarded the Nobel Peace Prize. At the same time, however, many Americans who disliked Gore's politics also came to reject his message on climate change.

In 2009, a hacker illegally broke into computers at the University of East Anglia, U.K., and made public several thousand documents, including over 1,000 private emails among a handful of climate scientists. A few of these messages appeared to show questionable behavior in the use of data and the treatment of other researchers. Climate-change deniers named the incident "Climategate" and used it to accuse the entire scientific establishment of wrongdoing and conspiracy. The news media disseminated the story widely. However, subsequent investigations into the affair by independent panels exonerated the climate scientists, concluding that there was no evidence of wrongdoing. Although some individuals may have exercised poor taste or judgment at times, and some practices could be improved, the panels also pointed out that many media accounts trumpeting the news had misrepresented the content of the emails. The panels agreed that the hacked emails among a few individuals in no way called into question the vast array of research results compiled by thousands of hard-working independent climate scientists over several decades.

Today, most of the world's people accept that our fossil fuel consumption is altering the planet that our children will inherit. As youth and grassroots activists spread this message, and as political leaders begin to respond, everyday people are searching for solutions. As a result of this shift in public perception, and in response to demand from their shareholders, many corporations and industries are looking

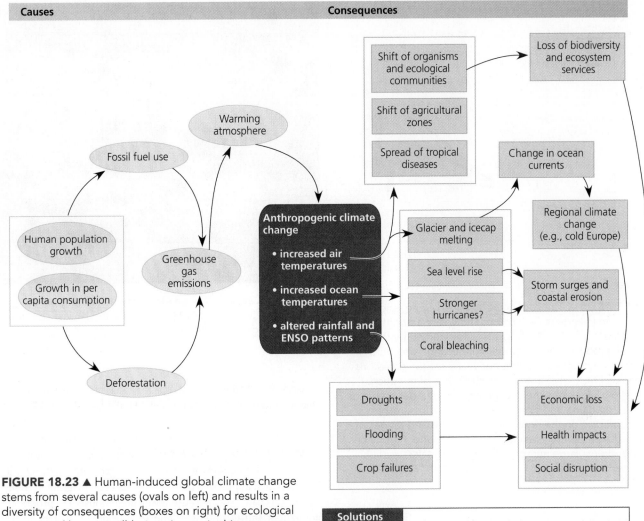

FIGURE 18.23 ▲ Human-induced global climate change stems from several causes (ovals on left) and results in a diversity of consequences (boxes on right) for ecological systems and human well-being. Arrows in this concept map lead from causes to consequences. Note that items grouped within outlined boxes do not necessarily share any special relationship; the outlined boxes are intended merely to streamline the figure.

Solutions

As you progress through this chapter, try to identify as many solutions to anthropogenic climate change as you can. What could you personally do to help address this issue? Consider how each action or solution might affect items in the concept map above.

for ways to reduce their greenhouse gas emissions and are supporting policies to reduce them. Today in the United States and worldwide, the mainstream debate focuses on how best to respond to the challenges of climate change.

RESPONDING TO CLIMATE CHANGE

Today we possess a new, broad consensus that climate change is a clear and present challenge to our society. How, exactly, we should respond to climate change is a difficult question, however, and one we will likely be wrestling with for decades.

Shall we pursue mitigation or adaptation?

We can respond to climate change in two fundamental ways. One is to pursue actions that reduce greenhouse gas emissions, so as to lessen the severity of climate change.

This strategy is called **mitigation** because the aim is to mitigate, or alleviate, the problem. Examples include improving energy efficiency, switching to clean and renewable energy sources, preventing deforestation, recovering landfill gas, and encouraging farm practices that protect soil quality.

The second type of response is to pursue strategies to minimize the impacts of climate change on us. This strategy is called **adaptation** because the goal is to adapt to change by finding ways to cushion oneself from its blows. Erecting a seawall as Maldives residents did with the Great Wall of Malé is an example of adaptation using technology and engineering. The people of Tuvalu also adapted, but with a behavioral choice—some chose to leave their island and make a new life in New Zealand (see *Weighing the Issues*, p. 511). Other examples of adaptation include restricting coastal development; adjusting farming practices to cope with drought; and modifying water management practices to deal with reduced river flows, glacial outburst floods, or salt contamination of groundwater.

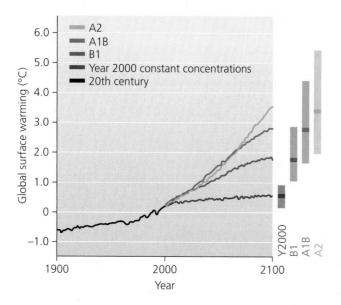

FIGURE 18.24 ▲ The sooner we stabilize our emissions, the less climate change we will cause. Shown are amounts of warming projected under four scenarios studied by the IPCC. The red line shows temperature change we could expect if we were able to limit our yearly carbon emissions to the level they were in the year 2000. The blue, green, and yellow lines show the change expected under scenarios of rapid, medium, and slow control of emissions, respectively. The bars on the right show the means and ranges of year-2100 temperature values for each scenario. Predictions are based on a large number of climate models. Data from Intergovernmental Panel on Climate Change, 2007. *Fourth assessment report.*

FIGURE 18.25 ▲ Coal-fired electricity-generating power plants, such as this one in Maryland, are the largest contributors to U.S. greenhouse emissions.

Both adaptation and mitigation are necessary. Adaptation is needed because even if we were to halt all our emissions now, global warming would continue until the planet's systems reach a new equilibrium, with temperature rising an estimated 0.6 °C (1.0 °F) more by the end of the century. Because we will face this change no matter what we do, it is wise to develop ways to minimize its impacts.

Mitigation is necessary because if we do nothing to diminish climate change, it will eventually overwhelm any efforts at adaptation we might make. To leave a sustainable future for our civilization and to safeguard the living planet that we know, we will need to pursue mitigation. The faster we begin reducing our emissions, the lower the level at which they will peak, and the less we will alter climate (**FIGURE 18.24**). We will spend the remainder of our chapter examining approaches for the mitigation of climate change.

Electricity generation is the largest source of U.S. greenhouse gases

The generation of electricity produces the largest portion (40%) of U.S. carbon dioxide emissions (**FIGURE 18.25**). Fossil fuel combustion generates 70% of U.S. electricity, and coal alone accounts for 50%, along with most of the emissions from electrical generation. There are two ways to reduce the amount of fossil fuels we burn to generate electricity: (1) encouraging

conservation and efficiency (pp. 555–557) and (2) switching to cleaner and renewable energy sources (Chapters 20 and 21).

Conservation and efficiency As individuals we all can make lifestyle choices to reduce electricity consumption. For nearly all of human history, people managed without the countless electrical appliances that most of us take for granted today. Each of us can choose to use fewer greenhouse-gas-producing appliances and technologies and to take practical steps to use electricity more efficiently.

New energy-efficient technologies make it easier to conserve. The U.S. Environmental Protection Agency's Energy Star Program rates household appliances, lights, windows, fans, office equipment, and heating and cooling systems by their energy efficiency. Replacing an old washing machine with an Energy Star washing machine can cut your CO_2 emissions by 200 kg (440 lb) annually. Replacing standard light bulbs with compact fluorescent lights reduces energy use for lighting by 40%. Energy Star homes use highly efficient windows, ducts, insulation, and heating and cooling systems to reduce energy use and emissions by 30% or more. Such technological solutions also save consumers money by reducing utility bills.

Sources of electricity We can also reduce greenhouse gas emissions by switching to clean energy sources. Alternatives to fossil fuels include nuclear power, biomass energy, hydroelectric power, geothermal power, photovoltaic cells, wind power, and ocean sources. These energy sources give off no net emissions during their use (but some in the production of their infrastructure). We will examine these clean and largely renewable energy sources in detail in Chapters 20 and 21.

Because our society is not ready to transition fully to these alternatives, we also need to consider the ways we use fossil fuels. Switching from coal to natural gas is a step in the right direction, because natural gas produces the same amount of energy as coal, with roughly one-half the emissions (pp. 536, 569). Approaches to boost the efficiency of fossil fuel use, such as cogeneration (p. 556), produce fewer emissions per unit energy generated.

Currently, interest in carbon capture and storage is intensifying. **Carbon capture** refers to technologies or approaches that remove carbon dioxide from power plant emissions. Successful carbon capture would allow facilities to continue using fossil fuels while cutting greenhouse gas pollution. The next step is **carbon sequestration** or **carbon storage**, in which the carbon is sequestered, or stored, underground under pressure in rock formations where it will not seep out. Depleted oil and gas deposits and deep salt mines are examples of the kinds of underground reservoirs being considered. However, we are still a long way from developing adequate technology and secure storage space to accomplish this without leakage. Moreover, some experts doubt that we will ever be able to sequester enough carbon to make a dent in our emissions. Carbon capture and storage is discussed in more detail in Chapter 19 (pp. 546–547).

Transportation is the second largest source of U.S. greenhouse gases

Can you imagine life without a car? Most Americans probably can't—a reason why transportation is the second-largest source of U.S. greenhouse emissions. One-third of the average American city—including roads, parking lots, garages, and gas stations—is devoted to use by the nation's 220 million registered automobiles. The average American family makes 10 trips by car each day, and governments across the nation spend $200 million per day on road construction and repairs.

Unfortunately, the typical automobile is highly inefficient. Over 85% of the fuel you pump into your gas tank does something other than move your car down the road (**FIGURE 18.26**). More aerodynamic designs, increased engine efficiency, and improved tire design can help reduce these losses, but gasoline-fueled automobiles may always remain somewhat inefficient.

Automotive technology The technology exists to make our vehicles far more fuel-efficient than they currently are. Indeed, the vehicles of many nations are more fuel-efficient than

those of the United States. Raising fuel efficiency (p. 555) for American-made vehicles will require government mandate and/or consumer demand, and as gasoline prices rise, demand for more fuel-efficient automobiles will intensify.

Advancing technology is also bringing us alternatives to the traditional combustion-engine automobile. These include hybrid vehicles that combine electric motors and gasoline-powered engines for greater efficiency (p. 556). They also include fully electric vehicles, alternative fuels such as compressed natural gas and biodiesel (pp. 580, 582), and hydrogen fuel cells that use oxygen and hydrogen and produce only water as a waste product (pp. 609–612).

Transportation choices We can also make lifestyle choices that reduce our reliance on cars. Some people are choosing to live nearer to their workplaces. Others use mass transit such as buses, subway trains, and light rail. Still others bike or walk to work or on errands (**FIGURE 18.27**). Public transportation in the United States currently serves 3–4% of passenger trips, reducing gasoline use by 4.2 billion gallons each year and saving 37 million metric tons of CO_2 emissions, the American Public Transportation Association estimates. (See Figure 13.11, p. 358, for data on how public transportation reduces energy use and costs relative to private car use.) If U.S. residents were to increase their use of mass transit to the levels of Canadians (7% of daily travel needs) or Europeans (10% of daily travel needs), the United States could cut its air pollution, its dependence on imported oil, and its contribution to climate change.

Unfortunately, reliable and convenient public transit is not yet available in many U.S. communities. Making automobile-based cities and suburbs more friendly to pedestrian and bicycle traffic and improving people's access to public transportation stand as central challenges for city and regional planners (pp. 357–359).

We can reduce emissions in other ways as well

Advances in agriculture, forestry, and waste management can help us mitigate climate change. In agriculture, sustainable land management on cropland and rangeland can enable soil to store more carbon. Techniques have been developed to reduce the emission of methane from rice cultivation and from

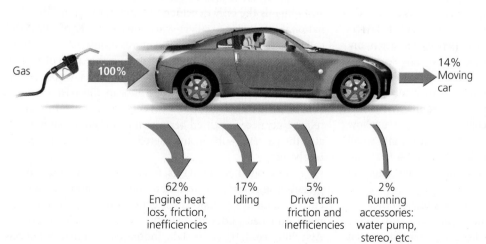

Gas 100% 14% Moving car

62% Engine heat loss, friction, inefficiencies

17% Idling

5% Drive train friction and inefficiencies

2% Running accessories: water pump, stereo, etc.

FIGURE 18.26 ◀ Conventional automobiles are inefficient. Only about 13–14% of the energy from a tank of gas actually moves the typical car down the road, while almost 85% of useful energy is lost, primarily as heat, according to the U.S. Department of Energy.

FIGURE 18.27 ▲ By choosing human-powered transportation methods, such as bicycles, we can greatly reduce our transportation-related greenhouse gas emissions. More people are choosing to live closer to their workplaces and to enjoy the dual benefits of exercise and reduced emissions by walking or cycling to work or school.

cattle and their manure, and to reduce nitrous oxide emissions from fertilizer. We can also grow renewable biofuel crops, although whether these decrease or increase emissions is an active area of research (pp. 579–583).

In forest management, the reforestation of cleared areas helps restore forests, which absorb carbon from the air. Sustainable forestry practices and preserving existing forests (Chapter 12) can help to reverse the carbon dioxide emissions resulting from deforestation.

Waste managers are doing their part to cut emissions by treating wastewater (pp. 426–427), generating energy from waste in incinerators (p. 624), and recovering methane seeping from landfills (p. 624). Individuals, communities, and waste haulers also help reduce emissions when they encourage recycling, composting, and the reuse of materials and products (pp. 625–627).

We will need to follow multiple strategies

We should not expect to find a single "magic bullet" for mitigating climate change. Reducing emissions will require many steps by many people and institutions across many sectors of our economy. The good news is that most reductions can be achieved using current technology and that we can begin implementing these changes right away. Environmental scientists Stephen Pacala and Robert Socolow advise that we follow some age-old wisdom: When the job is big, break it into small parts. Pacala and Socolow propose that we adopt a portfolio of strategies that together can stabilize our CO_2 emissions at current levels (**FIGURE 18.28**).

Pacala and Socolow began with graphs that predicted a doubling of emissions over the next 50 years and asked: What would we need to do to hold our emissions flat instead and avoid the additional future emissions represented by the triangular area of the graph above the flat trend line? The researchers subdivided their so-called *stabilization triangle* into seven equal wedges, like slices of a pie. To eliminate one wedge, a strategy would need to reduce emissions equivalent to 1 billion tons of carbon per year 50 years in the future. Pacala and Socolow identify not just 7, but a series of strategies (listed in Figure 18.28) that could each take care of one wedge if developed and deployed at a large scale.

In the long term, the stabilization-wedge approach will not be enough. To stop climate change, we will need to reduce emissions, not just stabilize them—and this may require us to develop new technology, modify our lifestyles, reduce our consumption, and/or reverse our population growth. However, there is plenty we can do in the meantime to mitigate climate change simply by scaling up technologies and approaches we already have developed.

The Kyoto Protocol sought to limit emissions

Climate change is a global problem, so global commitment is needed to forge effective solutions. This is why the world's policymakers have tried to tackle climate change by means of international treaties. In 1992 at the U.N. Conference on Environment and Development Earth Summit in Rio de Janeiro, Brazil, most of the world's nations signed the **U.N. Framework Convention on Climate Change (FCCC)**. This agreement outlined a plan for reducing greenhouse gas emissions to 1990 levels by the year 2000 through a voluntary, nation-by-nation approach.

By the late 1990s, it was clear that a voluntary approach was not likely to succeed. After watching the seas rise and observing the failure of most industrialized nations to cut their emissions, nations of the developing world—the Maldives and other island nations among them—helped create a binding international treaty that would *require* emissions reductions. An outgrowth of the FCCC drafted in 1997 in Kyoto, Japan, the **Kyoto Protocol** mandates signatory nations, by the period 2008–2012, to reduce emissions of six greenhouse gases to levels below those of 1990 (**TABLE 18.3**). The treaty took effect in 2005 after Russia became the 127th nation to ratify it.

The United States refused to ratify the Kyoto Protocol and remains the only developed nation not to join this international effort. U.S. leaders who oppose the Kyoto Protocol call the treaty unfair because it requires industrialized nations to reduce emissions but does not require the same of rapidly industrializing nations such as China and India, whose greenhouse emissions have risen over 50% in the past 15 years. Proponents of the Kyoto Protocol counter that the differential requirements are justified because industrialized nations created the current problem and therefore should take the lead in resolving it.

The United States emits fully one-fifth of the world's greenhouse gases, so its refusal to join international efforts to curb greenhouse emissions has generated widespread resentment and has undercut the effectiveness of global efforts. At a 2007 conference in Bali, Indonesia, where 190 nations

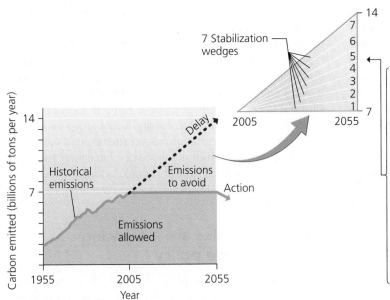

Ways to eliminate 1 "wedge" of emissions
- Double the fuel economy of cars
- Halve the miles driven by car
- Maximize efficiency in all buildings
- Double the efficiency of coal-powered plants
- Switch from coal to natural gas at 1,400 plants
- Capture and store carbon from 800 coal plants
- Capture and store carbon from 180 "synfuels" plants
- Increase hydrogen fuel production by 10 times
- Triple the world's nuclear capacity
- Increase wind power capacity by 50 times
- Increase solar power capacity by 700 times
- Increase ethanol production by 50 times
- Halt tropical deforestation and double reforestation
- Adopt conservation tillage on all croplands

FIGURE 18.28 ▲ We can accomplish the large job of stabilizing emissions by breaking it into multiple smaller steps. Environmental scientists Stephen Pacala and Robert Socolow began with a standard graph (left) of predicted carbon emissions from CO_2 showing the doubling of emissions that scientists expect to occur from 2005 to 2055. They added a flat line to represent the trend if emissions were held constant and separated the graph into emissions allowed (below the line) and emissions to be avoided (the triangular area above the flat line). They then divided this "stabilization triangle" into seven equal-sized portions, which they called "stabilization wedges" (center). Each stabilization wedge represents 1 billion tons of CO_2 emissions in 2055 to be avoided. Finally, they identified a series of strategies (box at right), each of which could take care of one wedge. If we accomplish just 7 of these strategies, we could halt our growth in emissions for the next half-century. Adapted from Pacala, S., and R. Socolow, 2004. Stabilization wedges: Solving the climate problem for the next 50 years with current technologies. *Science* 305: 968–972.

TABLE 18.3 Emissions Reductions Required and Achieved

Nation	Required change, 1990–2008/2012[1]	Observed change, 1990–2008[2]
Russia	0.0%	−32.9%[3]
Germany	−21.0%	−22.2%
United Kingdom	−12.5%	−18.4%
France	0.0%	−6.1%
Italy	−6.5%	+4.7%
Japan	−6.0%	+6.2%
United States[4]	−7.0%	+13.5%
Canada	−6.0%	+24.1%

[1]Percentage decrease in emissions (carbon-equivalents of six greenhouse gases) from 1990 to period 2008–2012, as mandated under the Kyoto Protocol.

[2]Actual percentage change in emissions (carbon-equivalents of six greenhouse gases) from 1990 to 2008. Negative values indicate decreases; positive values indicate increases. Values do not include influences of land use and forest cover.

[3]Russia's decrease was due mainly to economic contraction following the breakup of the Soviet Union.

[4]The United States has not ratified the Kyoto Protocol but was assigned a reduction requirement and reports its emissions, which are included here for comparison.

Data from U.N. Framework Convention on Climate Change, National Greenhouse Gas Inventory Reports, 2010.

strove to design a road map for future progress, the delegate from Papua New Guinea drew thunderous applause and cheers when he requested of the U.S. delegation, "If for some reason you are not willing to lead . . . please get out of the way."

As of 2008 (the most recent year with full international data), nations that signed the Kyoto Protocol had decreased their emissions by 5.2% from 1990 levels. However, much of this reduction was due to economic contraction in Russia and nations of the former Soviet Bloc following the breakup of the Soviet Union. When these nations are factored out, the remaining signatories showed a 7.9% *increase* in emissions from 1990 to 2008.

Kyoto Protocol critics and supporters alike acknowledge that even if every nation were to meet its treaty obligations, greenhouse gas emissions would continue to increase—albeit more slowly than they would without the treaty. Nations are now looking ahead and negotiating over what will come next to supercede Kyoto.

The Copenhagen conference failed to produce a treaty

An international conference in Copenhagen, Denmark, in December 2009 was intended to design a successor treaty to the Kyoto Protocol. Initially, hopes were high because the new administration of U.S. President Barack Obama favored international climate negotiations, and other nations hoped this would lead to U.S. participation in a full international agreement. However, Obama chose not to promise the international community more than the U.S. Congress had already agreed to. Although the House had passed legislation to mandate emissions reductions, the Senate had not, and Senate approval is needed for U.S. ratification of any treaty. Moreover, although China promised steep emissions cuts (45% by 2020!), it proved unwilling to allow international monitoring to confirm them.

With no agreements reached by the final day of the Copenhagen conference, Obama and a handful of other leaders of major nations put together a last-minute accord that fell far short of most nation's hopes. The most notable aspect of this Copenhagen Accord was an offer from developed nations to pay developing nations to help with their mitigation and adaptation efforts—$30 billion over three years and up to $100 billion per year by 2020. The accord also reflected widespread intent during the conference to begin counting deforestation in emissions calculations and rewarding nations that reduce forest loss. However, none of the accord's language was legally binding, and the conference ended without any specific targets being set or solid commitments being made.

In a dramatic final session, several nations denounced the accord, so that the conference could not even formally adopt it by consensus and had to merely "take note" of it. Tuvalu and the Maldives took opposite sides in the debate. Tuvalu insisted on protesting the accord on principle, whereas the Maldives concluded that a weak accord was better than no accord. Nations planned to try again at a conference in Mexico City in December 2010.

Will emissions cuts hurt the economy?

The U.S. Senate has steadfastly opposed emissions reductions out of fear that they will dampen the U.S. economy. Industrializing nations such as China and India have so far resisted emissions cuts under the same assumption. This assumption is understandable, given that so much of our economy depends on fossil fuels. Yet other nations have demonstrated that economic vitality does not require ever-higher emissions. For example, Germany has the third most technologically advanced economy in the world and is a leading producer of iron, steel, coal, chemicals, automobiles, machine tools, electronics, textiles, and other goods—yet it managed between 1990 and 2008 to reduce its greenhouse gas emissions by 22.2%. In the same period, the United Kingdom cut its emissions by 18.4%. Many nations' citizens enjoy standards of living comparable to that of U.S. citizens, but with lower per capita emissions.

Because resource use and per capita emissions are high in the United States and other industrialized nations, governments and industries there often feel they have more to lose economically from restrictions on emissions than developing nations do. However, industrialized nations are also the ones most likely to *gain* economically from major energy transitions, because they are best positioned to invent, develop, and market new technologies to power the world in a post-fossil-fuel era.

Germany has realized this and, along with Japan, is leading the world in production and deployment of solar energy technology (pp. 590–591). China also appears to realize this (**FIGURE 18.29**). China recently surpassed the United States to become the world's biggest greenhouse gas emitter (although it still emits far less per person than the United States). Yet China is also now embarking on a number of initiatives to develop and sell renewable energy technologies on a scale beyond what any other nation has yet attempted. If the United States does not fulfill its potential to develop energy technologies for the future, then the future could belong to nations like China, Germany, and Japan (see the *EnvisionIt* feature in Chapter 21, p. 598).

States and cities are advancing climate change policy

In the absence of action by the U.S. federal government to address climate change, state and local governments across the country are responding to popular sentiment and advancing policies to limit emissions. By 2010, mayors from over 1,000 cities from all 50 U.S. states had signed on to the U.S. Mayors Climate Protection Agreement, initiated by Seattle Mayor Greg Nickels (**FIGURE 18.30A**). Under this agreement, mayors commit their cities to pursue policies to "meet or beat" Kyoto Protocol guidelines.

At the state level, the boldest action so far has come in California, where in 2006 that state's legislature worked with Governor Arnold Schwarzenegger (**FIGURE 18.30B**) to pass the Global Warming Solutions Act, which aims to cut California's greenhouse gas emissions 25% by the year 2020. This law was the first state legislation with penalties for noncompliance and followed earlier efforts in California to mandate higher fuel efficiency for automobiles.

Action was also taken by 10 northeastern states that launched the Regional Greenhouse Gas Initiative (RGGI) in 2007. In this effort, Connecticut, Delaware, Maine, Maryland, Massachusetts, New Hampshire, New Jersey, New York, Rhode

FIGURE 18.29 ▼ China is racing to develop renewable energy technology and may soon surpass the United States in becoming a leader in green energy technology. Here, workers at a Chinese factory produce photovoltaic solar panels (p. 596).

(a) Greg Nickels

(b) Arnold Schwarzenegger

FIGURE 18.30 ▲ Seattle mayor Greg Nickels **(a)** convinced over 1,000 U.S. mayors to commit to fighting climate change, while California governor Arnold Schwarzenegger **(b)** promoted ambitious steps to lower his state's greenhouse emissions. In the absence of leadership at the federal level, elected officials at the state and local levels are taking charge.

Island, and Vermont set up a cap-and-trade program for carbon emissions from power plants. A similar effort, the Western Climate Initiative, involves Arizona, British Columbia, California, Manitoba, Montana, New Mexico, Ontario, Oregon, Quebec, Utah, and Washington. These emissions trading programs (pp. 183–184) show how government can engage the market economy to pursue public policy goals.

Market mechanisms are being used to address climate change

As we first discussed in Chapter 7 (pp. 183–184), permit trading programs aim to harness the economic efficiency of the free market to achieve public policy goals while allowing business,

industry, or utilities flexibility in how they meet those goals. Supporters of permit trading programs argue that they provide the fairest, least expensive, and most effective method of reducing emissions. Polluters choose how to cut their emissions and are given financial incentives for reducing emissions below the legally required amount (**FIGURE 18.31**).

As an example of how a *cap-and-trade* emissions trading program for carbon emissions can work, consider the approach of the Regional Greenhouse Gas Initiative:

1. Each state decided what polluting sources it would require to participate.

2. Each state set a cap on the total CO_2 emissions it would allow, equal to its 2009 levels.

3. Each state distributed to each emissions source one permit for each ton they emit, up to the amount of the cap.

4. Each state will lower its cap progressively, decreasing emissions by 10% by 2018.

5. Sources with too few permits to cover their emissions must find ways to reduce their emissions, buy permits from other sources, or pay for credits through a carbon offset project (p. 524). Sources with excess permits may keep them or sell them.

6. Any source emitting more than its permitted amount will face penalties.

Once up and running, it is hoped that the system will be self-sustaining. The price of a permit is meant to fluctuate freely in the market, creating the same kinds of financial incentives as any other commodity that is bought and sold in our capitalist system.

The world's first emissions trading program for greenhouse gas reduction (operating since 2003) is the Chicago Climate Exchange, which now involves over 350 corporations,

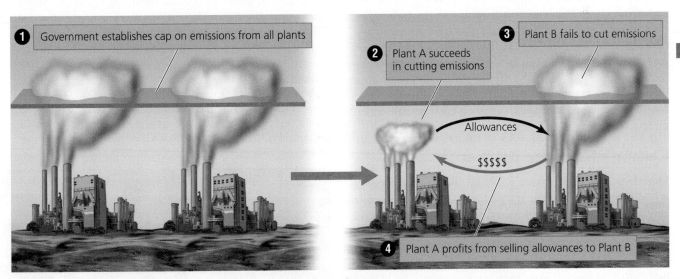

FIGURE 18.31 ▲ In a cap-and-trade emissions trading system, ❶ government first sets an overall cap on emissions. As polluting facilities respond, some will have better success reducing emissions than others. In this figure, ❷ Plant A succeeds in cutting its emissions well below the cap, whereas ❸ Plant B fails to cut its emissions at all. As a result, ❹ Plant B must pay money to Plant A to purchase allowances that Plant A is no longer using. Plant A profits from this sale, and the government cap is met, reducing pollution overall. Over time, the cap can be progressively lowered to achieve further emissions cuts.

institutions, and municipalities in North America and across the world. This voluntary but legally binding trading system has imposed a 6% reduction on overall emissions by 2010.

The world's largest cap-and-trade program is the European Union Emission Trading Scheme. This market got off to a successful start in 2005, but once investors discovered that national governments had allocated too many emissions permits to their industries, the price of carbon fell. The overallocation gave companies little incentive to reduce emissions, so permits lost their value, and prices in the market tanked to 1/100th of their high value. In 2008, Europeans tried to correct these problems by making emitters pay for permits and setting emissions caps across the entire European Union while expanding the program to include more greenhouse gases, more emissions sources, and additional members. In 2009 the market expanded, but prices fell. In the long run, permits will be valuable and the market will work only if government policies are in place to limit emissions.

Carbon taxes are another option

As the world's burgeoning carbon trading markets show mixed results early in their growth, a number of economists, scientists, and policymakers are saying that cap-and-trade systems are not effective enough, don't work quickly enough, or leave too much to chance. Many of these critics would prefer that governments enact a **carbon tax** instead. In this approach, governments charge polluters a fee for each unit of greenhouse gases they emit. This gives polluters a financial incentive to reduce emissions. Carbon taxes have so far been established in several European nations, in British Columbia, and in Boulder, Colorado.

The downside of a carbon tax is that most polluters simply pass the cost along to consumers by charging higher prices for the products or services they sell. Proponents of carbon taxes have responded by proposing an approach called **fee-and-dividend**. In this approach, funds from the carbon tax, or "fee," paid to government by polluters are transferred as a tax refund, or "dividend," to taxpayers. This way, if polluters pass their costs along to consumers, those consumers will be reimbursed for those costs by the tax refund they receive. In theory, the system should provide polluters a financial incentive to reduce emissions, while imposing no financial burden on taxpayers.

Carbon offsets are popular

Emissions trading programs generally allow participants to buy **carbon offsets**, voluntary payments intended to enable another entity to reduce emissions that one is unable to reduce oneself. The payment thus offsets one's own emissions. For example, a coal-burning power plant could pay a reforestation project to plant trees that will soak up as much carbon dioxide as the coal plant emits. Or a university could fund the development of clean and renewable energy projects to make up for fossil fuel energy the university uses.

Carbon offsets have fast become popular among utilities, businesses, universities, governments, and individuals trying to achieve **carbon-neutrality**, a state in which no net carbon is emitted. For busy people with enough wealth, offsets represent a simple and convenient way to reduce one's emissions without investing in efforts to change one's habits.

In principle, carbon offsets seem a great idea, but without rigorous oversight to make sure that the offset money actually accomplishes what it is intended for, carbon offsets risk being little more than a way for wealthy consumers to assuage a guilty conscience. Offsets are effective only if they fund emissions reductions that would not occur otherwise. And because trees can soak up only so much carbon dioxide, at some point our ability to reduce emissions by funding reforestation could reach its limit. Efforts to create a transparent and enforceable system for verifying the effectiveness of offsets are ongoing. If these efforts succeed, then carbon offsets could become an important means of mitigating climate change.

You can reduce your carbon footprint

Carbon taxes, carbon offsets, emissions trading schemes, national policies, international treaties, and technological innovations will all play roles in mitigating climate change. But in the end, the most influential factor may be the collective decisions of millions of regular people. In our everyday lives, each one of us can take steps to approach a carbon-neutral lifestyle by reducing greenhouse emissions that result from our decisions and activities. Just as we each have an ecological footprint (p. 5), we each have a **carbon footprint** that expresses the amount of carbon we are responsible for emitting.

College students are a vital key to driving the personal and societal changes needed to reduce carbon footprints and mitigate climate change—both through everyday lifestyle choices and through lobbying and activism (**FIGURE 18.32**). Today a groundswell of interest is sweeping across campuses (pp. 665–677). This was evident on January 31, 2008, when over 1,900 schools participated in the Focus the Nation teach-in on global warming. Young people have played a large part in subsequent grassroots events, including Focus the Nation's 2009 nationwide Town Hall on America's Energy Future, and 350.org's International Day of Climate Action on October 24, 2009. This last event—kicked off by the Maldives' underwater

FIGURE 18.32 ▼ Many college-aged people attended the Copenhagen climate conference in December 2009 to take in the proceedings and to express their views. Here, a group of activists shows delegates their support for island nations like Tuvalu and for the goal of bringing the atmospheric carbon dioxide level down to 350 parts per million.

cabinet meeting—featured 5,200 events in 181 nations and was called "the most widespread day of political action in the planet's history."

Global climate change may be the biggest challenge facing us and our children. Fortunately, it is still early enough that, with concerted action, we may be able to avert the most severe impacts. To help reduce emissions, we each can apply many of the strategies discussed in this chapter in our everyday lives—from deciding where to live and how to get to work to choosing appliances. You will encounter still more solutions in our discussions of energy sources, conservation, and renewable energy in Chapters 19–21 and elsewhere throughout this book.

➤ CONCLUSION

Many factors influence Earth's climate, and human activities have come to play a major role. Climate change is well underway, and further greenhouse gas emissions will increase global warming and cause increasingly severe and diverse impacts. Sea level rise and other consequences of global climate change will affect locations worldwide from the Maldives to Bangladesh to Alaska to Florida. As scientists and policymakers come to better understand anthropogenic climate change and its environmental, economic, and social consequences, more and more of them are urging immediate action. Reducing greenhouse gas emissions and taking other steps to mitigate and adapt to climate change represents the foremost challenge for our society in the coming years.

R E V I E W I N G   O B J E C T I V E S

You should now be able to:

DESCRIBE EARTH'S CLIMATE SYSTEM AND EXPLAIN THE MANY FACTORS INFLUENCING GLOBAL CLIMATE

- Earth's climate changes naturally over time, but it is now changing rapidly because of human influence. (p. 495)
- The sun provides most of Earth's energy. Earth absorbs about 70% of incoming solar radiation and reflects about 30% back into space. (pp. 495–496)
- Greenhouse gases such as carbon dioxide, methane, water vapor, nitrous oxide, ozone, and halocarbons warm the atmosphere by absorbing and re-emitting infrared radiation. (pp. 495–497)
- Earth is experiencing radiative forcing of 1.6 watts/m^2 of thermal energy above what it was experiencing 250 years ago. (p. 498)
- Milankovitch cycles influence climate in the long term. (pp. 498–499)
- Solar radiation, ocean absorption, and ocean circulation also influence climate. (p. 499)

CHARACTERIZE HUMAN INFLUENCES ON THE ATMOSPHERE AND ON CLIMATE

- Increased greenhouse gas emissions enhance the greenhouse effect. (p. 495)
- By burning fossil fuels, clearing forests, and manufacturing halocarbons, people are increasing atmospheric concentrations of many greenhouse gases. (p. 497)
- Human input of aerosols into the atmosphere exerts a variable but slight cooling effect. (p. 498)

SUMMARIZE HOW RESEARCHERS STUDY CLIMATE

- Proxy indicators, such as data from ice cores, sediment cores, tree rings, packrat middens, and coral reefs, reveal information about past climate. (pp. 499–501)

- Direct measurements of temperature, precipitation, and other conditions tell us about current climate. (p. 502)
- Climate models serve to predict future changes in climate. (pp. 502–503)

OUTLINE CURRENT AND FUTURE TRENDS AND IMPACTS OF GLOBAL CLIMATE CHANGE

- The Intergovernmental Panel on Climate Change (IPCC) synthesizes current climate research, and its periodic reports represent the consensus of the scientific community. (pp. 503–504)
- Temperatures on Earth have warmed by an average of 0.74 °C (1.33 °F) over the past century and are predicted to rise 1.8–4.0 °C (3.2–7.2 °F) over the next century. (pp. 504, 506)
- Changes in precipitation have varied by region. (pp. 505, 507)
- Melting glaciers will diminish water supplies, and melting ice sheets will add to sea level rise. (pp. 505–507)
- Sea level has risen an average of 17 cm (7 in.) over the past century. (pp. 507, 509)
- Other impacts include ocean acidification; extreme weather events; effects on organisms and ecosystems; and impacts on agriculture, forestry, health, and economics. (pp. 511–513)
- Climate change and its impacts will vary regionally. (pp. 513–515)
- Despite some remaining uncertainties, the scientific community feels that evidence for humans' role in influencing climate is strong enough to justify taking action to reduce greenhouse emissions. (p. 516)

SUGGEST WAYS WE MAY RESPOND TO CLIMATE CHANGE

- Both adaptation and mitigation are necessary. (pp. 517–518)
- Conservation, energy efficiency, and new clean and renewable energy sources will help reduce greenhouse gas emissions. (pp. 518–519)

- New automotive technologies and investment in public transportation will help reduce emissions. (pp. 519–520)
- Addressing climate change will require multiple strategies. (pp. 520–521)
- The Kyoto Protocol provided a first step for nations to begin addressing climate change. (pp. 520–521)
- Despite failure of the Copenhagen conference, nations are continuing efforts to design a treaty to follow the Kyoto Protocol. (pp. 521–522)
- U.S. states and cities are acting to address emissions because the federal government has not. (pp. 522–523)
- Emissions trading programs provide a way to harness the free market and engage industry in reducing emissions. (pp. 523–524)
- Many people feel that a carbon tax, specifically a fee-and-dividend approach, is a better option. (p. 524)
- Individuals are increasingly exploring carbon offsets and other means of reducing personal carbon footprints. (p. 524)

TESTING YOUR COMPREHENSION

1. What happens to solar radiation after it reaches Earth? How do greenhouse gases warm the lower atmosphere?

2. Why is carbon dioxide considered the main greenhouse gas? How could an increase in water vapor create either a positive or negative feedback effect?

3. How do scientists study the ancient atmosphere? Describe what a proxy indicator is, and give two examples.

4. Has simulating climate change with computer programs been effective in helping us predict climate? How do these programs work?

5. List five major trends in climate that scientists have documented so far. Now list five future trends or impacts that they are predicting.

6. Describe how rising sea levels, caused by global warming, can create problems for people. How may climate change affect marine ecosystems?

7. How might a warmer climate affect agriculture? How is it affecting distributions of plants and animals? How might it affect human health?

8. What are the largest two sources of greenhouse gas emissions in the United States? How can we reduce these emissions?

9. What roles have international treaties played in addressing climate change? Give two specific examples.

10. Describe one market-based approach for reducing greenhouse emissions. Explain one reason it may work well and one reason it may not work well.

SEEKING SOLUTIONS

1. To determine to what extent current climate change is the result of human activity versus natural processes, which type(s) of scientific research do you think is (are) most helpful? Why?

2. Some people argue that we need "more proof," or "better science," before we commit to substantial changes in our energy economy. How much certainty do you think we need before we should take action regarding climate change? How much certainty do you need in your own life before you make a major decision? Should nations and elected officials follow a different standard? Do you believe that the precautionary principle (pp. 265, 394) is an appropriate standard in the case of global climate change? Why or why not?

3. Describe several ways in which we can reduce greenhouse gas emissions from transportation. Which approach do you think is most realistic, which approach do you think is least realistic, and why?

4. Suppose that you would like to make your own lifestyle carbon-neutral and that you aim to begin by reducing the emissions you are responsible for by 25%. What three actions would you take first to achieve this reduction?

5. **THINK IT THROUGH** You have been appointed as the United States representative to an international conference to negotiate terms of a treaty to take hold after the Kyoto Protocol ends. All nations recognize that the Kyoto Protocol was not fully effective, and most are committed to creating a stronger agreement. The U.S. government has instructed you to take a leading role in designing the new treaty and to engage constructively with other nations' representatives while protecting your nation's economic and political interests. What type of agreement will you try to shape? Describe at least three components that you would propose or agree to, and at least one that you would oppose.

6. **THINK IT THROUGH** You have just been elected governor of a medium-sized U.S. state. Polls show that the public wants you to take bold action to reduce greenhouse gas emissions. However, polls also show that the public does not want prices of gasoline or electricity to rise very much. Carbon-emitting industries in your state are wary of emissions reductions being required of them but are willing to explore ideas with you. Your state legislature will support you in your efforts as long as you remain popular with voters. The state to your west has just passed ambitious legislation mandating steep greenhouse gas emissions reductions. The state to your east has just joined a new regional emissions trading consortium. What actions will you take in your first year as governor?

CALCULATING ECOLOGICAL FOOTPRINTS

Global climate change is something to which we all contribute, because fossil fuel combustion plays such a large role in supporting the lifestyles we lead. Conversely, as individuals, each one of us can contribute to mitigating global climate change through personal decisions and actions that affect the way we live our lives. Several online calculators enable you to calculate your own personal *carbon footprint*, the amount of carbon emissions for which you are responsible. Go to www.nature.org/initiatives/climatechange/calculator/, take the quiz, and enter the relevant data in the table.

	Carbon footprint (tons per person per year)
World average	
U.S. average	
Your footprint	
Your footprint with three changes	

1. How does your personal carbon footprint compare to that of the average U.S. resident? How does it compare to that of the average person in the world? Why do you think your footprint differs from these in the ways it does?

2. As you took the quiz and noted the impacts of various choices and activities, which one surprised you the most?

3. Think of three changes you could make in your lifestyle that would lower your carbon footprint. Now take the footprint quiz again, incorporating these three changes. Enter your resulting footprint in the table. By how much did you reduce your yearly emissions?

4. What do you think would be an admirable yet realistic goal for you to set as a target value for your own footprint? Would you choose to purchase carbon offsets to help reduce your impact? Why or why not?

Mastering ENVIRONMENTAL SCIENCE™

Go to **www.masteringenvironmentalscience.com** for practice quizzes, Pearson eText, videos, current events, and more.

Oil production facility at Prudhoe Bay, Alaska

19 FOSSIL FUELS, THEIR IMPACTS, AND ENERGY CONSERVATION

UPON COMPLETING THIS CHAPTER, YOU WILL BE ABLE TO:

- Identify the energy sources that we use
- Describe the nature and origin of coal and evaluate its extraction and use
- Describe the nature and origin of natural gas and evaluate its extraction and use
- Describe the nature and origin of crude oil and evaluate its extraction, use, and future depletion

- Describe the nature, origin, and potential of alternative fossil fuels
- Outline and assess environmental impacts of fossil fuel use and explore potential solutions
- Evaluate political, social, and economic aspects of fossil fuel use
- Specify strategies for conserving energy and enhancing efficiency

Oil or Wilderness on Alaska's North Slope?

"America is addicted to oil."

—U.S. President George W. Bush

"The era of easy oil is over."

—Chevron CEO David J. O'Reilly

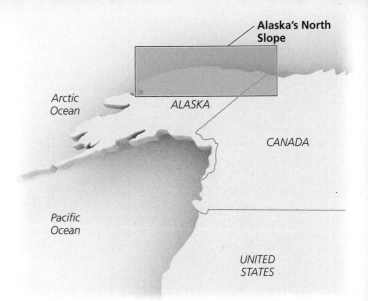

Alaska's North Slope

Arctic Ocean

ALASKA

CANADA

Pacific Ocean

UNITED STATES

Above the Arctic Circle, at the top of the North American continent, the land drops steeply down from the jagged mountains of Alaska's spectacular Brooks Range and stretches north in a vast, flat expanse of tundra until it meets the icy waters of the Arctic Ocean. Few Americans have been to this remote region, yet it has come to symbolize a struggle between two values in our modern life.

Polar bears walk past oil field infrastructure near Prudhoe Bay

For some U.S. citizens, Alaska's North Slope is the last great expanse of wilderness in their industrialized nation—a rare place that people have left untouched. For these millions of Americans, simply knowing that this wilderness still exists is of tremendous value. For millions of others, this land represents something else entirely—a source of petroleum, the natural resource that, more than any other, fuels our society and shapes our way of life. To these people, it seems wrong to leave such an important resource sitting unused in the ground. Those who propose drilling for oil here accuse wilderness preservationists of neglecting the country's economic interests. Advocates for wilderness maintain that drilling will sacrifice the nation's natural heritage for little gain.

These competing visions for Alaska's North Slope exist side by side across three regions of this vast swath of land (**FIGURE 19.1**). The westernmost

portion of the North Slope was set aside in 1923 by the U.S. government as an emergency reserve for petroleum. This parcel of land, the size of Indiana, is called the National Petroleum Reserve–Alaska (NPR–A) and was intended to remain untapped for oil unless the nation faced an emergency. In recent years, most of this region's 9.5 million ha (23.5 million acres) have been opened for development.

East of the National Petroleum Reserve are state lands that were developed after oil was discovered at Prudhoe Bay in 1968. Since drilling began in 1977, we have extracted over 14 billion barrels (1 barrel = 159 L, or 42 gal) of crude oil from 19 oil fields spread over 160,000 ha (395,000 acres) of this region. The oil is transported across Alaska by the 1,300-km (800-mi) trans-Alaska pipeline south to the port of Valdez, where it is loaded onto tankers.

East of the Prudhoe Bay region lies the Arctic National Wildlife Refuge, an area the size of South Carolina consisting of federal lands set aside in 1960 and 1980 mainly to protect wildlife and preserve pristine ecosystems of tundra, mountains, and seacoast. This scenic region is home to 160 nesting bird species, numerous fish and marine mammals, grizzly bears, polar bears, Arctic foxes, timber wolves, musk

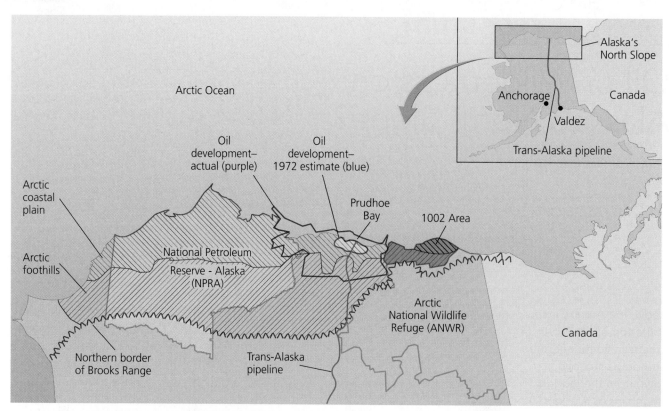

FIGURE 19.1 ▲ Alaska's North Slope is the site of both arctic wilderness and oil exploration. In the western portion of this region, the U.S. government established the National Petroleum Reserve–Alaska as an area in which to drill for oil if it is needed in an emergency. It is well explored but not yet widely developed. To the east of this area, the Prudhoe Bay region is the site of widespread oil extraction. Oil development has expanded much farther than experts estimated it would in 1972. Farther east lies the Arctic National Wildlife Refuge, home to untrammeled Arctic wilderness. Debate has focused on whether the 1002 Area of the coastal plain north of the Brooks Range should be opened to oil development.

oxen, and other animals. In summer, thousands of caribou arrive from the south and give birth to and raise their calves. Because of the vast caribou herd and the other large mammals, the Arctic National Wildlife Refuge has been called "the Serengeti of North America."

The Arctic Refuge has been the focus of debate for decades. Advocates of oil drilling have tried to open its lands for development, and proponents of wilderness have fought for its preservation. Scientists, oil industry experts, politicians, environmental groups, citizens, and Alaska residents have all been part of the debate. So have the two Native groups in the area, the Gwich'in and the Inupiat, who disagree over whether the refuge should be opened to oil development. The Gwich'in depend on hunting caribou and fear that oil industry activity will reduce caribou herds, whereas the Inupiat see oil extraction as one of the few opportunities for economic development in the area.

In a compromise in 1980, the U.S. Congress put most of the refuge off limits to development but reserved for future decision making a 600,000-ha (1.5-million-acre) area of coastal plain. This region,

called the 1002 Area (after Section 1002 of the bill that established it), can be opened for oil development by a vote of both houses of Congress. Its unsettled status has made it the center of the oil-versus-wilderness debate, and Congress has been caught between passionate feelings on both sides for a quarter of a century.

In recent years, Republican-controlled Congresses repeatedly tried to open the Arctic Refuge to drilling. Today's Democratic control of Congress and the presidency has taken the issue off the table in the short term, but as our society depletes the resource it depends on most, and as gasoline prices rise, calls to drill in the refuge will come up again and again.

We can also expect accelerated development of lands in the National Petroleum Reserve–Alaska, which have until now served as a *de facto* wildlife refuge. Stymied in its efforts to drill in the Arctic Refuge, the George W. Bush administration in 2006—ignoring objections from biologists, environmentalists, and hunters—opened for development an ecologically sensitive area of the NPR–A around Lake Teshekpuk, a key breeding ground for caribou and for many of the continent's migratory waterfowl.

Behind the noisy policy debate, scientists have attempted to inform the dialogue through research. Geologists have tried to ascertain how much oil lies underneath the refuge, and biologists have tried to assess the impacts of drilling on Arctic ecosystems. Moreover, scientists and nonscientists alike are debating whether we need the oil beneath the refuge for our security and prosperity, at a time when climate change is worsening and renewable energy sources remain underdeveloped.

SOURCES OF ENERGY

Humanity has devised many ways to harness the renewable and nonrenewable forms of energy available on our planet (**TABLE 19.1**). We use these energy sources to heat and light our homes; power our machinery; fuel our vehicles; and produce plastics, pharmaceuticals, synthetic fibers, and more. Our success in harnessing Earth's energy sources has provided us the comforts and conveniences to which we've grown accustomed in the industrial age.

We use a variety of energy sources

Most of Earth's energy comes ultimately from the sun. We can harness energy from the sun's radiation directly by using various solar power technologies. Solar radiation also helps drive wind patterns and the hydrologic cycle, making it possible for us to harness wind power and hydroelectric power. And of course, sunlight drives photosynthesis (pp. 31–32) and the growth of plants, from which we take wood and other biomass as a fuel source. Finally, when plants die and are preserved in sediments under particular conditions, they may impart their stored chemical energy to **fossil fuels**, highly combustible substances formed from the remains of organisms from past geologic ages. The three fossil fuels we use widely today are oil, coal, and natural gas.

In addition, a great deal of energy emanates from Earth's core, making geothermal power available for our use. Energy also results from the gravitational pull of the moon and sun, and we are just beginning to harness power from the ocean tides that these forces generate. An immense amount of energy resides within the bonds among protons and neutrons in atoms, and this energy provides us with nuclear power. We will explore all these energy sources in Chapters 20 and 21, after focusing on fossil fuels in the present chapter.

Since the industrial revolution, fossil fuels have replaced biomass as our society's dominant source of energy. Global consumption of the three main fossil fuels has risen steadily for years and is now at its highest level ever (**FIGURE 19.2**). The high energy content of fossil fuels makes them efficient to burn, ship, and store. Besides providing for transportation, heating, and cooking, these fuels are used to generate **electricity**, a secondary form of energy that is easier to transfer over long distances and apply to a variety of uses.

The energy stream of the United States is complex, but is dominated by coal, oil, and natural gas (**FIGURE 19.3**). As you can see by examining Figure 19.3 on the next page, we use different energy sources in different ways.

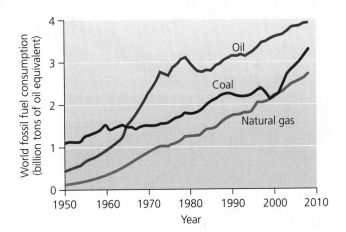

FIGURE 19.2 ▲ Global consumption of fossil fuels has risen greatly over the past half-century. Oil use rose steeply during the 1960s to overtake coal, and today it remains our leading energy source. Data from U.S. Energy Information Administration and International Energy Agency.

TABLE 19.1	Energy Sources We Use Today		
Energy source	**Description**	**Type of energy**	**Chapter**
Crude oil	Fossil fuel extracted from ground (liquid)	Nonrenewable	19
Natural gas	Fossil fuel extracted from ground (gas)	Nonrenewable	19
Coal	Fossil fuel extracted from ground (solid)	Nonrenewable	19
Nuclear energy	Energy from atomic nuclei of uranium	Nonrenewable	20
Biomass energy	Energy stored in plant matter from photosynthesis	Renewable	20
Hydropower	Energy from running water	Renewable	20
Solar energy	Energy from sunlight directly	Renewable	21
Wind energy	Energy from wind	Renewable	21
Geothermal energy	Earth's internal heat rising from core	Renewable	21
Tidal and wave energy	Energy from tidal forces and ocean waves	Renewable	21

As we first noted in Chapter 1 (pp. 3–4), energy sources such as sunlight, geothermal energy, and tidal energy are considered perpetually renewable because they will not be depleted by our use. Other sources, such as timber, are renewable if we do not harvest them too quickly. In contrast, energy sources such as oil, coal, and natural gas are considered nonrenewable, because at our current rates of consumption we will use up Earth's accessible store of them in decades to centuries. Nuclear power as currently harnessed through the fission of uranium (p. 565) is nonrenewable to the extent that uranium ore is in limited supply, although we can also reprocess some uranium and reuse it.

These nonrenewable fuels result from ongoing natural processes, but they are created extremely slowly. It takes so long for fossil fuels to form that, once depleted, they cannot be replaced within any time span useful to our civilization. It takes a thousand years for the biosphere to generate the amount of organic matter that must be buried to produce a single day's worth of fossil fuels for our society. To replenish

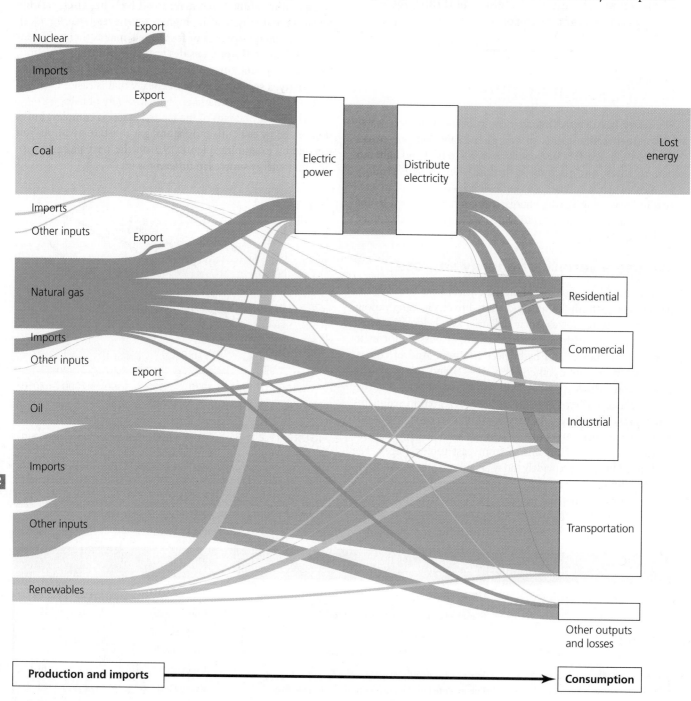

FIGURE 19.3 ▲ This diagram shows the total energy flow of the United States. Amounts are represented by the thickness of each bar; note the predominance of coal, oil, and natural gas. Domestic production and imports of fuels are shown on the left, and destinations of the energy are shown on the right. Some portions of each fuel are used directly to power the residential, commercial, industrial, and transportation sectors, while other portions are used to generate electricity, which in turn goes toward these sectors. The large amount of energy lost as waste heat is shown for electricity generation; large losses (not shown) also occur from energy use in each sector. Data are for 2008, from U.S. Energy Information Administration, 2009. *Annual Energy Review 2008.*

the fossil fuels we have depleted so far would take many millions of years. For this reason, and because fossil fuels exert severe environmental impacts, renewable energy sources increasingly are being developed as alternatives to fossil fuels, as we will see in Chapters 20 and 21.

Fossil fuels are indeed fuels created from "fossils"

The fossil fuels we burn today in our vehicles, homes, industries, and power plants were formed from the tissues of organisms that lived 100–500 million years ago. The energy these fuels contain came originally from the sun and was converted to chemical-bond energy as a result of photosynthesis. The

FIGURE 19.4 ▲ Fossil fuels begin to form when organisms die and end up in oxygen-poor conditions, such as when trees fall into lakes and are buried by sediment, or when phytoplankton and zooplankton drift to the seafloor and are buried (top diagram). Organic matter that undergoes slow anaerobic decomposition deep under sediments forms kerogen (middle diagram). Geothermal heating then acts on kerogen to create crude oil and natural gas (bottom diagram). Oil and gas come to reside in porous rock layers beneath dense, impervious layers. Coal is formed when plant matter is compacted so tightly that there is little decomposition.

chemical energy in these organisms' tissues then became concentrated as these tissues decomposed and their hydrocarbon compounds were altered and compressed (**FIGURE 19.4**).

Most organisms, after death, do not end up as part of a coal, gas, or oil deposit. A tree that falls and decays as a rotting log undergoes mostly **aerobic** decomposition; in the presence of air, bacteria and other organisms that use oxygen break down plant and animal remains into simpler carbon molecules that are recycled through the ecosystem. Fossil fuels are produced only when organic material is broken down in an **anaerobic** environment, one that has little or no oxygen. Such environments include the bottoms of deep lakes, swamps, and shallow seas. Over millions of years, organic matter that accumulates at the bottoms of such water bodies is converted into crude oil, natural gas, or coal. In any given location different fuels may form, depending on (1) the chemical composition of the starting material, (2) the temperatures and pressures to which the material is subjected, (3) the presence or absence of anaerobic decomposers, and (4) the passage of time.

Fossil fuel reserves are unevenly distributed

Fossil fuel deposits are localized and unevenly distributed over Earth's surface, so some regions have substantial reserves of fossil fuels whereas others have very few. How long each nation's fossil fuel reserves will last depends on how much the nation extracts, how much it consumes, and how much it imports from and exports to other nations. Nearly two-thirds of the world's proven reserves of crude oil lie in the Middle East. The Middle East is also rich in natural gas, but Russia holds more natural gas than any other country. Russia is also rich in coal, as is China, but the United States possesses the most coal of any nation (**TABLE 19.2**).

TABLE 19.2 Nations with the Largest Proven Reserves of Fossil Fuels

Oil (% world reserves)	Natural gas (% world reserves)	Coal (% world reserves)
Saudi Arabia, 19.8	Russia, 23.7	United States, 28.9
Venezuela, 12.9	Iran, 15.8	Russia, 19.0
Iran, 10.3	Qatar, 13.5	China, 13.9
Iraq, 8.6	Turkmenistan, 4.3	Australia, 9.2
Kuwait, 7.6	Saudi Arabia, 4.2	India, 7.1
United Arab Emirates, 7.3	United States, 3.7	Ukraine, 4.1
Russia, 5.6	United Arab Emirates, 3.4	Kazakhstan, 3.8
Libya, 3.3	Venezuela, 3.0	South Africa, 3.7
Kazakhstan, 3.0	Nigeria, 2.8	Poland, 0.9
Nigeria, 2.8	Algeria, 2.4	Brazil, 0.9

Data from U.S. Energy Information Administration; and BP plc. 2010. *Statistical review of world energy 2010*. Canada ranks second in oil reserves if oil sands are included.

Developed nations consume more energy than developing nations

Citizens of developed regions generally consume far more energy than do those of developing regions (**FIGURE 19.5**). Per person, the most-industrialized nations use up to 100 times more energy than do the least-industrialized nations. The United States has only 4.5% of the world's population, but it consumes over 20% of the world's energy.

Developed and developing nations also differ in how they apportion their energy use. Industrialized nations use roughly one-third of their energy on transportation, one-third on industry, and one-third on all other uses. Developing nations devote a greater proportion of energy to subsistence activities such as agriculture, food preparation, and home heating, and much less on transportation. In addition, people in developing countries often rely on manual or animal energy sources instead of automated ones. For instance, rice farmers in Bali plant rice by hand, but industrial rice growers in California use airplanes. Because industrialized nations rely more on equipment and technology, they use more fossil fuels. In the United States, oil, coal, and natural gas together supply 84% of energy needs.

It takes energy to make energy

We don't simply get energy for free. To harness, extract, process, and deliver the energy we use, we need to invest substantial inputs of energy. For instance, drilling for oil at Prudhoe Bay has required the construction of an immense infrastructure of roads, wells, vehicles, storage tanks, pipelines, housing for workers, and more—all requiring the use of energy. Piping and shipping the oil out of the North Slope, and then refining it into products we can use, requires further energy inputs. Thus, when evaluating an energy source, it is important to subtract costs in energy invested from benefits in energy received. **Net energy** expresses the difference between energy returned and energy invested:

$$\text{Net energy} = \text{Energy returned} - \text{Energy invested}$$

When assessing energy sources, it is useful to use a ratio often denoted as **EROI—energy returned on investment**. EROI ratios are calculated as follows:

$$\text{EROI} = \text{Energy returned}/\text{Energy invested}$$

Higher ratios mean that we receive more energy from each unit of energy that we invest. Fossil fuels are widely used because their EROI ratios have historically been high. However, EROI ratios can change over time. For instance, those for U.S. oil and natural gas declined from over 100:1 in the 1940s to about 30:1 in the 1970s, and today they hover around 5:1. The EROI ratios declined because we extracted the easiest deposits first and now must work harder and harder to extract the remaining amounts.

COAL, NATURAL GAS, AND OIL

The three major fossil fuels on which our modern industrial society was built, and on which we rely today, are coal, natural gas, and oil. We will consider each in turn, and afterwards will review some of the environmental and social impacts of their use.

534

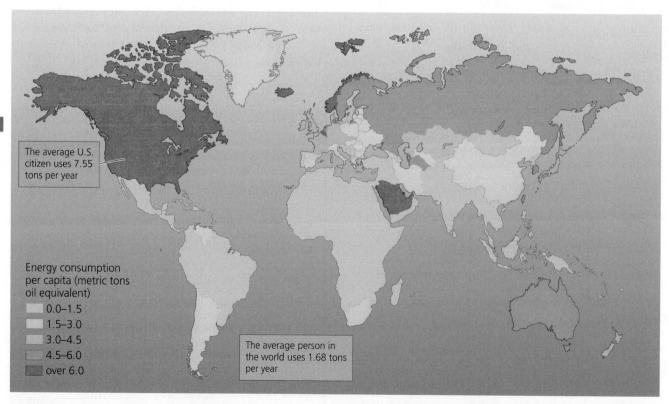

The average U.S. citizen uses 7.55 tons per year

The average person in the world uses 1.68 tons per year

Energy consumption per capita (metric tons oil equivalent)
- 0.0–1.5
- 1.5–3.0
- 3.0–4.5
- 4.5–6.0
- over 6.0

FIGURE 19.5 ▲ People in wealthy industrialized nations consume the most energy per person. This map combines all types of energy, standardized to metric tons of "oil equivalent," that is, the amount of fuel needed to produce the energy gained from combusting one metric ton of crude oil. Data from *Statistical review of world energy 2010*, BP p.l.c. By permission.

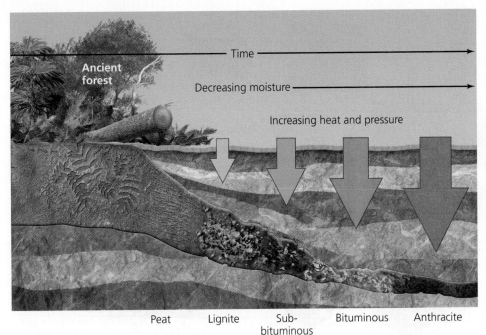

FIGURE 19.6 ◀ Coal forms as ancient plant matter is compressed underground. Scientists categorize coal into several types, depending on the amount of heat, pressure, and moisture involved in its formation. Anthracite coal is formed under the greatest pressure, where temperatures are high and moisture content is low. Lignite coal is formed under conditions of much less pressure and heat, but more moisture. Peat is also part of this continuum, representing plant matter that is minimally compacted. Anthracite coal has the densest carbon content and so contains the most potential energy.

Coal is most abundant

Coal is a hard blackish substance that forms from organic matter (generally woody plant material) that was compressed under very high pressure to form dense, solid carbon structures (**FIGURE 19.6**). Coal typically results when little decomposition takes place because the material cannot be digested or appropriate decomposers are not present. The proliferation 300–400 million years ago of swampy environments where organic material was buried has created coal deposits throughout the world. In fact, coal is the world's most abundant fossil fuel and provides 29% of global commercial energy consumption.

We mine coal and use it to generate electricity

We extract coal using two major methods. For deposits near the surface, we use *strip mining*, in which heavy machinery scrapes away huge amounts of earth to expose the coal. For deposits deep underground, we use *subsurface mining*, digging vertical shafts and blasting out networks of horizontal tunnels to follow seams, or layers, of coal. Recently, we have begun mining coal on immense scales, in some cases essentially lopping off entire mountaintops. This environmentally destructive process, called *mountaintop removal mining*, has now become common in the Appalachian Mountains. We will explore all these mining practices (and their impacts) more fully in our examination of minerals and mining in Chapter 23 (pp. 648–654).

People have extracted coal from the ground and burned it to cook food, heat homes, and fire pottery for thousands of years and in many cultures, from ancient China to the Roman Empire to the Hopi Nation. Coal-fired steam engines helped drive the industrial revolution, powering factories, agriculture, trains, and ships. The birth of the steel industry in 1875 increased demand further because coal fueled the furnaces used to produce steel. Then in the 1880s, people began to burn coal to generate electricity. Today coal provides half the electrical generating capacity of the United States, and powers China's surging economy. China and the United States are the primary producers and consumers of coal (**TABLE 19.3**). We generate electricity in modern coal-fired power plants by combusting coal to convert water to steam, which turns a turbine (**FIGURE 19.7**).

Coal varies in its qualities

Coal varies from deposit to deposit in its water content, carbon content, and the amount of potential energy it contains. Organic material that is broken down anaerobically but remains wet, near the surface, and not well compressed is called *peat*. As peat decomposes further, as it becomes buried more deeply under sediments, as pressure and heat increase, and as time passes, water is squeezed out of the material, and carbon compounds are packed more tightly together, forming coal. Scientists classify coal into four types: lignite, sub-bituminous, bituminous, and anthracite. Lignite is the least-compressed type of coal, and anthracite is the most-compressed type (see Figure 19.6). The more coal is compressed, the greater is its carbon content and the greater is the energy content per unit volume.

TABLE 19.3 Top Producers and Consumers of Coal

Production (% world production)	Consumption (% world consumption)
China, 39.2	China, 39.1
United States, 16.1	United States, 15.5
India, 7.8	India, 8.8
Australia, 6.0	Germany, 3.7
Russia, 4.9	Russia, 3.7

Data are for 2008, from U.S. Energy Information Administration.

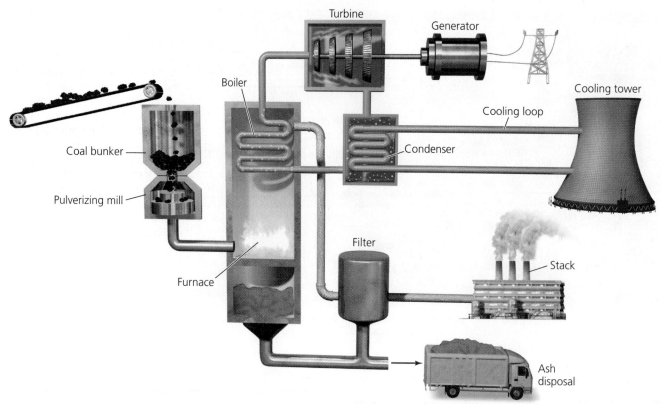

FIGURE 19.7 ▲ At a coal-fired power plant, coal is pulverized and blown into a high-temperature furnace. Heat from the combustion boils water, and the resulting steam turns a turbine, generating electricity by passing magnets past copper coils. The steam is then cooled and condensed in a cooling loop and returned to the furnace. "Clean coal" technologies help filter out pollutants from the combustion process, and toxic ash residue is taken to hazardous waste disposal sites.

Coal deposits also vary in the amount of impurities they contain, including sulfur, mercury, arsenic, and other trace metals. Sulfur content depends partly on whether the coal was formed in freshwater or saltwater sediments. Coal from the eastern United States tends to be high in sulfur because it was formed in marine sediments, where sulfur from seawater was present. Most coal in China is even more sulfur-rich. The impurities in coal are emitted in smoke from its combustion unless pollution control measures are in place. We will examine the many health and environmental impacts from coal combustion, along with solutions to these problems, later in this chapter (pp. 545–549). Reducing pollution from coal is important because society's demand for this relatively abundant fossil fuel may rise once supplies of oil and natural gas begin to decline. Scientists estimate that Earth holds enough coal to supply our society for perhaps a few hundred years more—far longer than oil or natural gas will remain available.

Natural gas burns more cleanly than coal

Natural gas today provides for one-quarter of global commercial energy consumption and is increasingly favored because it is versatile and clean-burning, emitting just half as much carbon dioxide per unit of energy produced as coal and two-thirds as much as oil. We use natural gas to generate electricity in power plants, to heat and cook in our homes, and for much else. Converted to a liquid at low temperatures (*liquefied natural gas,*

or *LNG*), it can be shipped long distances in refrigerated tankers, although this poses small risks of catastrophic explosions. Russia and the United States lead the world in gas production and gas consumption, respectively (**TABLE 19.4**). At predicted levels of use, world supplies of natural gas are projected to last perhaps 60 more years.

Natural gas is formed in two ways

Natural gas consists primarily of methane, CH_4, and typically includes varying amounts of other volatile hydrocarbons. Natural gas can arise from either of two processes. *Biogenic* gas is created at shallow depths by the anaerobic

TABLE 19.4 Top Producers and Consumers of Natural Gas	
Production (% world production)	Consumption (% world consumption)
Russia, 21.3	United States, 21.0
United States, 18.6	Russia, 15.2
Canada, 5.5	Iran, 3.8
Iran, 3.7	Japan, 3.2
Norway, 3.2	United Kingdom, 3.1

Data are for 2008, from U.S. Energy Information Administration.

decomposition of organic matter by bacteria. An example is the "swamp gas" you may smell when stepping into the muck of a swamp. In contrast, *thermogenic* gas results from compression and heat deep underground. As organic matter is buried more deeply, the pressure exerted by overlying sediments grows, and temperatures increase. Carbon bonds in the organic matter begin breaking, and the organic matter turns to a substance called *kerogen*, which acts as a source material for both natural gas and crude oil. Further heat and pressure act on the kerogen to degrade complex organic molecules into simpler hydrocarbon molecules. At depths below about 3 km (1.9 mi), the high temperatures and pressures tend to form natural gas. Whereas biogenic gas is nearly pure methane, thermogenic gas contains small amounts of other gases as well as methane.

Thermogenic gas may be formed directly, along with coal or crude oil, or from coal or oil that is altered by heating. Most gas extracted commercially is thermogenic and is found above deposits of crude oil or seams of coal, so its extraction often accompanies the extraction of those fossil fuels.

Often, natural gas goes to waste as it escapes from coal mines or oil wells. Methane from coal seams, called *coalbed methane*, commonly leaks to the atmosphere during mining. To avoid this waste, and because methane is a potent greenhouse gas that contributes to climate change (pp. 495–497), mining engineers now try to capture more of this gas for energy. Likewise, in remote oil-drilling areas, where transporting natural gas is prohibitively expensive, natural gas is flared—wasted by being simply burned off. In Alaska, gas captured during oil drilling requires too much expense to export, but is now being reinjected into the ground for potential future extraction.

One source of biogenic natural gas is the decay process in landfills, and many landfill operators are now capturing this gas to sell as fuel (p. 624). This practice decreases energy waste, can be profitable for the operator, and helps reduce the atmospheric release of methane.

Natural gas extraction becomes more challenging with time

To access some natural gas deposits, prospectors need only drill an opening, because pressure and low molecular weight drive the gas upward naturally. The first gas fields to be tapped were of this type. Most fields remaining today, however, require that gas be pumped to the surface. In Texas, Kansas, California, and other areas of the United States, it is common to see horsehead pumps (**FIGURE 19.8**). A horsehead pump moves a rod in and out of a shaft, creating pressure to pull both natural gas and crude oil to the surface.

As with oil and coal, many of the most accessible natural gas reserves have already been exhausted. The average U.S. well produces just one-quarter as much as it did in 1970. Thus, much extraction today makes use of "fracturing techniques" to break into rock formations and pump gas to the surface. One such technique is to pump salt water under high pressure into the rocks to crack them. Sand or small glass beads are inserted to hold the cracks open once the water is withdrawn.

FIGURE 19.8 ▲ Horsehead pumps are used to extract natural gas as well as oil. They are a common feature of the landscape in areas such as west Texas. The pumping motion of the machinery draws gas and oil upward from below ground. Gas and oil are then transported through pipelines (left foreground), ultimately to homes and businesses.

Offshore drilling produces much of our gas and oil

Drilling for natural gas, as well as for oil, takes place not just on land but also in the seafloor on the continental shelves (**FIGURE 19.9**). Offshore drilling has required developing technology that can withstand the forces of wind, waves, and ocean currents. Some drilling platforms are fixed, standing platforms built with unusual strength. Others are resilient floating platforms anchored in place above the drilling site. Roughly one-third of the oil and 13% of the natural gas extracted in the United States today comes from offshore sites, primarily in the Gulf of Mexico and secondarily off the

FIGURE 19.9 ▼ Most remaining deposits of oil and natural gas in the United States lie offshore under ocean water. Offshore drilling platforms like this one off Santa Barbara, California, allow the oil industry to drill for petroleum in the seafloor on the continental shelves.

southern California coast. Geologists estimate that most U.S. gas and oil remaining to be extracted occurs offshore.

In 2008 the U.S. Congress lifted a long-standing moratorium on offshore drilling along much of the U.S. coast. The Obama administration in 2010 then designated vast areas open for drilling that were formerly closed. These included most waters along the Atlantic coast from Delaware south to central Florida, a region of the eastern Gulf of Mexico, and most waters off Alaska's North Slope. However, just weeks after this announcement, British Petroleum's *Deepwater Horizon* oil drilling platform exploded and sank, causing the largest accidental oil spill in world history (pp. 547–548). The Obama administration was forced to backtrack, canceling offshore drilling projects it had approved and putting a hold on further approvals.

Oil is the world's most-used fuel

Our most-used fuel is oil, which today accounts for 35% of the world's commercial energy consumption. The modern extraction and use of oil for energy began in the mid-19th century. A Dartmouth College scholar named George Bissell started the Pennsylvania Rock Oil Company in 1854, and five years later Edwin Drake drilled the world's first oil well for that company, in Titusville, Pennsylvania. Over the next 40 years, Pennsylvania's oil fields produced half the world's oil supply and helped establish a fossil-fuel-based economy that would hold sway for decades to come.

Today our global society produces and consumes nearly 750 L (200 gal) of oil each year for every man, woman, and child. World oil consumption has risen 15% in the past decade and shows little sign of abating. The United States consumes 23% of the world's oil, but rapidly industrializing populous nations such as China and India are increasingly driving world demand. **TABLE 19.5** shows the top oil-producing and oil-consuming nations.

Heat and pressure act on organic matter to form oil

The sludgelike liquid we know as **oil**, or **crude oil**, tends to form within a window of temperature and pressure conditions often found 1.5–3 km (1–2 mi) below the surface. Oil is also known as **petroleum**, although this term is commonly used to refer to both oil and natural gas. Like natural gas, the crude oil we extract from places like Alaska's North Slope was formed

TABLE 19.5	Top Producers and Consumers of Oil
Production (% world production)	Consumption (% world consumption)
Russia, 11.8	United States, 22.3
Saudi Arabia, 11.6	China, 9.8
United States, 10.8	Japan, 5.2
Iran, 5.0	India, 3.6
China, 4.7	Russia, 3.4

Data are for 2009, from U.S. Energy Information Administration.

when dead plant material (and small amounts of animal material) drifted down through coastal marine waters millions of years ago and was buried in sediments on the ocean floor. These organic remains were then transformed by time, heat, and pressure into the petroleum of today. All crude oil is a mixture of hundreds of different types of hydrocarbon molecules (p. 28). The specific properties of crude oil depend on the chemistry of the organic starting materials, the geologic environment of its location, and the details of the formation process.

Petroleum geologists infer the location and size of deposits

Because petroleum forms only under certain conditions, it occurs in isolated deposits. Once geothermal heating separates hydrocarbons from their source material and produces crude oil, this liquid migrates upward through rock pores, sometimes assisted by seismic faulting. It tends to collect in porous layers beneath dense, impermeable layers.

Geologists searching for oil (or other fossil fuels) drill cores and conduct ground, air, and seismic surveys to map underground rock formations and predict where fossil fuel deposits might lie. Using these methods, geologists with the U.S. Geological Survey (USGS) have estimated that 11.6–31.5 billion barrels of oil lay beneath the Arctic National Wildlife Refuge's 1002 Area, of which 4.3–11.8 billion barrels are "technically recoverable" with current technology (see **THE SCIENCE BEHIND THE STORY**, pp. 542–543).

However, oil companies will not be willing to extract these entire amounts. Some oil would be so difficult to extract that the expense of doing so would exceed the income the company would receive from the oil's sale. Thus, the amount a company chooses to drill for will be determined by the costs of extraction (and transportation), balanced against the current price of oil on the world market. Because the price of oil fluctuates, the portion of oil from a given deposit that is "economically recoverable" fluctuates as well. At higher prices, economically recoverable amounts approach technically recoverable amounts.

Thus, technology sets a limit on the amount that *can* be extracted, whereas economics determines how much *will* be extracted. The amount of oil, or any other fossil fuel, in a deposit that is technologically and economically feasible to remove under current conditions is termed the **proven recoverable reserve** of that fuel.

We drill to extract oil

Once geologists have identified a promising location for an oil deposit, an oil company will typically conduct *exploratory drilling*. Exploratory holes are usually small in circumference but descend to great depths. If enough oil is encountered, extraction begins. Just as you would squeeze a sponge to remove its liquid, pressure is required to extract oil from porous rock. Oil is typically already under pressure—from above by rock or trapped gas, from below by groundwater, or internally from natural gas dissolved in the oil. All these forces are held in place by surrounding rock until drilling reaches the deposit, whereupon oil will often rise to the surface of its own accord.

Once pressure is relieved and some oil has risen to the surface, however, the remaining oil becomes more difficult to extract and may need to be pumped out. Even after pumping, a great deal of oil remains stuck to rock surfaces. As much as two-thirds of a deposit may remain in the ground after **primary extraction**, the initial drilling and pumping of available oil (**FIGURE 19.10A**). Companies may then begin **secondary extraction**, in which solvents are used or underground rocks are flushed with water or steam to remove additional oil (**FIGURE 19.10B**).

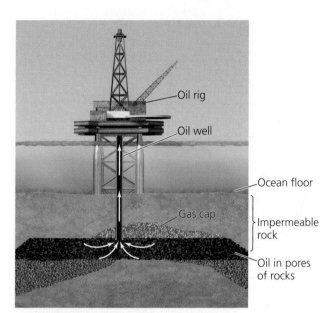

(a) Primary extraction of oil

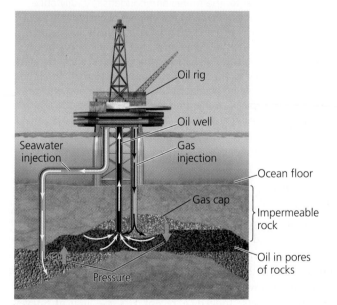

(b) Secondary extraction of oil

FIGURE 19.10 ▲ In primary extraction **(a)**, oil is drawn up through the well by keeping pressure at the top lower than pressure at the level of the oil deposit. Once the pressure in the deposit drops, however, material must be injected into the deposit to increase the pressure. Thus, secondary extraction **(b)** involves injecting seawater beneath the oil and/or injecting gases just above the oil to force more oil up and out of the deposit.

At Prudhoe Bay, seawater is piped in from the coast and pumped into wells to flush out the remaining oil. Even after secondary extraction, quite a bit of oil can remain; we lack the technology to remove every last drop. Secondary extraction is more expensive than primary extraction, so many U.S. deposits did not undergo secondary extraction when they were first drilled because the price of oil was too low to make the procedure economical. When oil prices rose in the 1970s, many drilling sites were reopened for secondary extraction. More are being reopened today.

Petroleum products have many uses

Because crude oil is a complex mix of hydrocarbons, we can create many types of petroleum products by separating its various components. This is what takes place at an oil refinery (**FIGURE 19.11**). The many types of hydrocarbon molecules in crude oil have carbon chains of different lengths (p. 28). A chain's length affects its chemical properties, and this has consequences for our use, such as whether a given fuel burns cleanly in a car engine. Through the process of **refining**, hydrocarbon molecules are separated into different size classes and are chemically transformed to create specialized fuels for heating, cooking, and transportation, and to create lubricating oils, asphalts, and the precursors of plastics and other petrochemical products.

Since the 1920s, refining techniques and chemical manufacturing have greatly expanded our uses of petroleum to include a wide array of products and applications, from lubricants to plastics to fabrics to pharmaceuticals. Today, petroleum-based products are all around us in our everyday lives (**FIGURE 19.12**). Because petroleum products have become so central to our lives, many fossil fuel experts today are voicing concern that oil production may soon decline as we continue to deplete the world's recoverable oil reserves.

We may have already depleted half our oil reserves

Some scientists and oil industry analysts calculate that we have already extracted half the world's oil reserves. So far we have used up about 1.1 trillion barrels of oil, and most estimates hold that somewhat more than 1 trillion barrels remain. To estimate how long this remaining oil will last, analysts calculate the **reserves-to-production ratio**, or **R/P ratio**, by dividing the amount of total remaining reserves by the annual rate of production (i.e., extraction and processing). At current levels of production (30 billion barrels globally per year), 1.2 trillion barrels would last about 40 more years.

Unfortunately, this does not mean that we have 40 years to figure out what to do once the oil runs out. A growing number of scientists and analysts insist that we will face a crisis not when the last drop of oil is pumped, but when the rate of production comes to a peak and then begins to decline—a point in time nicknamed **"peak oil."** They point out that as production declines following peak oil, if demand continues to increase (because of rising global population and consumption), then we will experience an oil shortage immediately. Because production tends to decline once reserves are depleted halfway, most of these experts calculate

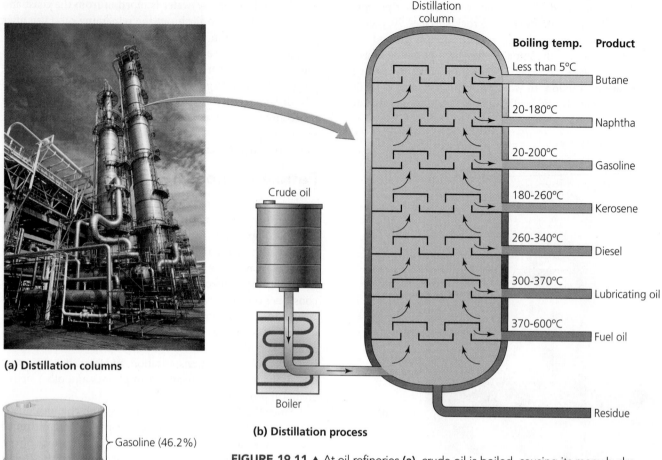

(a) Distillation columns

Distillation column

Boiling temp.	Product
Less than 5°C	Butane
20-180°C	Naphtha
20-200°C	Gasoline
180-260°C	Kerosene
260-340°C	Diesel
300-370°C	Lubricating oil
370-600°C	Fuel oil
	Residue

Crude oil

Boiler

(b) Distillation process

Gasoline (46.2%)

Diesel fuel and heating oil (20.3%)

Liquefied petroleum gases (10.0%)

Jet fuel (7.8%)

Heavy fuel oil (3.2%)

Other (12.5%)

(c) Typical composition of refined oil

FIGURE 19.11 ▲ At oil refineries **(a)**, crude oil is boiled, causing its many hydrocarbon constituents to volatilize and proceed upward through a distillation column **(b)**. Constituents that boil at the hottest temperatures and condense readily once the temperature cools will condense at low levels in the column. Constituents that volatilize at cooler temperatures will continue rising through the column and condense at higher levels, where temperatures are cooler. In this way, heavy oils (generally those with hydrocarbon molecules with long carbon chains) are separated from lighter oils (generally those with short-chain hydrocarbon molecules). The refining process produces a range of petroleum products. Shown in **(c)** are percentages of each major category of product typically generated from a barrel of crude oil. Data (c) from U.S. Energy Information Administration 2009. *Annual energy review* 2008.

that the peak oil crisis will likely begin within the next several years.

To understand the basis of these concerns, we need to turn back the clock to 1956. In that year, Shell Oil geologist M. King Hubbert calculated that U.S. oil production would peak around 1970. His prediction was ridiculed at the time, but it proved to be accurate; U.S. production peaked in that very year and has continued to fall since then (**FIGURE 19.13A**). The peak in production came to be known as **Hubbert's peak**.

In 1974, Hubbert analyzed data on technology, economics, and geology, predicting that global oil production would peak in 1995. It grew past 1995, but many scientists using newer, better data today predict that at some point in the coming decade, production will begin to decline (**FIGURE 19.13B**). Discoveries of new oil fields peaked 30 years ago, and since then we have been extracting and consuming more oil than we have been discovering.

Predicting an exact date for peak oil and the coming decline in production is difficult. Because of year-to-year variability in production, we will not be able to recognize that we have passed the peak of oil production until several years after it has happened. Many companies and governments do not reveal their true data on oil reserves, and estimates differ as to how much oil we can extract secondarily from existing deposits. Indeed, a recent U.S. Geological Survey report estimated 2 trillion barrels remaining in the world, rather than 1 trillion, and some estimates predict still greater amounts. A 2007 report by the U.S. General Accounting Office reviewed 21 studies and found that most estimates for the timing of the oil production peak ranged from now through 2040. Regardless of the exact timing, it seems certain that a peak in production of this limited and nonrenewable resource will occur. Meanwhile, global demand continues to rise, particularly as China and India industrialize rapidly.

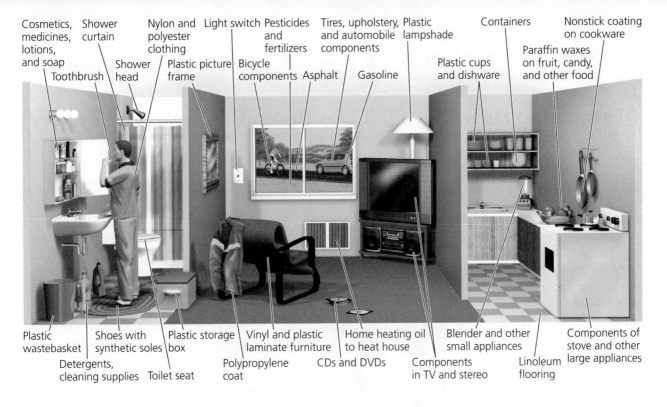

Cosmetics, medicines, lotions, and soap
Shower curtain
Nylon and polyester clothing
Light switch
Pesticides and fertilizers
Tires, upholstery, and automobile components
Plastic lampshade
Containers
Nonstick coating on cookware
Toothbrush
Shower head
Plastic picture frame
Bicycle components
Asphalt
Gasoline
Plastic cups and dishware
Paraffin waxes on fruit, candy, and other food

Plastic wastebasket
Shoes with synthetic soles
Plastic storage box
Vinyl and plastic laminate furniture
Home heating oil to heat house
Blender and other small appliances
Components of stove and other large appliances
Detergents, cleaning supplies
Toilet seat
Polypropylene coat
CDs and DVDs
Components in TV and stereo
Linoleum flooring

FIGURE 19.12 ▲ Besides the fuels we use for transportation and heating, petroleum products include many of the fabrics we wear and the plastics in countless items we use every day.

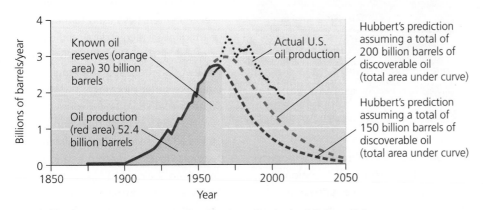

Known oil reserves (orange area) 30 billion barrels

Actual U.S. oil production

Hubbert's prediction assuming a total of 200 billion barrels of discoverable oil (total area under curve)

Oil production (red area) 52.4 billion barrels

Hubbert's prediction assuming a total of 150 billion barrels of discoverable oil (total area under curve)

(a) Hubbert's prediction of peak in U.S. oil production, with actual data

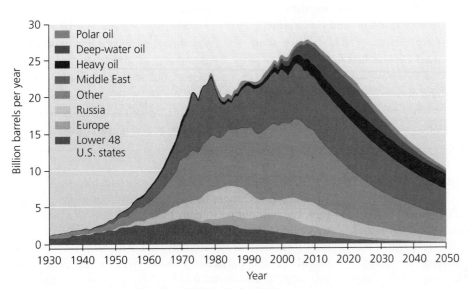

Polar oil
Deep-water oil
Heavy oil
Middle East
Other
Russia
Europe
Lower 48 U.S. states

(b) Modern prediction of peak in global oil production

FIGURE 19.13 ◄ Because fossil fuels are nonrenewable resources, supplies at some point pass the midway point of their depletion, and annual production begins to decline. U.S. oil production peaked in 1970 **(a)**, just as geologist M. King Hubbert had predicted. Success in Alaska, the Gulf of Mexico, and with secondary extraction increased production past his prediction during the decline. Today many analysts believe global oil production is about to peak. Shown in **(b)** is a recent projection, from a 2009 analysis by scientists at the Association for the Study of Peak Oil. Data from (a) Hubbert, M.K., 1956. Nuclear energy and the fossil fuels. Shell Development Co. Publ. No. 95, Houston, TX; and U.S. Energy Information Administration; and (b) Campbell, C.J., and Association for the Study of Peak Oil. By permission of Dr. Colin Campbell.

Estimating How Much Oil Lies Beneath the Arctic Refuge

To judge whether it is worth drilling for oil in the Arctic National Wildlife Refuge, most of us would want to know how much oil is there. How can we tell? For that matter, how do oil companies anywhere know where to drill for oil?

Oil companies employ petroleum geologists who study underground rock formations to predict where deposits of oil and natural gas might lie. Because the organic matter that gave rise to fossil fuels was buried in sediments, geologists know to look for sedimentary rock that may act as a source. They also know that oil and gas tends to seep upward through porous rock until being trapped by impermeable layers.

To map subsurface rock layers, petroleum geologists survey the landscape on the ground and from airplanes, studying rocks on the surface. Because rock layers often become tilted over geologic time, these strata may protrude at the surface, giving geologists an informative "side-on" view. Geologists also drill exploratory cores to take samples.

But to really understand what's deep beneath the surface, scientists need to conduct seismic surveys. Recall from an earlier *Science behind the Story* (pp. 38–39) that researchers are using seismic surveying to map subsurface geology across the entire

Petroleum geologists study maps of seismic data to determine where oil might likely be found

United States. Petroleum geologists have long used these techniques to look for fossil fuel deposits.

In seismic surveying, a base station first creates powerful vibrations at the surface by exploding dynamite, thumping the ground with a large weight, or using an electric vibrating machine (**see first figure**). As a result, seismic waves travel down and outward in all directions through the ground, just as ripples spread when a pebble is dropped into a

pond. As they travel, the waves encounter layers of different types of rock. The waves travel more quickly through denser rock. Each time a seismic wave encounters a new type of rock with a different density, some of the wave's energy is reflected off the boundary, and the rest passes through the boundary into the new layer. Some wave energy may be refracted, or bent, along the edge of the layer, sending refraction waves upward.

As reflected and refracted waves return to the surface, devices called seismometers (also used to measure earthquakes) record data on their strength and precise timing. Scientists collect data from seismometers at multiple surface locations and run the data through computer programs for analysis. By analyzing how long it takes all the reflected and refracted seismic waves to reach the various receiving stations, and how strong the waves are at each site, researchers can infer the densities, thicknesses, and locations of underlying geologic layers.

Seismic surveying is similar to the way we use sonar in water, or how bats use echolocation. It is also used for finding coal deposits, salt and mineral deposits, and geothermal energy hotspots, as well as for studying faults, aquifers, and engineering sites.

542

The coming divergence of demand and supply will likely have momentous economic, social, and political consequences that will profoundly affect the lives of each and every one of us. One prophet of peak oil, writer James Howard Kunstler, has sketched a frightening scenario of our post-peak world during what he calls "the long emergency": Lacking cheap oil with which to transport goods long distances, today's globalized economy would collapse, and our economies would become intensely localized. Large cities could no longer be supported without urban agriculture, and without petroleum-based fertilizers and pesticides we could feed only a fraction of the world's 7 billion people, even if we expand agricultural land. The American suburbs would be hit particularly hard because of their utter dependence on the automobile. Kunstler argues that the suburbs will become the slums of the future, a bleak and

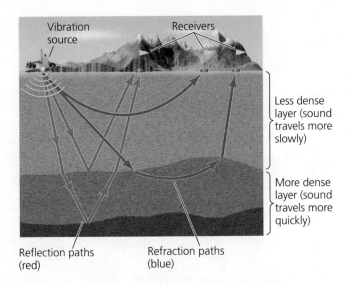

Vibration source

Receivers

Less dense layer (sound travels more slowly)

More dense layer (sound travels more quickly)

Reflection paths (red)

Refraction paths (blue)

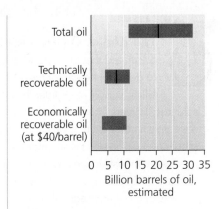

Total oil

Technically recoverable oil

Economically recoverable oil (at $40/barrel)

0 5 10 15 20 25 30 35
Billion barrels of oil, estimated

Of the total oil that scientists from the U.S. Geological Survey have estimated to exists in the 1002 Area of the Arctic Refuge, only a portion is technically recoverable, and not all of that is economically recoverable. Data from U.S. Geological Survey.

In seismic surveying, powerful vibrations are created in one location, and receivers measure how long it takes seismic waves to reach other locations. Waves travel more quickly through denser layers, and density differences cause waves to reflect or refract. Scientists interpret the patterns of wave reception to infer the densities, thicknesses, and locations of underlying geological layers, providing clues to the location and size of fuel deposits.

Although seismic surveys provide clues about the location and size of fuel deposits, even with good survey data, oil companies may still have only a 10% success rate in finding oil or gas. They need to conduct exploratory drilling to confirm whether oil or gas actually exists in any given location.

With data from such techniques, geologists from the U.S. Geological Survey (USGS) in 1998 assessed the subsurface geology of the Arctic National Wildlife Refuge to predict how much oil it may hold. Over three years, dozens of scientists conducted fieldwork and combined their results with a reanalysis of 2,300 km (1,400 mi) of seismic survey data that had been collected by industry in the 1980s.

After studying their resulting subsurface maps, USGS scientists concluded, with 95% certainty, that between 11.6 and 31.5 billion barrels

of oil lay underneath the Refuge's 1002 Area. The mean estimate of 20.7 billion barrels represents their best guess as to how much oil the 1002 Area holds. This amount is enough to supply the United States for 35 months at its current rate of consumption.

However, some portion of oil will be impossible to extract using current technology, and will need to wait for future advances in extraction equipment or methods. Thus, geologists generally estimate "technically recoverable" amounts of fuels. In its estimate for the 1002 Area, the USGS calculated technically recoverable oil to total 4.3–11.8 billion barrels, with a mean estimate of 7.7 billion barrels (just over 1 year of U.S. consumption).

The portion of this oil that is "economically recoverable" depends on the costs of extracting it and the price of oil

on the world market. USGS scientists calculated that at a price of $40 per barrel, 3.4–10.8 billion barrels would be economically worthwhile to recover from the 1002 Area (**see second figure**). At today's higher prices, the economically recoverable amount would be closer to the technically recoverable amount.

Because some Native lands and state-owned offshore areas can be developed if the 1002 Area is developed, the USGS also surveyed these areas and estimated that 1.4–4.2 billion additional barrels lie beneath them.

In 2002, the USGS conducted similar analyses for the National Petroleum Reserve–Alaska, and estimated that it contained 9.3 billion barrels of technically recoverable oil, spread over an area 15 times as large as the Arctic National Wildlife Refuge. Petroleum geologists worldwide are using similar methods to try to determine how much oil remains to exploit. Their work is of great importance as the world's nations try to set energy policy. ■

crime-ridden landscape littered with the hulls of rusted-out SUVs.

More optimistic observers argue that as oil supplies dwindle, rising prices will create powerful incentives for businesses, governments, and individuals to conserve energy and develop alternative energy sources (Chapters 20 and 21)—and that these developments will save us from major disruptions caused by the coming oil peak.

Indeed, to achieve a sustainable society, we will need to switch to renewable energy sources. Investments in energy efficiency and conservation (pp. 555–557) can extend the time we have to make this transition. However, the research and development needed to construct the infrastructure for a new energy economy depend on having cheap oil, and the time we will have to make this enormous transition will be quite limited.

The End of Oil Physicist David Goodstein has calculated that once world oil production begins to decline, the gap between rising demand and falling supply may widen by 5% per year. As a result, just 10 years after the production peak, we would have only half the oil that we had at the peak. In such a scenario, some experts worry that we may not be able to modify our infrastructure and institutions quickly enough to bring other energy sources online before the economic impacts of an oil shortage undermine our very ability to do so. How do you think your life would be affected if our society were to suffer a 50% drop in oil availability over the next decade? What steps would you take to adapt to these changes? What steps should our society take to deal with the coming depletion of oil?

OTHER FOSSIL FUELS

As oil production declines, we will rely more on natural gas and coal—yet these in turn will also peak and decline in future years. Are there other fossil fuels that can replace them and stave off our day of reckoning? At least three types of alternative fossil fuels exist in large amounts: oil sands, oil shale, and methane hydrate.

Canada is mining oil sands

Oil sands (also called **tar sands**) are deposits of moist sand and clay containing 1–20% *bitumen*, a thick and heavy form of petroleum that is rich in carbon and poor in hydrogen. Oil sands represent crude oil deposits that have been degraded and chemically altered by water erosion and bacterial decomposition.

Bitumen is too thick to extract by conventional oil drilling, so oil sands near the surface are generally removed by strip mining (**FIGURE 19.14**). For deposits 75 m (250 ft) or more below ground, a variety of extraction techniques are being devised. Most involve injecting steam or chemical solvents to liquefy the bitumen so it can be extracted through conventional wells. After extraction, bitumen may be sent to specialized refineries, where several types of chemical reactions that add hydrogen or remove carbon can upgrade it into more valuable synthetic crude oil.

Three-quarters of the world's oil sands lie in two areas: eastern Venezuela and northeastern Alberta. Oil sands in each region hold about 175 billion barrels of oil. In Alberta, strip mining began in 1967. Rising crude oil prices have made oil sands more profitable, and dozens of companies are now angling to begin mining projects in the region. Canadian oil sands are now producing well over 1 million barrels of oil per day, contributing half of Canada's petroleum production. If oil sands are included in proven reserves of petroleum, then Canada is second only to Saudi Arabia in total oil reserves.

Oil shale is abundant in the U.S. West

Oil shale is sedimentary rock filled with kerogen (organic matter) and can be processed to produce liquid petroleum. Oil shale is formed by the same processes that form crude oil

FIGURE 19.14 ▲ In Alberta, companies strip-mine oil sands with the world's largest dump trucks and power shovels. On average, 2 metric tons of oil sands are required to produce 1 barrel of synthetic crude oil.

but occurs when kerogen was not buried deeply enough or subjected to enough heat and pressure to form oil. We mine oil shale using strip mines or subsurface mines. Once mined, oil shale can be burned directly like coal, or can be baked in the presence of hydrogen and in the absence of air to extract liquid petroleum (a process called *pyrolysis*).

The world's known deposits of oil shale may be able to produce over 600 billion barrels of oil (roughly half as much as the conventional crude oil remaining in the world). About 40% of global oil shale reserves are in the United States, mostly on federally owned land in Colorado, Wyoming, and Utah. Historically, low prices for crude oil have kept investors away from the more costly oil shale, but today's higher crude prices mean that oil shale is again attracting attention.

Methane hydrate shows potential

Another novel potential source of fossil fuel energy occurs in sediments on the ocean floor. **Methane hydrate** (also called *methane clathrate* or *methane ice*) is an ice-like solid consisting of molecules of methane (CH_4, the main component of natural gas) embedded in a crystal lattice of water molecules. Methane hydrate is stable at temperature and pressure conditions found in many sediments on the Arctic seafloor and the continental shelves. Most methane in these gas hydrates was formed by bacterial decomposition in anaerobic environments, but some results from thermogenic formation deeper below the surface.

Scientists believe there to be immense amounts of methane hydrate on Earth. The USGS estimates that the world's methane hydrate deposits may hold twice as much carbon as all known deposits of oil, coal, and natural gas combined. However, we do not yet know how to extract these energy sources safely. Destabilizing a methane hydrate deposit during extraction could lead to a catastrophic release of gas. This could cause a massive landslide and tsunami and would also release huge amounts of methane, a potent greenhouse gas, into the atmosphere, worsening global climate change.

These alternative fuels have drawbacks

Oil sands, oil shale, and methane hydrate are abundant, but they are no panacea for our energy challenges. For one thing, their net energy values are low, because they are expensive to extract and process. Thus, the energy returned on energy invested (EROI) ratio is low. For instance, much of the energy content of oil shale is consumed in its production, and oil shale's EROI is only about 2:1 or 3:1, compared to a 5:1 or greater ratio for conventional crude oil.

Second, these fuels exert severe environmental impacts. Oil sands and oil shale require extensive strip mining, which utterly devastates landscapes over large areas and pollutes waterways running into still more areas. (We discuss impacts of strip-mining below with coal and again in Chapter 23, pp. 648–649). Although most governments require mining companies to restore mined areas to their original condition, regions denuded by the very first oil sand mine in Alberta 30 years ago have still not recovered.

Besides impacts from their extraction, our combustion of alternative fossil fuels would emit at least as much carbon dioxide, methane, and other air pollutants as our use of coal, oil, and gas. This would worsen the impacts fossil fuels are already causing, including air pollution and global climate change.

ENVIRONMENTAL IMPACTS AND SOLUTIONS

Our society's love affair with fossil fuels and the many petrochemical products we develop from them has eased constraints on travel, has helped lengthen our life spans, and has boosted our material standard of living beyond what our ancestors could have dreamed. However, it has also caused harm to the environment and human health. Concern over these impacts is a prime reason many scientists, environmental advocates, businesspeople, and policymakers are increasingly looking toward renewable sources of energy that exert less impact on natural systems.

Fossil fuel emissions cause pollution and drive climate change

When we burn fossil fuels, we alter fluxes in Earth's carbon cycle (pp. 124–128). We essentially take carbon that has been retired into a long-term reservoir underground and release it into the air. This occurs as carbon from the hydrocarbon molecules of fossil fuels unites with oxygen from the atmosphere during combustion, producing carbon dioxide (CO_2). Carbon dioxide is a greenhouse gas (p. 495), and CO_2 released from fossil fuel

combustion warms our planet and drives changes in global climate (Chapter 18). Because global climate change may have diverse, severe, and widespread ecological and socioeconomic impacts, carbon dioxide pollution (**FIGURE 19.15**) is becoming recognized as the greatest environmental impact of fossil fuel use. Moreover, methane is a potent greenhouse gas that drives climate warming. In Chapter 18 we explored the many avenues being considered to address climate change (pp. 517–525).

Besides modifying our climate, fossil fuel emissions affect human health directly. Gasoline combustion in automobiles releases pollutants that irritate the nose, throat, and lungs. Some hydrocarbons, such as benzene and toluene, are carcinogenic to laboratory animals and likely also to people. Gases such as hydrogen sulfide can evaporate from crude oil, irritate the eyes and throat, and cause asphyxiation. Crude oil also often contains trace amounts of known poisons such as lead and arsenic. As a result, workers at drilling operations, refineries, and other jobs that entail frequent exposure to oil or its products can develop serious health problems, including cancer.

The combustion of oil in our vehicles and coal in our power plants releases sulfur dioxide and nitrogen oxides, which contribute to industrial and photochemical smog (pp. 476–478) and to acid deposition (pp. 482–486). Combustion of coal high in mercury content emits mercury that can bioaccumulate in organisms' tissues, poisoning animals as it moves up food chains (pp. 385–386) and presenting health risks to people.

As we saw in Chapter 17, air pollution from fossil fuel combustion has been reduced in developed nations in recent decades as a result of legislation such as the U.S. Clean Air Act and the government regulations that ensued to protect public health. In contrast, fossil fuel pollution is increasing in developing nations that are industrializing rapidly as they grow in population. Policies such as the Clean Air Act regulations have encouraged industry to develop technological advances for pollution control. Technologies such as catalytic converters

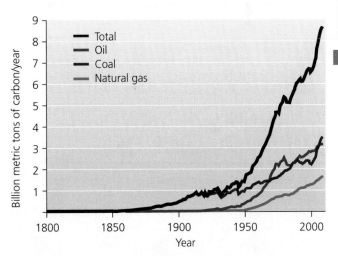

FIGURE 19.15 ▲ Emissions from fossil fuel combustion have risen dramatically as industrialization has proceeded and as population and consumption have grown. Here, global emissions of carbon from carbon dioxide are subdivided by their source (oil, coal, or natural gas). Other minor sources (such as cement production) are also included in the graphed total. Data from Carbon Dioxide Information Analysis Center, Oak Ridge National Laboratory, U.S. Department of Energy, Oak Ridge, TN.

that cut down on vehicle exhaust pollution (p. 682) have helped a great deal, and their wider adoption in the developing world would reduce pollution further. For pollution from coal-fired power plants, scientists and engineers are seeking ways to cleanse coal of its impurities so that it can continue to be used as an energy source while minimizing impact on health and the environment.

Clean coal technologies aim to reduce pollution from coal

Clean coal technologies refer to a wide array of techniques, equipment, and approaches that aim to remove chemical contaminants during the process of generating electricity from coal. Among these technologies are various types of scrubbers (p. 474), devices that chemically convert or physically remove pollutants. Some use minerals such as magnesium, sodium, or calcium in reactions to remove sulfur dioxide (SO_2) from smokestack emissions. Others use chemical reactions to strip away nitrogen oxides (NO_X), breaking them down into elemental nitrogen and water. Multilayered filtering devices are used to capture tiny ash particles.

Another type of clean coal approach is to dry coal with high water content, which makes it cleaner-burning. We can also gain more power from coal with less pollution through a process of *gasification*, in which coal is converted into a cleaner synthesis gas, or syngas, by reacting it with oxygen and steam at a high temperature. In an "integrated gasification combined cycle" process, syngas from coal is used to turn a gas turbine and also to heat water to turn a steam turbine.

The U.S. government and the coal industry have each invested billions of dollars in clean coal technologies, which have helped to reduce air pollution from sulfates, nitrogen oxides, mercury, and particulate matter, as we saw in Chapter 17 (pp. 473–474). If they were applied to the many older plants that still pollute our air, they could help still more. Still, many energy analysts and environmental advocates emphasize that these technologies may make for "cleaner" coal, but will never result in energy production that is completely clean. Coal, they maintain, is an inherently dirty way of generating power and should be replaced outright with cleaner energy sources.

Can we capture and store carbon?

Even if our clean coal technologies were able to remove every last chemical contaminant from power plant emissions, coal combustion would still pump huge amounts of carbon dioxide into the air, intensifying the greenhouse effect and worsening global climate change. This is why many current efforts focus on **carbon capture** and **carbon storage** or **sequestration** (p. 519). As mentioned in Chapter 18, this approach consists of capturing carbon dioxide emissions, converting the gas to a liquid form, and then sequestering (or storing) it in the ocean or underground in a geologically stable rock formation (**FIGURE 19.16**).

Carbon capture and storage (sometimes abbreviated as CCS) is now being attempted at a variety of new and retrofitted facilities, most of which also control other emissions from coal. The world's first coal-fired power plant to approach zero emissions opened in 2008 in Germany. This Swedish-built plant removes its sulfate pollutants and captures its carbon dioxide, then compresses the CO_2 into liquid form, trucks it 160 km (100 mi) away, and injects it 900 m (3,000 ft) underground into a depleted natural gas field.

In North Dakota, the Great Plains Synfuels Plant gasifies its coal, then sends half the CO_2 through a pipeline into Canada, where a Canadian oil company buys the gas to inject into an oilfield to help it pump out the remaining oil. The North Dakota plant also captures, isolates, and sells seven other types of gases for various purposes.

Currently the U.S. Department of Energy is teaming up with nine domestic and foreign energy companies to build a prototype of a near-zero-emissions coal-fired power plant. The $1.5-billion *FutureGen* project aims to design, construct, and operate a power plant that burns coal using gasification and

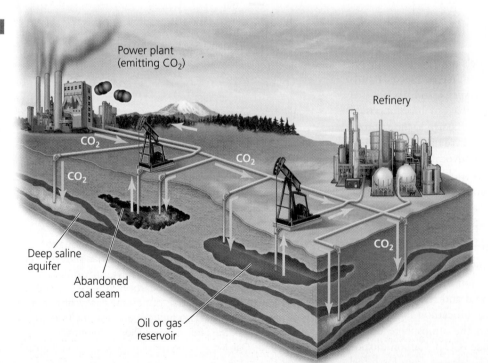

Power plant (emitting CO_2)

Refinery

CO_2

CO_2

CO_2

CO_2

CO_2

Deep saline aquifer

Abandoned coal seam

Oil or gas reservoir

FIGURE 19.16 ◀ Carbon capture and storage schemes propose to inject liquefied carbon dioxide emissions underground into depleted fossil fuel deposits, deep saline aquifers, or oil or gas deposits undergoing secondary extraction.

combined cycle generation, produces electricity and hydrogen, and captures its carbon dioxide emissions and sequesters the CO_2 underground. The project, located in Mattoon, Illinois, will pump CO_2 2,400 m (8,000 ft) underground into rock formations filled with saline water. Preliminary design work and cost estimates will be finished in 2010, and decisions on how to proceed made at that time. If this showcase project succeeds, it could be a model for a new generation of power plants across the world. Such facilities could enable us to keep on burning our cheapest and most abundant fossil fuel source in a way that does not contribute to climate change.

Energy experts are not ready to rely on carbon capture and storage quite yet, however. At this point the technology is too unproven to be the central focus of a clean energy strategy. We do not know whether we can ensure that carbon dioxide will stay underground once injected there, and we do not know whether these attempts might trigger earthquakes. Injection could in some cases contaminate groundwater supplies, and injecting carbon dioxide into the ocean would acidify its waters (pp. 440–441). Finally, many renewable energy advocates fear that the CCS approach takes the burden off emitters and prolongs our dependence on fossil fuels rather than facilitating a shift to renewables.

WEIGHING THE ISSUES

Clean Coal and Carbon Capture Do you think we should be spending billions of dollars to try to find ways to burn coal cleanly and to sequester carbon emissions from fossil fuels? Or is our money better spent on developing new clean and renewable energy sources that don't yet have enough infrastructure to produce power at the scale that coal can? What pros and cons do you see in each approach?

Fossil fuels pollute water as well as air

Even if we can clean up air pollution from coal, plenty of other pollution problems remain, ranging from effects of mining and extracting fossil fuels to impacts on water quality. Fossil fuels pollute water in many ways. What comes to most people's minds most readily is the pollution that occurs when massive oil spills from tanker ships or drilling platforms foul coastal waters and beaches. In 2010, British Petroleum's *Deepwater Horizon* offshore drilling platform exploded and sank off the coast of Louisiana (**FIGURE 19.17**). Eleven workers were killed, and oil began gushing out of the broken pipe on the ocean floor a mile beneath the surface at the rate of 62,000 barrels per day. Emergency shut-off systems had failed, and BP engineers tried one solution after another to stop the flow of oil and gas, which continued for three months.

The spill proved so difficult to control because we had never had to deal with a spill so deep underwater. It revealed that offshore drilling presents serious risks of environmental impact that may be difficult to address, even with our best engineering.

As the oil spread through the Gulf of Mexico and washed ashore, the region suffered a wide array of impacts (see **ENVISIONIT**, p. 548). The oil killed countless animals. Most conspicuous were birds, which die from hypothermia when

FIGURE 19.17 ▲ The explosion of British Petroleum's *Deepwater Horizon* drilling platform off the Louisiana coast in 2010 unleashed the world's biggest-ever accidental oil spill. Here vessels try to put out the blaze.

their feathers become coated with oil, ruining their ability to thermoregulate. However, the underwater nature of the BP spill meant that unknown numbers of fish, shrimp, and other marine animals were also killed, affecting coastal and ocean ecosystems in complex ways. Plants in coastal marshes were killed, leading to the erosion of marshes, which puts New Orleans and other coastal cities at greater risk of storm surges and flooding. Gulf Coast fisheries were hit hard by the spill, and these fisheries supply much of the nation's seafood. Thousands of fishermen and shrimpers were put out of work, and many were hired by BP to help clean up the spill. Beach tourism suffered, and many indirect economic and social impacts were expected to last for years. As this book went to press, Gulf coast residents were still dealing with the spill's cleanup and aftermath.

The *Deepwater Horizon* spill was the largest accidental oil spill in world history, far eclipsing the spill from the *Exxon Valdez* tanker in 1989 (p. 156). In that event, oil from Alaska's North Slope, piped to the port of Valdez through the trans-Alaska pipeline, exerted long-term damage to ecosystems and economies in Alaska's Prince William Sound when the tanker ran aground. Today, over two decades later, a layer of oil remains just inches beneath the sand of the region's beaches, and scientists debate whether it is better to try to remove it or to let it remain undisturbed.

Although large catastrophic oil spills have significant impacts on the marine environment, most water pollution from oil results from countless non-point sources (pp. 420–421) to which all of our actions contribute (see Figure 16.18b, p. 447). Oil from non-point sources such as automobiles, homes, industries, gas stations, and businesses runs off roadways and enters rivers and wastewater facilities, being discharged eventually into the ocean. Oil can also contaminate groundwater supplies, such as when leaks from oil operations penetrate deeply into soil. As we saw in Chapter 15 (p. 423), thousands of underground storage tanks containing petroleum products have leaked, threatening drinking water supplies. In addition, atmospheric deposition of pollutants from the combustion of fossil fuels exerts many impacts on freshwater ecosystems (p. 483).

The *Deepwater Horizon* spill disgorged over 200 million of gallons of oil into the Gulf of Mexico ...

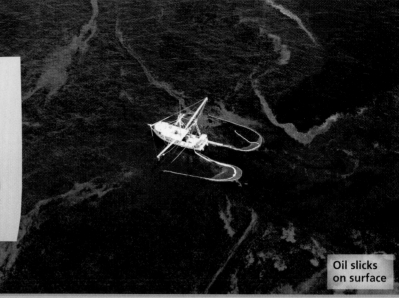

Oil slicks on surface

... where it fouled beaches, killed countless animals, and devastated fisheries and tourism.

Oiled brown pelican

Thousands of people threw themselves into the cleanup effort, but the spill's impacts will last for years.

Cleaning oiled beaches

YOU CAN MAKE A DIFFERENCE

➤ Volunteer for cleanups of oil spills in your region.

➤ Urge policymakers to strengthen regulations on offshore drilling.

➤ Limit your own oil consumption by driving less, driving a fuel-efficient car, eating local foods, reusing and recycling products, and supporting renewable energy.

Switching to alternative fossil fuel sources could worsen many of our environmental impacts. For instance, mining oil sands in Canada requires using an immense amount of water, and polluted wastewater is left to sit in gigantic reservoirs. The Syncrude company's massive tailings pond near Fort McMurray, Alberta, is so large that it is held back by the world's second-largest dam. Migratory waterfowl land on water bodies like this and are killed as the oily water gums up their feathers and impairs their ability to insulate themselves.

Coal mining devastates natural systems

The mining of coal has substantial impacts on natural systems and human well-being, as we will explore further in Chapter 23 (pp. 648–653). Strip mining destroys large swaths of habitat and causes extensive soil erosion. It also can cause chemical runoff into waterways through the process of **acid drainage**. This occurs when sulfide minerals in newly exposed rock surfaces react with oxygen and rainwater to produce sulfuric acid. As the sulfuric acid runs off, it leaches metals from the rocks, many of which are toxic to organisms. Acid drainage is a natural phenomenon, but mining greatly accelerates the rate at which it occurs by exposing many new rock surfaces at once. Regulations in the United States require mining companies to restore strip-mined land following mining, but complete restoration is impossible, and ecological modifications are severe and long-lasting (pp. 654–655). Most other nations exercise less oversight.

Mountaintop removal (**FIGURE 19.18** and pp. 650–653) has even greater impacts than conventional strip mining. When countless tons of rock and soil are removed from the top of a mountain, material slides downhill, where immense areas of habitat can be degraded or destroyed and creek beds can be clogged and polluted. Loosening of U.S. government regulations in 2002 enabled mining companies to legally dump mountaintop rock and soil into valleys and rivers below, regardless of the consequences for ecosystems, wildlife, and local residents.

The costs of alleviating all these environmental impacts—to say nothing of the health impacts on miners and on nearby residents (p. 653)—are high, and the public eventually pays them in an inefficient manner. The reason is that the costs are generally not internalized in the market prices of fossil fuels. Instead, we all pay these external costs (pp. 150–151) through medical expenses, costs of environmental cleanup, and impacts on our quality of life. Moreover, the prices we pay at the gas pump or on our monthly utility bill do not even cover the financial costs of fossil fuel production. Rather, fossil fuel prices have been kept inexpensive as a result of government subsidies to extraction companies. As we saw in Chapter 7 (Figure 7.14, p. 182), the profitable and well-established fossil fuel industries still receive far more financial support from taxpayers than do the young and struggling renewable energy sources. Thus, we all pay extra for our fossil fuel energy through a portion of our tax dollars, generally without even realizing it.

Oil and gas extraction modify the environment

Much more than drilling is involved in the development of an oil or gas field (see the photo that opens this chapter, p. 528). Road networks must be constructed, and many sites may be explored in the course of prospecting. The extensive infrastructure needed to support a full-scale drilling operation typically includes housing for workers, access roads, transport pipelines, and waste piles for removed soil. Ponds may be constructed for collecting the toxic sludge that remains after the useful components of oil have been removed. At extraction sites for coalbed methane (p. 537), groundwater is pumped out to free gas to rise, but salty groundwater dumped on the surface can contaminate soil and kill vegetation over large areas.

To predict the possible ecological effects of drilling in the Arctic National Wildlife Refuge, scientists have examined the effects of oil field development on arctic vegetation, air quality, water quality, and wildlife (including caribou, grizzly bears, and various birds) in Alaska locales where the environment is similar to that of the Refuge's coastal plain (see **THE SCIENCE BEHIND THE STORY**, pp. 550–551). Based on these studies, many scientists anticipate substantial and wide-ranging damage to vegetation and wildlife if drilling were to take place in the refuge.

Others contend that drilling operations in the Arctic National Wildlife Refuge would have little environmental impact because drilling technology is more environmentally sensitive than when sites like Prudhoe Bay were developed. **Directional drilling** allows oil wells to be drilled in directions outward from a drilling pad. Oil companies have long been able to slant their drills so as to bore in at an angle, but

FIGURE 19.18 ▶ Strip mining for coal in some areas is taking place on massive scales, such that entire mountain peaks are leveled and fill is dumped into adjacent valleys, as at this site in West Virginia. Such "mountaintop removal mining" can cause erosion and acid drainage into waterways that flow into surrounding valleys, affecting ecosystems over large areas, as well as the people who live there.

The SCIENCE behind the Story

Predicting Environmental Impacts of Oil Drilling on Alaska's North Slope

Just as geologists have tried to estimate the amount of oil we could extract if we were to drill in the Arctic National Wildlife Refuge, biologists have tried to estimate the ecological impacts such drilling would have. To predict the impacts of drilling in the Arctic Refuge's 1002 Area, scientists have studied the status of animals in the refuge and also the impacts of oil field development on the coastal plain in areas such as Prudhoe Bay.

Dozens of researchers, many with the U.S. Fish and Wildlife Service, have compared developed regions with undeveloped regions, or have monitored particular areas or populations before and after drilling. When possible, scientists have run small-scale manipulative experiments, to study, for example, the effects of ice roads and of secondary extraction methods such as seawater flushing. In one way or another they have examined the impacts of road building, oil pad construction, worker presence, oil spills, accidental fires, trash buildup, permafrost melting, off-road vehicle trails, and dust from roads.

These studies have led scientists to anticipate damage to air quality, water quality, vegetation, and wildlife if drilling were to take place in the Refuge. Air and water quality can be degraded by fumes from equipment and drilling operations, flaring of natural gas associated with oil extraction, sludge ponds, waste pits, and oil spills. Vegetation is killed when salt water pumped in for flushing deposits

Caribou cross a road on Alaska's North Slope

is spilled or when plants are buried under gravel pits or roads. Tundra ecosystems (p. 101) are especially vulnerable to human disturbance because plants grow slowly here, so even minor impacts can have long-lasting repercussions. For example, tundra vegetation at Prudhoe Bay still has not fully recovered from temporary roads last used 30 years ago.

The wildlife of the North Slope has been extensively studied. Researchers have found that road construction, prospecting, and drilling can produce enough noise and disruption to affect the behavior, habitat use, and reproductive success of many species. For instance, lesser snow geese are easily disturbed by aircraft. Repeated flushing reduces their energy intake, body condition, and reproductive success. Biologists conclude that these birds would be affected by the airplane and helicopter flights that oil development in the 1002 Area would entail.

Polar bears and musk oxen both live in the 1002 Area at high densities. Polar bears den and raise their cubs there, and musk oxen depend on river

habitats that would be disturbed for the gravel and water needed for road construction and maintenance.

Caribou have received more study than any other animal on Alaska's North Slope. The region is home to several major populations, or herds, each of which roams the interior during much of the year but migrates to the coastal plain in early summer, when females bear and raise their calves.

Since 1983, U.S. and Canadian wildlife biologists have placed radio-collars on female caribou and monitored their movements with radio receivers and by satellite. With the ability to track individual animals, biologists have been able to map the seasonal movements of herds and estimate the reproductive success of females each year.

Biologists with the Alaska Department of Fish and Game also conduct population censuses—and counting caribou is more difficult than you'd think. Because the animals are spread over such great distances, biologists need to wait until swarms of biting insects cause the caribou to aggregate tightly during a short period in the summer. Data from the radio-collars tell them when this is occurring and help them find the herds from airplanes. If the herds are packed tightly enough, and if weather conditions are safe for flying and offer good visibility, scientists take aerial photos from the planes of the herds, and then count individuals in the photos.

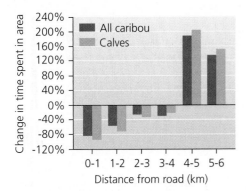

When a road was built through the middle of the Central Arctic Caribou Herd's traditional calving area, the herd split into two sub-areas, avoiding the road. Caribou and their calves decreased time spent close to the road (left side of graph) and moved to areas at least 4 km away from the road (right side of graph). Data from Cameron, R.D., et al., 1992. Redistribution of caribou in response to oil field developments on the Arctic Slope of Alaska. *Arctic* 45: 338–342, Fig 2. Used by permission from the Artic Institute of North America.

Whether Prudhoe Bay's oil operations have had a negative impact on that region's caribou is hotly debated. Surveys show that the Prudhoe Bay region's population—the Central Arctic Herd—increased during the 25 years after Prudhoe Bay was developed. Supporters of drilling have used this trend to argue that oil development does no harm to caribou. However, caribou populations in other parts of Alaska increased by even more during this period. Thus, natural conditions may have been favorable to them in all regions. Moreover, studies show that female caribou and their calves avoid all parts of the Prudhoe Bay oil complex, including its roads, sometimes detouring miles to do so (**see first figure**). Studies also show that female caribou in the Prudhoe Bay region suffer lower reproduction than those in undeveloped areas in Alaska.

A different population, the Porcupine Caribou Herd, uses parts of

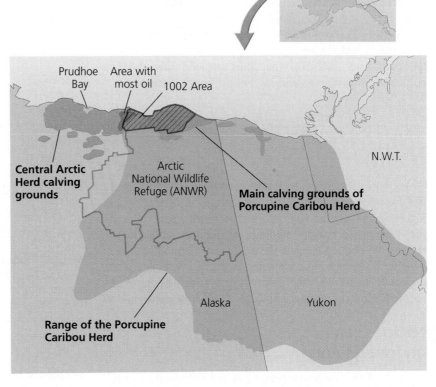

The Porcupine Caribou Herd's known range covers 260,000 km² in Alaska, Yukon, and the Northwest Territories. The caribou migrate to areas of the coastal plain, especially the 1002 Area of the Arctic Refuge, to bear and raise their calves.

the 1002 Area for calving (**see second figure**). This herd increased from 100,000 animals in the 1970s to 178,000 in 1989, declined to 123,000 in 2001, and today is estimated at 90,000–100,000. The reasons for its fluctuations are unknown, but because it has not increased like Alaska's other populations, wildlife managers are concerned that it would be vulnerable if the 1002 Area were to be developed.

Most oil is thought to lie in a portion of the 1002 Area that the caribou have not used much for calving, which suggests there could be little impact if only that portion is developed. However, caribou rely on different parts of the coastal plain each year as conditions vary, so putting some areas

off-limits could restrict their choices and lead to reproductive failure and population losses in certain years.

Proponents of drilling argue that impacts to caribou and other animals would be minimal. Roads would be built of ice that will melt in the summer, they point out, and most drilling activity would be confined to the winter, when caribou are elsewhere. Moreover, much of the technology used at Prudhoe Bay is now outdated, and the Refuge would be developed with more environmentally sensitive approaches, such as directional drilling. Nonetheless, research from biologists makes clear that some impacts are unavoidable. The question is how much impact we would be willing to accept in return for oil. ∎

today's technology also allows drillers to bore down vertically and then curve to drill horizontally. This allows extraction companies to follow the course of horizontal layered deposits to extract the most they can from them. It also allows drilling to reach a large underground area (up to several thousand meters in radius) around a drill pad. Thus, fewer drill pads are needed, and the surface footprint of drilling is smaller. Oil companies maintain that this advance substantially reduces the environmental impact of drilling operations.

POLITICAL, SOCIAL, AND ECONOMIC ASPECTS

The political, social, and economic consequences of fossil fuel use are numerous, varied, and far-reaching. Our discussion focuses on addressing several negative consequences of fossil fuel dependence, but it is important to bear in mind that fossil fuel use has enabled much of the world's population to achieve a higher material standard of living than ever before. It is also important to ask in each case whether switching to more renewable sources of energy would solve existing problems.

Nations can become dependent on foreign energy

Virtually all our modern technologies and services depend in some way on fossil fuels. Putting all of one's eggs in one basket is always a risky strategy. Because our economies rely so heavily on fossil fuels, we are vulnerable to supplies' becoming unavailable or costly. Nations that lack adequate fossil fuel reserves of their own are especially vulnerable. For instance, Germany, France, South Korea, and Japan consume far more energy than they produce and thus rely almost entirely on imports (**FIGURE 19.19**). Since its 1970 oil production peak, the United States has relied more and more on foreign energy, and today imports two-thirds of its crude oil.

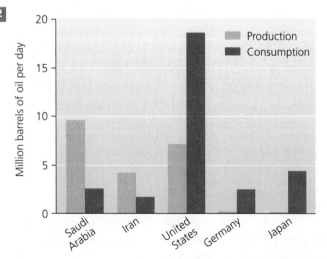

FIGURE 19.19 ▲ Japan, Germany, and the United States are among nations that consume far more oil than they produce. Iran and Saudi Arabia produce more oil than they consume and are able to export oil to high-consumption countries. Data from U.S. Energy Information Administration and BP p.l.c.

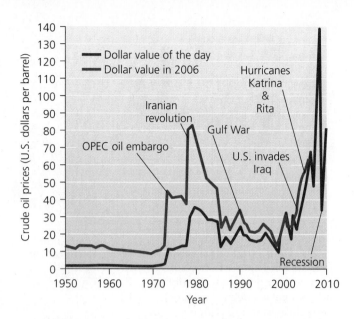

FIGURE 19.20 ▲ World oil prices have gyrated greatly over the decades, often because of political and economic events in oil-producing countries. The greatest price hikes in recent times have resulted from wars and unrest in the oil-rich Middle East. Data from U.S. Energy Information Administration.

Reliance on foreign oil means that seller nations can control energy prices, forcing buyer nations to pay more as supplies dwindle. This became clear in 1973, when the *Organization of Petroleum Exporting Countries (OPEC)* resolved to stop selling oil to the United States. The predominantly Arab nations of OPEC opposed U.S. support of Israel in the Arab-Israeli Yom Kippur War and sought to raise prices by restricting supply. The embargo created panic in the West and caused oil prices to skyrocket (**FIGURE 19.20**), spurring inflation. Fear of oil shortages drove American consumers to wait in long lines at gas pumps.

Oil supply and prices affect the economies of nations

More recently, when Hurricanes Katrina and Rita slammed into the Gulf Coast in 2005, they damaged offshore platforms and refineries, causing oil and gas prices to spike. The economic ripple effects served to remind us how much we rely on a steady supply of petroleum.

With the majority of world oil reserves located in the politically volatile Middle East, crises such as the 1973–1974 embargo, the Iranian Revolution, the Iran-Iraq War, and recent events in Iraq are a constant concern for U.S. policymakers. The United States has cultivated a close relationship with Saudi Arabia, the owner of 22% of world oil reserves, despite the fact that that country's political system allows for little of the democracy that U.S. leaders claim to cherish and promote. The world's third-largest holder of oil reserves, at 10%, is Iraq, which is why many people around the world believe the U.S.-led invasion of that nation in 2003 was motivated primarily to secure access to oil.

In response to the 1973 embargo, the U.S. government enacted a series of policies designed to reduce reliance on foreign oil. The government called for developing additional domestic sources, such as those on Alaska's North Slope. Since then, concern over reliance on foreign oil has repeatedly driven the proposal to open the Arctic Refuge to drilling, despite critics' charges that such drilling would do little to alleviate the nation's dependence.

The United States also diversified its sources of petroleum and today receives most of it from non–Middle Eastern nations, including Canada, Mexico, Venezuela, and Nigeria. However, there is no guarantee that all of these nations will continue exporting reliably to the United States over the long term. Current petroleum trade relations among nations and regions of the world are depicted in **FIGURE 19.21**.

The U.S. government also urged oil companies to pursue secondary extraction at sites shut down after primary extraction had ceased being cost-effective. It capped the price that domestic producers could charge for oil, funded research into renewable energy sources, and enacted conservation measures we will discuss below. It also established a stockpile of oil as a short-term buffer against future shortages. Stored deep underground in salt caverns in Louisiana, this is called the *Strategic Petroleum Reserve*. Currently the Reserve contains over 700 million barrels of oil; at present rates of U.S. consumption (20 million barrels/day), this equals just over one month's supply.

Residents may or may not benefit from their fossil fuel reserves

The extraction of fossil fuels can be extremely lucrative; many of the world's wealthiest corporations deal in fossil fuel energy or related industries. These industries provide jobs to millions of employees and pay dividends to millions of investors. Development can potentially yield economic benefits for people who live in petroleum-bearing areas, as well. Since the construction of the trans-Alaska pipeline in the 1970s, the state of Alaska has received over $65 billion in oil revenues. Alaska's state constitution requires that one-quarter of state oil revenues be placed in the Permanent Fund, which pays yearly dividends to all Alaska residents. Since 1982, each Alaska resident has received annual payouts ranging from $331 to $2,069.

Development of the Arctic Refuge would add to this fund and create jobs. Some estimates anticipate creation of many thousands of jobs and billions of dollars of income. For this reason, most Alaska residents support oil drilling in the Refuge. The Native people who live on the North Slope are split over the proposed drilling. The Inupiat want income for health care, police and fire protection, and other services that are scarce in this remote region. The Gwich'in oppose drilling because they fear it will threaten the caribou herds and Arctic ecosystems they depend on.

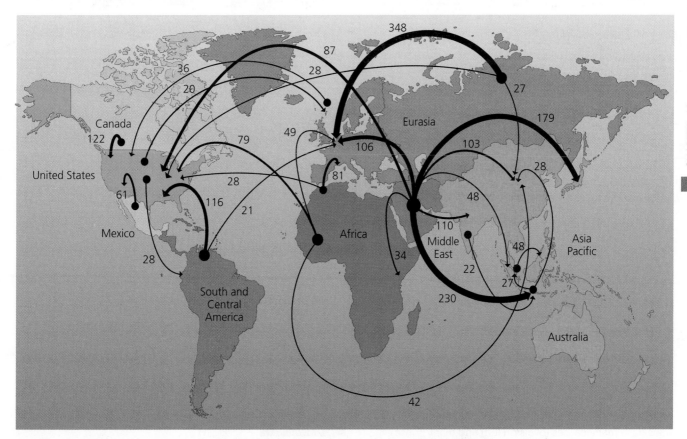

FIGURE 19.21 ▲ The global trade in oil is lopsided; relatively few nations account for most exports, and some nations are highly dependent on others for energy. The United States obtains most of its imported oil from Canada, Saudi Arabia, Venezuela, Nigeria, and Mexico. Numbers in the figure represent millions of metric tons. Adapted from *Statistical review of world energy 2010*, BP p.l.c. By permission.

Alaska's distribution of revenue among its citizenry is unusual. In most parts of the world where fossil fuels have been extracted, local residents have not seen benefits. When multinational corporations extract oil or gas in developing countries, paying those countries' governments for access, the money often does not trickle down to the people who live where the extraction takes place. Moreover, oil-rich developing nations such as Ecuador, Venezuela, and Nigeria tend to have few environmental regulations, and existing regulations may go unenforced if a government does not want to risk losing the large sums of money associated with oil development.

In Nigeria, oil was discovered in 1958 in the territory of the native Ogoni people, and the Shell Oil Company moved in to develop oil fields. Although Shell extracted $30 billion of oil from Ogoni land over the years, the Ogoni still live in poverty, with no running water or electricity. The profits from oil extraction on Ogoni land went to Shell and to the military dictatorships of Nigeria. The development resulted in oil spills, noise, and constantly burning gas flares, all of which caused illness among people living nearby. From 1962 until his death in 1995, Ogoni activist and leader Ken Saro-Wiwa worked for fair compensation to the Ogoni for oil extraction and environmental degradation on their land. After years of persecution by the Nigerian government, Saro-Wiwa was arrested in 1994, given a trial universally regarded as a sham, and put to death by military tribunal.

How will we convert to renewable energy?

Fossil fuels are not a sustainable long-term solution to our energy needs. Fossil fuel supplies are limited, and their use has health, environmental, political, and socioeconomic consequences (**FIGURE 19.22**). Given this, the world's nations have several policy options for guiding future energy use. One option is to continue relying on fossil fuels until they are no longer economically practical and to develop other energy sources only after supplies have dwindled. A second option is to immediately and dramatically increase funding to develop alternative energy sources in order to hasten a rapid shift to them. Third, we could steer a middle course and attempt to reduce our reliance on fossil fuels gradually.

Regardless of which course we take, it will benefit us to prolong the availability of fossil fuels as we make the transition to renewable sources. We can prolong our access to fossil fuels by instituting measures to conserve energy, primarily through lifestyle changes that reduce energy use and technological advances that improve efficiency.

554

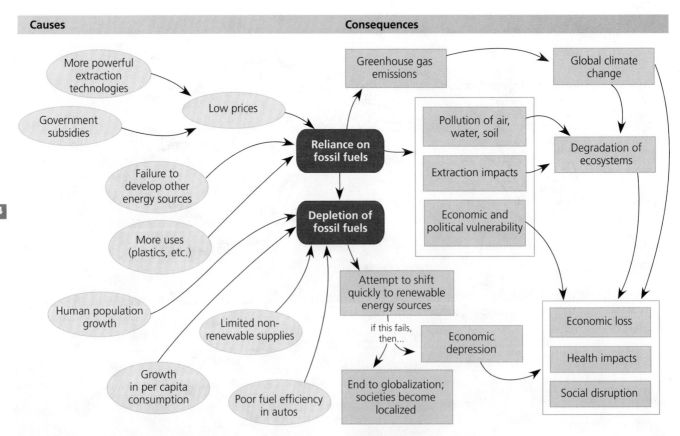

FIGURE 19.22 ▲ Our reliance on, and depletion of, fossil fuels have many causes (ovals on left) and many consequences (boxes on right). Arrows in this concept map lead from causes to consequences. Note that items grouped within outlined boxes do not necessarily share any special relationship; the outlined boxes are merely intended to streamline the figure.

Solutions

As you progress through this chapter, try to identify as many solutions to our reliance on and depletion of fossil fuels as you can. What could you personally do to help address this issue? Consider how each action or solution might affect items in the concept map above.

ENERGY EFFICIENCY AND CONSERVATION

Until our society makes the transition to renewable energy sources, we will need to find ways to minimize the expenditure of energy from our dwindling fossil fuel resources. **Energy efficiency** describes the ability to obtain a given result or amount of output while using less energy input. **Energy conservation** describes the practice of reducing energy use. In general, efficiency is a result of technological improvements, whereas conservation results from behavioral choices. Because greater efficiency also allows us to reduce energy use, efficiency is one primary means toward conservation. These efforts allow us to extend the lifetimes of our nonrenewable energy supplies, to be less wasteful, and to reduce our environmental impact.

Automobile fuel efficiency affects our ability to conserve

In the United States, many people first saw the value of conserving energy following the OPEC embargo of 1973–1974. Measures enacted by the U.S. government in response to that event included a mandated increase in the mile-per-gallon (mpg) fuel efficiency of automobiles and a reduction in the national speed limit to 55 miles per hour.

Over the next three decades, however, many of the conservation initiatives that followed the 1973–1974 oil crisis were abandoned. Without high market prices and an immediate threat of shortages, people lacked economic motivation to conserve. Government funding for research into alternative energy sources decreased, speed limits increased, and U.S. policymakers repeatedly failed to raise the *corporate average fuel efficiency (CAFE) standards*, which set benchmarks for auto manufacturers to meet. The average fuel efficiency of new vehicles fell from a high of 22.0 mpg in 1987 to 19.3 mpg in 2004, as sales of light trucks, including sport-utility vehicles,

increased relative to sales of cars. Average fuel efficiency then climbed up to 21.1 mpg in 2009 (**FIGURE 19.23**).

This recent rise in fuel efficiency occurred after Congress passed legislation in 2007 mandating that automakers raise average fuel efficiency to 35 mpg by the year 2020. This was a substantial advance, yet even after this boost, American automobiles will still lag behind the efficiency of the vehicles of most other developed nations. For instance, the fuel efficiency of European and Japanese cars is nearly twice that of U.S. cars and is slated to keep improving.

The United States has also kept its taxes on gasoline extremely low, relative to other nations. Americans pay two to three times *less* per gallon of gas than drivers in many European countries, for example. This means that gasoline prices do not account for the substantial external costs (pp. 150–151) that oil production and consumption impose on society, and it also diminishes our economic incentives to conserve.

In 2009, the U.S. government under President Barack Obama sought to improve automobile fuel efficiency while stimulating economic activity and saving jobs during a severe recession. The popular "Cash for Clunkers" program—formally named the Consumer Assistance to Recycle and Save (CARS) Act—paid Americans $3,500 or $4,500 each to turn in old vehicles and purchase newer, more fuel-efficient ones. The $3-billion program subsidized the sale or lease of 678,000 vehicles averaging 24.9 mpg that replaced vehicles averaging 15.8 mpg. It is estimated that 824 million gallons of gasoline will be saved as a result, preventing 9 million metric tons of greenhouse gas emissions and creating social benefits worth $278 million.

Many critics of oil drilling in the Arctic National Wildlife Refuge point out the vast amounts of oil wasted by our fuel-inefficient automobiles. They argue that a small amount of efficiency and conservation would save the nation far more oil than it would ever obtain from the Arctic Refuge. The U.S. Geological Survey's average estimate for recoverable oil in the 1002 Area, 7.7 billion barrels, represents just one year's supply for the United States at current consumption rates. Spread over a period of extraction of many years, the proportion of U.S. oil demand that the Refuge would fulfill appears strikingly small (**FIGURE 19.24**).

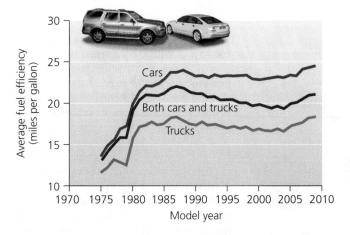

FIGURE 19.23 ▲ Fuel efficiency for automobiles in the United States rose dramatically in the late 1970s as a result of legislative mandates, but then stagnated because no further laws were enacted to improve fuel economy. Recent legislation is now improving it again. Data from U.S. Environmental Protection Agency, 2009. *Light-duty automotive technology, carbon dioxide emissions, and fuel economy trends: 1975 through 2009.*

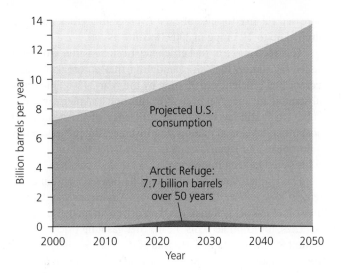

FIGURE 19.24 ▲ Opponents of oil drilling in the Arctic National Wildlife Refuge contend that its recoverable oil would make only a small contribution toward overall U.S. oil demand. In this graph, the best U.S. Geological Survey estimate of oil from the 1002 Area is shown in red, in the context of total U.S. oil consumption (in orange), assuming that current consumption trends are extrapolated into the future and that oil production takes place over many years. Adapted from Natural Resources Defense Council, 2002. *Oil and the Arctic National Wildlife Refuge*; and U.S. Geological Survey, 2001. *Arctic National Wildlife Refuge, 1002 Area, petroleum assessment, 1998, including economic analysis.*

WEIGHING THE ISSUES

Drilling in the Arctic Refuge Do you think the United States should open the Arctic National Wildlife Refuge to oil extraction? Why or why not? What would be gained? What would be lost?

Personal choice and increased efficiency are two routes to conservation

Energy conservation can be accomplished in two primary ways. As individuals, we can make conscious choices to reduce our own energy consumption by driving less, turning off lights when rooms are not being used, dialing down thermostats, and cutting back on the use of energy-intensive machines and appliances. For any given individual or business, reducing energy consumption can save money while also helping to conserve resources.

As a society, we can conserve energy by making our energy-consuming devices and processes more efficient. Currently, more than two-thirds of the fossil fuel energy we use is simply lost, as waste heat, in automobiles and power plants. The United States burns through twice as much energy per dollar of Gross Domestic Product (GDP) as do most other industrialized nations. However, the good news is that over the past three decades the United States has decreased its energy use per dollar of GDP by about 50%. Clearly, we have achieved tremendous gains in efficiency already, and we should be able to make still-greater progress in the future.

In the case of automobiles, we already possess the technology to increase fuel efficiency far above the current U.S. average of 21 mpg. We can accomplish this with more efficient gasoline engines; lightweight materials; continuously variable transmissions; and alternative technology vehicles such as electric cars, electric/gasoline hybrids (**FIGURE 19.25**), or vehicles that use hydrogen fuel cells (pp. 609–612).

We can also vastly improve the efficiency of our power plants. One way is to use **cogeneration**, in which excess heat produced during the generation of electricity is captured and used to heat nearby workplaces and homes and to produce other kinds of power. Cogeneration can almost double the efficiency of a power plant. The same is true of coal gasification and combined cycle generation (p. 546). In this process, coal is treated to create hot gases that turn a gas turbine, while the

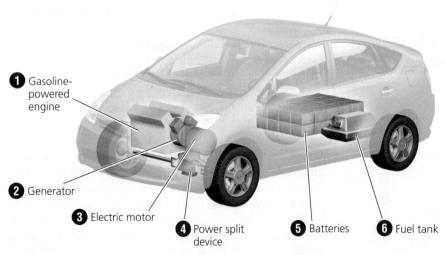

FIGURE 19.25 ◀ A hybrid car, such as the Toyota Prius diagrammed here, uses a small, clean, and efficient gasoline-powered engine ❶ to produce power that the generator ❷ can convert to electricity to drive the electric motor ❸. The power split device ❹ integrates the engine, generator, and motor, serving as a continuously variable transmission. The car automatically switches between all-electrical power, all-gas power, and a mix of the two, depending on the demands being placed on the engine. Typically, the motor provides power for low-speed city driving and adds extra power on hills. The motor and generator charge a pack of nickel-metal-hydride batteries ❺, which can in turn supply power to the motor. Energy for the engine comes from gasoline carried in a typical fuel tank ❻.

❶ Gasoline-powered engine
❷ Generator
❸ Electric motor
❹ Power split device
❺ Batteries
❻ Fuel tank

FIGURE 19.26 ▲ Many of our homes and offices could be made more energy-efficient. One way to determine how much heat a building is losing is to take a photograph that records energy in the infrared portion of the electromagnetic spectrum (p. 31). In such a photograph, or *thermogram* (shown here), white, yellow, and red signify hot and warm temperatures at the surface of the house, whereas blue and green shades signify cold and cool temperatures. The white, yellow, and red colors indicate areas where heat is escaping.

hot exhaust of this turbine heats water to drive a conventional steam turbine like that shown on p. 536.

In homes and public buildings, a significant amount of heat is lost in winter and gained in summer because of inadequate insulation (**FIGURE 19.26**). Improvements in the design of homes and offices can reduce the energy required to heat and cool them. Such design changes can involve the building's location, the color of its roof (light colors keep buildings cooler by reflecting the sun's rays), and its insulation.

Among consumer products, scores of appliances, from refrigerators to lightbulbs, have been reengineered through the years to increase energy efficiency. Energy-efficient lighting, for example, can reduce energy use by 80%. Compact fluorescent bulbs are much more efficient than incandescent light bulbs, and many governments are phasing out incandescent bulbs for this reason; the U.S. phaseout is scheduled to be complete in 2014.

Federal standards for energy-efficient appliances have already reduced per-person home electricity use below what it was in the 1970s. The EPA's Energy Star program (p. 518), which labels appliances for their energy efficiency, has been highly successful. Even so, there remains room for further improvement.

While manufacturers can improve the energy efficiency of appliances, consumers need to "vote with their wallets" by purchasing these energy-efficient appliances. Decisions by consumers to purchase energy-efficient products are crucial in keeping those products commercially available. The U.S. Environmental Protection Agency (EPA) estimates that if all U.S. households purchased energy-efficient appliances, the national annual energy expenditure would be reduced by $200 billion. For the individual consumer, studies show that the slightly higher cost of buying energy-efficient washing machines is rapidly offset by savings on water and electricity bills. On the national level, France, Great Britain, and many other developed countries have standards of living equal to that of the United States, but they use much less energy per capita. This disparity indicates that U.S. citizens could significantly reduce their energy consumption without decreasing their quality of life.

Both conservation and renewable energy are needed

It is often said that reducing our energy use is equivalent to finding a new oil reserve. Some estimates hold that effective energy conservation in the United States could save 6 million barrels of oil a day. Such a savings would likely far more than offset the energy produced by any oil under the Arctic National Wildlife Refuge while also reducing the negative impacts of fossil fuel extraction and use. Indeed, conserving energy is better than finding a new reserve, because it lessens impacts on the environment while extending our access to fossil fuels.

However, energy conservation does not add to our supply of available fuel. Regardless of how much we conserve, we will still need energy, and it will need to come from somewhere. The only sustainable way of guaranteeing ourselves a reliable long-term supply of energy is to ensure sufficiently rapid development of renewable energy sources—a challenge we will examine in our next two chapters.

➤ CONCLUSION

Over the past 200 years, fossil fuels have helped us build the complex industrialized societies we enjoy today. However, we are now approaching a turning point in history: Our production of fossil fuels will begin to decline. We can respond to this new challenge in creative ways, encouraging conservation and developing alternative energy sources. Or, we can continue our current dependence on fossil fuels and wait until they near depletion before we try to develop new technologies and ways of life. The path we choose will have far-reaching consequences for human health and well-being, for Earth's climate, for our environment, and for the stability and progress of our civilization.

The ongoing debate over the future of the Arctic National Wildlife Refuge is a microcosm of this debate over our energy future. Fortunately, there is not simply a trade-off between benefits of energy for us and harm to the environment, climate, and health. Instead, as evidence builds that renewable energy sources are becoming increasingly feasible and economical, it becomes easier to envision giving up our reliance on fossil fuels and charting a win-win future for humanity and our environment.

REVIEWING OBJECTIVES

You should now be able to:

IDENTIFY THE ENERGY SOURCES THAT WE USE

- Many renewable and nonrenewable energy sources are available to us. (pp. 531–532)
- Since the industrial revolution, nonrenewable fossil fuels—including coal, natural gas, and oil—have become our primary sources of energy. (pp. 531–532)
- Fossil fuels are formed very slowly as buried organic matter is chemically transformed by heat, pressure, and/or anaerobic decomposition. (p. 533)
- In evaluating energy sources, it is important to compare the amount of energy obtained from them with the amount invested in their extraction and production. (p. 534)

DESCRIBE THE NATURE AND ORIGIN OF COAL AND EVALUATE ITS EXTRACTION AND USE

- Coal is our most abundant fossil fuel. It results from organic matter that undergoes compression but little decomposition. (p. 535)
- Coal is mined underground and strip-mined from the land surface, and is used today principally to generate electricity. (pp. 535–536)
- Coal varies in its composition. Combustion of coal that is high in contaminants emits toxic air pollution. (pp. 535–536)

DESCRIBE THE NATURE AND ORIGIN OF NATURAL GAS AND EVALUATE ITS EXTRACTION AND USE

- Natural gas is cleaner-burning than coal or oil. (p. 536)
- Natural gas consists mostly of methane and can be formed in two ways. (pp. 536–537)
- Natural gas often occurs with oil and coal deposits, is extracted in similar ways, and becomes depleted in similar ways. (p. 537)
- Most remaining U.S. gas and oil deposits are located offshore. (pp. 537–538)

DESCRIBE THE NATURE AND ORIGIN OF CRUDE OIL AND EVALUATE ITS EXTRACTION, USE, AND FUTURE DEPLETION

- Crude oil is a thick, liquid mixture of hydrocarbons that is formed underground under high temperature and pressure. (p. 538)
- Scientists locate fossil fuel deposits by analyzing subterranean geology. We then estimate the technically and economically recoverable portions of those reserves. (pp. 538, 542–543)
- Primary extraction of oil may be followed by secondary extraction, in which gas or liquid is injected into the ground to help force up additional oil. (p. 539)
- Components of crude oil are separated in refineries to produce a wide variety of fuel types. (pp. 539–540)
- Petroleum-based products, from gasoline to clothing to plastics, are everywhere in our daily lives. (pp. 539, 541)

- We have depleted nearly half the world's oil. Once we pass the peak and production slows, the gap between rising demand and falling supply may pose immense economic and social challenges for our society. (pp. 539–544)

DESCRIBE THE NATURE, ORIGIN, AND POTENTIAL OF ALTERNATIVE FOSSIL FUELS

- Oil sands can be mined and processed into synthetic oil. (p. 544)
- Oil shale is abundant in the western United States. (p. 544)
- Methane hydrate could provide a source of methane gas. (pp. 544–545)
- These new fossil fuels exert environmental impacts. (p. 545)

OUTLINE AND ASSESS ENVIRONMENTAL IMPACTS OF FOSSIL FUEL USE AND EXPLORE POTENTIAL SOLUTIONS

- Emissions from fossil fuel combustion pollute air, pose human health risks, and drive global climate change. (pp. 545–546)
- Clean coal technologies aim to reduce pollution from coal combustion. (p. 546)
- If we can safely and effectively capture carbon dioxide and sequester it underground, we would mitigate a primary drawback of fossil fuels. Carbon capture and storage remains unproven, however. (pp. 546–547)
- Oil is a major contributor to water pollution. (pp. 547–548)
- Coal mining can devastate ecosystems and pollute waterways. (p. 549)
- Development for oil and gas extraction exerts various environmental impacts. (pp. 549–552)

EVALUATE POLITICAL, SOCIAL, AND ECONOMIC ASPECTS OF FOSSIL FUEL USE

- Today's societies are so reliant on fossil fuel energy that sudden restrictions in oil supplies can have major economic consequences. (pp. 552–553)
- Nations that consume far more fossil fuels than they produce are especially vulnerable to supply restrictions. (pp. 552–553)
- People living in areas of fossil fuel extraction do not always benefit from their extraction. (pp. 553–554)

SPECIFY STRATEGIES FOR CONSERVING ENERGY AND ENHANCING EFFICIENCY

- Energy conservation involves both personal choices and efficient technologies. (pp. 555–557)
- Automotive fuel efficiency plays a key role in conserving energy. (p. 555)
- Efficiency in power plant combustion, consumer appliances, and more can help us conserve. (pp. 556–557)
- Conservation lengthens our access to fossil fuels and reduces environmental impact, but to build a sustainable society we will also need to shift to renewable energy sources. (p. 557)

TESTING YOUR COMPREHENSION

1. Why are fossil fuels our most prevalent source of energy today? Why are they considered nonrenewable sources of energy?

2. How are fossil fuels formed? How do environmental conditions determine what type of fossil fuel is formed in a given location? Why are fossil fuels often concentrated in localized deposits?

3. Describe how *net energy* differs from *energy returned on investment (EROI)*. Why are these concepts important when evaluating energy sources?

4. Describe how coal is used to generate electricity.

5. How have geologists estimated the total amount of oil beneath the Arctic National Wildlife Refuge's 1002 Area? How does this amount differ from the "technically recoverable" and "economically recoverable" amounts of oil?

6. How do we create petroleum products? Provide examples of several of these products.

7. What is meant by "peak oil"? Why do many experts think we are about to pass the global production peak for oil? What consequences could there be for our society if we do not shift soon to renewable energy sources?

8. Describe three environmental impacts of fossil fuel production and consumption. Compare contrasting views regarding the environmental impacts of drilling for oil in the Arctic National Wildlife Refuge.

9. Give an example of a clean coal technology. Now describe how carbon capture and storage is intended to work.

10. Describe two main approaches to energy conservation, and give a specific example of each.

SEEKING SOLUTIONS

1. What impacts might you expect on your lifestyle once our society arrives at peak oil? How much oil is thought to remain in the world, and when do experts predict the supply will begin to decline? What lessons do you think we can take from the conservation methods adopted by the United States in response to the "energy crisis" of 1973–1974? What steps do you think we should take to avoid energy shortages in a post-peak-oil future?

2. Compare the effects of coal and oil consumption on the environment. Which process, our use of oil or our use of coal, do you think has ultimately had greater environmental impact, and why? What steps could governments, industries, and individuals take to reduce environmental impacts?

3. If the United States and other developed countries reduced dependence on foreign oil and on fossil fuels in general, do you think that their economies would benefit or suffer? Might your answer be different for the short term and the long term? What factors come into play in trying to make such a judgment?

4. Contrast the experiences of the Ogoni people of Nigeria with those of the citizens of Alaska. How have they been similar and different? Do you think businesses or governments should take steps to ensure that local people benefit from oil drilling operations? How could they do so?

5. **THINK IT THROUGH** You have been elected U.S. senator from the state of Alaska. Your colleague, the other senator from Alaska, has just introduced legislation to open the Arctic National Wildlife Refuge to oil drilling. Would you vote in favor of the legislation? Why or why not? Now imagine that you are instead a senator from another U.S. state. How would you vote on the bill, and why?

6. **THINK IT THROUGH** Throughout this book in these questions, we have asked you to imagine yourself in various roles. This time we ask you simply to be yourself. Given the information in this chapter on petroleum supplies, consumption, and depletion, what actions, if any, do you plan to take to prepare yourself for changes in our society that may come about as oil production declines? Describe in detail how you think your life may change, and suggest one thing you could do to help reduce negative impacts of oil depletion on our society.

CALCULATING ECOLOGICAL FOOTPRINTS

Scientists at the Global Footprint Network calculate the energy component of our ecological footprint by estimating the amount of ecologically productive land and sea required to absorb the carbon released from fossil fuel combustion. This translates into 6.5 ha of the average American's 9.4-ha ecological footprint. Another way to think about our footprint, however, is to estimate how much land would be needed to grow biomass with an energy content equal to that of the fossil fuel we burn.

Assume that you are an average American who burns about 6.7 metric tons of oil-equivalent in fossil fuels each year and that average terrestrial net primary productivity (p. 116) can be expressed as 0.0037 metric tons/ha/year. Calculate how many hectares of land it would take to supply our fuel use by present-day photosynthetic production.

	Hectares of land for fuel production H
You	1,811
Your class	
Your state	
United States	

1. Compare the energy component of your ecological footprint calculated in this way with the 6.5 ha calculated using the method of the Global Footprint Network. Explain why results from the two methods may differ.

2. Earth's total land area is approximately 15 billion hectares. Compare this to the hectares of land for fuel production from the table.

3. In the absence of stored energy from fossil fuels, how large a human population could Earth support at the level of consumption of the average American, if all of Earth's area were devoted to fuel production? Do you consider this realistic? Provide two reasons why or why not.

Mastering**ENVIRONMENTALSCIENCE**™

Go to **www.masteringenvironmentalscience.com** for practice quizzes, Pearson eText, videos, current events, and more.

Steam rising from cooling towers of a nuclear power plant, Shropshire, U.K

20 CONVENTIONAL ENERGY ALTERNATIVES

UPON COMPLETING THIS CHAPTER, YOU WILL BE ABLE TO:

- Discuss the reasons for seeking energy alternatives to fossil fuels
- Summarize the contributions to world energy supplies of conventional alternatives to fossil fuels
- Describe nuclear energy and how it is harnessed

- Outline the societal debate over nuclear power
- Describe the major sources, scale, and impacts of bioenergy
- Describe the scale, methods, and impacts of hydro-electric power

Sweden's Search for Alternative Energy

"Nowhere has the public debate over nuclear power plants been more severely contested than Sweden."

—Writer and analyst Michael Valenti

"If [Sweden] phases out nuclear power, then it will be virtually impossible for the country to keep its climate-change commitments."

—Yale University economist William Nordhaus

On the morning of April 28, 1986, workers at a nuclear power plant in Sweden detected suspiciously high radiation levels. Their concern turned to confusion when they determined that the radioactivity was coming not from their own plant, but through the atmosphere from the direction of the Soviet Union.

They had, in fact, discovered evidence of the disaster at Chernobyl, more than 1,200 km (750 mi) away in what is now the nation of Ukraine. Chernobyl's nuclear reactor had exploded two days earlier, but the Soviet government had not yet admitted it to the world.

As low levels of radioactive fallout rained down on the Swedish countryside in the days ahead, contaminating crops and cows' milk, many Swedes felt more certain than ever about the decision they had made collectively six years earlier. In a 1980 referendum, Sweden's electorate had voted to phase out their country's nuclear power program, shutting down all nuclear plants by the year 2010.

But trying to phase out nuclear power has proven difficult. As 2010 arrived, Sweden found itself relying on 10 nuclear reactors for 31% of its energy supply and 42% of its electricity. If nuclear plants are

Biomass power plant, Skellefteå, Sweden

shut down, something will have to take their place. Aware of the environmental impacts of fossil fuels, Sweden's government and citizens do not favor expanding fossil fuel use. In fact, Sweden is one of the few nations that have managed to *decrease* use of fossil fuels since the 1970s. It has done so largely by replacing them with nuclear power.

To make up for energy that would be lost with a nuclear phaseout, Sweden's government promoted research and development of renewable energy sources. Hydroelectric power from running water was already supplying most of the rest of the nation's electricity, but it could not be expanded much more. The government hoped that energy from biomass sources and wind power could fill the gap.

Sweden made itself an international leader in renewable energy alternatives, but because renewables were taking longer to develop than hoped, policymakers repeatedly postponed the nuclear phaseout. Then in 2009, the government's ruling coalition announced that it was reversing its long-standing policy and would not phase out nuclear power after all. Instead, new power plants would be built as existing ones needed replacing.

In announcing this decision, the government said it would be fiscally and socially irresponsible to dismantle the nation's nuclear program without a ready replacement. Without an abundance of clean renewable power at hand, a nuclear phaseout would mean a return to fossil fuels, or else would require cutting down extensive areas of forest to combust biomass. Policymakers also cited Sweden's international obligations to hold down its carbon emissions under the Kyoto Protocol (pp. 520–521). Nuclear energy is free of atmospheric pollution and seems, to many, the most effective way to minimize greenhouse gas emissions in the short term.

Many in the government's minority parties disagreed and continued to support bringing an end to nuclear power. "To rely on nuclear power to reduce CO_2 emissions," said Greenpeace spokeswoman Martina Kruger, "is like smoking to lose weight. It's not a good idea."

The government's policy shift reflected a gradual shift in public sentiment. As more and more Swedes had become convinced of the importance of fighting climate change and developing clean energy sources, support for nuclear power had grown. Public opinion polls from 2003 onward showed substantial majorities of Swedish citizens to be in favor of maintaining or increasing their nation's production of nuclear power.

Still greater majorities of Swedish citizens continued to support boosting the development of renewable energy sources, keeping their nation a world leader in the shift away from fossil fuels.

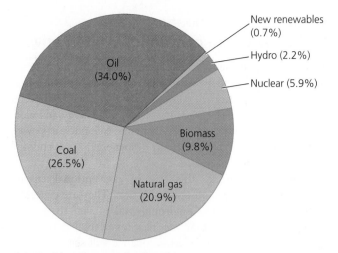

(a) World energy production, by source

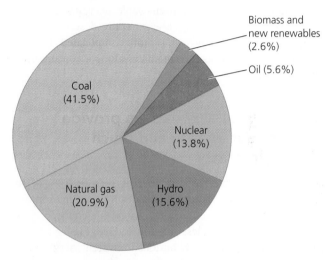

(b) World electricity generation, by source

FIGURE 20.1 ▲ Fossil fuels account for 81% **(a)** of the world's energy production. Nuclear power and hydroelectric power contribute substantially to global electricity generation **(b)**, but fossil fuels still power two-thirds of our electricity. Data are for 2008, from the International Energy Agency (IEA), 2009. *Key world energy statistics 2009.* Paris: IEA.

ALTERNATIVES TO FOSSIL FUELS

Fossil fuels helped to drive the industrial revolution and to create the unprecedented material prosperity we enjoy today. Our global economy is largely powered by fossil fuels; over 80% of our energy comes from oil, coal, and natural gas (**FIGURE 20.1A**). These three fuels also generate more than two-thirds of the world's electricity (**FIGURE 20.1B**). However, these nonrenewable energy sources will not last forever. As we saw in Chapter 19, oil production is widely expected to peak soon, and easily extractable supplies of oil and natural gas may not last half a century more (pp. 539–541). Moreover, the use of coal, oil, and natural gas entails substantial environmental impacts, as described in Chapters 17, 18, and 19.

For these reasons, most energy experts accept that we will need to shift from fossil fuels to energy sources that are less easily depleted and gentler on our environment.

Developing alternatives to fossil fuels has the added benefit of helping to diversify an economy's mix of energy, thus lessening price volatility and dependence on foreign fuel imports.

We have developed a range of alternatives to fossil fuels. Most of these energy sources are renewable, and most have less impact on health and the environment than oil, coal, or natural gas. However, at this time most remain more expensive than fossil fuels, at least in the short term and when external costs (pp. 150–151) are not included in market prices. As technologies are further developed and as we invest in infrastructure to better transmit power from renewable sources, prices will come down further and help us transition toward these new energy sources.

Nuclear power, bioenergy, and hydropower are conventional alternatives

Three alternative energy sources are currently the most developed and most widely used: nuclear power, hydroelectric power, and energy from biomass. Each of these well-established energy sources plays substantial roles in our energy and electricity budgets today. We can therefore call nuclear power, hydropower, and biomass energy (bioenergy) "conventional alternatives" to fossil fuels.

In some respects, this trio of conventional energy alternatives makes for an odd collection. They are generally considered to exert less environmental impact than fossil fuels, but more impact than the "new renewable" alternatives we will discuss in Chapter 21. Yet as we will see, they each involve a unique and complex mix of benefits and drawbacks for people and the environment. Nuclear power is commonly considered a nonrenewable energy source, and hydropower and bioenergy are generally described as renewable. The reality, however, is more complicated. They are perhaps best viewed as intermediates along a continuum of renewability.

Conventional alternatives provide some of our energy and much of our electricity

Fuelwood and other biomass sources provide 10% of the world's energy, nuclear power provides 6%, and hydropower provides 2%. The less-established renewable energy sources together account for less than 1% (see Figure 20.1A). Although their global contributions to overall energy supply are minor, alternatives to fossil fuels do contribute greatly to our generation of electricity. Nuclear energy and hydropower together account for nearly 30% of the world's electricity generation (see Figure 20.1B).

Energy consumption patterns in the United States (**FIGURE 20.2A**) are similar to those globally, except that the United States relies less on fuelwood and more on fossil fuels and nuclear power than most other countries. A graph showing trends in energy consumption in the United States over the past 60 years (**FIGURE 20.2B**) reveals two things. First, conventional alternatives play minor yet substantial roles in overall energy use. Second, use of conventional alternatives has been growing more slowly than use of fossil fuels.

Sweden and some other nations, however, have shown that it is possible to replace fossil fuels gradually with alternative sources. Since 1970, Sweden has decreased its fossil fuel use by 36%, and today nuclear power, bioenergy, and hydropower together provide Sweden with 60% of its total energy and virtually all of its electricity.

NUCLEAR POWER

Nuclear power occupies an odd and conflicted position in our modern debate over energy. It is free of the air pollution produced by fossil fuel combustion, so it has long been put

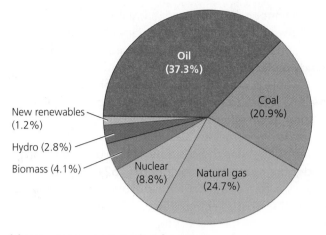

(a) U.S. energy consumption, by source

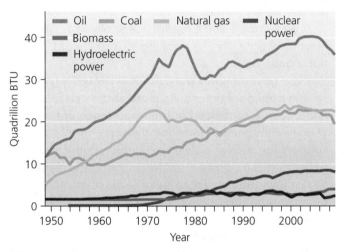

(b) U.S. energy consumption, 1949–2009

FIGURE 20.2 ▲ In the United States, fossil fuels account for 84% of energy consumption **(a)**, nuclear power for 8.5%, biomass for 3.9%, and hydropower for 2.5%. Over the past half century **(b)**, U.S. consumption of the three fossil fuels has grown faster than that of biomass or hydropower. Nuclear power grew considerably between 1970 and 2000.
Data are for 2009, from Energy Information Administration, 2010. *Annual energy review 2009*. U.S. Department of Energy.

forth as an environmentally friendly alternative to fossil fuels. Yet nuclear power's great promise has been clouded by nuclear weaponry, the dilemma of radioactive waste disposal, and the long shadow of Chernobyl and other power plant accidents. As such, public safety concerns and the costs of addressing them have constrained the development and spread of nuclear power in the United States, Sweden, and many other nations.

First developed commercially in the 1950s, nuclear power has expanded 15-fold worldwide since 1970, experiencing most of its growth during the 1970s and 1980s. Of all nations, the United States generates the most electricity from nuclear power, followed by France and Japan. However, only 20% of U.S. electricity comes from nuclear sources. A number of other nations rely more heavily on nuclear power (**TABLE 20.1**). France leads the list, receiving 76% of its electricity from nuclear power.

TABLE 20.1 Top Producers of Nuclear Power			
Nation	Nuclear power produced*	Number of plants*	Percentage of electricity from nuclear power†
United States	100.7	104	20.2
France	63.1	58	75.2
Japan	46.8	54	28.9
Russia	22.7	32	17.8
Germany	20.5	17	26.1
South Korea	17.7	20	34.8
Ukraine	13.1	15	48.6
Canada	12.6	18	14.9
United Kingdom	10.1	19	17.9
Sweden	9.3	10	37.4
Rest of world	55.4	91	8.0

*In gigawatts, 2010 data, from the European Nuclear Society.
†2009 data, from the International Atomic Energy Agency.

Fission releases nuclear energy

Strictly defined, **nuclear energy** is the energy that holds together protons and neutrons (p. 26) within the nucleus of an atom. We harness this energy by converting it to thermal energy, which can then be used to generate electricity. Several processes can convert the energy within an atom's nucleus into thermal energy, releasing it and making it available for use. Each process involves transforming isotopes (p. 26) of one element into isotopes of other elements by the addition or loss of neutrons.

The reaction that drives the release of nuclear energy in power plants is **nuclear fission**, the splitting apart of atomic nuclei (**FIGURE 20.3**). In fission, the nuclei of large, heavy

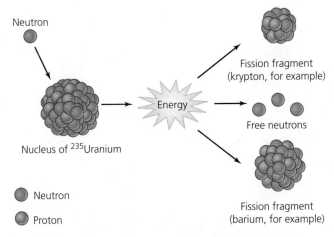

FIGURE 20.3 ▲ In nuclear fission, the nucleus of an atom of uranium-235 is bombarded with a neutron. The collision splits the uranium atom into smaller atoms and releases two or three neutrons, along with energy in the form of heat, light, and radiation. The neutrons can continue to split uranium atoms and set in motion a runaway chain reaction, so engineers at nuclear plants must absorb excess neutrons with control rods to regulate the rate of the reaction.

atoms, such as uranium or plutonium, are bombarded with neutrons. Ordinarily neutrons move too quickly to split nuclei when they collide with them, but if neutrons are slowed down they can break apart nuclei. Each split nucleus emits energy in the form of heat, light, and radiation, and also releases multiple neutrons. These neutrons (two to three in the case of uranium-235) can in turn bombard other nearby uranium-235 (^{235}U) atoms, resulting in a self-sustaining chain reaction.

If not controlled, this chain reaction becomes a runaway process of positive feedback that releases enormous amounts of energy. It is this process that creates the explosive power of a nuclear bomb. Inside a nuclear power plant, however, fission is controlled so that, on average, only one of the two or three neutrons emitted with each fission event goes on to induce another fission event. In this way, the chain reaction maintains a constant output of energy at a controlled rate.

Nuclear energy comes from processed and enriched uranium

We generate electricity from nuclear power by controlling fission in **nuclear reactors**, facilities contained within nuclear power plants. But this is just one step in a longer process sometimes called the *nuclear fuel cycle*. This process begins when the naturally occurring element uranium is mined from underground deposits, as we saw with the mines on Australian Aboriginal land in Chapter 6.

Uranium-containing minerals are uncommon, and uranium ore is in finite supply, which is why nuclear power is generally considered a nonrenewable energy source. Uranium is used for nuclear power because it is radioactive. Radioactive isotopes, or *radioisotopes* (p. 26), emit subatomic particles and high-energy radiation as they decay into lighter radioisotopes until they ultimately become stable isotopes. The isotope uranium-235 decays into a series of daughter isotopes, eventually forming lead-207. Each radioisotope decays at a rate determined by that isotope's *half-life* (p. 26), the time it takes for half of the atoms to give off radiation and decay. The half-life of ^{235}U is about 700 million years.

Over 99% of the uranium in nature occurs as the isotope uranium-238. Uranium-235 (with three fewer neutrons) makes up less than 1% of the total. Because ^{238}U does not emit enough neutrons to maintain a chain reaction when fissioned, we use ^{235}U for commercial nuclear power. Therefore, mined uranium ore must be processed to enrich the concentration of ^{235}U to at least 3%. The enriched uranium is formed into pellets of uranium dioxide (UO_2), which are incorporated into metallic tubes called *fuel rods* (**FIGURE 20.4**) that are used in nuclear reactors. After several years in a reactor, enough uranium has decayed so that the fuel no longer generates adequate energy, and it must be replaced with new fuel. In some countries, the spent fuel is reprocessed to recover the remaining usable energy. However, this process is costly relative to the low prices of uranium on the world market in recent years, so most spent fuel is disposed of as radioactive waste (pp. 573–575).

FIGURE 20.4 ▲ Enriched uranium fuel is packaged into fuel rods, which are encased in metal and used to power fission inside the cores of nuclear reactors. In this photo, the fuel rods are visible, arrayed in a circle within the blue glowing water.

Fission in reactors generates electricity in nuclear power plants

For fission to begin in a nuclear reactor, the neutrons bombarding uranium are slowed down with a substance called a *moderator*, most often water or graphite. As fission proceeds, it becomes necessary to soak up the excess neutrons produced when uranium nuclei divide, so that on average only a single neutron from each nucleus goes on to split another nucleus. For this purpose, *control rods*, made of a metallic alloy that absorbs neutrons, are placed into the reactor among the water-bathed fuel rods. Engineers move these control rods into and out of the water to maintain the fission reaction at the desired rate.

All this takes place within the reactor core and is the first step in the electricity-generating process of a nuclear power plant (**FIGURE 20.5**). The reactor core is housed within a reactor vessel, and the vessel, steam generator, and associated plumbing are often protected within a containment building. Containment buildings, with their meter-thick concrete and steel walls, are constructed to prevent leaks of radioactivity due to accidents or natural catastrophes such as earthquakes. Not all nations require containment buildings, which points out the key role that government regulation plays in protecting public safety.

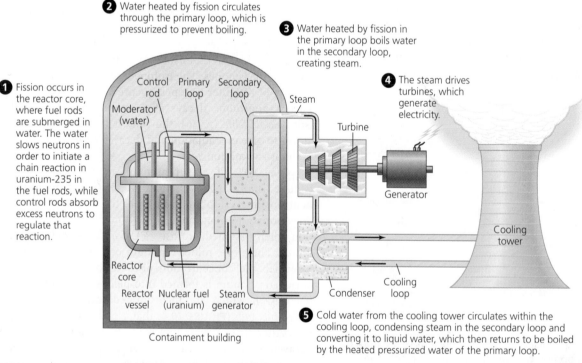

❷ Water heated by fission circulates through the primary loop, which is pressurized to prevent boiling.

❸ Water heated by fission in the primary loop boils water in the secondary loop, creating steam.

❶ Fission occurs in the reactor core, where fuel rods are submerged in water. The water slows neutrons in order to initiate a chain reaction in uranium-235 in the fuel rods, while control rods absorb excess neutrons to regulate that reaction.

❹ The steam drives turbines, which generate electricity.

❺ Cold water from the cooling tower circulates within the cooling loop, condensing steam in the secondary loop and converting it to liquid water, which then returns to be boiled by the heated pressurized water of the primary loop.

FIGURE 20.5 ▲ In a pressurized light water reactor, the most common type of nuclear reactor, uranium fuel rods are placed in water, which slows neutrons so that fission can occur **❶**. Control rods that can be moved into and out of the reactor core absorb excess neutrons to regulate the chain reaction. Water heated by fission circulates through the primary loop **❷** and warms water in the secondary loop, which turns to steam **❸**. Steam drives turbines, which generate electricity **❹**. The steam is then cooled in the cooling tower by water from an adjacent river or lake and returns to the containment building **❺**, to be heated again by heat from the primary loop.

Nuclear power delivers energy more cleanly than fossil fuels

Using fission, nuclear power plants generate electricity without creating air pollution from stack emissions. In contrast, combusting fossil fuels can emit sulfur dioxide, which contributes to acid deposition; particulate matter, which threatens human health; and carbon dioxide and other greenhouse gases, which drive global climate change. Even considering all the steps involved in building plants and generating power, researchers from the International Atomic Energy Agency (IAEA) have calculated that nuclear power reduces emissions 4–150 times below fossil fuel combustion (see **THE SCIENCE BEHIND THE STORY**, pp. 568–569). IAEA scientists estimate that at current global levels of use, nuclear power helps us avoid emitting 600 million metric tons of carbon each year, equivalent to 7% of global greenhouse gas emissions.

Nuclear power has additional advantages over fossil fuels—coal in particular. For residents living downwind from power plants, scientists calculate that nuclear power poses far fewer chronic health risks from pollution than does fossil fuel combustion. And because uranium generates far more power than coal by weight or volume, less of it needs to be mined, so uranium mining causes less damage to landscapes and generates less solid waste than coal mining. Moreover, in the course of normal operation, nuclear power plants are safer for workers than are coal-fired plants.

Nuclear power also has drawbacks. One is that the waste it produces is radioactive. Radioactive waste must be handled with great care and must be disposed of in a way that minimizes danger to present and future generations. The second main drawback is that if an accident occurs at a power plant, or if a plant is sabotaged, the consequences can potentially be catastrophic.

Given this mix of advantages and disadvantages (**FIGURE 20.6**), many governments (although not necessarily most citizens) have judged the good to outweigh the bad, and today the world has 439 operating nuclear plants in 30 nations.

WEIGHING THE ISSUES

Choose Your Risk Given the choice of living next to a nuclear power plant or living next to a coal-fired power plant, which would you choose? What would concern you most about each option?

Breeder reactors make better use of fuel, but have raised safety concerns

Using ^{235}U as fuel for fission is only one potential way to harness nuclear energy. **Breeder reactors** make use of ^{238}U, which in conventional fission goes unused as a waste product.

Environmental Impacts of Coal-fired and Nuclear Power		
Type of Impact	**Coal**	**Nuclear**
Land and ecosystem disturbance from mining	Extensive, on surface or underground	Less extensive
Greenhouse gas emissions	Considerable emissions	None from plant operation; much less than coal over the entire life cycle
Other air pollutants	Sulfur dioxide, nitrogen oxides, particulate matter, and other pollutants	No pollutant emissions
Radioactive emissions	No appreciable emissions	No appreciable emissions during normal operation; possibility of emissions during severe accident
Occupational health among workers	More known health problems and fatalities	Fewer known health problems and fatalities
Health impacts on nearby residents	Air pollution impairs health	No appreciable known health impacts under normal operation
Effects of accident or sabotage	No widespread effects	Potentially catastrophic widespread effects
Solid waste	More generated	Less generated
Radioactive waste	None	Radioactive waste generated
Fuel supplies remaining	Should last several hundred more years	Uncertain; supplies could last longer or shorter than coal supplies

FIGURE 20.6 ▲ Coal-fired power plants and nuclear power plants pose very different risks and impacts to human health and the environment. This chart compares the major impacts of each mode of electricity generation. For each type of impact, the more severe impact is indicated by a red box.

The SCIENCE behind the Story

Assessing Emissions from Power Sources

Combusting coal, oil, or natural gas emits carbon dioxide and other greenhouse gases into the atmosphere, where they contribute to global climate change (Chapter 18). Reducing greenhouse gas emissions is one of the main reasons so many people want to replace fossil fuels with alternative energy sources.

But determining how different energy alternatives compare in emissions is complicated. A number of studies have tried to quantify and compare emission rates of different energy sources, but the varied methods used have made it hard to synthesize this information into a coherent picture.

Researchers from the International Atomic Energy Agency (IAEA) attempted such a synthesis for the generation of electricity. Experts met and reviewed the scientific literature, together with data from industry and government. Their goal was to come up with a range of estimates of greenhouse gas emissions for nuclear power, each major fossil fuel type, and each major renewable energy

Coal-fired power plant in China

source. IAEA scientists Joseph Spadaro, Lucille Langlois, and Bruce Hamilton then published the results in the *IAEA Bulletin* in 2000.

The researchers had to decide how much of the total life cycle of electric power production to include in their estimates. Simply comparing the rotation of turbines at a wind farm to the operation of a coal-fired power plant might not be fair, because it would not reveal that greenhouse gases were emitted as a result of manufacturing the turbines, transporting them to the site, and erecting them there. Similarly, because uranium mining is part of the nuclear fuel cycle, and because we use oil-fueled machinery to mine uranium, perhaps emissions

from mining should be included in the estimate for nuclear power.

The researchers decided to conduct a "cradle-to-grave" analysis and include all sources of emissions throughout the entire life cycle of each energy source. This included not just power generation, but also the mining of fuel, preparation and transport of fuel, manufacturing of equipment, construction of power plants, disposal of wastes, and decommissioning of plants. They did, however, separate stack emissions from all other sources of emissions in the chain of steps so that these data could be analyzed independently.

Different greenhouse gases were then standardized to a unit of "carbon equivalence" according to their global warming potential (p. 496). The researchers then calculated rates of emission per unit of electric power produced, and presented figures in grams of carbon-equivalent emitted per kilowatt-hour (g C_{eq}/kWh).

The overall pattern they found was clear: Fossil fuels produce

---

In breeder reactors, the addition of a neutron to ^{238}U and its subsequent electron loss forms plutonium (^{239}Pu). When plutonium is bombarded by a neutron, it splits into fission products and releases more neutrons, which convert more of the remaining ^{238}U fuel into ^{239}Pu, continuing the process. Because 99% of all uranium is ^{238}U, fission in breeder reactors makes much better use of fuel, generates far more power, and produces far less waste.

However, breeder reactors are more dangerous than conventional reactors because highly reactive liquid sodium, rather than water, is used as a coolant, raising the risk of explosive accidents. Breeder reactors also are more expensive than conventional reactors. Finally, breeder reactors can be

used to supply plutonium to nuclear weapons programs. As a result, all but a handful of the world's breeder reactors have now been shut down.

Fusion remains a dream

For as long as scientists and engineers have generated power from nuclear fission, they have tried to figure out how they might use nuclear fusion instead. **Nuclear fusion**—the process that drives our sun's vast output of energy, and the force behind hydrogen bombs (thermonuclear bombs)—involves forcing together the small nuclei of lightweight elements under extremely high temperature and pressure. The hydrogen isotopes

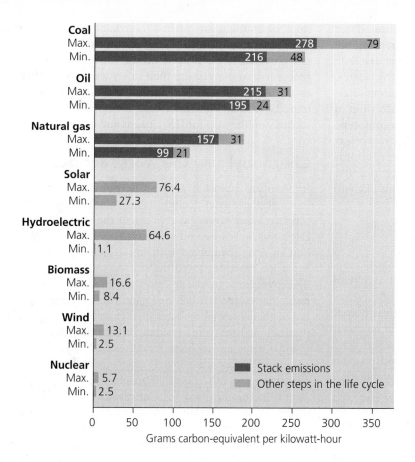

Coal, oil, and natural gas emit far more greenhouse gases than do renewable energy sources and nuclear energy. Red portions of bars represent stack emissions, and orange portions show emissions from other steps in the life cycle. Maximum and minimum values are given for each energy source. Adapted from Spadaro, J. V., et al., 2000. Greenhouse gas emissions of electricity generation chains: Assessing the difference. *IAEA Bulletin* 42(2). By permission of the International Atomic Energy Agency.

much higher emission rates than renewable energy sources and nuclear energy (**see figure**). Emissions decreased in the following order: coal, oil, natural gas, photovoltaic solar, hydroelectric, biomass, wind, and nuclear.

The maximum emission rate for fossil fuels (357 g C_{eq}/kWh for coal) was 4.7 times higher than the maximum emission rate for any renewable energy source (76.4 g C_{eq}/kWh for solar power). The minimum emission rate for any fossil fuel (120 g C_{eq}/kWh for

natural gas) was nearly 100 times greater than the minimum rate for renewables (1.1 g C_{eq}/ kWh for one form of hydropower). Most fossil fuel emissions were stack emissions directly from power generation, whereas the amounts due to other steps in the life cycle were roughly comparable to those from renewable sources.

Within each category, emissions values varied considerably. This variation was due to many factors, including the type of technology used, geographic location and transport costs, carbon content of the fuel, and the efficiency with which fuel was converted to electricity.

However, technology was expected to improve in the future, creating greater fuel-to-electricity conversion efficiency and lowering emissions rates. Thus, the researchers devised separate emissions estimates for newer technologies expected between 2005 and 2020. These estimates suggested that fossil fuels will improve but still will not approach the cleanliness of nuclear energy and most renewable sources.

Because the IAEA is charged with promoting nuclear energy, critics point out that the agency has clear motivation for conducting a study that shows nuclear power in a favorable light. However, few experts would quibble with the overall trend in the study's data: Nuclear and renewable energy sources are demonstrably cleaner than fossil fuels. ■

deuterium and tritium can be fused together to create helium, releasing a neutron and a tremendous amount of energy (**FIGURE 20.7**).

Overcoming the mutually repulsive forces of protons in a controlled manner is difficult, and fusion requires temperatures of many millions of degrees Celsius. Thus, researchers have not yet developed this process for commercial power generation. Despite billions of dollars of funding and decades of research, fusion experiments in the lab still require scientists to input more energy than they produce from the process. That is, they experience a loss in *net energy* (p. 534) and a ratio of *energy returned on investment* (EROI; p. 534) lower than 1.

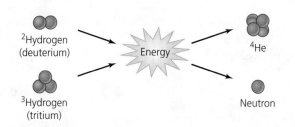

FIGURE 20.7 ▲ In nuclear fusion, two small atoms, such as the hydrogen isotopes deuterium and tritium, are fused together, releasing energy along with a helium nucleus and a free neutron. So far, however, scientists have not been able to fuse atoms without supplying far more energy than the reaction produces, so this process is not used commercially.

However, the potential payoffs are immense: If we were to develop a way to control fusion in a reactor, we could produce vast amounts of energy using water as a fuel. The process would create only low-level radioactive wastes, without pollutant emissions or the risk of dangerous accidents, sabotage, or weapons proliferation. A consortium of industrialized nations is collaborating to build a prototype fusion reactor called the International Thermonuclear Experimental Reactor (ITER) in southern France. It aims to achieve an EROI of 10:1. Even if this multi-billion-dollar effort succeeds, however, power from fusion seems likely to remain many years in the future.

Nuclear power poses small risks of large accidents

Although nuclear power delivers energy more cleanly than fossil fuels, the possibility of catastrophic accidents has spawned a great deal of public anxiety over nuclear power. Two events were influential in shaping public opinion about nuclear energy.

The first took place at the **Three Mile Island** plant in Pennsylvania in 1979 (**FIGURE 20.8**). Through a combination of mechanical failure and human error, coolant water began draining from the reactor vessel, temperatures rose inside the reactor core, and metal surrounding the uranium fuel rods began to melt, releasing radiation. This process is termed a **meltdown**, and it proceeded through half of one reactor core at Three Mile Island. Area residents stood ready to be evacuated as the nation held its breath, but fortunately most radiation remained trapped inside the containment building.

The accident was brought under control within days, the damaged reactor was shut down, and multi-billion-dollar cleanup efforts stretched on for years. Three Mile Island is best regarded as a near-miss; the emergency could have been far worse had the meltdown proceeded through the entire

FIGURE 20.8 ▼ The Three Mile Island nuclear power plant near Harrisburg, Pennsylvania, was the site of a partial meltdown in 1979. This emergency was a "near-miss"—radiation was released but was mostly contained, and no health impacts were confirmed. The incident put the world on notice, however, that a major accident could potentially occur.

stock of uranium fuel or had the containment building not contained the radiation. Moreover, residents of the area were not provided full and accurate information as the incident proceeded, so that risks to their safety were greater than necessary. Although residents have shown no significant health impacts in the years since, the event put safety concerns squarely on the map for U.S. citizens and policymakers.

Chernobyl saw the worst accident yet

In 1986 an explosion at the **Chernobyl** plant in Ukraine (part of the Soviet Union at the time) caused the most severe nuclear power plant accident the world has yet seen. Engineers had turned off safety systems to conduct tests, and human error, combined with unsafe reactor design, led to explosions that destroyed the reactor and sent clouds of radioactive debris billowing into the atmosphere. For 10 days radiation escaped from the plant while emergency crews risked their lives (some later died from radiation exposure) putting out fires. Most residents of the surrounding countryside remained at home for these 10 days, exposed to radiation, before the Soviet government belatedly began evacuating more than 100,000 people.

In the months and years afterwards, workers erected a gigantic concrete sarcophagus around the demolished reactor, scrubbed buildings and roads, and removed irradiated materials (**FIGURE 20.9**). However, the landscape for at least 30 km (19 mi) around the plant remains contaminated, the demolished reactor is still full of dangerous fuel and debris, and radioactivity leaks from the hastily built and quickly deteriorating sarcophagus. Today an international team is trying to build a larger sarcophagus around the original one to prevent a catastrophic re-release of radiation.

The accident killed 31 people directly and sickened or caused cancer in thousands more. Exact numbers are uncertain because of inadequate data and the difficulty of determining long-term radiation effects (see **THE SCIENCE BEHIND THE STORY**, pp. 572–573). Health authorities estimate that most of the over 4,000 cases of thyroid cancer diagnosed in people who were children at the time resulted from radioactive iodine spread by the accident. Estimates for the total number of cancer cases attributable to Chernobyl, past and future, vary widely, but an international consensus effort 20 years after the event estimated that the cancer rate among exposed people rose by up to a few percentage points, resulting in up to several thousand fatal cancer cases.

Atmospheric currents carried radioactive fallout from Chernobyl across much of the Northern Hemisphere, particularly Ukraine, Belarus, and parts of Russia and Europe (**FIGURE 20.10**). Fallout was greatest where rainstorms brought radioisotopes down from the radioactive cloud. Parts of Sweden received high amounts of fallout. The accident reinforced the Swedish public's fears about nuclear power. A survey taken after the event asked, "Do you think it was good or bad for the country to invest in nuclear energy?" The proportion of respondents answering "bad" jumped from 25% before Chernobyl to 47% afterward.

Fortunately, the world has not experienced another accident on the scale of Chernobyl. Moreover, the design

(a) The destroyed reactor at Chernobyl

(b) Technicians measuring radiation

FIGURE 20.9 ▲ The world's worst nuclear power plant accident unfolded in 1986 at Chernobyl, in present-day Ukraine (then part of the Soviet Union). As part of the extensive cleanup operation, the destroyed reactor **(a)** was encased in a massive concrete sarcophagus to contain further radiation leakage. Technicians scoured the landscape surrounding the plant **(b)**, measuring radiation levels, removing soil, and scrubbing roads and buildings.

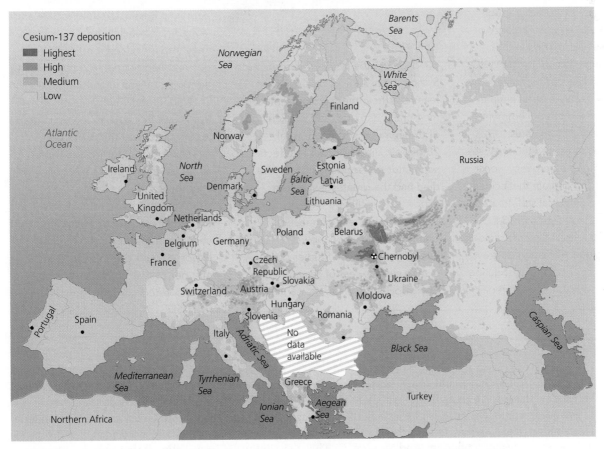

FIGURE 20.10 ▲ Radioactive fallout from the Chernobyl disaster was deposited across Europe in complex patterns resulting from atmospheric currents and rainstorms in the days following the accident. Darker colors in this map of cesium-137 deposition indicate higher levels of radioactivity. Although Chernobyl produced 100 times more fallout than the U.S. bombs dropped on Hiroshima and Nagasaki in World War II, it was distributed over a much wider area. Thus, levels of contamination in any given place outside of Ukraine, Belarus, and western Russia were relatively low; during these several days, the average European received less than the amount of radiation a person receives naturally in a year. Data from chernobyl.info, Swiss Agency for Development and Cooperation, Bern, 2005.

The SCIENCE behind the Story

Health Impacts of Chernobyl

In the wake of the nuclear power plant accident at Chernobyl in 1986, medical scientists from around the world rushed to study how the release of radiation would affect human health. Yet determining long-term health impacts of an event is difficult, so the hundreds of researchers trying to pin down Chernobyl's impacts sometimes came up with very different conclusions.

In an effort to reach consensus, the World Health Organization (WHO) engaged 100 experts from various nations to review all studies through 2006 and issue a report summarizing what scientists had learned in the 20 years since the accident. The WHO report was part of a collaborative 20-year review by an array of international agencies and the governments of Ukraine, Russia, and Belarus.

Doctors documented the most severe effects among emergency workers who battled to contain the incident in its initial days. Medical staff treated and recorded the progress of 134 workers hospitalized with acute radiation sickness (ARS). Radiation destroys cells in the body, and if the destruction outpaces the body's abilities to repair

Chernobyl-area boy and his mother after his surgery for thyroid cancer

the damage, the person will soon die. Symptoms of ARS include vomiting, fever, diarrhea, thermal burns, mucous membrane damage, and weakening of the immune system. In total, 28 people died from acute effects soon after the accident. Those who died had the greatest estimated exposure to radiation.

The major health impact of Chernobyl's radiation, however, has been thyroid cancer. Studies document a clear excess of cases among Chernobyl-area residents, particularly

children. The thyroid gland is where the human body concentrates iodine, and one of the most common radioactive isotopes released early in the disaster was iodine-131 (^{131}I). Children have large and active thyroid glands, so they are especially vulnerable to thyroid cancer induced by radio-isotopes.

Realizing that thyroid cancer might be a problem, medical workers measured iodine activity in the thyroid glands of several hundred thousand people in Russia, Ukraine, and Belarus in the months following the accident. They also measured food contamination and had people fill out questionnaires on their food consumption. These data showed that drinking milk from cows that had grazed on contaminated grass was the main route of exposure to ^{131}I for most people, although fresh vegetables also contributed.

As doctors had feared, rates of thyroid cancer began rising among children in the regions of highest exposure (**see graph**). The yearly number of thyroid cancer cases far exceeded numbers from years

of most reactors in the United States and other western nations is far safer than that of Chernobyl's. Yet, smaller-scale incidents have occurred; for instance, a 1999 accident at a plant in Tokaimura, Japan, killed two workers and exposed over 400 others to leaked radiation. And Sweden experienced a near-miss in 2006, when the Forsmark plant north of Stockholm narrowly avoided a meltdown after only two of four generators started up following a power outage.

As plants around the world age, they require more maintenance and are therefore less safe. New concerns have

also surfaced. The September 11, 2001, terrorist attacks raised fears that similar airplane attacks could be carried out against nuclear plants. Moreover, radioactive material could be stolen from plants and used in terrorist attacks. This possibility is especially worrisome in the cash-strapped nations of the former Soviet Union, where hundreds of former nuclear sites have gone without adequate security for years. In a cooperative international agreement, the U.S. government, through the "megatons to megawatts" program, has been buying up some of this material and diverting it to peaceful use in power generation.

before Chernobyl. Multiple studies found linear dose-response relationships (p. 390) in data from Ukraine and Belarus.

Fortunately, treatment of thyroid cancer has a high success rate, and as of 2002, only 15 of the 4,000 children diagnosed with thyroid cancer had died from it. By 2006, medical professionals were estimating that the number of cases had increased to 5,000 and was still rising.

Critics pointed out that any targeted search tends to turn up more of whatever medical problem is being looked for, but experts now agree that the increase in childhood thyroid cancer was attributable to Chernobyl. Thyroid cancer also appears to have risen markedly in adults in Belarus and the most contaminated regions of Russia.

Studies addressing other health impacts of the accident have turned up varying results. Some research has shown an increase in cataracts due to radiation. Other data have shown no rise in reproductive problems. Studies have also revealed psychological effects among exposed people, including heightened anxiety about health. The WHO report noted that much of this anxiety may result from constant portrayals of exposed people as victims and sufferers rather than as survivors—causing those people to think of themselves in that way.

Researchers have conducted many studies on leukemia and other

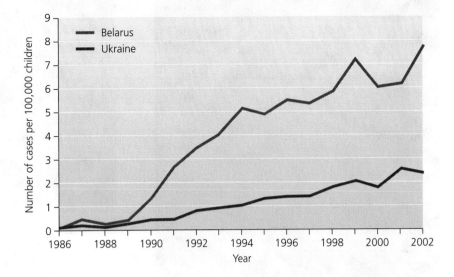

The incidence of thyroid cancer jumped in Belarus and Ukraine starting 4 years after the Chernobyl accident released radioactive iodine isotopes. Many babies and young children at the time of the accident developed thyroid cancer in later years. Most have undergone treatment and survived. Adapted from The Chernobyl Forum, 2006. *Chernobyl's legacy: Health, environmental, and socio-economic impacts.* The Chernobyl Forum: 2003–2005, 2nd rev. version. By permission of the IAEA, International Atomic Energy Agency, Vienna.

cancers, but the WHO report concluded that there is "no convincing evidence" so far that rates of any cancer (aside from thyroid cancer) have increased among people exposed to Chernobyl's radiation.

Based on previous studies on survivors of the atomic bombs dropped on Hiroshima and Nagasaki, researchers had expected increases in cancer mortality among the 600,000 people with significant exposure from Chernobyl to be "up to a few percent." This would translate into adding perhaps 4,000 cancer fatalities to a normally expected 100,000 cancer

deaths in the population from all other causes. However, conducting epidemiological studies (p. 387) with enough statistical power to detect minor increases in rare events is difficult, requiring that huge numbers of people be observed over many years.

Thus, the jury is still out on leukemia and other cancers. Moreover, many cancers do not generally appear for at least 10–15 years after exposure, so it is possible that most illnesses have yet to arise. For all these reasons, the WHO recommends continued monitoring to measure the full scope of health effects from Chernobyl. ■

Waste disposal remains a problem

Even if nuclear power generation could be made completely safe, we would still be left with the conundrum of what to do with spent fuel rods and other radioactive waste. Recall that fission utilizes ^{235}U as fuel, leaving as waste the 97% of uranium that is ^{238}U. This ^{238}U, as well as all irradiated material and equipment that is no longer being used, must be disposed of in a location where radiation will not escape. Because the half-lives of uranium, plutonium, and many other radioisotopes are far longer than

multiple human lifetimes, this waste will continue emitting radiation for thousands of years. Thus, radioactive waste must be placed in unusually stable and secure locations where radioactivity will not harm future generations.

For many years, the United States, the Soviet Union, and more than a dozen other nations routinely dumped radioactive waste into the oceans, believing that barrels would not leak or that any leaked radioactivity would be adequately diluted by seawater. Today we know that that is not the case and that ocean dumping poses threats to fisheries, people, and marine ecosystems, so the practice is banned.

(b) Dry storage

(a) Wet storage

FIGURE 20.11 ▲ Spent uranium fuel rods are currently stored at nuclear power plants and will likely remain at these scattered sites until a central repository for commercial radioactive waste is developed. Spent fuel rods are most often kept in "wet storage" in pools of water **(a)**, which keep them cool and reduce radiation release. Alternatively, the rods may be kept in "dry storage" in thick-walled casks layered with lead, concrete, and steel **(b)**.

Currently, nuclear waste from power generation is being held in temporary storage at nuclear power plants across the world. Spent fuel rods are sunken in pools of cooling water to minimize radiation leakage (**FIGURE 20.11A**). However, three-fourths of U.S. plants now have no room left for this type of storage. These plants are now storing waste in thick casks of steel, lead, and concrete (**FIGURE 20.11B**). In total, U.S. power plants are storing over 60,000 metric tons of high-level radioactive waste—enough to fill a football field to the depth of 6 m (20 ft)—as well as much more low-level radioactive waste. This waste is held at more than 120 sites spread across 39 states (**FIGURE 20.12**). A 2005 report from the National Academy of Sciences judged that most of these sites were vulnerable to terrorist attacks. Over 161 million U.S. citizens live within 125 km (75 mi) of temporarily stored waste.

Because storing waste at many dispersed sites creates a large number of potential hazards, nuclear waste managers would prefer to send all waste to a central repository that can be heavily guarded. In Sweden, that nation's nuclear industry conducted 20 years of research looking for a suitable location, and in 2009 selected the Forsmark power plant site as its single disposal location. If the site is approved by government agencies and constructed as planned, spent fuel rods and other high-level waste will be systematically buried in canisters about 500 m (1,650 ft) underground within stable bedrock.

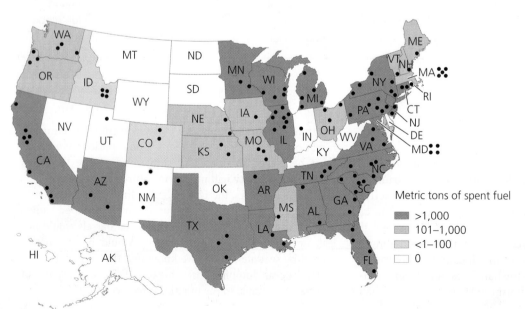

FIGURE 20.12 ◀ High-level radioactive waste from civilian reactors is currently stored at over 120 sites in 39 states across the United States. In this map, dots indicate storage sites, and the four shades of color indicate the total amount of waste stored in each state. Slightly different classifications of waste mean that some states shaded white show storage sites for certain types of waste. Data from Office of Civilian Radioactive Waste Management, U.S. Department of Energy; and Nuclear Energy Institute, Washington, D.C.

Metric tons of spent fuel
- \>1,000
- 101–1,000
- <1–100
- 0

(a) Yucca Mountain

(b) Scientific testing

1 **Canisters of radioactive waste are shipped to the site**

Yucca Mountain

Processing site

2 **Radioactive waste is placed in a multilayered steel storage container and sent underground**

Tunnel system

Storage container

300 m (1,000 ft)

300 m (1,000 ft)

Container

Ramp to tunnels

3 **Containers are stored along the tunnels**

Water table

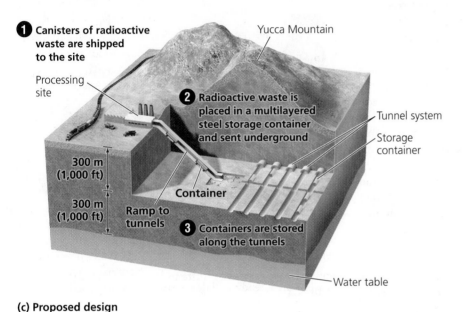

(c) Proposed design

FIGURE 20.13 ▲ Yucca Mountain **(a)**, in a remote part of Nevada, was being developed as the central repository site for all commercial nuclear waste in the United States until support was withdrawn in 2010. Here **(b)**, technicians are testing the effects of extreme heat from radioactive decay on the stability of rock. As seen in this cutaway view of the mountain **(c)**, waste was to be buried in a network of tunnels deep underground, yet still high above the water table.

In the United States, the multiyear search homed in on Yucca Mountain, a remote site in the desert of southern Nevada, 160 km (100 mi) from Las Vegas (**FIGURE 20.13A**). Choice of this site followed extensive study by government scientists (**FIGURE 20.13B**), and $13 billion was spent on its development, although Nevadans were not happy about the choice. In 2010, as the site was awaiting approval from the Nuclear Regulatory Commission, President Barack Obama's administration ended support for the project. Ironically this came just days after Obama had urged expanding nuclear power in his State of the Union address. Most political observers agree that the opposition of Senate Majority Leader Harry Reid, who represents Nevada, was a key reason for the change in policy. Without Yucca Mountain, the United States has no place designated to dispose of its radioactive waste from commercial nuclear power plants, so this waste will remain at its numerous current locations across the country.

Yucca Mountain had been expected to begin receiving radioactive waste from nuclear reactors and military installations in 2017. Waste was to be stored in a network of tunnels 300 m (1,000 ft) underground, yet 300 m (1,000 ft) above the water table (**FIGURE 20.13C**). Scientists and policymakers chose the Yucca Mountain site because they determined that it is remote and unpopulated, has minimal risk of earthquakes, receives little rain that could contaminate groundwater with radioactivity, has a deep water table atop an isolated aquifer, and is on federal land that can be protected from sabotage.

Some scientists, antinuclear activists, and concerned Nevadans challenged these conclusions. They argued that earthquakes or volcanic activity might destabilize the site's geology and that fissures in the mountain's rock could allow rainwater to seep into the caverns. Another concern with Yucca Mountain or any other centralized repository is that nuclear waste will need to be transported there from the 120-some current storage areas and from current and future nuclear plants and military installations. Because this would involve many thousands of shipments by rail and truck across hundreds of public highways through almost every state of the union, many people worry that the risk of an accident or of sabotage is unacceptably high.

WEIGHING THE ISSUES

How to Store Waste? Which do you think is a better option—to transport nuclear waste cross-country to a single repository or to store it permanently at numerous power plants and military bases scattered across the nation? Would your opinion be affected if you lived near the repository site? Near a power plant? On a highway route along which waste is transported?

Multiple dilemmas have slowed nuclear power's growth

Dogged by concerns over waste disposal, safety, and expensive cost overruns, nuclear power's growth has slowed. Since the late 1980s, nuclear power has grown by 2.5% per year worldwide, about the same rate as electricity generation overall. Public anxiety in the wake of Chernobyl made utilities less willing to invest in new plants. So did the enormous expense of building, maintaining, operating, and ensuring the safety of nuclear facilities. Almost every nuclear plant has turned out to be far more expensive than expected. In addition, plants have aged more quickly than expected because of problems that were underestimated, such as corrosion in coolant pipes. The plants that have been shut down—well over 100 around the world to date—have served on average less than half their expected lifetimes. Moreover, shutting down, or decommissioning, a plant can sometimes be more expensive than the original construction.

As a result of these economic issues, electricity from nuclear power today remains more expensive than electricity from coal and other sources. Governments are still subsidizing nuclear power to keep electricity costs to ratepayers down, but many private investors lost interest long ago. Nonetheless, nuclear power remains one of the few currently viable alternatives to fossil fuels with which we can generate large amounts of electricity in short order. For a nation wishing to cut its carbon emissions quickly and substantially, nuclear power is in many respects the leading option.

Many experts predict nuclear power will decline because three-quarters of Western Europe's plants will be retired by 2030. However, increased concern over climate change and carbon emissions has recently led several European nations that had planned to phase out nuclear power to reverse their decisions, just as Sweden did. Today Germany, Belgium, and Spain are the only European nations still considering nuclear phaseouts. Asian nations, meanwhile, are actively adding nuclear capacity. China, India, and South Korea are expanding their nuclear programs to help power their rapidly growing economies. Japan is so reliant on imported oil that it is eager to diversify its energy options. Altogether, Asia hosts two-thirds of the most recent nuclear plants to go into operation and most of the 61 plants now under construction.

In the United States, the nuclear industry stopped building plants following Three Mile Island, and public opposition scuttled many that were under construction. The $5.5 billion Shoreham Nuclear Power Plant on New York's Long Island was shut down just two months after being licensed because officials determined that evacuation would be impossible in this densely populated area should an accident ever occur. Of the 259 U.S. nuclear plants ordered since 1957, nearly half have been cancelled. At its peak in 1990, the United States had 112 operable plants; today it has 104.

Expanding U.S. nuclear capacity would decrease reliance on coal for electricity and would allow the nation to cut its greenhouse gas emissions that contribute to climate change. Proponents of expanding nuclear power stress that engineers are planning a new generation of reactors designed to be safer and less expensive. As a result, many U.S. policymakers are now urging construction of new nuclear power plants. Opponents, meanwhile, point out that issues of cost, security, and waste management have not been resolved.

WEIGHING THE ISSUES

More Nuclear Power? Sweden and many other European nations have reduced or slowed their carbon emissions by expanding their use of nuclear power to replace fossil fuels. Do you think the United States should expand its nuclear power program? Why or why not?

With slow growth expected for nuclear power, fossil fuels in limited supply, an oil production peak looming, and climate change worsening, where will our growing human population turn for clean and sustainable energy? Increasingly, people are turning to renewable sources of energy: energy sources that cannot be depleted by our use. Many promising renewable sources are still early in their development (Chapter 21), but two of them—bioenergy and hydroelectric power—are already well developed and widely used.

BIOENERGY

Bioenergy—also known as **biomass energy**—is energy obtained from biomass resources. **Biomass** (p. 85) consists of organic material derived from living or recently living organisms, and it contains chemical energy (p. 30) that originated ultimately with sunlight and photosynthesis. We harness bioenergy from many types of plant matter, including wood from trees, charcoal from wood charred in the absence of oxygen, and matter from agricultural crops, as well as from combustible animal waste products such as cattle manure. Indeed, bioenergy means different things to different people. To a poor farmer in Africa, bioenergy entails cutting wood from trees or collecting livestock manure by hand and burning it to heat and cook for her family. To an industrialized farmer in Iowa, bioenergy means shipping his grain to a hi-tech refinery that converts it to liquid fuel to run automobiles. The diversity of sources and approaches involved in bioenergy (**TABLE 20.2**) gives us many ways to address our energy challenges.

However, we must be careful in selecting the paths we follow. In principle, bioenergy is renewable and releases no net carbon dioxide into the atmosphere. For these reasons, it would seem an excellent replacement for our nonrenewable and polluting fossil fuels. However, the picture is not so simple. Judging the sustainability of any given bioenergy strategy

TABLE 20.2 Major Bioenergy Sources

Direct combustion for heating

▸ Wood cut from trees (fuelwood)
▸ Charcoal
▸ Manure from farm animals

Biofuels for powering vehicles

▸ Corn grown for ethanol
▸ Bagasse (sugarcane residue) grown for ethanol
▸ Soybeans, rapeseed, and other crops grown for biodiesel
▸ Used cooking oil for biodiesel
▸ Plant matter treated with enzymes to produce cellulosic ethanol
▸ Algae grown for biofuels

Biopower for generating electricity

▸ Crop residues (such as cornstalks) burned at power plants
▸ Forestry residues (such as wood waste from logging) burned at power plants
▸ Processing wastes (such as solid or liquid waste from sawmills, pulp mills, and paper mills) burned at power plants
▸ "Landfill gas" burned at power plants
▸ Livestock waste from feedlots for gas from anaerobic digesters
▸ Organic components of municipal solid waste from landfills

requires careful consideration of the type of biomass source we are using and the way we gain energy from it. In the following pages we survey the major bioenergy approaches and the advantages and disadvantages of each.

Traditional biomass sources are widely used in the developing world

Over 1 billion people use wood from trees as their principal energy source. In developing nations, especially in rural areas, families gather fuelwood to burn in their homes for heating, cooking, and lighting (**FIGURE 20.14**; also see Figure 17.28,

FIGURE 20.14 ▼ Well over a billion people in developing countries rely on fuelwood for heating and cooking. Wood cut from trees remains the major source of biomass energy used in the world today. In theory, biomass is renewable, but in practice it may not be if forests are overharvested.

p. 487). Although fossil fuels are replacing traditional energy sources as developing nations industrialize, fuelwood, charcoal, and manure still account for fully 35% of energy use in these nations—up to 90% in the poorest nations.

Fuelwood and other traditional biomass sources constitute nearly 80% of all renewable energy used worldwide. However, biomass is renewable only if it is not overharvested. Harvesting fuelwood at unsustainably rapid rates can lead to deforestation, soil erosion, and desertification (pp. 231, 233, 318), thereby damaging landscapes, diminishing biodiversity, and impoverishing human societies. Heavily populated arid regions that support meager woodlands are most vulnerable to overharvesting, and these include many regions of Africa and Asia. Another drawback of burning fuelwood and other biomass in traditional ways for cooking and heating is that it leads to health hazards from indoor air pollution (p. 487).

New bioenergy strategies are being developed in industrialized countries

Although much of the world still relies on fuelwood, charcoal, and manure, new bioenergy approaches are being developed using a variety of materials to provide innovative types of energy (see Table 20.2). Some of these biomass sources can be burned in power plants to produce **biopower** (also called *biomass power*), generating heat and electricity in the same way that we burn coal for power (p. 536). Other sources can be converted into **biofuels**, liquid fuels used primarily to power automobiles. Because many of these novel biofuels and biopower strategies depend on technologies resulting from extensive research and development, they are being developed primarily in wealthier industrialized nations, such as Sweden and the United States.

Biopower generates electricity from biomass

We harness biopower by combusting biomass to generate electricity. This can be done using a variety of sources and techniques.

Waste products Many sources used for biopower are the waste products of existing industries or processes. For instance, the forest products industry generates large amounts of woody debris in logging operations and at sawmills, pulp mills, and paper mills (**FIGURE 20.15**). Sweden's efforts to promote bioenergy have focused largely on using forestry residues. Because so much of the nation is forested and the timber industry is a major part of the Swedish economy, plenty of forestry waste is available.

Other waste sources used for biopower in nations from Sweden to the United States include residue from agricultural crops (such as cornstalks and corn husks), animal waste from feedlots, and organic waste from municipal landfills. The anaerobic bacterial breakdown of waste in landfills produces methane and other components, and this "landfill gas" is now being captured and sold as fuel (p. 624). Methane and other gases can also be produced in a more controlled way in

FIGURE 20.15 ▲ Forestry residues (here from a Swedish logging operation) are a major source material for biopower in some regions.

anaerobic digestion facilities. This *biogas* can then be burned in a power plant's boiler to generate electricity.

Bioenergy crops Besides using waste, we also are beginning to grow certain types of plants as crops to generate biopower. These include fast-growing grasses such as bamboo, fescue, and switchgrass, and fast-growing trees such as specially bred willows and poplars (**FIGURE 20.16**). Many of these plants are also being grown to produce liquid biofuels.

Combustion strategies Power plants built to combust biomass operate similarly to those fired by fossil fuels; combustion heats water, creating steam to turn turbines and generators, thereby generating electricity. Much of the biopower

FIGURE 20.16 ▼ These hybrid poplars are specially bred for fast growth in dense plantations, and they are harvested for use in biopower. This is renewable energy, but it also has environmental impacts: The vast monocultural plantations, which may replace natural systems over large areas, do not function ecologically as forests.

produced so far comes from power plants that use cogeneration (p. 556) to generate both electricity and heating. These plants are often located where they can take advantage of forestry waste.

Biomass is increasingly being combined with coal in coal-fired power plants in a process called *co-firing*. Wood chips, wood pellets, or other biomass is introduced with coal into a high-efficiency boiler that uses one of several technologies. We can substitute biomass for up to 15% of the coal with only minor equipment modification and no appreciable loss of efficiency. Co-firing is a relatively easy and inexpensive way for utilities to expand their use of renewable energy.

We also harness biopower through *gasification* (p. 546), in which biomass is vaporized at extremely high temperatures in the absence of oxygen, creating a gaseous mixture including hydrogen, carbon monoxide, carbon dioxide, and methane. This mixture can generate electricity when used to turn a gas turbine to propel a generator in a power plant. We can also treat gas from gasification in various ways to produce methanol, synthesize a type of diesel fuel, or isolate hydrogen for use in hydrogen fuel cells (pp. 609–612). An alternative method of heating biomass in the absence of oxygen results in *pyrolysis* (p. 544), which produces a mix of solids, gases, and liquids. This includes a liquid fuel called pyrolysis oil, which can be burned to generate electricity.

Scales of production At small scales, farmers, ranchers, or villages can operate modular biopower systems that use livestock manure to generate electricity. Small household biodigesters now provide portable and decentralized energy production for remote rural areas. At large scales, industries such as the forest products industry are using their waste to generate power, and industrialized farmers are growing bioenergy crops. In Sweden, nearly one-fifth of the nation's energy supply now comes from biomass, and biomass provides more fuel for electricity generation than coal, oil, or natural gas. Pulp mill liquors are the main source, but solid wood waste, municipal solid waste, and biogas from digestion are all used. In the United States, several dozen biomass-fueled power plants are now operating, and several dozen coal-fired plants are experimenting with co-firing.

Advantages Biomass power has several advantages over fossil fuel combustion. By enhancing energy efficiency and recycling waste products, biopower helps move our utilities and industries toward sustainability. The U.S. forest products industry now obtains over half its energy by combusting the waste it recycles, including woody waste and liquor from pulp mill processing. Biopower also helps mitigate climate change by reducing carbon dioxide emissions, and capturing landfill gas reduces emissions of methane, a potent greenhouse gas. Relative to fossil fuels, biopower also benefits human health. By replacing coal in co-firing and direct combustion, biopower reduces emissions of sulfur dioxide because plant matter, unlike coal, contains no appreciable sulfur content. In addition, biomass resources tend to be geographically widespread and well dispersed, so using them should help support rural economies and reduce many nations' dependence on imported fuels.

Disadvantages   A main disadvantage of biomass power is that when we burn crops or plant matter for power, we deprive the soil of the nutrients it would have gained from having the plant matter returned to the soil. We essentially draw fertility from the soil and never return it, so that the soil becomes progressively depleted. This also is the case when we burn crops or plant matter as a fuel, as discussed next. The depletion of soil fertility is a major long-term problem for bioenergy, and is one reason that relying solely on bioenergy is not a sustainable option.

(a) Corn grown for ethanol

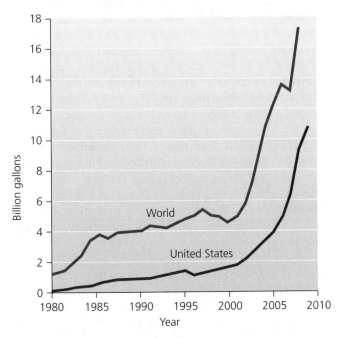

(b) Ethanol production, 1980–2009

FIGURE 20.17 ▲ Over 20% of the U.S. corn crop **(a)** is used to produce ethanol, a biofuel that is widely added to gasoline in the United States. Brazil produces most of the rest of the world's ethanol, from *bagasse* (sugarcane residue). Ethanol production **(b)** has grown rapidly in the last several years. Data (b) from Renewable Fuels Association.

Ethanol can power automobiles

Liquid fuels from biomass sources are powering millions of vehicles on today's roads. The two primary biofuels developed so far are ethanol (for gasoline engines) and biodiesel (for diesel engines).

Ethanol is the alcohol in beer, wine, and liquor. It is produced as a biofuel by fermenting biomass, generally from carbohydrate-rich crops, in a process similar to brewing beer. In fermentation, carbohydrates are converted to sugars and then to ethanol. Spurred by the 1990 Clean Air Act amendments and generous government subsidies, ethanol is widely added to gasoline in the United States to reduce automotive emissions. In 2009 in the United States, over 40 billion L (10.7 billion gal) of ethanol were produced, mostly from corn (**FIGURE 20.17**). This amount is growing rapidly, and nearly 200 U.S. ethanol production facilities are now operating. More growth is assured, as the Energy Independence and Security Act passed by Congress in 2007 mandates production and use of 136 billion L (36 billion gal) per year of ethanol by 2022.

Any vehicle with a gasoline engine runs well on gasoline blended with up to 10% ethanol, but more and more vehicles are being produced that can run primarily on ethanol. Sweden has many such public buses, and the U.S. "big three" automakers are now producing *flexible-fuel vehicles* that run on E-85, a mix of 85% ethanol and 15% gasoline. Over 8 million such cars are on U.S. roads today. Most gas stations do not yet offer E-85 (**FIGURE 20.18**), so drivers often fill these cars with conventional gasoline, but this situation is changing as infrastructure for ethanol increases.

FIGURE 20.18 ▼ More and more gas stations are offering biofuels. This station in New Mexico sells a 20% biodiesel blend (left pump), an 85% ethanol blend (center pump), and a standard 10% ethanol blend (right pump).

In Brazil, sugarcane residue is crushed to make *bagasse*, a material that is then used to make ethanol. Half of all new Brazilian cars are flexible-fuel vehicles, and ethanol from sugarcane accounts for 40% of all automotive fuel that Brazil's drivers use.

Ethanol may not be our most sustainable energy choice

The enthusiasm for corn-based ethanol shown by U.S policymakers is not widely shared by environmental scientists. One reason is that growing corn to produce ethanol exerts considerable impacts on ecosystems, including pesticide use, fertilizer use, fresh water depletion, and other consequences of monocultural industrialized agriculture (pp. 255–256). Corn ethanol crops take up precious land that might otherwise be left in its natural condition or developed for other purposes. If we were to try to produce all the automotive fuel currently used in the United States with ethanol from U.S. corn, the nation would need to expand its already immense corn acreage by five times, with no loss of productivity (see **ENVISIONIT**, p. 581).

Even at our current level of production, ethanol already competes with food production and drives up food prices. About 21% of the U.S. corn crop today is used to make ethanol. As farmers shifted more corn crops to ethanol in 2006–2008, corn supplies for food dropped, and international corn prices skyrocketed, making it difficult for the poor to afford food. Protests and riots ensued in many nations. In Mexico, for instance, where corn tortillas are emblematic of the national cuisine, working-class citizens found themselves struggling to buy their staple food, and protests erupted across the country.

Growing corn for ethanol also requires substantial inputs of energy. We currently operate farm equipment using oil, and farmers apply petroleum-based pesticides and fertilizers to increase yields. Moreover, oil is used to transport corn to processing plants, and fossil fuels are used in refineries to heat water so that we can distill ethanol. Thus, simply shifting from gasoline to corn ethanol for our transportation needs would not eliminate our reliance on fossil fuels.

Moreover, corn ethanol yields only a modest amount of energy relative to the energy that needs to be input. The EROI (*energy returned on investment*) ratio (p. 534) for corn-based ethanol is variable and controversial, but recent estimates place it around 1.5:1. This means that to gain 1.5 units of energy from ethanol, we need to expend 1 unit of energy. The EROI of Brazilian bagasse ethanol is much higher, but the low ratio for corn ethanol makes this fuel quite inefficient. For this reason, many critics do not view corn ethanol as an effective path to sustainable energy use.

Biodiesel powers diesel engines

Drivers of vehicles with diesel engines can use **biodiesel**, a fuel produced from vegetable oil, used cooking grease, or animal fat. The oil or fat is mixed with small amounts of ethanol or methanol (wood alcohol) in the presence of a chemical catalyst. In Europe, where most biodiesel is used, rapeseed oil is the oil of choice, whereas U.S. biodiesel producers use mostly soybean oil. Vehicles with diesel

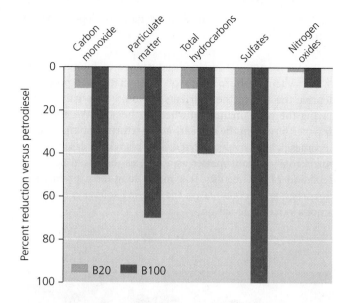

FIGURE 20.19 ▲ Burning biodiesel in a diesel engine emits less pollution than burning conventional petroleum-based diesel. Shown are the percentage reductions in several major automotive pollutants that one can attain by using B20 (a mix of 20% biodiesel and 80% petrodiesel) and B100 (pure biodiesel). Data from U.S. Environmental Protection Agency.

engines can run on 100% biodiesel, or biodiesel can be mixed with conventional petrodiesel; a 20% biodiesel mix (called B20) is common today.

Biodiesel cuts down on emissions compared with petrodiesel (**FIGURE 20.19**). Its fuel economy is almost as good, and it costs just slightly more at today's oil prices. It is also nontoxic and biodegradable. Increasing numbers of environmentally conscious individuals in North America and Europe are fueling their cars with biodiesel from waste oils, and some buses and recycling trucks are now running on biodiesel. Governments are encouraging its use, too, and many state and federal fleets are using biodiesel blends.

Some enthusiasts have taken biofuel use further. Eliminating the processing step that biodiesel requires, they use straight vegetable oil in their diesel engines (which requires modifying the engine). One notable effort is the BioTour, in which a group of students and young biofuel advocates drives across North America in buses fueled entirely by waste oil from restaurants (**FIGURE 20.20**). Each summer a group goes on tour, hosting festive events and seminars on environmental sustainability, spreading the word about nonpetroleum fuels.

Using waste oil as a biofuel is sustainable, but most biodiesel today, like most ethanol, comes from crops grown specifically for the purpose. As with ethanol, these crops have environmental impacts. Growing soybeans in Brazil (p. 320) and oil palms in Southeast Asia (p. 321) cause the deforestation of tropical rainforest. Growing soybeans in the United States and rapeseed in Europe takes up large areas of land as well. Because the major crops grown for biodiesel and for ethanol exert heavy impacts on the land, farmers and agricultural scientists are experimenting with a variety of other crops, from wheat, sorghum, cassava, and sugar beets to less-known plants such as hemp, jatropha, and the grass miscanthus.

In the quest for carbon-neutral alternatives to gasoline, what biofuel could be better, many people have thought, than ethanol made from America's #1 crop, corn?

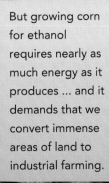

But growing corn for ethanol requires nearly as much energy as it produces ... and it demands that we convert immense areas of land to industrial farming.

Ethanol refinery

And when fuel crops compete with food crops, that drives food prices up.

Area of corn grown in the U.S. today

Area of corn that would need to be grown if ethanol were to replace all gasoline in U.S.

STOP!

NOT GASOLINE!
E85 for use in Flexible
Fuel Vehicles (FFVs) only.
Please consult your vehicle
owner's manual or store clerk.

E85
85% Ethanol

Ethanol at the pump

So, scientists are racing to find more efficient and sustainable biofuels.

YOU CAN MAKE A DIFFERENCE

➤ Ask your legislators to support university research on non-food-crop biofuels like switchgrass, algae, and cellulosic ethanol.

➤ Urge policymakers to create financial incentives for sustainable biofuels.

➤ Reduce your gasoline consumption by driving less and driving a more fuel-efficient vehicle. You'll save money, too!

FIGURE 20.20 ▲ Each summer, young alternative-fuel advocates go on tour across North America in buses fueled entirely on used vegetable oil from restaurants. Their "BioTours" sponsor events that include music and seminars on sustainability and alternative fuels. Their motto: "Solar-powered sound, veggie-powered bus."

Novel biofuels are being developed

Possibly the most exciting next-generation biofuel crop is algae (better known to most people as green pond scum)! Several species of these photosynthetic organisms produce large amounts of lipids that can be converted to biodiesel. Alternatively, carbohydrates in algae can be fermented to create ethanol. In fact, a variety of fuels, including jet fuel, can be produced from algae.

Algae can be grown in open ponds, including "raceway" ponds that circulate algae around a racetrack. More controlled and intensive production can be achieved in closed tanks or in closed transparent tubes called photobioreactors (**FIGURE 20.21**). Algae grow much faster than terrestrial crops, can be harvested every few days, and produce many

FIGURE 20.21 ▼ Algae is a leading candidate for a next-generation biofuel. Here it is seen growing in networks of lighted tubes at a demonstration facility. Algae grows quickly and productively, can be farmed in many places (including locations where other crops cannot be grown), and can produce biodiesel, ethanol, or other fuels.

times more oil than other biofuel crops. Algae farms can be set up just about anywhere and can use wastewater, ocean water, or saline water. Because algae need nutrients, wastewater from sewage treatment plants can actually be a good source of water. And because piping in carbon dioxide speeds their growth, algae can even make use of smokestack emissions. Indeed, placing algae farms next to power plants as a source of carbon capture (pp. 519, 546–547) is a promising combination. So far, producing biofuels from algae is expensive, but recent investment from the private sector will likely bring costs down soon.

Because depending on any monocultural crop for energy may not be a sustainable strategy, researchers are refining techniques to use enzymes to produce ethanol from the cellulose that gives structure to all plant material—a biofuel called **cellulosic ethanol**. This would be a substantial advance because ethanol as currently made from corn or sugarcane uses starch, which is valuable to us as food. Cellulose, in contrast, is of no food value to people yet is abundant in all plants. If we can produce cellulosic ethanol in commercially feasible ways, then ethanol could be made from low-value crop waste (residues such as corn stalks and husks), rather than from high-value crops.

In addition, certain plants hold promise as crops specifically for cellulosic ethanol. Switchgrass (**FIGURE 20.22**) could potentially be a sustainable choice for the United States because it is a native grass of the North American prairies. Thus it could provide wildlife habitat while serving as a crop, especially if planted in a polyculture mixed with other species. Cellulosic ethanol from switchgrass has an EROI ratio of 5:1 (much better than corn ethanol's), and growing this perennial grass could prevent the soil erosion and depletion of soil carbon that would occur as a result of removing too much crop waste for ethanol.

Is bioenergy carbon-neutral?

In principle, energy from biomass is carbon-neutral, releasing no net carbon into the atmosphere. Although burning biomass emits plenty of carbon, this is balanced by the fact that photosynthesis had pulled this amount of carbon from the atmosphere to create the biomass just years, months, or weeks before. Therefore, in theory, when we replace fossil fuels with bioenergy, we reduce net carbon flux to the atmosphere, helping to mitigate global climate change.

However, in practice things are not so simple. Burning biomass for energy is not carbon-neutral if natural forests are destroyed in order to plant bioenergy crops. Forests sequester more carbon (in vegetation and in soil) than do croplands, so deforesting land to plant crops will increase carbon flux to the atmosphere. In addition, bioenergy use is not carbon-neutral if we need to input fossil-fuel energy to produce the biomass (for instance, by driving tractors, using fertilizers, and applying pesticides to produce biofuel crops).

On top of this, international climate change policy so far has failed to encourage sustainable bioenergy approaches. The Kyoto Protocol (pp. 520–521) requires nations to submit data on emissions from energy use and from change in land use, but only the emissions from energy use are "counted" toward judging how well nations are controlling their

FIGURE 20.22 ▲ Switchgrass, a fast-growing plant native to the North American prairies, provides fuel for biopower now and is being studied as a crop to provide cellulosic ethanol.

emissions under the treaty. This creates loopholes that result in unintended, perverse incentives. For example, let's say an Asian nation clears its rainforest and replaces it with oil palm plantations (p. 321), then sells the palm oil to a European nation for use as biodiesel. The Asian nation reports carbon emissions from its land use change, and the European nation does not have to report emissions from its biodiesel use because this was already "counted" by the Asian nation. Because the Protocol only mandates reductions in emissions from energy use, this rewards Europe for consuming palm oil while giving Asia no incentive to preserve forest. As a result, there is an incentive to destroy forest for biofuel crops.

Scientists hope policymakers will fix this policy problem in future climate negotiations. In the meantime, researchers are busy trying to develop means of using bioenergy that are truly renewable and carbon-neutral. With continued research and careful decision making, our many bioenergy options may provide promising avenues for sustainably replacing fossil fuels.

WEIGHING THE ISSUES

Biofuels Do you think producing and using ethanol from corn is a good idea? Do the benefits outweigh the drawbacks? Do you think we should invest billions of dollars into developing next-generation biofuels like algae and cellulosic ethanol? Can you suggest ways of using biofuels that would minimize environmental impacts?

HYDROELECTRIC POWER

Next to biomass, we draw more renewable energy from the motion of water than from any other resource. In **hydroelectric power**, or **hydropower**, we use the kinetic energy of moving water to turn turbines and generate electricity. We examined hydropower and its environmental impacts in our discussion of freshwater resources in Chapter 15 (pp. 411–413). Now we will take a closer look at hydropower as an energy source.

Modern hydropower uses two approaches

Most of our hydroelectric power today comes from impounding water in reservoirs behind concrete dams that block the flow of river water and then letting that water pass through the dam. Because immense amounts of water are stored behind dams, this is called the **storage** technique. If you have ever seen Hoover Dam on the Colorado River (p. 400) or any other large dam, you have witnessed an impressive example of the storage approach to hydroelectric power.

As reservoir water passes through a dam, it turns the blades of turbines, which cause a generator to generate electricity (**FIGURE 20.23**). Electricity generated in the powerhouse of a dam is transmitted to the electric grid by transmission lines, and the water is allowed to flow into the riverbed below the dam to continue downriver. The amount of power generated depends on the distance the water falls and the volume of water released. By storing water in reservoirs, dam operators can ensure a steady and predictable supply of electricity at all times, even during periods of naturally low river flow.

An alternative approach is the **run-of-river** approach, which generates electricity without greatly disrupting the flow of river water. This approach sacrifices the reliability of water flow across seasons that the storage approach guarantees, but it minimizes many of the impacts of large dams. The run-of-river approach can use several methods, one of which is to divert a portion of a river's flow through a pipe or channel, passing it through a powerhouse and then returning it to the river (**FIGURE 20.24**). This can be done with or without a small reservoir that pools water temporarily, and the pipe or channel can be run along the surface or underground. Another method is to flow river water over a dam small enough not to impede fish passage, siphoning off water to turn turbines, and then returning the water to the river. Run-of-river systems are particularly useful in areas remote from established electrical grids and in regions without the economic resources to build and maintain large dams.

Hydroelectric power is widely used

Hydropower accounts for 2.2% of the world's energy supply and 15.6% of the world's electricity production (see Figure 20.1). For nations with large amounts of river water and the economic resources to build dams, hydroelectric power has been a keystone of their development and wealth. Canada, Brazil, Norway, Austria, Switzerland, Venezuela, and many other nations today obtain large amounts of their energy from hydropower (**TABLE 20.3**).

Sweden receives 10.6% of its total energy and nearly half its electricity from hydropower. In the wake of the nation's 1980 referendum to phase out nuclear power, many people had hoped hydropower could play a still larger role and compensate for the electrical capacity that would be lost. However, Sweden has already dammed so many of its rivers that it cannot gain much additional hydropower by erecting more dams. Moreover, Swedish citizens have made it clear that they want some rivers to remain undammed, preserved in their natural state.

(a) Ice Harbor Dam, Snake River, Washington

(b) Turbine generator inside McNary Dam, Columbia River

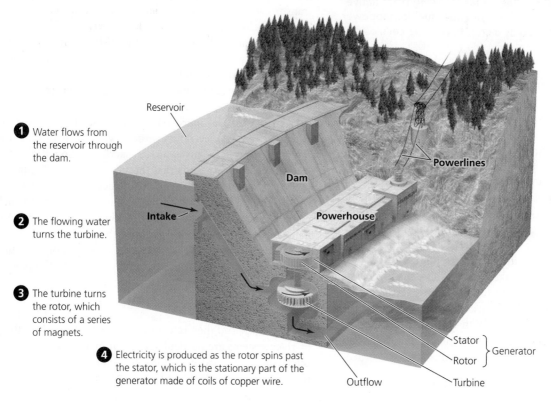

1 Water flows from the reservoir through the dam.

2 The flowing water turns the turbine.

3 The turbine turns the rotor, which consists of a series of magnets.

4 Electricity is produced as the rotor spins past the stator, which is the stationary part of the generator made of coils of copper wire.

Reservoir

Dam

Powerlines

Intake

Powerhouse

Stator ⎫
Rotor ⎬ Generator
Turbine

Outflow

(c) Hydroelectric power

584

FIGURE 20.23 ▲ Large dams **(a)** generate substantial amounts of hydroelectric power. Inside these dams, flowing water is used to turn turbines **(b)** and generate electricity. Water is funneled from the reservoir through a portion of the dam **(c)** to rotate turbines, which turn rotors containing magnets. The spinning rotors generate electricity as their magnets pass coils of copper wire. Electrical current is transmitted away through power lines, and the river's water flows out through the base of the dam.

The great age of dam building for hydroelectric power (as well as for flood control and irrigation) began in the 1930s in the United States, when the federal government constructed dams as public projects, partly to employ people and help end the economic depression of the time. U.S. dam construction peaked in 1960, when 3,123 dams were completed in a single year. American engineers subsequently exported their dam-building technologies, notably to nations of the developing world. In India, Prime Minister Jawaharlal Nehru commented in 1963 that "dams are the temples of modern India," referring to their central importance in the development of energy capacity in that nation.

Hydropower is clean and renewable

For producing electricity, hydropower has two clear advantages over fossil fuels. First, it is renewable; as long as precipitation falls from the sky and fills rivers and reservoirs, we can use water to turn turbines.

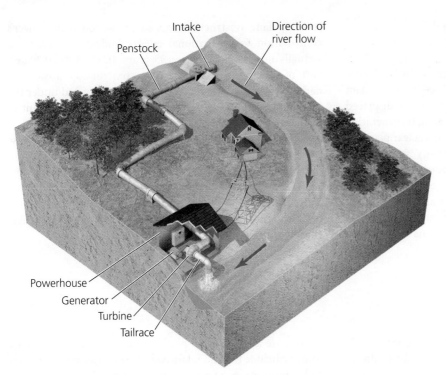

Penstock
Intake
Direction of river flow
Powerhouse
Generator
Turbine
Tailrace

FIGURE 20.24 ◄ Run-of-river systems divert a portion of a river's water and can be designed in various ways. Some designs, such as the one shown here, involve piping water downhill through a powerhouse and releasing it downriver, and some involve using water as it flows over shallow dams.

Second, no carbon compounds are burned in the production of hydropower, so no carbon dioxide or other pollutants are emitted into the atmosphere, which helps safeguard air quality, climate, and human health. Of course, fossil fuels *are* used in constructing and maintaining dams—and recent evidence indicates that large reservoirs release the greenhouse gas methane as a result of anaerobic decay in deep water. But overall, hydropower accounts for only a small fraction of the greenhouse gas emissions typical of fossil fuel combustion.

In addition, hydropower is efficient, for it is thought to have an EROI (energy returned on investment) ratio of 10:1 or more, at least as high as any other modern-day energy source. Fossil fuels had higher EROI values in the past, but as it has become more expensive to reach remaining deposits, their EROI values have dipped below those of hydropower.

Hydropower has negative impacts

Although it is renewable, efficient, and produces little air pollution, hydropower does create other impacts. Damming rivers destroys habitat for wildlife as riparian areas above dam sites are submerged and those below dam sites often are starved of water. Because water discharge is regulated to optimize electricity generation, the natural flooding cycles of rivers are disrupted. Suppressing flooding prevents river floodplains from receiving fresh nutrient-laden sediments. Instead, sediments become trapped behind dams, where they begin filling the reservoir.

Dams also cause thermal pollution (p. 421), because water downstream may become unusually warm if water levels are kept unnaturally shallow. Moreover, periodic flushes of cold water occur from the release of reservoir water. Such thermal shocks, together with habitat alteration, have diminished or eliminated many native fish populations in dammed waterways throughout the world. In addition, dams generally block the passage of fish and other aquatic creatures, effectively fragmenting the river and reducing biodiversity in each stretch.

The weight of water in a large reservoir has also been known to cause geologic impacts such as earthquakes, particularly where reservoir water seeps into fractures in the bedrock. And dams do collapse occasionally, sometimes as a result of earthquakes or landslides, resulting in deaths among the people downstream.

All these ecological impacts generally translate into negative social and economic impacts on local communities. We discussed the environmental, economic, and social impacts of dams, and their advantages and disadvantages, more fully in Chapter 15 (pp. 411–413).

TABLE 20.3 Top Producers of Hydropower

Nation	Hydropower produced (terawatt-hours)	Percentage of electricity generation from hydropower
China	485	14.8
Brazil	374	84.0
Canada	369	57.6
United States	276	6.3
Russia	179	17.6
Norway	135	98.2
India	124	15.4
Japan	84	7.4
Venezuela	83	72.3
Sweden	66	44.5
Rest of world	987	13.9

Data is for 2007, from the International Energy Agency.

Hydropower may not expand much more

Today the world is witnessing some gargantuan hydroelectric projects. China's recently completed Three Gorges Dam (pp. 412–413) is the world's largest. Its reservoir displaced over 1 million people, and the dam should soon be generating as much electricity as dozens of coal-fired or nuclear plants.

However, unlike other renewable energy sources, hydropower is not likely to expand much more. One reason is that, as in Sweden, most of the world's large rivers that offer excellent opportunities for hydropower are already dammed. Another reason is that people have grown more aware of the ecological impacts of dams, and in some regions residents are resisting dam construction. Indeed, in Sweden, hydropower's contribution to the national energy budget has remained virtually unchanged for over 35 years. In the United States, 98% of rivers appropriate for dam construction already are dammed, many of the remaining 2% are protected under the Wild and Scenic Rivers Act, and many people now want to dismantle some dams and restore river habitats (p. 413).

Overall, hydropower will likely continue to increase in developing nations that have yet to dam their rivers, but in developed nations hydropower growth will likely slow or stop. The International Energy Agency (IEA) forecasts that hydropower's share of electricity generation will decline between now and 2030, whereas the share of other renewable energy sources will triple, from 2% to 6%.

➤ CONCLUSION

Given limited supplies of fossil fuels and their considerable environmental impacts, many nations have sought to diversify their energy portfolios with alternative energy sources. The three most developed and widely used alternatives so far are nuclear power, bioenergy, and hydropower.

Nuclear power showed promise at the outset to be a pollution-free and highly efficient form of energy. But high costs and public fears over safety in the wake of accidents at Chernobyl and Three Mile Island stalled its growth, and a few nations are attempting to phase it out completely. Bioenergy sources include traditional fuelwood, as well as newer biofuels and various means of generating biopower. These sources can be carbon-neutral but are not all strictly renewable. Hydropower is a renewable, pollution-free alternative, but it is nearing its maximal use and can involve substantial ecological impacts. Although some nations, such as Sweden, already rely heavily on these three conventional alternatives, it appears that we will need further renewable sources of energy.

REVIEWING OBJECTIVES

You should now be able to:

DISCUSS THE REASONS FOR SEEKING ENERGY ALTERNATIVES TO FOSSIL FUELS

- Fossil fuels are nonrenewable resources, and we are gradually depleting them. (p. 563)
- Fossil fuel combustion causes air pollution that results in many environmental and health impacts and contributes to global climate change. (p. 563)

SUMMARIZE THE CONTRIBUTIONS TO WORLD ENERGY SUPPLIES OF CONVENTIONAL ALTERNATIVES TO FOSSIL FUELS

- "Conventional energy alternatives" are the alternatives to fossil fuels that are most widely used. They include nuclear power, bioenergy, and hydroelectric power. In their renewability and environmental impact, these sources fall between fossil fuels and the less widely used "new renewable" sources. (p. 564)
- Biomass provides 10% of global primary energy use, nuclear power provides 6%, and hydropower provides 2%. (pp. 563–564)
- Nuclear power generates 14% of the world's electricity, and hydropower generates 16%. (pp. 563–564)

DESCRIBE NUCLEAR ENERGY AND HOW IT IS HARNESSED

- Nuclear power comes from converting the energy of subatomic bonds into thermal energy, using uranium isotopes. (pp. 564–565)
- Uranium is mined, enriched, processed into pellets and fuel rods, and used in nuclear reactors. (pp. 565–566)
- By controlling the reaction rate of nuclear fission, nuclear power plant engineers produce heat that powers electricity generation. (pp. 565–566)

OUTLINE THE SOCIETAL DEBATE OVER NUCLEAR POWER

- Many advocates of clean energy support nuclear power because it does not emit the pollutants that fossil fuels do. (pp. 567–569)
- For many people, the risk of a major power plant accident, such as the one at Chernobyl, outweighs the benefits of clean energy. (pp. 570–573)
- The disposal of nuclear waste remains a major dilemma. Temporary storage and single-repository plans each involve health, security, and environmental risks. (pp. 573–575)
- Economic factors and cost overruns have slowed the nuclear industry's growth. (p. 576)

DESCRIBE THE MAJOR SOURCES, SCALE, AND IMPACTS OF BIOENERGY

- Bioenergy theoretically is renewable and adds no net carbon to the atmosphere. But overharvesting of wood can lead to deforestation, and growing crops solely for fuel production can be inefficient and ecologically damaging. (pp. 576–583)
- Fuelwood remains the major source of bioenergy today, especially in developing nations. (p. 577)
- We use biomass to generate electrical power (biopower) in several ways. Sources include special crops as well as waste products from agriculture and forestry. (pp. 577–579)
- Biofuels, including ethanol and biodiesel, are used to power automobiles. Some crops are grown specifically for this purpose, and waste oils are also used. (pp. 579–582)

- Next-generation biofuels such as algae and cellulosic ethanol need further research and development, but hold promise for more sustainable biofuel production. (p. 582)

DESCRIBE THE SCALE, METHODS, AND IMPACTS OF HYDROELECTRIC POWER

- Hydroelectric power is generated when water from a river runs through a powerhouse and turns turbines. (pp. 583–585)
- Dams and reservoirs and run-of-river systems are alternative approaches. (pp. 583–585)
- Hydropower produces little air pollution, but dams and reservoirs can greatly alter riverine ecology and local economies. (pp. 584–585)

TESTING YOUR COMPREHENSION

1. How much of our global energy supply do nuclear power, bioenergy, and hydroelectric power contribute? How much of our global electricity do these three conventional energy alternatives generate?

2. Describe how nuclear fission works. How do nuclear plant engineers control fission and prevent a runaway chain reaction?

3. In terms of greenhouse gas emissions, how does nuclear power compare to coal, oil, and natural gas? How do hydropower and bioenergy compare?

4. In what ways did the incident at Three Mile Island differ from that at Chernobyl? What consequences resulted from each of these incidents?

5. List several concerns about the disposal of radioactive waste. What has been done so far about disposing of radioactive waste?

6. List five sources of bioenergy. What is the world's most-used source of bioenergy? How does bioenergy use differ between developed and developing nations?

7. Describe two biofuels, where each comes from, and how each is used.

8. Evaluate two potential benefits and two potential drawbacks of bioenergy.

9. Contrast two major approaches to generating hydroelectric power.

10. Assess two benefits and two negative environmental impacts of hydroelectric power.

SEEKING SOLUTIONS

1. Given what you learned about fossil fuels in Chapter 19 and about some of the conventional alternatives discussed in this chapter, do you think it is important for us to minimize our use of fossil fuels and maximize our use of alternatives? If such a shift were to require massive public subsidies, how much should we subsidize this? What challenges or obstacles would we need to overcome in order to shift to such alternatives?

2. Nuclear power has by now been widely used for over four decades, and the world has experienced only one major accident (Chernobyl) responsible for any significant number of injuries or deaths. Would you call this a good safety record? Should we maintain, decrease, or increase our reliance on nuclear power? Why might safety at nuclear power plants be better in the future? Why might it be worse?

3. How serious a problem do you think the disposal of radioactive waste represents? How would you like to see this issue addressed?

4. There are many different sources of biomass and many ways of harnessing energy from biomass. Discuss one that seems particularly beneficial to you, and one with which you see problems. What bioenergy sources and strategies do you think our society should focus on investing in?

5. **THINK IT THROUGH** Imagine that you are the head of the national department of energy in a country that has just experienced a minor accident at one of its nuclear plants. A partial meltdown released radiation, but the radiation was fully contained inside the containment building, and there were no health impacts on area residents. However, citizens are terrified, and the media is playing up the dangers of nuclear power. Your country relies on its five nuclear plants for 25% of its energy and 50% of its electricity needs. It has no fossil fuel deposits and recently began a promising but still-young program to develop renewable energy options. What

will you tell the public at your next press conference, and what policy steps will you recommend taking to ensure a safe and reliable national energy supply?

6. **THINK IT THROUGH** You are an investor and would like to invest in alternative energy. You are considering buying stock in companies that (1) construct nuclear reactors, (2) build turbines for hydroelectric dams, (3) build corn ethanol refineries, (4) are ready to supply farm and forestry waste for cellulosic ethanol, and (5) are developing algae farms. What questions would you research about each of these companies before deciding how to invest your money? How do you expect you might apportion your investments, and why? Do you expect there may be ways in which you will feel torn between doing what seems financially wise and what seems best for the environment or for energy security? Explain.

CALCULATING ECOLOGICAL FOOTPRINTS

Each of the conventional energy alternatives releases considerably less net carbon dioxide (CO_2) to the atmosphere than do any of the fossil fuels.

For each fuel source in the table, calculate the net greenhouse gas emissions it produces while providing electricity to a typical U.S. household that uses 30 kilowatt-hours per day. Use data on emissions rates from each of the energy sources as provided in the figure from **The Science behind the Story** on p. 569. For each energy source, use the average of the maximum and minimum values given.

Energy source	Greenhouse gas emission rate (g C_{eq}/kWh)	Greenhouse gas emissions per day (g C_{eq})
Coal	311	9,330
Oil		
Natural gas		
Photovoltaic solar		
Hydroelectric		
Biomass		
Wind		
Nuclear		

1. According to the EPA, one acre of pine or fir forest sequesters 1,280,000 grams of carbon emissions each year. How much forest would be required to compensate for the emissions of a typical household using coal-fired power for one year? Using hydropower for one year? Using nuclear power for one year?

2. What is the ratio of greenhouse emissions from fossil fuels, on average, to greenhouse emissions from alternative energy sources, on average?

3. Which energy source has the lowest emission rate? Would you advocate that your nation further develop that source? Why or why not?

Mastering ENVIRONMENTAL SCIENCE™

Go to **www.masteringenvironmentalscience.com** for practice quizzes, Pearson eText, videos, current events, and more.

A wind turbine is reflected in solar panels near Mainz, Germany

21 NEW RENEWABLE ENERGY ALTERNATIVES

UPON COMPLETING THIS CHAPTER, YOU WILL BE ABLE TO:

- Outline the major sources of renewable energy and assess their potential for growth
- Describe solar energy and the ways it is harnessed, and evaluate its advantages and disadvantages
- Describe wind power and how we harness it, and evaluate its benefits and drawbacks

- Describe geothermal energy and the ways we make use of it, and assess its advantages and disadvantages
- Describe ocean energy sources and how we could harness them
- Explain hydrogen fuel cells and weigh options for energy storage and transportation

Germany Goes Solar

"Someday we will harness the rise and fall of the tides and imprison the rays of the sun."

—Thomas A. Edison, 1921

"[Renewable energy] will provide millions of new jobs. It will halt global warming. It will create a more fair and just world. It will clean our environment and make our lives healthier."

—Hermann Scheer, energy expert and German parliament member, 2009

RUSSIA

GERMANY

Atlantic
Ocean   EUROPE

AFRICA

When we think of solar energy, most of us envision a warm sunny place like Arizona or southern California. Yet the country that produces the most solar power in the world is Germany, a European nation at the latitude of Canada with a climate like Maine's. Germany is the world's top user of photovoltaic (PV) solar technology, which produces electricity from sunshine. In recent years Germany has installed nearly half the world's total of this technology. Germany now obtains more of its energy from solar power than any other nation, and the amount grows each year.

How is this happening in such a cool and cloudy country? A bold federal policy is using economic incentives to promote solar power and other forms of renewable energy. Germany has a **feed-in tariff** system whereby utilities are mandated to buy power from anyone who can generate power from renewable energy sources and feed it into the electric grid. Under this system, utilities must pay guaranteed premium prices for this power under long-term contract. As a result, German homeowners and businesses have rushed to install more and more PV panels each year, and they are selling their extra solar power to the utilities at a profit.

Homes in the Vauban neighborhood of Freiburg, Germany, which produce more solar power than they use, and sell it to the grid

It's not just solar—the feed-in tariffs apply to all forms of renewable energy. As a result of these incentives, Germany established itself as the world leader in wind power in the 1990s until being overtaken by the United States in 2008 and China in 2009. It ranks fourth in the world in using power from biomass, and third in solar water heating. In overall electrical power capacity from renewable sources, Germany ranks third in the world, trailing only China and the United States, which have far more people and businesses. Well over 1 million German households and 60,000 businesses purchase renewable power from their utilities, more than any other nation.

Boosted by domestic demand, German industries have become global leaders in "green tech," designing and selling renewable energy technologies around the world. Germany is second in PV production behind China, leads the world in production of biodiesel, and has recently developed several cellulosic ethanol (p. 582) facilities. Renewable energy industries in Germany today employ over 300,000 citizens.

Germany's push for renewable energy goes back to 1990. In the wake of the Chernobyl nuclear disaster (pp. 570–573), Germany decided to phase out its nuclear power plants. However, by shutting these down, the nation would lose virtually all its clean energy. With few domestic fossil fuel supplies, Germans would find their economy utterly dependent on oil, gas, and coal imported from producers in Russia and the Middle East.

Enter Hermann Scheer, a member of the German parliament and an expert on renewable energy. While everyone else assumed technologies like solar, wind, and geothermal energy were costly, risky, and not ready for prime time, Scheer saw them as a great economic opportunity—and as the only long-term answer. In 1990, Scheer helped push through a landmark law establishing feed-in tariffs. Ten years later, the law was revised and strengthened: The Renewable Energy Sources Act of 2000 aimed to promote renewable energy production and use, enhance the security of the energy supply, reduce carbon emissions, and lessen the many external costs (pp. 150–151) of fossil fuel use.

In 2004 and in 2009 the government revised the law, adjusting the amounts utilities were required to pay homeowners and businesses for their energy production. Each renewable source is assigned its own payment rate according to market considerations, and most rates are reduced year-by-year in order to encourage increasingly efficient means of producing power. Then in 2010 the German government slashed PV solar tariff rates by an additional 16%, because many Germans felt the feed-in-tariffs were less necessary in view of the fact that they had already helped PV market prices to fall by as much as 40%. In response, sales of PV modules skyrocketed in late 2009 and early 2010 as Germans rushed to lock in the old rates. In 2009 alone, Germans installed 3.8 gigawatts of PV solar capacity—more than twice the total cumulative capacity of the United States—and experts predicted still-greater installations in 2010.

The 2009 version of the law also set targets to meet, specifying that Germany should obtain 30% of its electricity and 14% of its heating energy from renewable sources by 2020. To back all this up, the German government has been allotting more public money to renewable energy than any other nation—close to $20 billion annually in recent years.

By developing renewable energy to replace some of its fossil fuel use, Germany has decreased its emissions of carbon dioxide by 122 million tons per year—equal to taking 21 million cars off the road. Nearly half of this total is due to energy paid for under the feed-in tariff system. Since 1990, carbon dioxide emissions from German energy sources have fallen by over 20%, and emissions of seven other major pollutants (CH_4, N_2O, SO_2, NO_x, CO, VOCs, and dust) have all been reduced by 12–95%.

Reducing its greenhouse gas emissions enables Germany to wield political clout in international climate negotiations. It can boast the greatest emissions reductions of any nation under the Kyoto Protocol. Most of this reduction was due to modernizing polluting industries in the former East Germany following reunification, but the move toward renewable energy also has played a role.

Germany's success is serving as a model for other nations in Europe and around the world. As of 2010, more than 60 nations had implemented some sort of feed-in tariffs. These policies are having effects similar to those in Germany; for instance, Spain ignited its wind and solar development as a result. In North America in 2009, Vermont and Ontario established feed-in tariff systems similar to Germany's, while California and Washington conduct more-limited programs. In 2010, Gainesville, Florida, became the first U.S. city to establish feed-in tariffs. In addition, utilities in 46 U.S. states now offer **net-metering**, in which utilities credit customers who produce renewable power and feed it into the grid. As more nations, states, and cities develop policies to encourage renewable energy, we may soon experience a historic transition in the way we meet our energy demands.

"NEW" RENEWABLE ENERGY SOURCES

Germany's bold support of solar and wind power is just one facet of a global shift toward renewable energy. Across the world, nations are searching for ways to move away from fossil fuels while ensuring a reliable and affordable supply of energy for their economies. This is because the economic and social costs, security risks, and health and environmental impacts of fossil fuel dependence (Chapter 19) are all continuing to intensify.

In Chapter 20 we explored the two renewable energy sources that are most developed and widely used: biomass energy, the energy from combustion of wood and other plant matter, and hydropower, the energy from running water. These "conventional" alternatives to fossil fuels are renewable, but they can be depleted with overuse and they exert some undesirable environmental impacts.

In this chapter we explore a group of alternative energy sources that are often called "new renewables." These include energy from the sun, from wind, from Earth's geothermal heat, and from ocean water. These energy sources are not truly new. In fact, they are as old as our planet, and people have used them for millennia. People commonly refer to them as "new" because (1) they are not yet used on a wide scale in our modern industrial society, (2) they are harnessed using technologies that are still in a rapid phase of development, and (3) they will likely come to play a much larger role in our energy use in the future.

New renewable sources currently provide little of our energy

Like the other energy sources we have discussed, the new renewables can provide energy for three types of applications: (1) power for electricity, (2) heating of air or water, and (3) fuel for vehicles. However, today we obtain less than 1% of our global energy from the new renewable energy sources. Fossil fuels provide 81% of the world's energy, nuclear power provides 6%, and renewable energy sources account for 13%, nearly all of which is provided by biomass and hydropower (see Figure 20.1a, p. 563). The new renewables make a similarly small contribution to our global generation of electricity (see Figure 20.1b, p. 563). Only 18% of our electricity comes from renewable energy, and of this amount, hydropower accounts for the vast majority.

Nations and regions vary in the renewable sources they use. In the United States, most renewable energy comes from biomass and hydropower. As of 2009, wind power accounted for 9.0%, geothermal energy for 4.8%, and solar energy for 1.4% (**FIGURE 21.1A**). Of electricity generated in the United States from renewables, hydropower accounts for 65%, and biomass 10%. As of 2009, wind power accounted for 17%, geothermal for 8%, and solar for just 0.2% (**FIGURE 21.1B**).

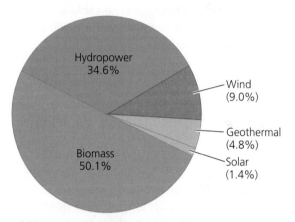

(a) U.S. consumption of renewable energy, by source

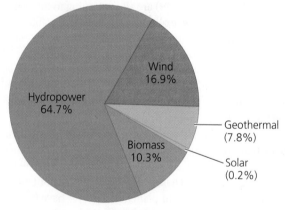

(b) U.S. electricity generation from renewable sources

FIGURE 21.1 ▲ Only 8% of the energy consumed in the United States each year comes from renewable sources. Of this amount **(a)**, most derives from biomass energy and hydropower. Wind power, geothermal energy, and solar energy together account for just 15% of this amount. Similarly, only 10% of electricity generated in the United States **(b)** comes from renewable energy sources, predominantly hydropower. Data are for 2009, from Energy Information Administration, U.S. Department of Energy, 2010. *Annual energy review, 2009.*

The new renewables are growing fast

Although they comprise a minuscule proportion of our energy budget, the new renewable energy sources are growing at much faster rates than are conventional energy sources. Over the past four decades, solar, wind, and geothermal energy sources have grown far faster than has the overall energy supply (**FIGURE 21.2**). The leader in growth is wind power, which has expanded by nearly 50% *each year* since the 1970s. Because these sources started from such low levels of use, however, it will take them some time to catch up to conventional sources. The absolute amount of energy added by a 50% increase in wind power today equals the amount added by just a 1% increase in oil, coal, or natural gas. In fact, 2008 was the first year in which our global society added more energy capacity from renewable energy sources than from fossil fuels and nuclear power.

Use of new renewables has been expanding quickly because of growing concerns over diminishing fossil fuel supplies (which cause both high prices and national security concerns) and because of the environmental and health impacts of fossil fuel combustion (Chapter 19). Advances in technology are also making it easier and less expensive to harness renewable energy sources.

The new renewables promise several benefits over fossil fuels. For one, they help alleviate air pollution and the greenhouse gas emissions that drive global climate change (Chapter 18). For another, renewable sources are inexhaustible on any time scale relevant to our society. Moreover, developing renewables can diversify an economy's mix of energy, helping to reduce price volatility and buffer us against supply restrictions. Finally, novel energy sources can create new employment opportunities and sources of income and property tax for communities, especially in rural areas passed over by other economic development. New and developing technologies generally require more labor per unit of energy output than do established technologies, so shifting to

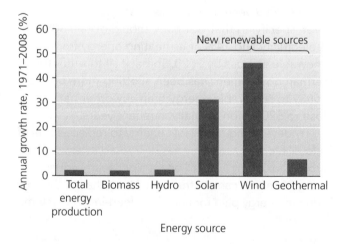

FIGURE 21.2 ▲ Globally, the "new renewable" energy sources are growing substantially faster than the total primary energy supply. Solar power has grown by 31% each year since 1971, and wind power has grown by 46% each year. Because these sources began from such low starting levels, however, their overall contribution to our energy supply remains small. Data from International Energy Agency Statistics.

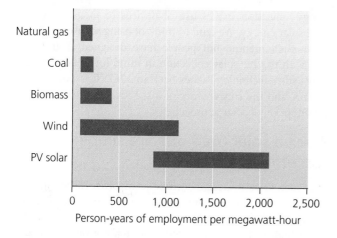

FIGURE 21.3 ▲ Manufacturing, installing, servicing, and operating new renewable energy technologies requires more labor than for established fossil fuel technologies, so shifting to renewable energy should increase employment by creating "green-collar jobs." Renewable energy sources such as photovoltaic (PV) solar, wind, and biomass tend to support more jobs than natural gas and coal. Data from Worldwatch Institute and Center for American Progress, 2006. *American energy: The renewable path to energy security.* p. 10. www.worldwatch.org. By permission.

renewable energy should generate employment (**FIGURE 21.3**). The design, installation, maintenance, and management required to develop technologies and rebuild and operate our society's energy infrastructure promises to be a major source of employment for young people today, through so-called **green-collar jobs**.

Rapid growth in renewable energy sectors seems likely to continue as population and consumption grow, global energy demand expands, fossil fuel supplies decline, and people demand cleaner environments. Renewable energy efforts received support when many national governments responded to the global financial downturn of 2008–2009 by enacting stimulus packages and boosting spending on green energy programs to help create jobs. As more governments, utilities, corporations, and consumers promote and use renewable energy, prices of renewables should continue to fall, further hastening their adoption.

Policy can accelerate our transition

If our civilization is to persist in the long term, it will need to shift to renewable energy sources. A key question is whether we will be able to shift soon enough and smoothly enough to avoid both (1) cataclysmic damage to our environment as a result of continued fossil fuel combustion and (2) economic depression, social unrest, and war as affordable supplies of fossil fuels diminish. The answer to this question will largely determine the quality of life for all of us in the coming decades.

We cannot switch completely to renewable energy sources overnight, because there are technological and economic barriers. Currently, most renewables lack adequate technology and infrastructure to transfer power on the required scale. However, rapid advances in recent years suggest that most remaining barriers are political. Renewable energy sources have received far less in subsidies and tax breaks from governments than have conventional sources. In the United States over the

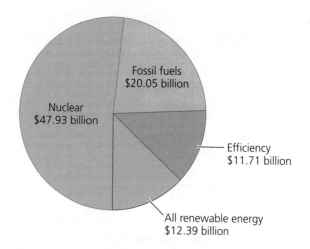

FIGURE 21.4 ▲ Most U.S. research and development funding for energy has gone toward nuclear power and fossil fuels. Between 1974 and 2005, only about 13% went toward all renewable energy sources combined. Data from International Energy Agency.

past three decades, renewable energy sources have been granted just one-sixth the public funding for research and development that nuclear energy and fossil fuels have received (**FIGURE 21.4**). Research and development of renewable sources have gone underfunded because fossil fuels continue to be available at inexpensive prices, even though these low prices are enabled in part by government policy that responds to lobbying from fossil fuel interests.

Most corporations in the fossil fuel, automobile, and transportation industries understand that they will eventually need to switch to renewable sources. They also know that when the time comes, they will need to act fast to stay ahead of their competitors. However, in light of continuing short-term profits and unclear policy signals, business and industry have not been eager to invest in shifting from fossil fuels to renewable energy. Under these circumstances, our best hope may be for a gradual shift driven largely by economic supply and demand. However, if we wait for market forces alone to drive the transition, while government continues to favor fossil fuels with subsidies, then renewable sources might not be fully ready in time to replace dwindling fossil fuel supplies or to avert severe climate change. Thus, encouraging the speedy development of renewable energy alternatives holds promise for bringing us a vigorous and sustainable energy economy without the environmental impacts of fossil fuels.

Germany's feed-in tariff policy—a prime example of an economic policy tool (p. 181)—would seem to be a good model. After intensive study, the European Commission concluded in 2008 that "well-adapted feed-in tariff regimes are generally the most efficient and effective support schemes for promoting renewable energy." Numerous other agencies and experts have concurred.

At the same time, we are learning lessons about how to apply these policies. Spain adopted feed-in tariffs that were so generous (58 cents per kilowatt-hour) that thousands of people rushed to set up solar facilities. Within a single year (2008), Spain increased its solar capacity fivefold, accounting for half the solar power installed worldwide that year, and becoming second to Germany in capacity. However, the generous payments attracted a flood of unqualified contractors and speculators, and many of the hastily built solar plants were poorly designed or badly located. The government realized that many of these new plants would never be economically self-sufficient, so it abruptly slashed the tariff rates. As more nations begin to use economic incentives to achieve policy goals, and to monitor what works and what does not, we may finally be able to safely and profitably accelerate our societal transition to renewable energy (see **ENVISIONIT**, p. 598).

SOLAR ENERGY

The sun releases astounding amounts of energy by converting hydrogen to helium through nuclear fusion (pp. 568–569). The tiny proportion of this energy that reaches Earth is enough to drive most of the processes in the biosphere, helping to make life possible on our planet. Each day in total, Earth receives enough energy from the sun to power human consumption for a quarter of a century. On average, each square meter of Earth's surface receives about 1 kilowatt of **solar energy**, or energy from the sun—17 times the energy of a light bulb. As a result, a typical home has enough roof area to meet all its power needs with rooftop panels that harness solar energy. However, we are still in the process of developing solar technologies and learning the most effective and cost-efficient ways to put the sun's energy to use.

The most commonly used way to harness solar energy is through **passive solar** energy collection. In this approach, buildings are designed and building materials are chosen to maximize absorption of sunlight in winter and keep the interior cool in the heat of summer. In contrast, **active solar**

energy collection makes use of devices to focus, move, or store solar energy. We tend to think of using solar energy as a novel phenomenon, but people have designed their living sites with passive solar collection in mind for millennia, and active solar technology dates back to 1767. It is policy, economics, and the convenience of fossil fuels that have limited our adoption of solar approaches.

Passive solar methods are simple and effective

One passive solar design technique involves installing low, south-facing windows to maximize the capture of sunlight in winter. Overhangs shade these windows in summer, when the sun is high in the sky and when cooling, not heating, is desired. Passive solar techniques also include the use of construction materials that absorb heat, store it, and release it later. Such *thermal mass* (of straw, brick, concrete, or other materials) often makes up floors, roofs, and walls, or can be used in portable blocks. Planting vegetation around a building to buffer the structure from temperature swings is another passive solar approach. By heating buildings in cold weather and cooling them in warm weather, passive solar methods conserve energy and reduce energy costs.

Active solar methods can heat air and water in buildings

We can use various active solar technologies to heat water and air in our homes and businesses. One common method involves installing *flat-plate solar collectors* on rooftops. These panels generally consist of dark-colored, heat-absorbing metal plates mounted in flat glass-covered boxes. Water, air, or antifreeze runs through tubes that pass through the collectors, transferring heat from the collectors to the building or its water tank (**FIGURE 21.5**). Heated water can be stored for later use and passed through pipes designed to release the heat into the building.

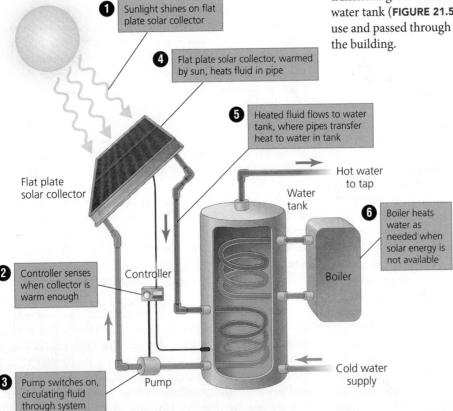

1 Sunlight shines on flat plate solar collector

4 Flat plate solar collector, warmed by sun, heats fluid in pipe

5 Heated fluid flows to water tank, where pipes transfer heat to water in tank

Hot water to tap

Flat plate solar collector

Water tank

6 Boiler heats water as needed when solar energy is not available

Boiler

2 Controller senses when collector is warm enough

Controller

3 Pump switches on, circulating fluid through system

Pump

Cold water supply

FIGURE 21.5 ◀ Solar systems for heating water vary in their designs, but typically **1** sunlight is gathered on a flat-plate solar collector, until a controller **2** switches on a pump **3** to circulate fluid through pipes to the collector. The sunlit collector heats the fluid **4**, which flows through pipes **5** to a water tank. The hot fluid in the pipes transfers heat to the water in the tank, and this heated water is available for the taps of the home or business. Generally an external boiler **6** kicks in to heat water when solar energy is not available.

Over 1.5 million homes and businesses in the United States heat water with solar panels, although most of this is water for swimming pools. Active solar heating is used more widely in China and in Europe, where Germans motivated by feed-in tariffs installed 200,000 new systems in 2008 alone. Because solar heating systems do not need to be connected to an electrical grid, they are useful in isolated locations. In Gaviotas, a remote town on the high plains of Colombia in South America, residents use active solar technology for heating, cooling, and water purification. This shows that solar energy need not be confined to wealthy communities or to regions that are always sunny.

Concentrating solar rays magnifies energy

We can magnify the intensity of solar energy by gathering sunlight from a wide area and focusing it on a single point. This is the principle behind *solar cookers*, simple portable ovens that use reflectors to focus sunlight onto food and cook it (**FIGURE 21.6A**). Such cookers are proving extremely useful in the developing world.

At much larger scales, utilities are using the principle behind solar cookers to generate electricity. **Concentrated solar power (CSP)** is being harnessed by several methods (**FIGURE 21.6B**) in sunny regions in Spain, the U.S. Southwest, and elsewhere. The dominant technology so far is the trough

approach (leftmost illustration in Figure 21.6B), in which curved mirrors focus sunlight on synthetic oil in pipes. The superheated oil is piped to an adjacent facility where it creates steam that drives turbines to generate electricity as in conventional power plants. In another approach, numerous mirrors concentrate sunlight onto a receiver atop a tall "power-tower" (**FIGURE 21.6C**). From this central receiver, heat is transported by air or fluids (often molten salts) and piped to a steam-driven generator to create electricity. CSP facilities can harness light from lenses or mirrors spread across large areas of land, and the lenses or mirrors may move to track the sun's movement across the sky.

CSP facilities need to be located in sunny areas, but they have great potential. The International Energy Agency estimates that just 260 km^2 (100 mi^2) of Nevada desert could generate enough electricity to power the entire U.S. economy. Another study estimated that CSP could fulfill one-quarter of global electricity demand by 2050 if we step up investment. Currently, German industrialists and investors are spearheading an ambitious effort to create an immense CSP facility in the Sahara Desert. In this planned $775 billion project, called Desertec, thousands of mirrors spread across vast areas of desert in Morocco would harness the Sahara's abundant sunlight and transmit electricity to Europe, the Middle East, and North Africa. Backers of the controversial project say it could one day supply 15% of Europe's electricity. Critics—including Hermann Scheer—call it "a mirage," saying the project would

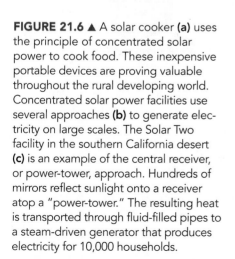

(a) Solar cooker in India

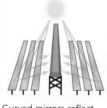

Curved reflectors heat liquid in horizontal tubes

Each curved reflector focuses light on its own small receiver

Curved mirrors reflect light onto absorber tube

Field of heliostats focus light on central power tower

(b) Four methods of concentrating solar power

FIGURE 21.6 ▲ A solar cooker (a) uses the principle of concentrated solar power to cook food. These inexpensive portable devices are proving valuable throughout the rural developing world. Concentrated solar power facilities use several approaches (b) to generate electricity on large scales. The Solar Two facility in the southern California desert (c) is an example of the central receiver, or power-tower, approach. Hundreds of mirrors reflect sunlight onto a receiver atop a "power-tower." The resulting heat is transported through fluid-filled pipes to a steam-driven generator that produces electricity for 10,000 households.

(c) The Solar Two power tower facility in California

be vulnerable to sandstorms and political disputes and would be less reliable and more expensive than the decentralized production from solar panels that feed-in tariffs are already promoting in Europe.

Photovoltaic cells generate electricity directly

The most direct way to produce electricity from sunlight involves photovoltaic (PV) systems. **Photovoltaic (PV) cells** convert sunlight to electrical energy by making use of the *photovoltaic effect*, or *photoelectric effect*. This effect occurs when light reaches the PV cell and strikes one of a pair of plates made primarily of silicon, a semiconductor that conducts electricity. The light causes one plate to release electrons, which are attracted by electrostatic forces to the opposing plate. Connecting the two plates with wires enables the electrons to flow back to the original plate, creating an electrical current (direct current, DC), which can be converted into alternating current (AC) and used for residential and commercial electrical power (**FIGURE 21.7**). Small PV cells may already power your watch or your calculator. Atop the roofs of homes and other buildings, PV cells are arranged

in modules, which comprise panels, which can be gathered together in arrays.

Researchers are experimenting with further variations on PV technology, and manufacturers today are already developing **thin-film solar cells**, photovoltaic materials compressed into ultra-thin sheets. Thin-film solar cells are lightweight and far less bulky than the standard crystalline silicon cells shown in Figure 21.7. Although less efficient at converting sunlight to electricity, they are cheaper to produce. Thin-film technologies can be incorporated into roofing shingles and potentially many other types of surfaces, even highways! For all these reasons, many people view thin-film solar technologies as a promising opportunity for the future.

Photovoltaic cells of all types can be connected to batteries that store the accumulated charge until it is needed. Or, producers of PV electricity can sell their power to their local utility if they are connected to the regional electric grid. In parts of 46 U.S. states, homeowners can sell power to their utility in a process called **net-metering**, in which the value of the power the consumer provides is subtracted from the consumer's monthly utility bill. Feed-in tariff systems like Germany's go a step further by paying producers more than

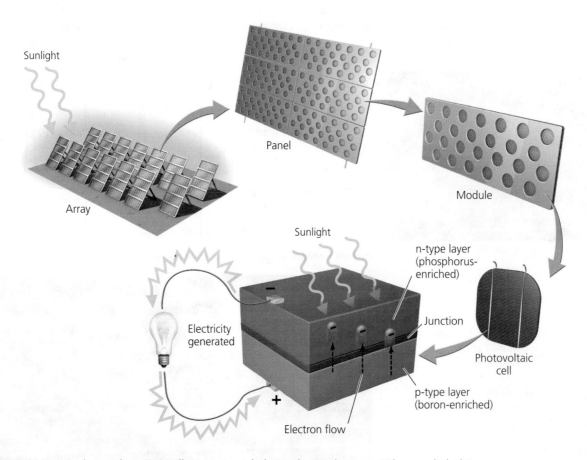

FIGURE 21.7 ▲ A photovoltaic (PV) cell converts sunlight to electrical energy. When sunlight hits the silicon layers of the cell, electrons are knocked loose from some of the silicon atoms and tend to move from the boron-enriched "p-type" layer toward the phosphorus-enriched "n-type" layer. Connecting the two layers with wiring remedies this imbalance as electrical current flows from the n-type layer back to the p-type layer. This direct current (DC) is converted to alternating current (AC) to produce usable electricity. PV cells are grouped in modules, which comprise panels, which can be erected in arrays.

the market price of the power. Feed-in tariffs thereby generally offer power-producers the hope of turning a profit.

Solar energy is fast growing

Although active solar technology dates from the 18th century, it was pushed to the sidelines as fossil fuels came to dominate our energy economy. Funding for research and development of solar technology has been erratic. After the 1973 oil embargo (p. 552), the U.S. Department of Energy funded the installation and testing of over 3,000 PV systems, providing a boost to companies in the solar industry. But later, as oil prices declined, so did government support for solar power, and funding for solar has remained far below that for fossil fuels.

Largely because of the lack of investment, solar energy contributes only a minuscule portion of today's energy production. In 2009, solar accounted for only 0.012%— 12 parts in 10,000—of the U.S. energy supply, and just 0.02% of U.S. electricity generation. Even in Germany, which gets more of its energy from solar than any other nation, its percentage is only 1.0%. However, solar energy use has grown by 31% annually worldwide since 1971, a growth rate second only to that of wind power. Solar energy is proving especially attractive in developing countries, many of which are rich in sun but poor in power infrastructure, and where hundreds of millions of people still live without electricity. Some multinational fossil fuel corporations are now investing in alternative energy as well. BP Solar, British Petroleum's solar energy wing, has sponsored multi-million-dollar projects to bring solar power to tens of thousands of homes in remote villages in Malaysia, Indonesia, and the Philippines.

PV technology is the fastest-growing power generation technology today, having recently doubled every two years (**FIGURE 21.8**). China leads the world in yearly production of PV cells, followed by Germany and Japan. Germany leads the world in installation of PV technology, and German rooftops host over half of all PV cells in the world. Its investment began in 1998 when Hermann Scheer spearheaded a "100,000 Rooftops" program to install PV panels atop 100,000 German roofs. The popular program ended up easily surpassing this goal.

The United States ranks fifth in production of PV cells. Recent federal tax credits and state-level initiatives may help the United States recover the leadership it lost to other nations in this technology, but China is moving faster and may soon dominate the market. The degree to which the United States lags in solar technology was highlighted when in 2007 the U.S. Department of Energy opened up its biannual Solar Decathlon house-building competition (p. 673) to foreign teams, and a German team promptly beat all the American contenders. The German team won again at the next contest in 2009.

As production of PV cells increases, prices are falling (see Figure 21.8). At the same time, efficiencies are increasing, making each unit more powerful. Some types of PV cells have doubled in efficiency during the period 1995–2008 shown in Figure 21.8. Throughout the world, use of solar technology should continue to increase as prices fall, technologies improve, and governments enact economic incentives to spur investment.

Solar energy offers many benefits

The fact that the sun will continue burning for another 4–5 billion years makes it practically inexhaustible as an energy source for human civilization. Moreover, the amount of solar energy reaching Earth should be enough to power our civilization once we develop technology adequate to harness it. These advantages of solar energy are clear, but the technologies themselves also provide benefits. PV cells and other solar technologies use no fuel, are quiet and safe, contain no moving parts, require little maintenance, and do not require a turbine or generator to create electricity. An average unit can produce energy for 20–30 years.

Solar systems also allow for local, decentralized control over power. Homes, businesses, and isolated communities can use solar power to produce electricity without being near a power plant or connected to a grid. This is especially helpful in developing nations, where solar cookers (see Figure 21.6a) enable families to cook food without gathering fuelwood. This lessens people's daily workload and helps reduce deforestation. In refugee camps, solar cookers are helping to relieve social and environmental stress. The low cost of solar cookers—many can be built locally for $2–10 each—has made them accessible in many impoverished areas. In developed nations, most PV systems are connected to the regional electric grid. As a result, homeowners with PV systems can sell their excess solar energy to their local utility through feed-in tariffs or net metering.

The development and deployment of solar systems is producing many new "green-collar" jobs. Currently, among major energy sources, PV technology employs the most people per unit energy output (see Figure 21.3).

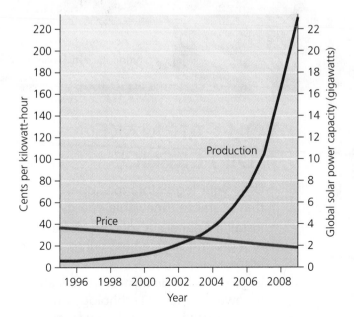

FIGURE 21.8 ▲ Global production of PV cells has been growing exponentially, and prices have fallen rapidly. Data from REN21, 2009. *Renewables global status report: 2009 update.* Paris: REN21 Secretariat; European Photovoltaic Industry Association; and U.S. Department of Energy.

"The nation that leads the clean energy economy will be the nation that leads the global economy." So said President Barack Obama in his 2010 State of the Union speech. ... But will the United States be that nation?

Silicon for PV panel, Germany

Denmark gets **10 times more** of its electricity from wind than the U.S does.

Per person, Germany surpasses the U.S. in wind power by **2.8 times**, photovoltaic (PV) solar by **21 times**, and total renewable energy by **3.2 times**.

Solar-powered homes are commonplace in Japan.

China has **50 times more** solar water heating capacity than the U.S., and **doubles** its wind power yearly.

Producing PV cells, China

YOU CAN MAKE A DIFFERENCE

➤ Urge your legislators to support bills that promote renewable energy.
➤ Install PV panels, a solar water heater, or a ground source heat pump at your home.
➤ Urge college administrators to install renewable energy technology on campus or to buy electricity from wind power.

Finally, a major advantage of solar power over fossil fuels is that it does not pollute the air with greenhouse gas emissions and other air pollutants. The manufacture of photovoltaic cells *does* currently require fossil fuel use, but once up and running, a PV system produces no emissions. Consumers can access online calculators offered by the U.S. Department of Energy and the U.S. Environmental Protection Agency to estimate the economic and environmental effects of installing a PV system. At the time of this writing, these calculators estimated that installing a 5-kilowatt PV system in a home in Fort Worth, Texas, to provide over half of its annual power needs would save the homeowners $681 per year on energy bills and would prevent over 5 tons of carbon dioxide emissions per year—as much CO_2 as results from burning 570 gallons of gasoline. Even in overcast Seattle, Washington, a 5-kilowatt system producing half a home's energy needs can save $310 per year and prevent over 3.5 tons of CO_2 emissions (equal to 390 gallons of gas).

Location and cost can be drawbacks

Solar energy currently has two major disadvantages. One is that not all regions are sunny enough to provide adequate power, given current technology. Although Earth as a whole receives vast amounts of sunlight, not every location on Earth does (**FIGURE 21.9**). People in cities such as Seattle might find it difficult to harness enough sunlight most of the year to rely on solar power. Daily or seasonal variation in sunlight can also pose problems for stand-alone solar systems if storage capacity in batteries or fuel cells is not adequate or if backup power is not available from a municipal electric grid.

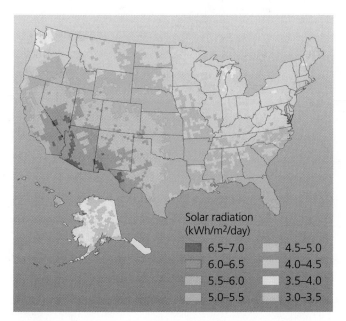

Solar radiation
(kWh/m²/day)

6.5–7.0	4.5–5.0
6.0–6.5	4.0–4.5
5.5–6.0	3.5–4.0
5.0–5.5	3.0–3.5

FIGURE 21.9 ▲ Some locations receive more sunlight than others, so harnessing solar energy is more profitable in some areas than in others. In the United States, many areas of Alaska and the Pacific Northwest receive only 3–4 kilowatt-hours per square meter per day, whereas most areas of the Southwest receive 6–7 kilowatt-hours per square meter per day. Data from National Renewable Energy Laboratory, U.S. Department of Energy.

The primary disadvantage of current solar technology is the up-front cost of the equipment. Because of the investment cost, solar power remains the most expensive way to produce electricity. For most U.S. homes, it may take 20 years or more for homeowners to break even on their investment. Proponents of solar power argue that decades of government support for fossil fuels and nuclear power have made solar power unable to compete. Because the external costs (pp. 150–151) of polluting and nonrenewable energy sources are not included in market prices, these energy sources have remained relatively cheap. Governments, businesses, and consumers thus have had little economic incentive to switch to solar energy.

However, declines in price and improvements in efficiency of solar technologies so far are encouraging, even in the absence of significant funding from government and industry. At their advent in the 1950s, solar technologies had efficiencies of around 6% while costing $600 per watt. Today, PV cells are showing up to 20% efficiency commercially and 40% efficiency in lab research, suggesting that future solar cells could be more efficient than any energy technologies we have today. Solar systems are becoming less expensive and now can sometimes pay for themselves in 10–20 years or less. After that time, they provide energy virtually for free as long as the equipment lasts.

WIND ENERGY

Wind energy—energy derived from the movement of air—is really an indirect form of solar energy, because it is the sun's differential heating of air masses on Earth that causes wind to blow. We can harness power from wind by using devices called **wind turbines**, mechanical assemblies that convert wind's kinetic energy (p. 29), or energy of motion, into electrical energy.

Today's wind turbines have their historical roots in Europe, where wooden windmills were used for 800 years to grind grain into flour and to pump water to drain wetlands and irrigate crops. In North America, ranchers use windmills to draw groundwater up for thirsty cattle. The first wind turbine built to generate electricity was constructed back in the late 1800s in Cleveland, Ohio. But it was not until after the 1973 oil embargo that governments and industry in North America and Europe began funding research and development for wind power. This moderate infusion of funding boosted technological progress, and today's wind turbines look more like airplane propellers or sleek new helicopters than romantic old Dutch paintings.

Wind turbines convert kinetic energy to electrical energy

Wind blowing into a turbine turns the blades of the rotor, which rotate machinery inside a compartment called a *nacelle*, which sits atop a tall tower (**FIGURE 21.10**). Inside the nacelle are a gearbox and a generator, as well as equipment to monitor and control the turbine's activity. Most of today's

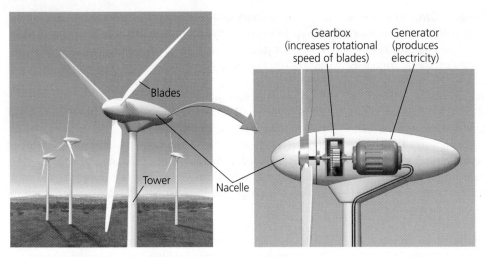

FIGURE 21.10 ◄ A wind turbine converts wind's energy of motion into electrical energy. Wind causes the blades of a wind turbine to spin, turning a shaft that extends into the nacelle that is perched atop the tower. Inside the nacelle, a gearbox converts the rotational speed of the blades, which can be up to 20 revolutions per minute (rpm) or more, into much higher rotational speeds (over 1500 rpm). These high speeds provide adequate motion for a generator inside the nacelle to produce electricity.

towers range from 45 to 105 m (148 to 344 ft) in height, so the largest are taller than a football field is long. Higher is generally better, to minimize turbulence (and potential damage) while maximizing wind speed. Most rotors consist of three blades and measure 57–99 m (187–325 ft) across. Turbines are often erected in groups called wind parks or *wind farms*. The world's largest wind farms contain hundreds of turbines spread across the landscape.

Engineers design turbines to yaw, or rotate back and forth in response to changes in wind direction, ensuring that the motor faces into the wind at all times. They also design them to begin turning at specific wind speeds to harness wind energy as efficiently as possible. Some turbines generate low levels of electricity by turning in light breezes. Others are programmed to rotate only in strong winds, operating less frequently but generating large amounts of electricity in short time periods. Slight differences in wind speed yield substantial differences in power output, for two reasons. First, the energy content of wind increases as the square of its velocity; thus if wind velocity doubles, energy quadruples. Second, an increase in wind speed causes more air molecules to pass through the wind turbine per unit time, making power output equal to wind velocity cubed. Thus a doubled wind velocity actually results in an eightfold increase in power output.

Wind power is growing fast

Like solar energy, wind provides just a small proportion of the world's power needs, but wind power is growing fast—doubling every three years in recent years (**FIGURE 21.11**). Five nations account for three-quarters of the world's wind power output (**FIGURE 21.12A**), but dozens of nations now produce wind power. Germany had long produced the most, but the United States overtook it as world leader in 2008.

Denmark is the global leader in obtaining the greatest percentage of its energy from wind power; in this small European nation, wind farms supply one-fifth of Danish electricity needs (**FIGURE 21.12B**). Germany is fifth in this respect, and the United States is 13th. Texas accounts for the most wind power generated of all U.S. states.

Wind power's growth in the United States has been haphazard because Congress has not committed to a long-term federal tax credit for wind development, but instead has passed a series of short-term renewals, leaving the industry uncertain about how much to invest. However, experts agree that wind power's rapid growth will continue, because only a small portion of this resource is currently being tapped and because wind power at favorable locations already generates electricity nearly as cheaply as do fossil fuels. A 2008 report by a consortium of government, industry, and environmental experts outlined how the United States could meet fully one-fifth of its electrical demands with wind power by 2030.

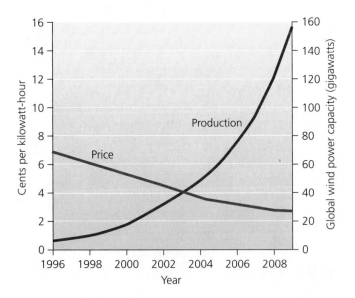

FIGURE 21.11 ▲ Global production of wind power has been doubling every three years in recent years, and prices have fallen rapidly. Data from REN21, 2009. *Renewables global status report: 2009 update.* Paris: REN21 Secretariat; Global Wind Energy Council, 2010. *Global wind 2009 report*; and U.S. Department of Energy. Price data is for Class 4 wind; stronger winds lead to cheaper prices.

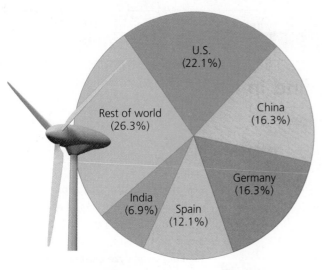

U.S.
(22.1%)

China
(16.3%)

Rest of world
(26.3%)

Germany
(16.3%)

India
(6.9%)

Spain
(12.1%)

(a) Percentage of global wind power in each nation

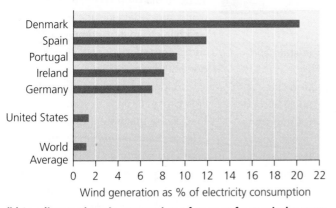

Denmark
Spain
Portugal
Ireland
Germany

United States

World
Average

0 2 4 6 8 10 12 14 16 18 20 22
Wind generation as % of electricity consumption

(b) Leading nations in proportion of energy from wind power

FIGURE 21.12 ▲ Most of the world's fast-growing wind-power generating capacity **(a)** is concentrated in a handful of countries, led by the United States, China, and Germany. Yet tiny Denmark **(b)** obtains the highest percentage of its energy needs from wind. Data from (a) Global Wind Energy Council, 2010. *Global wind 2009 report*. GWEC, Brussels, Belgium; and (b) Energy Efficiency and Renewable Energy, 2008. Annual report on U.S. wind power installation, cost, and performance trends: 2007. U.S. Department of Energy, Washington, D.C.

Offshore sites hold promise

Wind speeds on average are roughly 20% greater over water than over land. There is also less air turbulence over water. For these reasons, offshore wind turbines are becoming popular (**FIGURE 21.13**). Costs to erect and maintain turbines in water are higher, but the stronger, less turbulent winds produce more power and make offshore wind potentially more profitable. Currently, offshore wind farms are limited to shallow water, where towers are sunk into sediments singly or using a tripod configuration. In the future, towers may also be placed on floating pads anchored to the seafloor in deep water. At great distances from land, it may be best to store the generated electricity in hydrogen fuel and then ship or pipe this to land (instead of building submarine cables to carry electricity to shore), but further research is needed.

FIGURE 21.13 ▲ More and more wind farms are being developed offshore because offshore winds tend to be stronger yet less turbulent. Much of Denmark's wind power comes from offshore turbines. This Danish wind farm is one of several that provide over 20% of that nation's electricity.

Denmark erected the first offshore wind farm in 1991. Over the next decade, nine more came into operation across northern Europe, where the North Sea and Baltic Sea offer strong winds. Germany has no offshore wind farms yet, but as soon as Parliament raised the feed-in tariff rate for offshore wind from 9 cents to 15 cents per kilowatt-hour in 2009, one project began construction and many more initiated plans. In the United States, no offshore wind farms have yet been constructed, but development of the first was given U.S. government approval in 2010 after nine years of debate. The Cape Wind offshore wind farm, if constructed, will feature 130 turbines rising from Nantucket Sound 8 km (5 mi) off the coast of Cape Cod in Massachusetts. In announcing the government's approval, U.S. Interior Secretary Ken Salazar predicted that it would be "the first of many projects up and down the Atlantic coast."

Wind power has many benefits

Like solar energy, wind power produces no emissions once the equipment is manufactured and installed. As a replacement for fossil fuel combustion in the average U.S. power plant, the U.S. Environmental Protection Agency (EPA) has calculated that running a 1-megawatt wind turbine for 1 year prevents the release of more than 1,500 tons of carbon dioxide, 6.5 tons of sulfur dioxide, 3.2 tons of nitrogen oxides, and 60 lb of mercury. The amount of carbon pollution that all U.S. wind turbines together prevent from entering the atmosphere is greater than the emissions from 6.5 million cars, or from combusting the cargo of a 500-car freight train of coal each and every day.

Wind power, under optimal conditions, appears considerably more efficient than conventional power sources in its energy returned on investment (EROI; p. 534). One study found that wind turbines produce 23 times more energy than they consume. For nuclear energy, the ratio was 16:1; for coal it was 11:1; and for natural gas it was 5:1. Wind farms also use less water than do conventional power plants.

Wind turbine technology can be used on many scales, from a single tower for local use to fields of hundreds that supply large regions. Small-scale turbine development can

The SCIENCE behind the Story

Prospecting for Wind in Idaho

Installing a wind turbine

Where does wind translate into energy? In Idaho, where resource planners decided that the state's power future lies in generating electricity from wind. Now scores of Idaho residents, from small farmers to Native American tribes, have joined in the search for gusts with energy potential.

By handing out wind-measuring devices to interested landowners, Idaho turned its citizens into "wind prospectors" who pinpoint potential areas for wind farms. Idaho first launched the public wind prospecting program in 2001, after joining several other northwestern states in a regional research effort. People who join the program must collect data on wind speed and direction and share those data with the state and the public.

A promising wind farm site requires infrastructure, such as roads for erecting wind turbines and transmission lines for sending out power the turbines generate. But the most important factor is the speed and frequency of the wind. Effective commercial wind farms have a steady flow of wind just above ground level, with regular gusts of at least 21 km/hr (13 mi/hr) at a height of about 50 m (164 ft).

People analyze existing data to determine whether a site merits further study. In many parts of the developed world, these data include decades of weather information compiled into computerized wind maps that indicate general wind conditions.

In Idaho, energy planners provide prospectors with starter maps that divide the state into seven wind "classes" and reveal which areas might have enough wind to make a wind farm worthwhile. Areas listed as "Class 3" or higher, with wind speeds of about 23 km (14.3 mi)

---

help make local areas more self-sufficient, just as solar energy can. For instance, the Rosebud Sioux Tribe of Native Americans in 2003 set up a single turbine on its reservation in South Dakota. The turbine produces electricity for 200 homes and brings the tribe an estimated $15,000 per year in revenue. Wind resources are rich in this region, and the tribe is now developing a 30-megawatt wind farm nearby.

Another benefit of wind power is that farmers and ranchers can lease their land for wind development, which provides them extra revenue while also increasing property tax income for rural communities. A single large turbine can bring in $2,000 to $4,500 in annual royalties while occupying just a quarter-acre of land. Because each turbine takes up only a small area, most of the land can still be used for farming, ranching, or other uses.

Financially, wind power involves up-front costs for erecting turbines and expanding infrastructure to allow electricity distribution, but over the lifetime of a project it requires only maintenance costs. Unlike fossil fuel power plants, wind turbines incur no ongoing fuel costs. Startup costs of wind farms generally are higher than those of fossil-fuel plants, but wind farms incur fewer expenses once they are up and running. Moreover, advancing technology is driving down the costs, per unit of electricity produced, of wind farm construction.

Finally, just as solar energy creates job opportunities, so does wind power (**FIGURE 21.14**). The American Wind Energy Association estimates that 35,000 new U.S. jobs were created in 2008 alone, bringing the U.S. wind industry's total by that time to 85,000 employees. More than 100 colleges and universities now offer programs and degrees that train people in the skills needed for jobs in wind power and other renewable energy fields.

Wind power has some downsides

Wind is an intermittent resource; we have no control over when wind will occur, and this unpredictability is a major limitation in relying on it as an electricity source. However, this poses little problem if wind is one of several sources contributing to a utility's power generation. Moreover, batteries or hydrogen fuel (p. 609) can store energy generated by wind and release it later when needed.

Just as wind varies from time to time, it varies from place to place; some areas are windier than others. Global wind patterns combine with local topography—mountains, hills, water bodies, forests, cities—to create local wind patterns. Resource planners and wind power companies study these patterns closely before planning a wind farm (see **THE SCIENCE BEHIND THE STORY**, above). Meteorological research has given us information with which to judge prime areas for locating wind

per hour at 50 m (164 ft) above ground, offer the best possibilities.

Such maps, however, may not provide enough detail about a specific location. A piece of property, for example, may sit in a Class 3 area but be sheltered by a hill that blocks the wind. Knowing that kind of detail requires site-specific on-the-ground research.

To make such research possible, Idaho lends landowners in areas listed as Class 3 or higher devices called *anemometers* (**see figure**), which measure wind speed and direction. The Idaho program uses a common cup anemometer, with an array of three or four hollow cups set to catch the wind and rotate around a vertical rod. The greater the wind speed, the faster the cups rotate. Wind direction is measured by a vane that turns on a vertical axis to point into the wind. The cup wheel and wind vane are connected electrically to speed and direction dials, which relay wind data.

More than 80 landowners borrowed anemometers from the state in

To determine where to build wind farms, Idaho's wind prospectors use anemometers, which collect and relay wind data. Cup wheels rotate to indicate wind speed, and a vane turns to reveal wind direction.

the Idaho program's first year and sent data to the state every 60 days for review by energy planners and for subsequent posting online. The studies

generated new funding and wind farm plans in the state. In 2003, one farm near Idaho Falls won a $500,000 federal grant to help build a 1.5-megawatt wind farm to supply power for 500 homes. As of 2010, Idaho has a total of 147 megawatts of wind power in operation.

In eastern Idaho, five anemometers set up on Shoshone-Bannock tribal lands have revealed good prospects for a commercial wind farm on two Native American reservations. The research effort has shown that the lands are "world-class sites" for wind power, according to a state energy official. With average wind speeds in the 29 km/hr (18 mi/hr) range, further study of the tribal lands revealed possible sites for large-scale commercial wind farms, which could mean jobs and revenue for the reservations.
Similar wind prospecting programs are now underway on other reservations and in other states, including Utah, Oregon, Virginia, and Missouri. ■

FIGURE 21.14 ◄ As the wind power industry expands, it is becoming a major source of new "green-collar jobs." This technician is servicing a turbine at a wind farm in Culberson County, Texas, which provides electricity to customers in Houston and Austin.

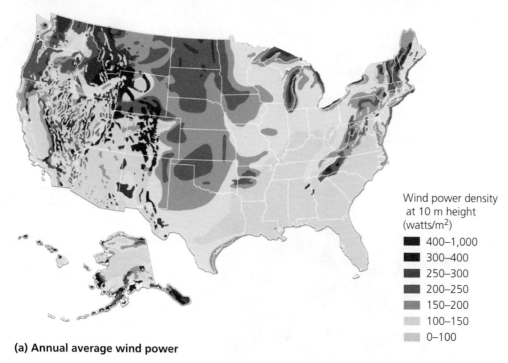

(a) Annual average wind power

Wind power density
at 10 m height
(watts/m²)

- 400–1,000
- 300–400
- 250–300
- 200–250
- 150–200
- 100–150
- 0–100

FIGURE 21.15 ◀ Wind's capacity to generate power varies according to wind speed. Meteorologists have measured wind speed to calculate the potential generating capacity from wind in different areas. The map in **(a)** shows average wind power across the United States, in watts per square meter at a height of 10 m (33 ft) above ground. Such maps are used to help guide placement of wind farms. The development of U.S. wind power so far is summarized in **(b)**, which shows the megawatts of generating capacity developed in each state through mid-2010. *Sources:* (a) Elliott, D. L., et al., 1987. *Wind energy resource atlas of the United States.* Golden, CO: Solar Energy Research Institute; (b) Data from 1st Quarter 2010 Market Report, American Wind Energy Association.

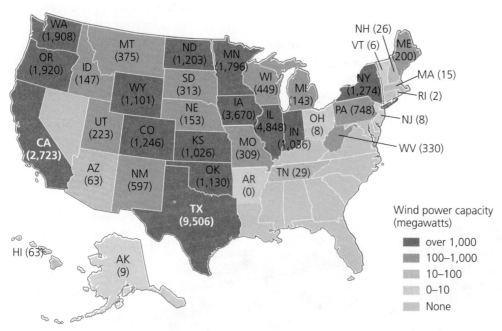

Wind power capacity
(megawatts)

- over 1,000
- 100–1,000
- 10–100
- 0–10
- None

(b) Wind generating capacity, 2010

farms. A map of average wind speeds across the United States (**FIGURE 21.15A**) reveals that mountainous regions are best, along with areas of the Great Plains. Based on such information, the wind power industry has located much of its generating capacity in states with high wind speeds (**FIGURE 21.15B**) and is seeking to expand in the Great Plains and mountain states. Provided that wind farms are strategically erected in optimal locations, an estimated 15% of U.S. energy demand could be met using only 43,000 km² (16,600 mi²) of land (with less than 5% of this land area actually occupied by turbines, equipment, and access roads).

Good wind resources, however, are not always near population centers that need the energy. Most of North America's people live near the coasts, far from the Great Plains and mountain regions that have the best wind resources. Thus, transmission networks would need to be greatly expanded to get wind power to where people live.

When wind farms *are* proposed near population centers, local residents often oppose them. Turbines are generally located in exposed, conspicuous sites, and some people object to wind farms for aesthetic reasons, feeling that the structures clutter the landscape. Although polls show wide public approval of existing wind projects and of the concept of wind power, newly proposed wind projects often elicit the so-called **not-in-my-backyard (NIMBY)** syndrome among people living nearby. For instance, the Cape Wind project has faced years of opposition from wealthy residents of Cape Cod, Nantucket, and Martha's Vineyard, even though many of these residents like to think of themselves as progressive environmentalists.

Wind turbines also pose a threat to birds and bats, which are killed when they fly into the rotating blades. At California's Altamont Pass wind farm, turbines killed many golden eagles and other raptors during the 1990s. Studies since then at other sites have suggested that bird deaths may be a less severe problem than was initially feared, but uncertainty remains. For instance, one European study indicated that migrating seabirds fly past offshore turbines without problem, but other data show that resident seabird densities decline near turbines. On land, an estimated 1–2 birds are killed per turbine per year—far fewer than the millions being killed annually by television, radio, and cell phone towers; tall buildings; automobiles; pesticides; and domestic cats. Bat mortality appears to be a more severe problem, and more research is urgently needed. One key for protecting birds and bats may be selecting sites that are not on migratory flyways or in the midst of prime habitat for species that are likely to fly into the blades.

WEIGHING THE ISSUES

Wind and NIMBY If you could choose to get your electricity from a wind farm or a coal-fired power plant, which would you choose? How would you react if the electric utility proposed to build the wind farm that would generate your electricity atop a ridge running in back of your neighborhood, such that the turbines would be clearly visible from your living room window? Would you support or oppose the development? Why? If you would oppose it, where would you suggest the farm be located? Do you think anyone might oppose it in that location?

GEOTHERMAL ENERGY

Geothermal energy is one form of renewable energy that does not originate from the sun. Instead, **geothermal energy** is thermal energy that arises from beneath Earth's surface. The radioactive decay of elements (p. 26) amid the extremely high pressures deep in the interior of our planet generates heat that rises to the surface through magma (molten rock, p. 36) and through cracks and fissures. Where this energy heats groundwater, natural spurts of heated water and steam are sent up from below and may erupt through the surface as terrestrial geysers (p. 24) or submarine hydrothermal vents (p. 33).

Geothermal energy manifests itself at the surface in these ways only in certain areas—such as The Geysers region of northern California that we visited in Chapter 2. The nation of Iceland has a wealth of geothermal energy resources because it is built from lava that extruded above the ocean's surface and cooled at the Mid-Atlantic Ridge (p. 36), along the spreading boundary of two tectonic plates. Because of the geothermal heat in this region, volcanoes and geysers are numerous in Iceland. The United States is the world leader in the use of geothermal power, but only some U.S. regions have adequate geothermal resources near the surface that can be readily used (**FIGURE 21.16**).

We harness geothermal energy for heating and electricity

Geothermal energy can be harnessed directly from geysers at the surface, but most often wells must be drilled down hundreds or thousands of meters toward heated groundwater. Hot groundwater can be used directly for heating homes, offices, and greenhouses; for driving industrial processes; and for drying crops. Iceland heats most of its homes through direct heating with piped hot water. Iceland began putting geothermal energy to use in the 1940s, and today 30 municipal district heating systems and 200 small private rural networks supply heat to 86% of the nation's residences. Such direct use of naturally heated water is efficient and inexpensive, but it is feasible only in places such as Iceland, where geothermal energy sources are available and near where the heat must be transported.

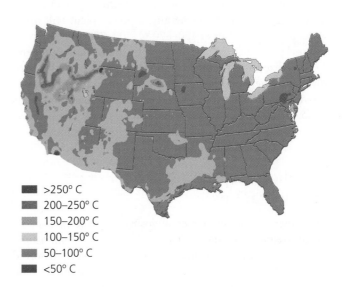

FIGURE 21.16 ▶ Geothermal resources in the United States are greatest in the western states. This map shows water temperatures 3 km (1.9 mi) below ground. Although deep subterranean temperatures are greatest in the West, ground-source heat pumps can be used anywhere in the country. Data from Idaho National Laboratory, 2007.

>250° C
200–250° C
150–200° C
100–150° C
50–100° C
<50° C

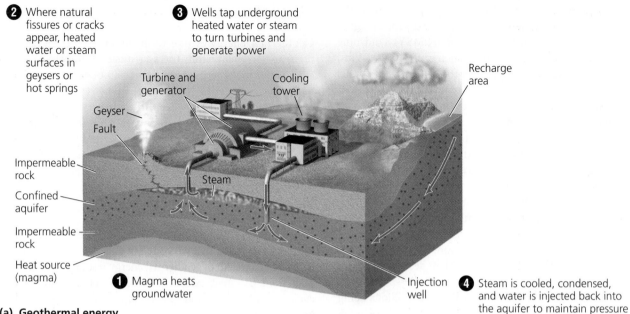

2 Where natural fissures or cracks appear, heated water or steam surfaces in geysers or hot springs

3 Wells tap underground heated water or steam to turn turbines and generate power

Recharge area

Turbine and generator

Cooling tower

Geyser

Fault

Impermeable rock

Confined aquifer

Impermeable rock

Heat source (magma)

Steam

1 Magma heats groundwater

Injection well

4 Steam is cooled, condensed, and water is injected back into the aquifer to maintain pressure

(a) Geothermal energy

(b) Nesjavellir geothermal power station, Iceland

FIGURE 21.17 ◄ With geothermal energy **(a)**, magma heats groundwater deep in the earth **1**, some of which is let off naturally through surface vents such as geysers **2**. Geothermal facilities tap into heated water below ground and channel steam through turbines in buildings to generate electricity **3**. After being used, the steam is often condensed, and the water is pumped back into the aquifer to maintain pressure **4**. At the Nesjavellir geothermal power station in Iceland **(b)**, steam is piped from wells to a condenser at the plant, where cold water pumped from lakeshore wells is heated. The heated water is sent through an insulated pipeline to the capital city, where residents use it for washing and space heating.

Geothermal power plants harness the energy of naturally heated underground water and steam to generate electricity (**FIGURE 21.17**). Generally, a power plant will bring water at temperatures of 150–370°C (300–700°F) or more to the surface and convert it to steam by lowering the pressure in specialized compartments. The steam is then used to turn turbines to generate electricity. Geothermal power currently provides more electricity than solar and a similar amount as wind. The world's largest geothermal power plants, The Geysers in northern California (p. 24), provide enough electricity to supply 750,000 homes.

Geothermal power has benefits and limitations

Like other renewable sources, all forms of geothermal energy greatly reduce emissions relative to fossil fuel combustion. Although geothermally heated water can release dissolved gases, including carbon dioxide, methane, ammonia, and hydrogen sulfide, these are generally in small quantities, and

facilities using the latest filtering technologies produce even fewer emissions. By one estimate, each megawatt of geothermal power prevents the emission of 7.0 million kg (15.5 million lb) of carbon dioxide emissions each year.

On the negative side, geothermal sources may not always be truly sustainable. Geothermal energy is renewable in principle (its use does not affect the amount of thermal energy produced underground), but the power plants we build to capture this energy may not all be able to operate indefinitely. If a geothermal plant uses heated water more quickly than groundwater is recharged, the plant will eventually run out of water. This was occurring at The Geysers in California, where the first generator was built in 1960. In response, operators began injecting municipal wastewater into the ground to replenish the supply. More geothermal plants worldwide are now injecting water back into aquifers to help maintain pressure and thereby sustain the resource.

A second reason geothermal energy may not always be renewable is that patterns of geothermal activity in Earth's crust shift naturally over time. This means that an area that produces hot groundwater now may not always do so. In addition, the water of many hot springs is laced with salts and minerals that corrode equipment and pollute the air. These factors may shorten the lifetime of plants, increase maintenance costs, and add to pollution. The greatest limitation of geothermal electric power, however, is that it is limited to regions where we can tap the energy from naturally heated groundwater. Places like Iceland, northern California, and Yellowstone National Park are rich in naturally heated groundwater, but most areas of the world are not.

Enhanced geothermal systems might widen our reach

Engineers are presently trying to overcome our limitation to areas where naturally heated groundwater occurs. As we saw in Chapter 2, efforts are being made to develop **enhanced geothermal systems** (**EGS**; pp. 25, 41). In the EGS approach, engineers drill extremely deeply into dry rock, fracture the rock, and pump in cold water. The water becomes heated deep underground and is then drawn up through an outlet well and used to generate power. EGS thereby uses natural thermal energy underground but supplies the water, which can be reused repeatedly. Because naturally occurring heated water is not needed, in theory we could use EGS widely in many locations. In Germany, for instance, which has little heated groundwater, feed-in-tariffs have enabled an EGS facility to operate profitably. For these reasons EGS technology shows great promise, and a 2006 report estimated that heat resources below the United States alone are enough to power the world's energy demands for several millennia. However, EGS also appears to trigger minor earthquakes—a problem that could render it undesirable. Unless we can develop ways to use EGS safely without causing earthquakes, our use of geothermal power will remain more localized than solar, wind, biomass, or hydropower.

Heat pumps make use of temperature differences above and below ground

Although heated groundwater is available only in certain areas, we can follow another strategy just about anywhere to use the earth's thermal energy: We can take advantage of the temperature differences that naturally exist between the soil and the air. Soil varies in temperature from season to season less than air does, because it absorbs and releases heat more slowly and because warmth and cold do not penetrate deeply below ground. Just several inches below the surface, temperatures are nearly constant year-round. Geothermal heat pumps, or **ground source heat pumps** (**GSHPs**), make use of this fact. These pumps heat buildings in the winter by transferring heat from the ground into buildings, and they cool buildings in the summer by transferring heat from buildings into the ground. Both types of heat transfer are accomplished by a single network of underground plastic pipes that circulate water and antifreeze (**FIGURE 21.18**). Because heat is simply moved from place to place rather than being produced using outside energy inputs, heat pumps can be highly energy-efficient.

More than 600,000 GSHPs are already used to heat U.S. homes. Compared to conventional electric heating and cooling systems, GSHPs heat spaces 50–70% more efficiently, cool them 20–40% more efficiently, can reduce electricity use by 25%–60%, and can reduce emissions by up to 70%.

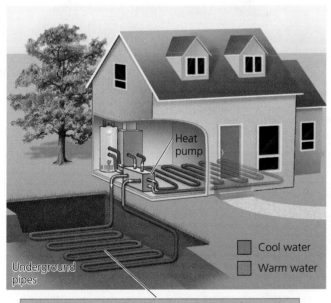

In summer, soil underground is cooler than surface air. Water flowing through the pipes transfers heat from the house to the ground, cooling the house.

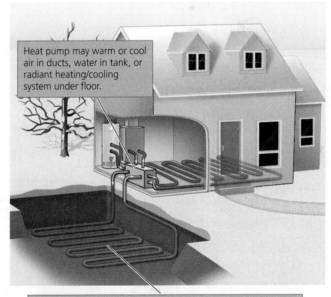

Heat pump may warm or cool air in ducts, water in tank, or radiant heating/cooling system under floor.

In winter, soil underground is warmer than surface air. Water flowing through the pipes transfers heat from the ground to the house, warming the house.

FIGURE 21.18 ▲ Ground-source heat pumps provide an efficient way to heat and cool air and water in one's home. A network of plastic pipes filled with water and antifreeze extend underground from the house. Soil is cooler than the air in the summer (left), and warmer than the air in the winter (right), so by running fluid between the house and the ground, these systems help adjust temperatures inside.

OCEAN ENERGY SOURCES

The oceans are home to several underexploited energy sources stemming from continuous natural processes. Of the four approaches being developed, three involve motion, and one involves temperature.

We can harness energy from tides, waves, and currents

Just as dams on rivers use flowing fresh water to generate hydroelectric power, we can use kinetic energy from the natural motion of ocean water to generate electrical power. The rise and fall of ocean tides (p. 442) twice each day moves large amounts of water past any given point on the world's coastlines. Differences in height between low and high tides are especially great in long, narrow bays such as Alaska's Cook Inlet or the Bay of Fundy between New Brunswick and Nova Scotia (**FIGURE 21.19**). Such locations are best for harnessing **tidal energy**, by erecting dams across the outlets of tidal basins. The incoming tide flows through sluice gates and is trapped behind them. Then, as the outgoing tide passes through the gates, it turns turbines to generate electricity (**FIGURE 21.20**). Some designs generate electricity from water moving in both directions.

FIGURE 21.19 ▶ Ocean tides are extreme at Canada's Bay of Fundy, where boats docked at high tide (**top photo**) become stranded on the mud at low tide (**bottom photo**) as the water recedes. Energy can be extracted from the movement of the tides at coastal sites where tidal flux is great enough.

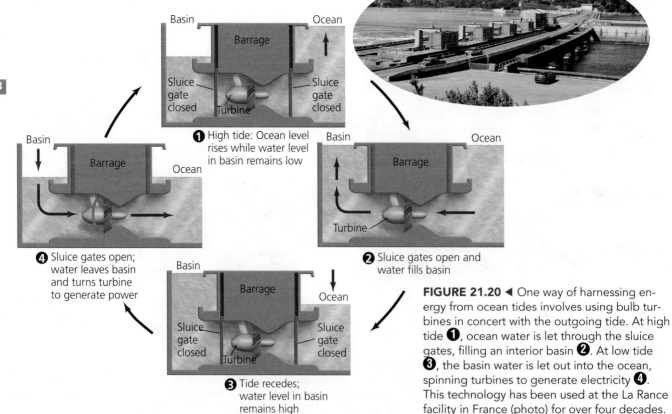

❶ High tide: Ocean level rises while water level in basin remains low

❷ Sluice gates open and water fills basin

❸ Tide recedes; water level in basin remains high

❹ Sluice gates open; water leaves basin and turns turbine to generate power

FIGURE 21.20 ◀ One way of harnessing energy from ocean tides involves using bulb turbines in concert with the outgoing tide. At high tide ❶, ocean water is let through the sluice gates, filling an interior basin ❷. At low tide ❸, the basin water is let out into the ocean, spinning turbines to generate electricity ❹. This technology has been used at the La Rance facility in France (photo) for over four decades.

The world's largest tidal generating facility is the La Rance facility in France (see Figure 21.20 inset photo), which has operated for over 40 years. Smaller facilities operate in China, Russia, and Canada. San Francisco is considering building a tidal energy station under the Golden Gate Bridge, and New York City is studying whether to establish one on the East River. Tidal stations release few or no pollutant emissions, but they can affect the ecology of estuaries and tidal basins.

People are also working to harness the motion of ocean waves and convert this mechanical energy into electricity. Many designs for machinery to harness **wave energy** have been invented, but few have been adequately tested. Some designs for offshore facilities involve floating devices that move up and down with the waves. Wave energy is greater at deep-ocean sites, but transmitting the electricity produced to shore would be expensive. Some designs for coastal on-shore facilities funnel waves from large areas into narrow channels and elevated reservoirs, from which water then flows out, generating electricity as hydroelectric dams do. Other coastal designs use rising and falling waves to push air into and out of chambers, turning turbines (**FIGURE 21.21**). No commercial wave energy facilities are operating yet, but demonstration projects exist in Europe, Japan, and Oregon.

A third way to harness marine kinetic energy is to use the motion of ocean currents (p. 436), such as the Gulf Stream. Devices that look like underwater wind turbines have been erected in European waters to test this idea.

The ocean stores thermal energy

Each day the tropical oceans absorb an amount of solar radiation equivalent to the heat content of 250 billion barrels of oil—enough to provide 20,000 times the electricity used daily in the United States. The ocean's sun-warmed surface is warmer than its deep water, and **ocean thermal energy conversion (OTEC)** approaches are based on this gradient in temperature.

In the *closed cycle* approach, warm surface water is piped into a facility to evaporate chemicals, such as ammonia, that boil at low temperatures. These evaporated gases spin turbines to generate electricity. Cold water piped in from ocean depths then condenses the gases so they can be reused. In the *open cycle* approach, warm surface water is evaporated in a vacuum, and its steam turns turbines and then is condensed by cold water. Because ocean water loses salts as it evaporates, the water can be recovered, condensed, and sold as desalinized fresh water for drinking or agriculture. Research on OTEC systems has been conducted in Hawaii and elsewhere, but costs remain high, and so far no facility is commercially operational.

HYDROGEN

All the renewable energy sources we have discussed can be used to generate electricity more cleanly than can fossil fuels. As useful as electricity is, however, it cannot be stored easily in large quantities for use when and where it is needed. This is why most vehicles rely on gasoline from oil for power. However, the development of fuel cells and of fuel consisting of hydrogen—the universe's simplest and most abundant element—shows promise to store energy conveniently and in considerable quantities, and to produce electricity at least as cleanly and efficiently as renewable energy sources.

Some yearn for a "hydrogen economy"

Some energy experts envision that hydrogen fuel, together with electricity, could serve as the basis for a clean, safe, and efficient energy system. In such a system, electricity generated from intermittent renewable sources, such as wind or solar energy, could be used to produce hydrogen. Fuel cells—essentially, hydrogen batteries (**FIGURE 21.22**)—could then use hydrogen to produce electrical energy as needed to power vehicles, computers, cell phones, home heating, and countless other applications. NASA's space flight programs have used fuel cell technology since the 1960s.

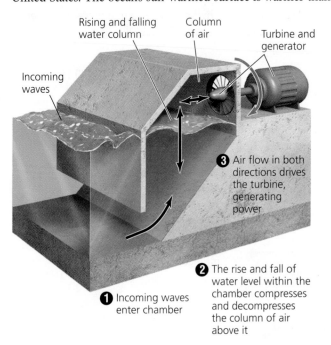

Rising and falling water column

Column of air

Turbine and generator

Incoming waves

❸ Air flow in both directions drives the turbine, generating power

❷ The rise and fall of water level within the chamber compresses and decompresses the column of air above it

❶ Incoming waves enter chamber

FIGURE 21.21 ▲ Coastal facilities can make use of energy from the motion of ocean waves. In one design, as waves are let into and out of a tightly sealed chamber ❶, the air inside is alternately compressed and decompressed ❷, creating air flow that rotates turbines ❸ to generate electricity.

The SCIENCE behind the Story

Algae as a Hydrogen Fuel Source

As scientists search for new ways to generate energy, some are looking to an unlikely power source—pond scum. Algae are being studied as an innovative way to generate large amounts of hydrogen to move society toward a more sustainable energy future.

Hydrogen's benefits hinge on how hydrogen fuel is produced. Some methods release substantial amounts of carbon dioxide, and other, nonpolluting, processes can be costly. These drawbacks have kept scientists searching for new hydrogen sources.

At the University of California at Berkeley, biologist Anastasios Melis thought one possible hydrogen source might be a single-celled aquatic plant. The alga *Chlamydomonas reinhardtii* was known to emit small amounts of hydrogen for brief periods of time when deprived of light.

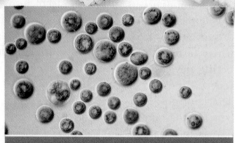

The green alga *Chlamydomonas reinhardtii*

So Melis set up an experiment with energy experts at the National Renewable Energy Laboratory in Colorado, aiming to develop ways to tweak the alga's basic biological functions so that the plant produced greater quantities of hydrogen.

Green algae photosynthesize, absorbing energy from light that converts carbon dioxide and water into food, and then expelling oxygen as a waste product. Additional nutrients from soil or water, and catalysts called

enzymes within the plant, keep this process running smoothly. To conduct photosynthesis effectively, *Chlamydomonas reinhardtii* needs the element sulfur as a nutrient. The alga also contains an enzyme called *hydrogenase*, which can trigger the alga to stop producing oxygen as a metabolic by-product and start releasing hydrogen instead.

Hydrogenase is normally active only after *Chlamydomonas reinhardtii* has been deprived of light. When deprived of light, the light reactions of photosynthesis (pp. 31–32) ebb, little oxygen is produced, and hydrogenase is activated. When light returns and the alga begins producing oxygen again, hydrogenase is deactivated, and the hydrogen release stops.

Melis's team wanted to activate hydrogenase so that more hydrogen

3 The electrons move from the negative electrode to the positive electrode, creating a current and generating electricity

Hydrogen fuel, H_2

Oxygen, O_2

1 Hydrogen molecules are stripped of electrons at the negative electrode, leaving hydrogen ions (protons, H^+)

H^+

4 Water is formed when oxygen combines with the protons and electrons that flow from the positive electrode

2 The protons traverse the membrane

Negative electrode

Proton (H^+) exchange membrane

Positive electrode

Water, H_2O

FIGURE 21.22 ◄ Hydrogen fuel drives electricity generation in a fuel cell, creating water as a waste product. First, atoms of hydrogen are split **1** into protons and electrons. The protons, or hydrogen ions **2**, pass through a proton exchange membrane. The electrons, meanwhile, move from a negative electrode to a positive one via an external circuit **3**, creating a current and generating electricity. The protons and electrons then combine with oxygen **4** to form water molecules.

would be produced. But simply keeping the algae in the dark would not escalate hydrogen production because the alga's metabolic functions slowed without light.

The researchers decided to try putting the algae on a sulfur-free, bright-light regimen. The lack of sulfur would hinder photosynthesis, limiting oxygen output enough to activate hydrogenase and trigger hydrogen production. The presence of light would keep the algae metabolically active and releasing large amounts of by-products.

The researchers cultured large quantities of the algae in bottles in labs. Then they deprived the cultures of sulfur but kept the algae exposed to light for long periods of time—up to 150 hours. After the sustained light exposure, gas and liquids were extracted from the bottles and analyzed.

The analysis supported the team's hypothesis. Without sulfur or photosynthesis, the algae were not producing oxygen. This low-oxygen, or anaerobic, environment had induced hydrogenase, which caused the algae to begin splitting water molecules and releasing

Dr. Anastasios Melis inspects flasks of algae in his laboratory.

hydrogen gas. Hydrogen dominated the alga's emissions—in gas collection analysis, approximately 87% of the gas was hydrogen, 1% was carbon dioxide, and the remaining 12% was nitrogen with traces of oxygen. The research teams published their findings in the journal *Plant Physiology* in 2000.

Since then, Melis and his collaborators have genetically engineered a strain of algae to give it enhanced energy efficiency. With further advances, they envision farms of algae pumping out hydrogen for our use in the future.

Questions remain about how much fuel can be harvested continuously using this *photobiological* process. But if we are ever to attain a future hydrogen economy, results from creative research like this will likely play an important role. ■

Basing an energy system on hydrogen could alleviate dependence on foreign fuels and help fight climate change. For these reasons, governments are funding research into hydrogen and fuel-cell technology, and automobile companies are investing in research and development to produce vehicles that run on hydrogen. The island nation of Iceland a decade ago decided to move toward a "hydrogen economy" and to set an example for the rest of the world to follow. Iceland has achieved several early steps in its 30–50-year plan to phase out fossil fuels, such as converting buses in the capital city of Reykjavik to run on hydrogen fuel. But the global economic downturn in 2008–2009 and delays in manufacturing hydrogen cars has delayed its progress. Meanwhile, Germany is one of several other nations with hydrogen-fueled city buses (**FIGURE 21.23**), and it plans by 2015 to launch a network of hydrogen filling stations for hydrogen cars that are being designed.

Hydrogen fuel may be produced from water or from other matter

Hydrogen gas (H_2) does not tend to exist freely on Earth; rather, hydrogen atoms bind to other molecules, becoming incorporated in everything from water to organic molecules. To obtain hydrogen gas for fuel, we must force these substances to release their hydrogen atoms, and this requires an input of energy. Several potential ways of producing hydrogen are being studied (see **THE SCIENCE BEHIND THE STORY**, above). In the process of **electrolysis**, electricity is input to split hydrogen atoms from the oxygen atoms of water molecules:

$$2H_2O \rightarrow 2H_2 + O_2$$

Electrolysis produces pure hydrogen, and it does so without emitting the carbon- or nitrogen-based pollutants of fossil fuel combustion. However, whether this strategy for producing hydrogen will cause pollution over its entire life cycle depends on the source of the electricity used for the electrolysis. If coal is burned to create the electricity, then the entire process will not reduce emissions compared with reliance on fossil fuels. If the electricity is produced by some less-polluting renewable source, then hydrogen production by electrolysis would create much less pollution and greenhouse warming than reliance on fossil fuels. The "cleanliness" of a future hydrogen economy would, therefore, depend largely on the source of electricity used in electrolysis.

The environmental impact of hydrogen production will also depend on the source material for the hydrogen. Besides water, hydrogen can be obtained from biomass and fossil fuels. Obtaining hydrogen from these sources generally requires less energy input but results in emissions of carbon-based pollutants.

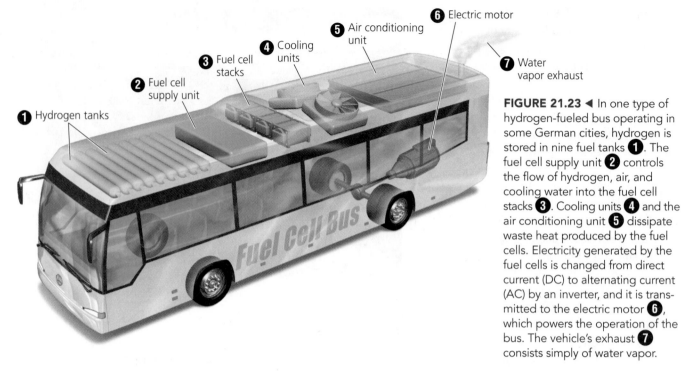

① Hydrogen tanks
② Fuel cell supply unit
③ Fuel cell stacks
④ Cooling units
⑤ Air conditioning unit
⑥ Electric motor
⑦ Water vapor exhaust

FIGURE 21.23 ◀ In one type of hydrogen-fueled bus operating in some German cities, hydrogen is stored in nine fuel tanks ①. The fuel cell supply unit ② controls the flow of hydrogen, air, and cooling water into the fuel cell stacks ③. Cooling units ④ and the air conditioning unit ⑤ dissipate waste heat produced by the fuel cells. Electricity generated by the fuel cells is changed from direct current (DC) to alternating current (AC) by an inverter, and it is transmitted to the electric motor ⑥, which powers the operation of the bus. The vehicle's exhaust ⑦ consists simply of water vapor.

For instance, extracting hydrogen from the methane (CH_4) in natural gas entails producing one molecule of the greenhouse gas carbon dioxide for every four molecules of hydrogen gas:

$$CH_4 + 2H_2O \rightarrow 4H_2 + CO_2$$

Thus, whether a hydrogen-based energy system is environmentally cleaner than a fossil fuel system depends on how the hydrogen is extracted.

Fuel cells produce electricity by joining hydrogen and oxygen

Once isolated, hydrogen gas can be used as a fuel to produce electricity within fuel cells. The chemical reaction involved in a fuel cell is simply the reverse of that shown for electrolysis; an oxygen molecule and two hydrogen molecules each split so that their atoms can bind and form two water molecules:

$$2H_2 + O_2 \rightarrow 2H_2O$$

Figure 21.22 shows how this occurs within one common type of fuel cell. Hydrogen gas (usually compressed and stored in an attached fuel tank) is allowed into one side of the cell, whose middle consists of two electrodes that sandwich a membrane that only protons (hydrogen ions) can move across. One electrode, helped by a chemical catalyst, strips the hydrogen gas of its electrons, creating two hydrogen ions that begin moving across the membrane. Meanwhile, on the other side of the cell, oxygen molecules from the open air are split into their component atoms along the other electrode. These oxygen ions soon bind to pairs of hydrogen ions traveling across the membrane, forming molecules of water that are expelled as waste, along with heat. While this is occurring, the electrons from the hydrogen atoms have traveled to a device that completes an electric current between the two electrodes. The movement of the hydrogen's electrons from one electrode to the other creates the output of electricity.

Hydrogen and fuel cells have costs and benefits

As a fuel, hydrogen offers advantages and disadvantages. One major drawback at this point is a lack of infrastructure. To convert a nation like Iceland or Germany to hydrogen would require massive and costly development of facilities to transport, store, and provide the fuel.

Another issue is that some research suggests that leakage of hydrogen from its production, transport, and use could potentially deplete stratospheric ozone (pp. 463, 479), and lengthen the atmospheric lifetime of the greenhouse gas methane. Research into these questions is ongoing, because scientists do not want society to switch from fossil fuels to hydrogen without first knowing the risks.

Hydrogen's benefits include the fact that we will never run out of it, because it is the most abundant element in the universe. Hydrogen can be clean and nontoxic to use, and—depending on its source and the source of electricity for its extraction—it may produce few greenhouse gases and other pollutants. Pure water and heat may be the only waste products from a hydrogen fuel cell, along with negligible traces of other compounds. In terms of safety for transport and storage, hydrogen can catch fire and explode, but if kept under pressure, it is probably no more dangerous than gasoline in tanks.

Hydrogen fuel cells are energy-efficient. Depending on the type of fuel cell, 35–70% of the energy released in the reaction can be used. If the system is designed to capture heat as well as electricity, then the energy efficiency of fuel cells can rise to 90%. These rates are comparable or superior to most nonrenewable alternatives.

Fuel cells are also silent and nonpolluting. Unlike batteries (which also produce electricity through chemical reactions), fuel cells will generate electricity whenever hydrogen fuel is supplied, without ever needing recharging. For all these reasons, hydrogen fuel cells may soon be used to power cars, much as they are already powering buses operating on the streets of some German cities.

➤ CONCLUSION

The increasing concern over air pollution, global climate change, health impacts, and security risks resulting from our dependence on fossil fuels—as well as concerns over dwindling supplies of oil and natural gas—have convinced many people that we need to shift to renewable energy sources that pollute far less and that will not run out. Renewable sources with promise for sustaining our civilization far into the future without greatly degrading our environment include solar energy, wind energy, geothermal energy, and ocean energy sources. Moreover, by using electricity from renewable sources to produce hydrogen fuel, we may be able to use fuel cells to produce electricity when and where it is needed,

helping to create a nonpolluting and renewable transportation sector.

Most renewable energy sources have been held back by inadequate funding for research and development and by artificially cheap market prices for nonrenewable fuels that do not include external costs. Despite these obstacles, renewable technologies have progressed far enough to offer hope that we can shift smoothly from fossil fuels to renewable energy. Whether we can also limit environmental impact will depend on how soon and how quickly we make the transition, and to what extent we put efficiency and conservation measures into place.

REVIEWING OBJECTIVES

You should now be able to:

OUTLINE THE MAJOR SOURCES OF RENEWABLE ENERGY AND ASSESS THEIR POTENTIAL FOR GROWTH

- The "new renewable" energy sources include solar, wind, geothermal, and ocean energy sources. They are not truly "new," but rather are in a stage of rapid development. (p. 591)
- The new renewables currently provide far less energy and electricity than we obtain from fossil fuels or other conventional energy sources. (pp. 591–592)
- Use of new renewables is growing quickly, and this growth is expected to continue as people seek to move away from fossil fuels. (pp. 592–593)
- Government subsidies have long favored fossil fuels, but public policies such as feed-in tariffs can help speed and smooth our transition to renewable sources. (pp. 593–594)

DESCRIBE SOLAR ENERGY AND THE WAYS IT IS HARNESSED, AND EVALUATE ITS ADVANTAGES AND DISADVANTAGES

- Energy from the sun's radiation can be harnessed using passive methods or by active methods involving powered technology. (p. 594)
- Solar technologies include flat-plate collectors for heating water and air, mirrors to concentrate solar rays, and photovoltaic (PV) cells to generate electricity. (pp. 594–596)
- PV solar and wind power are the fastest-growing energy sources today. (p. 597)
- Solar energy is perpetually renewable, creates no emissions, and enables decentralized power. (pp. 597, 599)
- Solar radiation varies in intensity from place to place and time to time, and harnessing solar energy remains expensive. (p. 599)

DESCRIBE WIND POWER AND HOW WE HARNESS IT, AND EVALUATE ITS BENEFITS AND DRAWBACKS

- Energy from wind is harnessed using wind turbines mounted on towers. (pp. 599–600)

- Turbines are often erected in arrays at wind farms located on land or offshore, in locations with optimal wind conditions. (pp. 600–603)
- Wind power and PV solar are the fastest-growing energy sources today. (p. 600)
- Wind energy is renewable, turbine operation creates no emissions, wind farms can generate economic benefits, and the cost of wind power is competitive with that of electricity from fossil fuels. (pp. 601–602)
- Wind is an intermittent resource and is adequate only in some locations. Turbines kill some birds and bats, and wind farms often face opposition from local residents. (pp. 602, 604–605)

DESCRIBE GEOTHERMAL ENERGY AND THE WAYS WE MAKE USE OF IT, AND ASSESS ITS ADVANTAGES AND DISADVANTAGES

- Thermal energy from radioactive decay in Earth's core rises toward the surface and heats groundwater. This energy is harnessed by geothermal power plants and used to directly heat water and air and to generate electricity. (pp. 605–606)
- Geothermal energy can be efficient, clean, and renewable, but naturally heated water occurs near the surface only in certain areas, and this water may be exhausted if it is overpumped. (pp. 605–607)
- Enhanced geothermal systems allow us to gain geothermal energy from more regions, but the approach can trigger earthquakes. (p. 607)
- Ground source heat pumps heat and cool homes and businesses by circulating water whose temperature is moderated underground. (p. 607)

DESCRIBE OCEAN ENERGY SOURCES AND HOW WE COULD HARNESS THEM

- Major ocean energy sources include the motion of tides, waves, and currents, and the thermal heat of ocean water. (pp. 608–609)

- Ocean energy is perpetually renewable and holds much promise, but so far technologies have seen limited development. (pp. 608–609)

EXPLAIN HYDROGEN FUEL CELLS AND WEIGH OPTIONS FOR ENERGY STORAGE AND TRANSPORTATION

- Hydrogen can serve as a fuel to store and transport energy, so that electricity generated by renewable sources can be made portable and used to power vehicles. (pp. 609–612)

- Hydrogen can be produced cleanly through electrolysis, but also by using fossil fuels—in which case its environmental benefits are reduced. (pp. 610–612)
- Fuel cells create electricity by controlling an interaction between hydrogen and oxygen, and they produce only water as a waste product. (pp. 610–612)
- Hydrogen infrastructure requires much more development, but hydrogen can be clean, safe, and efficient. Fuel cells are silent, are nonpolluting, and do not need recharging. (p. 612)

TESTING YOUR COMPREHENSION

1. About how much of our energy now comes from renewable sources? What is the most prevalent form of renewable energy we use? What form of renewable energy is most used to generate electricity?

2. What factors and concerns are causing renewable energy sectors to expand? Which two renewable sources are experiencing the most rapid growth?

3. Contrast passive and active solar heating. Describe how each works, and give examples of each.

4. Define the photoelectric effect. Explain how photovoltaic (PV) cells function and are used.

5. Describe several environmental and economic advantages of solar power. What are some disadvantages?

6. How do modern wind turbines generate electricity? How does wind speed affect the process? What factors affect where we place wind turbines?

7. Describe several environmental and economic benefits of wind power. What are some drawbacks?

8. Define *geothermal energy*, and explain several main ways in which it is obtained and used. In what ways is it renewable, and in what way is it not renewable?

9. List and describe four approaches for obtaining energy from ocean water.

10. How is hydrogen fuel produced? Is this a clean process? What factors determine the amount of pollutants hydrogen production will emit?

SEEKING SOLUTIONS

1. Explain how Germany accelerated its development of renewable energy by establishing a system of feed-in tariffs. Do you think the United States should adopt a similar system? Why or why not?

2. For each source of renewable energy discussed in this chapter, what factors are standing in the way of an expedient transition from fossil fuel use? What could be done in each case to ease a shift to these renewable sources?

3. Do you think we can develop and implement renewable energy resources to replace fossil fuels without great social, economic, and environmental disruption? What steps would we need to take? Will market forces alone suffice to bring about this transition? Do you think such a shift will be good for our economy? Why or why not?

4. Explain the circumstances under which using hydrogen fuel could be helpful in moving toward a low-emission future. If you could advise policymakers of Iceland or Germany on hydrogen, what would you tell them?

5. **THINK IT THROUGH** You have just graduated from college, gotten married, landed a good job, and purchased your first home. You and your spouse plan to stay in this home for the foreseeable future and are considering installing flat plate solar collectors and/or PV panels on your roof. What factors will you consider, and what questions will you ask before deciding whether to make the investment in solar energy?

6. **THINK IT THROUGH** You are the CEO of a company that develops wind farms. Your staff is presenting you with three options, listed below, for sites for your next development. Describe at least one likely advantage and at least one likely disadvantage you would expect to encounter with each option. What further information would you like to know before deciding which to pursue?

   ▶ Option A: A remote rural site in North Dakota
   ▶ Option B: A ridge-top site among the suburbs of Philadelphia
   ▶ Option C: An offshore site off the Florida coast

CALCULATING ECOLOGICAL FOOTPRINTS

Assume that average per capita residential consumption of electricity is 12 kilowatt-hours per day, that photovoltaic cells have an electrical output of 15% incident solar radiation, and that PV panels cost $1,000 per square meter. Now refer to Figure 21.9 on p. 599, and estimate the area and cost of the PV panels needed to provide all of the residential electricity used by each group in the table.

	Area of photovoltaic cells	Cost of photovoltaic cells
You		
A resident of Arizona		
A resident of Alaska		
Total for all U.S. residents		

1. What additional information would you need to increase the accuracy of your estimates for the areas in the table above?

2. Considering the distribution of solar radiation in the United States, where do you think it will be most feasible to greatly increase the percentage of electricity generated from photovoltaic solar cells?

3. The purchase price of a photovoltaic system is considerable. What other costs and benefits should you consider, in addition to the purchase price, when contemplating "going solar"?

Mastering**ENVIRONMENTALSCIENCE**™

Go to **www.masteringenvironmentalscience.com** for practice quizzes, Pearson eText, videos, current events, and more.

Containers en route to a recycling facility

22 MANAGING OUR WASTE

UPON COMPLETING THIS CHAPTER, YOU WILL BE ABLE TO:

- Summarize and compare the types of waste we generate
- List the major approaches to managing waste
- Delineate the scale of the waste dilemma
- Describe conventional waste disposal methods: landfills and incineration

- Evaluate approaches for reducing waste: source reduction, reuse, composting, and recycling
- Discuss management of industrial solid waste and principles of industrial ecology
- Assess issues in managing hazardous waste

Transforming New York's Fresh Kills Landfill

CANADA

UNITED STATES

New York

Long Island

New York City

Staten Island

Fresh Kills Landfill

Atlantic Ocean

"An extraterrestrial observer might conclude that conversion of raw materials to wastes is the real purpose of human economic activity."

—Gary Gardner and Payal Sampat, Worldwatch Institute

"Recycling is one of the best environmental success stories of the late 20th century."

—U.S. Environmental Protection Agency

The closure of a landfill is not the kind of event that normally draws politicians and the press, but the Fresh Kills Landfill was no ordinary dump. The largest landfill in the world (and said to be the largest human-made structure on Earth), Fresh Kills was the primary repository of New York City's garbage for half a century. On March 22, 2001, New York City Mayor Rudolph Giuliani and New York Governor George Pataki were on hand to celebrate as a barge arrived on the western shore of New York City's Staten Island and dumped the final load of 650 tons of trash at Fresh Kills.

The landfill's closure was a welcome event for Staten Island's 450,000 residents, who had long viewed the landfill as a bad-smelling eyesore, health threat, and civic blemish. The 890-ha (2,200-acre) landfill featured six gigantic mounds of trash and soil. The highest, at 69 m (225 ft), was higher than the nearby Statue of Liberty.

New York City had grandiose plans for the site. It planned to transform the old landfill into a world-class public park—a verdant landscape of rolling hills and wetlands teeming with wildlife, a mecca for recreation

THEN: Fresh Kills Landfill in operation.

NOW: Fresh Kills Landfill site today.

Fresh Kills Landfill, Staten Island, New York

for New York's residents. The site certainly had potential. It was almost three times bigger than Manhattan's Central Park. It was the region's largest remaining complex of saltwater tidal marshes and freshwater creeks and wetlands, which still attracted birds and wildlife. And the mounds offered panoramic views of the Manhattan skyline. The city sponsored an international competition to select a landscape architecture firm to design plans for the new park.

Meanwhile, with its only landfill closed, New York City began exporting its waste. The city began plans to develop an efficient network of stations to package and transfer the waste and ship it outward by barge and railroad. However, these plans soon fell apart amid neighborhood opposition, economic misjudgments, and accusations of political favoritism and mob influence. New York City instead found itself paying contractors exorbitant prices to haul its garbage away one truckload at a time. In the years following the Fresh Kills closure, trucks full of trash rumbled through neighborhood streets, carrying 12,000 tons of waste each day bound for 26 different landfills and incinerators in

New York, New Jersey, Virginia, Pennsylvania, and Ohio. The city sanitation department's budget nearly doubled, and budget woes caused the city to scale back its recycling program. Some New Yorkers suggested reopening Fresh Kills.

The landfill *was* reopened, but not for a reason anyone could have foreseen. After the September 11, 2001, terrorist attacks, the 1.8 million tons of rubble from the collapsed World Trade Center towers, including unrecoverable human remains, was taken by barge to Fresh Kills. A monument will be erected at the site as part of the new park.

Today, plans for the park are forging ahead, and public tours of the site are being offered. A draft master plan that incorporated suggestions from the public was released in 2006, and an environmental impact statement underwent public comment in 2008. The master plan involves everything from ecological restoration of the wetlands to construction of roads, ball fields, sculptures, and rollerblading rinks. People will be able to bicycle on trails paralleling tidal creeks of the region's largest estuary and reach stunning vistas atop the hills. The first parcels to be developed, recreation areas called Owl Hollow Fields and Schmul Park, should be completed by 2011, and work has begun on a parcel called North Park, which overlooks an adjacent wildlife refuge.

This immense undertaking—one of the largest public works projects in the world—will not be completed overnight. Designers and city officials expect to develop and open the new park one portion at a time over the next 30 years. In the end, they aim to transform a longtime symbol of waste into a world-class center for recreation and urban ecological restoration.

APPROACHES TO WASTE MANAGEMENT

As the world's human population rises, and as we produce and consume more material goods, we generate more waste. **Waste** refers to any unwanted material or substance that results from a human activity or process.

For management purposes, we divide waste into several main categories. **Municipal solid waste** is nonliquid waste that comes from homes, institutions, and small businesses. **Industrial solid waste** includes waste from production of consumer goods, mining, agriculture, and petroleum extraction and refining. **Hazardous waste** refers to solid or liquid waste that is toxic, chemically reactive, flammable, or corrosive. It can include everything from paint and household cleaners to medical waste to industrial solvents.

Another type of waste is *wastewater*, water we use in our households, businesses, industries, or public facilities and drain or flush down our pipes, as well as the polluted runoff from our streets and storm drains. We discussed wastewater in Chapter 15 (pp. 426–428).

We have several aims in managing waste

Waste can degrade water quality, soil quality, and air quality, thereby degrading human health and the environment. Waste is also a measure of inefficiency, so reducing waste can potentially save money and resources. Waste is also unpleasant aesthetically. For these and other reasons, waste management has become a vital pursuit.

There are three main components of **waste management**:

1. Minimizing the amount of waste we generate
2. Recovering waste materials and finding ways to recycle them
3. Disposing of waste safely and effectively

Minimizing waste at its source—called **source reduction**—is the preferred approach. There are several ways to reduce the amount of waste that enters the **waste stream**, the flow of waste as it moves from its sources toward disposal destinations (**FIGURE 22.1**). Manufacturers can use materials more efficiently. Consumers can buy fewer goods, buy goods with less packaging, and use those goods longer. Reusing goods you already own, purchasing used items, and donating your used items for others also help reduce the amount of material entering the waste stream.

The next-best strategy in waste management is **recovery**, which consists of recovering or removing waste from the waste stream. Recovery includes both recycling and composting. **Recycling** is the process of collecting used goods and sending them to facilities that extract and reprocess raw materials that can then be used to manufacture new goods. Our ability to recycle newspaper, white paper, cardboard, glass, metal cans, appliances, and some plastic containers has increased as technologies have been developed and as markets for recycled materials have grown. **Composting** is the practice of recovering organic waste (such as food and yard waste), by converting it into mulch or humus (p. 228) through natural biological processes of decomposition. Recycling and composting are not concepts that people invented. They are a fundamental feature of the way natural systems function. As we saw in Chapter 5, all materials in nature are broken down at some point, and matter cycles through ecosystems.

Regardless of how effectively we reduce our waste stream through source reduction and recovery, there will likely always be some waste left to dispose of. Disposal methods include burying waste in landfills and burning waste in incinerators. Waste managers attempt to find disposal methods that minimize impact to human health and environmental quality. In this chapter we first examine how the three major approaches are used to manage municipal solid waste, and then we address industrial solid waste and hazardous waste.

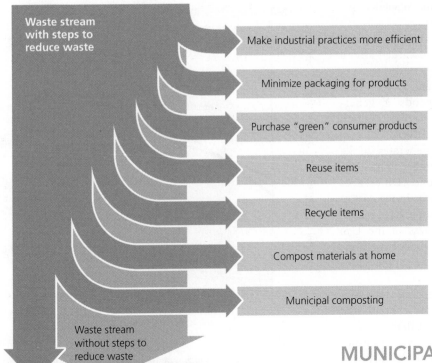

Waste stream with steps to reduce waste

Make industrial practices more efficient

Minimize packaging for products

Purchase "green" consumer products

Reuse items

Recycle items

Compost materials at home

Municipal composting

Waste stream without steps to reduce waste

Waste disposal (landfill, incinerator)

FIGURE 22.1 ◄ The most effective way to manage waste is to minimize the amount of material that enters the waste stream. To do this, manufacturers can increase efficiency, and consumers can buy "green" products that have minimal packaging or are produced in ways that minimize waste.

MUNICIPAL SOLID WASTE

Municipal solid waste is waste produced by consumers, public facilities, and small businesses. It is what we commonly refer to as "trash" or "garbage." Everything from paper to food scraps to roadside litter to old appliances and furniture is considered municipal solid waste.

In the United States, paper, yard debris, food scraps, and plastics are the principal components of municipal solid waste, together accounting for 69% of what enters the waste stream (**FIGURE 22.2A**). After recycling and composting reduce the waste stream, paper is still the largest component of U.S. municipal solid waste, followed by food scraps and plastics (**FIGURE 22.2B**). Patterns differ in developing countries; there,

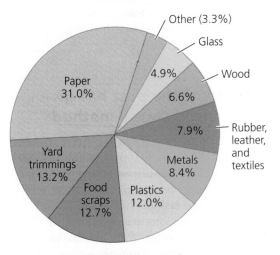

(a) Before recycling and composting

Other (3.3%)
Glass
Paper 31.0%
4.9%
Wood 6.6%
Rubber, leather, and textiles 7.9%
Yard trimmings 13.2%
Food scraps 12.7%
Plastics 12.0%
Metals 8.4%

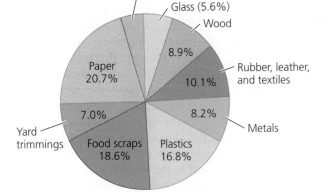

(b) After recycling and composting

Other (4.3%)
Glass (5.6%)
Wood
Paper 20.7%
8.9%
Rubber, leather, and textiles 10.1%
Metals 8.2%
Yard trimmings 7.0%
Food scraps 18.6%
Plastics 16.8%

FIGURE 22.2 ▲ Paper products comprise the largest component of the municipal solid waste stream in the United States by weight **(a)**, followed by yard trimmings, food scraps, and plastics. After recycling and composting removes many items **(b)**, the waste stream becomes one-third smaller (as shown by the smaller size of the pie chart). Paper products are still the largest contributor but are followed by food scraps and plastics, because most yard waste is composted. Data from U.S. Environmental Protection Agency, 2009. *Municipal solid waste generation, recycling, and disposal in the United States: Facts and figures for 2008.*

food scraps are often the primary contributor to solid waste, and paper makes up a smaller proportion.

Most municipal solid waste comes from packaging and nondurable goods (products meant to be discarded after a short period of use). In addition, consumers throw away old durable goods and outdated equipment as they purchase new products. As we acquire more goods, we generate more waste. In 2008, U.S. citizens produced 250 million tons of municipal solid waste (before recovery), almost 1 ton per person. The average American generates 2.0 kg (4.5 lb) of trash per day. This is considerably more than people in most other developed nations. Differences among nations result in part from differences in the cost of waste disposal. If disposal is expensive, people have incentive to waste less. The relative wastefulness of the U.S. lifestyle, with its excess packaging and reliance on nondurable goods, has caused critics to label the United States "the throwaway society."

In developing nations, people consume less and generate considerably less waste. One study found that people of high-income nations waste more than twice as much as people of low-income nations. However, wealthier nations also invest more in waste collection and disposal, so they are often better able to manage their waste proliferation and minimize impacts on human health and the environment.

More consumption creates more waste

In the United States since 1960, waste generation (before recovery) has increased (**FIGURE 22.3**) by 2.8 times, and per capita waste generation has risen by 67%. Plastics, which came into wide consumer use only after 1970, have accounted for the greatest relative increase in the waste stream during the last several decades.

The intensive consumption that has long characterized the United States and other wealthy nations is now spreading rapidly in developing nations as they become more affluent, and these nations are now creating increasing amounts of waste. To some extent, the increase in waste reflects rising material standards of living. However, it also results from an increase in packaging, manufacturing of nondurable goods, and production of inexpensive, poor-quality goods that wear out quickly. As a result, trash is piling up and littering the landscapes of countries from Mexico to Kenya to Indonesia. Over the past three decades, per capita waste generation rates have more than doubled in Latin American nations and have risen more than fivefold in the Middle East. Like U.S. consumers in the "throwaway society," wealthy consumers in developing nations often discard items that can still be used. In fact, at many dumps and landfills in the developing world, poor people support themselves by selling items they scavenge (**FIGURE 22.4**).

In many industrialized nations, per capita generation rates have leveled off or declined in recent years. For instance, note in Figure 22.3 that per capita waste production in the United States has been essentially stable since 1990. This

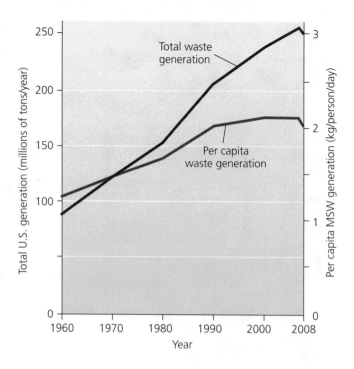

FIGURE 22.3 ▲ Total U.S. waste generation before recycling (blue line) has nearly tripled since 1960, whereas U.S. per capita waste generation before recycling (red line) has risen by 67%. Our society continues to produce more waste overall, but per person, our waste generation has leveled off in recent years, largely because of source-reduction efforts. Data from U.S. Environmental Protection Agency, 2009. *Municipal solid waste generation, recycling, and disposal in the United States: Facts and figures for 2008.*

leveling off is due largely to source reduction and reuse, especially by businesses. Moreover, increased recycling and composting has reduced the amount of waste we need to dispose of. We will examine reduction, reuse, recycling, and composting shortly, but let's first assess how we dispose of waste.

Open dumping of the past has given way to improved disposal methods

Historically, people dumped their garbage wherever it suited them. For example, until the mid-19th century, New York City's official method of garbage disposal was to dump it off piers into the East River. As population densities increased, municipalities took on the task of consolidating trash into open dumps at specified locations to keep other areas clean. To decrease the volume of trash, these dumps would be burned from time to time. Open dumping and burning still occur throughout much of the world.

As population and consumption rose in developed nations, waste increased, and dumps grew larger. At the same time, expanding cities and suburbs forced more people into the vicinity of dumps and exposed them to the noxious smoke of dump burning. In response to outcry from citizens living near dumps, and to a rising awareness of health

FIGURE 22.4 ▶ Tens of thousands of people used to scavenge each day from the dump at Payatas, outside Manila in the Philippines, finding items for themselves and selling material to junk dealers for 100–200 pesos (U.S. $2–$4) per day. That so many people could support themselves this way testifies to the immense amount of usable material needlessly discarded by wealthier portions of the population. The dump was closed in 2000 after an avalanche of trash killed hundreds of people.

and environmental threats posed by unregulated dumping and burning, many nations improved their methods of waste disposal. Most industrialized nations now bury waste in lined and covered landfills and burn waste in incineration facilities.

In the 1980s in the United States, waste generation increased while incineration was restricted, and recycling was neither economically feasible nor widely popular. As a result, landfill space became limited, and there was much talk of a "solid waste crisis." New York and other East Coast urban areas felt this situation most acutely; Fresh Kills Landfill accepted its all-time annual peak of trash, 29,000 tons, in 1986–1987. Since the late 1980s, however, recovery of materials for recycling has expanded, decreasing the pressure on landfills (**FIGURE 22.5**). As of 2008, U.S. waste managers were landfilling 54% of municipal solid waste, incinerating 13%, and recovering 33% for composting and recycling.

Sanitary landfills are regulated with health and environmental guidelines

In modern **sanitary landfills**, waste is buried in the ground or piled up in large, carefully engineered mounds. In contrast to open dumps, sanitary landfills are designed to prevent waste from contaminating the environment and threatening public health (**FIGURE 22.6**). Most municipal landfills in the United States are regulated locally or by the states, but they must meet national standards set by the U.S. Environmental Protection Agency (EPA), under the federal **Resource Conservation and Recovery Act (RCRA)**, enacted in 1976 and amended in 1984.

In a sanitary landfill, waste is partially decomposed by bacteria and compresses under its own weight to take up less space. Soil is layered along with the waste to speed decomposition, reduce odor, and lessen infestation by pests. Some infiltration of rainwater into the landfill is good, because it

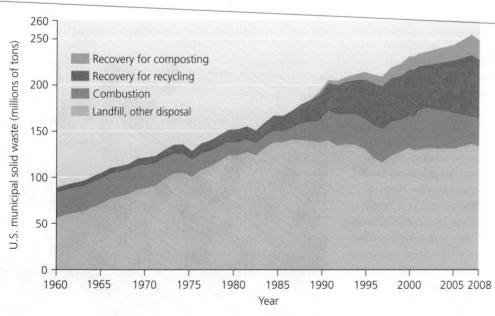

FIGURE 22.5 ◀ Since the 1980s, recycling and composting have grown in the United States, allowing a smaller proportion of waste to go to landfills. As of 2008, 54% of U.S. municipal solid waste was going to landfills and 13% to incinerators, whereas 33% was being recovered for composting and recycling. Data from U.S. Environmental Protection Agency, 2009. *Municipal solid waste generation, recycling, and disposal in the United States: Facts and figures for 2008.*

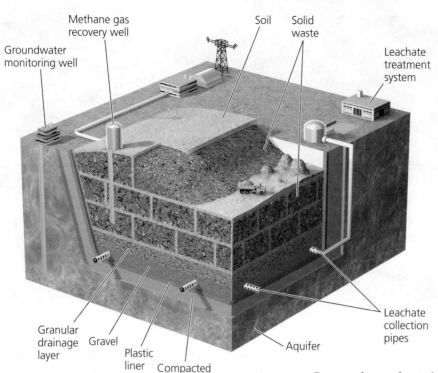

Methane gas recovery well

Groundwater monitoring well

Soil

Solid waste

Leachate treatment system

Granular drainage layer

Gravel

Plastic liner

Compacted impermeable clay

Aquifer

Leachate collection pipes

FIGURE 22.6 ◄ Sanitary landfills are engineered to prevent waste from contaminating soil and groundwater. Waste is laid in a large depression lined with plastic and impervious clay designed to prevent liquids from leaching out. Pipes of a leachate collection system draw out these liquids from the bottom of the landfill. Waste is layered along with soil until the depression is filled, and it continues to be built up until the landfill is capped. Landfill gas produced by anaerobic bacteria may be recovered, and waste managers monitor groundwater for contamination.

encourages biodegradation by aerobic and anaerobic bacteria— yet too much is not, because contaminants can escape if water carries them out.

To protect against environmental contamination, U.S. regulations require that landfills be located away from wetlands and earthquake-prone faults and be at least 6 m (20 ft) above the water table. The bottoms and sides of sanitary landfills must be lined with heavy-duty plastic and 60–120 cm (2–4 ft) of impermeable clay to help prevent contaminants from seeping into aquifers. Sanitary landfills also have systems of pipes, collection ponds, and treatment facilities to collect and treat **leachate**, liquid that results when substances from the trash dissolve in water as rainwater percolates downward. Landfill managers are required to maintain leachate collection systems for 30 years after a landfill has closed. Regulations also require that area groundwater be monitored regularly for contamination.

Once a landfill is closed, it is capped with an engineered cover that must be maintained. This cap consists of a hydraulic barrier of plastic, which prevents water from seeping down and gas from seeping up; a gravel layer above the hydraulic barrier, which drains water, lessening pressure on the hydraulic barrier; a soil barrier of at least 60 cm (24 in.), which stores water and protects the hydraulic layer from weather extremes; and a topsoil layer of at least 15 cm (6 in.), which encourages plant growth to minimize erosion.

Although it was considered a model for advanced landfill technology at the time of its construction, the Fresh Kills Landfill predated most of the EPA guidelines. As a result, it caused some environmental contamination. However, engineers have retrofitted the landfill with clay liners and a sophisticated leachate collection system. Three of the six mounds have been capped with a "final cover," and the remaining mounds will soon be capped. Indeed, the city's investment of several hundred million dollars in an existing landfill to bring environmental protection measures up to modern standards was unprecedented.

Because these safeguards need to be maintained and monitored for 30 years after closure, designs for a public park at Fresh Kills have had to work around these constraints.

Landfills have drawbacks

Despite improvements in liner technology and landfill siting, liners can be punctured, and leachate collection systems eventually cease to be maintained. Moreover, landfills are kept dry to reduce leachate, but the bacteria that break down material thrive in wet conditions. Dryness, therefore, slows waste decomposition. In fact, it is surprising how slowly some materials biodegrade when they are tightly compressed in a landfill. This was one key insight revealed by self-styled "garbologist" William Rathje, a recently retired University of Arizona archaeologist known as "the Indiana Jones of solid waste" (**FIGURE 22.7**). Since 1987, Rathje and

FIGURE 22.7 ▼ "Garbologist" William Rathje pioneered the study of our culture through the waste we generate. Here he is shown at a California landfill.

his research teams have drilled and burrowed into landfills across the United States, sorting through tons of trash to glean information about what we throw out and what happens to it. One major finding of the research is that waste doesn't decay much. Instead, in the low-oxygen conditions of most landfills, trash turns into a sort of time capsule. Rathje's team would routinely come across whole hot dogs, intact pastries that were decades old, and grass clippings that were still green. Newspapers up to 40 years old were often still legible, so the researchers used them to date layers of trash.

A second challenge with landfills is finding suitable areas to locate them, because most communities do not want them nearby. This not-in-my-backyard (NIMBY) reaction (p. 605) is one reason why New York City decided to export its waste and why residents of states receiving that waste are increasingly protesting. As a result of the NIMBY syndrome, landfills are rarely sited in neighborhoods that are home to wealthy and educated people with the political clout to keep them out. Instead, they are disproportionately sited in poor and minority communities, as environmental justice advocates (pp. 144–146) have frequently pointed out.

The unwillingness of most communities to accept waste became apparent with the famed case of the "garbage barge." In Islip, New York, in 1987, the town's landfills were full, prompting town administrators to ship waste by barge to a methane production plant in North Carolina. Prior to the barge's arrival, however, it became known that the shipment was contaminated with 16 bags of medical waste, including syringes, hospital gowns, and diapers. Because of the medical waste, the methane plant rejected the entire load. The barge sat in a North Carolina harbor for 11 days before heading for Louisiana. However, Louisiana would not permit the barge to dock. The barge traveled toward Mexico, but the Mexican navy prevented it from entering that nation's waters. In the end, the barge traveled 9,700 km (6,000 mi) before eventually returning to New York, where, after several court battles, the waste was finally incinerated at a facility in Queens.

Landfills can be transformed after closure

Today thousands of landfills lie abandoned. One reason is that waste managers have closed many smaller landfills and consolidated the trash stream into fewer, much larger, landfills. In 1988 the United States had nearly 8,000 landfills, but today it has fewer than 1,800.

Growing numbers of cities have converted closed landfills into public parks (**FIGURE 22.8**). The Fresh Kills redevelopment endeavor will be the world's largest landfill conversion project, but such efforts date back at least to 1938, when an ash landfill at Flushing Meadows, in Queens, was redeveloped for the 1939 World's Fair. The site subsequently hosted the United Nations and the 1964–1965 World's Fair. Designated a park in 1967, today the site hosts Shea Stadium, the Queens Museum of Art, the New York Hall of Science, and the Queens Botanical Garden, as well as playgrounds, wetlands, and festival events.

FIGURE 22.8 ▲ Old landfills, once properly capped, can serve other purposes. A number of them, such as Cesar Chavez Park in Berkeley, California, shown here, have been developed into areas for recreation.

Incinerating trash reduces pressure on landfills

Just as sanitary landfills are an improvement over open dumping, incineration in specially constructed facilities can be an improvement over open-air burning of trash. **Incineration**, or combustion, is a controlled process in which mixed garbage is burned at very high temperatures (**FIGURE 22.9**). At incineration facilities, waste is generally sorted and metals removed. Metal-free waste is chopped into small pieces to aid combustion and then is burned in a furnace. Incinerating waste reduces its weight by up to 75% and its volume by up to 90%.

However, the ash remaining after trash is incinerated contains toxic components and therefore must be disposed of in hazardous waste landfills (p. 638). Moreover, when trash is burned, hazardous chemicals—including dioxins, heavy metals, and polychlorinated biphenyls (PCBs) (Chapter 14)—can be created and released into the atmosphere. Such releases caused a backlash against incineration from citizens concerned about health hazards. Most developed nations now regulate incinerator emissions, and some have banned incineration outright.

As a result of real and perceived health threats from incinerator emissions—and of community opposition to these plants—engineers have developed technologies to mitigate emissions. *Scrubbers* chemically treat the gases produced in combustion to remove hazardous components and neutralize acidic gases, such as sulfur dioxide and hydrochloric acid, turning them into water and salt. Scrubbers generally do this either by spraying liquids formulated to neutralize the gases or by passing the gases through dry lime. A liquid-spraying scrubber was illustrated in Chapter 17 as an example of technology to mitigate air pollution (see Figure 17.16, p. 474).

Particulate matter is physically removed from incinerator emissions in a system of huge filters known as a *baghouse*. These tiny particles, called fly ash, often contain some of the worst dioxin and heavy metal pollutants. In addition, burning

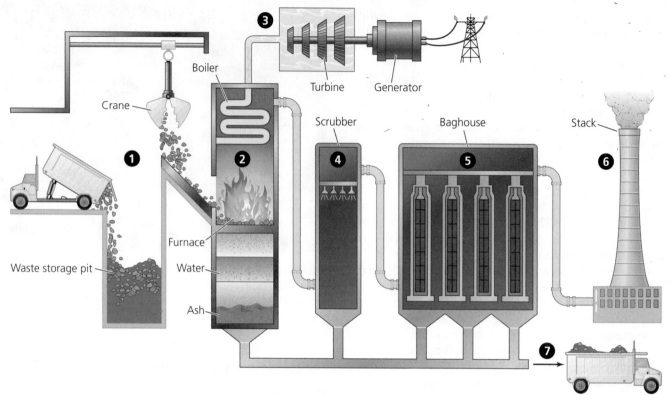

Boiler

Crane

Waste storage pit

Furnace

Water

Ash

Turbine   Generator

Scrubber

Baghouse

Stack

Wastewater and ash
for treatment or
disposal in landfill

FIGURE 22.9 ▲ Incinerators reduce the volume of solid waste by burning it but may emit toxic compounds into the air. Many incinerators are waste-to-energy (WTE) facilities that use the heat of combustion to generate electricity. In a WTE facility, solid waste ❶ is burned at extremely high temperatures ❷, heating water, which turns to steam. The steam turns a turbine ❸, which powers a generator to create electricity. In an incinerator outfitted with pollution control technology, toxic gases produced by combustion are mitigated chemically by a scrubber ❹, and airborne particulate matter is filtered physically in a baghouse ❺ before air is emitted from the stack ❻. Ash remaining from the combustion process is disposed of ❼ in a landfill.

garbage at especially high temperatures can destroy certain pollutants, such as PCBs. Even all these measures, however, do not fully eliminate toxic emissions.

624

WEIGHING THE ISSUES

Environmental Justice? Do you know where your trash goes? Where is your landfill or incinerator located? Are the people who live closest to the facility wealthy, poor, or middle class? What race or ethnicity are they? Do you know whether the people of this neighborhood protested against the introduction of the landfill or incinerator?

We can gain energy from incineration

Incineration was initially practiced simply to reduce the volume of waste, but today it often serves to generate electricity as well. Most North American incinerators now are **waste-to-energy (WTE)** facilities that use the heat produced by waste combustion to boil water, creating steam that drives electricity generation or that fuels heating systems. When burned, waste generates approximately 35% of the energy generated by burning coal. Over 100 WTE facilities are operating across the United States (mostly in the Northeast and South), with a total capacity to process nearly 100,000 tons of waste per day.

Revenues from power generation, however, are usually not enough to offset the considerable financial cost of building and running incinerators. Because it can take many years for a WTE facility to become profitable, many companies that build and operate these facilities require communities contracting with them to guarantee the facility a minimum amount of garbage. On occasion, such long-term commitments have interfered with communities' later efforts to reduce their waste through recycling and source reduction.

Landfills can produce gas for energy

Combustion in WTE plants is not the only way to gain energy from waste. Deep inside landfills, bacteria decompose waste in an oxygen-deficient environment. This anaerobic decomposition produces **landfill gas**, a mix of gases consisting of roughly half methane (pp. 28, 495–497). Landfill gas can be collected, processed, and used in the same way as natural gas (pp. 536–537).

Today hundreds of landfills in the United States and other nations are collecting landfill gas and selling it for energy. At Fresh Kills, collection wells pull landfill gas upward through a network of pipes by vacuum pressure. Landfill gas collected from Fresh Kills is sold by the city for $11 million per year and provides energy for 22,000 Staten Island homes. Where gas is not collected for commercial use, it is burned off in flares to reduce odors and greenhouse emissions.

Reducing waste is a better option

Reducing the amount of material entering the waste stream avoids costs of disposal and recycling, helps conserve resources, minimizes pollution, and can often save consumers and businesses money. Recall that preventing waste generation in this way is known as source reduction.

Much of our waste stream consists of materials used to package goods. Packaging serves worthwhile purposes—preserving freshness, preventing breakage, protecting against tampering, and providing information—but much packaging is extraneous. Consumers can give manufacturers incentive to reduce packaging by choosing minimally packaged goods, buying unwrapped fruit and vegetables, and buying food in bulk. In addition, manufacturers can use packaging that is more recyclable. They can also reduce the size or weight of goods and materials, as they already have with many items, such as aluminum cans, plastic soft drink bottles, and personal computers.

WEIGHING THE ISSUES

Reducing Packaging: Is It a Wrap?

Reducing packaging cuts down on the waste stream, but how, when, and how much should we reduce? Packaging can serve very worthwhile purposes, such as safeguarding consumer health and safety. Can you think of three products for which you would not want to see less packaging? Why? Can you name three products for which packaging could easily be reduced without ill effect to the consumer? Would you be any more or less likely to buy these products if they had less packaging?

Some governments have recently taken aim at a major source of waste and litter—plastic grocery bags. These lightweight polyethylene bags can persist for centuries in the environment, choking and entangling wildlife and littering the landscape—yet Americans discard 100 billion of them each year. A number of major cities and over 20 nations have now enacted bans or limits on their use. In 2007, San Francisco became the first U.S. city to ban nonbiodegradable plastic bags. The city's action saves 14,000 bags every day, and California is now poised to enact a statewide ban. Financial incentives are also effective. When Ireland began taxing these bags, their use dropped 90%. The Ikea company began charging for them and saw similar drops in usage. Increasing numbers of stores now give discounts if you bring your own reusable bags.

Increasing the longevity of goods also helps reduce waste. Consumers generally choose goods that last longer, all else being equal. To maximize sales, however, companies often produce short-lived goods that need to be replaced frequently. Thus, increasing the longevity of goods is largely up to the consumer. If demand is great enough, manufacturers will respond.

Reuse is a main strategy to reduce waste

To reduce waste, you can save items to use again or substitute disposable goods with durable ones. Habits as simple as bringing your own coffee cup to coffee shops or bringing

---

TABLE 22.1 Some Everyday Things You Can Do to Reduce and Reuse

- ▶ Donate used items to charity
- ▶ Reuse boxes, paper, plastic wrap, plastic containers, aluminum foil, bags, wrapping paper, fabric, packing material, etc.
- ▶ Rent or borrow items instead of buying them, when possible . . . and lend your items to friends
- ▶ Buy groceries in bulk
- ▶ Decline bags at stores when you don't need them
- ▶ Bring reusable cloth bags shopping
- ▶ Make double-sided photocopies
- ▶ Bring your own coffee cup to coffee shops
- ▶ Pay a bit extra for durable, long-lasting, reusable goods rather than disposable ones
- ▶ Buy rechargeable batteries
- ▶ Select goods with less packaging
- ▶ Compost kitchen and yard wastes in a compost bin or worm bin (often available from your community or waste hauler)
- ▶ Buy clothing and other items at resale stores and garage sales
- ▶ Use cloth napkins and rags rather than paper napkins and towels
- ▶ Write to companies to tell them what you think about their packaging and products
- ▶ When solid waste policy is being debated, let your government representatives know your thoughts
- ▶ Support organizations that promote waste reduction

Data from U.S. Environmental Protection Agency, 2005.

---

sturdy reusable cloth bags to the grocery store can, over time, have substantial impact. You can also donate unwanted items and shop for used items yourself at yard sales and resale centers. Over 6,000 reuse centers exist in the United States, including stores run by organizations that resell donated items, such as Goodwill Industries and the Salvation Army. Besides being sustainable, reusing items is often economically advantageous. Used items are often every bit as functional as new ones, and much cheaper. **TABLE 22.1** presents a sampling of actions we all can take to reduce the waste we generate.

Composting recovers organic waste

Composting is the conversion of organic waste into mulch or humus (p. 228) through natural decomposition. The compost can then be used to enrich soil. People can place waste in compost piles, underground pits, or specially constructed containers. As wastes are added, heat from microbial action builds in the interior, and decomposition proceeds. Banana peels, coffee grounds, grass clippings, autumn leaves, and countless other organic items can be converted into rich, high-quality compost through the actions of earthworms, bacteria, soil mites, sow bugs, and other detritivores and decomposers (p. 84). Home composting is a prime example of how we can live more sustainably by mimicking natural cycles and incorporating them into our daily lives.

Municipal composting programs—3,500 across the United States at last count—divert food and yard waste from the waste stream to central composting facilities, where they decompose into mulch that community residents can use for gardens and landscaping. Nearly half of U.S. states now ban yard waste from the municipal waste stream, helping accelerate the drive toward composting. Approximately one-fifth of the U.S. waste stream is made up of materials that can easily be composted. Composting reduces landfill waste, enriches soil and enhances soil biodiversity, helps soil resist erosion, makes for healthier plants and more pleasing gardens, and reduces the need for chemical fertilizers.

Recycling consists of three steps

Recycling, too, offers many benefits. Recycling involves collecting used materials and breaking them down so that they can be reprocessed to manufacture new items. Recycling diverted 61 million tons of materials away from incinerators and landfills in the United States in 2008.

The recycling loop contains three basic steps (**FIGURE 22.10**). The first step is collecting and processing used goods and materials. Communities may designate locations where residents can drop off recyclables or receive money for them. Many of these have now been replaced by the more convenient option of curbside recycling, in which trucks pick up recyclable items in front of homes, usually in conjunction with municipal trash pickup. Curbside recycling has grown rapidly, and its convenience has helped boost recycling rates. Nearly 9,000 curbside recycling programs across all 50 U.S. states now serve nearly half of all Americans.

Items collected are taken to **materials recovery facilities (MRFs)**, where workers and machines sort items, using automated processes including magnetic pulleys, optical sensors, water currents, and air classifiers that separate items by weight and size. The facilities clean the materials, shred them, and prepare them for reprocessing.

Once readied, these materials are used in manufacturing new goods. Newspapers and many other paper products use recycled paper, many glass and metal containers are now made from recycled materials, and some plastic containers are of recycled origin. Some large objects, such as benches and bridges in city parks, are now made from recycled plastics, and glass is sometimes mixed with asphalt (creating "glassphalt") for paving roads and paths. The paper used in this textbook, besides being FSC-certified (Forest Stewardship Council; pp. 314–315), also contains 10% recycled post-consumer waste.

If the recycling loop is to function, consumers and businesses must complete the third step in the cycle by purchasing products made from recycled materials. Buying recycled goods provides economic incentive for industries to recycle materials and for new recycling facilities to open or existing ones to expand. In this arena, individual consumers have power to encourage environmentally friendly options through the free market. Many businesses now advertise their use of recycled materials, a widespread instance of ecolabeling (p. 161). As markets for products made with recycled materials expand, prices continue to fall.

Recycling has grown rapidly and can expand further

The thousands of curbside recycling programs and the 500 MRFs in operation today have sprung up only in the last 25 years. Recycling in the United States has risen from 6.4% of the waste stream in 1960 to 24.4% in 2008 (and 33.2% if you include composting), according to EPA data (**FIGURE 22.11**). The EPA calls the growth of recycling "one of the best environmental success stories of the late 20th century."

Recycling rates vary greatly from one product or material type to another and from one location to another. Rates for different types of materials and products range from nearly

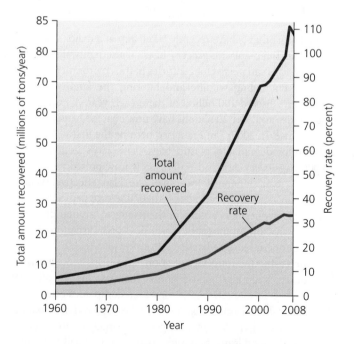

FIGURE 22.10 ▲ The familiar recycling symbol consists of three arrows to represent the three components of a sustainable recycling strategy: collection and processing of recyclable materials, use of the materials in making new products, and consumer purchase of these products.

FIGURE 22.11 ▲ Recovery has risen sharply in the United States over the past 50 years. Today over 80 million tons of material are recovered (61 million tons through recycling and 22 million tons through municipal composting), comprising one-third of the waste stream. Data from U.S. Environmental Protection Agency, 2005.

TABLE 22.2 Recovery Rates for Various Materials in the United States

Material	Percentage that is recycled or composted
Lead-acid batteries	99
Newspapers	88
Major appliances	67
Yard trimmings	65
Paper and paperboard	55
Aluminum cans	48
Glass containers	28
Total plastics	7

Data from U.S. Environmental Protection Agency.

zero to almost 100% (**TABLE 22.2**). Recycling rates among U.S. states also vary greatly, from 2% to 43% (**FIGURE 22.12**).

Recycling's growth has been propelled in part by economic forces as established businesses see opportunities to save money and as entrepreneurs see opportunities to start new businesses. It has also been driven by the desire of municipalities to reduce waste and by the satisfaction people take in recycling. These latter two forces have driven the rise of recycling even when it has not been financially profitable. In fact, many of the increasingly popular municipal recycling programs are run at an economic loss. The expense required to collect, sort, and process recycled goods is often more than recyclables are worth in the market. Furthermore, the more people recycle, the more glass, paper, and plastic is available to manufacturers for purchase, driving down prices. And transporting items to recycling facilities can sometimes involve surprisingly long distances (see **THE SCIENCE BEHIND THE STORY**, pp. 628–629).

Recycling advocates, however, point out that market prices do not take into account external costs (pp. 150–151)—in particular, the environmental and health impacts of *not* recycling. For instance, it has been estimated that globally, recycling saves enough energy to power 6 million households per year. Each

year in the United States, recycling and composting together prevent greenhouse gas emissions equal to 10 billion gallons of gasoline or 33 million cars. And recycling aluminum cans saves 95% of the energy required to make the same amount of aluminum from mined virgin bauxite, its source material.

As more manufacturers use recycled products and as more technologies and methods are developed to use recycled materials in new ways, markets should continue to expand, and new business opportunities may arise. We are still at an early stage in the shift from an economy that moves linearly from raw materials to products to waste, to an economy that moves circularly, using waste products as raw materials for new manufacturing processes. The steps we have taken in recycling so far are central to this transition, which many analysts view as key to building a sustainable economy.

WEIGHING THE ISSUES

Costs of Recycling and Not Recycling

Should recycling programs be subsidized by governments even if they are run at an economic loss? What types of external costs—costs not reflected in market prices—do you think would be involved in not recycling, say, aluminum cans? Do you feel these costs justify sponsoring recycling programs even when they are not financially self-supporting? Why or why not?

We can recycle material from landfills

With improved technology for sorting rubbish and recyclables, many businesses and entrepreneurs are weighing the economic benefits and costs of rummaging through landfills and salvaging materials of value that can be recycled. Steel, aluminum, copper, and other metals are abundant enough in some landfills to make such salvage operations profitable when market prices for the metals are high enough. For instance, Americans throw out so many aluminum cans that at 2010 prices for aluminum, the nation buries $1.8 billion of this metal in landfills each year. If we could retrieve all the aluminum

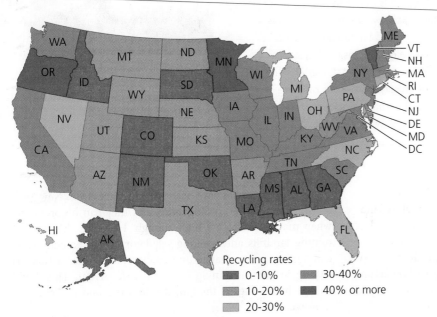

Recycling rates
- 0-10%
- 10-20%
- 20-30%
- 30-40%
- 40% or more

FIGURE 22.12 ◀ U.S. states vary greatly in the rates at which their citizens recycle. Data are for 2006 (with earlier data for several states), from Arsova, L., et al., 2008. The state of garbage in America. *BioCycle* 49 (12): 22.

The SCIENCE behind the Story

Tracking Trash

Where does your trash go once you throw it away? Where does your recycling go? How far might it travel, and how much energy might it take to get rid of it?

With the help of the latest tracking technology, we can find out. Researchers from the SENSEable City Lab at the Massachusetts Institute of Technology (MIT) are affixing tiny sensors to everyday items in our trash and monitoring them to reveal their hidden travels. By documenting what actually happens to trash and to recyclables, they hope to help make the trash removal process more effective and to encourage better recycling.

The Trash Track project was launched in 2009 in New York City and in Seattle, and it is now expanding to other cities. Inspired by PlaNYC (p. 363), which aims to raise New York City's recycling rate from 30% to 100% by 2030, the project was supported by the Architectural League of New York. Early results were unveiled in real-time at public exhibitions put on in New York and Seattle in late 2009.

Here's how trash-tracking works: Project director Carlo Ratti and associate director Assaf Biderman, both architects at MIT, organize research teams and local volunteers in the target city to affix tiny electronic tags (see **first figure**) to hundreds of different items being

Where will this cup from a New York City Starbucks shop end up?

thrown away. As each item makes its final journey through the waste stream, its tag calculates its location every few hours and relays the information via the cell phone network to a central server at the MIT lab. A computer plots the movements atop satellite maps, helping the researchers to visualize and interpret the migration of trash.

Each trash tag calculates its position by a method of triangulation using the nearest cell phone towers. The tag measures the signal strength from each cell tower within its range and compares this to a map of tower locations, allowing it to estimate its position relative to the towers and thus its exact position globally. The tags used in New York City and Seattle are not as accurate as a global positioning system (GPS), but their signals carry better through such barriers as building roofs, walls of garbage trucks, and deep piles of trash. New second-generation tags

the project is now using are more accurate, combining GPS technology with better-quality cell network triangulation.

To extend battery life to two months or more, the tags are programmed to "sleep" when they are motionless and to "wake up" and report their location frequently when they are moving (when they sense new cell towers coming into range).

As an example, a plastic container of liquid soap was tagged on September 5, 2009, and placed in the trash at 457 Madison Avenue in Manhattan (see **second figure**). Mapping reveals that the truck that picked it up looped through the city's streets a few times on its route, crossed

Tags used in New York City and Seattle used the cell phone network to calculate their location. Tags now being used in other cities use GPS technology as well.

from U.S. landfills, it would exceed the amount the world produces from a year's worth of mining ore. Besides metals, landfills also offer organic waste that can be mined and sold as premium compost. Old landfill waste can also be incinerated in newer, cleaner-burning WTE facilities to produce energy. Some companies are even looking into gaining carbon credits (p. 524) by harvesting methane leaking from huge open dumps in developing nations in Asia and Africa.

Such approaches have been tried in places from New York to Israel to Sweden to Singapore, and they can be profitable when market prices are high enough. However, the costs of mining landfills and meeting regulatory requirements while commodity prices change unpredictably have meant that investing in landfill mining has been risky so far. This could change in the future, though, if prices rise and technologies improve.

the Hudson River through the Lincoln Tunnel, then headed to Rutherford, New Jersey. Here it turned south and continued on its way, and was in transit along the Bellevue Turnpike in Kearny, New Jersey, three days later when the tag's battery gave out.

As of summer 2010, the project had posted mapped data of all its Seattle items online for the public to view, and was preparing to do the same for its New York items.

The results from Seattle reveal some expected patterns but also some odd surprises. Of the 760 trash items tagged, about 200 ended up at the city's Allied Waste Recycling Center and Transfer Station after brief journeys. Smaller numbers were transported to other city or regional landfills and recycling centers. But some items followed surprisingly circuitous routes, being transferred from one waste center to another, or back and forth between cities. Some ended up in seemingly random places, perhaps having fallen off a garbage truck and blown down a roadside.

A few items made very long journeys. Two printer cartridges were driven down Interstate 5 to California's border with Mexico, perhaps to be disassembled at a *maquiladora* (pp. 190–191) facility along the border. Two cell phones were transported halfway across the country to Dallas (one flown directly and one making apparent stops at three other cities). A compact fluorescent bulb went to St. Louis after traveling to Portland, Oregon, and back to Seattle. Chicago received a coffee cup that apparently made its way east on the nation's interstates. Shipments of batteries were flown 1,500 miles to Minneapolis, 2,500 miles to Pittsburgh, and 2,600 miles to Atlanta.

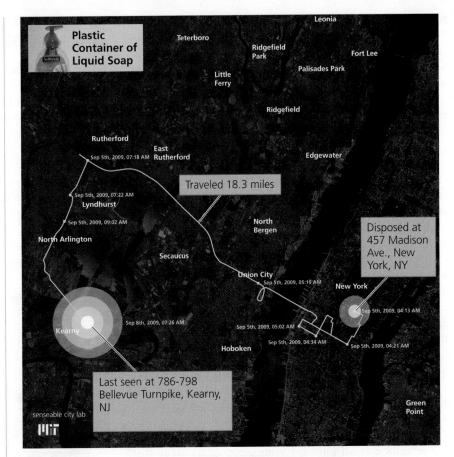

A plastic container of liquid soap put out with the trash on Madison Avenue in New York City looped through midtown Manhattan, crossed the Hudson River, and was last detected traveling down the Bellevue Turnpike in New Jersey.

The longest-traveling piece of trash was a cell phone that was transported over 3,000 miles to the other corner of the United States, ending up near Ocala, Florida.

Why did each of these trash items migrate so far? How much gasoline was used to transport them? When we move items such great distances to dispose of them or recycle their parts, does that reflect efficiency from an economy of scale—or does it indicate an excessive waste of resources? Do we need more recycling and disassembly facilities nearer to each major city? Questions like these are what MIT's researchers and the waste managers who will use their data will try to address.

A vice president for Waste Management, a company that funded the project, says he hopes the results can help the company find ways to improve the logistics of how it handles waste and recycling. And associate project director Biderman hopes that being able to see where our trash goes will encourage all of us to make more sustainable decisions about how much we consume. ■

Financial incentives can help address waste

Waste managers have employed economic incentives to reduce the waste stream. The "pay-as-you-throw" approach to garbage collection uses a financial incentive to influence consumer behavior. In these programs, municipalities charge residents for home trash pickup according to the amount of trash they put out. The less waste the household generates, the less the resident has to pay. Over 7,000 of these programs now exist in the United States, serving more than one out of every four communities.

Bottle bills represent another approach that hinges on financial incentive. Eleven U.S. states have these laws, which allow consumers to return bottles and cans to stores after use and receive a refund—generally 5 cents per bottle or can. The first bottle bills were passed in the 1970s to cut down on litter, but they have also served to decrease the waste stream. In states where they have been enacted, these laws have proved profoundly effective and resoundingly popular; they are recognized as among the most successful state legislation of recent decades (**FIGURE 22.13**). States with bottle bills have reported that their beverage container litter has decreased by 69–84%, their total litter has decreased by 30–64%, and their per capita container recycling rates have risen 2.6-fold.

As of 2010, 5 of the 11 bottle bill states were seeking to expand their programs, and 7 other states were considering establishing programs. It is a testament to the lobbying influence of the beverage industries and grocery retailers, which have traditionally opposed passage of bottle bills, that more states do not have such legislation.

States with bottle bills now face two challenges. One is to amend these laws to include new kinds of containers. In New York State, where polls showed that 70% of the public favored expanding their state's law to include more types of containers, the legislature in 2009 added bottled water containers to the law's coverage. It also mandated that most of the $140 million in unclaimed deposit money (for unreturned containers) go to the state rather than to the bottling industry.

The second challenge for bottle-bill states is to adjust refunds for inflation. In the 39 years since Oregon passed the nation's first bottle bill, the value of a nickel has dropped such that today, the refund would need to be 27 cents to reflect the refund's original intended value. Proponents argue that increasing refund amounts will raise return rates, and available data support this view (see Figure 22.13).

One Canadian city showcases reduction and recycling

Edmonton, Alberta, has created one of the world's most advanced waste management programs. Just 40% of the city's waste stream goes to its sanitary landfill, whereas 15% is recycled and an impressive 45% is composted. Almost 9 in 10 Edmonton citizens participate in its curbside recycling program.

When Edmonton's residents put out their trash, city trucks take it to their new co-composting plant—at the size of eight football fields, the largest in North America (**FIGURE 22.14A**). The waste is dumped on the floor of the facility, and large items, such as furniture, are removed and landfilled. The bulk of the waste is mixed with dried sewage sludge for 1–2 days in five large rotating drums, each the length of six buses. The resulting

(a) Composting facility, Edmonton, Alberta

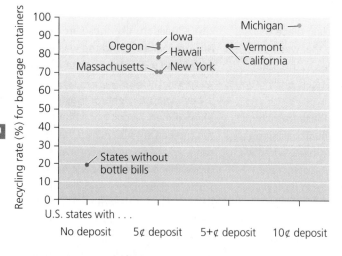

FIGURE 22.13 ▲ Data suggest that bottle bills increase recycling rates and that higher redemption amounts boost recycling rates further. The eight states with bottle bills shown in this graph all have much higher recycling rates for beverage containers than states without bottle bills. California and Vermont offer redemption rates of 5 cents for most containers and more than 5 cents for some, and these states have higher recycling rates than most states that offer only 5-cent redemption rates. Michigan, the only state with a 10-cent deposit, has the highest recycling rate of all. Data from Container Recycling Institute, Arlington, VA, 2010, using most recent data from the states, mostly from 2009. Maine, Connecticut, and Delaware also have bottle bills but have not kept detailed data on recycling rates.

(b) Aeration building, Edmonton composting facility

FIGURE 22.14 ▲ Edmonton, Alberta, boasts one of North America's most successful waste management programs. Edmonton's gigantic composting facility **(a)** is the size of eight football fields. Inside the aeration building **(b)**, which is the size of 14 professional hockey rinks, mixtures of solid waste and sewage sludge are exposed to oxygen and composted for 14–21 days.

mix travels on a conveyor to a screen that removes non-biodegradable items. It is aerated for several weeks in the largest stainless steel building in North America (**FIGURE 22.14B**). The mix is then passed through a finer screen and finally is left outside for 4–6 months. The resulting compost—80,000 tons annually—is made available to area farmers and residents. The facility even filters the air it emits with a layer of compost, bark, and wood chips to eliminate odors.

Edmonton's program also includes a state-of-the-art MRF that handles 30,000–40,000 tons of waste annually, a leachate treatment plant, a research center, public education programs, and a wetland and landfill revegetation program. In addition, 100 pipes collect enough landfill gas to power 4,600 homes, bringing thousands of dollars to the city and helping power the new waste management center. Five area businesses reprocess the city's recycled items. Newsprint and magazines are turned into new newsprint and cellulose insulation, and cardboard and paper are converted into building paper and shingles. Household metal is made into rebar and blades for tractors and graders, and recycled glass is used for reflective paint and signs.

Moreover, Edmonton has just built a new integrated processing and transfer facility to handle both compostable and recyclable waste from homes and businesses, and the city will soon complete a biofuels facility to create ethanol (p. 579) from waste that cannot be recycled or composted. With these new facilities, Edmonton hopes to achieve 90% recovery by 2013.

INDUSTRIAL SOLID WASTE

Each year, U.S. industrial facilities generate about 7.6 billion tons of waste, according to the EPA, about 97% of which is wastewater. Thus, very roughly, 230 million or so tons of solid waste are generated by 60,000 facilities each year—an amount about equal to that of municipal solid waste. In the United States, *industrial solid waste* is defined as solid waste that is considered neither municipal solid waste nor hazardous waste under the Resource Conservation and Recovery Act.

Whereas the federal government regulates municipal solid waste, state or local governments regulate industrial solid waste (with federal guidance). Industrial waste includes waste from factories, mining activities, agriculture, petroleum extraction, and more. Waste is generated at various points along the process from raw materials extraction to manufacturing to sale and distribution (**FIGURE 22.15**).

Regulation and economics each influence industrial waste generation

Most methods and strategies of waste disposal, reduction, and recycling by industry are similar to those for municipal solid waste. For instance, businesses that manage their own waste on site generally dispose of it in landfills, and companies must design and manage their landfills in ways that meet state, local, or tribal guidelines. Other businesses pay to have their waste disposed of at municipal disposal sites. Regulation varies greatly from state to state and area to area, but in most

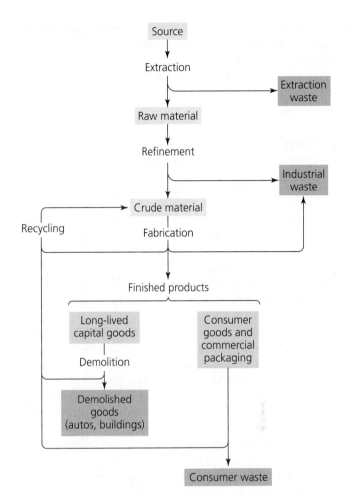

FIGURE 22.15 ▲ Industrial and municipal waste is generated at a number of stages throughout the life cycles of products. Waste is first generated when raw materials needed for production are extracted. Further industrial waste is produced as raw materials are processed and as products are manufactured. Waste results from the demolition or disposal of products after businesses and individuals use them. Each stage often presents opportunities for efficiency improvements, waste reduction, or recycling.

cases, state and local regulation of industrial solid waste is less strict than federal regulation of municipal solid waste. In many areas, industries are not required to have permits, install landfill liners or leachate collection systems, or monitor groundwater for contamination.

The amount of waste generated by a manufacturing process is a good measure of its efficiency; the less waste produced per unit or volume of product, the more efficient that process is, from a physical standpoint. However, physical efficiency is not always reflected in economic efficiency. Often it is cheaper for industry to manufacture its products or perform its services quickly but messily. That is, it can be cheaper to generate waste than to avoid generating waste. In such cases, economic efficiency is maximized, but physical efficiency is not. Because our market system awards only economic efficiency, all too often industry has no financial incentive to achieve physical efficiency. The frequent mismatch between these two types of efficiency is a major reason why the output of industrial waste is so great.

Rising costs of waste disposal, however, enhance the financial incentive to decrease waste and increase physical efficiency. Once either government or the market makes the physically efficient use of raw materials also economically efficient, businesses have financial incentives to reduce their own waste.

Industrial ecology seeks to make industry more sustainable

To reduce waste, growing numbers of industries today are experimenting with industrial ecology. A holistic approach that integrates principles from engineering, chemistry, ecology, and economics, **industrial ecology** seeks to redesign industrial systems to reduce resource inputs and to maximize both physical and economic efficiency. Industrial ecologists would reshape industry so that nearly everything produced in a manufacturing process is used, either within that process or in a different one.

The larger idea behind industrial ecology is that industrial systems should function more like ecological systems, in which almost everything produced is used by some organism, with very little being wasted. This principle brings industry closer to the ideal of ecological economists, in which human economies attain sustainability by functioning in a circular fashion rather than a linear one (p. 155).

Industrial ecologists pursue their goals in several ways. For one, they examine the entire life cycle of a given product—from its origins in raw materials, through its manufacturing, to its use, and finally its disposal—and look for ways to make the process more efficient. This strategy is called **life-cycle analysis**.

Industrial ecologists also try to identify how waste products from one manufacturing process might be used as raw materials for a different process. For instance, used plastic beverage containers cannot be refilled because of the potential for contamination, but they can be shredded and reprocessed to make other plastic items, such as benches, tables, and decks. In addition, industrial ecologists examine industrial processes with an eye toward eliminating environmentally harmful products and materials. Finally, they study the flow of materials through industrial systems to look for ways to create products that are more durable, recyclable, or reusable. Goods that are currently thrown away when they become obsolete, such as computers, automobiles, and some appliances, could be designed to be more easily disassembled so more of their component parts can be easily reused or recycled.

Businesses are adopting industrial ecology

Attentive businesses are taking advantage of the insights of industrial ecology to save money while reducing waste. For example, American Airlines switched from hazardous to nonhazardous materials in its Chicago facility, decreasing its need to secure permits from the EPA. The company used over 50,000 reusable plastic containers to ship goods, reducing packaging waste by 90%. Its Dallas–Fort Worth headquarters recycled enough aluminum cans and white paper in five years to save $205,000 and recycled 3,000 broken baggage containers into lawn furniture. A program to gather suggestions from employees brought over 700 ideas to reduce waste—and 15 of these ideas saved the company over $8 million in the first year of implementation.

The Swiss Zero Emissions Research and Initiatives (ZERI) Foundation sponsors dozens of innovative projects worldwide that attempt to create goods and services without generating waste. One example involves breweries, currently being pursued in Canada, Sweden, Japan, and Namibia (**FIGURE 22.16**). Brewers in these projects take waste from the

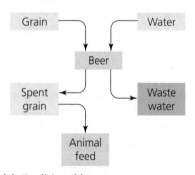

(a) Traditional brewery process

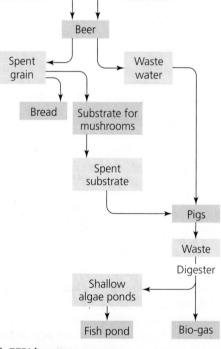

(b) ZERI brewery process

FIGURE 22.16 ◀ Traditional breweries **(a)** produce only beer while generating much waste, some of which goes toward animal feed. Breweries sponsored by the Zero Emissions Research and Initiatives (ZERI) Foundation **(b)** use their waste grain to make bread and to farm mushrooms. Waste from the mushroom farming, along with brewery wastewater, goes to feed pigs. The pigs' waste is digested in containers that capture natural gas and collect nutrients used to nourish algae for growing fish in fish farms. The brewer derives income from bread, mushrooms, pigs, gas, and fish, as well as beer.

beer-brewing process and use it to fuel other processes. As a result, the brewer can make money from bread, mushrooms, pigs, gas, and fish, as well as beer, all while producing little waste. Although most ZERI projects are not fully closed-loop systems, they attempt to approach this ideal. In so doing, they cut down on waste while increasing output and income, and often they generate new jobs as well.

Few businesses have taken industrial ecology to heart as much as the carpet tile company Interface, which founder Ray Anderson set on the road to sustainability over a decade ago (p. 162). Interface asks customers to return used tiles for recycling and for reuse as backing for new carpet. It modified its tile design and its production methods to reduce waste. It adapted its boilers to use landfill gas for its energy needs. Through such steps, Anderson's company has cut its waste generation by 80%, its fossil fuel use by 45%, and its water use by 70%—all while saving $30 million per year, holding prices steady for its customers, and raising profits by 49%.

For businesses, governments, and individuals alike, there are plenty of ways to reduce waste and mitigate the impacts of our waste generation—and quite often, doing so brings economic benefits. This is true both for solid waste and for hazardous waste.

HAZARDOUS WASTE

Hazardous wastes are diverse in their chemical composition and may be liquid, solid, or gaseous. By EPA definition, **hazardous waste** is waste that is one of the following:

▸ *Ignitable.* Substances that easily catch fire (for example, natural gas or alcohol).

▸ *Corrosive.* Substances that corrode metals in storage tanks or equipment.

▸ *Reactive.* Substances that are chemically unstable and readily react with other compounds, often explosively or by producing noxious fumes.

▸ *Toxic.* Substances that harm human health when they are inhaled, are ingested, or contact human skin.

Materials with these characteristics can harm human health and environmental quality. Flammable and explosive materials can cause ecological damage and atmospheric pollution. For instance, fires at large tire dumps in California's Central Valley have caused air pollution and highway closures. Toxic wastes in lakes and rivers have caused fish die-offs and closed important domestic fisheries, such as those in Chesapeake Bay.

Hazardous wastes are diverse

Industry, mining, households, small businesses, agriculture, utilities, and building demolition all create hazardous waste. Industry produces the largest amounts of hazardous waste, but in most developed nations industrial waste generation and disposal is highly regulated. This regulation has reduced the amount of hazardous waste entering the environment

from industrial activities. As a result, households currently are the largest source of unregulated hazardous waste.

Household hazardous waste includes a wide range of items, such as paints, batteries, oils, solvents, cleaning agents, lubricants, and pesticides. U.S. citizens generate 1.6 million tons of household hazardous waste annually, and the average home contains close to 45 kg (100 lb) of it in sheds, basements, closets, and garages.

Although many hazardous substances become less hazardous over time as they degrade chemically, two classes of chemicals are particularly hazardous because their toxicity persists over time: organic compounds and heavy metals.

Organic compounds and heavy metals can be hazardous

In our day-to-day lives, we rely on the capacity of synthetic organic compounds and petroleum-derived compounds to resist bacterial, fungal, and insect activity. Items such as plastic containers, rubber tires, pesticides, solvents, and wood preservatives are useful to us precisely because they resist decomposition. We use these substances to protect our buildings from decay, kill pests that attack crops, and keep stored goods intact. However, the resistance of these compounds to decay is a double-edged sword, for it also makes them persistent pollutants. Many synthetic organic compounds are toxic because they can be readily absorbed through the skin and can act as mutagens, carcinogens, teratogens, and endocrine disruptors (pp. 380–381).

Heavy metals such as lead, chromium, mercury, arsenic, cadmium, tin, and copper are used widely in industry for wiring, electronics, metal plating, metal fabrication, pigments, and dyes. Heavy metals enter the environment when paints, electronic devices, batteries, and other materials are disposed of improperly. Lead from fishing weights and from hunters' lead shot has accumulated in many rivers, lakes, and forests. In older homes, lead from pipes contaminates drinking water, and lead paint remains a problem, especially for infants. Heavy metals that are fat-soluble and break down slowly are prone to bioaccumulate and biomagnify (p. 385). In California's Coast Range, for instance, mercury washed downstream from abandoned mercury mines enters low-elevation lakes and rivers, is consumed by bacteria and invertebrates, and accumulates in increasingly larger quantities up the food chain, poisoning organisms at higher trophic levels and making fish unsafe to eat.

"E-waste" is growing

Today's proliferation of computers, printers, cell phones, handheld devices, TVs, DVD players, fax machines, MP3 players, and other electronic technology has created a substantial new source of waste (see **ENVISIONIT**, page 634). These products have short lifetimes before people judge them obsolete, and most are discarded after only a few years. The amount of this **electronic waste**—often called **e-waste**—is growing rapidly, and now comprises 2% of the U.S. solid waste stream. Over 3 billion electronic devices have been sold in the United States since 1980. Of these, half have been disposed of, about 40% are still being used (or reused), and 10% are in storage. American

ENVISION IT

Every five minutes, Americans throw away the number of cell phones shown on this page.

The 426,000 cell phones entering the U.S. waste stream daily can leach toxic heavy metals into the environment ...

... or they can be recycled for reuse and for the recovery of valuable metals.

YOU CAN MAKE A DIFFERENCE

➤ Recycle your old phone with an approved e-waste recycling service.

➤ Donate your phone to a person or a charity that can reuse it.

➤ Think twice before buying yet another new electronic gadget that you don't really need.

households discard close to 400 million electronic devices per year—two-thirds of them still in working order.

Of the electronic items we discard, roughly four of five go to landfills and incinerators, where they have traditionally been treated as conventional solid waste. However, most electronic products contain heavy metals and toxic flame retardants, and recent research suggests that e-waste should instead be treated as hazardous waste (see **THE SCIENCE BEHIND THE STORY**, pp. 636–637). The EPA and a number of states are now taking steps to keep e-waste out of conventional sanitary landfills and incinerators and instead treat it as hazardous waste.

More and more e-waste today is being recycled. The devices are taken apart, and parts and materials are refurbished and reused in new products. According to EPA estimates, Americans were recycling 15% of e-waste in 1999, and this rose to 18% by 2007. However, so many more items have been manufactured each year that the amount of e-waste we sent to landfills and incinerators in that time period increased by a greater amount. **FIGURE 22.17** shows how disposal has risen faster than recycling: In 2007 we recycled 45 million more tons of e-waste than in 1999, but we also disposed of 169 million more tons of e-waste in landfills and incinerators.

Besides keeping toxic substances out of our environment, e-waste recycling is beneficial because a number of

FIGURE 22.18 ▲ The gold, silver, and bronze medals awarded to athletes at the 2010 Winter Olympic Games in Vancouver were manufactured in part from precious metals recycled from discarded electronic waste items.

trace metals used in electronics are globally rare, so they can be lucrative to recover. As we will see in the next chapter (p. 661), a typical cell phone contains close to a dollar's worth of precious metals. Every bit of metal we can recycle from a manufactured item is a bit of metal we don't need to mine from the ground, so "mining" e-waste for precious metals helps reduce the environmental impacts that mining exerts. By one estimate, 1 ton of computer scrap contains more gold than 16 tons of mined ore from a gold mine. In one of the more intriguing efforts to promote sustainability through such recycling, the 2010 Winter Olympic Games in Vancouver produced its stylish gold, silver, and bronze medals (**FIGURE 22.18**) from metals recovered from recycled and processed e-waste!

There are serious concerns, however, about the health risks that recycling may pose to workers doing the disassembly. Wealthy nations ship much of their e-waste to developing countries, where low-income workers disassemble the devices and handle toxic materials with minimal safety regulations. These environmental justice concerns need to be resolved, but if electronics recycling can be done responsibly, it seems likely to be the way of the future.

In many North American cities, used electronics are collected by businesses, nonprofit organizations, or municipal services, and are processed for reuse or recycling. So next time you upgrade to a new computer, TV, DVD player, cell phone, or handheld device, find out what opportunities exist in your area to recycle your old ones.

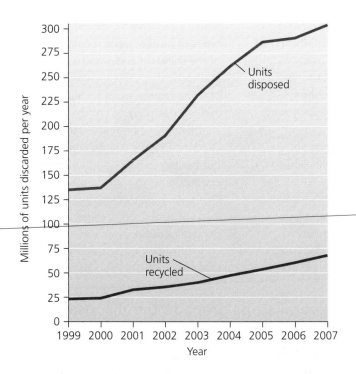

FIGURE 22.17 ▲ More electronic items are being recycled in the United States each year (red line), but for every e-waste item recycled, more than four are disposed of in landfills or incinerators (blue line). Thus, even though the amount and percentage of recycled items has grown, the amount of e-waste we are sending to landfills and incinerators has increased by a greater amount. Data from U.S. Environmental Protection Agency, 2008. *Electronic waste management in the United States: Approach 1.*

The SCIENCE behind the Story

Testing the Toxicity of "E-Waste"

Most electronic waste, or "e-waste," is disposed of in conventional sanitary landfills. However, most electronic appliances contain heavy metals that can cause environmental contamination and public health risks. For instance, over 6% of a typical computer is composed of lead.

The EPA funded Timothy Townsend's lab at the University of Florida at Gainesville to determine whether e-waste is toxic enough to be classified as hazardous waste under the Resource Conservation and Recovery Act.

With students and colleagues, Townsend determined in 1999–2000 that cathode ray tubes (CRTs) from computer monitors and color televisions leach an average of 18.5 mg/L of lead, far above the regulatory threshold of 5 mg/L. Following this research, the EPA proposed classifying CRTs as hazardous waste, and several U.S. states banned these items from conventional landfills.

Then in 2004, Townsend's lab group completed experiments on 12 other types of electronic devices. To measure their toxicity, Townsend's group used the EPA's standard test,

Dr. Brajesh Dubey (L) and Dr. Timothy Townsend (R) preparing the TCLP test

the Toxicity Characteristic Leaching Procedure (TCLP), designed to mimic the process by which chemicals leach out of solid waste in landfills. In the TCLP, waste is ground up into fine pieces, and 100 g (3.5 oz) of it is put in a container with 2 L (0.53 gal) of an acidic leaching fluid. The container is rotated for 18 hours, after which the leachate is analyzed for its chemical content. Researchers look for eight heavy metals—arsenic, barium, cadmium, chromium, lead, mercury, selenium, and silver—and determine for each whether their concentration in the leachate exceeds that allowed by EPA regulations. Of these eight elements, electronic devices contain notable amounts of four: cadmium, chromium, lead, and mercury.

To conduct the standard TCLP, Townsend's team ground up the central processing units (CPUs) of personal computers, creating a mix made up by weight of 15.8% circuit board, 7.5% plastic, 68.2% ferrous metal, 5.4% nonferrous metal, and 3.1% wire and cable. However, grinding up a computer into small bits is no easy task, and it is hard to obtain a sample that accurately represents all components and materials. So the researchers also designed a modified TCLP test in which they placed whole CPUs—with the parts disassembled but not ground up—in a rotating 55-gallon drum full of leaching liquid. Then they tested their 12 types of devices using a combination of the standard and modified TCLP methods.

The team's results are summarized in the accompanying bar chart. Lead was the only heavy metal found to exceed the EPA's regulatory threshold, but this threshold (5 mg/L) was exceeded in the majority of trials. Computer monitors leached the most lead (47.7 mg/L on average), as expected, because monitors include the

WEIGHING THE ISSUES

Toxic Computers? The cathode ray tubes in televisions and computer screens can hold up to 5 kg (8 lb) of heavy metals, such as lead and cadmium. These represent the second-largest source of lead in U.S. landfills today, behind auto batteries. With more computer screens being purchased, the transition to high-definition television, and the rapid turnover of computers, what future waste problems and environmental health issues might you expect? How do you think we should handle the reuse, recycling, and disposal of these products?

Several steps precede the disposal of hazardous waste

For many years we discarded hazardous waste without special treatment. In many cases, people did not know that certain substances were harmful to human health. In other cases, the danger posed by these substances was known or suspected, but it was assumed that the substances would disappear or be sufficiently diluted in the environment. The resurfacing of toxic chemicals in a residential area years after their burial at Love Canal (p. 639) in upstate New York demonstrated to the public that hazardous waste deserves special attention and treatment.

Discarded electronic waste can leach heavy metals and should be considered hazardous waste, researchers say.

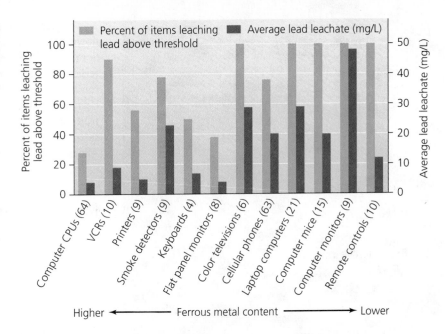

Some proportion of all 12 devices tested exceeded the EPA regulatory standard for lead leachate. Devices with higher ferrous metal content tended to leach less lead. Where both standard and modified Toxicity Characteristic Leaching Procedures (TCLPs) were used, results are averaged. Data from Townsend, T.G., et al. 2004. *RCRA toxicity characterization of computer CPUs and other discarded electronic devices.* July 15, 2004, report to the U.S. EPA.

cathode ray tubes already known to be a problem. However, laptops, color TVs, smoke detectors, cell phones, and computer mice also leached high levels of lead. Next came remote controls, VCRs, keyboards, and printers, all of which leached more lead on average than the EPA threshold, and did so in 50% or more of the trials. Whole CPUs and flat panel monitors were the only devices to leach less than 5 mg/L of lead on average, but even these exceeded the threshold more than one-quarter of the time.

The researchers found that items containing more ferrous metals (such as iron) tended to leach less lead. For instance, CPUs contain 68% ferrous metals (compared to only 7% in laptops), and laptops leached seven times as much lead as CPUs. Further experiments confirmed that ferrous metals were chemically reacting with lead and stopping it from leaching.

Townsend says the work suggests that many electronic devices have the potential to be classified as hazardous waste because they frequently surpass the toxicity criterion for lead. However, EPA scientists must decide how to judge results from the modified TCLP methods, and must evaluate other research, before determining whether to alter regulatory standards.

Furthermore, lab tests may or may not accurately reflect what actually happens in landfills. So Townsend's team is filling columns measuring 24 cm (2 ft) wide by 4.9 m (16 ft) long with e-waste and municipal solid waste, burying them in a Florida landfill, and then testing the leachate that results. The results from such research should help regulators decide how best to dispose of the e-waste that is not reused or recycled. ■

Since the 1980s, many communities have designated sites or special collection days to gather household hazardous waste, or have designated facilities for the exchange and reuse of substances (**FIGURE 22.19**). Once consolidated in such sites, the waste is transported for treatment and ultimate disposal.

Under the Resource Conservation and Recovery Act, the EPA sets standards by which states are to manage hazardous waste. RCRA also requires large generators of hazardous waste to obtain permits and mandates that hazardous materials be tracked "from cradle to grave." As hazardous waste is generated, transported, and disposed of, the producer, carrier, and disposal facility must each report to the EPA the type and amount of material generated; its location, origin, and destination; and the way it is being handled. This process is intended to prevent illegal dumping and to encourage the use of reputable waste carriers and disposal facilities.

Because current U.S. law makes disposing of hazardous waste quite costly, irresponsible companies sometimes illegally dump waste, creating health risks for residents and financial headaches for local governments forced to deal with the mess (**FIGURE 22.20**).

Hazardous waste from industrialized nations is also sometimes dumped illegally in developing nations—a major environmental justice issue (p. 146). This occurs despite the Basel Convention, an international treaty to prevent this practice. In 2006, a ship secretly dumped toxic wastes in

FIGURE 22.19 ▲ Many communities designate collection sites or collection days for household hazardous waste. Here, workers handle waste from an Earth Day collection event near Los Angeles.

Abidjan, the capital of the Ivory Coast, after being told by Dutch authorities that the Netherlands would charge it money to dispose of the waste in Amsterdam. The waste caused several deaths and thousands of illnesses in Abidjan, and street protests forced the government to resign over the scandal. Because of the difficulty of tracking deliveries in the international shipping industry, the responsible parties have not yet been brought to justice.

High costs of disposal, however, have also encouraged conscientious businesses to invest in reducing their hazardous

FIGURE 22.20 ▼ Unscrupulous individuals or businesses sometimes dump hazardous waste illegally to avoid disposal costs.

waste. Many biologically hazardous materials can be broken down by incineration at high temperatures in cement kilns. Some hazardous materials can be treated by exposure to bacteria that break down harmful components and synthesize them into new compounds. Additionally, various plants have been bred or engineered to take up specific contaminants from soil and then break down organic contaminants into safer compounds or concentrate heavy metals in their tissues. The plants are eventually harvested and disposed of.

We have three disposal methods for hazardous waste

We have developed three primary means of hazardous waste disposal: landfills, surface impoundments, and injection wells. These do nothing to lessen the hazards of the substances, but they help keep the waste isolated from people, wildlife, and ecosystems. Design and construction standards for landfills that receive hazardous waste are stricter than those for ordinary sanitary landfills. Hazardous waste landfills must have several impervious liners and leachate removal systems and must be located far from aquifers. Dumping of hazardous waste in ordinary landfills has long been a problem. In New York City, Fresh Kills largely managed to keep hazardous waste out, but most of the city's older landfills were declared to be hazardous sites because of past toxic waste dumping.

Liquid hazardous waste, or waste in dissolved form, may be stored in ponds or **surface impoundments**, shallow depressions lined with plastic and an impervious material, such as clay. Water containing dilute hazardous waste is placed in the pond and allowed to evaporate, leaving a residue of solid hazardous waste on the bottom (**FIGURE 22.21**). This process is repeated until the dry material is removed and transported elsewhere for permanent disposal. Impoundments are not ideal. The underlying layer can crack and leak waste. Some material may evaporate or blow into surrounding areas. Rainstorms may cause waste to overflow and contaminate nearby areas. For these reasons, surface impoundments are used only for temporary storage.

The third method is intended for long-term disposal. In **deep-well injection**, a well is drilled deep beneath the water table into porous rock, and wastes are injected into it (**FIGURE 22.22**). The waste is meant to remain deep underground, isolated from groundwater and human contact. However, wells can corrode and can leak wastes into soil, contaminating aquifers. Roughly 34 billion L (9 billion gal) of hazardous waste continue to be placed in U.S. injection wells each year.

Radioactive waste is especially hazardous

Radioactive waste is particularly dangerous to human health and is persistent in the environment. The dilemma of disposal has dogged the nuclear energy industry and the U.S. military for decades. As we saw in our discussion of radioactive waste disposal in Chapter 20 (pp. 573–576), the United States has no designated single site to dispose of its commercial nuclear waste if Yucca Mountain in Nevada is removed from consideration. Instead, waste will continue to accumulate at the many nuclear power plants spread through the nation.

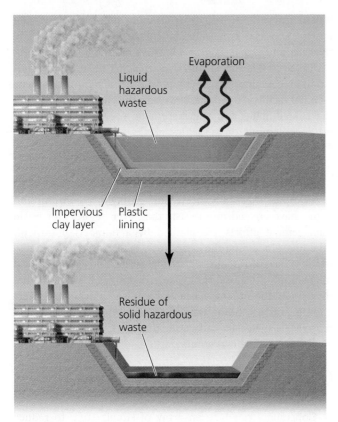

FIGURE 22.21 ▲ Surface impoundments are a strategy for temporarily disposing of liquid hazardous waste. The waste, mixed with water, is poured into a shallow depression lined with plastic and clay to prevent leakage. When the water evaporates, leaving a crust of the hazardous substance, new liquid is poured in and the process repeated. This method alone is not satisfactory, because waste can potentially leak, overflow, evaporate, or blow away.

Currently, a site in the Chihuahuan Desert in southeastern New Mexico serves as a permanent disposal site for radioactive waste. The Waste Isolation Pilot Plant (WIPP) is the world's first underground repository for transuranic waste from nuclear weapons development. The mined caverns holding the waste are located 655 m (2,150 ft) below ground in a huge salt formation thought to be geologically stable. Twenty years in the planning, WIPP became operational in 1999 and is receiving thousands of shipments of waste from 23 other locations over the next three decades.

Contaminated sites are being cleaned up, slowly

Many thousands of former military and industrial sites remain contaminated with hazardous waste in the United States and virtually every other nation on Earth. For most nations, dealing with these messes is simply too difficult, time-consuming, and expensive. In 1980, however, the U.S. Congress passed the Comprehensive Environmental Response Compensation and Liability Act (CERCLA). This legislation established a federal program to clean up U.S. sites polluted with hazardous waste from past activities. The EPA administers this cleanup program, called the **Superfund**. Under EPA auspices, experts identify sites polluted with hazardous chemicals, take action to

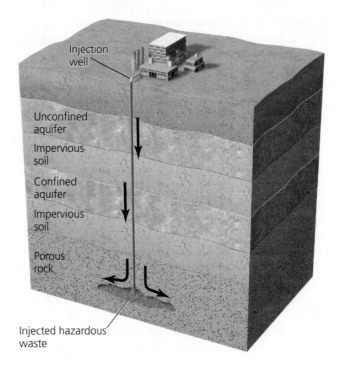

FIGURE 22.22 ▲ Liquid hazardous waste may be pumped deep underground, by deep-well injection. The well must be drilled below any aquifers, into porous rock separated by impervious clay. The technique is expensive, and waste may leak from the well shaft into groundwater.

protect groundwater near these sites, and clean up the pollution. Later laws also charged the EPA with cleaning up **brownfields**, lands whose reuse or development are complicated by the presence of hazardous materials.

Two well-publicized events spurred creation of the Superfund legislation. In *Love Canal*, a residential neighborhood in Niagara Falls, New York, families were evacuated in 1978–1980 after toxic chemicals buried by a company and the city in past decades rose to the surface, contaminating homes and an elementary school. In Missouri, the entire town of *Times Beach* was evacuated and its buildings demolished after being contaminated by dioxin (p. 396) from waste oil sprayed on its roads.

Once a Superfund site is identified, EPA scientists evaluate how close the site is to human habitation, whether wastes are currently confined or likely to spread, and whether the site threatens drinking water supplies. Sites that appear harmful are placed on the EPA's National Priority List, ranked according to the level of risk to human health that they pose. Cleanup proceeds on a site-by-site basis as funds are available. Throughout the process, the EPA is required to hold public hearings to inform area residents of its findings and to receive feedback.

The objective of CERCLA was to charge the polluting parties for cleanup of their sites, according to the *polluter-pays principle* (p. 170). For many sites, however, the responsible parties cannot be found or held liable, and in such cases—roughly one of four so far—Superfund activities have been covered by taxpayers' funds and from a trust fund established by a federal tax on industries producing petroleum and chemical raw materials. However, Congress let the tax expire, the trust fund went bankrupt in 2004, and neither the president nor Congress has moved to restore it, so taxpayers are now shouldering the entire burden of the program. As funding dwindles and the remaining cleanup jobs become more expensive, fewer cleanups are being completed.

As of mid-2010, 1,277 Superfund sites remained on the National Priorities List, and only 343 have been cleaned up or otherwise deleted from the list. The average cleanup has cost over $25 million and has taken nearly 15 years. Many sites are contaminated with hazardous chemicals we have no effective way to deal with. In such cases, cleanups simply involve trying to isolate waste from human contact, either by building trenches and clay or concrete barriers around a site or by excavating contaminated material, placing it in industrial-strength containers, and shipping it to a hazardous waste disposal facility. For all these reasons, the current emphasis in the United States and elsewhere is on preventing hazardous waste contamination in the first place.

➤ CONCLUSION

Our societies have made great strides in addressing our waste problems. Modern methods of waste management are far safer for people and gentler on the environment than past practices of open dumping and open burning. In many countries, recycling and composting efforts are making rapid strides. The United States has gone in a few decades from a country that did virtually no recycling to a nation in which one-third of all solid waste is diverted from disposal. The continuing growth of recycling, driven by market forces, government policy, and consumer behavior, shows potential to further alleviate our waste problems.

Despite these advances, our prodigious consumption habits have created more waste than ever before. Our waste management efforts are marked by a number of difficult dilemmas, including the cleanup of Superfund sites, safe disposal of hazardous and radioactive waste, and frequent local opposition to disposal sites. These dilemmas make clear that the best solution to our waste problem is to reduce our generation of waste. Finding ways to reduce, reuse, and efficiently recycle the materials and goods that we use stands as a key challenge for the new century.

REVIEWING OBJECTIVES

You should now be able to:

SUMMARIZE AND COMPARE THE TYPES OF WASTE WE GENERATE

- Municipal and industrial solid waste, hazardous waste, and wastewater are major types of waste. (p. 618)

LIST THE MAJOR APPROACHES TO MANAGING WASTE

- Source reduction, recovery, and disposal are the three main components of waste management. (pp. 618–619)

DELINEATE THE SCALE OF THE WASTE DILEMMA

- Developed nations generate far more waste than developing nations. (p. 620)
- Waste everywhere is increasing as a result of growth in population and consumption. (p. 620)

DESCRIBE CONVENTIONAL WASTE DISPOSAL METHODS: LANDFILLS AND INCINERATION

- Sanitary landfills guard against contamination of groundwater, air, and soil. Nonetheless, such contamination can occur. (pp. 621–623)
- Incinerators reduce waste volume by burning it. Pollution control technology removes most pollutants from emissions, but some escape, and highly toxic ash needs to be disposed of in landfills. (pp. 623–624)
- We are harnessing energy from landfill gas and generating electricity from incineration. (p. 624)

EVALUATE APPROACHES FOR REDUCING WASTE: SOURCE REDUCTION, REUSE, COMPOSTING, AND RECYCLING

- Reducing waste before it is generated is the best waste management approach. Recovery is the next-best option. (p. 625)

- Consumers can take an array of simple steps to reduce their waste output. (p. 625)
- Composting reduces waste while creating organic matter for gardening and agriculture. (pp. 625–626)
- Recycling has grown in recent years and now removes 24% of the U.S. waste stream. (pp. 626–627)
- We can recycle materials from landfills, given the right technology and high-enough market prices. (pp. 627–628)
- Bottle bills are one way in which financial incentives can be used to reduce waste. (pp. 629–630)

DISCUSS MANAGEMENT OF INDUSTRIAL SOLID WASTE AND PRINCIPLES OF INDUSTRIAL ECOLOGY

- Regulations differ, but industrial waste management is similar to that for municipal solid waste. (p. 631)
- Industrial ecology urges industrial systems to mimic ecological systems and provides ways for industry to increase its efficiency. (pp. 632–633)

ASSESS ISSUES IN MANAGING HAZARDOUS WASTE

- Hazardous waste is ignitable, corrosive, reactive, or toxic. (p. 633)
- Electronic waste may be considered hazardous, and constitutes a growing waste source. (pp. 633–637)
- Hazardous waste is regulated and monitored, yet illegal dumping remains a problem. (pp. 636–638)
- No fully satisfactory method of disposing of hazardous waste has yet been devised. (pp. 638–639)
- The Superfund program cleans up hazardous waste sites, but cleanup is a long and expensive process. (pp. 639–640)

TESTING YOUR COMPREHENSION

1. Describe five major methods of managing waste. Why do we practice waste management?

2. Why have some people labeled the United States "the throwaway society"? How much solid waste do Americans generate, and how does this amount compare to that of people from other countries?

3. Name several guidelines by which sanitary landfills are regulated. Describe three problems with landfills.

4. Describe the process of incineration or combustion. What happens to the resulting ash? What is one drawback of incineration?

5. What is composting, and how does it help reduce input to the waste stream?

6. What are the three elements of a sustainable process of recycling?

7. What are the goals of industrial ecology?

8. What four criteria are used to define hazardous waste? Why are heavy metals and synthetic organic compounds particularly hazardous?

9. What are the largest sources of hazardous waste? Describe three ways to dispose of hazardous waste.

10. What is the Superfund program? How does it work?

SEEKING SOLUTIONS

1. How much waste do you generate? Look into your waste bin at the end of the day, and categorize and measure the waste there. List all other waste you may have generated in other places throughout the day. How much of this waste could you have avoided generating? How much could have been reused or recycled?

2. Some people have criticized current waste management practices as merely moving waste from one medium to another. How might this criticism apply to the methods now in practice? What are some potential solutions?

3. Of the various waste management approaches covered in this chapter, which ones are your community or campus pursuing, and which are they not pursuing? Would you suggest that your community or campus start pursuing any new approaches? If so, which ones, and why?

4. Can manufacturers and businesses benefit from source reduction if consumers were to buy fewer products as a result? How? Given what you know about industrial ecology, what do you think the future of sustainable manufacturing may look like?

5. **THINK IT THROUGH** You are the CEO of a major corporation that produces containers for soft drinks and a wide variety of other consumer products. Your company's shareholders are asking that you improve the company's image—while not cutting into profits—by taking steps to reduce waste. What steps would you consider taking?

6. **THINK IT THROUGH** You are the president of your college or university. Your trustees want you to engage with local businesses and industries in ways that benefit both the school and the community. Your faculty and students want you to make the school a leader in waste reduction and industrial ecology. Consider the industries and businesses in your community and the ways they interact with facilities on your campus. Bearing in mind the principles of industrial ecology, can you think of any novel ways that your school and local businesses might mutually benefit from one another's services, products, or waste materials? Are there waste products from one business, industry, or campus facility that another might put to good use? Can you design an eco-industrial park that might work on your campus? What steps would you propose to take as president?

CALCULATING ECOLOGICAL FOOTPRINTS

The 16th biennial "State of Garbage in America" survey documents the ability of U.S. residents to generate prodigious amounts of municipal solid waste (MSW). According to the survey, on a per capita basis, Idaho residents generate the least MSW (4.66 lb/day), and Indiana residents generate the most (11.78 lb/day). The average for the entire country is 7.56 lb MSW per person per day. Calculate the amount of MSW generated in 1 day and in 1 year by each of the groups indicated, at each of the rates shown in the accompanying table.

Groups generating municipal solid waste	Per capita MSW generation rates					
	U.S. average (7.56 lb/day)		Idaho (4.66 lb/day)		Indiana (11.78 lb/day)	
	Day	Year	Day	Year	Day	Year
You	7.25	2,647				
Your class						
Your state						
United States						
World						

Data from Arsova, L., et al., 2008. The state of garbage in America. BioCycle 49 (12): 22.

1. Suppose your town of 50,000 people has just approved construction of a landfill nearby. Estimates are that it will accommodate 1 million tons of MSW. Assuming the landfill is serving only your town, and that your town's residents generate waste at the U.S. average rate, for how many years will it accept waste before filling up? How much longer would a landfill of the same capacity serve a town of the same size in Idaho?

2. One recent study estimated that the average world citizen generates 1.47 pounds of trash per day. How many times more does the average U.S. citizen generate?

3. The same study showed that the average resident of a low-income nation generates 1.17 pounds of waste per day and the average resident of a high-income nation generates 2.64 pounds per day. Why do you think U.S. residents generate so much more MSW than people in other "high-income" countries, when standards of living in those countries are comparable?

Go to **www.masteringenvironmentalscience.com** for practice quizzes, Pearson eText, videos, current events, and more.

Processing coltan ore mined in central Africa

23 MINERALS AND MINING

UPON COMPLETING THIS CHAPTER, YOU WILL BE ABLE TO:

- Outline types of mineral resources and how they contribute to our products and society
- Describe the major methods of mining
- Characterize the environmental and social impacts of mining
- Assess reclamation efforts and mining policy
- Evaluate ways to encourage sustainable use of mineral resources

Mining for . . . Cell Phones?

"The conflict in the Democratic Republic of the Congo has become mainly about access, control, and trade of five key mineral resources: coltan, diamonds, copper, cobalt, and gold."

—Report to the United Nations Security Council, April 2001

"Coltan . . . is not helping the local people. In fact, it is the curse of the Congo."

—African journalist Kofi Akosah-Sarpong

Atlantic Ocean

AFRICA

Region of coltan mining

Democratic Republic of the Congo

Indian Ocean

Pulling a cell phone from her pocket, a student on a college campus in the United States dials a friend. Inside her phone is a little-known metal called tantalum—just a tiny amount, but no cell phone could operate without it.

Half a world away, a dirt-poor miner in the heart of Africa toils all day in a jungle streamed, sifting sediment for nuggets of coltan ore, which contain tantalum. At nightfall, rebel soldiers take most of his ore, leaving him to sell what little remains to buy food for his family at the squalid mining camp where they live.

Coltan miners in eastern Congo

In bedeviling ways, tantalum links our glossy global high-tech economy with one of the most badly wrecked regions on Earth. The Democratic Republic of the Congo has been embroiled in a sprawling conflict that has involved six nations and various rebel militias. Over 5 million people have lost their lives in this war since 1998. For its population size, this is as though Congo were being hit by three September 11 terrorist attacks every day for a decade. It is the latest chapter in the sad history of a nation rich in natural resources—copper, cobalt, gold, diamonds, uranium, and timber—whose impoverished people keep losing control of those resources to others.

At the center of the recent conflict is tantalum (Ta), element number 73 on the Periodic Table (**APPENDIX C**). We rely on this metal for our cell phones, computer chips, DVD players, game consoles, and digital cameras.

Tantalum powder is ideal for capacitors (the components that store energy and regulate current in miniature circuit boards) because it is highly heat resistant and readily conducts electricity.

Tantalum comes from a dull blackish mineral called tantalite, which often occurs with a mineral called columbite—so the ore is referred to as columbite-tantalite, or *coltan* for short. In eastern Congo, men dig craters in rainforest streambeds, panning for coltan much as early California miners panned for gold.

As information technology boomed in the late 1990s, global demand for tantalum rose, and market prices for the metal shot up to $500/kg ($230/lb) in 2001. High prices led some Congolese men to mine coltan by choice, but many more were forced into it. As the war began in 1998, local militias, supported by forces from neighboring Rwanda and Uganda, overran eastern Congo. Farmers were chased off their land, villages were burned, and civilians were raped, tortured, and killed. Soldiers from each army seized control of mining operations. They forced farmers, refugees, prisoners, and children to work, and the soldiers skimmed profits from the coltan the people mined. Children and teachers abandoned school and worked in the mines, while prostitution spread AIDS and sexually transmitted disease through the mining camps. The turmoil also caused ecological havoc as miners and soldiers streamed into national parks, clearing rainforests and killing wildlife for food, including forest elephants, hippopotamuses, endangered gorillas, and the okapi, a rare relative of the giraffe.

Most miners ended up with little, while rebels, soldiers, and bandits enriched themselves selling coltan to traders, who sold it to processing companies in Europe and the United States. These companies refine and sell tantalum powder to capacitor manufacturers, which in turn sell capacitors to Nokia, Motorola, Sony, Intel, Compaq, Dell, and other high-tech corporations.

In 2001, an expert panel commissioned by the United Nations Security Council concluded that coltan riches were fueling, financing, and prolonging the war. The panel urged a U.N. embargo on coltan and other minerals smuggled from Congo and exported by neighboring nations. A grass-roots activist movement advanced the slogan, "No blood on my cell phone!"

Sony, Nokia, Ericsson, and other corporations rushed to assure consumers that they were not using tantalum from eastern Congo—and the region was in fact producing less than 10% of the world's supply. Meanwhile, some observers felt an embargo could hurt the long-suffering Congolese people, rather than help them. The mining life may be miserable, but it pays better than most alternatives in a land where the average income is 20 cents a day.

Soon, however, the high-tech boom went bust, and global demand for tantalum diminished. This occurred just as Australia and other countries were ramping up industrial-scale tantalite mining. As supply outpaced demand, the market price of tantalum fell, and several major producers quit mining tantalum. But nations began to work through their stockpiles, and by 2010 demand had grown, driving prices up once again.

Today, the war is declared over and foreign troops are out of Congo, but internal factions continue to fight, and thousands of people continue to die or to flee their homes. Western electronics companies avoid knowingly purchasing tantalum from Congo, but as a result, much of it ends up being sold to China. In 2010 the U.S. Congress included in its financial reform bill an amendment requiring all electronics companies to report the origin of the tantalum in the products they sell. Yet the trade has so many middlemen and so little transparency that a company like Apple Computer will find it very difficult to determine where its tantalum actually comes from.

In the meantime, some Congolese men are returning to the coltan mines, while others mine for tin, copper, or cobalt. Similar stories are playing out with these and other "conflict minerals" that we in more wealthy nations put to use in our products every day.

EARTH'S MINERAL RESOURCES

Coltan provides just one example of how we extract raw materials from beneath our planet's surface and turn them into products we use in our everyday lives. In Chapter 2 we examined the science of geology and some of the physical processes that take place in the lithosphere (p. 34), the region that includes the uppermost layers of rock near Earth's surface. We saw how the rock cycle (p. 37) creates new rock and alters existing rock. We saw how plate tectonics (pp. 35–37) builds mountains; shapes the geography of oceans, islands, and continents; and gives rise to earthquakes and volcanoes. The coltan mining areas of eastern Congo are situated along the western edge of Africa's Great Rift Valley system, a region where the African tectonic plate is slowly pulling itself apart. Some of the world's largest lakes have formed in the immense valley floors, far below towering volcanoes such as Mount Kilimanjaro. Here and throughout the world, geological processes are fundamental to shaping the world around us.

We will now take a closer look at how rock and the resources of the lithosphere contribute directly to our economies and our lives. We will first examine the mineral resources we mine and the products they provide us. Next we will study the various ways we extract minerals from the earth. We will then examine the many social and environmental impacts that our mining efforts exert, and see what we can do to mitigate these impacts. Finally, because mineral resources are nonrenewable on human time scales, we need to be attentive to finite and decreasing supplies of economically important minerals. Thus we will examine solutions we can pursue to make our mineral use more sustainable.

Rocks provide the minerals we use

As we first learned in Chapter 2, a **rock** is a solid aggregation of minerals, and a **mineral** is a naturally occurring solid chemical element or inorganic compound with a crystal structure, a specific chemical composition, and distinct physical properties. (See the Periodic Table in **APPENDIX C** for chemical elements.) For instance, the mineral tantalite consists of the elements tantalum, oxygen, iron, and manganese. Tantalite occurs most commonly in pegmatite, a type of igneous rock (pp. 37–38) similar to granite. In addition to tantalite, pegmatite generally contains the minerals feldspar, quartz, and mica, and occasionally even includes gemstones and other rare minerals. Geologic processes influence the distribution of rocks and minerals in the lithosphere and their availability to us.

We depend on a wide array of mineral resources as raw materials for our products, so we mine and process these resources. Just consider a typical scene from a student lounge at a college or university (**FIGURE 23.1**), and note how many items are made with elements from the minerals we take from Earth. Without the resources from beneath the ground that we use to make building materials, wiring, clothing, appliances, fertilizers for crops, and so much more, civilization as we know it could not exist.

We obtain minerals by mining

We obtain the minerals we use in all these ways through the process of mining. The term *mining* in the broad sense describes

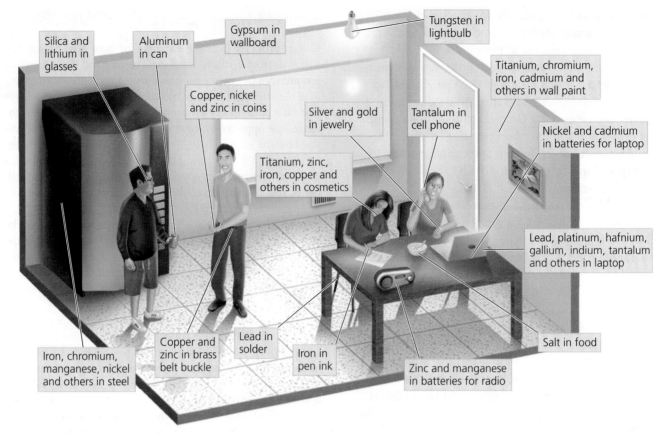

FIGURE 23.1 ▲ Elements from minerals that we mine are everywhere in the products we use in our everyday lives. This scene from a typical college student lounge points out just a few of the many elements from minerals that surround us.

the extraction of any resource that is nonrenewable on the timescale of our society. In this sense, we mine fossil fuels and groundwater, as well as minerals. When used specifically in relation to minerals, **mining** refers to the systematic removal of rock, soil, or other material for the purpose of extracting minerals of economic interest. Because most minerals of interest are widely spread but in low concentrations, miners and mining geologists first try to locate concentrated sources of minerals before mining begins.

Metals are extracted from ores

Some minerals can be mined for metals. A **metal** is a type of chemical element, or a mass of such an element, that typically is lustrous, opaque, malleable, and can conduct heat and electricity. Most metals are not found in a pure state in Earth's crust, but instead are present within **ore**, a mineral or grouping of minerals from which we extract metals.

Copper, iron, lead, gold, and aluminum are among the many economically valuable metals we extract from mined ore. These metals and others serve so many purposes that our modern lives would be impossible without them. The tantalum used in the electronic components of computers, cell phones, DVD players, and other devices is a metal that comes from the mineral tantalite (**FIGURE 23.2**). In nature, tantalite is often found with the mineral columbite within the ore called coltan. Columbite contains the metal niobium, formerly known as columbium, which is utilized in ways similar to tantalum.

We process metals after mining ore

Extracting minerals from the ground is the first step in putting them to use. However, most minerals need to be processed in some way to become useful for our products. For example, after ores are mined, the rock is crushed and pulverized, and the desired metals are isolated by chemical or physical means. The material is then processed to purify the metals we desire. With coltan, processing facilities use acid solvents to separate tantalite from columbite. Other chemicals are then used to produce metallic tantalum powder. This powder can be consolidated by various melting techniques and can be shaped into wire, sheets, or other forms.

Sometimes we mix, melt, and fuse a metal with another metal or a nonmetal substance to form an *alloy*. For example, steel is an alloy of the metal iron that has been fused with a small quantity of carbon. The strength and malleability of this particular alloy make steel ideal for its many applications in buildings, vehicles, appliances, and more. To make steel, we first mine iron ore, which consists of iron-containing compounds such as iron oxide. Steelmakers then heat the ore and chemically extract the iron with carbon in a process known as **smelting** (heating ore beyond its melting point and combining it with other metals or chemicals). They then melt and reprocess the mixture, removing precise amounts of carbon and shaping the product into rods, sheets, or wires. During this melting process, certain other metals may be added to modify the strength, malleability, or other characteristics of the steel, as desired.

(a) Tantalite ore

(b) Purified tantalum

(c) Capacitor containing tantalum

FIGURE 23.2 ▲ Tantalite ore **(a)** is mined from the ground and then processed to extract the pure metal tantalum **(b)**. This metal is used in capacitors **(c)** and other electronic components in computer chips, cell phones, and many other devices.

Processing minerals exerts environmental impacts. Most methods are water-intensive and energy-intensive. Moreover, many chemical reactions and heating processes used for extracting metals from ore emit air pollution, and smelting plants in particular have long been hotspots of toxic air pollution. In addition, soil and water commonly become polluted by **tailings**, portions of ore left over after metals have been extracted. Tailings may leach heavy metals present in the ore waste as well as chemicals applied in the extraction process. For instance, we use cyanide to extract gold from ore, and we use sulfuric acid to extract copper. Mining operations often pump a toxic slurry of tailings to a tailings pond, from which water can evaporate, and it is difficult to line such ponds well enough to prevent leaching into the environment.

We also mine nonmetallic minerals and fuels

We also mine and use many minerals that do not contain metals. **FIGURE 23.3** illustrates a selection of economically useful mineral resources, both metallic and nonmetallic. For each one, its major nation of origin and several main uses are shown.

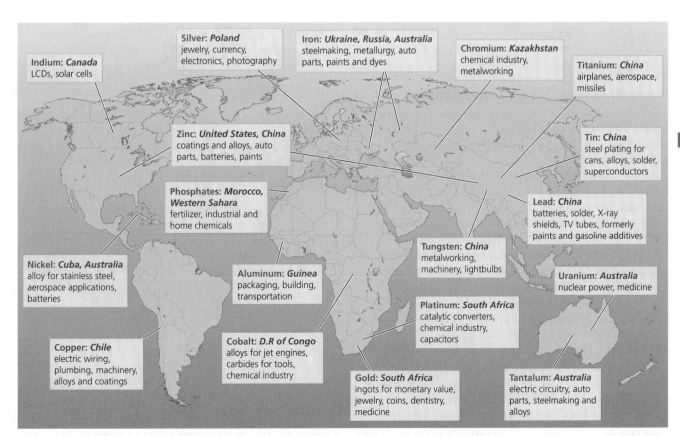

Silver: *Poland*
jewelry, currency, electronics, photography

Iron: *Ukraine, Russia, Australia*
steelmaking, metallurgy, auto parts, paints and dyes

Chromium: *Kazakhstan*
chemical industry, metalworking

Titanium: *China*
airplanes, aerospace, missiles

Indium: *Canada*
LCDs, solar cells

Zinc: *United States, China*
coatings and alloys, auto parts, batteries, paints

Tin: *China*
steel plating for cans, alloys, solder, superconductors

Phosphates: *Morocco, Western Sahara*
fertilizer, industrial and home chemicals

Lead: *China*
batteries, solder, X-ray shields, TV tubes, formerly paints and gasoline additives

Nickel: *Cuba, Australia*
alloy for stainless steel, aerospace applications, batteries

Aluminum: *Guinea*
packaging, building, transportation

Tungsten: *China*
metalworking, machinery, lightbulbs

Uranium: *Australia*
nuclear power, medicine

Platinum: *South Africa*
catalytic converters, chemical industry, capacitors

Copper: *Chile*
electric wiring, plumbing, machinery, alloys and coatings

Cobalt: *D.R of Congo*
alloys for jet engines, carbides for tools, chemical industry

Gold: *South Africa*
ingots for monetary value, jewelry, coins, dentistry, medicine

Tantalum: *Australia*
electric circuitry, auto parts, steelmaking and alloys

FIGURE 23.3 ▲ The minerals we use come from all over the world. Shown is a selection of economically important minerals (mostly metals, with several nonmetals), together with their major uses and their main nation of origin. Only a minority of minerals, uses, and origins is shown.

Sand and gravel (the most commonly mined mineral resources) provide fill and construction materials, including for the manufacture of concrete. Each year over $7 billion of sand and gravel are mined in the United States. Phosphates provide us fertilizer. We mine limestone, salt, potash, and other minerals for a number of diverse purposes.

Gemstones are treasured for their rarity and beauty. For instance, diamonds have long been prized—and like coltan, they have fueled resource wars. Besides the conflict in eastern Congo, the diamond trade has acted to fund, prolong, and intensify wars in Angola, Sierra Leone, Liberia, and elsewhere, as armies exploit local people for mine labor, then sell the diamonds for profit. This is why you may hear the phrase "blood diamonds," just as coltan has been called a "conflict mineral." The more that we in developed nations consume, the more our economic demand affects people elsewhere. Throughout the world, people in developing regions from Africa to Asia to the Amazon often suffer unintended consequences of the developed world's appetite for mineral resources.

We also mine substances we use for fuel, as we have seen in previous chapters. Uranium ore is a mineral from which we extract the metal uranium, which we use in nuclear power (pp. 564–576). One of the most common fuels we mine is coal. Coal (p. 535) is not a mineral, because it consists of organic matter, but we consider coal mining below because it has relevance for many general mining issues. Other fossil fuels—petroleum, natural gas, and alternative fossil fuels such as oil sands, oil shale, and methane hydrates—are also organic and are extracted from the earth, as we saw in Chapter 19.

MINING METHODS AND THEIR IMPACTS

Mining for minerals is an important industry that provides jobs for people and revenue for communities in many regions. Mining supplies us raw materials for countless products we use daily, so it is necessary for the lives we lead. In 2009, raw materials from mining contributed $57 billion to the U.S. economy, and after processing, mineral materials contributed $454 billion. About 28,000 Americans were employed directly in mining for metals in 2009, and the mining industry, together with processors and manufacturers of products from mined materials, employed nearly 1.2 million people.

At the same time, mining also exerts a price in environmental and social impacts. Because minerals of interest often make up only a small portion of the rock in a given area, typically very large amounts of material must be removed in order to obtain the desired minerals. This frequently means that mining disturbs large areas of land, thereby exerting severe impacts on the environment and on people living nearby.

Depending on the nature of the mineral deposit, any of several mining methods may be employed to extract the resource from the ground. Mining companies select which method to use based largely on its economic efficiency. We will examine seven major mining approaches commonly used throughout the world, and will also take note of the impacts of each approach as we proceed.

Strip mining removes surface layers of soil and rock

When a resource occurs in shallow horizontal deposits near the surface, the most effective mining method is often **strip mining**, whereby layers of surface soil and rock are removed from large areas to expose the resource. Heavy machinery removes the overlying soil and rock (termed *overburden*) from a strip of land, and the resource is extracted. This strip is then refilled with the overburden that had been removed, and miners proceed to an adjacent strip of land and repeat the process. Strip mining is commonly used for coal (p. 535; **FIGURE 23.4A**) and oil sands (p. 544), and sometimes for sand and gravel.

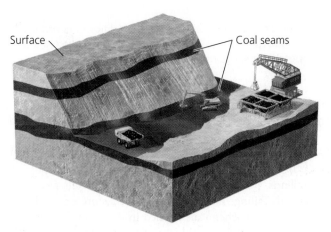

(a) Strip mining

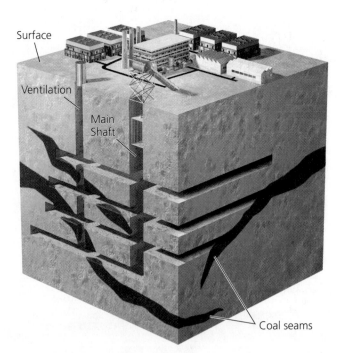

(b) Subsurface mining

FIGURE 23.4 ▲ Coal mining illustrates two types of mining approaches. In strip mining **(a)**, soil is removed from the surface in strips, exposing seams from which coal is mined. In subsurface mining **(b)**, miners work below ground in shafts and tunnels blasted through the rock. These passageways provide access to underground seams of coal or minerals.

Strip mining for coal and oil sands can be economically efficient, but it causes severe environmental impacts. By completely removing vegetative cover and nutrient-rich topsoil, strip mining obliterates natural communities over large areas. Soil from refilled areas easily erodes away. Strip mining also pollutes waterways through the process of **acid drainage**, which occurs when sulfide minerals in newly exposed rock surfaces react with oxygen and rainwater to produce sulfuric acid. As the sulfuric acid runs off, it leaches metals from the rocks, many of which are toxic to organisms (**FIGURE 23.5**). This toxic liquid is called *leachate*, a term used in Chapter 22 (p. 622) to describe the toxic liquids that form in landfills. Whereas leachates from landfills are caused by anaerobic reactions in organic substances, leachates from mining sites result from aerobic reactions in inorganic substances. Acid drainage is a natural phenomenon, but mining greatly accelerates this process by exposing many new rock surfaces at once.

FIGURE 23.5 ▲ The discolored water in this stream downstream from a mine site is a sign of acid drainage. Acid drainage can make stream water toxic and reduce aquatic biodiversity. The site shown here is near the Iron Mountain Mine in northern California. A century of mining for iron, gold, silver, zinc, and copper has made the area a Superfund (p. 639) hazardous waste cleanup site.

In subsurface mining, miners work underground

When a resource occurs in concentrated pockets or seams deep underground, and the earth allows for safe tunneling, then mining companies pursue **subsurface mining**. In this approach, shafts are excavated deep into the ground, and networks of tunnels are dug or blasted out to follow deposits of the mineral (**FIGURE 23.4B**). Miners remove the resource systematically and ship it to the surface.

We use subsurface mining for metals such as zinc, lead, nickel, tin, gold, copper, and uranium, as well as for diamonds, phosphate, salt, and potash. In addition, a great deal of coal is mined using the subsurface technique. The scale of subsurface mining can be mind-boggling; the world's deepest mines (certain gold mines in South Africa) extend nearly 4 km (2.5 mi) underground.

Subsurface mining is the most dangerous form of mining and indeed one of society's most dangerous occupations. Fatal accidents are not unusual; for instance, 29 coal miners died underground after an explosion in the Upper Big Branch mine in Montcoal, West Virginia, in 2010. In China, coal-mining conditions are so dangerous that in 2009 over 2,600 miners lost their lives. Besides risking injury or death from dynamite blasts, natural gas explosions, and collapsing shafts and tunnels, miners inhale toxic fumes and coal dust, which can lead to respiratory diseases, including fatal black lung disease.

Occasionally subsurface mines can affect people years after they are closed, if tunnels collapse. The collapse of tunnels at the Retsof Salt Mine in Genessee Valley, New York, after a minor earthquake in 1994 created sinkholes at the surface that damaged roads, bridges, and homes and sucked groundwater from neighborhood wells. In terms of environmental impact, subsurface mining creates acid drainage just as surface mining does, and toxic leachate can make its way down into groundwater. Abandoned mine sites can continue polluting groundwater long after mining has ceased.

Open pit mining creates immense holes in the ground

When a mineral is spread widely and evenly throughout a rock formation, or when the earth is unsuitable for tunneling, the method of choice is **open pit mining**. This essentially involves digging a gigantic hole and removing the desired ore, along with waste rock that surrounds the ore. Some open pit mines are inconceivably enormous. The world's largest, the Bingham Canyon Mine near Salt Lake City, Utah, is 4 km (2.5 mi) across and 1.2 km (0.75 mi) deep (**FIGURE 23.6**). Conveyor systems and immense trucks with tires taller than a person carry out nearly half a million tons of ore and waste rock each day.

Open pit mines are terraced so that men and machinery can move about, and waste rock is left in massive heaps outside the pit. The pit is expanded until the resource runs out or becomes unprofitable to mine. Open pit mining is used for copper, iron, gold, diamonds, and coal, among other resources. We also use this technique to extract clay, gravel, sand, and stone such as limestone, granite, marble, and slate, but we generally call these pits *quarries*.

FIGURE 23.6 ▲ The Bingham Canyon open pit mine outside Salt Lake City, Utah, is the world's largest human-made hole in the ground. This immense mine produces mostly copper.

FIGURE 23.7 ▲ Miners in eastern Congo find coltan by placer mining. Sediment is placed in plastic tubs and water is run through them. A mixing motion allows the sediment to be poured off while the heavy coltan settles to the bottom.

Open pit mines are so large because huge volumes of waste rock need to be removed in order to extract relatively small amounts of ore, which in turn contain still smaller traces of valuable minerals. The sheer size of these mines means that the degree of habitat loss and aesthetic degradation is considerable. Another impact is chemical contamination from acid drainage as water runs off the waste heaps or collects in the pit. Once mining is complete, abandoned pits generally fill up with groundwater, which soon becomes toxic as water and oxygen react with sulfides from the ore and produce sulfuric acid. Acidic water from the pit can harm wildlife and can percolate into aquifers and spread through the region. Regulations in developed nations require that waste heaps be capped with clay, then soil, and then planted with vegetation once mines are closed. However, many dumps will likely leach acid for hundreds or thousands of years.

The Berkeley Pit, a former copper mine near Butte, Montana, is today one of the largest Superfund toxic waste cleanup sites (p. 639) in the United States. After its closure in 1982, it filled with groundwater and became so acidic (pH of 2.2) and concentrated with toxic metals that microbiologists discovered new species of microbes in the water—the harsh conditions were so rare in nature that scientists had never before encountered microbes adapted to them!

Placer mining uses running water to isolate minerals

Some metals and gems accumulate in riverbed deposits, having been displaced from elsewhere and carried along by flowing water. To search for these metals and gems, miners sift through material in modern or ancient riverbed deposits, generally using running water to separate lightweight mud and gravel from heavier minerals of value (**FIGURE 23.7**). This technique is called **placer mining** (pronounced "plasser").

Placer mining is the method used by Congo's coltan miners, who wade through streambeds, sifting through large amounts of debris by hand with a pan or simple tools, searching for high-density tantalite that settles to the bottom while low-density material washes away. Today's African miners practice small-scale placer mining similar to the method used by American miners long ago who ventured to California in

the Gold Rush of 1849, and later to Alaska in the Klondike Gold Rush of 1896–1899. Indeed, placer mining for gold is still practiced in areas of Alaska and Canada, although today it uses large dredges and heavy machinery.

Besides the many social impacts of placer mining in places like Congo, placer mining is environmentally destructive because most methods wash large amounts of debris into streams, making them uninhabitable for fish and other life for many miles downstream. Gold mining in northern California's rivers in the decades following the Gold Rush washed so much debris all the way to San Francisco Bay that a U.S. district court ruling in 1884 finally halted this mining practice. Placer mining also disturbs stream banks, causing erosion and harming ecologically important riparian plant communities.

Mountaintop mining reshapes ridges and can fill valleys

When a resource occurs in underground seams near the tops of ridges or mountains, we may practice **mountaintop removal mining**, in which several hundred vertical feet of mountaintop may be removed to allow recovery of entire seams of the resource (**FIGURE 23.8A**). This method of mining is used primarily for coal in the Appalachian Mountains of the eastern United States. In mountaintop removal mining, a mountain's forests are clear-cut and the timber is sold, topsoil is removed, and then rock is repeatedly blasted away to expose the coal for extraction. Overburden is placed back onto the mountaintop, but this waste rock is unstable and typically takes up more volume than the original rock, so generally a great deal of waste rock is dumped into adjacent valleys (a practice called "valley filling").

(a) Mountaintop mining in eastern Kentucky

(b) Train hauls coal past homes in West Virginia

(c) Flood damage below a West Virginia mine site

FIGURE 23.8 ▲ In the Appalachians, mountaintop mining for coal **(a)** takes place on massive scales. Communities near these sites **(b)** experience environmental and social impacts. For example, mountaintop mining contributed to flash floods that damaged the homes of Bernice Waugh **(c)** and other people living along Scrabble Creek in Fayette County, West Virginia, in 2001. Mountaintop removal may be out of sight and out of mind for most of us, but it is our demand for electricity **(d)** that drives the demand for coal extraction.

(d) We all contribute

So far, mountaintop removal has blasted away an area the size of Delaware and has buried nearly 3,200 km (2,000 mi) of streams.

Mountaintop removal mining has expanded in recent years because it is an economically efficient way for companies to extract coal. In addition, 1990 Clean Air Act amendments that encouraged clean-burning low-sulfur coal led to more mining in Appalachia, where low-sulfur coal predominates. Moreover, most easily accessible coal in level areas near the surface has already been mined. Thus, the mining industry says mountaintop removal is necessary if we are to obtain coal from the Appalachian region in order to provide the electricity we all use. As a result, the Clinton administration opposed strengthening regulations on valley filling, and the Bush administration loosened regulations to allow coal companies to dump waste rock directly into valleys and streams.

Scientists are finding that dumping tons of debris into valleys degrades or destroys immense areas of habitat, clogs streams and rivers, and pollutes waterways with acid drainage. With slopes deforested and valleys filled with debris, erosion intensifies, mudslides become frequent, and flash floods ravage the lower valleys (see **THE SCIENCE BEHIND THE STORY**, pp. 652–653).

People living in communities near the sites experience social and health impacts (**FIGURE 23.8B**). Blasts from mines crack house foundations and wells, loose rock tumbles down into yards and homes, overloaded coal trucks speed down once-peaceful rural roads, and floods tear through properties (**FIGURE 23.8C**). Coal dust causes respiratory ailments, and contaminated water unleashes a variety of health problems. Some landowners are pressured to sell their land to coal companies, while local politicians and judges do little to help. All around them, people lose the forests and landscapes they have lived with since childhood.

Although the people of Appalachia have long relied on the coal industry for employment, mountaintop removal is so efficient that fewer workers are needed for mining. As a result, employment has declined in recent years even as coal extraction has risen. In all these ways, the people of Appalachia—already among the poorest in the United States—suffer substantial external costs (pp. 150–151). The rest of us benefit from the electricity that we produce with their coal (**FIGURE 23.8D**).

Critics of mountaintop removal mining argue that valley filling violates the Clean Water Act. In 2010 the Environmental Protection Agency introduced new regulations to limit damage from mountaintop mining and valley filling. As of this writing, Congress is considering two bills that would effectively halt these practices.

The SCIENCE behind the Story

Mountaintop Removal Mining: Assessing the Environmental Impacts

Mountaintop removal mining has attracted a great deal of criticism from environmental activists. But what do scientific studies tell us about the impacts of this mining method? Is the outcry justified?

Stream ecologist Margaret Palmer of the University of Maryland and a team of 11 other scientists aimed to find out. They reviewed all the scientific literature on mountaintop mining impacts. Writing in the journal *Science* in 2010, the team concluded that this method of mining causes "serious environmental impacts that mitigation practices cannot successfully address." The team also concluded that published studies indicate "a high potential for human health impacts."

The Appalachian forests that are cleared in mountaintop mining are some of the richest forests for biodiversity in the nation. Yet forest clearance—and the loss of biodiversity and ecosystem services—is merely the most obvious impact. Palmer's team stressed that many more consequences result when waste material from mountaintop removal is dumped into adjacent valleys, burying streams and trees.

Once buried, headwater streams are lost as ecosystems. Gone with them are certain rare endemic species, as well as the ecosystem's ability to cycle nutrients and produce organic matter for downstream systems. Flash floods also result: Because mining removes vegetation and topsoil, alters topography, and

Dr. Margaret Palmer, University of Maryland

compacts soil, less rain infiltrates the ground, and instead rain runs off quickly, causing flooding downstream. Moreover, when a valley is filled with mined material, water that runs through the fill emerges at the bottom carrying a brew of toxic substances. Through the process of acid drainage (p. 649), the acidic water carries dissolved heavy metals leached from the rock.

Researchers with the United States Geological Survey (USGS) carried out extensive water quality surveys in Appalachian streams in the late 1990s. For example, separate teams led by Katherine Paybins and by James Sams took hundreds of samples from diverse areas, chemically analyzed the samples, and then mapped their data, looking for geographic patterns. Their data clearly showed that concentrations of pollutants such as sulfates were strongly linked to upstream mining activities. They also showed that this pollution and its impacts (such as biodiversity loss) are long-lasting. Indeed, many studies show biodiversity declines in streams disturbed by mining, and no

study has yet documented a recovery of stream life in the years after mining pollution.

Palmer's team tested for water quality impacts by tapping into a large database of measurements made by staff of the West Virginia Department of Environmental Protection. These workers had measured concentrations of pollutants in the waters of 1,058 streams in West Virginia and had recorded numbers and types of aquatic insects in the streams.

Palmer's team first confirmed that sulfate concentrations were tightly linked with upstream mining activity. They then looked for statistical correlations between sulfate concentrations and other water quality indicators, such as concentrations of selenium, iron, aluminum, and manganese. They found that concentrations of all these minerals rose with concentrations of sulfates (**see first graph**).

The researchers then graphed their insect data against sulfate concentrations (**see second graph**). All types of insects decreased as sulfate concentrations rose, suggesting that the changes in water chemistry brought about by mining were diminishing insect diversity in the streams.

Water pollution from mining can affect organisms higher on the food chain as well. Selenium is known to bioaccumulate (p. 385) in organisms and to cause birth defects (**see photo**). EPA scientists surveying 78 streams

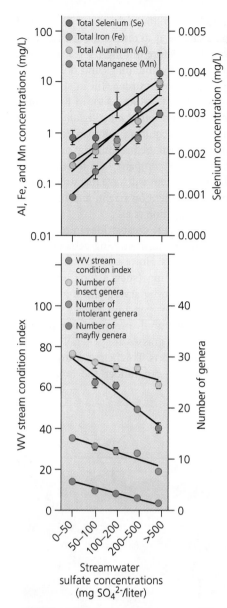

Data from over 1,000 West Virginia streams show that (**top**) concentrations of four pollutants (selenium, iron, aluminum, and manganese) rise along with the increased sulfate concentrations that result from mining, and also that (**bottom**) insect diversity decreases with rising sulfate concentrations. *Source:* Palmer, M.A., et al., 2010. Mountaintop mining consequences. *Science* 327: 148–149. Reprinted with permission of AAAS.

near mines in 2002 found that 73 had selenium concentrations above the level at which studies have documented risks to insects, fish, and the birds that eat them.

High selenium concentrations in stream water can lead to birth defects. This embryo of a fish has a severely curved spine as a result of selenium downstream from a mine. *Source:* Lemly, A.D. 2008. Aquatic hazard of selenium pollution from coal mining, pp. 167–183, in G.B. Fosdyke, ed., *Coal mining: research, technology, and safety.* Nova Science Publishers, Inc.

Scientific research also documents health impacts on people apparently due to mountaintop mining. Health impacts come from inhaling air pollution and dust, drinking contaminated groundwater, and eating fish contaminated with selenium and other toxic substances. A 2009 study documented that people in mountaintop mining areas show elevated levels of lung cancer, heart disease, kidney disease, pulmonary disorders, hypertension, and mortality. The public health researchers who carried out this study, Michael Hendryx and Melissa Ahern, found that these health problems occurred in women as well as men, so they could not be due to direct occupational exposure among coal miners.

How much damage is too much, when it comes to a watershed? Various studies over the years have found that when over 5–10% of a watershed's area is disturbed by some sort of human land alteration, water quality and biodiversity in a stream decline. Mountaintop mining typically disturbs a greater percentage than this. For instance, of eight mountaintop mining permits issued

in West Virginia in 2008, mining was allowed to cover 17–51% of each watershed.

Can mined mountaintops, filled valleys, and human health be restored to their original condition after mining? The science so far says no. A 2006 USGS study showed that even after reclamation efforts, groundwater from people's wells still contains higher levels of mine-derived chemicals than water from wells in unmined areas.

Reclamation efforts have traditionally focused on planting grasses and herbs—a far less diverse plant assemblage than was destroyed. Moreover, the degraded, compacted soils do not hold water, nutrients, or organic matter well, so trees have a hard time establishing. A 2008 study found little or no regrowth of trees and shrubs a full 15 years after reclamation was completed. Another study projected that even 60 years later, the soil would hold only 77% as much carbon as it did originally.

Palmer's team concluded that the U.S. government is failing to enforce the Clean Water Act and the Surface Mining Control and Reclamation Act and that these laws may not be strong enough to prevent severe impacts from mountaintop mining. The researchers urged the U.S. government to strengthen regulation of mountaintop mining practices. "Regulators," they wrote, "should no longer ignore rigorous science."

Just months after publication of this paper, the U.S. Environmental Protection Agency announced that it was strengthening its regulations on mountaintop mining. After reviewing EPA-sponsored and independent scientific studies, the agency adopted rules aimed at protecting 95% of aquatic life in Appalachian streams. Congress in 2010 was also considering bills to limit mountaintop mining and valley filling. ∎

Solution mining dissolves and extracts resources in place

When a deposit is especially deep and the resource can be dissolved in a liquid, miners may use a technique called *solution mining* or *in-situ recovery*. In this technique, a narrow borehole is drilled deep into the ground to reach the deposit, and water, acid, or another liquid is injected down the borehole to leach the resource from the surrounding rock and dissolve it in the liquid. The resulting solution is then sucked out, and the desired resource is isolated. Salts can be mined in this way; water is pumped into deep salt caverns, the salt dissolves in the water, and the salty solution is extracted. Besides sodium chloride (table salt), such salts include lithium, boron, bromine, magnesium, and potash. In-situ recovery also is sometimes used for copper (dissolved with acids) and uranium (dissolved with acids or carbonates).

Solution mining generally exerts less environmental impact than other mining techniques, because less area at the surface is disturbed. The main potential impacts involve accidental leakage of acids into groundwater surrounding the borehole, and the contamination of aquifers with acids, heavy metals, or uranium leached from the rock.

Some mining occurs in the ocean

The oceans hold many minerals useful to our society. We extract some minerals from seawater, such as magnesium from salts held in solution. We extract other minerals from the ocean floor. Using large vacuum-cleaner-like hydraulic dredges, miners collect sand and gravel from beneath the sea. They extract sulfur from salt deposits in the Gulf of Mexico, and phosphate from offshore areas near the California coast and elsewhere. Other valuable minerals found on or beneath the seafloor include calcium carbonate (used in making cement) and silica (used as fire-resistant insulation and in manufacturing glass), as well as copper, zinc, silver, and gold ore. Many minerals are concentrated in manganese nodules, small ball-shaped accretions that are scattered across parts of the ocean floor. Over 1.5 trillion tons of manganese nodules may exist in the Pacific Ocean alone, and their reserves of metal may exceed all terrestrial reserves. The logistical difficulty of mining them, however, has kept their extraction uneconomical so far.

As land resources become scarcer and as undersea mining technology develops, mining companies may turn increasingly to the seas. Some companies already are exploring hydrothermal vents (p. 33) as potentially concentrated sources of metals such as gold, silver, and zinc, because these vents emit dissolved metals resulting from underground volcanic activity.

Impacts of undersea mining are largely unknown, but such mining would undoubtedly destroy marine habitats and organisms that have not yet been studied. It would also likely cause some metals to diffuse into the water column at toxic concentrations and enter the food chain.

Restoration of mined sites is often only partly effective

Because of the environmental impacts of mining (see **ENVISIONIT**, p. 655), governments of the United States and other developed nations now require that mining companies restore, or reclaim, surface-mined sites following mining. The aim of such restoration, or **reclamation**, is to restore the site to a condition similar to its condition before mining. To restore a site, companies are required to remove buildings and other structures used for mining, replace overburden, fill in shafts, and replant the area with vegetation (**FIGURE 23.9**). In the United States, the 1977 *Surface Mining Control and Reclamation Act* mandates restoration efforts, requiring companies to post bonds to cover reclamation costs before mining can be approved. This ensures that if the company fails to restore the land for any reason, the government will have the money to do so. Most other nations exercise less oversight, and in nations such as Congo, there is no regulation at all.

The mining industry has made great strides in reclaiming mined land and employs many hard-working ecologists and engineers to conduct these efforts. However, even on sites that are restored, impacts from mining (such as soil and water damage from acid drainage) can be severe and long-lasting. Moreover, reclaimed sites do not generally regain the same biotic communities that were naturally present before mining. One reason is that fast-growing grasses are generally used to initiate and anchor restoration efforts. This helps control erosion quickly from the outset, but it can hinder the longer-term

FIGURE 23.9 ▲ More mine sites are being restored today, but restoration rarely is able to recreate the natural community present before mining. Here, reclamation workers in Ghana, West Africa, plant trees on the benches and floor of an abandoned gold-mining pit.

The metals and the coal that mining brings us have ecological and social costs. Mining affects land, water, wildlife, and people near mine sites.

Acid drainage pollution

Flash flood below a mine

Protest in Frankfort, Kentucky

Citizens have pushed government to make mining safer and cleaner.

Today old mine sites are reclaimed, and in their place we create fields, farms, and golf courses.

Golf course built on reclaimed mine

YOU CAN MAKE A DIFFERENCE

➤ Reuse and recycle metal items and electronics, to reduce the need to mine virgin ore.

➤ Reduce electricity use and enhance energy efficiency, to reduce the need for coal mining.

➤ Urge policymakers to promote clean and safe mining and reclamation practices.

The SCIENCE behind the Story

Using Bacteria to Clean Mine Water and Recover Metals

Toxic waters of the Berkeley Pit near Butte, Montana

Mining poses two dilemmas: It exerts environmental impacts, and it can deplete nonrenewable minerals. Now, some scientists are seeking to address both these drawbacks in one fell swoop. They're aiming to clean up polluted mine sites and amass valuable minerals at the same time. Their secret weapon? Bacteria.

Let's set the stage by traveling to western Montana, where in 1982 the Atlantic Richfield Company ceased mining at the Berkeley Pit, a massive open pit copper mine covering just over 1 square mile. Once the mine was shut down and pumps were turned off, groundwater started to fill the 540-m (1,780-ft)-deep crater.

As water bathed the rock walls, it mixed with oxygen from the air and reacted with sulfide minerals in the freshly exposed rock, producing acid drainage (p. 649). Sulfuric acid formed, and water accumulating in the Berkeley Pit became as acidic as lemon juice, reaching pH values as low as 2.2. The wet rock began leaching metals into the acidic water, including ions of iron, zinc, aluminum, manganese, nickel, cadmium, cobalt, and arsenic, as well as copper. These metals reacted with sulfuric acid to form metal sulfates that

dissolved in the water. For instance, zinc formed zinc sulfate ($ZnSO_4$) and iron formed iron sulfate ($FeSO_4$).

High concentrations of these dissolved metals made the water toxic to wildlife, and waterfowl landing at the site were later found dead. The water here was so unusually acidic and metal-rich that biologists interested in microbes that thrive in extreme environments came here searching for odd species never before encountered. As mentioned on p. 650, they did in fact discover novel species of bacteria that lived in the harsh conditions.

As groundwater continued to fill the pit at a rate of 13 vertical feet per year, experts estimated that by the year 2015 water would overflow into the Clark Fork River drainage, polluting ecosystems and poisoning people's

drinking water. The race was on to find a solution.

Traditionally, mining engineers have tried to treat acid mine drainage by adding a strongly alkaline substance such as lime or sodium hydroxide to raise the water's pH and precipitate the dissolved heavy metals. In recent years, researchers have begun to put sulfate-reducing bacteria to work in converting metal sulfates in acid mine drainage to metal sulfides, harmless compounds that allow the metals to be removed and processed.

Sulfate-reducing bacteria are microbes that thrive in the absence of oxygen by chemically reducing sulfates to obtain energy and creating sulfides that they expel as waste. In nature, some such bacteria degrade organic materials in the mud of swamps and produce hydrogen sulfide, the gas that gives swamps and mudflats their distinctive rotten-egg odor. Hydrogen sulfide in turn reacts with metal sulfates to create metal sulfides, which give swamp mud its blackish color.

At the Berkeley Pit, EPA scientist Henry Tabak, University of Cincinnati chemical engineer Rakesh Govind, and three colleagues examined whether they could use sulfate-reducing bacteria

establishment of forests, wetlands, or other complex natural communities. Instead, grasses may outcompete slower-growing native plants in the acidic, compacted, nutrient-poor soils that usually result from mining. Moreover, many inconspicuous but vital symbiotic relationships (pp. 82–83) that maintain ecosystems—such as specialized relationships between plants and fungi and plants and insects—are eliminated by mining and are very difficult to restore.

Water polluted by mining and acid drainage can also be reclaimed, if pH can be moderated and if toxic heavy metals can be removed. Like the reclamation of land, this is a

challenging and imperfect process, but researchers and the mining industry are making progress in improving techniques. **THE SCIENCE BEHIND THE STORY** (above) examines an example of one such method to clean up acid drainage and recover excess metals at the same time.

WEIGHING THE ISSUES

Restoring Mined Areas Mining has severe environmental impacts, but restoring mined sites to their pre-mining condition is

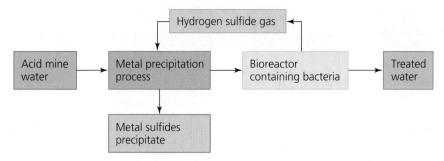

Acidic mine water rich in sulfates is fed into two tanks. Bacteria in the bioreactor produce hydrogen sulfide (H_2S), which circulates to the metal precipitation tanks. In the presence of hydrogen sulfide, metal ions bond to sulfur to form metal sulfides. The sulfides are not soluble, so they precipitate out of the water. The sulfides are collected and may be processed further to harvest minerals of economic value. The treated water is discharged from the system, clean enough to use for agriculture.

Percent Recovery of Metals from Mine Water Using Sulfate-Reducing Bacteria

Metal	Percent recovery of metal
Aluminum	99.8
Cadmium	99.7
Cobalt	99.1
Copper	99.8
Iron	97.1
Manganese	87.4
Nickel	47.8
Zinc	100.0

Adapted with kind permission from Springer Science and Business Media from Tabak, H.H., et al., 2003. Advances in biotreatment of acid mine drainage and biorecovery of metals: 1. Metal precipitation for recovery and recycle. Biodegradation 14: 423–436.

to clean up the water and recover valuable metals.

Highly acidic water harms most bacteria, so Tabak and Govind's team designed a two-step process. They grew their bacteria in water kept at a neutral pH in a tank called a bioreactor and fed them sulfates from the mine water, along with carbon-containing nutrients, letting them produce hydrogen sulfide. They then funneled this hydrogen sulfide into a series of tanks, where it reacted with the acid mine drainage from the Berkeley Pit. The reactions in these tanks caused the metals to precipitate out as insoluble metal sulfides.

A simple schematic diagram (**see figure**) shows this process. In the bioreactor, the bacteria take sulfate ions (SO_4^{2-}) from the mine water and carbon-based nutrients (CH_2O) and create hydrogen sulfide (H_2S) and bicarbonate (HCO_3^-):

$$SO_4^{2-} + 2CH_2O \rightarrow H_2S + 2HCO_3^-$$

The hydrogen sulfide is fed back to the metal precipitation tanks, where it drives reactions in which metal ions bond to sulfur atoms to form metal sulfides, which are insoluble in water and thus precipitate into a solid and settle to the bottom. For example, with the metal zinc (Zn), the equation would be:

$$H_2S + Zn^{2+} \rightarrow ZnS + 2H^+$$

Thus, overall, soluble metal sulfates, such as zinc sulfate ($ZnSO_4$), are converted to insoluble metal sulfides, such as zinc sulfide (ZnS).

The metal sulfides are then removed from the precipitation tanks. The research team measured its success recovering metals by this process by calculating the average percent recovery for each of several metals (**see table**). Precipitates of the various metals ranged in purity from 75% to 98%.

Besides recovering high percentages of metals that can be processed and recycled, the process cleaned heavy metals from the mine water and also made it less acidic with help from the bicarbonate. After full treatment, the researchers determined that the water was pure enough to meet safety standards for using to irrigate agricultural crops.

This research, published in 2003, was done at the lab-bench scale, and more work remains to scale it up to treat large amounts of water from the mine. Today, a wide variety of similar efforts are ongoing around the world, from Montana to South Africa, as scientists develop better ways of removing metals from acid mine drainage and as engineers work to scale these systems up. These approaches are showing promise to reclaim mine water and recycle metals at the same time. ∎

costly and difficult. How much do you think we should require mining companies to restore after a mine is shut down, and what criteria should we use to guide restoration? Should we require complete restoration? No restoration? What should our priorities be—to minimize water pollution, health impacts, biodiversity loss, soil damage, or other factors? Should the amount of restoration we require depend on how much money the company made from the mine? Explain your recommendations.

An 1872 law still guides U.S. mining policy

Government policy plays a role in the ways that mining companies stake claims and use land. In the United States, this has been controversial because policy is still guided by a law that is well over a century old. The **General Mining Act of 1872** encourages people and companies to prospect for minerals on federally owned land by allowing any U.S. citizen or any company with permission to do business in the United States to stake a claim on any plot of public land open to mining. The person or company owning the claim gains the sole right to take minerals from the area. The claim-holder can also patent the claim (i.e., buy the land) for only about $5 per acre. Regardless of the profits they might make on minerals they extract, the law requires no payments of any kind to the public, and until recently no restoration of the land after mining was required.

The General Mining Act of 1872 was enacted partly in response to the chaos of the California Gold Rush and other episodes, and it was designed to bring some order to mining activities. It also aimed to promote mining at a time when the government was trying to hasten settlement of the West in an orderly way. The law may have made good sense in 1872, but the United States has changed a great deal since then, and many question the law's suitability for today's nation.

Supporters of the policy say that it is appropriate and desirable to continue encouraging the domestic mining industry, which must undertake substantial financial risk and investment to locate resources that are vital to our economy. Critics counter that the policy gives valuable public resources away to private interests nearly for free. They also point out that many claims made under this law have eventually led to lucrative land development schemes (such as condominium development) that have nothing to do with mining.

Critics have tried to amend the law many times over the years, mostly without success. In 2007, the U.S. House of Representatives passed a bill that would largely end the patenting process, put some public lands off-limits to mining, mandate that mined sites be restored to some semblance of their former condition, and require miners to pay the government royalties of 4% of profits from new mines and 8% from existing mines. The money would help to fund cleanups of mined sites and reimburse communities affected by mining. This 2007 bill was never brought to a vote in the Senate, however. A new version of this bill was reintroduced in 2009, named the Hardrock Mining and Reclamation Act of 2009, and at the time of this writing was in committee in both houses of Congress.

The General Mining Act of 1872 covers a wide variety of metals, gemstones, uranium, and minerals used for building materials. In contrast, fossil fuels, phosphates, sodium, and sulfur are governed by the Mineral Leasing Act of 1920. This law sets terms for leasing public lands that vary according to the resource being mined, but in all cases the terms include the payment of rents for the use of the land and the payment of royalties on profits.

WEIGHING THE ISSUES

Reforming the 1872 Mining Law You are a legislator in the U.S. Congress, and your colleagues are asking you to help prepare a bill to reform the General Mining Act of 1872. Would you join this effort to reform the law? Why or why not? If you would, then what would you seek to include in the bill?

TOWARD SUSTAINABLE MINERAL USE

Mining exerts plenty of environmental impacts, but we also have another concern to keep in mind: Minerals are nonrenewable resources (pp. 3–4) in finite supply. Like fossil fuels, they form far more slowly than we use them, and if we continue to mine them, they will eventually be depleted. As a result, it will benefit us to find ways to conserve the supplies we have left and to make them last. Reducing waste and developing means of recovering and recycling used mineral resources are ways we can pursue the use of mineral resources more sustainably. We will likely never achieve 100% recovery but we can do much better than we are doing today.

Minerals are nonrenewable resources in limited supply

Unlike sunlight or water or forests, minerals do not regenerate fast enough to provide us a new supply once we have mined all known reserves. They are therefore considered nonrenewable resources. Some minerals we use are abundant in their supply and will likely never run out, but others are rare enough that they could soon become unavailable.

For instance, geologists in 2008 calculated that the world's known reserves of tantalum will last about 129 more years at today's rate of consumption. If demand for tantalum increases, it could run out faster. And if everyone in the world began consuming tantalum at the rate of U.S. citizens, then it would last for only 18 years! Most pressing may be dwindling supplies of indium. This obscure metal, which is used for LCD screens, might last only another 32 years. Because of these supply concerns and price volatility, industries now are working hard to develop ways of substituting other materials for indium. A lack of indium and gallium would threaten the production of high-efficiency cells for solar power. Platinum is dwindling too, and if it became unavailable this would make it harder to develop fuel cells for vehicles. However, platinum's high market price encourages recycling, which may keep it available, albeit as an expensive metal.

FIGURE 23.10 shows estimated years remaining for several selected minerals at today's consumption rates. Note that each bar in the figure consists of two parts. The left (maroon) portion of the bar shows the number of years researchers calculate that we will have the mineral available under today's economic conditions. As minerals become scarcer, demand for them increases and price rises. Higher market prices make it more profitable for companies to mine the resource, so they become willing to spend more to reach further deposits that were not economically worthwhile originally. Thus the entire length of each bar in Figure 23.10 shows the number of years researchers calculate that we will have the mineral available in total, as prices rise. These two categories used by the U.S. Geological Survey (USGS) are similar to the way the USGS classifies fossil fuel deposits as being either "economically recoverable" or "technically recoverable" (pp. 538, 543).

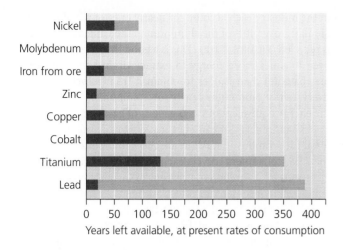

FIGURE 23.10 ▲ Minerals are nonrenewable resources, so supplies of metals are limited. Shown in red are the numbers of remaining years that certain metals are estimated to be economically recoverable at current prices, given known global reserves and assuming current rates of consumption. The entire lengths of the bars (red plus orange) show the numbers of remaining years that certain metals are estimated to be available using current technology on all known deposits, whether economically recoverable or not. All these time periods could increase if more reserves are found, or decrease if consumption rates rise. Data are for 2009, from U.S. Geological Survey, 2010. *Mineral Commodity Summaries 2010*. USGS, Washington, D.C.

Several factors affect how long mineral deposits may last

Calculating how long a given mineral resource will be available to us, as shown by the data in Figure 23.10, is beset by a great deal of uncertainty. There are several major reasons why such estimates may increase or decrease over time.

Discovery of new reserves   As we discover new deposits of a mineral, the known reserves—and thus the years this mineral is available to us—increase. For this reason, some previously predicted shortages have not come to pass, and we may have access to these minerals for longer than currently estimated. As one recent example, in 2010 geologists associated with the U.S. military in Afghanistan discovered that Afghanistan holds immense mineral riches that were previously unknown. The newly discovered reserves of iron, copper, niobium, lithium, and many other metals are estimated to be worth over $900 billion—enough to realign the entire Afghan economy around mining. (Note however, that such riches are not guaranteed to make Afghanistan a wealthy nation; history teaches us that regions rich in nonrenewable resources, such as Congo and Appalachia, have often been unable to prosper from them.)

New extraction technologies   Just as rising prices of scarce minerals encourage companies to expend more effort to reach difficult deposits, rising prices also may favor the development of enhanced mining technologies that can reach more minerals at less expense. If more powerful technologies

are developed, then these may increase the amounts of minerals that are technically feasible for us to mine.

Changing social and technological dynamics   New societal developments and new technologies in the marketplace can modify demand for minerals in unpredictable ways. Just as cell phones and computer chips boosted demand for tantalum, fiber-optic cables decreased demand for copper as they replaced copper wiring in communications applications. Today lithium-ion batteries are replacing nickel-cadmium batteries in many devices. Synthetically made diamonds are driving down prices of natural diamonds and extending their availability. Additionally, health concerns sometimes motivate change: We have replaced toxic substances such as lead and mercury with safer materials in many applications, for example.

Changing consumption patterns   Changes in the rates and patterns of consumption also alter the speed with which we exploit mineral resources. For instance, economic recession depressed demand and caused production and consumption of most minerals to decrease in 2008 and 2009. However, over the long term, demand has been rising. This is especially true today as China, India, and other major industrializing nations rapidly increase their consumption.

Recycling   Advances in recycling technologies and the extent of recycling have been helping us to extend the lifetimes of many mineral resources. Further progress in recycling will likely continue to do so.

Despite these sources of uncertainty, we would be wise to be concerned about Earth's finite supplies of mineral resources and to try to use them more sustainably. Sustainable use will benefit future generations by conserving resources for them to use. It will also benefit us today, because conserving mineral resources through reuse and recycling can prevent price hikes that result from reduced supply. And as we saw with fossil fuels (Chapter 19), domestic conservation of resources helps make a national economy less vulnerable in instances when other nations decide to withhold resources. Currently, the United States is 100% dependent on imports from other nations for 19 of the 63 major minerals on which the USGS reports annually. For 19 more of these minerals, the United States relies on imports for 60% or more of its supply.

We can make our mineral use more sustainable

We can address both major challenges facing us regarding mineral resources—finite supply and environmental damage—by encouraging recycling of these resources. Metal-processing industries regularly save resources and money by reusing some of the waste products produced during their refining processes. In addition, municipal recycling programs help provide metals by handling used items that we as consumers place in recycling bins and divert from the waste stream (**FIGURE 23.11**). In 2008, fully 35% of metals in the U.S. municipal solid waste stream were diverted for recycling. Programs for recycling car batteries have resulted in enough recovery of lead that fully 80% of the lead we consume today comes from recycled materials. Similarly, 35% of our copper comes from recycled copper

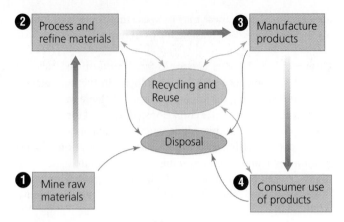

FIGURE 23.11 ▲ After we ❶ mine ore or other raw materials, we ❷ process and refine the minerals of interest and then ❸ manufacture products from them, which are then sold to and used by consumers ❹. At each of these stages, some waste is disposed of, but at each stage there are also opportunities for reusing and recycling minerals.

sources such as pipes and wires. We recycle steel, iron, platinum, and other metals from auto parts. Altogether, we have found ways to recycle much of our gold, lead, iron and steel scrap, chromium, zinc, aluminum, and nickel. **TABLE 23.1** shows minerals that currently boast the highest recycling rates in the United States.

In many cases, recycling can decrease energy use substantially. For instance, making steel by remelting recycled iron and steel scrap requires much less energy than producing steel from virgin iron ore. Because this practice saves money, the steel industry today is designed to make efficient use of iron and steel scrap. Over half its scrap comes from discarded consumer items like cars, cans, and appliances; scrap produced within their plants and scrap produced by other types of plants

in the industry each account for nearly one-fourth of the total. Similarly, over 40% of the aluminum in the United States today is recycled (**FIGURE 23.12**). This is a good thing because it takes over 20 times more energy to extract virgin aluminum from ore (bauxite) than it does to obtain it from recycled sources. Every ton of aluminum cans your community recycles saves the energy equivalent of 1,665 gallons of gasoline.

Saving energy means cutting down on pollution from fossil fuel combustion. All in all, the 7 million tons of metals that U.S. consumers recycle each year reduce greenhouse gas emissions by 25 million metric tons of carbon dioxide equivalent, the Environmental Protection Agency estimates. This is like taking more than 4.5 million cars off the road for a year.

Tantalum is recycled from scrap by-products generated during the manufacture of electronic components and also from scrap from tantalum-containing alloys and manufactured materials. Currently the industry estimates that recycling accounts for 20–25% of the tantalum available for use in products. This percentage has been growing quickly, but its future growth will depend on how quickly we expand recycling efforts for used cell phones and other electronic waste (pp. 634–635) and on how well we enable recycling facilities to recover metals from these products.

We can recycle metals from e-waste

We saw in Chapter 22 (pp. 633–637) that electronic waste, or e-waste, from discarded computers, printers, cell phones, handheld devices, and other electronic products is rising fast—and that e-waste contains hazardous substances. Recycling old electronic devices helps keep them out of landfills and also helps us conserve valuable minerals such as tantalum. As Dr. Timothy Townsend (featured in Chapter 22's **Science behind the Story** on pp. 636–637) has put it, "If we know these metals are, overall, bad for us, it doesn't make

TABLE 23.1	Top recycled minerals in the United States
Mineral	**U.S. recycling rate**
▶ Gold	More is recycled than is consumed.
▶ Lead	80% consumed comes from recycled post-consumer items.
▶ Iron and steel scrap	100% for autos, 90% for appliances, 70–98% for construction materials, 65% for cans.
▶ Chromium	61% is recycled in stainless steel production.
▶ Zinc	54% produced is recovered, mostly from recycled materials used in processing.
▶ Aluminum	42% produced comes from recycled post-consumer items.
▶ Nickel	42% consumed is from recycled nickel.
▶ Tungsten	37% consumed is from recycled scrap.
▶ Copper	35% of supply comes from various recycled sources.
▶ Germanium	30% consumed worldwide is recycled. Optical device manufacturing recycles over 60%.
▶ Molybdenum	About 30% gets recycled as part of steel scrap that is recycled.
▶ Silver	28% consumed is from recycled silver. U.S. recovers as much as it produces.
▶ Cobalt	24% consumed comes from recycled scrap.
▶ Tin	24% consumed is from recycled tin.
▶ Bismuth	All scrap metal containing bismuth is recycled, providing 20% of consumption.
▶ Niobium (Columbium)	Perhaps 20% gets recycled as part of steel scrap that is recycled.
▶ Diamond (Industrial)	12% of production is from recycled diamond dust, grit, and stone.
▶ Platinum-group metals	About 12% consumed is recycled.

Data are for 2009, from U.S. Geological Survey, 2010. *Mineral Commodity Summaries 2010.* USGS, Washington, D.C.

TABLE 23.2 Metal Content of 500 Million Unused Cell Phones

Metal	Amount (metric tons)	Value
Copper	7,900	$17 million
Silver	178	$31 million
Gold	17	$199 million
Palladium	7.4	$63 million
Platinum	0.18	$3.9 million
Total	**8,102**	**$314 million**

Data are for 2005, from U.S. Geological Survey, 2006. *Recycled cell phones—A treasure trove of valuable metals.* USGS Fact Sheet 2006-3097. Values have risen since 2005 with the prices of gold and other metals.

FIGURE 23.12 ▲ When you recycle aluminum cans, you contribute to valuable efforts to save mineral resources, money, and energy.

sense to keep digging them up from the earth's crust and bringing them into the biosphere while—at the same time—we're taking the ones we've already got and burying them."

In fact, each of the 1.2 billion cell phones sold each year contains about 200 chemical compounds and close to a dollar's worth of precious metals. In 2005, researchers estimated that about 500 million old cell phones were left inactive in people's homes and offices. **TABLE 23.2** shows estimates of the value of minerals they contain.

When you turn in your old phone to a recycling and reuse center rather than discarding it, the phone may be refurbished and resold in a developing country. People in African nations in particular readily buy used cell phones, because they are inexpensive and because land-line phone service does not always exist in poor and rural areas. Alternatively, the phone may be dismantled in a developing country and the various parts refurbished and reused, or recycled for their metals. Either way, you are helping to extend the availability of resources through reuse and recycling and to decrease waste of valuable minerals.

Today only about 10% percent of old cell phones are recycled. That leaves a long way to go! As more of us recycle our phones, computers, and other electronic items, more tantalum and other metals may be recovered and reused. By recycling more, we can reduce demand for virgin ore and decrease pressure on African people and ecosystems where coltan is mined. Throughout the world, using recycling to make better use of the mineral resources we have already mined will help minimize the impacts of mining and assure us access to resources farther into the future.

➤ CONCLUSION

We depend on a diversity of minerals and metals to help manufacture products widely used in our society. We mine these nonrenewable resources by various methods, according to how the minerals are distributed. Economically efficient mining methods have greatly contributed to our material wealth, but they have also resulted in extensive environmental impacts, ranging from habitat loss to acid drainage. Restoration efforts and enhanced regulation help to minimize the environmental and social impacts of mining, although to some extent these impacts will always exist. We can lengthen our access to mineral resources and make our mineral use more sustainable by maximizing the recovery and recycling of key minerals.

R E V I E W I N G   O B J E C T I V E S

You should now be able to:

OUTLINE TYPES OF MINERAL RESOURCES AND HOW THEY CONTRIBUTE TO OUR PRODUCTS AND SOCIETY

- Minerals we mine from the earth provide raw materials for most of the products we use every day. (pp. 645–646)
- We mine ore for metals, as well as nonmetallic minerals and fuels. (pp. 646–648)

- Processing and refining metals through methods such as smelting is an important step between mining ore and manufacturing products. (pp. 646–647)

DESCRIBE THE MAJOR METHODS OF MINING

- Strip mining removes surface layers of soil and rock to expose resources. (p. 648)
- In subsurface mining, miners tunnel underground. (pp. 648–649)

- Open pit mining involves digging gigantic holes. (pp. 649–650)
- Placer mining uses running water to isolate minerals. (p. 650)
- Mountaintop removal mining removes immense amounts of rock from mountaintops and dumps it into valleys below. (pp. 650–651)
- Solution mining uses water to dissolve minerals in place and extract them. (p. 654)
- A great deal of mineral wealth exists in the oceans but is mostly uneconomical (so far) to reach. (p. 654)

CHARACTERIZE THE ENVIRONMENTAL AND SOCIAL IMPACTS OF MINING

- Many methods of mining completely remove vegetation, soil, and habitat. (pp. 649–653)
- Acid drainage occurs when water leaches compounds from freshly exposed waste rock. It is often toxic to aquatic organisms. (p. 649)
- Erosion, sediment disturbance, and other impacts add to water pollution. (pp. 649–654)

- Mountaintop removal for coal destroys forests, mountaintops, and adjacent valleys and streams. (pp. 650–653)
- Mining may have health impacts on miners, and diverse social impacts on people living near mines. (pp. 649–653)

ASSESS RECLAMATION EFFORTS AND MINING POLICY

- Reclamation efforts generally fall short of effective ecological restoration. (pp. 654, 656)
- U.S. mining policy remains dominated by a law passed in 1872. (pp. 657–658)

EVALUATE WAYS TO ENCOURAGE SUSTAINABLE USE OF MINERAL RESOURCES

- Minerals are nonrenewable resources, and some are limited in supply. (pp. 658–659)
- Several factors affect how long a given mineral resource will last. (p. 659)
- Reuse and recycling by industry and consumers are the keys to more sustainable practices of mineral use. (pp. 659–660)
- E-waste can be a source of recovered metals. (p. 661)

TESTING YOUR COMPREHENSION

1. Define each of the following and contrast them with one another: (1) mineral, (2) metal, (3) ore, (4) alloy.

2. A mining geologist locates a horizontal seam of coal very near the surface of the land. What type of mining method will the mining company use to extract it? What is one common environmental impact of this type of mining?

3. How does strip mining differ from subsurface mining? How does each of these approaches differ from open pit mining?

4. What type of mining is used for both coltan and gold? What does a miner do to conduct this type of mining?

5. Describe and contrast how water is used in placer mining and solution mining.

6. What is acid drainage, and why can it be toxic to fish?

7. Describe three major environmental or social impacts of mountaintop removal mining.

8. Explain why reclamation efforts after mining frequently fail to effectively restore natural communities. Include reference to both soil and vegetation in your answer.

9. List five factors that can influence how long global supplies of a given mineral will last, and explain how each might increase or decrease the time span the mineral will be available to us.

10. Name three types of metal that we currently recycle, and identify the products or materials that are recycled to recover these metals.

SEEKING SOLUTIONS

1. List three impacts of mining on the natural environment, and describe how particular mining practices can lead to each of these impacts. How are these impacts being addressed? Can you think of additional solutions to prevent, reduce, or mitigate these impacts?

2. List three impacts of mining on people's health, lifestyles, or well-being, and describe how particular mining practices can lead to each of these impacts. How are these impacts being addressed? Can you think of additional solutions to prevent, reduce, or mitigate these impacts?

3. You have won a grant from the EPA to work with a mining company to develop a more effective way of restoring a mine site that is about to be abandoned. Describe a few preliminary ideas for carrying out restoration better than it is typically being done. Now

describe a field experiment you would like to run to test one of your ideas.

4. Present one common argument in favor of retaining the General Mining Act of 1872. Now present one common argument in favor of repealing or reforming the law. Describe how you think the United States today should balance our need for minerals with our need to protect public lands and public interests. Would you retain, repeal, or reform the 1872 law?

5. **THINK IT THROUGH** The story of coltan in the Congo is just one example of how an abundance of exploitable resources can often worsen or prolong military conflicts in nations that are too poor or ineffectively governed to protect these resources. In such "resource wars," civilians often suffer the most as civil society breaks down. Suppose you

are the head of an international aid agency that has ear-marked $10 million to help address conflicts related to mining in the Democratic Republic of the Congo. You have access to government and rebel leaders in Congo and neighboring countries, to ambassadors of the world's nations in the United Nations, and to representatives of international mining corporations. Based on what you know from this chapter, what steps would you consider taking to help improve the situation in the Congo?

6. **THINK IT THROUGH** As you finish your college degree, you learn that the mountains behind your childhood home in the hills of Kentucky are slated to be mined for coal using the mountaintop removal method. Your parents, who still live there, are worried for their health and safety and do not want to lose the beautiful forested creek and ravine behind their property. However, your brother is out of work and could use a mining job. What would you attempt to do in this situation?

CALCULATING ECOLOGICAL FOOTPRINTS

As we saw in Figure 23.10, some metals are in limited enough supply that, at today's prices, they could be available to us for only a few more decades. After that, prices will rise as they become scarcer. The number of years of total availability (at all prices) depends on a number of factors: On the one hand, metals will be available for longer if new deposits are discovered, new mining technologies are developed, or recycling efforts are improved. On the other hand, if our consumption of metals increases, this will decrease the number of years we have left to use them.

Currently the United States consumes metals at a much higher per-person rate than the world does as a whole. If one goal of humanity is to lift the rest of the world up to U.S. living

standards, then this will sharply increase pressures on mineral supplies.

The chart shows currently known economically recoverable global reserves for several metals, together with the amount used per year (each figure in thousands of metric tons). For each metal, calculate and enter in the fourth column the years of supply left at current prices by dividing the reserves by the amount used annually.

The fifth column shows the amount that the world would use if everyone in the world consumed the metal at the rate that Americans do. Now calculate the years of supply left at current prices for each metal if the world were to consume the metals at the U.S. rate, and enter these values in the sixth column.

Metal	Known economic reserves	Amount used per year	Years of economic supply left	Amount used per year if everyone consumed at U.S. rate	Years of economic supply left if everyone consumed at U.S. rate
Titanium	730,000	5,720		24,420	
Copper	540,000	15,800		36,850	
Nickel	71,000	1,430		3,370	
Tin	5,600	307		1,100	
Tungsten	2,800	58		240	
Antimony	2,100	187		497	
Silver	400	21.4		118	
Gold	47	2.35		3.77	

Data are for 2009, from U.S. Geological Survey, 2010. *Mineral Commodity Summaries 2010*. USGS, Washington, D.C. All numbers are in thousands of metric tons. World consumption data are assumed equal to world production data. "Known economic reserves" include extractable amounts under current economic conditions. Additional reserves exist that could be mined at greater cost.

1. Which of these eight metals will last the longest under current economic conditions and at current rates of global consumption? For which of these metals will economic reserves be depleted fastest?

2. If the average citizen of the world consumed metals at the rate that the average U.S. citizen does, which of these eight metals' economic reserves would last the longest? Which would be depleted fastest?

3. In this chart, our calculations of years of supply left do not factor in population growth. All else being equal, how do you think population growth will affect these numbers?

4. Describe two general ways that we could increase the years of supply left for these metals. What do you think it will take to accomplish this?

Mastering ENVIRONMENTAL SCIENCE™

Sri Lankan children planting tree seedlings on deforested hillsides around their village

24 SUSTAINABLE SOLUTIONS

UPON COMPLETING THIS CHAPTER, YOU WILL BE ABLE TO:

- List and describe approaches being taken on college and university campuses to promote sustainability
- Explain the concept of sustainable development
- Discuss how protecting the environment can promote economic well-being

- Describe and assess key approaches to designing sustainable solutions
- Explain how time is limited but how human potential to solve problems is tremendous

De Anza College Strives for a Sustainable Campus

CANADA

UNITED STATES

San Francisco
De Anza College
California

Pacific Ocean

MEXICO

Atlantic Ocean

"Sustainability considers our impact not just on air, land, and water, but includes our impact on community vibrancy, environmental stewardship, social equity, and financial responsibility."

—De Anza College's Sustainability Management Plan

"Sustainability isn't a distant, unattainable concept to be discussed in the abstract. It is a very real way of making decisions that can be integrated into every single person's daily actions."

—Yale University student Jacquelyn Maitram Truong

California's Silicon Valley has long been a hub for innovation. So perhaps it's no surprise that a college in the heart of the region is a leader in the burgeoning movement for campus sustainability. De Anza College, located in Cupertino, California (home of Apple Computer, Inc.), serves 24,000 students, making it one of North America's largest community colleges. Today De Anza has become one of the "greenest" community college campuses as well, thanks to the ongoing commitment of students, faculty, staff, and administrators.

We can view colleges and universities as microcosms of society at large. These institutions of learning consume resources, emit pollution, and exert a variety of environmental impacts. Proponents of campus sustainability seek ways to help these institutions reduce their ecological footprints. Student-run environmental organizations often play a key role in initiating recycling programs, finding ways to reduce energy use, and agitating for new courses or majors in environmental science or environmental studies.

At De Anza College, sustainability efforts reach back to 1990, when faculty member Julie Phillips began incorporating concepts of sustainability into

De Anza College students in front of their LEED-certified Science Center.

the curriculum and energized students with the idea of a green building project. Soon the College Environmental Advisory Group (CEAG) was created, bringing together students, faculty, staff, administrators, and community members to encourage green building and other sustainable practices on campus. Eventually, those early students brought their classroom experiences, sustainable practices, and green building vision to reality by urging the De Anza Associated Student Body to allocate $180,000 for the conceptual design of what would become the Kirsch Center for Environmental Studies.

In 2005, the Kirsch Center opened for instruction, as the nation's first LEED-Platinum sustainable building (pp. 363, 366) at a community college. The Kirsch Center shows students and the public how to merge energy efficiency, pollution prevention, and biodiversity protection. Showcasing hi-tech and innovative labs and classrooms that promote visual and hands-on learning, the Kirsch Center aims to be "a building that teaches about energy, resources, and stewardship."

Indeed, the building does teach. Displays inside the entrance show real-time data on the energy generated by the 36.5-kilowatt photovoltaic energy system on the rooftop. Other monitors show current temperature readings that guide the advanced radiant heating and cooling system, while red and green lights advise when windows should be opened and closed. Students use labs, classrooms, and open study stations that are bathed in natural lighting from outdoors, which saves on electricity while creating a bright and pleasant environment that helps students learn better. Everything from carpeting to furniture to structural steel to toilet seats is made with recycled materials, and features abound to conserve water and energy.

Located adjacent to the Kirsch Center is a 1.5-acre arboretum called the Cheeseman Environmental Study Area. Here, 12 California native plant communities are represented, with over 400 species of native plants. Five other LEED-certified green buildings grace De Anza's campus, as well.

In 2006, De Anza's administrators signed a sustainability policy, committing to green building renovation and construction, and to the selection of vendors experienced with sustainability.

The school took another major step in 2007 when it adopted the Sustainability Management Plan that CEAG had spearheaded. This plan helps to identify environmental and health risks and to prioritize opportunities for addressing them. It specifies six focus areas:

- ▶ Reducing solid and hazardous waste
- ▶ Conserving energy and reducing carbon emissions
- ▶ Conserving water
- ▶ Making sustainable purchasing decisions
- ▶ Pursuing ecologically responsible landscaping and maintenance
- ▶ Undertaking green building practices in construction and renovation

This plan guides De Anza's greening efforts, while CEAG monitors progress and makes policy recommendations.

Challenges abound, and the main ones today are financial as California and other states suffer severe budget cutbacks in higher education. In tough financial times, more and more young people are enrolling in colleges, yet as state budgets and private endowments shrink, schools receive fewer resources. Sustainability initiatives often cost money, and their proponents generally have to argue their case to penny-conscious administrators time after time. But De Anza's administration, like increasing numbers of others, has recognized that many of the short-term costs associated with sustainability efforts are actually investments that can save substantial amounts of money in the long term.

Despite these challenges, De Anza continues to make strides. Today, 60% of the campus's waste is diverted from landfills. Ninety percent of the chemicals used by the custodial staff are environmentally friendly. The dining services are using biodegradable

FIGURE 24.1 ▲ Sustainability-oriented curricula extend far beyond the classroom. In California's Coyote Valley, De Anza College students are trained in tracking and surveying wildlife as they help perform research into the importance of habitat corridors.

containers and local and organic food. Students are getting transit passes from the local bus service. De Anza's sustainability proponents have produced a policy handbook on energy efficiency and are leading workshops for other community colleges.

Moreover, thousands of students enroll in De Anza's environmental studies and environmental sciences courses each year. The learning extends far beyond the classroom. In one new program, students are trained in wildlife tracking and help conduct research to benefit their region as they learn the science of conservation biology (**FIGURE 24.1**). All in all, De Anza is providing an inspiring model for other community colleges in California and for colleges and universities nationwide as they strive for campus sustainability.

SUSTAINABILITY ON CAMPUS

Whether on campus or around the world, *sustainability* means living in a way that can be lived far into the future. As we saw in Chapter 1 (p. 15) and throughout this book, sustainability involves conserving resources to prevent their depletion, protecting ecological processes, and eliminating waste and pollution, so as to ensure that our society's practices can continue and our civilization can endure.

A *sustainable solution* is one that results in truly renewable resource use, whereby whatever natural capital we take from Earth can be replenished by the planet's systems, so that resources are not depleted. A sustainable solution preserves ecosystem services so that they continue to function and provide us their many benefits. A sustainable solution eliminates pollution, being carbon-neutral and emitting no toxic substances. A sustainable solution enables us to reuse or recycle waste so as to truly close the loop in our processes of production. And a sustainable solution will satisfy all three pillars of sustainability: environmental quality, economic well-being, and social justice.

In most cases, the solutions that are immediately viable today will fall short of true sustainability. Yet we need to start somewhere, and we need to take steps that society can agree upon. True sustainability may be our goal, but we may need to take interim steps to approach this ideal. Moving toward sustainability is an ongoing process that we will be engaged in for the remainder of our lifetimes. It may take a while, but it is our one necessary journey.

If we are to attain a sustainable civilization, we will need to make efforts at every level, from the individual to the household to the community to the nation to the world. Governments, corporations, and organizations must all encourage and pursue sustainable practices. Among the institutions that can contribute to sustainability efforts are colleges and universities.

Why strive for campus sustainability?

We tend to think of colleges and universities as enlightened and progressive institutions that benefit society. However, colleges and universities are also centers of lavish resource consumption. Institutions of higher education feature extensive infrastructure including classrooms, offices, research labs, and residential housing. Most have dining establishments, sports arenas, vehicle fleets, and road networks. The 4,300 campuses in the United States interact with thousands of businesses and spend nearly $400 billion each year on products and services. The ecological footprint of a typical college or university is substantial, and together these institutions generate perhaps 2% of U.S. carbon emissions.

Reducing the size of this footprint is challenging. Colleges and universities tend to be bastions of tradition, where institutional habits are deeply ingrained and where bureaucratic inertia can block the best intentions for positive change. Nonetheless, faculty, staff, administrators, and students are progressing on a variety of fronts to make the operations of educational institutions more sustainable (see **ENVISIONIT**, p. 668). Students are often the ones who initiate change, although support from faculty, staff, and administrators is crucial for success. Students often feel freest to express themselves. Students also arrive on campus with new ideas and perspectives, and they generally are less attached to traditional ways of doing things.

Students who promote more sustainable practices on their campuses accomplish several things at once. First, reducing the ecological footprint of a campus really can make a difference, because the consumptive impact of educating, feeding, and housing hundreds or thousands of students is immense. Second, students who act to advance campus sustainability serve as models for their peers, helping to make them aware of the need to address problems. Finally, the student who engages in sustainability efforts learns and grows as a result. The challenges, successes, and failures that you encounter while working with others as part of a team can serve as valuable preparation for similar efforts in our broader society.

An audit is a useful way to begin

Campus sustainability efforts often begin with a quantitative assessment of the institution's operations. An audit provides baseline information on what an institution is doing or how much it is consuming. Audits also help set priorities and goals. Students can conduct an audit themselves, as when University of Vermont graduate student Erika Swahn measured heating, transportation, electricity, waste, food, and water use to calculate her school's ecological footprint (it turned out to be 4.5 acres for each student, instructor, and staff member). If you set out to do an audit yourself, one useful tool is a handheld device called a "Kill-A-Watt" meter (**FIGURE 24.2**) which can help measure energy use plug-by-plug and room-by-room.

Sometimes an audit can be done as part of a class, as when students at Stetson University in Florida conducted a greenhouse gas inventory of their campus. Alternatively, staff or hired consultants may conduct an audit. Harford Community College in Maryland hired specialists to audit its energy use and pollutant emissions, which then served as the basis for setting reduction goals. At De Anza College, students helped conduct a comprehensive audit of various buildings on campus, describing the buildings' inefficiencies and identifying opportunities for improvement. The information from this audit was then included in the college's Sustainability Management Plan.

It is most useful in an audit to target items that can lead directly to specific recommendations. For instance, an audit should quantify the performance of individual appliances so that decision makers can identify particular ones to replace. Once changes are implemented, the institution can monitor progress by comparing future measurements to the audit's baseline data.

Recycling and waste reduction are common campus efforts

Campus sustainability efforts frequently involve waste reduction, recycling, or composting. According to the most recent comprehensive survey of campus sustainability efforts (**TABLE 24.1**), most schools recycle or compost at least some waste, and the average recycling rate reported was 29%. Waste management

FIGURE 24.2 ▲ A valuable tool for conducting an audit of energy use on campus is a Kill-a-Watt meter. This device measures the electrical current drawn from outlets by various appliances and fixtures.

Students everywhere are helping to make their campuses more sustainable.

Installing solar panels at Arizona State

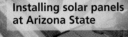

CARTRIDGES BATTERIES PHONES
INK   CELL

Bike-sharing at U. of Rhode Island

URIde
CELS Sustainability Initiative
401-874-4947

Recycling at Davidson

Every item you recycle, each light you turn off, and each trip made by bike instead of car makes a difference — and serves as a positive example for your peers.

E-waste drive at UT-Austin

18 million American college students are a powerful force for change!

Biodiesel at U. of Virginia

YOU CAN MAKE A DIFFERENCE

➤ Join the student environmental club on your campus . . . or start one!

➤ Register to vote, and help with voter registration drives on campus.

➤ Talk to your institution's president about signing the American College and University Presidents' Climate Commitment.

TABLE 24.1 Frequency of Campus Sustainability Efforts

Activity	Percentage of campuses performing activity[*]
Waste management	
▶ Recycling and composting	50–89[†]
▶ Diverting surplus materials from the waste stream through an exchange program	77
▶ Reducing the need for paper	68
Campus buildings	
▶ Constructing or renovating green buildings with LEED (pp. 363, 366) certification	35
▶ Creating green roofs	13
Water efficiency upgrades	76
Energy efficiency upgrades	
▶ Upgrading lighting efficiency	81
▶ Upgrading HVAC systems	73
▶ Reducing energy load in IT systems	66
Energy use and policy	
▶ Creating plans to reduce greenhouse gases	27
▶ Purchasing carbon offsets	8
▶ Generating clean renewable energy	12
▶ Using clean renewable energy from off-campus	32
Sustainable purchasing policies	61
Transportation	
▶ Offering free or discounted mass transit	31
▶ Providing adequate and protected bike racks	61
▶ Using alternative fuels in fleet vehicles	27
Landscaping	
▶ Landscaping to help wildlife	39
▶ Restoring habitat	40
▶ Using integrated pest management (pp. 259–260)	61
▶ Removing invasive species	38
Professional development in sustainability for faculty	36
Staff coordinator for sustainability issues	51

[*]Data are for 2008, based on voluntary responses from 1,086 schools. Most numbers are likely underestimates, as some campuses did not respond to some questions.

[†]50–89%, depending on type of material.

Source: McIntosh, M., et al., 2008. *Campus environment 2008: A national report card on sustainability in higher education.* National Wildlife Federation Campus Ecology; survey conducted by Princeton Survey Research Associates International.

initiatives are relatively easy to start and maintain because they offer many opportunities for small-scale improvements and because people generally enjoy recycling and reducing waste.

The best-known collegiate waste management event is *RecycleMania,* a 10-week competition among schools to see which can recycle the most. This annual competition, which was started by students, involved 607 schools in its 10th year in 2010, when California State University at San Marcos won the overall award, recycling 72% of its waste. Students at the United States Coast Guard Academy won the title for per capita recycling, with 37 kg (82 lb) of material recycled per student. Other top-performing schools in 2010 included North Lake College, Rutgers University, New Mexico State University, American University, Kalamazoo College, and Stetson University. In earlier years Colorado State University students had done so well

in this event that they were recruited to run a recycling program at the 2008 Democratic National Convention in Denver, and they succeeded in recycling 80% of the waste.

"Trash audits" or "landfill on the lawn" events involve tipping dumpsters onto a campus open space and sorting out recyclable items (**FIGURE 24.3**). When students at Ashland University in Ohio audited their waste, they found that 70% was recyclable, and they used this data to press their administration to support recycling programs. Louisiana State University students initiated recycling efforts at home football games, and over three seasons they recycled 68 tons of refuse that otherwise would have gone to the landfill.

Composting is becoming popular as well. Ball State University in Indiana composts bulky wood waste, shredding surplus furniture and wood pallets and making them into mulch to nourish campus plantings. A composting program at Texas State University designed by a graduate student combines food scraps, paper waste, and waste from a nearby chicken feedlot—and then adds water hyacinth plants, an invasive aquatic plant unwanted in the area. The program aims to determine whether the heat generated by composting can kill seeds of the water hyacinth while producing top-quality compost. At Ithaca College in New York, 44% of the food waste generated annually on campus is composted. Disposal fees at the local landfill are $60 per ton, so composting saves the college $11,500 each year. The compost is used on campus plantings, and experiments showed that the plantings grew better with the compost mix than with chemical soil amendments.

Reuse is more sustainable than recycling, so students at some campuses systematically collect unwanted items and donate them to charity or resell them to returning students in the fall. Students at the University of Texas at Austin run a "Trash to Treasure" program. Each May, they collect 40–50 tons of items that students discard as they leave and then resell them at low prices in August to arriving students. This keeps waste out of the landfill, provides arriving students with items they need at low cost, and raises $10,000–20,000 per year that gets plowed back into campus sustainability efforts.

FIGURE 24.3 ▼ In "landfill on the lawn" events, volunteers sort through rubbish, separating out recyclables, as these students at the University of North Carolina at Greensboro are doing. These events can demonstrate just how many recyclable items are needlessly thrown away.

Green building design is a key to sustainable campuses

Buildings are responsible for 70–90% of a campus's carbon emissions, so making them more efficient can make a big difference. Many campuses now boast "green" buildings that are constructed from sustainable building materials and whose design and technologies reduce pollution, use renewable energy, and encourage efficiency in energy and water use. As with any type of ecolabeling, there need to be agreed-upon standards, and for sustainable buildings these are the *Leadership in Energy and Environmental Design (LEED)* standards (pp. 363, 366). Developed and maintained by the nonprofit U.S. Green Building Council, LEED standards guide the design and certification of new construction and the renovation of existing structures.

De Anza College's Kirsch Center for Environmental Studies (**FIGURE 24.4**) has achieved a "platinum" LEED ranking (the highest possible). Besides the features described earlier (pp. 665–666), the building contains recycled steel in its beams, cabinets made with FSC-certified (Forest Stewardship Council) wood, tiles made from car windshields, toilet seat lids made from recycled water bottles, concrete floors containing fly ash from coal-fired power plants, and carpeting made from recycled materials with no PVCs, vinyl, or toxic adhesives. The building's layout and orientation utilize passive solar design concepts (p. 594), natural daylighting, and energy-efficiency strategies. South-facing windows let in sunlight in the winter while the high solar angle, exterior sun shades, and deciduous trees effectively block the sun in the summer.

Students helped launch the Kirsch Center when the De Anza Associated Student Body contributed $180,000 for the conceptual design of the building. The Steve and Michele Kirsch Foundation contributed $2 million toward construction, and local bond funds provided $8 million. Its many energy- and water-saving features help the college save money in the long run, relative to a conventional building. Arup, the architectural firm responsible for the energy and mechanical design of the building, has calculated that the Kirsch Center's energy efficiency features save $65,000 per year (the total regulated energy cost is reduced by 88%) when compared to a typical building.

One of the first green buildings on a college campus was the Adam Joseph Lewis Center for Environmental Studies at Oberlin College in Ohio (**FIGURE 24.5A**). This building was constructed using materials that were recycled or reused, took little energy to produce, or were locally harvested, produced, or distributed. Carpeting materials are leased and then returned to the company for recycling when they wear out. The Lewis Center contains energy-efficient lighting, heating, and appliances, and it maximizes indoor air quality with a state-of-the-art ventilation system, as well as paints, adhesives, and carpeting that emit few volatile organic compounds. The structure is powered largely by solar energy from photovoltaic (PV) panels on the roof, active solar heating, and passive solar heating from south-facing walls of glass and a tiled slate floor that acts as a thermal mass (p. 594). Over 150 sensors throughout the building monitor conditions such as temperature and air quality.

Bren Hall at the University of Santa Barbara in California (**FIGURE 24.5B**) uses an innovative and efficient heating and cooling system integrated with solar panels and white roofing material to reflect sunlight. The structural steel and other materials in the building are composed of 40% recycled content, and over 90% of construction waste was recycled. The building contains no formaldehyde, asbestos, or CFCs, and other toxic substances are kept to a minimum. Bren Hall conserves water with low-flow fixtures, waterless urinals, reclaimed water for toilets, and automatic sensors. Sustainable materials and approaches added only 2% to the $26 million construction cost, and this added cost is easily being recovered through energy savings.

The demand for "green buildings" is growing fast. More than one-third of schools responding to a 2008 survey had at least one LEED-certified building or retrofit, and over half planned to pursue them in the future. Butler University in Indiana even has a LEED-certified fraternity house. The University of Florida has built or started construction on 18 green buildings since 2003. Pacific Union College in California is seeking to build an entire "ecovillage" of dorms designed for sustainability. At Catawba College in North

FIGURE 24.4 ▼ The Kirsch Center for Environmental Studies at De Anza College is a climate-responsive and energy-efficient building that embraces water conservation and other sustainable practices. It uses recycled and nontoxic materials, generates renewable energy including solar thermal and solar electric power, and features special outdoor learning spaces and labs. The spacious interior gathers natural daylight (bottom photo), providing a welcoming learning environment.

(a) Lewis Center at Oberlin

(b) Bren Hall at UC Santa Barbara

FIGURE 24.5 ▲ Oberlin College's Adam Joseph Lewis Center for Environmental Studies **(a)** and Bren Hall at the University of Santa Barbara, California **(b)**, are two of the best-known green buildings on American campuses.

Carolina, students convinced their board of trustees to commit to construct only sustainable buildings.

Sustainable architecture doesn't stop at a building's walls. The landscaping around the Kirsch Center and around Bren Hall each features drought-tolerant native plants that require little watering. Oberlin's Lewis Center is set among orchards, gardens, and a restored wetland that helps filter wastewater, and with urban agriculture and lawns of grass specially bred to require less chemical care. Careful design of campus landscaping can create livable spaces that promote social interaction and where plantings supply shade, prevent soil erosion, create attractive settings, and provide wildlife habitat. It has been said, in fact, that groundskeepers are more vital to colleges' recruiting efforts than are vice presidents!

Water conservation is important

Conserving water is a key element of sustainable campuses—especially in arid regions, as students at the University of Arizona in Tucson know. Despite Tucson's dry desert climate, rain falls in torrents during the late-summer monsoon season, when water pours off paved surfaces and surges down riverbeds, causing erosion, carrying pollution, and flowing too swiftly to sink into the ground and recharge aquifers. So UA students sought to redirect these floodwaters and put them to use. With the help of a grant written by Dr. James Riley and student Chet Phillips, the group created an independent study course to design and implement a rainwater harvesting project on campus (**FIGURE 24.6**). Students surveyed sites, researched their hydrology, and worked with staff to design and engineer channels, dams, berms, and basins to slow the water down and direct it into swales, where it can nourish plants and sink in to recharge the aquifer.

Water conservation is just as important indoors. A student organization called Greeks Going Green pursues sustainable solutions for students in fraternities and sororities. In its water conservation program at the University of Washington, students installed 300 5-minute shower timers, 300 low-flow showerheads, and 500 low-flow aerators for sink faucets. At Denison University in Ohio, students staged a month-long "water wars" competition among dorms. The dorm that reduced its water use the most was given a financial reward, which was donated to a community charity voted on by residents of the dorm. Students at Reading Area Community College in Pennsylvania installed water-bottle fillers at drinking fountains in college buildings and mounted an information campaign urging their peers to fill used bottles with tap water instead of buying bottled water, which involves more environmental impacts (pp. 416–418).

Water-saving technologies such as waterless urinals and "living machines" to treat wastewater are being installed on many campuses. The University of British Columbia in Vancouver, Canada, sank $35 million into retrofitting 300 campus buildings with water- and energy-efficient upgrades that now save the school $2.6 million annually. The upgrades reduce water use by 30% each year—enough water for 12,000 homes.

FIGURE 24.6 ▼ University of Arizona students reengineer the landscape on their campus to prevent flooding and harvest rainwater.

Energy efficiency is easy to improve

Students are finding many ways to conserve energy. At Colorado College, students launched a four-month public awareness campaign, using "eco-reps" in dorms to give advice on saving energy, as well as e-mails, murals, sculptures, and energy-saving equipment. In those four months, the college saved $100,000 in utility costs and reduced its greenhouse gas emissions by 10%. Students at SUNY-Purchase in New York saved their school $86,000 per year simply by turning down hot water temperatures by 5 °F (2.8 °C). Students at California State University, Chico, voted to urge the university to adjust building thermostats by 3 °F (1.7 °C). This simple adjustment saves the institution $151,000 each year and prevents the emission of 1,100 tons of carbon dioxide.

Campuses can harness large energy savings simply by not powering unused buildings. Central Florida Community College booked its summer classes into a minimum number of efficient buildings so that other buildings could be shut down. Central Florida students also surveyed staff and then removed light bulbs the staff said they did not need. The college also installed motion detectors to turn off lights when people are not in rooms. At Augusta State University in Georgia, students studied lighting in the science building and concluded that installing automated sensors would save $1,200 per year in electricity bills and save 6.4 tons of CO_2 emissions. At this rate, the cost of the sensors would be recouped in just four years.

At a number of schools, students have distributed thousands of energy-efficient compact fluorescent bulbs to their peers and to community members, in exchange for standard incandescent lightbulbs (**FIGURE 24.7**). University of Vermont students have been swapping bulbs since 2004, and they estimate that through 2009 they saved $16,000 and 22 tons of carbon dioxide emissions. Morehouse College in Atlanta is trying to outdo all these efforts: Its ambitious "Let's Raise a Million" project aims to raise money to install 1 million CFL bulbs in low-income houses in the Atlanta area.

672

FIGURE 24.7 ▲ Replacing traditional incandescent lightbulbs with compact fluorescent bulbs (CFLs) is an easy way to reduce electricity use. Here student Brandon Geller of the Harvard University Green Campus Initiative holds a CFL.

One way to get people to conserve energy is to challenge them with a competition. Many schools now stage contests among residence halls to see how little electricity (or water) each can use over the course of a day, month, or term. Students at Williams College in Massachusetts were some of the first to transform energy conservation into a kind of intramural sport. Their "Do It in the Dark" competition pitted residence halls against one another, and it soon was producing a 13% cut in energy consumption as students got caught up in the fun. At Connecticut College, students added an extra incentive: They took 25% of the money saved by energy conservation during the winter and used it to fund a concert for the whole student body in the spring.

At Oberlin College, students built on this by developing monitoring systems that display in real time the energy use in a dorm and compare that to energy use in other dorms. The students who invented the technology went on to found a company that markets similar devices for various uses. Dartmouth College students, meanwhile, adapted this idea and made it visual, creating online monitors that show a cartoon polar bear happy when energy use is low and increasingly distressed as energy use rises.

In 2007, students at 16 Minnesota colleges and universities took things to the next level, waging a month-long "Campus Energy Wars" competition among schools. Carleton College won by reducing energy use by 21% in its dorms and 10% overall. This inspired a national contest the next year. The National Campus Energy Challenge in 2008 was won by St. John's University of Minnesota.

Students are promoting renewable energy

Campuses can reduce fossil fuel consumption and emissions by altering the type of energy they use. Ball State University replaced its coal-fired heating plant with cleaner and more efficient fluidized-bed combustion technology. Middlebury College in Vermont is switching its central plant from fuel oil to carbon-neutral wood chips, which should reduce emissions by 40%.

Solar power plays a role on many campuses. Schools in sunny climates such as De Anza College are a natural fit for solar power, including solar thermal and photovoltaic (PV) systems. These applications provide a portion of De Anza's energy, and the Foothill-De Anza District plans to increase the PV-generated electricity to over 1 megawatt over the next few years. Butte College in California gets 28% of its electricity from its solar array. California State University, Long Beach, is installing a PV solar system that will decrease emissions equal to taking 50,000 cars off the road. Yet as we saw in Chapter 21, solar power also works in cooler and cloudier climes. At the University of Vermont, solar PV panels provide enough electricity for nine desktop computers or 95 energy-efficient lightbulbs for 10 hours each day. William Paterson University in New Jersey is embarking on the installation of what could become the largest single-campus PV solar installation in the nation, which could provide 14% of campus electricity needs and save $4.4 million over 15 years.

Middlebury College and Minnesota's Macalester College were pioneers in installing wind turbines to help meet their

FIGURE 24.8 ▲ A wind turbine at the University of Maine at Presque Isle saves $100,000 per year in electricity costs while reducing CO_2 emissions and serving as an educational resource for the campus.

energy needs, but today more colleges are doing so. The University of Maine at Presque Isle installed a wind turbine that will save $100,000 each year in electricity expenses with a corresponding reduction in carbon emissions (**FIGURE 24.8**). St. Olaf College in Minnesota gets one-third of its electricity from its wind turbine. Massachusetts Maritime Academy is erecting a wind turbine expected to generate 25% of the school's electricity and save $300,000 per year.

Campuses that do not generate renewable energy themselves can still invest in renewable energy by purchasing "green tags," or carbon offsets (p. 524), which subsidize renewable energy sources. Western Washington University, College of the Atlantic in Maine, Georgian Court University in New Jersey, and other schools now offset 100% of their greenhouse gas emissions by buying green tags for renewable energy. The California State University system buys 20% of its power from renewable sources, and students in the University of California system were integral in convincing the UC regents to pass a system-wide policy back in 2003 to encourage renewable energy. Renewable power generally costs more than power from fossil fuels, but on more than 30 campuses so far, students have voted to increase student fees—to tax themselves—to help fund the purchase of renewable energy. At the University of Tennessee, Knoxville, green fees brought in $1.4 million to buy green power and make energy efficiency upgrades.

Some students even design renewable energy technology! In 2009, teams of students from 20 universities competed in the fourth Solar Decathlon. In this remarkable biennial event, teams of students travel to the National Mall in Washington, D.C., bringing material for solar-powered homes they have spent months designing. The teams erect their homes on the Mall, where they stand for three weeks (**FIGURE 24.9**). The homes are judged on 10 criteria, and prizes are awarded to winners in each category. In the 2009 competition, a team from Germany beat all the American teams to win the event. Apparently Germany's students brought their nation's prowess in solar technology (pp. 590–591) with them! Some of the solar houses have been brought back to campus and put to use: The 2007 Georgia Institute of Technology team re-erected its house on the Georgia Tech campus, where it serves as a solar-powered lab and classroom.

Carbon-neutrality is a major goal

Now that global climate change has vaulted to the forefront of society's concerns, reducing greenhouse gas emissions from fossil fuel combustion has become a top priority for campus sustainability proponents. Today many campuses are aiming to become carbon-neutral.

Students at Lewis and Clark College in Oregon began the trend in 2002, when they made their school the first in the nation to comply with the Kyoto Protocol (pp. 520–521). After conducting a campus audit, student leaders reduced greenhouse emissions by the percentage required under the Protocol, largely by purchasing carbon offsets from a non-profit that funds energy efficiency and revegetation projects. While U.S. leaders were citing economic expense in refusing to cut emissions at the national level, Lewis and Clark students found that becoming Kyoto-compliant on their campus cost only $10 per student per year. Similarly, Western

FIGURE 24.9 ▶ In October 2009, 20 teams from colleges and universities across the United States and the world converged on the Mall in Washington, D.C., for the fourth Solar Decathlon. Each team erected an entire house of the students' own design fully powered by solar energy.

Washington University's carbon offset program costs just $10.50 per student per year—an amount that students voted overwhelmingly to pay.

The University of Minnesota, Morris, plans to be carbon-neutral after 2010, without relying on offsets. It already gets half its electricity from a wind turbine on campus and is constructing a second turbine to cover the other half. A newly built biomass gasification plant should offset nearly all campus fossil fuel use with local biomass feedstocks.

Today, student pressure and petitions at many campuses are nudging administrators and trustees to set targets for reducing greenhouse emissions. As of 2010, nearly 700 university presidents had signed onto the American College and University Presidents' Climate Commitment. Presidents taking this pledge commit to inventory emissions, set target dates and milestones for becoming carbon-neutral, and take immediate steps to lower emissions with short-term actions, while also integrating sustainability into the curriculum.

Education and advocacy on climate change is spilling out from the campus into the general society. On January 31, 2008, members of the public joined an estimated 240,000 students on 1,900 American campuses as part of *Focus the Nation*, a national "teach-in" on solutions to global climate change. The next year, Focus the Nation sponsored a Nationwide Town Hall on America's Energy Future that aimed to educate the public and influence lawmakers.

In 2009, 12,000 students converged on Washington, D.C., for the second Power Shift conference, networking with one another and urging legislators to promote clean renewable energy and green-collar jobs. College students were also integral in 350.org's International Day of Climate Action on October 24, 2009 (p. 494), a massive outreach and activism effort to urge action on climate change. Other events are continuing into 2010 and beyond, powered by the energy and commitment of students eager to address the climate change that looms over our future.

Dining services and campus farms can promote more sustainable food

Activism at landmark events is one way to promote sustainability concerns, but we all can make a difference three times a day, each time we eat. One way is by cutting down on waste, estimated at 25% of food that students take. Composting food scraps is an effective method of recycling waste once it is created, but trayless dining can reduce waste at its source. Coe College in Iowa is one of many schools where dining halls are eliminating trays and asking students to carry plates individually. Not having to wash trays saves on water, detergent, and energy, and Coe's dining manager calculated that the trayless system prevents 200–300 pounds of food waste per week.

Campus food services also can buy organic produce, purchase food in bulk or with less packaging, and buy locally grown or produced food. At Sterling College in Vermont, many foods are organic, grown by local farmers, or produced by Vermont-based companies. Some foods are grown and breads are baked on the Sterling campus, and food shipped in is purchased in bulk. Dish soap is biodegradable, and kitchen scraps are composted along with unbleached paper products. Some college campuses even have gardens or farms where students help to grow food that is eaten on

(a) Winona State University

(b) Yale University

FIGURE 24.10 ▲ Some campuses have gardens where students can grow organic produce for their dining halls. At Winona State University in Minnesota **(a)**, Dr. Bruno Borsari shows students herbs grown on campus. At Yale University **(b)**, student Claire Bucholz harvests beets from a campus farm.

campus (**FIGURE 24.10**). University of Montana students tend a 10-acre farm that produces 20,000 pounds of organic produce that is given to a local food bank. Students help run sizeable organic farms at Colorado College, Boston College, University of Nevada–Reno, Berea College in Kentucky, Augustana College in Illinois, Michigan State University, and others, all of which provide food for the dining halls and often for local communities. At Dickinson College in Pennsylvania, a small garden originating with a class project grew to three-fourths of an acre and began supplying a local food bank, and then moved off-campus to a 20-acre site, where it now produces $28,000 worth of food for community-supported agriculture (CSA) members (p. 274), as well as for the campus food service.

Loyola University Chicago is going still further. Besides raising bees to produce honey on campus, Loyola students and staff are going into the community and tending community gardens to feed the homeless. They also are planting fruit- and nut-bearing trees in urban neighborhoods in

Chicago to help establish homegrown food sources for working-class residents.

Purchasing decisions wield influence

The kinds of purchasing decisions made in dining halls favoring local food, organic food, and biodegradable products can be applied across the entire spectrum of a campus's needs. When campus purchasing departments buy recycled paper, certified sustainable wood, energy-efficient appliances, goods with less packaging, and other ecolabeled products, they send signals to manufacturers and increase the demand for such items.

At Chatham College in Pennsylvania, students chose to honor their school's best-known alumnus, Rachel Carson (pp. 175, 379), by seeking to eliminate toxic chemicals on campus. Administrators agreed, provided that alternative products to replace the toxic ones worked just as well and were not more expensive. Students brought in the CEO of a company that produces nontoxic cleaning products, who demonstrated to the janitorial staff that his company's products were superior. The university switched to the nontoxic products, which were also cheaper, and proceeded to save $10,000 per year. Chatham students then found a company offering paint without volatile organic compounds and negotiated with it for a free paint job and discounted prices on later purchases. Students also worked with grounds staff to eliminate herbicides and fertilizers used on campus lawns and to find alternative treatments.

Transportation alternatives are many

Many campuses struggle with traffic congestion, parking shortages, commuting delays, and pollution from vehicle exhaust. Indeed, commuting to and from campuses in vehicles accounts for over half of the carbon emissions of the average college or university (**FIGURE 24.11**). Some are addressing these issues by establishing or expanding bus and shuttle systems; introducing alternative vehicles to university fleets; and encouraging walking, carpooling, and bicycling (**FIGURE 24.12**).

Juniata College in Pennsylvania runs a bike-sharing program in which students can borrow bikes from a fleet.

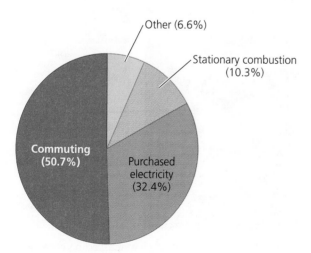

FIGURE 24.11 ▲ Commuting to and from campuses in automobiles accounts for over 50% of the greenhouse gas emissions of the average college or university, as measured in CO_2-equivalents. Data from American College and University Presidents' Climate Commitment.

Other (6.6%)

Stationary combustion (10.3%)

Commuting (50.7%)

Purchased electricity (32.4%)

FIGURE 24.12 ▲ Emma Edmondson is one of many Oberlin College students who bike to class on this bicycle-oriented campus.

University of Texas at Austin students refurbish donated bicycles and then provide them free to new students. Passing bikes from one generation to the next reduces traffic congestion, pollution, and parking costs while promoting a healthy mode of transportation.

Humboldt State University in California grants its students free unlimited bus service. As a result, enough students shifted from cars to buses that a planned parking garage did not need to be built. At the University of Montana, students actually run a mass transit system. In 1999, students took over a small van service, and it has steadily grown, fed by student demand and funded by self-imposed student fees. The popular service gave 343,000 free rides on six buses in 2008.

Alternative vehicles and alternative fuels are playing larger and larger roles. Butte College in California operates three natural gas buses and 10 biodiesel buses that keep 1,100 cars off campus each day. These buses were funded through a student-approved fee. At University of California–Irvine, students pushed for the bus system to be converted to biodiesel, and the 20 converted buses today save 480 tons of carbon emissions per year. The State University of New York College of Environmental Science and Forestry in Syracuse acquired electric vehicles, a gas-electric hybrid car, and a delivery van that runs on compressed natural gas while also converting its buses to biodiesel.

Middlebury College students began Project Bio Bus, which has crisscrossed North America each summer in a biodiesel bus spreading the gospel of this alternative fuel. Dartmouth College students are now taking their own Big Green Bus on the road in a similar effort. Meanwhile, students at Rice University, MIT, Loyola University Chicago, and the University of Central Florida are taking biodiesel initiatives to the next level—they are producing biodiesel (**FIGURE 24.13**). These students are collecting waste cooking oil from dining halls and restaurants, and brewing biodiesel for campus bus fleets.

The University of Washington is a leader in transportation efforts. Its transportation program provides unlimited mass transit access, discounts on bicycle equipment, rides home in

FIGURE 24.13 ▲ At Loyola University Chicago, students and staff produce biodiesel from waste vegetable oil from the dining halls, and use it to fuel this biodiesel van. A grant from the U.S. EPA funds them to transport this mini-biodiesel reactor to local high schools to teach students about alternative fuels.

emergencies, merchant discounts, subsidies for carpooling, rentals of hybrid vehicles, and more. As a result, each day, three-quarters of the population commutes by some other means than driving alone in a car. The program has kept peak traffic below 1990 levels, despite 23% growth in the campus population.

Campuses are restoring native plants, habitats, and landscapes

No campus sustainability program would be complete without some effort to enhance the campus's natural environment. Such efforts remove invasive species, restore native plants and communities, improve habitat for wildlife, enhance soil and water quality, and create healthier, more attractive surroundings.

These efforts are diverse in their scale and methods. Seattle University in Washington landscapes its grounds with native plants and has not used pesticides since 1986. Its campus includes an ethno-botanical garden and areas for wildlife. At Warren Wilson College in North Carolina, students and the landscaping supervisor built a greenhouse, expanded an arboretum, and are propagating local grasses and wildflowers. Ohio State University and the New College of California in San Francisco have rooftop gardens. At De Anza College, the Cheeseman Environmental Study Area maintains small-scale examples of entire plant communities native to California. Docents lead schoolchildren on field trips here, thousands of environmental studies students gain knowledge of California native plants, and the demonstration garden receives thousands of visitors interested in learning about native landscaping.

Some schools have embarked on ambitious projects of ecological restoration (pp. 95, 308–309). The University of Central Florida manages 500 of its 1,400 acres for biodiversity conservation. It is using prescribed fire (p. 329) and has so far restored 100 acres for fire-dependent species while protecting the campus against out-of-control wildfires (**FIGURE 24.14**). In Washington, Cascadia Community College and the University of Washington–Bothell together restored a 58-acre wetland, dismantling levees and ditches to restore natural water flow and reintroducing native vegetation. At Northland College in Wisconsin, sustainability advocates re-landscaped half their campus, replacing invasive plants with native ones and planting meadows that capture stormwater runoff and filter pollution. At North Hennepin Community College in Minnesota, students have planted over 35,000 seedlings and are transforming 2.8 ha (7 acres) of lawn and channelized ponds into marsh, prairie/savanna, and native forest. Besides providing wildlife habitat, the restored area reduces erosion and maintenance costs and provides opportunities for biological and environmental education.

Sustainability efforts include curricular changes

Along with this diversity of efforts to make campus operations more sustainable, colleges and universities have also been transforming their academic curricula and course offerings (**FIGURE 24.15**). The course for which you are using this book right now likely did not exist a generation ago. As our society comes to appreciate the looming challenge of sustainability, colleges and universities are attempting to train students to confront this challenge more effectively. For instance, since 1992, De Anza College has been committed to institutionalizing Environmental Studies as a recognized department and discipline, along with Math, English, and other sciences and social sciences. Its curriculum has grown to over 70 courses and four certificate and degree tracks.

However, the latest nationwide survey of campus sustainability efforts (in 2008) found that curriculum offerings across American colleges and universities overall had actually not risen between 2001 and 2008. Surprisingly, the

FIGURE 24.14 ▲ Many schools have embarked on ecological restoration projects to beautify their campuses, provide wildlife habitat, restore native plants, and filter water runoff. Here, staff and students at the University of Central Florida conduct prescribed burns to restore healthy pine woodland on campus.

FIGURE 24.15 ▲ One way to get sustainability into the curriculum is to get out of the classroom. At Davidson College in North Carolina, students and professors use lands near campus as living laboratories. Sampling streams for insect larvae that reveal water quality is one of many learning activities in which they engage.

percentage of schools that require all students to take at least one course related to environmental science or sustainability decreased from 8% in 2001 to just 4% in 2008. At most schools, fewer than half of students take even one course on the basic functions of Earth's natural systems, and still fewer take courses on the links between human activity and sustainability. As a result, the report states, "students are slightly less likely to be environmentally literate when they graduate in 2008 than in 2001."

This surprising finding suggests that students like you are in a privileged minority, benefiting from a valuable education that most of your peers are missing. You will come away from your college years with a better understanding of how the world works. You will be better prepared for a future dominated by sustainability concerns and better qualified for green-collar job opportunities.

WEIGHING THE ISSUES

Sustainability on Your Campus Find out what sustainability efforts are being made on your campus. What results have these achieved so far? What further efforts would you like to see pursued on your campus? Do you foresee any obstacles to these efforts? How could these obstacles be overcome? How could you become involved?

Organizations assist campus efforts

Many campus sustainability initiatives are supported by organizations such as the Association for the Advancement of Sustainability in Higher Education and the Campus Ecology program of the National Wildlife Federation. These organizations act as information clearinghouses for campus sustainability efforts. Each year the Campus Ecology program recognizes the most successful campus sustainability initiatives with awards. In addition, national and international conferences are growing, such as the biennial Greening of the Campus conferences at Ball State University. With the assistance of these organizations and events, it is easier than ever to

start sustainability efforts on your own campus and obtain the support to carry them through to completion.

SUSTAINABILITY AND SUSTAINABLE DEVELOPMENT

Efforts toward sustainability on college and university campuses parallel efforts in the world at large. As more people come to appreciate Earth's limited capacity to accommodate our rising population and consumption, they are voicing concern that we will need to modify our behaviors, institutions, and technologies if we wish to sustain our civilization and the natural environment on which it depends. In the quest for sustainability, the strategies pursued on campuses reflect those pursued in the wider society, and they also can serve as models.

When people speak of *sustainability*, what precisely do they mean to sustain? Generally they mean to sustain human institutions in a healthy and functional state—and also to sustain ecological systems in a healthy and functional state. The contributions of biodiversity (pp. 296–300) and ecosystem goods and services (pp. 3, 121–122, 160–161) to human welfare are tremendous. Indeed, they are so fundamental (some would say infinitely valuable, thus literally priceless) that we have long taken them for granted.

Sustainable development aims to achieve a triple bottom line

We first explored sustainable development in our opening chapter (p. 19) and offered the definition put forth by the United Nations: "Development that meets the needs of the present without compromising the ability of future generations to meet their own needs." Today, it is widely recognized that sustainability does not mean simply protecting the environment against the ravages of human development. Instead, it means finding ways to promote social justice, economic well-being, and environmental quality at the same time (**FIGURE 24.16**). Meeting

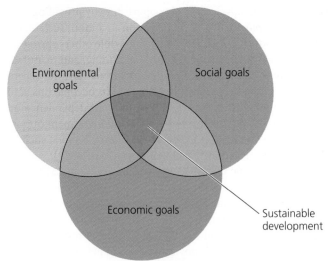

FIGURE 24.16 ▲ Modern conceptions of sustainable development hold that sustainability occurs where three sets of goals overlap: social, economic, and environmental goals. Adapted from Millennium Ecosystem Assessment, 2005. Ecosystems and Human Well-being: Biodiversity Synthesis. World Resources Institute, Washington, DC. By permission.

this *triple bottom line* (p. 19) is the goal of modern sustainable development. It is our primary challenge for this century and likely for the rest of our species' time on Earth.

"Environmental sustainability" is one of the United Nations' *Millennium Development Goals* (pp. 219–220) set by the international community at the turn of this century, and the drive for sustainability overall meshes with all eight goals. In the past few years, the Millennium Project and the Millennium Ecosystem Assessment (pp. 16, 18) each have determined that:

▶ Environmental degradation is a major barrier to achieving the Millennium Development Goals.

▶ Investing in environmental assets and management is vital to relieving poverty, hunger, and disease.

▶ Reaching environmental goals requires progress in eradicating poverty.

The many actions being taken today at campuses such as De Anza's and by governments, businesses, industries, organizations, and individuals across the globe are giving people optimism that achieving these goals and developing in sustainable ways is within reach.

Environmental protection can enhance economic opportunity

Environmental and social concerns are intimately tied to economic ones, yet for too long we have labored under the misconception that economic well-being and environmental protection are in conflict. In reality, the opposite is true: Our well-being depends on a healthy environment, and protecting environmental quality can improve our economic bottom line.

For individuals, businesses, and institutions, reducing resource consumption and waste often saves money, as many colleges and universities discover when they embark on sustainability initiatives. Sometimes savings accrue immediately, and other times an up-front investment brings long-term savings. For society as a whole, promoting environmental quality can enhance economic opportunity by providing new types of employment. As we transition to a more sustainable economy, some industries will decline while others spring up to take their place. As jobs in logging, mining, and manufacturing have disappeared in developed nations in recent decades, jobs have proliferated in service occupations and high-technology sectors. As we decrease our dependence on fossil fuels, green-collar jobs (p. 593) and investment opportunities are opening up in renewable energy sectors, such as wind power and fuel-cell technology (**FIGURE 24.17**).

Moreover, people desire to live in areas that have clean air, clean water, intact forests, and parks and open space. Environmental protection increases a region's attractiveness, drawing residents and increasing property values and the tax revenues that help fund social services. As a result, regions that act to protect their environments are generally the ones that retain and increase their wealth and quality of life.

Thus, environmental protection is likely to enhance economic opportunity. Indeed, a recent U.S. government review concluded that the economic benefits of environmental regulations greatly exceed their economic costs (see *The Science behind the Story*, Chapter 7, pp. 180–181). Consider also that

FIGURE 24.17 ▲ As we progress toward a sustainable economy, new job opportunities open up. Green-collar jobs such as employment as a wind power technician are already increasing in the United States and other nations.

the U.S. and global economies have each expanded rapidly in the past 40 years, the very period during which environmental protection measures have proliferated.

In addition, if we look beyond conventional economic accounting (which measures only private economic gain and loss) and include external costs and benefits (pp. 150–151) that affect people at large, then environmental protection becomes still more valuable. Take several studies reviewed by the Millennium Ecosystem Assessment: They each show how overall economic value is maximized by conserving natural resources rather than exploiting them for short-term private gain (**FIGURE 24.18**).

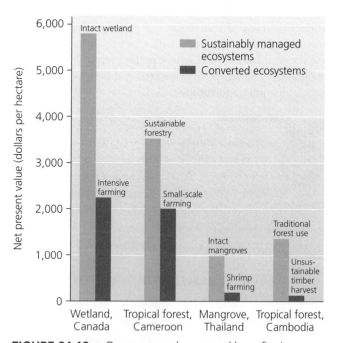

FIGURE 24.18 ▲ Once external costs and benefits (pp. 150–151) are factored in, the economic value of sustainably managed ecosystems generally exceeds the economic value of ecosystems that have been converted for intensive private resource harvesting. Shown are land values calculated by researchers in four such comparisons from sites around the world. Adapted from Millennium Ecosystem Assessment, 2005. *Ecosystems and human well-being: biodiversity synthesis.* World Resources Institute, Washington, DC. By permission.

We are part of our environment

Several factors may account for the widespread assumption that we cannot simultaneously protect the environment and provide for people's needs. For one thing, economic development since the industrial revolution has clearly diminished biodiversity, destroyed habitat, and degraded ecological systems. For another, many people believe that command-and-control environmental policy (pp. 179–180) poses excessive costs for industry and restricts rights of private citizens.

Ultimately, the explanation may lie in our species' long history. For the thousands of years that we lived as nomadic hunter-gatherers, population densities were low, and consumption and environmental impact were limited. With natural resources in little danger of running out, people were free to exploit them limitlessly and had little reason to adopt a conservation ethic. However, our establishment of sedentary agricultural societies, and then cities, increased our impact—while at the same time causing us to overlook the connections between our economies and our environments.

On a day-to-day basis, it is easy to feel disconnected from the natural environment, particularly in industrialized nations and large cities. We live inside houses, work in shuttered buildings, travel in enclosed vehicles, and generally know little about the plants and animals around us. Millions of urban citizens have never set foot in an undeveloped area. Just a few centuries or even decades ago, most of the world's people were able to name and describe the habits of the plants and animals that lived nearby. They knew exactly where their food, water, and clothing came from. Today it seems that water comes from the faucet, clothing from the mall, and food from the grocery store. It's little wonder we have lost track of the connections that tie us to our natural environment.

However, this doesn't make our connections to the environment any less real. Consider a thoroughly un-"natural" (yet delicious!) invention of the human species: the banana split (**FIGURE 24.19**). Even in this triumph of human creation, seemingly concocted *de novo* at an ice cream shop, each and every element has ties to the resources of the natural environment, and each exerts environmental impacts.

Once we learn to consider where the things we use and value each day actually come from, it becomes easier to see how we are part of our environment. And once we reestablish this connection, it becomes readily apparent that our own interests are best served by preservation or responsible stewardship of the natural systems around us. Because what is good for the environment can also be good for people, win-win solutions are very much within reach, if we learn from what science can teach us, think creatively, and act on our ideas.

STRATEGIES FOR SUSTAINABILITY

Sustainable solutions to environmental problems are numerous, and we have seen specific examples throughout this book. Yet several challenges confront us when we seek sustainable solutions to problems. One lies in being imaginative enough to think of creative solutions that have not yet been tried. Another is being shrewd and dogged enough to overcome political or economic obstacles that may lie in the path of their implementation. Another challenge—and a clear role for research in environmental science—lies in being able to measure the effect of a given change to determine whether it is truly sustainable. Does the change result in

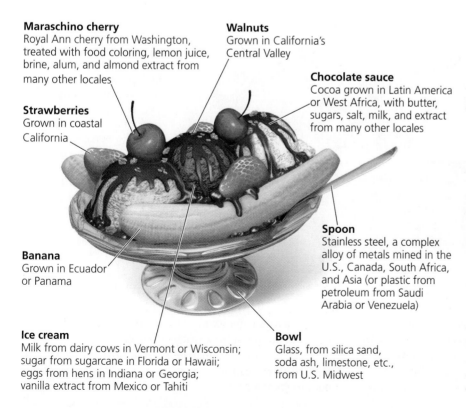

Maraschino cherry
Royal Ann cherry from Washington, treated with food coloring, lemon juice, brine, alum, and almond extract from many other locales

Walnuts
Grown in California's Central Valley

Chocolate sauce
Cocoa grown in Latin America or West Africa, with butter, sugars, salt, milk, and extract from many other locales

Strawberries
Grown in coastal California

Banana
Grown in Ecuador or Panama

Spoon
Stainless steel, a complex alloy of metals mined in the U.S., Canada, South Africa, and Asia (or plastic from petroleum from Saudi Arabia or Venezuela)

Ice cream
Milk from dairy cows in Vermont or Wisconsin; sugar from sugarcane in Florida or Hawaii; eggs from hens in Indiana or Georgia; vanilla extract from Mexico or Tahiti

Bowl
Glass, from silica sand, soda ash, limestone, etc., from U.S. Midwest

FIGURE 24.19 ◄ A banana split eaten at an ice cream shop in Tulsa, Denver, or Des Moines consists of ingredients from around the world, whose production has impacts on the environments of many far-away locations. Ice cream requires milk from dairy cows that graze pastures or are raised in feedlots on grain grown in industrial monocultures. Ice cream is sweetened with sugar from sugar beet farms or sugarcane plantations. The banana was shipped thousands of miles by oil-fueled transport from a tropical country, where it grew on a plantation that displaced rainforest and where it was liberally treated with fertilizers and fungicides. Fruits and nuts grown in California's Central Valley were irrigated generously with water piped in from the Sierras. The spoon originated with metal ores mined along with thousands of tons of soil and processed into stainless steel using energy from fossil fuels.

The SCIENCE behind the Story

Rating the Environmental Performance of Nations

Dr. Daniel Esty, Yale University

How effectively is our world moving toward environmental sustainability? That depends on where in the world you are, because the governments of some nations are performing better than others.

To measure the progress of nations toward environmental sustainability, researchers devised the *Environmental Performance Index (EPI)*, which rates countries by using data from 25 indicators of environmental conditions for which governments can be held accountable. The EPI analyses, first introduced in 2006, are revised every two years by researchers from the Yale Center for Environmental Law and Policy and Columbia University's Center for International Earth Science Information Network, in collaboration with the World Economic Forum and the Joint Research Centre of the European Commission. Daniel Esty of the Yale Center led the initial report, and the third report was led in 2010 by Jay Emerson of the Yale Center, along with seven colleagues.

These researchers created the EPI because the U.N. Millennium Development Goals (pp. 219–220) did not define how to quantify progress on environmental measures. The lack of quantitative measures, these researchers said, had stymied progress on how to implement effective environmental policy. To address this problem, this team aimed to track the performance of environmental policy with the same quantitative rigor as statistics are tracked for health, poverty reduction, and other development goals. Giving nations scores and ranking them would reveal "leaders and laggards," showing which nations are on the right track and which are not.

In the 2010 report, the researchers gathered internationally available data from U.N. and other sources on 25 indicators, which funneled into 10 categories: environmental disease burden, air pollution impacts on people, water resource impacts on people, air pollution impacts on ecosystems, water resource impacts on people, biodiversity

EPI score
(by quintile)

- 87.4–93.5
- 81.3–87.3
- 75.2–81.2
- 69.0–75.1
- 62.9–68.9
- 56.8–62.8
- 50.6–56.7
- 44.5–50.5
- 38.4–44.4
- 32.1–38.3
- No data

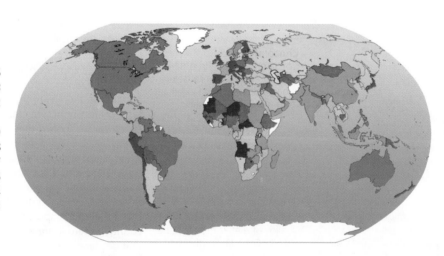

Researchers scored nations for their performance in approaching environmental sustainability goals. Higher Environmental Performance Index (EPI) values (green is highest in this map) indicate better performance. Data from Emerson, J., et al., 2010. *2010 Environmental Performance Index.* By permission of Yale Center for Environmental Law and Policy. V.

truly renewable resource use? Does it prevent pollution completely? Does it enable us to reuse or recycle waste so as to truly close the loop in a production process? In most cases, our solutions will fall short. Yet by taking steps toward the ultimate goal, by measuring our successes and failures along the way, and by improving our efforts, we may ultimately succeed in our quest for a sustainable future.

We will now summarize several broad strategies or approaches that can help move us toward truly sustainable solutions (**TABLE 24.2**).

Top-Ranked Nations	
Nation	EPI Score
Iceland	93.5
Switzerland	89.1
Costa Rica	86.4
Sweden	86.0
Norway	81.1
Mauritius	80.6
France	78.2
Austria	78.1
Cuba	78.1
Colombia	76.8

Bottom-Ranked Nations	
Nation	EPI Score
Benin	39.6
Haiti	39.5
Mali	39.4
Turkmenistan	38.4
Niger	37.6
Togo	36.4
Angola	36.3
Mauritania	33.7
Central African Republic	33.3
Sierra Leone	32.1

Data for both tables from Emerson, J. et al. 2010. *Environmental performance index.* By permission of Yale Center for Environmental Law and Policy.

and habitat, forestry, fisheries, agriculture, and climate change. The first three categories were considered components of "environmental health," and the latter seven categories were considered components of "ecosystem vitality." Esty's team scored nations on their performance in each category and gave them overall scores, based 50% on their environmental health score and 50% on their ecosystem vitality scores. Scores for each category ranged from 0 to 100, with 100 representing a target value established by international consensus. The researchers included 163 nations in their 2010 report (**see map**).

The main pattern in the results was conspicuous: Nations with the highest scores (**first table**) are wealthy, industrialized nations with the capacity to commit substantial resources toward environmental protection. Nations with the lowest scores (**second table**) are largely developing nations with few resources to invest. They include countries with dense populations and stressed ecosystems that are trying to industrialize, those with arid environments and limited natural resources, and those facing extreme poverty.

This correlation between economic vitality and environmental protection was expected, but the researchers were more interested in the variation around this trend. For any given level of income or development, some nations outperformed others. The researchers say this demonstrates that other factors are at work—namely, the political choices that national leaders make. Indeed, their data show a strong correlation between EPI scores and "good governance," which involves aspects such as rule of law, open political debate, and lack of corruption.

For instance, neighboring Chile and Argentina are similar economically, but Chile ranked 16th while Argentina ranked 70th—a difference the researchers attribute to Chile's superior investments in environmental protection policies. Likewise, the United States placed 61st—far below most of its economic peers, and behind more than 20 European nations. Yet this is perhaps not surprising, as U.S. environmental policy has been considerably weaker than policies of other developed nations for years now.

Other trends were apparent with deeper examination of the data. For instance, wealthy industrialized nations tended to score highest in environmental health and lowest in ecosystem vitality indicators, especially biodiversity and habitat. This is presumably because they possess money to invest in protecting the health of their citizens but achieved economic development by exploiting their natural resources and degrading their natural environment.

Conversely, poorer nations tended to score low on environmental health and more highly on indicators that reflected the availability of remaining unexploited natural resources. Overall across all nations, performance was worst for the indicators of renewable energy and wilderness protection.

Some critics say the rankings have little meaning because only certain data sets were measured, and these were then weighted subjectively. Supporters counter that although no one will ever agree on exact numbers, the EPI provides a useful tool for evaluating actions and investments, highlighting policies that work, and identifying future priorities.

The EPI scores are only as solid as the data that go into them, and not every nation provides reliably accurate data on every issue. Moreover, the research team acknowledges that there are plenty of ways in which their formulas might be improved in the future. Nonetheless, the Environmental Performance Index provides a unique way to assess what is working in environmental policy. As the methodology is refined year by year, the EPI promises to become a more accurate performance indicator and a more helpful policy tool. ■

Citizens exert political influence

Policy is an essential tool for pursuing sustainability (see **THE SCIENCE BEHIND THE STORY**, above). Politically open democracies offer a compelling route for pursuing sustainability: the power of the vote. Many sustainable solutions require policymakers to usher them through, and policymakers respond to whoever exerts influence. Corporations and interest groups employ lobbyists to influence politicians all the time. Citizens in a democratic republic have the same power, *if* they choose to exercise it.

TABLE 24.2	Some Major Approaches to Sustainability

▶ Be politically active
▶ Vote with our wallets
▶ Rethink our assumptions about economic growth
▶ Consume less while maintaining quality of life
▶ Limit population growth
▶ Encourage green technologies
▶ Mimic natural systems by promoting closed-loop industrial processes
▶ Enhance local self-sufficiency, yet embrace some aspects of globalization
▶ Think in the long term
▶ Promote research and education

You can exercise your power at the ballot box, by attending public hearings, by donating to advocacy groups that promote positions you favor, and by writing letters and making phone calls to office-holders.

Today's environmental and consumer protection laws came about because citizens pressured their representatives to act. The raft of legislation enacted in the 1960s and 1970s in the United States and other nations might never have come about had ordinary citizens not stepped up and demanded action. We owe it to future generations to be engaged and to act responsibly now so that they have a better world in which to live. The words of anthropologist Margaret Mead are worth repeating: "Never doubt that a small group of thoughtful, committed people can change the world. Indeed, it's the only thing that ever has."

Consumers vote with their wallets

Expressing our preferences through the political system is vital, but each of us also wields influence in the choices we make as consumers. When products produced sustainably are ecolabeled (p. 161), consumers can "vote with their wallets" by purchasing these products. Consumer choice has helped drive sales of everything from recycled paper to organic produce to "dolphin-safe" tuna.

Individuals can multiply their own influence by promoting "green" purchasing habits at their school or workplace. We saw how purchasing power at colleges and universities has spurred sales of certified sustainable wood, organic food, energy-efficient appliances, and more. Employees in businesses and government agencies can often promote change in those institutions by voicing their preferences in purchasing decisions.

We can rethink our assumptions about economic growth

It is conventional among economists and policymakers to speak of economic growth as an ultimate goal. Many politicians view nurturing economic expansion as their prime responsibility while in office. Yet economic growth is merely a tool with which we try to attain the real goal of maximizing human happiness. If economic growth depends on ever-increasing consumption and depletion of nonrenewable resources, then trying to endlessly expand the size of our economy will not bring us long-term happiness.

Moreover, the data that economists use to calculate economic growth do not incorporate external costs (pp. 150–151), those social, environmental, and personal costs not included in the market prices of goods and services. Currently, goods and services are priced as though pollution and resource extraction involved no costs to society. If, instead, we can make our accounting practices reflect indirect consequences to the public, then we can provide a clearer view of the full costs and benefits of a given action, decision, or product—and the free market could become the optimal tool for improving environmental quality, our economy, and our quality of life. Implementing green taxes (p. 182) and phasing out harmful subsidies could further hasten our attainment of prosperous and sustainable economies. The political obstacles to all these changes are considerable, and such changes will require educated citizens to push for them and courageous policymakers to implement them.

Good quality of life does not require intensive consumption

Economic growth is largely driven by consumption, the purchase of material goods and services (and thus the use of resources involved in their manufacture) by consumers (**FIGURE 24.20**). Our tendency to believe that more, bigger,

FIGURE 24.20 ◀ Citizens of the United States consume more than the people of any other nation. Unless we find ways to increase the sustainability of our manufacturing processes, our rising rate of consumption cannot be sustained in the long run.

and faster are always better is reinforced by advertisers seeking to sell us more goods more quickly. Consumption has grown tremendously, with the wealthiest nations leading the way. The United States, with less than 5% of the world's population, consumes 30% of world energy resources and 40% of total global resources. U.S. homes are larger than ever, sports-utility vehicles remain popular, and many citizens have more material belongings than they know what to do with. We think nothing today of having home computers and wireless handheld devices with high-speed Internet access, let alone the televisions, telephones, refrigerators, and dishwashers that were marvels just decades ago.

Because many of Earth's natural resources are limited and nonrenewable, consumption cannot continue growing forever. Eventually, if we do not shift to sustainable resource use, per capita consumption will drop for rich and poor alike as resources dwindle. We are enjoying the greatest material prosperity in all of history, but if we do not find ways to make our wealth sustainable, the party may not last much longer.

Fortunately, material consumption alone does not reflect a person's quality of life, and the accumulation of possessions does not necessarily bring contentment. Observing how affluent people often fail to find happiness in their material wealth, social critics have given this phenomenon a name like a dread disease: *affluenza* (p. 151). It may not be a disease, but scientific research does back up the contention that money cannot buy nearly as much happiness as people typically believe (**FIGURE 24.21**).

We can reduce our consumption while enhancing our quality of life—squeezing more from less—in at least three ways. One way is to improve the technology of materials and the efficiency of manufacturing processes, so that industry produces goods using fewer natural resources. Another way is to develop sustainable manufacturing systems—ones that are circular and recycle, in which the waste from a process becomes raw material for input into that process or others (pp. 632–633). A third way is to modify our behavior, attitudes, and lifestyles to minimize consumption. At the outset, such choices may seem like sacrifices, but people who have slowed down the pace of their busy lives and freed themselves of an attachment to material possessions say it can feel tremendously liberating.

Population growth must eventually cease

Just as continued growth in consumption is not sustainable, neither is growth in the human population. We have seen (pp. 66–67) that populations may grow exponentially for a time but eventually encounter limiting factors and level off or decline. We have used technology to increase Earth's carrying capacity for our species, but our population cannot continue

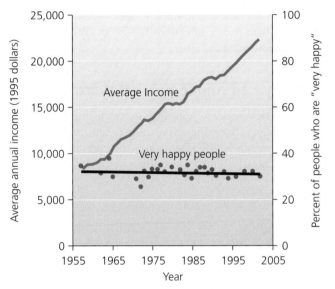

(a) Happiness versus income, through time

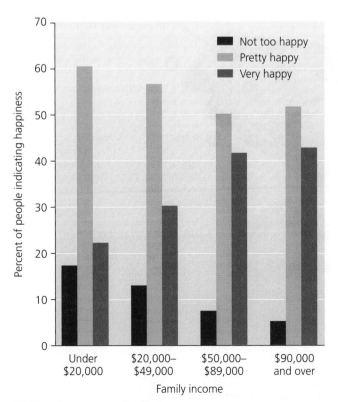

(b) Happiness versus family income

FIGURE 24.21 ◄ Although average income in the United States has risen steadily in the past half-century, the percentage of people reporting themselves as being "very happy" has remained stable or declined slightly **(a)**. A different study found that Americans in higher income brackets were more likely to report themselves as "very happy" and less likely to report themselves as "not too happy." **(b)**. However, the difference was far less than people generally expect, and the researchers showed this finding, too. For instance, when asked how much a fivefold increase in income improves a person's mood day to day, respondents guessed on average that such an increase would improve mood by 32%, but the actual improvement, from respondents' self-reporting, was only 12%. Data in (a) from Gardner, G. and E. Assadourian. 2004. Rethinking the good life. Worldwatch Institute, *State of the World* 2004. www.worldwatch.org. By permission. Data in (b) from Kahneman, D., et al., 2006. Would you be happier if you were richer? A focusing illusion. *Science* 312: 1908–1910.

growing forever; sooner or later, human population growth will cease. The question is how: through war, plagues, and famine, or through voluntary means as a result of wealth and education?

The demographic transition (pp. 210–211) is already far along in many developed nations thanks to urbanization, wealth, education, and the empowerment of women. If today's developing nations also pass through the demographic transition, then there is hope that humanity may halt its population growth while creating a more prosperous and equitable society.

Technology can help us toward sustainability

It is largely technology—developed with the agricultural revolution, the industrial revolution, and advances in medicine and health—that has spurred our population increase. Technology has magnified our impacts on Earth's environmental systems, yet it can also give us ways to reduce our impact. Recall the IPAT equation (p. 201), which summarizes human environmental impact (I) as the interaction of population (P), consumption or affluence (A), and technology (T). Technology can exert either a positive or negative value in this equation. The shortsighted use of

technology may have gotten us into this mess, but wiser use of environmentally friendly, or "green," technology can help get us out.

In recent years, we have intensified environmental impacts in developing countries by exporting industrial technologies from the developed world poorer nations eager to industrialize. In developed nations, meanwhile, we have begun using green technologies to mitigate our impacts. Catalytic converters on cars have reduced emissions (**FIGURE 24.22**), as have scrubbers on smokestacks (Figure 17.16, p. 474). Recycling technology and wastewater treatment are reducing our waste output. Solar, wind, and geothermal energy technologies are producing cleaner, renewable energy. Technological advances such as these help explain why people of the United States and western Europe today enjoy cleaner environments—although they consume far more—than people of eastern Europe or rapidly industrializing nations such as China.

Industry can mimic natural systems

As industries seek to develop green technologies and sustainable practices, they have an excellent model: nature itself. As we saw in Chapter 5 and throughout this book, environmental systems tend to operate in cycles featuring

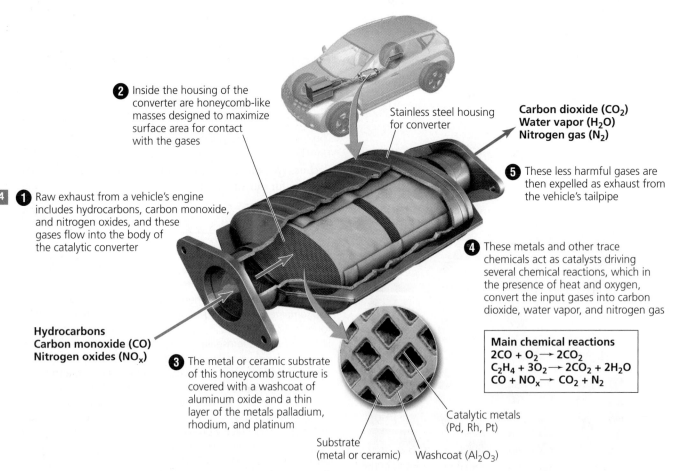

2 Inside the housing of the converter are honeycomb-like masses designed to maximize surface area for contact with the gases

Stainless steel housing for converter

Carbon dioxide (CO_2)
Water vapor (H_2O)
Nitrogen gas (N_2)

5 These less harmful gases are then expelled as exhaust from the vehicle's tailpipe

684

1 Raw exhaust from a vehicle's engine includes hydrocarbons, carbon monoxide, and nitrogen oxides, and these gases flow into the body of the catalytic converter

4 These metals and other trace chemicals act as catalysts driving several chemical reactions, which in the presence of heat and oxygen, convert the input gases into carbon dioxide, water vapor, and nitrogen gas

Hydrocarbons
Carbon monoxide (CO)
Nitrogen oxides (NO_x)

3 The metal or ceramic substrate of this honeycomb structure is covered with a washcoat of aluminum oxide and a thin layer of the metals palladium, rhodium, and platinum

Catalytic metals
(Pd, Rh, Pt)

Substrate
(metal or ceramic)   Washcoat (Al_2O_3)

Main chemical reactions
$$2CO + O_2 \rightarrow 2CO_2$$
$$C_2H_4 + 3O_2 \rightarrow 2CO_2 + 2H_2O$$
$$CO + NO_x \rightarrow CO_2 + N_2$$

FIGURE 24.22 ▲ The catalytic converter is a classic example of green technology. This device filters air pollutants from vehicle exhaust and has helped bring about the improvement of air quality in the United States and other nations.

feedback loops and the circular flow of materials. In natural systems, output is recycled into input. In contrast, human manufacturing processes have run on a linear model in which raw materials are input and processed to create a product while by-products and waste are generated and discarded. Some forward-thinking industrialists are making their processes more sustainable by transforming linear pathways into circular ones, in which waste is recycled and reused (pp. 632–633). For instance, several companies now produce carpets that can be retrieved from the consumer when they wear out, and these materials are then recycled to create new carpeting (p. 633). Some automobile manufacturers are planning cars that can be disassembled and recycled into new cars. Proponents of this industrial model see little reason why virtually all products cannot be recycled, given the right technology. Their ultimate vision is to create truly closed-loop industrial processes, generating no waste.

We can promote local self-sufficiency yet embrace some aspects of globalization

As our societies become more globally interconnected, we experience a diversity of impacts, positive and negative. Encouraging local self-sufficiency is one important element of building sustainable societies. When people feel closely tied to the area in which they live, they tend to value the area and seek to sustain its environment and its human community. Moreover, relying on locally made products cuts down on fossil fuel use from long-distance transportation. This argument is frequently made in relation to the cultivation and distribution of food, in encouraging local organic or sustainable agriculture (pp. 273, 276).

Many advocates of local self-sufficiency criticize globalization. They are troubled by the homogenization of the world's societies, in which a few cultures and worldviews displace many others. For instance, many of the world's languages are going extinct as large and powerful industrialized societies displace smaller traditional societies. Traditional ways of life in many areas are being abandoned as more people take up the material and cultural trappings of a few dominant cultures, particularly those of the United States.

The growing power of large multinational corporations is a key driver of these trends. In today's globalizing world, multinational corporations are attaining greater and greater power over global trade while governments retain less and less. Critics of globalization consider corporations more likely than governments to promote a high-consumption lifestyle and less likely than governments to support environmental protection, so they feel that globalization will hinder progress toward sustainability.

In recent years, many people have reacted against the power of multinational corporations and the homogenizing effects of globalization. In Seattle in 1999, thousands of protestors picketed a meeting of the World Trade Organization, in what was called the "Battle in Seattle" (**FIGURE 24.23**). Since then, protestors have picketed every WTO meeting.

However, globalization is a complicated phenomenon with diverse consequences. One positive aspect is the way that people of the world's diverse cultures are increasingly communicating and learning about one another. Air travel, books, television, and the Internet have made us more aware of one another's cultures and more likely to respect and celebrate, rather than fear, differences among cultures. Moreover, globalization may foster sustainability because Western democracy, as imperfect as it is, serves as a model and a beacon for people living under repressive governments. Open societies allow for entrepreneurship and the flowering of creativity in business, research, and academia. Millions of free minds thinking about issues are more likely to come up with sustainable solutions than the minds of a few holding authoritarian power.

WEIGHING THE ISSUES

Globalization From your own experience, what advantages and disadvantages do you see in globalization? Have you personally benefited or been hurt by it in any way? In what ways might promoting local self-sufficiency be helpful for the pursuit of global sustainability? In what ways might it not?

Now consider this: Many people enjoy eating at Vietnamese restaurants in the United States, yet many who criticize globalization would frown on the presence of McDonald's restaurants in Vietnam. Do you think this represents a double standard, or are there reasons the two are not comparable?

FIGURE 24.23 ▲ Thousands of protesters picketed the World Trade Organization's meeting in Seattle in 1999, criticizing the homogenizing effects of globalization, as well as relaxations in labor and environmental protections brought about by free trade.

We can think in the long term

Whatever solutions we pursue, we must base our decisions on long-term thinking, because to be sustainable, a solution must work in the long term. Often the best long-term solution is not the best short-term solution, which explains why much of what we currently do is not sustainable. Policymakers in democracies often act for short-term good because they aim to produce quick results that will help them be reelected. Yet many environmental dilemmas are cumulative, worsen gradually, and can be resolved only over long periods. Often the costs of addressing an environmental problem are short term, whereas the benefits are long term, giving politicians little incentive to tackle the problem. In such a situation, citizen pressure on policymakers is especially vital.

Students can be especially compelling advocates for long-term thinking. At the Copenhagen climate conference in December 2009, 17-year-old student Christina Ora of the Solomon Islands spoke for the youth climate movement when she told the assembled delegates:

> "I was born in 1992. You have been negotiating all my life. You cannot tell us that you need more time. Please commit to this decision now, because you hold our future in our hands, and survival is not negotiable."

Promoting research and education is vital

Finally, we each can magnify our influence by educating others and by serving as role models through our actions. The campus sustainability efforts at De Anza College and so many other colleges and universities accomplish both approaches. The discipline of environmental science plays a key role in providing information that people can use to make wise decisions about environmental issues. By promoting scientific research and by educating the public about environmental science, we can all assist in the pursuit of sustainable solutions.

686

PRECIOUS TIME

By shifting our conventional patterns of thinking and following the approaches outlined above, we can bring sustainable solutions within reach. However, the natural systems we depend on are changing quickly. Human impacts continue to intensify, including deforestation, overfishing, wetland loss, resource extraction, and climate change. Our window of opportunity for turning these trends around is getting short. Even if we can visualize sustainable solutions to our many problems, how can we find the time to implement them before we do irreparable damage to our environment and our future?

We need to reach again for the moon

In 1961, U.S. President John F. Kennedy announced that within a decade the United States would be "landing a man on the moon and returning him safely to the Earth" (**FIGURE 24.24**). It was a bold and astonishing statement; the technology to achieve this unprecedented, almost unimaginable, feat did not

FIGURE 24.24 ▲ In 1961, U.S. President John F. Kennedy called on Congress to fund a space program to send men to the moon by 1970. Addressing our environmental problems and shifting our political, economic, and social institutions to a paradigm of sustainable development will require even more vision, resolve, and commitment. The fact that astronauts reached the moon just eight years after Kennedy's speech demonstrates the power of human ingenuity in meeting a challenge, and provides hope that we will be able to meet the larger challenge of living sustainably on Earth.

yet exist. Kennedy's directive had powerful motivation behind it, however. The United States was dueling the Soviet Union in the Cold War. In this competition for global hegemony, the two nations, held mutually at bay with nuclear weapons, tried to prove their mettle by other means, and the race for dominance in space became the centerpiece of the rivalry. The prospect of "losing" the space race prompted Kennedy's administration to set a national goal on an ambitious timeline. Congress supplied funding, NASA performed the science and engineering, and in 1969 astronauts walked on the moon.

The United States accomplished this milestone in human history by building public support for a goal and giving its scientists and engineers the wherewithal to develop technology and strategies to meet the goal. Great challenges were also met when the United States confronted the Great Depression, when it threw itself into World War II, and when it conducted the Marshall Plan after that war to help rebuild western Europe.

Today humanity faces a challenge more important than any previous one—achieving sustainability. Attaining sustainability is a larger and more complex process than traveling to the moon. However, it is one to which every person on Earth can contribute; in which government, industry, and citizens can all cooperate; and toward which all nations can work together. If America was able to reach the moon in a mere eight years, then certainly humanity can begin down the road to sustainability with comparable speed. Human ingenuity is capable of it; we merely need to rally public resolve and engage our governments, institutions, and entrepreneurs in the race.

FIGURE 24.25 ▲ This photo of Earth, taken by astronauts orbiting the moon, shows our planet as it truly is—an island in space. Everything we know, need, love, and value comes from and resides on this small sphere, so we had best treat it well.

We must think of Earth as an island

We began this book with the vision of Earth as an island, and indeed that is what it is (**FIGURE 24.25**). Islands can be paradise, as Easter Island (pp. 6–7) likely was when the Polynesians first reached it. But when Europeans arrived there, they witnessed the aftermath of a civilization that had depleted its island's resources, degraded its environment, and collapsed as a result. For the few people who remained of the once-mighty culture, life was difficult and unrewarding. They had lost even the knowledge of the history of their ancestors, who had cut trees unsustainably, kicking the base out from beneath their prosperous civilization.

As Easter Island's trees disappeared, some individuals must have spoken out for conservation and for finding ways to live sustainably amid dwindling resources. Others likely ignored those calls and went on extracting more than the land could bear, assuming that somehow things would turn out all right. Indeed, whoever cut the last tree atop the most remote mountaintop could have looked out across the island and seen that it was the last tree. And yet that person cut it down.

It would be tragic folly to let such a fate occur to our planet as a whole. By recognizing this, by deciding to modify our individual behavior and our cultural institutions in ways that encourage sustainable practices, and by employing science to help us achieve these ends, we may yet be able to live happily and sustainably on our wondrous island, Earth.

➤ CONCLUSION

In any society facing dwindling resources and environmental degradation, there will be those who raise alarms and those who ignore them. Fortunately, in our global society today we have many thousands of scientists who study Earth's processes and resources. For this reason, we are amassing a detailed knowledge and an ever-developing understanding of our dynamic planet, what it offers us, and what impacts it can bear. The challenge for our global society today, our one-world island of humanity, is to support that science so that we may judge false alarms from real problems and distinguish legitimate concerns from thoughtless denial. This science, this study of Earth and of ourselves, offers us hope for our future.

REVIEWING OBJECTIVES

You should now be able to:

LIST AND DESCRIBE APPROACHES BEING TAKEN ON COLLEGE AND UNIVERSITY CAMPUSES TO PROMOTE SUSTAINABILITY

- Audits produce baseline data on how much a campus consumes and pollutes. (p. 667)
- Recycling and waste reduction are common campus sustainability efforts. (pp. 667–669)
- Green buildings are being constructed on a growing number of campuses. (pp. 670–671)
- There are many ways to reduce water use, and these efforts often save money. (p. 671)
- Students have many feasible ways to conserve energy and to promote renewable energy sources. (pp. 672–673)
- To address climate change, a current drive is to make campuses "carbon-neutral." (pp. 673–674)
- Dining services and campus farms and gardens can help provide local food and reduce waste. (p. 674)
- Colleges and universities can favor sustainable products in institutional purchasing. (p. 675)
- Campuses can use alternative fuels and vehicles and encourage bicycling, walking, and public transportation. (pp. 675–676)
- Restoration of plants, habitat, and landscapes are popular campus sustainability activities. (p. 676)
- Curricula have added courses related to issues of sustainability and environmental science. (pp. 676–677)

EXPLAIN THE CONCEPT OF SUSTAINABLE DEVELOPMENT

- Sustainable development entails environmental protection, economic development, and social justice. (pp. 677–678)
- Proponents of sustainable development feel that economic development and environmental quality can enhance one another. (p. 678)

DISCUSS HOW PROTECTING THE ENVIRONMENT CAN PROMOTE ECONOMIC WELL-BEING

- Environmental protection and green technologies and industries can create rich sources of new jobs. (p. 678)
- Safeguarding environmental quality enhances a community's desirability and economy. (p. 678)

DESCRIBE AND ASSESS KEY APPROACHES TO DESIGNING SUSTAINABLE SOLUTIONS

- We discuss ten general approaches that can inspire specific sustainable solutions. (pp. 679–686)
- Growth in population and per capita consumption will likely need to be halted if we are to create a sustainable society. (pp. 682–684)
- Technology has traditionally increased environmental impact, but new "green" technologies can help reduce impact. (p. 684)

EXPLAIN HOW TIME IS LIMITED BUT HOW HUMAN POTENTIAL TO SOLVE PROBLEMS IS TREMENDOUS

- Time for turning around our increasing environmental impacts is running short. (p. 686)
- The United States and other nations have met tremendous challenges before, so we have reason to hope that we will be able to attain a sustainable society. (p. 686)

TESTING YOUR COMPREHENSION

1. In what ways are campus sustainability efforts relevant to sustainability efforts in the broader society?

2. Describe one way in which campus sustainability proponents have addressed each of the following areas: (1) recycling and waste reduction, (2) "green" building, and (3) water conservation.

3. Describe one way in which campus sustainability proponents have addressed each of the following areas: (1) energy efficiency, (2) renewable energy, and (3) global climate change.

4. Describe one way in which campus sustainability proponents have addressed each of the following areas: (1) dining services, (2) institutional purchasing, (3) transportation, and (4) habitat restoration.

5. What do environmental scientists mean by *sustainable development*?

6. Describe three ways in which environmental protection can enhance economic well-being.

7. Why are many people now living at the highest level of material prosperity in history? Is this level of consumption sustainable? How can it feel good to consume less?

8. In what ways can technology help us achieve sustainability? How do natural processes provide good models of sustainability for manufacturing? Provide examples.

9. Why do many people feel that local self-sufficiency is important? What consequences of globalization may threaten sustainability? How can open democratic societies help to promote sustainability?

10. How can thinking of Earth as an island help prevent us from repeating the mistakes of previous civilizations?

SEEKING SOLUTIONS

1. What sustainability initiatives would you like to see attempted on your campus? If you were to take the lead in promoting such initiatives, how would you go about it? What obstacles would you expect to face, and how would you deal with them?

2. Choose one item or product that you enjoy, and consider how it came to be. Think of as many components of the item or product as you can, and determine how each of them was obtained or created. Now refer to Figure 24.18 as a guide. What steps were involved in creating your item's components, and where did the raw materials come from? How was your item manufactured? How was it delivered to you?

3. Do you think that we can increase our quality of life through development while also protecting the integrity of the environment? Discuss examples from your course or from other chapters of this book that illustrate possible win-win solutions. Are you familiar with any cases in your community or at your college that bear on this issue? Describe such a case, and state what lessons you would draw from it.

4. Reflect on the experiences of prior human civilizations and how they came to an end. What is your prognosis for our current human civilization? Do you see a vast world of independent cultures all individually responsible for themselves, or do you consider human civilization to be one great entity? If we accept that all people depend on the same environmental systems for sustenance, what resources and strategies do we have to ensure that the actions of a few do not determine the outcome for all and that sustainable solutions are a common global goal?

5. **THINK IT THROUGH** You have been elected president of your college class, and your school's administrators promise to be responsive to student concerns. Many of your fellow students are asking you to promote sustainability initiatives on your campus. Consider the many approaches and activities pursued by the colleges and universities mentioned in this chapter, and now think about your own school. Which of these approaches and activities are most needed at your school? Which might be most effective? What ideas would you prioritize and promote during your term as president of your class?

6. **THINK IT THROUGH** In our final "Think It Through" question, you are . . . *you!* In this chapter and throughout this book, you have encountered a diversity of ideas for sustainable solutions to environmental problems. Many of these are approaches you can pursue in your own life. Name at least five ways in which you think you can make a difference—and would most like to make a difference—in helping to attain a more sustainable society. For each approach, describe one specific thing you could do today or tomorrow or next week to begin.

CALCULATING ECOLOGICAL FOOTPRINTS

As we have seen throughout this book, individuals can contribute to sustainable solutions for our society and our planet in many ways. Some of these involve advocating for change at high levels of government or business or academia. But plenty of others involve the countless small choices we make in how we live our lives day to day. Where we live, what we buy, how we travel—these types of choices we each make as citizens, consumers, and human beings determine how we affect the environment and the people around us. As you know, such personal choices are summarized (crudely, but usefully) in an ecological footprint.

Turn back to the Calculating Ecological Footprints exercise in Chapter 1 (pp. 21–22), and recover the numerical value of your own personal ecological footprint that you calculated at the beginning of your course. Enter it in the table. Now return to the same online ecological footprint calculator that you used for Chapter 1's Calculating Ecological Footprints exercise. For many of you, this will have been www .myfootprint.org or http://www.footprintnetwork.org/en/index .php/GFN/page/personal_footprint. Take the footprint quiz again, and calculate your current footprint.

	Footprint value (hectares per person)
World average	2.7
U.S. average	9.4
Your footprint from Chapter 1	
Your footprint now	
Your footprint with three more changes	

1. Enter your current footprint, as determined by the online calculator, in the table. How does this value compare to your footprint at the beginning of your course? By what percentage did your footprint decrease or increase? If it changed, why do you think it changed? What changes have you made in your lifestyle since beginning this course that influence your environmental impact?

2. How does your personal footprint compare to the average footprint of a U.S. resident? How does it compare to that of the average person in the world? What do you think would be an admirable yet realistic goal for you to set as a target value for your own footprint?

3. Now think of three changes in your lifestyle that would lower your footprint. These should be changes that you would like to make and that you believe you could reasonably make. Take the footprint quiz again, incorporating these three changes. Enter the resulting footprint in the table.

4. Now set as a goal reducing your footprint by 25%, and experiment by changing various answers in your footprint quiz. What changes would allow you to attain a 25% reduction in your footprint? What changes would be needed to reduce your footprint to the hypothetical target value you set in Question 2?

Mastering**ENVIRONMENTALSCIENCE**™

APPENDIX A
Some Basics on Graphs

Presenting data in ways that help make trends and patterns visually apparent is a vital part of the scientific endeavor. For scientists, businesspeople, policymakers, and others, the primary tool for expressing patterns in data is the graph. Thus, the ability to interpret graphs is a skill that you will want to cultivate. This appendix guides you in how to read graphs, introduces a few vital conceptual points, and surveys the most common types of graphs, giving rationales for their use.

Navigating a Graph

A graph is a diagram that shows relationships among *variables*, which are factors that can change in value. The most common types of graphs relate values of a *dependent variable* to those of an *independent variable*. As explained in Chapter 1 (p. 11), a dependent variable is so named because its values "depend on" the values of an independent variable. In other words, as the values of an independent variable change, the values of the dependent variable change in response. In a manipulative experiment (pp. 12–13), changes that a researcher specifies in the value of the independent variable *cause* changes in the value of the dependent variable. In observational studies, there may be no causal relationship, and scientists may plot a correlation (pp. 12–13). In a positive correlation, values of one variable go up or down along with values of another. In a negative correlation, values of one variable go up when values of the other go down, or go down when the others go up. Whether we are graphing a correlation or a causal relationship, the values of the independent variable are known or specified by the researcher, and the values of the dependent variable are unknown until the research has taken place, and are what we are interested in observing or measuring.

By convention, independent variables are generally represented on the horizontal axis, or *x axis*, of a graph, while dependent variables are represented on the vertical axis, or *y axis*. Numerical values of variables generally become larger as one proceeds rightward on the *x* axis or upward on the *y* axis. Note that the tick marks along the axes must be uniformly spaced so that when the data are plotted, the graph gives an accurate visual representation of the scale of quantitative change in the data.

In many cases, independent variables are not numbers, but categories. For example, in a graph that presents population sizes of several nations, the nations would comprise a categorical independent variable, whereas population size would be a numerical dependent variable.

As a simple example, **FIGURE A.1** shows data from a classic early lab experiment that measured population growth among yeast cells. The *x* axis shows values of the independent variable, which in this case was time, expressed in units of hours. The researcher was interested in how many yeast cells would propagate over time, so the dependent variable, presented on the *y* axis, is the number of yeast cells present. For each hour at which data were measured during the experiment, a data point on the graph is plotted to show the number of yeast cells present. In this particular graph, a line (red curve) was then drawn through the actual data points (orange dots), showing how closely the empirical data matched the logistic growth curve (p. 67), a theoretical phenomenon of importance in ecology.

Now that you're familiar with the basic building blocks of a graph, let's survey the most common types of graphs you'll see, and examine a few vital concepts in graphing.

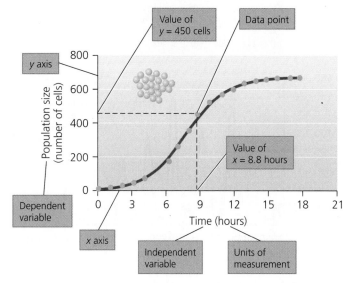

FIGURE A.1 ▲ Logistic population growth, demonstrated by the growth of yeast cells over time in a classic lab experiment. (Figure 3.17a, p. 70)

A-1

GRAPH TYPE: Line Graph

A line graph is used when a data set involves a sequence of some kind, such as a series of values that occur one by one and change through time or across distance. In a line graph, a line runs from one data point to the next. Line graphs are most appropriate when the *y* axis expresses a continuous numerical variable, and the *x* axis expresses either continuous numerical data or discrete sequential categories (such as years). **FIGURE A.2** shows values for the size of the ozone hole over Antarctica in recent years. Note how the data show that the size increases until 1987, when the Montreal Protocol (pp. 481–482) came into force, and then begins to stabilize afterwards.

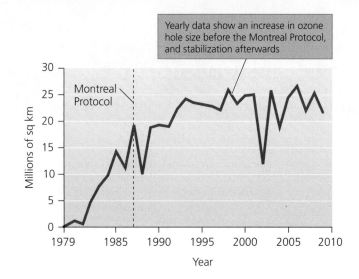

FIGURE A.2 ▲ Size of the Antarctic ozone hole before and after a treaty intended to address it. (Figure 17.24, p. 482)

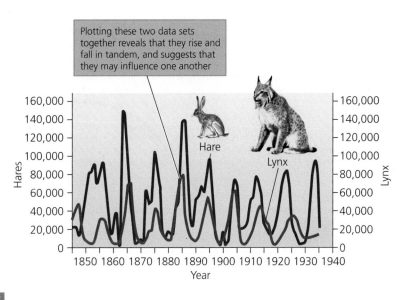

FIGURE A.3 ▲ Fluctuations in recorded numbers of hare and lynx. (Figure 4.5, p. 81)

One useful technique is to plot two or more data sets together on the same graph. This allows us to compare trends in the data sets to see whether and how they may be related. In **FIGURE A.3**, recorded numbers of a predator species rise and fall immediately following those of its prey, suggesting a possible connection.

KEY CONCEPT: Projections

Besides showing observed data, we can use graphs to show data that are predicted for the future. Such *projections* of data are based on models, simulations, or extrapolations from past data, but they are only as good as the information that goes into them—and future trends may not hold if conditions change in unforeseen ways. Thus, in this textbook, projected future data on a line graph are shown with dashed lines, as in **FIGURE A.4**, to indicate that they are less certain than data that have already been observed. Be careful when interpreting graphs in the popular media, however; often newspapers, magazines, and ads will show projected future data in the same way as known past data!

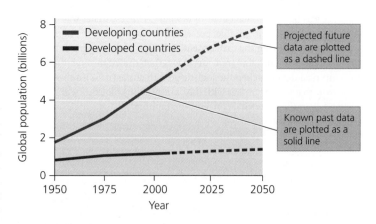

FIGURE A.4 ▲ Past and projected population growth for developing and developed countries. (Figure 8.22, p. 217)

GRAPH TYPE: Bar Chart

A bar chart is most often used when one variable is a category and the other is a number. In such a chart, the height (or length) of each bar represents the numerical value of a given category. Higher or longer bars mean larger values. In **FIGURE A.5**, the bar for the category "Automobile" is higher than that for "Light rail," indicating that automobiles use more energy per passenger-mile (the numerical variable on the *y* axis) than light rail systems do.

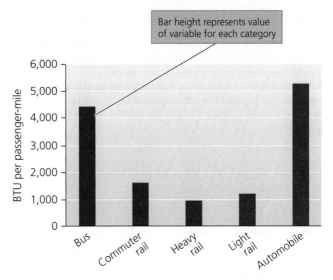

FIGURE A.5 ▲ Energy consumption for different modes of transit. (Figure 13.11a, p. 358)

It is often instructive to graph two or more data sets together to reveal patterns and relationships. A bar chart such as **FIGURE A.6** lets us compare two data sets (oil production and oil consumption) both within and among nations. A graph that does double duty in this way allows for higher-level analysis (in this case, suggesting which nations depend on others for petroleum imports). Most bar charts in this book illustrate multiple types of information at once in this manner.

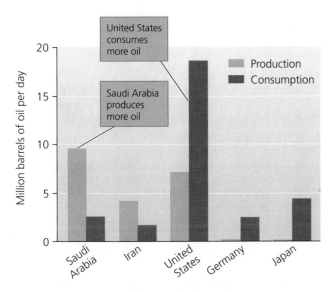

FIGURE A.6 ▲ Oil production and consumption by selected nations. (Figure 19.18, p. 552)

FIGURE A.7 illustrates two ways in which a bar chart can be modified. First, note that the orientation of bars is horizontal instead of vertical. In this configuration, the categorical variable is along the *y* axis and the numerical variable is along the *x* axis. Second, the bars extend in either direction from a central *x*-axis value of zero, representing either positive values (right) or negative values (left). Sometimes such arrangements can make for a clearer presentation. In this example, "Total forest" has been decreasing by over 5 million hectares per year, while "Productive plantations" have been increasing by 5 million hectares per year.

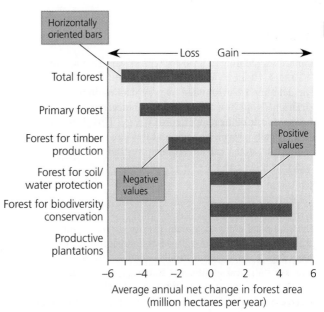

FIGURE A.7 ▲ Loss or gain of forests in recent years, by forest type. (Figure 12.19, p. 332)

KEY CONCEPT: Statistical Uncertainty

Most data sets involve some degree of uncertainty. When a graphed value represents the *mean* (average) of many measurements, the researcher may want to show the degree to which the raw data vary around this mean. Mathematical techniques are used to obtain statistically precise degrees of variation around a mean. Results from such statistical analyses may be expressed in a number of ways, and the three graphs at right show methods used in this book.

In an error-bar plot (**FIGURE A.8**), each mean is shown as a dot, and thin black lines called *error bars* extend in each direction from the dot, representing the degree of variation around the mean. Longer error bars indicate more uncertainty or variation, whereas short error bars mean we can place high confidence in the mean value. In this example, a researcher measured the number of bird species each year in plots of forest. The mean numbers decreased after the forest was broken apart into fragments, yet the error bars reveal considerable variation among plots. Thus the researcher needed to perform further statistical analyses to determine how much confidence could be placed in the apparent decrease in bird species numbers.

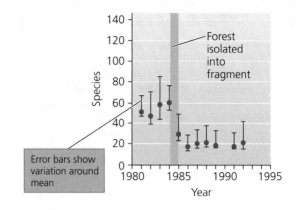

FIGURE A.8 ▲ Species richness of Amazon rainforest birds before and after habitat fragmentation. (Figure 12.SBS2, p. 337)

In a bar chart (**FIGURE A.9**), error bars may be shown extending above and below the tops of the bars, or simply above them. In this example of woody debris remaining after salvage logging, error bars show the most variation in measurements at "Burned and logged" sites.

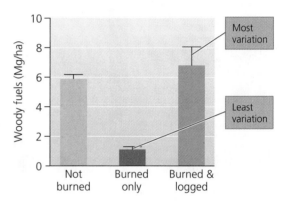

FIGURE A.9 ▲ Fine-scale woody debris left after treatments in salvage logging study. (Figure 12.SBS1, p. 331)

Sometimes shading is used to express variation around a mean. **FIGURE A.10** shows mean global temperature readings since 1850. The data line is surrounded by gray shading indicating statistical variation. Note how the amount of uncertainty is exceeded by the sheer scale of the temperature increase. This gives us confidence that globally warming temperatures are a real phenomenon, despite the statistical uncertainty we find around mean values each year.

The statistical analysis of data is critically important in science. In this book we provide a broad and streamlined introduction to many topics, so we often omit error bars from our graphs and details of statistical significance from our discussions. Bear in mind that this is for clarity of presentation only; the research we discuss analyzes its data in far more depth than any textbook could possibly cover.

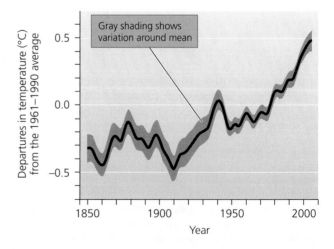

FIGURE A.10 ▲ Change in global temperature, measured since 1850. (Figure 18.11, p. 506)

GRAPH TYPE: Scatter Plot

A scatter plot is often used when data are not sequential, and when a given *x*-axis value could have multiple *y*-axis values. A scatter plot allows us to visualize a broad positive or negative correlation between variables. **FIGURE A.11** shows a negative correlation (that is, one value goes up while the other goes down): Nations with higher rates of school enrollment for girls tend to have lower fertility rates. Jamaica has high enrollment and low fertility, whereas Ethiopia has low enrollment and high fertility.

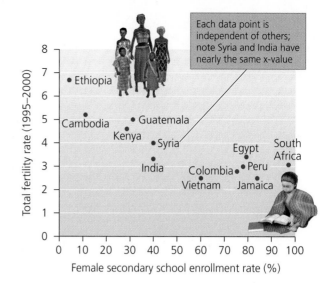

FIGURE A.11 ▲ Fertility rate and female education. (Figure 8.18, p. 213)

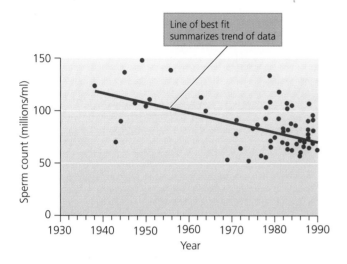

FIGURE A.12 ▲ Declining sperm count in men across the world. (Figure 14.20, p. 391)

A "line of best fit" may be drawn through a scatter plot in order to make a trend in the data more clear to the eye. The placement of such a line is determined by precise mathematical analysis through a statistical technique called linear regression. In general, the closer the data points are to the line, the more reliably the line reflects any true correlation in the data. **FIGURE A.12** shows an apparent decline in human sperm counts during the 20th century. The data varies greatly around the line of best fit, however, and so this research conclusion has remained controversial.

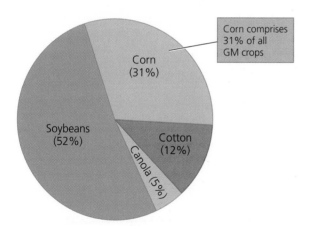

FIGURE A.13 ▲ Genetically modified crops grown worldwide, by type. (Figure 10.15, p. 266)

GRAPH TYPE: Pie Chart

A pie chart is used when we wish to compare the numerical proportions of some whole that are taken up by each of several categories. Each category is represented visually like a slice from a pie, with the size of the slice reflecting the percentage of the whole that is taken up by that category. For example, **FIGURE A.13** shows the percentages of genetically modified crops worldwide that are soybeans, corn, cotton, and canola.

Mastering ENVIRONMENTAL SCIENCE™

Don't stop here. Take advantage of the graphing resources at **www.masteringenvironmentalscience.com**. The **GRAPHit!** tutorials allow you to plot your own data, and the **Interpreting Graphs and Data** exercises guide you through critical-thinking questions on graphed data from recent research in environmental science. Both features will help you further expand your comprehension and use of graphs.

Measurement	Unit and Abbreviation	Metric Equivalent	Metric to English Conversion Factor	English to Metric Conversion Factor
Length	1 kilometer (km)	$= 1{,}000 \ (10^3)$ meters	1 km $=$ 0.62 mile	1 mile $=$ 1.61 km
	1 meter (m)	$= 100 \ (10^2)$ centimeters	1 m $=$ 1.09 yards	1 yard $=$ 0.914 m
		$= 1{,}000$ millimeters	1 m $=$ 3.28 feet	1 foot $=$ 0.305 m
			1 m $=$ 39.37 inches	
	1 centimeter (cm)	$= 0.01 \ (10^{-2})$ meter	1 cm $=$ 0.394 inch	1 foot $=$ 30.5 cm
				1 inch $=$ 2.54 cm
	1 millimeter (mm)	$= 0.001 \ (10^{23})$ meter	1 mm $=$ 0.039 inch	
Area	1 square meter (m^2)	$= 10{,}000$ square centimeters	1 m^2 $=$ 1.1960 square yards	1 square yard $=$ 0.8361 m^2
			1 m^2 $=$ 10.764 square feet	1 square foot $=$ 0.0929 m^2
	1 square centimeter (cm^2)	$= 100$ square millimeters	1 cm^2 $=$ 0.155 square inch	1 square inch $=$ 6.4516 cm^2
Mass	1 metric ton (t)	$= 1{,}000$ kilograms	1 t $=$ 1.103 ton	1 ton $=$ 0.907 t
	1 kilogram (kg)	$= 1{,}000$ grams	1 kg $=$ 2.205 pounds	1 pound $=$ 0.4536 kg
	1 gram (g)	$= 1{,}000$ milligrams	1 g $=$ 0.0353 ounce	1 ounce $=$ 28.35 g
	1 milligram (mg)	$= 0.001$ gram		
Volume (solids)	1 cubic meter (m^3)	$= 1{,}000{,}000$ cubic centimeters	1 m^3 $=$ 1.3080 cubic yards	1 cubic yard $=$ 0.7646 m^3
			1 m^3 $=$ 35.315 cubic feet	1 cubic foot $=$ 0.0283 m^3
	1 cubic centimeter (cm^3 or cc)	$= 0.000001$ cubic meter	1 cm^3 $=$ 0.0610 cubic inch	1 cubic inch $=$ 16.387 cm^3
		$= 1$ milliliter		
	1 cubic millimeter (mm^3)	$= 0.000000001$ cubic meter		
Volume (liquids and gases)	1 kiloliter (kl or kL)	$= 1{,}000$ liters	1 kL $=$ 264.17 gallons	1 gallon $=$ 3.785 L
	1 liter (l or L)	$= 1{,}000$ milliliters	1 L $=$ 0.264 gallons	1 quart $=$ 0.946 L
			1 L $=$ 1.057 quarts	
	1 milliliter (ml or mL)	$= 0.001$ liter	1 ml $=$ 0.034 fluid ounce	1 quart $=$ 946 ml
		$= 1$ cubic centimeter	1 ml $=$ approximately $\frac{1}{4}$ teaspoon	1 pint $=$ 473 ml
				1 fluid ounce $=$ 29.57 ml
				1 teaspoon $=$ approx. 5 ml
Time	1 millisecond (ms)	$= 0.001$ second		
Temperature	Degrees Celsius (°C)		$°C = \frac{5}{9} (°F - 32)$	$°F = \frac{9}{5} °C + 32$
Energy and Power	1 kilowatt-hour	$= 34{,}113$ BTUs $= 860{,}421$ calories		
	1 watt	$= 3.413$ BTU/hr		
		$= 14.34$ calorie/min		
	1 calorie	$=$ the amount of heat necessary to raise the temperature of 1 gram (1 cm^3) of water 1 degree Celsius		
	1 horsepower	$= 7.457 \times 102$ watts		
	1 joule	$= 9.481 \times 10^{-4})$ BTU		
		$= 0.239$ cal		
		$= 2.778 \times 10^{-7}$ kilowatt-hour		
Pressure	1 pound per square inch (psi)	$= 6894.757$ pascal (Pa)		
		$= 0.068045961$ atmosphere (atm)		
		$= 51.71493$ millimeters of mercury (mm hg $=$ Torr)		
		$= 68.94757$ millibars (mbar)		
		$= 6.894757$ kilopascal (kPa)		
	1 atmosphere (atm)	$= 101.325$ kilopascal (kPa)		

Periodic Table of the Elements

Periodic Table

Representative (main group) elements

IA	IIA	IIIB	IVB	VB	VIB	VIIB	VIIIB			IB	IIB	IIIA	IVA	VA	VIA	VIIA	VIIIA
1 **H** 1.0079 Hydrogen																	2 **He** 4.003 Helium
3 **Li** 6.941 Lithium	4 **Be** 9.012 Beryllium											5 **B** 10.811 Boron	6 **C** 12.011 Carbon	7 **N** 14.007 Nitrogen	8 **O** 15.999 Oxygen	9 **F** 18.998 Fluorine	10 **Ne** 20.180 Neon
11 **Na** 22.990 Sodium	12 **Mg** 24.305 Magnesium											13 **Al** 26.982 Aluminum	14 **Si** 28.086 Silicon	15 **P** 30.974 Phosphorus	16 **S** 32.066 Sulfur	17 **Cl** 35.453 Chlorine	18 **Ar** 39.948 Argon
19 **K** 39.098 Potassium	20 **Ca** 40.078 Calcium	21 **Sc** 44.956 Scandium	22 **Ti** 47.88 Titanium	23 **V** 50.942 Vanadium	24 **Cr** 51.996 Chromium	25 **Mn** 54.938 Manganese	26 **Fe** 55.845 Iron	27 **Co** 58.933 Cobalt	28 **Ni** 58.69 Nickel	29 **Cu** 63.546 Copper	30 **Zn** 65.39 Zinc	31 **Ga** 69.723 Gallium	32 **Ge** 72.61 Germanium	33 **As** 74.922 Arsenic	34 **Se** 78.96 Selenium	35 **Br** 79.904 Bromine	36 **Kr** 83.8 Krypton
37 **Rb** 85.468 Rubidium	38 **Sr** 87.62 Strontium	39 **Y** 88.906 Yttrium	40 **Zr** 91.224 Zirconium	41 **Nb** 92.906 Niobium	42 **Mo** 95.94 Molybdenum	43 **Tc** 98 Technetium	44 **Ru** 101.07 Ruthenium	45 **Rh** 102.906 Rhodium	46 **Pd** 106.42 Palladium	47 **Ag** 107.868 Silver	48 **Cd** 112.411 Cadmium	49 **In** 114.82 Indium	50 **Sn** 118.71 Tin	51 **Sb** 121.76 Antimony	52 **Te** 127.60 Tellurium	53 **I** 126.905 Iodine	54 **Xe** 131.29 Xenon
55 **Cs** 132.905 Cesium	56 **Ba** 137.327 Barium	57 **La** 138.906 Lanthanum	72 **Hf** 178.49 Hafnium	73 **Ta** 180.948 Tantalum	74 **W** 183.84 Tungsten	75 **Re** 186.207 Rhenium	76 **Os** 190.23 Osmium	77 **Ir** 192.22 Iridium	78 **Pt** 195.08 Platinum	79 **Au** 196.967 Gold	80 **Hg** 200.59 Mercury	81 **Tl** 204.383 Thallium	82 **Pb** 207.2 Lead	83 **Bi** 208.980 Bismuth	84 **Po** 209 Polonium	85 **At** 210 Astatine	86 **Rn** 222 Radon
87 **Fr** 223 Francium	88 **Ra** 226.025 Radium	89 **Ac** 227.028 Actinium	104 **Rf** 261 Unnilquadium	105 **Db** 262 Unnilpentium	106 **Sg** 263 Unnilhexium	107 **Bh** 262 Unnilseptium	108 **Hs** 265 Unniloctium	109 **Mt** 266 Unnilennium	110 **Uun** 269 Ununnilium	111 **Uuu** 272 Unununium	112 **Uub** 277 Ununbium		114		116		

Transition metals

Rare earth elements

Lanthanides

58 **Ce** 140.115 Cerium	59 **Pr** 140.908 Praseodymium	60 **Nd** 144.24 Neodymium	61 **Pm** 145 Promethium	62 **Sm** 150.36 Samarium	63 **Eu** 151.964 Europium	64 **Gd** 157.25 Gadolinium	65 **Tb** 158.925 Terbium	66 **Dy** 162.5 Dysprosium	67 **Ho** 164.93 Holmium	68 **Er** 167.26 Erbium	69 **Tm** 168.934 Thulium	70 **Yb** 173.04 Ytterbium	71 **Lu** 174.967 Lutetium

Actinides

90 **Th** 232.038 Thorium	91 **Pa** 231.036 Protactinium	92 **U** 238.029 Uranium	93 **Np** 237.048 Neptunium	94 **Pu** 244 Plutonium	95 **Am** 243 Americium	96 **Cm** 247 Curium	97 **Bk** 247 Berkelium	98 **Cf** 251 Californium	99 **Es** 252 Einsteinium	100 **Fm** 257 Fermium	101 **Md** 258 Mendelevium	102 **No** 259 Nobelium	103 **Lr** 262 Lawrencium

The periodic table arranges elements according to atomic number and atomic weight into horizontal rows called periods and vertical columns called groups.

Elements of each group in Class A have similar chemical and physical properties. This reflects the fact that members of a particular group have the same number of valence shell electrons, which is indicated by the group's number. For example, group IA elements have one valence shell electron, group IIA elements have two, and group VA elements have five. In contrast, as you progress across a period from left to right, properties of the elements change, varying from the very metallic properties of groups IA and IIA to the nonmetallic properties of group VIIA to the inert elements (noble gases) in group VIIA. This reflects changes in the number of valence shell electrons.

Class B elements, or transition elements, are metals, and generally have one or two valence shell electrons. In these elements, some electrons occupy more distant electron shells before the deeper shells are filled.

In this periodic table, elements with symbols printed in black exist as solids under standard conditions (25°C and 1 atmosphere of pressure), while elements in red exist as gases, and those in dark blue as liquids. Elements with symbols in green do not exist in nature and must be created by some type of nuclear reaction.

Geologic Time Scale

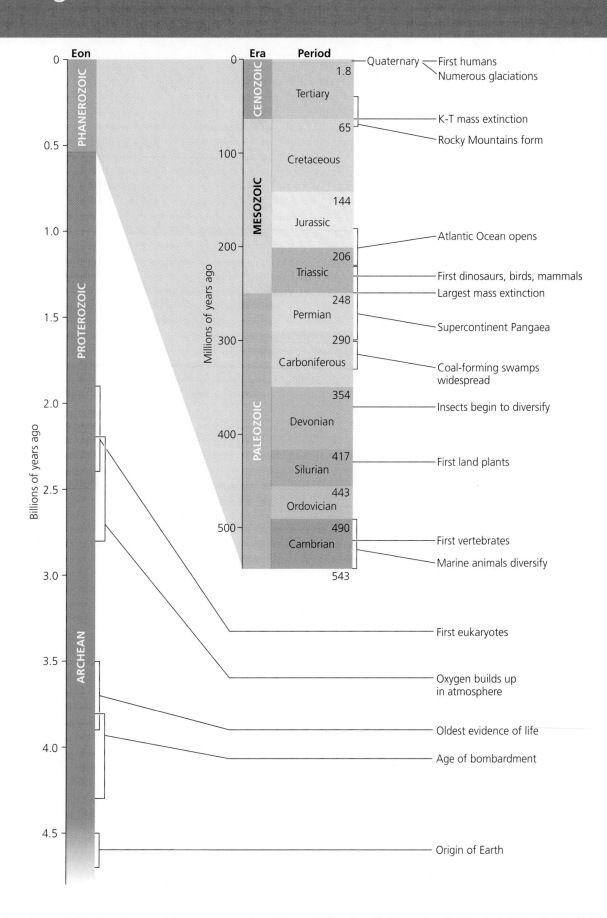

GLOSSARY

abiotic   Non-living. Compare *biotic*.

acid deposition   The settling of acidic or acid-forming pollutants from the *atmosphere* onto Earth's surface. This can take place by precipitation, fog, gases, or the deposition of dry particles. Compare *acid rain*; *acid precipitation*.

acid drainage   A process in which sulfide minerals in newly exposed rock surfaces react with *oxygen* and rainwater to produce sulfuric acid, which causes chemical *runoff* as it *leaches* metals from the rocks. Acid drainage is a natural phenomenon, but mining greatly accelerates it by exposing many new surfaces.

acidic   The property of a *solution* in which the concentration of *hydrogen* (H⁺) *ions* is greater than the concentration of hydroxide (OH⁻) ions. Compare *basic*.

acid-neutralizing capacity   The capacity of soil, rock, or water to resist *pH* change from *acid deposition*, due to alkaline chemistry.

acid precipitation   The deposition of *acidic* or acid-forming pollutants from the *atmosphere* onto Earth's surface by *precipitation*. Compare *acid rain*.

acid rain   *Acid deposition* that takes place through rain. Compare *acid precipitation*.

active solar energy collection   An approach in which technological devices are used to focus, move, or store solar energy. Compare *passive solar energy collection*.

acute exposure   Exposure to a *toxicant* occurring in high amounts for short periods of time. Compare *chronic exposure*.

adaptive management   The systematic testing of different management approaches to improve methods over time.

adaptive radiation   A burst of *species* formation due to natural selection.

adaptive trait (adaptation)   A trait that confers greater likelihood that an individual will reproduce.

aerobic   Occurring in an *environment* where *oxygen* is present. For example, the decay of a rotting log proceeds by aerobic decomposition. Compare *anaerobic*.

aerosols   Very fine liquid droplets or solid particles aloft in the atmosphere.

affluenza   Term coined by social critics to describe the failure of material goods to bring happiness to people who have the financial means to afford them.

age distribution   The relative numbers of organisms of each age within a *population*. Age distributions can have a strong effect on rates of population growth or decline and are often expressed as a ratio of age classes, consisting of organisms (1) not yet mature enough to reproduce, (2) capable of reproduction, and (3) beyond their reproductive years.

age structure   See *age distribution*.

age structure diagram (population pyramid)   A diagram demographers use to show the age structure of a population. The width of each horizontal bar represents the relative number of individuals in each age class.

agricultural extension agent   An expert from a university or government agency who assists farmers by providing information on new research and helping them apply this knowledge with new techniques.

agricultural revolution   The shift around 10,000 years ago from a hunter-gatherer lifestyle to an agricultural way of life in which people began to grow their own crops and raise domestic animals. Compare *industrial revolution*.

agriculture   The practice of cultivating *soil*, producing crops, and raising livestock for human use and consumption.

A horizon   A layer of *soil* found in a typical *soil profile*. It forms the top layer or lies below the *O horizon* (if one exists). It consists of mostly inorganic mineral components such as *weathered* substrate, with some organic matter and *humus* from above mixed in. The A horizon is often referred to as *topsoil*. Compare *B horizon; C horizon; E horizon; R horizon*.

air pollutants   Gases and particulate material added to the atmosphere that can affect *climate* or harm people or other organisms.

air pollution   The act of polluting the air, or the condition of being polluted by *air pollutants*.

albedo   The capacity of a surface to reflect light. Higher albedo values refer to greater reflectivity.

allelopathy   A phenomenon whereby certain plants release poisonous chemicals that harm others nearby.

allergen   A *toxicant* that overactivates the immune system, causing an immune response when one is not necessary.

allopatric speciation   Species formation due to the physical separation of populations over some geographic distance. Compare *sympatric speciation*.

alpine tundra   *Tundra* that occurs at the tops of mountains.

alternative hypothesis   A *hypothesis* that proposes a particular association among variables, and is tested against a *null hypothesis* that postulates no association.

ambient air pollution   See *outdoor air pollution*.

amensalism   A relationship between members of different *species* in which one organism is harmed and the other is unaffected. Compare *commensalism*.

amino acids   Organic molecules that join in long chains to form *proteins*.

anaerobic   Occurring in an *environment* that has little or no *oxygen*. The conversion of organic matter to *fossil fuels (crude oil, coal, natural gas)* at the bottom of a deep lake, swamp, or shallow sea is an example of anaerobic decomposition. Compare *aerobic*.

anemometer   A device that measures wind speed and direction.

Anthropocene   A term proposed by some geoscientists as a formal name for the most recent 200 or so years of historical time, delineating the period of Earth's history in which human beings are having the most extreme impact on the planet's systems.

anthropocentrism   A human-centered view of our relationship with the *environment*.

application   An applied use of science, such as a new technology, policy decision, or resource management strategy.

aquaculture   The raising of aquatic organisms for food in controlled *environments*.

aquifer   An underground water reservoir.

area effect   In *island biogeography theory*, the pattern that large islands host more species than smaller islands, because larger islands provide larger targets for immigration and because extinction rates are reduced.

artesian aquifer   See *confined aquifer*.

artificial selection   *Natural selection* conducted under human direction. Examples include the *selective breeding* of crop plants, pets, and livestock.

asbestos   Any of several types of *mineral* that form long, thin microscopic fibers—a structure that allows asbestos to insulate buildings for heat, muffle sound, and resist fire. When inhaled and lodged in lung tissue, asbestos scars the tissue and may eventually lead to lung cancer or *asbestosis*.

asbestosis   A disorder resulting from lung tissue scarred by acid following prolonged inhalation of *asbestos*.

Asian Brown Cloud   A persistent 2-mile-thick layer of *air pollution* from southern Asia that hangs over the Indian subcontinent throughout the dry season, each December through April. It is estimated to reduce sunlight reaching Earth's surface in that region by 10–15%, decrease rice productivity by 5–10%, and cause thousands of deaths each year.

asthenosphere   A layer of the upper *mantle*, just below the *lithosphere*, consisting of especially soft rock.

atmosphere   The thin layer of gases surrounding planet Earth. Compare *biosphere; hydrosphere; lithosphere*.

Atmospheric Brown Cloud   See *Asian Brown Cloud*.

atmospheric deposition   The wet or dry deposition on land of a wide variety of pollutants, including mercury, nitrates, organochlorines, and others. *Acid deposition* is one type of atmospheric deposition.

atmospheric pressure   The weight per unit area produced by a column of air.

atoll   A ring-shaped island (generally of *coral reef*) surrounding an older submerged island area.

atom   The smallest component of an *element* that maintains the chemical properties of that element.

atomic number   The number of *protons* in a given *atom*.

autotroph (primary producer)   An organism that can use the energy from sunlight to produce its own food. Includes green plants, algae, and cyanobacteria.

***Bacillus thuringiensis* (Bt)**   A naturally occurring *soil* bacterium that produces a protein that kills many pests, including caterpillars and the larvae of some flies and beetles.

background rate of extinction The average rate of *extinction* that occurred before the appearance of humans. For example, the *fossil record* indicates that for both birds and mammals, one *species* in the world typically became extinct every 500–1,000 years. Compare *mass extinction event*.

bagasse Crushed sugarcane residue, whose sugars are used in Brazil to make *ethanol* that helps powers millions of vehicles.

baghouse A system of large filters that physically removes *particulate matter* from *incinerator emissions*.

barrier island A long thin island that parallels a shoreline. Generally of sand or *coral reef*, barrier islands protect coasts from storms.

basic The property of a *solution* in which the concentration of hydroxide (OH⁻) *ions* is greater than the concentration of *hydrogen* (H⁺) ions. Compare *acidic*.

bathymetry The study of ocean depths.

bedrock The continuous mass of solid rock that makes up Earth's *crust*.

benthic Of, relating to, or living on the bottom of a water body. Compare *pelagic*.

benthic zone The bottom layer of a water body. Compare *littoral zone*; *limnetic zone*; *profundal zone*.

B horizon The layer of *soil* that lies below the *E horizon* and above the *C horizon*. *Minerals* that leach out of the E horizon are carried down into the B horizon (or subsoil) and accumulate there. Sometimes called the "zone of accumulation" or "zone of deposition." Compare *A horizon; O horizon; R horizon*.

bioaccumulation The buildup of *toxicants* in the tissues of an animal.

biocapacity A term in *ecological footprint* accounting meaning the amount of biologically productive land and sea available to us.

biocentrism A philosophy that ascribes relative values to actions, entities, or properties on the basis of their effects on all living things or on the integrity of the *biotic* realm in general. The biocentrist evaluates an action in terms of its overall impact on living things, including—but not exclusively focusing on—human beings.

biocontrol See *biological control*.

biodiesel Diesel fuel produced by mixing vegetable oil, used cooking grease, or animal fat with small amounts of *ethanol* or methanol in the presence of a chemical catalyst.

biodiversity (biological diversity) The variety of life across all levels of biological organization, including the diversity of *species*, their *genes*, their *populations*, and their *communities*.

biodiversity hotspot An area that supports an especially great diversity of *species*, particularly species that are *endemic* to the area.

bioenergy (biomass energy) *Energy* harnessed from plant and animal matter, including wood from trees, charcoal from burned wood, and combustible animal waste products, such as cattle manure. *Fossil fuels* are not considered biomass energy sources because their organic matter has not been part of living organisms for millions of years and has undergone considerable chemical alteration since that time.

biofuel Fuel produced from *biomass energy* sources and used primarily to power automobiles.

biogas Methane-rich gas produced by bacterial action in anaerobic digestion facilities. Can be burned in a power plant to generate electricity.

biogenic Type of *natural gas* created at shallow depths by the anaerobic decomposition of organic matter by bacteria. Consists of nearly pure *methane*. Compare *thermogenic*.

biogeochemical cycle See *nutrient cycle*.

biological control (biocontrol) Control of pests and weeds with organisms that prey on or parasitize them, rather than with *pesticides*.

biological diversity See *biodiversity*.

biological hazard Human health hazards that result from ecological interactions among organisms. These include *parasitism* by viruses, bacteria, or other *pathogens*. Compare *infectious disease*; *chemical hazard*; *cultural hazard*; *physical hazard*.

biological weathering *Weathering* that occurs when living things break down *parent material* by physical or chemical means. Compare *chemical weathering*; *physical weathering*.

biomagnification The magnification of the concentration of *toxicants* in an organism caused by its consumption of other organisms in which toxicants have *bioaccumulated*.

biomass (1) In ecology, organic material that makes up living organisms; the collective mass of living matter in a given place and time. (2) In energy, organic material derived from living or recently living organisms, containing chemical energy that originated with photosynthesis.

biomass energy See *bioenergy*.

biomass power See *biopower*.

biome A major regional complex of similar plant *communities*; a large *ecological* unit defined by its dominant plant type and vegetation structure.

biophilia A phenomenon proposed by E. O. Wilson as "the connections that human beings subconsciously seek with the rest of life."

biopower The burning of *biomass energy* sources to generate electricity.

bioremediation The attempt to clean up *pollution* by enhancing natural processes of biodegradation by living organisms.

biosphere The sum total of all the planet's living organisms and the *abiotic* portions of the *environment* with which they interact.

biosphere reserve A tract of land with exceptional *biodiversity* that couples preservation with *sustainable development* to benefit local people. Biosphere reserves are designated by UNESCO (the *United Nations Educational, Scientific, and Cultural Organization*) following application by local stakeholders.

biotechnology The material application of biological *science* to create products derived from organisms. The creation of *transgenic* organisms is one type of biotechnology.

biotic Living. Compare *abiotic*.

biotic potential An organism's capacity to produce offspring.

birth control The effort to control the number of children one bears, particularly by reducing the frequency of pregnancy. Compare *contraception*; *family planning*.

bisphenol A (BPA) A substance widely used in plastics and to line food and drink cans, which has raised health concerns because it is an estrogen mimic.

bitumen A thick and heavy form of *petroleum* rich in *carbon* and poor in *hydrogen*.

bog A type of *wetland* in which a pond is thoroughly covered with a thick floating mat of vegetation. Compare *freshwater marsh*; *swamp*.

boreal forest A *biome* of northern coniferous forest that stretches in a broad band across much of Canada, Alaska, Russia, and Scandinavia. Also known as taiga, boreal forest consists of a limited number of *species* of evergreen trees, such as black spruce, that dominate large regions of forests interspersed with occasional bogs and lakes.

Borlaug, Norman (1914–2009) American agricultural scientist who introduced specially bred crops to developing nations in the 20th century, helping to spur the *Green Revolution*.

bottleneck A step in a process that limits the progress of the overall process.

bottom-trawling Fishing practice that involves dragging weighted nets across the seafloor to catch *benthic* organisms. Trawling crushes many organisms in its path and leaves long swaths of damaged sea bottom.

braided river A river that flows as an interconnected series of watercourses because it runs through a steeply sloped region or carries a great deal of sediment. Compare *meandering river*.

breakdown product A *compound* that results from the degradation of a toxicant.

breeder reactor A nuclear reactor that creates more fissile material than it consumes, and uses primarily uranium-238 and plutonium-239. Although breeder reactors make better use of fuel, generate more power, and produce less waste than conventional reactors, most have been closed because of concerns over nuclear weapons proliferation.

brownfield An area of land whose redevelopment or reuse is complicated by the presence or potential presence of hazardous material.

building-related illness Any sickness caused by indoor pollution.

Bureau of Land Management (BLM) Federal agency that owns and manages most U.S. rangelands. The BLM is the nation's single largest landowner; its 106 million ha (261 million acres) are spread across 12 western states.

by-catch That portion of a commercial fishing catch consisting of animals caught unintentionally. By-catch kills many thousands of fish, sharks, marine mammals, and birds each year.

Calvin cycle In *photosynthesis*, a series of chemical reactions in which *carbon atoms* from *carbon dioxide* are linked together to manufacture sugars.

canopy The upper level of tree leaves and branches in a *forest*.

cap-and-trade A *permit-trading* system in which government determines an acceptable level of *pollution* and then issues polluting parties permits to pollute. A company receives credit for amounts it does not emit and can then sell this credit to other companies. A type of *emissions trading system*.

capitalist market economy An *economy* in which buyers and sellers interact to determine which *goods* and *services* to produce, how much of them to produce, and how to distribute them. Compare *centrally planned economy*.

captive breeding The practice of capturing members of threatened and endangered *species* so that their young can be bred and raised in controlled *environments* and subsequently reintroduced into the wild.

carbohydrate An *organic compound* consisting of *atoms* of *carbon*, *hydrogen*, and *oxygen*.

carbon The chemical *element* with six protons and six neutrons. A key element in *organic compounds*.

carbon capture Technologies or approaches that remove *carbon dioxide* from power plant or other emissions, in an effort to mitigate *global climate change*.

carbon cycle A major *nutrient cycle* consisting of the routes that *carbon atoms* take through the nested networks of environmental *systems*.

carbon dioxide (CO_2) A colorless gas used by plants for *photosynthesis*, given off by *respiration*, and released by burning *fossil fuels*. A primary *greenhouse gas* whose buildup contributes to *global climate change*.

carbon footprint The cumulative amount of carbon, or *carbon dioxide*, that a person or institution emits, and is indirectly responsible for emitting, into the *atmosphere*, contributing to *global climate change*. Compare *ecological footprint*.

carbon monoxide (CO) A colorless, odorless gas produced primarily by the incomplete combustion of fuel. An EPA *criteria pollutant*.

carbon neutrality The state in which an individual, business, or institution emits no net carbon to the atmosphere. This may be achieved by reducing carbon emissions and/or employing *carbon offsets* to offset emissions.

carbon offset A voluntary payment to another entity intended to enable that entity to reduce the *greenhouse gas* emissions that one is unable or unwilling to reduce oneself. The payment thus offsets one's own emissions.

carbon sequestration Technologies or approaches to sequester, or store, *carbon dioxide* from industrial emissions, e.g., underground under pressure in locations where it will not seep out, in an effort to mitigate *global climate change*. We are still a long way from developing adequate technology and secure storage space to accomplish this.

carbon storage See *carbon sequestration*.

carbon tax A fee charged to entities that pollute by emitting *carbon dioxide*. A carbon tax gives polluters a financial incentive to reduce pollution, and is thus foreseen as a way to address *global climate change*.

carcinogen A chemical or type of radiation that causes cancer.

carnivore An organism that consumes animals. Compare *herbivore*; *omnivore*.

carrying capacity The maximum *population size* that a given *environment* can sustain.

case history Medical approach involving the observation and analysis of individual patients.

case law A body of law made up of cumulative decisions rendered by courts.

Cassandra A worldview (or a person holding the worldview) that predicts doom and disaster as a result of our environmental impacts. In Greek mythology, Cassandra was the princess of Troy with the gift of prophecy, whose dire predictions were not believed. Compare *Cornucopian*.

categorical imperative An *ethical standard* described by Immanuel Kant, which roughly approximates Christianity's "golden rule": to treat others as you would prefer to be treated yourself.

cation exchange Process by which plants' roots donate *hydrogen ions* to the soil in exchange for cations (positively charged ions) such as those of calcium, magnesium, and potassium, which plants use as *nutrients*. The soil particles then replenish these cations by exchange with soil water.

cation exchange capacity A soil's ability to hold cations, preventing them from *leaching* and thus making them available to plants. A useful measure of soil fertility.

cell The most basic organizational unit of organisms.

cellular respiration The process by which a *cell* uses the chemical reactivity of *oxygen* to split glucose into its constituent parts, water and *carbon dioxide*, and thereby release chemical energy that can be used to form chemical bonds or to perform other tasks within the cell. Compare *photosynthesis*.

cellulosic ethanol *Ethanol* produced from the cellulose in plant tissues, by treating it with enzymes. Techniques for producing cellulosic ethanol are under development because of the desire to make ethanol from low-value crop waste (residues such as corn stalks and husks), rather than from the sugars of high-value crops.

centrally planned economy An *economy* in which a nation's government determines how to allocate resources in a top-down manner. Also called a "state socialist economy." Compare *capitalist market economy*.

chaparral A *biome* consisting mostly of densely thicketed evergreen shrubs occurring in limited small patches. Its "Mediterranean" *climate* of mild, wet winters and warm, dry summers is induced by oceanic influences. In addition to ringing the Mediterranean Sea, chaparral occurs along the coasts of California, Chile, and southern Australia.

character displacement A phenomenon resulting from *competition* among *species* in which competing species evolve characteristics that better adapt them to specialize on the portion of the resource they use. The species essentially become more different from one another, reducing their competition.

chemical hazard Chemicals that pose human health hazards. These include *toxins* produced naturally, as well as many of the disinfectants, *pesticides*, and other synthetic chemicals that our society produces. Compare *biological hazard*; *cultural hazard*; *physical hazard*.

chemical weathering *Weathering* that results when water or other substances chemically interact with *parent material*. Compare *biological weathering*; *physical weathering*.

chemosynthesis The process by which bacteria in *hydrothermal vents* use the chemical energy of hydrogen sulfide (H_2S) to transform inorganic *carbon* into *organic compounds*. Compare *photosynthesis*.

Chernobyl Site of a nuclear power plant in Ukraine (then part of the Soviet Union), where in 1986 an explosion caused the most severe *nuclear reactor* accident the world has yet seen. As with *Three Mile Island*, the term is often used to denote the accident itself.

chlorofluorocarbon (CFC) One of a group of human-made *organic compounds* derived from simple *hydrocarbons*, such as ethane and methane, in which *hydrogen atoms* are replaced by chlorine, bromine, or fluorine. CFCs deplete the protective *ozone layer* in the *stratosphere*.

chlorophyll The light-absorbing pigment that enables *photosynthesis* and makes plants green.

chloroplast A cell organelle containing *chlorophyll* in which *photosynthesis* occurs.

C horizon The layer of *soil* that lies below the *B horizon* and above the *R horizon*. It contains rock particles that are larger and less *weathered* than the layers above. It consists of *parent material* that has been altered only slightly or not at all by the process of *soil* formation. Compare *A horizon*; *E horizon*; *O horizon*.

chronic exposure Exposure for long periods of time to a *toxicant* occurring in low amounts. Compare *acute exposure*.

city planning The professional pursuit that attempts to design cities in such a way as to maximize their efficiency, functionality, and beauty.

classical economics Founded by *Adam Smith*, the study of the behavior of buyers and sellers in a free-market *economy*. Holds that individuals acting in their own self-interest may benefit society, provided that their behavior is constrained by the rule of law and by private property rights and operates within competitive markets. See also *neoclassical economics*.

clay *Sediment* consisting of particles less than 0.002 mm in diameter. Compare *sand*; *silt*.

Clean Air Act of 1970 Revision of prior Congressional *legislation* to control *air pollution* that set stricter standards for air quality, imposed limits on emissions from new stationary and mobile sources, provided new funds for *pollution*-control research, and enabled citizens to sue parties violating the standards.

Clean Air Act of 1990 Congressional *legislation* that strengthened *regulations* pertaining to air quality standards, auto emissions, toxic *air pollution*, *acid deposition*, and depletion of the *ozone layer*, while also introducing market-based incentives to reduce *pollution*.

clean coal technologies A wide array of techniques, equipment, and approaches that seek to remove chemical contaminants (such as sulfur) during the process of generating electricity from coal.

clear-cutting The harvesting of timber by cutting all the trees in an area, leaving only stumps. Although it is the most cost-efficient method, clear-cutting is also the most damaging to the *environment*.

Clear Skies A George W. Bush administration initiative that aimed to abandon a command-and-control policy approach to air pollution and establish a market-based cap-and-trade program for sulfur dioxide, nitrogen oxides, and mercury. The Clear Skies legislation was stopped in 2005 by senators who concluded that it would increase pollution, relative to existing Clean Air Act policy.

climate The pattern of atmospheric conditions found across large geographic regions over long periods of time. Compare *weather*.

climate diagram (climatograph) A visual representation of a region's average monthly temperature and *precipitation*.

climate model A computer program that combines what is known about weather patterns, atmospheric circulation, atmosphere-ocean interactions, and feedback mechanisms to simulate *climate* processes.

climax community In the traditional view of ecological *succession*, a *community* that remains in place with little modification until disturbance restarts the successional process. Today, ecologists recognize that community change is more variable and less predictable than originally thought, and that assemblages of species may instead form complex mosaics in space and time.

closed cycle An approach in *ocean thermal energy conversion* in which warm surface water is used to evaporate chemicals that boil at low temperatures. These evaporated gases spin turbines to generate electricity. Cold water piped in from ocean depths then condenses the gases so they can be reused.

cloud forests Moist *forests*, generally at high elevations in the tropics and subtropics, that derive much of their moisture from low-moving clouds.

clumped distribution Distribution pattern in which organisms arrange themselves in patches, generally according to the availability of the resources they need.

coal A *fossil fuel* composed of organic matter that was compressed under very high pressure to form a dense, solid *carbon* structure.

coal gasification The chemical breakdown of *coal* by treatment with steam and heat to form a mix of gases.

coalbed methane *Methane* that emanates from *coal* seams, which commonly leaks to the *atmosphere* during coal mining. To avoid this waste and reduce methane emissions, engineers are trying to capture more of this gas for energy.

coevolution Process by which two or more species evolve in response to one another. Parasites and hosts may coevolve, as may flowers and their pollinators.

co-firing A process in which *biomass* is combined with *coal* in coal-fired power plants. Can be a relatively easy and inexpensive way for *fossil-fuel*-based utilities to expand their use of *renewable energy*.

cogeneration A practice in which the extra heat generated in the production of electricity is captured and put to use heating workplaces and homes, as well as producing other kinds of power.

cold front The boundary where a mass of cold air displaces a mass of warmer air. Compare *warm front*.

colony collapse disorder An undiagnosed cause of mass die-offs of honeybees (*Apis mellifera*) in recent years.

combined cycle A process in which coal is treated to create hot gases that turn a gas turbine, while the hot exhaust of this turbine heats water to drive a conventional steam turbine. Compare *gasification*.

command and control An approach to protecting the *environment* that sets strict legal limits and threatens punishment for violations of those limits.

commensalism A relationship between members of different *species* in which one organism benefits and the other is unaffected. Compare *amensalism*.

communicable disease See *infectious disease*.

community An assemblage of *populations* of organisms that live in the same place at the same time.

community-based conservation The practice of engaging local people to protect land and wildlife in their own region.

community ecology The study of the interactions among *species*, from one-to-one interactions to complex interrelationships involving entire *communities*.

community-supported agriculture A system in which consumers pay farmers in advance for a share of their yield, usually in the form of weekly deliveries of produce.

competition A relationship in which multiple organisms seek the same limited resource.

competitive exclusion An outcome of *interspecific competition* in which one *species* excludes another species from resource use entirely.

compost A mixture produced when decomposers break down organic matter, including food and crop waste, in a controlled environment.

composting The conversion of organic *waste* into mulch or *humus* by encouraging, in a controlled manner, the natural biological processes of decomposition.

compound A *molecule* whose *atoms* are composed of two or more *elements*.

concentrated animal feeding operation See *feedlot*.

concession The right to extract a resource, granted by a government to a corporation. Compare *conservation concession*.

confined (artesian) aquifer A water-bearing, porous layer of rock, *sand*, or gravel that is trapped between an upper and lower layer of less permeable substrate, such as *clay*. The water in a confined aquifer is under pressure because it is trapped between two impermeable layers. Compare *unconfined aquifer*.

conservation biology A scientific discipline devoted to understanding the factors, forces, and processes that influence the loss, protection, and restoration of *biological diversity* within and among *ecosystems*.

conservation concession A type of *concession* in which a conservation organization purchases the right to prevent resource extraction in an area of land, generally to preserve habitat in developing nations.

conservation district One of many county-based entities created by the Soil Conservation Service (now the Natural Resources Conservation Service) to promote practices that conserve *soil*.

conservation ethic An *ethic* holding that humans should put *natural resources* to use but also have a responsibility to manage them wisely. Compare *preservation ethic*.

conservation geneticist A scientist who studies genetic attributes of organisms, generally to infer the status of their *populations* in order to help conserve them.

Conservation Reserve Program U.S. policy in farm bills since 1985 that pays farmers to stop cultivating highly erodible cropland and instead place it in conservation reserves planted with grasses and trees.

conservation tillage *Agriculture* that limits the amount of tilling (plowing, disking, harrowing, or chiseling) of *soil*. Compare *no-till*.

consumer See *heterotroph*.

consumptive use *Freshwater* use in which water is removed from a particular *aquifer* or surface water body and is not returned to it. *Irrigation* for *agriculture* is an example of consumptive use. Compare *nonconsumptive use*.

continental collision The meeting of two tectonic plates of continental *lithosphere* at a *convergent plate boundary*, wherein the continental *crust* on both sides resists *subduction* and instead crushes together, bending, buckling, and deforming layers of rock and forcing portions of the buckled crust upward, often creating mountain ranges.

continental shelf The gently sloping underwater edge of a continent, varying in width from 100 m (330 ft) to 1,300 km (800 mi), with an average slope of 1.9 m/km (10 ft/mi).

continental slope The portion of the ocean floor that angles somewhat steeply downward, connecting the *continental shelf* to the deep ocean basin below.

contingent valuation A technique that uses surveys to determine how much people would be willing to pay to protect a resource or to restore it after damage has been done. Compare *expressed preference*.

contour farming The practice of plowing furrows sideways across a hillside, perpendicular to its slope, to help prevent the formation of rills and gullies. The technique is so named because the furrows follow the natural contours of the land.

contraception The deliberate attempt to prevent pregnancy despite sexual intercourse. Compare *birth control*.

control The portion of an *experiment* in which a *variable* has been left unmanipulated, to serve as a point of comparison with the *treatment*.

controlled burn See *prescribed burn*.

controlled experiment An *experiment* in which the effects of all *variables* are held constant, except the one whose effect is being tested by comparison of *treatment* and *control* conditions.

control rods Rods made of a metallic alloy that absorbs *neutrons*, which are placed in a *nuclear reactor* among the water-bathed *fuel rods* of uranium. Engineers move these control rods into and out of the water to maintain the *fission* reaction at the desired rate.

convective circulation A circular *current* (of air, water, magma, etc.) driven by temperature differences. In the atmosphere, warm air rises into regions of lower *atmospheric pressure*, where it expands and cools and then descends and becomes denser, replacing warm air that is rising. The air picks up heat and moisture near ground level and prepares to rise again, continuing the process.

convention A *treaty* or binding agreement among national governments.

conventional law International law that arises from *conventions*, or treaties, that nations agree to enter into. Compare *customary law*.

Convention on Biological Diversity An international treaty that aims to conserve *biodiversity*, use biodiversity in a *sustainable* manner, and ensure the fair distribution of biodiversity's benefits. Although many nations have agreed to the treaty (as of 2007, 188 nations had become parties to it), several others, including the United States, have not.

Convention on International Trade in Endangered Species of Wild Fauna and Flora (CITES) A 1973 treaty facilitated by the *United Nations* that protects endangered *species* by banning the international transport of their body parts.

convergent plate boundary Area where tectonic plates converge or come together. Can result in *subduction* or *continental collision*. Compare *divergent plate boundary* and *transform plate boundary*.

coral Tiny marine animals that build *coral reefs*. Corals attach to rock or existing reef and capture passing food with stinging tentacles. They also derive nourishment from photosynthetic symbiotic algae known as *zooxanthellae*.

coral reef A mass of calcium carbonate composed of the skeletons of tiny colonial marine organisms called *corals*.

core The innermost part of the Earth, made up mostly of iron, that lies beneath the *crust* and *mantle*.

Coriolis effect The apparent deflection of north-south air *currents* to a partly east-west direction, caused by the faster spin of regions near the equator than of regions near the poles as a result of Earth's rotation.

Cornucopian A worldview (or a person holding the worldview) that we will find ways to make Earth's natural resources meet all of our needs indefinitely and that human ingenuity will see us through any difficulty. In Greek mythology, cornucopia—literally "horn of plenty"—is the name for a magical goat's horn that overflowed with grain, fruit, and flowers. Compare *Cassandra*.

corporate average fuel efficiency (CAFE) standards   Miles-per-gallon fuel efficiency standards set by the U.S. Congress for auto manufacturers to meet, by a sales-weighted average of all models of the manufacturer's fleet.

correlation   A relationship among *variables*.

corridor   A passageway of protected land established to allow animals to travel between islands of protected *habitat*.

corrosive   Able to corrode metals. One criterion for defining *hazardous waste*.

cost-benefit analysis   A method commonly used by *neoclassical economists*, in which estimated costs for a proposed action are totaled and then compared to the sum of benefits estimated to result from the action.

cover crop   A crop that covers and anchors the *soil* during times between main crops, intended to reduce *erosion*.

criteria pollutants   Six *air pollutants*—*carbon monoxide, sulfur dioxide, nitrogen dioxide, tropospheric ozone, particulate matter,* and *lead*—for which the *Environmental Protection Agency* has established maximum allowable concentrations in ambient outdoor air because of the threats they pose to human health.

cropland   Land that humans use to raise plants for food and fiber.

crop rotation   The practice of alternating the kind of crop grown in a particular field from one season or year to the next.

crude birth rate   The number of births per 1,000 individuals for a given time period.

crude death rate   The number of deaths per 1,000 individuals for a given time period.

crude oil (petroleum)   A *fossil fuel* produced by the conversion of *organic compounds* by heat and pressure. Crude oil is a mixture of hundreds of different types of *hydrocarbon* molecules characterized by *carbon* chains of different lengths.

crust   The lightweight outer layer of the Earth, consisting of rock that floats atop the malleable *mantle*, which in turn surrounds a mostly iron *core*.

cultural hazard   Human health hazards that result from the place we live, our socioeconomic status, our occupation, or our behavioral choices. These include choosing to smoke cigarettes, or living or working with people who do. Compare *biological hazard; chemical hazard; physical hazard*.

culture   The overall ensemble of knowledge, beliefs, values, and learned ways of life shared by a group of people.

current   The flow of a liquid or gas in a certain direction.

customary law   International law that arises from long-standing practices, or customs, held in common by most *cultures*. Compare *conventional law*.

cyclone   A cyclonic storm that forms over the ocean but can do damage upon its arrival on land.

Daly, Herman   Contemporary American economist and well-known proponent of a *steady-state economy*.

dam   Any obstruction placed in a river or stream to block the flow of water so that water can be stored in a reservoir. Dams are built to prevent floods, provide drinking water, facilitate *irrigation*, and generate electricity.

Darwin, Charles (1809–1882)   English naturalist who proposed the concept of *natural selection* as a mechanism for *evolution* and as a way to explain the great variety of living things. See also *Wallace, Alfred Russel*.

data   Information, generally quantitative information.

debt-for-nature swap   A transaction in which a conservation organization pays off a portion of a developing nation's international debt in exchange for the nation's promise to set aside reserves, fund environmental education, and better manage protected areas.

deciduous   Term describing trees that lose their leaves each fall and remain dormant during winter, when hard freezes would endanger leaves.

decomposer   An organism, such as a fungus or bacterium, that breaks down leaf litter and other nonliving matter into simple constituents that can be taken up and used by plants. Compare *detritivore*.

deep ecology   A philosophy established in the 1970s based on principles of self-realization (the awareness that humans are inseparable from nature) and *biocentric* equality (the precept that all living beings have equal value). Holds that because we are truly inseparable from our *environment*, we must protect all other living things as we would protect ourselves.

Deepwater Horizon   The British Petroleum offshore drilling platform that sank in 2010, creating the largest oil spill in U.S. history.

deep-well injection   A *hazardous waste* disposal method in which a well is drilled deep beneath an area's *water table* into porous rock below an impervious *soil* layer. Wastes are then injected into the well, so that they will be absorbed into the porous rock and remain deep underground, isolated from *groundwater* and human contact. Compare *surface impoundment*.

deforestation   The clearing and loss of *forests*.

demand   The amount of a product people will buy at a given price if free to do so. Compare *supply*.

demographer   A scientist who studies human populations.

demographic fatigue   An inability on the part of governments to address overwhelming challenges related to population growth.

demographic transition   A theoretical *model* of economic and cultural change that explains the declining death rates and birth rates that occurred in Western nations as they became industrialized. The model holds that industrialization caused these rates to fall naturally by decreasing mortality and by lessening the need for large families. Parents would thereafter choose to invest in quality of life rather than quantity of children.

demography   A *social science* that applies the principles of *population ecology* to the study of statistical change in human *populations*.

denitrifying bacteria   Bacteria that convert the nitrates in *soil* or water to gaseous *nitrogen* and release it back into the *atmosphere*.

density-dependent factor   A *limiting factor* whose effects on a *population* increase or decrease depending on the *population density*. Compare *density-independent factor*.

density-independent factor   A *limiting factor* whose effects on a *population* are constant regardless of *population density*. Compare *density-dependent factor*.

deoxyribonucleic acid   See *DNA*.

dependent variable   The *variable* that is affected by manipulation of the *independent variable*.

deposition   The arrival of eroded *soil* at a new location. Compare *erosion*.

desalination (desalinization)   The removal of salt from seawater.

desert   The driest *biome* on Earth, with annual *precipitation* of less than 25 cm. Because deserts have relatively little vegetation to insulate them from temperature extremes, sunlight readily heats them in the daytime, but daytime heat is quickly lost at night, so temperatures vary widely from day to night and in different seasons.

desertification   A form of *land degradation* in which more than 10% of a land's productivity is lost due to *erosion, soil* compaction, forest removal, *overgrazing*, drought, *salinization, climate* change, water depletion, or other factors. Severe desertification can result in the expansion of desert areas or creation of new ones. Compare *land degradation; soil degradation*.

detritivore   An organism, such as a millipede or soil insect, that scavenges the waste products or dead bodies of other community members. Compare *decomposer*.

development   The use of natural resources for economic advancement (as opposed to simple subsistence, or survival).

dike   A long raised mound of earth erected along a river bank to protect against floods by holding rising water in the main channel.

directional drilling   Modern methods of drilling underground for *oil* and *natural gas* that enable the drill to be bent and curved as it descends, allowing many areas to be reached from a single drill pad and thus reducing the environmental impact of drilling on the surface.

directional selection   Mode of *natural selection* in which selection drives a feature in one direction rather than another—for example, toward larger or smaller, or faster or slower. Compare *disruptive selection; stabilizing selection*.

discounting   A practice in *neoclassical economics* by which short-term costs and benefits are granted more importance than long-term costs and benefits. Future effects are thereby "discounted," under the notion that an event far in the future should count much less than one in the present.

disruptive selection   Mode of *natural selection* in which a trait diverges from its starting condition in two or more directions. Compare *directional selection; stabilizing selection*.

distance effect   In *island biogeography theory*, the pattern that islands far from a mainland host fewer *species* because fewer species tend to find and colonize it.

divergent plate boundary   Area where tectonic plates push apart from one another as *magma* rises upward to the surface, creating new *lithosphere* as it cools and spreads. A prime example is the Mid-Atlantic Ridge. Compare *convergent plate boundary* and *transform plate boundary*.

DNA (deoxyribonucleic acid)   A double-stranded *nucleic acid* composed of four nucleotides, each of which contains a sugar (deoxyribose), a phosphate group, and a nitrogenous base. DNA carries the hereditary information for living organisms and is responsible for passing traits from parents to offspring. Compare *RNA*.

doldrums A region near the equator with little wind activity.

dose The amount of *toxicant* a test animal receives in a dose-response test. Compare *response*.

dose–response analysis A set of experiments that measure the *response* of test animals to different *doses* of a *toxicant*. The response is generally quantified by measuring the proportion of animals exhibiting negative effects.

dose–response curve A curve that plots the *response* of test animals to different *doses* of a *toxicant*, as a result of *dose–response analysis*.

downwelling In the ocean, the flow of warm surface water toward the ocean floor. Downwelling occurs where surface *currents* converge. Compare *upwelling*.

drainage basin See *watershed*.

driftnet Fishing net that spans large expanses of water, arrayed strategically to drift with currents so as to capture passing fish, and held vertical by floats at the top and weights at the bottom. Driftnetting captures substantial *by-catch* of dolphins, seals, sea turtles, and nontarget fish.

Dust Bowl An area that loses huge amounts of *topsoil* to wind *erosion* as a result of drought and/or human impact; first used to name the region in the North American Great Plains severely affected by drought and topsoil loss in the 1930s. The term is now also used to describe that historical event and others like it.

dynamic equilibrium The state reached when processes within a *system* are moving in opposing directions at equivalent rates so that their effects balance out.

earthquake A release of energy occurring as Earth relieves accumulated pressure between masses of lithosphere, and which results in shaking at the surface.

ecocentrism A philosophy that considers actions in terms of their damage or benefit to the integrity of whole ecological *systems*, including both *biotic* and *abiotic* elements. For an ecocentrist, the well-being of an individual organism— human or otherwise—is less important than the long-term well-being of a larger integrated ecological system.

ecofeminism A philosophy holding that the patriarchal (male-dominated) structure of society is a root cause of both social and environmental problems. Ecofeminists hold that a *worldview* traditionally associated with women, which interprets the world in terms of interrelationships and cooperation, is more in tune with nature than a worldview traditionally associated with men, which interprets the world in terms of hierarchies and competition.

ecolabeling The practice of designating on a product's label how the product was grown, harvested, or manufactured, so that consumers buying it are aware of the processes involved and can differentiate between brands that use processes believed to be *environmentally* beneficial (or less harmful than others) and those that do not.

ecological economics A developing school of *economics* that applies the principles of *ecology* and *systems* thinking to the description and analysis of *economies*. Compare *environmental economics; neoclassical economics*.

ecological footprint The cumulative amount of land and water required to provide the raw materials a person or *population* consumes and to dispose of or *recycle* the *waste* that is produced.

ecological modeling The practice of constructing and testing *models* that aim to explain and predict how ecological systems function.

ecological restoration Efforts to reverse the effects of human disruption of ecological systems and to restore *communities* to their condition before the disruption. The practice that applies principles of *restoration ecology*.

ecology The *science* that deals with the distribution and abundance of organisms, the interactions among them, and the interactions between organisms and their *abiotic environments*.

economically recoverable Extractable such that income from a resource's sale exceeds the costs of extracting it. Applied to *fossil fuel* deposits. Compare *technically recoverable*.

economic growth An increase in an economy's activity; that is, an increase in the production and consumption of goods and services.

economics The study of how we decide to use scarce resources to satisfy the demand for *goods* and *services*.

economy A social *system* that converts resources into *goods* and *services*.

ecosystem All organisms and nonliving entities that occur and interact in a particular area at the same time.

ecosystem-based management The attempt to manage the harvesting of resources in ways that minimize impact on the *ecosystems* and ecological processes that provide the resources.

ecosystem diversity The number and variety of ecosystems in a particular area. One way to express *biodiversity*. Related concepts consider the geographic arrangement of *habitats*, *communities*, or *ecosystems* at the landscape level, including the sizes, shapes, and interconnectedness of patches of these entities.

ecosystem ecology The study of how the living and nonliving components of *ecosystems* interact.

ecosystem service An essential service an *ecosystem* provides that supports life and makes *economic* activity possible. For example, ecosystems naturally purify air and water, cycle *nutrients*, provide for plants to be *pollinated* by animals, and serve as receptacles and *recycling* systems for the *waste* generated by our economic activity.

ecotone A transitional zone where *ecosystems* meet.

ecotourism Visitation of natural areas for tourism and recreation. Most often involves tourism by more-affluent people, which may generate *economic* benefits for less-affluent communities near natural areas and thus provide economic incentives for conservation of natural areas.

ED$_{50}$ (effective dose–50%) The amount of a *toxicant* it takes to affect 50% of a *population* of test animals. Compare *threshold dose; LD$_{50}$*.

edge effect An impact on organisms, populations, or communities that results because conditions along the edge of a habitat fragment differ from conditions in the interior.

effluent Water that flows out of a facility such as a wastewater treatment plant or power plant.

E horizon The layer of *soil* that lies below the *A horizon* and above the *B horizon*. The letter E stands for *eluviation*, meaning "loss," and the E horizon is characterized by the loss of certain *minerals* through *leaching*. It is sometimes called the "zone of leaching." Compare *C horizon; O horizon; R horizon*.

electricity A secondary form of energy that can be transferred over long distances and applied for a variety of uses.

electrolysis A process in which electrical current is passed through a *compound* to release *ions*. Electrolysis offers one way to produce *hydrogen* for use as fuel: Electrical current is passed through water, splitting the water *molecules* into hydrogen and *oxygen atoms*.

electron A negatively charged particle that surrounds the nucleus of an *atom*.

electronic waste (e-waste) Discarded electronic products such as computers, monitors, printers, DVD players, cell phones, and other devices. *Heavy metals* in these products mean that this waste may be judged hazardous.

element A fundamental type of matter; a chemical substance with a given set of properties, which cannot be broken down into substances with other properties. Chemists currently recognize 92 elements that occur in nature, as well as more than 20 others that have been artificially created.

El Niño The exceptionally strong warming of the eastern Pacific Ocean that occurs every 2 to 7 years and depresses local fish and bird *populations* by altering the marine *food web* in the area. Originally, the name that Spanish-speaking fishermen gave to an unusually warm surface *current* that sometimes arrived near the Pacific coast of South America around Christmas time. Compare *La Niña*.

El Niño–Southern Oscillation (ENSO) A systematic shift in atmospheric pressure, sea-surface temperature, and ocean circulation in the tropical Pacific Ocean. ENSO cycles give rise to *El Niño* and *La Niña*.

emergent property A characteristic that is not evident in a *system*'s components.

emergent tree An especially tall tree that protrudes above the *canopy* of a *tropical rainforest*.

Emerson, Ralph Waldo (1803–1882) American author, poet, and philosopher who espoused transcendentalism, a philosophy that views nature as a direct manifestation of the divine, and who promoted a holistic view of nature among the public.

emigration The departure of individuals from a *population*.

emissions trading system A *permit-trading* system for emissions in which a government issues marketable emissions permits to conduct environmentally harmful activities. Under a *cap-and-trade* system, the government determines an acceptable level of *pollution* and then issues permits to pollute. A company receives credit for amounts it does not emit and can then sell this credit to other companies. Compare *cap-and-trade*.

Endangered Species Act (ESA) The primary *legislation*, enacted in 1973, for protecting *biodiversity* in the United States. It forbids the government and private citizens from taking actions (such as developing land) that would destroy endangered *species* or their *habitats*, and it prohibits trade in products made from endangered species.

endemic Native or restricted to a particular geographic region. An endemic species occurs in one area and nowhere else on Earth.

endocrine disruptor A *toxicant* that interferes with the *endocrine (hormone) system*.

endocrine system The body's *hormone* system.

energy The capacity to change the position, physical composition, or temperature of matter; a force that can accomplish work.

energy conservation The practice of reducing *energy* use as a way of extending the lifetime of our *fossil fuel* supplies, of being less wasteful, and of reducing our impact on the *environment*. Conservation can result from behavioral decisions or from technologies that demonstrate *energy efficiency*.

energy conversion efficiency The ratio of useful output of *energy* to the amount that needs to be input. See also *EROI; net energy*.

energy efficiency The ability to obtain a given result or amount of output while using less energy input. Technologies permitting greater energy efficiency are one main route to *energy conservation*.

enhanced geothermal systems A new approach whereby engineers drill deeply into rock, fracture it, pump in water, and then pump it out once it is heated below ground. This approach would enable us to obtain geothermal energy in any location.

entropy The degree of disorder in a substance, *system*, or process. See *second law of thermodynamics*.

environment The sum total of our surroundings, including all of the living things and nonliving things with which we interact.

environmental economics A developing school of *economics* that modifies the principles of *neoclassical economics* to address environmental challenges. An environmental economist believes that we can attain *sustainability* within our current economic *systems*. Whereas ecological economists call for revolution, environmental economists call for reform. Compare *ecological economics; neoclassical economics*.

environmental ethics The application of *ethical standards* to environmental questions.

environmental health The study of environmental factors that influence human health and quality of life and the health of *ecological* systems essential to environmental quality and long-term human well-being.

environmental impact statement (EIS) A report of results from detailed studies that assess the potential effects on the *environment* that would likely result from development projects or other actions undertaken by the government.

environmentalism A social movement dedicated to protecting the natural world.

environmental justice The fair and equitable treatment of all people with respect to environmental policy and practice, regardless of their income, race, or ethnicity. Responds to the perception that minorities and the poor suffer more pollution than whites and the rich.

Environmental Performance Index (EPI) An index that rates nations on their progress toward environmental sustainability, using data from 16 indicators of environmental conditions for which governments can be held accountable.

environmental policy *Public policy* that pertains to human interactions with the *environment*. It generally aims to regulate resource use or reduce *pollution* to promote human welfare and/or protect natural systems.

Environmental Protection Agency (EPA) An administrative agency created by executive order in 1970. The EPA is charged with conducting and evaluating research, monitoring environmental quality, setting standards, enforcing those standards, assisting the states in meeting standards and goals, and educating the public.

environmental resistance The collective force of *limiting factors*, which together stabilize a *population size* at its *carrying capacity*.

environmental science The study of how the natural world functions and how humans and the *environment* interact.

environmental studies An academic *environmental science* program that heavily incorporates the social sciences as well as the natural sciences.

environmental toxicology The study of *toxicants* that come from or are discharged into the *environment*, including the study of health effects on humans, other animals, and *ecosystems*.

enzyme A chemical that catalyzes a chemical reaction.

epidemiological study A study that involves large-scale comparisons among groups of people, usually contrasting a group known to have been exposed to some *toxicant* and a group that has not.

EROI (Energy Returned on Investment) The ratio determined by dividing the quantity of *energy* returned from a process by the quantity of energy invested in the process. Higher EROI ratios mean that more energy is produced from each unit of energy invested. See also *energy conversion efficiency; net energy*.

erosion The removal of material from one place and its transport to another by the action of wind or water.

estuary An area where a river flows into the ocean, mixing *fresh water* with salt water.

ethanol The alcohol in beer, wine, and liquor, produced as a *biofuel* by fermenting biomass, generally from *carbohydrate*-rich crops such as corn.

ethical standard A criterion that helps differentiate right from wrong.

ethics The study of good and bad, right and wrong. The term can also refer to a person's or group's set of moral principles or values.

European Union (EU) Political and economic organization formed after World War II to promote Europe's economic and social progress. As of 2007, the EU consisted of 27 member nations.

eutrophic Term describing a water body that has high-nutrient and low-oxygen conditions. Compare *oligotrophic*.

eutrophication The process of *nutrient* enrichment, increased production of organic matter, and subsequent *ecosystem* degradation in a water body.

evaporation The conversion of a substance from a liquid to a gaseous form.

even-aged Condition of timber plantations—generally *monocultures* of a single *species*—in which all trees are of the same age. Most *ecologists* view plantations of even-aged stands more as crop *agriculture* than as ecologically functional *forests*. Compare *uneven-aged*.

evenness See *relative abundance*.

evolution Genetically based change in the appearance, functioning, and/or behavior of organisms across generations, often by the process of *natural selection*.

evolutionary arms race A duel of escalating adaptations between species. Like rival nations racing to stay ahead of one another in military technology, *host* and *parasite* may repeatedly evolve new responses to the other's latest advance.

e-waste See *electronic waste*.

executive branch The branch of the U.S. government that is headed by the president and that includes administrative agencies. Among other powers, the president may approve (enact) or reject (veto) legislation and issue executive orders. Compare *judicial branch; legislative branch*.

executive order A specific legal instruction for government agencies ordered by the U.S. president.

exotic Non-native to an area (as, an exotic organism).

experiment An activity designed to test the validity of a *hypothesis* by manipulating *variables*. See *manipulative experiment* and *natural experiment*.

exploitative interaction A species interaction in which one participant benefits while another is harmed; that is, one species exploits the other. Such interactions include *predation, parasitism*, and *herbivory*.

exploratory drilling Drilling that takes place after a *fossil fuel* deposit has been identified, in order to gauge how much of the fuel exists and whether extraction will prove worthwhile. Involves drilling small holes that descend to great depths.

exponential growth The increase of a *population* (or of anything) by a fixed percentage each year.

expressed preference An approach in environmental economics that quantifies nonmarket values by asking people how much they would pay for specified items. *Contingent valuation* is an example. Compare *revealed preference*.

external cost A negative *externality*; a cost borne by someone not involved in an economic transaction. Examples include harm to citizens from *water pollution* or *air pollution* discharged by nearby factories.

externality A cost or benefit of a transaction that affects people other than the buyer or seller.

extinction The disappearance of an entire *species* from the face of the Earth. Compare *extirpation*.

extirpation The disappearance of a particular *population* from a given area, but not the entire *species* globally. Compare *extinction*.

extrusive Term for igneous rock formed when magma is ejected from a *volcano* and cools quickly (e.g., basalt).

factory farm See *feedlot*.

factory fishing A highly industrialized approach to commercial fishing, employing fossil fuels, huge vessels, and powerful new technologies to capture fish in immense volumes. Factory fishing vessels even process and freeze their catches while at sea.

family planning The effort to plan the number and spacing of one's children, so as to offer children and parents the best quality of life possible.

farmers' markets Markets at which local farmers and food producers sell fresh locally grown items.

fee-and-dividend A program of *carbon taxes* in which proceeds from the taxes are paid to consumers as a tax refund or "dividend." This way, if polluters pass their costs along to consumers, consumers will not lose money.

feedback loop A circular process in which a *system*'s output serves as input to that same system. See *negative feedback loop; positive feedback loop*.

feed-in tariff A system whereby utilities are mandated to buy power from anyone who can generate power from renewable energy sources and feed it into the electric grid. When prices are set at a premium, this can act as a powerful mechanism to promote the spread of renewable energy technologies. Compare *net metering*.

feedlot A huge barn or outdoor pen designed to deliver *energy*-rich food to animals living at extremely high densities. Also called a *factory farm* or *concentrated animal feeding operation (CAFO)*.

Ferrel cell One of a pair of cells of *convective circulation* between 30° and 60° north and south latitude that influence global *climate* patterns. Compare *Hadley cell; polar cell*.

fertilizer A substance that promotes plant growth by supplying essential *nutrients* such as *nitrogen* or *phosphorus*.

first law of thermodynamics Physical law stating that *energy* can change from one form to another but cannot be created or lost. The total energy in the universe remains constant and is said to be conserved.

flagship species A *species* that has wide appeal with the public and that can be used to promote conservation efforts that also benefit other less charismatic species.

flat-plate solar collectors See *solar panels*.

flexible fuel vehicle A vehicle that runs on fuel that is a mixture of *ethanol* and gasoline, such as E-85, a mix of 85% ethanol and 15% gasoline.

flooding The spillage of water over a river's banks due to heavy rain or snowmelt.

floodplain The region of land over which a river has historically wandered and periodically floods.

flux The movement of nutrients among *pools* or *reservoirs* in a *nutrient cycle*.

food chain A linear series of feeding relationships. As organisms feed on one another, energy is transferred from lower to higher *trophic levels*. Compare *food web*.

food security An adequate, reliable, and available food supply to all people at all times.

food web A visual representation of feeding interactions within an *ecological community* that shows an array of relationships between organisms at different *trophic levels*. Compare *food chain*.

forensics See *forensic science*.

forensic science The scientific analysis of evidence to make an identification or answer a question, most often relating to a crime or accident. *Conservation biologists* are now employing forensic science to protect *species* from illegal harvesting.

forest Any ecosystem characterized by a high density of trees.

forester Professionals who manage *forests* through the practice of *forestry*.

forestry The professional management of *forests*.

forest type Categories of *forest* defined by the predominant tree *species*.

formaldehyde A common synthetically made *volatile organic compound* that causes a variety of health impacts.

fossil The remains, impression, or trace of an animal or plant of past geological ages that has been preserved in rock or *sediments*.

fossil fuel A *nonrenewable natural resource*, such as *crude oil, natural gas*, or *coal*, produced by the decomposition and compression of organic matter from ancient life.

fossil record The cumulative body of *fossils* worldwide, which paleontologists study to infer the history of past life on Earth.

Fourth Assessment Report A 2007 report from the *Intergovernmental Panel on Climate Change (IPCC)* that summarizes thousands of scientific studies, documenting observed trends in surface temperature, precipitation patterns, snow and ice cover, sea levels, storm intensity, and other factors. It also predicts future changes in these phenomena under a range of emission scenarios; addresses impacts of *climate change* on wildlife, ecosystems, and human societies; and discusses strategies we might pursue in response. The Fourth Assessment represents the consensus of scientific climate research from around the world.

Framework Convention on Climate Change (FCCC) International agreement to reduce *greenhouse gas* emissions to 1990 levels by the year 2000, signed by nations represented at the 1992 Earth Summit convened in Rio de Janeiro by the *United Nations*. The FCCC called for a voluntary, nation-by-nation approach, but by the late 1990s it had become apparent that it would not succeed. Its imminent failure sparked introduction of the *Kyoto Protocol*.

free rider A party that fails to invest in controlling *pollution* or carrying out other *environmentally* responsible activities and instead relies on the efforts of other parties to do so. For example, a factory that fails to control its emissions gets a "free ride" on the efforts of other factories that do make the sacrifices necessary to reduce emissions.

fresh water Water that is relatively pure, holding very few dissolved salts.

freshwater marsh A type of *wetland* in which shallow water allows plants such as cattails to grow above the water surface. Compare *swamp; bog*.

front The boundary between air masses that differ in temperature and moisture (and therefore density). See *warm front; cold front*.

fuel rods Rods of uranium that supply the fuel for nuclear *fission*, and are kept bathed in a *moderator* in a *nuclear reactor*.

fumarole A natural spray of heated steam pressurized below ground that erupts through the surface. Compare *geyser*.

fundamental niche The full *niche* of a *species*. Compare *realized niche*.

fungicide A type of chemical *pesticide* that kills fungi.

gasification A process in which *biomass* is vaporized at extremely high temperatures in the absence of *oxygen*, creating a gaseous mixture including *hydrogen, carbon monoxide*, and *methane*, in order to produce *biopower* or *biofuels*. In coal gasification, *coal* is converted to a syngas by reacting it with oxygen and steam at a high temperature.

geology The scientific study of Earth's physical features, processes, and history.

gene A stretch of *DNA* that represents a unit of hereditary information.

genera Plural of *genus*.

General Land Ordinances of 1785 and 1787 Laws that gave the U.S. government the right to manage Western lands and created a grid system for surveying them and readying them for private ownership.

General Mining Act of 1872 U.S. law that legalized and promoted mining by private individuals on public lands for just $5 per acre, subject to local customs, with no government oversight.

generalist A *species* that can survive in a wide array of *habitats* or use a wide array of resources. Compare *specialist*.

genetically modified (GM) organism An organism that has been *genetically engineered* using a technique called *recombinant DNA* technology.

genetic diversity A measurement of the differences in *DNA* composition among individuals within a given *species*.

genetic engineering Any process scientists use to manipulate an organism's genetic material in the lab by adding, deleting, or changing segments of its *DNA*.

genome The entirety of all an organism's genes.

Genuine Progress Indicator (GPI) An *economic* indicator introduced in 1995 that attempts to differentiate between desirable and undesirable economic activity. The GPI accounts for benefits such as *volunteerism* and for costs such as *environmental* degradation and social upheaval. Compare *Gross Domestic Product (GDP)*.

genus A taxonomic level in the Linnaean classification system that is above *species* and below family. A genus is made up of one or more closely related species.

geographic information system (GIS) Computer software that takes multiple types of data (for instance, on geology, hydrology, vegetation, animal species, and human development) and overlays them on a common set of geographic coordinates. The idea is to create a complete picture of a landscape and to analyze how elements of the different datasets are arrayed spatially and how they may be correlated. A common tool of geographers, landscape ecologists, resource managers, and conservation biologists.

geothermal energy Renewable *energy* that is generated deep within Earth. The radioactive decay of elements amid the extremely high pressures and temperatures at depth generate heat that rises to the surface in magma and through fissures and cracks. Where this energy heats *groundwater*, natural eruptions of heated water and steam are sent up from below.

geyser A natural spurt of heated groundwater and steam pressurized and sent up from below ground that erupts through the surface.

glaciation The extension of ice sheets from the polar regions far into Earth's temperate zones during cold periods of Earth's history.

global climate change Any change in aspects of Earth's *climate*, such as temperature, *precipitation*, and storm intensity. Generally refers today to the current warming trend in global temperatures and associated climatic changes. Compare *global warming*.

global warming An increase in Earth's average surface temperature. The term is most frequently used in reference to the pronounced warming trend of recent years and decades. Global warming is one aspect of *global climate change*, and in turn drives other components of climate change.

global warming potential A quantity that specifies the ability of one *molecule* of a given *greenhouse gas* to contribute to atmospheric warming, relative to *carbon dioxide*.

good A material commodity manufactured for and bought by individuals and businesses.

gray water Wastewater from showers and sinks.

greenbelt A long and wide corridor of parkland, often encircling an entire urban area.

green building (1) A structure that minimizes the ecological footprint of its construction and operation by using sustainable materials, using minimal energy and water, reducing health impacts, limiting pollution, and recycling waste.

(2) The pursuit of constructing or renovating such buildings.

green-collar jobs Jobs resulting from new employment opportunities in a more sustainably oriented economy, such as jobs in renewable energy.

greenhouse effect The warming of Earth's surface and *atmosphere* (especially the *troposphere*) caused by the *energy* emitted by *greenhouse gases*.

greenhouse gas A gas that absorbs infrared radiation released by Earth's surface and then warms the surface and *troposphere* by emitting *energy*, thus giving rise to the *greenhouse effect*. Greenhouse gases include *carbon dioxide* (CO_2), water vapor, ozone (O_3), nitrous oxide (N_2O), halocarbon gases, and *methane* (CH_4).

green manure Organic *fertilizer* composed of freshly dead plant material.

Green Revolution An intensification of the industrialization of *agriculture* in the developing world in the latter half of the 20th century that has dramatically increased crop yields produced per unit area of farmland. Practices include devoting large areas to *monocultures* of crops specially bred for high yields and rapid growth; heavy use of *fertilizers, pesticides,* and *irrigation* water; and sowing and harvesting on the same piece of land more than once per year or per season.

green tax A levy on *environmentally* harmful activities and products aimed at providing a market-based incentive to correct for *market failure*. Compare *subsidy*.

greenwashing A public-relations effort by a corporation or institution to mislead customers or the public into thinking it is acting more sustainably than it actually is.

greenway A strip of park land that connects parks or neighborhoods; often located along rivers, streams, or canals.

Gross Domestic Product (GDP) The total monetary value of final *goods* and *services* produced in a country each year. The GDP sums all *economic* activity, whether good or bad, and does not account for benefits such as volunteerism or for *external costs* such as *environmental* degradation and social upheaval. Compare *Genuine Progress Indicator (GPI)*.

gross primary production The *energy* that results when *autotrophs* convert solar energy (sunlight) to energy of chemical bonds in sugars through *photosynthesis*. Autotrophs use a portion of this production to power their own metabolism, which entails oxidizing *organic compounds* by *cellular respiration*. Compare *net primary production*.

groundfish A name given to *benthic* fish that live or eat along the bottom, such as cod, halibut, pollock, haddock, and flounder.

ground-level ozone See *tropospheric ozone*.

ground source heat pump A pump that harnesses *geothermal energy* from near-surface sources of earth and water, in order to heat and cool buildings. Operates on the principle that temperatures below ground are more stable than temperatures above ground.

groundwater Water held in aquifers underground. Compare *surface water*.

growth rate The net change in a *population's* size, per 1,000 individuals. Calculated by adding the *crude birth rate* to the *immigration* rate and then subtracting the *crude death rate* and the *emigration* rate, each expressed as the number per 1,000 individuals per year.

Haber-Bosch process A process to synthesize ammonia on an industrial scale. Developed by German chemists Fritz Haber and Carl Bosch, the process has enabled humans to double the natural rate of *nitrogen fixation* on Earth and thereby increase *agricultural* productivity, but it has also dramatically altered the *nitrogen cycle*.

habitat The specific *environment* in which an organism lives, including both *biotic* and *abiotic factors*.

habitat conservation plan A cooperative agreement that allows landowners to harm threatened or endangered species in some ways if they voluntarily improve habitat for the species in others.

habitat fragmentation The process by which an expanse of natural *habitat* becomes broken up into discontinuous fragments, often as a result of farming, logging, road-building, and other types of human development and land use.

habitat heterogeneity The diversity of habitats in a given area.

habitat selection The process by which organisms select *habitats* from among the range of options they encounter.

habitat use The process by which organisms use *habitats* from among the range of options they encounter.

Hadley cell One of a pair of cells of *convective circulation* between the equator and 30° north and south latitude that influence global *climate* patterns. Compare *Ferrel cell; polar cell*.

half-life The amount of time it takes for one-half the atoms of a *radioisotope* to emit radiation and decay. Different radioisotopes have different half-lives, ranging from fractions of a second to billions of years.

halocarbon A class of human-made chemical *compounds* derived from simple *hydrocarbons* in which *hydrogen* atoms are replaced by halogen atoms such as bromine, fluorine, or chlorine. Many halocarbons are *ozone-depleting substances* and/or *greenhouse gases*.

harmful algal bloom A *population* explosion of toxic algae caused by excessive *nutrient* concentrations.

hazardous waste *Waste* that is toxic, chemically reactive, flammable, or corrosive. Compare *industrial solid waste; municipal solid waste*.

Healthy Forests Restoration Act Controversial legislation initiated by the George W. Bush administration and passed by Congress in 2003 that promotes the physical removal of small trees, underbrush, and dead trees on *national forests* by timber companies in order to reduce the chance of fire. Critics contend that it focuses too little on *prescribed burning*, encourages *salvage logging*, subsidizes private companies, and reduces citizen participation in forest management decisions.

heat capacity A measure of the heat energy required to increase the temperature of a given substance by a given amount.

herbicide A type of chemical *pesticide* that kills plants.

herbivore An organism that consumes plants. Compare *carnivore; omnivore*.

herbivory The consumption of plants by animals.

heterotroph (consumer) An organism that consumes other organisms. Includes most animals, as well as fungi and microbes that decompose organic matter.

high-pressure system An air mass with elevated *atmospheric pressure*, containing air that descends, typically bringing fair *weather*. Compare *low-pressure system*.

historical sciences Sciences, such as cosmology or paleontology, that study phenomena that change across historical time.

homeostasis The tendency of a *system* to maintain constant or stable internal conditions.

Homestead Act of 1862 U.S. law that allowed any citizen to claim, for a $16 fee, 65 ha (160 acres) of public land by living there for 5 years and cultivating the land or building a home. A waiting period of only 14 months was available to those who could pay $176 for the land.

horizon A distinct layer of *soil*. See *A horizon; B horizon; C horizon; E horizon; O horizon; R horizon*.

hormone A chemical messenger that travels though the bloodstream to stimulate growth, development, and sexual maturity; and regulate brain function, appetite, sexual drive, and many other aspects of physiology and behavior.

host The organism in a parasitic relationship that suffers harm while providing the *parasite* nourishment or some other benefit.

hotspot A localized area where plugs of molten rock from the *mantle* erupt through Earth's *crust*.

Hubbert's peak The peak in production of *crude oil* in the United States, which occurred in 1970 just as Shell Oil geologist M. King Hubbert had predicted in 1956.

humus A dark, spongy, crumbly mass of material made up of complex organic compounds, resulting from the partial decomposition of organic matter.

hurricane A cyclonic storm that forms over the ocean but can do damage upon its arrival on land. A type of *cyclone* or *typhoon* that usually forms over the Atlantic Ocean.

hydrocarbon An *organic compound* consisting solely of *hydrogen* and *carbon atoms*.

hydroelectric power (hydropower) The generation of electricity using the *kinetic energy* of moving water.

hydrogen The chemical *element* with 1 proton. The most abundant element in the universe. Also a possible fuel for our future economy.

hydrogenase An enzyme that can trigger algae to stop producing *oxygen* as a metabolic by-product and start releasing *hydrogen* instead.

hydrogen bond A special type of interaction in which the oxygen atom of one water molecule is weakly attracted to one or two hydrogen atoms of another.

hydrologic cycle The flow of water—in liquid, gaseous, and solid forms—through our *biotic* and *abiotic environment*.

hydropower See *hydroelectric power*.

hydrosphere All water—salt or fresh, liquid, ice, or vapor—in surface bodies, underground, and in the *atmosphere*. Compare *biosphere; lithosphere*.

hydrothermal vent Location in the deep ocean where heated water spurts from the seafloor, carrying *minerals* that precipitate to form rocky structures. Unique and recently discovered *ecosystems* cluster around these vents; tubeworms, shrimp, and other creatures here use symbiotic bacteria to derive their energy from chemicals in the heated water rather than from sunlight.

hypothesis A statement that attempts to explain a phenomenon or answer a *scientific* question. Compare *theory*.

hypoxia The condition of extremely low dissolved *oxygen* concentrations in a body of water.

igneous rock   One of the three main categories of rock. Formed from cooling *magma*. Granite and basalt are examples of igneous rock. Compare *metamorphic rock; sedimentary rock.*

ignitable   Easily able to catch fire. One criterion for defining *hazardous waste.*

immigration   The arrival of individuals from outside a *population.*

inbreeding depression   A state that occurs in a *population* when genetically similar parents mate and produce weak or defective offspring as a result.

incineration   A controlled process of burning solid waste for disposal in which mixed garbage is combusted at very high temperatures. Compare *sanitary landfill.*

independent variable   The *variable* that the scientist manipulates in a *manipulative experiment.*

indoor air pollution   *Air pollution* that occurs indoors.

industrial agriculture   A form of *agriculture* that uses large-scale mechanization and *fossil fuel* combustion, enabling farmers to replace horses and oxen with faster and more powerful means of cultivating, harvesting, transporting, and processing crops. Other aspects include *irrigation* and the use of *inorganic fertilizers.* Use of chemical herbicides and *pesticides* reduces *competition* from weeds and *herbivory* by insects. Compare *traditional agriculture.*

industrial ecology   A holistic approach to industry that integrates principles from engineering, chemistry, *ecology, economics,* and other disciplines and seeks to redesign industrial *systems* in order to reduce resource inputs and minimize inefficiency.

industrial revolution   The shift in the mid-1700s from rural life, animal-powered agriculture, and manufacturing by craftsmen to an urban society powered by *fossil fuels* such as *coal* and *crude oil.* Compare *agricultural revolution.*

industrial smog   Gray-air smog caused by the incomplete combustion of *coal* or oil when burned. Compare *photochemical smog.*

industrial solid waste   Nonliquid *waste* that is not especially hazardous and that comes from production of consumer goods, mining, *petroleum* extraction and *refining,* and *agriculture.* Compare *hazardous waste; municipal solid waste.*

industrial stage   The third stage of the *demographic transition* model, characterized by falling birth rates that close the gap with falling death rates and reduce the rate of *population* growth. Compare *pre-industrial stage; post-industrial stage; transitional stage.*

infectious disease   A disease in which a pathogen attacks a host.

inherent value   See *intrinsic value.*

inorganic fertilizer   A *fertilizer* that consists of mined or synthetically manufactured mineral supplements. Inorganic fertilizers are generally more susceptible than *organic fertilizers* to *leaching* and *runoff* and may be more likely to cause unintended off-site impacts.

insecticide   A type of chemical *pesticide* that kills insects.

instrumental value   Value ascribed to something for the pragmatic benefits it brings us if we put it to use. Synonymous with *utilitarian value.*

integrated pest management (IPM)   The use of multiple techniques in combination to achieve long-term suppression of pests, including *biocontrol,* use of *pesticides,* close monitoring of *populations, habitat* alteration, *crop rotation, transgenic* crops, alternative tillage methods, and mechanical pest removal.

intercropping   Planting different types of crops in alternating bands or other spatially mixed arrangements.

interdisciplinary field   A field that borrows techniques from several more traditional fields of study and brings together research results from these fields into a broad synthesis.

interest group   A small group of people seeking private gain that may work against the larger public interest.

interglacial   Term describing relatively warm periods in Earth history that occur between periods of *glaciation.*

Intergovernmental Panel on Climate Change (IPCC)   An international panel of *climate* scientists and government officials established in 1988 by the *United Nations Environment Programme* and the World Meteorological Organization. The IPCC's mission is to assess and synthesize scientific research on *global climate change* and to offer guidance to the world's policymakers. The IPCC's 2007 *Fourth Assessment Report* summarizes current and projected future global trends in climate, and represents the consensus of climate scientists around the world.

International Polar Year   A large international scientific program coordinating research in the Arctic and Antarctic in 2007–2009.

interspecific competition   *Competition* that takes place among members of two or more different *species.* Compare *intraspecific competition.*

intertidal   Of, relating to, or living along shorelines between the highest reach of the highest *tide* and the lowest reach of the lowest tide.

intraspecific competition   *Competition* that takes place among members of the same *species.* Compare *interspecific competition.*

intrinsic value   Value ascribed to something for its intrinsic worth; the notion that the thing has a right to exist and is valuable for its own sake.

intrusive   Term for igneous rock formed when magma cools slowly while it is well below Earth's surface (e.g., granite).

invasive   Spreading widely, quickly, and becoming dominant in a *community* (as, an *invasive species*).

invasive species   A *species* that spreads widely and rapidly becomes dominant in a *community,* interfering with the community's normal functioning.

inversion layer   In a *temperature inversion,* the band of air in which temperature rises with altitude (instead of falling with altitude, as temperature does normally).

ion   An electrically charged *atom* or combination of atoms.

ionic compound (salt)   An association of *ions* that are bonded electrically in an ionic bond.

IPAT model   A formula that represents how humans' total impact (I) on the *environment* results from the interaction among three factors: *population* (P), affluence (A), and technology (T).

irrigation   The artificial provision of water to support *agriculture.*

island biogeography theory   A *theory* that was initially applied to oceanic islands to explain how *species* come to be distributed among them. Since its development, researchers have increasingly applied the theory to islands of *habitat* (patches of one type of habitat isolated within vast "seas" of others). Aspects of the theory include *immigration* and *extinction* rates, the effect of island size, and the effect of distance from the mainland. Full name is the equilibrium theory of island biogeography.

isotope   One of several forms of an *element* having differing numbers of *neutrons* in the nucleus of its *atoms.* Chemically, isotopes of an element behave almost identically, but they have different physical properties because they differ in mass.

judicial branch   The branch of the U.S. government, consisting of the Supreme Court and various lower courts, that is charged with interpreting the law. Compare *executive branch; legislative branch.*

kelp   Large brown algae or seaweed that can form underwater "forests," providing habitat for marine organisms.

kerogen   A substance derived from deeply buried organic matter that acts as a source material for both *natural gas* and *crude oil.*

keystone species   A *species* that has an especially far-reaching effect on a *community.*

kinetic energy   *Energy* of motion. Compare *potential energy.*

K–selected   Term denoting a *species* with low *biotic potential* whose members produce a small number of offspring and take a long time to gestate and raise each of their young, but invest heavily in promoting the survival and growth of these few offspring. *Populations* of K–selected species are generally regulated by *density-dependent factors.* Compare *r–selected.*

kwashiorkor   A form of *malnutrition* that results from a high-*starch* diet with inadequate *protein* or *amino acids.* In children, causes bloating of the abdomen, deterioration and discoloration of hair, mental disability, immune suppression, developmental delays, anemia, and reduced growth.

Kyoto Protocol   An agreement drafted in 1997 that calls for reducing, by 2012, emissions of six *greenhouse gases* to levels lower than their levels in 1990. Although the United States has refused to ratify the protocol, it came into force in 2005 when Russia ratified it, the 127th nation to do so.

lahar   A large flow of destabilized mud racing downslope.

land degradation   A general deterioration of land that diminishes its productivity and biodiversity, impairs the functioning of its ecosystems, and reduces the ecosystem services the land can offer us. Compare *desertification; soil degradation.*

landfill gas   A mix of gases that consists of roughly half *methane* produced by anaerobic decomposition deep inside *landfills.*

landrace   A locally adapted domesticated variety of an agricultural crop native to a particular area.

landscape ecology   The study of how landscape structure affects the abundance, distribution, and interaction of organisms. This approach to the study of organisms and their *environments* at the landscape scale focuses on broad geographical areas that include multiple *ecosystems.*

landslide   The collapse and downhill flow of large amounts of rock or soil. A severe and sudden form of *mass wasting.*

land trust   Local or regional organization that preserves lands valued by its members. In most cases, land trusts purchase land outright with the aim of preserving it in its natural condition. The Nature Conservancy may be considered the world's largest land trust.

La Niña   An exceptionally strong cooling of surface water in the equatorial Pacific Ocean that occurs every 2 to 7 years and has widespread climatic consequences. Compare *El Niño*.

latitudinal gradient The increase in *species richness* as one approaches the equator. This pattern of variation with latitude has been one of the most obvious patterns in *ecology*, but one of the most difficult ones for scientists to explain.

lava *Magma* that is released from the *lithosphere* and flows or spatters across Earth's surface.

law of conservation of matter Physical law stating that *matter* may be transformed from one type of substance into others, but that it cannot be created or destroyed.

LD$_{50}$ (lethal dose–50%) The amount of a *toxicant* it takes to kill 50% of a *population* of test animals. Compare *ED$_{50}$; threshold dose*.

leachate Liquids that seep through liners of a *sanitary landfill* and leach into the *soil* underneath.

leaching The process by which solid materials such as *minerals* are dissolved in a liquid (usually water) and transported to another location.

lead A heavy metal that may be ingested through water or paint, or that may enter the *atmosphere* as a particulate pollutant through combustion of leaded gasoline or other processes. Atmospheric lead deposited on land and water can enter the *food chain*, accumulate within body tissues, and cause *lead poisoning* in animals and people. An EPA *criteria pollutant*.

Leadership in Energy and Environmental Design (LEED) The leading set of standards for sustainable building. Compare *green building*.

lead poisoning Poisoning by ingestion or inhalation of the heavy metal *lead*, causing an array of maladies including damage to the brain, liver, kidney, and stomach; learning problems and behavioral abnormalities; anemia; hearing loss; and even death. Lead poisoning can result from drinking water that passes through old lead pipes or ingesting dust or chips of old lead-based paint.

legislation Statutory law.

legislative branch The branch of the U.S. government that passes laws; it consists of Congress, which includes the House of Representatives and the Senate. Compare *executive branch; judicial branch*.

Leopold, Aldo (1887–1949) American scientist, scholar, philosopher, and author. His book *The Land Ethic* argued that humans should view themselves and the land itself as members of the same *community* and that humans are obligated to treat the land *ethically*.

levee See *dike*.

life-cycle analysis In *industrial ecology*, the examination of the entire life cycle of a given product—from its origins in raw materials, through its manufacturing, to its use, and finally its disposal—in an attempt to identify ways to make the process more *ecologically* efficient.

life expectancy The average number of years that individuals in particular age groups are likely to continue to live.

lifestyle hazard See *cultural hazard*.

light pollution Pollution from city lights that obscures the night sky, impairing people's visibility of stars.

light rail Public transit systems consisting of rail cars powered by electricity.

light reactions In *photosynthesis*, a series of chemical reactions in which water molecules are split and react to form *hydrogen ions* (H^+), molecular *oxygen* (O_2), and small, high-energy molecules used to fuel the *Calvin cycle*.

limiting factor A physical, chemical, or biological characteristic of the *environment* that restrains *population* growth.

limnetic zone In a water body, the layer of open water through which sunlight penetrates. Compare *littoral zone; benthic zone; profundal zone*.

liquefied natural gas (LNG) *Natural gas* that has been converted to a liquid at low temperatures and that can be shipped long distances in refrigerated tankers.

lithification The formation of rock through the processes of compaction, binding, and crystallization.

lithosphere The outer layer of Earth, consisting of *crust* and uppermost *mantle*, and located just above the *asthenosphere*. More generally, the solid part of the Earth, including the rocks, *sediment*, and *soil* at the surface and extending down many miles underground. Compare *atmosphere; biosphere; hydrosphere*.

littoral See *intertidal*.

littoral zone The region ringing the edge of a water body. Compare *benthic zone; limnetic zone; profundal zone*.

Living Planet Index A metric that summarizes trends in the *populations* of over 1,100 *species* that are well enough monitored to provide reliable data. Developed by scientists at the World Wildlife Fund and the *United Nations Environment Programme* to give an overall idea of how natural populations are faring. Between 1970 and 2003, this index fell by roughly 30%.

loam *Soil* with a relatively even mixture of *clay-*, *silt-*, and *sand*-sized particles.

lobbying The expenditure of time or money in an attempt to influence an elected official.

logistic growth curve A plot that shows how the initial *exponential growth* of a *population* is slowed and finally brought to a standstill by *limiting factors*.

longline fishing Fishing practice that involves setting out extremely long lines with up to several thousand baited hooks spaced along their lengths. Kills turtles, sharks, and an estimated 300,000 seabirds each year in *by-catch*.

Love Canal A residential neighborhood in Niagara Falls, New York, from which families were evacuated after buried toxic chemicals rose to the surface, contaminating homes and an elementary school. The well-publicized event helped spark passage of the *Superfund* legislation. Compare *Times Beach*.

low-input agriculture *Agriculture* that uses smaller amounts of *pesticides, fertilizers, growth hormones, water*, and *fossil fuel* energy than are used in *industrial agriculture*. Compare *sustainable agriculture; organic agriculture*.

low-pressure system An air mass in which the air moves toward the low *atmospheric pressure* at the center of the system and spirals upward, typically bringing clouds and *precipitation*. Compare *high-pressure system*.

macromolecule A very large molecule, such as a *protein, nucleic acid, carbohydrate*, or lipid.

macronutrient A *nutrient* that organisms require in relatively large amounts. Compare *micronutrient*.

magma Molten, liquid rock.

malnutrition The condition of lacking *nutrients* the body needs, including a complete complement of vitamins and minerals.

Malthus, Thomas (1766–1834) British economist who maintained that increasing human *population* would eventually deplete the available food supply until starvation, war, or disease arose and reduced the population.

mangrove A tree with a unique type of roots that curve upward to obtain *oxygen*, which is lacking in the mud in which they grow, and that serve as stilts to support the tree in changing water levels. Mangrove forests grow on the coastlines of the tropics and subtropics.

manipulative experiment An *experiment* in which the researcher actively chooses and manipulates the *independent variable*. Compare *natural experiment*.

mantle The malleable layer of rock that lies beneath Earth's *crust* and surrounds a mostly iron *core*.

maquiladora A U.S.-owned factory on the Mexican side of the U.S.-Mexico border.

marasmus A form of *malnutrition* that results from *protein* deficiency together with a lack of calories, causing wasting or shriveling among millions of children in the developing world.

marine protected area (MPA) An area of the ocean set aside to protect marine life from fishing pressures. An MPA may be protected from some human activities but be open to others. Compare *marine reserve*.

marine reserve An area of the ocean designated as a "no-fishing" zone, allowing no extractive activities. Compare *marine protected area*.

market failure The failure of markets to take into account the *environment*'s positive effects on *economies* (for example, *ecosystem services*) or to reflect the negative effects of economic activity on the environment and thereby on people (*external costs*).

mass extinction event The extinction of a large proportion of the world's *species* in a very short time period due to some extreme and rapid change or catastrophic event. Earth has seen five mass extinction events in the past half-billion years.

mass number The combined number of *protons* and *neutrons* in an *atom*.

mass wasting The downslope movement of soil and rock due to gravity. Compare *landslide*.

materials recovery facility (MRF) A *recycling* facility where items are sorted, cleaned, shredded, and prepared for reprocessing into new items.

matter All material in the universe that has mass and occupies space. See *law of conservation of matter*.

maximum sustainable yield The maximal harvest of a particular *renewable natural resource* that can be accomplished while still keeping the resource available for the future.

meandering river A river that winds in a single course through a relatively flat landscape, shaping the landscape of its floodplain. Compare *braided river*.

mechanical weathering See *physical weathering*.

meltdown The accidental melting of the uranium fuel rods inside the core of a *nuclear reactor*, causing the release of radiation.

mesosphere The atmospheric layer above the *stratosphere*, extending 50–80 km (31–56 mi) above sea level.

metal A type of chemical element, or a mass of such an element, that typically is lustrous, opaque, malleable, and can conduct heat and electricity.

metamorphic rock One of the three main categories of rock. Formed by great heat and/or pressure that reshapes crystals within the rock and changes its appearance and physical properties. Common metamorphic rocks include marble and slate. Compare *igneous rock; sedimentary rock*.

metapopulation A network of subpopulations, most of whose members stay within their respective landscape *patches*, but some of whom move among patches or mate with members of other patches.

methane (CH$_4$) A colorless gas produced primarily by *anaerobic* decomposition. The major constituent of *natural gas*, and a *greenhouse gas* that is molecule-for-molecule more potent than *carbon dioxide*.

methane hydrate (methane clathrate or **methane ice)** An ice-like solid consisting of molecules of *methane* embedded in a crystal lattice of water molecules. Methane hydrates are being investigated as a potential new source of *energy* from *fossil fuels*.

microclimate A small-scale localized pattern of weather conditions.

micronutrient A *nutrient* that organisms require in relatively small amounts. Compare *macronutrient*.

Milankovitch cycle One of three types of variations in Earth's rotation and orbit around the sun that result in slight changes in the relative amount of solar radiation reaching Earth's surface at different latitudes. As the cycles proceed, they change the way solar radiation is distributed over Earth's surface and contribute to changes in *atmospheric* heating and circulation that have triggered the ice ages and other *climate* changes.

Mill, John Stuart (1806–1873) British philosopher who believed that as resources become harder to find and extract, *economic growth* will slow and eventually stabilize into a *steady-state economy*.

Millennium Development Goals A program of targets for *sustainable development* set by the international community through the *United Nations* at the turn of this century.

Millennium Ecosystem Assessment The most comprehensive scientific assessment of the present condition of the world's ecological systems and their ability to continue supporting our civilization. Prepared by over 2,000 of the world's leading environmental scientists from nearly 100 nations, and completed in 2005.

mineral A naturally occurring solid *element* or inorganic *compound* with a crystal structure, a specific chemical composition, and distinct physical properties. Compare *ore* and *rock*.

minimum viable population size The *population* size for a given population of a *species* that should assure its continued existence indefinitely into the future. Calculated and used by *conservation geneticists*, population biologists, and wildlife managers.

mining (1) In the broad sense, the extraction of any resource that is nonrenewable on the timescale of our society (such as fossil fuels or groundwater). (2) In relation to mineral resources, the systematic removal of rock, soil, or other material for the purpose of extracting minerals of economic interest.

mixed economy An economy that combines elements of a *capitalist market economy* and state socialism.

model A simplified representation of a complex natural process, designed by scientists to help understand how the process occurs and to make predictions.

moderator Within a *nuclear reactor*, a substance, most often water or graphite, that slows the *neutrons* bombarding uranium so that *fission* can begin.

molecule A combination of two or more *atoms*.

monoculture The uniform planting of a single crop over a large area. Characterizes *industrial agriculture*. Compare *polyculture*.

Montreal Protocol International treaty ratified in 1987 in which 180 signatory nations agreed to restrict production of *chlorofluorocarbons (CFCs)* in order to forestall stratospheric ozone depletion. Because of its effectiveness in decreasing global CFC emissions, the Montreal Protocol is considered the most successful effort to date in addressing a global *environmental* problem.

mortality Rate of death within a *population*.

mosaic In *landscape ecology*, a spatial configuration of *patches* arrayed across a landscape.

moulin A tunnel extending downward from the top surface of a glacier, created by meltwater and through which water flows.

mountaintop removal mining A large-scale form of *coal* mining in which entire mountaintops are leveled. The technique exerts extreme environmental impact on surrounding ecosystems and human residents.

Muir, John (1838–1914) Scottish immigrant to the United States who eventually settled in California and made the Yosemite Valley his wilderness home. Today, he is most strongly associated with the *preservation ethic*. He argued that nature deserved protection for its own inherent values (an *ecocentrist* argument) but also claimed that nature played a large role in human happiness and fulfillment (an *anthropocentrist* argument).

multiple use A principle that has nominally guided management policy for *national forests* over the past half century. The multiple use principle specifies that the forests be managed for recreation, wildlife habitat, *mineral* extraction, and various other uses.

municipal solid waste Nonliquid *waste* that is not especially hazardous and that comes from homes, institutions, and small businesses. Compare *hazardous waste; industrial solid waste*.

mutagen A *toxicant* that causes *mutations* in the *DNA* of organisms.

mutation An accidental change in *DNA* that may range in magnitude from the deletion, substitution, or addition of a single nucleotide to a change affecting entire sets of chromosomes. Mutations provide the raw material for evolutionary change.

mutualism A relationship in which all participating organisms benefit from their interaction. Compare *parasitism*.

nacelle Compartment in a *wind turbine* containing machinery for generating power.

natality Rate of birth within a *population*.

national ambient air quality standards Maximum allowable concentrations of *criteria pollutants* in ambient outdoor air, set by the U.S. *EPA*.

National Environmental Policy Act (NEPA) A U.S. law enacted on January 1, 1970, that created an agency called the Council on Environmental Quality and required that an *environmental impact statement* be prepared for any major federal action.

national forest Public lands consisting of 191 million acres (more than 8% of the nation's land area) in many tracts spread across all but a few states.

National Forest Management Act *Legislation* passed by the U.S. Congress in 1976, mandating that plans for renewable resource management be drawn up for every *national forest*. These plans were to be explicitly based on the concepts of *multiple use* and *sustainable development* and be subject to broad public participation.

national monument An area of designated public land that may later become a *national park*.

national park A scenic area set aside for recreation and enjoyment by the public. The national park system today numbers 392 sites totaling 84 million acres and includes national historic sites, national recreation areas, national wild and scenic rivers, and other types of areas.

national wildlife refuge An area set aside to serve as a haven for wildlife and also sometimes to encourage hunting, fishing, wildlife observation, photography, environmental education, and other public uses.

natural experiment An *experiment* in which the researcher cannot directly manipulate the *variables* and therefore must observe nature, comparing conditions in which variables differ, and interpret the results. Compare *manipulative experiment*.

natural gas A *fossil fuel* composed primarily of methane (CH$_4$), produced as a by-product when bacteria decompose organic material under *anaerobic* conditions.

natural rate of population change The rate of change in a *population's* size resulting from birth and death rates alone, excluding migration.

natural resource Any of the various substances and *energy* sources we need in order to survive.

Natural Resources Conservation Service U.S. agency that promotes soil conservation, as well as water quality protection and pollution control. Prior to 1994, known as the Soil Conservation Service.

natural science An academic discipline that studies the natural world. Compare *social science*.

natural selection The process by which traits that enhance survival and reproduction are passed on more frequently to future generations of organisms than those that do not, thus altering the genetic makeup of populations through time. Natural selection acts on genetic variation and is a primary driver of *evolution*.

negative feedback loop A *feedback loop* in which output of one type acts as input that moves the *system* in the opposite direction. The input and output essentially neutralize each other's effects, stabilizing the system. Compare *positive feedback loop*.

neoclassical economics A *theory* of *economics* that explains market prices in terms of consumer preferences for units of particular commodities. Buyers desire the lowest possible price, whereas sellers desire the highest possible price. This conflict between buyers and sellers results in a compromise price being reached and the "right" quantity of commodities being bought and sold. Compare *ecological economics; environmental economics*.

net energy The quantitative difference between *energy* returned from a process and energy invested in the process. Positive net energy values mean that a process produces more energy than is invested. See also *energy conversion efficiency*; *EROI*.

net metering Process by which owners of houses with *photovoltaic* systems or *wind turbines* can sell their excess *solar energy* or *wind energy* to their local power utility. Whereas *feed-in tariffs* award producers with prices above market rates, net metering offers market-rate prices.

net primary production The *energy* or biomass that remains in an ecosystem after *autotrophs* have metabolized enough for their own maintenance through *cellular respiration*. Net primary production is the energy or biomass available for consumption by *heterotrophs*. Compare *gross primary production; secondary production*.

net primary productivity The rate at which *net primary production* is produced. See *productivity; gross primary production; net primary production; secondary production*.

neurotoxin A *toxicant* that assaults the nervous system. Neurotoxins include heavy metals, *pesticides*, and some chemical weapons developed for use in war.

neutron An electrically neutral (uncharged) particle in the nucleus of an *atom*.

new forestry A set of *ecosystem-based management* approaches for harvesting timber that explicitly mimic natural disturbances. For instance, "sloppy clear-cuts" that leave a variety of trees standing mimic the changes a *forest* might experience if hit by a severe windstorm.

new source review A policy whereby old polluting facilities were exempted from pollution requirements introduced in 1977, as long as they installed the "best available" current technology for pollution control when upgrading their plants in the future.

new urbanism A school of thought among architects, planners, and developers that seeks to design neighborhoods in which homes, businesses, schools, and other amenities are within walking distance of one another. In a direct rebuttal to *sprawl*, proponents of new urbanism aim to create functional neighborhoods in which families can meet most of their needs close to home without the use of a car.

niche The functional role of a *species* in a *community*. See *fundamental niche; realized niche*.

NIMBY See *not-in-my-backyard*.

nitrification The conversion by bacteria of ammonium ions (NH_4^+) first into nitrite ions (NO_2^-) and then into nitrate ions (NO_3^-).

nitrogen The chemical *element* with seven *protons* and seven *neutrons*. The most abundant element in the *atmosphere*, a key element in *macromolecules*, and a crucial plant *nutrient*.

nitrogen cycle A major *nutrient cycle* consisting of the routes that *nitrogen atoms* take through the nested networks of environmental *systems*.

nitrogen dioxide (NO_2) A foul-smelling reddish brown gas that contributes to *smog* and *acid deposition*. It results when atmospheric *nitrogen* and *oxygen* react at the high temperatures created by combustion engines. An EPA *criteria pollutant*.

nitrogen fixation The process by which inert *nitrogen* gas combines with *hydrogen* to form ammonium ions (NH_4^+), which are chemically and biologically active and can be taken up by plants.

nitrogen-fixing Term describing bacteria that live in a *mutualistic* relationship with many types of plants and provide *nutrients* to the plants by converting *nitrogen* to a usable form.

noise pollution Undesired ambient sound.

nonconsumptive use *Fresh water* use in which the water from a particular *aquifer* or surface water body either is not removed or is removed only temporarily and then returned. The use of water to generate electricity in hydroelectric *dams* is an example. Compare *consumptive use*.

nongovernmental organization (NGO) An organization unaffiliated with any government or corporation that exists to promote an issue or agenda. Many operate internationally, and some exert influence over environmental policy.

nonmarket value A value that is not usually included in the price of a *good* or *service*.

non-point source A diffuse source of *pollutants*, often consisting of many small sources. Compare *point source*.

nonrenewable natural resource A *natural resource* that is in limited supply and is formed much more slowly than we use it. Compare *renewable natural resource*.

normative Prescribing norms or standards, as in *ethics*.

North Atlantic Deep Water (NADW) The deep portion of the *thermohaline circulation* in the northern Atlantic Ocean.

Northwest Forest Plan A 1994 plan developed by the Clinton administration to allow logging of *forests* of western Washington, Oregon, and northwestern California with increased protection for *species* and *ecosystems*. The Northwest Forest Plan represented one of the first large-scale applications of *adaptive management*.

no-till *Agriculture* that does not involve tilling (plowing, disking, harrowing, or chiseling) the *soil*. The most extreme form of *conservation tillage*.

not-in-my-backyard (NIMBY) Syndrome in which people do not want something (e.g., a polluting facility) near where they live, even if they may want or need the thing to exist somewhere.

n-type layer The silicon layer in a *photovoltaic cell* that is enriched with *phosphorus* and is rich in *electrons*. Compare *p-type layer*.

nuclear energy The *energy* that holds together *protons* and *neutrons* within the nucleus of an *atom*. Several processes, each of which involves transforming *isotopes* of one *element* into isotopes of other elements, can convert nuclear energy into thermal energy, which is then used to generate electricity. See also *nuclear fission; nuclear reactor*.

nuclear fission The conversion of the *energy* within an *atom*'s nucleus to usable thermal energy by splitting apart atomic nuclei. Compare *nuclear fusion*.

nuclear fuel cycle The process from the mining of *uranium* through its use as fuel in a *nuclear reactor* through its disposal as *waste*.

nuclear fusion The conversion of the *energy* within an *atom*'s nucleus to usable thermal energy by forcing together the small nuclei of lightweight *elements* under extremely high temperature and pressure. Developing a commercially viable method of nuclear fusion remains an elusive goal.

nuclear reactor A facility within a nuclear power plant that initiates and controls the process of *nuclear fission* in order to generate electricity.

nucleic acid A *macromolecule* that directs the production of *proteins*. Includes *DNA* and *RNA*.

null hypothesis A *hypothesis* that postulates no association among variables, and is tested against *alternative hypotheses* that propose particular associations.

nutrient An *element* or *compound* that organisms consume and require for survival.

nutrient cycle The comprehensive set of cyclical pathways by which a given *nutrient* moves through the *environment*.

ocean acidification The process by which today's oceans are become more *acidic* (attaining lower *pH*), as a result of increased *carbon dioxide* concentrations in the atmosphere. Ocean acidification occurs as ocean water absorbs CO_2 from the air and forms carbonic acid. Ocean acidification threatens to dissolve *coral reefs*, which would have strong impacts on marine biodiversity.

oceanography The study of the physics, chemistry, biology, and geology of the oceans.

ocean thermal energy conversion (OTEC) A potential *energy* source that involves harnessing the solar radiation absorbed by tropical oceans in the tropics. See *closed cycle; open cycle*.

O horizon The top layer of *soil* in some *soil profiles*, made up of organic matter, such as decomposing branches, leaves, crop residue, and animal waste. Compare *A horizon; B horizon; C horizon; E horizon; R horizon*.

oil See *petroleum*.

oil sands (tar sands) Deposits that can be mined from the ground, consisting of moist sand and clay containing 1–20% *bitumen*, that some envision as a replacement for *crude oil* as this resource is depleted. Oil sands represent crude oil deposits that have been degraded and chemically altered by water *erosion* and bacterial decomposition.

oil shale *Sedimentary rock* filled with *kerogen* that can be processed to produce liquid *petroleum*. Oil shale is formed by the same processes that form *crude oil*, but occurs when kerogen was not buried deeply enough or subjected to enough heat and pressure to form oil.

oligotrophic Term describing a water body that has low-nutrient and high-oxygen conditions. Compare *eutrophic*.

omnivore An organism that consumes both plants and animals. Compare *carnivore; herbivore*.

open cycle An approach in *ocean thermal energy conversion* whereby warm surface water is evaporated in a vacuum, its steam turns turbines, and it is then condensed by cold water.

open pit mining A *mining* technique that involves digging a gigantic hole and removing the desired *ore*, along with waste rock that surrounds the ore.

ore A *mineral* or grouping of minerals from which we extract *metals*.

organ In an organism, a collection of *tissues*, which performs a particular bodily function.

organic agriculture *Agriculture* that uses no synthetic *fertilizers* or *pesticides* but instead relies on biological approaches such as *composting* and *biocontrol*. Compare *low-input agriculture; sustainable agriculture*.

organic compound A *compound* made up of *carbon atoms* (and, generally, *hydrogen* atoms) joined by covalent bonds and sometimes including other *elements*, such as *nitrogen*, *oxygen*, sulfur, or *phosphorus*. The unusual ability of carbon to build elaborate *molecules* has resulted in millions of different organic compounds showing various degrees of complexity.

organic fertilizer A *fertilizer* made up of natural materials (largely the remains or *wastes* of organisms), including animal manure, crop residues, fresh vegetation, and *compost*. Compare *inorganic fertilizer*.

Organization of Petroleum Exporting Countries (OPEC) Cartel of predominantly Arab nations that in 1973 embargoed oil shipments to the United States and other nations supporting Israel, setting off the nation's first oil shortage.

organ system In an organism, a system of *organs* coordinated to perform a particular set of bodily functions.

outdoor air pollution *Air pollution* that occurs outdoors.

outlet glacier A glacier, or slow-moving stretch of ice, that extends outward from the edge of an ice sheet or ice cap, and through which water from melting ice flows.

overgrazing The consumption by too many animals of plant cover, impeding plant regrowth and the replacement of *biomass*. Overgrazing can exacerbate damage to *soils*, natural *communities*, and the land's productivity for further grazing.

overnutrition A condition of excessive food intake in which people receive more than their daily caloric needs.

overshoot The amount by which humanity has surpassed Earth's long-term carrying capacity for our species.

oxbow A U-shaped loop in a river.

oxbow lake A U-shaped water body that becomes isolated from a river when river water erodes a shortcut from one end of an *oxbow* to the other, so that the river pursues a direct course.

oxygen The chemical *element* with eight *protons* and eight *neutrons*. A key element in the *atmosphere* that is produced by *photosynthesis*.

ozone A *molecule* consisting of three atoms of *oxygen*. Absorbs ultraviolet radiation in the *stratosphere*. Compare *ozone layer; tropospheric ozone*.

ozone-depleting substances Airborne chemicals, such as *halocarbons*, that destroy *ozone* molecules and thin the *ozone layer* in the *stratosphere*.

ozone hole Term popularly used to describe the thinning of the stratospheric *ozone layer* that occurs over Antarctica each year, as a result of *chlorofluorocarbons* and other *ozone*-destroying *pollutants*.

ozone layer A portion of the *stratosphere*, roughly 17–30 km (10–19 mi) above sea level, that contains most of the *ozone* in the *atmosphere*.

paleoclimate Climate in the geologic past.

Pangaea A "supercontinent" 225 million years ago consisting of all Earth's landmasses joined together.

paper park A term referring to parks that are protected "on paper" but not in reality.

paradigm A dominant philosophical and theoretical framework within a scientific discipline.

parasite The organism in a parasitic relationship that extracts nourishment or some other benefit from the *host*.

parasitism A relationship in which one organism, the *parasite*, depends on another, the *host*, for nourishment or some other benefit while simultaneously doing the host harm. Compare *mutualism*.

parasitoid An insect that parasitizes other insects, generally causing eventual death of the *host*. Compare *parasite*.

parent material The base geological material in a particular location.

particulate matter Solid or liquid particles small enough to be suspended in the *atmosphere* and able to damage respiratory tissues when inhaled. Includes *primary pollutants* such as dust and soot as well as *secondary pollutants* such as sulfates and nitrates. An EPA *criteria pollutant*.

passive solar energy collection An approach in which buildings are designed and building materials are chosen to maximize their direct absorption of sunlight in winter, even as they keep the interior cool in the summer. Compare *active solar energy collection*.

patch In *landscape ecology*, spatial areas within a landscape. Depending on a researcher's perspective, patches may consist of habitat for a particular organism, or communities, or ecosystems. An array of patches forms a *mosaic*.

pathogen A microbe that causes disease.

peace park A *transboundary park* intended to help ease tensions by acting as a buffer between nations that have quarreled over boundary disputes.

peak oil Term used to describe the point of maximum production of petroleum in the world (or for a given nation), after which oil production declines. This is also expected to be roughly the midway point of extraction of the world's oil supplies. Compare *Hubbert's peak*.

peat A kind of precursor stage to *coal*, produced when organic material that is broken down by *anaerobic* decomposition remains wet, near the surface, and poorly compressed.

peer review The process by which a manuscript submitted for publication in an academic journal is examined by other specialists in the field, who provide comments and criticism (generally anonymously) and judge whether the work merits publication in the journal.

pelagic Of, relating to, or living between the surface and floor of the ocean. Compare *benthic*.

periodic table of the elements (see Appendix C.) Standard table in chemistry that summarizes information on the *elements* in a comprehensive and elegant way.

permafrost In *tundra*, underground soil that remains more or less permanently frozen.

permanent gas A gas that remains at a stable concentration in the *atmosphere*.

permit-trading The practice of buying and selling government-issued marketable emissions permits to conduct environmentally harmful activities. Under a *cap-and-trade* system, the government determines an acceptable level of *pollution* and then issues permits to pollute. A company receives credit for amounts it does not emit and can then sell this credit to other companies. Compare *emissions trading system*.

peroxyacyl nitrate A chemical created by the reaction of NO_2 with hydrocarbons that can induce further reactions that damage living tissues in animals and plants.

pest A pejorative term for any organism that damages crops that are valuable to us. The term is subjective and defined by our own economic interest, and is not biologically meaningful. Compare *weed*.

pesticide An artificial chemical used to kill insects (insecticide), plants (herbicide), or fungi (fungicide).

pesticide drift Airborne transport of *pesticides*.

petroleum See *crude oil*. However, the term is also used to refer to both *oil* and *natural gas* together.

pH A measure of the concentration of *hydrogen ions* in a *solution*. The pH scale ranges from 0 to 14: A solution with a pH of 7 is neutral; solutions with a pH below 7 are *acidic*, and those with a pH higher than 7 are *basic*. Because the pH scale is logarithmic, each step on the scale represents a tenfold difference in hydrogen ion concentration.

phase shift A fundamental shift in the overall character of an ecological community, generally occurring after some extreme disturbance, and after which the community may not return to its original state. Also known as a *regime shift*.

phosphorus The chemical *element* with 15 *protons* and 15 *neutrons*. An abundant element in the *lithosphere*, a key element in *macromolecules*, and a crucial plant *nutrient*.

phosphorus cycle A major *nutrient cycle* consisting of the routes that *phosphorus* atoms take through the nested networks of environmental *systems*.

photic zone In the ocean or a freshwater body, the well-lit top layer of water where *photosynthesis* occurs.

photobiological Term describing a process that produces fuel by utilizing light and living organisms.

photochemical smog Brown-air smog caused by light-driven reactions of *primary pollutants* with normal atmospheric *compounds* that produce a mix of over 100 different chemicals, ground-level ozone often being the most abundant among them. Compare *industrial smog*.

photoelectric effect Effect that occurs when light strikes one of a pair of metal plates in a *photovoltaic cell*, causing the release of *electrons*, which are attracted by electrostatic forces to the opposing plate. The flow of electrons from one plate to the other creates an electrical current.

photosynthesis The process by which *autotrophs* produce their own food. Sunlight powers a series of chemical reactions that convert *carbon dioxide* and water into sugar (glucose), thus transforming low-quality *energy* from the sun into high-quality energy the organism can use. Compare *cellular respiration*.

photovoltaic (PV) cell A device designed to collect sunlight and directly convert it to electrical *energy* by making use of the *photoelectric effect*.

photovoltaic effect See *photoelectric effect*.

phylogenetic tree A treelike diagram that represents the history of divergence of *species* or other taxonomic groups of organisms.

physical hazard Physical processes that occur naturally in our environment and pose human health hazards. These include discrete events such as *earthquakes*, *volcanic eruptions*, fires, floods, blizzards, *landslides*, hurricanes, and droughts, as well as ongoing natural phenomena such as ultraviolet radiation from sunlight. Compare *biological hazard; chemical hazard; cultural hazard*.

physical weathering *Weathering* that breaks rocks down without triggering a chemical change in the *parent material*. Wind and rain are two main forces. Compare *biological weathering; chemical weathering*.

phytoplankton Microscopic photosynthetic algae, protists, and cyanobacteria that drift near the surface of water bodies and generally form the first *trophic level* in an aquatic *food chain*. Compare *zooplankton*.

Pinchot, Gifford (1865–1946) The first professionally trained American *forester*, Pinchot helped establish the U.S. Forest Service. Today, he is the person most closely associated with the *conservation ethic*.

pioneer species A *species* that arrives earliest, beginning the ecological process of *succession* in a terrestrial or aquatic *community*.

placer mining A *mining* technique that involves sifting through material in modern or ancient riverbed deposits, generally using running water to separate lightweight mud and gravel from heavier minerals of value.

plastics Synthetic (human-made) *polymers* used in numerous manufactured products.

plate tectonics The process by which Earth's surface is shaped by the extremely slow movement of tectonic plates, or sections of *crust*. Earth's surface includes about 15 major tectonic plates. Their interaction gives rise to processes that build mountains, cause *earthquakes*, and otherwise influence the landscape.

point source A specific spot—such as a factory's smokestacks—where large quantities of *pollutants* are discharged. Compare *non-point source*.

polar cell One of a pair of cells of *convective circulation* between the poles and 60° north and south latitude that influence global *climate* patterns. Compare *Ferrel cell; Hadley cell*.

polar stratospheric clouds High-altitude icy clouds containing condensed nitric acid that enhance the destruction of stratospheric ozone in the spring.

polar vortex Circular wind currents that trap air over Antarctica during winter, worsening ozone depletion over the continent.

policy A rule or guideline that directs individual, organizational, or societal behavior.

pollination An interaction in which one organism (for example, bees) transfers pollen (male sex cells) from one flower to the ova (female cells) of another, fertilizing the female flower, which subsequently grows into a fruit.

polluter-pays principle Principle specifying that the party responsible for producing *pollution* should pay the costs of cleaning up the pollution or mitigating its impacts.

pollution Any matter or *energy* released into the *environment* that causes undesirable impacts on the health and well-being of humans or other organisms. Pollution can be physical, chemical, or biological, and can affect water, air, or soil.

polybrominated diphenyl ethers (PBDEs) Synthetic compounds that provide fire-retardant properties and are used in a diverse array of consumer products, including computers, televisions, plastics, and furniture. Released during production, disposal, and use of products, these chemicals persist and accumulate in living tissue, and appear to be *endocrine disruptors*.

polyculture The planting of multiple crops in a mixed arrangement or in close proximity. An example is some traditional Native American farming that mixed maize, beans, squash, and peppers. Compare *monoculture*.

polymer A chemical *compound* or mixture of compounds consisting of long chains of repeated *molecules*. Some polymers play key roles in the building blocks of life.

pool A location in which nutrients in a *biogeochemical cycle* remain for a period of time before moving to another pool. Can be living or nonliving entities. Synonymous with *reservoir*. Compare *flux; residence time*.

population A group of organisms of the same *species* that live in the same area. Species are often composed of multiple populations.

population density The number of individuals within a *population* per unit area. Compare *population size*.

population dispersion See *population distribution*.

population distribution The spatial arrangement of organisms within a particular area.

population ecology The study of the quantitative dynamics of population change and the factors that affect the distribution and abundance of members of a population.

population pyramid See *age structure diagram*.

population size The number of individual organisms present at a given time.

positive feedback loop A *feedback loop* in which output of one type acts as input that moves the *system* in the same direction. The input and output drive the system further toward one extreme or another. Compare *negative feedback loop*.

post-industrial stage The fourth and final stage of the *demographic transition* model, in which both birth and death rates have fallen to a low level and remain stable there, and *populations* may even decline slightly. Compare *industrial stage; pre-industrial stage; transition stage*.

potential energy *Energy* of position. Compare *kinetic energy*.

precautionary principle The idea that one should not undertake a new action until the ramifications of that action are well understood.

precedent A legal ruling that serves as a guide for later cases, steering judicial decisions through time.

precipitation Water that condenses out of the *atmosphere* and falls to Earth in droplets or crystals.

predation The process in which one *species* (the *predator*) hunts, tracks, captures, and ultimately kills its *prey*.

predator An organism that hunts, captures, kills, and consumes individuals of another species, the *prey*.

prediction A specific statement, generally arising from a *hypothesis*, that can be tested directly and unequivocally.

pre-industrial stage The first stage of the *demographic transition* model, characterized by conditions that defined most of human history. In pre-industrial societies, both death rates and birth rates are high. Compare *industrial stage; post-industrial stage; transitional stage*.

prescribed (controlled) burns The practice of burning areas of *forest* or grassland under carefully controlled conditions to improve the health of *ecosystems*, return them to a more natural state, and help prevent uncontrolled catastrophic fires.

prescriptive Guiding or directing how one should behave, as in *ethics*.

preservation ethic An ethic holding that we should protect the natural *environment* in a pristine, unaltered state. Compare *conservation ethic*.

prey An organism that is killed and consumed by a *predator*.

primary consumer An organism that consumes *producers* and feeds at the second *trophic level*.

primary extraction The initial drilling and pumping of the most easily accessible *crude oil*. Compare *secondary extraction*.

primary forest *Forest* uncut by people. Compare *secondary forest*.

primary pollutant A hazardous substance, such as soot or *carbon monoxide*, that is emitted into the *troposphere* in a form that is directly harmful. Compare *secondary pollutant*.

primary producer See *autotroph*.

primary production The conversion of solar energy to the energy of chemical bonds in sugars during *photosynthesis*, performed by *autotrophs*. Compare *secondary production*.

primary succession A stereotypical series of changes as an *ecological community* develops over time, beginning with a lifeless substrate. In terrestrial *systems*, primary succession begins when a bare expanse of rock, *sand*, or *sediment* becomes newly exposed to the atmosphere and *pioneer species* arrive. Compare *secondary succession*.

primary treatment A stage of *wastewater* treatment in which contaminants are physically removed. Wastewater flows into tanks in which sewage solids, grit, and particulate matter settle to the bottom. Greases and oils float to the surface and can be skimmed off. Compare *secondary treatment*.

probability A quantitative description of the likelihood of a certain outcome.

productivity The rate at which plants convert solar *energy* (sunlight) to *biomass*. Ecosystems whose plants convert solar energy to biomass rapidly are said to have high productivity. See *net primary productivity; gross primary production; net primary production*.

profundal zone In a water body, the volume of open water that sunlight does not reach. Compare *littoral zone; benthic zone; limnetic zone*.

protein A *macromolecule* made up of long chains of amino acids.

proton A positively charged particle in the nucleus of an *atom*.

proven recoverable reserve The amount of a given *fossil fuel* in a deposit that is technologically and economically feasible to remove under current conditions.

proxy indicator A type of indirect evidence that serves as a proxy, or substitute, for direct measurement, and that sheds light on conditions of the past. For example, pollen from *sediment* cores and air bubbles from ice cores provide data on the past *climate*.

p-type layer The silicon layer in a *photovoltaic cell* that is enriched with boron and is electron-poor. Compare *n-type layer*.

public policy *Policy* that is made by governments, including those at the local, state, federal, and international levels; it consists of *legislation*, *regulations*, orders, incentives, and practices intended to advance societal welfare. See also *environmental policy*.

public-private partnership A combined effort of government and a for-profit entity, generally intended to use the efficiency of the private sector to help achieve a public policy goal.

pycnocline A zone of the ocean beneath the surface in which density increases rapidly with depth.

pyroclastic flow A fast-moving cloud of toxic gas, ash, and rock fragments that races down the slopes of a *volcano*.

pyrolysis The chemical breakdown of organic matter (such as *biomass*, *oil shale*, and so on) by heating in the absence of *oxygen*, which often produces materials that can be more easily converted to useable energy. A number of variations on this process exist.

quarry A pit used to extract mineral resources such as clay, gravel, sand, and stone such as limestone, granite, marble, and slate.

radiative forcing The amount of change in *energy* that a given factor (such as *aerosols*, *albedo*, or *greenhouse gases*) exerts over Earth's energy balance. Positive radiative forcing warms the surface, whereas negative radiative forcing cools it.

radioactive The quality by which some *isotopes* "decay," changing their chemical identity as they shed atomic particles and emit high-energy radiation.

radiocarbon dating A technique to date preserved organic materials by measuring the percentage of carbon that is the *radioisotope* carbon-14, and matching this value against carbon-14's clocklike progression of decay.

radioisotope A radioactive *isotope* that emits subatomic particles and high-*energy* radiation as it "decays" into progressively lighter isotopes until becoming a stable isotope.

radon A highly *toxic*, radioactive, colorless gas that seeps up from the ground in areas with certain types of bedrock and can build up inside basements and homes with poor air circulation.

rainshadow A region on one side of a mountain or mountain range that experiences arid climate. This occurs because moisture-laden air rising over the terrain from the opposite direction releases precipitation on the windward slope as it cools, leaving the air's humidity low as it descends over the peak and into the rainshadow region.

rainwater harvesting The act of capturing rainwater *runoff* from a rooftop in a barrel, in order to conserve water.

Ramsar Convention on Wetlands of International Importance A 1971 international treaty to inventory and protect important wetlands around the world.

random distribution Distribution pattern in which individuals are located haphazardly in space in no particular pattern (often when needed resources are spread throughout an area and other organisms do not strongly influence where individuals settle).

rangeland Land used for grazing livestock.

REACH Program of the European Union that shifts the burden of proof for testing chemical safety from national governments to industry, and requires that chemical substances produced or imported in amounts of over 1 metric ton per year be registered with a new European Chemicals Agency. *REACH* stands for Registration, Evaluation, Authorisation and Restriction of Chemicals, and went into effect in 2007.

reactive Chemically unstable and readily able to react with other compounds, often explosively or by producing noxious fumes. One criterion for defining *hazardous waste*.

realized niche The portion of the *fundamental niche* that is fully realized (used) by a *species*.

recharge zone An area where water infiltrates Earth's surface and reaches an *aquifer* below.

reclamation The act of restoring a *mining* site to an approximation of its pre-mining condition. To reclaim a site, companies are required to remove buildings and other structures used for mining, replace overburden, fill in shafts, and replant the area with vegetation.

recombinant DNA *DNA* that has been patched together from the DNA of multiple organisms in an attempt to produce desirable traits (such as rapid growth, disease and pest resistance, or higher nutritional content) in organisms lacking those traits.

recovery Waste management strategy composed of *recycling* and *composting*.

recycling The collection of materials that can be broken down and reprocessed to manufacture new items.

Red List An updated list of *species* facing unusually high risks of *extinction*. The list is maintained by the World Conservation Union.

red tide A *harmful algal bloom* consisting of algae that produce reddish pigments that discolor surface waters.

Reducing Emissions from Deforestation and Forest Degradation (REDD) A program to fight *climate change*, whereby wealthy industrialized nations would pay poorer developing nations to conserve *forest*, and the wealthy nations would receive credits for *carbon offsets*.

refining Process of separating the *molecules* of the various *hydrocarbons* in *crude oil* into different-sized classes and transforming them into various fuels and other petrochemical products.

regime shift A fundamental shift in the overall character of an ecological community, generally occurring after some extreme disturbance, and after which the community may not return to its original state. Also known as a *phase shift*.

regional planning *City planning* done on broader geographic scales, generally involving multiple municipal governments.

regulation A specific rule issued by an administrative agency, based on the more broadly written statutory law passed by Congress and enacted by the president.

regulatory taking The deprivation of a property's owner, by means of a law or *regulation*, of most or all economic uses of that property.

relative abundance The extent to which numbers of individuals of different species are equal or skewed. One way to express species diversity. See *evenness*; compare *species richness*.

relative humidity The ratio of the water vapor contained in a given volume of air to the maximum amount the air could contain, for a given temperature.

relativist An ethicist who maintains that *ethics* do and should vary with social context. Compare *universalist*.

renewable natural resource A *natural resource* that is virtually unlimited or that is replenished by the *environment* over relatively short periods of hours to weeks to years. Compare *nonrenewable natural resource*.

replacement fertility The *total fertility rate (TFR)* that maintains a stable *population* size.

reserves-to-production ratio (R/P ratio) The total remaining reserves of a *fossil fuel* divided by the annual rate of production (extraction and processing).

reservoir (1) An artificial water body behind a dam that stores water for human use. (2) See *pool*.

residence time In a biogeochemical cycle, the amount of time a nutrient remains in a given *pool* or *reservoir* before moving to another. Compare *flux*.

resilience The ability of an ecological *community* to change in response to disturbance but later return to its original state. Compare *resistance*.

resistance The ability of an ecological *community* to remain stable in the presence of a disturbance. Compare *resilience*.

Resource Conservation and Recovery Act (RCRA) Congressional *legislation* (enacted in 1976 and amended in 1984) that specifies, among other things, how to manage *sanitary landfills* to protect against environmental contamination.

resource management Strategic decision making about who should extract resources and in what ways, so that resources are used wisely and not wasted.

resource partitioning The process by which *species* adapt to *competition* by evolving to use slightly different resources, or to use their shared resources in different ways, thus minimizing interference with one another.

response The type or magnitude of negative effects an animal exhibits in response to a *dose* of *toxicant* in a *dose-response analysis*. Compare *dose*.

restoration ecology The study of the historical conditions of *ecological communities* as they existed before humans altered them. Principles of restoration ecology are applied in the practice of *ecological restoration*.

revealed preference An approach in environmental economics that quantifies nonmarket values by measuring people's preferences as revealed by data on their actual behavior. Compare *expressed preference*.

revolving door The movement of powerful officials between the private sector and government agencies.

R horizon The bottommost layer of *soil* in a typical *soil profile*. Also called *bedrock*. Compare *A horizon; B horizon; C horizon; E horizon; O horizon*.

ribonucleic acid See *RNA*.

riparian Relating to a river or the area along a river.

risk The mathematical probability that some harmful outcome (for instance, injury, death, *environmental* damage, or *economic* loss) will result from a given action, event, or substance.

risk assessment The quantitative measurement of *risk*, together with the comparison of risks involved in different activities or substances.

risk management The process of considering information from scientific *risk assessment* in light of economic, social, and political needs and values, in order to make decisions and design strategies to minimize *risk*.

river system A river and all its tributaries. River systems drain *watersheds*.

RNA (ribonucleic acid) A usually single-stranded *nucleic acid* composed of four nucleotides, each of which contains a sugar (ribose), a phosphate group, and a nitrogenous base. RNA carries the hereditary information for living organisms and is responsible for passing traits from parents to offspring. Compare *DNA*.

roadless rule A 2001 Clinton Administration executive order that put 31% of national forest land off-limits to road construction or maintenance.

rock A solid aggregation of *minerals*.

rock cycle The very slow process in which *rocks* and the *minerals* that make them up are heated, melted, cooled, broken, and reassembled, forming *igneous*, *sedimentary*, and *metamorphic* rocks.

rotation time The number of years that pass between the time a *forest* stand is cut for timber and the next time it is cut.

r–selected Term denoting a *species* with high *biotic potential* whose members produce a large number of offspring in a relatively short time but do not care for their young after birth. *Populations* of r–selected species are generally regulated by *density-independent factors*. Compare *K–selected*.

runoff The water from *precipitation* that flows into streams, rivers, lakes, and ponds, and (in many cases) eventually to the ocean.

run-of-river Any of several methods used to generate *hydroelectric power* without greatly disrupting the flow of river water. Run-of-river approaches eliminate much of the *environmental* impact of large *dams*. Compare *storage*.

Ruskin, John (1819–1900) British art critic, poet, and writer who criticized industrialized cities and their *pollution*, and who believed that people no longer appreciated the *environment*'s spiritual or aesthetic benefits.

safe harbor agreement A cooperative agreement that allows landowners to harm threatened or endangered species in some ways if they voluntarily improve habitat for the species in others.

salinization The buildup of salts in surface *soil* layers.

salt See *ionic compound*.

salt marsh Flat land that is intermittently flooded by the ocean where the *tide* reaches inland. Salt marshes occur along temperate coastlines and are thickly vegetated with grasses, rushes, shrubs, and other herbaceous plants.

salvage logging The removal of dead trees following a natural disturbance. Although it may be economically beneficial, salvage logging can be ecologically destructive, because the dead trees provide food and shelter for a variety of insects and wildlife and because removing timber from recently burned land can cause severe *erosion* and damage to *soil*.

sand *Sediment* consisting of particles 0.005–2.0 mm in diameter. Compare *clay*; *silt*.

sanitary landfill A site at which solid waste is buried in the ground or piled up in large mounds for disposal, designed to prevent the waste from contaminating the *environment*. Compare *incineration*.

savanna A *biome* characterized by grassland interspersed with clusters of acacias and other trees. Savanna is found across parts of Africa (where it was the ancestral home of our *species*), South America, Australia, India, and other dry tropical regions.

science A systematic process for learning about the world and testing our understanding of it.

scientific method A formalized method for testing ideas with observations that involves several assumptions and a more or less consistent series of interrelated steps.

scrubber Technology to chemically treat gases produced in combustion to remove hazardous components and neutralize acidic gases, such as *sulfur dioxide* and hydrochloric acid, turning them into water and salt, in order to reduce smokestack emissions.

secondary consumer An organism that consumes *primary consumers* and feeds at the third *trophic level*.

secondary extraction The extraction of *crude oil* remaining after *primary extraction* by using solvents or by flushing underground rocks with water or steam. Compare *primary extraction*.

secondary forest *Forest* that has grown back after *primary forest* has been cut. Consists of *second-growth* trees.

secondary pollutant A hazardous substance produced through the reaction of substances added to the *atmosphere* with chemicals normally found in the atmosphere. Compare *primary pollutant*.

secondary production The total *biomass* that *heterotrophs* generate by consuming *autotrophs*. Compare *primary production*.

secondary succession A stereotypical series of changes as an *ecological community* develops over time, beginning when some event disrupts or dramatically alters an existing community. Compare *primary succession*.

secondary treatment A stage of *wastewater* treatment in which biological means are used to remove contaminants remaining after *primary treatment*. Wastewater is stirred up in the presence of *aerobic* bacteria, which degrade organic pollutants in the water. The wastewater then passes to another settling tank, where remaining solids drift to the bottom. Compare *primary treatment*.

second-growth Term describing trees that have sprouted and grown to partial maturity after virgin timber has been cut.

second law of thermodynamics Physical law stating that the nature of *energy* tends to change from a more-ordered state to a less-ordered state; that is, *entropy* increases.

sediment The eroded remains of rocks.

sedimentary rock One of the three main categories of rock. Formed when dissolved *minerals* seep through *sediment* layers and act as a kind of glue, crystallizing and binding sediment particles together. Sandstone and shale are examples of sedimentary rock. Compare *igneous rock*; *metamorphic rock*.

seed bank A storehouse for samples of the world's crop diversity.

seed-tree Timber harvesting approach that leaves small numbers of mature and vigorous seed-producing trees standing so that they can reseed a logged area.

seismometer An instrument that measures and records seismic waves generated by *earthquakes*.

selection system Method of timber harvesting whereby single trees or groups of trees are selectively cut while others are left, creating an *uneven-aged* stand.

selective breeding See *artificial selection*.

septic system A *wastewater* disposal method, common in rural areas, consisting of an underground tank and series of drainpipes. Wastewater runs from the house to the tank, where solids precipitate out. The water proceeds downhill to a drain field of perforated pipes laid horizontally in gravel-filled trenches, where microbes decompose the remaining waste.

service Work done for others as a form of business.

sex ratio The proportion of males to females in a *population*.

shale oil *Sedimentary rock* filled with organic matter that was not buried deeply enough to form *oil*. Some envision this as a replacement for current *petroleum* as this resource declines.

shelf-slope break The portion of the ocean floor where the *continental shelf* drops off with relative suddenness.

shelterbelt A row of trees or other tall perennial plants that are planted along the edges of farm fields to break the wind and thereby minimize wind *erosion*.

shelterwood Timber harvesting approach that leaves small numbers of mature trees in place to provide shelter for seedlings as they grow.

sick-building syndrome A *building-related illness* produced by indoor *pollution* in which the specific cause is not identifiable.

silt *Sediment* consisting of particles 0.002–0.005 mm in diameter. Compare *clay*; *sand*.

silviculture See *forestry*.

sink In a *nutrient cycle*, a *pool* that accepts more *nutrients* than it releases.

sinkhole An area where the ground has given way with little warning as a result of subsidence caused by depletion of water from an *aquifer*.

SLOSS (Single Large or Several Small) dilemma The debate over whether it is better to make reserves large in size and few in number or many in number but small in size.

sludge Solid material that is removed during the *wastewater* treatment process.

smart growth A *city planning* concept in which a community's growth is managed in ways that limit *sprawl* and maintain or improve residents' quality of life. It involves guiding the rate, placement, and style of development such that it serves the *environment*, the *economy*, and the community.

smelting A process in which ore is heated beyond its melting point and combined with other metals or chemicals, in order to form metal with desired characteristics. Steel is created by smelting iron ore with carbon, for example.

Smith, Adam (1723–1790) Scottish philosopher known today as the father of *classical economics*. He believed that when people are free to pursue their own economic self-interest in a competitive marketplace, the marketplace will behave as if guided by "an invisible hand" that ensures that their actions will benefit society as a whole.

smog Term popularly used to describe unhealthy mixtures of air *pollutants* that often form over urban areas. See *industrial smog*; *photochemical smog*.

snag A dead or dying tree that is still standing. Snags are valuable for wildlife.

socially responsible investing Investing in companies that have met certain criteria for environmental or social sustainability.

social science An academic discipline that studies human interactions and institutions. Compare *natural science*.

soil A complex plant-supporting *system* consisting of disintegrated rock, organic matter, air, water, *nutrients*, and microorganisms.

soil degradation A deterioration of soil quality and decline in soil productivity, resulting primarily from forest removal, cropland agriculture, and *overgrazing* of livestock. Compare *desertification*; *land degradation*.

soil profile The cross-section of a *soil* as a whole, from the surface to the *bedrock*.

solar cooker A simple portable oven that uses reflectors to focus sunlight onto food and cook it.

solar energy Energy from the sun. It is perpetually renewable and may be harnessed in several ways.

solution A homogenous mixture of substances in which elements, molecules, or compounds come together without chemically bonding. Most often liquids, but sometimes gases or solids.

solution mining A *mining* technique in which a narrow borehole is drilled deep into the ground to reach a *mineral* deposit, and water, acid, or another liquid is injected down the borehole to leach the resource from the surrounding rock and dissolve it in the liquid. The resulting solution is then sucked out, and the desired resource is isolated.

source In a *nutrient cycle*, a *pool* that releases more *nutrients* than it accepts.

source reduction The reduction of the amount of material that enters the *waste stream* to avoid the costs of disposal and *recycling*, help conserve resources, minimize *pollution*, and save consumers and businesses money.

specialist A *species* that can survive only in a narrow range of *habitats* that contain very specific resources. Compare *generalist*.

speciation The process by which new *species* are generated.

species A *population* or group of populations of a particular type of organism whose members share certain characteristics and can breed freely with one another and produce fertile offspring. Different biologists may have different approaches to diagnosing species boundaries.

species-area curve A graph showing how number of *species* varies with the geographic area of a landmass or water body. *Species richness* commonly doubles as area increases tenfold.

Species at Risk Act (SARA) Canada's endangered species protection law, enacted in 2002.

species coexistence An outcome of *interspecific competition* in which no competing *species* fully excludes others, and the species continue to live side by side.

species diversity The number and variety of *species* in the world or in a particular region.

species richness The number of species in a particular region. One way to express species diversity. Compare *evenness; relative abundance*.

sprawl The unrestrained spread of urban or *suburban* development outward from a city center and across the landscape. Sometimes specified as growth in which the area of development outpaces *population* growth.

stabilizing selection Mode of *natural selection* in which selection produces intermediate traits, in essence preserving the status quo. Compare *directional selection; disruptive selection*.

stable isotope An *isotope* that is not *radioactive*.

steady-state economy An *economy* that does not grow or shrink but remains stable.

stock In fisheries management, a *population* of a particular *species* of fish.

Stockholm Convention on Persistent Organic Pollutants A 2004 international *treaty* that aims to end the use of 12 persistent organic *pollutants* nicknamed the "dirty dozen."

storage Technique used to generate *hydroelectric power*, in which large amounts of water are impounded in a reservoir behind a concrete *dam* and then passed through the dam to turn *turbines* that generate electricity. Compare *run-of-river*.

storm surge A temporary and localized rise in sea level brought on by the high tides and winds associated with storms.

Strategic Petroleum Reserve A stockpile of *oil* stored deep underground in salt caverns in Louisiana by the U.S. government as a short-term buffer against future shortages.

stratosphere The layer of the *atmosphere* above the *troposphere* and below the *mesosphere*; it extends from 11 km (7 mi) to 50 km (31 mi) above sea level.

strip-mining The use of heavy machinery to remove huge amounts of earth to expose *coal* or *minerals*, which are mined out directly. Compare *subsurface mining*.

subcanopy The middle and lower levels of trees in a *forest*, beneath the *canopy*.

subduction The *plate tectonic* process by which denser *crust* slides beneath lighter crust at a *convergent plate boundary*. Often results in *volcanism*.

subsidy A government incentive (a giveaway of cash or publicly owned resources, or a *tax break*) intended to encourage a particular industry or activity. Compare *green tax*.

subsistence agriculture The oldest form of *traditional agriculture*, in which farming families produce only enough food for themselves.

subsistence economy A survival *economy*, one in which people meet most or all of their daily needs directly from nature and do not purchase or trade for most of life's necessities.

subspecies *Populations* of a *species* that occur in different geographic areas and vary from one another in some characteristics. Subspecies are formed by the same processes that drive *speciation* but result when divergence does not proceed far enough to create separate species.

subsurface mining Method of mining underground *coal* deposits, in which shafts are dug deeply into the ground and networks of tunnels are dug or blasted out to follow coal seams. Compare *strip-mining*.

suburb A smaller community that rings a city.

succession A stereotypical series of changes in the composition and structure of an *ecological community* through time. See *primary succession; secondary succession*.

sulfur dioxide (SO₂) A colorless gas resulting in part from the combustion of *coal*. In the *atmosphere*, it may react to form sulfur trioxide and sulfuric acid, which may return to Earth in *acid deposition*. An EPA *criteria pollutant*.

Superfund A program administered by the *Environmental Protection Agency* in which experts identify sites polluted with hazardous chemicals, protect *groundwater* near these sites, and clean up the *pollution*.

supply The amount of a product offered for sale at a given price. Compare *demand*.

surface impoundment A *hazardous waste* disposal method in which a shallow depression is dug and lined with impervious material, such as *clay*. Water containing small amounts of hazardous waste is placed in the pond and allowed to evaporate, leaving a residue of solid hazardous waste on the bottom. Compare *deep-well injection*.

Surface Mining Control and Reclamation Act A 1977 U.S. law that mandates efforts to reclaim (restore) *mining* sites after use, requiring companies to post bonds to cover *reclamation* costs before mining can be approved.

surface water Water located atop Earth's surface. Compare *groundwater*.

survivorship curve A graph that shows how the likelihood of death for members of a *population* varies with age.

sustainability A guiding principle of *environmental science* that requires us to live in such a way as to maintain Earth's systems and its *natural resources* for the foreseeable future.

sustainable agriculture *Agriculture* that does not deplete *soils* faster than they form, or reduce the clean water and genetic diversity essential to long-term crop and livestock production.

sustainable development Development that satisfies our current needs without compromising the future availability of *natural resources* or our future quality of life.

sustainable forest certification A form of *ecolabeling* that identifies timber products that have been produced using *sustainable* methods. The Forest Stewardship Council and several other organizations issue such certification.

swamp A type of *wetland* consisting of shallow water rich with vegetation, occurring in a forested area. Compare *bog; freshwater marsh*.

swidden The traditional form of *agriculture* in tropical forested areas, in which the farmer cultivates a plot for 1 to a few years and then moves on to clear another plot, leaving the first to grow back to *forest*. When the forest is burned this may be called "slash-and-burn" agriculture.

symbiosis A *parasitic* or *mutualistic* relationship between different *species* of organisms that live in close physical proximity.

sympatric speciation *Species* formation that occurs when *populations* become reproductively isolated within the same geographic area. Compare *allopatric speciation*.

synergistic effect An interactive effect (as of *toxicants*) that is more than or different from the simple sum of their constituent effects.

system A network of relationships among a group of parts, elements, or components that interact with and influence one another through the exchange of *energy*, matter, and/or information.

taiga See *boreal forest*.

tailings Portions of *ore* left over after *metals* have been extracted in *mining*.

takings clause A clause from the Fifth Amendment to the U.S. Constitution that ensures, in part, that private property shall not "be taken for public use without just compensation." Courts have interpreted this clause to ban not only the literal taking of private property but also *regulatory takings*.

Talloires Declaration A document composed in Talloires, France, in 1990 that commits university leaders to pursue *sustainability* on their campuses. It has been signed by over 300 university presidents and chancellors from more than 40 nations.

tar sands See *oil sands*.

tax break A common form of *subsidy* in which government reduces or eliminates taxes required of a business or an individual, for the purpose of promoting industries or activities deemed desirable.

taxonomist A scientist who classifies species, using an organism's physical appearance and/or genetic makeup, and who groups species by their similarity into a hierarchy of categories meant to reflect evolutionary relationships.

technically recoverable Extractable using current technology. Applied to *fossil fuel* deposits. Compare *economically recoverable*.

temperate deciduous forest A *biome* consisting of midlatitude *forests* characterized by broad-leafed trees that lose their leaves each fall and remain dormant during winter. These forests occur in areas where *precipitation* is spread relatively evenly throughout the year: much of Europe, eastern China, and eastern North America.

temperate grassland A *biome* whose vegetation is dominated by grasses and features more extreme temperature differences between winter and summer and less *precipitation* than *temperate deciduous forests*.

temperate rainforest A *biome* consisting of tall coniferous trees, cooler and less *species*-rich than *tropical rainforest* and milder and wetter than *temperate deciduous forest*.

temperature (thermal) inversion A departure from the normal temperature distribution in the *atmosphere*, in which a pocket of relatively cold air occurs near the ground, with warmer air above it. The cold air, denser than the air above it, traps *pollutants* near the ground and causes a buildup of *smog*.

teratogen A *toxicant* that causes harm to the unborn, resulting in birth defects.

terracing The cutting of level platforms, sometimes with raised edges, into steep hillsides to contain water from *irrigation* and *precipitation*. Terracing transforms slopes into series of steps like a staircase, enabling farmers to cultivate hilly land while minimizing their loss of *soil* to water *erosion*.

tertiary consumer An organism that consumes *secondary consumers* and feeds at the fourth *trophic level*.

theory A widely accepted, well-tested explanation of one or more cause-and-effect relationships that has been extensively validated by a great amount of research. Compare *hypothesis*.

thermal inversion See *temperature inversion*.

thermal mass Construction materials that absorb heat, store it, and release it later, for use in *passive solar energy* approaches.

thermogenic Type of *natural gas* created by compression and heat deep underground. Contains *methane* and small amounts of other *hydrocarbon* gases. Compare *biogenic*.

thermohaline circulation A worldwide system of ocean currents in which warmer, fresher water moves along the surface and colder, saltier water (which is more dense) moves deep beneath the surface.

thermosphere The *atmosphere*'s top layer, extending upward to an altitude of 500 km (300 mi).

Thoreau, Henry David (1817–1862) American transcendentalist author, poet, and philosopher. His book *Walden*, recording his observations and thoughts while he lived at Walden Pond away from the bustle of urban Massachusetts, remains a classic of American literature.

Three Mile Island Nuclear power plant in Pennsylvania that in 1979 experienced a partial *meltdown*. The term is often used to denote the accident itself, the most serious *nuclear reactor* malfunction that the United States has thus far experienced.

threshold dose The amount of a *toxicant* at which it begins to affect a *population* of test animals. Compare ED_{50}; LD_{50}.

tidal creek A channel in a *salt marsh* through which the *tide* flows in and out.

tidal energy *Energy* harnessed by erecting a *dam* across the outlet of a tidal basin. Water flowing with the incoming or outgoing *tide* through sluices in the dam turns turbines to generate *electricity*.

tide The periodic rise and fall of the ocean's height at a given location, caused by the gravitational pull of the moon and sun.

Timber Culture Act of 1873 U.S. law that granted 65 ha (160 acres) to any citizen promising to cultivate trees on one-quarter of that area.

Times Beach A town in Missouri whose residents were evacuated and whose buildings were demolished after being contaminated by *dioxin* from waste oil sprayed on its roads. The well-publicized event helped spark passage of the *Superfund* legislation. Compare *Love Canal*.

tissue In an organism, a collection of *cells* that perform the same function.

topsoil That portion of the *soil* that is most nutritive for plants and is thus of the most direct importance to *ecosystems* and to *agriculture*. Also known as the *A horizon*.

tornado A type of cyclonic storm in which funnel clouds pick up soil and objects and can do great damage to structures.

tort law A system of law addressing harm caused by one entity to another, which operates primarily through lawsuits.

total fertility rate (TFR) The average number of children born per female member of a *population* during her lifetime.

toxic Poisonous; able to harm health of people or other organisms when a substance is inhaled, ingested, or touched. One criterion for defining *hazardous waste*.

toxic air pollutant *Air pollutant* that is known to cause cancer, reproductive defects, or neurological, developmental, immune system, or respiratory problems in humans, and/or to cause substantial *ecological* harm by affecting the health of nonhuman animals and plants. The *Clean Air Act of 1990* identifies 188 toxic air pollutants, ranging from the heavy metal mercury to *volatile organic compounds* such as benzene and methylene chloride.

toxicant A substance that acts as a poison to humans or wildlife.

toxicity The degree of harm a chemical substance can inflict.

toxicology The scientific field that examines the effects of poisonous chemicals and other agents on humans and other organisms.

Toxic Substances Control Act A 1976 U.S. law that directs the *Environmental Protection Agency* to monitor thousands of industrial chemicals and gives the EPA authority to regulate and ban substances found to pose excessive risk.

toxin A *toxic* chemical stored or manufactured in the tissues of living organisms. For example, a chemical that plants use to ward off *herbivores* or that insects use to deter *predators*.

trade winds Prevailing winds between the equator and 30° latitude that blow from east to west.

traditional agriculture Biologically powered *agriculture*, in which human and animal muscle power, along with hand tools and simple machines, perform the work of cultivating, harvesting, storing, and distributing crops. Compare *industrial agriculture*.

transboundary Crossing a political boundary such as a national border.

transboundary park A reserve of protected land that overlaps national borders.

transform plate boundary Area where two tectonic plates meet and slip and grind alongside one another, creating *earthquakes*. For example, the Pacific Plate and the North American Plate rub against each other along California's San Andreas Fault. Compare *convergent plate boundary* and *divergent plate boundary*.

transgene A *gene* that has been extracted from the *DNA* of one organism and transferred into the DNA of an organism of another *species*.

transgenic Term describing an organism that contains *DNA* from another *species*.

transitional stage The second stage of the *demographic transition* model, which occurs during the transition from the *pre-industrial stage* to the *industrial stage*. It is characterized by declining death rates but continued high birth rates. See also *post-industrial stage*. Compare *industrial stage*; *post-industrial stage*; *pre-industrial stage*.

transit-oriented development A development approach in which compact communities in the *new urbanist* style are arrayed around stops on a major rail transit line.

transmissible disease See *infectious disease*.

transpiration The release of water vapor by plants through their leaves.

trawling Fishing method that entails dragging immense cone-shaped nets through the water, with weights at the bottom and floats at the top to keep the nets open. Compare *bottom-trawling*.

treatment The portion of an *experiment* in which a *variable* has been manipulated in order to test its effect. Compare *control*.

treaty See *convention*.

tributary A smaller river that flows into a larger one.

triple bottom line An approach to sustainability that attempts to meet environmental, economic, and social goals simultaneously.

trophic cascade A series of changes in the *population* sizes of organisms at different *trophic levels* in a *food chain*, occurring when *predators* at high trophic levels indirectly promote populations of organisms at low trophic levels by keeping species at intermediate trophic levels in check. Trophic cascades may become apparent when a top predator is eliminated from a system.

trophic level Rank in the feeding hierarchy of a *food chain*. Organisms at higher trophic levels consume those at lower trophic levels.

tropical dry forest A *biome* that consists of deciduous trees and occurs at tropical and subtropical latitudes where wet and dry seasons each span about half the year. Widespread in India, Africa, South America, and northern Australia.

tropical rainforest A *biome* characterized by year-round rain and uniformly warm temperatures. Found in Central America, South America, southeast Asia, west Africa, and other tropical regions. Tropical rainforests have dark, damp interiors; lush vegetation; and highly diverse *biotic communities*.

tropopause The boundary between the *troposphere* and the *stratosphere*. Acts like a cap, limiting mixing between these atmospheric layers.

troposphere The bottommost layer of the *atmosphere*; it extends to 11 km (7 mi) above sea level. See also *stratosphere*.

tropospheric ozone *Ozone* that occurs in the *troposphere*, where it is a *secondary pollutant* created by the interaction of sunlight, heat, nitrogen oxides, and volatile *carbon*-containing chemicals. A major component of *smog*, it can injure living tissues and cause respiratory problems. An EPA *criteria pollutant*.

tsunami An immense swell, or wave, of ocean water triggered by an *earthquake*, *volcano*, or *landslide*, that can travel long distances across oceans and inundate coasts.

tundra A *biome* that is nearly as dry as *desert* but is located at very high latitudes along the northern edges of Russia, Canada, and Scandinavia. Extremely cold winters with little daylight and moderately cool summers with lengthy days characterize this landscape of lichens and low, scrubby vegetation.

turbine A rotary device that converts the *kinetic energy* of a moving substance, such as steam, into mechanical energy. Used widely in commercial power generation from various types of energy sources.

type I A *survivorship curve* that shows higher death rates at older ages.

type II A *survivorship curve* that shows equal rates of death at all ages.

type III A *survivorship curve* that shows highest death rates at young ages.

typhoon See *cyclone*.

umbrella species A *species* for which meeting its *habitat* needs automatically helps meet those of many other species. Umbrella species generally are species that require large areas of habitat.

unconfined aquifer A water-bearing, porous layer of rock, *sand*, or gravel that lies atop a less-permeable substrate. The water in an unconfined aquifer is not under pressure because there is no impermeable upper layer to confine it. Compare *confined aquifer*.

undernutrition A condition of insufficient *nutrition* in which people receive less than 90% of their daily caloric needs.

understory The layer of a *forest* consisting of small shrubs and trees, above the forest floor and below the *subcanopy*.

uneven-aged Term describing stands of trees in timber plantations that are of different ages. Uneven-aged stands more closely approximate a natural *forest* than do *even-aged* stands.

uniform distribution Distribution pattern in which individuals are evenly spaced (as when individuals hold territories or otherwise compete for space).

United Nations (U.N.) Organization founded in 1945 to promote international peace and to cooperate in solving international economic, social, cultural, and humanitarian problems.

United Nations Environment Programme (UNEP) Agency within the United Nations that deals with *environmental policy*. Created in 1972.

United Nations Framework Convention on Climate Change See *Framework Convention on Climate Change*.

universalist An *ethicist* who maintains that there exist objective notions of right and wrong that hold across cultures and situations. Compare *relativist*.

upwelling In the ocean, the flow of cold, deep water toward the surface. Upwelling occurs in areas where surface *currents* diverge. Compare *downwelling*.

urban ecology A scientific field that views cities explicitly as *ecosystems*. Researchers in this field seek to apply the fundamentals of *ecosystem ecology* and *systems* science to urban areas.

urban growth boundary (UGB) In *city planning*, a geographic boundary intended to separate areas desired to be urban from areas desired to remain rural. Development for housing, commerce, and industry are encouraged within urban growth boundaries, but beyond them such development is severely restricted.

urban heat island effect The phenomenon whereby a city becomes warmer than outlying areas because of the concentration of heat-generating buildings, vehicles, and people, and because buildings and dark paved surfaces absorb heat.

urbanization The shift from rural to city and *suburban* living.

urban planning See *city planning*.

utilitarian value See *instrumental value*.

utility An *ethical standard*, elaborated by British philosophers Jeremy Bentham and *John Stuart Mill*, holding that something is right when it produces the greatest practical benefits for the most people.

variable In an *experiment*, a condition that can change. See *dependent variable* and *independent variable*.

variable gas An atmospheric gas that varies in concentration from time to time or place to place, as a result of natural processes or human activities.

vector An organism that transfers a *pathogen* to its *host*. An example is a mosquito that transfers the malaria pathogen to humans.

vernal pool A type of seasonal wetland that forms in spring from rain and snowmelt, then dries up later in the year.

vested interest A strong personal interest in the outcome of a decision that may result in one's private gain or loss.

virtue An *ethical standard* that, as the ancient Greek philosopher Aristotle held, involves the

personal achievement of moral excellence in character through reasoning and moderation.

volatile organic compound (VOC) One of a large group of potentially harmful organic chemicals used in industrial processes.

volcano A site where molten rock, hot gas, or ash erupt through Earth's surface, often creating a mountain over time as cooled *lava* accumulates.

Wallace, Alfred Russel (1823–1913) English naturalist who proposed, independently of *Charles Darwin*, the concept of *natural selection* as a mechanism for *evolution* and as a way to explain the great variety of living things.

warm front The boundary where a mass of warm air displaces a mass of colder air. Compare *cold front*.

waste Any unwanted product that results from a human activity or process.

waste management Strategic decision making to minimize the amount of *waste* generated and to dispose of waste safely and effectively.

waste stream The flow of *waste* as it moves from its sources toward disposal destinations.

waste-to-energy (WTE) facility An incinerator that uses heat from its furnace to boil water to create steam that drives electricity generation or that fuels heating systems.

wastewater Any water that is used in households, businesses, industries, or public facilities and is drained or flushed down pipes, as well as the polluted *runoff* from streets and storm drains.

waterlogging The saturation of *soil* by water, in which the *water table* is raised to the point that water bathes plant roots. Waterlogging deprives roots of access to gases, essentially suffocating them and eventually damaging or killing the plants.

water mining The withdrawal of water at a rate faster than it can be replenished.

water pollution The act of polluting water, or the condition of being polluted by water pollutants.

watershed The entire area of land from which water drains into a given river.

water table The upper limit of *groundwater* held in an *aquifer*.

wave energy *Energy* harnessed from the motion of wind-driven waves at the ocean's surface. Many designs for machinery to harness wave energy have been invented, but few have been adequately tested.

weather The local physical properties of the *troposphere*, such as temperature, pressure, humidity, cloudiness, and wind, over relatively short time periods. Compare *climate*.

weathering The physical, chemical, and biological processes that break down rocks and *minerals*, turning large particles into smaller particles.

weed A pejorative term for any plant that competes with our crops. The term is subjective and defined by our own economic interest, and is not biologically meaningful. Compare *pest*.

westerlies Prevailing winds from 30° to 60° latitude that blow from west to east.

wetland A system in which the soil is saturated with water and which generally features shallow standing water with ample vegetation. These biologically productive systems include *freshwater marshes*, *swamps*, *bogs*, and seasonal wetlands such as *vernal pools*.

Wetlands Reserve Program U.S. policy in recent farm bills that pays landowners who

protect, restore, or enhance wetland areas on their property.

Whitman, Walt (1819–1892) American poet who espoused transcendentalism. See also *Emerson, Ralph Waldo* and *Thoreau, Henry David*.

wilderness area Federal land that is designated off-limits to development of any kind but is open to public recreation, such as hiking, nature study, and other activities that have minimal impact on the land.

wildland-urban interface A region where urban or suburban development meets forested or undeveloped lands.

windbreak See *shelterbelt*.

wind energy *Energy* from the motion of wind. In this source of renewable energy, the passage of wind through *wind turbines* is used to generate *electricity*.

wind farm A development involving a group of *wind turbines*.

wind turbine A mechanical assembly that converts the wind's *kinetic energy*, or energy of motion, into electrical energy.

wise-use movement A loose confederation of individuals and groups that coalesced in the 1980s and 1990s as a response to the increasing success of environmental advocacy. The movement favors extracting more resources from public lands, obtaining greater local control of lands, and obtaining greater motorized recreational access to public lands.

woodland An ecosystem consisting of trees with open areas among them, similar to a *forest* but with a lower density of trees.

World Bank Institution founded in 1944 that serves as one of the globe's largest sources of funding for *economic* development, including such major projects as *dams*, *irrigation* infrastructure, and other undertakings.

world heritage site A location internationally designated by the *United Nations* for its cultural or natural value. There are over 890 such sites worldwide.

World Trade Organization (WTO) Organization based in Geneva, Switzerland, that represents multinational corporations and promotes free trade by reducing obstacles to international commerce and enforcing fairness among nations in trading practices.

worldview A way of looking at the world that reflects a person's (or a group's) beliefs about the meaning, purpose, operation, and essence of the world.

xeriscaping Landscaping using plants adapted to arid conditions.

zone of aeration The upper layer of an aquifer, containing pore spaces partly filled with water. Compare *zone of saturation*.

zone of saturation The lower layer of an aquifer, containing pore spaces completely filled with water. Compare *zone of aeration*.

zoning The practice of classifying areas for different types of development and land use.

zooplankton Tiny aquatic animals that feed on *phytoplankton* and generally comprise the second *trophic level* in an aquatic *food chain*. Compare *phytoplankton*.

zooxanthellae Symbiotic algae that inhabit the bodies of *corals* and produce food through *photosynthesis*.

PHOTO CREDITS

Getty Images **8.15** David Turnley/Corbis **8.19** Getty Images, Inc.-Agence France Presse **The Science behind the Story: Fertility Decline in Bangladesh miniphoto** Karen Tweedy-Holmes **main** Mark Edwards/Still Pictures/Peter Arnold **8.23** Tiziana and Gianni Baldizzone/Corbis **8.25a** Peter Menzel Photography **8.25b** Peter Ginter **8.27** AP Photo/George Osodi

Chapter 9 Opening Photo © Jim Wark/Agstockusa/Age fotostock **Case Study a** USDA NRCS **Case Study b** USDA NRCS **Case Study** © Curt Maas/age fotostock **9.1a** © Xinhua/Photoshot **9.7a** Ron Giling/Peter Arnold Inc **9.7** USGS **9.7** Fletcher & Baylis/Photo Researchers, Inc **9.7b** © Philip Gould/CORBIS **9.9** Photo Courtesy of USDA NRCS **EnvisionIt photos** George Steinmetz/Corbis, STRDEL/ AFP/Getty Image, © GE Gong/Reuters/Corbis, Adrian Bradshaw/Liaison/Getty Images **9.12a** Finney County Historical Society **The Science behind the Story: Measuring Erosion with Pins and . . . Nuclear Fallout? miniphoto** Courtesy of Jerry Ritchie. USDA Agricultural Research Service **9.13a** Photo Courtesy of USDA NRCS **9.13b** Ted Spiegel/Corbis **9.14a** © Scott Sinklier/Agstockusa/age fotosock **9.14b** Kevin Horan/Stone/Getty Images **9.14c** Keren Su/Stone/Getty Images **9.14d** Ron Giling/Lineair/Peter Arnold **9.14e** Yann Arthus-Bertrand/Corbis **9.14f** U.S. Department of Agriculture **9.15** AP Photo/J.D. Pooley **9.17** China Photos/China Photos **9.18a** Getty Images, Inc.- Photodisc. **9.18c** Carol Cohen/Corbis **9.19a** Nigel Cattlin/Photo Researchers, Inc **9.19b** © Phil Klein/Corbis **9.23** USDA **The Science behind the Story: Restoring the Malpai Borderlands miniphoto** Malpai Borderlands Group **main 1, 2** Richard Hamilton Smith/AgStockUSA **9.24** Richard Hamilton Smith/AgStockUSA

Chapter 10 Opening Photo Steve Satushek/Getty Images Inc.-Image Bank **Case Study** Josef Polleross/The Image Works **10.3** Peter Turnley/CORBIS **10.4** Art Rickerby/Time Life Pictures/Getty Images/Time Life Pictures **10.6a** Jack Dykinga/Getty Images Inc.-Image Bank **10.6b** © Nigel Cattlin/Alamy **10.7a** Native Seeds/SEARCH **10.7b** Native Seeds/SEARCH **10.8** sipaphotostwo355230/Newscom **10.10a** Australia Department of Lands **10.10b** Australia Department of Lands **10.12a** Getty Images, Inc.–Photodisc **10.12b** Bob Rowan/Progressive Image/Corbis **The Science behind the Story: Transgenic Contamination of Native Maize? miniphoto** Peg Skorpinski Photography **10.16** John Schmeiser **10.18** © GlowImages Cuisine/ SuperStock **10.21a** © IndexStock /SuperStock **10.21b** FLPA/Mark Newman/age.fotostock **10.22** © Bertrand Rieger/Hemis/Corbis **10.24** AP Wide World Photos **10.25a** Ambient Images Inc./Alamy **10.25b** USDA **10-27** Danita Delimont/Alamy **The Science behind the Story: Does Organic Farming Work? miniphoto** Paul Maeder **main** Andreas Fliessbach, FiBL Switzerland, Andreas Fliessbach, FiBL Switzerland **EnvisionIt photos** Lyza Danger Gardner, ©Miichael Siluk/The Image Works, David R. Frazier Photolibrary, Inc./Alamy, Alamy **10.28** AP Wide World Photos **10.29** Peter Baker/International Photobank/age fotostock

Chapter 11 Opening Photo Christian Ziegler **Case Study** Andywork/iStockphoto.com **11.09** Elusive Ivory" by Larry Chandler/ www.ivory-bill-woodpecker.com **11.12** Photobank/Shutterstock **11.15** Konrad Wothe/Photolibrary.com **The Science behind the Story: Amphibian Diversity and Decline miniphoto** Asanga Ratnaweera and Madhava Meegaskumbura **main** Ant/NHPH/Photoshot **11.17** Jan Martin Will/Shutterstock **11.22** Matthew Cavenaugh/AP **11.23** Jeremy Holden **11.24** © Drnickburton/Dreamstime **11.25** Galen Rowell/Corbis **The Science behind the Story: Using Forensics to Uncover Illegal Whaling miniphoto** Courtesy of Dr. Scott Baker **main** © Neil Beer/CORBIS **11.26** Corbis **11.27** Josephine Marsden/Alamy **11.29** © Nik Wheeler/CORBIS **EnvisionIt photos** Martin Harvey/Alamy, Jordi Bas Casas/NHPA/Photoshot, frans Lanting//NGS Image Collection, Marvin Dembinsky Photo Associates/Alamy, AP Photo/Hong Kong's Agriculture, Fisheries and Conservation Department (AFCD), Philippe Psaila/Photo Researchers Inc.

Chapter 12 Opening Photo Carl R. Sams II/Peter Arnold Images/Photolibrary **Case Study** NewPage Corporation **12.2a** © NHPA/

Photoshot **12.2b** © Danita Delimont/Alamy **12.2c** © Clint Farlinger/Alamy **12.2d** © Tetra Images/Alamy **12.4** © Ron Austing; Frank Lane Picture Agency/CORBIS **EnvisionIt photos** NASA/UNEP, NASA/UNEP, MICHAEL NICHOLS/NGS Image Collection, Photo by Rodrigo Baleia/LatinContent/Getty Images, Ron Giling/APIS/Peter Arnold **12.8** © Frans Lanting/Corbis **12.12** Weyerhaeuser Company **12.13** Rob Badger/Getty Images, Inc.-Taxi **12.15a** Scott J. Ferrell/Congressional Quarterly/Alamy **12.15b** © William Leaman/Alamy **12.16** Karl Mondon/MCT/Landov **12.18b** © Tracy Ferrero/Alamy **12.18b1** Photo by Dion Manastyrski/Ministry of Forests, southern Interior Forest Region **The Science behind the Story: Fighting over Fire and Forests miniphoto** The Oregonian Publishing Co. **12.20** LAIF/Redux **12.21** SashaBuzko/iStockphoto.com **12.22** © aaron peterson.net/Alamy **12.23b** © Frans Lanting/Corbis **The Science behind the Story: Forest Fragmentation in the Amazon miniphoto** David S. Holloway/Getty Images **main** Courtesy of Dr. Rob Bierregaard, Jr **12.24a** Gary Braasch/Corbis **12.24c** James Zipp/Photo Researchers, Inc.

Chapter 13 Opening Photo Rich Iwaski/AGE Fotostock America, Inc. **Case Study** © E Nugent/ AGE Fotostock **13.2** © Noah Addis/Corbis **13.3a** Chad Palmer/Shutterstock **13.3b** Alamy Images Royalty Free **13.5** NASA **The Science behind the Story: Measuring the Impacts of Sprawl miniphoto** Reid Ewing **main** © Bjeayes | Dreamstime.com **13.6a** Lester Lefkowitz/Corbis **13.6b** Bob Krist/Corbis **13.6c** David R. Frazier Photolibrary, Inc./Getty Images Inc.-Stone Allstock **13.6d** Aldo Torelli/Getty Images **13.7a** Library of Congress **13.7b** Stock Connection Blue/Alamy **EnvisionIt photos** © William Manning/Corbis, Elena Elisseeva/Shutterstock, Alex Segre/Alamy, PHILIPPE DESMAZES/AFP/Getty Images **13.1** Cooper Carry and Association **13.12a** Steve Semler/Alamy **13.13** © WU HONG/epa/Corbis **13.14** Getty Images/Robert Harding World Imagery **13.15** © Bohemian Nomad Picturemakers/CORBIS **13.19** Courtesy of DOE/NREL/Katrin Scholz-Barth **The Science behind the Story: Baltimore and Phoenix Showcase Urban Ecologyminiphoto** Steward Pickett/BES LTER

Chapter 14 Opening Photo © Wendy Stone/Corbis **Case Study** © Lester Lefkowitz/CORBIS **14.1** Gary Porter/Milwaukee Journal Sentinel/MCT)/Newscome **14.2a** Alamy Images Royalty Free **14.2b** Spencer Grant/Photo Edit **14.2c** © Knorre | Dreamstime.com **14.2d** iStockphoto/Thinkstock **14.3** SciMAT/ Photo Researchers, /inc **14.3a** Photofusion Picture Library/Alamy **14.7** © DONG YANJUN/epa/Corbis **EnvisionIt photos** © Remi Benali/Corbis, Regina Bermes/laif/Redux, zumaltwo/newscom, Du Cane Medical Imaging Ltd./Photo Researchers, Inc., AP photo Kearney Hub, Brad Norton **14.9** Joel W. Rogers/Corbis **14.11** Bettmann/Corbis **14-13a** Photolibrary.com-Royalty Free **14.13b** Rosanne Olson/Getty Images Inc.-Stone Allstock **The Science behind the Story: Testing the Safety of Bisphenol A miniphoto** Patricia Hunt **main** Reproduced by permission from Hunt, PA, KE Koehler, M Susiarjo, CA Hodges, A Ilagan, RC Voight, S Thomas, BF Thomas and TJ Hassold.. Bispenol A exposure causes meiotic aneuploidy in the female mouse. Current Biology 13, 546-553. Copyright (c)2003 by Elsevier Science Ltd. **main** Reproduced by permission from Hunt, PA, KE Koehler, M Susiarjo, CA Hodges, A Ilagan, RC Voight, S Thomas, BF Thomas and TJ Hassold.. Bispenol A exposure causes meiotic aneuploidy in the female mouse. Current Biology 13, 546-553. Copyright (c)2003 by Elsevier Science Ltd. **14.17a** Howard K. Suzuki **14.17b** Peg Skorpinski Photography **The Science behind the Story: Pesticides and Child Development in Mexico's Yaqui Valley miniphoto** Jeff Conant **main** Elizabeth A. Guillette **14.18** Masterfile

Chapter 15 Opening Photo Lester Lefkowitz/Corbis **Case Study** © Photoquest | Dreamstime.com **15.3** Alexey Stiop/Shutterstock **15.5a** Michael p. Gadomski/Photo Researchers, Inc. **15.5b** © Thomas Kitchin & Victoria Hurst/First Light/Corbis **15.9** Laura Rauch/AP **15.10** © Tony Hertz/Alamy **15.11a** Courtesy of U.S. Department of the Interior/Bureau of Reclamation, Lower Colorado Region **15.11b** High Country News

SELECTED SOURCES AND REFERENCES FOR FURTHER READING

Chapter 1

Bahn, Paul, and John Flenley. 1992. *Easter Island, Earth island*. Thames and Hudson, London.

Bowler, Peter J. 1993. *The Norton history of the environmental sciences*. W.W. Norton, New York.

Brown, Lester R. 2009. *Plan B 4.0: Mobilizing to save civilization*. Earth Policy Institute and W.W. Norton, New York.

Diamond, Jared. 2005. *Collapse: How societies choose to fail or succeed*. Viking, New York.

Flenley, John, and Paul Bahn. 2003. *The enigmas of Easter Island*. Oxford Univ. Press.

Global Footprint Network, 2009. *How we can bend the curve: Global Footprint Network annual report*. Oakland, Calif. http://www.footprintnetwork.org.

Global Footprint Network, 2010. *The ecological wealth of nations*. Oakland, Calif.

Goudie, Andrew. 2005. *The human impact on the natural environment*, 6th ed. Blackwell Publishing, London.

Hardin, Garrett. 1968. The tragedy of the commons. *Science* 162: 1243–1248.

Harrison, Paul, and Fred Pearce, eds. 2000. *AAAS atlas of population & environment*. Univ. of California Press, Berkeley.

Hunt, Terry L., and Carl P. Lipo. 2006. Late colonization of Easter Island. *Science* 311: 1603–1606.

Kuhn, Thomas S. 1962. *The structure of scientific revolutions*, 2nd ed., 1970. Univ. of Chicago Press, Chicago.

Lomborg, Bjorn. 2001. *The skeptical environmentalist: Measuring the real state of the world*. Cambridge Univ. Press.

Millennium Ecosystem Assessment. 2005. *Ecosystems and human well-being: General synthesis*. Millennium Ecosystem Assessment and World Resources Institute.

Musser, George. 2005. The climax of humanity. *Scientific American* 293(3): 44–47.

Ponting, Clive. 1991. *A green history of the world: The environment and the collapse of great civilizations*. Penguin Books, New York.

Popper, Karl R. 1959. *The logic of scientific discovery*. Hutchinson, London.

Redman, Charles R. 1999. *Human impact on ancient environments*. Univ. of Arizona Press, Tucson.

Sagan, Carl. 1997. *The demon-haunted world: Science as a candle in the dark*. Ballantine Books, New York.

Siever, Raymond. 1968. Science: Observational, experimental, historical. *American Scientist* 56: 70–77.

U.N. Environment Programme. 2007. *Global environment outlook 4 (GEO-4)*. UNEP and Progress Press, Nairobi and Malta.

Valiela, Ivan. 2001. *Doing science: Design, analysis, and communication of scientific research*. Oxford Univ. Press.

Van Tilburg, Jo Anne. 1994. *Easter Island: Archaeology, ecology, and culture*. Smithsonian Institution Press, Washington, D.C.

Wackernagel, Mathis, and William Rees. 1996. *Our ecological footprint: Reducing human impact on the earth*. New Society Publishers, Gabriola Island, British Columbia, Canada.

Wackernagel, Mathis, et al. 2002. Tracking the ecological overshoot of the human economy. *PNAS* 99: 9266–9271.

Worldwatch Institute. 2010. *State of the world 2010: Transforming cultures*. Worldwatch Institute, Washington, D.C.

Worldwatch Institute. 2010. *Vital signs 2010*. Worldwatch Institute, Washington, D.C. http://vitalsigns.worldwatch.org.

WWF–World Wide Fund for Nature. 2008. *Living planet report 2008*. WWF, Gland, Switzerland.

Young, Emma. 2006. A monumental collapse? *New Scientist*, 29 July 2006: 30–34.

Chapter 2

Berardelli, Phil. 2008. Human-driven planet: Time to make it official? *ScienceNOW*. 24 Jan. 2008. http://sciencenow.sciencemag.org/cgi/content/full/2008/124/1.

Berry, R. Stephen. 1991. *Understanding energy: Energy, entropy and thermodynamics for every man*. World Scientific Publishing Co.

Calpine. The Geysers. http://www.geysers.com.

Christopherson, Robert W. 2008. *Geosystems: An introduction to physical geography*, 7th ed. Prentice Hall, Upper Saddle River, N.J.

Craig, James R., David J. Vaughan, and Brian J. Skinner. 2010. *Earth resources and the environment*, 4th ed. Benjamin Cummings, San Francisco.

Keller, Edward A. 2008. *Introduction to environmental geology*, 4th ed. Prentice Hall, Upper Saddle River, N.J.

Keller, Edward A., and Robert H. Blodgett. 2008. *Natural hazards: Earth's processes as hazards, disasters, and catastrophes*, 2nd ed. Prentice Hall, Upper Saddle River, N.J.

Keller, Edward A., and Nicholas Pinter. 2002. *Active tectonics: Earthquakes, uplift, and landscape*, 2nd ed. Prentice Hall, Upper Saddle River, N.J.

Lancaster, Mike, et al. 2010. *Green chemistry: An introductory text*. Royal Society of Chemistry, London.

Lund, John W. 2007. Characteristics, development, and utilization of geothermal resources. *GHC Bulletin*, June 2007: 1–9.

Manahan, Stanley E. 2009. *Environmental chemistry*, 9th ed. Lewis Publishers, CRC Press, Boca Raton, Fla.

Massachusetts Institute of Technology. 2006. *The future of geothermal energy: Impact of enhanced geothermal systems (EGS) on the United States in the 21st century*. Idaho National Laboratory, Idaho Falls.

McMurry, John E. 2007. *Organic chemistry*, 7th ed. Brooks/Cole, San Francisco.

Montgomery, Carla. 2010. *Environmental geology*, 9th ed. McGraw-Hill, New York.

Skinner, Brian J., and Stephen C. Porter. 2003. *The dynamic earth: An introduction to physical geology*, 5th ed. Wiley and Sons, Hoboken, N.J.

Tarbuck, Edward J., Frederick K. Lutgens, and Dennis Tasa. 2009. *Earth science*, 12th ed. Prentice Hall, Upper Saddle River, N.J.

Timberlake, Karen C. 2007. *General, organic, and biological chemistry*, 2nd ed. Benjamin Cummings, San Francisco.

Van Dover, Cindy Lee. 2000. *The ecology of deep-sea hydrothermal vents*. Princeton Univ. Press.

Van Ness, H.C. 1983. *Understanding thermodynamics*. Dover Publications, Mineola, New York.

Zalasiewicz, Jan, et al. 2008. Are we now living in the Anthropocene? *GSA Today* 18(2) (Feb. 2008): 4–8.

Chapter 3

Alvarez, Luis W., et al. 1980. Extraterrestrial cause for the Cretaceous-Tertiary extinction. *Science* 208: 1095–1108.

Begon, Michael, Colin R. Townsend, and John L. Harper. 2005. *Ecology: From individuals to ecosystems*, 4th ed. Blackwell Publishing, London.

Campbell, Neil A., and Jane B. Reece. 2008. *Biology*, 8th ed. Benjamin Cummings, San Francisco.

Clark K.L., et al. 1998. Cloud water and precipitation chemistry in a tropical montane forest, Monteverde, Costa Rica. *Atmospheric Environment* 32: 1595–1603.

Crump, L. Martha, et al. 1992. Apparent decline of the golden toad: Underground or extinct? *Copeia* 1992: 413–420.

Darwin, Charles. 1859. *The origin of species by means of natural selection*. John Murray, London.

Endler, John A. 1986. *Natural selection in the wild*. Monographs in Population Biology 21, Princeton Univ. Press.

Freeman, Scott, and Jon C. Herron. 2006. *Evolutionary analysis*, 4th ed. Prentice Hall, Upper Saddle River, N.J.

Futuyma, Douglas J. 2009. *Evolution*, 2nd ed. Sinauer Associates, Sunderland, Mass.

Krebs, Charles J. 2009. *Ecology: The experimental analysis of distribution and abundance*, 6th ed. Benjamin Cummings, San Francisco.

Lawton, Robert O., et al. 2001. Climatic impact of tropical lowland deforestation on nearby montane cloud forests. *Science* 294: 584–587.

Lips, Karen R., et al. 2008. Riding the wave: Climate change, emerging infectious disease and amphibian declines. *PLOS Biology* 6: e72.

Molles, Manuel C. 2010. *Ecology: Concepts and applications*, 5th ed. McGraw-Hill, Boston.

Nadkarni, Nalini M., and Nathaniel T. Wheelwright, eds. 2000. *Monteverde: Ecology and conservation of a tropical cloud forest*. Oxford Univ. Press.

Pounds, J. Alan. 2001. Climate and amphibian declines. *Nature* 410: 639.

Pounds, J. Alan, and Martha L. Crump. 1994. Amphibian declines and climate disturbance: The case of the golden toad and the harlequin frog. *Conservation Biology* 8: 72–85.

Pounds, J. Alan, et al. 1997. Tests of null models for amphibian declines on a tropical mountain. *Conservation Biology* 11: 1307–1322.

Pounds, J. Alan, et al. 1999. Biological response to climate change on a tropical mountain. *Nature* 398: 611–615.

Pounds, J. Alan, et al. 2006. Widespread amphibian extinctions from epidemic disease driven by global warming. *Nature* 439: 161–167.

Powell, James L. 1998. *Night comes to the Cretaceous: Dinosaur extinction and the transformation of modern geology*. W.H. Freeman, New York.

Raup, David M. 1991. *Extinction: Bad genes or bad luck?* W.W. Norton, New York.

Ricklefs, Robert E., and Gary L. Miller. 2000. *Ecology*, 4th ed. W.H. Freeman, New York.

Ricklefs, Robert E., and Dolph Schluter, eds. 1993. *Species diversity in ecological communities*. Univ. of Chicago Press, Chicago.

Rohr, Jason R., et al. 2008. Evaluating the links between climate, disease spread, and amphibian declines. *PNAS* 105:17436–17441.

Savage, Jay M. 1966. An extraordinary new toad (*Bufo*) from Costa Rica. *Revista de Biologia Tropical* 14: 153–167.

Savage, Jay M. 1998. The "brilliant toad" was telling us something. *Christian Science Monitor*, 14 September 1998: 19.

Smith, Thomas M., and Robert L. Smith. 2009. *Elements of ecology*, 7th ed. Benjamin Cummings, San Francisco.

Wilson, Edward O. 1992. *The diversity of life*. Harvard Univ. Press.

Chapter 4

Baskin, Yvonne. 2002. *A plague of rats and rubbervines: The growing threat of species invasions*. Island Press, Washington, D.C.

Breckle, Siegmar-Walter. 2002. *Walter's vegetation of the Earth: The ecological systems of the geo-biosphere*, 4th ed. Springer-Verlag, Berlin.

Bright, Chris. 1998. *Life out of bounds: Bioinvasion in a borderless world*. Worldwatch Institute and W.W. Norton, Washington, D.C., and New York.

Bronstein, Judith L. 1994. Our current understanding of mutualism. *Quarterly Journal of Biology* 69: 31–51.

Clewell, Andre F., and James Aronson. 2008. *Ecological restoration: Principles, values, and structure of an emerging profession*. Island Press.

Connell, Joseph H., and Ralph O. Slatyer, 1977. Mechanisms of succession in natural communities. *American Naturalist* 111: 1119–1144.

Drake, John M., and Jonathan M. Bossenbroek. 2004. The potential distribution of zebra mussels in the United States. *BioScience* 54: 931–941.

Estes, J.A., et al. 1998. Killer whale predation on sea otters linking oceanic and nearshore ecosystems. *Science* 282: 473–476.

Estes, J.A., et al. 2004. Complex trophic interactions in kelp forest ecosystems. *Bull. Marine Science* 74: 621–638.

Falk, Donald A., et al., eds. 2005. *Foundations of restoration ecology*. Island Press.

Hobbs, Richard J., et al. 2006. *Foundations of restoration ecology: The science and practice of ecological restoration*. Island Press, Washington, D.C.

Menge, Bruce A., et al. 1994. The keystone species concept: Variation in interaction strength in a rocky intertidal habitat. *Ecological Monographs* 64: 249–286.

Molles, Manuel C. 2010. *Ecology: Concepts and applications*, 5th ed. McGraw-Hill, Boston.

Nijhuis, Michelle. 2007. Wish you weren't here. [quagga mussels.] *High Country News*, 5 Mar. 2007.

Pimentel, David, et al. 2005. Update on the environmental and economic costs associated with alien-invasive species in the United States. *Ecological Economics* 52: 273–288.

Power, Mary E., et al. 1996. Challenges in the quest for keystones. *BioScience* 46: 609–620.

Ricklefs, Robert E. 2010. *The economy of nature*, 6th ed. W.H. Freeman and Co., New York.

Schrope, Mark. 2007. Killer in the kelp. *Nature* 445: 703–705.

Shea, Katriona, and Peter Chesson. 2002. Community ecology theory as a framework for biological invasions. *Trends in Ecology and Evolutionary Biology* 17: 170–176.

Smith, Robert L., and Thomas M. Smith. 2001. *Ecology and field biology*, 6th ed. Benjamin Cummings, San Francisco.

Springer, A.M., et al. 2003. Sequential megafaunal collapse in the North Pacific Ocean: An ongoing legacy of industrial whaling? *PNAS* 100: 12223–12228.

Stokstad, Erik. 2007. Feared quagga mussel turns up in western United States. *Science* 315: 453.

Strayer, David L. 2009. Twenty years of zebra mussels: Lessons from the mollusk that made headlines. *Frontiers in Ecology and the Environment* 7: 135–141.

Strayer, David L., et al. 1999. Transformation of freshwater ecosystems by bivalves: A case study of zebra mussels in the Hudson River. *BioScience* 49: 19–27.

Strayer, David L., et al. 2004. Effects of an invasive bivalve (*Dreissena polymorpha*) on fish in the Hudson River estuary. *Canadian Journal of Fisheries and Aquatic Sciences* 61: 924–941.

Thompson, John N. 1999. The evolution of species interactions. *Science* 284: 2116–2118.

U.S. Geological Survey. Zebra and quagga mussel information resource page. http://nas.er.usgs.gov/taxgroup/mollusks/zebramussel.

Van Andel, Jelte, and James Aronson. 2005. *Restoration ecology: The new frontier*. Blackwell Publishing, London.

Weigel, Marlene, ed. 1999. *Encyclopedia of biomes*. UXL, Farmington Hills, Michigan.

Whittaker, Robert H., and William A. Niering. 1965. Vegetation of the Santa Catalina Mountains, Arizona: A gradient analysis of the south slope. *Ecology* 46: 429–452.

Woodward, Susan L. 2003. *Biomes of Earth: Terrestrial, aquatic, and human-dominated*. Greenwood Publishing, Westport, Conn.

Chapter 5

Alexander, Richard B., et al. 2008. Differences in phosphorus and nitrogen delivery to the Gulf of Mexico from the Mississippi River Basin. *Env. Sci. Technol.* 42: 822–830.

Appenzeller, Tim. 2004. The case of the missing carbon. *National Geographic*, Feb. 2004: 88–117.

Carpenter, Edward J., and Douglas G. Capone, eds. 1983. *Nitrogen in the marine environment*. Academic Press, New York.

Committee on Environment and Natural Resources. 2000. *An integrated assessment: Hypoxia in the northern Gulf of Mexico*. CENR, National Science and Technology Council, Washington, D.C.

Committee on the Mississippi River and the Clean Water Act. 2009. *Nutrient control actions for improving water quality in the Mississippi River basin and northern Gulf of Mexico*. National Academies Press, Washington, D.C.

Diaz, Robert J., and Rutger Rosenberg. 2008. Spreading dead zones and consequences for marine ecosystems. *Science* 321: 926–929.

Ferber, Dan. 2004. Dead zone fix not a dead issue. *Science* 305: 1557.

Field, Christopher B., et al. 1998. Primary production of the biosphere: Integrating terrestrial and oceanic components. *Science* 281: 237–240.

Gruber, Nicolas, and James N. Galloway. 2008. An Earth-system perspective on the global nitrogen cycle. *Nature* 451: 293–296.

Jacobson, Michael, et al. 2000. *Earth system science from biogeochemical cycles to global changes*. Academic Press.

Larsen, Janet. 2004. Dead zones increasing in world's coastal waters. *Eco-economy Update #41*, 16 June 2004. Earth Policy Institute. www.earth-policy.org/Updates/Update41.htm.

Mississippi River/Gulf of Mexico Watershed Nutrient Task Force. 2008. *Gulf hypoxia action plan 2008*. U.S. EPA, Washington, D.C.

Mississippi River/Gulf of Mexico Watershed Nutrient Task Force. 2009. *Moving forward on Gulf hypoxia: Annual report 2009*. U.S. EPA, Washington, D.C.

Mitsch, William J., et al. 2001. Reducing nitrogen loading to the Gulf of Mexico from the Mississippi River Basin: Strategies to counter a persistent ecological problem. *BioScience* 51: 373–388.

National Science and Technology Council, Committee on Environment and Natural Resources. 2003. *An assessment of coastal hypoxia and eutrophication in U.S. waters*. National Science and Technology Council, Washington, D.C.

Rabalais, Nancy N., et al. 2002. Beyond science into policy: Gulf of Mexico hypoxia and the Mississippi River. *BioScience* 52: 129–142.

Rabalais, Nancy N., et al. 2002. Hypoxia in the Gulf of Mexico, a.k.a. "The dead zone." *Annual Review of Ecology and Systematics* 33: 235–263.

Raloff, Janet. 2004. Dead waters: Massive oxygen-starved zones are developing along the world's coasts. *Science News* 165: 360–362.

Raloff, Janet. 2004. Limiting dead zones: How to curb river pollution and save the Gulf of Mexico. *Science News* 165: 378–380.

Ricklefs, Robert E. 2010. *The economy of nature*, 6th ed. W.H. Freeman and Co., New York.

Scavia, Donald, and Kristina A. Donnelly. 2007. Reassessing hypoxia forecasts for the Gulf of Mexico. *Env. Sci. Technol.* 41: 8111–8117.

Schlesinger, William H. 1997. *Biogeochemistry: An analysis of global change*, 2nd ed. Academic Press, London.

Smith, Thomas M., and Robert L. Smith. 2009. *Elements of ecology*, 7th ed. Benjamin Cummings, San Francisco.

Takahashi, Taro. 2004. The fate of industrial carbon dioxide. *Science* 305: 352–353.

Turner, Monica, et al. 2003. *Landscape ecology in theory and practice: Pattern and process*. Springer.

Turner, R. Eugene, and Nancy N. Rabalais. 2003. Linking landscape and water quality in the Mississippi River Basin for 200 years. *BioScience* 53: 563–572.

U.S. Department of Energy. 2002. *An evaluation of the Department of Energy's free-air carbon dioxide enrichment (FACE) experiments as scientific user facilities*. U.S. DOE, Washington, D.C.

Vitousek, Peter M., et al. 1997. Human alteration of the global nitrogen cycle: Sources and consequences. *Ecological Applications* 7: 737–750.

Whittaker, Robert H. 1975. *Communities and ecosystems*, 2nd ed. Macmillan, New York.

Wu, Jianguo, and Richard J. Hobbs, eds. 2007. *Key topics in landscape ecology*. Cambridge Univ. Press.

Chapter 6

Balmford, Andrew, et al. 2002. Economic reasons for conserving wild nature. *Science* 297: 950–953.

Barbour, Ian G. 1992. *Ethics in an age of technology*. Harper Collins, San Francisco.

Beddoe, Rachael, et al. 2009. Overcoming systemic roadblocks to sustainability: The evolutionary redesign of worldviews, institutions, and technologies. *PNAS* 106: 2483–2489.

Brown, Lester. 2001. *Eco-economy: Building an economy for the Earth*. Earth Policy Institute and W.W. Norton, New York.

Carson, Richard T., et al. 1994. Valuing the preservation of Australia's Kakadu Conservation Zone. *Oxford Economic Papers* 46: 727–749.

Cole, Luke W., and Sheila R. Foster. 2001. *From the ground up: Environmental racism and the rise of the environmental justice movement*. New York Univ. Press.

Costanza, Robert, et al. 1997. *An introduction to ecological economics*. St. Lucie Press, Boca Raton, Fla.

Costanza, Robert, et al. 1997. The value of the world's ecosystem services and natural capital. *Nature* 387: 253–260.

Daily, Gretchen. 1997. *Nature's services: Societal dependence on natural ecosystems*. Island Press, Washington, D.C.

Daly, Herman E. 1996. *Beyond growth*. Beacon Press, Boston.

Daly, Herman E. 2005. Economics in a full world. *Scientific American* 293(3): 100–107.

De Graaf, John, et al. 2002. *Affluenza: The all-consuming epidemic*. Berrett-Koehler Publishers, San Francisco.

Esty, Daniel C., and Andrew S. Winston. 2006. *Green to gold: How smart companies use environmental strategy to innovate, create value, and build competitive advantage*. Yale Univ. Press, New Haven, Conn.

Fox, Stephen. 1985. *The American conservation movement: John Muir and his legacy*. Univ. of Wisconsin Press, Madison.

Gardner, Gary, et al. 2004. The state of consumption today. Pp. 3–23 in *State of the world 2004*. Worldwatch Institute, Washington, D.C.

Goodstein, Eban. 1999. *The tradeoff myth: Fact and fiction about jobs and the environment*. Island Press, Washington, D.C.

Goodstein, Eban. 2010. *Economics and the environment*, 6th ed. Wiley & Sons, Hoboken, N.J.

Gundjeihmi Aboriginal Corporation. Welcome to the Mirarr site. www.mirarr.net.

Hawken, Paul, Amory Lovins, and L. Hunter Lovins. 1999. *Natural capitalism*. Little, Brown, and Co., Boston.

HM Treasury. 2006. *Stern review on the economics of climate change*. HM Treasury and Cambridge Univ. Press.

Kolstad, Charles D. 2010. *Environmental economics*, 2nd ed. Oxford Univ. Press.

Leopold, Aldo. 1949. *A Sand County almanac, and sketches here and there*. Oxford Univ. Press.

Meadows, Donella, Jørgen Randers, and Dennis Meadows. 2004. *Limits to growth: The 30-year update.* Chelsea Green Publishing Co., White River Junction, Vermont.

Millennium Ecosystem Assessment. 2005. *Ecosystems and human well-being: Opportunities and challenges for business and industry.* Millennium Ecosystem Assessment and World Resources Institute.

Nash, Roderick F. 1989. *The rights of nature.* Univ. of Wisconsin Press, Madison.

Nash, Roderick F. 1990. *American environmentalism: Readings in conservation history,* 3rd ed. McGraw-Hill, New York.

Nordhaus, William. 2007. Critical assumptions in the Stern Review on Climate Change. *Science* 317: 201–202.

O'Neill, John O., et al., eds. 2002. *Environmental ethics and philosophy.* Edward Elgar, Cheltenham, U.K.

Renner, Michael. 2008. *Green jobs: Working for people and the environment.* Worldwatch Report 177. Worldwatch Institute, Washington, D.C.

Ricketts, Taylor, et al. 2004. Economic value of tropical forest to coffee production. *PNAS* 101: 12579–12582.

Sachs, Jeffrey. 2005. Can extreme poverty be eliminated? *Scientific American* 293(3): 56–65.

Singer, Peter, ed. 1993. *A companion to ethics.* Blackwell Publishers, Oxford.

Smith, Adam. 1776. *An inquiry into the nature and causes of the wealth of nations.* 1993 ed., Oxford Univ. Press.

Stone, Christopher D. 1972. Should trees have standing? Towards legal rights for natural objects. *Southern California Law Review* 1972: 450–501.

Talberth, John, et al. 2007. *The genuine progress indicator 2006: A tool for sustainable development.* Redefining Progress, Oakland, Calif.

Tietenberg, Tom, and Lynne Lewis. 2010. *Environmental economics and policy,* 6th ed. Prentice Hall, Upper Saddle River, N.J.

Turner, R. Kerry, et al. 1993. *Environmental economics: An elementary introduction.* Johns Hopkins Univ. Press, Baltimore.

Wenz, Peter S. 2001. *Environmental ethics today.* Oxford Univ. Press.

White, Lynn. 1967. The historic roots of our ecologic crisis. *Science* 155: 1203–1207.

Worldwatch Institute. 2008. *State of the world 2008: Innovations for a sustainable economy.* Worldwatch Institute, Washington, D.C.

Chapter 7

Bae, Chang-Hee Christine. 2003. Tijuana–San Diego: Globalization and the transborder metropolis. *Annals of Regional Science* 37: 463–477.

Carpentier, Chantal Line. 2006. NAFTA Commission for Environmental Cooperation: Ongoing assessment of trade liberalization in North America. *Impact Assessment and Project Appraisal* 24: 259–272.

Clark, Ray, and Larry Canter. 1997. *Environmental policy and NEPA: Past, present, and future.* St. Lucie Press, Boca Raton, Fla.

Dietz, Thomas, et al. 2003. The struggle to govern the global commons. *Science* 302: 1907–1912.

Environmental Law Institute. 2009. *Estimating U.S. government subsidies to energy sources: 2002–2008.* ELI, Washington, D.C.

Fogleman, Valerie M. 1990. *Guide to the National Environmental Policy Act.* Quorum Books, New York.

Green Scissors. 2010. *Green Scissors 2010: More than $200 billion in cuts to wasteful and environmentally harmful spending.* Friends of the Earth, Taxpayers for Common Sense, Environment America, and Public Citizen.

Gruben, William C. 2001. Was NAFTA behind Mexico's high maquiladora growth? *Economic and Financial Review* Third Quarter 2001: 11–21.

Houck, Oliver. 2003. Tales from a troubled marriage: Science and law in environmental policy. *Science* 302: 1926–1928.

Kubasek, Nancy K., and Gary S. Silverman. 2008. *Environmental law,* 6th ed. Prentice Hall, Upper Saddle River, N.J.

Myers, Norman, and Jennifer Kent. 2001. *Perverse subsidies: How misused tax dollars harm the environment and the economy.* Island Press, Washington, D.C.

The National Environmental Policy Act of 1969, as amended. www.nepa.gov/nepa/regs/nepa/nepaeqia.htm.

Nordhaus, Ted, and Michael Shellenberger. 2007. *Break Through: From the death of environmentalism to the politics of possibility.* Houghton Mifflin, Boston.

Office of Management and Budget, Executive Office of the President of the United States, Washington, D.C. 2003. *Informing regulatory decisions: 2003 report to Congress on the costs and benefits of federal and unfunded mandates on state, local, and tribal entities.* Washington, D.C.

Percival, Robert V., et al. 2009. *Environmental regulation: Law, science, and policy,* 6th ed. Aspen Publishers, New York.

Rosenbaum, Walter. 2010. *Environmental politics and policy,* 8th ed. CQ Press, Congressional Quarterly, Inc., Washington, D.C.

Shellenberger, Michael, and Ted Nordhaus. 2004. *The death of environmentalism: Global warming politics in a post-environmental world.* Presented at the Environmental Grantmakers Association meeting, Oct. 2004.

Tietenberg, Tom, and Lynne Lewis. 2010. *Environmental economics and policy,* 6th ed. Prentice Hall, Upper Saddle River, N.J.

U.S. Congress. House. H.R. 3378. 2000. The Tijuana River Valley Estuary and Beach Sewage Cleanup Act of 2000.

U.S. Environmental Protection Agency. 2000. *Regulatory impact analysis: Heavy-duty engine and vehicle standards and highway diesel fuel sulfur control requirements.* EPA420-R-00-026. EPA, Washington, D.C.

U.S. Environmental Protection Agency. 2009. *Acid rain and related programs: 2008 highlights.* U.S. EPA, Washington, D.C.

U.S. Environmental Protection Agency. Summary of the Clean Water Act. www.epa.gov/regulations/laws/cwa.html.

U.S. Government Accountability Office. 2008. *Border wastewater treatment.* GAO-08-595R.

Vig, Norman J., and Michael E. Kraft, eds. 2009. *Environmental policy: New directions for the twenty-first century,* 7th ed. CQ Press, Congressional Quarterly, Inc., Washington, D.C.

Chapter 8

Ausubel, Jesse. H. 1996. Can technology spare the earth? *American Scientist* 84: 166–178.

Ball, Philip. 2008. Where have all the flowers gone? [China's one-child legacy.] *Nature* 454: 374–375.

Balter, Michael. 2006. The baby deficit. *Science* 312: 1894–1897.

Cohen, Joel E. 1995. *How many people can the Earth support?* W.W. Norton, New York.

Cohen, Joel E. 2003. Human population: The next half century. *Science* 302: 1172–1175.

Cohen, Joel E. 2005. Human population grows up. *Scientific American* 293(3): 48–55.

De Souza, Roger-Mark, et al. 2003. Critical links: Population, health, and the environment. *Population Bulletin* 58(3). Population Reference Bureau, Washington, D.C.

Eberstadt, Nicholas. 2000. China's population prospects: Problems ahead. *Problems of Post-Communism* 47: 28.

Ehrlich, Paul. 1968. *The population bomb.* 1997 reprint, Buccaneer Books, Cutchogue, New York.

Ehrlich, Paul R., and Anne H. Ehrlich. 1990. *The population explosion.* Touchstone, New York.

Ehrlich, Paul R., and John P. Holdren. 1971. Impact of population growth: Complacency concerning this component of man's predicament is unjustified and counterproductive. *Science* 171: 1212–1217.

Greenhalgh, Susan. 2001. Fresh winds in Beijing: Chinese feminists speak out on the one-child policy and women's lives. *Signs: Journal of Women in Culture & Society* 26: 847–887.

Haberl, Helmut, et al. 2007. Quantifying and mapping the human appropriation of net primary production in earth's terrestrial ecosystems. *PNAS* 104: 12942–12947.

Harrison, Paul, and Fred Pearce, eds. 2000. *AAAS atlas of population & environment.* Univ. of California Press, Berkeley.

Haub, Carl, and O.P. Sharma. 2006. India's population reality: Reconciling change and tradition. *Population Bulletin* 61(3). Population Reference Bureau, Washington, D.C.

Holdren, John P., and Paul R. Ehrlich. 1974. Human population and the global environment. *American Scientist* 62: 282–292.

Imhoff, Marc L., et al. 2004. Global patterns in human consumption of net primary production. *Nature* 429: 870–873.

Kane, Penny. 1987. *The second billion: Population and family planning in China.* Penguin Books, Australia, Ringwood, Victoria.

Kane, Penny, and Ching Y. Choi. 1999. China's one child family policy. *British Medical Journal* 319: 992.

Kent, Mary M., and Carl Haub. 2005. Global demographic divide. *Population Bulletin* 60(4). Population Reference Bureau, Washington, D.C.

Lamptey, Peter R., et al. 2006. The global challenge of HIV and AIDS. *Population Bulletin* 61(1). Population Reference Bureau, Washington, D.C.

Malthus, Thomas R. *An essay on the principle of population.* 1983 ed. Penguin USA, New York.

Notestein, Frank. 1953. Economic problems of population change. Pp. 13–31 in *Proceedings of the Eighth International Conference of Agricultural Economists.* Oxford Univ. Press.

Population Reference Bureau. 2010. *2010 World population data sheet.* Population Reference Bureau, Washington, D.C.

Riley, Nancy E. 2004. *China's population: New trends and challenges. Population Bulletin* 59(2). Population Reference Bureau, Washington, D.C.

UNAIDS and World Health Organization. 2009. *09 AIDS epidemic update.* UNAIDS and WHO, New York.

U.N. Population Division. 2009. *World population ageing 2009.* UNPD, New York.

U.N. Population Division. 2009. *World population prospects: The 2008 revision.* UNPD, New York.

U.N. Population Fund. 2009. *State of world population 2009.* UNFPA, New York.

U.N. Population Fund. 2010. *The millennium development goals report 2010.* UNFPA.

U.S. Census Bureau. www.census.gov.

Wackernagel, Mathis, and William Rees. 1996. *Our ecological footprint: Reducing human impact on the earth.* New Society Publishers, Gabriola Island, British Columbia, Canada.

Chapter 9

Ashman, Mark R., and Geeta Puri. 2002. *Essential soil science: A clear and concise introduction to soil science.* Wiley-Blackwell.

Brown, Lester R. 2002. World's rangelands deteriorating under mounting pressure. *Eco-Economy Update #6,* 5 Feb. 2002. Earth Policy Institute. www.earth-policy.org/Updates/Update6.htm.

Curtin, Charles G. 2002. Integration of science and community-based conservation in the Mexico/U.S. borderlands. *Conservation Biology* 16: 880–886.

Diamond, Jared. 1999. *Guns, germs, and steel: The fates of human societies.* W.W. Norton, New York.

Diamond, Jared, and Peter Bellwood. 2003. Farmers and their languages: The first expansions. *Science* 300: 597–603.

Food and Agriculture Organization. 2001. Conservation agriculture: Case studies in Latin America and Africa. *FAO Soils Bulletin No. 78.* FAO, Rome.

Glanz, James. 1995. *Saving our soil: Solutions for sustaining Earth's vital resource.* Johnson Books, Boulder, Colorado.

Huggins, David R., and John P. Reganold. 2008. No-till: How farmers are saving the soil by parking their plows. *Scientific American,* 30 June 2008.

Iowa State University Extension. 2009. *Considerations in selecting no-till.* Iowa State University Extension and NRCS.

Jenny, Hans. 1941. *Factors of soil formation: A system of quantitative pedology.* McGraw-Hill, New York.

Kaiser, Jocelyn. 2004. Wounding Earth's fragile skin. *Science* 304: 1616–1618.

Malpai Borderlands Group. Malpai Borderlands Group. www.malpaiborderlandsgroup.org.

Millennium Ecosystem Assessment. 2005. *Ecosystems and human well-being: Desertification synthesis.* Millennium Ecosystem Assessment and World Resources Institute.

Montgomery, David R. 2007. Soil erosion and agricultural sustainability. *PNAS* 104: 13268–13272.

Morgan, R.P.C. 2005. *Soil erosion and conservation,* 3rd ed. Blackwell, London.

Natural Resources Conservation Service. 2010. *2007 national resources inventory: Soil erosion on cropland.* NRCS, USDA, Washington, D.C.

Natural Resources Conservation Service. Soils. NRCS, USDA. www.soils.usda.gov.

Pieri, Christian, et al. 2002. *No-till farming for sustainable rural development.* Agriculture & Rural Development Working Paper. International Bank for Reconstruction and Development, Washington, D.C.

Pierzynski, Gary M., et al. 2005. *Soils and environmental quality,* 3rd ed. CRC Press, Boca Raton, Fla.

Ritchie, Jerry C. 2000. Combining cesium-137 and topographic surveys for measuring soil erosion/deposition patterns in a rapidly accreting area. TEKTRAN, USDA Division of Agricultural Research.

Soil Science Society of America. About soils. https://www.soils.org/about-soils.

Stocking, M.A. 2003. Tropical soils and food security: The next 50 years. *Science* 302: 1356–1359.

Trimble, Stanley W., and Pierre Crosson. 2000. U.S. soil erosion rates—myth and reality. *Science* 289: 248–250.

Troeh, Frederick R., and Louis M. Thompson. 2004. *Soil and soil fertility,* 6th ed. Blackwell Publishing, London.

U.N. Convention to Combat Desertification. 2001. *Global alarm: Dust and sandstorms from the world's drylands.* UNCCD and others, Bangkok, Thailand.

U.N. Environment Programme. 2007. Land. Pp. 81–114 in *Global environment outlook 4 (GEO-4).* UNEP and Progress Press, Nairobi and Malta.

Uri, Noel D. 2001. The environmental implications of soil erosion in the United States. *Env. Monitoring and Assessment* 66: 293–312.

Wilkinson, Bruce H. 2005. Humans as geologic agents: A deep-time perspective. *Geology* 33: 161–164.

Chapter 10

Bazzaz, Fakhri A. 2001. Plant biology in the future. *PNAS* 98: 5441–5445.

Brown, Lester R. 2004. *Outgrowing the Earth: The food security challenge in an age of falling water tables and rising temperatures.* Earth Policy Institute, Washington, D.C.

Buchmann, Stephen L., and Gary Paul Nabhan. 1996. *The forgotten pollinators.* Island Press/Shearwater Books, Washington, D.C./Covelo, California.

Center for Food Safety. 2007. *Monsanto vs. U.S. farmers: November 2007 update.* Center for Food Safety, Washington, D.C.

Cerdeira, Antonio L., and Stephen O. Duke. 2006. The current status and environmental impacts of glyphosate-resistant crops: A review. *J. Environ. Quality* 35: 1633–1658.

Commission for Environmental Cooperation. 2004. *Maize and biodiversity: The effects of transgenic maize in Mexico.* CEC Secretariat.

[Correspondence to *Nature*, various authors]. 2002. *Nature* 416: 600–602, and 417: 897–898.

Dimitri, Carolyn, and Lydia Oberholtzer. 2009. *Marketing U.S. organic foods.* Economic Research Service, USDA, Washington, D.C.

Dyer, George A., et al. 2009. Dispersal of transgenes through maize seed systems in Mexico. *PLoS One* 4: e5734, pp. 1–9.

The Farm Scale Evaluations of spring-sown genetically modified crops. 2003. A themed issue from *Philosophical Transactions of the Royal Society of London B: Biological Sciences* 358 (1439), 29 Nov. 2003.

Fedoroff, Nina, and Nancy Marie Brown, 2004. *Mendel in the kitchen: A scientist's view of genetically modified foods.* National Academies Press, Washington, D.C.

Food and Agriculture Organization. 2006. *Livestock's long shadow: Environmental issues and options.* FAO, Rome.

Food and Agriculture Organization. 2009. *The state of food insecurity in the world.* FAO, Rome.

Food and Agriculture Organization. 2009. *The state of world fisheries and aquaculture 2008.* FAO Fisheries and Aquaculture Department, Rome.

Gardner, Gary, and Brian Halweil. 2000. *Underfed and overfed: The global epidemic of malnutrition.* Worldwatch Paper #150. Worldwatch Institute, Washington, D.C.

Halweil, Brian. 2004. *Eat here: Reclaiming homegrown pleasures in a global supermarket.* Worldwatch Institute, Washington, D.C.

Halweil, Brian. 2008. *Farming fish for the future.* Worldwatch Report 176. Worldwatch Institute, Washington, D.C.

International Food Information Council. http://www.foodinsight.org.

James, Clive. 2010. *Global status of commercialized biotech/GM crops: 2009.* International Service for the Acquisition of Agri-biotech Applications.

Kristiansen, P., et al., eds. 2006. *Organic agriculture: A global perspective.* CABI Publishing, Oxfordshire, U.K.

Liebig, Mark A., and John W. Doran. 1999. Impact of organic production practices on soil quality indicators. *J. Env. Qual.* 28: 1601–1609.

Losey, John E., et al. 1999. Transgenic pollen harms monarch larvae. *Nature* 399: 214.

Maeder, Paul, et al. 2002. Soil fertility and biodiversity in organic farming. *Science* 296: 1694–1697.

Mann, Charles C. 2002. Transgene data deemed unconvincing. *Science* 296: 236–237.

Manning, Richard. 2000. *Food's frontier: The next green revolution.* North Point Press, New York.

Miller, Henry I., and Gregory Conko. 2004. *The frankenfood myth: How protest and politics threaten the biotech revolution.* Praeger Publishers, Westport, Connecticut.

Nierenberg, Danielle. 2005. *Happier meals: Rethinking the global meat industry.* Worldwatch Paper #171. Worldwatch Institute, Washington, D.C.

Nierenberg, Danielle, and Brian Halweil. 2005. Cultivating food security. Pp. 62–79 in *State of the world 2005.* Worldwatch Institute, Washington, D.C.

Norris, Robert F., et al. 2003. *Concepts in integrated pest management.* Prentice Hall, Upper Saddle River, N.J.

Organic Trade Association. 2009. *The Organic Trade Association's 2009 organic industry survey.* OTA, Greenfield, Mass.

Ortiz-García, Sol, et al. 2005. Absence of detectable transgenes in local landraces of maize in Oaxaca, Mexico (2003–2004). *PNAS* 102: 12338–12343.

Paoletti, Maurizio G., and David Pimentel. 1996. Genetic engineering in agriculture and the environment: Assessing risks and benefits. *BioScience* 46: 665–673.

Pearce, Fred. 2002. The great Mexican maize scandal. *New Scientist* 174: 14.

Pedigo, Larry P., and Marlin Rice. 2009. *Entomology and pest management,* 6th ed. Prentice Hall, Upper Saddle River, N.J.

Piñeyro-Nelson, A., et al. 2009. Transgenes in Mexican maize: Molecular evidence and methodological considerations for GMO detection in landrace populations. *Molecular Ecology* 18: 750–761.

Polak, Paul. 2005. The big potential of small farms. *Scientific American* 293(3): 84–91.

Pringle, Peter. 2003. *Food, Inc.: Mendel to Monsanto—The promises and perils of the biotech harvest.* Simon and Schuster, New York.

Quist, David, and Ignacio H. Chapela. 2001. Transgenic DNA introgressed into traditional maize landraces in Oaxaca, Mexico. *Nature* 414: 541–543.

Roberts, Paul. 2008. *The end of food.* Houghton Mifflin, Boston.

Ruse, Michael, and David Castle, eds. 2002. *Genetically modified foods: Debating technology.* Prometheus Books, Amherst, N.Y.

Schmeiser, Percy. Monsanto vs. Schmeiser. www.percyschmeiser.com.

Shiva, Vandana. 2000. *Stolen harvest: The hijacking of the global food supply.* South End Press, Cambridge, Mass.

Smil, Vaclav. 2001. *Feeding the world: A challenge for the twenty-first century.* MIT Press, Cambridge, Mass.

Soleri, Daniela, et al. 2006. Transgenic crops and crop varietal diversity: The case of maize in Mexico. *BioScience* 56.

Stewart, C. Neal. 2004. *Genetically modified planet: Environmental impacts of genetically engineered plants.* Oxford Univ. Press.

Sustainable Agriculture Network. 2010. *The new American farmer: Profiles of agricultural innovation,* 2nd ed. Sustainable Agriculture Network, Beltsville, MD.

Sustainable Agriculture Research and Education (SARE). 2010. *Exploring sustainability in agriculture.* SARE.

U.S. Department of Agriculture. 2010. *2008 Census of agriculture.* USDA, Washington, D.C.

Weasel, Lisa. 2008. *Food fray: Inside the controversy over genetically modified food.* AMACOM Books.

Willer, Helga. 2010. *Organic agriculture worldwide: The main results of the FiBL–IFOAM survey 2010.* www.fibl.org.

Willer, H., and L. Klicher, eds. 2009. *The world of organic agriculture: Statistics and emerging trends 2009.* FiBL and IFOAM, Bonn and Frick.

Wolfenbarger, L. La Reesa. 2000. The ecological risks and benefits of genetically engineered plants. *Science* 290: 2088.

Chapter 11

Balmford, Andrew, et al. 2002. Economic reasons for conserving wild nature. *Science* 297: 950–953.

Barnosky, Anthony D., et al. 2004. Assessing the causes of late Pleistocene extinctions on the continents. *Science* 306: 70–75.

Baskin, Yvonne. 1997. *The work of nature: How the diversity of life sustains us.* Island Press, Washington, D.C.

CITES Secretariat. Convention on International Trade in Endangered Species of Wild Fauna and Flora. www.cites.org.

Convention on Biological Diversity. www.biodiv.org.

Daily, Gretchen C., ed. 1997. *Nature's services: Societal dependence on natural ecosystems.* Island Press, Washington, D.C.

Ehrenfeld, David W. 1970. *Biological conservation.* Holt, Rinehart, and Winston, New York.

Gascon, Claude, et al., eds. 2007. Amphibian conservation action plan. IUCN, Gland, Switzerland.

Gaston, Kevin J., and John I. Spicer. 2004. *Biodiversity: An introduction,* 2nd ed. Blackwell, London.

Groom, Martha J., et al. 2005. *Principles of conservation biology,* 3rd ed. Sinauer Associates, Sunderland, Mass.

Groombridge, Brian, and Martin D. Jenkins. 2002. *Global biodiversity: Earth's living resources in the 21st century.* UNEP, World Conservation Monitoring Centre, and Aventis Foundation; World Conservation Press, Cambridge, U.K.

Groombridge, Brian, and Martin D. Jenkins. 2002. *World atlas of biodiversity: Earth's living resources in the 21st century.* Univ. of California Press, Berkeley.

Hanken, James. 1999. Why are there so many new amphibian species when amphibians are declining? *Trends in Ecology and Evolution* 14: 7–8.

Harris, Larry D. 1984. *The fragmented forest: Island biogeography theory and the preservation of biotic diversity.* Univ. of Chicago Press, Chicago.

Jenkins, Martin. 2003. Prospects for biodiversity. *Science* 302: 1175–1177.

Laurance, William F., et al. 2002. Ecosystem decay of Amazonian forest fragments: A 22-year investigation. *Conservation Biology* 16: 605–618.

Louv, Richard. 2005. *Last child in the woods: Saving our children from nature-deficit disorder.* Algonquin Books, Chapel Hill, North Carolina.

Lovejoy, Thomas E., and Lee Hannah, eds. 2006. *Climate change and biodiversity.* Yale Univ. Press, New Haven, Conn.

Mac Arthur, Robert H., and Edward O. Wilson. 1967. *The theory of island biogeography.* Princeton Univ. Press.

Meegaskumbura, Madhava, et al. 2002. Sri Lanka: An amphibian hot spot. *Science* 298: 379.

Millennium Ecosystem Assessment. 2005. *Ecosystems and human well-being: Biodiversity synthesis.* Millennium Ecosystem Assessment and World Resources Institute.

Mooney, Harold A., and Richard J. Hobbs, eds. 2000. *Invasive species in a changing world.* Island Press, Washington, D.C.

Newmark, William D. 1987. A land-bridge perspective on mammal extinctions in western North American parks. *Nature* 325: 430.

O'Neill, Elizabeth. 2008. Tigers: Worth more dead than alive. *World Watch* Jul/Aug. 2008, pp. 6–11.

Pimm, Stuart L. 1998. The forest fragment classic. *Nature* 393: 23–24.

Pisupati, Balakrishna, and Renata Rubian. 2008. *MDG on reducing biodiversity loss and the CBD's 2010 target.* UNU-IAS report, Yokohama, Japan.

Primack, Richard B. 2010. *Essentials of conservation biology,* 5th ed. Sinauer Associates, Sunderland, Mass.

Quammen, David. 1996. *The song of the dodo: Island biogeography in an age of extinction.* Touchstone, New York.

Relyea, Rick, and Nathan Mills. 2001. Predator-induced stress makes the pesticide carbaryl more deadly to gray treefrog tadpoles. *PNAS* 98: 2491–2496.

Rosenzweig, Michael L. 1995. *Species diversity in space and time.* Cambridge Univ. Press.

Sepkoski, John J. 1984. A kinetic model of Phanerozoic taxonomic diversity. *Paleobiology* 10: 246–267.

Simberloff, Daniel. 1998. Flagships, umbrellas, and keystones: Is single-species management passé in the landscape era? *Biological Conservation* 83: 247–257.

Simberloff, Daniel S., and Edward O. Wilson. 1970. Experimental zoogeography of islands: A two-year record of colonization. *Ecology* 51: 934–937.

Smithsonian Tropical Research Institute. Biological Dynamics of Forest Fragments Project. http://www.stri.org/english/research/facilities/affiliated_stations/bdffp/index.php.

Soulé, Michael E. 1986. *Conservation biology: The science of scarcity and diversity.* Sinauer Associates, Sunderland, Mass.

Takacs, David. 1996. *The idea of biodiversity: Philosophies of paradise.* Johns Hopkins Univ. Press, Baltimore.

U.N. Environment Programme. 2007. Biodiversity. Pp. 157–194 in *Global environment outlook 4 (GEO-4).* UNEP and Progress Press, Nairobi and Malta.

U.S. Fish and Wildlife Service. Endangered species program. www.fws.gov/endangered.

Wilson, Edward O. 1984. *Biophilia.* Harvard Univ. Press.

Wilson, Edward O. 1992. *The diversity of life.* Harvard Univ. Press.

Wilson, Edward O. 2002. *The future of life.* Alfred A. Knopf, New York.

World Conservation Union. IUCN Red List. www.iucnredlist.org.

WWF–World Wide Fund for Nature. 2008. *Living planet report 2008.* WWF, Gland, Switzerland.

Chapter 12

British Columbia Ministry of Forests. Introduction to Silvicultural Systems. www.for.gov.bc.ca/hfd/pubs/SSIntroworkbook/index.htm. B.C. Ministry of Forests, Victoria, B.C.

Clary, David. 1986. *Timber and the Forest Service.* Univ. Press of Kansas, Lawrence.

Donato, Daniel C., et al. 2006. Post-wildfire logging hinders regeneration and increases fire risk. *Science* 311: 352.

Food and Agriculture Organization. 2009. *State of the world's forests 2009.* FAO Forestry Department, Rome.

Food and Agriculture Organization. 2010. *Global forest resources assessment.* FAO Forestry Department, Rome.

Forest Stewardship Council. www.fsc.org.

Myers, Norman, and Jennifer Kent. 2001. *Perverse subsidies: How misused tax dollars harm the environment and the economy.* Island Press, Washington, D.C.

National Forest Management Act of 1976. October 22, 1976 (P.O. 94–588, 90 Stat. 2949, as amended; 16 U.S.C.).

NewPage Corporation. www.newpagecorp.com.

Runte, Alfred. 1979. *National parks and the American experience.* Univ. of Nebraska Press, Lincoln.

Sedjo, Robert A. 2000. *A vision for the U.S. Forest Service.* Resources for the Future, Washington, D.C.

Shatford, Jeffrey, et al. 2007. Conifer regeneration after forest fire in the Klamath-Siskiyous: How much, how soon? *J. Forestry* 105: 139–146.

Smith, David M., et al. 1996. *The practice of silviculture: Applied forest ecology,* 9th ed. Wiley, New York.

Soulé, Michael E., and John Terborgh, eds. 1999. *Continental conservation.* Island Press, Washington, D.C.

Stegner, Wallace. 1954. *Beyond the hundredth meridian: John Wesley Powell and the second opening of the West.* Houghton Mifflin, Boston.

Thompson, Jonathan R., et al. 2007. Re-burn severity in managed and unmanaged vegetation in a large wildfire. *PNAS* 104: 10743–10748.

U.N. Environment Programme. 2007. Land. Pp. 81–114 in *Global environment outlook 4 (GEO-4).* UNEP and Progress Press, Nairobi and Malta.

USDA Forest Service. 2008. Forest resources of the United States, 2007. U.S. Department of Agriculture, Washington, D.C.

Westerling, A.L., et al. 2006. Warming and earlier spring increase western U.S. forest wildfire activity. *Science* 313: 940–943.

Chapter 13

Abbott, Carl. 2001. *Greater Portland: Urban life and landscape in the Pacific Northwest.* Univ. of Pennsylvania Press.

Abbott, Carl. 2002. Planning a sustainable city. Pp. 207–235 in Squires, Gregory D., ed. *Urban sprawl: Causes, consequences, and policy responses.* Urban Institute Press, Washington, D.C.

Breuste, Jurgen, et al. 1998. *Urban ecology.* Springer-Verlag, Berlin.

Brockerhoff, Martin P. 2000. An urbanizing world. *Population Bulletin* 55(3). Population Reference Bureau, Washington, D.C.

Cronon, William. 1991. *Nature's metropolis: Chicago and the great West*. W.W. Norton, New York.

Duany, Andres, et al. 2001. *Suburban nation: The rise of sprawl and the decline of the American dream*. North Point Press, New York.

Ewing, Reid, et al. 2002. *Measuring sprawl and its impact*. Smart Growth America.

Ewing, Reid, et al. 2003. Measuring sprawl and its transportation impacts. *Transportation Research Record* 1831: 175–183.

Girardet, Herbert. 2004. *Cities people planet: Livable cities for a sustainable world*. Academy Press.

Glaeser, Edward L., and Matthew E. Kahn. 2003. Sprawl and urban growth. Harvard Institute of Economic Research Discussion Paper No. 2004.

Hall, Kenneth B., and Gerald A. Porterfield. 2001. *Community by design: New urbanism for suburbs and small communities*. McGraw-Hill, New York.

Jacobs, Jane. 1992. *The death and life of great American cities*. Vintage.

Kalnay, Eugenia, and Ming Cai. 2003. Impact of urbanization and land-use change on climate. *Nature* 423: 528–531.

Kirdar, Uner, ed. 1997. *Cities fit for people*. United Nations, New York.

Litman, Todd. 2004. *Rail transit in America: A comprehensive evaluation of benefits*. Victoria Transport Policy Institute and American Public Transportation Association.

Logan, Michael F. 1995. *Fighting sprawl and city hall*. Univ. of Arizona Press, Tucson.

Metro. www.metro-region.org.

New Urbanism. www.newurbanism.org.

Northwest Environment Watch. 2004. *The Portland exception: A comparison of sprawl, smart growth, and rural land loss in 15 U.S. cities*. Northwest Environment Watch, Seattle.

Oppenheimer, Laura. 2006. Measure 37 changed state, but how much? *Oregonian*, 3 Dec. 2006: 1 and A15.

Pearce, Fred. 2005. Cities lead the way to a greener world. *New Scientist*, 4 June 2005: 8–9.

Portney, Kent. E. 2003. *Taking sustainable cities seriously: Economic development, the environment, and quality of life in American cities (American and comparative environmental policy)*. MIT Press, Cambridge, Mass.

Pugh, Cedric, ed. 1996. *Sustainability, the environment, and urbanization*. Earthscan Publications, London.

Sheehan, Molly O'Meara. 2001. *City limits: Putting the brakes on sprawl*. Worldwatch Paper #156. Worldwatch Institute, Washington, D.C.

U.N. Population Division. 2010. *World urbanization prospects: The 2009 revision*. UNPD, New York.

U.S. Environmental Protection Agency. Smart growth. www.epa.gov/smartgrowth.

U.S. Green Building Council. www.usgbc.org.

Wiewel, Wim, and Joseph J. Persky, eds. 2002. *Suburban sprawl: Private decisions and public policy*. M.E. Sharpe, Armond, New York.

Worldwatch Institute. 2007. *State of the world 2007: Our urban future*. Worldwatch Institute, Washington, D.C.

Chapter 14

Ames, Bruce N., et al. 1990. Nature's chemicals and synthetic chemicals: Comparative toxicology. *PNAS* 87: 7782–7786.

Bloom, Barry. 2005. Public health in transition. *Scientific American* 293(3): 92–99.

Carlsen, Elisabeth, et al. 1992. Evidence for decreasing quality of semen during past 50 years. *British Medical Journal* 305: 609–613.

Carson, Rachel. 1962. *Silent spring*. Houghton Mifflin, Boston.

Colburn, Theo, Dianne Dumanoski, and John P. Myers. 1996. *Our stolen future*. Penguin USA, New York.

Consumer Reports. 2009. Concern over canned foods: Our tests find wide range of bisphenol A in soups, juice, and more. *Consumer Reports* Dec. 2009.

Curtis, Kathleen, and Bobbi Chase Wilding. 2010. *Is it in us? Chemical contamination in our bodies*. Body Burden Working Group and Commonweal Biomonitoring Resource Center.

Environmental Defence. 2008. *Toxic baby bottles in Canada*. Environmental Defence, Toronto, Ontario.

European Commission, Environment Directorate General. 2007. *REACH in brief*. European Commission.

Gilliom, Robert J., et al. 2006. *Pesticides in the nation's streams and ground water, 1992–2001—A summary*. USGS National Water-Quality Assessment Program Circular 1291.

Gross, Liza. 2007. The toxic origins of disease. *PloS Biology* 5(7): e193. doi:10.1371/journal.pbio.0050193.

Guillette, Elizabeth A., et al. 1998. An anthropological approach to the evaluation of preschool children exposed to pesticides in Mexico. *Env. Health Perspectives* 106: 347–353.

Guillette, Louis J. Jr., et al. 2000. Alligators and endocrine disrupting contaminants: A current perspective. *American Zoologist* 40: 438–452.

Hayes, Tyrone, et al. 2003. Atrazine-induced hermaphroditism at 0.1 PPB in American leopard frogs (*Rana pipiens*): Laboratory and field evidence. *Env. Health Perspectives* 111: 568–575.

Hunt, Patricia A., et al. 2003. Bisphenol A exposure causes meiotic aneuploidy in the female mouse. *Current Biology* 13: 546–553.

Kent, Mary M., and Sandra Yin, 2006. Controlling infectious diseases. *Population Bulletin* 61(2), 24 pp. Population Reference Bureau, Washington, D.C.

Kolpin, Dana W., et al. 2002. Pharmaceuticals, hormones, and other organic wastewater contaminants in U.S. streams, 1999–2000: A national reconnaissance. *Env. Sci. Technol.* 36: 1202–1211.

Landis, Wayne G., and Ming-Ho Yu. 2004. *Introduction to environmental toxicology*, 3rd ed. Lewis Press, Boca Raton, Fla.

Lang, Iain A., et al. 2008. Association of urinary bisphenol A concentration with medical disorders and laboratory abnormalities in adults. *JAMA* 300: 1303–1310.

Loewenberg, Samuel. 2003. E.U. starts a chemical reaction. *Science* 300: 405.

Manahan, Stanley E. 2009. *Environmental chemistry*, 9th ed. Lewis Publishers, CRC Press, Boca Raton, Fla.

McGinn, Anne Platt. 2000. *Why poison ourselves? A precautionary approach to synthetic chemicals*. Worldwatch Paper #153. Worldwatch Institute, Washington, D.C.

Millennium Ecosystem Assessment. 2005. *Ecosystems and human well-being: Health synthesis*. World Health Organization.

Milwaukee Journal-Sentinel. 2009–2010. *Chemical fallout* [series of articles on bisphenol A]. *Milwaukee Journal-Sentinel*, http://www.jsonline.com/watchdog/34405049.html.

Moeller, Dade. 2004. *Environmental health*, 3rd ed. Harvard Univ. Press.

Myers, Samuel S. 2009. *Global environmental change: The threat to human health*. Worldwatch Report 181. Worldwatch Institute, Washington, D.C.

National Center for Environmental Health; U.S. Centers for Disease Control and Prevention. 2005. *Third national report on human exposure to environmental chemicals*. NCEH Pub. No. 05-0570, Atlanta.

National Center for Health Statistics. 2009. *Health, United States, 2009, with special feature on medical technology*. Hyattsville, Maryland.

NOW. 2008. Efforts to ban phthalates. http://www.pbs.org/now/shows/412/ban-phthalates.html.

Our Stolen Future. http://www.ourstolenfuture.org/.

Pirages, Dennis. 2005. Containing infectious disease. Pp. 42–61 in *State of the world 2005*. Worldwatch Institute, Washington, D.C.

President's Cancer Panel. 2010. *Reducing environmental cancer risk*. President's Cancer Panel 2008–2009 annual report. U.S. Department of Health and Human Services, Washington, D.C.

Renner, Rebecca. 2002. Conflict brewing over herbicide's link to frog deformities. *Science* 298: 938–939.

Rodricks, Joseph V. 1994. *Calculated risks: Understanding the toxicity of chemicals in our environment.* Cambridge Univ. Press.

Stancel, George, et al. 2001. Report of the bisphenol A sub-panel. Chapter 1 in *National Toxicology Program's report of the endocrine disruptors low-dose peer review.* U.S. EPA and NIEHS, NIH.

Stockholm Convention on Persistent Organic Pollutants. www.pops.int.

United Health Foundation. 2009. *America's health rankings: A call to action for individuals and their communities.* 20th anniversary ed. United Health Foundation, Minnetonka, Minn.

U.S. Environmental Protection Agency. Summary of the Toxic Substances Control Act. www.epa.gov/regulations/laws/tsca.html.

Vogel, Sarah. 2009. The politics of plastic: The making and unmaking of bisphenol A "safety". *Framing Health Matters, Am. J. Public Health* Suppl 3, vol. 99 #S3, S559–S566.

World Health Organization. 2008. *The global burden of disease: 2004 update.* WHO, Geneva, Switzerland.

World Health Organization. 2009. *Global health risks: Mortality and burden of disease attributable to selected major risks.* WHO, Geneva, Switzerland.

World Health Organization. 2009. *World health statistics 2009.* WHO, Geneva, Switzerland.

Zogorski, John S., et al. 2006. *Volatile organic compounds in the nation's ground water and drinking-water supply wells.* USGS National Water-Quality Assessment Program Circular 1292.

Chapter 15

American Rivers. 2002. *The ecology of dam removal: A summary of benefits and impacts.* American Rivers, Washington, D.C.

British Geographical Society and Bangladesh Department of Public Health Engineering. 2001. *Arsenic contamination of groundwater in Bangladesh.* Technical report WC/00/19, volume 1: Summary.

De Villiers, Marq. 2000. *Water: The fate of our most precious resource.* Mariner Books.

Gleick, P.H., and H.S. Cooley. 2009. Energy implications of bottled water. *Environ. Res. Lett.* 4: 014009 (6 pp).

Gleick, Peter H., et al. 2009. *The world's water 2008–2009: The biennial report on freshwater resources.* Island Press, Washington, D.C.

Harvey, Charles F., et al. 2002. Arsenic mobility and groundwater extraction in Bangladesh. *Science* 298: 1602–1606.

Jenkins, Matt. 2005. Squeezing water from a stone. *High Country News* 37(17), 19 Sept. 2005.

Jenkins, Matt. 2006. Running on empty in Sin City. *High Country News* 38(17), 18 Sept. 2006.

Marston, Ed. 2001. Quenching the big thirst. *High Country News* 33(10), 21 May 2001.

Millennium Ecosystem Assessment. 2005. *Ecosystems and human well-being: Wetlands and water synthesis.* Millennium Ecosystem Assessment and World Resources Institute.

Naidenko, Olga, et al. 2008. *Bottled water quality investigation: 10 major brands, 38 pollutants.* Environmental Working Group. www.ewg.org.

New York Times. 2009–2010. *Toxic Waters: A series about the worsening pollution in American waters and regulators' response. New York Times,* http://projects.nytimes.com/toxic-waters.

Nickson, Ross, et al. 1998. Arsenic poisoning of Bangladesh groundwater. *Nature* 395: 338.

Pala, Christopher. 2006. Once a terminal case, the North Aral Sea shows new signs of life. *Science* 312: 183.

Postel, Sandra. 1999. *Pillar of sand: Can the irrigation miracle last?* W.W. Norton, New York.

Postel, Sandra. 2005. *Liquid assets: The critical need to safeguard freshwater ecosystems.* Worldwatch Paper #170. Worldwatch Institute, Washington, D.C.

Reisner, Marc. 1986. *Cadillac desert: The American West and its disappearing water.* Viking Penguin, New York.

Smith, Lingas Rahman. 2000. Contamination of drinking water by arsenic in Bangladesh: A public health emergency. *Bull. World Health Organization* 78(9).

Stone, Richard. 1999. Coming to grips with the Aral Sea's grim legacy. *Science* 284: 30–33.

U.N. Environment Programme. 2007. Water. Pp. 115–156 in *Global environment outlook 4 (GEO-4).* UNEP and Progress Press, Nairobi and Malta.

U.N. Environment Programme. 2008. *Water quality for ecosystem and human health,* 2nd ed. UNEP Global Environment Monitoring System (GEMS)/Water Programme, Burlington, Ontario.

U.N. World Water Assessment Programme. 2009. *U.N. world water development report: Water in a changing world.* Paris, New York, and Oxford, UNESCO and Berghahn Books.

U.S. Bureau of Reclamation, Lower Colorado Regional Office. www.usbr.gov/lc/region.

U.S. Environmental Protection Agency. 2004. *Primer for municipal wastewater treatment systems.* EPA 832-R-04-001. Office of Wastewater Management, Washington D.C.

U.S. Environmental Protection Agency. 2009. *Water on tap: What you need to know.* EPA 816-K-09-002. Office of Water, Washington, D.C.

U.S. Geological Survey. 2004. *Climatic fluctuations, drought, and flow in the Colorado River Basin.* USGS Fact Sheet 2004–3062.

U.S. Government Accountability Office (GAO). 2009. *Bottled water: FDA safety and consumer protections are often less stringent than comparable EPA protections for tap water.* GAO Report to Congressional Requesters, GAO-09-610.

Wagner, Martin, and Jörg Oehlmann. 2009. Endocrine disruptors in bottled mineral water: Total estrogenic burden and migration from plastic bottles. *Env. Sci. Pollut. Res.* 16: 278–286.

Chapter 16

Allsopp, Michelle, et al. 2007. *Oceans in peril: Protecting marine biodiversity.* Worldwatch Report 174. Worldwatch Institute, Washington, D.C.

Bellwood, David R., et al. 2004. Confronting the coral reef crisis. *Nature* 429: 827–833.

[Correspondence to *Science*, various authors]. 2001. *Science* 295: 1233–1235.

Food and Agriculture Organization. 2009. *The state of world fisheries and aquaculture 2008.* FAO Fisheries and Aquaculture Department, Rome.

Frank, Kenneth T., et al. 2005. Trophic cascades in a formerly cod-dominated ecosystem. *Science* 308: 1621–1623.

Garrison, Tom. 2009. *Oceanography: An invitation to marine science,* 7th ed. Brooks/Cole, San Francisco.

Gell, Fiona R., and Callum M. Roberts. 2003. Benefits beyond boundaries: The fishery effects of marine reserves. *Trends in Ecology and Evolution* 18: 448–455.

Halpern, Benjamin S., and Robert R. Warner. 2002. Marine reserves have rapid and lasting effects. *Ecology Letters* 5: 361–366.

Halpern, Benjamin S., and Robert R. Warner. 2003. Matching marine reserve design to reserve objectives. *Proceedings of the Royal Society of London B:* 270: 1871–1878.

Halweil, Brian. 2006. *Catch of the day: Choosing seafood for healthier oceans.* Worldwatch Paper #172. Worldwatch Institute, Washington, D.C.

Henderson, Caspar. 2006. Ocean acidification: The *other* CO_2 problem. *New Scientist* 5 Aug. 2006.

Hoegh-Guldberg, O., et al. 2007. Coral reefs under rapid climate change and ocean acidification. *Science* 318: 1737–1742.

Jackson, Jeremy B.C., et al. 2001. Historical overfishing and the recent collapse of coastal ecosystems. *Science* 293: 629–638.

Kurlansky, Mark. 1998. *Cod: A biography of the fish that changed the world*. Penguin Books, New York.

Lotze, Heike L., et al. 2006. Depletion, degradation, and recovery potential of estuaries and coastal seas. *Science* 312: 1806–1809.

Mayo, Ralph, and Loretta O'Brien. 2006. Status of fishery resources off the northeastern U.S. NEFSC. http://www.nefsc.noaa.gov/sos/spsyn/pg/cod/.

Moore, Charles James. 2008. Synthetic polymers in the marine environment: A rapidly increasing, long-term threat. *Environmental Research* 108: 131–139.

Myers, Ransom A., and Boris Worm. 2003. Rapid worldwide depletion of predatory fish communities. *Nature* 423: 280–283.

National Center for Ecological Analysis and Synthesis (NCEAS) and Communication Partnership for Science and the Sea (COMPASS), sponsors. 2001. *Scientific consensus statement on marine reserves and marine protected areas.* www.nceas.ucsb.edu/consensus.

National Research Council. 2003. *Oil in the sea III: Inputs, fates, and effects*. National Academies Press, Washington, D.C.

Norse, Elliott, and Larry B. Crowder, eds. 2005. *Marine conservation biology: The science of maintaining the sea's biodiversity*. Island Press, Washington, D.C.

Nybakken, James W., and Mark D. Bertness. 2004. *Marine biology: An ecological approach*, 6th ed. Benjamin Cummings, San Francisco.

Orr, James C. 2005. Anthropogenic ocean acidification over the twenty-first century and its impact on calcifying organisms. *Nature* 437: 681–686.

Palumbi, Stephen. 2003. *Marine reserves: A tool for ecosystem management and conservation*. Pew Oceans Commission.

Pauly, Daniel, et al. 2002. Towards sustainability in world fisheries. *Nature* 418: 689–695.

Pauly, Daniel, et al. 2003. The future for fisheries. *Science* 302: 1359–1361.

Pew Oceans Commission. 2003. *America's living oceans: Charting a course for sea change*. A report to the nation. Pew Oceans Commission, Arlington, Va.

Roberts, Callum M., et al. 2001. Effects of marine reserves on adjacent fisheries. *Science* 294: 1920–1923.

Rosenberg, A., et al. 2006. Rebuilding U.S. fisheries: Progress and problems. *Frontiers in Ecology and the Environment* 4(6).

Sumich, James L., and John F. Morrissey. 2008. *Introduction to the biology of marine life*, 9th ed. Jones & Bartlett, Boston.

Trujillo, Alan P., and Harold V. Thurman. 2008. *Essentials of oceanography*, 9th ed. Prentice Hall, Upper Saddle River, N.J.

U.N. Environment Programme. 2007. Water. Pp. 115–156 in *Global environment outlook 4 (GEO-4)*. UNEP and Progress Press, Nairobi and Malta.

U.S. Commission on Ocean Policy. 2004. *An ocean blueprint for the 21st century*. Final report. Washington, D.C.

U.S. Department of Commerce and U.S. Department of the Interior. Marine protected areas of the United States. www.mpa.gov.

Weber, Michael L. 2001. *From abundance to scarcity: A history of U.S. marine fisheries policy*. Island Press, Washington, D.C.

Weiss, Kenneth R., and Usha Lee McFarling. 2006. Altered oceans (a special five-part series). *Los Angeles Times*, 30 July–3 August, 2006. Available at: http://www.latimes.com/news/local/la-oceans-series,0,7783938.special.

Worm, Boris, et al. 2006. Impacts of biodiversity loss on ocean ecosystem services. *Science* 314: 787–790.

Chapter 17

Ahrens, C. Donald. 2008. *Meteorology today*, 9th ed. Brooks/Cole, San Francisco.

Akimoto, Hajime. 2003. Global air quality and pollution. *Science* 302: 1716–1719.

Bernard, Susan M., et al. 2001. The potential impacts of climate variability and change on air pollution-related health effects in the United States. *Env. Health Perspectives* 109(Suppl 2): 199–209.

Bruce, Nigel, Rogelio Perez-Padilla, and Rachel Albalak. 2000. Indoor air pollution in developing countries: A major environmental and public health challenge. *Bull. World Health Organization* 78: 1078–1092.

Cooper, C. David, and F. C. Alley. 2002. *Air pollution control*, 3rd ed. Waveland Press.

Davis, Devra. 2002. *When smoke ran like water: Tales of environmental deception and the battle against pollution*. Basic Books, New York.

Davis, Devra L., et al. 2002. A look back at the London smog of 1952 and the half century since. *Env. Health Perspectives* 110: A734.

Driscoll, Charles T., et al. 2001. *Acid rain revisited: Advances in scientific understanding since the passage of the 1970 and 1990 Clean Air Act Amendments*. Hubbard Brook Research Foundation.

Godish, Thad. 2003. *Air quality*, 4th ed. CRC Press, Boca Raton, Fla.

Hall, Jane V., et al. 2008. *The benefits of meeting federal clean air standards in the South Coast and San Joaquin Valley air basins*. William and Flora Hewlett Foundation.

Hoffman, Matthew J. 2005. *Ozone depletion and climate change: Constructing a global response*. SUNY Press, New York.

Jacobson, Mark Z. 2002. *Atmospheric pollution: History, science, and regulation*. Cambridge Univ. Press.

Likens, Gene E. 2004. Some perspectives on long-term biogeochemical research from the Hubbard Brook ecosystem study. *Ecology* 85: 2355–2362.

Lutgens, Frederick K., Edward J. Tarbuck, and Dennis Tasa. 2010. *The atmosphere: An introduction to meteorology*, 11th ed. Prentice Hall. Upper Saddle River, N.J.

Molina, Mario J., and F. Sherwood Rowland. 1974. Stratospheric sink for chlorofluoromethanes: Chlorine atom catalyzed destruction of ozone. *Nature* 249: 810–812.

Parson, Edward A. 2003. *Protecting the ozone layer: Science and strategy*. Oxford Univ. Press.

U.N. Environment Programme. 2007. Atmosphere. Pp. 39–80 in *Global environment outlook 4 (GEO-4)*. UNEP and Progress Press, Nairobi and Malta.

U.N. Environment Programme, Ozone Secretariat. Montreal Protocol. http://unep.org/ozone/Treaties_and_Ratification/2B_montreal_protocol.asp.

U.S. Environmental Protection Agency. 2008. *National air quality: Status and trends through 2007*. EPA 454/R-08-006. Washington, D.C.

U.S. Environmental Protection Agency. 2009. *Acid rain and related programs: 2008 highlights*. EPA, Washington, D.C.

U.S. Environmental Protection Agency, Office of Air and Radiation. www.epa.gov/air.

World Health Organization. Indoor air pollution. www.who.int/indoorair/en/index.html.

Chapter 18

Alley, Richard B. 2000. *The two-mile time machine: Ice cores, abrupt climate change, and our future*. Princeton Univ. Press.

Appenzeller, Tim. 2007. The big thaw. *National Geographic*, June 2007: 56–71.

Arndt, D.S., et al., eds. 2010. State of the climate in 2009. *Bull. Amer. Meteor. Soc.* 91(6):S1–S224.

Bianco, Nicholas M., and Franz T. Litz. 2010. *Reducing greenhouse gas emissions in the United States using existing federal authorities and state action*. WRI Report, World Resources Institute, Washington, D.C.

Bloom, Arnold J. 2009. *Global climate change: Convergence of disciplines*. Sinauer Associates, Sunderland, Mass.

Burroughs, William James. 2007. *Climate change: A multidisciplinary approach*, 2nd ed. Cambridge Univ. Press.

Caldeira, Kenneth, and Michael E. Wickett. 2003. Anthropogenic carbon and ocean pH. *Nature* 425: 365.

EPICA community members. 2004. Eight glacial cycles from an Antarctic ice core. *Nature* 429: 623–628.

Flannery, Tim. 2005. *The weather makers: The history and future impact of climate change*. Text Publishing, Melbourne, Australia.

Gelbspan, Ross. 1997. *The heat is on: The climate crisis, the cover-up, the prescription*. Perseus Books, New York.

Gelbspan, Ross. 2004. *Boiling point: How politicians, big oil and coal, journalists, and activists are fueling the climate crisis—and what we can do to avert disaster*. Basic Books, New York.

Goodstein, Eban, 2007. *Fighting for love in the century of extinction: How passion and politics can stop global warming*. Univ. Press of New England, Lebanon, N.H.

Gore, Al. 2006. *An inconvenient truth: The planetary emergency of global warming and what we can do about it*. Rodale Press and Melcher Media, New York.

Intergovernmental Panel on Climate Change. 2001. *IPCC third assessment report—Climate change 2001: Synthesis report*. World Meteorological Organization and U.N. Environment Programme, Geneva, Switzerland.

Intergovernmental Panel on Climate Change. 2007. *IPCC fourth assessment report—Climate change 2007: The AR4 synthesis report*. World Meteorological Organization and U.N. Environment Programme, Geneva, Switzerland.

Intergovernmental Panel on Climate Change. 2007. *Climate change 2007: The physical science basis*. Contribution of Working Group I to the fourth assessment report of the IPCC. WMO and UNEP.

Intergovernmental Panel on Climate Change. 2007. *Climate change 2007: Impacts, adaptation, and vulnerability*. Contribution of Working Group II to the fourth assessment report of the IPCC. WMO and UNEP.

Intergovernmental Panel on Climate Change. 2007. *Climate change 2007: Mitigation of climate change*. Contribution of Working Group III to the fourth assessment report of the IPCC. WMO and UNEP.

Intergovernmental Panel on Climate Change. www.ipcc.ch.

Jonzén, Niclas, et al. 2006. Rapid advance of spring arrival dates in long-distance migratory birds. *Science* 312: 1959–1961.

Karl, Thomas R., and Kevin E. Trenberth. 2003. Modern global climate change. *Science* 302: 1719–1723.

Kerr, Richard A. 2006. A worrying trend of less ice, higher seas. *Science* 311: 1698–1701.

Kerr, Richard A. 2006. A tempestuous birth for hurricane climatology. *Science* 312: 676–678.

Mann, Michael, and Lee R. Kump. 2008. *Dire Predictions: Understanding global warming*. DK Publishing and Pearson Education, New York.

Mayewski, Paul A., and Frank White. 2002. *The ice chronicles: The quest to understand global climate change*. Univ. Press of New England, Hanover, N.H.

National Geographic. 2008. Changing climate. *National Geographic special report*, 22 June 2008.

Pacala, Stephen, and Robert Socolow. 2004. Stabilization wedges: Solving the climate problem for the next 50 years with current technologies. *Science* 305: 968–972.

Parmesan, Camille, and Gary Yohe. 2003. A globally coherent fingerprint of climate change impacts across natural systems. *Nature* 421: 37–42.

Pew Center on Global Climate Change. www.pewclimate.org.

Real Climate. www.realclimate.org.

Rignot, Eric, and Pannir Kanagaratnam. 2006. Changes in the velocity structure of the Greenland Ice Sheet. *Science* 311: 986–990.

Root, Terry L., et al. 2003. Fingerprints of global warming on wild animals and plants. *Nature* 421: 57–60.

The Royal Society. 2005. *Ocean acidification due to increasing atmospheric carbon dioxide*. The Royal Society, London, U.K., June 2005.

Scherr, Sara J., and Sajal Sthapit. 2009. *Mitigating climate change through food and land use*. Worldwatch Report 179. Worldwatch Institute, Washington, D.C.

Schiermeier, Quirin. 2006. Climate credits. *Nature* 444: 976–977.

Schneider, Stephen H., and Terry L. Root, eds. 2002. *Wildlife responses to climate change: North American case studies*. Island Press, Washington, D.C.

Siegenthaler, Urs, et al. 2005. Stable carbon cycle–climate relationship during the Late Pleistocene. *Science* 310: 1313–1317.

Spahni, Renato, et al. 2005. Atmospheric methane and nitrous oxide of the Late Pleistocene from Antarctic ice cores. *Science* 310: 1317–1321.

Speth, James Gustave. 2004. *Red sky at morning: America and the crisis of the global environment*. Yale Univ. Press, New Haven, Conn.

Taylor, David. 2003. Small islands threatened by sea level rise. Pp. 84–85 in *Vital signs 2003*. Worldwatch Institute, Washington D.C.

Time. 2006. Special report: Global warming. *Time*, 3 Apr. 2006: 28–62.

U.S. Climate Change Science Program. 2008. *Scientific assessment of the effects of global change on the United States*. Committee on Environment and Natural Resources, National Science and Technology Council, Washington, D.C.

U.S. Climate Change Science Program. 2009. *Coastal sensitivity to sea-level rise: A focus on the mid-Atlantic region*. Synthesis and Assessment Product 4.1, Washington, D.C.

U.S. Global Change Research Program (Karl, Thomas R., Jerry M. Melillo, and Thomas C. Peterson, eds.). 2009. *Global climate change impacts in the United States*. U.S. Global Change Research Program and Cambridge Univ. Press.

United Nations. U.N. Framework Convention on Climate Change. http://unfccc.int/2860.php.

United Nations. Kyoto Protocol. http://unfccc.int/kyoto_protocol/items/2830.php.

Victor, David G., et al. 2005. A Madisonian approach to climate policy. *Science* 309: 1820–1821.

Walsh, Bryan. 2008. How to win the war on global warming. *Time*, 28 Apr. 2008.

World Meteorological Organization. 2009. *WMO statement on the status of the global climate in 2008*. WMO No. 1039. Geneva, Switzerland.

Worldwatch Institute. 2009. *State of the world 2009: Into a warming world*. Worldwatch Institute, Washington, D.C.

Zwally, H. Jay, et al. 2002. Surface melt–induced acceleration of Greenland Ice-Sheet flow. *Science* 297: 218–222.

Chapter 19

American Petroleum Institute. 2009. *Offshore access to oil and natural gas reserves*. American Petroleum Institute, Washington, D.C.

Appenzeller, Tim. The end of cheap oil. *National Geographic*, June 2004: 80–109.

Association for the Study of Peak Oil and Gas. www.peakoil.net.

Ayres, Robert, et al., eds. 2004. *Encyclopedia of Energy*. Elsevier.

British Petroleum. 2010. *BP statistical review of world energy 2010*. BP, London.

Campbell, Colin J. 1997. *The coming oil crisis*. Multi-Science Publishing Co., Essex, U.K.

Deffeyes, Kenneth S. 2001. *Hubbert's peak: The impending world oil shortage*. Princeton Univ. Press.

Deffeyes, Kenneth S. 2005. *Beyond oil: The view from Hubbert's peak*. Farrar, Straus, and Giroux, New York.

Douglas, D.C., et al., eds. 2002. *Arctic Refuge coastal plain terrestrial wildlife research summaries. Biological science report.* USGS/BRD/BSR-2002-0001. U.S. Geological Survey, Washington, D.C.

Energy Information Administration, U.S. Department of Energy. www.eia.doe.gov.

Energy Information Administration, U.S. Department of Energy. 1999. *Petroleum: An energy profile, 1999.* DOE/EIA-0545(99).

Energy Information Administration, U.S. Department of Energy. 2010. *Annual energy review 2009.* DOE/EIA, Washington, D.C.

Hubbert, M. King. 1956. *Nuclear energy and the fossil fuels.* Publication No. 95, Shell Development Company, Houston, Texas.

International Energy Agency. 2005. *Resources to reserves: Oil and gas technologies for the energy markets of the future.* IEA, Paris.

International Energy Agency. 2009. *Key world energy statistics 2009.* IEA, Paris.

Kunstler, James H. 2005. *The long emergency.* Atlantic Monthly Press, New York.

Lovins, Amory B. 2005. More profit with less carbon. *Scientific American* 293(3): 74–83.

Lovins, Amory B., et al. 2004. *Winning the oil endgame: Innovation for profits, jobs, and security.* Rocky Mountain Institute, Snowmass, Colorado.

Martinot, Eric, and Li Junfeng. 2007. *Powering China's development.* Worldwatch Report 175. Worldwatch Institute, Washington, D.C.

Nellemann, Christian, and Raymond D. Cameron. 1998. Cumulative impacts of an evolving oil-field complex on the distribution of calving caribou. *Canadian Journal of Zoology* 76: 1425–1430.

Pelley, Janet. 2001. Will drilling for oil disrupt the Arctic National Wildlife Refuge? *Env. Sci. Technol.* 35: 240–247.

Powell, Stephen G. 1990. Arctic National Wildlife Refuge: How much oil can we expect? *Resources Policy,* Sept. 1990: 225–240.

Ristinen, Robert A., and Jack J. Kraushaar, 2005. *Energy and the environment,* 2nd ed. Wiley and Sons, New York.

Roberts, Paul. 2004. *The end of oil: On the edge of a perilous new world.* Houghton Mifflin, Boston.

Roberts, Paul. 2008. Tapped out. *National Geographic,* June 2008, pp. 84–91.

Russell, D.E., and P. McNeil. 2005. *Summer ecology of the Porcupine caribou herd.* Porcupine Caribou Management Board, Whitehorse, Yukon.

U.S. Environmental Protection Agency. 2009. *Light-duty automotive technology, carbon dioxide emissions, and fuel economy trends: 1975 through 2009.* EPA420-R-09-014. EPA Office of Transportation and Air Quality, Washington, D.C.

U.S. Fish and Wildlife Service. 2001. Potential impacts of proposed oil and gas development on the Arctic Refuge's coastal plain: Historical overview and issues of concern. http://arctic.fws.gov/issues1.htm.

U.S. Geological Survey. 2001. *The National Petroleum Reserve–Alaska (NPRA) data archive.* USGS Fact Sheet FS-024-01.

U.S. Geological Survey. 2001. *Arctic National Wildlife Refuge, 1002 Area, petroleum assessment, 1998, including economic analysis.* USGS Fact Sheet FS-028-01.

U.S. Geological Survey. 2002. *Petroleum resource assessment of the National Petroleum Reserve Alaska (NPRA).* USGS, Washington, D.C.

U.S. Government Accountability Office (GAO). 2007. *Crude oil: Uncertainty about future oil supply makes it important to develop a strategy for addressing a peak and decline in oil production.* Report to Congressional Requesters.

Walker, Donald A. 1997. Arctic Alaskan vegetation disturbance and recovery. Pp. 457–479 in R.M.M. Crawford, ed., *Disturbance and recovery in Arctic lands.* Kluwer Academic Publishers, Dordrecht, Netherlands.

Chapter 20

Ayres, Robert, et al., eds. 2004. *Encyclopedia of Energy.* Elsevier.

Biomass Research and Development Board. 2008. *National biofuels action plan.* Biomass Research and Development Initiative, U.S. DOE and USDA.

British Petroleum. 2010. *BP statistical review of world energy 2010.* BP, London.

The Chernobyl Forum. 2006. *Chernobyl's legacy: Health, environmental and socio-economic impacts* and *recommendations to the governments of Belarus, the Russian Federation and Ukraine. The Chernobyl Forum: 2003–2005.* Second revised version. World Health Organization and International Atomic Energy Agency, Vienna.

Earley, Jane, and Alice McKeown. 2009. *Red, white, and green: Transforming U.S. biofuels.* Worldwatch Report 180. Worldwatch Institute, Washington, D.C.

Energy Efficiency and Renewable Energy, U.S. Department of Energy. www.eere.energy.gov.

Energy Information Administration, U.S. Department of Energy. www.eia.doe.gov.

Energy Information Administration, U.S. Department of Energy. 2010. *Annual energy outlook 2010.* Washington, D.C.

Energy Information Administration, U.S. Department of Energy. 2010. *Annual energy review 2009.* DOE/EIA, Washington, D.C.

Energy Information Administration, U.S. Department of Energy. International energy statistics. http://tonto.eia.doe.gov/cfapps/ipdbproject/IEDIndex3.cfm.

European Commission/International Atomic Energy Agency/World Health Organization. 1996. One decade after Chernobyl: Summing up the consequences of the accident. Summary of the conference results. Vienna, Austria, 8–12 April 1996. EC/IAEA/WHO.

Flavin, Christopher. 2008. *Low-carbon energy: A roadmap.* Worldwatch Report 178. Worldwatch Institute, Washington, D.C.

International Atomic Energy Agency. *Nuclear power and sustainable development.* IAEA Information Series 02-01574/FS Series 3/01/E/Rev.1. Vienna, Austria.

International Atomic Energy Agency. 2006. *Environmental consequences of the Chernobyl accident and their remediation: Twenty years of experience.* Report of the U.N. Chernobyl Forum Expert Group "Environment." IAEA, Vienna.

International Energy Agency. 2007. *Biomass for power generation and CHP.* IEA, Paris.

International Energy Agency. 2009. *Key world energy statistics 2009.* IEA, Paris.

International Energy Agency. 2009. *World energy outlook 2009.* IEA, Paris.

National Renewable Energy Lab, U.S. Department of Energy. www.nrel.gov.

Nature. 2006. Special report: Chernobyl and the future. *Nature* 440: 982–989.

Nuclear Energy Agency. 2002. *Chernobyl: Assessment of radiological and health impacts.* 2002 update of *Chernobyl: Ten years on.* OECD, Paris.

Pearce, Fred. 2006. Fuels gold: Are biofuels really the greenhouse-busting answer to our energy woes? *New Scientist,* 23 Sept. 2006: 36–41.

REN21 Renewable Energy Policy Network. 2009. *Renewables global status report: 2009 update.* REN21 Secretariat, Paris.

Renewable Fuels Association. 2010. *Climate of opportunity: 2010 ethanol industry outlook.* RFA, Washington, D.C.

Sawin, Janet L., and William R. Moomaw. 2009. *Renewable revolution: Low-carbon energy by 2030.* Worldwatch Report 182. Worldwatch Institute, Washington, D.C.

Science. 2005. News Focus: Rethinking nuclear power. *Science* 309: 1168–1179.

Spadaro, Joseph V., et al. 2000. Greenhouse gas emissions of electricity generation chains: Assessing the difference. *IAEA Bulletin* 42(2).

Swedish Bioenergy Association (SVEBIO). 2003. *Focus: Bioenergy.* Nos. 1–10. SVEBIO, Stockholm.

Swedish Energy Agency. 2004. *Renewable electricity is the future's electricity.* Swedish Energy Agency, Eskilstuna, Sweden.

Swedish Energy Agency. 2009. *Energy in Sweden 2009.* Swedish Energy Agency, Eskilstuna, Sweden.

Swedish Energy Agency. 2010. *Energy indicators 2009.* Swedish Energy Agency, Eskilstuna, Sweden.

U.N. Environment Programme. 2009. *Assessing biofuels.* UNEP, Nairobi, Kenya.

World Health Organization. 2006. *Health effects of the Chernobyl accident and special health care programmes.* Report of the U.N. Chernobyl Forum Expert Group "Health." WHO, Geneva.

World Nuclear Association. 2009. *Nuclear power in Sweden.* www. world-nuclear.org/info/inf42.html.

Worldwatch Institute. 2007. *Biofuels for transport: Global potential and implications for sustainable agriculture and energy in the 21st century.* Worldwatch Institute, Washington, D.C.

Worldwatch Institute and Center for American Progress. 2006. *American energy: The renewable path to energy security.* Washington, D.C.

Chapter 21

American Wind Energy Association. 2009. *Annual wind industry report.* AWEA, Washington, D.C.

Ananthaswamy, Anil. 2003. Reality bites for the dream of a hydrogen economy. *New Scientist*, 15 Nov. 2003: 6–7.

Arnason, Bragi, and Thorsteinn I. Sigfusson. 2000. Iceland—a future hydrogen economy. *Intl. J. Hydrogen Energy* 25: 389–394.

Ayres, Robert, et al., eds. 2004. *Encyclopedia of Energy.* Elsevier.

Chow, Jeffrey, et al. 2003. Energy resources and global development. *Science* 302: 1528–1531.

Dunn, Seth. 2000. The hydrogen experiment. *World Watch* 13: 14–25.

Energy Efficiency and Renewable Energy, U.S. Department of Energy. www.eere.energy.gov.

Energy Efficiency and Renewable Energy, U.S. Department of Energy. 2008. *Annual report on U.S. wind power installation, cost, and performance trends: 2007.* EERE, Washington, D.C.

Energy Information Administration, U.S. Department of Energy. www.eia.doe.gov.

Energy Information Administration, U.S. Department of Energy. 2009. *Renewable energy annual 2007.* Washington, D.C.

Energy Information Administration, U.S. Department of Energy. 2010. *Annual energy outlook 2010.* Washington, D.C.

Energy Information Administration, U.S. Department of Energy. 2010. *Annual energy review 2009.* DOE/EIA, Washington, D.C.

Federal Ministry for the Environment, Nature Conservation, and Nuclear Safety [Germany]. http://www.erneuerbare-energien. de/inhalt/3860.

Federal Ministry for the Environment, Nature Conservation, and Nuclear Safety [Germany]. 2009. *Electricity from renewable energy sources: What does it cost?* Berlin.

Federal Ministry for the Environment, Nature Conservation, and Nuclear Safety [Germany]. 2009. *Renewable energy sources in figures: National and international development.* Berlin.

Federal Ministry of Economics and Technology [Germany]. http:// www.renewables-made-in-germany.com/index.php?id=50&L=1.

Flavin, Christopher. 2008. *Low-carbon energy: A roadmap.* Worldwatch Report 178. Worldwatch Institute, Washington, D.C.

Global Wind Energy Council. 2010. *Global wind 2009 report.* GWEC, Brussels, Belgium.

Grant, Paul M., et al. 2006. A power grid for the hydrogen economy. *Scientific American*, July 2006: 77–83.

Idaho National Laboratory. Wind power: Idaho wind data. http://www.inl.gov/wind/idaho/.

International Energy Agency. 2007. *Renewables in global energy supply: An IEA fact sheet.* IEA, Paris.

International Energy Agency. 2009. *Key world energy statistics 2009.* IEA, Paris.

International Energy Agency. 2009. *Technology roadmap: Wind energy.* IEA, Paris

International Energy Agency. 2009. *World energy outlook 2009.* IEA, Paris.

International Energy Agency. 2010. *Renewables information 2010.* IEA, Paris.

Jacobson, Mark Z., et al. 2005. Cleaning the air and improving health with hydrogen fuel-cell vehicles. *Science* 308: 1901–1905.

Knott, Michelle. 2003. Power from the waves. *New Scientist*, 20 Sept. 2003: 33–35.

Martinot, Eric, et al. 2005. *Renewable energy markets and policies in the United States.* Center for Resource Solutions, San Francisco. www.martinot.info/Martinot_et_al_CRS.pdf.

Melis, Anastasios, et al. 2000. Sustained photobiological hydrogen gas production upon reversible inactivation of oxygen evolution in the green alga *Chlamydomonas reinhardtii. Plant Physiology* 122: 127–135.

National Renewable Energy Lab, U.S. Department of Energy. www. nrel.gov.

REN21 Renewable Energy Policy Network. 2009. *Renewables global status report: 2009 update.* REN21 Secretariat, Paris.

Ristinen, Robert A., and Jack J. Kraushaar, 2005. *Energy and the environment*, 2nd ed. Wiley and Sons, New York.

Sawin, Janet. 2004. *Mainstreaming renewable energy in the 21st century.* Worldwatch Paper 169. Worldwatch Institute, Washington, D.C.

Sawin, Janet L. 2008. Another sunny year for solar power. In *Vital signs 2008–2009.* Worldwatch Institute, Washington, D.C.

Sawin, Janet L., and William R. Moomaw. 2009. *Renewable revolution: Low-carbon energy by 2030.* Worldwatch Report 182. Worldwatch Institute, Washington, D.C.

Weisman, Alan. 1998. *Gaviotas: A village to reinvent the world.* Chelsea Green Publishing Co., White River Junction, Vermont.

World Alliance for Decentralized Energy. http://www.localpower. org.

World Future Council. 2010. *Feed-in tariffs—Boosting energy for our future.* Earthscan Publications.

Worldwatch Institute and Center for American Progress. 2006. *American energy: The renewable path to energy security.* Washington, D.C.

Chapter 22

Arsova, L., et al. 2008. The state of garbage in America. *BioCycle* 49 (12): 22.

Ayres, Robert U., and Leslie W. Ayres. 1996. *Industrial ecology: Towards closing the materials cycle.* Edward Elgar Press, Cheltenham, U.K.

Beede, David N., and David E. Bloom. 1995. The economics of municipal solid waste. *World Bank Research Observer* 10: 113–150.

Container Recycling Institute. http://www.container-recycling.org.

Douglas, Ed. 2009. There's gold in them there landfills. *New Scientist*, 1 Oct. 2008.

Edmonton, City of. Edmonton Waste Management Centre. http://www.edmonton.ca/for_residents/garbage_recycling/ edmonton-waste-management-centre.aspx.

Graedel, Thomas E., and Braden R. Allenby. 2002. *Industrial ecology*, 2nd ed. Prentice Hall, Upper Saddle River, N.J.

Lilienfeld, Robert, and William Rathje. 1998. *Use less stuff: Environmental solutions for who we really are.* Ballantine, New York.

Manahan, Stanley E. 1999. *Industrial ecology: Environmental chemistry and hazardous waste.* Lewis Publishers, CRC Press, Boca Raton, Fla.

McDonough, William, and Michael Braungart. 2002. *Cradle to cradle: Remaking the way we make things.* North Point Press, New York.

New York City Department of Parks and Recreation. Fresh Kills Park. www.nycgovparks.org/sub_your_park/fresh_kills_park/ html/ fresh_kills_park.html.

New York City Department of Planning. Fresh Kills Park Project. www.nyc.gov/html/dcp/html/fkl/fkl3.shtml.

Rathje, William, and Colleen Murphy. 2001. *Rubbish! The archeology of garbage.* Univ. of Arizona Press.

Trash Track. http://senseable.mit.edu/trashtrack/.

U.S. Environmental Protection Agency. 2007. *Fact Sheet: Management of electronic waste in the United States.* EPA 530-F-08-014.

U.S. Environmental Protection Agency. 2008. *Electronic waste management in the United States Approach I.* Office of Solid Waste, Washington, D.C.

U.S. Environmental Protection Agency. 2009. *Municipal solid waste generation, recycling, and disposal in the United States: Facts and figures for 2008.* EPA 530-F-009-021. Office of Solid Waste and Emergency Response.

U.S. Environmental Protection Agency. Summary of the Comprehensive Environmental Response, Compensation, and Liability Act (Superfund). www.epa.gov/regulations/laws/cercla.html.

U.S. Environmental Protection Agency. Summary of the Resource Conservation and Recovery Act. www.epa.gov/regulations/laws/ rcra.html.

Chapter 23

Brugge, Doug, and Rob Goble. 2002. The history of uranium mining and the Navajo people. *Am. J. Public Health* 92(9): 1410–1419.

Christopherson, Robert W. 2008. *Geosystems: An introduction to physical geography*, 7th ed. Prentice Hall, Upper Saddle River, N.J.

Cohen, David. 2007. Earth audit. *New Scientist*, 26 May 2007: pp. 34–41.

Craig, James R., David J. Vaughan, and Brian J. Skinner. 2010. *Earth resources and the environment*, 4th ed. Benjamin Cummings, San Francisco.

Essick, Kristi. 2001. Guns, money, and cell phones. *The Industry Standard Magazine*, 11 Jun. 2001.

Gordon, R.B., et al. 2006. Metal stocks and sustainability. *PNAS* 103(5): 1209–1214.

Harden, Blaine. 2001. The dirt in the new machine. *New York Times Magazine* 12 Aug. 2001.

Hendryx, Michael, and Melissa M. Ahern. 2009. Mortality in Appalachian coal mining regions: The value of statistical life lost. *Public Health Reports* 124: 541–550.

Keller, Edward A. 2008. *Introduction to environmental geology*, 4th ed. Prentice Hall, Upper Saddle River, N.J.

Lovgren, Stefan. 2006. Can cell-phone recycling help African gorillas? *National Geographic News*, 20 Jan. 2006.

Mineral Information Institute. Mine Reclamation [extensive examples of mine reclamation efforts with photographs.] http://www. mii.org/recl.html.

Mooallem, Jon. 2008. The afterlife of cellphones. *New York Times Magazine* 13 Jan. 2008.

Mountain Justice. What is mountaintop removal mining? http:// www.mountainjusticesummer.org/facts/steps.php.

Palmer, Margaret, et al. 2010. Mountaintop mining consequences. *Science* 327: 148–149.

Perkins, Dexter. 2010. *Mineralogy*, 3rd ed. Prentice Hall, Upper Saddle River, N.J.

Rogich, D.G., and G.R. Matos. 2008. *The global flows of metals and minerals.* U.S. Geological Survey Open-File Report 2008-1355, 11 pp.

Sibley, Scott F., ed. 2004. *Flow studies for recycling metal commodities in the United States.* USGS Circular 1196-A-M. U.S. Geological Survey, Reston, Va.

Skinner, Brian J., and Stephen C. Porter. 2003. *The dynamic earth: An introduction to physical geology*, 5th ed. Wiley and Sons, Hoboken, N.J.

Sullivan, Daniel E. 2006. *Recycled cell phones—A treasure trove of valuable metals.* USGS Fact Sheet 2006-3097. U.S. Geological Survey, Denver, Colo.

Tabak, Henry H., and Rakesh Govind. 2003. Advances in biotreatment of acid mine drainage and biorecovery of metals: 2. Membrane bioreactor system for sulfate reduction. *Biodegradation* 14: 437–452.

Tabak, Henry H., et al. 2003. Advances in biotreatment of acid mine drainage and biorecovery of metals: 1. Metal precipitation for recovery and recycle. *Biodegradation* 14: 423–436.

Tantalum-Niobium International Study Center. Tantalum: Raw materials and processing. http://tanb.org/tantalum.

Tarbuck, Edward J., Frederick K. Lutgens, and Dennis Tasa. 2009. *Earth science*, 12th ed. Prentice Hall, Upper Saddle River, N.J.

U.N. Security Council. 2001. *Report of the panel of experts on the illegal exploitation of natural resources and other forms of wealth of the Democratic Republic of the Congo.* U.N. Security Council, 12 Apr. 2001.

U.N. Security Council. 2007. *Interim report of the group of experts on the Democratic Republic of the Congo, pursuant to Security Council resolution 1698 (2006).* U.N. Security Council, 25 Jan. 2007.

U.S. Department of the Interior. 2003. *Surface coal mining reclamation: 25 years of progress, 1977-2002.* Office of Surface Mining, Washington D.C.

U.S. Geological Survey, 2010. *Mineral commodity summaries.* USGS, 193 pp.

U.S. Geological Survey. Minerals information. http://minerals.usgs. gov/minerals.

Chapter 24

Association for the Advancement of Sustainability in Higher Education. http://www.aashe.org.

Bartlett, Peggy, and Geoffrey W. Chase, eds. 2004. *Sustainability on campus: Stories and strategies for change.* MIT Press, Cambridge, Mass.

Brower, Michael, and Warren Leon. 1999. *The consumer's guide to effective environmental choices: Practical advice from the Union of Concerned Scientists.* Three Rivers Press, New York.

Brown, Lester. 2001. *Eco-economy: Building an economy for the Earth.* Earth Policy Institute and W.W. Norton, New York.

Brown, Lester R. 2009. *Plan B 4.0: Mobilizing to save civilization.* Earth Policy Institute and W.W. Norton, New York.

Campus Ecology. National Wildlife Federation. www.nwf.org/ campusecology.

Campus Ecology. 2008. *Campus environment 2008: A national report card on sustainability in higher education.* Campus Ecology program of the National Wildlife Federation.

Carlson, Scott. 2006. In search of the sustainable campus: With eyes on the future, universities try to clean up their acts. *Chronicle of Higher Education* 53: A10.

Creighton, Sarah Hammond. 1998. *Greening the ivory tower: Improving the environmental track record of universities, colleges, and other institutions.* MIT Press, Cambridge, Mass.

Daly, Herman E. 1996. *Beyond growth.* Beacon Press, Boston.

Dasgupta, Partha, et al. 2000. Economic pathways to ecological sustainability. *BioScience* 50: 339–345.

De Anza College. Sustainability. http://www.deanza.edu/sustainability.

Durning, Alan. 1992. *How much is enough? The consumer society and the future of the Earth.* Worldwatch Institute, Washington, D.C.

Emerson, Jay, et al. 2010. *2010 Environmental Performance Index.* Yale Center for Environmental Law & Policy, New Haven, Conn.

Erickson, Christina, and David J. Eagan. 2009. *Generation E: Students leading for a sustainable, clean energy future.* Campus Ecology program of the National Wildlife Federation.

French, Hilary. 2004. Linking globalization, consumption, and governance. Pp. 144–163 in *State of the world 2004.* Worldwatch Institute, Washington, D.C.

Hawken, Paul. 1994. *The ecology of commerce: A declaration of sustainability.* Harper Business, New York.

Kahneman, Daniel, et al. 2006. Would you be happier if you were richer? A focusing illusion. *Science* 312: 1908–1910.

Keniry, Julian. 1995. *Ecodemia: Campus environmental stewardship at the turn of the 21st century.* National Wildlife Federation, Washington, D.C.

McMichael, A.J., et al. 2003. New visions for addressing sustainability. *Science* 302: 1919–1921.

Meadows, Donella, Jørgen Randers, and Dennis Meadows. 2004. *Limits to growth: The 30-year update.* Chelsea Green Publishing Co., White River Junction, Vermont.

Millennium Ecosystem Assessment. 2005. *Ecosystems and human well-being: General synthesis.* Millennium Ecosystem Assessment and World Resources Institute.

National Research Council, Board on Sustainable Development. 1999. *Our common journey: A transition toward sustainability.* National Academies Press, Washington, D.C.

Renner, Michael. 2008. *Green jobs: Working for people and the environment.* Worldwatch Report 177. Worldwatch Institute, Washington, D.C.

Sanderson, Eric W., et al. 2002. The human footprint and the last of the wild. *BioScience* 52: 891–904.

Schor, Juliet B., and Betsy Taylor, eds. 2002. *Sustainable planet: Solutions for the twenty-first century.* The Center for a New American Dream. Beacon Press, Boston.

Toor, Will, and Spenser W. Havlick. 2004. *Transportation and sustainable campus communities: Issues, examples, solutions.* Island Press, Washington, D.C.

United Nations. 2002. *Report of the World Summit on Sustainable Development, Johannesburg, South Africa, 26 August–4 September 2002.* U.N., New York.

U.N. Department of Economic and Social Affairs. 2010. *The Millennium Development Goals Report 2010.* U.N. DESA, New York.

U.N. Development Programme. 2009. *Human development report 2009.* Oxford Univ. Press.

U.N. Environment Programme. 2007. *Global environment outlook 4 (GEO-4).* UNEP and Progress Press, Nairobi and Malta.

University Leaders for a Sustainable Future. www.ulsf.org.

World Bank. 2010. *World development indicators 2010.* World Bank, Washington, D.C. http://data.worldbank.org.

World Commission on Environment and Development. 1987. *Our common future.* Oxford Univ. Press.

Worldwatch Institute. 2008. *State of the world 2008: Innovations for a sustainable economy.* Worldwatch Institute, Washington, D.C.

Worldwatch Institute. 2010. *State of the world 2010: Transforming cultures.* Worldwatch Institute, Washington, D.C.